This book provides the first self-contained comprehensive exposition of the theory of dynamical systems as a core mathematical discipline closely intertwined with most of the main areas of mathematics. The authors introduce and rigorously develop the theory while providing researchers interested in applications with fundamental tools and paradigms.

The book begins with a discussion of several elementary but fundamental examples. These are used to formulate a program for the general study of asymptotic properties and to introduce the principal theoretical concepts and methods. The main theme of the second part of the book is the interplay between local analysis near individual orbits and the global complexity of the orbit structure. The third and fourth parts develop in depth the theories of low-dimensional dynamical systems and hyperbolic dynamical systems.

The book is aimed at students and researchers in mathematics at all levels from advanced undergraduate up. Scientists and engineers working in applied dynamics, nonlinear science, and chaos will also find many fresh insights in this concrete and clear presentation. It contains more than four hundred systematic exercises.

ENCYCLOPEDIA OF MATHEMATICS AND ITS APPLICATIONS

EDITED BY G.-C. ROTA

Volume 54

Introduction to the Modern Theory of Dynamical Systems

ENCYCLOPEDIA OF MATHEMATICS AND ITS APPLICATIONS

18 H. O. Fattorini *The Cauchy problem*
19 G. G. Lorentz, K. Jetter, and S. D. Riemenschneider *Birkhoff interpolation*
21 W. T. Tutte *Graph theory*
22 J. R. Bastida *Field extensions and Galois theory*
23 J. R. Cannon *The one-dimensional heat equation*
25 A. Salomaa *Computation and automata*
26 N. White (ed.) *Theory of matroids*
27 N. H. Bingham, C. M. Goldie, and J. L. Teugels *Regular variation*
28 P. P. Petrushev and V. A. Popov *Rational approximation of real functions*
29 N. White (ed.) *Combinatorial geometries*
30 M. Pohst and H. Zassenhaus *Algorithmic algebraic number theory*
31 J. Aczel and J. Dhombres *Functional equations containing several variables*
32 M. Kuczma, B. Chozewski, and R. Ger *Iterative functional equations*
33 R. V. Ambartzumian *Factorization calculus and geometric probability*
34 G. Gripenberg, S.-O. Londen, and O. Staffans *Volterra integral and functional equations*
35 G. Gasper and M. Rahman *Basic hypergeometric series*
36 E. Torgersen *Comparison of statistical experiments*
37 A. Neumaier *Interval methods for systems of equations*
38 N. Korneichuk *Exact constants in approximation theory*
39 R. A. Brualdi and H. J. Ryser *Combinatorial matrix theory*
40 N. White (ed.) *Matroid applications*
41 S. Sakai *Operator algebras in dynamical systems*
42 W. Hodges *Model theory*
43 H. Stahl and V. Totik *General orthogonal polynomials*
44 R. Schneider *Convex bodies*
45 G. Da Prato and J. Zabczyk *Stochastic equations in infinite dimensions*
46 A. Bjorner, M. Las Vergnas, B. Sturmfels, N. White, and G. Ziegler *Oriented matroids*
47 E. A. Edgar and L. Sucheston *Stopping times and directed processes*
48 C. Sims *Computation with finitely presented groups*
49 T. Palmer *Banach algebras and the general theory of *-algebras*
50 F. Borceux *Handbook of categorical algebra I*
51 F. Borceux *Handbook of categorical algebra II*
52 F. Borceux *Handbook of categorical algebra III*

ENCYCLOPEDIA OF MATHEMATICS AND ITS APPLICATIONS

Introduction to the Modern Theory of Dynamical Systems

ANATOLE KATOK
Pennsylvania State University

BORIS HASSELBLATT
Tufts University

With a supplement by Anatole Katok and Leonardo Mendoza

CAMBRIDGE UNIVERSITY PRESS
Cambridge, New York, Melbourne, Madrid, Cape Town, Singapore,
São Paulo, Delhi, Dubai, Tokyo, Mexico City

Cambridge University Press
The Edinburgh Building, Cambridge CB2 8RU, UK

Published in the United States of America by Cambridge University Press, New York

www.cambridge.org
Information on this title: www.cambridge.org/9780521575577

First published 1995
Reprinted 1996
First paperback edition 1997
Reprinted 1998, 1999

A catalogue record for this publication is available from the British Library

ISBN 978-0-521-34187-5 Hardback
ISBN 978-0-521-57557-7 Paperback

To Sveta and Kathy

Contents

Preface

The theory of dynamical systems is a major mathematical discipline closely intertwined with most of the main areas of mathematics. Its mathematical core is the study of the global orbit structure of maps and flows with emphasis on properties invariant under coordinate changes. Its concepts, methods, and paradigms greatly stimulate research in many sciences and have given rise to the vast new area of applied dynamics (also called nonlinear science or chaos theory). The field of dynamical systems comprises several major disciplines, but we are interested mainly in finite-dimensional differentiable dynamics. This theory is inseparably connected with several other areas, primarily ergodic theory, symbolic dynamics, and topological dynamics. So far there has been no account that treats differentiable dynamics from a sufficiently comprehensive point of view encompassing the relations with these areas. This book attempts to fill this gap. It provides a self-contained coherent comprehensive exposition of the fundamentals of the theory of smooth dynamical systems together with the related areas of other fields of dynamics as a core mathematical discipline while providing researchers interested in applications with fundamental tools and paradigms. It introduces and rigorously develops the central concepts and methods in dynamical systems and their applications to a wide variety of topics.

What this book contains. We begin with a detailed discussion of a series of elementary but fundamental examples. These are used to formulate the general program of the study of asymptotic properties as well as to introduce the principal notions (differentiable and topological equivalence, moduli, structural stability, asymptotic orbit growth, entropies, ergodicity, etc.) and, in a simplified way, a number of important methods (fixed-point methods, coding, KAM-type Newton method, local normal forms, homotopy trick, etc.).

The main theme of the second part is the interplay between local analysis near individual (e.g., periodic) orbits and the global complexity of the orbit structure. This is achieved by exploring hyperbolicity, transversality, global topological invariants, and variational methods. The methods include the study of stable and unstable manifolds, bifurcations, index and degree, and construction of orbits as minima and minimaxes of action functionals.

In the third and fourth parts the general program outlined in the first part is carried out to considerable depth for low-dimensional and hyperbolic dynamical systems which are particularly amenable to such analysis. Hyperbolic systems are the prime example of well-understood complexity. This manifests itself in an orbit structure that is rich both from the topological and statistical point of view and stable under perturbation. At the same time the principal features can be described qualitatively and quantitatively with great precision. In low-dimensional dynamical systems on the other hand there are two situations. In the "very low-dimensional" case the orbit structure is simplified and admits only a limited amount of complexity. In the "low-dimensional" case some complexity is possible, yet additional major aspects of the orbit structure can be understood via hyperbolicity or related types of behavior.

Although we develop most themes related to differentiable dynamics in some depth we have not tried to write an encyclopedia of differentiable dynamics. Even if this were possible, the resulting work would be strictly a reference source and not useful as an introduction or a text. Consequently we also do not strive to present the most definitive results available but rather to provide organizing principles for methods and results. This is also not a book on applied dynamics and the examples are not chosen from those models that are widely studied in various disciplines. Instead our examples arise naturally from the internal structure of the subject and contribute to its understanding. The emphasis placed on various areas in the field is not dictated by the relative amount of published work or research activity in those areas, but reflects our understanding of what is basic and fundamental in the subject. An obvious disparity appears in the area of one-dimensional (real and especially complex) dynamics, which witnessed a great surge of activity in the past 15 years producing a number of brilliant results. It plays a relatively modest role in this book. Real one-dimensional dynamics is used mainly as an easy model situation in which various methods can be applied with considerable success. Complex dynamics, which is in our view a fascinating but rather specialized area, appears only as a source of examples of hyperbolic sets. On the other hand we try to point out and emphasize the interactions of dynamics with other areas of mathematics (probability theory, algebraic and differential topology, geometry, calculus of variations, etc.) even in some situations where the current state of knowledge is somewhat tentative.

How to use this book. This book can be used both as a text for a course or for self-study and as a reference book. As a text it would most naturally be used as the primary source for graduate students with background equivalent to one year of graduate study at a major U.S. university who are interested in becoming specialists in dynamical systems or want to acquire solid general knowledge of the field. Some portions of this book do not assume as much background and can be used by advanced undergraduate students or graduate students in science and engineering who want to learn about the subject without becoming experts. Those portions include Chapter 1, most of Chapters 2, 3, and 5, parts of Chapters 4, 6, 8, and 9, Chapters 10 and 11, and most of Chapters 12, 14, 15, and 16. The 472 exercises are a very important part of the book. They fall into several categories. Some of them directly illustrate the use of results or methods from the text; others explore examples that are not discussed in the text or indicate further developments. Sometimes an important side topic is developed in a series of exercises. Those 317 that we do not consider routine have been provided with hints or brief solutions in the back of the book. An asterisk indicates our subjective assessment of higher difficulty, due to the need for either inventiveness or familiarity with material not obviously related to the subject at hand.

Each of the four parts of the book can be the basis of a course roughly at the second-year graduate level running one semester or longer. From this book one can tailor many courses dedicated to more specialized topics, such as variational methods in classical mechanics, hyperbolic dynamical systems, twist maps and applications, an introduction to ergodic theory and smooth ergodic theory, and the mathematical theory of entropy. In order to assist both students and teachers in selecting material for a course we summarize the principal interrelations between the chapters in Figure F.1. A solid arrow A⟶B indicates that a major portion of the material from Chapter A is used in Chapter B (this relation is transitive). A dashed arrow A--→B indicates that material from Chapter A is used in some parts of Chapter B.

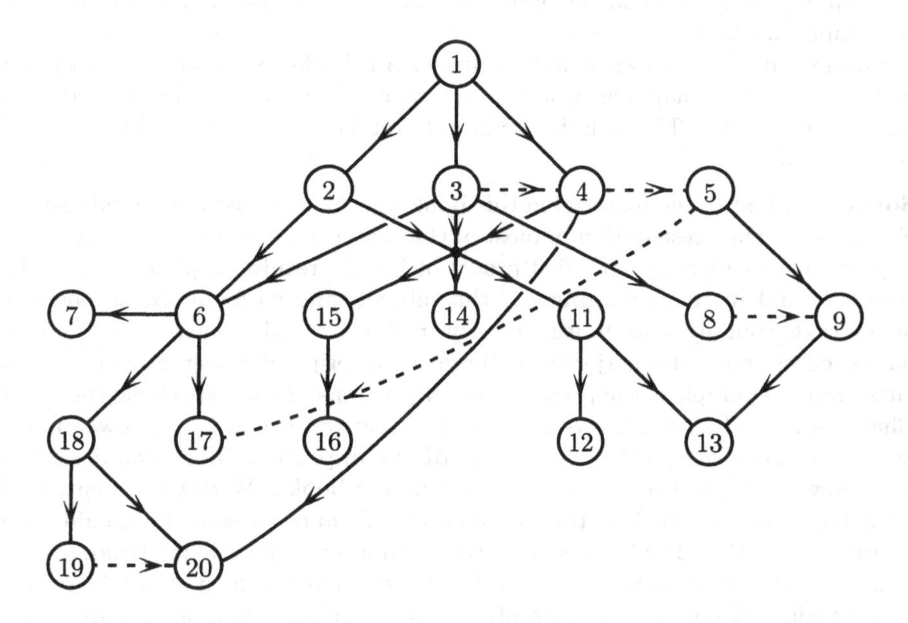

FIGURE F.1.

With the exception of Chapters 1–4, which form a common basis for the rest of the book, generally the material in the left part of the diagram deals with hyperbolic dynamics, that in the middle with low-dimensional dynamics, and that in the right with aspects of differentiable dynamics related to topology and classical mechanics.

There are various kinds of material used in this book. First of all we tacitly use, and assume familiarity with, the results of linear algebra (including Jordan normal forms), calculus of several variables, the basic theory of ordinary differential equations (including systems), elementary complex analysis, basic set theory, elementary Lebesgue integration, basic group theory, and some Fourier series. There is a next higher level of essential background material which is reviewed in the appendix. Most of the material in the appendix is of this nature, namely, the standard theory of topological, metric, and Banach spaces, elementary homotopy theory, the basic theory of differentiable manifolds including vector fields, bundles and differential forms, and the definition and basic properties of Riemannian metrics. Some topics are used on isolated occasions only. This last level of material includes the basic topology and geometry of surfaces and the general theory of measures, σ-algebras, and Lebesgue spaces, homology theory, material related to Lie groups and symmetric spaces, curvature and connections on manifolds, transversality, and normal families of complex functions. Most, but not all, of this material is also reviewed in the appendix, usually in a less detailed fashion. Either such material can be taken on faith without loss to the application in the text, or otherwise the pertinent portion of the text can be skipped without great loss.

On several occasions we include an important background fact without proof in the text. This happens when a certain result is organically related to a particular section. The Lefschetz Fixed-Point Formula is a good example of such a result.

Sources. Most of the material in this book does not consist of original results. Nevertheless the presentation of most of the material is our own and consists of original or considerably modified proofs of known results, explanations of the structure and interconnectedness of the subject, and so forth. Some portions of the text, roughly a sixth of it, mostly in Parts 3 and 4, closely follow other published sources, the majority of these being original research articles. An outstanding example is the presentation of portions of the hyperbolic theory in Chapters 18 and 20 which was given such a clear treatment by R. Bowen in his articles in the seventies that it could hardly be improved. On several occasions we follow the exposition of a subject in existing books. With the exception of some basic subjects such as Hamiltonian formalism or variational calculus this occurs only in Part 3. The reason for this is that low-dimensional dynamics has a much better developed expository literature than the field as a whole. We acknowledge all borrowings of proofs and presentations of which we are aware in the notes near the end of the book.

Since we aim to present the subject by developing it from first principles and in a self-contained way, rather than to give an exhaustive account of the development and current state of the field, we do not attempt a comprehensive listing of relevant references which would easily increase our bibliography to a thousand items or more. In particular, not all theorems are attributed to the original authors, especially if the results are part of the broader developments in the field, rather than landmark results or of a rather special nature. Most of the attributions are relegated to the notes. These consist of general comments arranged by section and some numbered remarks to particular points in the main text. Furthermore, in order not to interrupt the logical flow of the text all bibliographical references for the main part of the book are also relegated to the notes. Our historical comments, in both the introduction and the notes, do not aim to present a coherent account of the development of the subject, but to subjectively select some of its major moments.

We have included several types of literature in the bibliography. First, we have tried to list all major monographs and representative textbooks and surveys covering the principal branches of dynamics. Next there are landmark papers that introduce and develop various branches of our subject, define principal notions, or contain proofs of major results. We try to list all the sources on which the presentation in various parts of the book is based, or that inspired our presentation in other places and many (but not all) of the original sources for specific results presented in the text. Finally there is a sample of references to important work, both original and surveys, in some areas touched upon but not treated in the text. According to our principle of selecting models for their intrinsic interest rather than their value for concrete scientific problems we omit works by nonmathematicians (even important ones) that are dedicated to the study of models motivated by scientific problems so long as these contain only hypotheses and numerical results. References to such works are widely available in many of the books and surveys that we quote.

History and acknowledgments. The general idea of writing a broad introduction to the theory of dynamical systems first occurred to the first author when he taught a graduate course at the California Institute of Technology in 1984–5. This course resulted in two sets of lecture notes prepared by the second author and by his fellow graduate student John Lindner to whom we are deeply grateful. The key idea of introducing the principal notions and methods via a presentation of a series of basic examples crystallized when the first author was preparing and teaching an intensive four-week course in July 1986 at the Summer Mathematics Institute for graduate students at Fudan University in Shanghai. The summary and notes from that course became the germ for major parts of Chapters 1–4. Further progress was made during another graduate course at the California Institute of Technology in 1986–7 after which it became clear that the original project of a book of 300–350 pages would result in too sketchy and incomplete an account of the subject. In the summer of 1989 we developed a detailed plan of the book which has been carried out with some substantial later modification. Another graduate course during the first author's first year at the Pennsylvania State University (1990–1) helped to test some existing parts of the book and develop some new material.

We feel deep gratitude to the California Institute of Technology, Tufts University, and the Pennsylvania State University for providing excellent working conditions and supporting several mutual visits. Special thanks are due to the Mathematical Sciences Research Institute in Berkeley, California, where we worked together on major portions of the book in the summer of 1992. During this period our project was transformed from a collection of drafts into an incomplete but coherent product.

We also owe thanks and gratitude to numerous individuals for providing various kinds of help and inspiration during this project. We apologize for any omissions of people whose comments and suggestions may have been incorporated and forgotten.

Jessica Madow, the technical typist at the California Institute of Technology, typed major portions of the then existing manuscript in Exp. Kathy Wyland and Pat Snare at the Pennsylvania State University typed the first drafts of many chapters in TeX. Several people helped with computer support or typesetting advice. David Glaubman at the Mathematical Sciences Research Institute was very helpful, Michael Downes at the Technical Support Department of the American Mathematical Society helped to make the running heads come out right on every single page, and our colleague Uwe Schmock at the ETH wrote the overbar macro for TeX and made other useful comments. Boris Katok made the majority of the illustrations for the book. Bill Schlesinger gave us the initial tutoring that enabled us to make numerous pictures using Matlab. We are deeply grateful to the editors of Cambridge University Press: David Tranah who encouraged and prodded us during the earlier stages of the project and Lauren Cowles who patiently guided us through the process of finishing the book and getting it ready for production. This book was typeset in TeX using AMS-TeX, the TeX macro package of the American Mathematical Society.

Viorel Niţică and Alexej Kononenko wrote solutions to the majority of the exercises. Their work helped to correct some flawed exercises, and we used their solutions to write many of our hints.

The following people made numerous suggestions, including pointing out mathematical and stylistical errors, misprints, and the need for better explanations: The greatest amount of this kind of help came from Howie Weiss at the Pennsylvania State University. Further comments were given to us by Luis Barreira, Misha Brin, Mirko Degli-Esposti, David DeLatte, Serge Ferleger, Eugene Gutkin, Moisey Guysinsky, Miaohua Jiang, Tasso Kaper, Alexej Kononenko, Viorel Niţică, Ralf Spatzier, Garrett Stuck, Andrew Török, and Chengbo Yue.

In particular Howie Weiss, Tasso Kaper, Garrett Stuck, Ralf Spatzier, and Misha Brin taught from parts of the book and were very helpful in polishing it.

We had fruitful discussions with Michael Yakobson, Welington de Melo, Mikhael Lyubich, and Zbigniew Nitecki concerning one-dimensional maps and with Eduard Zehnder on variational methods. These were useful in crystallizing the content and presentation of those respective chapters. Gene Wayne helped by providing references concerning infinite-dimensional dynamical systems and Mike Boyle gave some useful guidance for sources in symbolic dynamics.

A number of corrections were made between printings. We would like to thank Luis Barreira, Marlies Gerber, Karl Friedrich Siburg, Garrett Stuck and Andrew Török who pointed out many small errors. Peter Walters found inaccuracies in Lemma 4.5.2 and Lemma 20.2.3. Robert McKay pointed out that some results of Section 14.2 needed recurrence hypotheses and Jonathan Robbins noted problems with the first version of Step 5 in the proof of the Hadamard-Perron Theorem 6.2.8. Tim Hunt corrected the DA construction. Corrections are listed at `http://www.tufts.edu/~bhasselb/thebook.html`.

A serious omission survived three printings: Section 20.6 is entirely due to Charles Toll, an attribution we inadvertently failed to make. Our sincere apologies.

Last, and most importantly, we wish to thank Svetlana Katok and Kathleen Hasselblatt for constant support and inspiration.

0

Introduction

1. Principal branches of dynamics

The most general and somewhat vague notion of a dynamical system includes the following ingredients:

(i). A "phase space" X, whose elements or "points" represent possible states of the system.

(ii). "Time", which may be discrete or continuous. It may extend either only into the future (irreversible or noninvertible processes) or into the past as well as the future (reversible or invertible processes). The sequence of time moments for a reversible discrete-time process is in a natural correspondence to the set of all integers; irreversibility corresponds to considering only nonnegative integers. Similarly, for a continuous-time process, time is represented by the set of all real numbers in the reversible case and by the set of nonnegative real numbers for the irreversible case.

(iii). The time-evolution law. In the most general setting this is a rule that allows us to determine the state of the system at each moment of time t from its states at all previous times. Thus, the most general time-evolution law is time dependent and has infinite memory. In the course of this book, however, we will consider only those evolution laws that allow us to define all future (and for reversible systems also past) states given a state at any particular moment. Furthermore we will assume that the law of time evolution itself does not change with time. In other words, the result of time evolution will depend only on the initial position of the system and on the length of the evolution but not on the moment when the state of the system was initially registered. Thus, if our system was initially at a state $x \in X$, it will find itself after time t at a new state, which is uniquely determined by x and t, and thus can be denoted by $F(x,t)$. Fixing t, we obtain a transformation $\varphi^t\colon x \mapsto F(x,t)$ of the phase space into itself. These transformations for different t are related to each other. Namely, the evolution of the state x for time $s+t$ can be accomplished by first applying the transformation φ^t to x and then by applying φ^s to the new state $\varphi^t(x)$. Thus, we have $F(x, t+s) = F(\varphi^t(x), s)$ or equivalently, the transformation φ^{t+s} is equal to the composition of φ^t and φ^s. In other words, the transformations φ^t form a semigroup. For a reversible system the transformations φ^t are defined for both positive and negative values of t and each φ^t is invertible. Thus, a reversible discrete-time dynamical system is represented by a cyclic group $\{F^n = (\varphi^1)^n \mid n \in \mathbb{Z}\}$ of one-to-one transformations of the phase space onto itself. Similarly, a reversible continuous-time dynamical system determines a one-parameter group $\{\varphi^t \mid t \in \mathbb{R}\}$ of one-to-one transformations of X onto itself.

The most characteristic feature of dynamical theories, which distinguishes them from other areas of mathematics dealing with groups of automorphisms of various mathematical structures, is the emphasis on asymptotic behavior, especially in the presence of nontrivial recurrence, that is, properties related with the behavior as time goes to infinity. The best way to explain what significant asymptotic properties are is to examine specific examples of dynamical systems and to determine the most characteristic features of their behavior. We will do that in Chapter 1 and then we will summarize some of our findings and present a list of interesting properties in Sections 3.1, 3.3, 4.1, 4.2d, and 4.3. This summary is preceded by an examination of natural equivalence relations for dynamical systems in Chapter 2 which sets the stage for treating asymptotic properties as invariants of those equivalence relations.

Historically, smooth continuous-time dynamical systems appeared first because of Newton's discovery that the motions of mechanical objects can be described by second-order ordinary differential equations. More generally, many other natural and social phenomena, such as radioactive decay, chemical reactions, population growth, or dynamics of prices on the market, may be modeled with various degrees of accuracy by systems of ordinary differential equations. These situations fit into the domain of our investigation if there is no explicit time-dependence in the coefficients and right-hand parts of the equations.

In virtually all situations of interest the phase space of a dynamical system possesses a certain structure which the evolution law respects. Different structures give rise to theories dealing with dynamical systems that preserve those structures. Let us mention the most important of those theories.

1. Ergodic theory. Here the phase space X is a "good" measure space, that is, a Lebesgue space (cf. Section 6 of the Appendix) with a finite or σ-finite measure μ. We can consider as a structure in X either the measure μ itself or its equivalence class which is determined by the collection of all sets of measure zero. Accordingly, ergodic theory concerns groups or semigroups of measurable transformations of X that either preserve μ or transform it into an equivalent measure. In the latter case the measure μ is called *quasi-invariant.* In this book ergodic theory plays an important but auxiliary role. It provides the appropriate paradigms and tools for studying asymptotic distribution and statistical behavior of orbits for smooth dynamical systems. Some central concepts and results of ergodic theory are introduced and discussed in Chapter 4.

The origins of ergodic theory go back to the famous ergodic hypothesis of Boltzmann who postulated equality of time averages and space averages for systems in statistical mechanics. Within mathematics the notions of ergodic theory arose from the study of uniform distributions of sequences. The Kronecker–Weyl Equidistribution Theorem (Proposition 4.2.1) is an early example of such a result. H. Poincaré observed that the preservation of a finite invariant measure forces strong conclusions about recurrence which are encapsulated in his Recurrence Theorem (Theorem 4.1.19). The systematic development of ergodic theory as a mathematical subject started around 1930 by von Neumann who looked at the subject primarily from a functional-analytic viewpoint. Among the early major contributors to the subject were G. D. Birkhoff, E. Hopf, and S. Kakutani. The critical point in the development of ergodic theory which forever changed the emphasis from the functional-analytic to the probabilistic and later geometric and combinatorial viewpoints was the introduction of entropy by A. Kolmogorov around 1958. It built upon C. Shannon's seminal development of information theory which was given the appropriate mathematical treatment by A. Khinchin. Kolmogorov's work was quickly followed by the development of an entropy theory based on the probabilistic viewpoint primarily by Y. Sinai and V. Rokhlin which culminated in Sinai's weak isomorphism theorem. The next crucial juncture was the first proof of the isomorphism of Bernoulli shifts of equal entropy which was obtained by D. Ornstein via combinatorial constructions. This work was followed by the development of the isomorphism theory which in particular gave necessary and sufficient conditions for metric isomorphism to a Bernoulli shift. Among later major developments one should note the Kakutani (monotone) equivalence theory, H. Furstenberg's theory of multiple recurrence, and the finitary isomorphism theory.

2. Topological dynamics. The phase space in this theory is a good topological space, usually a metrizable compact or locally compact space (see Section 1 of the Appendix). Topological dynamics concerns itself with groups of homeomorphisms and semigroups of continuous transformations of such spaces. Sometimes these objects are called topological dynamical systems. Similarly to the case of ergodic theory we use in this book notions and results from topological dynamics primarily as a framework and a tool for studying smooth dynamical systems. Though we are not making an attempt to provide a comprehensive introduction to the field, a fair amount of material from topological dynamics appears in this book, beginning with our first survey of examples in Chapter 1 and then in Chapter 3. Sections 4.1, 4.5 and later 20.1 and 20.2 provide crucial links between topological dynamics and ergodic theory. Some material in Chapter 8 (for example, Theorem 8.3.1) as well as all of Chapters 11 and 15 deal with particular classes of dynamical systems without any differentiability assumptions and thus belong to topological dynamics.

Topological dynamics was founded by Poincaré when he introduced the idea of qualitatively describing the solutions of differential equations that could not be solved analytically. One of his early achievements was the classification of circle maps (Theorem 11.2.7). M. Morse and G. D. Birkhoff made major contributions to topological dynamics in the process of trying to understand more classical systems (behavior of geodesics and Hamiltonian systems). Later a more intrinsic approach was developed by G. Hedlund, J. Oxtoby, and others. An important subject in topological dynamics is H. Furstenberg's theory of distal extensions which was further developed by R. Ellis.

3. The theory of smooth dynamical systems or differentiable dynamics. As the name suggests, the phase space here possesses the structure of a smooth manifold, for example, a domain or a closed surface in a Euclidean space (see Section 3 of the Appendix for a more detailed description). This theory, which is the prime subject of this book, concerns diffeomorphisms and flows (smooth one-parameter groups of diffeomorphisms) of such manifolds and iterates of noninvertible differentiable maps. In this book we will deal mostly with finite-dimensional situations. Interest in infinite-dimensional dynamical systems has been growing steadily during the past two decades, to a large extent stimulated by problems in fluid dynamics, statistical mechanics, and other fields of mathematical physics. Several directions in infinite-dimensional dynamics have been developed to a considerable extent starting from analogies with various branches in finite-dimensional dynamics.

Since a finite-dimensional smooth manifold possesses a natural locally compact topology, the theory of smooth dynamical systems naturally draws upon notions and results from topological dynamics. Another deeper reason for these interrelations arises from the fact that in dealing with asymptotic behavior of smooth dynamical systems one is likely to encounter very complicated nonsmooth phenomena, which in other contexts would be dismissed as pathological. In particular, some important invariant sets for a smooth system, for example, attractors (Definition 3.3.1), may not have any smooth structure and consequently, such sets should be studied from a different, nonsmooth, point of view. *Symbolic dynamics*, the study of a specific class of topological dynamical systems which occur as closed invariant subsets of the shift transformation in a sequence space (cf. Section 1.9), is particularly important in that respect. For further motivation of the relationships between topological and smooth dynamics see Section 2.3.

Relations with ergodic theory are also intimate, both because invariant measures provide a powerful tool for the study of asymptotic properties of smooth dynamical systems and because the smooth structure on a finite-dimensional manifold determines a natural class of quasi-invariant measures for differentiable dynamical systems (see Section 5.1).

Sometimes the part of the theory of smooth dynamical systems that concerns measure-theoretic properties of such systems is given the separate name *smooth ergodic theory.* One might also say that smooth ergodic theory is the study of automorphisms of a composite structure formed by a smooth manifold and a reasonable measure on it. Chapter 20 and the Supplement are dedicated to this subject. A number of results belonging to smooth ergodic theory are scattered among the earlier chapters.

Poincaré is also the father of differentiable dynamics. His main contribution was to emphasize the qualitative approach as opposed to the traditional emphasis on explicit solutions of differential equations of mechanics. His other achievement was the founding of the local theory of maps and vector fields near fixed and periodic orbits (cf. Sections 2.1, 6.3, 6.6). Other principal figures in the early stages of the field were A. M. Lyapunov and J. Hadamard who introduced various concepts of stability and developed major analytic tools (for example, the Hadamard–Perron Theorem 6.2.8). Part of Poincaré's program was carried out by G. D. Birkhoff who proved, among other things, Poincaré's celebrated "Last Geometric Theorem" which gives a mechanism responsible for dynamical complexity in mechanical systems with two degrees of freedom. Another aspect of Poincaré's program was developed by A. Denjoy who introduced some key new ideas in the process of completing Poincaré's theory of circle maps and flows on the two-dimensional torus. Symbolic dynamics appeared as a very useful tool beginning with a seminal paper by E. Artin and it was greatly developed by Morse and Hedlund. E. Hopf was the first to realize that hyperbolicity is a key mechanism that produces complicated behavior in nonlinear dynamical systems. His proof of ergodicity of the geodesic flow of surfaces of negative curvature can be viewed as the first major result in smooth ergodic theory.

Another principal root of the modern global approach to the study of smooth dynamical systems was the notion of structural stability which was first introduced by A. Andronov and L. S. Pontryagin in the study of flows on surfaces and later developed in that setting by Peixoto. It was given a second life by Smale who discovered that systems with complicated orbit behavior (the "horseshoe", Section 2.5) can be structurally stable. Subsequently Smale, Anosov, Sinai, and Bowen developed the core of the theory of hyperbolic dynamical systems. They greatly developed methods from ergodic theory and topological dynamics due to Hopf and Hedlund as well as more classical ideas going back to Hadamard, Perron, and Lyapunov. Identifying a certain hyperbolicity as sufficient (J. Robbin, C. Robinson) and necessary (R. Mañé) for structural stability was one of the crowning achievements of the theory of smooth dynamical systems. A major impetus to smooth ergodic theory was given by D. Ruelle and Y. Sinai who introduced ideas and methods from statistical mechanics to the theory of smooth dynamical systems. The next important step was made by Y. B. Pesin who developed the general structural theory of smooth measure-preserving systems based on the concept of nonuniform hyperbolicity. We should also mention the work of M. Herman and J.-C. Yoccoz on smooth classification of circle diffeomorphisms and the work of D. V. Anosov and A. Katok on constructions of smooth dynamical systems with various often unexpected properties.

4. Hamiltonian or symplectic dynamics. This theory is a natural generalization of a study of differential equations of classical mechanics. The phase space here is an even-dimensional smooth manifold with a nondegenerate closed differential 2-form Ω. One-parameter groups of Ω-preserving diffeomorphisms correspond to Hamiltonian differential equations in classical mechanics. An individual Ω-preserving diffeomorphism generalizes the notion of a canonical transformation. We first encounter such systems in Section 1.5 and return to this field in a more systematic way in Section 5.5.

The origin of Hamiltonian dynamics as an object of study from the point of view of dynamical systems is largely in the questions of celestial mechanics. Again Poincaré introduced the fundamental approach of the qualitative study of the n-body problem. Later two distinct directions of study emerged: (i) the investigation of dynamical complexity in this problem due to some hyperbolicity (Alekseev, Conley) and (ii) the study of integrable systems and their perturbations which led to the KAM theory. Though both the hyperbolic and integrable paradigm were available since Poincaré, it was Kolmogorov's profound contribution to realize that many qualitative features of (the very exceptional) integrable systems persist to some extent under perturbations and appear also in generic situations (for example, near an elliptic fixed point). Both of these lines of thought were influenced by the question of the stability of the solar system which was addressed by the hyperbolic approach in terms of the stability of an n-body system and by the KAM approach by considering perturbations, for example, of the (integrable) central force problem without interactions between planets. The work of Conley and Zehnder established a synthesis of topological and variational methods which became the cornerstone of modern global symplectic geometry. A renaissance of the study of completely integrable systems started with a seminal paper by Gardner, Greene, Kruskal, and Miura and the discovery by P. Lax of new mechanisms for producing integrable systems. It led both to a proliferation of new interesting examples of finite-dimensional integrable systems as well as to the theory of infinite-dimensional Hamiltonian systems whose applications to nonlinear partial differential equations were a major breakthrough by providing for the first time means for a complete qualitative analysis in situations other than those with the most simple asymptotic behavior.

2. Flows, vector fields, differential equations

The description of a dynamical system is somewhat easier when time is discrete, because the map generating a discrete-time system often can be given explicitly, usually by means of some formulas. In contrast, a continuous-time dynamical system is usually given infinitesimally (for example, by means of differential equations) and the reconstruction of the dynamics from this infinitesimal description involves some kind of integration process. In this and the next section we will very briefly discuss this local (in time) aspect of the theory of continuous-time dynamical systems and some simple relations between the discrete-time and the continuous-time situations.

We assume that the phase space is a smooth manifold of dimension m which we will usually denote by M, and thus our time evolution is given by a smooth function $F(x,t) = \varphi^t(x)$, $x \in M$, $t \in \mathbb{R}$, which satisfies the group (composition) property $\varphi^t \circ \varphi^s = \varphi^{t+s}$ and may or may not be defined for all x and t. Let us consider first the local aspect of the situation. When we fix $x \in M$ and vary t we obtain a parameterized smooth curve on M. Let $\xi(x)$ be the tangent vector to this curve at $t = 0$, that is, at the point x. Properly speaking, the vector $\xi(x)$ belongs to the tangent space T_xM which is an m-dimensional linear space "attached" to M at the point x. The map $x \mapsto \xi(x)$ forms a section of the tangent bundle $TM = \bigcup_{x\in M} T_xM$ or a *vector field* on M (see Section 3 of the Appendix for more details). Of course, the local version of this construction is familiar to everybody who completed a standard course of advanced calculus. Namely, let $U \subset M$ be a coordinate neighborhood with coordinates $(s_1, \dots, s_m)$. Then the tangent bundle TU is simply a direct product $U \times \mathbb{R}^m$ and a vector field is determined by a map from U to $\mathbb{R}^m$, that is, by m real-valued functions $v_1, \dots, v_m$, as follows. Denoting by $\frac{\partial}{\partial s_i}$ the basic vector fields which associate to every point the ith vector of the standard basis in $\mathbb{R}^m$ we can represent every vector field locally as $\sum_{i=1}^m v_i(s_1, \dots, s_n)\frac{\partial}{\partial s_i}$. If our initial point x is represented by coordinates $s_1^0, \dots, s_m^0$ then the evolution of this point is obtained by solving the system of first-order ordinary differential equations

$$\frac{ds_i}{dt} = v_i(s_1, \dots, s_m)$$

with initial conditions $s_i(0) = s_i^0$ for $i = 1, \dots, m$.

We know from the standard theory of ordinary differential equations that under very moderate smoothness assumptions, for example, if the functions v_i are continuously differentiable, the solution for sufficiently small time exists, is unique, and depends smoothly on the initial condition.

Thus, at least for small values of t, the transformation φ^t can be recovered from the vector field. For larger t one should take compositions of maps defined in local coordinates. If solutions exist for all real values of t, the vector field is called *complete*. We should keep in mind that on a manifold we have to work in different local coordinate systems if t is large, but this does not present any difficulties. If the manifold M is compact and has no boundary then it can be covered by a finite number of coordinate charts. Inside any chart the solutions exist for a fixed length of time. Since every point $x \in M$ belongs to a coordinate neighborhood which is not very small, this implies that any C^1 vector field on a closed compact manifold without boundary is complete and thus defines a *smooth flow*, that is, a one-parameter group of diffeomorphisms of M.

This is one of the reasons why we will often prefer to consider dynamical systems on compact manifolds. This preference will not be universal because in many situations such as local and semilocal problems (cf. Section 0.4 and Chapter 6) or systems of differential equations associated to many concrete mechanical and other problems, this assumption would be too restrictive.

Exercise

0.2.1. *Show (in detail) that a smooth vector field on a compact manifold is complete.*

3. Time-one map, section, suspension

There are several useful relations between continuous-time and discrete-time dynamical systems.

The most obvious way to associate a discrete-time system to a flow $\{\varphi^t\}_{t\in\mathbb{R}}$ is to take the iterates of the map φ^{t_0} for some value of t_0, say, $t_0 = 1$. However, only very few diffeomorphisms may be obtained that way. For example, let $f = \varphi^{t_0}$ and assume that $f^k(x) = x$, where $k > 1$, but $f(x) \neq x$ so that the orbit of x is periodic, but not fixed. But then for every $t \in \mathbb{R}$

$$f^k\varphi^t(x) = \varphi^{kt_0+t}(x) = \varphi^t(\varphi^{kt_0}(x)) = \varphi^t(f^k(x)) = \varphi^t(x).$$

Hence every point $\varphi^t(x)$ is also a periodic point of period k for f. Thus if f has an isolated periodic point of period greater than one, the map f cannot be obtained as the time-t map of any flow.

Another more local but also more useful method is the construction of a *Poincaré (first-return) map*. Let us take a point $x \in M$ such that $\xi(x) \neq 0$ and an $(m-1)$-dimensional (codimension-one) submanifold N containing x and transversal to the vector field. The latter property simply means that for every point $y \in N$ the vector $\xi(y)$ is not tangent to N. If we assume that the point x is periodic for the flow, that is, $\varphi^{t_0}(x) = x$ for some $t_0 > 0$, then every nearby orbit of the flow intersects the surface N at a time close to t_0 so we have defined for a neighborhood U of x on N a map $F_N: U \to N$ such that $F_N(x) = x$. This map is called a *section map* or *first-return map* or *Poincaré map* for the flow. This construction (also called inducing) also works if x is not periodic but comes sufficiently close to itself (see below).

Finally, for any diffeomorphism $f: M \to M$ one can construct a *suspension flow* on the *suspension manifold* M_f which is obtained from the direct product $M \times [0,1]$ by identifying pairs of points of the form $(x,1)$ and $(f(x),0)$ for $x \in M$. The *suspension flow* σ_f^t is determined by the "vertical" vector field $\dfrac{\partial}{\partial t}$ on M_f.

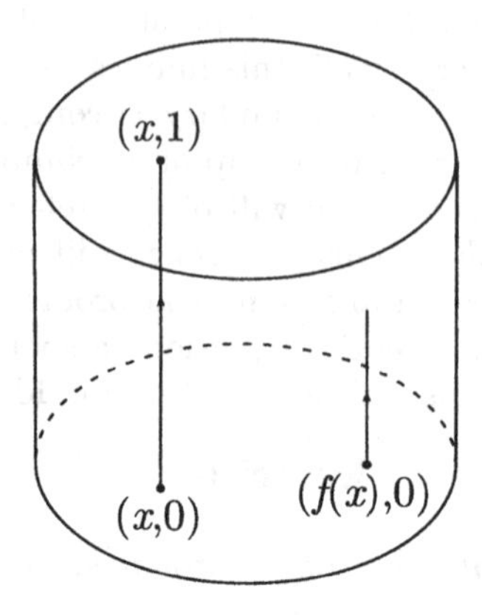

FIGURE 0.3.1. A suspension

This construction is closely related to the solution of a system of ordinary differential equations with periodic coefficients. First recall that a time-dependent system of ordinary differential equations is given by a family of vector fields v_t and thus determines a family of time evolutions $\varphi^{t,s}$ from the moment t to the moment $t+s$ which is not a group. It can, however, be interpreted as a single vector field $w(x,t) = v_t(x) + \frac{\partial}{\partial t}$ in the extended phase space $M \times \mathbb{R}$. The time evolution $\Phi^s(x,t) = (\varphi^{t,s}(x), t+s)$ in $M \times \mathbb{R}$ does have the group property. Of course, the space $M \times \mathbb{R}$ is never compact.

The situation changes, however, if the system of ordinary differential equations is periodic in time with period τ, say. Then $v_{t+\tau} = v_t$ and $\varphi^{t+k\tau,s} = \varphi^{t,s}$ for $k \in \mathbb{Z}$. In this case one can reduce the time evolution in $M \times \mathbb{R}$ to one in a factor space by identifying (x,t) with $(x, t+\tau)$. The factor space thus obtained is compact if M is compact and the projection of the flow Φ^s to that space is diffeomorphic to the suspension flow over the map $\varphi^{0,\tau}$ by the map $h\colon (\varphi^{0,t}(x), t) \mapsto (\varphi^{0,\tau}(x), t)$ $(0 \le t \le \tau)$ to $M_{\varphi^{0,\tau}}$.

The suspension construction is generalized to the construction of *the flow under a function* or *special flow*. Namely, add to our data a smooth positive function φ on M and consider the manifold $M_{f,\varphi}$ obtained from the subset $M_\varphi = \{(x,t) \mid x \in M,\ t \in \mathbb{R},\ 0 \le t \le \varphi(x)\}$ of the direct product $M \times \mathbb{R}$ by identifying pairs $(x, \varphi(x))$ and $(f(x), 0)$. Of course, topologically $M_{f,\varphi}$ is the same as M_f, but the "vertical" vector field $\frac{\partial}{\partial t}$ on $M_{f,\varphi}$ determines a new flow $\sigma^t_{f,\varphi}$ (the special flow under φ built over f) which differs from the suspension by a time change (see Definition 2.2.3).

Exercises

0.3.1. *Let $M = [0,1]$ and $f(x) = 1 - x$. Show that the manifold M_f is homeomorphic to the Möbius strip. The suspension flow has one orbit of period one and infinitely many orbits of period two. Show that the period-one orbit does not separate M_f and that any period-two orbit except the one that forms the boundary separates it into two pieces, one homeomorphic to the Möbius strip and the other to the cylinder $[0,1] \times S^1$.*

0.3.2. *Let $M = S^1 = \{z \in \mathbb{C} \mid |z| = 1\}$ and $f(z) = -z$. Show that the manifold M_f is homeomorphic to the two-torus $\mathbb{T}^2 = S^1 \times S^1$.*

0.3.3. *Let $M = S^1$ and $f(z) = \bar{z}$. Show that the manifold M_f is homeomorphic to the Klein bottle. The suspension flow has two orbits of period one and infinitely many orbits of period two. Show that period-one orbits do not separate M_f and that any period-two orbit separates it into two pieces homeomorphic to the Möbius strip.*

0.3.4. *Describe the smooth structure on the suspension manifold M_f and, more generally, on the manifold $M_{f,\varphi}$.*

4. Linearization and localization

We will see in the next three chapters that a large number of useful concepts related to the asymptotic behavior of smooth dynamical systems in fact belong to topological dynamics, that is, they are defined only in terms of topology, not the differentiable structure. We already mentioned some reasons for that in Section 0.1. Now we would like to point out some specific features that distinguish the theory of smooth dynamical systems from topological dynamics.

Already in elementary calculus one learns how useful it is to represent a function $\varphi(t)$ of one real variable t near a point t_0 as the main linear part $\varphi(t_0) + \varphi'(t_0)(t - t_0)$ plus an "infinitesimal of higher order", $o(t - t_0)$. A less elementary version of the same idea plays a central role in the theory of smooth dynamical systems. First, if $U \subset \mathbb{R}^m$ is an open neighborhood of x_0 and $f: U \to \mathbb{R}^m$ is a differentiable map, we can represent f near the point x_0 as the constant part $f(x_0)$ plus the linear part $Df_{x_0}(x - x_0)$ plus higher-order terms. The differential Df is a linear operator in $\mathbb{R}^n$ that is represented in coordinate form by the matrix of partial derivatives. If $f(t_1, \dots, t_m) = (f_1(t_1, \dots, t_m), \dots, f_m(t_1, \dots, t_m))$, then

$$Df_{x_0}(t_1, \dots, t_m) = \left(\frac{\partial f_i}{\partial t_j}\right)_{i,j=1,\dots,m},$$

where the partial derivatives are calculated at the values of the coordinates corresponding to the point x_0. If the map is regular at x_0 this operator is invertible.

The picture remains essentially the same for differentiable maps of smooth manifolds with the only difference that instead of the standard coordinate system in $\mathbb{R}^m$ one should use appropriate local coordinate systems near a point and its image. A more invariant way to express the same idea is to describe the differential Df_{x_0} of the map $f: M \to M$ as a linear map of the tangent space $T_{x_0}M$ into the space $T_{f(x_0)}M$. If f is a diffeomorphism then the differential is invertible. This construction can be globalized by considering the tangent bundle $TM = \bigcup_{x \in M} T_xM$ which can be provided with the structure of a differentiable manifold whose dimension is twice the dimension of M (see Section 3 of the Appendix). Any local coordinate system on M induces a coordinate system in TM which is global in the tangent direction. Namely, tangent vectors to the coordinate curves form a basis in each tangent space and the $2n$ coordinates of a tangent vector include n coordinates of its base point plus the coordinates of the vector with respect to that basis.

When we consider iterates of a map f, the differential $Df_x^n: T_xM \to T_{f_x^n}M$ of the nth iterate is a composition of the differentials $Df_{f^i(x)}: T_{f^i(x)} \to T_{f^{i+1}(x)}$, $i = 0, \dots, n-1$:

$$\underbrace{T_xM \xrightarrow{Df_x} T_{f(x)}M \xrightarrow{Df_{f(x)}} T_{f^2(x)}M \xrightarrow{Df_{f^2(x)}} \dots \xrightarrow{Df_{f^{n-1}(x)}} T_{f^n(x)}M}_{Df_x^n}.$$

In this localized picture the asymptotic properties of f correspond to the properties of products of linear maps thus obtained, when the number of factors goes to infinity. Once the behavior of such products is understood, the question arises as to what extent this behavior reflects the properties of the original nonlinear system. The crucial point here is that the differential at any given point approximates well the behavior of points near to the point at which the differential has been calculated. The quality of this approximation depends on the nonlinear terms, for example, on the size of the second derivatives of the function representing our map in a neighborhood of the original point. When we pass to the iterates of a map, as a rule, the size of second derivatives grows (by the chain rule), so, a priori, the quality of the linear approximation should deteriorate. Under certain conditions the influence of nonlinear terms can be controlled, so that we obtain a picture of the behavior of those orbits that stay near the original orbit for sufficiently long time. Considerations of this kind represent the content of what is usually called the local analysis of smooth dynamical systems. This is the central theme of Chapter 6 and of the first three sections of the Supplement.

An ideal setting for the local approach appears when the original orbit is periodic, say, $f^n(x_0) = x_0$. Then the sequence of differentials is also periodic and the main role in the local analysis is played by the iterates of a single linear operator, $Df^n_{x_0}$, which represents the infinitesimal behavior of nearby orbits for the period. In particular, the eigenvalues of that operator play a crucial role in the local analysis near the point x_0. See Section 1.2 for an analysis of linear maps and Sections 6.3 and 6.6 for a local analysis of nonlinear maps near a periodic point. For continuous-time dynamical systems the role of the differential is played by the variational equation whose right-hand side represents the infinitesimal generator for the one-parameter group of differentials of the maps forming the flow.

Though the local analysis concerns itself with the relative behavior of nearby orbits or, in the case of a neighborhood of a periodic orbit, with the behavior of orbits or orbit segments as long as they stay near the periodic orbit, the main goal of the theory of smooth dynamical systems is to understand the global behavior of nonlinear maps. Sometimes local analysis plays a crucial role in the global consideration. This happens, for example, if a periodic point represents an attractor, that is, if neighboring orbits approach it asymptotically with time (cf. Sections 1.1 and 3.3). More generally, we may try to localize certain parts of the phase space that play a particularly important role for the asymptotic behavior and to study the orbits inside and nearby this part. It is also possible that the behavior of orbits with certain initial conditions is particularly important due to the nature of a particular scientific problem which is represented by the dynamical system.

All this reasoning leads to a "semilocal" approach which lies between the local analysis and the global study of a system as a whole. Namely, let us assume that M is a smooth manifold (not necessarily compact), $U \subset M$ is an open subset of M, and $\Lambda \subset U$ is a compact set. Let furthermore $f\colon U \hookrightarrow M$ be a smooth map that leaves Λ invariant. We may be interested in the study of orbits of f that lie inside Λ or stay nearby. The local tool of this study is the differential Df localized to the restricted tangent bundle $T_\Lambda M = \bigcup_{x\in\Lambda} T_x M$.

Let us illustrate this approach by an example. Consider the hyperbolic linear map $f: \mathbb{R}^2 \to \mathbb{R}^2$, $f(x, y) = (2x, y/2)$ in a neighborhood of the origin:

FIGURE 0.4.1. The map $(x, y) \mapsto (2x, y/2)$

The segment of the y-axis consists of points asymptotic to the origin in positive time and the segment of the x-axis consists of points asymptotic to the origin in negative time, while all other points move along hyperbolas $xy = \text{const}$. (cf. Section 1.2 for a more detailed description). Suppose we extend our map nonlinearly to a larger area such that the preimage of the y-axis and the image of the x-axis intersect at a point p and form a nonzero angle (see Figure 6.5.1).

Such a point p is called a *transverse homoclinic point* for the fixed point 0. Obviously $f^n(p) \to 0$ as $|n| \to \infty$ so $\Lambda = \{0\} \cup (\bigcup_{n=-\infty}^{\infty} f^n(p))$ is a closed f-invariant set. A theory by G. D. Birkhoff asserts that any neighborhood of Λ contains periodic points of arbitrarily high period. S. Smale gave a complete analysis of the structure of orbits staying in a sufficiently small neighborhood of the set Λ (see Section 6.5). His work played a crucial role in the development of the modern theory of dynamical systems.

Another version of the semilocal analysis involves the study of orbits which stay inside a certain usually open noninvariant set. Of course, there may not be such orbits at all, but under certain conditions the existence of such orbits can be guaranteed. The constructions of an invariant Cantor set in the quadratic map and of Smale's "horseshoe" discussed in Section 2.5 are simple but nontrivial examples of such analysis.

Exercise

0.4.1. *Give an example of a diffeomorphism $f: \mathbb{R} \to \mathbb{R}$ such that $f(0) = (0)$, every orbit of the differential $f'(0)$ is bounded, and every orbit of f different from the origin is unbounded.*

Part 1

Examples and fundamental concepts

1

First examples

This chapter is intended to illustrate the concept of a dynamical system and the notion of asymptotic behavior by an assortment of examples. In the course of our survey we proceed from simple to more complicated types of asymptotic behavior and identify certain properties for a more systematic analysis in the future.

1. Maps with stable asymptotic behavior

a. Contracting maps. The simplest imaginable kind of asymptotic behavior is represented by the convergence of iterates of any given state to a particular state.

Definition 1.1.1. Let (X, d) be a metric space. A map $f: X \to X$ is called *contracting* if there exists $\lambda < 1$ such that for any $x, y \in X$

$$d(f(x), f(y)) \leq \lambda d(x, y). \tag{1.1.1}$$

The inequality (1.1.1) implies that the map f is continuous and therefore its positive iterates form a discrete-time topological dynamical system.

By iterating (1.1.1), one sees that for any positive integer n

$$d(f^n(x), f^n(y)) \leq \lambda^n d(x, y) \tag{1.1.2}$$

so

$$d(f^n(x), f^n(y)) \to 0 \quad \text{as } n \to \infty.$$

This means that the asymptotic behavior of all points is the same. On the other hand, for any $x \in X$ the sequence $\{f^n(x)\}_{n \in \mathbb{N}}$ is a Cauchy sequence because for $m \geq n$

$$\begin{aligned} d(f^m(x), f^n(x)) &\leq \sum_{k=0}^{m-n-1} d(f^{n+k+1}(x), f^{n+k}(x)) \\ &\leq \sum_{k=0}^{m-n-1} \lambda^{n+k} d(f(x), x) \leq \frac{\lambda^n}{1-\lambda} d(f(x), x) \xrightarrow[n\to\infty]{} 0. \end{aligned} \tag{1.1.3}$$

Thus for any $x \in X$ the limit of $f^n(x)$ as $n \to \infty$ exists if the space is complete, and by (1.1.2) this limit is the same for all x. Let us denote this limit by p and show that p is a fixed point for f. For any $x \in X$ and any integer n one has

$$\begin{aligned} d(p, f(p)) &\leq d(p, f^n(x)) + d(f^n(x), f^{n+1}(x)) + d(f^{n+1}(x), f(p)) \\ &\leq (1+\lambda) d(p, f^n(x)) + \lambda^n d(x, f(x)). \end{aligned}$$

Since $d(p, f^n(x)) \to 0$ as $n \to \infty$, we have $f(p) = p$. Taking the limit in (1.1.3) as $m \to \infty$ we obtain that $d(f^n(x), p) \leq \frac{\lambda^n}{1-\lambda} d(f(x), x)$. We will say that two sequences $\{x_n\}_{n\in\mathbb{N}}$ and $\{y_n\}_{n\in\mathbb{N}}$ of points in a metric space *converge exponentially* (or *with exponential speed*) *to each other* if $d(x_n, y_n) < c\lambda^n$ for some $c > 0$, $\lambda < 1$. In particular, if one of the sequences is constant, that is, $y_n = y$, we will say that x_n *converges exponentially* to y.

The above argument contains the proof of the following fundamental result which gives a complete and very simple picture of asymptotic behavior for the dynamical system generated by a contracting map.

Proposition 1.1.2. (Contraction Mapping Principle) *Let X be a complete metric space. Under the action of iterates of a contracting map $f: X \to X$ all points converge with exponential speed to the unique fixed point of f.*

Definition 1.1.3. If X is a topological space, $f: X \to X$, $f(p) = p$, and $f^n(x) \to p$ as $n \to \infty$ then we say that x is *(positively) asymptotic* to p. If f is invertible and $f^{-n}(x) \to p$ as $n \to \infty$ then we say that x is *negatively asymptotic* to p.

We denote the set of fixed points of any map f by $\text{Fix}(f)$.

Thus for a contracting map all points are asymptotic to a unique fixed point. This result will often be used in the course of our study of dynamical systems with more complicated behavior. Typically, we will be applying it not to the original dynamical system in the phase space but to a certain map in a functional space associated to the dynamical system. Right now, however, we give a straightforward and simple illustration of the use of the *Contraction Mapping Principle* in dynamics where the principle is applied to a certain derived system in the same space.

Proposition 1.1.4. *If p is a periodic point of period m for a C^1 map f and the differential Df_p^m does not have one as an eigenvalue then for every map g sufficiently close to f in the C^1 topology there exists a unique periodic point of period m close to p.*

Proof. Let us introduce local coordinates near p with p as the origin. In these coordinates Df_0^m becomes a matrix. Since 1 is not among its eigenvalues the map $F = f^m - \mathrm{Id}$ defined locally in these coordinates is locally invertible by the Inverse Function Theorem. Now let g be a map C^1-close to f. Near 0 one can write $g^m = f^m - H$ where H is small together with its first derivatives. A fixed point for g^m can be found from the equation $x = g^m(x) = (f^m - H)(x) = (F + \mathrm{Id} - H)(x)$ or $(F - H)(x) = 0$ or

$$x = F^{-1}H(x).$$

Since F^{-1} has bounded derivatives and H has very small first derivatives one can show that $F^{-1}H$ is a contracting map. More precisely, let $\|\cdot\|_0$ denote the C^0-norm, $\|dF^{-1}\|_0 = L$, and

$$\max\left(\|H\|_0, \|dH\|_0\right) \leq \epsilon.$$

Then, since $F(0) = 0$, we get $\|F^{-1}H(x) - F^{-1}H(y)\| \leq \epsilon L\|x - y\|$ for every x, y close to 0 and $\|F^{-1}H(0)\| \leq L\|H(0)\| \leq \epsilon L$, and hence $\|F^{-1}H(x)\| \leq \|F^{-1}H(x) - F^{-1}H(0)\| + \|F^{-1}H(0)\| \leq \epsilon L\|x\| + \epsilon L$. Thus if $\epsilon \leq \dfrac{R}{L(1+R)}$ the disc $X = \{x \mid \|x\| \leq R\}$ is mapped by $F^{-1}H$ into itself and the map $F^{-1}H\colon X \to X$ is contracting. By the Contraction Mapping Principle it has a unique fixed point in X which is thus a unique fixed point for g^m near 0. □

Let us illustrate the notion of a contracting map with an elementary example. Consider the real line with the metric induced by the absolute value. Suppose $f\colon \mathbb{R} \to \mathbb{R}$ is a continuously differentiable function whose derivative is bounded by some $\lambda < 1$. If $x, y \in \mathbb{R}$ then by the Mean Value Theorem there exists some ξ between x and y such that $f(x) - f(y) = f'(\xi)(x - y)$. Thus $|f(x) - f(y)| = |f'(\xi)||x - y| \leq \lambda|x - y|$ and f is a contracting map according to Definition 1.1.1. Thus any such map has a unique fixed point. Exercise 1.1.3 contains a generalization of this example.

The next section contains a few additional examples of contracting maps.

b. Stability of contractions. We now make an observation about the orbit structure of contraction that is interesting in itself, but also of utility when we apply the Contraction Mapping Principle to operators associated to a dynamical system under study: namely, that changing the contracting map slightly does not move the fixed point very much.

Proposition 1.1.5. *If $f\colon X \to X$ is a contraction of a complete metric space X with fixed point x_0 and contraction constant λ as in Definition 1.1.1 then for every $\epsilon > 0$ there exists a $\delta \in (0, 1 - \lambda)$ such that for any map $g\colon X \to X$ with*

(1) *$d(f(x), g(x)) < \delta$ for all $x \in X$ and*
(2) *$d(g(x), g(y)) \leq (\lambda + \delta)d(x, y)$ for all $x, y \in X$*

the fixed point y_0 of g satisfies $d(x_0, y_0) < \epsilon$.

Proof. Take $\delta = \dfrac{\epsilon(1-\lambda)}{1+\epsilon}$. Since $g^n(x_0) \to y_0$ we have

$$d(x_0, y_0) \le \sum_{n=0}^{\infty} d(g^n(x_0), g^{n+1}(x_0)) < d(x_0, g(x_0)) \sum_{n=0}^{\infty} (\lambda+\delta)^n$$
$$< \frac{\delta}{1-\lambda-\delta} = \frac{\epsilon(1-\lambda)}{(1+\epsilon)(1-\lambda-\frac{\epsilon(1-\lambda)}{1+\epsilon})} = \epsilon.$$

□

c. Increasing interval maps. The next simple asymptotic behavior is convergence of every orbit to a fixed point but in the presence of more than one fixed point. This occurs in the case of increasing functions of a real variable viewed as maps. This example is instructive because it demonstrates an important method in low-dimensional dynamics, the systematic use of the Intermediate Value Theorem.

Proposition 1.1.6. *If $I \subset \mathbb{R}$ is a closed interval and $f: I \to I$ is a nondecreasing continuous map then all $x \in I$ are asymptotic to a fixed point of f. If f is increasing (hence invertible) then all $x \in I$ are either fixed or positively and negatively asymptotic to adjacent fixed points.*

Proof. Note that the set $\mathrm{Fix}(f)$ is closed by continuity and nonempty by the Intermediate Value Theorem. If $\mathrm{Fix}(f) = I$ then there is nothing to show. Otherwise consider $x \in I \setminus \mathrm{Fix}(f)$ and let (a, b) be the maximal open interval of $I \setminus \mathrm{Fix}(f)$ containing x. Since f is nondecreasing we have $f(a, b) \subset [a, b]$ and by the Intermediate Value Theorem $f - \mathrm{Id}$ does not change sign on (a, b). To be specific suppose $f(y) > y$ for $y \in (a, b)$ (the other case is similar). Then $x_n := f^n(x)$ defines a nondecreasing sequence bounded by b, hence convergent to some $x_0 \in (a, b]$. But $f(x_0) = f(\lim_{n\to\infty} x_n) = \lim_{n\to\infty} f(x_n) = \lim_{n\to\infty} x_{n+1} = x_0$, so $x_0 \in \mathrm{Fix}(f)$ and in fact $x_0 = b$. Note that for the case $f(x) < x$ on (a, b) we would likewise obtain $f^n(x) \to a$ for all $x \in (a, b)$ as $n \to \infty$.

In case f is increasing note that the sign of $f^{-1} - \mathrm{Id}$ on such an interval (a, b) is opposite to that of $f - \mathrm{Id}$, so every $x \in (a, b)$ is positively and negatively asymptotic to opposite ends of $[a, b]$. □

Exercises

Problems 1.1.1 and 1.1.2 investigate the effect of replacing the uniform contraction condition (1.1.1) by weaker assumptions. As in the text X is a complete metric space and $f: X \to X$ a map of X into itself.

1.1.1. *Construct an example of a map f such that $d(f(x), f(y)) < d(x, y)$ for $x \neq y$, f has no fixed point, and $d(f^n(x), f^n(y))$ does not converge to zero for some x, y.*

1.1.2. *Suppose that $d(f(x), f(y)) < d(x,y)$ for $x \neq y$ and in addition the space X is compact. Then the iterates of every point $x \in X$ converge to a single fixed point of f as $n \to \infty$. Give an example showing that convergence need not be exponential.*

1.1.3. *Let $f: M \to M$ be a C^1 map of a complete Riemannian manifold to itself. Then f is a contraction if and only if the norm of the differential is bounded by a constant $\lambda < 1$.*

1.1.4. *Use Proposition 1.1.5 to show that the fixed point of a contraction depends continuously on the contraction with respect to the C^1 topology.*

1.1.5. *Prove that no contracting map of a compact metric space with more than one point is invertible.*

2. Linear maps

Next let us consider the dynamical system defined by the iterates of a linear map A of the Euclidean space $\mathbb{R}^n$. If A is invertible, this system is reversible.

Definition 1.2.1. Let $A: \mathbb{R}^n \to \mathbb{R}^n$ be a linear map. We call the set of eigenvalues the *spectrum* of A and denote it by $\operatorname{sp} A$. We denote the maximal absolute value of an eigenvalue of A by $r(A)$ and call it the *spectral radius* of A.

Given any norm on $\mathbb{R}^n$ we define the norm of a linear map A by $\|A\| := \sup_{\|v\|=1} \|Av\|$. Clearly $\|A\| \geq r(A)$ and with respect to the Euclidean norm we have $\|A\| = r(A)$ whenever A is diagonal. The following fact from linear algebra is useful for the understanding of dynamics of linear maps even if they cannot be diagonalized.

Proposition 1.2.2. *For every $\delta > 0$ there exists a norm in $\mathbb{R}^n$ such that $\|A\| < r(A) + \delta$.*

Proof. Using the Jordan normal form we can find a basis in $\mathbb{R}^n$ such that the matrix of our map has the block diagonal form

$$\begin{pmatrix} A_1 & & 0 \\ & \ddots & \\ 0 & & A_k \end{pmatrix}$$

where each block is either a Jordan block corresponding to a real eigenvalue λ:

$$\begin{pmatrix} \lambda & 1 & & & 0 \\ 0 & \lambda & 1 & & 0 \\ & & \ddots & \ddots & \\ 0 & & & \lambda & 1 \\ 0 & & & & \lambda \end{pmatrix} \tag{1.2.1}$$

or a combination of two blocks corresponding to a pair of complex conjugate eigenvalues $\lambda = \rho e^{i\varphi}$ and $\bar{\lambda} = \rho e^{-i\varphi}$:

$$\begin{pmatrix} \rho R_\varphi & \mathrm{Id} & \dots & 0 \\ & \rho R_\varphi & \mathrm{Id} & 0 \\ & & \ddots & \\ 0 & & \dots & \rho R_\varphi \end{pmatrix}, \tag{1.2.2}$$

where Id is the 2×2 identity matrix $\begin{pmatrix} 1 & 0 \\ 0 & 1 \end{pmatrix}$ and $R_\varphi := \begin{pmatrix} \cos\varphi & \sin\varphi \\ -\sin\varphi & \cos\varphi \end{pmatrix}$ is the 2×2 matrix corresponding to the rotation of the plane by the angle φ. Let us fix $\delta > 0$. By making an extra diagonal coordinate change of the form

$$\begin{pmatrix} 1 & & & 0 \\ & \delta^{-1} & & \\ & & \ddots & \\ & 0 & & \delta^{-m+1} \end{pmatrix}$$

for an m-block of the form (1.2.1) and of the form

$$\begin{pmatrix} \mathrm{Id} & & & 0 \\ & \delta^{-1}\,\mathrm{Id} & & \\ & & \ddots & \\ 0 & & & \delta^{-m+1}\,\mathrm{Id} \end{pmatrix}$$

for a $2m$-block of the form (1.2.2) one can make the off-diagonal entries in (1.2.1) and (1.2.2) equal to δ. Now for the standard Euclidean norm with respect to this new basis we have (after a slightly messy calculation)

$$\|A\| := \sup_{\|v\|=1} \|Av\| \le r(A) + \delta. \tag{1.2.3}$$

□

Remark. In fact, since all norms in $\mathbb{R}^n$ are equivalent up to a bounded multiple, one can conclude that for any norm and for every $\epsilon > 0$ there exists C_ϵ such that for any $v \in \mathbb{R}^n$

$$\|A^n v\| \le C_\epsilon (r(A) + \epsilon)^n \|v\|.$$

We begin our study of asymptotic behavior of linear maps with an important particular case.

Corollary 1.2.3. *Assume all eigenvalues of a linear map A have absolute value less than one. Then there exists a norm in $\mathbb{R}^n$ such that A is a contracting map with respect to the metric generated by that norm.*

Proof. If δ is chosen small enough we can conclude from Proposition 1.2.2 that $\|A\| < 1$ and since $d(x, y) = \|x - y\|$, A is a contracting map. □

The concept of exponential convergence does not depend on a particular choice of a norm. Thus Proposition 1.1.2 and Corollary 1.2.3 immediately imply the following statement.

Corollary 1.2.4. *If all eigenvalues of a linear map $A: \mathbb{R}^n \to \mathbb{R}^n$ have absolute values less than one, then the positive iterates of every point converge to the origin with exponential speed. If in addition A is an invertible map, that is, if zero is not an eigenvalue for A, then negative iterates of every point go to infinity exponentially.*

Let us look at a few examples of linear maps of $\mathbb{R}^2$ of this kind. The first example that comes to mind is the map given by the matrix $\begin{pmatrix} \lambda & 0 \\ 0 & \lambda \end{pmatrix}$ when $\lambda \in (0,1)$. Every vector v is contracted by the factor λ, so the iterates of any vector move toward the origin along a line through 0. Since every line through 0 is mapped to itself (while being contracted), we can draw the orbit picture as follows.

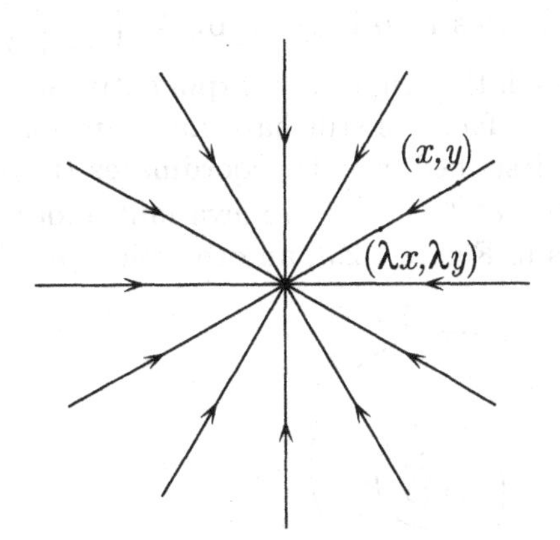

FIGURE 1.2.1. Orbits of contracting homothety

The next example to try is the map $\begin{pmatrix} \lambda & 0 \\ 0 & \mu \end{pmatrix}$ when $\lambda, \mu \in (0,1)$. Let us suppose that $\mu < \lambda$. Then every point still moves toward the origin, but not on a given straight line. Nevertheless, the orbits move along curves that are invariant under $\begin{pmatrix} \lambda & 0 \\ 0 & \mu \end{pmatrix}$. One can easily verify that these are the axes and the curves given by $x^{\log \mu} = \text{const.}\, y^{\log \lambda}$. Thus the corresponding orbit picture is as shown in Figure 1.2.2. A fixed point of a map of this kind is referred to as a *node*.

The next example is that of a map with two complex eigenvalues of absolute value less than one. To be specific consider the map given by $\lambda \begin{pmatrix} \cos\theta & \sin\theta \\ -\sin\theta & \cos\theta \end{pmatrix}$. Observe that this is the composition of the rotation by θ and contraction by

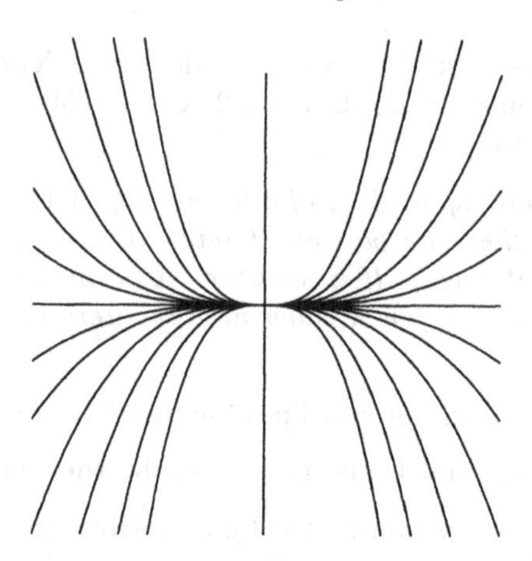

FIGURE 1.2.2. A node

λ and the nth iterate of this map is given by $\lambda^n \begin{pmatrix} \cos n\theta & \sin n\theta \\ -\sin n\theta & \cos n\theta \end{pmatrix}$. Thus, while points still approach the origin with exponential speed, at the same time they rotate around 0. In fact, we still have invariant curves, namely, spirals, which are most easily described in polar coordinates (r, ϕ). One checks easily that the curves $r = \text{const.}\, e^{-(\theta^{-1} \log \lambda)\phi}$ are invariant under this map. Thus here we obtain a portrait as in Figure 1.2.3, an orbit picture called a *focus.*

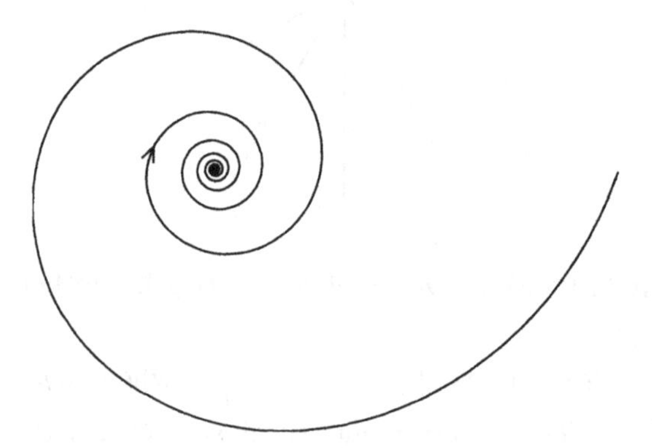

FIGURE 1.2.3. A focus

The appearance of invariant curves in the above examples is not accidental. The linear maps we described above arise from solutions of ordinary differential equations, whose flows interpolate iterates of the maps above. Consider first the ordinary differential equation

$$\begin{pmatrix} \dot{x} \\ \dot{y} \end{pmatrix} = \begin{pmatrix} \log \lambda & 0 \\ 0 & \log \lambda \end{pmatrix} \begin{pmatrix} x \\ y \end{pmatrix}$$

whose solutions are given by

$$\begin{pmatrix} x(t) \\ y(t) \end{pmatrix} = \begin{pmatrix} x(0)e^{t\log\lambda} \\ y(0)e^{t\log\lambda} \end{pmatrix} = \begin{pmatrix} x(0)\lambda^t \\ y(t)\lambda^t \end{pmatrix}.$$

Thus for $t = 1$ we get the first contracting map. The second one is evidently obtained from the ordinary differential equation with coefficient matrix $\begin{pmatrix} \log\lambda & 0 \\ 0 & \log\mu \end{pmatrix}$. Thus the first two figures above show the orbit structures of two *linear flows*. The second of these is also called a node in the context of ordinary differential equations.

Finally the focus is obtained from the linear ordinary differential equation with coefficient matrix

$$\begin{pmatrix} \log\lambda & \theta \\ -\theta & \log\lambda \end{pmatrix}.$$

Again taking $t = 1$ gives the above map with a focus.

Next we proceed to a somewhat more general case which will play an important role in our future considerations of nonlinear dynamical systems.

Definition 1.2.5. A linear map of $\mathbb{R}^n$ is called *hyperbolic* if all of its eigenvalues have absolute values different from one.

For every linear map $A: \mathbb{R}^n \to \mathbb{R}^n$ and for a real eigenvalue λ of A let us denote by E_λ the root space corresponding to λ, that is, the space of all vectors $v \in \mathbb{R}^n$ such that $(A - \lambda\,\mathrm{Id})^k v = 0$ for some k.

Similarly, for a pair of complex conjugate eigenvalues λ, $\bar\lambda$ let $E_{\lambda,\bar\lambda}$ be the intersection of $\mathbb{R}^n$ with the sum of root spaces corresponding to E_λ and $E_{\bar\lambda}$ for the complexification of A (that is, the extension to the complex space $\mathbb{C}^n$). For brevity we will call $E_{\lambda,\bar\lambda}$ a root space, too. Let

$$E^- = E^-(A) = \bigoplus_{|\lambda|<1} E_\lambda \oplus \bigoplus_{|\lambda|<1} E_{\lambda,\bar\lambda} \tag{1.2.4}$$

and similarly

$$E^+ = E^+(A) = \bigoplus_{|\lambda|>1} E_\lambda \oplus \bigoplus_{|\lambda|>1} E_{\lambda,\bar\lambda}. \tag{1.2.5}$$

If the map A is invertible, then $E^+(A) = E^-(A^{-1})$. Finally, let

$$E^0 = E^0(A) = E_1 \oplus E_{-1} \oplus \bigoplus_{|\lambda|=1} E_{\lambda,\bar\lambda}. \tag{1.2.6}$$

The spaces E^-, E^+, E^0 are obviously invariant with respect to A and $\mathbb{R}^n = E^- \oplus E^+ \oplus E^0$.

An equivalent way to describe hyperbolic linear maps is to say that A is hyperbolic if $E^0 = \{0\}$ or, equivalently, if $\mathbb{R}^n = E^+ \oplus E^-$

Since the restriction of A to the space $E^-(A)$ is a linear operator with all eigenvalues of absolute value less than one, one obtains immediately from Corollary 1.2.4

Corollary 1.2.6. *There exists a norm such that the restriction of a linear map A to the space $E^-(A)$ is a contracting map. If A is invertible then in addition the restriction of A^{-1} to the space $E^+(A)$ is a contracting map.*

Definition 1.2.7. The space $E^-(A)$ above is called the *contracting* subspace, the space $E^+(A)$ the *expanding* subspace.

Remark. Note that the expanding subspace is not characterized by the fact that vectors in it expand under iterates of the map—all vectors outside the contracting subspace are expanded by a sufficiently large iterate of the map. The characterization of E^+ is given by the description of Corollary 1.2.6, namely that preimages contract.

Our next statement describes the asymptotic behavior of iterates of a hyperbolic linear map.

Proposition 1.2.8. *Let $A: \mathbb{R}^n \to \mathbb{R}^n$ be a hyperbolic linear map. Then:*

(1) *For every $v \in E^-$, the positive iterates $A^n v$ converge to the origin with exponential speed as $n \to \infty$ and if A is invertible then the negative iterates $A^n v$ go to infinity with exponential speed as $n \to -\infty$.*
(2) *For every $v \in E^+$ the positive iterates of v go to infinity exponentially and if A is invertible then the negative iterates converge exponentially to the origin.*
(3) *For every $v \in \mathbb{R}^n \setminus (E^- \cup E^+)$ the iterates $A^n v$ go to infinity exponentially as $n \to \infty$ and if A is invertible also as $n \to -\infty$.*

Proof. Statement (1) follows from Corollary 1.2.6 and Proposition 1.1.2; (2) reduces to (1) when A is replaced by A^{-1}. Finally, if $v \in \mathbb{R}^n \setminus (E^- \cup E^+)$ then $v = v^- + v^+$ where $v^- \in E^-$, $v^+ \in E^+$ and $v^-, v^+ \neq 0$.

Then for large positive n one has

$$\|A^n v\| = \|A^n(v^- + v^+)\| \geq \|A^n v^+\| - \|A^n v^-\| \geq \lambda^n c\|v^+\| - \lambda^{-n} c'\|v^-\| \geq \lambda^n c'',$$

where $\lambda > 1$ and $c, c', c'' > 0$ do not depend on n.

The argument for negative iterates is the same with v^+ and v^- exchanged. □

Let us quickly illustrate the behavior of the hyperbolic linear maps by considering the two-dimensional case. In this case the subspaces E^+ and E^- are of course one-dimensional. In the example where these are the x and y axes (which can be arranged by a coordinate change) we can easily draw a picture. The map is given by $\begin{pmatrix} \lambda & 0 \\ 0 & \mu \end{pmatrix}$ with $0 < \mu < 1 < \lambda$. As for the node one checks that the axes and the curves given by $x^{\log \mu} = \text{const.}\, y^{\log \lambda}$ are invariant.

Note that the two exponents here now have different sign, so we do not get curves through the origin. Indeed the picture is as in Figure 0.4.1; it is usually referred to as a *saddle*. As in the previous linear examples this one also arises from a linear ordinary differential equation, this time with coefficient matrix $\begin{pmatrix} \log\lambda & 0 \\ 0 & \log\mu \end{pmatrix}$.

An interesting special case is that of a map which is also *area preserving*. In this case we must have $\lambda\mu = 1$, so the invariant curves are the standard hyperbolas $xy = \text{const}$. This is the reason for using the word "hyperbolic" in Definition 1.2.5 and in many other contexts later in this book. There is also an interesting case of a hyperbolic linear map which does *not* come from an ordinary differential equation. It is given by a matrix $\begin{pmatrix} \lambda & 0 \\ 0 & \mu \end{pmatrix}$ where $\lambda < -1 < \mu < 0$, that is, where λ and μ are negative numbers on opposite sides of -1. These *inverted saddles* play an interesting role in some global problems (cf. Section 8.4 and Exercise 9.2.7).

In order to describe the behavior of iterates for a nonhyperbolic linear map we should first understand what happens inside the subspace E^0. This subspace splits into root subspaces of E_1, E_{-1}, and $E_{\lambda,\bar{\lambda}}$ for $|\lambda| = 1$, $\lambda \neq \pm 1$. Inside each of those subspaces there is a corresponding invariant eigenspace which we will denote by $\tilde{E}_1$, $\tilde{E}_{-1}$, and $\tilde{E}_{\lambda,\bar{\lambda}}$, correspondingly. The first two of those spaces yield a fairly trivial behavior; namely, all points inside $\tilde{E}_1$ are fixed and those in $\tilde{E}_{-1} \setminus \{0\}$ are periodic with period two. A more interesting situation appears in $\tilde{E}_{\lambda,\bar{\lambda}}$ if λ is not real, say $\lambda = e^{2\pi i\varphi}$. If one of those spaces is not empty then A has an invariant plane such that in an appropriate coordinate system the map acts in that plane as a rotation by the angle φ about the origin.

Every circle with center at the origin is invariant under the map. Thus in order to continue our analysis of linear maps we should first understand the behavior of iterates for a rotation of the circle. This is the first time in our survey when we will encounter the phenomenon of nontrivial recurrence, that is, iterates of a point coming back arbitrary close to the initial position without returning exactly to that position. A detailed study of rotations is our next task. The structure of general linear maps is discussed in Exercises 1.2.4 and 1.2.5.

Exercises

1.2.1. *Prove (1.2.3).*

1.2.2. *Show that the eigenvalues of a linear map A depend continuously on A. Do they depend smoothly?*

1.2.3. *Show that hyperbolic linear maps are an open dense subset of the set of linear maps $\mathbb{R}^n \to \mathbb{R}^n$.*

1.2.4. *Suppose all eigenvalues of a linear map $A: \mathbb{R}^n \to \mathbb{R}^n$ have absolute value one. Then there exists an invariant subspace $C = C(A) \subset \mathbb{R}^n$ and a norm in $\mathbb{R}^n$ such that A acts in C as an isometry and for every vector $v \in \mathbb{R}^n \setminus C$ the norm $\|A^n v\|$ grows polynomially as $|n| \to \infty$, that is, for some positive integer k and $c > 0$*

$$\lim_{|n|\to\infty} \frac{\|A^n v\|}{\|v\|\,|n|^k} = c$$

(k and c may depend on v). Show how to determine the maximal value of k for a given map.

1.2.5. *Use Proposition 1.2.8 and Exercise 1.2.4 to describe the asymptotic behavior of points for an arbitrary invertible linear map in terms of the decomposition*

$$\mathbb{R}^n = E^+(A) \oplus E^-(A) \oplus E^0(A).$$

3. Rotations of the circle

We can use either multiplicative notation so that the circle is represented as the unit circle in the complex plane

$$S^1 = \{z \in \mathbb{C} \mid |z| = 1\} = \{e^{2\pi i\varphi} \mid \varphi \in \mathbb{R}\}$$

or additive notation

$$S^1 = \mathbb{R}/\mathbb{Z},$$

the factor group of the additive group of real numbers modulo the subgroup of integers. The logarithm map

$$e^{2\pi i\varphi} \mapsto \varphi$$

establishes an isomorphism between these representations. We will use the symbol R_α to denote the rotation by angle $2\pi\alpha$. In multiplicative notation

$$R_\alpha z = z_0 z \text{ with } z_0 = e^{2\pi i\alpha}.$$

Not surprisingly, in additive notation we have

$$R_\alpha x = x + \alpha \pmod 1,$$

where (mod 1) means that numbers which differ by an integer are identified. The iterates of the rotation are correspondingly

$$R_\alpha^n z = R_{n\alpha} z = z_0^n z \text{ or } R_\alpha^n x = x + n\alpha \pmod 1. \tag{1.3.1}$$

A crucial distinction appears between the cases of rational and irrational α.

In the former case, write $\alpha = p/q$, where p, q are relatively prime integers. Then $R_\alpha^q x = x$ for all x so R_α^q is the identity map and after q iterates the transformation simply repeats itself.

The latter case is much more interesting. We begin with two general definitions which belong to topological dynamics.

Definition 1.3.1. A topological dynamical system $f: X \to X$ is called *topologically transitive* if there exists a point $x \in X$ such that its orbit $\mathcal{O}_f(x) := \{f^n(x)\}_{n\in\mathbb{Z}}$ is dense in X.

The definitions for noninvertible and continuous-time systems are similar.

Definition 1.3.2. A topological dynamical system $f: X \to X$ is called *minimal* if the orbit of every point $x \in X$ is dense in X, or, equivalently, if f has no proper closed invariant sets.

Proposition 1.3.3. *If α is irrational then the rotation R_α is minimal.*

Proof. Let $A \subset S^1$ be the closure of an orbit. If the orbit is not dense, the complement $S^1 \smallsetminus A$ is a nonempty open invariant set which consists of disjoint intervals. Let I be the longest of those intervals (or one of the longest, if there are several of the same length). Since rotation preserves the length of any interval, the iterates $R_\alpha^n I$ do not overlap. Otherwise $S^1 \smallsetminus A$ would contain an interval longer than I. Since α is irrational, no iterates of I can coincide; because then an endpoint x of an iterate of I would come back to itself and we would have $x + k\alpha = x \pmod 1$ with $k\alpha = l$ an integer and $\alpha = l/k$ a rational number. Thus the intervals $R_\alpha^n I$ are all of equal length and all disjoint, but this is impossible because the circle has finite length and the sum of lengths of disjoint intervals can not exceed the length of the circle. □

Irrational rotations serve as the starting point for a number of very fruitful generalizations. Let us discuss one of them. The circle is a compact abelian group and the rotation can be represented in group terms as the group multiplication or left translation

$$L_{g_0}: G \to G, \quad L_{g_0} g = g_0 g. \tag{1.3.2}$$

The orbit of the unit element $e \in G$ is the cyclic subgroup $\{g_0^n\}_{n\in\mathbb{Z}}$ and it is easy to deduce from Proposition 1.3.3 that the circle does not have proper infinite closed subgroups.

Proposition 1.3.4. *If the translation L_{g_0} on a topological group G is topologically transitive then it is minimal.*

Proof. For $g, g' \in G$ denote by $A, A' \subset G$ the closures of the orbits of g and g', respectively. Now $g_0^n g' = g_0^n g(g^{-1}g')$, so $A' = Ag^{-1}g'$ and $A' = G$ if and only if $A = G$. □

Exercises

1.3.1. *Prove that the decimal expansion of the number 2^n may begin with any finite combination of digits.*

1.3.2. *Let G be a metrizable compact topological group. Suppose for some $g_0 \in G$ the translation L_{g_0} is topologically transitive. Prove that G is abelian.*

1.3.3. *Define the following metric d_2 on the group $\mathbb{Z}$ of all integers: $d_2(m,n) = \|m-n\|_2$ where*

$$\|n\|_2 = 2^{-k} \quad \textit{if } n = 2^k l \textit{ with an odd number } l.$$

The completion of $\mathbb{Z}$ with respect to that metric is called the group of 2-adic integers and is usually denoted by $\mathbb{Z}_2$. It is a compact topological group. Let $\mathbb{Z}_2^+$ be the closure of the even integers with respect to the metric d_2. $\mathbb{Z}_2^+$ is a subgroup of $\mathbb{Z}_2$ of index two.

Prove that for $g_0 \in \mathbb{Z}_2$ the translation $L_{g_0}\colon \mathbb{Z}_2 \to \mathbb{Z}_2$ is topologically transitive if and only if $g_0 \in \mathbb{Z}_2 \smallsetminus \mathbb{Z}_2^+$.

This is an example of a class of systems called *adding machines.* We will encounter them again in Section 15.4.

4. Translations on the torus

This is a generalization of rotations and a special case of group translations. This example plays the central role in the theory of completely integrable Hamiltonian systems, which we will touch upon at the end of the next section. The phase space is the n-dimensional torus

$$\mathbb{T}^n = \underbrace{S^1 \times \cdots \times S^1}_{n \text{ times}} = \mathbb{R}^n/\mathbb{Z}^n = \underbrace{\mathbb{R}/\mathbb{Z} \times \cdots \times \mathbb{R}/\mathbb{Z}}_{n \text{ times}}.$$

A natural fundamental domain for $\mathbb{R}^n/\mathbb{Z}^n$ is the unit cube:

$$I^n = \{(x_1,\dots,x_n) \in \mathbb{R}^n \mid 0 \le x_i \le 1 \text{ for } i = 1,\dots,n\}.$$

In order to represent the torus, one should identify opposite faces of I^n so that the point $(x_1,\dots,x_{i-1},0,x_{i+1},\dots,x_n)$ is identified with $(x_1,\dots,x_{i-1},1,x_{i+1},\dots,x_n)$ because these two points represent the same element of the factor group.

Similar to the case of the circle there are two convenient coordinate systems on $\mathbb{T}^n$, namely,

(1) multiplicative, where the elements of $\mathbb{T}^n$ are represented as $(z_1,\dots,z_n)$ with $z_i \in \mathbb{C}$ and $|z_i| = 1$ for $i = 1,\dots,n$, and
(2) additive, when they are represented by n-vectors $(x_1,\dots,x_n)$, where each coordinate is defined mod 1.

The correspondence $(x_1,\dots,x_n) \mapsto (e^{2\pi i x_1},\dots,e^{2\pi i x_n})$ establishes an isomorphism between these two representations. In additive notation let $\gamma = (\gamma_1,\dots,\gamma_n) \in \mathbb{T}^n$. The translation T_γ has the form

$$T_\gamma(x_1,\dots,x_n) = (x_1+\gamma_1,\dots,x_n+\gamma_n) \pmod 1.$$

If all coordinates of the vector γ are rational numbers, then T_γ is periodic. However, unlike the circle case, it is not true any more that minimality is the only alternative to periodicity. For example, if $n=2$ and $\gamma=(\alpha,0)$ where α is an irrational number then the torus $\mathbb{T}^2$ splits into a family of invariant circles $x_2 =$ const. and every orbit stays on one of these circles and fills it densely.

Proposition 1.4.1. *The translation T_γ is minimal if and only if the numbers $\gamma_1,\dots,\gamma_n$ and 1 are rationally independent, that is, if $\sum_{i=1}^n k_i\gamma_i$ is not an integer for any collection of integers $k_1,\dots,k_n$ except for $k_1 = k_2 = \cdots = k_n = 0$.*

Remark. One may give an algebraic proof of this proposition which involves the classification of all closed subgroups of $\mathbb{T}^n$ and induction on dimension. We prefer an analytical approach which anticipates some of the most fruitful methods used for the study of smooth dynamical systems and will be developed further in Section 4.2.

Before we proceed to the proof of this proposition we need to establish some general criteria for topological transitivity.

Lemma 1.4.2. *Let $f\colon X \to X$ be a continuous map of a locally compact separable metric space X into itself. The map f is topologically transitive if and only if for any two nonempty open sets $U, V \subset X$ there exists an integer $N = N(U,V)$ such that $f^N(U)\cap V$ is nonempty.*

Proof. Let f be topologically transitive and suppose the orbit of $x \in X$ is dense. Then, in particular, this orbit intersects both U and V so $f^n(x) \in U$, $f^m(x) \in V$, where, say $m \geq n$. Consequently $f^{m-n}(U) \cap V$ is nonempty. (Recall that $f^{-1}(A) := \{x \in X \mid f(x) \in A\}$.)

Now let us assume that the intersection condition holds. Let $U_1, U_2, \dots$ be a countable base of open subsets of X. This means that for any $x \in X$ and every open set with $x \in U \subset X$ there is an n such that $x \in U_n \subset U$. Let us furthermore choose U_1 in such a way that its closure $\bar{U}_1$ is compact. In order to prove topological transitivity, it is enough to construct an orbit which intersects every U_n. By assumption, there exists an integer N_1 such that $f^{N_1}(U_1) \cap U_2$ is nonempty. Let V_1 be a nonempty open set such that $\bar{V}_1 \subset U_1 \cap f^{-N_1}(U_2)$. Obviously $\bar{V}_1$ is compact. There exists an integer N_2 such that $f^{N_2}(V_1) \cap U_3$ is nonempty. Again, take an open set V_2 such that $\bar{V}_2 \subset V_1 \cap f^{-N_2}(U_3)$. By induction, we construct a nested sequence of open sets V_n such that $\bar{V}_{n+1} \subset V_n \cap f^{-N_{n+1}}(U_{n+2})$. The intersection $V = \bigcap_{n=1}^\infty \bar{V}_n = \bigcap_{n=1}^\infty V_n$ is nonempty because the $\bar{V}_n$ are compact. If $x \in V$ then $f^{N_{n-1}}(x) \in U_n$ for every $n \in \mathbb{N}$. $\square$

Corollary 1.4.3. *A continuous open map f of a locally compact separable metric space is topologically transitive if and only if there are no two disjoint open nonempty f-invariant sets.*

Proof. If $U, V \subset X$ are open then the invariant open sets $\tilde{U} := \bigcup_{n\in\mathbb{Z}} f^n(U)$ and $\tilde{V} := \bigcup_{n\in\mathbb{Z}} f^n(V)$ are not disjoint, so that $f^n(U) \cap f^m(V) \neq \varnothing$ for some $n, m \in \mathbb{Z}$, and $f^{n-m}(U) \cap V \neq \varnothing$. □

Corollary 1.4.4. *If $f: X \to X$ is topologically transitive then there is no f-invariant nonconstant continuous function $\varphi: X \to \mathbb{R}$.*

Proof. Let $\varphi: X \to \mathbb{R}$ be f-invariant, that is, $\varphi(f(x)) = \varphi(x)$ for all $x \in X$. Since it is not a constant, there exists $t \in \mathbb{R}$ such that both $\{x \in X \mid \varphi(x) > t\}$ and $\{x \in X \mid \varphi(x) < t\}$ are nonempty. Since φ is invariant, these sets are also invariant. Since φ is continuous, they are open. □

Proof of Proposition 1.4.1. First let us show that if $\sum_{i=1}^n k_i\gamma_i = k$ and not all integers $k_1, \dots, k_n$ are zero, then T_γ is not topologically transitive. We will construct a continuous T_γ-invariant function and then use Corollary 1.4.4. Our function is $\varphi(x) = \sin 2\pi \left(\sum k_i x_i\right)$. It is defined on $\mathbb{T}^n$ by the periodicity of $\sin(x)$ and it is not constant by our assumption. On the other hand φ is invariant because

$$\begin{aligned}\varphi(T_\gamma x) &= \sin(2\pi \sum k_i(x_i + \gamma_i)) \\ &= \sin(2\pi \sum k_i x_i + 2\pi k) = \sin(2\pi \sum k_i x_i) = \varphi(x).\end{aligned}$$

To prove the converse it is enough to show that rational independence of $\gamma_1, \dots, \gamma_n, 1$ implies topological transitivity of T_γ. Since T_γ is a translation on a group, this will imply minimality by Proposition 1.3.4. We will use Corollary 1.4.3 and prove the contrapositive. Let U, V be two disjoint nonempty open T_γ-invariant sets. Let χ be the characteristic function of U. By invariance of U we have

$$\chi(T_\gamma x) = \chi(x).$$

Take the Fourier expansion

$$\chi(x_1, \dots, x_n) = \sum_{(k_1,\dots,k_n)\in\mathbb{Z}^n} \chi_{k_1,\dots,k_n} \exp\left(2\pi i \sum_{j=1}^n k_j x_j\right)$$

of χ. Then

$$\begin{aligned}\chi(T_\gamma x) &= \chi(x_1 + \gamma_1, \dots, x_n + \gamma_n) \\ &= \sum_{(k_1,\dots,k_n)\in\mathbb{Z}^n} \chi_{k_1,\dots,k_n} \exp\left(2\pi i \sum_{j=1}^n k_j(x_j + \gamma_j)\right) \\ &= \sum_{(k_1,\dots,k_n)\in\mathbb{Z}^n} \chi_{k_1,\dots,k_n} \exp\left(2\pi i \sum_{j=1}^n k_j\gamma_j\right) \exp\left(2\pi i \sum_{j=1}^n k_j x_j\right).\end{aligned}$$

Invariance of χ and uniqueness of the Fourier expansion imply that for every $k_1, \dots, k_n$ we have $\chi_{k_1,\dots,k_n} = \chi_{k_1,\dots,k_n} \exp\left(2\pi i \sum_{j=1}^n k_j\gamma_j\right)$ or

$$\chi_{k_1,\dots,k_n}\left(1 - \exp 2\pi i \sum_{j=1}^{n} k_j\gamma_j\right) = 0,$$

which means that either $\chi_{k_1,\dots,k_n} = 0$ or $\exp(2\pi i \sum k_i\gamma_i) = 1$, that is, $\sum k_i\gamma_i$ is an integer. Since both U and its complement contain nonempty open sets, which have positive Lebesgue measure, χ is not constant almost everywhere. Therefore there is some $(k_1, \dots, k_n) \neq 0$ such that $\chi_{k_1,\dots,k_n} \neq 0$ and hence $\sum k_i\gamma_i$ is an integer. □

Let us point out a relation between toral translations and linear maps. Let $A: \mathbb{R}^{2n} \to \mathbb{R}^{2n}$ be given by the matrix

$$\begin{pmatrix} R_{\varphi_1} & & 0 \\ & \ddots & \\ 0 & & R_{\varphi_n} \end{pmatrix}.$$

Using a complex coordinate $z_k = x_{2k-1} + ix_{2k}$ in each two-plane $0 = x_1 = \dots = x_{2k-2} = x_{2k+1} = \dots = x_{2n}$ for $k = 1, \dots, n$ we can write the map A as $A(z_1, \dots, z_n) = (e^{i\varphi_1}z_1, \dots, e^{i\varphi_n}z_n)$. Let $\rho = (\rho_1, \dots, \rho_n)$ be a vector with nonnegative coordinates. The torus $\mathbb{T}^k_\rho = \{|z_1| = \rho_1, \dots, |z_n| = \rho_n\}$ is invariant with respect to A and its dimension k is equal to the number of nonzero coordinates in ρ. The restriction of A to that torus is obviously just the translation T_γ where γ is the k-dimensional vector composed of all φ_i's for which $\rho_i \neq 0$.

Exercises

1.4.1. *Prove that for any translation T_γ and any $x \in \mathbb{T}^n$ the closure $C(x)$ of the orbit of x is a finite union of tori of dimension k, $0 \le k \le n$, and that the restriction of T_γ to $C(x)$ is minimal.*

1.4.2. *Let X be a compact metrizable space which is perfect, that is, does not have isolated points. Prove that if a homeomorphism $f: X \to X$ is topologically transitive, that is, for some point $x \in X$ the whole orbit $\mathcal{O}(x) = \{f^n(x) \mid n \in \mathbb{Z}\}$ is dense, then there exists a point $y \in X$ whose* positive semiorbit *$\mathcal{O}^+(y) = \{f^n(y) \mid n = 0, 1, 2, \dots\}$ is dense.*

1.4.3. *Construct an example showing that the assertion of Exercise 1.4.2 is not true if X is an arbitrary compact metrizable space.*

1.4.4. *Prove that the map $A_\alpha: \mathbb{T}^2 \to \mathbb{T}^2$, $A_\alpha(x, y) = (x + \alpha, y + x) \pmod 1$ is topologically transitive if and only if α is irrational.*

5. Linear flow on the torus and completely integrable systems

In this section we consider continuous-time analogs of examples from the past two sections. We begin with the following system of differential equations on the 2-torus (we use additive notation)

$$\frac{dx_1}{dt} = \omega_1 , \quad \frac{dx_2}{dt} = \omega_2. \tag{1.5.1}$$

This system of differential equations can be easily integrated explicitly. The resulting flow $\{T^t_\omega\}_{t\in\mathbb{R}}$ has the form

$$T^t_\omega(x_1, x_2) = (x_1 + \omega_1 t,\ x_2 + \omega_2 t) \pmod 1. \tag{1.5.2}$$

We will present a geometric picture of this flow. As we already mentioned, the torus $\mathbb{T}^2 = \mathbb{R}^2/\mathbb{Z}^2$ can be represented as the unit square $I^2 = \{(x_1, x_2) \mid 0 \le x_1 \le 1,\ 0 \le x_2 \le 1\}$ with pairs of opposite sides identified: $(x, 0) \sim (x, 1)$ and $(0, x) \sim (1, x)$. In this representation the integral curves of the system (1.5.1) are pieces of straight lines with slope $\gamma = \omega_2/\omega_1$. The motion along the orbits is uniform with instantaneous "jumps" to the corresponding points when the orbit reaches the boundary of the square (compare with the suspension construction in Section 0.3). If we consider the successive moments when an orbit intersects the circle $C_1 = \{x_1 = 0\}$, the x_2 coordinate changes by exactly $\gamma \pmod 1$ between two such returns. Thus by Proposition 1.3.3 if γ is irrational, the closure of every orbit contains the circle C_1 and since the images of this circle under the flow $\{T^t_\omega\}$ cover the whole torus, the flow is minimal in the sense similar to that of Definition 1.3.2, that is, every orbit is dense in $\mathbb{T}^2$. If γ is rational, then every orbit is closed, as becomes immediately clear from (1.5.2).

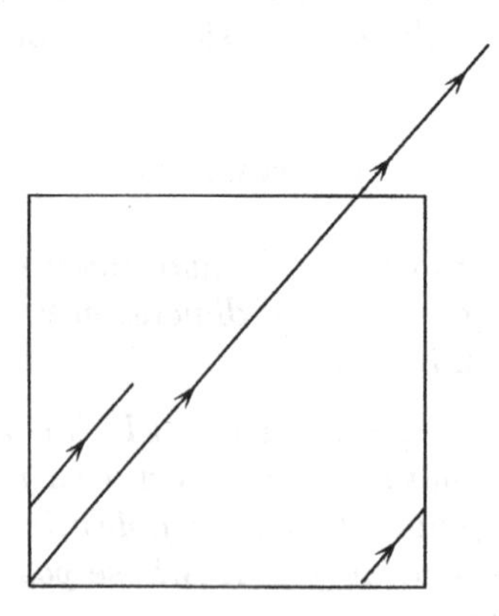

FIGURE 1.5.1. Linear flow on the torus

This example has a natural generalization to a torus of arbitrary dimension. Namely, let us consider the following system of differential equations on $\mathbb{T}^n$

$$\frac{dx_i}{dt} = \omega_i \text{ for } i = 1, \ldots, n.$$

Again integration produces a one-parameter group of translations

$$T_\omega^t(x_1,\dots,x_n)=(x_1+t\omega_1,\dots,x_n+t\omega_n)\pmod 1. \tag{1.5.3}$$

Obviously the flow $\{T_\omega^t\}$ is minimal if for some t_0 the transformation $T_\omega^{t_0}$ is minimal. This remark together with Proposition 1.4.1 allows us to establish the criterion of minimality for this case.

Proposition 1.5.1. *The flow $\{T_\omega^t\}$ is minimal if and only if the numbers $\omega_1,\dots,\omega_n$ are rationally independent, that is, if $\sum_{i=1}^n k_i\omega_i\neq 0$ for any integers $k_1,\dots,k_n$ unless $k_1=\dots=k_n=0$.*

Proof. Since $T_\omega^t=T_{t\omega}$ minimality follows from Proposition 1.4.1 once we show that for some real t and for any nonzero integer vector $(k_1,\dots,k_n)$ the sum $\sum_{i=1}^n tk_i\omega_i$ is never an integer. But for any collection of integers $k_1,\dots,k_n,k$ there is only one value of t such that

$$t\sum_{i=1}^n k_i\omega_i=k,$$

namely, $t=k/\sum k_i\omega_i$, unless $\sum_{i=1}^n k_i\omega_i=0$, which cannot be the case by our assumption. The proof in one direction is finished because there are only countably many different integer vectors $k_j,\dots,k_n,k$ and uncountably many values of t to take care of.

On the other hand, if $\sum_{i=1}^n k_i\omega_i=0$ for some nonzero vector $(k_1,\dots,k_n)$, then the function $\sin 2\pi\left(\sum_{i=1}^n k_ix_i\right)$ is continuous, nonconstant, and invariant under the flow $\{T_\omega^t\}$. □

Similarly to the discrete-time case, there is a close connection between linear flows on tori and solutions of certain linear systems of ordinary differential equations with constant coefficients. Let A be a $2n\times 2n$ real matrix with n pairs of distinct purely imaginary eigenvalues $\pm i\alpha_i$, $i=1,\dots,n$. Consider the system of ordinary differential equations

$$\frac{dx}{dt}=Ax. \tag{1.5.4}$$

The solution of (1.5.4) is

$$x(t)=e^{tA}x(0).$$

By a coordinate change the matrix A can be transformed into

$$\begin{pmatrix} 0 & \alpha_1 & & & \\ -\alpha_1 & 0 & & 0 & \\ & & \ddots & & \\ & 0 & & 0 & \alpha_n \\ & & & -\alpha_n & 0 \end{pmatrix}$$

so that the solution for the time t is the linear transformation given by the matrix

$$\begin{pmatrix} R_{t\alpha_1} & & & \\ & \ddots & & 0 \\ & & & \\ 0 & & \ddots & \\ & & & R_{t\alpha_n} \end{pmatrix}.$$

Arguing exactly as at the end of the previous section we split $\mathbb{R}^{2n}$ into invariant tori where the flow defined by (1.5.4) acts by translations.

A more important class of dynamical systems related to linear flows on the torus comes from Hamiltonian mechanics. Let us recall the most classical definition of a Hamiltonian system. Let H be a smooth function defined in an open subset U of Euclidean space $\mathbb{R}^{2n}$. The Hamiltonian equations for the Hamiltonian function H are

$$\begin{aligned} \frac{dx_i}{dt} &= \frac{\partial H}{\partial x_{i+n}}, \qquad i = 1,\dots,n, \\ \frac{dx_i}{dt} &= -\frac{\partial H}{\partial x_{i-n}}, \qquad i = n+1,\dots,2n. \end{aligned} \tag{1.5.5}$$

The more general definition involves a $2n$-dimensional smooth manifold M, a closed nondegenerate differential 2-form Ω on TM, that is, a form such that the exterior derivative $d\Omega = 0$ and the n-fold wedge product $\Omega^n \neq 0$, and a smooth function $H: M \to \mathbb{R}$. Then the Hamiltonian vector field V_H is defined by the condition that

$$\Omega(\xi, V_H(x)) = dH(\xi) \tag{1.5.6}$$

for $x \in M$, $\xi \in T_xM$. The Euclidean case (1.5.5) corresponds to $\Omega = \sum_{i=1}^{n} dx_i \wedge dx_{i+n}$. A more motivated and detailed description of Hamiltonian systems will be given in Section 5.5c.

It is not appropriate now to discuss a general notion of complete integrability, with its historical development and implications. It is enough for us to say that a Hamiltonian system is *completely integrable in the open set* $V \subset M$ if one can introduce coordinates $(I, \varphi) = (I_1, \dots, I_n; \varphi_1, \dots, \varphi_n)$ in V such that $\varphi_1, \dots, \varphi_n$ are defined mod 1, $I \in U \subset \mathbb{R}^n$, $\Omega = \sum_{i=1}^{n} d\varphi_i \wedge dI_i$ in these coordinates, and the Hamiltonian H depends only on I. Such coordinates are usually called *action–angle coordinates* for H. One immediately sees that in action–angle coordinates (1.5.6) implies that $V_H(I,\varphi) = (0, \dots, 0, -\partial H/\partial I_1, \dots, -\partial H/\partial I_n)$. Thus the action variables $I_1, \dots, I_n$ are preserved by the Hamiltonian flow and on each torus $\mathbb{T}_n^c = \{I_i = c_i, i = 1, \dots, n\}$ the flow is linear. However, unlike the case of the linear ordinary differential equation, the *frequency vector* $\omega(I) = (\partial H/\partial I_1, \dots, \partial H/\partial I_n)$ is usually different for different values of $c = (c_1, \dots, c_n)$. Thus in general, a completely integrable system in V looks like a collection of invariant tori with linear flows whose frequency vector and hence the type of recurrence changes from one torus to another. We will return to completely integrable Hamiltonian systems in Section 5.5c. In particular the Liouville Theorem 5.5.21 explains why the above notion of complete integrability is natural.

Exercise

1.5.1. *Consider a Lissajous figure on the plane*

$$x(t) = A\sin(t+\varphi)\ ,\ y(t) = B\sin(\omega t+\psi) \quad (t \in \mathbb{R}).$$

Prove that if ω *is irrational then for any phases* φ, ψ *the set* $\{(x(t), y(t))\}_{t\in\mathbb{R}}$ *is dense in the rectangle* $|x| \leq A$, $|y| \leq B$.

6. Gradient flows

Let $S^2 = \{(x, y, z) \mid x^2+y^2+z^2 = 1\}$ be the standard unit sphere in $\mathbb{R}^3$. We will consider the flow that moves every point downward (or "southward", if we think of S^2 as the surface of the globe and take the earth's axis to be vertical) along a great circle (meridian) connecting the point $(0,0,1)$ ("the north pole") and $(0,0,-1)$ ("the south pole"). The speed of the motion is equal to the derivative of the vertical coordinate along the meridian. In other words, our flow is generated by integrating the following vector field v on the sphere:

$$v(x, y, z) = (xz, yz, -x^2 - y^2). \tag{1.6.1}$$

To see this note that the downward unit vector tangent to the sphere at (x, y, z) is given by $(xz, yz, -(x^2+y^2))/\sqrt{x^2+y^2}$. The absolute value of its z-coordinate $\sqrt{x^2+y^2}$ gives the norm of the gradient vector, which is hence given by (1.6.1). The two poles are the only zeroes of this vector field and consequently they are fixed points for the flow. It is almost obvious that every point except for the north pole asymptotically approaches the south pole as time goes to plus infinity. In fact this convergence is exponential. Similarly, every point except for the south pole exponentially approaches the north pole as time goes to minus infinity.

To generalize this construction, let us consider a Riemannian metric on a compact smooth manifold M and a real-valued function F on M. At each point $x \in M$ that is not a critical point for F one can define the unique direction of fastest increase for F, that is, the unit tangent vector $\zeta(x) \in T_xM$ such that $\mathcal{L}_{\zeta(x)}F = \max_{\eta\in T_xM} \mathcal{L}_\eta F/\|\eta\|$, where $\mathcal{L}_\eta F$ denotes the Lie (directional) derivative of the function F along the vector η.

We define the gradient vector field ∇F by

$$\nabla F(x) = \begin{cases} \mathcal{L}_{\zeta(x)}F \cdot \zeta(x) & \text{if } x \text{ is noncritical,} \\ 0 & \text{if } x \text{ is critical.} \end{cases}$$

Suppose that in local coordinates $(x_1, \ldots, x_n)$ the Riemannian metric has the form $ds^2 = \sum g_{ij}(x_1, \ldots, x_n)dx_i dx_j$. Then

$$\nabla F(x_1, \ldots, x_n) = G^{-1}(x)\left(\frac{\partial F}{\partial x_1}, \ldots, \frac{\partial F}{\partial x_n}\right),$$

where $G(x) = \{g_{ij}(x)\}$ and G^{-1} is the inverse matrix, so it is a smooth vector field on M. The flow generated by the gradient vector field ∇F is called the *gradient flow* of F.

From calculus we know that the gradient is orthogonal to level sets of the function. Via coordinate calculations this is still true in this setting:

Lemma 1.6.1. *The gradient vector field is orthogonal to the level sets.*

Our first example was the gradient flow on the two-sphere provided with the Riemannian metric induced from the standard Euclidean metric in $\mathbb{R}^3$ for the function $F(x, y, z) = -z$. Let us consider two more examples.

Let M be the two-dimensional torus embedded into $\mathbb{R}^3$ as a doughnut or bagel standing up, that is, in the position of a doughnut being dunked, and F as before be the negative of the height function z, $F(x, y, z) = -z$. The function F has four critical points on the torus, a maximum A, two saddles B and C, and a minimum D. All orbits of the gradient flow other than those fixed points and six special orbits described below approach the minimum D as time goes to $+\infty$ and the maximum A as it goes to $-\infty$. Two special orbits connect A with B, two more connect B with C, and the last two connect C with D.

Now let us tilt this torus a little bit, that is, change our embedding but keep the function F the same. Equivalently we can consider the same embedding but the function $F = -z + \epsilon x$ for small $\epsilon > 0$. Four critical points remain, as well as the special orbits connecting the maximum with the upper saddle and the lower saddle with the minimum. However, the orbits connecting the two saddles disappear. Instead of these two orbits we will have four: two connecting the maximum with the lower saddle and two connecting the upper saddle with the minimum.

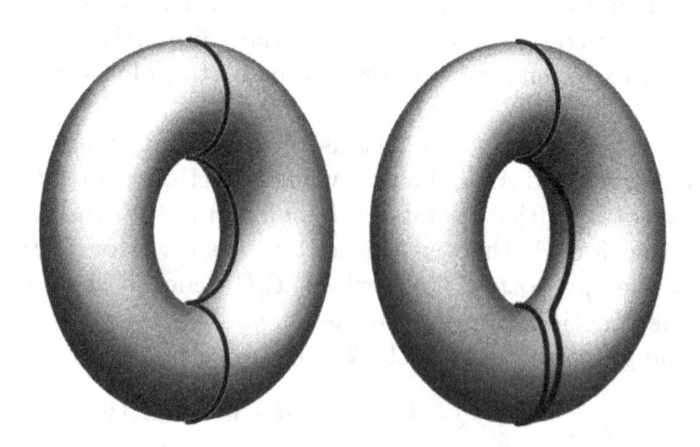

FIGURE 1.6.1. Gradient flows on the torus

Some features of the asymptotic behavior observed in our three examples remain true for general gradient flows. In order to describe those features we need some general definitions from topological dynamics. We will consider a topological dynamical system defined on the phase space X with discrete or continuous time.

Definition 1.6.2. A point $y \in X$ is called an *ω-limit point* (correspondingly, an *α-limit point*) for a point $x \in X$ if there exists a sequence of moments of time going to $+\infty$ (correspondingly, to $-\infty$) such that the images of x converge to y.

The set of all ω-limit (α-limit) points for x is denoted by $\omega(x)$ (correspondingly, $\alpha(x)$) and is called its *ω-limit (α-limit) set.*

$\omega(x)$ and $\alpha(x)$ are obviously closed and invariant. It follows from the definition that for the dynamical system $\{\varphi^t\}$

$$\begin{aligned} \omega(x) &= \bigcap_{T=0}^{\infty} (\overline{\bigcup_{t\geq T} \varphi^t x}), \\ \alpha(x) &= \bigcap_{T=0}^{-\infty} (\overline{\bigcup_{t\leq T} \varphi^t x}). \end{aligned} \tag{1.6.2}$$

Thus if X is compact, the sets $\omega(x)$ and $\alpha(x)$ are nonempty and every point sooner or later comes to any given neighborhood of its ω-limit set and stays there.

Let us denote by $\omega_F(x)$ (correspondingly, $\alpha_F(x)$) the ω-limit (α-limit) set of the point $x \in M$ with respect to the gradient flow for the function F.

Proposition 1.6.3. *The sets $\omega_F(x)$ and $\alpha_F(x)$ consist of critical points of F, that is, fixed points of the gradient flow.*

Proof. Let $\{\varphi^t\}_{t\in\mathbb{R}}$ be the gradient flow of the function F. Note that $F \circ \varphi^t$ is nondecreasing (in t) and increases at noncritical points. Thus if $y \in X$ is a noncritical point for F then $F(\varphi^t(y)) > F(y)$ for any $t > 0$. Assume $y \in \omega_F(x)$. Fix $t_0 > 0$ and let $\delta_0 = F(\varphi^{t_0}(y)) - F(y)$. If $x_n \to y$ then by continuity of the gradient flow $F(\varphi^{t_0}(x_n)) \to F(y) + \delta_0$. In particular, if $y \in \omega(x)$ then there is a sequence $t_n \to \infty$ such that $\varphi^{t_n}(x) \to y$ and hence $F(\varphi^{t_0+t_n}(x)) > F(y) + \delta_0/2$ for sufficiently large n, and since F does not decrease along the orbits, $F(\varphi^t(x)) > F(y) + \delta_0/2$ for sufficiently large t. But this contradicts the convergence $\varphi^{t_n}(x) \to y$. □

Proposition 1.6.4. *For any $x \in M$ and any F the set $\omega_F(x)$ is either a single point or an infinite set.*

Proof. Since M is compact, the set $\omega_F(x)$ is nonempty. Suppose $\omega_F(x)$ is finite and $y, z \in \omega_F(x)$, $y \neq z$. We have $y = \lim \varphi^{t_n}(x)$ and $z = \lim \varphi^{s_n}(x)$, where as before $\{\varphi^t\}$ is the gradient flow. Let B be a ball around y and S the boundary of B such that $(B \cup S) \cap \omega_f(x) = \{y\}$. Since the orbit of x enters and leaves B infinitely many times the intersection $\mathcal{O}(x) \cap S$ is an infinite set and by compactness of S it contains a limit point which must belong to $\omega(x)$. □

Corollary 1.6.5. *If the function F has only isolated critical points then every orbit of the gradient flow of F converges to a critical point of F as $t \to +\infty$.*

We will see in Section 9.3 how this property of a gradient flow constructed in an auxiliary space helps in finding special orbits for certain dynamical systems.

From a certain formal point of view there is a duality between gradient flows and Hamiltonian dynamical systems. Let us mention it only for the most elementary Euclidean case. In $\mathbb{R}^{2n}$ provided with the standard Euclidean metric the standard Hamiltonian form $\Omega = \sum_{i=1}^{n} dx_i \wedge dx_{i+n}$ can be expressed via the metric and the operator

$$I = \begin{pmatrix} 0 & \mathrm{Id} \\ -\mathrm{Id} & 0 \end{pmatrix}.$$

Namely, for any two tangent vectors $\xi, \eta \in T_x\mathbb{R}^{2n}$

$$\Omega(\xi, \eta) = \langle \xi, I\eta \rangle,$$

where $\langle \cdot, \cdot \rangle$ is the Euclidean scalar product. Accordingly the Hamiltonian vector fields $V_H = (\partial H/\partial x_{n+1}, \ldots, \partial H/\partial x_{2n}, -\partial H/\partial x_1, \ldots, -\partial H/\partial x_n)$ can be written as $V_H = I\,\nabla H$. It is curious to point out that the types of asymptotic behavior for a Hamiltonian vector field on a compact energy manifold $H = \text{const.}$ and for a gradient flow are in a sense diametrically opposite. Whereas for gradient vector fields the only recurrent behavior is represented by fixed points, for Hamiltonian systems nontrivial recurrence is a rule. We have seen this already for completely integrable Hamiltonian systems in the previous section. In general this fact follows from the Liouville Theorem (Proposition 5.5.12) and the Poincaré Recurrence Theorem (Theorem 4.1.19).

Exercises

1.6.1. *Prove that the ω-limit set of any point with respect to a gradient flow is connected.*

1.6.2. *Give an example of a C^∞ function F on a compact manifold and a point $x \in M$ such that $\omega_F(x)$ contains more than one point.*

1.6.3. *For every $g \geq 1$ construct a C^∞ function on the compact orientable surface of genus g with exactly 3 critical points. Describe the dynamics of the gradient flow for that function.*

7. Expanding maps

Consider the following noninvertible map E_2 of the circle: in multiplicative notation

$$E_2(z) = z^2 , \qquad |z| = 1,$$

or in additive notation

$$E_2(x) = 2x \pmod 1). \tag{1.7.1}$$

Algebraically this map represents an endomorphism of the group $S^1 = \mathbb{R}/\mathbb{Z}$ onto itself. Geometrically it is a double cover of S^1.

This is the first example where we will encounter simultaneously and in an essential way nontrivial recurrence as in Sections 1.3–1.5 and different asymptotic behavior for different orbits as in Sections 1.2. and 1.6. The combination of these two phenomena makes the orbit structure for this seemingly very simple transformation much more complicated than everything we have seen so far.

Definition 1.7.1. For a transformation $f: X \to X$ we let $P_n(f)$ be the number of periodic points of f with (not necessarily minimal) period n, that is, the number of fixed points for f^n.

The following proposition uncovers some of the features of the complicated orbit structure for E_2.

Proposition 1.7.2. $P_n(E_2) = 2^n - 1$, *periodic points for* E_2 *are dense in* S^1, *and* E_2 *is topologically transitive.*

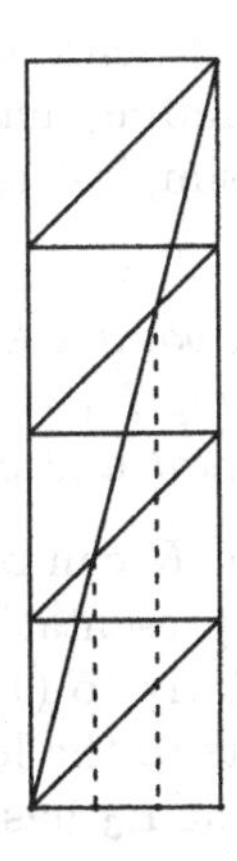

FIGURE 1.7.1. Periodic points for an expanding map

Proof. If $E_2^n(z) = z$, then $z^{2^n} = z$, and $z^{2^n - 1} = 1$. Thus every root of unity of order $2^n - 1$ is a periodic point for E_2 of period n. There are exactly $2^n - 1$ such roots of unity. Furthermore, they are uniformly spread over the circle with equal intervals and when n becomes large these intervals become small.

To prove topological transitivity we will consider the binary intervals

$$\Delta_n^k = \left[\frac{k}{2^n}, \frac{k+1}{2^n}\right] \quad \text{for } n = 1, \dots \quad \text{and} \quad k = 0, 1, \dots, 2^n - 1.$$

Let $x = 0.x_1x_2\dots$ be the binary representation of the number $x \in [0,1]$. Then $2x = x_1.x_2x_3\dots = 0.x_2x_3\dots \pmod 1$. Thus

$$E_2(x) = 0.x_2x_3\dots \pmod 1. \tag{1.7.2}$$

Let $k_1\dots k_n$ be the binary representation of the integer k, maybe with several zeroes at the beginning. Then $x \in \Delta_n^k$ if and only if $x_i = k_i$ for $i = 1, \dots n$. Thus $E_2^n(\Delta_n^k) = S^1$ and since every interval $I \subset S^1$ contains a binary interval, $E_2^n(I) = S^1$ for some n. Thus for any nonempty open sets U, V there is an $n \in \mathbb{N}$ such that $E_2^n(U) \cap V$ is nonempty and by Lemma 1.4.2 E_2 is topologically transitive. □

The maps

$$E_m: x \mapsto mx \pmod 1,$$

where m is an integer of absolute value greater than one, represent a straightforward generalization of the map E_2. Not surprisingly these maps are also topologically transitive and have dense sets of periodic orbits. The proof of Proposition 1.7.2 holds verbatim with the replacement of the binary representation by the representation with base m for positive m and with a very minor modification for negative m.

Furthermore, besides periodic and dense orbits there are other types of asymptotic behavior for orbits of expanding maps. One can construct such orbits for E_2 (cf. Exercise 1.7.5) but the simplest and most elegant example appears for the map E_3.

Proposition 1.7.3. *There exists a point $x \in S^1$ such that in additive notation its ω-limit set with respect to the map E_3 is the standard middle-third Cantor set K. In particular K is E_3-invariant and contains a dense orbit.*

Proof. The middle-third Cantor set K can be described as the set of all points on the unit interval which have a representation in base 3 with only 0's and 2's as digits (see Exercise 1.7.4). Similarly to (1.7.2) in the base 3 representation the map E_3 acts as the shift of digits to the left. This implies that the set K is E_3-invariant. It remains to show that E_3 has a dense orbit in K.

Every point in K has a unique representation in base three without ones. Let $x \in K$ and

$$0.x_1x_2x_3\dots \tag{1.7.3}$$

be such a representation. Let $h(x)$ be the number whose representation in base two is $0.\frac{x_1}{2}\frac{x_2}{2}\frac{x_3}{2}\dots$, that is, it is obtained from (1.7.3) by replacing twos by ones. Thus we have constructed a map $h: K \to [0,1]$ which is continuous, monotone (that is, $x > y$ implies $h(x) > h(y)$), and one-to-one except for the fact that binary rationals have two preimages each. Furthermore $h \circ E_3 = E_2 \circ h$. Let $D \subset [0,1]$ be a dense set of points which does not contain binary rationals. Then $h^{-1}(D)$ is dense in K. This immediately follows from the fact that if Δ is an open interval such that $\Delta \cap K \neq \varnothing$ then $h(\Delta)$ is a nonempty interval, open, closed, or semiclosed. Now take any $x \in [0,1]$ whose E_2 orbit is dense; the E_3 orbit of $h^{-1}(x) \in K$ is dense in K. □

Let us emphasize again a crucial difference between all our earlier examples and the expanding maps. In most of the previous examples either the recurrent behavior was very simple, that is, only fixed points, as for contracting maps, hyperbolic linear maps, and gradient flows, or, if nontrivial recurrence was present, all recurrent orbits behaved similarly as for translations and linear flows on the torus. True, for general completely integrable systems different orbits behave differently and nontrivial recurrence takes place. But the phase space of such a system splits into invariant pieces (tori) and all orbits on each torus have the same structure. By contrast, for expanding maps orbits with different behavior (such as periodic, dense, or with a Cantor closure) are interspersed and cannot be separated. This makes the orbit structure very complicated and asymptotic behavior of an *individual* orbit very sensitive to the initial condition and unstable. Furthermore any two orbits will diverge from each other exponentially until they are separated some distance δ. Consequently it is impossible to predict the behavior of an orbit for a long time if the initial position is known only with limited accuracy. For example, iterating E_2 on a computer will clearly yield only as many useful iterations as there are significant binary digits in the initial data. Furthermore any increase in accuracy will only yield a very modest increase in the time over which one can make reasonable predictions: Although doubling the number of significant digits in the initial data and the computation is likely to double the time span of possible prediction, the required improvement in the measurement of the initial state is of astronomical (and illusory) magnitude. Likewise cutting the initial error in half yields only one more step of valid iteration.

Rather surprisingly we will see later in Section 2.4 that the orbit structure *as a whole* is remarkably stable in a certain sense.

There are also important examples of one-dimensional maps which are not expanding. Here is one that we will encounter in the exercises and many times later. For $\lambda \in \mathbb{R}$ let $f_\lambda : \mathbb{R} \to \mathbb{R}$, $f_\lambda(x) := \lambda x(1-x)$. For $0 \le \lambda \le 4$ the f_λ map the unit interval $I = [0,1]$ into itself. The family f_λ, $\lambda \in [0,4]$, is referred to as the quadratic family. It is by far the most popular model in one-dimensional dynamics, both real and complex (in the latter case the maps are extended to $\mathbb{C}$).[1]

Exercises

1.7.1. *Calculate the number of periodic points of period n for the map E_m.*

1.7.2. *Consider the family f_λ.*

(1) *Calculate $P_n(f_4)$.*
(2) *Calculate $P_n(f_3)$,*
(3) *Show that for $\lambda > 3$ there is a period-two orbit.*
(4) *Show that for $\lambda \in (3, 1+\sqrt{6}]$ there are no points of period higher than two.*

1.7.3. *Show that for any point $x \in S^1$ the set*

$$P_x = \{y \in S^1 \mid \exists n \in \mathbb{N} \text{ such that } E_m^n(y) = x\}$$

is dense in S^1.

1.7.4. *Show that the middle-third Cantor set is the set of points on the unit interval which have a representation in base 3 with only 0's and 2's as digits.*

1.7.5*. *Consider the set T of all points on the unit interval which have a binary representation without two successive zeroes. Prove that T is a perfect nowhere dense set invariant with respect to the map E_2. Prove that there is a point $x \in T$ whose orbit with respect to E_2 is dense in T.*

1.7.6*. *Find a point $x \in S^1$ such that $E_3^n(x) \in S^1 \setminus K$ for $n = 0, 1, \dots$ and the orbit closure consists of the orbit itself and the middle-third Cantor set K. In other words, the points of the orbit are isolated in the orbit closure but accumulate to the perfect set K.*

8. Hyperbolic toral automorphisms

Hyperbolic toral automorphisms are an invertible analog of the expanding maps E_m. They have very similar properties and analyzing them will give us a chance to preview some of the methods used in the theory of hyperbolic dynamical systems.

We consider the following linear map of $\mathbb{R}^2$:

$$L(x, y) = (2x + y, x + y).$$

If two vectors (x, y) and (x', y') represent the same element of $\mathbb{T}^2$, that is, if $(x - x', y - y') \in \mathbb{Z}^2$, then $L(x, y)$ and $L(x', y')$ also represent the same element of $\mathbb{T}^2$. Thus L defines a map $F_L \colon \mathbb{T}^2 \to \mathbb{T}^2$:

$$F_L(x, y) = (2x + y, x + y) \pmod 1.$$

The map F_L is invertible because the matrix $\begin{pmatrix} 2 & 1 \\ 1 & 1 \end{pmatrix}$ has determinant one, so L^{-1} also has integer entries. Moreover F_L is an automorphism of the abelian group $\mathbb{T}^2 = \mathbb{R}^2/\mathbb{Z}^2$. Before we proceed to the study of the map F_L let us discuss some properties of the linear map L.

First, the eigenvalues of L are

$$\lambda_1 = \frac{3+\sqrt{5}}{2} > 1 \text{ and } \lambda_1^{-1} = \lambda_2 = \frac{3-\sqrt{5}}{2} < 1.$$

Since the matrix L is symmetric, the eigenvectors are orthogonal. The eigenvectors corresponding to the first eigenvalue belong to the line $y = \frac{\sqrt{5}-1}{2}x$. The family of lines parallel to it is invariant under L and L uniformly expands distances on those lines by a factor λ_1. Similarly, there is an invariant family of contracting lines $y = \frac{-\sqrt{5}-1}{2}x + \text{const}$.

Figure 1.8.1 gives an idea of the action of F_L on the fundamental square $I = \{(x,y) \mid 0 \le x < 1,\ 0 \le y \le 1\}$. The lines with arrows show the eigendirections.

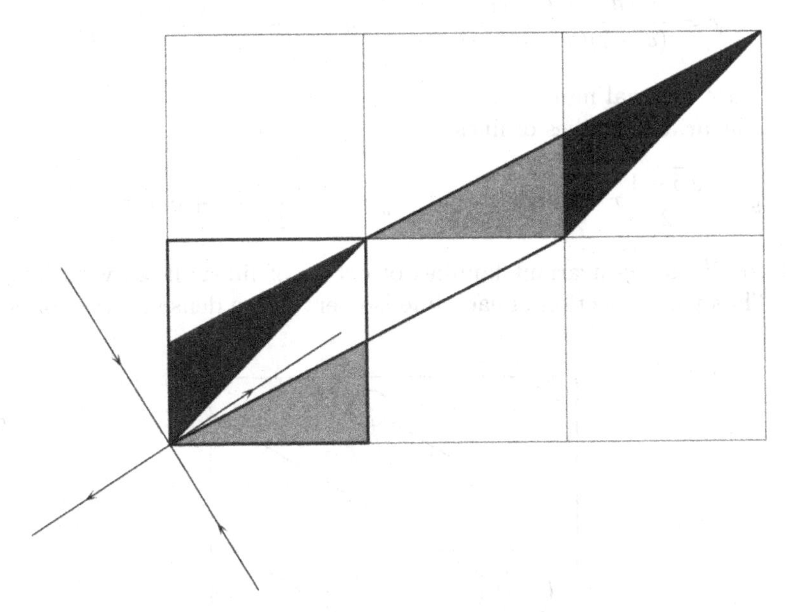

FIGURE 1.8.1. The hyperbolic toral map

Proposition 1.8.1. *Periodic points of F_L are dense, F_L is topologically transitive, and $P_n(F_L) = \lambda_1^n + \lambda_1^{-n} - 2$, where $P_n(F_L)$ is as in Definition 1.7.1.*

Proof. First, let us show that points with rational coordinates are periodic points of F_L. Let $x = s/q$, $y = t/q$, where s, t, q are integers. Then $F_L\left(\frac{s}{q}, \frac{t}{q}\right) = \left(\frac{2s+t}{q}, \frac{s+t}{q}\right)$, that is, it is a rational point whose coordinates also have denominator q. But there are only q^2 different points on $\mathbb{T}^2$ whose coordinates can be represented as rational numbers with denominator q and all iterates $F_L^n(s/q, t/q)$, $n = 0, 1, 2 \dots$, belong to that finite set. Thus they must repeat, that is, $F_L^n(s/q, t/q) = F_L^m(s/q, t/q)$ for some integers n, m. But since F_L is invertible, $F_L^{n-m}(s/q, t/q) = (s/q, t/q)$. Thus we have proved the density of periodic orbits. Before we proceed further, let us show that points with rational coordinates are the only periodic points for F_L.

Assume $F_L^n(x, y) = (x, y)$. But $F_L^n(x, y) = (ax + by,\ cx + dy) \pmod 1$ where a, b, c, d are integers. Thus we have

$$ax + by = x + k,$$
$$cx + dy = y + l$$

for some integers k, l. Since 1 is not an eigenvalue for L^n we determine (x, y) uniquely from a, b, c, d, k, l.

$$x = \frac{(d-1)k - bl}{(a-1)(d-1) - cb}, \qquad y = \frac{(a-1)l - ck}{(a-1)(d-1) - cb}.$$

Thus x, y are rational numbers.

The L-invariant families of lines

$$y = \frac{\sqrt{5} - 1}{2} x + \text{const.} \quad \text{and} \quad y = \frac{-\sqrt{5} - 1}{2} x + \text{const.} \tag{1.8.1}$$

project to $\mathbb{T}^2$ as F_L-invariant families of orbits of linear flows with irrational slopes. Thus the projection of each line is everywhere dense on the torus.

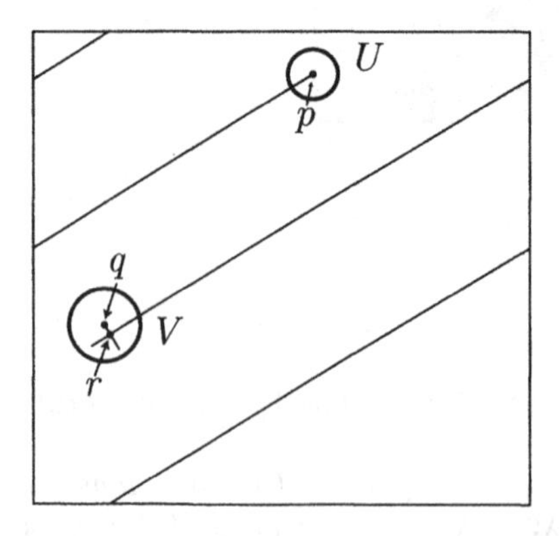

FIGURE 1.8.2. Topological transitivity

Now we are ready to prove that F_L is topologically transitive. Take arbitrary nonempty open subsets U, V of $\mathbb{T}^2$. Let $p \in U$, $q \in V$ be two periodic points and let n be their common period. The line from the first family passing through the point p is invariant under F_L^n and F_L^n expands it with coefficient $\lambda_1^n > 1$. Similarly, the line from the second family passing through the point q is F_L^n invariant and contracting. Let r be a point of intersection of these two lines. $F_L^{kn}(r) \to q$ as $k \to +\infty$ and $F_L^{kn}(r) \to p$ as $k \to -\infty$. Thus if k is large and positive then $F_L^{-kn}(r) \in U$, $F_L^{kn}(r) \in V$ and $F_L^{2kn}(U) \cap V$ is nonempty.

Finally, let us calculate $P_n(F_L)$. As before, if $F_L^n(x,y) = (x,y)$ then $(a-1)x + by$ and $cx + (d-1)y$ are integers. The map $G = F_L^n - \mathrm{Id}\colon (x,y) \mapsto ((a-1)x + by,\ cx + (d-1)y) \pmod 1$ is a well-defined noninvertible map of the torus onto itself. Every point of the torus, including the integer point $(0,0)$, has the same number of preimages which is $|\det(L^n - \mathrm{Id})| = |(\lambda_1^n - 1)(\lambda_1^{-n} - 1)| = \lambda_1^n + \lambda_1^{-n} - 2$. (We have used the fact that the number of points of $\mathbb{Z}^2$ in $G([0,1] \times [0,1])$ is the area of $G([0,1] \times [0,1])$.) This is exactly the number of different points $(x,y) \in \mathbb{T}^2$ for which both numbers $(a-1)x + by$ and $cx + (d-1)y$ are integers, that is, the number of fixed points for F_L^n. □

Let us compare the asymptotic properties of the map F_L with those of the toral translations discussed in Section 1.4.

According to Proposition 1.4.1 if the coordinates of the vector γ are rationally independent from 1, then the translation T_γ is topologically transitive; as we have just proved, the automorphism F_L possesses the same property. On the other hand, all orbits of T_γ are dense whereas for F_L dense orbits coexist with a dense set of periodic orbits, each of the latter obviously not being dense. Thus the recurrence of orbits represented by their density is uniform with respect to initial conditions for a translation and is highly sensitive to initial conditions for F_L.

Another aspect of asymptotic behavior is related to regularity of recurrence with respect to time. Topological transitivity implies that iterates of any open set from time to time intersect any other open set. A stronger version of recurrence is reflected by the following property.

Definition 1.8.2. A topological dynamical system $f\colon X \to X$ is called *topologically mixing* if for any two open nonempty sets $U, V \subset X$ there exists a positive integer $N = N(U,V)$ such that for every $n > N$ the intersection $f^n(U) \cap V$ is nonempty.

By Lemma 1.4.2 every topologically mixing map is topologically transitive. On the other hand, no translation T_γ is topologically mixing. This follows from the fact that translations preserve the natural metric on the torus induced by the standard Euclidean metric on $\mathbb{R}^n$ and from the following general criterion.

Lemma 1.8.3. *If a topological dynamical system $f\colon X \to X$ preserves a metric on X which generates the topology of X, then f is not topologically mixing.*

Proof. Fix an invariant metric on X, take three different points $x, y, z \in X$, and let δ be one tenth of the minimum of the pairwise distances between those points. Let U, V_1, V_2 be δ-balls around x, y, z correspondingly. Since f preserves the diameter of any set, the diameter of $f^n(U)$ does not exceed 2δ whereas the distance between any two points $p \in V_1$, $q \in V_2$ is greater than 7δ. Thus for each n either $f^n(U) \cap V_1$ or $f^n(U) \cap V_2$ is empty. □

By contrast, we have the following statement.

Proposition 1.8.4. *The automorphism F_L is topologically mixing.*

Proof. The expanding lines $y = \frac{\sqrt{5}-1}{2}x + \text{const.}$ are orbits of the linear flow T^t_ω, where $\omega = \left(1, \frac{\sqrt{5}-1}{2}\right)$. By Proposition 1.5.1 this flow is minimal. Any open set U contains a piece J of an expanding line. Let us fix $\epsilon > 0$. Then there exists $T = T(\epsilon)$ such that every segment of an expanding line of length T intersects any ϵ-ball on the torus. The existence of at least one such segment follows from the topological transitivity of the flow T^t_ω. But since all segments of a given length are translations of one another this property holds for all segments. Thus for any fixed open set V for any sufficiently large n, $f^n(J)$ intersects V and so does $f^n(U)$. □

Similarly to the expanding maps of the previous section the behavior of orbits of the map F_L depends on the initial point in a very sensitive way. Any two orbits will diverge from each other exponentially in the future or in the past and often in both past and future (always up to a certain distance). Again this presents difficulties for numerical calculations, for example, if an orbit is computed numerically when the initial condition is known to three decimal places, for example, to within 1/2000. Then, since $(\frac{3+\sqrt{5}}{2})^8 > 2000$, the error may be of order one already after eight iterations, rendering the calculation useless. Again, cutting the initial error in half barely yields one more step of valid iteration.

The construction of the example from this section can be generalized.

Let $L: \mathbb{R}^m \to \mathbb{R}^m$ be an $m \times m$ matrix with integer entries, with determinant $+1$ or -1, and without eigenvalues of absolute value 1, that is, a hyperbolic matrix. Then $L\mathbb{Z}^m = \mathbb{Z}^m$ and L is invertible on $\mathbb{Z}^m$, so L determines an invertible map of the m-torus $\mathbb{R}^m/\mathbb{Z}^m$ which has properties very similar to those of the map F_L discussed earlier. We will call such a map a *hyperbolic toral automorphism.* Furthermore if one drops the determinant condition, the resulting map still can be defined on the torus although it ceases to be invertible. Those maps are called *hyperbolic toral endomorphisms.* For $m = 1$ these are the expanding maps of the circle.

Exercises

1.8.1. *Calculate the number $P_n(H)$ of periodic points of period n for a general hyperbolic automorphism H of the torus. Show that* $\lim_{n\to\infty} P_n(H)/\lambda^n$ *exists for some* $\lambda > 0$.

1.8.2. *Every integer matrix L with determinant ± 1 defines a map of the torus. Show that the resulting map has finitely many periodic points of each period if and only if no eigenvalue of L is a root of unity.*

1.8.3*. *Show that periodic orbits of any hyperbolic toral endomorphism F_L are dense.*

9. Symbolic dynamical systems

a. Sequence spaces. We now introduce a class of topological dynamical systems of particular importance for the theory of smooth dynamical systems. One reason is that in many respects symbolic systems serve as models for smooth ones; it is often easier to see many properties in the symbolic case first and then to carry them over to the smooth case. Second, restrictions of some smooth dynamical systems to various (often important) invariant sets look very much like symbolic systems. Furthermore, symbolic systems can be used to "code" some smooth systems.

Let us consider for each natural number $N \geq 2$ the space

$$\Omega_N = \{\omega = (\dots, \omega_{-1}, \omega_0, \omega_1, \dots) \mid \omega_i \in \{0, 1, \dots, N-1\} \text{ for } i \in \mathbb{Z}\}$$

of two-sided sequences of N symbols and a similar one-sided space

$$\Omega_N^R = \{\omega = (\omega_0, \omega_1, \omega_2, \dots) \mid \omega_i \in \{0, 1, \dots, N-1\} \text{ for } i \in \mathbb{N}_0\}.$$

We can define a topology by noting that Ω_N is the direct product of $\mathbb{Z}$ copies of the finite set $\{0, 1, \dots, N-1\}$, each with the discrete topology, and using the product topology.

Notice that if we consider the finite set $\{0, 1, \dots, N-1\}$ as the finite group $\mathbb{Z}/n\mathbb{Z}$ then this product is naturally a compact abelian topological group.

Fix integers $n_1 < n_2 < \dots < n_k$ and numbers $\alpha_1, \dots, \alpha_k \in \{0, 1, \dots, N-1\}$ and call the subset

$$C_{\alpha_1, \dots, \alpha_k}^{n_1, \dots, n_k} = \{\omega \in \Omega_N \mid \omega_{n_i} = \alpha_i \text{ for } i = 1, \dots, k\} \tag{1.9.1}$$

a *cylinder* and the number k of fixed digits the *rank* of that cylinder. Cylinders in the space Ω_N^R are defined similarly.

An alternative way to define the topology in the space Ω_N (and similarly in Ω_N^R) is by declaring that all cylinders are open sets and that they form a base for the topology. Then every cylinder is also closed because the complement to a cylinder is a finite union of cylinders. The most general open set is a countable union of cylinders.

One more way is to introduce a metric

$$d_\lambda(\omega, \omega') = \sum_{n=-\infty}^{\infty} \frac{|\omega_n - \omega'_n|}{\lambda^{|n|}}$$

for any fixed $\lambda > 1$. These metrics are particularly convenient for large λ, say for $\lambda = 10N$, because any symmetric cylinder $C^{-n,\dots,n}_{\alpha_{-n},\dots,\alpha_n}$ of rank $2n+1$ is a ball with respect to such a metric. Ω_N is a perfect, totally disconnected compact space so it is homeomorphic to a Cantor set.

The different metrics d_λ not only define the same topology on Ω_N (although they are not equivalent as metrics) but also determine a Hölder structure. This means that the notion of Hölder-continuous function with respect to the metric d_λ does not depend on λ. That class of Hölder-continuous functions plays an extremely important role in applications to differentiable dynamics (see Chapters 19 and 20) and can be described as follows.

Let φ be a continuous complex-valued function defined on Ω_N or on a closed subset and write $\omega = (\dots,\omega_{-1},\omega_0,\omega_1,\dots)$ and $\omega' = (\dots,\omega'_{-1},\omega'_0,\omega'_1,\dots)$. Then for $n = 0, 1, \dots$ let

$$V_n(\varphi) := \max\{|\varphi(\omega) - \varphi(\omega')| \mid \omega_k = \omega'_k \text{ for } |k| \le n\}.$$

Since Ω_N is compact, φ is uniformly continuous and $V_n(\varphi) \to 0$ as $n \to \infty$. We will say that φ has *exponential type* if for some $a, c > 0$

$$V_n(\varphi) \le ce^{-an}.$$

It is not difficult to see that φ has exponential type if and only if it is Hölder continuous with respect to some (and hence any) metric d_λ. (See Exercise 1.9.1.)

An equivalent way to express independence of the class of Hölder-continuous functions of λ is to point out that for any λ, μ the identity map $\mathrm{Id}\colon (\Omega_N, d_\lambda) \to (\Omega_N, d_\mu)$ is Hölder continuous, that is, there exist $a, c > 0$ such that for any $\omega, \omega' \in \Omega_N$ we have

$$d_\mu(\omega,\omega') < c\, d_\lambda(\omega,\omega')^a. \tag{1.9.2}$$

All of the above discussion translates with obvious changes to the spaces Ω_N^R.

b. The shift transformation. Let us consider the left shift in Ω_N

$$\sigma_N\colon \Omega_N \to \Omega_N, \quad \sigma_N(\omega) = \omega' = (\dots,\omega'_0,\omega'_1,\dots), \tag{1.9.3}$$

where $\omega'_n = \omega_{n+1}$.

σ_N is a one-to-one map and takes cylinders into cylinders. Thus it is a homeomorphism of Ω_N. Sometimes the shift σ_N is called a *topological Bernoulli shift.*

Similarly let us define the *one-sided* N-shift$\sigma_N^R\colon \Omega_N^R \to \Omega_N^R$ by

$$\sigma_N^R(\omega_0,\omega_1,\omega_2,\dots) = (\omega_1,\omega_2,\omega_3,\dots).$$

This is a continuous noninvertible transformation of the space Ω_N^R onto itself.

The shifts σ_N^R and σ_N possess a number of properties already familiar to us from Sections 1.7 and 1.8.

Proposition 1.9.1. *Periodic points for the shifts σ_N and σ_N^R are dense in Ω_N and Ω_N^R, correspondingly, $P_n(\sigma_N) = P_n(\sigma_N^R) = N^n$, and both transformations σ_N and σ_N^R are topologically mixing.*

Proof. Periodic orbits for a shift are periodic sequences, that is, $(\sigma_N)^m \omega = \omega$ if and only if $\omega_{n+m} = \omega_n$ for all $n \in \mathbb{Z}$ and similarly for σ_N^R. In order to prove the density of periodic points it is enough to find a periodic point in every cylinder. Since any cylinder in Ω_N contains a symmetric cylinder of rank $2m+1$ for some m such as

$$C^{-m,\dots,m}_{\alpha_{-m},\dots,\alpha_m} =: C^m_\alpha,$$

where $\alpha = \alpha_{-m}, \dots, \alpha_m$, it is enough to consider only such cylinders. But the sequence obtained by simply repeating the finite sequence $\alpha_{-m}, \dots, \alpha_m$, that is, ω where $\omega_n = \alpha_{n'}$ for $|n'| \leq m$, $n' = n \pmod{2m+1}$, obviously lies in our cylinder and has period $2m+1$.

Every periodic sequence ω of period n is uniquely determined by its coordinates $\omega_0, \dots, \omega_{n-1}$. There are N^n different finite sequences $(\omega_0, \dots, \omega_{n-1})$.

Finally, in order to prove topological mixing, it is enough to show that for any $\alpha = \alpha_{-m}, \dots, \alpha_m$ and $\beta = \beta_{-m}, \dots, \beta_m$ and n sufficiently large $\sigma_N^n(C^m_\alpha)$ intersects C^m_β. We take $n > 2m+1$, say $n = 2m+k+1$ with $k > 0$. Consider any sequence ω such that

$$\omega_i = \alpha_i \text{ for } |i| \leq m, \quad \omega_i = \beta_{i-n} \text{ for } i = m+k+1, \dots, 3m+k+1.$$

Obviously, $\omega \in C^m_\alpha$ and $\sigma_N^n(\omega) \in C^m_\beta$.

The arguments for the one-sided shift are completely similar. □

Remark. The map $\pi\colon \Omega_2^R \to K$, $\pi(\omega_0, \omega_1, \dots) = 0.\beta(\omega_0)\beta(\omega_1)\dots$, where K is the middle-third Cantor set and $\beta(0) = 0$, $\beta(1) = 2$, is a homeomorphism and obviously $\pi \circ \sigma_2^R = E_3 \circ \pi$. Thus Proposition 1.7.3 implies topological transitivity of σ_2^R. This is the simplest example of the situation where a restriction of a smooth system to an invariant set looks like a shift. Accordingly, the correspondence h, described in the proof of Proposition 1.7.3, is the simplest example of coding. We will discuss this topic in greater detail in Sections 2.4 and 2.5.

Definition 1.9.2. The restriction of the shifts σ_N or σ_N^R to any closed, shift-invariant subset of Ω_N or Ω_N^R, respectively, is called a *symbolic dynamical system.*

Properties of symbolic dynamical systems vary widely. These systems provide a rich source of examples and counterexamples for topological dynamics and ergodic theory.

c. Topological Markov chains. Here we will consider only one special (although probably the most important) class of symbolic dynamical systems.

Let $A = (a_{ij})_{i,j=0}^{N-1}$ be an $N \times N$ matrix whose entries a_{ij} are either zeroes or ones. (We call such a matrix a 0-1 matrix.) Let

$$\Omega_A := \{\omega \in \Omega_N \mid a_{\omega_n \omega_{n+1}} = 1 \text{ for } n \in \mathbb{Z}\}. \tag{1.9.4}$$

In other words, the matrix A determines all admissible transitions between the symbols $0, 1, \dots, N-1$. The set Ω_A is obviously shift invariant.

Definition 1.9.3. The restriction

$$\sigma_N \restriction_{\Omega_A} =: \sigma_A$$

is called the *topological Markov chain*[1] determined by the matrix A. Sometimes σ_A is also called a *subshift of finite type.*

There is a useful geometric representation for topological Markov chains. Let us identify the symbols $0, 1, \dots, N-1$ with points $x_0, \dots, x_{N-1}$ and connect x_i with x_j by an arrow if $a_{ij} = 1$. This way we obtain a graph G_A with N vertices and a certain number of oriented edges. We will call a finite or infinite sequence of vertices of G_A an *admissible path* or *admissible sequence* if any two consecutive vertices in the sequence are connected by an oriented arrow. A point of Ω_A corresponds to a doubly infinite path in G_A with a marked origin; the topological Markov chain σ_A corresponds to moving the origin to the next vertex.

$$\begin{array}{ccc} x_1 & \longrightarrow & x_2 \\ \upharpoonleft\downharpoonright & \searrow & \downarrow \\ x_4 & \longrightarrow & x_3 \\ & & \circlearrowleft \end{array}$$

FIGURE 1.9.1. A Markov graph

The following simple combinatorial lemma is a key to the study of topological Markov chains:

Lemma 1.9.4. *For every $i, j \in \{0, 1, \dots, N-1\}$, the number N_{ij}^m of admissible paths of length $m+1$ that begin at x_i and end at x_j is equal to the entry a_{ij}^m of the matrix A^m.*

Proof. We will use induction on m. First, it follows immediately from the definition of the graph G_A that $N_{ij}^1 = a_{ij}$. Let us show that

$$N_{ij}^{m+1} = \sum_{k=0}^{N-1} N_{ik}^m a_{kj}. \tag{1.9.5}$$

For every $k \in \{0, \dots, N-1\}$ every admissible path of length $m+1$ connecting x_i and x_k produces exactly one admissible path of length $m+2$ connecting x_i and x_j by adding x_j to it, if and only if $a_{kj} = 1$. This proves (1.9.5). Now, assuming by induction that $N_{ij}^m = a_{ij}^m$ for all ij, we obtain from (1.9.5) that $N_{ij}^{m+1} = a_{ij}^{m+1}$. □

Every admissible closed path of length $m+1$ with marked origin, that is, a path that begins and ends at the same vertex of G_A, produces exactly one periodic point of σ_A of period m. Thus we have

Corollary 1.9.5. $P_n(\sigma_A) = \operatorname{tr} A^n$.

Topological Markov chains can be classified according to recurrence properties of various orbits they contain. Some principal elements of this classification are given in Exercises 1.9.4–1.9.9. Now we will concentrate on the most interesting special class of topological Markov chains that possess the strongest recurrence properties.

Definition 1.9.6. A 0-1 matrix A is called *transitive* if for some positive m all entries of the matrix A^m are positive numbers. We will call a topological Markov chain σ_A *transitive* if A is a transitive matrix.

Lemma 1.9.7. *If all entries of A^m are positive then for any $n \geq m$ all entries of A^n are positive too.*

Proof. First notice that if $a_{ij}^n > 0$ for all i, j, then for each j there is a k such that $a_{kj} = 1$. Otherwise $a_{ij}^n = 0$ for every n and i. Now use induction. Assume that $a_{ij}^n > 0$ for all i, j; then $a_{ij}^{n+1} = \sum_{k=0}^{N-1} a_{ik}^n a_{kj} > 0$ because $a_{kj} = 1$ for at least one k. □

Lemma 1.9.8. *If A is transitive and $\alpha = (\alpha_{-k}, \dots, \alpha_k)$ is admissible, that is, $a_{\alpha_i \alpha_{i+1}} = 1$ for $i = -k, \dots, k-1$, then the intersection $\Omega_A \cap C_\alpha^k =: C_{\alpha,A}^k$ is nonempty and contains a periodic point.*

Proof. Take m such that $a_{\alpha_k, \alpha_{-k}}^m > 0$. Then one can extend the sequence α to an admissible sequence of length $2k+m+1$ which begins and ends with α_{-k}. Repeating this sequence periodically we obtain a periodic point in $C_{\alpha,A}^k$. □

Proposition 1.9.9. *If A is a transitive matrix then the topological Markov chain σ_A is topologically mixing and its periodic orbits are dense in Ω_A.*

Proof. The density of periodic orbits follows immediately from Lemma 1.9.8.

In order to establish topological mixing it is enough to show that if for two sequences $\alpha = (\alpha_{-k}, \dots, \alpha_k)$ and $\beta = (\beta_{-k}, \dots, \beta_k)$ the cylinders $C_{\alpha,A}^k$ and $C_{\beta,A}^k$ are nonempty, then for any sufficiently large n the set $\sigma_A^n(C_{\alpha,A}^k) \cap C_{\beta,A}^k$ is also nonempty. Take $n \geq 2k+1+m$, where m is taken from Definition 1.9.6, say $n = 2k+1+m+l$ with $l \geq 0$. Then $a_{\alpha_k \beta_{-k}}^{m+l} > 0$ by Lemma 1.9.7, so one can construct an admissible sequence of length $4k+2+m+l$ whose first $2k+1$ symbols are identical to α and the last $2k+1$ symbols to β. By Lemma 1.9.8 this sequence can be extended to a periodic element of Ω_A which obviously belongs to $\sigma_A^n(C_{\alpha,A}^k) \cap C_{\beta,A}^k$. □

There is a natural class of symbolic systems more general than Markov chains.

Definition 1.9.10. Let $\mathcal{A}: \{1, \dots, N\}^{n+1} \to \{0,1\}$ and $\Omega_{\mathcal{A}} := \{\omega \in \Omega_N \mid \mathcal{A}(\omega_m, \dots, \omega_{m+n}) = 1 \text{ for } m \in \mathbb{Z}\}$. Then the restriction $\sigma_{\mathcal{A}}$ of σ_N to $\Omega_{\mathcal{A}}$ is called an *n-step topological Markov chain.*

From the point of view of their intrinsic dynamics n-step topological Markov chains are the same as topological Markov chains, since they can be described as topological Markov chains over the alphabet $\{1, \dots, N\}^n$ by taking the matrix A given by $A_{(i_1,\dots,i_n),(j_1,\dots,j_n)} = 1$ if $j_k = i_{k+1}$ for $k = 1, \dots, n-1$ and $\mathcal{A}(i_1, \dots, i_n, j_n) = 1$.

Let $\lambda_1, \dots, \lambda_N$ be the eigenvalues of the matrix A taken with their multiplicities and ordered in decreasing order of their absolute values. By Corollary 1.9.5 we have $P_n(\sigma_A) = \sum_{i=1}^n \lambda_i^n$. For a transitive matrix A a very precise asymptotic of the last expression can be found. It is based on some results about positive matrices which we will present here both for the sake of completeness and with an eye on future uses.

d. The Perron–Frobenius theorem for positive matrices.

Theorem 1.9.11. (Perron–Frobenius Theorem)[2] *Let L be an $N \times N$ matrix with nonnegative entries such that for a certain power L^n all entries are positive. Then L has one (up to a scalar) eigenvector e with positive coordinates and no other eigenvectors with nonnegative coordinates. Moreover, the eigenvalue corresponding to e is simple, positive, and greater than the absolute values of all other eigenvalues.*

Corollary 1.9.12. $P_n(\sigma_A) = \lambda_{\max}^n + \mu_n$ *for a transitive* 0-1 *matrix A, where $\lambda_{\max} > 1$ is the eigenvalue corresponding to the positive eigenvector and $|\mu_n| < C\lambda^n$ for some $C > 0$ and $\lambda < \lambda_{\max}$.*

Proof of Corollary 1.9.12. All statements except for $\lambda_{\max} > 1$ follow immediately from Corollary 1.9.5 and Theorem 1.9.11. Let $x = (x_0, \dots, x_{N-1})$, $x_i > 0$ for $i = 0, \dots, N-1$, and $Ax = \lambda_{\max} x$. Then $A^n x = \lambda_{\max}^n x$, that is, $\lambda_{\max}^n x_i = \sum_{j=0}^{N-1} a_{ij}^n x_j$ with $a_{ij}^n \geq 1$. Thus $\lambda_{\max}^n x_i \geq \sum_{j=0}^{N-1} x_j > x_i$; hence $\lambda_{\max}^n > 1$ and $\lambda_{\max} > 1$. □

Proof of Theorem 1.9.11. Let us denote by P the set of all vectors in $\mathbb{R}^N$ with nonnegative coordinates and by σ the unit simplex in P, that is, $\sigma = \{(x_1, \dots, x_N) \mid x_i \geq 0, \sum_{i=1}^N x_i = 1\}$. By assumption $LP \subset P$. Thus for every $x \in \sigma$ there exists a unique vector $Tx \in \sigma$ proportional to Lx. Thus we have defined a map $T: \sigma \to \sigma$. Obviously for each convex subset $S \subset \sigma$ both the image TS and the preimage are convex. By assumption $L^n P \subset \operatorname{Int} P$; hence $T^n \sigma \subset \operatorname{Int} \sigma$.

For the map T the extreme points of the image of any closed convex set are among images of its extreme points (see Definition A.2.8).

The set $\sigma_0 = \bigcap_{n=0}^{\infty} T^n \sigma \subset \operatorname{Int} \sigma$ is closed, convex, and strictly T-invariant (that is, $T\sigma_0 = \sigma_0$).

Let us show that σ_0 has no more than N extreme points. Let $x \in \sigma_0 \subset T^n\sigma$. Then x is a convex linear combination of extreme points of $T^n\sigma$, but as we pointed out all extreme points of $T^n\sigma$ are among the images of the vertices $e_1, \dots, e_N$ of σ. Thus $x = \sum_{i=1}^N \lambda_i^{(n)} T^n e_i$, where $\lambda_i^{(n)} \geq 0$ and $\sum_{i=1}^N \lambda_i^{(n)} = 1$. One can find a subsequence $n_k \to \infty$ such that $T^{n_k} e_i$ and $\lambda_i^{(n_k)}$ converge for all $i = 1, \dots, N$. Let $\lim T^{n_k} e_i = p_i$, $\lim \lambda_i^{(n_k)} = \lambda_i$. We have $x = \sum_{i=1}^N \lambda_i p_i$. If x is different from p_i, $i = 1, \dots, N$, it is not an extreme point of σ_0.

The set of extreme points of σ_0 is thus finite and T-invariant, so all those points are fixed points for T^m for some m. Each of these points corresponds to an eigenvector of L^m with positive coordinates. We are going to show that L^m may have only one (up to a scalar multiple) such eigenvector. Assuming that there are at least two, we consider two cases.

Case 1: All eigenvectors have the same eigenvalue. Thus we have $e, f \in \operatorname{Int} P$, $L^m e = \lambda e$, $L^m f = \lambda f$. It is actually enough to assume only that $f \in \operatorname{Int} P$, because then there exists a positive number α such that the vector $e - \alpha f$ belongs to the boundary of P. Since $T^m(e - \alpha f) = e - \alpha f$, this contradicts the assumption that for large n, $L^n P \subset \operatorname{Int} P$.

Case 2: There are two eigenvectors with different eigenvalues

$$e, f \in P, \quad L^m e = \lambda e, \quad L^m f = \mu f.$$

Obviously λ and μ are positive numbers, so we can assume that $\lambda > \mu$. Consider the plane generated by e and f. The lines containing e and f divide it into four sectors.

Consider the sector S bounded by the half-lines containing f and $-e$. If $x \in S$ then $x = \alpha f - \beta e$, $\alpha, \beta > 0$, so $L^{nm} x = \alpha \mu^n f - \beta \lambda^n e$, that is, the direction of that vector approaches that of $-e$ as $n \to \infty$. In particular, for a large n, $L^{nm} x$ does not belong to P. However, since $f \in \operatorname{Int} P$, $f - \epsilon e \in P$ for small $\epsilon > 0$. Thus we have found a vector in P which eventually leaves P, contradicting our assumption. This completes the proof of uniqueness of an eigenvector for L because our argument implies that σ_0 consists of a single point which is then fixed for L and thus generates an eigenvector $e \in P$ for L.

Let us show that if $Le = \lambda e$ and μ is another eigenvalue of L, then $\lambda > |\mu|$.

If μ is real, consider an eigenvector f with eigenvalue μ and the plane generated by e and f. We already proved that no other vectors with eigenvalues $\pm\lambda$ exist. Assume $|\mu| > \lambda$. Then as before for $\alpha, \beta > 0$ the direction of the vector $L^n(\alpha e + \beta f)$ approaches the direction of f as $n \to \infty$ and hence for large n these vectors are outside P. But if ϵ is small then $e + \epsilon f \in P$, a contradiction. Similarly if μ is complex, say $\mu = \rho \cdot e^{i2\pi\varphi}$, one finds an invariant two-plane where L acts as a multiplication by ρ and rotation by φ. If $\rho > \lambda$ we obtain a similar contradiction by considering the action of L on the 3-dimensional space generated by e and that plane.

If $\rho = \lambda$ we take a vector in that 3-space which lies on the boundary of P. This vector either eventually returns to ∂P (if φ is rational) or comes arbitrary

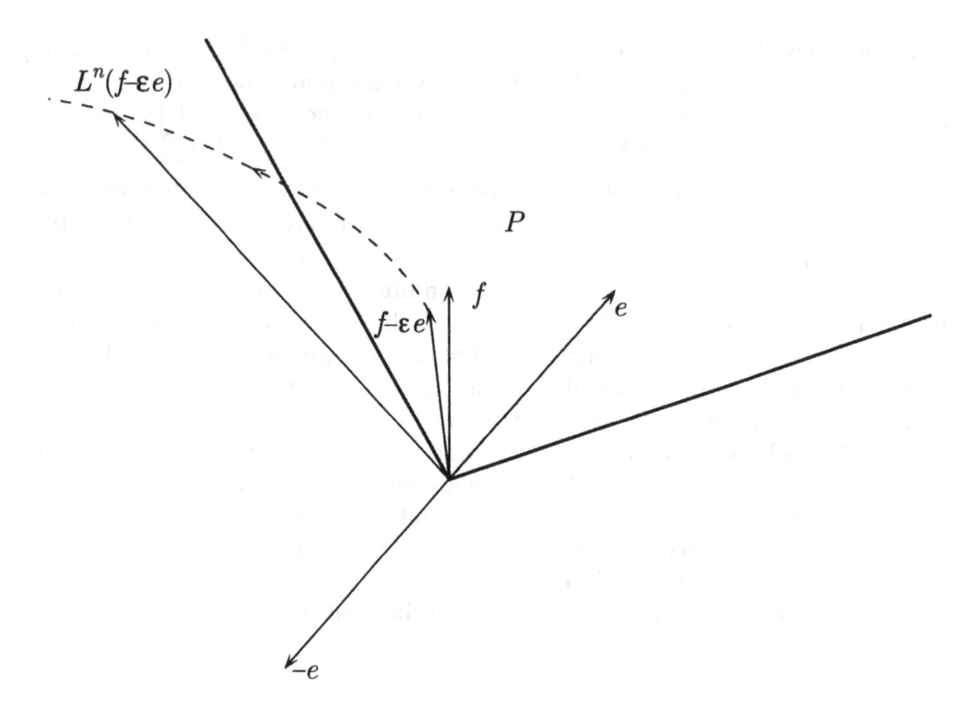

FIGURE 1.9.2. The case of two eigenvalues

close to it (if φ is irrational) contradicting the fact that $L^m P$ lies strictly inside P for large m.

It remains to prove that λ is a simple eigenvalue. We already proved that the space of eigenvectors with eigenvalue λ is one-dimensional. The remaining possibility is that the root space corresponding to λ is more than one-dimensional. But then there exists $f \in \mathbb{R}^N$ such that $Lf = \lambda(f+e)$. Then for small ϵ

$$e - \epsilon f \in P \quad \text{and} \quad L^n(e-\epsilon f) = \lambda^n(-\epsilon f + (1-\epsilon n)e).$$

For large positive n the direction of the latter vector approaches that of $-e$, that is, it leaves P, which is impossible. □

Exercises

1.9.1. *Justify the statements made at the end of subsection a. Namely:*

(1) *Show that a function $\varphi: \Omega_N \to \mathbb{C}$ is of exponential type if and only if it is Hölder continuous with respect to the metric d_λ for some $\lambda > 1$.*

(2) *Prove (1.9.2).*

1.9.2. *Let d_* be the following metric on Ω_N*

$$d_*(\omega, \omega') = \sum_{n=-\infty}^{\infty} \frac{|\omega_n - \omega'_n|}{n^2}.$$

Show that it determines the same topology as any of the metrics d_λ. Show that every function of exponential type is Hölder continuous with respect to d_* but there are also some functions not of exponential type that are Hölder continuous with respect to that metric.

1.9.3. *Consider the natural homeomorphism $H: \Omega_2^R \to K$, where K is the middle-third Cantor set:*

$$H(\omega_0, \omega_1, \omega_2, \dots) := 0.\alpha(\omega_0)\alpha(\omega_1)\dots,$$

where $\alpha(0) = 0$, $\alpha(1) = 2$, and the number on the right-hand side is written in base 3.

Prove that the class of functions of exponential type on Ω_2^R becomes the class of functions which are Hölder continuous with respect to the metric on K induced by the Euclidean metric on $[0, 1]$.

Let us assume from now on that A is a 0-1 $m \times m$ matrix which has at least one 1 in each row and each column.

1.9.4. *Prove that for every $i \in \{0, \dots, m-1\}$ the set $\Omega_{A,i} = \{\omega \in \Omega_A \mid \omega_0 = i\}$ is nonempty.*

1.9.5. *Prove that if there is an element $\omega \in \Omega_A$ that contains the symbol i at least twice then there is a periodic element $\omega' \in \Omega_A$ such that $\omega'_0 = i$.*

1.9.6. *Let us call symbols i satisfying the condition of the previous problem essential. Prove that any ω-limit point (see Definition 1.6.2) of any element of Ω_A contains only essential symbols.*

1.9.7. *Let us call two essential symbols i and j equivalent if there exist $\omega, \omega' \in \Omega_A$, $k_1 < k_2$, $l_1 < l_2$ such that*

$$\omega_{k_1} = \omega'_{l_2} = i, \quad \omega_{k_2} = \omega'_{l_1} = j.$$

Prove that the set of all essential symbols splits into disjoint subsets of mutually equivalent symbols (that is, this is an equivalence relation).

1.9.8. *Prove that σ_A has a dense positive semiorbit if and only if all symbols are essential and equivalent.*

1.9.9. *Show that under the condition of the previous problem there exists a positive integer N and a decomposition of Ω_A into closed disjoint subsets $\Lambda_1, \dots, \Lambda_N = \Lambda_0$ such that $\sigma_A \Lambda_i = \Lambda_{i+1}$ for $i = 0, 1, \dots, N-1$ and the restriction of $(\sigma_A)^N$ to each Λ_i is topologically mixing. Moreover the decomposition* *of Ω_A into Λ_i's corresponds to a decomposition of the set $\{1,\ldots,m\}$ into N equal groups such that every element $\omega \in \Omega_A$ has only symbols from one group in positions equal modulo N.*

This is called the *spectral decomposition.*[3]

1.9.10. *Let $B_k = \{\omega \in \Omega_2 \mid \forall m,n \in \mathbb{Z}, m > n, \ \left|\sum_{i=n}^{m}(-1)^{\omega_i}\right| \leq k\}$. Prove that B_k is a closed σ_2-invariant subset of Ω_2. Denote $S_k = \sigma_2\restriction_{B_k}$. Prove that S_k is topologically transitive but not topologically mixing.*

1.9.11. *Prove that there exists a 0-1 matrix A and a continuous map $H: \Omega_A \to B_2$ such that the diagram*

$$\begin{array}{ccc} \Omega_A & \xrightarrow{\sigma_A} & \Omega_A \\ {\scriptstyle H}\downarrow & & \downarrow{\scriptstyle H} \\ B_2 & \xrightarrow{S_2} & B_2 \end{array} \tag{1.9.6}$$

is commutative and all but two points in B_2 have exactly one preimage.

1.9.12*. *Prove that there is no homeomorphism satisfying (1.9.6), that is, the map S_2 is not C^0 equivalent or topologically conjugate (see Definition 2.1.1 and Definition 2.3.1 in the sequel) to any topological Markov chain.*

2

Equivalence, classification, and invariants

In the previous chapter in the process of studying various examples we encountered a number of useful concepts related to the asymptotic behavior of dynamical systems. Our list includes so far the growth of the number of periodic orbits, their spatial distribution (for example, density), topological transitivity, minimality, α- and ω-limit sets, and topological mixing. This list will be considerably extended and systematized in Chapters 3 and 4. Before doing that, we are going to look into the problem of studying the asymptotic behavior of smooth dynamical systems from a different angle.

1. Smooth conjugacy and moduli for maps

a. Equivalence and moduli. We can consider the properties in question as some features of the global orbit structure independent of a particular choice of coordinates. From the global point of view a coordinate change is given by a diffeomorphism (in the case of a smooth structure) or a homeomorphism (for the topological situation) between the phase spaces. Thus we can introduce natural equivalence relations between dynamical systems associated with various classes of coordinate changes and interpret the problem of the description of the orbit structure as the classification of dynamical systems with respect to those equivalence relations.

We begin our discussion with the discrete-time case.

Definition 2.1.1. Two C^r maps $f: M \to M$ and $g: N \to N$ are said to be C^m *equivalent* or C^m *conjugate* $(m \leq r)$ if there exists a C^m diffeomorphism $h: M \to N$ such that $f = h^{-1} \circ g \circ h$. h, or the existence of such h, is referred to as a *(smooth) conjugacy.*

In other words, C^m equivalence means that f differs from g by a C^m coordinate change. This certainly looks like a natural equivalence relation in differentiable dynamics both from the general structural point of view and with regard to applications.

Definition 2.1.2. Let U be an open subset in the space $C^r(M,M)$ of C^r maps of M into itself with the C^r topology. A continuous functional $F: U \to \mathbb{R}$ is called a C^m *modulus* if there exists a $\delta > 0$ such that $F(f) = F(g)$ for any two maps $f, g \in U$ that are C^m equivalent via a diffeomorphism h with $\text{dist}_{C^m}(h, id) < \delta$.

The condition of closeness to the identity becomes particularly important for continuous-time systems. We will discuss appropriate examples in the next section. Right now we will show that at least in many interesting cases with nontrivial recurrence such as those described in Sections 1.7 and 1.8, there are many C^1 moduli.

Let p be a periodic point of period n for f. Obviously for any g that is C^m equivalent to f we have $g^n h(p) = hf^n(p) = h(p)$ so $q = h(p)$ is a periodic point for g of the same period. Thus, $P_n(f) = P_n(g)$ for $n \in \mathbb{N}$. Furthermore, if $m \geq 1$, one has for every (not necessarily periodic) point x and for every n

$$Df_x^n = Dh^{-1}_{g^n hx} \circ Dg^n_{hx} \circ Dh_x.$$

In particular if $f^n p = p$ then

$$Df_p^n = (Dh_p)^{-1} Dg^n_{hp} Dh_p$$

because in this case $g^n h(p) = h(p)$ and $(Dh^{-1})_{hp} = (Dh_p)^{-1}$. Thus the linear operators Df_p^n and Dg^n_{hp} are conjugate and in particular they have the same eigenvalues. Let us call the set of eigenvalues of Df_p^n the *spectrum* of the periodic point p.

Invoking Proposition 1.1.4 one sees that every periodic point p of f, whose spectrum does not contain one, determines several C^1 moduli. If we assume for simplicity that the eigenvalues of Df_p^n are simple, those eigenvalues can serve as the moduli. Since such periodic orbits are isolated, their spectra can be perturbed separately, at least for any finite collection of points. Thus, the moduli obtained from different periodic orbits are in a certain sense independent.

b. Local analytic linearization. On the other hand, at least in some cases *locally* the spectrum is a complete invariant of smooth conjugacy. Right now we will give a very simple example of such a situation which also sheds some light on certain methods used in more analytical aspects of the theory of smooth dynamical systems. Different approaches to the problem of local smooth conjugacy will be discussed in Sections 2.8 and 6.6.

Proposition 2.1.3. *Let $I = [-\delta, \delta]$, $f: I \to I$ be a real analytic contracting map, $f(0) = 0$, and $0 \neq \mu := f'(0)$. Then there are an interval $J_1 \subset I$ containing 0 and a real-analytic diffeomorphism $h: J_1 \to J_2 \subset \mathbb{R}$ preserving the origin and conjugating f with the linear map $x \mapsto \mu x$.*

There are also versions of Proposition 2.1.3 for C^∞ and C^r maps. The C^∞ version is contained in Theorem 6.6.6.

Proof. We will show how to find a formal power series for the conjugacy and then prove convergence of this series. This is a very simple example of the majorization method which is widely used in many local and semilocal problems concerning conjugacy of dynamical systems.

By a formal power series we mean an expression $u = \sum_{i=0}^{\infty} u_i x^i$ which we do not assume to converge anywhere except for $x = 0$. Given two formal power series $u = \sum_{i=0}^{\infty} u_i x^i$ and $v = \sum_{i=0}^{\infty} v_i x^i$ we say that v *majorizes* u if $|u_i| \leq v_i$ for all i. We write this as $u \prec v$ and call v a *majorant* for u. We use the same notation when u and v are analytic functions. Thus, for example, $1 \prec (1-x)^{-1}$ and $1 + 2x \prec (1-x)^{-2}$.

Write $f(x) = \sum_{i=1}^{\infty} f_i x^i$ and let $\epsilon = \min_{n \geq 2} |f_i|^{-1/(n-1)} > 0$ (since f is defined on a neighborhood of 0). Then

$$f'(x) := f(\epsilon x)/\epsilon = \lambda x + \sum_{i=2}^{\infty} f_i \epsilon^{i-1} x^i \prec |\lambda| x + \frac{x^2}{1-x} =: |\lambda| x + F(x).$$

If $h'(\lambda x) = f'(h'(x))$ and $h := h'/\epsilon$ then

$$f(h(x)) = \epsilon f'(h(x)/\epsilon) = \epsilon f'(h'(x)) = \epsilon h'(\lambda x) = h(\lambda x),$$

so we may assume $f \prec |\lambda| x + F(x)$. If $h(x) = x + \tilde{h}(x)$ and $f(x) = \lambda x + \tilde{f}(x)$ then $h(\lambda x) = f(h(x))$ becomes $\lambda x + \tilde{h}(\lambda x) = \lambda h(x) + \tilde{f}(h(x))$ or $\tilde{h}(\lambda x) - \lambda \tilde{h}(x) = \tilde{f}(h(x))$. For the coefficients of $h(x) = x + \sum_{i=2}^{\infty} h_n x^n$ this means

$$(\lambda^k - \lambda) h_k = (\tilde{f} \circ h)_k. \tag{2.1.1}$$

The right hand side involves only coefficients of h of order lower than k because $\tilde{f}$ begins with quadratic terms, so this determines the coefficients of h uniquely by recursion starting from $h_0 = 0$ and $h_1 = 1$. It remains to show that this power series converges. To that end let

$$c = \max_{k \geq 2} \frac{1}{|\lambda^k - \lambda|} = \frac{1}{|\lambda|} \max_{k \geq 2} \frac{1}{|1 - \lambda^{k-1}|} = \frac{1}{|\lambda|} \frac{1}{1 - |\lambda|}$$

and $\bar{h}$ the power series with coefficients $\bar{h}_k = |h_k|$. Then for $k \geq 2$

$$\bar{h}_k = |h_k| = \frac{1}{|\lambda^k - \lambda|} (\tilde{f} \circ h)_k \leq \frac{1}{|\lambda^k - \lambda|} (F \circ \bar{h})_k \leq c (F \circ \bar{h})_k.$$

Here we used that if $u \prec U$ and $v \prec V$ with $V(0) = 0$ then $u \circ v \prec U \circ V$. To see this, note that the kth order coefficient on either side of $u \circ v \prec U \circ V$ is a polynomial with positive coefficients in the coefficients of u and v on the one hand and U and V on the other hand. The polynomials are the same on either side and the arguments on the right side are larger in absolute value, giving larger values.

Thus $\bar{h} \prec x+cF\circ\bar{h} = x+c\dfrac{\bar{h}^2}{1-\bar{h}} \prec \dfrac{x+c\bar{h}^2}{1-\bar{h}}$, because $x \prec x(1+\bar{h}+\bar{h}^2+\dots) = x/1-\bar{h}$. Note that in the expression $\dfrac{x+cu^2}{1-u} = (x+cu^2)(1+u+u^2+\dots)$, where u is any formal power series, the kth order coefficient is given by a polynomial $P_k(u_0,\dots,u_{k-1})$ in lower order coefficients of u all of whose coefficients are positive. Thus

$$H = \frac{x+cH^2}{1-H} \tag{2.1.2}$$

defines a formal power series H by $H_k = P_k(H_0,\dots,H_{k-1})$ and inductively

$$|h_k| = \bar{h}_k \le P_k(h_0,\dots,h_{k-1}) \le P_k(H_0,\dots,H_{k-1}) = H_k, \tag{2.1.3}$$

or $h \prec H$. H converges because $(c+1)H^2 - H + x = 0$ or $a^2H^2 - 2aH + 2ax = 0$, where $a = 2(c+1)$, so $(1-aH)^2 = 1 - 2aH + a^2H^2 = 1 - 2ax$ and

$$1 + 2aH \prec 1 + 2aH + 2a^2H^2 + \dots = (1-aH)^{-2} = \frac{1}{1-2ax},$$

hence $2aH \prec \dfrac{1}{1-2ax} - 1 = \dfrac{2ax}{1-2ax}$ and $H \prec \dfrac{x}{1-2ax}$ converges for $|2ax| < 1$.

□

Proposition 2.1.3 is a particular case of Theorem 2.8.2. (Although Theorem 2.8.2 deals with transformations in a complex domain one easily sees from the proof that if f has real coefficients then so does the conjugating map h.) The proof of Theorem 2.8.2 will serve as an illustration of the fast-converging iteration method, sometimes called the Newton method. The method, described in a general way in Section 2.7, is one of the most powerful and versatile tools in the theory of smooth dynamical systems, especially for problems related to smooth equivalence. Its special importance is due to the fact that it is applicable to situations where, unlike in our case, no hyperbolicity is present.

c. Various types of moduli. Let us come back to the discussion of the global orbit structure. Our construction of independent moduli associated with periodic orbits shows that in the case of infinitely many periodic orbits as for the expanding maps E_m (Section 1.7) and the hyperbolic toral automorphism F_L (Section 1.8) there are infinitely many invariants of local C^1 equivalence. It turns out that for both those cases which represent the simplest examples of hyperbolic systems (cf. Section 6.4 and Part 4 of this book), the spectra of periodic orbits form a complete system of invariants for C^1 and even C^∞ equivalence in a neighborhood of the maps E_m and F, correspondingly. For C^1 conjugacy of toral maps this is contained in Theorem 20.4.3. A reasonable description of the set of possible values for eigenvalues of *all* periodic points remains an open problem.

A modulus of a different kind gives substantial although not complete information about smooth equivalence in a neighborhood of the circle rotation

R_α. The rotation number (see Definition 11.1.2) is a C^0 modulus and for some irrational values of α its levels determine the smooth equivalence class (see Theorem 12.3.1).

On the other hand, in many situations the set of all C^r equivalence classes for $r \geq 1$ is both too big and does not admit any reasonable structure. The case of $r = 0$, that is, the *topological* equivalence of *smooth* dynamical systems, is strikingly different and will be discussed in Section 2.3.

Now we will give an idea of how C^r classification might look for systems with a very simple recurrence pattern. For that purpose we consider a monotone analytic map φ of the unit interval $I = [0,1]$ to itself, which fixes the endpoints, has no other fixed points, and is such that $\varphi'(0) > 1$, $\varphi'(1) < 1$, for example,

$$\varphi(x) = -\frac{x^2}{2} + \frac{3}{2}x. \tag{2.1.4}$$

Thus $x = 0$ and $x = 1$ are attracting fixed points for negative and positive iterates of the map φ, respectively, that is, any point in between tends to 0 and 1 for negative and positive iterates accordingly (cf. Proposition 1.1.6).

First we will show that such a map φ intrinsically defines two smooth affine structures on the open interval $(0,1)$.

Lemma 2.1.4. *Any C^1 map defined in a neighborhood of the origin on the real line and commuting with a linear contraction $\Lambda\colon x \to \lambda x$, $|\lambda| < 1$, is linear.*

Proof. Let $f\colon [-\epsilon, \epsilon] \to \mathbb{R}$ be such a map. First, f must preserve the origin because $f(0)$ is a fixed point for Λ which is unique. Furthermore, the commutation condition implies $f(\lambda x) = \lambda(f(x))$ and inductively

$$f(\lambda^n x) = \lambda^n f(x). \tag{2.1.5}$$

Since f is differentiable at 0, $f(\lambda^n x)/\lambda^n x$ has a limit as $n \to \infty$ that is independent of x and will be denoted by a. By (2.1.5)

$$a = \lim_{n\to\infty} \frac{f(\lambda^n x)}{\lambda^n x} = \frac{f(x)}{x},$$

that is, $f(x) = ax$. □

Remark. The differentiability condition is important. By contrast there are many nonlinear Lipschitz maps commuting with Λ.

Corollary 2.1.5. *Let h_1, h_2 be two diffeomorphisms satisfying the assertion of Proposition 2.1.3. Then there exists a real number μ such that $h_2(x) = h_1(\mu x)$.*

In other words, each real-analytic contracting map preserves a uniquely defined smooth affine structure. For the map φ we have been discussing, the two structures, defined near the endpoints of the interval, meet in the middle. The transition function between the two structures at any fundamental domain $J = [a, \varphi(a)]$ provides an infinite-dimensional space of moduli for φ. Let us describe the last statement in more detail. Using Proposition 2.1.3 one can change coordinates so that φ will be affine as a map from $[0, \varphi^{-1}(a)]$ to $[0, a]$ and from $[\varphi(a), 1]$ to $[\varphi^2(a), 1]$. Those coordinates by Corollary 2.1.5 are defined uniquely up to two multiplicative constants, one at each end. Then the map $\varphi\restriction_{[a,\varphi(a)]}: [a, \varphi(a)] \to [\varphi(a), \varphi^2(a)]$ can be normalized to a diffeomorphism $\varphi_a: [0,1] \to [0,1]$

$$\varphi_a(t) = \frac{\varphi(a + t(\varphi(a) - a)) - \varphi(a)}{\varphi^2(a) - \varphi(a)} \tag{2.1.6}$$

and extended to the whole real line by the formula

$$\varphi_a(t + k) = \varphi_a(t) + k, \tag{2.1.7}$$

where $0 \le t \le 1$ and $k \in \mathbb{Z}$. We will call two diffeomorphisms ψ_1, ψ_2 of the real line with $\psi_1(0) = \psi_2(0) = 0$ *equivalent* if $\psi_2(t) = \psi_1(t + s) - \psi_1(s)$ for some $s \in [0,1]$. We will call $\varphi_a: \mathbb{R} \to \mathbb{R}$ defined by φ via (2.1.6) and (2.1.7) a *transition map* for φ. Such a map depends on the multiplicative constants determining the linearization of φ near the ends of the interval and on the choice of the base point a. It is obvious from (2.1.6) that a change in the linearization does not change the transition map if a is changed appropriately. A change in the base point replaces the transition map by an equivalent one. Thus we have described the moduli for our map φ; they include the eigenvalues at the endpoints and the equivalence class of the transition map.

Corollary 2.1.6. *Let $\tilde{\varphi}$ be a small analytic perturbation of the map φ. Then φ and $\tilde{\varphi}$ are analytically conjugate if and only if they are C^1 conjugate or if the following is true: $\tilde{\varphi}'(0) = \varphi'(0)$, $\tilde{\varphi}'(1) = \varphi'(1)$, and the transition maps for φ and $\tilde{\varphi}$ are equivalent.*

In contrast let us point out that the C^0 orbit structure of the map φ is stable:

Proposition 2.1.7. *The map φ is C^0 conjugate to any map ψ that is C^1-close to φ and such that $\psi(0) = 0$ and $\psi(1) = 1$.*

Proof. If $\psi'(0) > 1$, $\psi'(1) < 1$, and $\psi' > 0$ then $\lim_{n\to\infty} \psi^n(t) = 1$ for $t \neq 0$ and $\lim_{n\to-\infty} \psi^n(t) = 0$ for $t \neq 1$. Thus if we take any monotone continuous map H between the intervals $[a, \varphi(a)]$ and $[a, \psi(a)]$ and extend it to a map $h: [0,1] \to [0,1]$ by

$$h(0) = 0, h(1) = 1, h = \psi^n \circ H \circ \varphi^{-n} \text{ on } [\varphi^n(a), \varphi^{n+1}(a)]$$

for any $n \in \mathbb{Z}$, then h is a homeomorphism of $[0,1]$ and $\psi = h \circ \varphi \circ h^{-1}$. □

This argument provides a prototype of proofs of the stability of the orbit structure for systems with highly dissipative behavior. See Exercises 2.1.1, 2.3.3, and 2.3.4 as well as Section 6.3 for further applications of this method.

Exercises

2.1.1. *Let f, g be C^1 maps defined in a neighborhood of the origin on the real line, $f(0) = g(0) = 0$, and $0 < f'(0) < 1$, $0 < g'(0) < 1$. Prove that f and g are locally topologically conjugate near the origin, that is, there exists an open interval $I \ni \{0\}$ and a homeomorphic embedding $h: I \to \mathbb{R}$ such that $h(0) = 0$ and $f(h(x)) = h(g(x))$ for $x \in I$.*

2.1.2. *Prove that under the assumption of the previous problem h can be chosen Lipschitz continuous if and only if $f'(0) \le g'(0)$. Thus in particular h can be taken biLipschitz if and only if $f'(0) = g'(0)$.*

2.1.3.

(1) *Prove that the map $f: x \mapsto x + \dfrac{x^2}{1+x^2}$ of the real line is not locally conjugate near the origin to any linear map.*
(2) *Prove that the map $f: x \mapsto x + x^3$ is topologically conjugate to a linear map l given by $l(x) = \lambda x$, but the solution h of the equation $f = h \circ l \circ h^{-1}$ cannot be chosen Hölder continuous in any neighborhood of the origin.*

2.1.4. *Let $f_0 = R$ be a rotation of $\mathbb{R}^n$ and $f_{\pm 1}(x) = R(x \pm x\|x\|)$. Show that no two of the maps f_0, f_1, f_{-1} are topologically conjugate near 0.*

2.1.5. *Consider the family of quadratic maps $f_\lambda(x) = \lambda x(1-x)$ with $(0 \le \lambda \le 4)$ of the unit interval $[0,1]$ studied in Exercise 1.7.2. Prove that:*

(1) *The map f_4 is not topologically conjugate to any f_λ for $\lambda < 4$.*
(2) *All f_λ for $0 < \lambda \le 1$ are topologically conjugate to each other.*
(3) *If $\lambda \neq \mu$ then f_λ is not C^1 conjugate to f_μ.*
(4) *If $0 < \lambda \le 1$ and $\mu > 1$ then f_λ and f_μ are not topologically conjugate.*
(5) *If $1 < \lambda < 3$ then $p_\lambda = (\lambda - 1)/\lambda$ is an attracting fixed point, 0 is a repelling fixed point, and $\lim_{n\to\infty} f_\lambda^n(x) = p_\lambda$ whenever $0 < x < 1$.*
(6) *All maps f_λ for $1 < \lambda < 2$ are topologically conjugate to each other.*
(7)* *Show that for $\lambda_1, \lambda_2 \in (3, 1+\sqrt{6}]$ the maps f_{λ_1} and f_{λ_2} are topologically conjugate but not C^1 equivalent.*

2.1.6. *Prove that both the topological conjugacy h constructed in Proposition 2.1.7 and its inverse h^{-1} can be chosen to be Lipschitz continuous if and only if $\psi'(0) = \varphi'(0)$ and $\psi'(1) = \varphi'(1)$.*

2.1.7. *Consider a complex analytic map $f: U \to \mathbb{C}$ defined in a neighborhood of the origin on the complex plane such that $f(0) = 0$ and $|f'(0)| \neq 1$. Prove that f is locally analytically conjugate in a neighborhood of 0 to the linear map $Lz = f'(0)z$.*

2.1.8. *Let v be a real-analytic vector field in a neighborhood of the origin on the real line such that $v(0) = 0$, $v'(0) \neq 0$. Prove that the local flow φ^t generated by v in a neighborhood of 0 is locally analytically conjugate to the linear flow $\Phi^t x = e^{v'(0)t}x$.*

2.1.9*. *Show that for $a \in (0,1)$ the map $(x,y) \mapsto (a^2x + y^2, ay)$ can be C^1-linearized near 0, but not C^2-linearized.*

2.1.10. *Let $L \in SL(m,\mathbb{Z})$ be an integer matrix with determinant one that does not have one as an eigenvalue, and let $a \in \mathbb{R}^m$. Consider the affine map of the m-torus*

$$A_{L,a}(x) = Lx + a \pmod 1$$

Show that $A_{L,a}$ is C^∞ conjugate to the automorphism $F_L: x \mapsto Lx \pmod 1$.

2. Smooth conjugacy and time change for flows

The notion of smooth equivalence can be extended to the continuous-time case in a straightforward way.

Definition 2.2.1. Two C^r flows $\varphi^t: M \to M$ and $\psi^t: N \to N$ are said to be C^m *flow equivalent* $(m \leq r)$ if there exists a C^m diffeomorphism $h: M \to N$ such that $\varphi^t = h \circ \psi^t \circ h^{-1}$ for all $t \in \mathbb{R}$.

In other words, flow equivalence is the conjugacy of flows as differentiable actions of the group $\mathbb{R}$ of real numbers.

Let us consider an interesting simple example of flow equivalence.

Proposition 2.2.2. *Let $\omega = (\omega_1, \dots, \omega_{n-1}, 1)$. The linear flow T^t_ω on the n-torus $\mathbb{T}^n$ is C^∞ flow equivalent to the suspension of the translation T_γ on the $(n-1)$-torus, where $\gamma = (\omega_1, \dots, \omega_{n-1})$.*

Proof. Consider the map H from the suspension manifold $M = \mathbb{T}^{n-1}_{T_\gamma}$ to the torus $\mathbb{T}^n$ given by

$$H(x_1, \dots, x_{n-1}, t) = (x_1 + \omega_1 t, x_2 + \omega_2 t, \dots, x_{n-1} + \omega_{n-1} t, x_n + t).$$

It is obviously differentiable for $t \neq 0$. Differentiability at $t = 0$ follows from the definition of the smooth structure on the suspension manifold. The differential of H carries the upward vector field $\frac{\partial}{\partial t}$ to the vector field $\omega_1 \frac{\partial}{\partial x_1} + \omega_2 \frac{\partial}{\partial x_2} + \dots + \omega_n \frac{\partial}{\partial x_n}$ and hence conjugates the flows generated by those vector fields, which are exactly the suspension flow and the linear flow, respectively. □

The orbit structure of flows, unlike that for discrete-time systems, has two distinct aspects: (i) the relative behavior of points on different orbits and (ii) the evolution of an initial condition along its orbit with time. There is a natural way to change a given flow while preserving the first aspect of its orbit structure, namely, not to change orbits at all.

Definition 2.2.3. A flow ψ^t on M is a *time change* of another flow φ^t if for each $x \in M$ the orbits $\mathcal{O}_\varphi(x) = \{\varphi^t x\}_{t\in\mathbb{R}}$ and $\mathcal{O}_\psi(x) = \{\psi^t x\}_{t\in\mathbb{R}}$ coincide and the orientations given by the change of t in the positive direction are the same.

If ψ^t is a time change for φ^t then $\varphi^t x = \psi^{\alpha(t,x)} x$ for every $x \in M$, where α is a real-valued function. The group properties $\varphi^{t+s} = \varphi^s \circ \varphi^t$ and $\varphi^{-t} = (\varphi^t)^{-1}$ translate into the following equations

$$\begin{aligned} \alpha(t+s,x) &= \alpha(t,x) + \alpha(s,\varphi^t x), \\ \alpha(-t,x) &= -\alpha(t,\varphi^{-t}x). \end{aligned} \tag{2.2.1}$$

Such a function $\alpha(t,x)$ is called an *(untwisted) one-cocycle* over φ^t. Cocycles are discussed in Section 2.9 in a more general framework. The preservation of orientation means that

$$\alpha(t,x) \geq 0 \quad \text{if } t \geq 0 \tag{2.2.2}$$

and the fact that orbits cannot collapse means that either x is a fixed point for φ^t, that is, $\varphi^t x = x$ for all $t \in \mathbb{R}$, or $\alpha(t,x) > 0$ if $t > 0$.

Obviously, the fixed points for φ^t and ψ^t coincide.

If both φ^t and ψ^t are C^r and x is not a fixed point, then the Implicit Function Theorem immediately implies that $\alpha(t,x)$ is a C^r function in both variables. Simple examples show that this is not true for fixed points. $\alpha(t,x)$ for a fixed t may not even have a limit as $x \to x_0$, a fixed point of φ^t.

Another description of time change can be given through the vector fields $\xi = \frac{d\varphi^t}{dt}\big|_{t=0}$ and $\eta = \frac{d\psi^t}{dt}\big|_{t=0}$. The uniqueness of solutions for differential equations implies that zeros of a vector field are fixed points of the corresponding flow. Thus, we have

$$\xi(x) = 0 \text{ if and only if } \eta(x) = 0. \tag{2.2.3}$$

Furthermore if x is not a fixed point, tangent vectors to the curves $\{\varphi^t x\}$ and $\{\psi^t x\}$ are collinear, do not vanish, and have the same direction. Thus $\eta(x) = a(x)\xi(x)$ where a is a scalar function defined and positive at all nonfixed x, namely, $a(x) = \frac{\partial}{\partial t}\alpha(t,x)\big|_{t=0}$. If both φ^t and ψ^t are C^r flows and $\xi(x) \neq 0$ then $a(x)$ is C^{r-1} at x. Sometimes we will use the term time change for the flow generated by a scalar multiple of the vector field ξ even if it vanishes at some points where $\xi \neq 0$.

There is a simple case when a time change produces a flow that is flow equivalent to the original one, namely, if the conjugating diffeomorphism itself preserves each orbit of the original flow. In other words, one chooses

$$h(x) = \varphi^{\beta(x)}(x), \tag{2.2.4}$$

where β is a differentiable function and its derivative in the direction of the flow

$$(\xi\beta)(x) = \frac{d\beta(\varphi^t(x))}{dt}\Big|_{t=0}$$

is positive if $\xi(x) \neq 0$. One easily sees that

$$(h \circ \varphi^t \circ h^{-1})(x) = \varphi^{\beta(x)+t-\beta(\varphi^t x)}(x), \tag{2.2.5}$$

where

$$\alpha(t,x) - t = \beta(x) - \beta(\varphi^t x), \tag{2.2.6}$$

so α obviously satisfies (2.2.1)–(2.2.2). Differentiating with respect to t at $t = 0$ shows that the flow $h \circ \varphi^t \circ h^{-1}$ is generated by the vector field $(\xi\beta) \cdot \xi$, that is, we have

$$a - 1 = \xi\beta. \tag{2.2.7}$$

We will call time changes of the form (2.2.5) *trivial.* It is natural to try to describe all time changes of a given flow modulo trivial ones. This problem is essentially equivalent to describing the space of all sufficiently smooth positive functions up to an addition of a function that is a derivative of another smooth function in the direction of the flow. The general framework for studying this and similar problems is described in Section 2.9. In some cases this problem is solved using natural moduli, that is, periods of periodic orbits. This is true, for example, for special flows built over a hyperbolic toral automorphism (cf. Section 19.2c).

Definition 2.2.4. Two C^r flows φ^t on M and ψ^t on N are said to be C^m *orbit equivalent* ($m \leq r$) if there exists a C^m diffeomorphism $h\colon M \to N$ such that the flow $\chi^t = h^{-1} \circ \psi^t \circ h$ is a time change of the flow φ^t.

Equivalently one can say that h maps orbits of the flow ψ^t onto orbits of φ^t preserving the orientation given by the positive-time direction.

An interesting example of orbit equivalence is provided by the construction of the flow under a function in Section 1.3.

Proposition 2.2.5. *Let M be a compact differentiable manifold, $f\colon M \to M$ a C^m diffeomorphism, and $\varphi\colon M \to \mathbb{R}_+$ a C^m function on M. Then the special flow on the manifold M_f^φ is C^m orbit equivalent to the suspension flow on M_f.*

Proof. Let $k := \min \varphi$ and $K := \max \varphi$. Consider a C^∞ function $g\colon [0,1] \times [k,K] \to \mathbb{R}$ such that

(1) $g(t,s) = t$ for $t \in [0, k/4]$,
(2) $g(t,s) = t + s - 1$ for $t \in [1 - k/4, 1]$,
(3) $\frac{\partial}{\partial t} g(t,s) > 0$.

Then the map $(x,t) \mapsto (x, g(t, \varphi(x)))$ is a diffeomorphism between M_f and M_f^φ which takes the vertical vector field $\frac{\partial}{\partial t}$ on M_f to a vertical vector field on M_f^φ, and hence conjugates the suspension flow with a time change of the special flow under φ. □

The definition of a C^m modulus can be extended to the case of flows in two ways assuming preservation of its values either for flow-equivalent or for orbit-equivalent perturbations.

In order to emphasize the difference between flow equivalence and orbit equivalence for flows let us point out that in both cases the image of any periodic orbit of the first flow is a periodic orbit for the second flow. However, even C^0 flow equivalence implies that the period of such an orbit is preserved, whereas an orbit equivalence may lead to a change of period. In fact, a typical time change does change the period of a given orbit and those changes can be made independently for different orbits. Thus, periods of periodic orbits are moduli of even C^0 flow equivalence and in cases such as the suspension of the hyperbolic toral automorphism F there are infinitely many such moduli. Relations between those moduli and the ones coming from the eigenvalues of the linearized map are far from trivial.

Let us discuss a simple example showing that the requirement that the conjugating map is close to identity in the definition of moduli is reasonable. Consider a linear flow (1.5.2) on the two-torus. It is generated by the constant vector field $\omega_1 \frac{\partial}{\partial x} + \omega_2 \frac{\partial}{\partial y}$. Obviously if the slopes of two such flows are equal they have the same orbits and if the ω's are positive they are obtained by a time change from each other. A linear unimodular map

$$G(x, y) = (ax + by, cx + dy),$$

where $a, b, c, d \in \mathbb{Z}$, $ad - bc = \pm 1$, takes the vector field (ω_1, ω_2) into the vector field $(a\omega_1 + b\omega_2, c\omega_1 + d\omega_2)$ with slope $\frac{c\omega_1 + d\omega_2}{a\omega_1 + b\omega_2} = \frac{c + d\gamma}{a + b\gamma}$, where $\gamma = \omega_2/\omega_1$ is the slope of the vector field (ω_1, ω_2).

It is easy to see that for any γ the numbers of the form $\frac{c + d\gamma}{a + b\gamma}$ are dense. Thus, the set of all vectors (ω_1, ω_2) is divided into dense equivalence classes which determine orbit-equivalent flows. In fact, those flows are almost flow equivalent since the time change in each case is constant. On the other hand, if one requires that the conjugating map is close to the identity then these linear flows are orbit equivalent only if their vector fields have the same slope. (See Exercise 2.2.1.) The notion of rotation vector (see Section 14.7) provides a C^0 modulus for that kind of equivalence.

Exercises

2.2.1. *Prove that if two linear flows $\{T^t_\omega\}$ and $\{T^t_{\omega'}\}$ on the n-torus are orbit equivalent via a homeomorphism homotopic to the identity (for example, close to the identity) then the vectors ω and ω' are proportional.*

2.2.2.

(1) *Consider a hyperbolic toral automorphism F_L from Section 1.8. Each periodic orbit of F_L determines a unique periodic orbit of any special flow built over F_L. Pick any finite collection $\mathcal{O}_1, \dots, \mathcal{O}_m$ of periodic orbits of F_L. Show that for any positive real numbers $t_1, \dots, t_m$ there exists a positive C^∞ function φ on $\mathbb{T}^2$ such that the orbit of the special flow on $(\mathbb{T}^2)^{\varphi}_{F_L}$ determined by the orbit $\mathcal{O}_i$ has period t_i, $i = 1, \dots, m$.*

(2) **Show that the function φ in (1) can in fact be chosen to be a trigonometric polynomial.*

2.2.3. *Prove that two flows $\sigma^t_{f,\varphi}$, $\sigma^t_{f,\psi}$ under functions φ and ψ built over the same homeomorphism $f: X \to X$ are flow equivalent if there exists a continuous function $\Phi: X \to \mathbb{R}$ such that*

$$\varphi(x) - \psi(x) = \Phi(f(x)) - \Phi(x). \tag{2.2.8}$$

3. Topological conjugacy, factors, and structural stability

Differentiable conjugacy discussed in the previous sections seems to be a very natural basis for a classification of dynamical systems. However, even our fragmentary analysis indicates that at least for global problems the notion is usually far too subtle. The cases where a C^r classification for $r \geq 1$ can be carried out in a reasonable way deserve very careful consideration, but they are vastly outnumbered by situations where such classification is impossible. Besides, all important asymptotic properties identified in Chapter 1 are in fact invariants of C^0 equivalence which we will also call topological conjugacy. Let us state separately this particular case of Definition 2.1.1, which will play a very important role in later parts of this book.

Definition 2.3.1. For $r \geq 0$ two C^r maps $f: M \to M$ and $g: N \to N$ are said to be *topologically conjugate* if there exists a homeomorphism $h: M \to N$ such that $f = h^{-1} \circ g \circ h$.

There is a useful related notion of semiconjugacy:

Definition 2.3.2. A map $g: N \to N$ is a *factor* (or *topological factor*) of $f: M \to M$ if there exists a surjective continuous map $h: M \to N$ such that $h \circ f = g \circ h$. The map h is called a *semiconjugacy*.

Let us point out that the construction of moduli associated with generic periodic orbits in Section 2.1 does not work for topological conjugacy. Furthermore, we will soon see (Section 2.4 and 2.6) that the complete orbit structure of a differentiable map may be stable in the topological sense. This possibility is reflected in the following definition.

Definition 2.3.3. A C^r map f is C^m *structurally stable* $(1 \leq m \leq r)$ if there exists a neighborhood U of f in the C^m topology such that every map $g \in U$ is topologically conjugate to f.

A somewhat stronger version of structural stability is both natural and practical.

Definition 2.3.4. A C^r map is C^m *strongly structurally stable* if it is structurally stable and in addition for any $g \in U$ one can choose a conjugating homeomorphism $h = h_g$ in such a way that both h_g and h_g^{-1} uniformly converge to the identity as g converges to f in C^m topology.

It is important to point out that the previous notions involve *topological* conjugacy of *differentiable* maps. Attempts to either replace topological conjugacy by smooth equivalence or to allow arbitrary continuous maps or even arbitrary homeomorphisms as perturbations lead to vacuous notions. The former statement is supported by the discussion in Section 2.1. The latter follows from the observation that the topological structure of any map can be made more complicated by an arbitrarily small C^0 perturbation. For example any isolated periodic point can be "blown up" into an uncountable set of such points. However, there is a notion of topological stability that is substantial and in certain respects supplementary to structural stability.

Definition 2.3.5. A C^r diffeomorphism is *topologically stable* if it is a factor of any *homeomorphism* sufficiently close to it in the uniform (C^0) topology.

This definition remains substantial for local diffeomorphisms (that is, covering maps) but not for arbitrary differentiable maps.

As before, there are two ways to extend the notion of structural stability to flows. We will not formulate the first, straightforward one involving flow equivalence of all perturbations. Although not completely vacuous, it is rarely fulfilled; for example, in the presence of periodic orbits, their periods are moduli. We will reserve the name of structural stability for the second version which is encountered much more often.

Definition 2.3.6. A C^r flow φ^t is C^m *structurally stable* $(1 \leq m \leq r)$ (correspondingly, C^m *strongly structurally stable*) if any flow sufficiently close to φ^t in the C^m topology is C^0 orbit equivalent to it (correspondingly, if in addition the homeomorphism in question can be chosen close enough to the identity for small perturbations).

The notions of factor and topological stability are modified accordingly.

Definition 2.3.7. A flow $\psi^t\colon N \to N$ is an *orbit factor* of $\varphi^t\colon M \to M$ if there exists a surjective continuous map $h\colon M \to N$ that takes orbits of φ^t onto orbits of ψ^t. A C^r flow φ^t is *topologically stable* if it is an orbit factor of any continuous flow sufficiently close to it in the uniform topology.

Certainly in all definitions of this section compactness of the phase spaces involved is not necessary. Furthermore, one can naturally extend the definitions to the case where for some points the dynamical system is defined only for a finite segment in time, for example, a neighborhood of a hyperbolic fixed point for a linear map. This generalization leads to the concepts of local and semilocal (in a neighborhood of an invariant set) structural stability in the spirit of Section 0.4.

Let us examine some of the examples discussed earlier from the point of view of structural stability and its localized versions.

For a general contracting map the phase space may not have a smooth structure, so our concepts are not directly applicable. However, a contracting map defined in a small disc in Euclidean space is structurally stable, as is a hyperbolic linear map in a neighborhood of its fixed point. In the one-dimensional case this follows from Exercise 2.1.1. In the general situation this will be shown in Section 6.3b. Notice also that Proposition 2.1.7 and Exercise 2.3.3 give examples of structural stability for a global problem with very simple recurrence behavior of the kind discussed in Section 1.1c.

The examples discussed in Sections 1.3–1.5 are not structurally stable. Since topological conjugacy preserves periodic orbits, the rotation R_α with irrational α can not be conjugate to one for which this number is rational. But since both rational and irrational numbers are dense, among arbitrarily small perturbations of a rational rotation one can find irrational ones and vice versa. Similarly for a toral shift T_γ: If all γ_i's are rational numbers then all orbits are periodic, so the vectors with periodic and minimal shifts are intermingled. The same argument works for linear flows on $\mathbb{T}^n$ which have only periodic orbits if $\omega_1, \dots, \omega_n$ are all rational.

Of the three examples of gradient flows discussed in Section 1.6 the second one (vertical torus) is evidently not structurally stable. This follows from the fact that the number of orbits whose α- or ω-limit sets are saddle points changes when the torus is tilted. Those numbers are invariants of topological orbit equivalence. The other two examples (round sphere and tilted torus) are in fact C^1 strongly structurally stable. For the first of these examples an even stronger stability property is true (Exercise 2.3.4). Anyway, the global orbit structures of those flows are rather simple and stability is not surprising.

The really interesting and maybe somewhat surprising examples are the expanding maps from Section 1.7 and hyperbolic toral automorphisms from Section 1.8. These maps have complicated orbit structure (Propositions 1.7.2, 1.7.3, 1.8.1, and 1.8.4) and preservation of such structure under perturbation is certainly revealing. We proceed now to the examination of stability for those examples as well as some relations between them and symbolic systems. In fact, in the next section we will do more than establish structural stability of the maps E_m: we will show that the degree gives a complete topological classification for a large class of circle maps which includes C^1 perturbations of E_m.

Exercises

2.3.1. *Let $f\colon M \to M$ be a C^1 diffeomorphism of a compact manifold. Prove that if f^n has infinitely many fixed points for some n then f is not C^1 strongly structurally stable.*

2.3.2. *Let $f\colon S^1 \to S^1$ be a C^∞ map. Let x be a periodic point of period n such that $|(f^n)'(x)| = 1$. Prove that f is not C^k strongly structurally stable for any k.*

2.3.3. *Let $f\colon S^1 \to S^1$ be a map that has one attracting and one repelling fixed point and no other fixed points, for example, $f(x) = x + \lambda \sin 2\pi x$, where $-1 < \lambda < 1$. Prove that f is strongly structurally stable.*

2.3.4. *Prove that any C^1 flow C^1-close to the gradient flow φ^t on the round sphere is C^0 conjugate to φ^t via a homeomorphism close to the identity.*

4. Topological classification of expanding maps on a circle

a. Expanding maps. Expanding maps in general can be defined similarly to contracting maps from Section 1.1 with the important caveat that distances increase only locally. Let X be a metric space with distance function d.

Definition 2.4.1. A continuous map $f\colon X \to X$ is called *expanding* if for some $\mu > 1, \epsilon_0 > 0$ and every $x, y \in X$ with $x \neq y$ and $d(x, y) < \epsilon_0$ one has

$$d(f(x), f(y)) > \mu d(x, y). \tag{2.4.1}$$

The maps $E_m\colon S^1 \to S^1, |m| \geq 2$,

$$E_m z = z^m \tag{2.4.2}$$

provide examples of expanding maps on the circle. Existence of an expanding map is a nontrivial topological restriction. In fact, among orientable compact surfaces without boundary only the torus possesses expanding maps (because one can show without difficulty that expanding maps are covering maps and the torus is the only compact surface that admits covering maps onto itself; see Section 5 of the Appendix). There are expanding maps among the hyperbolic endomorphisms of the torus described at the end of Section 1.8.

Besides Cartesian products $E_m \times E_k(z_1, z_2) = (z_1^m, z_2^k)$ of circle maps one can take any matrix $L = \begin{pmatrix} a & b \\ c & d \end{pmatrix}$ with integer entries and both eigenvalues of absolute value greater than one and project L to the torus as in Section 1.8,

$$F_L(x, y) = (ax + by,\ cz + dy) \pmod 1.$$

Here we have switched to additive notation. Since $|\det L| > 1$, the map F_L is not invertible. This construction obviously generalizes to arbitrary dimension. An example of an expanding map on a differentiable manifold other than a torus is constructed in Section 17.3.

It follows immediately from (2.4.1) that if M is a Riemannian manifold and $f: M \to M$ is a differentiable expanding map, then for any $x \in M$ the linearized map $D_x f: T_x M \to T_{f(x)} M$ is expanding with respect to the norm generated by the Riemannian metric: There exists $\mu > 1$ such that for every $v \in T_x M \setminus \{0\}$ we have

$$\|D_x f v\| > \mu \|v\|. \tag{2.4.3}$$

On compact manifolds this is sufficient:

Proposition 2.4.2. *If M is compact, condition (2.4.3) is sufficient for the map f to be expanding.*

Proof. First notice that by the Implicit Function Theorem f is a local diffeomorphism. By compactness one can choose $\delta_0 > 0$ such that every ball of radius δ_0 is mapped diffeomorphically onto its image and $\delta_1 > 0$ such that every connected component of the preimage of a δ_1-ball has a diameter less than δ_0. Finally let ϵ_0 be such that $d(x, y) \le \epsilon_0$ implies $d(f(x), f(y)) < \delta_1/2$. Let $\gamma: [0,1] \to M$ with $\gamma(0) = f(x), \gamma(1) = f(y)$ be a smooth curve connecting $f(x)$ and $f(y)$ and lying inside a δ_1-ball. Then $\widetilde{\gamma}$ uniquely defined by $\widetilde{\gamma}(0) = x, \widetilde{\gamma}(1) = y, f(\widetilde{\gamma}(t)) = \gamma(t)$ is a smooth curve connecting x, y and

$$\operatorname{length} \gamma = \int_0^1 \| D_{\widetilde{\gamma}(t)} f \, \widetilde{\gamma}'(t) \| \, dt > \mu \int_0^1 \| \widetilde{\gamma}'(t) \| \, dt = \mu \operatorname{length} \widetilde{\gamma}.$$

Since $d(f(x), f(y)) = \inf_\gamma \operatorname{length} \gamma$, where the infimum is taken over smooth curves connecting $f(x)$ and $f(y)$ inside the δ_1-ball around $f(x)$, we have

$$d(f(x), f(y)) > \mu d(x, y).$$

□

Thus in the case of the circle, using additive notation, we can say that a C^1-map f is expanding if $|f'| > 1$. Since the circle is compact, $\mu := \min_{x \in S^1} |f'(x)| > 1$. By the chain rule $|(f^n)'(x)| > \mu^n$ for any iterate f^n and, if the restriction of f^n to an interval $I \subset S^1$ is an injection, then $\operatorname{length}(f^n I) > \mu^n \operatorname{length} I$. Otherwise the image of I covers S^1. Let us introduce the notion of degree which will be very useful later.

Lemma 2.4.3. *Let $f: S^1 \to S^1$ be an arbitrary continuous map. Consider the universal covering $\pi: \mathbb{R} \to S^1$ of $S^1 = \mathbb{R}/\mathbb{Z}$ and lift f to $\mathbb{R}$ (that is, consider a map $F: \mathbb{R} \to \mathbb{R}$ such that $f \circ \pi = \pi \circ F$). Then $F(x+1) - F(x)$ is an integer which is independent of x and the lift.*

Proof. $\pi(F(x+1)) = f(\pi(x+1)) = f(\pi(x)) = \pi(F(x))$ so $F(x+1) - F(x) \in \mathbb{Z}$ and hence is also independent of x by continuity. If $\tilde{F}$ is another lift of f then $\pi(\tilde{F}(x)) = f(\pi(x)) = \pi(F(x))$, so $\tilde{F} - F$ is a continuous integer-valued function and hence constant, so $\tilde{F}(x+1) - \tilde{F}(x) = F(x+1) - F(x)$. □

Definition 2.4.4. If $f: S^1 \to S^1$ and F is any lift of f then $F(x+1) - F(x)$ is called the *degree* of f and denoted by $\deg(f)$.

Lemma 2.4.5. *The degree is continuous and hence locally constant in the C^0 (uniform) topology.*

Proof. Let $g: S^1 \to S^1$ be a map uniformly close to f. We may assume that $\operatorname{dist}(g(x), f(x)) < 1/4$. Lift both f and g to the real line and denote the lifts by F and G, respectively. Let $\varphi(x) = G(x) - F(x)$. For $x \in [0,1]$ we have

$$G(x+1) - \varphi(x+1) = F(x+1) = F(x) + \deg(f) = G(x) + \deg(f) - \varphi(x)$$

so

$$-1/2 < |G(x+1) - G(x) - \deg(f)| < 1/2. \tag{2.4.4}$$

Since g is a map of the circle, $G(x+1) - G(x)$ is an integer and from (2.4.4) it must be equal to $\deg(f)$. □

If f is an expanding map then

$$|f(x+1) - f(x)| = \left| \int_0^1 f'(x+t)dt \right| > \lambda > 1.$$

Thus the degree of any expanding map is greater than one in absolute value. On the other hand the map $E_k: x \mapsto kx \pmod 1$ is expanding for any integer k with $|k| > 1$.

We will return to the discussion of the degree of a circle map in Section 8.2 when it will serve as a motivation for a more general and less obvious concept of degree.

b. Conjugacy via coding.

Theorem 2.4.6. *Every expanding map f of the circle of degree k is topologically conjugate to the map E_k. Furthermore, if f is close enough to E_k in the C^0 (uniform) topology, then the conjugating homeomorphism can be chosen close to the identity.*

Lemma 2.4.7. *Every continuous circle map f of degree k, $|k| \neq 1$, has a lift with a fixed point $p \in [-\frac{1}{2}, \frac{1}{2}]$. If f is C^0-close to E_k then p is near 0.*

Proof. Let F be a lift of f and $H(x) := F(x) - x$. The interval with endpoints $H(-\frac{1}{2})$ and $H(\frac{1}{2}) = F(\frac{1}{2}) - \frac{1}{2} = F(-\frac{1}{2}) + k - \frac{1}{2} = H(-\frac{1}{2}) + k - 1$ has length $|k-1| \geq 1$ and hence contains some $m \in \mathbb{Z}$. By the Intermediate Value Theorem there is a $p \in [-\frac{1}{2}, \frac{1}{2}]$ such that $H(p) = m$. Replacing F with $F - m$ then yields $F(p) = p$. If f is close to E_k then $F(x) - x$ is close to $kx - x$. Hence it changes sign near 0. □

Proof of Theorem 2.4.6. We will give the proof for positive k and mention at the end the fairly obvious modifications for negative values of the degree.

Let $\Delta_0^0 = [0,1]$ and $\Delta_n^m = [m/k^n, (m+1)/k^n]$ for $n \in \mathbb{N}, 0 \leq m \leq k^n - 1$. Then

$$E_k \Delta_n^m = \Delta_{n-1}^{m'} \tag{2.4.5}$$

for $n \in \mathbb{N}$, where m' is the unique integer between 0 and $k^{n-1} - 1$ such that $m' = m \pmod{k^{n-1}}$.

Let $\xi_n = \{\Delta_n^0, \dots, \Delta_n^{k^n-1}\}$ be the partition of S^1 into the intervals Δ_n^m. We slightly abuse the word "partition" since the ends of successive intervals overlap. For a given expanding map f of degree k we will construct a nested sequence of partitions

$$\eta_n = \{\Gamma_n^0, \dots, \Gamma_n^{k^n-1}\}$$

into intervals which will be in a natural, order-preserving correspondence with the standard sequence ξ_n. Let p be a fixed point for the lift F of f as in Lemma 2.4.7. If f is close to E_k, pick p close to 0. Since $F(p) = p$, $F(p+1) = p+k$, and F is a strictly monotone function, there are uniquely defined points $a_1^0 = p < a_1^1 < a_1^2 < \cdots < a_1^{k-1} < p+1 = a_1^k$ such that $F(a_1^m) = p+m$ for $m = 0, 1, \dots, k$. Let us write $\Gamma_1^m = \pi([a_1^m, a_1^{m+1}])$ for $m = 0, \dots, k-1$. Obviously $f(\Gamma_1^m) = S^1$ and f is injective on the interval Γ_1^m except for the identification of the ends. If f is close to E_k then clearly each number a_1^m is close to m/k.

Furthermore, on each interval Γ_1^m one can find uniquely defined points $a_1^m = a_2^{km} < a_2^{km+1} < \cdots < a_2^{km+k-1} < a_2^{k(m+1)} = a_1^{m+1}$ such that $F(a_2^{km+i}) = a_1^i \pmod 1$ for $i = 0, \dots, k$ and again a_2^{km+i} is close to $(km+i)/k^2$ if f is close to E_k. Let $\Gamma_2^m = \pi([a_2^m, a_2^{m+1}])$ for $m = 0, \dots, k^2-1$, so that $f(\Gamma_2^m) = \Gamma_1^{m'}$, where m' is the unique integer between 0 and $k-1$ such that $m = m' \pmod k$.

We continue inductively and for each $n \in \mathbb{N}$ define points a_n^{km+i} for $m = 0, \dots, k^{n-1} - 1$ and $i = 0, \dots, k$ such that $a_{n-1}^m = a_n^{km} < a_n^{km+1} < \cdots < a_n^{km+k-1} < a_n^{k(m+1)} = a_{n-1}^{m+1}$ and

$$F(a_n^{km+i}) = a_{n-1}^{m'} \pmod 1, \tag{2.4.6}$$

where, as before, $0 \leq m' \leq k^{n-1} - 1$ and $m' = km + i \pmod{k^{n-1}}$. We let $\pi([a_n^m, a_n^{m+1}]) =: \Gamma_n^m$ for $m = 0, \dots, k^n - 1$ so that $f(\Gamma_n^m) = \Gamma_{n-1}^{m'}$, where again $0 \leq m' \leq k^{n-1} - 1$ and $m = m' \pmod{k^{n-1}}$.

So far we have used only the fact that f is strictly monotone and has degree k. By induction $f^n(\Gamma_n^m) = S^1$ and f^n is injective on Γ_n^m except for its ends. If f is expanding, that is, $|f'| \geq \mu > 1$, this implies that the length of each interval Γ_n^m does not exceed μ^{-n}, so the set of all points $\{a_n^m\}_{n \in \mathbb{N}, m=0,\dots,k^n-1}$ is dense on S^1. This is the only place in the proof where we are using the fact that f is an expanding map. (In fact, the use of differentiability could be easily avoided.)

Furthermore, for any N and ϵ one can find $\delta > 0$ such that if f is δ-close to E_k in the uniform topology, then

$$|a_n^m - \frac{m}{k^n}| < \frac{\epsilon}{3} \qquad \text{for } n = 1, \dots, N, \quad m = 0, 1, \dots, k^n - 1. \tag{2.4.7}$$

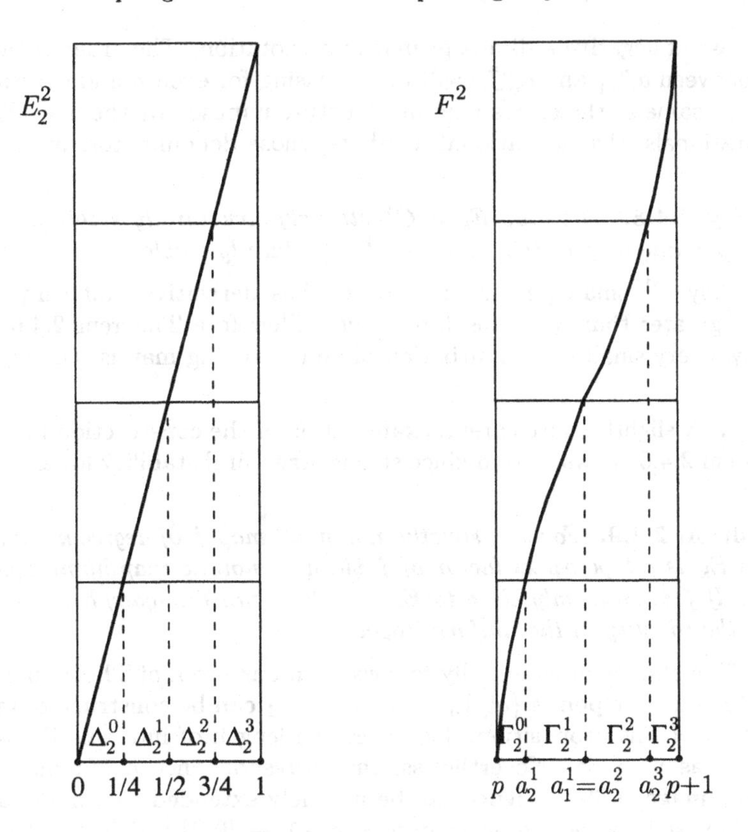

FIGURE 2.4.1. Conjugacy via coding

We define the correspondence h between the set $\{a_n^m\}$ and all rational numbers on the interval $[0,1]$ whose denominators are powers of k by

$$h(a_n^m) = \frac{m}{k^n}$$

This correspondence is monotone and since the set of a's is dense in the interval $[p, p+1]$ it can be uniquely extended to a continuous strictly monotone map (hence, a homeomorphism) $h: [p, p+1] \to [0,1]$. Since $h(\Gamma_n^m) = \Delta_n^m$ for all n, m, (2.4.5) and (2.4.6) imply that

$$E_k \circ h = h \circ f \tag{2.4.8}$$

Assuming under the conditions of (2.4.7) that N and ϵ are chosen such that $1/k^n < \epsilon/3$, one sees in addition that $|a_n^m - h(a_n^m)| < \epsilon$ for any $n \in \mathbb{N}, m = 0, 1, \dots, k^n - 1$, and hence $|h(x) - x| < \epsilon$ for all x. A similar argument shows that $|h^{-1}(x) - x| < \epsilon$.

The case of negative k differs primarily by notation. The order of the points a_n^{km+i} between a_{n-1}^m and a_{n-1}^{m+1} will be increasing for even n and decreasing for odd n, the same as the corresponding structure imposed by the map E_k on the k-nary rationals, that is, rational numbers whose denominators are powers of k. □

Corollary 2.4.8. *The map E_k is C^1 strongly structurally stable for $|k| > 1$. Every expanding map of the circle is C^1-structurally stable.*

Proof. Any C^1 small perturbation of E_k has derivative uniformly close to k, hence greater than 1 in absolute value. Therefore Theorem 2.4.6 applies. Similarly, every small C^1 perturbation of an expanding map is still expanding. □

Remark. A slightly more careful examination of the construction in the proof of Theorem 2.4.6 allows us to deduce strong structural stability for any expanding map.

Proposition 2.4.9. *For any strictly monotone map f of degree k with $|k| > 1$ the map E_k is a topological factor of f via a monotone map homotopic to the identity. If f is uniformly close to E_k then the semiconjugacy h can be chosen close to the identity in the uniform topology.*

Proof. The statement essentially follows from our proof of Theorem 2.4.6. As we pointed out, the points $\{a_n^m\}_{n\in\mathbb{N},m=0,1,\dots,k^n-1}$ can be constructed under the assumption of this proposition. However, the length of intervals Γ_n^m may not go to zero as $n \to \infty$. Nevertheless, the correspondence h is monotone and has dense image on $[0,1]$, hence can be uniquely extended to a monotone, but possibly not strictly monotone, map $h\colon [p, p+1] \to [0,1]$ satisfying (2.4.8). If f is C^0 close to E_k then (2.4.7) ensures that the map h is uniformly close to the identity. □

Remark. Since the map h constructed in the last proof is monotone, it may be constant on no more than countably many intervals. This implies that the set of points with more than one preimage is at most countable.

Let us go back to the construction of Section 1.7. In the course of the proof of Proposition 1.7.2 we effectively constructed a semiconjugacy between the one-sided 2-shift σ_2^R and the map E_2 (cf. (1.7.2)). Thus, E_2 is a factor of σ_2^R. This construction obviously extends to $\sigma_{|k|}^R$ and E_k for any k with $|k| > 1$, and by Theorem 2.4.6 we can replace E_k by any expanding map of the circle of degree k. The noninvertibility of the semiconjugacy has to do with the fact that any binary rational $m/2^n$ has two different binary representations, with 0's and 1's at the end. The semiconjugacy $h\colon \Omega_2^R \to S^1$, assigning to a sequence ω of 0's and 1's the number between 0 and 1 for which the sequence ω provides a binary representation, is obviously not a homeomorphism; it has a dense countable set of points of noninvertibility. This is the simplest case of a "natural" semiconjugacy between a symbolic and a smooth system. Another not so self-evident case will be discussed in the next section. It involves hyperbolic automorphisms of the two-torus.

c. The fixed-point method. We will finish this section by outlining another, seemingly less constructive proof of Theorem 2.4.6 and Proposition 2.4.9 which presages in our simple setting some general methods used in the theory of hyperbolic dynamical systems (cf. Sections 2.6, 6.2, and 18.1).

We want to solve the functional equation (2.4.8)

$$E_k \circ h = h \circ f$$

for h. Let us try to reformulate the problem of finding such a map h as a fixed-point problem for a contracting operator in a function space. First, let us assume that 0 is a fixed point of f. The difficulty lies in the noninvertibility of the map E_k. We bypass it by rewriting (2.4.8). Let $\mathfrak{C}$ be the space of all continuous maps h from the interval $[0,1]$ such that $h(0) = 0, h(1) = 1$, endowed with the uniform metric. These are exactly the maps projecting to maps on S^1 of degree one, that is, those homotopic to the identity. We then restate (2.4.8) as

$$h = \mathcal{F}(h), \tag{2.4.9}$$

where $\mathcal{F}$ is the operator on $\mathfrak{C}$ defined by

$$(\mathcal{F}h)(x) = \begin{cases} \frac{1}{k}h(\{F(x)\}) + \frac{m}{k} \pmod 1 & \text{for } a_1^m \leq x < a_1^{m+1},\ 0 \leq m < k, \\ 1 & \text{for } x = 1. \end{cases} \tag{2.4.10}$$

Here $\{x\}$ means the fractional part. In other words, we apply the mth branch $(x+m)/k$ of the inverse to E_k on the interval Γ_1^m. Notice that $\mathcal{F}$ maps $\mathfrak{C}$ to itself: Substituting (2.4.10) into (2.4.9) and acting by E_k from the left we obtain (2.4.8). The fact that $\mathcal{F}(h)$ is continuous is obvious at all points except for $x = a_1^m$ $(m = 0, \dots, n-1)$; but since all those points are mapped to 0 by F, changing branches does not destroy continuity. Thus $\mathcal{F}$ maps the space $\mathfrak{C}$ into itself.

Let $h_1, h_2 \in \mathfrak{C}$. Then

$$\operatorname{dist}(\mathcal{F}(h_1), \mathcal{F}(h_2)) = \sup_{x\in[0,1]} \left|\frac{1}{k} h_1(\{F(x)\}) - \frac{1}{k}h_2(\{F(x)\})\right| = \frac{1}{|k|} \operatorname{dist}(h_1, h_2). \tag{2.4.11}$$

Hence $\mathcal{F}$ is a contracting map of the space $\mathfrak{C}$. By Proposition 1.1.2 it has a unique fixed point h_0. This gives the first statement of Theorem 2.4.6.

Closeness of the solution to the identity can be derived by noticing that

$$\operatorname{dist}(\mathrm{Id}, \mathcal{F}(\mathrm{Id})) = \sup_{\substack{0\leq m\leq k-1 \\ a_1^m \leq x \leq a_1^{m+1}}} \left|\frac{F(x)+m}{k} - x\right|$$

is obviously small if f is close to E_k and using the fact that the fixed point can be obtained as the limit of iterates under $\mathcal{F}$ of any initial map, for example, the identity. By (2.4.11),

$$\operatorname{dist}(\mathrm{Id}, h_0) \leq \sum_{i=0}^{\infty} \operatorname{dist}(\mathcal{F}^{i+1}(\mathrm{Id}), \mathcal{F}^i(\mathrm{Id})) = \frac{|k|}{|k|-1} \operatorname{dist}(\mathrm{Id}, \mathcal{F}(\mathrm{Id})).$$

Finally, in order to prove that h_0 is a homeomorphism if f is an expanding map, one should consider the functional equation for its inverse

$$h = \tilde{\mathcal{F}}(h),$$

where

$$\tilde{\mathcal{F}}(h) = f_m^{-1}(h(kx)) \text{ for } \frac{m}{k} \leq x \leq \frac{m+1}{k}, m = 0, \dots, k-1, \tag{2.4.12}$$

and f_m^{-1} is the mth branch of f^{-1}, which maps the whole circle onto the interval $[a_1^m, a_1^{m+1}]$. One can see that $|(f_m^{-1})|' < 1$ and thus derive an estimate similar to (2.4.11), although not so explicit. Thus, $\tilde{\mathcal{F}}$ is a contracting operator with a unique fixed point $\tilde{h}_0$. Finally one can deduce that $h_0 \circ \tilde{h}_0 \circ E_k = h_0 \circ f \circ \tilde{h}_0 = E_k \circ h_0 \circ \tilde{h}_0$. Since $\mathfrak{C}$ is closed under composition we utilize the following

Lemma 2.4.10. *The identity is the only map $g: S^1 \to S^1$ of degree one which commutes with some E_k with $|k| > 1$.*

Proof. Lift E_k and g to $\mathbb{R}$ and denote the lift of g by $\mathrm{Id} + \tilde{G}$ with $\tilde{G}$ periodic since $\deg g = 1$. From $E_k \circ g = g \circ E_k$ we obtain

$$kx + k\tilde{G}(x) = kx + \tilde{G}(kx)$$

and

$$\tilde{G}(k^n x) = k^n \tilde{G}(x) \tag{2.4.13}$$

for all $n \in \mathbb{Z}$. But the right-hand side in (2.4.13) goes to infinity if $\tilde{G}(x) \neq 0$. Since $\tilde{G}$ is periodic and hence bounded we have $\tilde{G} = 0$. □

Since both h_0 and $\tilde{h}_0$ have degree one, so does $h_0 \circ \tilde{h}_0$. By Lemma 2.4.10 we thus have $h_0 \circ \tilde{h}_0 = \mathrm{Id}$ and h_0 and $\tilde{h}_0$ are homeomorphisms.

Exercises

2.4.1. *Let $g: [0,1] \to [0,1]$ be the "tent" map*

$$g(x) = \begin{cases} 2x, & 0 \leq x \leq 1/2, \\ 2 - 2x, & 1/2 \leq x \leq 1. \end{cases}$$

Prove the following restricted and modified version of C^1 strong structural stability for g: For any continuous map $g_1: [0,1] \to [0,1]$ such that $g_1(0) = g_1(1) = 0$, $g_1(1/2) = 1$, $g - g_1$ is C^1 on $[0,1] \setminus \{1/2\}$, and $|g - g_1|$ is small enough together with its derivative, there exists a homeomorphism $h: [0,1] \to [0,1]$ that is C^0-close to the identity and such that $g_1 = h \circ g \circ h^{-1}$.

2.4.2*. *Prove that the map g from the previous problem is topologically conjugate to the quadratic map $f_4: x \mapsto 4x(1-x)$.*

2.4.3. *Prove that f_4 is topologically transitive.*

2.4.4*. *Let k, l be two positive integers such that $k^m \neq l^n$ for any positive integers m, n, $f: S^1 \to S^1$ an analytic expanding map of degree k, and g an analytic map of degree l which commutes with f. Prove that there exists an analytic diffeomorphism $h: S^1 \to S^1$ such that*

$$h^{-1} \circ f \circ h = E_k, \quad h^{-1} \circ g \circ h = E_l.$$

2.4.5. *Let $f: S^1 \to S^1$ be a strictly monotone continuous map of degree k, $|k| \geq 2$, such that $f(0) = 0$ and let $F: \mathbb{R} \to \mathbb{R}$ be the lift of f for which $F(0) = 0$. Prove that the semiconjugacy h between f and the linear map E_k with $h(0) = 0$ is given by $h(x) = \lim_{n\to\infty} F^n(x)/k^n$.*

2.4.6. *Show that if a map $f: S^1 \to S^1$ is expanding with respect to some Riemannian metric on S^1 then there exist $\lambda > 1$ and $C > 0$ such that for all $n \in \mathbb{N}$ we have $|f^{n\prime}| > C\lambda^n$.*

5. Coding, horseshoes, and Markov partitions

a. Markov partitions. Semiconjugacies between one-sided shifts and expanding maps of the circle represent a simple but very effective example of coding of orbits for a smooth system, a concept alluded to at the beginning of Section 1.9. In general one can try to code orbits of, say, a diffeomorphism, or even a homeomorphism f, by dividing its phase space X into finitely many pieces $X_0, \ldots, X_{N-1}$ and registering to which of the pieces successive iterates of a point $x \in X$ belong. Two basic difficulties appear in this construction:

(1) If the pieces overlap (as the intervals Δ_n^m do in the previous section) one point is coded by more than one sequence and
(2) all points in the intersection

$$\bigcap_{n\in\mathbb{Z}} f^{-n}(X_{\omega_n}) \tag{2.5.1}$$

are coded by the same sequence $\omega = \{\omega_n\}_{n\in\mathbb{Z}}$.

Thus in general our procedure does not produce any map in either direction between the space X and a subset of the sequence space Ω_N. In order to have a reasonable relation between the topologies in the phase space and in the sequence space the pieces should be closed sets. Thus if X is, say, a connected manifold, the first difficulty cannot be avoided. This statement should be qualified in two ways. First, one can sometimes avoid overlaps in the case of semilocal analysis, as we will see later in this section. Second, overlaps of measure zero are not significant if one studies statistical properties of typical orbits with respect to a measure invariant under f (cf. Section 4.1), since in that setting sets of measure zero can be neglected.

If every intersection (2.5.1) contains no more than one point one can define a continuous map h from a closed subset $\Lambda \subset \Omega_N$ onto X such that $f \circ h = h \circ \sigma_N$. Thus in this case the map f is a factor of some symbolic system. This construction is particularly informative when the set Λ has a well-understood structure, for example, if $\Lambda = \Omega_A$ for a 0-1 matrix A (cf. Definition 1.9.3), and, in addition, the sets of different codings for the same points are not very massive and can be described reasonably well. For example it is good if on a "large" set the map h is one-to-one. Clearly, the semiconjugacy described at the end of Subsection 2.4b satisfies all those conditions, modified in an obvious way for noninvertible systems. For it to be a semiconjugacy, however, we had to deviate slightly from the prescription (2.5.1) because the sets Δ_n^k in the proof of Proposition 1.7.2 (or the sets Δ_n^m in (2.4.5)) are not of the form $\bigcap_{i=0}^n f^{-n}\Delta_1^{\omega_i}$, for example, $E_2^{-1}(\Delta_1^0) \cap \Delta_1^0 = \Delta_2^0 \cup \{1/2\}$. Thus, instead of (2.5.1), we consider expressions like

$$\bigcap_{n \in \mathbb{Z}} \overline{\operatorname{Int}(\bigcap_{|k| \leq n} f^{-k}(X_{\omega_k}))} \tag{2.5.2}$$

to ensure that despite overlaps on the boundaries we obtain intersections consisting of a single point.

One would like to call the decomposition $(X_0, \dots, X_N)$, which provides a semiconjugacy between a topological Markov chain σ_A and f, is one-to-one on a large set, and has identifications describable in a finite "Markov" way, a *Markov partition.* We postpone a more technical discussion to Sections 15.1 and 18.7. Now we describe several specific situations (other than expanding circle maps) where the phenomenon of Markov partition occurs in an unambiguous way.

b. Quadratic maps. We first describe a class of maps that are somewhat similar to the expanding maps. But unlike in the case of expanding maps we obtain a conjugacy rather than semiconjugacy, to a symbolic system. Consider the quadratic map

$$f_\lambda : \mathbb{R} \to \mathbb{R}, \quad x \to \lambda x(1-x),$$

where $\lambda > 2 + \sqrt{5}$[1]. For notational convenience we write f instead of f_λ. Note first that for $x < 0$ we have $f(x) < x$ and $f'(x) > \lambda > 4$, so $f^n(x) \to -\infty$ when $x < 0$. Likewise one sees that $f^n(x) \to -\infty$ when $x > 1$. Thus the collection Λ of points with bounded orbits coincides with the collection of points whose orbits are in $[0,1]$, that is, $\Lambda = \bigcap_{n \in \mathbb{N}_0} f^{-n}([0,1])$.

Let

$$\Delta^0 = \left[0, \frac{1}{2} - \sqrt{\frac{1}{4} - \frac{1}{\lambda}}\right] \text{ and } \Delta^1 = \left[\frac{1}{2} + \sqrt{\frac{1}{4} - \frac{1}{\lambda}}, 1\right].$$

Then $f^{-1}([0,1]) = \Delta^0 \cup \Delta^1$ by solving the quadratic equation $f(x) = 1$. Next one sees that $f^{-2}([0,1]) = \Delta^{00} \cup \Delta^{01} \cup \Delta^{11} \cup \Delta^{10}$ consists of four intervals, and so forth.

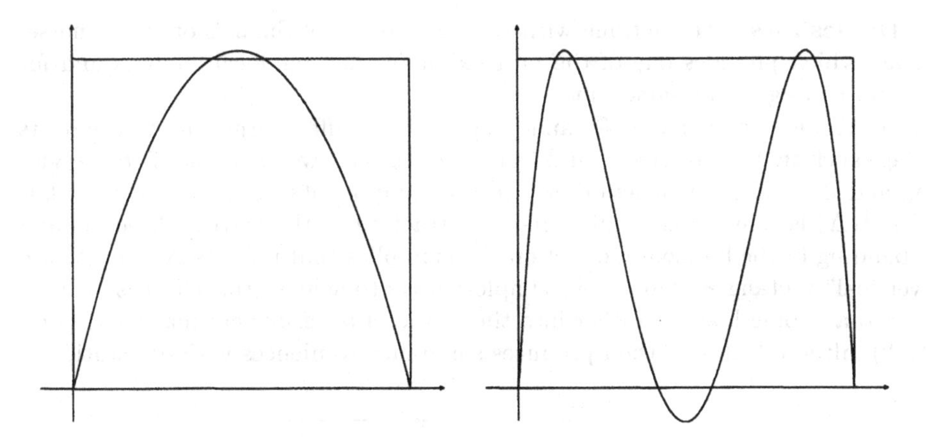

FIGURE 2.5.1. A quadratic map and its second iterate

Consider the partition of Λ by Δ^0 and Δ^1. These pieces do not overlap. Note furthermore that $|f'(x)| \geq \sqrt{\lambda^2 - 4\lambda} > 1$ on $\Delta^0 \cup \Delta^1$:

$$\begin{aligned}|f'(x)| &= |\lambda(1-2x)| = 2\lambda|x - \frac{1}{2}| \\ &\geq 2\lambda\sqrt{\frac{1}{4} - \frac{1}{\lambda}} = \sqrt{\lambda^2 - 4\lambda} > \sqrt{(2+\sqrt{5})^2 - 4(2+\sqrt{5})} = 1.\end{aligned}$$

This shows that for any sequence $\omega = (\omega_0, \omega_1, \dots)$ the diameter of the intersections

$$\bigcap_{n=0}^{N} f^{-n}(\Delta^{\omega_n})$$

decreases (exponentially) as $N \to \infty$. Thus $\Lambda = \bigcap_{n\in\mathbb{N}} f^{-n}([0,1])$ is a Cantor set and for a sequence $\omega = (\omega_0, \omega_1, \dots)$ the intersection

$$h(\{\omega\}) = \bigcap_{n\in\mathbb{N}_0} f^{-n}(\Delta^{\omega_n}) \tag{2.5.3}$$

consists of exactly one point. Furthermore the map

$$h\colon \Omega_2^R \to \Lambda$$

defined by (2.5.3) is a bijection. Since two sequences ω and ω' which are close to each other in Ω_2 have a long initial segment in common, their images x and x' under h are close (their distance is exponentially small as a function of the length of the common segment), and hence h is continuous. Conversely, two nearby points x, x' arise from nearby sequences; thus h is a homeomorphism.

Thus f restricted to the set of points with bounded orbits is topologically conjugate to the full one-sided 2-shift σ_2^R.

c. Horseshoes. We continue with the description of Smale's original "horseshoe" which provides one of the best examples for semilocal analysis and for perfect coding at the same time.

Let Δ be a rectangle in $\mathbb{R}^2$ and $f: \Delta \to \mathbb{R}^2$ a diffeomorphism of Δ onto its image such that the intersection $\Delta \cap f(\Delta)$ consists of two "horizontal" rectangles Δ_0 and Δ_1 and the restriction of f to the components $\Delta^i \subset f^{-1}(\Delta)$, $i = 0, 1$, of $f^{-1}(\Delta)$ is a hyperbolic affine map, contracting in the vertical direction and expanding in the horizontal direction. This implies that the sets Δ^0 and Δ^1 are "vertical" rectangles. One of the simplest ways to achieve this effect is to bend Δ into a "horseshoe", or rather into the shape of a permanent magnet (Figure 2.5.2), although this method produces some inconveniences with orientation.

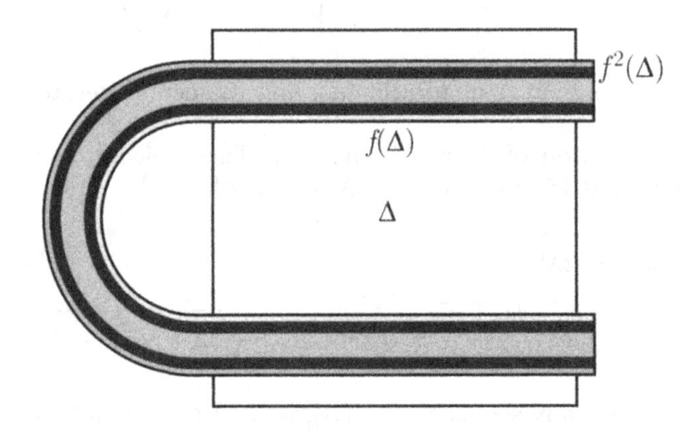

FIGURE 2.5.2. The horseshoe

Another way, which is better from the point of view of orientation, is to bend Δ roughly into a "G" shape (Figure 2.5.3).

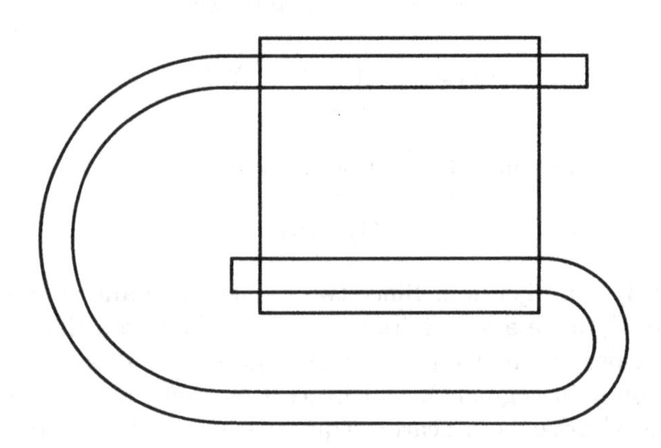

FIGURE 2.5.3. The alternate horseshoe

Let us study the maximal invariant subset of Δ. Clearly $\Lambda = \bigcap_{n=-\infty}^{\infty} f^n(\Delta)$, but it is not clear whether Λ is contained in the interior of Δ. We will use Δ^0 and Δ^1 as the "pieces" in the coding construction and will start with positive iterates. The intersection $\Delta \cap f(\Delta) \cap f^2(\Delta)$ consists of four thin horizontal rectangles: $\Delta_{ij} = \Delta_i \cap f(\Delta_j) = f(\Delta^i) \cap f^2(\Delta^j)$, $i, j \in \{0, 1\}$ (cf. Figure 2.5.2). Continuing inductively one sees that $\bigcap_{i=0}^{n} f^i(\Delta)$ consists of 2^n thin disjoint horizontal rectangles whose heights are exponentially decreasing with n. Each such rectangle has the form $\bigcap_{i=1}^{n} f^i(\Delta^{\omega_i})$, where $\omega_i \in \{0, 1\}$ for $i = 1, \dots, n$, and we will denote it by $\Delta_{\omega_1, \dots, \omega_n}$. Each infinite intersection $\bigcap_{n=1}^{\infty} f^n(\Delta^{\omega_n})$, $\omega_n \in \{0, 1\}$, is a horizontal segment and the intersection $\bigcap_{n=1}^{\infty} f^n(\Delta)$ is the product of the horizontal segment with a Cantor set in the vertical direction. Similarly, one defines and studies vertical rectangles $\Delta^{\omega_0, \dots, \omega_{-n}} = \bigcap_{i=0}^{n} f^{-i}(\Delta^{\omega_{-i}})$, the vertical segments $\bigcap_{n=0}^{\infty} f^{-n}(\Delta^{\omega_{-n}})$, and the set $\bigcap_{n=0}^{\infty} f^{-n}(\Delta)$, which is the product of a segment in the vertical direction with a Cantor set in horizontal direction. Finally, the desired invariant set $\Lambda = \bigcap_{n=-\infty}^{\infty} f^n(\Delta)$ is the product of two Cantor sets, hence a Cantor set itself, and the map

$$h \colon \Omega_2 \to \Lambda, \qquad h(\{\omega\}) = \bigcap_{n=-\infty}^{\infty} f^{-n}(\Delta^{\omega_n}) \tag{2.5.4}$$

is a homeomorphism conjugating the shift σ_2 and the restriction of the diffeomorphism f to the set Λ. Incidentally, we see that the invariant set Λ belongs to the interior of the rectangle Δ. Since periodic points and topological mixing are invariants of topological conjugacy, an application of Proposition 1.9.1 immediately gives substantial information about the behavior of f on Λ.

Corollary 2.5.1. *Periodic points of f are dense in Λ, $P_n(f\restriction_{\Lambda}) = 2^n$, and the restriction of f to the set Λ is topologically mixing.*

Naturally, the construction of our "horseshoe" can be modified and generalized in several ways. First, instead of a single rectangle Δ whose image intersects it twice, one can start from a collection of disjoint rectangles $\Delta^{(1)}, \dots, \Delta^{(N)}$ with parallel sides, which we will call "vertical" and "horizontal", and map them in such a way that every connected component of $f^{-1}(\Delta^{(j)}) \cap \Delta^{(i)}$ is a "vertical" subrectangle of $\Delta^{(i)}$ mapped affinely onto a "horizontal" subrectangle of $\Delta^{(j)}$. Three important conditions must be observed. The former component has full "height", its image has full "length", and all components are contracted in the vertical direction and expanded in the horizontal one. It is rather clear that the action of f on the maximal invariant subset Λ of $\bigcup_{i=1}^{N} \Delta^{(i)}$ is topologically conjugate to the topological Markov chain (see Definition 1.9.3) whose states are identified with the connected components of $f^{-1}(\Delta^{(j)}) \cap \Delta^{(i)}$ and whose transition matrix A has ones exactly where the image of the component corresponding to the row intersects the component corresponding to the column.

Secondly, one can consider multidimensional "horseshoes" where rectangles (that is, the products of intervals) are replaced by products of "nice" subsets of higher-dimensional spaces.

Finally, strict linearity of the map on the components of intersection is not necessary. For example, any C^1 small perturbation of the situation described above still produces an invariant set topologically equivalent to a topological Markov chain—this is a particular case of Theorem 18.2.1 dealing with structural stability. More general sufficient conditions for the existence of nonlinear horseshoes will be given in Section 6.5 (see Definition 6.5.2 and Theorem 6.5.5). A far-reaching example of the use of nonlinear horseshoes in the general structural theory of smooth dynamical systems is Theorem S.5.9 of the supplement and its corollaries.

d. Coding of the toral automorphism. Next we will show how the idea of coding can be applied in a natural way to hyperbolic toral automorphisms. In order to simplify notations and keep the construction more visual we will consider the specific map F_L of the two-torus from Section 1.8

$$F_L(x, y) = (2x + y,\ x + y) \qquad (\text{mod } 1).$$

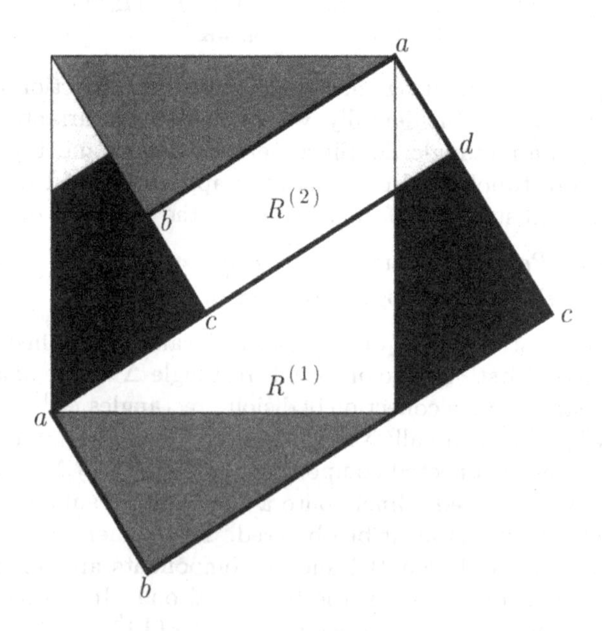

FIGURE 2.5.4. Partitioning the torus

Let us draw segments of the two eigenlines at the origin until they cross sufficiently many times and separate the torus into disjoint rectangles. Although this prescription contains an ambiguity, direct inspection shows that it can be effected by taking a segment of the contracting line in the fourth quadrant until it intersects the segment of the expanding line twice in the first quadrant and once in the third quadrant (see Figure 2.5.4). The resulting configuration is a decomposition of the torus into two rectangles $R^{(1)}$ and $R^{(2)}$. Three pairs among the seven vertices of the plane configuration are identified, so there are only four different points on the torus which serve as vertices of the rectangles. This agrees with our description: those vertices are exactly the origin and three intersection points. Although $R^{(1)}$ and $R^{(2)}$ are not disjoint, let us try to apply the prescription for the construction of a generalized horseshoe described previously, using $R^{(1)}$ and $R^{(2)}$ as basic rectangles. Naturally, the expanding and contracting eigendirections will play the role of the "horizontal" and "vertical" directions correspondingly. It is rather easy to see even without an explicit calculation that the image $F(R^{(i)})$ $(i = 1,2)$ consists of several "horizontal" rectangles of full length. The union of the boundaries $\partial R^{(1)} \cup \partial R^{(2)}$ consists of the segments of the two eigenlines at the origin just described. The image of the contracting segment is a part of that segment. Thus, the images of $R^{(1)}$ and $R^{(2)}$ have to be "anchored" at parts of their "vertical" sides, that is, once one of the images "enters" either $R^{(1)}$ or $R^{(2)}$ it has to stretch all the way through it. An explicit calculation shows that $F(R^{(1)})$ consists of three components, two in $R^{(1)}$ and one in $R^{(2)}$. The image of $R^{(2)}$ has two components, one in each rectangle (see Figure 2.5.5).

We can use these five components $\Delta_0, \Delta_1, \Delta_2, \Delta_3, \Delta_4$ (or their preimages) as the pieces in our coding construction. Due to the contraction of F in the "vertical" direction and contraction of F^{-1} in the "horizontal" direction each intersection (2.5.1) contains no more than one point. On the other hand, because of the "Markov" property described previously, that is, the images going full length through rectangles, the following is true: If $\omega \in \Omega_5$ and $F(\Delta_{\omega_n}) \cap \Delta_{\omega_{n+1}} \neq \varnothing$ for all integers $n \in \mathbb{Z}$, then $\bigcap_{n\in\mathbb{Z}} F^n(\Delta_{\omega_n}) \neq \varnothing$. In other words, we have a coding, that is, a semiconjugacy $h\colon \Omega_A \to \mathbb{T}^2$, where

$$A = \begin{pmatrix} 1 & 1 & 0 & 1 & 0 \\ 1 & 1 & 0 & 1 & 0 \\ 1 & 1 & 0 & 1 & 0 \\ 0 & 0 & 1 & 0 & 1 \\ 0 & 0 & 1 & 0 & 1 \end{pmatrix} \tag{2.5.5}$$

such that

$$F \circ h = h \circ \sigma_A. \tag{2.5.6}$$

Let us try to describe the identifications arising from that semiconjugacy, that is, what points on the torus have more than one preimage. First, obviously, the topological Markov chain σ_A has three fixed points, namely, the constant sequences of 0's, 1's, and 4's whereas the toral automorphism F has only one, the origin. It is easy to see that all three fixed points are indeed mapped to the origin. This explains the difference in the calculation of the number of periodic points: $P_n(F) = \lambda_1^n + \lambda_1^{-n} - 2$ (Proposition 1.8.1), whereas $P_n(\sigma_A) = \operatorname{tr} A^n = \lambda_1^n + \lambda_1^{-n} = P_n(F) + 2$ (Proposition 1.9.1), where $\lambda_1 = (3+\sqrt{5})/2$ is

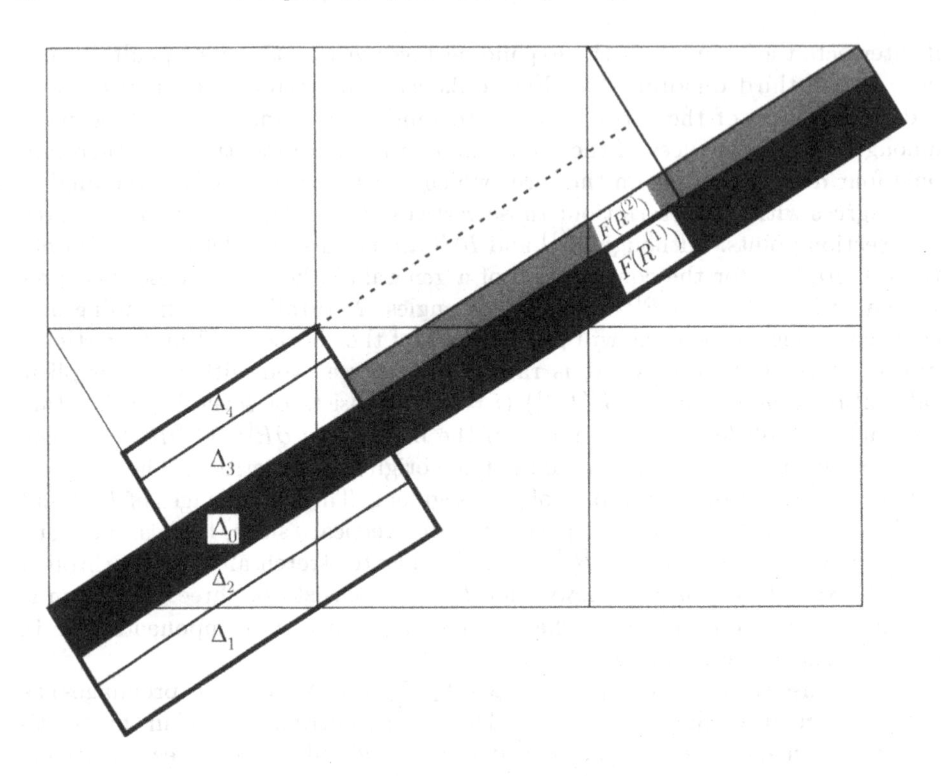

FIGURE 2.5.5. The image of the partition

the maximal eigenvalue for both the 2×2 matrix $\begin{pmatrix} 2 & 1 \\ 1 & 1 \end{pmatrix}$ and for the 5×5 matrix (2.5.5).

Furthermore, one can see that every point $q \in \mathbb{T}^2$ whose positive and negative iterates avoid the boundaries $\partial R^{(1)}$ and $\partial R^{(2)}$ has a unique preimage and vice versa. The points of Ω_A whose images are on those boundaries or their iterates under F fall into three categories corresponding to the three segments of stable and unstable manifolds through 0 which define parts of the boundary. Thus sequences are identified in the following cases: They have a constant infinite right (future) tail consisting of 0's or 4's, and agree otherwise, or else an infinite left (past) tail (of 0's and 1's, or of 4's) and agree otherwise. Let us summarize some of the properties of the coding.

Corollary 2.5.2. *The semiconjugacy between σ_A and F is one-to-one on all periodic points except for the fixed points. The number of preimages of any point not negatively asymptotic to the fixed point is bounded.*

Exercises

2.5.1. *Prove that for $\lambda \geq 1$ every bounded orbit of the quadratic map f_λ is in* $[0, 1]$.

2.5.2. *Prove the assertion of Corollary 2.5.2 for some 0-1 matrix A for any automorphism*

$$F_L: \mathbb{T}^2 \to \mathbb{T}^2, x \mapsto Lx \pmod 1,$$

where L is an integer 2×2 matrix with determinant $+1$ or -1 and with real eigenvalues different from ± 1.

2.5.3. *Construct a Markov partition and describe the corresponding topological Markov chain for the automorphism F_L, where $L = \begin{pmatrix} 1 & 1 \\ 2 & 1 \end{pmatrix}$.*

2.5.4. *Given a 0-1 $n \times n$-matrix A describe a system of n rectangles $\Delta_1, \dots, \Delta_n$ in R^2 and map $f: \Delta := \bigcup_{i=1}^n \Delta_i \to \mathbb{R}^2$ such that the restriction of f to the set of points that stay inside Δ for all iterates of f is topologically equivalent to the topological Markov chain σ_A.*

6. Stability of hyperbolic toral automorphisms

At the end of Section 2.4 we showed that the map effecting the semiconjugacy between a monotone map of the circle of degree k with $|k| \geq 2$ and the linear expanding map E_k can be found as the fixed point of a certain contracting operator in a space of continuous functions. Now we are going to use a similar method on the torus.

Theorem 2.6.1. *Any hyperbolic linear automorphism F_L of the two-torus is a factor of any homeomorphism g in the same homotopy class via a uniquely determined semiconjugacy homotopic to the identity. If g is C^0-close to F_L, then the semiconjugacy is close to the identity in the C^0 topology.*

Proof. Let $g: \mathbb{T}^2 \to \mathbb{T}^2$ be a homeomorphism homotopic to F_L. We are going to prove existence of a continuous map $h: \mathbb{T}^2 \to \mathbb{T}^2$ homotopic to the identity (and hence surjective since its degree is nonzero, cf. Section 8.2) such that

$$h \circ g = F_L \circ h \quad \text{or} \quad h = F_L^{-1} \circ h \circ g. \tag{2.6.1}$$

Let us develop a convenient way of writing our equations. Any map of the torus into itself can be lifted to the universal cover $\mathbb{R}^2$; furthermore a map $S: \mathbb{R}^2 \to \mathbb{R}^2$ is the lift of a map of $\mathbb{T}^2$ if and only if there exists an endomorphism $A: \mathbb{Z}^2 \to \mathbb{Z}^2$ such that $S(x+m) = Sx + Am$ for any $x \in \mathbb{R}^2, m \in \mathbb{Z}^2$. In particular for the lift of a map homotopic to the identity we have $A = \mathrm{Id}$, that is, $S - \mathrm{Id}$ is a doubly periodic map.

The lift of F_L is the hyperbolic linear map L; let us denote the lift of g by $L + \tilde{g}$ where $\tilde{g}$ is doubly periodic, that is, $\tilde{g}(x+m) = \tilde{g}(x)$ for $m \in \mathbb{Z}^2$, and the lift of h by $\mathrm{Id} + \tilde{h}$ with a doubly periodic $\tilde{h}$.

The second equation in (2.6.1) is equivalent to

$$\mathrm{Id} + \tilde{h} = L^{-1} \circ (\mathrm{Id} + \tilde{h}) \circ (L + \tilde{g}) \quad \text{or}$$

$$\tilde{h} = L^{-1}\tilde{g} + L^{-1} \circ \tilde{h} \circ (L + \tilde{g}). \tag{2.6.2}$$

Unlike (2.4.9) the right-hand side of (2.6.2) can not be viewed as a contracting operator acting on $\tilde{h}$. However, solving (2.6.2) can be reduced to finding fixed points of contracting operators using the decomposition of $\mathbb{R}^2$ into eigenspaces of the matrix L. Let e_1, e_2 be eigenvectors of L so that $Le_1 = \lambda_1 e_1, Le_2 = \lambda_2 e_2, |\lambda_1| = |\lambda_2^{-1}| > 1$. Let us decompose the vector functions $\tilde{h}$ and $\tilde{g}$ as

$$\tilde{h} = h_1 e_1 + h_2 e_2, \quad \tilde{g} = g_1 e_1 + g_2 e_2. \tag{2.6.3}$$

Then (2.6.2) is equivalent to two equations involving unknown scalar doubly periodic continuous functions h_1 and h_2:

$$h_1 = \lambda_1^{-1} g_1 + \lambda_1^{-1} h_1 \circ (L + \tilde{g}), \tag{2.6.4}$$

$$h_2 = \lambda_2^{-1} g_2 + \lambda_2^{-1} h_2 \circ (L + \tilde{g}). \tag{2.6.5}$$

Let us denote the right-hand side of (2.6.4) by $\mathcal{F}_1(h_1)$ and consider $\mathcal{F}_1$ as an operator on the space of doubly periodic continuous functions on $\mathbb{R}^2$ with the uniform topology. One easily sees that $\mathcal{F}_1$ is contracting:

$$\begin{aligned}\|\mathcal{F}_1(h) - \mathcal{F}_1(h')\| &= |\lambda_1^{-1}| \sup_{x \in \mathbb{T}^2} |h(Lx + \tilde{g}(x)) - h'(Lx + \tilde{g}(x))| \\ &\leq |\lambda_1^{-1}| \sup_{y \in \mathbb{T}^2} |h(y) - h'(y)| = |\lambda_1|^{-1} \|h - h'\|.\end{aligned}$$

Thus by the Contraction Mapping Principle (Proposition 1.1.2) $\mathcal{F}_1$ has a unique fixed point h_1 whose norm can be estimated by iterating the zero map:

$$\|h_1\| \leq \sum_{n=0}^{\infty} \|\mathcal{F}_1^{n+1}(0) - \mathcal{F}_1^n(0)\| = \frac{1}{1 - |\lambda_1|^{-1}} \|\mathcal{F}_1(0)\| = \frac{|\lambda_1|}{|\lambda_1| - 1} \|g_1\|.$$

Equation (2.6.5) should be slightly rewritten in order to represent it as the fixed-point equation for a contracting operator. Using the fact that g and hence $L + \tilde{g}$ is a homeomorphism we invert the latter map and denote the inverse by S. Then (2.6.5) becomes

$$h_2 = \lambda_2 h_2 \circ S - g_2 \circ S =: \mathcal{F}_2(h_2). \tag{2.6.6}$$

The same calculation as above shows that $\mathcal{F}_2$ is a contracting operator whose fixed point h_2 satisfies the estimate

$$\|h_2\| \leq \frac{\|g_2\|}{1 - |\lambda_2|}.$$

Substituting solutions of (2.6.4) and (2.6.6) into (2.6.3) and projecting $\mathrm{Id} + \tilde{h}$ to the torus one obtains a solution of (2.6.1) which is in fact unique among continuous maps of $\mathbb{R}^2$ homotopic to the identity. □

If g is a small perturbation of F_L in the C^1 topology, one can reverse the roles played by F_L and g in our argument.

Proposition 2.6.2. *Any C^1 map g sufficiently close to F_L in the C^1 topology is a factor of F_L.*

Proof. We need to solve the equation

$$g \circ h = h \circ F_L. \tag{2.6.7}$$

Using the same notation as in the proof of Theorem 2.6.1 we reduce (2.6.7) to

$$\mathcal{L}(\tilde{h}) = \tilde{g} \circ (\mathrm{Id} + \tilde{h}), \tag{2.6.8}$$

where $\mathcal{L}(\tilde{h}) := \tilde{h} \circ L - L \circ \tilde{h}$. The operator $\mathcal{L}$ is linear on the space of all continuous doubly periodic mappings of $\mathbb{R}^2$ onto itself. In fact $\mathcal{L}$ has bounded inverse: Using (2.6.3) one can represent $\mathcal{L}$ as

$$\mathcal{L}(\tilde{h}) = \mathcal{L}_1(h_1)e_1 + \mathcal{L}_2(h_2)e_2,$$

where

$$\mathcal{L}_1(h_1) = h_1 \circ L - \lambda_1 h_1,$$
$$\mathcal{L}_2(h_2) = h_2 \circ L - \lambda_2 h_2$$

and both $\mathcal{L}_1$ and $\mathcal{L}_2$ can be inverted explicitly:

$$\mathcal{L}_1^{-1}(h_1) = -\sum_{n=0}^{\infty} \lambda_1^{-(n+1)} h_1 \circ L^n,$$
$$\mathcal{L}_2^{-1}(h_2) = \sum_{n=0}^{\infty} \lambda_2^n h_2 \circ L^{-(n+1)}.$$

Note next that (2.6.8) is equivalent to the fixed-point equation

$$\tilde{h} = (\mathcal{L}^{-1}\mathcal{T})\tilde{h},$$

where $\mathcal{T}(\tilde{h}) = \tilde{g} \circ (\mathrm{Id} + \tilde{h})$.

Since

$$\begin{aligned} \|\mathcal{T}(h) - \mathcal{T}(h')\| &= \sup_{x \in \mathbb{R}^2} |\tilde{g}(x + h(x)) - \tilde{g}(x + h'(x))| \\ &\leq \|D\tilde{g}\| \sup_{x \in \mathbb{R}^2} |h(x) - h'(x)| = \|D\tilde{g}\| \cdot \|h - h'\|, \end{aligned}$$

we have $\|\mathcal{L}^{-1}\mathcal{T}\| \leq \|D\tilde{g}\| \cdot \|\mathcal{L}^{-1}\|$.

The second factor depends only on L. Hence if $\|D\tilde{g}\| < \|\mathcal{L}^{-1}\|^{-1}$, then $\mathcal{L}^{-1}\mathcal{T}$ is a contracting operator which has a unique fixed point $\tilde{h}$ by the Contraction Mapping Principle. □

Equation (2.6.7) can be rewritten as

$$h = g^{-1} \circ h \circ F_L. \tag{2.6.9}$$

If $h' = F_L^{-1} \circ h' \circ g$ is a solution of (2.6.1) and $h'' = g^{-1} \circ h'' \circ F_L$ is a solution of (2.6.9) then

$$h' \circ h'' = F_L^{-1} \circ h' \circ h'' \circ F_L$$

by taking the composition of both sides of the equation. Thus the map $h' \circ h''$ commutes with F_L; it is also close to the identity. Later (Proposition 3.2.15) we will see that the only such map is the identity. Thus we obtain the following result.

Theorem 2.6.3. *Any hyperbolic linear automorphism of the two-torus is C^1 strongly structurally stable.*

Exercises

2.6.1. *Show that the proof of C^1 structural stability for hyperbolic linear automorphisms of the two-torus can be generalized to any hyperbolic linear automorphism of the m-torus for $m \geq 2$.*

2.6.2. *Prove the following semilocal version of C^1 structural stability for the "horseshoe" map $f: \Delta \to \mathbb{R}^2$ described in Section 2.5: Let $g: \Delta \to \mathbb{R}^2$ be any map sufficiently C^1-close to g. Then there exists an injective continuous map $h = h_g: \Lambda \to \Delta$ such that $g \circ h_g = h_g \circ f$. Naturally then $\Lambda_g := h_g(\Lambda)$ is a closed g-invariant set and $g_{\restriction \Lambda_g}$ is topologically conjugate to the full shift σ_2.*

7. The fast-converging iteration method (Newton method) for the conjugacy problem

a. Methods for finding conjugacies. On several occasions we constructed solutions of the conjugacy equation

$$f = h \circ g \circ h^{-1} \tag{2.7.1}$$

between given transformations f and g in local, semilocal, and global situations. So far we have used the following methods:

1. The *fundamental-domain method* used in Section 2.1 for proving structural stability of an interval map with attracting and repelling points at the ends as well as describing the moduli for smooth conjugacy. See also Exercises 2.1.1, 2.1.3(2), 2.3.3, and 2.3.4. It works for some systems with highly *dissipative* behavior, that is, where most orbits do not recur and one can find nice fundamental domains for the action. It has applications beyond the one-dimensional situation, for example, in the proof of the Hartman–Grobman Theorem 6.3.1, and the Sternberg wedge method gives an alternative proof of Theorem 6.6.6 in Section 6.6d. However, it cannot be used for systems with nontrivial recurrence behavior (see the discussion at the end of Section 3.3).

2. The *majorization method* for the local linearization problem in Section 2.1b. Recall that we first constructed the formal solutions (2.1.1), that is, we determined a formal power series for the conjugating transformation h at the origin. Then we proved convergence of that series and thus showed that the analytic function determined by those coefficients solves the conjugacy equation. This method heavily depends on the local character of the problem.

3. The *coding method*, which we first used in the proof of the topological conjugacy of an arbitrary expanding circle map with the linear map of the same degree (Theorem 2.4.6). It appeared on three more occasions: In the semilocal situation in Sections 2.5b, 2.5c in the construction of the topological conjugacy between the full 2-shift and the invariant sets of the quadratic and the "horseshoe" map and then in Section 2.5d when we constructed the semiconjugacy between a topological Markov chain and a toral automorphism. This method is very powerful in global and semilocal hyperbolic problems, that is, when nearby orbits exhibit exponential divergence as in those examples (cf. Chapter 6, especially Definition 6.4.1 and Definition 6.4.2 for more details). One of the main features is its direct character. In particular it does not require considering an auxiliary space of "candidate" conjugacies. On the other hand it applies only to the problem of topological (as opposed to smooth) conjugacy and semiconjugacy. However, it is particularly effective in low-dimensional situations where it often works without hyperbolicity assumptions (see Sections 14.5, 14.6, 15.4).

4. The *contraction mapping method.* It has been used in the second proof of conjugacy for expanding maps of the circle (Section 2.4c) and in the proofs of the global version of topological stability (Theorem 2.6.1) and of structural stability (Theorem 2.6.3) for hyperbolic automorphisms of the 2-torus. In this approach one rewrites the conjugacy equation as a fixed-point equation for the conjugating map $h = f^{-1} \circ h \circ g$ by considering the operator

$$\mathcal{L}h = f^{-1} \circ h \circ g$$

on an appropriate space of transformations. If f is noninvertible a trick involving the branches of f^{-1} makes it possible to construct a similar operator (cf. (2.4.10)). Then one shows that the operator $\mathcal{L}$ inherits hyperbolicity from the map f and this allows us to construct from $\mathcal{L}$ a contraction map whose fixed points satisfy the conjugacy equation. This is an infinite-dimensional counterpart of the construction used in the proof of Proposition 1.1.4. This method will be used extensively in the hyperbolic situation (Chapters 6 and 18). Its particular strength is that it works in the situation where f and g are not necessarily close to each other. However, the necessary hyperbolic behavior of $\mathcal{L}$ can only be established with respect to a sufficiently coarse topology, such as the C^0 or Hölder topology, and hence this method is also restricted to finding topological conjugacies.

b. Construction of the iteration process. We will now describe another method which also makes use of a related operator in a function space. This method is commonly referred to as the "Newton method". Indeed it can be viewed as a far-reaching generalization of the elementary Newton method of finding zeros of functions. It is often called the *KAM method* after Kolmogorov, Arnold, and Moser. We will reduce the conjugacy equation to an implicit-function problem rather than to a fixed-point problem as in the contraction mapping method. Unlike the latter the Newton method is suitable for finding smooth and analytic conjugacies and it works in many nonhyperbolic problems. However, its applicability is restricted to perturbation problems, that is, situations where f and g are close to each other.

Consider the following operator depending on two (functional) variables f and h

$$\mathcal{F}(f,h) = h^{-1} \circ f \circ h$$

and write the conjugacy equation as

$$\mathcal{F}(f,h) = g.$$

The main feature of this operator is the following "group property":

$$\begin{aligned} \mathcal{F}(f, \varphi \circ \psi) &= \mathcal{F}(\mathcal{F}(f,\varphi),\psi), \\ \mathcal{F}(f, \mathrm{Id}) &= f. \end{aligned} \tag{2.7.2}$$

As in the elementary Newton method we will want to linearize the operator and hence we need to assume that there is a linear structure on a neighborhood of (g, Id) in the functional space and that f is close to g. Then one can linearize $\mathcal{F}$ on this neighborhood. We write $D_1\mathcal{F}$ and $D_2\mathcal{F}$ for the partial differentials with respect to f and h, respectively, and we will look for an "approximate solution" $h = \mathrm{Id} + w$ of the conjugacy equation linearized at (g, Id). Thus we write

$$\mathcal{F}(f,h) = \mathcal{F}(g,\mathrm{Id}) + D_1\mathcal{F}(g,\mathrm{Id})(f-g) + D_2\mathcal{F}(g,\mathrm{Id})(h-\mathrm{Id}) + \mathcal{R}(f,h),$$

where $\mathcal{R}(f,h)$ is of second order in $(f-g, h-\mathrm{Id})$. In other words, if h solves the linearized equation (obtained by dropping $\mathcal{R}$), then $w = h - \mathrm{Id}$ is a solution of the following equation:

$$\mathcal{F}(g,\mathrm{Id}) + D_1\mathcal{F}(g,\mathrm{Id})(f-g) + D_2\mathcal{F}(g,\mathrm{Id})w = g. \tag{2.7.3}$$

Note that $D_1\mathcal{F}(g,\mathrm{Id}) = \mathrm{Id}$ since $\mathcal{F}(\cdot,\mathrm{Id}) = \mathrm{Id}(\cdot)$ by (2.7.2). Thus (2.7.3) simplifies to

$$(f-g) + D_2\mathcal{F}(g,\mathrm{Id})w = 0. \tag{2.7.4}$$

If we set $u = f - g$ and assume that $D_2\mathcal{F}(g,\mathrm{Id})$ is invertible, this is solved by

$$w = -\left(D_2\mathcal{F}(g,\mathrm{Id})\right)^{-1} u.$$

In this case w is of the same order as u, and substituting $h = \mathrm{Id} + w$ into $\mathcal{F}(f,h)$ we obtain a function $f_1 = h^{-1} \circ f \circ h = \mathcal{F}(f,h) = g + \mathcal{R}(f,h)$, so the size of $u_1 = f_1 - g$ should be of second order in the size of $u = f - g$. In order to justify this, we will need to estimate the difference between $\mathcal{F}$ and its linearization near (g, Id).

Thus we consider an iterative process as follows. Assuming that $f_1, \dots, f_n$ have been constructed we solve the equation

$$f_n - g + D_2\mathcal{F}(g, \mathrm{Id})w_{n+1} = 0$$

and set

$$h_{n+1} = h_n \circ (\mathrm{Id} + w_{n+1}) \text{ and } f_{n+1} = (\mathrm{Id} + w_{n+1})^{-1} \circ f_n \circ (\mathrm{Id} + w_n).$$

The last step of the construction is the proof of convergence of the sequence h_n in an appropriate topology. It follows from the same estimates that provide the fast decrease of the size of the $f_n - g$.

Notice that at every step we invert the linear part at (g, Id), rather than at the intermediate points as in the elementary Newton method. This is precisely the reason this method can be applied in nonhyperbolic situations.

Exercises

2.7.1. *Set up iterative processes similar to the one described in Subsection b. for solving the functional equations $f \circ h = g$ or $h \circ f = g$ with respect to an unknown transformation h, where f and g are known maps of a space X to a space Y, so that the process involves inverting the linearized operator only at (g, Id).*

2.7.2. *Use the previous exercise to derive the formula*

$$(I - A)^{-1} = \prod_{n=0}^{\infty} (I + A^{2^n}),$$

where I is the identity matrix and A is a matrix of norm less than one.

8. The Poincaré–Siegel Theorem

As we mentioned in Section 2.1b we are going to use the Newton method to give another proof of local analytic conjugacy between an analytic contracting map on the line and its linear part (Proposition 2.1.3). In fact, almost the same proof will work for another problem which analytically looks very similar, but is dynamically very different.

The conjugacy constructed in (2.1.1) works for the map f extended to a *complex* neighborhood of 0 since both f and the conjugacy are defined via power series (cf. Exercise 2.1.7). Furthermore one need not assume that the map f preserves the real line, that is, that its Taylor coefficients, including the first term λ, are real. The only hypothesis on λ will be that $|1-\lambda^n|$ is uniformly bounded away from 0 for all n. This is equivalent to $|\lambda| \neq 1$ and this situation is referred to as the *Poincaré case.*

However, we can also consider a holomorphic map $f: U \to \mathbb{C}$ on a neighborhood of 0 such that $f(0) = 0$ and $|f'(0)| = 1$. The linearized map $\Lambda z := \lambda z$ is the rotation around the origin by an angle $\arg \lambda$. If this angle is a rational multiple of 2π then the linear map is periodic, although this is usually not the case for f, for example, the quadratic map $z \mapsto \exp 2\pi i p/q z + az^2$ is not periodic. Suppose, however, that $(1/2\pi)\arg\lambda$ is not only irrational, but *not too well approximable by rationals.* (see Definition 2.8.1). This is called the *Siegel case.* In this situation the Newton method allows us to construct a holomorphic conjugacy between f and Λ in a certain neighborhood of 0. Since every circle $|z| = \text{const.}$ is invariant under Λ, its image is invariant under f. Thus the conjugacy is defined on an *invariant* disk and its existence in the Siegel case is not simply a local, but a *semilocal* fact.

Definition 2.8.1. A number α is called *Diophantine* of type (c, d) if for any nonzero $p, q \in \mathbb{Z}$ we have $|q\alpha - p| > cq^{-d}$. α is called Diophantine if there exist $c > 0, d > 1$ such that α is Diophantine of type (c, d).

Theorem 2.8.2. (Poincaré–Siegel Theorem) *Let $f(z) = \lambda z + \sum_{n=2}^{\infty} f_n z^n$ be a holomorphic map in a neighborhood of 0, where either $|\lambda| \neq 1$ or $\lambda = e^{2\pi i\alpha}$ for some Diophantine number α. Then there exist $\delta > 0$ and a holomorphic map $h(z) = z + \sum_{n=2}^{\infty} h_n z^n$ such that*

$$h^{-1} \circ f \circ h = \Lambda \text{ for } |z| < \delta, \tag{2.8.1}$$

where $\Lambda(z) = \lambda z$.

Remark. The *majorization method* of Section 2.1b can be made to work in the Siegel case as well,[1] but the estimates are much more formidable than in the Poincaré case and the method does not work in some other problems in which the Newton method is applicable.

Proof. (Moser) For the estimates needed in the Newton method we use the fact that for analytic functions C^0 estimates and estimates of Taylor coefficients are equivalent in the following way: Given a C^0 estimate on a disk, the Cauchy integral formula gives estimates for the derivatives in a smaller disk, and these in turn give a C^0 estimate on the smaller disk.

Lemma 2.8.3.

(1) *Suppose a function* $\varphi = \sum_{k=0}^{\infty} \varphi_k z^k$ *is analytic on* $B_r := \{z \mid |z| < r\}$ *and continuous on* $\overline{B_r}$ *and* $|\varphi| < \epsilon$ *on* B_r*; then* $|\varphi_k| < \epsilon r^{-k}$ *for* $k \in \mathbb{N}$.

(2) *Suppose* $|\varphi_k| < K r^{-k}$ *for* $k \in \mathbb{N}$. *Then the function* $\varphi = \sum_{k=0}^{\infty} \varphi_k z^k$ *is analytic on* B_r *and* $|\varphi| \le Kr/\delta$ *on* $B_{r-\delta}$.

Proof. (1) $|\varphi_k| = \left|\frac{1}{2\pi i}\int_{|z|=r} \frac{\varphi(z)}{z^{k+1}} dz\right| \le \frac{1}{2\pi}\int_{|z|=r}\left|\frac{\varphi(z)}{z^{k+1}}\right| dz \le \epsilon/r^k$.

(2) $|\varphi(z)| \le K \sum_{k=0}^{\infty} r^{-k}(r-\delta)^k = Kr/\delta$. □

We note that the hypothesis on λ implies the Diophantine condition

$$|\lambda^q - 1| \ge \frac{q^{-d}}{c_0|\lambda|} \tag{2.8.2}$$

for some $c_0, d \in \mathbb{N}$. This is trivial in the hyperbolic case when $|\lambda| \ne 1$ and hence $|\lambda^q - 1|$ is bounded from below. In fact, in this case one can take d as small as desired by adjusting c_0. If $\lambda = e^{2\pi i\alpha}$, where α is as in Definition 2.8.1, let us take $p \in \mathbb{Z}$ to be the integer closest to $q\alpha$. Then we find

$$|\lambda^q - 1| = |e^{2\pi i q\alpha} - e^{2\pi i p}| = |e^{2\pi i(q\alpha - p)} - 1| \ge c'|q\alpha - p| \ge c'cq^{-d},$$

where d is as in Definition 2.8.1, proving (2.8.2).

We now invert $D_2\mathcal{F}(g, \mathrm{Id})$. First note that by systematically discarding terms of higher order in t we get

$$\begin{aligned} D_2\mathcal{F}(\Lambda, \mathrm{Id})w &= \lim_{t\to 0}\frac{1}{t}\left(\mathcal{F}(\Lambda, \mathrm{Id}+tw) - \mathcal{F}(\Lambda, \mathrm{Id})\right)\\ &= \lim_{t\to 0}\frac{1}{t}\left((\mathrm{Id}+tw)^{-1}\circ\Lambda\circ(\mathrm{Id}+tw) - \Lambda\right)\\ &= \lim_{t\to 0}\frac{1}{t}\left((\mathrm{Id}-tw)\circ\Lambda\circ(\mathrm{Id}+tw) - \Lambda\right)\\ &= \lim_{t\to 0}\frac{1}{t}\left(t\Lambda\circ w - tw\circ\Lambda\right) = \Lambda\circ w - w\circ\Lambda. \end{aligned}$$

Thus to solve (2.7.4) we have to solve

$$u = w\circ\Lambda - \Lambda\circ w \tag{2.8.3}$$

for w, given $u = \sum_{k=2}^{\infty} f_k z^k$. If we write $w = \sum_{k=2}^{\infty} w_k z^k$, then we must have

$$\begin{aligned} \sum_{k=2}^{\infty} f_k z^k &= u = w\circ\Lambda - \Lambda\circ w\\ &= \sum_{k=2}^{\infty}(w_k\lambda^k - \lambda w_k)z^k = \sum_{k=2}^{\infty}(\lambda^k - \lambda)w_k z^k, \end{aligned} \tag{2.8.4}$$

so the power series for w must have coefficients

$$w_k = \frac{f_k}{\lambda^k - \lambda} \tag{2.8.5}$$

for $k \geq 2$. We now show that this series converges on B_r if u does, and give a useful estimate.

Lemma 2.8.4. *Let $\varphi = \sum_{k=0}^{\infty} \varphi_k z^k$ be analytic on B_ρ, $|\varphi| < \delta$ on B_ρ, and λ be as in (2.8.2). Then*

$$\psi(z) := \sum_{k=2}^{\infty} \frac{\varphi_k}{\lambda^k - \lambda} z^k$$

is analytic on B_ρ and there exists a $c(d) > 0$ such that

$$|\psi| < \delta c_0 c(d) \Delta^{-(d+1)}$$

on $\overline{B_{\rho(1-\Delta)}}$.

Proof. Convergence of ψ follows from the Cauchy estimate $|\varphi_k| < \delta \rho^{-k}$ and (2.8.2), since

$$|\psi_k| = \left| \frac{\varphi_k}{\lambda^k - \lambda} \right| \leq \delta c_0 (k-1)^d \rho^{-k} \leq \delta c_0 k^d \rho^{-k}.$$

Notice that $\sum_{k=0}^{\infty} k^d x^k$ is a linear combination of derivatives of $\sum_{k=0}^{\infty} x^k = 1/1 - x$ up to order d with polynomial coefficients, so for $|x| < 1$ we have $\sum_{k=0}^{\infty} k^d x^k \leq \dfrac{c(d)}{(1-x)^{d+1}}$ for some $c(d) > 0$. Thus on $B_{\rho(1-\Delta)}$ we have

$$|\psi(z)| < \delta c_0 \sum_{k=2}^{\infty} k^d \rho^{-k} z^k \leq \delta c_0 \sum_{k=2}^{\infty} k^d (1-\Delta)^k \leq \delta c_0 c(d) \Delta^{-(d+1)}.$$

□

In particular, $D_2\mathcal{F}(\Lambda, \mathrm{Id})$ is indeed invertible.

Now we begin the iterative procedure. To begin with, recall that $u = f - \Lambda$ vanishes to second order at 0 since Λ is the linear part of f. Consequently we can choose r so that

$$|u'| < \epsilon \text{ on } B_r, \tag{2.8.6}$$

where ϵ will be specified later.

We also assume that $|\lambda| \leq 1$, which is not a restriction in the Siegel case and in the Poincaré case can be achieved by possibly considering the inverse of f instead of f. Apply Lemma 2.8.4 to u with $\rho = r$ and $\delta = \epsilon\rho$ (since $|u'| < \epsilon$ on B_r) to get $|w(z)| < \epsilon\rho c_0 c(d) \Delta^{-(d+1)}$ for $|z| \leq \rho(1-\Delta)$. Thus for $|z| = \rho(1-\Delta)$ we have

$$|w(z)| < \epsilon\rho c_0 c(d) \Delta^{-(d+1)} = \epsilon \frac{c_0 c(d)}{1-\Delta} \Delta^{-(d+1)} |z|.$$

This is independent of ρ so

$$|w| < \epsilon \frac{c_0 c(d)}{1-\Delta} \Delta^{-(d+1)} r \text{ on } \overline{B_{r(1-\Delta)}}. \tag{2.8.7}$$

The same argument applied to $zu'(z)$ yields

$$|w'| < \epsilon \frac{c_0 c(d)}{1-\Delta} \Delta^{-(d+1)} \text{ on } \overline{B_{r(1-\Delta)}}. \tag{2.8.8}$$

The next two lemmas show that the new map

$$f_1 = h^{-1} \circ f \circ h,$$

where $h = \mathrm{Id} + w$, is defined on $B_{r(1-4\Delta)}$ if ϵ was chosen small enough to begin with.

Lemma 2.8.5. *If*

$$\epsilon c_0 c(d) < \Delta^{d+2}(1-\Delta) \textit{ and } 0 < \Delta < \frac{1}{4} \tag{2.8.9}$$

then

$$h(B_{r(1-4\Delta)}) \subset B_{r(1-3\Delta)} \textit{ and } B_{r(1-2\Delta)} \subset h(B_{r(1-\Delta)}). \tag{2.8.10}$$

Proof. To prove the first inclusion in (2.8.10) take $|z| < r(1-4\Delta)$ and use (2.8.7) to get

$$\begin{aligned} |h(z)| \le |z| + |w(z)| &< r\big(1 - 4\Delta + \frac{\epsilon c_0 c(d)}{1-\Delta} \Delta^{-(d+1)}\big) \\ &< r(1 - 4\Delta + \Delta) = r(1-3\Delta). \end{aligned}$$

To show the second inclusion of (2.8.10) note that $|h(z) - z| = |w(z)| < r\Delta$ by (2.8.7) and (2.8.9), while for $|z| = r(1-\Delta)$ we clearly have $|z| - r(1-2\Delta) = r\Delta > |w(z)|$. Thus $|h(z)| = |z - (h(z) - z)| \ge |z| - |w(z)| > r(1-2\Delta)$. Since $h(0) = 0$ this implies the claim. □

Now we prove that $f_1 = h^{-1} \circ f \circ h$ is defined on $B_{r(1-4\Delta)}$ and give the quadratic estimate for its nonlinear part. We write

$$f_1 = \Lambda + u_1.$$

Lemma 2.8.6. *If*

$$\epsilon c_0 c(d) < \Delta^{d+2}(1-\Delta) \textit{ and } 0 < \epsilon < \Delta < \frac{1}{5} \tag{2.8.11}$$

then f_1 *is defined on* $B_{r(1-4\Delta)}$ *and*

$$|u_1'| \le \epsilon^2 \frac{5c_0c(d)}{4(1-\Delta)\Delta^{d+2}} \text{ on } B_{r(1-5\Delta)}.$$

Proof. By Lemma 2.8.5 we have $h(B_{r(1-4\Delta)}) \subset B_{r(1-3\Delta)}$. Since $|\lambda| < 1$, (2.8.6) yields $|f(z)| \le r(1-3\Delta)+r\epsilon < r(1-2\Delta)$ on $B_{r(1-3\Delta)}$. Since $B_{r(1-2\Delta)} \subset h(B_{r(1-\Delta)})$ by Lemma 2.8.5, h^{-1} is defined on $B_{r(1-2\Delta)}$; hence $f_1 = h^{-1} \circ f \circ h$ is defined on $B_{r(1-4\Delta)}$.

To estimate u_1' rewrite $h \circ f_1 = f \circ h$ as

$$\lambda z + u_1(z) + w(\lambda z + u_1(z)) = \lambda(z + w(z)) + u(h(z)),$$

or, using (2.8.3),

$$u_1(z) = w(\lambda z) - w(\lambda z + u_1(z)) + u(h(z)) - u(z). \tag{2.8.12}$$

Now by the Mean Value Theorem and (2.8.8)

$$|w(\lambda z)-w(\lambda z+u_1(z))| \le \sup|w'|\sup|u_1| \le \epsilon\frac{c_0c(d)}{1-\Delta}\Delta^{-(d+1)}\sup|u_1| < \frac{\sup|u_1|}{5},$$

using (2.8.11). Thus (2.8.7) and (2.8.12) yield

$$\frac{4}{5}\sup|u_1| \le |u(h(z)) - u(z)| \le \sup|u'||w| < \epsilon^2\frac{c_0c(d)}{1-\Delta}\Delta^{-(d+1)}r$$

on $B_{r(1-4\Delta)}$. Thus by Lemma 2.8.3(1)

$$|u_1'| \le \epsilon^2\frac{5c_0c(d)}{4(1-\Delta)}\Delta^{-(d+2)}$$

on $B_{r(1-5\Delta)}$. □

Now we apply these estimates inductively. If f_n is given by (2.8.15) and $u_n = f_n - \Lambda$ and we have

$$|u_n'| \le \epsilon_n \text{ on } B_{r_n}$$

then

$$|u_{n+1}'| \le \epsilon_n^2 \frac{5c_0c(d)}{4(1-\Delta_n)\Delta_n^{d+2}} =: \epsilon_{n+1} \le c_1\frac{\epsilon_n^2}{\Delta_n^{d+2}} \text{ on } B_{r_{n+1}}. \tag{2.8.13}$$

If we take $r_n = r(1+2^{-n})/2 > r/2$ and

$$\Delta_n = \frac{1}{10(2^n+1)} \tag{2.8.14}$$

then $r_{n+1} = r_n(1 - 5\Delta_n)$. To show that $\epsilon_n \to 0$ we note from (2.8.13) and (2.8.14) that $\epsilon_{n+1} = 10^{d+2}c_1\epsilon_n^2(2^n+1)^{d+2} \le c_2^{n+1}\epsilon_n^2$ for some $c_2 > 0$, so $\epsilon'_n := c_2^n\epsilon_n$ satisfies

$$\epsilon'_{n+1} = c_2^{n+1}\epsilon_{n+1} \le c_2^{n+1}c_2^{n+1}\epsilon_n{}^2 = \epsilon'_n{}^2,$$

and hence ϵ'_n and ϵ_n tend to 0 superexponentially if we take $\epsilon_0 < 1$. Note also that (2.8.11) is satisfied for all n if we take ϵ_0 small enough.

Consider now the maps

$$k_n := h_0 \circ \cdots \circ h_{n-1}$$

which by Lemma 2.8.6 are defined on $B_{r_{n-1}}$. Then

$$f_n = k_n^{-1} \circ f \circ k_n \tag{2.8.15}$$

is defined on $B_{r_{n-1}} \supset B_{r/2}$ as well.

Let us show that $\{k_n\}_{n\in\mathbb{N}}$ converges uniformly on $B_{r/2}$. By the chain rule

$$|k'_n| = \prod_{l=0}^{n-1} h'_l = \prod_{l=0}^{n-1}(1 + w'_l),$$

where $|w'_l| \le c_3^l\epsilon_l$ by (2.8.8). Thus $\prod_{l=0}^{n-1}(1 + |w'_l|) \le c_4$ on $B_{r/2}$, so

$$|k_{n+1} - k_n| \le c_4 \sup|h_n - \mathrm{Id}\,| \le c_4|w_n| \le c_4c_5^n\epsilon'_n.$$

Therefore

$$k_n \to h \text{ and } f_n \to \Lambda,$$

which yields (2.8.1). □

Exercises

2.8.1. *Prove that the set of Diophantine numbers with a given $d > 1$ and arbitrary c has full Lebesgue measure.*

2.8.2. *Prove that the number $\alpha = \sum_{n=0}^{\infty} 2^{-n!}$ is not Diophantine.*

2.8.3. *Prove that for any holomorphic function w which is not a polynomial there exists a number $\lambda = \exp 2\pi i\alpha$, where α is irrational, such that the linearized equation (2.8.3) does not have a holomorphic solution.*

9. Cocycles and cohomological equations

In the course of this chapter we have found that several different problems reduce to solving a linear functional equation of a particular kind. These occurred in (2.2.6), (2.2.7), and (2.2.8) in the study of time change for flows; in (2.6.4), (2.6.5) in the proof of topological stability of hyperbolic toral automorphisms (Theorem 2.6.1, see also the proof of Proposition 2.6.2); and in the linearized equation (2.8.3) for conjugacy when using the Newton method. In the discrete-time case all of these equations can be expressed in the form

$$g(x) = \lambda\varphi(f(x)) - \varphi(x), \tag{2.9.1}$$

where $f: X \to X$ is a given map, g is a given scalar function on X, λ is a given constant, and φ is an unknown scalar function. One can also view equation (2.6.2) as a vector equation of the same form, where λ is a linear operator acting on a vector. Equations of the form (2.9.1) are called *cohomological equations* and appear in various problems other than the aforementioned ones.

Let us describe in general terms the way in which cohomological equations appear. Let G denote either $\mathbb{N}$, $\mathbb{Z}$, or $\mathbb{R}$, that is, time, $\rho: G \to GL(k, \mathbb{R})$ a homomorphism, that is, a linear representation of G, and $T: G \times X \to X$ a dynamical system with phase space X and time G.

Definition 2.9.1. A *one-cocycle* twisted by ρ is a map $\alpha: G \times X \to \mathbb{R}^k$ such that

$$\alpha(g_2 + g_1, x) = \rho(g_1)\alpha(g_2, T(g_1)x) + \alpha(g_1, x).$$

If ρ is the identity representation then such an α is called an *untwisted cocycle* or just a cocycle.

Note that cocycles form a linear space and that $\alpha(0, x) = 0$ from the definition. Any function $\varphi: X \to \mathbb{R}^k$ defines a cocycle via

$$\alpha(g, x) := \rho(g)\varphi(T(g)x) - \varphi(x). \tag{2.9.2}$$

Cocycles of this form are called *coboundaries*. Two cocycles are called *cohomologous* if their differe' ce is a coboundary.

In the discrete-time case there is a natural bijection between cocycles and functions on X. Namely, every cocycle is determined by the function $a(x) := \alpha(1, x)$. Solving a cohomological equation is equivalent to showing that a given cocycle is a coboundary: Let $f := T(1)$, $R = \rho(1)$, and $a(x) = \alpha(1, x)$. Then (2.9.2) becomes

$$a(x) = R\varphi(f(x)) - \varphi(x), \tag{2.9.3}$$

that is, a vector form of (2.9.1).

It turns out that there is a great difference between the case of hyperbolic ρ (that is, $\rho(1)$ hyperbolic) and nonhyperbolic ρ (in particular the untwisted case). When we solved (2.6.2) and (2.6.8) we essentially used the following basic result for the hyperbolic case:

Theorem 2.9.2. *Suppose R is hyperbolic. Then for bounded a the equation (2.9.3) has a unique bounded solution φ. If a is continuous then the solution of (2.9.3) is continuous also.*

Proof. As in the proof of Theorem 2.6.1 we decompose R into contracting and expanding parts R_- and R_+ and use the norm given by Proposition 1.2.2. Then we obtain two equations

$$a_+(x) = R_+\varphi_+(f(x)) - \varphi_+(x),$$
$$a_-(x) = R_-\varphi_-(f(x)) - \varphi_-(x),$$

where $\|R_-\|, \|R_+^{-1}\| < 1$. They are solved by the uniformly convergent series $\varphi_+(x) = \sum_{k=1}^{\infty} R_+^{-k} a_+(f^{-k}(x))$ and $\varphi_-(x) = -\sum_{k=0}^{\infty} R_+^k a_-(f^k(x))$. Thus boundedness and continuity are evident. Uniqueness follows from the fact that the difference ξ of two solutions solves (2.9.2) with $a = 0$, that is, $R\xi(f(x)) = \xi(x)$. The only bounded solution of this equation is zero since R is hyperbolic. □

It is important to point out that even if the data involved in (2.9.3), that is, a, f, and ρ, are smooth the solution φ cannot be expected to be very regular. The best general result in this direction is that one obtains a Hölder continuous solution if a is Hölder continuous and f is Lipschitz continuous. This lack of regularity is related to the fact that systems with hyperbolic behavior are structurally stable but the conjugacy is not smooth (see Section 19.1b, where we show that the conjugacy is Hölder continuous).

As an example of the nonhyperbolic case of a cohomological equation we consider an untwisted cohomological equation which already appeared in (2.2.8). In this case there are obvious obstructions for solving (2.9.2), namely, the values of α over periodic points must vanish. To see what this implies for a note that in this case

$$\alpha(n,x) = \sum_{i=0}^{n-1} a(f^i(x)) \text{ for } n \geq 0 \text{ and } \alpha(n,x) = -\sum_{i=n}^{-1} a(f^i(x)) \text{ for } n < 0. \tag{2.9.4}$$

If $f^n(x) = x$ and α is a coboundary then $\sum_{i=0}^{n-1} a(f^i(x)) = \alpha(n,x) = \varphi(f^n(x)) - \varphi(x) = 0$. We call the sum of the function over a periodic orbit the *periodic obstruction.*

If one does not require any structure in the solution of the untwisted cohomological equation, then there is no problem, as long as these obstructions vanish since we may choose a point x from each orbit and then set $\varphi(f^n(x)) = \sum_{i=0}^{n-1} a(f^i(x))$. If, however, there is any structure on X, this may be very unsatisfactory. For example, in the case of an irrational rotation of the circle, this construction would necessarily give a nonmeasurable collection of points x (the standard construction of a nonmeasurable set in analysis), and thus most likely a nonmeasurable (and highly unbounded) solution.

There is, however, a natural condition for existence of a bounded solution of (2.9.4).

Theorem 2.9.3. *If $\alpha(n,x)$ is bounded uniformly in n and x then $a(x) = \varphi(f(x)) - \varphi(x)$ has the solution*

$$\varphi(x) = \sup_{n\in\mathbb{N}} \Big[-\sum_{i=0}^{n} a(f^i(x))\Big]. \tag{2.9.5}$$

Proof. φ is well defined and

$$\begin{aligned}\varphi(f(x)) - \varphi(x) &= \sup_{n\in\mathbb{N}} \Big[-\sum_{i=0}^{n} a(f^{i+1}(x))\Big] - \sup_{n\in\mathbb{N}} \Big[-\sum_{i=0}^{n} a(f^i(x))\Big] \\ &= \sup_{n\in\mathbb{N}} \Big[-\sum_{i=0}^{n} a(f^{i+1}(x))\Big] - \Big(\sup_{n\in\mathbb{N}} \Big[-\big(a(x) + \sum_{i=0}^{n} a(f^{i+1}(x))\big)\Big]\Big) = a(x).\end{aligned}$$
□

Instead of the supremum we could have taken any other intrinsic characteristic of the sequence $\alpha(n,x)$, such as the infimum, the upper or lower limit, any linear combination of the above, or even more sophisticated expressions of a "center of mass" type, which take into account not just the set of values but their distribution.

It is easy to see that vanishing of the periodic obstructions alone may not suffice for the existence of reasonable solutions of an untwisted cohomological equation. For example, since irrational circle rotations R_α have no periodic points there are no periodic obstructions although there is another obvious necessary condition: If $g(x) = \varphi(R_\alpha(x)) - \varphi(x)$ and φ is integrable then $\int_{S^1} g = 0$.

Theorem 2.9.3 does not guarantee continuity of the solution of an untwisted cohomological equation even if the data are continuous (see Exercise 2.9.2). Furthermore, even if there is a continuous solution of an untwisted cohomological equation the solution given by (2.9.5) may not be continuous (see Exercise 2.9.3). There is, however, an interesting case where continuity is guaranteed.

Theorem 2.9.4. (Gottschalk, Hedlund) *Let X be a compact metric space, $f: X \to X$ a minimal continuous map (see Definition 1.3.2), and $g: X \to \mathbb{R}$ continuous such that $M := \sup_{n\in\mathbb{N}} \big|\sum_{i=0}^{n} g\circ f^i(x_0)\big| =: M < \infty$ for some $x_0 \in X$. Then there is a continuous $\varphi: X \to \mathbb{R}$ such that*

$$\varphi \circ f - \varphi = g. \tag{2.9.6}$$

Proof. First note that $\big|\sum_{i=0}^{n} g\circ f^i(x)\big|$ is bounded for all x. This follows from minimality: If $\big|\sum_{i=0}^{n} g\circ f^i(y)\big| > 2M$ then the same inequality holds for any z sufficiently close to y, in particular for some iterate $f^{n_0}(x_0)$ of x_0. But then $2M < \big|\sum_{i=n_0}^{n_0+N} g\circ f^i(x_0)\big| \le \big|\sum_{i=0}^{n_0+N} g\circ f^i(x_0)\big| + \big|\sum_{i=0}^{n_0-1} g\circ f^i(x_0)\big|$, contrary to the definition of M. Thus the assumptions of Theorem 2.9.3 are satisfied and we can take

$$\varphi(x) := \sup_{n\in\mathbb{N}} \Big[-\sum_{i=0}^{n} g(f^{-i}(x))\Big].$$

The *oscillation* of a function ψ at a point x is defined as

$$\mathrm{Osc}_\psi(x) := \lim_{\delta\to 0}(\sup\{\psi(y) \mid |x-y|<\delta\} - \inf\{\psi(y) \mid |x-y|<\delta\}).$$

Notice that the oscillation of a function vanishes if and only if the function is continuous and that the oscillation of any φ as in (2.9.6) is therefore f-invariant by continuity of g. Thus the set $\{x \mid \mathrm{Osc}_\varphi(x) \geq \epsilon\}$ is closed by definition and by minimality it is either X or empty. Thus it suffices to show that it is not X for any $\epsilon > 0$.

Notice that φ is the pointwise, in fact monotone, limit of the sequence $\varphi_k(x) = \sup_{n\leq k} - \sum_{i=0}^{n} g(f^{-i}(x))$ of continuous functions. Let $O_{\epsilon,n}$ be the set on which $\varphi - \varphi_n \leq \epsilon/2$. Then $\mathrm{Osc}_\varphi(x) \leq \epsilon$ for $x \in O_{\epsilon,n}$ and this set is nonempty for sufficiently large n. □

Exercise 2.9.1 shows that the above continuous solution is unique up to an additive constant.

Nonhyperbolic twisted cohomological equations appear in the proof of the Poincaré–Siegel Theorem 2.8.2 (namely, (2.8.3) in the Siegel case) and in very much the same way later in (12.3.3). In these cases they can be solved in the analytic category using Fourier analysis and under a crucial arithmetic assumption (2.8.2). A very similar result using the same method, but in the C^∞ category, is the following result about flows on the torus.

Proposition 2.9.5. *Suppose ρ is a Diophantine number and φ a positive C^∞ function on the circle. Then the flow under φ built over R_ρ is C^∞ flow equivalent to the linear flow T_t^ω, where $\omega = (\rho\int\varphi, \int\varphi)$.*

Proof. It suffices to show equivalence with the flow under the constant function $\varphi_0 := \int\varphi$, since the latter is a constant rescaling of the suspension, which is smoothly flow equivalent to $T_t^{(\rho,1)}$ by Proposition 2.2.2.

For that purpose we construct a new section of the special flow such that the return time is constant. The section will be homotopic to the original base section given by $t = 0$. Moreover it will have the following form. If the special flow is denoted by ψ^t then the section is given by moving the base section along the orbits for some variable time, that is, sending each point $(x, 0)$ to $\psi^{h(x)}(x, 0)$. In order for this to be a section we need some monotonicity, namely, we need to have $h(x) < \varphi(x) + h(R_\rho(x))$, so that the order of returns to the transversal is preserved. In order to achieve constant return time we will try to find h such that in fact $h(x) + \varphi_0 = \varphi(x) + h(x + \rho)$, that is,

$$h(x+\rho) - h(x) = \varphi_0 - \varphi(x). \tag{2.9.7}$$

To solve this cohomological equation we use Fourier expansion both for the known function $\varphi(x) = \sum_{k\in\mathbb{Z}} \varphi_k \exp(2\pi ikx)$ and for the unknown function $h(x) = \sum_{k\in\mathbb{Z}} h_k \exp(2\pi ikx)$. Then by (2.9.7) the Fourier coefficients of h for $k \neq 0$ are given by

$$h_k = \frac{\varphi_k}{1 - \exp(2\pi ik\rho)}.$$

In order to show that these numbers are indeed the Fourier coefficients of a C^∞ function we use the Diophantine condition. Recall that a necessary and sufficient condition for a function to be C^∞ is that the Fourier coefficients decrease faster than any power of k. The Diophantine condition gives $|1 - \exp(2\pi i k\rho)| > \text{const.}\, k^{-r}$. Since the φ_k decay faster than any power of k, so do the h_k. □

Exercises

2.9.1. *Let $f: X \to X$ be a topologically transitive homeomorphism of a compact metric space and ψ a continuous function on X. Prove that any two continuous solutions φ of the cohomological equation*

$$\varphi(f(x)) - \varphi(x) = \psi(x)$$

differ by a constant.

2.9.2. *Consider the symbolic system S_k from Exercise 1.9.10. Prove that the cohomological equation*

$$\varphi(S_k\omega) - \varphi(\omega) = (-1)^{\omega_0}$$

has a bounded Borel solution but no continuous solution.

2.9.3. *For $\omega \in \Omega_2$ let $\Phi(\omega) = \sum_{n\in\mathbb{Z}} \omega_n 2^{-|n|}$ and $a(\omega) = \Phi(\sigma_2\omega) - \Phi(\omega)$. Show that the solution φ of the cohomological equation $a(\omega) = \varphi(\sigma_2\omega) - \varphi(\omega)$ given by (2.9.5) has a dense set of discontinuity points.*

3

Principal classes of asymptotic topological invariants

In this chapter we will embark upon the task of systematically identifying important specific phenomena associated with the asymptotic behavior of smooth dynamical systems. We will build upon the results of our survey of specific examples in Chapter 1 as well as on the insights gained from the general structural approach outlined and illustrated in Chapter 2.

Most of the properties discussed in the present chapter are in fact topological invariants and can be defined for broad classes of topological dynamical systems, including symbolic ones. The predominance of topological invariants fits well with the picture that emerges from the considerations of Sections 2.1, 2.3, 2.4, and 2.6. The considerations of the previous chapter make it very plausible that smooth dynamical systems are virtually never differentiably stable and can only rarely be classified locally up to smooth conjugacy. In contrast, structural and the related topological stability seem to be fairly widespread phenomena.

We will consider three broad classes of asymptotic invariants: (i) growth of the numbers of orbits of various kinds and of the complexity of orbit families, (ii) types of recurrence, and (iii) asymptotic distribution and statistical behavior of orbits. The first two classes are of a purely topological nature; they are discussed in the present chapter. The last class is naturally related to ergodic theory and hence we will provide an introduction to key aspects of that subject. This will require some space so we put that material into a separate chapter. The two chapters are intimately connected. Our motivation for introducing invariant measures and ergodic theory in the next chapter will be an attempt to understand in a more quantitative way qualitative recurrence properties in the topological and smooth cases discussed in the present chapter.

1. Growth of orbits

a. Periodic orbits and the ζ-function. Periodic orbits represent the most distinctive special class of orbits. In Definition 1.7.1 we introduced the numbers $P_n(f)$ of periodic points for a map f. Those numbers are obviously topological invariants. Let us point out that $P_n(f)$ gives the total number of points for which the positive integer n is *a* period, not necessarily the smallest possible period. If n is a prime number, then $P_n(f) - P_1(f)$ gives exactly the number of points with smallest period n, and hence $\frac{1}{n}(P_n(f) - P_1(f))$ is the number of periodic *orbits* of period exactly n. In general, the connection between the number of points of period n and the number of orbits of period n is more complicated. If one denotes by $\tilde{P}_n(f)$ the number of points for which n is the minimal period then the number of orbits of period n is naturally $\tilde{P}_n(f)/n$. The numbers $\tilde{P}_n(f)$ are also topological invariants of f; they can be expressed through $P_n(f)$ and vice versa via some number-theoretic functions. In general, it is more convenient to work with $P_n(f)$ than with $\tilde{P}_n(f)$.

The most natural measure of asymptotic growth of the number of periodic points is the exponential growth rate $p(f)$ for the sequence $P_n(f)$:

$$p(f) = \overline{\lim_{n\to\infty}} \frac{\log(\max(P_n(f), 1))}{n}. \tag{3.1.1}$$

We write $\max(P_n(f), 1)$ instead of $P_n(f)$ in order to avoid taking $\log 0$.

If $p(f) = 0$ it is sometimes useful to consider the polynomial growth rate of periodic points given by

$$\overline{\lim_{n\to\infty}} \frac{\log(\max(P_n(f), 1))}{\log n}. \tag{3.1.2}$$

If $p(f) < \infty$, that is, if the growth rate of periodic points is at most exponential, one can incorporate all the information about the numbers of periodic points into a nice analytical object, the *ζ-function* of f defined by

$$\zeta_f(z) = \exp \sum_{n=1}^{\infty} \frac{P_n(f)}{n} z^n, \tag{3.1.3}$$

where z is a complex number.[1] This series converges for $|z| < \exp(-p(f))$ and always has singularities on the circle $|z| = \exp(-p(f))$. In many cases the function $\zeta_f(z)$ admits an analytic continuation, often to a meromorphic function in the whole complex plane. Naturally, poles, zeroes, and residues of the extended ζ-function provide additional topological invariants for f which are determined by the numbers of periodic points but often provide nontrivial insights into the orbit structure.

Let us calculate the ζ-function for several of our examples.

For the linear expanding map E_m with $m > 1$ (and hence, by Theorem 2.4.6, for every expanding map of the circle of degree m) one has $P_n(E_m) = m^n - 1$, so

$$\begin{aligned} \zeta_{E_m}(z) &= \exp\left(\sum_{n=1}^{\infty} \frac{m^n - 1}{n} z^n\right) \\ &= \exp\left(-\log(1 - mz) + \log(1 - z)\right) = \frac{1 - z}{1 - mz}. \end{aligned} \tag{3.1.4}$$

Thus, this function indeed has a meromorphic continuation into the whole plane with a single pole exactly at the point $1/m = \exp(-p(E_m))$.

Similarly for the N-shift σ_N and the one-sided N-shift σ_N^R we have

$$\zeta_{\sigma_N}(z) = \zeta_{\sigma_N^R}(z) = \exp\sum_{n=1}^{\infty}\frac{N^n}{n}z^n = \frac{1}{(1-Nz)}. \tag{3.1.5}$$

The difference between the two expressions (3.1.4) and (3.1.5) is mostly the presence of the zero at $z = 1$ for the first function. It is related to the fact that the semiconjugacy described in the proof of Proposition 1.7.2 maps m fixed points of the shift σ_m^R into the same fixed point $x = 0$ for E_m.

Finally, for the hyperbolic toral automorphism $F_L\colon \mathbb{T}^2 \to \mathbb{T}^2$, $F_L(x,y) = (2x+y, x+y) \pmod 1$ we have

$$\begin{aligned}\zeta_{F_L}(z) &= \exp\sum_{n=1}^{\infty}\left(\left(\frac{3+\sqrt{5}}{2}\right)^n + \left(\frac{3-\sqrt{5}}{2}\right)^n - 2\right)\frac{z^n}{n}\\ &= \frac{(1-z)^2}{\left(1-\left(\frac{3+\sqrt{5}}{2}\right)z\right)\left(1-\left(\frac{3-\sqrt{5}}{2}\right)z\right)}.\end{aligned} \tag{3.1.6}$$

Again, the ζ-function has a meromorphic extension to a rational function on the complex plane.

For flows one should count periodic orbits instead of periodic points. This can be done in two different ways. It would be closest to the discrete-time case to count periodic orbits weighted by their length, which is what counting of periodic points amounts to in the discrete-time case. (Compare with the description of the Bowen measure for flows at the end of Section 20.1.) On the other hand one can count just the number of periodic orbits without weighting by their lengths. If, however, the number of periodic orbits grows exponentially then the distinction is immaterial because most orbits of length up to T will have length close to T, so the growth rate is the same.

Now we consider the invariants given by the numbers $P_T(\varphi^t)$ of all periodic orbits of period less than or equal to T. Correspondingly, one defines the exponential growth rate for the number of periodic orbits for a flow

$$p(\varphi^t) := \overline{\lim_{T\to\infty}}\frac{\log(\max(P_T(\varphi^t),1))}{T} \tag{3.1.7}$$

and the ζ-function

$$\zeta_{\varphi^t}(z) = \prod_{\gamma}(1-\exp(-z\,l(\gamma)))^{-1}, \tag{3.1.8}$$

where the product is taken over all nonfixed periodic orbits γ of the flow and $l(\gamma)$ is the smallest positive period of γ. Assuming that the flow has finitely for infinite products that the product (3.1.8) converges for $\mathrm{Re}\, z > p(\varphi^t)$ and has singularities on the critical line $\mathrm{Re}\, z = p(\varphi^t)$.

The numbers $P_T(\varphi^t)$ and the ζ-function are obviously invariants of flow equivalence. In general, they are *not* invariant under an orbit equivalence. Although an orbit equivalence takes periodic orbits into periodic orbits, it may change their period. However, a cruder property survives a time change.

Proposition 3.1.1. *Let X, Y be compact metric spaces and $\varphi^t : X \to X$ and $\psi^t : Y \to Y$ continuous flows without fixed points. Suppose that the flows are orbit equivalent and $p(\psi^t) = 0$. Then $p(\varphi^t) = 0$.*

Proof. Let $h: X \to Y$ be a homeomorphism that maps orbits of the flow φ^t onto orbits of ψ^t. Then $h(\varphi^1(x)) = \psi^{\alpha(x)}h(x)$. The function α is continuous and positive; hence it is bounded from above, say $\alpha(x) \leq M$. Thus the image of any orbit segment for φ^t of length one is an orbit segment of ψ^t of length at most M. Hence the image of a segment of length at most T has length less than $(T+1)M$ which for $T \geq 1$ is less than $2MT$. That means that the image of any periodic orbit of period $\leq T$ has period $\leq 2MT$, that is,

$$P_{2MT}(\psi^t) \geq P_T(\varphi^t),$$

and $p(\psi^t) \geq p(\varphi^t)/2M$. □

b. Topological entropy. The most important numerical invariant related to the orbit growth is topological entropy. It represents the exponential growth rate for the number of orbit segments distinguishable with arbitrarily fine but finite precision. In a sense, the topological entropy describes in a crude but suggestive way the total exponential complexity of the orbit structure with a single number.

As usual, we begin our discussion with the discrete-time case. Let $f: X \to X$ be a continuous map of a compact metric space X with distance function d. We define an increasing sequence of metrics d_n^f, $n = 1, 2, \ldots$, starting from $d_1^f = d$ by

$$d_n^f(x, y) = \max_{0 \leq i \leq n-1} d(f^i(x), f^i(y)) \tag{3.1.9}$$

In other words, d_n^f measures the distance between the orbit segments $I_x^n = \{x, \ldots, f^{n-1}x\}$ and I_y^n. We will denote the open ball $\{y \in X \mid d_n^f(x, y) < \epsilon\}$ by $B_f(x, \epsilon, n)$.

A set $E \subset X$ is said to be *(n, ϵ)-spanning* if $X \subset \bigcup_{x \in E} B_f(x, \epsilon, n)$. Let $S_d(f, \epsilon, n)$ be the minimal cardinality of an (n, ϵ)-spanning set, or equivalently the cardinality of a *minimal (n, ϵ)-spanning set.* One can verbally express the meaning of that quantity by saying that it is equal to the minimal number of initial conditions whose behavior up to time n approximates the behavior of *any* initial condition up to ϵ. Consider the exponential growth rate for that quantity

$$h_d(f, \epsilon) := \overline{\lim_{n \to \infty}} \frac{1}{n} \log S_d(f, \epsilon, n). \tag{3.1.10}$$

Obviously $h_d(f,\epsilon)$ does not decrease with ϵ. We define the *topological entropy* $h_d(f)$ as

$$h_d(f) = \lim_{\epsilon \to 0} h_d(f,\epsilon).$$

A priori, this quantity might depend on the metric d. We will show that actually it does not.

Proposition 3.1.2. *If d' is another metric on X which defines the same topology as d, then $h_{d'}(f) = h_d(f)$.*

Proof. Consider the set D_ϵ of all pairs $(x_1,x_2) \in X \times X$ for which $d(x_1,x_2) \geq \epsilon$. This is a compact subset of $X \times X$ with the product topology. The function d' is continuous on $X \times X$ in that topology and consequently it reaches its minimum $\delta(\epsilon)$ on D_ϵ. This minimum is positive; otherwise there would be points $x_1 \neq x_2$ such that $d'(x_1,x_2) = 0$. Thus, if $d'(x_1,x_2) < \delta(\epsilon)$, then $d(x_1,x_2) < \epsilon$, that is, any $\delta(\epsilon)$-ball in the metric d' is contained in an ϵ-ball in the metric d. This argument extends immediately to the metrics d'^f_n and d^f_n. Thus for every n we have $S_{d'}(f,\delta(\epsilon),n) \geq S_d(f,\epsilon,n)$ so $h_{d'}(f,\delta(\epsilon)) \geq h_d(f,\epsilon)$ and $h_{d'}(f) \geq \lim_{\epsilon\to 0} h_{d'}(f,\delta(\epsilon)) \geq \lim_{\epsilon \to 0} h_d(f,\epsilon) = h_d(f)$. Interchanging the metrics d and d' one obtains $h_d(f) \geq h_{d'}(f)$. □

Definition 3.1.3. The quantity $h_d(f)$ calculated for any metric generating the given topology in X is called the *topological entropy* of f and is denoted by $h(f)$ or sometimes $h_{\text{top}}(f)$.[2]

Corollary 3.1.4. *The topological entropy is an invariant of topological conjugacy.*

Proof. Let $f: X \to X$, $g: Y \to Y$ be topologically conjugate via a homeomorphism $h: X \to Y$. Fix a metric d on X and define d' on Y as the pullback of d, that is, $d'(y_1,y_2) = d(h^{-1}(y_1), h^{-1}(y_2))$. Then h becomes an isometry so $h_d(f) = h_{d'}(g)$. □

There are several quantities similar to $S_d(f,\epsilon,n)$ that can be used to define topological entropy. For example, let $D_d(f,\epsilon,n)$ be the minimal number of sets whose diameter in the metric d^f_n is at most ϵ and whose union covers X. Obviously, the diameter of an ϵ-ball is less than or equal to 2ϵ so every covering by ϵ-balls is a covering by sets of diameter $\leq 2\epsilon$, that is,

$$D_d(f,2\epsilon,n) \leq S_d(f,\epsilon,n). \tag{3.1.11}$$

On the other hand, any set of diameter $\leq \epsilon$ is contained in the ϵ-ball around each of its points so

$$S_d(f,\epsilon,n) \leq D_d(f,\epsilon,n). \tag{3.1.12}$$

Lemma 3.1.5. *For any $\epsilon > 0$ the limit $\lim_{n\to\infty}(1/n)\log D_d(f,\epsilon,n)$ exists.*

Remark. The corresponding limit for $S_d(f,\epsilon,n)$ may not exist.

Proof. The statement will follow from the inequality

$$D_d(f,\epsilon,m+n) \leq D_d(f,\epsilon,n) \cdot D_d(f,\epsilon,m)$$

for all m, n. For then the sequence $a_n = \log D_d(f,\epsilon,n)$ is subadditive, that is, $a_{m+n} \leq a_n + a_m$, and hence $\lim_{n\to\infty} a_n/n$ exists. This is a very useful elementary fact which will be used numerous times later (for example, (3.1.20), Proposition 4.3.6). Proposition 9.6.4 is a stronger version of this fact.

To prove this inequality, let us notice that if A is a set of d_n^f-diameter less than ϵ and B is a set of d_m^f-diameter less than ϵ, then $A \cap f^{-n}(B)$ is a set of d_{m+n}^f-diameter less than ϵ. Thus if $\mathfrak{A}$ is a cover of X by $D_d(f,\epsilon,n)$ sets of d_n^f-diameter less than ϵ and $\mathfrak{B}$ is a cover of X by $D_d(f,\epsilon,m)$ sets of d_m^f-diameter less than ϵ, the cover by all sets $A \cap f^{-n}(B)$, where $A \in \mathfrak{A}$, $B \in \mathfrak{B}$, which contains not more than $D_d(f,\epsilon,n) \cdot D_d(f,\epsilon,m)$ sets, is a cover by sets of d_{m+n}^f-diameter less than ϵ. □

Let us define $\lim_{n\to\infty}(1/n)\log D_d(f,\epsilon,n) = \tilde{h}_d(f,\epsilon)$. From (3.1.11) and (3.1.12) we derive

$$\tilde{h}_d(f,\epsilon) \geq h_d(f,\epsilon) \geq \tilde{h}_d(f,2\epsilon)$$

and the analogous inequality holds for $\underline{h}_d(f,\epsilon) := \underline{\lim}_{n\to\infty}(1/n)\log S_d(f,\epsilon,n)$ instead of $h_d(f,\epsilon)$.

Thus we conclude that $\lim_{\epsilon\to 0}\tilde{h}_d(f,\epsilon) = \lim_{\epsilon\to 0}\underline{h}_d(f,\epsilon) = h(f)$ and

$$\lim_{\epsilon\to 0}\left(h_d(f,\epsilon) - \underline{h}_d(f,\epsilon)\right) = 0.$$

One more way to define topological entropy is via the numbers $N_d(f,\epsilon,n)$, the maximal number of points in X with pairwise d_n^f-distances at least ϵ. We will call such a set of points (n,ϵ)-*separated.* Such points generate the maximal number of orbit segments of length n that are distinguishable with precision ϵ. A maximal (n,ϵ)-separated set is an (n,ϵ)-spanning set, that is, for any such set of points the ϵ-balls around them cover X, because otherwise it would be possible to increase the set by adding any point not covered. Thus

$$N_d(f,\epsilon,n) \geq S_d(f,\epsilon,n). \tag{3.1.13}$$

On the other hand, no ϵ-ball can contain two points 2ϵ apart. Thus

$$S_d(f,\epsilon,n) \geq N_d(f,2\epsilon,n). \tag{3.1.14}$$

Using (3.1.13) and (3.1.14) we obtain

$$\begin{aligned} &\overline{\lim_{n\to\infty}}\frac{1}{n}\log N_d(f,\epsilon,n) \geq h_d(f,\epsilon), \\ &\underline{\lim_{n\to\infty}}\frac{1}{n}\log N_d(f,2\epsilon,n) \leq \overline{\lim_{n\to\infty}}\frac{1}{n}\log N_d(f,2\epsilon,n) \leq h_d(f,\epsilon) \end{aligned} \tag{3.1.15}$$

and hence

$$h_{\text{top}}(f) = \lim_{\epsilon\to 0}\overline{\lim_{n\to\infty}}\frac{1}{n}\log N_d(f,\epsilon,n) = \lim_{\epsilon\to 0}\underline{\lim_{n\to\infty}}\frac{1}{n}\log N_d(f,\epsilon,n), \tag{3.1.16}$$

justifying the verbal description we gave at the beginning of this subsection.

Proposition 3.1.6. *If the map g is a factor of f, then $h_{\text{top}}(g) \leq h_{\text{top}}(f)$.*

Proof. Let $f: X \to X$, $g: Y \to Y$, $h: X \to Y$, $h \circ f = g \circ h$, $h(X) = Y$, and d_X, d_Y be the distance functions in X and Y, correspondingly.

h is uniformly continuous, so for any $\epsilon > 0$ there is $\delta(\epsilon) > 0$ such that if $d_X(x_1, x_2) < \delta(\epsilon)$, then $d_Y(h(x_1), h(x_2)) < \epsilon$. Thus the image of any $(d_X)_n^f$ ball of radius $\delta(\epsilon)$ lies inside a $(d_Y)_n^g$ ball of radius ϵ, that is,

$$S_{d_X}(f, \delta(\epsilon), n) \geq S_{d_Y}(g, \epsilon, n).$$

Taking logarithms and limits, we obtain the result. □

The following proposition contains an incomplete list of standard elementary properties of topological entropy. The proofs demonstrate the usefulness of switching back and forth from one of the three definitions to another.

Proposition 3.1.7.

(1) *If Λ is a closed f-invariant set, then $h_{\text{top}}(f\restriction_\Lambda) \leq h_{\text{top}}(f)$.*

(2) *If $X = \bigcup_{i=1}^m \Lambda_i$, where Λ_i, $(i = 1, \ldots, m)$ are closed f-invariant sets, then $h_{\text{top}}(f) = \max_{1 \leq i \leq m} h_{\text{top}}(f\restriction_{\Lambda_i})$.*

(3) $h_{\text{top}}(f^m) = |m| h_{\text{top}}(f)$.

(4) $h_{\text{top}}(f \times g) = h_{\text{top}}(f) + h_{\text{top}}(g)$.

Here if $f: X \to X$, $g: Y \to Y$, then $f \times g: X \times Y \to X \times Y$ is defined by $(f \times g)(x, y) = (f(x), g(y))$.

Proof. (1) is obvious since every cover of X by sets of d_n^f-diameter less than ϵ is at the same time a cover of Λ.

To prove (2) let us point out that the union of covers of $\Lambda_1, \ldots, \Lambda_m$ by sets of diameter less than ϵ is a cover of X, so

$$D_d(f, \epsilon, n) \leq \sum_{i=1}^m D_d(f\restriction_{\Lambda_i}, \epsilon, n),$$

that is, for at least one i

$$D_d(f\restriction_{\Lambda_i}, \epsilon, n) \geq \frac{1}{m} D_d(f, \epsilon, n).$$

Since there are only finitely many i, at least one i works for infinitely many n. For this $i \in \{1, \ldots, m\}$

$$\overline{\lim_{n\to\infty}} \frac{\log D_d(f\restriction_{\Lambda_i}, \epsilon, n)}{n} \geq \overline{\lim_{n\to\infty}} \frac{\log D_d(f, \epsilon, n) - \log m}{n} = \tilde{h}_d(f, \epsilon).$$

This proves (2).

For positive m (3) follows from two remarks. First

$$d_n^{f^m}(x, y) = \max_{0 \leq i < n} d(f^{im}(x), f^{im}(y)) \leq \max_{0 \leq i \leq m(n-1)} d(f^i(x), f^i(y)) = d_{nm-m+1}^f(x, y)$$

so $B_f(x,\epsilon,mn-m+1) \subset B_{f^m}(x,\epsilon,n)$ for any $x \in X$ and

$$S_d(f^m,\epsilon,n) \leq S_d(f,\epsilon,mn). \tag{3.1.17}$$

Hence $h_{\text{top}}(f^m) \leq m h_{\text{top}}(f)$. On the other hand, for every $\epsilon > 0$ there is a $\delta(\epsilon) > 0$ such that $B(x,\delta(\epsilon)) \subset B_f(x,\epsilon,m)$ for all $x \in X$. Thus

$$\begin{aligned} B_{f^m}(x,\delta(\epsilon),n) &= \bigcap_{i=0}^{n-1} f^{-im} B(f^{im}(x),\delta(\epsilon)) \\ &\subset \bigcap_{i=0}^{n-1} f^{-im} B_f(f^{im}(x),\epsilon,m) = B_f(x,\epsilon,mn). \end{aligned}$$

Consequently,

$$S_d(f,\epsilon,mn) \leq S_d(f^m,\delta(\epsilon),n) \tag{3.1.18}$$

and $m h_{\text{top}}(f) \leq h_{\text{top}}(f^m)$. Now let f be invertible. Then $B_f(x,\epsilon,n) = B_{f^{-1}}(f^{n-1}(x),\epsilon,n)$ and $S_d(f,\epsilon,n) = S_d(f^{-1},\epsilon,n)$, so $h_{\text{top}}(f) = h_{\text{top}}(f^{-1})$.

For negative m (3) follows from the statement for $m > 0$ and $n = -1$.

Finally, in order to prove (4), one introduces the product metric in $X \times Y$, $d((x_1,y_1),(x_2,y_2)) = \max(d_X(x_1,x_2), d_Y(y_1,y_2))$. Then balls in the product metric are products of balls on X and Y. The same is true for balls in $d_n^{f\times g}$. Thus $S_d(f\times g,\epsilon,n) \leq S_{d_X}(f,\epsilon,n) S_{d_Y}(g,\epsilon,n)$ and $h_{\text{top}}(f\times g) \leq h_{\text{top}}(f) + h_{\text{top}}(g)$. On the other hand, the product of any (n,ϵ)-separated set in X for f and any (n,ϵ)-separated set in Y for g is an (n,ϵ)-separated set for $f \times g$. Thus

$$N_d(f \times g,\epsilon,n) \geq N_{d_X}(f,\epsilon,n) \times N_{d_Y}(g,\epsilon,n)$$

and hence $h_{\text{top}}(f \times g) \geq h_{\text{top}}(f) + h_{\text{top}}(g)$. □

The definition of topological entropy $h_{\text{top}}(\Phi)$ for a flow $\Phi = \{\varphi^t\}_{t\in\mathbb{R}}$ is completely parallel to that for the discrete-time case. The counterpart of (3.1.9) is the following nondecreasing family of metrics

$$d_T^\Phi(x,y) = \max_{0\leq t\leq T} d(\varphi^t(x),\varphi^t(y)).$$

The only property worth special notice is the following counterpart of Proposition 3.1.7(3).

Proposition 3.1.8. $h_{\text{top}}(\Phi) = h_{\text{top}}(\varphi^1)$.

Proof. By compactness and continuity for $\epsilon > 0$ one can find $\delta(\epsilon) > 0$ such that if $d(x,y) \leq \delta(\epsilon)$ then $\max_{0\leq t\leq 1} d(\varphi^t(x),\varphi^t(y)) < \epsilon$. This implies that any ϵ-ball in the metric d_T^Φ contains a $\delta(\epsilon)$-ball in the metric $d_{[T]}^{\varphi^1}$. On the other hand, obviously $d_n^\Phi \geq d_n^{\varphi^1}$. These two remarks imply the statement. □

The topological entropy for a flow is obviously invariant under flow equivalence. It changes under time change and hence under orbit equivalence in a rather complicated way. Arguing similarly to the proof of Proposition 3.1.1 one can show that for a flow without fixed points any time change preserves vanishing of the topological entropy (cf. Exercise 3.1.5). If the topological entropy for a map or a flow vanishes, the subexponential asymptotic of any of the quantities involved in its definition may provide a useful insight into the complexity of the orbit structure.

c. Volume growth. In addition to considering the growth of discrete families of orbits as in Subsections a and b, one can try to measure the growth of continuous families of orbits. For example, for a smooth dynamical system one can take a compact smooth arc γ in the phase space and measure the length of its successive images as a function of time. This length grows at most exponentially and, as before, its exponential growth rate is the most natural and suggestive measure of growth. It does not depend on the choice of Riemannian metric used to measure length. In order to obtain an invariant of at least smooth conjugacy one should consider the supremum of such growth rates for all arcs. The quantity thus obtained in general is not invariant under topological conjugacy. Surprisingly, for C^∞ diffeomorphisms of a two-dimensional manifold it is equal to the topological entropy and hence is a topological invariant.

More generally, one can consider a smooth compact k-dimensional cell on an m-dimensional compact manifold M, that is, the image of the standard closed ball in $\mathbb{R}^k$ under an embedding into M, and calculate the exponential growth rate of the volume of its images under a given smooth dynamical system on M. If the latter is noninvertible the volume should be taken with appropriate multiplicities. Taking the supremum over all k-dimensional cells one obtains an invariant of smooth conjugacy which for a fixed k in general is not an invariant of topological conjugacy. In fact, for a C^∞ map the maximum of these quantities over k, $0 \le k \le m$, is equal to the topological entropy.[3]

d. Topological complexity: Growth in the fundamental group. An alternative to measuring the growth of families of orbits by their volumes is to consider their topological complexity. This idea leads to several algebraic counterparts of entropy. An obvious attraction of that approach is that it automatically produces invariants of topological conjugacy.

The first invariant of this kind is related to the growth of homotopical complexity of iterates for a closed loop.

To that end we first define a purely algebraic notion of entropy for an endomorphism of a finitely generated group. Let π be such a group, $\Gamma = \{\gamma_1, \dots, \gamma_s\}$ a system of generators for π, and $F: \pi \to \pi$ an endomorphism. Let us consider for an element $\gamma \in \pi$ all possible representations

$$\gamma = \gamma_1^{i_1} \gamma_2^{i_2} \cdots \gamma_s^{i_s} \gamma_1^{i_{s+1}} \cdots \gamma_s^{i_{2s}} \cdots \gamma_s^{i_{ks}}, \quad i_j \in \mathbb{Z}, \tag{3.1.19}$$

and let $L(\gamma,\Gamma)$ be the minimum of $\sum_{j=1}^{ks} |i_j|$ over all such representations. Furthermore let $L_n(F,\Gamma) = \max_{1\le i\le s} L(F^n\gamma_i,\Gamma)$. One can concatenate representations of the form (3.1.19), so

$$L(\gamma\cdot\gamma',\Gamma) \le L(\gamma,\Gamma) + L(\gamma',\Gamma)$$

and

$$L(F^n\gamma,\Gamma) \le L(\gamma,\Gamma)\cdot L_n(F,\Gamma).$$

Since F is an endomorphism one can obtain a representation of $F^{m+n}\gamma$ by substituting the representation for $F^n\gamma_i$ into any representation for $F^m\gamma$ of the form (3.1.19). Thus $L(F^{m+n}\gamma,\Gamma) \le L(F^m\gamma,\Gamma)L_n(F,\Gamma)$, hence

$$L_{m+n}(F,\Gamma) \le L_m(F,\Gamma)L_n(F,\Gamma)$$

and consequently

$$\lim_{n\to\infty} \frac{1}{n} \log L_n(F,\Gamma) =: h_A^\Gamma(F) \tag{3.1.20}$$

exists.

Now let $\Gamma' = \{\gamma_1',\dots,\gamma_t'\}$ be another system of generators for π. Each γ_i, $i=1,\dots,s$, has a representation through Γ' and vice versa. Let K be such that all those representations can be chosen to have length at most K. Then $F^n\gamma_i'$ has a representation of the form (3.1.19) of length at most $KL_n(F,\Gamma)$ and hence a representation of the form

$$(\gamma_1')^{i_1}(\gamma_2')^{i_2}\cdots(\gamma_t')^{i_t}\cdots(\gamma_t')^{i_{kt}}$$

of length at most $K^2L_n(F,\Gamma)$, so

$$L_n(F,\Gamma') \le K^2L_n(F,\Gamma)$$

and $h_A^{\Gamma'}(F) \le h_A^\Gamma(F)$. Interchanging Γ and Γ' one obtains $h_A^\Gamma(F) \le h_A^{\Gamma'}(F)$ so $h_A^\Gamma(F)$ is independent of the choice of a system of generators.

Definition 3.1.9. The quantity $h_A(F) = h_A^\Gamma(F)$ for any Γ is called the *algebraic entropy* of the endomorphism F.

It is immediate from the definition that the algebraic entropy is an invariant of conjugacy of group endomorphisms, that is, if $S\colon \pi\to\pi'$ is an isomorphism, and $F\colon\pi\to\pi$ an endomorphism, then

$$h_A(SFS^{-1}) = h_A(F). \tag{3.1.21}$$

For nonabelian groups there is a class of automorphisms that for our purposes should be considered as trivial, namely, the *inner automorphisms* which are of the form $I_{\gamma_0}\gamma = \gamma_0^{-1}\gamma\,\gamma_0$ for fixed $\gamma_0\in\pi$.

Proposition 3.1.10. *For any endomorphism F of a finitely generated group π and any $\gamma_0 \in \pi$*

$$h_A(I_{\gamma_0}F) = h_A(F).$$

Proof. Since $F = I_{\gamma_0^{-1}}(I_{\gamma_0}F)$ it is sufficient to prove that $h_A(I_{\gamma_0}F) \leq h_A(F)$. For $\gamma \in \pi$ one has

$$(I_{\gamma_0}F)^n\gamma = \gamma_0^{-1}F(\gamma_0)^{-1}\cdots(F^{n-1}(\gamma_0))^{-1}F^n(\gamma)F^{n-1}(\gamma_0)\cdots\gamma_0$$

and hence

$$\begin{aligned} L\left((I_{\gamma_0}F)^n\gamma,\Gamma\right) &\leq L(F^n\gamma,\Gamma) + 2\sum_{k=0}^{n-1} L(F^k\gamma_0,\Gamma) \\ &\leq L_n(F,\Gamma)\cdot L(\gamma,\Gamma) + 2\sum_{k=0}^{n-1} L_k(F,\Gamma)\cdot L(\gamma_0,\Gamma). \end{aligned} \tag{3.1.22}$$

Let $\bar{L}_n(F,\Gamma) = \max_{0\leq k\leq n} L_k(F,\Gamma)$. The existence of the limit (3.1.20) implies that

$$\lim_{n\to\infty} \frac{\log \bar{L}_n(F,\Gamma)}{n} = h_A(F),$$

and since

$$\bar{L}_n(I_{\gamma_0}F,\Gamma) \leq (1 + 2nL(\gamma_0,\Gamma))\bar{L}_n(F,\Gamma)$$

from (3.1.22), we obtain $h_A(I_{\gamma_0}F) \leq h_A(F)$. □

Now let us consider a continuous map f of a compact connected manifold M and let $p \in M$. Let us fix a continuous path α connecting p with its image $f(p)$, that is, a map $\alpha\colon [0,1] \to M$ such that $\alpha(0) = p$, $\alpha(1) = f(p)$. Then we can define an endomorphism f_*^α of the fundamental group $\pi = \pi_1(M,p)$, where the image of an element represented by a closed loop $\gamma\colon [0,1] \to M$, $\gamma(0) = \gamma(1) = p$, is represented by the loop $\alpha f(\gamma)\alpha^{-1}$ consisting of the path α followed by the loop $f \circ \gamma$ and then by α taken in the opposite direction.

Let us define the fundamental-group entropy of f (with respect to p and α) as

$$h_*^{p,\alpha}(f) := h_A(f_*^\alpha). \tag{3.1.23}$$

In what follows we will show that this definition is independent of the choice of α and p.

First, let α' be another path connecting p with $f(p)$ and let us denote by σ the element of $\pi_1(M,p)$ represented by the loop $\alpha(\alpha')^{-1}$. For $\gamma \in \pi_1(M,p)$ we have

$$f_*^{\alpha'}(\gamma) = \alpha' f(\gamma)(\alpha')^{-1} = \sigma^{-1}\alpha f(\gamma)\alpha^{-1}\sigma = I_\sigma f_*^\alpha(\gamma).$$

Thus by Proposition 3.1.10

$$h_A(f_*^\alpha) = h_A(f_*^{\alpha'}). \tag{3.1.24}$$

Now let $q \in M$ be another point and let β be a continuous path that connects q and p. Then the path $\gamma = \beta\alpha(f(\beta))^{-1}$ connects q with its image $f(q)$ and one can define the map $f_*^\gamma : \pi_1(M, q) \to \pi_1(M, q)$. The map $S_\beta : \pi_1(M, q) \to \pi_1(M, p)$, $\xi \mapsto \beta^{-1}\xi\beta$ is an isomorphism and direct calculation shows that

$$f_*^\gamma = S_\beta^{-1} f_*^\alpha S_\beta.$$

Thus from (3.1.24)

$$h_A(f_*^\gamma) = h_A(f_*^\alpha). \tag{3.1.25}$$

Then using (3.1.24) and (3.1.25) we see that $h_*^{p,\alpha}(f)$ is, in fact, independent of p and α. From now on we shall denote this quantity by $h_*(f)$ and we will call it the *fundamental-group entropy* of the map f.

Clearly $h_*(f)$ is a topological invariant of f. In Section 8.1b we will show that $h_*(f) \le h_{\text{top}}(f)$ for any continuous map f.[4]

e. Homological growth. Other useful topological growth invariants come from considering the linear maps f_{*i} induced by f on the homology groups $H_i(M, \mathbb{R})$. The most natural numerical measure of growth for a linear map is its spectral radius, which in the finite-dimensional case coincides with the maximum of absolute values of eigenvalues (cf. Definition 1.2.1 and the subsequent discussion). The spectral radii $r(f_{*i})$ are topological invariants of f.

Two particularly interesting cases of homology growth are $i = 1$ and $i = \dim M$. If $i = 1$ we use the Hurewicz isomorphism

$$\mathcal{H}: \pi_1(M, p)/[\pi_1(M, p), \pi_1(M, p)] \to H_1(M, \mathbb{Z})$$

between the abelianization of $\pi_1(M, p)$ and the first homology group (Theorem A.7.5). Any two $S_\beta : \pi_1(M, q) \to \pi_1(M, p)$ differ by an inner automorphism, so all S_β define the same isomorphism between $\pi_1(M, q)/[\pi_1(M, q), \pi_1(M, q)]$ and $\pi_1(M, p)/[\pi_1(M, p), \pi_1(M, p)]$. If

$$P: \pi_1(M, p) \to \pi_1(M, p)/[\pi_1(M, p), \pi_1(M, p)]$$

denotes the projection and $\mathcal{P}: H_1(M, \mathbb{Z}) \to H_1(M, \mathbb{R})$ the canonical homomorphism onto the integer lattice in $H_1(M, \mathbb{R})$ whose kernel consists of the elements of finite order then we let

$$\mathfrak{H}_p := \mathcal{P} \circ \mathcal{H} \circ P : \pi_1(M, p) \to H_1(M, \mathbb{R}).$$

The Hurewicz isomorphism is equivariant with respect to continuous maps, so for any continuous $f: M \to M$, $p \in M$ and any path α connecting p with $f(p)$ we have

$$f_{*1}\mathfrak{H}_p = \mathfrak{H}_p f_*^\alpha. \tag{3.1.26}$$

Thus f_{*1} restricted to the integer lattice is a factor map of f_*^α. Given generators Γ in $\pi_1(M, p)$ and a norm in $H_1(M, \mathbb{R})$ there is a $C > 0$ such that for any $\gamma \in \pi_1(M, p)$ one has

$$\|\mathfrak{H}_p(\gamma)\| \le CL(\gamma, \Gamma). \tag{3.1.27}$$

Equations (3.1.26) and (3.1.27) immediately yield

$$\log r(f_{*1}) \leq h_*(f). \tag{3.1.28}$$

The case $i = \dim M$ is of interest only for noninvertible maps of orientable manifolds. If M is orientable then $H_1(M, \mathbb{R}) = \mathbb{R}$ and $f_{*\dim M}$ is the multiplication by an integer $\deg(f)$ called the degree of f (see Section 8.2). For an invertible map the degree is always ± 1 since $f_{*\dim M}$ is invertible. In Section 8.3 we will show that

$$h_{\text{top}}(f) \geq \log |\deg(f)| = \log |r(f_{*\dim M})| \tag{3.1.29}$$

for every C^1 map $f: M \to M$ (Theorem 8.3.1). Thus any smooth map f with $|\deg(f)| \geq 2$ has complicated asymptotic behavior.

Unlike the relation between h_{top} and $r(f_{*1})$, however, (3.1.29) does not necessarily hold for continuous maps. In fact, continuous maps of degree greater than one may have relatively simple asymptotic behavior for individual orbits. We will discuss an appropriate example at the end of Section 8.2c.[5]

For continuous-time dynamical systems the invariants defined above are vacuous since every element of the flow is homotopic to the identity and hence induces trivial maps of the fundamental group and homology groups. There are, however, different ways to measure the growth of topological complexity. For example, on a compact connected manifold X one can fix a point $p \in X$ and a family of arcs $\Gamma = \{\gamma_x \mid x \in X\}$ of bounded length connecting p with various points of X. Then for a flow $\Phi = \varphi^t: X \to X$ one fixes T and considers for each $x \in X$ the closed loop $l(x, T)$ consisting of the arc γ_x, the orbit segment $\{\varphi^t x\}_{t=1}^T$, and the reverse of the arc $\gamma_{f_T x}$. Those loops represent different elements of the fundamental group $\pi_1(X, p)$. Their number, which we may denote by $\Pi(\Phi, p, \Gamma, T)$, grows at most exponentially and one can define the exponential growth rate

$$\overline{\lim_{T \to \infty}} \frac{\log \Pi(\Phi, p, \Gamma, T)}{T},$$

which is, in fact, independent of p and of the choice of a family of arcs Γ and hence can be simply denoted $h_{\text{hom}}(\Phi)$. We shall call $h_{\text{hom}}(\Phi)$ the *homotopical entropy* of the flow Φ. It is obviously invariant under flow equivalence.

We will develop this idea for a special case of dynamical systems, geodesic flows on Riemannian manifolds (see Definition 5.3.4) in Section 9.6.

Exercises

3.1.1. *Let $f: S^1 \to S^1$ be a C^1 map of degree $m \geq 2$. Suppose f has an attracting periodic point of period n, that is, $f^n(x) = x$ and $|(f^n)'(x)| < 1$. Prove that $P_n(f) \geq m^n + 1$.*

3.1.2. *Prove that for any nonnegative real number t there exists a homeomorphism $f: X \to X$ of a compact metric space such that $p(f) = t$.*

3.1.3. *Prove that the ζ-function of any topological Markov chain is rational.*

3.1.4. *Does there exist a map f with the ζ-function $\zeta_f(z) = \exp \exp z$?*

3.1.5. *Consider the free group on two generators a, b and calculate the algebraic entropy of the automorphism that sends a to aba and b to ab.*

3.1.6*. *Show that*

$$h_{\text{hom}}(\Phi) \leq h_{\text{top}}(\Phi).$$

The next three exercises introduce the original definition of topological entropy by Adler, Konheim, and McAndrew.

3.1.7. *Let $\mathfrak{A}$ be an open cover of a compact space X and $N(\mathfrak{A})$ the minimal number of elements in a subcover of $\mathfrak{A}$. For two covers $\mathfrak{A}, \mathfrak{B}$ let $\mathfrak{A} \vee \mathfrak{B}$ be the cover formed by all intersections $A \cap B$ where $A \in \mathfrak{A}$, $B \in \mathfrak{B}$. Show that*

$$N(\mathfrak{A} \vee \mathfrak{B}) \leq N(\mathfrak{A}) \cdot N(\mathfrak{B}).$$

3.1.8. *Prove the following: For a continuous map $f: X \to X$ and an open cover $\mathfrak{A}$ let $f^{-1}(\mathfrak{A})$ be the cover by all sets $f^{-1}(A)$ where $A \in \mathfrak{A}$. Then*

$$h(f, \mathfrak{A}) := \lim_{n \to \infty} \frac{1}{n} \log N(\mathfrak{A} \vee f^{-1}(\mathfrak{A}) \vee \cdots \vee f^{1-n}(\mathfrak{A}))$$

exists.

3.1.9*. *Show that*

$$h_{\text{top}}(f) = \sup_{\mathfrak{A}} h(f, \mathfrak{A}),$$

where the supremum is taken over all open covers of X.

2. Examples of calculation of topological entropy

In this section we will go through the list of examples discussed in Chapter 1 and calculate topological entropy for those examples. We will naturally restrict ourselves to the cases where the phase space is compact since we only defined topological entropy for that situation.

a. Isometries. Assume that X is a compact metric space and $f: X \to X$ is an isometry. Then $d_n^f = d$ for all n and consequently the quantities $S_n(f, \epsilon, n)$ do not depend on n and do not grow at all. In particular, $h(f) = 0$. This case of absence of any growth is most removed from the case of positive topological entropy. As we mentioned at the end of Section 3.1b, between these two extreme cases there is a variety of situations of "moderate", that is, subexponential, growth for those quantities. The situation with isometric flows is completely similar. Thus, we obtain the following result.

Proposition 3.2.1. *The topological entropy for any translation T_γ of the torus or any linear flow T_ω^t on the torus is equal to zero.*

b. Gradient flows. Let us consider the gradient flow for the round sphere as in Section 1.6. Fix an $\epsilon > 0$ and let N_ϵ and S_ϵ be ϵ-neighborhoods of of the two fixed points $N = (0, 0, 1)$ and $S = (0, 0, -1)$, respectively. Let $K := S^2 \smallsetminus (N_\epsilon \cup S_\epsilon)$. Any orbit spends only a bounded amount of time in K, and by compactness one can find a fixed number $M(\epsilon)$ of orbit segments of length at most $T(\epsilon)$ such that every orbit segment that begins and ends in the set K is ϵ-close to one of those segments all the time.

A typical orbit segment of very large length T stays in N_ϵ for a long time, then spends at most time $T(\epsilon)$ in K and the rest of the time in S_ϵ. Naturally all orbit segments inside N_ϵ and S_ϵ are ϵ-close to each other, whereas the part in K is ϵ-close to one of the chosen fixed number of segments. Thus, the divergence of orbit segments may occur only due to different times of entering the set K. On the other hand, by compactness for any given flow $\Phi = \{\varphi^t\}$ and for every $\epsilon > 0$ there exists $\delta(\epsilon) > 0$ such that for all T we have $d_T^\Phi(x, \varphi^s x) < \epsilon$ if $0 \le s \le \delta(\epsilon)$. Thus every orbit segment of length T is ϵ-close to a shift of one of the $M(\epsilon)$ chosen segments in K, where the shifts are by multiples of $\delta(\epsilon)$. The number of all such shifts grows linearly with T and hence the topological entropy of the flow is equal to zero. This is, in a way, an example of the slowest growth rate for the d_T^Φ balls possible, if any growth is present at all. Thus we have proved the following fact.

Proposition 3.2.2. *The topological entropy of the gradient flow on the round sphere is equal to zero.*

It is not difficult to see that our argument is general enough. In particular, it applies to the two examples from Section 1.6 of gradient flows on the torus. With a little effort we can apply it to the gradient flow for any function with isolated critical points. We will not do that because the resulting statement is superseded by (3.3.1) which implies that the topological entropy of any gradient flow is equal to zero.

c. Expanding maps. The expanding maps E_m introduced in Section 1.7 represent the first situation in our survey where a really complicated orbit structure appears. Since one of the features of this structure is exponential growth of periodic orbits (Proposition 1.7.2) it is natural to expect the total exponential orbit complexity, measured by the topological entropy, to be positive too.

Proposition 3.2.3. *If $m \in \mathbb{N}$, $|m| \geq 2$, then $h_{\text{top}}(E_m) = \log |m| = p(E_m)$.*

Proof. For the map E_m, and in fact for any expanding map, the distance between iterates of any two points grows until it becomes greater than a certain constant depending on the map ($1/2|m|$ for the map E_m). To simplify notations assume $m > 0$. Let $x, y \in S^1$, $d(x, y) < m^{-n}/2$. Then $d_n^{E_m}(x, y) = d(E_m^{n-1}(x), E_m^{n-1}(y))$. Thus if $d_n^{E_m}(x, y) \geq \epsilon$ then $d(x, y) \geq \epsilon m^{-n}$. Pick $\epsilon = m^{-k}$. Then the set of points $\{im^{-n-k} \mid i = 0, \ldots, m^{n+k} - 1\}$ is a maximal set of points whose pairwise $d_n^{E_m}$ distances are greater or equal than m^{-k}. Thus

$$N_d(E_m, m^{-k}, n) = m^{n+k}$$

and consequently

$$h(E_m) = \lim_{k\to\infty} \lim_{n\to\infty} \frac{\log N_d(E_m, m^{-k}, n)}{n} = \lim_{k\to\infty} \lim_{n\to\infty} \frac{n+k}{n} \log m = \log m.$$

The case $m < 0$ is completely parallel. □

For any map $f\colon S^1 \to S^1$ the induced map f_* of the fundamental group $\mathbb{Z}$ is simply the multiplication by $\deg f$. Using the natural generator of $\mathbb{Z}$ we immediately see that $h_*(f) = \log |\deg f|$.

Since topological entropy is invariant under topological conjugacy (Corollary 3.1.4) and every expanding map of degree m is topologically conjugate to the map E_m (Theorem 2.4.6) we obtain from Proposition 3.2.3

Corollary 3.2.4. *If $f\colon S^1 \to S^1$ is an expanding map of degree m then*

$$h_{\text{top}}(f) = h_*(F) = p(f) = \log |m|.$$

d. Shifts and topological Markov chains. As in Section 1.9a consider the space Ω_N with the metric $d = d_{10N}$ given by

$$d_{10N}(\omega, \omega') = \sum_{n=-\infty}^{\infty} \frac{|\omega_n - \omega'_n|}{(10N)^{|n|}}.$$

Then for $\alpha = (\alpha_{-n}, \ldots, \alpha_n)$ the symmetric cylinder $C_\alpha^m = \{\omega \in \Omega_n \mid \omega_i = \alpha_i \text{ for } |i| \leq m\}$ is at the same time the ball of radius $\epsilon_m = (10N)^{-m}/2$ around each of its points. Similarly if we fix numbers $\alpha_{-m}, \ldots, \alpha_{m+n}$, the cylinder

$$C_{\alpha_{-m},\ldots,\alpha_{n+m}}^{-m,\ldots,n+m} = \{\omega \in \Omega_N \mid \omega_i = \alpha_i \text{ for } -m \leq i \leq m+n\} \tag{3.2.1}$$

is at the same time the ball of radius ϵ_m around each of its points with respect to the metric $d_n^{\sigma_N}$ associated with the shift σ_N. Thus, any two $d_n^{\sigma_N}$ balls of radius ϵ_m are either identical or disjoint and there are exactly N^{n+2m+1} different ones of the form (3.2.1). The covering of Ω_N by those balls is obviously minimal, so $S_{d_{10N}}(\sigma_N, \epsilon_m, n) = N^{n+2m+1}$ and

$$h(\sigma_N) = \lim_{m\to\infty} \lim_{n\to\infty} \frac{1}{n} \log N^{n+2m+1} = \log N.$$

Similarly, for the topological Markov chain σ_A, $S_d(\sigma_A, \epsilon_m, n)$ equals the number of those cylinders (3.2.1) which have nonempty intersection with the set Ω_A. Let us assume that each row of the matrix A contains at least one 1. Since the number of admissible paths of length n that begin with the symbol i and end with the symbol j is equal to the entry a_{ij}^n of the matrix A^n (see Lemma 1.9.4), the number of nonempty cylinders of rank $n+1$ in Ω_A is equal to $\sum_{i,j=0}^{N-1} a_{ij}^n < C \cdot \|A^n\|$ for some constant C. On the other hand, since all numbers a_{ij}^n are nonnegative, $\sum_{i,j=0}^{N-1} a_{ij}^n > c\|A^n\|$ for another constant $c > 0$. Thus, we have

$$S_{d_{10N}}(\sigma_A, \epsilon_m, n) = \sum_{i,j=0}^{N-1} a_{ij}^{n+2m} \tag{3.2.2}$$

and

$$\begin{aligned} \log|\lambda_A^{\max}| = \log r(A) &= \lim_{n\to\infty} \frac{1}{n} \log \|A^n\| = \lim_{n\to\infty} \frac{1}{n} \log \|A^{n+2m}\| \\ &= \lim_{n\to\infty} \frac{1}{n} \log S_{d_{10N}}(\sigma_A, \epsilon_m, n) = h(\sigma_A), \end{aligned} \tag{3.2.3}$$

where $r(A)$ is the spectral radius of the matrix A (Definition 1.2.1). Comparing (3.2.2) and (3.2.3) with Corollary 1.9.5 we obtain

Proposition 3.2.5. $h_{\text{top}}(\sigma_A) = p(\sigma_A) = \log|\lambda_A^{\max}|$ *for any topological Markov chain* σ_A.

e. The hyperbolic toral automorphism. Finally, let us consider the toral automorphism

$$F_L(x, y) = (2x + y, x + y) \pmod 1.$$

In Section 2.5 we showed that F_L is a factor of the topological Markov chain σ_A, where

$$A = \begin{pmatrix} 1 & 1 & 0 & 1 & 0 \\ 1 & 1 & 0 & 1 & 0 \\ 1 & 1 & 0 & 1 & 0 \\ 0 & 0 & 1 & 0 & 1 \\ 0 & 0 & 1 & 0 & 1 \end{pmatrix}.$$

We have seen already that $\lambda_A^{\max} = \dfrac{3+\sqrt{5}}{2}$ is the same as the maximal eigenvalue of the matrix $L = \begin{pmatrix} 2 & 1 \\ 1 & 1 \end{pmatrix}$ that defines F_L. Thus by Proposition 3.1.6

$$h(F_L) \le h(\sigma_A) = \log \frac{3+\sqrt{5}}{2}. \tag{3.2.4}$$

On the other hand, let us consider all periodic points of the map F_L with period n. If p, q are two such points and if they are close enough to each other, say $\operatorname{dist}(p,q) < 1/4$, then there is a uniquely defined minimal rectangle $psqt$ formed by segments of expanding and contracting lines passing through p and q.

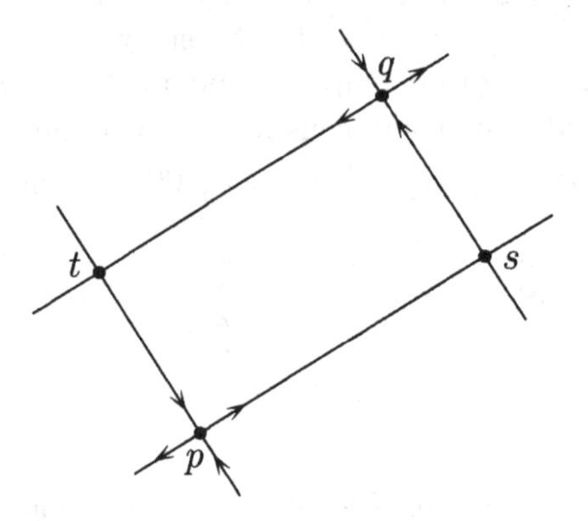

FIGURE 3.2.1. Heteroclinic points

Under the action of F_L the sides pt and sq expand with coefficient $\lambda_1 = \dfrac{3+\sqrt{5}}{2} > 2$ while the other two sides contract with coefficient λ_1^{-1}. Thus, either under the action of F_L or F_L^{-1}, the perimeter of the rectangle $psqt$ increases and then continues to increase under the iterates until the iterates of p and q are far enough from each other that the image of our rectangle is no longer minimal. Since p and q are periodic points this must happen because the nth image or preimage of the minimal rectangle cannot be minimal; its perimeter will be too big. Thus at some moment k the distance between $F_L^k(p)$ and $F_L^k(q)$ becomes greater than $1/4$, that is, all periodic points of period n form an $(n,\ 1/4)$-separated set in the metric $d_n^{F_L}$. Thus $N_d(F_L, 1/4, n) \ge P_n(F_L)$ and by Proposition 1.8.1

$$h(F_L) \ge p(F_L) = \log \frac{3+\sqrt{5}}{2}. \tag{3.2.5}$$

Again as for transitive topological Markov chains we obtain from (3.2.4) and (3.2.5)

Proposition 3.2.6. *If $F_L: \mathbb{T}^2 \to \mathbb{T}^2$ is given by $F_L(x,y) = (2x+y, x+y)$ (mod 1), then*

$$h(F_L) = p(F_L) = \frac{3+\sqrt{5}}{2}.$$

Remark. In the case of expanding maps E_m and for topological Markov chains σ_A one can also show that periodic points form (n, ϵ_0)-separated sets for some ϵ_0. This allows us to produce the inequality $h_{\text{top}} \geq p$ in a uniform way for all three cases.

Another similarity between expanding maps and the toral automorphism has to do with the fundamental-group entropy h_*. The fundamental group $\pi_1(\mathbb{T}^2, (0,0)) = \mathbb{Z}^2$ and the map F_{L*} acts on the natural generators $\gamma_1 = (1,0)$ and $\gamma_2 = (0,1)$ as $F_{L*}(\gamma_1) = 2\gamma_1 + \gamma_2$, $F_{L*}(\gamma_2) = \gamma_1 + \gamma_2$. Since the group is abelian and free, the representations for $F_{L*}^n \gamma_i$, $i = 1,2$, of the form (3.1.22) can be brought into a canonical form corresponding to the rows of the matrix L^n. This immediately implies that $h_*(F_L) = \log \frac{3+\sqrt{5}}{2}$.

Thus we have encountered an interesting pattern. For both smooth examples with complicated exponentially growing orbit structure, expanding maps and hyperbolic toral automorphisms, all three natural measures of the exponential orbit growth, the growth rate p of periodic points, the topological entropy h_{top}, and the fundamental-group entropy h_*, coincide. The coincidence of the first two quantities is a rather widespread although by far not universal phenomenon. It, as well as structural stability, is related to the local hyperbolic structure (see Sections 6.4 (Theorem 6.4.15) and 18.5 (Theorem 18.5.1)). The coincidence of h_* with the other two quantities is more accidental and depends both on hyperbolicity and on the low dimension. One can see that already for higher-dimensional toral automorphisms it may not take place (see Exercise 3.2.8). However, Theorem 8.1.1 will show that $h_* \leq h_{\text{top}}$. For topological Markov chains the growth rate of periodic points and topological entropy also coincide. Hyperbolicity is a relevant explanation here, too, since by the constructions of Section 2.5c topological Markov chains are topologically conjugate to the restriction of some smooth systems to special invariant sets that possess hyperbolic behavior.

f. Finiteness of entropy of Lipschitz maps. In all our examples the topological entropy of the map under consideration turned out to be finite. Now we will show that this always holds for Lipschitz continuous maps of compact metric spaces, in particular for differentiable maps and shifts.

Definition 3.2.7. Let (X,d) be a compact metric space and $b(\epsilon)$ the minimum cardinality of a covering of X by ϵ-balls. Then

$$D(X) := \overline{\lim_{\epsilon \to 0}} \frac{\log b(\epsilon)}{|\log \epsilon|} \in \mathbb{R} \cup \{\infty\}$$

is called the ball dimension of X.

Remark. $D(X)$ is clearly invariant under passing to a biLipschitz-equivalent metric. It is also not hard to see that $D([0,1]^n) = n$ and that for a differentiable manifold the ball dimension is the topological dimension. Finally we obviously have $D(Y) \le D(X)$ whenever $Y \subset X$. However, the ball dimension is not a topological invariant.

Definition 3.2.8. Let (X, d) be a metric space, $f: X \to X$ a Lipschitz continuous map. Then the *Lipschitz constant* $L(f)$ of f is defined by

$$L(f) := \sup_{x \neq y} \frac{d(f(x), f(y))}{d(x, y)}.$$

Theorem 3.2.9. *Let (X, d) be a compact metric space of finite ball dimension $D(X)$ and $f: X \to X$ a Lipschitz continuous map. Then*

$$h_{\text{top}}(f) \le D(X) \max(0, \log L(f)).$$

Proof. Let $L > \max(1, L(f))$. Then $d(f(x), f(y)) \le L d(x, y)$ for all $x, y \in X$. In particular $f^m(B_d(x, L^{-n}\epsilon)) \subset B_d(f(x), \epsilon)$ when $0 \le m \le n$ and hence $B_d(x, L^{-n}\epsilon) \subset B_{d_n^f}(x, \epsilon)$ for all $x \in X$, $n \in \mathbb{N}$, and $\epsilon > 0$. Thus

$$S(f, \epsilon, n) \le b(L^{-n}\epsilon).$$

Since $|\log(L^{-n}\epsilon)| = n \log L - \log \epsilon$ we have

$$n = \left|\frac{\log(L^{-n}\epsilon)}{\log L}\right| \left(1 + \frac{\log \epsilon}{|\log(L^{-n}\epsilon)|}\right) = \frac{1}{\log L} |\log(L^{-n}\epsilon)| \left(1 + O\left(\frac{1}{n}\right)\right),$$

so

$$\overline{\lim_{n\to\infty}} \frac{\log S(f, \epsilon, n)}{n} \le \overline{\lim_{n\to\infty}} \frac{\log b(L^{-n}\epsilon)}{n} = \log L \overline{\lim_{n\to\infty}} \frac{\log b(L^{-n}\epsilon)}{|\log(L^{-n}\epsilon)|} = D(X) \log L$$

and $h_{\text{top}}(f) \le D(X) \log L < \infty$. □

Since for a manifold the ball dimension is equal to the usual dimension and for C^1 maps f of a compact Riemannian manifold we have $L(f) = \sup_x \|Df_x\|$, we obtain

Corollary 3.2.10. *For a C^1 map $f: M \to M$ of a compact Riemannian manifold*

$$h_{\text{top}}(f) \le \max(0, \dim M \log \sup_x \|Df_x\|) < \infty.$$

Of course the latter estimate works for any Riemannian metric on M. It is natural to try to minimize the norm of $\|Df\|$. In fact, instead of that norm one can use the constant $l(f)$ defined after (3.2.6) which is independent of choice of Riemannian metric (Exercises 3.2.10 and 3.2.11).

g. Expansive maps. During the arguments in Subsections c, d, and e we noticed that in the calculation of topological entropy taking the limit as $\epsilon \to 0$ was not necessary because the quantity $h_d(f, \epsilon)$ stabilized for sufficiently small ϵ. This is not true universally but due to a certain topological property which is also interesting in itself. As before let X be a compact metric space.

Definition 3.2.11. A homeomorphism (correspondingly, a continuous map) $f: X \to X$ is called *expansive* if there exists a constant $\delta > 0$ such that if $d(f^n(x), f^n(y)) < \delta$ for all $n \in \mathbb{Z}$ (correspondingly, $n \in \mathbb{N}_0$) then $x = y$. A continuous flow φ^t is called *expansive* if given any continuous function $s: \mathbb{R} \to \mathbb{R}$ with $s(0) = 0$ and $d(\varphi^t(x), \varphi^{s(t)}(x)) < \delta$ then $d(\varphi^t(x), \varphi^{s(t)}(y)) < \delta$ implies $y = \varphi^\tau(x)$ for some $\tau \in \mathbb{R}$.

The maximal number δ_0 satisfying this property is usually called the *expansivity constant* for the dynamical system. By compactness, the property of being expansive does not depend on the particular choice of a metric on X defining a given topology, and hence is an invariant of topological conjugacy. However, the expansivity constant does depend on the choice of metric.

It is easy to see that none of the examples from the first part of our survey (rotations on the circle, translations on the torus, linear flows on the torus, completely integrable Hamiltonian systems, gradient flows) is expansive. On the other hand, the examples from the second group (expanding maps on the circle, topological Markov chains, hyperbolic toral automorphisms) are all expansive.

The proof of expansiveness for expanding maps is contained in the proof of Proposition 3.2.3. For topological Markov chains as well as for arbitrary symbolic dynamical systems the property is self-evident: $\omega \neq \omega'$ means that $\omega_n \neq \omega'_n$ for some n. But then applying the nth iterate of the shift we obtain elements whose zeroth coordinates are different. The distance between any two such elements is greater than a fixed constant.

Finally for the hyperbolic toral automorphism F we will use the same trick as in our proofs of topological transitivity (Proposition 1.8.1) and of the separation of periodic points (Subsection e of the present section). Namely, if $x, y \in \mathbb{T}^2$ are close, one of three possibilities holds. Either the two points lie on the same short arc of an expanding line or on the same short arc of a contracting line, or one can construct a unique minimal rectangle from the arcs of expanding and contracting lines of x and y. In the first two cases the situation is very similar to that for expanding maps. Namely, the distance between positive (correspondingly, negative) iterates of the points x, y is equal to the distance along the expanding (correspondingly, contracting) line and grows exponentially as long as it is less than $1/(3+\sqrt{5}) = 1/2\lambda$. In the last case one can apply the previous argument, substituting y by z and the intersection of the unstable arc of x with the stable arc of y, and use the triangle inequality. The distance between the positive iterates of x and z grows until it reaches at least $1/(2\lambda)$ while the distance between positive iterates of z and y is equal to $\mathrm{dist}(y, z) \cdot \lambda^{-n}$.

The principal reason we are discussing expansiveness now is its usefulness in the computation of topological entropy.

Lemma 3.2.12. *Let X be a compact metric space and $f: X \to X$ an expansive homeomorphism with expansivity constant δ_0. Then for $0 < \epsilon < \delta_0/2$ and $\delta > 0$ there exists $C_{\delta,\epsilon}$ such that for all $n \in \mathbb{N}$ we have*

$$N_d(f, \delta, n) \leq C_{\delta,\epsilon} N_d(f, \epsilon, n).$$

Proof. For $0 < \epsilon < \delta_0/2$ let $N \in \mathbb{N}$ and $\alpha > 0$ be such that

$$d^f_{2N+1}(f^{-N}(x), f^{-N}(y)) \leq 2\epsilon \Rightarrow d(x, y) < \delta$$

and $d(x, y) \leq \alpha \Rightarrow d^f_{2N+1}(f^{-N}(x), f^{-N}(y)) \leq \delta$. Existence of such an N follows from expansivity, and of α from uniform continuity. If E is a maximal (n, δ)-separated set and F a maximal (n, ϵ)-separated set, then for $x \in E$ there is a $z(x) \in F$ such that $d^f_n(x, z(x)) < \epsilon$. Thus $\operatorname{card}(E) \leq \sum_{z \in F} \operatorname{card}(E_z)$, where $E_z := \{x \in E \mid z(x) = z\}$, and the claim follows once we find a bound on $\operatorname{card}(E_z)$ that does not depend on n.

But if $x, y \in E_z$ then $d^f_n(x, y) \leq 2\epsilon$ by definition of E_z, hence $d(f^i(x), f^i(y)) \leq \delta$ for $i \in [N, n - N)$ by choice of N, and thus, by choice of α and since $\{x, y\}$ is (n, δ)-separated, $d(x, y) > \alpha$ or $d(f^n(x), f^n(y)) > \alpha$.

Therefore

$$\begin{aligned} \operatorname{card}(E_z) &= \operatorname{card}\{(x, f^n(x)) \mid x \in E_z\} \\ &\leq \max\{\operatorname{card} A \mid A \subset X \times X \text{ and } \forall a, b \in A \quad d(a, b) > \alpha\} =: M \end{aligned}$$

since the $(x, f^n(x))$ form just such an α-separated set. □

Remark. Obviously we can take $C_{\delta,\epsilon} = 1$ for $\delta > \epsilon$.

Corollary 3.2.13. *If $f: X \to X$ is expansive and δ is less than half its expansivity constant δ_0, then $h_d(f, \delta) = h(f)$. The same is true for flows.*

An equally easy property relates the topological entropy and periodic orbit growth (3.1.1) for expansive maps:

Proposition 3.2.14. *$p(f) \leq h_{\text{top}}(f)$ for an expansive homeomorphism f of a compact metric space.*

Proof. If δ_0 is the expansivity constant then $\operatorname{Fix}(f^n)$ is (n, δ_0)-separated for all $n \in \mathbb{N}$ since if $x \neq y \in \operatorname{Fix}(f^n)$ and $\delta := \max\{d(f^i(x), f^i(y)) \mid 0 \leq i < n\}$ then $d(f^i(x), f^i(y)) \leq \delta$ for $i \in \mathbb{Z}$ and hence $\delta > \delta_0$. Thus $P_n(f) \leq N(f, \epsilon, n)$ for $\epsilon < \delta_0$, implying the claim. □

Here is another property of expansive maps which is used among other things in the proofs of structural stability.

Proposition 3.2.15. *If $f: X \to X$ is expansive with expansivity constant δ_0, $h: X \to X$ is a continuous map such that $d(x, h(x)) < \delta_0$ for all x, and $f \circ h = h \circ f$, then h is equal to the identity.*

Proof. Let $h(x) = y \neq x$ and $d(f^n(x), f^n(y)) \geq \delta_0$. Since $f^n(y) = f^n(h(x)) = h(f^n(x))$ we have $d(f^n(x), h(f^n(x))) \geq \delta_0$, contradicting our assumption. □

Exercises

3.2.1. *Let $f: [0,1] \to [0,1]$ be a homeomorphism. Prove that $h(f) = 0$.*

3.2.2. *Let $f: S^1 \to S^1$ be a continuous map. Prove that $h(f) \geq \log|\deg f|$.*

3.2.3. *Prove that if $f: S^1 \to S^1$ is a monotone map (that is, its lift to $\mathbb{R}$ is a monotone function) then $h(f) = \log|\deg f|$.*

3.2.4. *Calculate the minimal positive value of topological entropy for a topological Markov chain σ_A, where A is a 3×3 matrix.*

3.2.5. *Calculate the topological entropy for the symbolic dynamical systems S_2 and S_3 defined in Problem 1.9.10.*

3.2.6. *Calculate the topological entropy for the following affine map A_α of the two-dimensional torus*

$$A_\alpha(x,y) = (x+\alpha,\ y+x) \pmod 1$$

(cf. Exercise 1.4.4).

3.2.7. *Let L be an $m \times m$ matrix with integer entries and determinant $+1$ or -1. Let $\lambda_1, \dots, \lambda_m$ be the eigenvalues of L counted with their multiplicities. Prove that for the automorphism $F_L: \mathbb{T}^m \to \mathbb{T}^m$*

$$h_{\text{top}}(F_L) \geq \sum_{i:|\lambda_i|>1} \log|\lambda_i|.$$

3.2.8. *Find a hyperbolic matrix $L \in SL(4,\mathbb{Z})$ such that $h_{\text{top}}(F_L) > h_*(F_L)$.*

3.2.9. *Prove that there exists a positive constant l such that for any Riemannian metric σ on M there exists another constant $c = c(f,\sigma)$ such that for every $x, y \in M$ and every $n \in \mathbb{N}$ we have*

$$d_\sigma(f^n(x), f^n(y)) \leq cl^n d_\sigma(x,y), \tag{3.2.6}$$

where d_σ is the distance generated by the Riemannian metric σ.

Denote the smallest constant l satisfying (3.2.6) by $l(f)$.

3.2.10. *Under the assumptions of Corollary 3.2.10 prove $h_{\text{top}}(f) \leq l(f) \dim M$.*

3.2.11. *Prove the counterpart of Corollary 3.2.13 for noninvertible maps.*

3. Recurrence properties

We have already encountered properties related to the character of recurrent behavior, including topological transitivity (Definition 1.3.1), minimality (Definition 1.3.2), and topological mixing (Definition 1.8.2). The topological type of the closure of the set $\mathrm{Per}(f)$ of all periodic points is another invariant of that type. Furthermore, for each point x, the f-invariant ω-limit and α-limit sets of x were defined by Definition 1.6.2. Some invariants of topological conjugacy can be produced from the collections of homeomorphism types of α- and ω-limit sets; for example, topological transitivity is equivalent to the fact that one of those collections contains the whole space. The union of all α- or ω-limit sets may not be closed. The homeomorphism type of the closure of each of those unions is, naturally, an invariant of topological conjugacy. We will denote those closures by $\alpha(f)$ and $\omega(f)$.

The notion of ω-limit point is related to a very important class of invariant sets for topological dynamical systems. As usual we give a definition only for the discrete-time case, since the continuous-time counterpart is obvious.

Definition 3.3.1. A compact set $A \subset X$ is called an *attractor* for f if there exists a neighborhood V of A and $N \in \mathbb{N}$ such that $f^N(V) \subset V$ and $A = \bigcap_{n\in\mathbb{N}} f^n(V)$.

Remark. Considering $V' = \bigcap_{n=0}^{N-1} f^n(V)$ we may take $N = 1$ in the definition.

If A is an attractor then there exists an open neighborhood V of A such that $\omega(x) \subset A$ for all $x \in V$. The converse, however, is false, as shown by the example of the diffeomorphism $f\colon S^1 \to S^1$ induced by the map

$$x \mapsto x + \frac{1}{10}\sin^2 \pi x \quad (\mathrm{mod}\ 1).$$

The point $x = 0$ is not an attractor but $\alpha(f) = \omega(f) = \{0\}$. Points that attract one one side and repel on the other are called *semistable*.

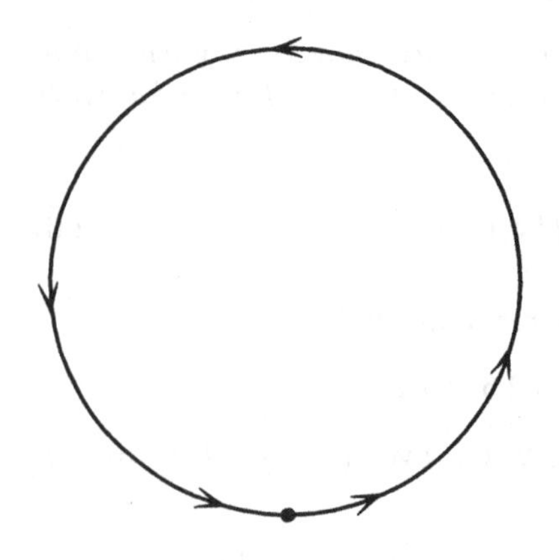

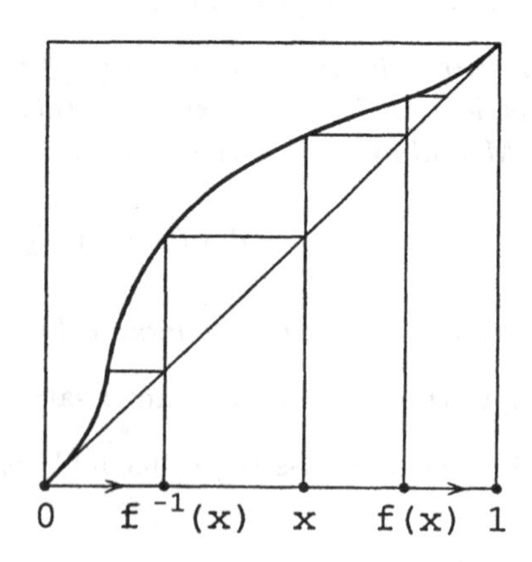

FIGURE 3.3.1. Not an attractor

Let us introduce two types of recurrence behavior for individual orbits which appear very often. Since there is no difference between discrete and continuous time we will formally write down the definitions only for the former case.

Definition 3.3.2. *A point* $x \in X$ *is positively recurrent if* $x \in \omega(x)$, that is, $x = \lim f^{n_k}(x)$ for some sequence $n_k \to \infty$.

If f is invertible, x is *negatively recurrent* if $x \in \alpha(x)$. Finally, x is *recurrent* if it is both positively and negatively recurrent.

The notion of recurrence, being very natural and straightforward, is not free from some deficiencies; for example, positive recurrence does not necessarily imply negative recurrence (cf. Exercise 3.3.1) and the sets of all positively recurrent, negatively recurrent, and recurrent points are often not closed since, for example, in the full shift there are points whose positive and negative semiorbits are dense as well as points that are neither positively nor negatively recurrent. We will denote the closures of the three sets by $R^+(f)$, $R^-(f)$, and $R(f)$.

Definition 3.3.3. A point $x \in X$ is *nonwandering* with respect to the map $f: X \to X$ if for any open set $U \ni x$ there is an $N > 0$ such that $f^N(U) \cap U \neq \varnothing$. The set of all nonwandering points of f is denoted by $NW(f)$.

Equivalently, one may assume that N is arbitrarily big. For, if for every U, $f^N(U) \cap U = \varnothing$ for $N \geq N_0$, then x is not periodic. Hence one can find a neighborhood $V \ni x$ such that $f^i(V) \cap V = \varnothing$ for $i = 0, 1, \ldots, N_0$, and x can not be nonwandering. In the definition for flows one should require from the beginning that the return time is not too small. Another simple comment is that for an invertible f, a nonwandering point x, and open $V \ni x$ there are arbitrarily large negative N such that $f^N(V) \cap V \neq \varnothing$.

Proposition 3.3.4. *The set* $NW(f)$ *is closed and* f*-invariant. It contains all* ω*- and* α*-limit points for all points.*

Proof. Let $x_n \in NW(f)$, $x_n \to x \in U$, which is open. Then $x_n \in U$ for a large enough n, so $f^N(U) \cap U$ is nonempty for some N and $x \in NW(f)$. f-invariance follows immediately: Let $f(x) \in U$ and $V = f^{-1}(U)$ be the complete preimage of U. If $x \in NW(f)$ then $V \cap f^N(V) \neq \varnothing$ for some $N > 0$. But then $f(V \cap f^N(V)) = U \cap f^N(U) \neq \varnothing$. Finally, let $x = \lim_{n_k \to \infty} f^{n_k}(y) \in U$, U open. Assume that n_k is an increasing sequence. Then $f^{n_k}(y) \in U$ for any large enough k and $f^{n_{k+1}}(y) \in U$ so $U \cap f^{n_{k+1}-n_k}(U) \neq \varnothing$. The argument for α-limit points is completely similar. □

Since for every $x \in X$ the ω-limit set $\omega(x)$ is nonempty we have

Corollary 3.3.5. *If* X *is compact and* $f: X \to X$ *then* $NW(f) \neq \varnothing$.

The construction of the nonwandering set can be iterated. Namely, let $NW_1(f) = NW(f)$ and by induction $NW_{n+1}(f) = NW(f\restriction_{NW_n(f)})$. This way we obtain a nested sequence of closed f-invariant sets whose intersection is denoted by $NW_\omega(f)$ and then the construction can be started again. In short, the process uses transfinite induction which stops at a countable ordinal. The final set is called the *center* of the dynamical system. Since recurrent points are defined intrinsically using only their own orbits, Proposition 3.3.4 can be applied to them inductively with the conclusion that the closure $R(f)$ of the set of all recurrent points (which we do not yet know to be nonempty) is contained in the center. In fact, in virtually all interesting examples this construction stabilizes very quickly, at most after one or two steps so the center is either $NW(f)$ or $NW_2(f)$, although it is not so difficult to construct examples where this is not so. (cf. Exercises 3.3.2 and 3.3.3).

It is thus clear that the set $NW(f)$ is the hub of recurrence behavior, since it contains all α- and ω-limit points and recurrent points, including, naturally, all periodic points. In view of Proposition 3.2.3, Proposition 3.2.5, and Proposition 3.2.6 it seems plausible that it should also contain enough orbits to account for the topological entropy. This is indeed true:

$$h_{\text{top}}(f) = h_{\text{top}}(f\restriction_{NW(f)}). \tag{3.3.1}$$

A direct proof of this fact is not very simple and is based on a rather far-reaching development of the simple idea used in Section 3.2b. We will derive (3.3.1) from considerations of invariant measures (see Section 4.1) and the Variational Principle Theorem 4.5.3, which establishes relationships between the topological entropy and entropies with respect to invariant measures. Since the support of any invariant measure belongs to $NW(f)$ (see Proposition 4.1.18) the Variational Principle implies (3.3.1).

Equation (3.3.1) is useful to see that h_{top} is not continuous in general. To that end consider the family of maps f_λ of the closed unit disk in $\mathbb{C}$ given by $f_\lambda(z) = (1-\lambda)z^2$ for $0 \le \lambda \le 1$. For $\lambda = 0$ the restriction $f_0\restriction_{NW(f_0)} = f_0\restriction_{S^1\cup\{0\}}$ and since $f_0\restriction_{S^1} = E_2$ and E_2 has topological entropy $\log 2$ we have $h_{\text{top}}(f_0) = \log 2$, whereas for $\lambda > 0$ the nonwandering set is the origin, so $h_{\text{top}}(f_\lambda) = 0$.

Periodic points represent the most perfect case of recurrence. However, not every dynamical system with a compact phase space has periodic orbits. The next case of very strong and uniform recurrence is represented by minimal sets.

Proposition 3.3.6. *Every continuous map f of a compact metric space X has an invariant minimal subset.*

Proof. Consider the collection $\mathcal{C}$ of all closed f-invariant subsets of f, partially ordered by inclusion. Since the intersection of any number of closed invariant subsets is still closed and invariant, any totally ordered subset of $\mathcal{C}$ has a lower bound. Then by Zorn's Lemma $\mathcal{C}$ has a minimal element, that is, a closed f-invariant set A that has no closed f-invariant subsets. Thus, the orbit of every point $x \in A$ is dense in A, that is, A is minimal. $\square$

Since every point of a minimal set is obviously recurrent we obtain

Corollary 3.3.7. $R(f) \neq \varnothing$.

Let us denote the closure of the union of all invariant minimal sets for f by $M(f)$.

Let us summarize the relationships among the different kinds of recurrence behavior that we have discussed so far:

$$\operatorname{Per}(f) \subset M(f) \subset R(f) \subset R^{+}(f) \cup R^{-}(f) \subset NW(f). \tag{3.3.2}$$

The homeomorphism types of all these sets are invariants of topological conjugacy. For flows the same is true for orbit equivalence. All sets in (3.3.2), except possibly $\operatorname{Per}(f)$, are nonempty. Finally, each of the five inclusions in (3.3.2) may be proper. Some of the appropriate examples are provided in Exercises 3.3.1, 3.3.2, and 3.3.5.

The presence of nonperiodic recurrent points is often referred to as *nontrivial recurrence*, especially in the literature on ordinary differential equations. It is the first indication of complicated asymptotic behavior. In certain low-dimensional situations such as homeomorphisms of the circle (Chapter 11) and flows on surfaces (Chapter 14) it is possible to give a comprehensive description of the nontrivial recurrence that can appear.

Exercises

3.3.1. *Construct a point in the space Ω_2 that is positively recurrent but not negatively recurrent with respect to the full 2-shift σ_2. Show that points satisfying that property are dense in Ω_2.*

3.3.2. *For any given natural number n construct a symbolic dynamical system f for which $NW_k(f)$ are all different for $k = 1, \dots, n$ but $NW_n(f) = NW_{n+1}(f)$.*

3.3.3. *Construct a homeomorphism f of a compact metric space for which the sets $NW_n(f)$ are all pairwise different for $n \in \mathbb{N}$.*

3.3.4*. (S. Simpson) *Prove existence of an invariant minimal set (Proposition 3.3.6) without using Zorn's Lemma or the Axiom of Choice.*

3.3.5. *Construct an example of a topologically transitive dynamical system of a compact manifold for which the only minimal set is a fixed point.*

3.3.6. *Prove that if f is a homeomorphism of a perfect connected compact metric space X and the set $\operatorname{Per}(f)$ is dense in X and $f^n \neq \operatorname{Id}$ for any $n \in \mathbb{N}$, then f has a nonperiodic recurrent point.*

3.3.7. *To give an alternative proof of Theorem 2.9.4 consider $F: X \times \mathbb{R} \to X \times \mathbb{R}$, $(x, t) \mapsto (f(x), t + g(x))$ and $T_\lambda: X \times \mathbb{R} \to X \times \mathbb{R}$, $(x, t) \mapsto (x, t + \lambda)$ and show that the assumptions of Theorem 2.9.4 guarantee the existence of a compact minimal set for F.*

4

Statistical behavior of orbits and introduction to ergodic theory

We begin by showing how various recurrence properties such as recurrence of an orbit (Definition 3.3.2), topological transitivity (Definition 1.3.1), minimality (Definition 1.3.2), and topological mixing (Definition 1.8.2) can be sharpened in a quantitative way by considering *asymptotic frequencies* with which corresponding types of recurrence appear. In order to show that for some orbits such asymptotic frequencies exist we have to appeal to measure theory. Later we will find out that the growth invariant, topological entropy, although already quantitative, also has statistical counterparts with which it is intimately connected.

1. Asymptotic distribution and statistical behavior of orbits

a. Asymptotic distribution, invariant measures. Let f be a continuous self-map of a metrizable space X and for $x \in X$ and a "sufficiently good" set $U \subset X$ let $F_U(f,x,n)$ be the number of integers $k \in [0, n-1]$ such that $f^k(x) \in U$, that is, the number of visits to the set U under the first n iterates of the point x. The limit

$$F_U(f,x) := \lim_{n\to\infty} \frac{F_U(f,x,n)}{n}, \tag{4.1.1}$$

if it exists, gives the *asymptotic density of the distribution* of the iterates between the set U and its complement $X \setminus U$. Unlike the orbit closure of x, which just tells what parts of the phase space are visited by the iterates of x, the asymptotic densities for various sets, for small balls, say, give quantitative information about frequencies of such visits. One can reformulate the definition of the asymptotic frequency $F_U(f,x)$ by pointing out that $f^k(x) \in U$ if and only if the value of the characteristic function χ_U of the set U at the point $f^k(x)$ is equal to one. Therefore

$$F_U(f,x,n) = \sum_{k=0}^{n-1} \chi_U(f^k(x)) \quad \text{and}$$
$$F_U(f,x) = \lim_{n\to\infty} \frac{1}{n} \sum_{k=0}^{n-1} \chi_U(f^k(x)). \tag{4.1.2}$$

The expression (4.1.2) is called the *time average* or *Birkhoff average* of the function χ_U. It is convenient to consider time averages of functions other than characteristic functions; continuous functions are the most natural candidates. Thus let us assume that for a given $x \in X$ and every continuous function φ on X the time average

$$I_x(\varphi) := \lim_{n\to\infty} \frac{1}{n} \sum_{k=0}^{n-1} \varphi(f^k(x)) \tag{4.1.3}$$

exists. We suppressed the dependence on f in the notation since in the subsequent discussion f will be fixed and we will concentrate on the dependence on x and φ. As usual let $C(X)$ be the space of all continuous functions on X with the uniform topology. Then the time average $I_x: C(X) \to \mathbb{R}$ satisfies the following properties:

(1) Linearity: $I_x(\alpha\varphi + \beta\psi) = \alpha I_x(\varphi) + \beta I_x(\psi)$, for α, $\beta \in \mathbb{R}$;
(2) Boundedness: $|I_x(\varphi)| \le \sup_{y\in X} |\varphi(y)|$;
(3) Positivity: $I_x(\varphi) \ge 0$ if $\varphi \ge 0$ and $I_x 1 = 1$;
(4) Invariance under f: $I_x(\varphi \circ f) = I_x(\varphi)$ or, equivalently, $I_{f(x)}(\varphi) = I_x(\varphi)$.

Properties (1) and (2) imply that I_x is continuous. Property (4) follows from a simple calculation:

$$\begin{aligned} I_x(\varphi\circ f) - I_x(\varphi) &= \lim_{n\to\infty} \frac{1}{n} \left(\sum_{k=0}^{n-1} \varphi(f^{k+1}(x)) - \sum_{k=0}^{n-1} \varphi(f^k(x)) \right) \\ &= \lim_{n\to\infty} \frac{1}{n} (\varphi(f^n(x)) - \varphi(x)) = 0. \end{aligned}$$

Now we can use the Riesz Representation Theorem A.2.7 which says that for any positive bounded linear functional on $C(X)$, that is, a functional $I: C(X) \to \mathbb{R}$ satisfying (1)–(3), there is a unique Borel probability measure μ such that $I(\varphi) = \int_X \varphi \, d\mu$. Property (4) is equivalent to f-invariance of μ, that is, $\mu(f^{-1}(A)) = \mu(A)$ for any measurable set. (The inverse appears because $\chi_U \circ f = \chi_{f^{-1}(U)}$.) This can be shown by considering first open sets U for which $\mu(\partial U) = 0$, approximating the characteristic function χ_U from above by continuous functions f_n such that $\int f_n \, d\mu \to \mu(U)$, and then approximating more general sets by such open sets. Thus we have shown that

$$I_x(\varphi) = \int_X \varphi \, d\mu_x, \tag{4.1.4}$$

where μ_x is an f-invariant Borel probability measure on X.

Two main questions arise:

(A) Are there points for which asymptotic distributions I_x exist?

(B) When does an invariant measure determine any asymptotic distribution of orbits, that is, given an f-invariant probability measure μ, can one find a point x such that for all $\varphi \in C(X)$ one gets $\int \varphi \, d\mu = I_x(\varphi)$?

These questions are answered by a combination of two fundamental theorems: one from topological dynamics and one from ergodic theory.

b. Existence of invariant measures.

Theorem 4.1.1. (Krylov–Bogolubov Theorem)[1] *Any continuous map on a metrizable compact space has an invariant Borel probability measure.*

Proof. Let $f: X \to X$ and $x \in X$. Take a dense countable set $\{\varphi_1, \varphi_2, \dots\}$ in the space $C(X)$. For each m the sequence $(1/n)\sum_{k=0}^{n-1} \varphi_m(f^k(x))$ is bounded; hence it contains a convergent subsequence. By the diagonal process one can find a subsequence n_k, $k = 1, 2, \dots$, such that for every $m = 1, 2, \dots$

$$\lim_{k\to\infty} \frac{1}{n_k} \sum_{l=0}^{n_k-1} \varphi_m(f^l(x)) =: J(\varphi_m)$$

exists. Now let φ be an arbitrary continuous function. Fix $\epsilon > 0$ and take φ_m such that $\sup_{x\in X} |\varphi(x) - \varphi_m(x)| < \epsilon$. Then

$$\frac{1}{n_k} \sum_{l=0}^{n_k-1} \varphi(f^l(x)) = \frac{1}{n_k} \sum_{l=0}^{n_k-1} \varphi_m(f^l(x)) + \frac{1}{n_k} \sum_{l=0}^{n_k-1} \left(\varphi(f^l(x)) - \varphi_m(f^l(x))\right)$$

The first term converges to $J(\varphi_m)$; the second is bounded in absolute value by ϵ. Thus all limit points for the sum on the left-hand side differ by no more than ϵ. Since ϵ is arbitrary the following limit exists:

$$\lim_{k\to\infty} \frac{1}{n_k} \sum_{l=0}^{n_k-1} \varphi(f^l(x)) =: J(\varphi). \tag{4.1.5}$$

The functional $J: C(X) \to \mathbb{R}$ defined by (4.1.5) is obviously linear, bounded, and positive. It is also invariant. That follows from the same calculation as in (4). Thus $J(\varphi) = \int \varphi \, d\mu_x$ by the Riesz Representation Theorem A.2.7, where μ_x is an f-invariant Borel probability measure. □

Remark. If f is a homeomorphism, μ an f-invariant measure, and $A \subset X$ measurable then $\mu(f(A)) = \mu(A)$.

For some applications the appropriate notion of a measure space is a *Lebesgue space* (Definition A.6.4), that is, a space isomorphic up to a set of measure zero to an interval with Lebesgue measure with at most countably many "atoms", that is, points of positive measure, added. Surprisingly, this notion is not too restrictive; virtually every probability space in analysis or geometry has this property. For example, any Borel measure on a complete separable metric space produces a Lebesgue space. See Definition A.6.4 and the discussion preceding it for more details. A (not necessarily invertible) transformation $T: X \to X$ of a measure space onto itself is called measure-preserving if for any measurable set A the full preimage $T^{-1}(A)$ is also measurable and $\mu(T^{-1}(A)) = \mu(A)$. Thus, the Krylov–Bogolubov Theorem implies that every continuous map of a metrizable compact space can be viewed as a measure-preserving transformation of a Lebesgue space generated by a Borel measure on X.

c. The Birkhoff Ergodic Theorem. The Birkhoff Ergodic Theorem deals with transformations of abstract spaces preserving a probability measure.

Theorem 4.1.2. (Birkhoff Ergodic Theorem)[2] *Let $T: (X,\mu) \to (X,\mu)$ be a measure-preserving transformation of a probability space, $\varphi \in L^1(X,\mu)$. Then for μ-almost every $x \in X$ the following time average exists:*

$$\lim_{n\to\infty} \frac{1}{n} \sum_{k=0}^{n-1} \varphi(T^k(x)) =: \varphi_T(x). \tag{4.1.6}$$

Proof. Let $f \in L^1(\mu)$, $\mathcal{I} := \{A \in \mathcal{B} \mid T^{-1}(A) = A\}$ the invariant σ-algebra, $f_{\mathcal{I}} := \left[\frac{(f\mu)\restriction_{\mathcal{I}}}{\mu\restriction_{\mathcal{I}}}\right]$ the (invariant) Radon–Nikodym derivative, $F_n := \max_{k\le n} \sum_{i=0}^{k-1} f \circ T^i$.

$$\overline{\lim_{n\to\infty}} \frac{1}{n} \sum_{k=0}^{n-1} f \circ T^k \le \overline{\lim_{n\to\infty}} \frac{F_n}{n} \le 0 \text{ off } A := \{x \mid F_n(x) \to \infty\} \in \mathcal{I}, \tag{4.1.7}$$

but $F_{n+1} - F_n \circ T = f - \min(0, F_n \circ T) \downarrow f$ on A, so by Dominated Convergence $0 \le \int_A (F_{n+1} - F_n)\, d\mu = \int_A (F_{n+1} - F_n \circ T)\, d\mu \to \int_A f\, d\mu = \int_A f_{\mathcal{I}}\, d\mu\restriction_{\mathcal{I}}$ and if $f_{\mathcal{I}} < 0$ then $\mu(A) = 0$. Now take $f = \varphi - \varphi_{\mathcal{I}} - \epsilon$. Then $f_{\mathcal{I}} = -\epsilon < 0$ and $\overline{\lim}_{n\to\infty} \sum_{k=0}^{n-1} (\varphi \circ T^k)/n - \varphi_{\mathcal{I}} - \epsilon \le 0$ μ-a.e. by (4.1.7). Replacing φ by $-\varphi$ gives $\underline{\lim}_{n\to\infty} \sum_{k=0}^{n-1} \varphi \circ T^k/n \ge \varphi_{\mathcal{I}} - \epsilon$ μ-a.e. Thus $\varphi_T = \varphi_{\mathcal{I}}$ μ-a.e. ☐

So φ_T is measurable and T-invariant and $\varphi_T = \varphi_{\mathcal{I}}$ implies

$$\int \varphi_T \, d\mu = \int \varphi \, d\mu. \tag{4.1.8}$$

If T is invertible, the Birkhoff Ergodic Theorem applies to T^{-1} and implies almost-everywhere convergence of negative time averages

$$\frac{1}{n} \sum_{k=0}^{n-1} \varphi(T^{-k}(x)) \to \bar{\varphi}_T(x) \tag{4.1.9}$$

and hence also for the two-sided time average

$$\frac{1}{2n-1}\sum_{k=-n+1}^{n-1}\varphi(T^k(x)).$$

Proposition 4.1.3. *$\varphi_T(x) = \bar{\varphi}_T(x)$ almost everywhere.*

Proof. If not, there exist $\epsilon > 0$ and a T-invariant set E of positive measure such that $\varphi_T(x) > \bar{\varphi}_T(x) + \epsilon$ for $x \in E$ (or $\varphi_T(x) < \bar{\varphi}_T(x) - \epsilon$ for $x \in E$). But applying (4.1.8) to $\varphi \cdot \chi_E$ gives $\int_E \varphi_T \, d\mu = \int_E \varphi \, d\mu = \int_E \bar{\varphi}_T \, d\mu$. □

d. Existence of asymptotic distribution. The exceptional set where the positive or negative time averages do not exist may, of course, depend on the function φ. However, since the union of countably many sets of measure zero still has measure zero one can, for any countable set of functions, for example, for a dense set of continuous functions, find a common set of full measure where the averages converge. Then by the same argument as in the proof of the Krylov–Bogolubov Theorem we establish convergence on the same set of full measure for all continuous functions. Thus we have

Corollary 4.1.4. *Let X be compact metrizable, f a continuous map. Then $\{x \in X \mid \lim_{n\to\infty}(1/n)\sum_{k=0}^{n-1}\varphi(f^k(x))$ exists for all continuous functions $\varphi\}$ has full measure with respect to any f-invariant Borel probability measure. If f is a homeomorphism then the same is also true for the set $\{x \in X \mid \lim_{n\to\infty}(1/n)\sum_{k=0}^{n-1}\varphi(f^k(x)) = \lim_{n\to\infty}(1/n)\sum_{k=0}^{n-1}\varphi(f^{-k}(x))$ for $\varphi \in C(X)\}$.*

Combining this corollary of the Birkhoff Ergodic Theorem 4.1.2 with the Krylov–Bogolubov Theorem 4.1.1 we obtain a positive answer to question (A):[3]

Corollary 4.1.5. *For any continuous map $f: X \to X$ of a compact metric space there exists a point $x \in X$ such that the time average $(1/n)\sum_{k=0}^{n-1}\varphi(f^k(x))$ has a limit for every continuous function φ on X and such that if f is a homeomorphism, then in addition $(1/n)\sum_{k=0}^{n-1}\varphi(f^{-k}(x))$ converges to the same limit.*

e. Ergodicity and unique ergodicity.

Definition 4.1.6. An f-invariant measure μ is called *ergodic* with respect to f if for any measurable f-invariant set $A \subset X$ either $\mu(A) = 0$ or $\mu(X \setminus A) = 0$.

Sometimes one says that f is ergodic with respect to μ, meaning the same thing.

Definition 4.1.7. A continuous map $f: X \to X$ of a metrizable compact space X is called *uniquely ergodic* if it has only one invariant Borel probability measure.

Proposition 4.1.8. *The only invariant probability measure μ of a uniquely ergodic map f is ergodic.*

Proof. For $A \subset X$ measurable with $\mu(A) > 0$ denote by μ_A the *conditional measure* defined by

$$\mu_A(B) := \frac{\mu(B \cap A)}{\mu(A)}. \tag{4.1.10}$$

If μ is not ergodic one can find an f-invariant measurable set A such that $0 < \mu(A) < 1$. Then μ_A and $\mu_{X \setminus A}$ are f-invariant probability measures. They are different since $\mu_A(A) = 1$, $\mu_{X \setminus A}(A) = 0$. □

Ergodicity can be reformulated in functional language: A μ-preserving $f: X \to X$ is ergodic if any measurable f-invariant real-valued function is constant outside of a set of measure zero.

Now we have an important corollary of the Birkhoff Ergodic Theorem 4.1.2 which states that for an ergodic transformation the time average is equal to the space average almost everywhere.

Corollary 4.1.9. *If $f: X \to X$ is an ergodic μ-preserving transformation, $\mu(X) = 1$, and $\varphi \in L^1(X, \mu)$ then for every x outside of a set of measure zero*

$$\varphi_f(x) = \lim_{n \to \infty} \frac{1}{n} \sum_{k=0}^{n-1} \varphi(f^k(x)) = \int_X \varphi \, d\mu.$$

Proof. Since φ_f is f-invariant it is constant almost everywhere. But by (4.1.8) that constant must be $\int_X \varphi \, d\mu$. □

Thus we have answered question (B) from Subsection a. An invariant measure determines the asymptotic distribution of μ-almost every point if it is ergodic. Let us point out that a nonergodic invariant measure μ may also determine the asymptotic distribution of *some* orbits, but such orbits will always be a set of μ-measure zero (cf. Exercise 4.1.3).

Naturally Corollary 4.1.9 leads to the question of whether every continuous map has an ergodic invariant measure. The answer turns out to be affirmative. Moreover, every invariant measure for a measure-preserving transformation can be decomposed into "ergodic components". For continuous maps of compact metrizable spaces the latter fact is a consequence of a powerful result from convex analysis, the Choquet Theorem A.2.10. We first prove existence of an ergodic measure using only the Krylov–Bogolubov Theorem 4.1.1 and some elementary functional analysis. We begin with a description of ergodicity in functional-analytic terms. The set $\mathfrak{M}$ of all Borel probability measures on a compact metrizable space possesses a natural convex structure and a natural topology, called the *weak* topology*. Namely, $\mu_n \to \mu$ if $\int_X \varphi \, d\mu_n \to \int \varphi \, d\mu$ for every continuous function φ. $\mathfrak{M}$ is compact with respect to this topology; this can be easily seen from a usual diagonal argument.

For $f: X \to X$ let $\mathfrak{M}(f)$ be the set of all f-invariant Borel probability measures. $\mathfrak{M}(f)$ is a convex, closed, and hence compact subset of $\mathfrak{M}$ (see Definition A.2.8).

Lemma 4.1.10. *If $\mu \in \mathfrak{M}(f)$ is not ergodic there exist $\mu_1, \mu_2 \in \mathfrak{M}(f)$ such that $\mu_1 \neq \mu_2$ and $0 < \lambda < 1$ such that $\mu = \lambda\mu_1 + (1-\lambda)\mu_2$.*

Proof. If A is f-invariant and $0 < \mu(A) < 1$, define $\mu_1 = \mu_A$ and $\mu_2 = \mu_{X \setminus A}$ by (4.1.10) to get $\mu = \mu(A)\mu_1 + (1-\mu(A))\mu_2$. □

Thus *extreme points* of $\mathfrak{M}(f)$ are ergodic measures. We already used the notion of extreme points in the proof of the Perron–Frobenius Theorem 1.9.11 in a finite-dimensional context. The set $\mathfrak{M}(f)$ is in general infinite-dimensional. We now prove existence of extreme points.

Theorem 4.1.11. *Every continuous map f on a metrizable compact space X has an ergodic invariant Borel probability measure.*

Proof. Let $\{\varphi_i\}_{i\in\mathbb{N}}$ be a dense set of continuous functions and define a nested sequence of sets $\mathfrak{M}_0 \supset \cdots \supset \mathfrak{M}_n \supset \cdots$ by $\mathfrak{M}_0 := \mathfrak{M}(f)$,

$$\mathfrak{M}_{i+1} := \Big\{\mu \in \mathfrak{M}_i \,\Big|\, \int \varphi_{i+1}\, d\mu = \max_{\nu\in\mathfrak{M}_i} \int \varphi_{i+1}\, d\nu\Big\}.$$

Since $\nu \mapsto \int \varphi_{i+1} d\nu$ is continuous on $\mathfrak{M}_i$ these are nonempty compact convex sets and their intersection $\mathfrak{E}$ is nonempty. To show that this set consists of extreme points take $\mu \in \mathfrak{E}$ and assume $\mu = \lambda\mu_1 + (1-\lambda)\mu_2$ with $\mu_1, \mu_2 \in \mathfrak{M}(f)$. Then $\int \varphi\, d\mu = \lambda \int \varphi\, d\mu_1 + (1-\lambda)\int \varphi\, d\mu_2$ for all continuous functions φ. Inductively one then sees that $\int \varphi_i\, d\mu_1 = \int \varphi_i\, d\mu_2 = \int \varphi_i\, d\mu$ and $\mu_1, \mu_2 \in \mathfrak{M}_i$ for every $i \in \mathbb{N}$. Density of $\{\varphi_i\}_{i\in\mathbb{N}}$ thus implies that $\int \varphi\, d\mu_1 = \int \varphi\, d\mu_2 = \int \varphi\, d\mu$ for all continuous φ, which by uniqueness of the representation in the Riesz Representation Theorem A.2.7 implies $\mu_1 = \mu_2 = \mu$ as required. Thus there exist extreme points μ of $\mathfrak{M}(f)$ and by Lemma 4.1.10 these are ergodic measures. □

Lemma 4.1.10 and the Choquet Theorem A.2.10 together in fact imply the following stronger *ergodic decomposition* theorem:

Theorem 4.1.12. *Every invariant Borel probability measure for a continuous map f of a metrizable compact space X can be decomposed into an integral of ergodic invariant Borel probability measures in the following sense: There is a partition (modulo null sets) of X into invariant subsets X_α, $\alpha \in A$ with A a Lebesgue space, and each X_α carrying an f-invariant ergodic measure μ_α such that for any function φ we have $\int \varphi\, d\mu = \iint \varphi\, d\mu_\alpha\, d\alpha$.*

Proposition 4.1.13. *If $f: X \to X$ is uniquely ergodic then for every continuous function φ the time averages $(1/n)\sum_{k=0}^{n-1} \varphi(f^k(x))$ converge uniformly.*

Proof. Suppose for some continuous function φ there is no uniform convergence. Then one can find numbers $a < b$, sequences of points x_k, $y_k \in X$,

$k = 1, 2, \ldots$, and a sequence $n_k \to \infty$ such that

$$\frac{1}{n_k}\sum_{l=0}^{n_k-1}\varphi(f^l(x_k)) < a, \qquad \frac{1}{n_k}\sum_{l=0}^{n_k-1}\varphi(f^l(y_k)) > b.$$

By a diagonal argument one can find a subsequence n_{k_j} such that for every $\psi \in C(X)$ both

$$J_1(\psi) = \lim_{j\to\infty}\frac{1}{n_{k_j}}\sum_{l=0}^{n_{k_j}-1}\psi(f^l(x_{k_j})) \text{ and } J_2(\psi) = \lim_{j\to\infty}\frac{1}{n_{k_j}}\sum_{l=0}^{n_{k_j}-1}\psi(f^l(y_{k_j}))$$

exist. Both J_1 and J_2 are bounded linear positive f-invariant functionals; thus $J_1(\psi) = \int \psi\, d\mu_1$, $J_2(\psi) = \int \psi\, d\mu_2$, where both μ_1 and μ_2 are f-invariant probability measures. Since $J_1(\varphi) \le a < b \le J_2(\varphi)$ we have $\mu_1 \ne \mu_2$ so f is not uniquely ergodic. □

Remark. The converse to this proposition is not true (see Exercise 4.2.2). However, it is true if f is topologically transitive (Exercise 4.1.5).

Corollary 4.1.14. *Let μ be the invariant probability measure for a uniquely ergodic map $f\colon X \to X$. Let $U \subset X$ be open and $\mu(\partial U) = 0$. Then the time averages $(1/n)\sum_{k=0}^{n-1}\chi_U(f^k(x))$ converge uniformly to $\mu(U)$.*

Proof. Let $\bar{\varphi}_m \ge \chi_U$ and $\underline{\varphi}_m \le \chi_U$ ($m \in \mathbb{N}$) be sequences of continuous functions such that $\int \bar{\varphi}_m\, d\mu \to \mu(U)$ and $\int \underline{\varphi}_m\, d\mu \to \mu(U)$. For each $n \in \mathbb{N}$ and $x \in X$ one has

$$\frac{1}{n}\sum_{k=0}^{n-1}\underline{\varphi}_m(f^k(x)) \le \frac{1}{n}\sum_{k=0}^{n-1}\chi_U(f^k(x)) \le \frac{1}{n}\sum \bar{\varphi}_m(f^k(x)). \tag{4.1.11}$$

Fix $\delta > 0$ and find m such that $\int \underline{\varphi}_m\, d\mu > \mu(U) - \delta/2$ and $\int \bar{\varphi}_m\, d\mu < \mu(U) + \delta/2$. By Proposition 4.1.13 we have

$$\mu(U) - \delta \le \frac{1}{n}\sum_{k=0}^{n-1}\chi_U(f^k(x)) \le \mu(U) + \delta$$

from (4.1.11) for sufficiently large n. Since δ is arbitrary, the claim follows. □

Remark. The condition $\mu(\partial U) = 0$ is needed for uniform convergence of time averages. In fact, on a nowhere dense set of positive measure one cannot have uniform convergence of averages; see Exercise 4.1.7.

Proposition 4.1.15. *If for every continuous function φ from a dense set Φ in the space $C(X)$ the time averages $(1/n)\sum_{k=0}^{n-1}\varphi(f^k(x))$ converge uniformly to a constant then f is uniquely ergodic.*

Proof. First, let us show that for every continuous function ψ the time averages also converge uniformly. This is essentially another repetition of the argument used in the proof of the Krylov–Bogolubov Theorem 4.1.1. Fix $\epsilon > 0$ and find $\varphi \in \Phi$ such that $\max_{x\in X}|\varphi(x) - \psi(x)| < \epsilon$. Let φ_0 be the limit value of the averages for φ. Then $|(1/n)\sum_{k=0}^{n-1}\psi(f^n(x)) - (1/n)\sum_{k=0}^{n-1}\varphi(f^n(x))| < \epsilon$ for all n and therefore $|\sup_{x\in X}\sup_{n\to\infty}(1/n)\sum_{k=0}^{n-1}\psi(f^k(x)) - \varphi_0| < \epsilon$ and $|\varphi_0 - \inf_{x\in X}\inf_{n\to\infty}(1/n)\sum_{k=0}^{n-1}\psi(f^k(x))| < \epsilon$.

Since ϵ is arbitrary, the time averages of ψ converge uniformly to a constant ψ_0. If μ is an f-invariant probability measure then $\int\psi_0\,d\mu = \int\psi\,d\mu$. Hence μ is unique. □

f. Statistical behavior and recurrence.

Definition 4.1.16. For a Borel measure μ on a separable metrizable space X define the *support* of μ to be the following set

$$\operatorname{supp}\mu := \{x \in X \mid \mu(U) > 0 \text{ whenever } x \in U, U \text{ open}\}.$$

Proposition 4.1.17. (1) *$\operatorname{supp}\mu$ is a closed set.*

(2) $\mu(X \smallsetminus \operatorname{supp}\mu) = 0$.

(3) *Any set of full measure is dense in* $\operatorname{supp}\mu$.

Proof. (1) Let $\{x_n\}_{n\in\mathbb{N}} \subset \operatorname{supp}\mu$, $x_n \to x$, and U be an open set with $x \in U$. Then $x_n \in U$ for sufficiently large n, and hence $\mu(U) > 0$.

(2) Every point $y \in X \smallsetminus \operatorname{supp}\mu$ has an open neighborhood U such that $\mu(U) = 0$. Since X is separable, the set $X \smallsetminus \operatorname{supp}\mu$ can be covered by at most countably many such neighborhoods; hence $\mu(X \smallsetminus \operatorname{supp}\mu) = 0$ by σ-additivity of μ.

(3) If $A \subset X$ and $x \in U := \operatorname{supp}\mu \smallsetminus \bar{A}$ then $\mu(X \smallsetminus A) \geq \mu(U) > 0$. □

Proposition 4.1.18. *Let f be a continuous map of a complete separable metrizable space X. Then:*

(1) *$\operatorname{supp}\mu \subset R(f)$ for any f-invariant Borel probability measure μ.*
(2) *If μ is ergodic then $f\restriction_{\operatorname{supp}\mu}$ has a dense orbit.*
(3) *If X is compact and $f\restriction_{\operatorname{supp}\mu}$ is uniquely ergodic, then $\operatorname{supp}\mu$ is a minimal set.*

In the proof of (1) we will use an important fact from ergodic theory which has many other applications.

Theorem 4.1.19. (Poincaré Recurrence Theorem) *Let T be a measure-preserving transformation of a probability space (X, μ) and let $A \subset X$ be a measurable set. Then for any $N \in \mathbb{N}$*

$$\mu(\{x \in A \mid \{T^n(x)\}_{n \geq N} \subset X \smallsetminus A\}) = 0.$$

Proof. Replacing T by T^N one sees that it is enough to prove the statement for $N = 1$. The set

$$\tilde{A} := \{x \in A \mid \{T^n(x)\}_{n \in \mathbb{N}} \subset X \smallsetminus A\} = A \cap \left(\bigcap_{n=1}^{\infty} T^{-n}(X \smallsetminus A) \right) \qquad (4.1.12)$$

is measurable. $T^{-n}(\tilde{A}) \cap \tilde{A} = \varnothing$ for every n and hence

$$T^{-n}(\tilde{A}) \cap T^{-m}(\tilde{A}) = \varnothing$$

for all $m, n \in \mathbb{N}$. $\mu(T^{-n}(\tilde{A})) = \mu(\tilde{A})$ since T preserves μ. Thus $\mu(\tilde{A}) = 0$ since $1 = \mu(X) \geq \mu\left(\bigcup_{n=0}^{\infty} T^{-n}(\tilde{A})\right) = \sum_{n=0}^{\infty} \mu(T^{-n}(\tilde{A})) = \sum_{n=0}^{\infty} \mu(\tilde{A})$. □

Proof of Proposition 4.1.18. (1) Take a countable base $\{U_1, U_2, \dots\}$ of open subsets of X and let R_+ be the set of all points x such that if $x \in U_m$ then infinitely many positive iterates of x also belong to U_m. Apply the Poincaré Recurrence Theorem 4.1.19 to each of the U_i to deduce that R_+ has full measure. If f is invertible then by the same argument the set R_- constructed similarly to R_+ but with negative iterates also has full measure. Hence $R := R_- \cap R_+$ has full measure and by (3) of Proposition 4.1.17, R is dense in $\operatorname{supp} \mu$. On the other hand, if $x \in R$ and $U \ni x$ is an open set then $U_m \subset U$ for some m; hence infinitely many positive and negative iterates of n lie in U, that is, R consists of positively recurrent points. Hence $\bar{R} = \operatorname{supp} \mu \subset R(f)$.

(2) Take a countable base $\{U_1, U_2, \dots\}$ of open sets for the induced topology on $\operatorname{supp} \mu$. By definition $\mu(U_m) > 0$ for $m \in \mathbb{N}$. Applying Corollary 4.1.9 simultaneously to the characteristic functions χ_{U_n} we obtain a set R of full measure such that for $x \in R$, $m \in \mathbb{N}$

$$\lim_{n \to \infty} \frac{1}{n} \sum_{i=0}^{n-1} \chi_{U_m}(f^i(x)) = \mu(U_m) > 0.$$

Thus the orbit of any $x \in R$ intersects all U_m and hence is dense.

(3) If there is a proper closed f-invariant subset $\Lambda \subset \operatorname{supp} \mu$, then by the Krylov–Bogolubov Theorem 4.1.1 there is an invariant Borel probability measure ν for $f_{\restriction \Lambda}$. But then $\operatorname{supp} \nu \subset \Lambda$. Hence $\mu \neq \nu$, a contradiction. □

Thus, we have found that having typical behavior with respect to an invariant measure is a statistical counterpart for recurrence, ergodicity is such for topological transitivity, and unique ergodicity is likewise for minimality. It is very important to point out that the converse to any of the statements of Proposition 4.1.18 is not true, even if we assume in addition that f is a diffeomorphism of a compact manifold. In other words, in general the closure of the union of the supports of all f-invariant measures may be smaller than the closure of the set of recurrent points of the map, a topologically transitive map may fail to have an ergodic measure with full support (that is, positive on all nonempty open sets), and a minimal set may support more than one invariant measure. However, though the corresponding counterexamples cannot be dismissed as pathological, they still should be considered as somewhat atypical. (See, for example, Exercise 4.1.9 and Corollary 12.6.4.) By examining the examples from Chapter 1 we will show that for all of them a natural correspondence between topological properties and their statistical analogs holds.

g. Measure-theoretic isomorphism and factors. We already encountered some properties (ergodicity) and results (Poincaré Recurrence Theorem 4.1.19, Birkhoff Ergodic Theorem 4.1.2) that deal with measure-preserving transformations in measure spaces and do not use topology or any other extra structure. This naturally brings us into the realm of ergodic theory which studies such transformations. Similarly to the theory of smooth dynamical systems and topological dynamics it has a dual agenda: the classification of various classes of measure-preserving transformations up to natural equivalence relations and the study of various asymptotic properties invariant under those relations. Ergodicity is an example of such an invariant which is a counterpart of topological transitivity; mixing, another recurrence-type invariant, will be discussed in Section 4.2d.; the most important growth-type invariant, measure-theoretic entropy, is the subject of Sections 4.3 and 4.4. Right now we will define and discuss the most natural equivalence relation in ergodic theory.

Definition 4.1.20. Let $T: X \to X$ and $S: Y \to Y$ be measure-preserving transformations of spaces (X,μ) and (Y,ν), correspondingly. T and S are called *metrically isomorphic* if there exists an isomorphism $R: (X,\mu) \to (Y,\nu)$, that is, an injective (mod 0) transformation such that $R_*\mu = \nu$ and

$$S = R \circ T \circ R^{-1}.$$

Definition 4.1.21. In the same notation S is called a (metric) *factor* of T if there exists a measure-preserving map $R: X \to Y$ (in general noninvertible) such that $R_*\mu = \nu$ and

$$S \circ R = R \circ T.$$

All properties of measure-preserving transformations that we are going to discuss are invariants of metric isomorphism; ergodicity quite obviously is. Furthermore, a factor of an ergodic transformation is also ergodic: If S is a factor of T and $A \subset Y$ is S-invariant, $0 < \nu(A) < 1$, then $B := R^{-1}(A)$ is T-invariant and $\mu(B) = \mu(A)$.

In certain cases invariants of metric isomorphism provide insights into properties of smooth or topological dynamical systems. For example, the metric isomorphism class of a uniquely ergodic map is an important invariant of topological conjugacy. The Variational Principle Theorem 4.5.3 provides another connection of that kind.

A powerful tool for the study of measure-preserving transformations is given by spectral analysis. One associates to $T: (X, \mu) \to (X, \mu)$ an isometric operator $U_T: L^2(X, \mu) \to L^2(X, \mu)$ by

$$(U_T f)(x) := f(T(x)).$$

If T is invertible then so is U_T and in this case U_T is a unitary operator. If T and S are metrically isomorphic via R then U_T and U_S are unitarily equivalent, namely,

$$U_S = U_R^{-1} \circ U_T \circ U_R,$$

where $U_R: L^2(X, \mu) \to L^2(Y, \nu)$,

$$(U_R f)(x) := f(R(x)).$$

Thus spectral invariants of U_T, for example, eigenvalues with their multiplicities, the spectrum, or spectral measures, are invariants of metric isomorphism of T.

However, it is possible even for invertible metrically nonisomorphic ergodic measure-preserving transformations T and S to have unitarily equivalent associated unitary operators (see Exercises 4.4.3 and 4.4.4).

Exercises

4.1.1. *Show that $f: [0,1] \to [0,1]$ given by*

$$f(x) = \begin{cases} x/2 & \text{when } 0 < x \le 1, \\ 1 & \text{when } x = 0 \end{cases}$$

has no invariant Borel probability measure.

4.1.2. *Show that different ergodic probability measures are mutually singular.*

4.1.3. *Let $\bar{0}$, $\bar{1}$ be the sequences in the space Ω_2 that consist entirely of zeros and ones, respectively. Prove that there exists a point $\omega \in \Omega_2$ such that for every continuous function φ*

$$\lim_{n \to \pm\infty} \frac{1}{n} \sum_{k=0}^{n-1} \varphi(\sigma_2^k \omega) = \frac{1}{2}\varphi(\bar{0}) + \frac{1}{2}\varphi(\bar{1}) = \int \varphi \, d\mu,$$

where μ is the probability measure with $\mu(\{\bar{0}\}) = \mu(\{\bar{1}\}) = 1/2$.

4.1.4. *Let (X,μ) be a measure space, $A \subset X$ a measurable set with $\mu(A) > 0$, $T: X \to X$ a measure-preserving transformation, and μ_A the conditional measure on A defined by (4.1.10). For $x \in A$ let $n(x) := \min\{n \in \mathbb{N} \mid T^n(x) \in A\}$. Prove that the formula $T_A(x) := T^{n(x)}(x)$ defines a transformation of A which preserves μ_A. The map T_A is called the first return map induced by T on the set A.*

4.1.5. *Suppose f is a topologically transitive continuous map of a compact space X and for every continuous function φ the averages $\frac{1}{n}\sum_{k=0}^{n-1}\varphi(f^k(x))$ converge uniformly. Prove that f is uniquely ergodic.*

4.1.6. *Show that unique ergodicity is a topological invariant.*

4.1.7. *Show that if $f: X \to X$ is ergodic and $N \subset X$ is nowhere dense with positive measure then the time averages $1/n\sum_{k=0}^{n-1}\chi_N(f^k(x))$ cannot converge uniformly.*

4.1.8. *Prove that the diffeomorphism $f: S^1 \to S^1$ defined by $f(x) = x + \frac{1}{10}\sin^2(\pi x) \pmod 1$ and mentioned in Section 3.3 is uniquely ergodic but not topologically transitive.*

4.1.9. *Show that by making an appropriate "time change" vanishing at one point in an irrational linear flow $\{T_\omega^t\}$ on the two-torus one can construct a real-analytic topologically transitive flow whose only invariant Borel probability measure is the δ-measure concentrated on the fixed point.*

4.1.10. (M. Boshernitzan) *Improve the Poincaré Recurrence Theorem 4.1.19 for a measurable transformation T of the interval $[0,1]$ that preserves Lebesgue measure as follows:*

$$\varliminf_{n\to\infty} n \cdot |T^n(x) - x| \le 1$$

for almost every $x \in [0,1]$.

4.1.11. *Prove that a measure-preserving transformation T is ergodic if and only if one is a simple eigenvalue of U_T.*

4.1.12. *Prove that for an ergodic measure-preserving transformation T every eigenvalue of the associated isometric operator U_T is simple.*

4.1.13. *Prove that the eigenvalues of the isometric operator associated to an ergodic measure-preserving transformation form a group.*

2. Examples of ergodicity; mixing

We have mentioned several times that our examples of dynamical systems with nontrivial recurrence split into two distinct groups:

(1) Systems with similar behavior for different orbits and low complexity of the global orbit structure. This group includes rotations on the circle (Section 1.3), translations (Section 1.4) and linear flows (Section 1.5) on the torus, and, to a large extent, completely integrable Hamiltonian systems (Section 1.5).

(2) Systems with different asymptotic behavior for different initial conditions, instability of asymptotic behavior with respect to initial conditions, and high (exponential) complexity of the global orbit structure, represented, for example by exponential growth of periodic orbits and positive topological entropy. This group includes expanding maps of the circle (Section 1.7), hyperbolic automorphisms of the torus (Section 1.8), and transitive topological Markov chains including the full shift (Section 1.9).

In this section we are going to examine the statistical behavior of orbits in both groups of examples.

a. Rotations. Let us begin with the irrational rotation R_α on the circle. Every rotation preserves Lebesgue measure.

Proposition 4.2.1. (Kronecker–Weyl Equidistribution Theorem) *Any irrational rotation is uniquely ergodic.*

Proof. By Proposition 4.1.15 it is sufficient to check that time averages for every continuous function from a dense set of continuous functions uniformly converge to a constant. By the Weierstrass Theorem, trigonometric polynomials form a dense set among all continuous functions in the uniform topology. Furthermore, uniform convergence to a constant is a linear property; if it takes place for φ and ψ, it also holds for $a\varphi + b\psi$, where a and b are constants. Thus, it is enough to check uniform convergence for any complete system of functions, for example, for the characters $\chi_m(x) = e^{2\pi imx}$. For $m \neq 0$ one has $\chi_m(R_\alpha x) = e^{2\pi im(x+\alpha)} = e^{2\pi im\alpha}e^{2\pi imx} = e^{2\pi im\alpha}\chi_m(x)$ and

$$\left|\frac{1}{n}\sum_{k=0}^{n-1}\chi_m(R_\alpha^k(x))\right| = \left|\frac{1}{n}\sum_{k=0}^{n-1}e^{2\pi imk\alpha}\right| = \frac{|1-e^{2\pi imn\alpha}|}{n|1-e^{2\pi im\alpha}|} \leq \frac{2}{n|1-e^{2\pi im\alpha}|} \to 0$$

as $n \to \infty$. □

This argument extends rather straightforwardly to any translation T_γ on the torus where $\gamma = (\gamma_1, \ldots, \gamma_n)$ is such that $\gamma_1, \ldots, \gamma_n$ and 1 are rationally independent. By Proposition 1.4.1 this condition is necessary and sufficient for the minimality of T_γ. In fact, in Section 1.4 we also showed that the same condition is necessary for topological transitivity and hence, since the support of

Lebesgue measure is the whole torus, by Proposition 4.1.18(2) is also necessary for ergodicity with respect to Lebesgue measure.

The same proof can also be carried out for the linear flow T^t_ω if ω satisfies the condition of Proposition 1.5.1. Again, this condition is necessary for topological transitivity and ergodicity with respect to Lebesgue measure and sufficient for minimality and unique ergodicity.

Now we will present an alternative proof of unique ergodicity for translations on the torus. It consists of two parts. First, using a Fourier analysis argument, we prove ergodicity. This kind of argument is very useful for many dynamical systems of an algebraic nature, including the expanding maps E_m and hyperbolic toral automorphisms. In the case of translations the proof of ergodicity is essentially contained in our proof of topological transitivity (Proposition 1.4.1).

Proposition 4.2.2. *If $\gamma_1, \dots, \gamma_n, 1$ are rationally independent, then the translation T_γ is ergodic with respect to Lebesgue measure.*

Proof. As in the proof of Proposition 1.4.1 let χ be a bounded measurable function invariant under T_γ, for example, the characteristic function of an invariant set. Consider the Fourier expansion

$$\chi(x_1, \dots, x_n) = \sum_{(k_1,\dots,k_n)\in\mathbb{Z}^n} \chi_{k_1,\dots,k_n} \exp\left(2\pi i \sum_{j=1}^{n} k_j x_j\right).$$

Using the calculation in the proof of Proposition 1.4.1 we deduce from T_γ-invariance of χ that for any $k_1, \dots, k_n$

$$\chi_{k_1,\dots,k_n}\left(1 - \exp\sum_{j=1}^{n} k_j\gamma_j\right) = 0.$$

The rational-independence condition implies that $\chi_{k_1,\dots,k_n} = 0$ except for possibly $k_1 = k_2 = \cdots = k_n = 0$. Hence χ is a constant outside of a set measure zero and T_γ is ergodic. □

The second step consists of showing that ergodicity with respect to Lebesgue measure implies unique ergodicity. The special property of Lebesgue measure is that it is invariant with respect to *all* translations. The natural context for the argument is thus the multiplication transformation on compact abelian groups (see end of Section 1.3). However, the method employed in the proof has implications beyond that case (see Exercises 4.2.3–4.2.7 where it is used to prove equidistribution of fractional parts of a polynomial).

Let G be a compact metrizable abelian group. There is a unique Borel probability measure λ_G invariant under all multiplications $L_{g_0}: g \to g_0 g$. This measure is called the *Haar measure.* For the torus $\mathbb{T}^n$ Haar measure is the usual Lebesgue measure. We will prove the following statistical counterpart of Proposition 1.3.4.

Proposition 4.2.3. *If a translation L_{g_0} on a compact metrizable abelian group G is ergodic with respect to the Haar measure λ_G, then it is uniquely ergodic.*

Proof. Let μ be any L_{g_0}-invariant Borel probability measure. Since g_0 commutes with any other element of G we see that for any $g \in G$ the pullback measure μ_g defined by

$$\mu_g(A) := \mu(L_g A)$$

is also L_{g_0}-invariant. Since the set $\mathfrak{M}(L_{g_0})$ is weak*-closed and convex, we can define for any measurable set E of positive Haar measure an L_{g_0}-invariant measure μ_E by

$$\mu_E(A) = \frac{1}{\lambda_G(E)} \int_E \mu_g(A) d\lambda_G(g). \tag{4.2.1}$$

If $E \cap F = \varnothing$ then

$$\lambda_G(E \cup F)\mu_{E\cup F} = \lambda_G(E)\mu_E + \lambda_G(F)\mu_F. \tag{4.2.2}$$

A change of variables in (4.2.1) shows that the measure μ_G is L_g-invariant for any $g \in G$; hence

$$\mu_G = \lambda_G \tag{4.2.3}$$

by uniqueness of Haar measure.

Suppose $\mu \neq \lambda_G$. Then there exists a continuous function φ such that $\int_G \varphi\, d\mu \neq \int \varphi d\lambda_G$. But since $\mu_G = \lambda_G$ one has

$$\int \varphi\, d\lambda_G = \int_G \Big(\int \varphi\, d\mu_g\Big)\, d\lambda_G = \int_G \Big(\int (\varphi \circ L_g)\, d\mu\Big)\, d\lambda_G.$$

The function $\bar{\varphi}_g = \int \varphi \circ L_g\, d\mu$ is continuous in g and since we assume that $\bar{\varphi}_{\mathrm{Id}} \neq \int \bar{\varphi}_g d\lambda_G$, $\bar{\varphi}_g$ is not constant. Thus we can find a number a such that $\lambda_G(E) > 0$ and $\lambda_G(F) > 0$, where $E = \{g \mid \varphi_g \geq a\}$, $F = G \setminus E$.

Then $\int \varphi\, d\mu_E \geq a$ and $\int \varphi\, d\mu_F < a$ so

$$\mu_E \neq \mu_F. \tag{4.2.4}$$

But by (4.2.2) and (4.2.3)

$$\lambda_G(E)\mu_E + \lambda_G(F)\mu_F = \mu_{E\cup F} = \mu_G = \lambda_G. \tag{4.2.5}$$

To finish the proof we need the converse to Lemma 4.1.10 which will also be used later.

Lemma 4.2.4. *If an f-invariant probability measure μ can be represented as $\lambda\mu_1 + (1-\lambda)\mu_2$ where $0 < \lambda < 1$, μ_1, μ_2 are f-invariant probability measures, and $\mu_1 \neq \mu_2$, then f is not ergodic with respect to μ.*

Proof of Lemma 4.2.4. Since every set of μ-measure zero has μ_1- and μ_2-measures zero, by the Radon–Nikodym Theorem the measures μ_1 and μ_2 can be represented by densities ρ_1 and ρ_2 with respect to μ, that is, $\int \varphi\, d\mu_i = \int \rho_i \varphi\, d\mu$ for $i = 1, 2$. By assumption $\lambda\rho_1 + (1-\lambda)\rho_2 = 1$, $\int \rho_1\, d\mu = \int \rho_2\, d\mu = 1$, and $\rho_1 \neq \rho_2$. Hence $\rho_1 \not\equiv$ const. Since ρ_1 is an f-invariant L^1-function, μ is not ergodic. □

Proposition 4.2.3 follows by contradiction from (4.2.4), (4.2.5), and Lemma 4.2.4. □

Remark. Lemmas 4.2.4 and 4.1.10 say that ergodic measures can be characterized as extreme points of the set $\mathfrak{M}(f)$ of all f-invariant Borel probability measures.

Although we have given already two proofs of ergodicity for translations on the torus, we present a sketch of one more proof that topological transitivity implies ergodicity. This proof is very geometric and introduces some ideas which are useful in the study of broad classes of dynamical systems, including those without an apparent algebraic structure (see, for example, Proposition 5.1.24 and Theorem 12.7.2).

Every measurable set on a small scale is densely concentrated; it fills some small balls or cubes almost completely and almost misses others because it can be approximated arbitrarily well (in measure) by finite collections of cubes. Fix an invariant set A and $\epsilon > 0$ and find a small cube Δ such that $\lambda(A \cap \Delta) > (1-\epsilon)\lambda(\Delta)$. Images of Δ under the iterates of our map have the same property since both λ and A are invariant. Since our map is an isometry, any image of Δ is again a cube of the same size. By topological transitivity one can find a collection of images that cover the whole phase space almost uniformly, without much overlap. In fact it is sufficient to assume that every point is covered no more than N times, where N is independent of ϵ, because then the measure of A must be greater than $1 - \epsilon N$. Since ϵ can be chosen arbitrarily small, this implies that A has full measure.

b. Extensions of rotations. We now describe a class of examples closely connected to rotations which will be shown in Section 12.4 to contain some minimal nonergodic examples. Let $\alpha \in \mathbb{R} \setminus \mathbb{Q}$.

Proposition 4.2.5. *Consider the torus $\mathbb{T}^2$, a function $\varphi: S^1 \to \mathbb{R}$, and a map $f: (x, y) \mapsto (x + \alpha, y + \varphi(x))$ of $\mathbb{T}^2$. If $\varphi(x) = \Phi(x + \alpha) - \Phi(x)$ for some Lebesgue measurable function $\Phi: S^1 \to \mathbb{R}$ then for any ergodic invariant measure f is metrically isomorphic to the rotation R_α and there are uncountably many different ergodic invariant measures.*

Proof. Take $h(x, y) = (x, y + \Phi(x))$. Then $h^{-1} \circ f \circ h(x, y) = (x + \alpha, y)$. Since the rotation is uniquely ergodic any invariant measure for f projects to Lebesgue measure on the circle and hence h defines a metric isomorphism for any such measure. Thus the invariant ergodic measures for f are exactly the measures induced from measures on circles. There are uncountably many of these because the graph of $\Phi + c$ for any $c \in \mathbb{R}$ supports such a measure. □

Proposition 4.2.6. *Consider the torus $\mathbb{T}^2$, a function $\varphi: S^1 \to \mathbb{R}$, and a map $f: (x, y) \mapsto (x+\alpha, y+\varphi(x))$ of $\mathbb{T}^2$. Then either $\varphi(x) = \Phi(x+\alpha) - \Phi(x) + r_1\alpha + r_2$ for some continuous $\Phi: S^1 \to \mathbb{R}$ and $r_1, r_2 \in \mathbb{Q}$ or f is minimal.*

Remark. For the case where Φ is continuous and $r = 0$ the map h above provides a topological conjugacy to $\mathbb{R}_\alpha \times \mathrm{Id}$ on $\mathbb{T}^2$. Notice that this is another instance where an untwisted cohomological equation arises (cf. Section 2.9).

Proof. By Proposition 3.3.6 there is an invariant minimal set M for f and the projection of this set to the first coordinate is invariant, hence is S^1. Consider the intersection of M with the fiber $\{x\}\times S^1$. We will show that if it contains two points y and $y+\tau$ then it is invariant under translation by τ in the fiber. Namely, by minimality there exist points $z = f^N(x,y)$ arbitrarily close to $(x, y+\tau)$, so the points $f^{kN}(x,y) \in M$ accumulate on $(x, y+k\tau)$, which are hence in M. Thus the closed set $M \cap (\{x\} \times S^1)$ is either a finite subgroup of $\{x\} \times S^1$ generated by $r \in \mathbb{Q}$ or equal to $\{x\} \times S^1$. Since M is closed the same case occurs for all x and by continuity we obtain the same subgroup for all x, hence giving either minimality or a collection of invariant closed curves for f. To complete the proof we need to show that those curves are graphs of functions of the first coordinate. If necessary we can reduce this case to the case of a single such curve by factoring the second coordinate modulo $1/q$. Thus M intersects every vertical exactly once and is hence the graph of a continuous function. On the universal cover it lifts to the graph of a function Φ' with $\Phi'(x+1) = \Phi'(x) + k$ and all its integer translates. Invariance under the lift F yields $(x+\alpha, \Phi'(x) + \varphi(x)) = F(x, \Phi'(x)) = (x+\alpha, \Phi'(x+\alpha) + n)$ for some $n \in \mathbb{Z}$. Thus we obtain $\varphi(x) = \Phi'(x+\alpha) - \Phi'(x) + n$. Recalling that we factored the second coordinate by $1/q$ and writing $\Phi'(x) = \Phi(x) + kx$ we obtain Proposition 4.2.6 with $r_1 = k/q$ and $r_2 = n/q$. □

One interesting application of the preceding two results is that a circle extension of the above form where one can write $\varphi(x) = \Phi(x+\alpha) - \Phi(x)$ for some measurable Φ but not $\varphi(x) = \Phi(x+\alpha) - \Phi(x) + r$ for any continuous Φ is an example of a minimal nonergodic diffeomorphism. In Corollary 12.6.4 we will construct functions φ with precisely this property.

c. Expanding maps. Now let us proceed to the second group of our examples beginning with the linear expanding map E_m. This map preserves Lebesgue measure. λ because the preimage of any interval of length l consists of $|m|$ disjoint intervals of length l/m.

Proposition 4.2.7. *The map E_m, $|m| \geq 2$, is ergodic with respect to Lebesgue measure.*

We will give two proofs of this fact which are related to our second and third proofs for the toral translations.

First proof. Let φ be a measurable bounded E_m-invariant function. Using the Fourier expansion

$$\varphi(x) = \sum_{k=-\infty}^{\infty} \varphi_k \exp(2\pi i k x)$$

we obtain $\varphi(E_m(x)) = \sum_{k=-\infty}^{\infty} \varphi_k \exp(2\pi i k m x)$.

Thus since $\varphi(x) = \varphi(E_m(x))$ almost everywhere we have

$$\varphi_k = \varphi_{k\cdot m}, \qquad m \in \mathbb{N}. \tag{4.2.6}$$

$\varphi \in L^1$, so $|\varphi_k| \to 0$ as $k \to \infty$. Hence (4.2.6) implies $\varphi_k = 0$ for $k \neq 0$ and $\varphi \equiv \varphi_0$ almost everywhere. □

Second proof. Let $A \subset S^1$ be a measurable E_m-invariant set of positive Lebesgue measure. $f^{-1}(A) = A$ implies forward-invariance of $S^1 \setminus A = f(S^1 \setminus A)$. As in the third proof for the toral translations, fix $\epsilon > 0$ and find an open interval Δ of length $|m|^{-n}$ for some n such that

$$\lambda(\Delta \setminus A) > (1-\epsilon)\lambda(\Delta) = (1-\epsilon)|m|^{-n}.$$

Since E_m has constant derivative it expands the Lebesgue measure of any set exactly $|m|$ times as long as E_m is injective on that set. Thus $\lambda(E_m^n(\Delta) \setminus A) = |m|^n \lambda(\Delta \setminus A) > 1 - \epsilon$. □

d. Mixing. Let us try to understand the difference in the statistical behavior of orbits between an irrational rotation R_α and a linear expanding map E_m. First of all, the former is uniquely ergodic whereas the latter has many different ergodic invariant measures; besides Lebesgue measure each periodic orbit carries a uniform δ-measure which is obviously ergodic. The Cantor set K carries the E_3-invariant pullback of Lebesgue measure. for E_2 under the natural conjugacy between $E_3\restriction_K$ and E_2 and there are many more ergodic invariant measures. In addition to that we would like to see the difference intrinsically in terms of behavior with respect to Lebesgue measure itself. We would also like this difference not to be a direct consequence of the obvious fact that expanding maps are noninvertible but also to be traceable to the case of hyperbolic toral automorphisms. The clue is provided by the notion of topological mixing (Definition 1.8.2) which distinguishes expanding maps and toral automorphisms from translations. Let us try to find a statistical property similar to topological mixing in the same sense as ergodicity is similar to topological transitivity.

Definition 4.2.8. A measure-preserving transformation $T\colon (X,\mu) \to (X,\mu)$ is called *mixing*[1] if for any two measurable sets A, B

$$\mu(T^{-n}(A) \cap B) \to \mu(A) \cdot \mu(B) \quad \text{as} \quad n \to \infty. \tag{4.2.7}$$

Obviously mixing is an invariant of metric isomorphism (Definition 4.1.20). Furthermore any factor (Definition 4.1.21) of a mixing map is mixing by an argument similar to that for ergodicity in Section 4.1g.

Since for every T-invariant set A and every n we have $\mu(T^{-n}(A) \cap (X \setminus A)) = 0$ we immediately see that mixing implies that $\mu(A) \cdot \mu(X \setminus A) = 0$ for such a set, that is, T is ergodic.

Remark. Obviously if a continuous map f has a mixing invariant measure μ then $f\restriction_{\operatorname{supp}\mu}$ is topologically mixing since if $A, B \subset \operatorname{supp}\mu$ are open subsets and n is sufficiently large then $\mu(T^{-n}(A) \cap B)$ is positive and hence the intersection is nonempty. However, the converse is not true: A topologically mixing map, even a minimal one, may fail to have a mixing invariant measure with full support. This phenomenon is, however, atypical similarly to the situation with other properties such as topological transitivity and ergodicity which we discussed at the end of Subsection a. As we will soon see, our topologically mixing examples are mixing with respect to natural invariant measures.

Definition 4.2.9. A collection $\mathfrak{U}$ of measurable sets in a measure space (X, μ) is called *dense* if for any measurable set A and any $\epsilon > 0$ one can find $A' \in \mathfrak{U}$ such that

$$\mu(A \bigtriangleup A') < \epsilon.$$

A collection of sets $\mathcal{C}$ of measurable sets is called *sufficient* if finite disjoint unions of elements of $\mathcal{C}$ form a dense collection.

Proposition 4.2.10. (1) *If (4.2.7) holds for any $A, B \in \mathcal{C}$, where $\mathcal{C}$ is a sufficient collection of sets, then T is mixing;*

(2) *T is mixing if and only if for a given complete system Φ of functions in L^2 and any $\varphi, \psi \in \Phi$*

$$\int_X \varphi(T^n x)\bar{\psi}(x)\, d\mu \to \left(\int \varphi\, d\mu\right) \cdot \left(\int \bar{\psi}\, d\mu\right) \quad \text{as } n \to \infty. \tag{4.2.8}$$

Proof. (1) Let $A_1, \dots, A_k$, $B_1, \dots, B_l \in \mathcal{C}$, $A = \bigcup_{i=1}^k A_i$, $B = \bigcup_{j=1}^l B_j$, $A_i \cap A_{i'} = \varnothing$ for $i \neq i'$, $B_j \cap B_{j'} = \varnothing$ for $j \neq j'$. Then $\mu(A) = \sum_{i=1}^k \mu(A_i)$, $\mu(B) = \sum_{j=1}^l \mu(B_j)$. By assumption

$$\mu(T^{-n}(A) \cap B) = \sum_{i=1}^{k}\sum_{j=1}^{l} \mu(T^{-n}(A_i) \cap B_j) \to \sum_{i=1}^{k}\sum_{j=1}^{l} \mu(A_i) \cdot \mu(B_j) = \mu(A) \cdot \mu(B).$$

Thus (4.2.7) holds for any elements of the dense collection $\mathfrak{U}$ formed by finite disjoint unions of elements of $\mathcal{C}$. Now let A, B be arbitrary measurable sets. Find A', $B' \in \mathfrak{U}$ such that $\mu(A \bigtriangleup A') < \epsilon/4$, $\mu(B \bigtriangleup B') < \epsilon/4$. Then using the triangle inequality

$$\begin{aligned} |\mu(T^{-n}(A) \cap B) - \mu(A)\mu(B)| &\leq \mu(T^{-n}(A \bigtriangleup A') \cap B) + \mu(T^{-n}(A') \cap (B \bigtriangleup B')) \\ &\quad + |\mu(T^{-n}(A') \cap B') - \mu(A')\mu(B')| \\ &\quad + \mu(A) \cdot \mu(B \bigtriangleup B') + \mu(B') \cdot \mu(A \bigtriangleup A') \\ &\leq |\mu(T^{-n}(A') \cap B') - \mu(A') \cdot \mu(B')| + \epsilon. \end{aligned}$$

Since $\epsilon > 0$ can be chosen arbitrarily small, this implies (4.2.7).

(2) First, let us show that if (4.2.8) holds for φ, $\psi \in \Phi$ it also holds for any two functions from $L^2(X, \mu)$. Since both sides of (4.2.8) are linear in φ and antilinear in ψ, it holds for any finite linear combination of functions from Φ, that is, for φ and ψ from a dense subset $L(\Phi) \subset L^2(X_\mu)$.

To show that it is true for arbitrary functions we use essentially the same approximation argument as in (1). Let $\varphi, \psi \in L^2(X, \mu)$. Fix $\epsilon > 0$ and find φ', $\psi' \in L(\Phi)$ such that $\|\varphi - \varphi'\| < \epsilon$, $\|\psi - \psi'\| < \epsilon$. Then, using the Schwarz inequality and the invariance of μ under T, we have

$$
\begin{aligned}
&\Big|\int \varphi(T^n(x))\bar{\psi}(x)\,d\mu - \int \varphi\,d\mu \int \bar{\psi}\,d\mu\Big| \\
&\quad = \Big|\int \varphi(T^n(x))(\overline{\psi(x) - \psi'(x)})\,d\mu + \int (\varphi(T^n(x)) - \varphi'(T^n(x)))\bar{\psi'}(x)\,d\mu \\
&\qquad + \int \varphi'(T^n(x))\bar{\psi'}(x)\,d\mu - \int \varphi'\,d\mu \int \bar{\psi'}\,d\mu \\
&\qquad + \int \varphi'\,d\mu \int \overline{\psi' - \psi}\,d\mu + \int (\varphi' - \varphi)\,d\mu \int \bar{\psi}\,d\mu\Big| \\
&\quad \le \|\varphi \circ T^n\| \cdot \|\psi - \psi'\| + \|(\varphi - \varphi') \circ T^n\| \\
&\qquad + \Big|\int \varphi'(T^n(x))\bar{\psi'}(x)\,d\mu - \int \varphi'\,d\mu \int \bar{\psi'}\,d\mu\Big| \\
&\qquad + \Big|\int \varphi'\,d\mu\Big| \cdot \|\psi - \psi'\| + \|\varphi - \varphi'\| \cdot \Big|\int \bar{\psi}\,d\mu\Big| \\
&\quad \le \Big|\int \varphi'(T^n(x))\bar{\psi'}(x)\,d\mu - \int \varphi'\,d\mu \int \bar{\psi'}\,d\mu\Big| \\
&\qquad + \epsilon\Big(\|\varphi\| + 1 + \Big|\int \varphi'\,d\mu\Big| + \Big|\int \bar{\psi}\,d\mu\Big|\Big).
\end{aligned}
$$

Since ϵ can be chosen arbitrarily small, (4.2.8) follows. In particular if $\varphi = \chi_A$ and $\psi = \chi_B$ then (4.2.8) becomes (4.2.7), thus proving mixing.

On the other hand, characteristic functions of measurable sets form a complete system in $L^2(X, \mu)$ and by the previous argument mixing implies (4.2.8) for all L^2 functions. □

Proposition 4.2.11. (1) *No translation T_γ of the torus is mixing with respect to Lebesgue measure.*
(2) *Every expanding endomorphism E_m, $|m| \ge 2$, is mixing with respect to Lebesgue measure.*

Proof. (1) Since mixing implies ergodicity it is enough to consider ergodic T_γ. If Δ is a small ball then, since T_γ is a topologically transitive isometry, there are infinitely many iterates n_k such that $T_\gamma^{-n_k}(\Delta) \cap \Delta = \varnothing$. Hence $\lambda(T^{-n_k}(\Delta) \cap \Delta) = 0$.

(2) By (1) of Proposition 4.2.10 it is enough to establish (4.2.7) for intervals A and B. The preimage $E_m^{-n}(A)$ consists of $|m|^n$ intervals of length $|m|^{-n} \cdot \lambda(A)$ uniformly spread on S^1, that is, every interval $\Delta = (i/|m|^n, (i+1)/|m|^n)$ contains exactly one component of the preimage $E_m^{-n}(A)$. Hence for n large B contains approximately $|m|^n \cdot \lambda(B)$ components of $E_m^{-n}(A)$ and $\lambda(E_m^{-n}(A) \cap B)$ converges to $|m|^n \lambda(B) |m|^{-n} \lambda(A) = \lambda(A) \cdot \lambda(B)$. □

e. Hyperbolic toral automorphisms. A linear map $L\colon \mathbb{R}^m \to \mathbb{R}^m$ preserves Lebesgue measure. λ if and only if $|\det L| = 1$. This extends to nonlinear maps. Namely, let $U \subset \mathbb{R}^m$ be an open set and $f: U \to \mathbb{R}^m$ be an injective differentiable map. For any domain $V \subset U$

$$\lambda(f(V)) = \int_V |\det Df(x)|\, d\lambda(x).$$

Hence $|\det Df| \equiv 1$ implies that Lebesgue measure is preserved. Conversely if $|\det Df(x_0)| > 1$, say, then $\lambda(f(V)) > \lambda(V)$ for a sufficiently small domain V containing x_0. The same argument works for an injective C^1 map $f: \mathbb{T}^m \to \mathbb{T}^m$. In particular, any linear automorphism is injective and its determinant is either $+1$ or -1. Hence it preserves Lebesgue measure.

In Section 5.1 we will discuss conditions for preservation of a measure in a more general context and more systematically.

Proposition 4.2.12. *Any hyperbolic automorphism of the two-torus is ergodic and mixing with respect to Lebesgue measure.*

Proof. We will first prove separately ergodicity and mixing using Fourier analysis arguments similar to our proof of Proposition 4.2.2 and the first proof of Proposition 4.2.7. In addition, we will present another, more geometric and visual proof of mixing, which is similar to our proof of Proposition 4.2.11(2) and is less dependent on the algebraic structure of the map.

Let $F_L: \mathbb{T}^2 \to \mathbb{T}^2$ be given by

$$F_L(x, y) = (ax + by, cx + dy) \pmod 1,$$

where the determinant of the matrix $L = \begin{pmatrix} a & b \\ c & d \end{pmatrix}$ is either $+1$ or -1 and its eigenvalues are real and different from ± 1. Consider the action of F_L on the characters

$$\chi_{m,n}(x, y) = \exp(2\pi i(mx + ny));$$
$$\chi_{m,n}(F_L(x, y)) = \exp(2\pi((am + cn)x + (bm + dn)y)) = \chi_{am+cn,bm+dn}(x, y).$$

Thus, if one identifies the character $\chi_{m,n}$ with the vector (m, n) in the integer lattice $\mathbb{Z}^2$, the map induced by F_L acts on that lattice by the transpose matrix L^t. All orbits of that action are infinite except for the fixed point at the origin. Hence if

$$\varphi(x, y) = \sum_{(m,n)\in\mathbb{Z}^2} \varphi_{m,n}\chi_{m,n}(x, y)$$

is a bounded F_L-invariant function then

$$\varphi_{m,n} = \varphi_{am+cn,\, bm+dn},$$

that is, the Fourier coefficients of φ are constant on the orbits of the action induced by F_L on the lattice $\mathbb{Z}^2$ (compare with (4.2.6)). This implies that either $\varphi_{m,n} = 0$ for $(m,n) \neq (0,0)$ and hence $\varphi \equiv$ const. (mod 0), or φ has infinitely many nonzero Fourier coefficients equal to each other. The latter possibility contradicts the well-known fact that $|\varphi_{m,n}| \to 0$ as $m^2 + n^2 \to \infty$ for any L^1-function φ. This proves ergodicity.

Since the characters $\chi_{m,n}$ for $(m,n) \in \mathbb{Z}^2$ form a complete system of functions, by Proposition 4.2.10(2) mixing follows once we prove that

$$\int \chi_{m,n}(F_L^N(x,y))\overline{\chi_{k,l}(x,y)}\, d\lambda \xrightarrow[N\to\infty]{} \int \chi_{m,n}\, d\lambda \int \overline{\chi_{k,l}}\, d\lambda. \tag{4.2.9}$$

If $m = n = k = l = 0$, both sides in (4.2.9) are equal to 1 for all N; thus we can assume that $(m,n) \neq (0,0)$. Then the right-hand side is equal to 0 since all characters except the constant have zero integrals. For any $(m,n) \neq (0,0)$ the norm of the vector $(L^t)^N(m,n)$ tends to infinity as $N \to \infty$ and hence for any sufficiently large N

$$(L^t)^N(m,n) \neq (k,l). \tag{4.2.10}$$

The left-hand side in (4.2.9) is equal to $\int \chi_{(L^t)^N(m,n)-(k,l)}(x,y)\, d\lambda$ and hence it is equal to zero as soon as (4.2.10) holds. □

Second proof of mixing. Let us consider the collection of all parallelograms in $\mathbb{T}^2$ whose sides are parallel to the eigenvectors of the matrix L. Since such parallelograms form a sufficient collection, according to Proposition 4.2.10(1) it is sufficient to establish (4.2.7) for any two such parallelograms A and B. For n large $F_L^{-n}(A)$ is the projection to $\mathbb{T}^2$ of a very long and thin parallelogram in R^2. Its long sides are orbit segments of length $l_n = l_0\lambda_1^n$ of an irrational linear flow $\{T_\omega^t\}$, where λ_1 is the larger of the absolute values of the eigenvalues of L and $\omega = (\omega_1, \omega_2)$ is an eigenvector corresponding to the smaller (contracting) eigenvalue of L. Let $J = \{T_\omega^t(x)\}_{t=0}^{l_n}$ be any such segment. The proportion of its length which lies inside B is equal to

$$\frac{1}{l_n}\int_0^{l_n} \chi_B(T_\omega^{-t}(x))\, dt. \tag{4.2.11}$$

Since the flow T_ω^t is uniquely ergodic by the counterpart of Corollary 4.1.14 for flows, (4.2.11) converges to $\lambda(B)$ uniformly in x as $n \to \infty$. Thus for large n and for every segment J comprising $F_L^{-n}(A)$

$$\left|\frac{\text{length }(J \cap B)}{l_n} - \lambda(B)\right| < \epsilon. \tag{4.2.12}$$

Integrating (4.2.12) along the short side of $F_L^{-n}(A)$, one obtains (4.2.7). □

f. Symbolic systems. We conclude our survey of statistical properties for specific examples introduced in Chapter 1 by looking at symbolic dynamical systems. In contrast with the smooth examples just discussed, for a symbolic dynamical system no "natural" invariant measure, similar to Lebesgue measure, is present. We will see later (Proposition 4.4.2, Exercise 4.5.3) that at least for a transitive topological Markov chain a certain invariant measure can be singled out as providing, in a sense, the most typical statistics for the asymptotic behavior of orbits. We begin with some fairly apparent invariant measures for the full shifts σ_N and σ_N^R.

Every probability distribution $p = (p_0, \dots, p_{N-1})$, where $0 \le p_i \le 1$ for $i = 0, \dots, N-1$ and $\sum_{i=0}^{N-1} p_i = 1$, determines the *product measures* μ_p and μ_p^R on the spaces Ω_n and Ω_N^R, correspondingly. Namely, for any cylinder $C_{\alpha_1,\dots,\alpha_k}^{n_1,\dots,n_k} = \{\omega \in \Omega_N \mid \omega_{n_i} = \alpha_i,\ i = 1,\dots,k\}$ we set $\mu_p\left(C_{\alpha_1,\dots,\alpha_k}^{n_1,\dots,n_k}\right) = \prod_{i=1}^k p_{\alpha_i}$ and then extend μ_p in the standard fashion to the σ-algebra of all Borel sets. The measures μ_p^R are defined similarly. The measures μ_p and μ_p^R thus defined are obviously shift invariant. Sometimes the product measures are called *Bernoulli measures* and the shifts considered as measure-preserving transformations with respect to such a measure are often called *Bernoulli shifts*. The term "*topological Bernoulli shift*" for topological dynamical systems was coined as an imitation of the more common term used in ergodic theory. Note that when only one component of p is nonzero the measures μ_p and μ_p^R are atomic, so we usually exclude this case.

Proposition 4.2.13. *σ_N and σ_N^R are mixing with respect to μ_p and μ_p^R.*

Proof. We will consider only σ_N; the other case is completely similar. Since symmetric cylinders $C_\alpha^m = \{\omega \in \Omega_N \mid \omega_i = \alpha_i,\ i = -m,\dots,m\}$ form a sufficient collection of μ_p-measurable sets it is enough to prove that

$$\mu_p\left(\sigma_N^{-n}(C_\alpha^k) \cap C_\beta^l\right) \xrightarrow[n\to\infty]{} \mu_p\left(C_\alpha^k\right) \cdot \mu_p\left(C_\beta^l\right) = \prod_{i=-k}^{k} p_{\alpha_i} \prod_{j=-l}^{l} p_{\beta_j}.$$

Since $\sigma_N^{-n}(C_\alpha^k) = C_{\alpha_{-k},\dots,\alpha_k}^{n-k,\dots,n+k}$, we have $\sigma_N^{-n}(C_\alpha^k) \cap C_\beta^l = C_{\beta_{-l},\dots,\beta_l,\alpha_{-k},\dots,\alpha_k}^{-l,\dots,l,n-k,\dots,n+k}$ for $n \ge 2k + 2l + 2$, and by definition of the measure μ_p

$$\mu\left(C_{\beta_{-l},\dots,\beta_l,\alpha_{-k},\dots,\alpha_k}^{-l,\dots,l,n-k,\dots,n+k}\right) = \prod_{i=-k}^{k} p_{\alpha_i} \cdot \prod_{j=-l}^{l} p_{\beta_j} = \mu\left(C_\alpha^k\right) \cdot \mu\left(C_\beta^l\right).$$

□

The product measure on Ω_N corresponding to the uniform distribution $(1/N, \dots, 1/N)$ possesses some special properties. Recall from Section 1.9a that one can introduce the structure of an abelian group in Ω_N by using coordinatewise addition modulo N. In other words if $\omega = (\dots, \omega_{-1}, \omega_0, \omega_1, \dots)$, $\alpha = (\dots, \alpha_{-1}, \alpha_0, \alpha_1, \dots)$ then $\omega + \alpha = (\dots, \beta_{-1}, \beta_0, \beta_1, \dots)$, where $\beta_n = \alpha_n + \omega_n \pmod N$ and $\beta_n \in \{0, 1, \dots, N-1\}$.

This operation is continuous with respect to the product topology. The group structure in Ω_N^R is defined similarly. The shifts σ_N and σ_N^R become a group automorphism and an endomorphism, correspondingly, strengthening the similarity with the toral automorphisms and the linear expanding map.

The Haar measures on Ω_N and Ω_N^R coincide with the product measures $\mu_{(1/N,\dots,1/N)}$ and $\mu^R_{(1/N,\dots,1/N)}$. Slightly modifying our previous notations for Haar measures we will denote those measures by λ_N and λ_N^R.

Let $\mu_{N,n}$ be the uniform δ-measure on the set of all periodic points of period n for the shift σ_N. Obviously

$$\mu_{N,n}(C_\alpha^m) = N^{-2m-1} = \lambda_N(C_\alpha^m)$$

for any symmetric cylinder C_α^m and $n \geq 2m+1$. Thus, the sequence $\mu_{N,n}$ converges to the measure λ_N in the weak* topology. This means that the measure λ_N in a sense reflects the asymptotic distribution of periodic points for the shift σ_N. The same construction with an obvious modification works for the one-sided shift σ_N^R.

A more general class of invariant measures for the N-shift and topological Markov chains are *Markov measures.* Let $\Pi := \{\pi_{ij}\}_{i,j=0,\dots,N-1}$ be an $N \times N$ matrix with nonnegative entries such that $\sum_{i=0}^{N-1} \pi_{ij} = 1$ for $j = 0,\dots,N-1$. Such matrices are called *stochastic.* Similarly to the case of 0-1 matrices (see Definition 1.9.6) we will call a stochastic matrix Π *transitive* if for some m all entries of Π^m are positive. The following fact is an easy consequence of the Perron–Frobenius Theorem 1.9.11 and some arguments from its proof.

Proposition 4.2.14. *Every stochastic matrix Π has an invariant vector p with nonnegative coordinates. If Π is transitive, such a vector is unique (up to rescaling), one is a simple eigenvalue, and all other eigenvalues of Π have absolute values less than one.*

Proof. Every stochastic matrix preserves the hyperplane $x_1 + \cdots + x_N = 1$ and the cone P of vectors with nonnegative coordinates. Hence it preserves the simplex

$$\Sigma := \Big\{(x_1,\dots,x_N)\Big| x_i \geq 0, \sum_{i=1}^{N} x_i = 1\Big\}.$$

The map $T\colon \sigma \to \sigma$ defined in the proof of the Perron–Frobenius Theorem 1.9.11 in this case simply coincides with the restriction of Π to the simplex σ. Following the proof of the Perron–Frobenius Theorem we conclude that Π preserves the convex set

$$\sigma_0 := \bigcap_{n=0}^{\infty} \Pi^n \sigma$$

which has at most N extreme points. Let us denote these extreme points by $p_1,\dots,p_S$ and their average $(p_1+\cdots+p_S)/S$ by p. Since Π permutes $(p_1,\dots,p_S)$ we deduce that $\Pi p = p$.

Now assume that Π is transitive. In this case the Perron–Frobenius Theorem applies directly and produces the unique invariant vector $p \in \sigma$ with positive coordinates. □

Given a stochastic matrix Π and a vector $p \in \sigma$ we define the Markov measure $\mu_{\Pi,p}$ on Ω_N by

$$\mu_{\Pi,p}(C_\alpha^m) = \Big(\prod_{i=-m}^{m-1} \pi_{\alpha_i \alpha_{i+1}}\Big) p_{\alpha_m}. \tag{4.2.13}$$

Let us emphasize that π_{ij} represents the proportion of the measure of the cylinder C_j^0 (whose measure is p_j) that is transported to C_i^0. (Compare with $\sum_j \pi_{ij} p_j = p_i$.) This makes stochasticity of the matrix an obvious necessary condition for invariance of the measure. An immediate calculation shows that $\Pi p = p$ guarantees σ_N-invariance of $\mu_{\Pi,p}$.

Suppose now A is a 0-1 matrix and suppose that a stochastic matrix Π is such that $\pi_{ij} = 0$ if $a_{ij} = 0$. Then obviously $\operatorname{supp} \mu_{\Pi,p} \subset \Omega_A$ and hence $\mu_{\Pi,p}$ can be viewed as an invariant measure for the topological Markov chain σ_A. If Π is a transitive matrix we will denote the measure $\mu_{\Pi,p}$ simply by μ_Π since the vector p is unique in this case.

Proposition 4.2.15. *If Π is a transitive stochastic matrix then the shift σ_N is mixing with respect to the measure μ_Π.*

Lemma 4.2.16. *If $\Pi^n = \{\pi_{ij}^{(n)}\}$ then $\pi_{ij}^{(n)} \to p_i$ as $n \to \infty$.*

Proof. It follows from Proposition 4.2.14 that Π^n is a sequence of stochastic matrices which converges to a projection to the line l generated by the vector p. Since stochasticity is a closed property the limit is also a stochastic matrix. But the only stochastic matrix that projects $\mathbb{R}^N$ to l is the matrix whose columns are copies of the vector p. □

Proof of Proposition 4.2.15. For $n > 2m+2$ the intersection $\sigma_N^{-n}(C_\alpha^m) \cap C_\beta^m$ is the union of cylinders of the form $C^{-m,\dots,m,m+1,\dots,n-m-1,n-m,\dots,n+m}_{\beta_{-m},\dots,\beta_m,\gamma_{m+1},\dots,\gamma_{n-m-1},\alpha_{-m},\dots,\alpha_m}$ for all $\gamma = (\gamma_{m+1},\dots,\gamma_{n-m-1})$. The μ_Π-measure of such a cylinder is equal to

$$\mu_\Pi(C_\beta^m) \cdot p_{\beta_m}^{-1} \cdot \mu_\Pi(C_\alpha^m) \cdot \pi_{\beta_m \gamma_{m+1}} \prod_{k=1}^{n-m-2} \pi_{\gamma_{m+k} \gamma_{m+k+1}} \cdot \pi_{\gamma_n \alpha_{-m}}.$$

Summing these expressions over all γ we obtain

$$\mu_\Pi(\sigma_N^{-n}(C_\alpha^m) \cap C_\beta^m) = \mu_\Pi(C_\beta^m) \cdot \mu_\Pi(C_\alpha^m) \cdot p_{\beta_m}^{-1} \cdot \pi_{\beta_m \alpha_{-m}}^{(n-m)}.$$

By Lemma 4.2.16 $\pi_{\beta_m \alpha_{-m}}^{(n-m)} \to p_{\beta_m}$ as $n \to \infty$. Hence $\mu_\Pi(\sigma_N^{-n}(C_\alpha^m) \cap C_\beta^m) \to \mu_\Pi(C_\beta^m) \cdot \mu_\Pi(C_\alpha^m)$ and since the system of symmetric cylinders for various m is sufficient, mixing follows from Proposition 4.2.10(1). □

Exercises

4.2.1. *Prove that every translation of the torus has a pure point spectrum, that is, the associated unitary operator has a complete orthogonal system of eigenfunctions.*

4.2.2. *Consider the map $f: S^1 \times [0,1] \to S^1 \times [0,1]$, $f(x,t) = (x+\alpha, t)$, where α is an irrational number. Prove that the averages of every continuous function φ converge uniformly but that f is not uniquely ergodic.*

4.2.3. *Prove that the affine map A_α of the two-dimensional torus that appeared in Exercises 1.4.4 and 3.2.6,*

$$A_\alpha(x,y) = (x+\alpha, y+x) \pmod 1,$$

is ergodic with respect to Lebesgue measure if α is an irrational number.

4.2.4. *Prove that the map A_α with α irrational is uniquely ergodic. Find all invariant Borel probability measures for A_α when α is rational.*

4.2.5. *Prove that the fractional part of any quadratic polynomial $\alpha n^2 + \beta n + \gamma$, $n = 0, 1, \dots$, where α is an irrational number, is uniformly distributed on the interval $[0,1]$, that is, if $x_n = \alpha n^2 + \beta n + \gamma - [\alpha n^2 + \beta n + \gamma]$, $0 \le a < b < 1$, and*

$$N_n(a,b) = \operatorname{card}\{i \in [0,\dots,n-1] \mid a \le x_n < b\}$$

then $(1/n)N_n(a,b) \to b - a$.

4.2.6. *Prove unique ergodicity of the following affine map $A_{\alpha,m}$ of $\mathbb{T}^m$:*

$$A_{\alpha,m}(x_1,\dots,x_m) = (x_1+\alpha, x_2+x_1, x_3+x_2, \dots, x_m+x_{m-1}) \pmod 1,$$

where α is an irrational number.

4.2.7. *Prove that the fractional parts of any polynomial*

$$\alpha n^m + \alpha_1 n^{m-1} + \cdots + \alpha_m$$

for $n \in \mathbb{N}_0$, where α is irrational, are uniformly distributed in the interval $[0,1]$.[2]

4.2.8. *Prove that the "tent" map $g: [0,1] \to [0,1]$ that appeared in Exercise 2.5.1 and is given by*

$$g(x) = \begin{cases} 2x, & 0 \le x \le \frac{1}{2}, \\ 2-2x, & \frac{1}{2} \le x \le 1, \end{cases}$$

preserves Lebesgue measure and is ergodic with respect to Lebesgue measure.

4.2.9. *Prove that the map $f_4\colon [0,1] \to [0,1]$ given by $f_4(x) = 4x(1-x)$ has an ergodic invariant measure equivalent to Lebesgue measure.*

4.2.10*. *Let $0 < \alpha < \beta < 1$ and consider the following piecewise-continuous transformation $I_{\alpha,\beta}$ of the interval $[0,1)$ onto itself:*

$$I_{\alpha,\beta}(x) := \begin{cases} x+1-\alpha, & 0 \le x \le \alpha, \\ x-\alpha+1-\beta, & \alpha \le x < \beta, \\ x-\beta, & \beta \le x < 1. \end{cases}$$

Prove that $I_{\alpha,\beta}$ is an injective map which preserves Lebesgue measure. Prove that $I_{\alpha,\beta}$ is ergodic with respect to Lebesgue measure if and only if $\beta/(1+\beta-\alpha)$ is irrational.

4.2.11. *Prove that the automorphism $F_L\colon \mathbb{T}^m \to \mathbb{T}^m$ determined by an integer $m \times m$ matrix with determinant ± 1 is ergodic with respect to Lebesgue measure if and only if the matrix L does not have an eigenvalue that is a root of unity. Prove that any ergodic F_L is in fact mixing.*

4.2.12. *In the presence of a root of unity as an eigenvalue show that there is a nonconstant F_L-invariant trigonometric polynomial so F_L is not topologically transitive and hence not ergodic.*

4.2.13. *Prove that a mixing transformation has continuous spectrum, that is, the associated isometric operator has no nonconstant eigenfunctions.*

4.2.14. *Suppose that Π is a nontransitive stochastic matrix. Show that among the measures $\mu_{\Pi,p}$ there are finitely many ergodic ones and every measure $\mu_{\Pi,p}$ is a convex linear combination of these measures.*

4.2.15. *Show that every ergodic measure $\mu_{\Pi,p}$ has the following structure. There exists an integer m such that $\mu_{\Pi,p}$ is not ergodic for the mth power σ_N^m of σ_N, namely, $\mu_{\Pi,p} = \sum_{i=1}^m \mu^{(i)}$, where each measure $\mu^{(i)}$ is invariant and mixing for σ_N^m and σ_N interchanges the $\mu^{(i)}$ cyclically.*

4.2.16. *Prove that for the measure μ_Π, where Π is a transitive stochastic matrix, mixing is exponential on cylinders, that is, for any cylinders C, C' we have*

$$|\mu_\Pi(\sigma_N^{-n}(C) \cap C') - \mu_\Pi(C) \cdot \mu_\Pi(C')| < c\exp(-\alpha n)$$

for some $c > 0$, $\alpha > 0$.

3. Measure-theoretic entropy

We saw in the previous sections that basic properties related to statistical behavior of orbits, namely, typical recurrent behavior with respect to an invariant measure, ergodicity, unique ergodicity, and mixing, can be viewed as stronger "quantitative" counterparts of "qualitative" recurrence properties, namely, orbit recurrence, topological transitivity, minimality, and topological mixing, correspondingly (see Proposition 4.1.18 and the remark after Definition 4.2.8). Now we will consider a statistical counterpart to a global orbit growth invariant, the topological entropy. This quantity is called entropy of a measure-preserving transformation or entropy with respect to an invariant measure (see Definition 4.3.7 and Definition 4.3.9). In the case of an ergodic measure entropy can be described similarly to its topological counterpart as the exponential growth rate of the number of statistically significant orbit segments distinguishable with arbitrarily fine but finite precision.[1]

The connection between the concepts of topological entropy and entropy with respect to an invariant measure is more complete and precise than in the case of the pairs orbit recurrence–typical recurrence with respect to an invariant measure, topological transitivity–ergodicity, minimality–unique ergodicity, and topological mixing–mixing. For those cases it is one-sided: The statistical property implies its topological counterpart, but not in general vice versa. In the case of entropies the connection is provided by the Variational Principle (Theorem 4.5.3) which asserts that the topological entropy of a continuous map is equal to the supremum of the entropies of that map with respect to all its invariant measures. Thus not only does a statistical property (say, positivity of entropy with respect to an invariant measure) guarantee its topological counterpart (in this case positivity of topological entropy), but conversely, positivity of topological entropy implies existence of an invariant measure with positive entropy, thus providing a quantitative extension of the Krylov–Bogolubov Theorem 4.1.1 in the case of maps with positive topological entropy.

Before proceeding to the Variational Principle we will develop in this section the general theory of entropy with respect to an invariant measure and then will test it in the next section on our standard array of examples.

a. Entropy and conditional entropy of partitions. Let $(X, \mathcal{B}, \mu)$ or (X, μ) be a probability space and I a finite or countable set of indices. A collection of measurable subsets $\xi = \{C_\alpha \in \mathcal{B} \mid \alpha \in I\}$ is called a *measurable partition* of X if $\mu(X \smallsetminus \bigcup_{\alpha \in I} C_\alpha) = 0$ and $\mu(C_{\alpha_1} \cap C_{\alpha_2}) = 0$ for $\alpha_1 \neq \alpha_2$. We will be considering mostly finite measurable partitions of X into sets of positive measure. As is natural and customary in measure theory we will consider measurable partitions mod 0. This means that two partitions ξ and η are identified if one can find a set A of measure zero such that the restrictions of ξ and η to $X \smallsetminus A$ coincide. Equivalently $\xi = \eta \pmod 0$ if for any element $C \in \xi$ of positive measure one can find an element $D \in \eta$ such that $\mu(C \,\Delta\, D) = 0$. Here Δ means symmetric difference: $A \,\Delta\, B = (A \cup B) \smallsetminus (A \cap B) = (A \smallsetminus B) \cup (B \smallsetminus A)$.

Definition 4.3.1. The *entropy* of a measurable partition ξ is given by

$$H(\xi) := H_\mu(\xi) := -\sum_{\substack{\alpha\in I\\ \mu(C_\alpha)>0}} \mu(C_\alpha)\log\mu(C_\alpha) \geq 0.$$

Alternatively we may agree that $0\log 0 = 0$ and write

$$H_\mu(\xi) = -\sum_{\alpha\in I}\mu(C_\alpha)\log\mu(C_\alpha).$$

For countable ξ the entropy may be infinite.

In most cases we will suppress the dependence of entropy on the measure, but where more than one measure is involved in a discussion we will use the lower index.

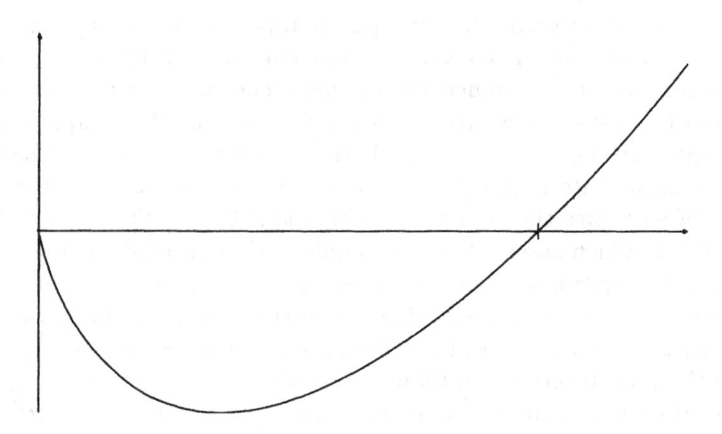

FIGURE 4.3.1. The function $x\log x$

If $T: X \to X$ is a measure-preserving transformation, ξ a measurable partition of X, and $T^{-1}(\xi) := \{T^{-1}(C_\alpha) \mid \alpha\in I\}$ then obviously

$$H(T^{-1}(\xi)) = H(\xi). \tag{4.3.1}$$

For a measurable partition ξ and $x\in X$ let $C_\xi(x)$ be the element of ξ that contains x. The function $I_\xi: X\to\mathbb{R}$,

$$I_\xi(x) = -\log\mu(C_\xi(x)) \tag{4.3.2}$$

is called the *information function* of the partition ξ. In this definition we disregard the set of measure zero for which $\mu(C_\xi(x)) = 0$. Using (4.3.2) one can write

$$H_\mu(\xi) = \int_X I_\xi\, d\mu. \tag{4.3.3}$$

This representation of entropy illuminates and makes natural the following notion of conditional entropy of a partition with respect to another partition which plays the central role in the entropy theory for measure-preserving transformations.

When considering a partition it is convenient to use an alternative notation for the conditional measures μ_A corresponding to elements of the partition by setting $\mu(A \mid B) := \mu(A \cap B)/\mu(B)$.

Definition 4.3.2. Let $\xi = \{C_\alpha \mid \alpha \in I\}$, $\eta = \{D_\alpha \mid \alpha \in J\}$ be two measurable partitions of (X, μ). The *conditional entropy* of ξ with respect to η is

$$H(\xi \mid \eta) := -\sum_{\beta \in J} \mu(D_\beta) \sum_{\alpha \in I} \mu(C_\alpha \mid D_\beta) \log \mu(C_\alpha \mid D_\beta).$$

Remark. If $\nu = \{X\}$ is the trivial partition then $H(\xi) = H(\xi \mid \nu)$. Similarly to (4.3.3) one can rewrite the definition of conditional entropy as

$$H(\xi \mid \eta) = \int_X I_{\xi,\eta} \, d\mu, \tag{4.3.4}$$

where $I_{\xi,\eta}$ is the *conditional information function* defined by

$$I_{\xi,\eta}(x) = -\log \mu(C_\xi(x) \mid C_\eta(x)).$$

Formula (4.3.4) allows us to define conditional entropy even in certain cases when ξ is a continuous partition. Without going into a general discussion about measurability and conditional measures for a continuous partition we will illustrate this case by an example. Let X be the unit square $D = [0,1] \times [0,1]$ with Lebesgue measure, ξ the partition into vertical intervals $\{x\} \times [0, f(x)]$ and $\{x\} \times (f(x), 1]$, where $f\colon [0,1] \to [0,1]$ is a measurable function, and η the partition into vertical intervals $\{x\} \times [0,1]$. Then

$$H(\xi \mid \eta) = -\int_0^1 [f(x) \log f(x) + (1 - f(x)) \log(1 - f(x))] \, dx.$$

Returning to the case of a finite or countable measurable partition let us note that if we denote by ξ_{D_β} the partition of the set D_β into the intersections $D_\beta \cap C_\alpha$, $\alpha \in I$, such that $\mu(D_\beta \cap C_\alpha) > 0$ then

$$H(\xi \mid \eta) = \sum_{\beta \in J} \mu(D_\beta) H(\xi_{D_\beta}),$$

where $H(\xi_{D_\beta})$ is calculated with respect to the conditional measure μ_β.

There is an obvious partial-ordering relation between partitions: $\xi \le \eta$ if and only if for all $D \in \eta$ there exists a $C \in \xi$ such that $D \subset C$. If $\xi \le \eta$ we will say that η is a *refinement* of ξ and that ξ is *subordinate* to η. An obvious construction of a simultaneous refinement of two partitions ξ and η is the *joint partition*

$$\xi \vee \eta := \{C \cap D \mid C \in \xi, D \in \eta, \mu(C \cap D) > 0\}$$

(compare with Exercise 3.1.7).

Finally we say that two partitions ξ and η are *independent* if

$$\mu(C \cap D) = \mu(C) \cdot \mu(D)$$

for all $C \in \xi$, $D \in \eta$.

The following proposition summarizes basic properties of entropy and conditional entropy which will be systematically used later.

Proposition 4.3.3. *Let $(X, \mathcal{B}, \mu)$ be a probability space and let $\xi = \{C_\alpha \mid \alpha \in I\}$, $\eta = \{D_\alpha \mid \alpha \in J\}$, $\zeta = \{E_\alpha \mid \alpha \in K\}$ be finite or countable measurable partitions of X and $\nu = \{X\}$. Then:*

(1) $0 < -\log(\sup_{\alpha\in I} \mu(C_\alpha)) \leq H(\xi) \leq \log \operatorname{card} \xi$; *furthermore if ξ is finite then $H(\xi) = \log \operatorname{card} \xi$ if and only if all elements of ξ have equal measure.*

(2) $0 \leq H(\xi \mid \eta) \leq H(\xi)$; *$H(\xi \mid \eta) = H(\xi)$ if and only if ξ and η are independent; $H(\xi \mid \eta) = 0$ if and only if $\xi \leq \eta$ (mod 0). If $\zeta \geq \eta$ then $H(\xi \mid \zeta) \leq H(\xi \mid \eta)$.*

(3) $H(\xi \vee \eta \mid \zeta) = H(\xi \mid \zeta) + H(\eta \mid \xi \vee \zeta)$. *In particular, for $\zeta = \nu$ we obtain*

$$H(\xi \vee \eta) = H(\xi) + H(\eta \mid \xi). \tag{4.3.5}$$

(4) $H(\xi \vee \eta \mid \zeta) \leq H(\xi \mid \zeta) + H(\eta \mid \zeta)$; *in particular $H(\xi \vee \eta) \leq H(\xi) + H(\eta)$.*

(5) $H(\xi \mid \eta) + H(\eta \mid \zeta) \geq H(\xi \mid \zeta)$.

(6) *If λ is another measure on X then for every partition ξ measurable for both μ and λ and for any $p \in [0, 1]$*

$$pH_\mu(\xi) + (1-p)H_\lambda(\xi) \leq H_{p\mu+(1-p)\lambda}(\xi).$$

Corollary 4.3.4. *For two measurable partitions ξ, η with $H(\xi) < \infty$, $H(\eta) < \infty$ let*

$$d_R(\xi, \eta) := H(\xi \mid \eta) + H(\eta \mid \xi). \tag{4.3.6}$$

Then d_R is a metric on the set of (all equivalence classes mod 0 of) measurable partitions with finite entropy. It is called the Rokhlin metric.

Proof of the corollary. $d_R(\xi, \eta) \geq 0$ by (2). If $d_R(\xi, \eta) = 0$ then $H(\xi \mid \eta) = H(\eta \mid \xi) = 0$. By (2) $\xi \geq \eta$ and $\eta \geq \xi$. But this immediately implies that $\xi = \eta$ (mod 0). The symmetry of d_R is immediate from (4.3.6). Finally, from (5)

$$\begin{aligned} d_R(\xi, \zeta) &= H(\xi \mid \zeta) + H(\zeta \mid \xi) \\ &\leq H(\xi \mid \eta) + H(\eta \mid \zeta) + H(\zeta \mid \eta) + H(\eta \mid \xi) = d_R(\xi, \eta) + d_R(\eta, \zeta). \end{aligned}$$ □

Proof of the proposition. (1) $H(\xi) \geq 0$ by definition; if ξ contains at least two elements of positive measure then $H(\xi) > 0$; thus $H(\xi) = 0$ implies that $\xi = \nu \pmod 0$. $-\log(\sup_{\alpha\in I}\mu(C_\alpha)) = \inf I(\xi)$; hence by (4.3.3) $H(\xi) \geq -\log(\sup_{\alpha\in I}\mu(C_\alpha))$.

In order to prove $H(\xi) \leq \log \operatorname{card} \xi$, we may assume that ξ is a finite partition. Consider the function

$$\phi(x) = \begin{cases} x\log x & \text{if } x \geq 0, \\ 0 & \text{if } x = 0. \end{cases}$$

$\phi''(x) = 1/x > 0$, so ϕ is strictly convex, that is, $\phi(\sum_{i=1}^{\infty} a_i x_i) \leq \sum_{i=1}^{\infty} a_i\phi(x_i)$ for nonnegative $a_1, \dots, a_n$ such that $\sum_{i=1}^{\infty} a_i = 1$, with equality only if $x_i = x_j$ when $a_i \neq 0 \neq a_j$. Now let $\xi = (C_1, \dots, C_k)$ and take $a_i = 1/k$, $x_i = \mu(C_i)$ for $i = 1, \dots, k$. Convexity of ϕ implies

$$-\frac{1}{k}\log k = \phi\Big(\frac{1}{k}\Big) = \phi\Big(\frac{1}{k}\sum_{i=1}^{k}\mu(C_i)\Big) \leq \sum_{i=1}^{k}\frac{1}{k}\phi(\mu(C_i)) = -\frac{1}{k}H(\xi), \qquad (4.3.7)$$

that is, $H(\xi) \leq \log k$. Strict convexity of ϕ implies that equality in (4.3.7) is equivalent to $\mu(C_i) = 1/k$ for all i.

(2) The inequality follows again from convexity of ϕ:

$$\begin{aligned} 0 \leq H(\xi \mid \eta) &= -\sum_{\beta\in J}\mu(D_\beta)\sum_{\alpha\in I}\phi(\mu(C_\alpha \mid D_\beta)) \\ &= -\sum_{\alpha\in I}\sum_{\beta\in J}\mu(D_\beta)\phi(\mu(C_\alpha \mid D_\beta)) \\ &\leq -\sum_{\alpha\in I}\phi\Big(\sum_{\beta\in J}\mu(D_\beta)\mu(C_\alpha \mid D_\beta)\Big) \\ &= -\sum_{\alpha\in I}\phi(\mu(C_\alpha)) = H(\xi). \end{aligned} \qquad (4.3.8)$$

Now $\phi(x) < 0$ for $0 < x < 1$, so if $H(\xi \mid \eta) = 0$ then for every β with $\mu(D_\beta) > 0$ we have $\phi(\mu(C_\alpha \mid D_\beta)) = 0$ for all $\alpha \in I$ and consequently $\xi \leq \eta$. If $H(\xi \mid \eta) = H(\xi)$ then we must have equality in (4.3.8) for each term of the summation over α, that is,

$$\phi(\mu(C_\alpha)) = \phi\Big(\sum_{\substack{\beta\in J \\ \mu(D_\beta)>0}}\mu(D_\beta)\mu(C_\alpha \mid D_\beta)\Big) = \sum_{\substack{\beta\in J \\ \mu(D_\beta)>0}}\mu(D_\beta)\phi(\mu(C_\alpha \mid D_\beta)).$$

By strict convexity of the function ϕ this implies that if $\mu(D_\beta) > 0$ and $\mu(C_\alpha) > 0$ then $\mu(C_\alpha \mid D_\beta) = \mu(C_\alpha)$, that is, $\mu(C_\alpha \cap D_\beta) = \mu(C_\alpha)\cdot\mu(D_\beta)$.

Applying the inequality $H_{\mu_C}(\xi \mid \zeta) \leq H_{\mu_C}(\xi)$ to the conditional measures μ_C on each element C of the partition η and integrating over that partition we obtain $H(\xi \mid \zeta) = H(\xi \mid \zeta \vee \eta) \leq H(\xi \mid \eta)$.

$$(3)\ H(\xi \vee \eta \mid \zeta) = -\sum_{(\alpha,\beta,\gamma)\in I\times J\times K}\mu(C_\alpha \cap D_\beta \cap E_\gamma)\log\frac{\mu(C_\alpha \cap D_\beta \cap E_\gamma)}{\mu(E_\gamma)}$$

$$
\begin{aligned}
&= -\sum_{\alpha,\beta,\gamma} \mu(C_\alpha \cap D_\beta \cap E_\gamma) \log \frac{\mu(C_\alpha \cap E_\gamma)}{\mu(E_\gamma)} \\
&\quad - \sum_{\alpha,\beta,\gamma} \mu(C_\alpha \cap D_\beta \cap E_\gamma) \log \frac{\mu(C_\alpha \cap D_\beta \cap E_\gamma)}{\mu(C_\alpha \cap E_\gamma)} \\
&= -\sum_{\alpha,\gamma} \mu(C_\alpha \cap E_\gamma) \log \frac{\mu(C_\alpha \cap E_\gamma)}{\mu(E_\gamma)} + H(\eta \mid \xi \vee \zeta) \\
&= H(\xi \mid \zeta) + H(\eta \mid \xi \vee \zeta).
\end{aligned}
$$

(4) This follows from (3) and the inequality $H(\eta \mid \xi \vee \zeta) \leq H(\eta \mid \zeta)$ which in turn follows from (2) since $\xi \vee \zeta \geq \zeta$.

(5) By (3) and (4) we have $H(\zeta \mid \xi \vee \eta) = H(\xi \vee \zeta \mid \eta) - H(\xi \mid \eta) \leq H(\zeta \mid \eta)$. Using (3) several times we obtain

$$
\begin{aligned}
H(\zeta \mid \eta) + H(\eta \mid \zeta) &= H(\xi \vee \eta) + H(\eta \vee \zeta) - H(\eta) - H(\zeta) \\
&= H(\xi \vee \eta) + H(\zeta \mid \eta) - H(\zeta) \\
&= H(\xi \vee \eta \vee \zeta) - H(\zeta \mid \xi \vee \eta) + H(\zeta \mid \eta) - H(\zeta) \\
&\geq H(\xi \vee \eta \vee \zeta) - H(\zeta) \geq H(\xi \vee \zeta) - H(\zeta) = H(\xi \mid \zeta).
\end{aligned}
$$

(6) This follows immediately from convexity of the function ϕ:

$$
\begin{aligned}
pH_\mu(\xi) + (1-p)H_\lambda(\xi) &= -p\sum_{\alpha \in I} \phi(\mu(C_\alpha)) - (1-p)\sum_{\alpha \in I} \phi(\lambda(C_\alpha)) \\
&\leq -\sum_{\alpha \in 1} \phi((p\mu + (1-p)\lambda)(C_\alpha)) = H_{p\mu+(1-p)\lambda}(\xi).
\end{aligned}
$$

□

For a measure space (X, μ) and $m \in \mathbb{N}$ consider the space $\mathcal{P}_m$ of all equivalence classes mod 0 of partitions of X into at most m measurable sets. By adding null sets if necessary, we may assume that every partition in $\mathcal{P}_m$ has exactly m elements. For $\xi, \eta \in \mathcal{P}_m$ consider now the set of bijections σ between the elements of ξ and η and set

$$
\mathcal{D}(\xi, \eta) := \min_\sigma \sum_{C \in \xi} \mu(C \vartriangle \sigma(C)). \tag{4.3.9}
$$

Obviously $\mathcal{D}$ is a metric. We will need the fact that convergence in this metric guarantees convergence in the Rokhlin metric.

Proposition 4.3.5. *Given $\epsilon > 0$ there exists $\delta > 0$ such that $\mathcal{D}(\xi, \eta) < \delta$ implies $d_R(\xi, \eta) < \epsilon$.*

Remark. In fact, the metrics $\mathcal{D}$ and d_R are equivalent in the space $\mathcal{P}_m$ (Exercise 4.3.1).

Proof. By symmetry it suffices to estimate $H(\eta \mid \xi)$. If $\mathcal{D}(\xi, \eta) = \delta$ write $\xi = (A_1, \ldots, A_m)$, $\eta = (B_1, \ldots, B_m)$ in such a way that $\sum_{i=1}^m \mu(A_i \vartriangle B_i) = \delta$. For $i \in \{1, \ldots, m\}$ such that $\mu(A_i) > 0$ let $\alpha_i := \mu(A_i \setminus B_i)/\mu(A_i)$. Then the contribution of A_i to the expression for $H(\eta \mid \xi)$ in Definition 4.3.2 is

$$\begin{aligned} &- \mu(B_i \cap A_i) \log \frac{\mu(B_i \cap A_i)}{\mu(A_i)} - \sum_{j \neq i} \mu(B_j \cap A_i) \log \frac{\mu(B_j \cap A_i)}{\mu(A_i)} \\ &\qquad \leq \mu(A_i)[-(1 - \alpha_i)\log(1 - \alpha_i) - \alpha_i \log \alpha_i + \alpha_i \log(m-1)] \\ &\qquad = \mu(A_i)\Big[(1 - \alpha_i)\log \frac{1}{1 - \alpha_i} + \alpha_i \log \frac{m-1}{\alpha_i}\Big] \leq \mu(A_i) \log m. \end{aligned}$$

Here the first inequality follows from Proposition 4.3.3(1) by considering the measure induced on $A_i \setminus B_i = \bigcup_{j \neq i}(A_i \cap B_j)$ and estimating the entropy of η with respect to that measure. The last inequality uses convexity of $-\log x$. Thus

$$\begin{aligned} H(\eta \mid \xi) \leq &\sum_{\mu(A_i) \geq \sqrt{\delta}} \mu(A_i)[-(1 - \alpha_i)\log(1 - \alpha_i) - \alpha_i \log \alpha_i + \alpha_i \log(m-1)] \\ &+ \sum_{\mu(A_i) < \sqrt{\delta}} \mu(A_i) \log m. \end{aligned}$$

The second term does not exceed $m \log m \sqrt{\delta}$. To estimate the first note that

$$\alpha_i \mu(A_i) = \mu(A_i \setminus B_i) = \sum_{j \neq i} \mu(B_j \cap A_i) \leq \sum_{j=1}^m \mu(A_j \vartriangle B_j) = \delta,$$

so for $\mu(A_i) \geq \sqrt{\delta}$ we get $\alpha_i \leq \sqrt{\delta}$. Now $\varphi(x) := -x \log x - (1 - x)\log(1 - x)$ is increasing on $(0, 1/2)$, so for $\delta < 1/4$ the first sum is dominated by $\varphi(\sqrt{\delta}) + \sqrt{\delta} \log(m-1)$ and hence $H(\eta \mid \xi) \leq \varphi(\sqrt{\delta}) + \sqrt{\delta}(m \log m + \log(m-1))$. Since $\varphi(x) \to 0$ as $x \to 0$ the statement follows. □

The notions of entropy and conditional entropy have counterparts in the topological case. Namely, let $\mathcal{A}$ be a finite open cover of a compact metrizable space X and $N(\mathcal{A})$ the minimal cardinality of a subcover. Then $H(\mathcal{A}) := \log N(\mathcal{A})$ (see Exercise 3.1.7) is a natural counterpart of entropy in that context. Furthermore for a subset $Y \subset X$ let $N_Y(\mathcal{A})$ be the minimal number of elements of $\mathcal{A}$ that still cover Y. Then for two covers $\mathcal{A}$ and $\mathcal{B}$

$$H(\mathcal{A} \mid \mathcal{B}) := \log(\max_{B \in \mathcal{B}} N_B(\mathcal{A}))$$

is an analog of the conditional entropy. However, this analogy is only partial. Though inequalities such as

$$H(\mathcal{A} \mid \mathcal{B} \vee \mathcal{C}) \leq H(\mathcal{A} \mid \mathcal{B}) \leq H(\mathcal{A})$$

hold, there are no topological counterparts to Proposition 4.3.3(3) and to the independence part of (2). This disparity explains why measure-theoretic entropy is a much more powerful and quantitative tool in ergodic theory than topological entropy is in topological dynamics.

b. Entropy of a measure-preserving transformation. For a measurable partition ξ and a measure-preserving (not necessarily invertible) transformation T we define the *joint partition* by

$$\xi_{-n}^{T} := \xi \vee T^{-1}(\xi) \vee \cdots \vee T^{-n+1}(\xi).$$

From now on, unless stated otherwise, we will assume that all partitions are finite or countable measurable partitions with finite entropy.

Proposition 4.3.6. *$\lim_{n\to\infty}(1/n)H(\xi_{-n}^{T})$ exists.*

Proof. One immediately sees from (4.3.5) and (4.3.1) that $H(\xi_{-n-m}^{T}) \leq H(\xi_{-n}^{T}) + H(\xi_{-m}^{T})$ and the statement follows. □

Remark. This is the third occasion after Lemma 3.1.5 and the proof of (3.1.20) when subadditivity was used to prove existence of the limit for a quantity related to exponential growth. We will have numerous opportunities later to use various modifications of this simple but powerful argument. (See Proposition 9.6.4.)

Definition 4.3.7. $h(T,\xi) := h_{\mu}(T,\xi) := \lim_{n\to\infty}(1/n)H(\xi_{-n}^{T})$ is called the *metric entropy* of the transformation T *relative* to the partition ξ.

The following proposition gives an alternative proof of existence of the limit $h(T,\xi)$ as well as another expression for it.

Proposition 4.3.8. *$h(T,\xi) = \lim_{n\to\infty} H(\xi \mid T^{-1}(\xi_{-n}^{T}))$ and $H(\xi \mid T^{-1}(\xi_{-n}^{T}))$ is nonincreasing in n.*

Proof. Using (4.3.5) inductively and taking into account the invariance property (4.3.1) we obtain

$$\begin{aligned} H(\xi_{-n}^{T}) &= H(T^{-1}(\xi_{-n+1}^{T})) + H(\xi \mid T^{-1}(\xi_{-n+1}^{T})) \\ &= H(\xi_{-n+1}^{T}) + H(\xi \mid T^{-1}(\xi_{-n+1}^{T})) = H(\xi_{0}^{T}) + \sum_{k=0}^{n-1} H(\xi \mid T^{-1}(\xi_{-k}^{T})). \end{aligned}$$

Since the partition $T^{-1}(\xi_{-k}^{T})$ in the "denominator" is refined as k increases, by Proposition 4.3.3(2) the sequence

$$b_n := H(\xi \mid T^{-1}(\xi_{-n}^{T}))$$

is nonincreasing. Thus $h_{\mu}(T,\xi) = \lim_{n\to\infty}(1/n)\sum_{k=0}^{n-1} b_k = \lim_{n\to\infty} b_n$. □

Definition 4.3.9. The *entropy* of T with respect to μ (or the entropy of μ) is

$$h(T) := h_\mu(T) := \sup\{h_\mu(T,\xi) \mid \xi \text{ is a measurable partition with } H(\xi) < \infty\}.$$

Obviously entropy is invariant under metric isomorphism. We will see soon that this definition is more constructive than it seems; in many cases $h_\mu(T) = h_\mu(T,\xi)$ for an appropriately chosen ξ. (See, for example, Corollary 4.3.14.)

Recalling the definition of the partition entropy through the information function (4.3.2)–(4.3.3) we can interpret the entropy $h_\mu(T,\xi)$ as the average amount of information provided by the knowledge of the "present state" in addition to the knowledge of an arbitrarily long past. Thus, a system with zero entropy can be viewed as strongly deterministic in the sense that an approximate knowledge of the entire past (that is, the past itinerary with respect to a finite partition) precisely determines the future itinerary.

c. Properties of entropy. The following proposition summarizes basic properties of the entropy $h(T,\xi)$ as a function of the partition ξ. It prepares the way for subsequent criteria which allow one to calculate the transformation entropy $h(T)$.

Proposition 4.3.10. *Let $T\colon (X,\mu) \to (X,\mu)$ be a measure-preserving transformation of a probability space and η, ξ be measurable partitions with finite entropy. Then:*

(1) $0 \le \overline{\lim}_{n\to\infty} -(1/n)\log(\sup_{C\in\xi^T_{-n}} \mu(C)) \le h(T,\xi) \le H(\xi)$.
(2) $h(T,\xi \vee \eta) \le h(T,\xi) + h(T,\eta)$.
(3) $h(T,\eta) \le h(T,\xi) + H(\eta \mid \xi)$*; in particular if* $\xi \le \eta$ *then* $h(T,\xi) \le h(T,\eta)$.
(4) $|h(T,\xi) - h(T,\eta)| \le H(\xi \mid \eta) + H(\eta \mid \xi)$ *(the Rokhlin inequality).*
(5) $h(T,T^{-1}(\xi)) = h(T,\xi)$ *and if* T *is invertible then* $h(T,\xi) = h(T,T(\xi))$.
(6) $h(T,\xi) = h(T,\bigvee_{i=0}^k T^{-i}(\xi))$ *for* $k \in \mathbb{N}$ *and if* T *is invertible then* $h(T,\xi) = h(T,\bigvee_{i=-k}^k T^i(\xi))$ *for* $k \in \mathbb{N}$.
(7) *If* ν *is another measure and* $p \in [0,1]$ *then*

$$ph_\mu(T,\xi) + (1-p)h_\nu(T,\xi) \le h_{p\mu+(1-p)\nu}(T,\xi).$$

Remark. Property (4) means that $h(T,\cdot)$ is a Lipschitz function with Lipschitz constant 1 on the space of partitions with finite entropy provided with the Rokhlin metric (4.3.6).

Proof. The middle inequality in (1) follows directly from Proposition 4.3.3(1) and the right inequality follows from Proposition 4.3.8 and Proposition 4.3.3(2).

(2) Since $(\xi \vee \eta)^T_{-n} = \xi^T_{-n} \vee \eta^T_{-n}$, this statement follows from (4.3.5) which is a particular case of Proposition 4.3.3(3).

(3) By (4.3.5) $H(\xi_{-n}^T) \le H(\xi_{-n}^T \vee \eta_{-n}^T) = H(\eta_{-n}^T) + H(\xi_{-n}^T \mid \eta_{-n}^T)$ and by using Proposition 4.3.3(3) inductively, we obtain

$$\begin{aligned} H(\xi_{-n}^T \mid \eta_{-n}^T) &= H(\xi \mid \eta_{-n}^T) + H(T^{-1}(\xi_{1-n}^T) \mid \xi \vee \eta_{-n}^T) \\ &\le H(\xi \mid \eta) + H(T^{-1}(\xi_{1-n}^T) \mid \eta_{-n}^T) \\ &\le H(\xi \mid \eta) + H(T^{-1}(\xi) \mid T^{-1}(\eta)) + H(T^{-2}(\xi_{2-n}^T) \mid \eta_{-n}^T) \\ &\le n\, H(\xi \mid \eta). \end{aligned}$$

Property (4) follows directly from (3).

Property (5) follows from the invariance property (4.3.1) since

$$H((T^{-1}(\xi))_{-n}^T) = H(T^{-1}(\xi_{-n}^T)) = H(\xi_{-n}^T)$$

and similarly if T is invertible

$$H((T(\xi))_{-n}^T) = H(T(\xi_{-n}^T)) = H(\xi_{-n}^T).$$

(6) $(\bigvee_{i=0}^k T^{-i}(\xi))_{-n}^T = \xi_{-n-k}^T$ and hence

$$H\Big(T, \bigvee_{i=0}^{k} T^{-i}(\xi)\Big) = \lim_{n\to\infty} \frac{1}{n} H(\xi_{-n-k}^T) = \lim_{n\to\infty} \frac{1}{n+k} H(\xi_{-n-k}^T) = h(T,\xi).$$

The argument for invertible T is completely similar.

Property (7) follows directly from Proposition 4.3.3(6). □

We can now formulate some criteria for calculating the entropy of a measure-preserving transformation. It is interesting to point out that invertible and noninvertible transformations have to be treated separately.

Definition 4.3.11. A family Ξ of measurable partitions with finite entropy is called *sufficient* with respect to the measure-preserving transformation T if

(1) for noninvertible T, partitions subordinate to partitions of the form $\bigvee_{i=0}^k T^{-i}(\xi)$ ($\xi \in \Xi$, $k \in \mathbb{N}$) form a dense subset in the space of all partitions with finite entropy provided with the Rokhlin metric (4.3.6);

(2) for invertible T the same holds for partitions subordinate to $\bigvee_{i=-l}^{l} T^i(\xi)$ ($\xi \in \Xi$, $l \in \mathbb{N}$).

Remark. Proposition 4.3.5 allows us to replace the Rokhlin metric by the metric $\mathcal{D}$ from (4.3.9) in this definition. In the case of a nonatomic Borel measure on a compact metric space a more obvious condition that guarantees sufficiency of a family $\Xi = \{\xi_n\}_{n\in\mathbb{N}}$ is $\operatorname{diam}(\xi_n) \to 0$, where $\operatorname{diam}(\xi) := \sup_{C\in\xi}(\operatorname{diam}(C))$.

Theorem 4.3.12. *$h_\mu(T) = \sup_{\xi\in\Xi} h_\mu(T,\xi)$ for any sufficient family Ξ of partitions.*

Proof. Let η be an arbitrary measurable partition of X with $H_\mu(\eta) < \infty$. Fix $\epsilon > 0$ and find $\xi \in \Xi$ and $k \in \mathbb{N}$ such that

$$d_R(\eta,\zeta) = H(\eta \mid \zeta) + H(\zeta \mid \eta) < \epsilon$$

for some partition $\zeta \le \bigvee_{i=0}^{k} T^{-i}(\xi)$ if T is noninvertible and $\zeta \le \bigvee_{i=-k}^{k} T^{i}(\xi)$ if T is invertible. Using consecutively Proposition 4.3.10(4), (3), and (6) we obtain in the noninvertible case $h_\mu(T,\eta) \le h_\mu(T,\zeta) + \epsilon \le h_\mu(T, \bigvee_{i=0}^{k} T^{-i}(\xi)) + \epsilon = h_\mu(T,\xi) + \epsilon$ (and similarly if T is invertible). Since ϵ is arbitrary, the statement follows. □

Definition 4.3.13. A partition ξ is called a *generator* for T if $\Xi = \{\xi\}$ is a sufficient family.

The following corollary is the best-known and simplest-sounding criterion for calculating entropy.

Corollary 4.3.14. *If ξ is a generator for T then $h_\mu(T) = h_\mu(T,\xi)$.*

At this point it is useful to stress the difference between the invertible and the noninvertible case. Let us call a partition ξ a *one-sided generator* for an invertible measure-preserving transformation T if partitions subordinate to partitions of the form $\bigvee_{i=0}^{k} T^{-i}(\xi)$ $(k \in \mathbb{N})$ are dense in the metric d_R.

Proposition 4.3.15. *If an invertible measure-preserving transformation possesses a one-sided generator then $h_\mu(T) = 0$.*

Proof. A one-sided generator ξ is obviously a generator for T so by Corollary 4.3.14 it is sufficient to prove that $h_\mu(T,\xi) = 0$. By Proposition 4.3.10(5) this is equivalent to $h_\mu(T,T\xi) = 0$. Suppose ξ is a one-sided generator and $\epsilon > 0$. Then take $k \in \mathbb{N}$ and $\zeta \le \bigvee_{i=0}^{k} T^{-i}(\xi)$ such that $d(T(\xi),\zeta) < \epsilon$ and hence $H(T(\xi) \mid \bigvee_{i=0}^{k} T^{-i}(\xi)) \le H(T(\xi) \mid \zeta) < \epsilon$. Thus since the sequence $a_n := H(T(\xi) \mid \bigvee_{i=0}^{n} T^{-i}(\xi))$ is nonincreasing, $h_\mu(T,T(\xi)) < \epsilon$ by Proposition 4.3.8. Since ϵ is arbitrary we have $h(T,T(\xi)) = 0$. □

The following proposition is a counterpart for measure-preserving transformations to Proposition 3.1.6 and Proposition 3.1.7.

Proposition 4.3.16.

(1) *If $S\colon (Y,\nu) \to (Y,\nu)$ is a factor (see Definition 4.1.21) of $T\colon (X,\mu) \to (X,\mu)$ then $h_\nu(S) \le h_\mu(T)$.*
(2) *If A is invariant for T and $\mu(A) > 0$ then $h_\mu(T) = \mu(A)h_{\mu_A}(T) + \mu(X \smallsetminus A)h_{\mu_{X\smallsetminus A}}(T)$.*
(3) *If μ, λ are two invariant probability measures for T then for any $p \in [0,1]$*

$$h_{p\mu+(1-p)\lambda}(T) \ge p h_\mu(T) + (1-p)h_\lambda(T).$$

(4) $h_\mu(T^k) = kh_\mu(T)$ *for any* $k \in \mathbb{N}$. *If* T *is invertible then* $h_\mu(T^{-1}) = h_\mu(T)$ *and hence* $h_\mu(T^k) = |k|h_\mu(T)$ *for any* $k \in \mathbb{Z}$.
(5) $h_{\mu\times\lambda}(T \times S) = h_\mu(T) + h_\lambda(S)$.

Proof. (1) For any measurable partition η of Y, the preimage

$$R^{-1}(\eta) = \{R^{-1}D \mid D \in \eta\}$$

under the factor map R is a measurable partition of X and by definition $H_\mu(R^{-1}\eta) = H_\nu(\eta)$ and $h_\mu(T, R^{-1}\eta) = h_\nu(S,\eta)$. Thus

$$\begin{aligned} h_\mu(T) = \sup\{h_\mu(T,\xi) \mid H_\mu(\xi) < \infty\} &\geq \sup\{h_\mu(T, R^{-1}(\eta)) \mid H_\mu(R^{-1}(\eta)) < \infty\} \\ &= \sup\{h_\nu(S,\eta) \mid H_\nu(\eta) < \infty\} = h_\nu(S). \end{aligned}$$

(2) Let ξ be a measurable partition of X, $H_\mu(\xi) < \infty$, and $\zeta = \{A, X \setminus A\}$. By replacing ξ by $\xi \vee \zeta$ if necessary, we may assume that $\xi \geq \zeta$. Then $H_\mu(\xi^T_{-n}) = \mu(A)H_{\mu_A}(\xi^T_{-n}) + \mu(X \setminus A)H_{\mu_{X\setminus A}}(\xi^T_{-n})$ since A is T-invariant.

(3) This follows from Proposition 4.3.3(6) applied to the partition ξ^T_{-n}.

(4) If $k \in \mathbb{N}$ then

$$\frac{1}{n}H_\mu\Big(\bigvee_{j=0}^{n-1} T^{-kj}\big(\bigvee_{i=0}^{k-1} T^{-i}(\xi)\big)\Big) = \frac{k}{nk}H_\mu\Big(\bigvee_{i=0}^{nk-1} T^{-i}(\xi)\Big)$$

and $h_\mu\Big(T^k, \bigvee_{i=0}^{k-1} T^{-i}(\xi)\Big) = k\, h_\mu(T,\xi)$. Furthermore $h_\mu(T,\xi) = h_\mu(T^{-1},\xi)$ since $\xi^T_{-n} = T^{-n+1}(\xi^{T^{-1}}_{-n})$.

(5) Let ξ and η be measurable partitions of X and Y, correspondingly. Consider the trivial partitions $\nu_X = \{X\}$ and $\nu_Y = \{Y\}$ of X and Y. Then $\xi \times \eta = (\xi \times \nu_Y) \vee (\nu_X \times \eta)$, where $\zeta \times \nu_Y$ and $\nu_X \times \eta$ are independent as partitions of $X \times Y$.

It follows from Proposition 4.3.3(2) and (3) that $H(\alpha \vee \beta) = H(\alpha) + H(\beta)$ for independent partitions α and β. Thus

$$H_{\mu\times\lambda}(\zeta \times \eta) = H_{\mu\times\lambda}(\xi \times \nu_Y) + H_{\mu\times\lambda}(\nu_Y \times \eta) = H_\mu(\xi) + H_\lambda(\eta).$$

Since $(\xi \times \eta)^{T\times S}_{-n} = \xi^T_{-n} \times \eta^S_{-n}$, this implies that $h_{\mu\times\lambda}(T \times S, \xi \times \eta) = h_\mu(T,\xi) + h_\lambda(S,\eta)$ and hence $h_{\mu\times\lambda}(T \times S) \leq h_\mu(T) + h_\lambda(S)$. But the family of partitions of $X \times Y$ of the form $\xi \times \eta$ where $H_\mu(\xi) < \infty$ and $H_\lambda(\eta) < \infty$ is sufficient with respect to any measure-preserving transformation of $X \times Y$. Hence $h_{\mu\times\lambda}(T \times S) = h_\mu(T) + h_\lambda(S)$ by Theorem 4.3.12. □

Corollary 4.3.17. *If* μ, ν *are two mutually singular invariant probability measures for* f *and* $\alpha \in [0,1]$ *then* $h_{\alpha\mu+(1-\alpha)\nu}(f) = \alpha h_\mu(f) + (1-\alpha)h_\nu(f)$.

Proof. Let $A \subset X$ be such that $\mu(A) = \nu(X \setminus A) = 1$. Then for $\alpha \in (0,1)$ $\mu = (\alpha\mu + (1-\alpha)\nu)_A$ and $\nu = (\alpha\mu + (1-\alpha)\nu)_{X\setminus A}$ and Proposition 4.3.16(2) applies. □

Corollary 4.3.17 is, in fact, true even if the two measures are not mutually singular. This can be proved using the *ergodic decomposition*, Theorem 4.1.12, of a measure into ergodic measures which gives a unique representation of an invariant measure as an integral over ergodic measures (thus the set of invariant measures is effectively a simplex). Nevertheless the behavior of metric entropy as a function of the measure is rather subtle because it is often not continuous (with respect to the weak topology). The coexistence of this "linearity" with discontinuity is related to the fact that even on the set of ergodic measures entropy is not continuous; for example, a weak limit of periodic δ-measures may have positive entropy.

Exercises

4.3.1. *Let $\xi, \eta \in \mathcal{P}_m$. Prove that for any $\epsilon > 0$ there is a $\delta > 0$ such that $d_R(\xi, \eta) < \delta$ implies $\mathcal{D}(\xi, \eta) < \epsilon$.*

4.3.2. *Prove that if a measure-preserving transformation $T: (X, \mu) \to (X, \mu)$ has a generator ξ with k elements then $h_\mu(T) \leq \log k$.*

4.3.3. *Prove that if under the assumption of the previous exercise T is invertible and $h_\mu(T) = \log k$ then T is metrically isomorphic to the full k-shift σ_k with the uniform Bernoulli measure $\mu_{(1/k,\ldots,1/k)}$*

4. Examples of calculation of measure-theoretic entropy

a. Rotations and translations. It is natural to expect that the entropy of a rotation with respect to Lebesgue measure λ is zero. There are several ways to establish this fact. Most straightforwardly one can take the partition ξ_N into N equal intervals. The family $\{\xi_N\}_{n\in\mathbb{N}}$ is obviously sufficient by the remark after Definition 4.3.11. The iterated (or joint) partition $(\xi_N)_n^{R_\alpha} = \bigvee_{i=0}^{n-1} R_\alpha^{-i}(\xi_N)$ has no more than $N \cdot n$ elements (in fact exactly that many if α is irrational). Hence by Proposition 4.3.3(1)

$$H\Big(\bigvee_{i=0}^{n-1} R_\alpha^{-i}(\xi_N)\Big) \leq \log Nn = \log N + \log n$$

and $h_\lambda(R_\alpha, \xi_N) \leq \lim_{n\to\infty} \dfrac{1}{n}(\log N + \log n) = 0.$

Another proof uses the fact that for an irrational α any partition ξ of S^1 into intervals is in fact a one-sided generator. That follows from the fact that the iterated partition

$$\xi_{-n}^{R_\alpha} := \bigvee_{i=0}^{n-1} R_\alpha^{-i}(\xi)$$

consists of intervals whose endpoints are preimages of the endpoints of the intervals that comprise ξ. Since every semiorbit is dense the maximal length of intervals in $\xi_{-n}^{R_\alpha}$ goes to zero as $n \to \infty$. Thus Proposition 4.3.15 implies that $h_\lambda(R_\alpha) = 0$ for irrational α. For $\alpha = \frac{p}{q}$ rational we have $R_{p/q}^q = \text{Id}$ and since obviously $h_\lambda(\text{Id}) = 0$ Proposition 4.3.16(4) implies that $h_\lambda(R_{p/q}) = \frac{1}{q} h_\lambda(\text{Id}) = 0$.

For a translation T_γ on a torus the easiest way to show that $h_\lambda(T_\gamma) = 0$ is to use Proposition 4.3.16(5). For, if $\gamma = (\gamma_1, \dots, \gamma_m)$, then $T_\gamma = R_{\gamma_1} \times R_{\gamma_2} \times \dots \times R_{\gamma_m}$ and hence $h_\lambda(T_\gamma) = \sum_{i=1}^m h(R_{\gamma_i}) = 0$. Alternatively one can repeat the first proof for the rotation and show that for the partition $\eta = \xi_N \times \dots \times \xi_N$ of the torus into equal cubes with sides $1/N$ the number of elements in the iterated partition $\eta_n^{T_\gamma}$ grows no faster than $(nN)^m$. Hence by Proposition 4.3.3(1)

$$h_\lambda(T_\gamma, \eta) = \lim_{n\to\infty} \frac{1}{n} H(\eta_{-n}^{T_\gamma}) \le \lim_{n\to\infty} \frac{1}{n}(m \log n + m \log N) = 0.$$

Since $\{\xi_N \times \dots \times \xi_N\}_{N\in\mathbb{N}}$ is a sufficient family of partitions, Theorem 4.3.12 implies that $h_\lambda(T_\gamma) = 0$.

b. Expanding maps. For the map E_k the standard partition ξ_1 into intervals $\left[\frac{m}{|k|}, \frac{m+1}{|k|}\right]$, $m = 0, \dots, k-1$, is a generator since the iterated partition $\bigvee_{i=0}^{n-1} E_k^{-i}(\xi_1) =: \xi_n$ is the partition into equal intervals of length $|k|^{-n}$ whose entropy is $n \log |k|$. Hence by Corollary 4.3.14

$$h_\lambda(E_k) = h_\lambda(E_k, \xi_1) = \lim_{n\to\infty} \frac{1}{n} H(\xi_n) = \log |k|.$$

Comparing this calculation with Corollary 3.2.4 we conclude that $h_\lambda(E_k) = h_{\text{top}}(E_k)$.

c. Bernoulli and Markov measures. Consider the N-shift σ_N provided with the *Bernoulli measure* μ_p. The standard partition ξ_0 into 1-cylinders

$$C_\alpha^0 = \{\omega \in \Omega_N \mid \omega_0 = \alpha\},$$

$\alpha = 0, \dots, N-1$, is a generator since $\bigvee_{i=-m}^m \sigma_N^i(\xi_0)$ is the partition into symmetric m-cylinders and the diameter of those partitions in any metric d_λ defined in Section 1.9a goes to 0 as $m \to \infty$. Since the partitions $\sigma_N^i(\xi_0)$ are independent for different i we have by Proposition 4.3.3(2) and (3)

$$h_{\mu_p}(\sigma_N) = h_{\mu_p}(\sigma_N, \xi_0) = \lim_{n\to\infty} \frac{1}{n} H\Big(\bigvee_{i=0}^{n-1} \sigma_N^{-i}(\xi_0)\Big) = \frac{nH(\xi_0)}{n} = -\sum_{i=1}^N p_i \log p_i.$$

Thus we see that in this case the sequence $\frac{1}{n} H(\bigvee_{i=0}^{n-1} \sigma_N^{-i}(\xi_0))$ in fact stabilizes.

By Proposition 4.3.3(1) and Section 3.2d we have

Corollary 4.4.1. *$h_{\mu_p}(\sigma_N) \le h_{\mathrm{top}}(\sigma_N)$ and equality takes place if and only if $p = (1/N, \dots, 1/N)$.*

For the Markov measures $\mu_{\Pi,p}$ the easiest way to calculate the entropy is through Proposition 4.3.8 applied to the partition ξ_0 into 1-cylinders. By a straightforward calculation one sees from the definition of the measure $\mu_{\Pi,p}$ on the cylinders (4.2.13) that for any $n \ge 1$

$$H(\xi_0 \mid \sigma_N^{-1} \vee \dots \vee \sigma_N^{-n}(\xi_0)) = -\sum_{i,j=1}^{N} p_j \pi_{ij} \log \pi_{ij}.$$

Since ξ_0 is a generator we obtain by Corollary 4.3.14

$$h_{\mu_{\Pi,p}}(\sigma_N) = -\sum_{i,j=1}^{N} p_j \pi_{ij} \log \pi_{ij}. \tag{4.4.1}$$

There is an important particular case of a Markov measure supported on the subspace Ω_A, where A is a transitive 0-1 matrix. Let $q = (q_1, \dots, q_N)$ be a positive eigenvector of A and $v = (v_1, \dots, v_N)$ the positive eigenvector of the transpose matrix A^T normalized in such a way that

$$\sum_{i=1}^{N} q_i v_i = 1. \tag{4.4.2}$$

Both of those vectors are unique up to a positive scalar multiple due to the Perron–Frobenius Theorem 1.9.11. Since the maximum eigenvalues for A and A^T are the same we have

$$\sum_{j=1}^{N} a_{ij} q_j = \lambda q_i \text{ for } i = 1, \dots, N \tag{4.4.3}$$

and

$$\sum_{i=1}^{N} a_{ij} v_i = \lambda v_j \text{ for } j = 1, \dots, N. \tag{4.4.4}$$

Let

$$\pi_{ij} = \frac{a_{ij} v_i}{\lambda v_j}. \tag{4.4.5}$$

$\Pi = \{\pi_{ij}\}_{i,j=1,\dots,N}$ is a stochastic matrix because

$$\sum_{i=1}^{N} \pi_{ij} = \sum_{i=1}^{N} \frac{a_{ij} v_i}{\lambda v_j} = \frac{\lambda v_j}{\lambda v_j} = 1$$

for all j. Since the vector v has positive coordinates the matrix Π has nonzero elements for all i, j for which $a_{ij} = 1$. Since A is a transitive 0-1 matrix this immediately implies that Π is a transitive matrix, since for the mth power Π^m the element $\pi_{ij}^{(m)} > 0$ if and only if $a_{ij}^{(m)} > 0$.

The vector $p = (p_1, \dots, p_N)$ with

$$p_i := q_i v_i \tag{4.4.6}$$

is invariant under Π:

$$\sum_{j=1}^{N} \pi_{ij} p_j = \sum_{j=1}^{N} \frac{a_{ij} v_i}{\lambda v_j} q_j v_j = \frac{\lambda v_i q_i}{\lambda} = p_i.$$

Since $\operatorname{supp} \mu_{\Pi,p} = \Omega_A$ by Proposition 4.2.15 we deduce that the topological Markov chain σ_A is mixing with respect to the measure $\mu_{\Pi,p} = \mu_\Pi$. This measure is called the *Parry measure* for the topological Markov chain σ_A.

Finally let us calculate the measure-theoretic entropy $h_{\mu_\Pi}(\sigma_A)$. By substituting (4.4.6) and (4.4.5) into the formula (4.4.1) for entropy we obtain

$$\begin{aligned} h_{\mu_\Pi}(\sigma_A) &= -\sum_{i,j=1}^{N} \frac{q_j a_{ij} v_i}{\lambda} \log \frac{a_{ij} v_i}{\lambda v_j} \\ &= \sum_{i,j=1}^{N} \frac{q_j a_{ij} v_i}{\lambda} \log \lambda + \Big(\sum_{i,j=1}^{N} \frac{q_j a_{ij} v_i}{\lambda} (\log v_j - \log a_{ij} v_i) \Big). \end{aligned}$$

It follows from (4.4.2), (4.4.3), and (4.4.4) that the first sum is equal to $\log \lambda$, that is, the topological entropy of σ_A (see Proposition 3.2.5), and the second sum is equal to zero because using (4.4.3) and (4.4.4) again we obtain

$$\sum_{i,j=1}^{N} \frac{a_{ij} q_j v_i}{\lambda} \log v_j = \sum_{j=1}^{N} v_j q_j \log v_j$$

and

$$\sum_{i,j=1}^{N} \frac{a_{ij} q_j v_i}{\lambda} \log a_{ij} v_i = \sum_{i,j=1}^{N} \frac{a_{ij} q_j v_i}{\lambda} \log v_i = \sum_{i,j=1}^{N} v_i q_i \log v_i,$$

where the first equality follows because if $a_{ij} = 0$ then the corresponding term in both parts is equal to zero anyway.

Thus we have proved

Proposition 4.4.2. *The entropy of a transitive topological Markov chain σ_A with respect to the measure μ_Π, where $\Pi = \{\pi_{ij}\}_{i,j=1,\dots,N}$ is given by (4.4.5), is equal to the topological entropy of σ_A and hence to the periodic orbit growth $p(\sigma_A)$.*

Let us point out that this property of the measure μ_Π is similar to that of Lebesgue measure for the linear expanding maps E_k and the uniform Bernoulli measure for the full shift σ_N (cf. Corollary 4.4.1). We will see in the next section that all those measures maximize the value of entropy among all invariant measures for the given transformation. Corollary 20.1.5 shows that μ_Π is the unique measure of maximal entropy.

The construction of the stochastic matrix Π out of a 0-1 matrix A by (4.4.5) may look somewhat mysterious but it indeed admits a natural interpretation. The measure μ_Π is nothing but the *asymptotic distribution* of *periodic orbits* of the topological Markov chain σ_A. To see that we should go back to the discussion of Section 1.9c which implies that the number of different periodic orbits of period n in the basic 0-cylinder C_i^0 is equal to the diagonal element $a_{ii}^{(n)}$ of the matrix A^n. It follows from the Perron–Frobenius Theorem 1.9.11 that $a_{ij}^{(n)}/\lambda^n \to v_i q_j$, where q and v are defined by (4.4.3) and (4.4.4). Thus the *proportion* of the periodic points of period n that belongs to C_i^0 is

$$\frac{a_{ii}^{(n)}}{\operatorname{tr} A^n} \to \frac{q_i v_i}{\sum_{k=1}^N q_k v_k} = q_i v_i,$$

by (4.4.2), justifying (4.4.6). Now the number of periodic points in $C_i^0 \cap \sigma_A(C_j^0)$ is $a_{ij} a_{ji}^{(n-1)}$. The ratio of that number to $a_{jj}^{(n)}$ converges to

$$a_{ij} \frac{\lambda^{n-1} v_i q_j}{\lambda^n q_j v_j} = a_{ij} \frac{v_i}{\lambda v_j},$$

justifying (4.4.5). See Exercise 4.4.2 for more details.

d. Hyperbolic toral automorphisms. To simplify notation we write F for a hyperbolic toral automorphism F_L in this section and denote the maximal eigenvalue by λ rather than λ_1. Let ξ be a finite partition of $\mathbb{T}^2$ into elements of diameter less than 1/10. We will estimate $H(\bigvee_{k=-n}^{n} F^k(\xi)) = H(\xi_{-2n-1}^F)$ from below by estimating from above the diameter and hence the Lebesgue measure of the elements of ξ_{-2n-1}^F. Let $C \in \bigvee_{k=-n}^{n} F^k(\xi)$ and $x, y \in C$. Consider the line parallel to the eigenvector with eigenvalue $\lambda > 1$ passing through the point x and the line parallel to the second eigenvector passing through y. As in Section 3.2e we define z as the first point of intersection of these lines. Then $d(F^k(x), F^k(y)) \le d(F^k(x), F^k(z)) + d(F^k(z), F^k(y))$. First, let $k > 0$. Then $d(F^k(z), F^k(y)) = \lambda^{-k} d(z, y) \le \lambda^{-k} d(x, y) < \lambda^{-k}/10$. Since for $k = 1, \dots, n$ the points $F^k(x)$, $F^k(y)$ belong to the same element of the partition ξ we have $d(F^k(x), F^k(y)) < 1/10$ and hence $d(F^k(x), F^k(z)) < 1/10 + \lambda^{-k}/10 < 1/5$. This implies by induction that the length of the line segment connecting $F^k(x)$ and $F^k(z)$ is also less than 1/5. Hence $d(x, z) = \lambda^{-n} d(F^n(x), F^n(z)) < \lambda^{-n}/5$. A similar argument for negative k shows that $d(y, z) < \lambda^{-n}/5$ and hence we have $d(x, y) < 2\lambda^{-n}/5$. Thus the diameter of any element of $\bigvee_{-n}^{n} F^{-k}(\xi)$ is at most $2\lambda^{-n}/5$ and hence by the isoperimetric inequality its Lebesgue measure is at most $2\pi\lambda^{-2n}/5$. Thus the left inequality in Proposition 4.3.10(1) gives

$h(F,\xi) \geq \log\lambda$. Since the family of partitions into sets of diameter at most $1/10$ is obviously sufficient we obtain for Lebesgue measure m

$$h_m(F) \geq \log\lambda = h_{\text{top}}(F).$$

Now we will obtain the reverse inequality $h_m(F) \leq \log\lambda$ to conclude

$$h_m(F) = \log\lambda. \tag{4.4.7}$$

We use an elementary version of a general method that allows us to obtain an upper estimate of entropy of a smooth dynamical system with respect to an invariant measure. See Section 2 of the Supplement for the general argument.

First, by Proposition 4.3.16(4)

$$h_m(F) = \frac{1}{n} h_m(F^n)$$

for any $n \in \mathbb{N}$. We fix $\epsilon > 0$ and pick n such that $\lambda^{-n} < \epsilon/10$. Now we estimate $h_m(F^n,\xi)$ for a sufficient family of partitions, namely, for any partition into squares with sides $1/N$, where N is sufficiently large. By Proposition 4.3.8 $h_m(F^n,\xi) \leq H(\xi \mid F^{-n}(\xi))$. In order to estimate $H(\xi \mid F^{-n}(\xi))$ we estimate for any given $C \in F^{-n}(\xi)$ the number $\mathcal{N}(C)$ of elements $D \in \xi$ that have nonempty intersection with C. By Proposition 4.3.3(1) and Definition 4.3.2 of conditional entropy we have

$$H(\xi \mid F^{-n}(\xi)) < \max \log \mathcal{N}(C)$$

for $C \in F^{-n}(\xi)$.

Any element C of $F^{-n}(\xi)$ is a parallelogram whose sides are images of vertical and horizontal segments under F^n. The diameter of C is less than $\sqrt{2}\lambda^n/N$; its "width", that is, the distance between the lines containing each pair of parallel sides, is at most $\lambda^{-n}/N < \epsilon/10N$ by assumption. If an element $D \in \xi$ intersects C it lies in the $\sqrt{2}/N$-neighborhood of C. but the total area of such a neighborhood is less than

$$\frac{\sqrt{2}\lambda^n + 2\sqrt{2}}{N} \times \frac{2\sqrt{2} + \epsilon}{N} < \frac{10\lambda^n}{N^2}$$

if n is large enough and $\epsilon < 1$. Let us emphasize that N is much greater than n and even λ^n. Since the area of any element of ξ is $1/N^2$ we obtain $\mathcal{N}(C) < 10\lambda^n$ and

$$H(\xi \mid F^{-n}(\xi)) < \log\lambda + \log 10.$$

Thus $h_m(F) = \dfrac{1}{n} h_m(F^n) < \log\lambda + \dfrac{\log 10}{n}$.

The inequality $h_m(F) \leq \log\lambda$ will also follow from the Variational Principle, Theorem 4.5.3. Another way of calculating the entropy of a hyperbolic toral automorphism is presented in Exercises 4.4.6 and 4.4.7.

Exercises

4.4.1. *Prove that any topologically transitive translation on a torus has a one-sided generator which consists of two elements.*

4.4.2. *Let A be a transitive $N \times N$ 0-1 matrix, $C \subset \Omega_A$ a cylinder, and $\mathcal{N}(n, C)$ the number of periodic orbits of σ_A of period n in C. Prove that*

$$\lim_{n\to\infty} \frac{\mathcal{N}(n,C)}{\operatorname{tr}(A^n)} = \mu_\Pi(C),$$

where the matrix Π is given by (4.4.5).

4.4.3. *Prove that the measure-preserving transformations $(\sigma_2, \mu_{(1/2,1/2)})$ and $(\sigma_3, \mu_{(1/3,1/3,1/3)})$ are not metrically isomorphic.*

4.4.4. *Prove that U_{σ_2} and U_{σ_3} (with the Bernoulli measures as before) are unitarily equivalent.*

4.4.5*. *For any probability distribution $p = (p_0, \dots, p_{N-1})$ with at least two nonzero coordinates consider the operator U_{σ_N} associated to the Bernoulli measure. Prove that any two of these operators are unitarily equivalent.*

4.4.6. *Show that the image of Lebesgue measure under the semiconjugacy of the hyperbolic toral automorphism F with the topological Markov chain σ_A given by the Markov partition of Section 2.5d is a Markov measure μ_Π.*

4.4.7*. *Show that the transition matrix Π for the measure μ_Π from the previous exercise is given by (4.4.5). Use this to give a new proof for $h_m(F) = \log \lambda$.*

4.4.8. *Use Exercises 4.4.2 and 4.4.7 to rigorously formulate and prove the statement "periodic points of the hyperbolic toral automorphism F are uniformly distributed with respect to Lebesgue measure".*

5. The Variational Principle

Topological entropy, which we introduced in Section 3.1, was actually discovered later than measure-theoretic entropy. Measure-theoretic entropy gives a quantitative measure of the complexity of a dynamical system as seen via an invariant measure. Topological entropy was found by extracting from the same concept an invariant of the topological dynamics only. Though there are some analogies in the definitions, the absence of a natural measure of the size of sets in topological dynamics leads to some differences between the two notions. Notably the measure-theoretic entropy of the union of two invariant sets is the sum of the entropies of the invariant sets, weighted by their measures by Proposition 4.3.16, whereas for topological entropy the entropy of a union is the maximum of the entropies of the two components by Proposition 3.1.7(2). Thus topological entropy measures the maximal dynamical complexity versus an average complexity reflected by measure-theoretic entropy. Therefore one expects measure-theoretic entropy to be no greater than topological entropy. Moreover, measures assigning most weight to regions of high complexity should have measure-theoretic entropy close to the topological entropy. This is indeed true, that is, the topological entropy is the supremum of the measure-theoretic entropies.

To study the measurable structure of homeomorphisms f of compact metric spaces X recall first from Section 4.1 that the set $\mathfrak{M}(f)$ of f-invariant Borel probability measures on X is a nonempty compact convex subset (in the weak* topology) of the compact set $\mathfrak{M}$ of Borel probability measures on X. (The extreme points are the ergodic measures.) For our discussion we need to find partitions with nice metric properties. Let us denote by $\bar{A}$ the closure of $A \subset X$ and by ∂A the boundary of A. Let $\partial\{A_1, \dots, A_k\} := \bigcup_{i=1}^k \partial A_i$.

Lemma 4.5.1. *Let X be a compact metric space, $\mu \in \mathfrak{M}$.*

(1) *For $x \in X$, $\delta > 0$ there exists $\delta' \in (0, \delta)$ such that $\mu(\partial B(x, \delta')) = 0$.*
(2) *For $\delta > 0$ there exists a finite measurable partition $\xi = \{C_1, \dots, C_k\}$ with $\operatorname{diam}(C_i) < \delta$ for all i and $\mu(\partial\xi) = 0$.*

Proof. (1) $\bigcup_{\delta' \in (0,\delta)} \partial B(x, \delta')$ is an uncountable disjoint union with finite measure.
(2) Let $\{B_1, \dots, B_k\}$ be a cover of X by balls of radius less than $\delta/2$ and with $\mu(\partial B_i) = 0$ for $0 < i \le k$. Let $C_1 = \overline{B_1}$, $C_i = \overline{B_i} \setminus \bigcup_{j=1}^{i-1} \overline{B_j}$ for $i > 1$ and $\xi := \{C_1, \dots, C_k\}$. Then ξ is as desired since $\partial\xi \subset \bigcup_{i=1}^k \partial B_i$. □

Remark. Note that if $\mu(\partial\xi) = 0$ and f is measure preserving then $\mu(\partial\xi_{-n}^f) = 0$ for any $n \in \mathbb{N}$. An important property of such partitions, which we will use later is that for a weak*-convergent sequence $\mu_n \to \mu$ we get $H_{\mu_n}(\xi) \to H_\mu(\xi)$.

We now describe a method of constructing measures of large entropy which we will use several times. Let us denote by δ_x the probability measure supported on $\{x\}$ and let $f_*\mu(A) := \mu(f^{-1}(A))$ whenever μ is a Borel measure, f measurable, and A a Borel set.

Lemma 4.5.2. *Let (X, d) be a compact metric space, $f\colon X \to X$ a homeomorphism, $E_n \subset X$ an (n, ϵ)-separated set, $\nu_n := (1/\operatorname{card}(E_n)) \sum_{x \in E_n} \delta_x$ the uniform δ-measure on E_n, and $\mu_n := (1/n) \sum_{i=0}^{n-1} f_*^i \nu_n$. Then there is an accumulation point μ of $\{\mu_n\}_{n \in \mathbb{N}}$ (in the weak* topology) that is f-invariant and satisfies*

$$\overline{\lim_{n\to\infty}} \frac{1}{n} \log \operatorname{card}(E_n) \le h_\mu(f).$$

Proof. Take any accumulation point μ of the sequence μ_{n_k}, where n_k is a sequence with $\lim_{k\to\infty} \log \operatorname{card}(E_{n_k})/n_k = \overline{\lim}_{n\to\infty} \log \operatorname{card}(E_n)/n$. Existence of an accumulation point follows from weak*-compactness of $\mathfrak{M}$, and f-invariance of μ is clear since $f_*\mu_n - \mu_n = (f_*^n \nu_n - \nu_n)/n$ and ν_n are probability measures. Consider thus a partition ξ with elements of diameter less than ϵ and

$\mu(\partial\xi) = 0$ as in Lemma 4.5.1. Note first that $\log\operatorname{card}(E_n) = H_{\nu_n}(\xi^f_{-n})$ since each $C \in \xi^f_{-n}$ contains at most one $x \in E_n$, so there are $\operatorname{card}(E_n)$ elements of ξ^f_{-n} with ν_n-measure $1/\operatorname{card}(E_n)$. Now suppose $0 < q < n$ and let $a(k) := [(n-k)/q]$ be the integer part of $(n-k)/q$ whenever $0 \le k < q$. Then $\{0, 1, \dots, n-1\} = \{k + rq + i \mid 0 \le r < a(k), 0 < i \le q\} \cup S$, where $S = \{0, 1, \dots, k, k+a(k)q+1, \dots, n-1\}$, and $\operatorname{card}(S) \le 2q$ since $k+a(k)q \ge n-q$ by definition of $a(k)$. Consequently

$$\xi^f_{-n} = \Big(\bigvee_{r=0}^{a(k)-1} f^{-(rq+k)}(\xi^f_{-q})\Big) \vee \Big(\bigvee_{i\in S} f^{-i}(\xi)\Big)$$

and

$$\begin{aligned}\log\operatorname{card}(E_n) = H_{\nu_n}(\xi^f_{-n}) &\le \sum_{r=0}^{a(k)-1} H_{\nu_n}(f^{-(rq+k)}(\xi^f_{-q})) + \sum_{i\in S} H_{\nu_n}(f^{-i}(\xi))\\ &\le \sum_{r=0}^{a(k)-1} H_{f_*^{rq+k}\nu_n}(\xi^f_{-q}) + 2q\log\operatorname{card}(\xi)\end{aligned} \tag{4.5.1}$$

by Proposition 4.3.3(1) and (4). Thus by Proposition 4.3.3(6)

$$\begin{aligned} q\log\operatorname{card}(E_n) = \sum_{k=0}^{q-1} H_{\nu_n}(\xi^f_{-n}) &\le \sum_{k=0}^{q-1}\Big(\sum_{r=0}^{a(k)-1} H_{f_*^{rq+k}\nu_n}(\xi^f_{-q}) + 2q\log\operatorname{card}(\xi)\Big)\\ &\le nH_{\mu_n}(\xi^f_{-q}) + 2q^2\log\operatorname{card}(\xi).\end{aligned}$$

Thus $\overline{\lim}_{n\to\infty}(1/n)\log\operatorname{card}(E_n) \le \lim_{k\to\infty} H_{\mu_{n_k}}(\xi^f_{-q})/q = H_\mu(\xi^f_{-q})/q$, so that $\overline{\lim}_{n\to\infty}(1/n)\log\operatorname{card}(E_n) \le h_\mu(f,\xi) \le h_\mu(f)$. □

We now prove the Variational Principle, namely, that topological entropy is the supremum of metric entropies.

Theorem 4.5.3. (Variational Principle) *If $f\colon X \to X$ is a homeomorphism of a compact metric space (X, d), then $h_{\mathrm{top}}(f) = \sup\{h_\mu(f) \mid \mu \in \mathfrak{M}(f)\}$.*

Proof. If $\xi = \{C_1, \dots, C_k\}$ is a measurable partition of X then, since μ is a Borel measure (cf. Definition A.6.6), $\mu(C_i) = \sup\{\mu(B) \mid B \subset C_i \text{ closed}\}$. Thus we can choose compact sets $B_i \subset C_i$ such that for $\beta = \{B_0, B_1, \dots, B_k\}$ with $B_0 = X \setminus \bigcup_{i=1}^k B_i$ we have $H(\xi \mid \beta) < 1$. (Think of $\{B_1, \dots, B_k\}$ as "islands" and B_0 as the "sea".) By Proposition 4.3.10(3)

$$h_\mu(f,\xi) \le h_\mu(f,\beta) + H_\mu(\xi \mid \beta) \le h_\mu(f,\beta) + 1.$$

Now $\mathcal{B} := \{B_0 \cup B_1, \dots, B_0 \cup B_k\}$ is an open cover of X. By Proposition 4.3.3(1) $H_\mu(\beta^f_{-n}) \le \log\operatorname{card}\beta^f_{-n} \le \log(2^n \operatorname{card}\mathcal{B}^f_{-n})$. If δ_0 is the Lebesgue number of

$\mathcal{B}$, that is, the supremum of $\delta > 0$ such that every δ-ball is contained in an element of $\mathcal{B}$, then δ_0 is also the Lebesgue number of $\mathcal{B}^f_{-n}$ with respect to the metric d^f_n. Since $\mathcal{B}^f_{-n}$ is a minimal cover every $C \in \mathcal{B}^f_{-n}$ contains a point x_C not in any other element of $\mathcal{B}^f_{-n}$. The x_C form a δ_0-separated set. Consequently $h_\mu(f, \beta) \le h_{\text{top}}(f) + \log 2$ and $h_\mu(f, \xi) \le h_\mu(f, \beta) + 1 \le h_{\text{top}}(f) + \log 2 + 1$. Therefore by using Proposition 4.3.16(4) and Proposition 3.1.7(3) we obtain $h_\mu(f) = h_\mu(f^n)/n \le (h_{\text{top}}(f^n) + \log 2 + 1)/n = h_{\text{top}}(f) + (\log 2 + 1)/n$ for all $n \in \mathbb{N}$, and hence $h_\mu(f) \le h_{\text{top}}(f)$.

On the other hand applying Lemma 4.5.2 to maximal (n, ϵ)-separated sets in X yields

$$\overline{\lim_{n\to\infty}} \frac{1}{n} \log N_X(f, \epsilon, n) \le h_\mu(f)$$

for a corresponding accumulation point $\mu \in \mathfrak{M}(f)$. Thus

$$\overline{\lim_{n\to\infty}} \frac{1}{n} \log N_X(f, \epsilon, n) \le \sup_\mu h_\mu(f)$$

and letting $\epsilon \to 0$ proves the Variational Principle. □

Remark. Note that if f is expansive and δ_0 is the expansivity constant then $\overline{\lim}_{n\to\infty}(1/n) \log N_X(f, \epsilon, n) = h_{\text{top}}(f)$ for $\epsilon < \delta_0$ by Corollary 3.2.13. Consequently the proof of the Variational Principle immediately yields:

Theorem 4.5.4. *Expansive maps of compact metric spaces have a measure of maximal entropy.*

A very natural question arises concerning conditions for *uniqueness* of such a measure. Obviously one can take the union of several disjoint copies of the same expansive system, which is obviously expansive, or of several different systems with the same entropy and by Proposition 3.1.7(2) and Proposition 4.3.16(2) there will be more than one measure with maximal entropy. Adding topological transitivity does not help (Exercise 4.5.2). However, as we will see in Section 20.1 for a large natural class of expansive dynamical systems, which in particular includes all transitive topological Markov chains, hyperbolic toral automorphisms, horseshoes, expanding maps, and so forth, the invariant measure with maximal entropy is indeed unique.

Exercises

4.5.1. *Construct an example of a homeomorphism of a compact metric space with finite topological entropy that does not have a measure with maximal entropy.*

4.5.2. *Construct an example of a topologically transitive expansive homeomorphism of a compact metric space that has more than one measure with maximal entropy.*

5

Systems with smooth invariant measures and more examples

Certain classes of invariant measures are natural for smooth systems. Those are absolutely continuous measures, that is, measures that are given by densities in local coordinate charts. In the first section we establish general criteria for the existence of such measures in three classes of dynamical systems: discrete-time invertible and noninvertible and continuous-time invertible systems. We demonstrate how those criteria can be used to show existence and uniqueness for smooth invariant measures for expanding maps. In the rest of this chapter we describe several classes of dynamical systems that arise from classical mechanics and differential geometry. Due to the presence of an extra structure all these systems preserve a naturally defined smooth invariant measure. Along the way we enrich our collection of standard examples by a few interesting items.

1. Existence of smooth invariant measures

a. The smooth measure class. The Krylov–Bogolubov Theorem 4.1.1 establishes the existence of an invariant measure for any topological dynamical system on a compact metrizable space. Corollary 4.1.4 says that an invariant measure reflects the asymptotic behavior of almost all points with respect to that measure. Yet it is unclear just how much information this really gives. For example, if there is a periodic point, then one can consider the measure concentrated on the periodic orbit. It is invariant and trivially reflects the asymptotics of the periodic point, but gives no information about any other point. Although in the next chapter we will obtain asymptotic information about the behavior of some orbits in a neighborhood of a periodic point, this is still not sufficient to make conclusions about the asymptotic behavior of "most" orbits. On the other hand, we naturally picked Lebesgue measure when studying the smooth examples of Sections 4.2a, c, e. In general, a dynamical system may have a great abundance of invariant measures, many of which possess some complex behavior. This was the case in the example of the full shift in Section 4.2f, which possesses Bernoulli and Markov measures, measures concentrated on periodic orbits, and, in fact, many others. In that case one may be interested in being able to select some of these as being more relevant for understanding the dynamics. We would like to be able to decide whether an invariant measure reflects the asymptotic behavior of orbits in a significant way, and to be able to choose for consideration the more significant measures among the invariant measures present. The special Markov measure (Parry measure) for a transitive topological Markov chain constructed in Section 4.4c is an example of such a dynamically significant measure.[1]

On the other hand, for a smooth dynamical system on a manifold M there is at least one natural meaning for the notion of a significant measure. Namely, the smooth structure defines a *measure class* which is invariant under diffeomorphisms and many other differentiable maps. Equivalently, there is naturally a collection of sets of measure zero that is invariant under such maps. This is simply the collection of sets A such that for any smooth local chart (U, φ) the set $\varphi(U \cap A) \subset \mathbb{R}^n$ has Lebesgue measure zero.

Definition 5.1.1. A measure μ on a differentiable manifold is called *absolutely continuous* if in any smooth local chart it is given by integrating a density. Such a measure is called *positive* if the density is almost everywhere positive in any chart. It is called *smooth* positive if the density is a smooth positive function.

All positive measures are equivalent, that is, they have the same collection of null sets. Any absolutely continuous measure is absolutely continuous with respect to any positive measure. The class of positive measures is invariant under diffeomorphisms as well as under surjective differentiable maps that are not too degenerate, that is, for maps whose Jacobian, the determinant of the matrix of partial derivatives in local coordinates, vanishes only on a null set.

If the manifold in question is orientable then these classes of measures may be introduced via differential n-forms. In particular any smooth positive measure is obtained by integrating a smooth nondegenerate n-form and vice versa. On a nonorientable manifold M there are no nondegenerate n-forms. There are two ways to use infinitesimal language in this case. One can either introduce the oriented double cover $\pi: M_0 \to M$ with the involution I which commutes with π and consider nondegenerate n-forms ω on M_0 such that $I_*\omega = -\omega$. Alternatively one can introduce a notion of an *odd* n-form which is a function on the nth exterior power of the tangent bundle TM such that a linear change of coordinates by a matrix A multiplies the value of the form by $|\det A|$. See Exercises 5.1.1–5.1.3 for a justification of these procedures. An odd n-form is called *nondegenerate* if it is nonzero on a basis.

There is a simple connection between uniqueness of an invariant measure in a certain class and ergodicity. Namely, let $T: (M, \mu) \to (M, \mu)$ be a measure-preserving transformation and ν another measure which is absolutely continuous with respect to μ with density ρ. If ν is T-invariant and f is measurable then

$$\int \rho f \, d\mu = \int f \, d\nu = \int f \circ T \, d\nu = \int \rho (f \circ T) \, d\mu = \int \rho \circ T^{-1} f \, d\mu,$$

that is, $\rho \circ T^{-1} = \rho$. If μ is ergodic then this implies that $\rho = \text{const.}$ and if both μ and ν are probability measures then this means that $\rho = 1$. Thus we have

Proposition 5.1.2. *If μ is an ergodic T-invariant probability measure then T has no other invariant probability measure that is absolutely continuous with respect to μ.*

For smooth dynamical systems on a smooth manifold it is natural to ask the following question: Does a given smooth dynamical system have an absolutely continuous or smooth positive invariant measure?

In this section we give a framework for investigating this question as well as some conditions for existence of such measures and present one nontrivial application. We will restrict our discussion to the case of orientable manifolds and orientation-preserving maps and leave the nonorientable case to the exercises.

b. The Perron–Frobenius operator and divergence. For a smooth measure, invariance under a diffeomorphism f becomes a differential (infinitesimal) condition. If Ω is an invariant volume, that is, an invariant nondegenerate n-form, then

$$\int_A \Omega = \mu(A) = \mu(f^{-1}(A)) = \int_{f^{-1}(A)} \Omega = \int_A f_* \Omega,$$

that is, $f_*\Omega = \Omega$, where $f_*\Omega(v_1, \dots, v_n) = \Omega(Df^{-1}v_1, \dots, Df^{-1}v_n)$ is the *pushforward* of Ω under f. One may write this equivalently as follows. Suppose Ω is a volume form, but not necessarily invariant. In analogy to calculus we would like to define the *Jacobian* $Jf \colon M \to \mathbb{R}$ of f with respect to Ω by the volume distortion, namely, by setting

$$\int_A (Jf) \, \Omega = \int_{f(A)} \Omega,$$

which amounts to saying that $(Jf)\Omega = f^*\Omega$, where $f^*\Omega$ is the *pullback* of Ω by f defined by $f^*\Omega(v_1, \dots, v_n) = \Omega(Dfv_1, \dots, Dfv_n)$. This is reasonable since we have:

Proposition 5.1.3. *Let $\Omega = dx_1 \wedge \cdots \wedge dx_n$ be the standard volume form on $\mathbb{R}^n$. If f is a diffeomorphism, then $f^*\Omega = (\det Df)\Omega$.*

Proof.

$$f^*\Omega = \Omega(Df(\cdot), Df(\cdot), \dots, Df(\cdot), Df(\cdot)) = f^* dx_1 \wedge f^* dx_2 \wedge \dots \wedge f^* dx_n$$
$$= \Big(\sum_{i=1}^n \frac{\partial f_1}{\partial x_i} dx_i\Big) \wedge \dots \wedge \Big(\sum_{i=1}^n \frac{\partial f_n}{\partial x_i} dx_i\Big)$$
$$= \sum_{i_1,\dots,i_n=1}^n \frac{\partial f_1}{\partial x_{i_1}} \cdots \frac{\partial f_n}{\partial x_{i_n}} dx_{i_1} \wedge \dots \wedge dx_{i_n}$$
$$= \Omega \sum_{\sigma \in S(n)} \frac{\partial f_1}{\partial x_{\sigma(1)}} \frac{\partial f_2}{\partial x_{\sigma(2)}} \cdots \frac{\partial f_n}{\partial x_{\sigma(n)}} = (\det Df)\Omega.$$

□

Thus we now define the Jacobian accordingly:

Definition 5.1.4. Let M be a manifold, Ω a volume form, and $f: M \to M$ differentiable (not necessarily a diffeomorphism). Then the *Jacobian* Jf of f with respect to Ω is the unique function on M defined by $Jf\Omega = f^*\Omega$, that is,

$$Jf(x) \cdot \Omega_x = \Omega_{f(x)}(Df(\cdot), \dots, Df(\cdot)).$$

Invariance of a volume form Ω is thus described by requiring $f^*\Omega = \Omega$, that is, $Jf = 1$.

Suppose now that f is a diffeomorphism and $\rho: M \to \mathbb{R}_+$ is the density of an f-invariant measure with respect to Ω. Then the preceding calculation, repeated with $Jf\rho\Omega$ instead of Ω, shows that we must have $Jf\rho = \rho \circ f^{-1}$ or

$$\rho \circ f = \frac{\rho}{Jf}.$$

In the noninvertible case one sees similarly that an invariant density must satisfy

$$\rho(x) = \sum_{y \in f^{-1}(\{x\})} \frac{\rho(y)}{Jf(y)}.$$

We have thus proved

Proposition 5.1.5. *Let M be a smooth manifold, Ω a volume form, $f: M \to M$ differentiable, and $\rho: M \to \mathbb{R}$ the density of an absolutely continuous f-invariant measure. Then $\rho(x) = \sum_{y \in f^{-1}(\{x\})} \dfrac{\rho(y)}{Jf(y)}$ for all $x \in M$. In particular if f is a diffeomorphism then $\rho \circ f = \rho / J(f)$.*

If ρ does not vanish then the latter relation can be rewritten as $Jf = \rho/\rho \circ f$ and by iterating this condition over a periodic orbit we obtain

Proposition 5.1.6. *Let M be a smooth manifold, Ω a volume form, and $f: M \to M$ a diffeomorphism with an absolutely continuous f-invariant measure with density $\rho: M \to \mathbb{R}_+$ which is defined everywhere and positive. Then $Jf^n(x) = 1$ for every $x \in \mathrm{Fix}(f^n)$.*

Proposition 5.1.5 motivates the following definition:

Definition 5.1.7. Let M be a smooth manifold with a volume form Ω, and $f: M \to M$ a surjective differentiable map. The *Perron–Frobenius operator* associated with f is the operator on nonnegative measurable functions defined by

$$(\mathcal{F}\rho)(x) = \sum_{y \in f^{-1}(\{x\})} \frac{\rho(y)}{Jf(y)}.$$

We have thus observed that invariant densities are fixed points of the Perron–Frobenius operator.

An even simpler condition describes invariance of a volume form under a flow. Since a smooth flow φ^t is a 1-parameter family of diffeomorphisms, we can clearly adopt the previous discussion of invariant forms to say that a volume form Ω is φ^t-invariant if and only if we have $(\varphi^t)^*\Omega = \Omega$ for all t. By differentiating this expression at $t = 0$ we obtain $\frac{d}{dt}(\varphi^t)^*\Omega = 0$, that is,

$$\mathcal{L}_v \Omega = 0,$$

where $v = \frac{d}{dt}\varphi^t\big|_{t=0}$ is the vector field generating φ and $\mathcal{L}$ is the Lie derivative.

Definition 5.1.8. The *divergence* of a vector field v with respect to a volume form Ω is defined to be the unique function $\operatorname{div} v$ satisfying

$$\mathcal{L}_v \Omega = \operatorname{div} v \cdot \Omega.$$

Proposition 5.1.9. *Let M be a manifold with a volume form Ω, φ^t a flow, and $v := \dot{\varphi}^t$. Then $\operatorname{div} v = 0$ if and only if φ^t preserves Ω.*

This proposition is true essentially by definition. Let us show that in $\mathbb{R}^n$ the definition of divergence coincides with the familiar one by partial derivatives.

Proposition 5.1.10. *Let $\Omega = dx_1 \wedge \cdots \wedge dx_n$ be the standard volume form on $\mathbb{R}^n$. If v is a vector field on $\mathbb{R}^n$ then $\mathcal{L}_v \Omega = \operatorname{div} v \cdot \Omega$.*

Proof.

$$\mathcal{L}_v \Omega = d(\Omega \lrcorner v) + d\Omega \lrcorner v = d(\Omega \lrcorner v) = d(\sum_{i=1}^{n} v_i dx_1 \wedge \cdots \wedge \widehat{dx_i} \wedge \cdots \wedge dx_n)$$

$$= \sum_{i=1}^{n} \frac{\partial v_i}{\partial x_i} dx_1 \wedge \cdots \wedge dx_i \wedge \cdots \wedge dx_n = \Omega \sum_{i=1}^{n} \frac{\partial v_i}{\partial x_i} = (\operatorname{div} v)\Omega.$$

□

From now on suppose M is a compact orientable manifold. For a smooth volume form Ω we denote by $\mathrm{Diff}^r(M,\Omega)$ the space of Ω-preserving C^r diffeomorphisms with the induced C^r topology. It is evidently a closed subset of $\mathrm{Diff}^r(M)$. With a certain effort one can also show that it possesses a local Banach structure. In the case of flows let $\Gamma^r(TM,\Omega) \subset \Gamma^r(TM)$ be the set of vector fields whose flows preserve Ω. Since these are exactly the divergence-free C^r vector fields they form a closed linear subspace of $\Gamma^r(TM)$ and hence a Banach space.

We now establish a connection between invariant volumes for a flow and invariant volumes for the return map to a section for the flow (cf. Section 0.3). This is a local fact in that we need only consider a local section and obtain an invariant volume form ω, also called a *flux* on the local section from a local construction involving the volume Ω on the manifold.

Proposition 5.1.11. *Let M be a manifold, $\varphi^t: M \to M$ a differentiable flow, and Ω an invariant volume form. If τ is a disk transversal to the flow then the volume form $\omega = \Omega \lrcorner X$ on τ, where $X = \dot{\varphi}^t$ is the vector field generating φ^t, is invariant under the return map.*

Proof. For $p \in \tau$ consider a frame $(X, v_2, \dots, v_n)$ at p with $(v_2, \dots, v_n)$ a frame for $T_p\tau$. The image $(w_2, \dots, w_n)$ of $(v_2, \dots, v_n)$ under the differential of the return map to τ is related to the image $(X, w_2', \dots, w_n')$ of $(X, v_2, \dots, v_n)$ under the flow by $w_i' - w_i = c_i \cdot X$. Thus $\Omega(X, w_2', \dots, w_n') = \Omega(X, v_2, \dots, v_n)$ using multilinearity and the fact that Ω is an alternating form. Since X and Ω are φ^t-invariant, this shows that $\Omega \lrcorner X$ is preserved by the return map. □

c. Criteria for existence of smooth invariant measures. The considerations of the previous subsection also yield necessary conditions for existence of a smooth invariant measure beyond Proposition 5.1.6. Since we observed earlier that for the density ρ with respect to Ω of an invariant measure we have $Jf = \rho/\rho \circ f$, we immediately see that we have

Proposition 5.1.12. *Let M be a smooth manifold, Ω a volume form, and $f: M \to M$ a diffeomorphism. If there exists an absolutely continuous f-invariant measure with positive continuous density then $\{Jf^n(x) \mid n \in \mathbb{Z}, x \in M\}$ is bounded.*

Proof. If ρ is the density then $Jf^n(x) = \dfrac{\rho(x)}{\rho(f^n(x))} \le \dfrac{\max_{x\in M}\rho(x)}{\min_{x\in M}\rho(x)}$. □

Having found necessary conditions for existence of smooth invariant measures, we would like to see whether there are sufficient conditions as well. If one relaxes the assumptions on the density to it being a positive Borel function, then there is a sufficient condition:

Theorem 5.1.13. *Suppose (M,Ω) is a manifold with volume form Ω and $f: M \to M$ an orientation-preserving diffeomorphism. If $\{Jf^n(x) \mid n \in \mathbb{Z}\}$ is bounded for almost every $x \in M$ then there exists a Borel function $\omega: M \to \mathbb{R}_+$*

such that $\omega \geq 1/Jf$ and $\omega\Omega$ is f-invariant. If $\{Jf^n(x) \mid n \in \mathbb{Z}\}$ is uniformly bounded for all x and n then ω is bounded.

Proof. Write $\Phi(x) := -\log Jf(x)$ and suppose $e^{\varphi}\Omega$ is f-invariant. Then

$$\begin{aligned} 0 = (f^* e^{\varphi}\Omega)_x - (e^{\varphi}\Omega)_x &= e^{\varphi(f^{-1}(x))}\Omega_{f^{-1}(x)}(Df^{-1}(\cdot), \ldots, Df^{-1}(\cdot)) - e^{\varphi(x)}\Omega_x \\ &= e^{\varphi(f^{-1}(x))}(Jf(x))^{-1}\Omega_x - e^{\varphi(x)}\Omega_x = (e^{\varphi(f^{-1}(x))}e^{\Phi(x)} - e^{\varphi(x)})\Omega. \end{aligned}$$

So we need to solve the cohomological equation

$$\varphi(f^{-1}(x)) - \varphi(x) = -\Phi(x). \tag{5.1.1}$$

From the proof of Theorem 2.9.3 we know that

$$\varphi(x) := \sup_{n \in \mathbb{N}} \sum_{i=0}^{n} \Phi(f^{-i}(x)) \geq \Phi(x), \tag{5.1.2}$$

which is a well-defined Borel function and finite almost everywhere, solves (5.1.1). Thus $\omega := e^{\varphi} \geq e^{\Phi} = 1/Jf$ is the desired density. In general it may not be L^1, hence the invariant measure may be infinite. If, however, $Jf^n(x)$ is uniformly bounded in both x and n then Theorem 2.9.3 applies directly, so the function φ defined by (5.1.2) is bounded and hence the invariant measure is finite. □

Remark. The sufficient condition of Theorem 5.1.13 implies the necessary condition of Proposition 5.1.6. For, if $Jf^n(x) = \lambda \neq 1$ for some $x = f^n(x)$ then $Jf^{mn}(x) = \lambda^m$ which goes to ∞ as $m \to \infty$ or $m \to -\infty$.

Theorem 19.2.7 asserts that for a certain class of smooth dynamical systems (Anosov systems; see Definition 6.4.2 and Definition 17.4.2) the latter condition is, in fact, sufficient for the existence of a positive smooth invariant measure and hence by Theorem 5.1.13 implies the former.

Next we derive a similar, albeit considerably more complicated-looking, condition for noninvertible maps and use it to prove existence of absolutely continuous measures for expanding maps of the circle.

Consider a differentiable map $f: M \to M$ such that every point $x \in M$ has only finitely many preimages and a volume form Ω on M.

Definition 5.1.14. A *section of preimages* of x is a finite subset $S \subset M$ such that any infinite semiorbit $\{x_{-n}\}_{n \in \mathbb{N}_0}$, where $x_0 = x$, $f(x_n) = x_{n+1}$, contains exactly one element of S. Let $S = \{y_1, \ldots, y_k\}$ be a section of preimages of x and $n_1, \ldots, n_k$ be defined by $f^{n_i}(y_i) = x$. We shall call the maximum of the n_i the *rank* and the minimum of the n_i the *depth* of S. We define a function Φ by

$$\Phi(S) = \sum_{i=1}^{k} (Jf^{n_i}(y_i))^{-1}, \tag{5.1.3}$$

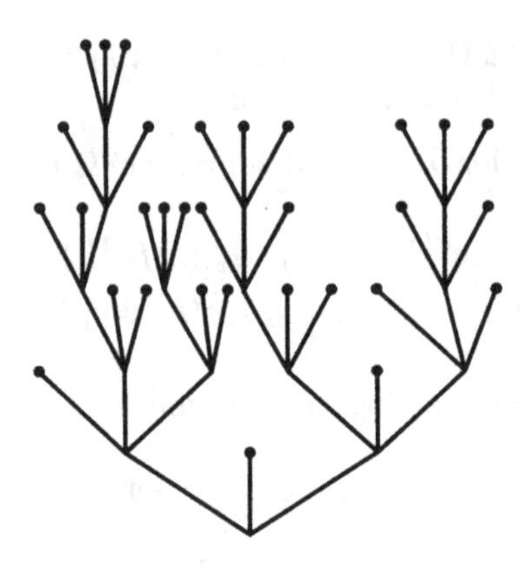

FIGURE 5.1.1. Section of preimages

where the Jacobian is taken with respect to Ω.

Let $x_1, \dots, x_m$ be the preimages of x and let $S_1, \dots, S_m$ be any sections of preimages of the points $x_1, \dots, x_m$, correspondingly. Then $S = \bigcup_{i=1}^m S_i$ is a section of preimages of x. Furthermore an immediate calculation shows that

$$\Phi(S) = \sum_{i=1}^m \frac{\Phi(S_i)}{Jf(x_i)} \tag{5.1.4}$$

and $\operatorname{rank} S = 1 + \max_{1 \le i \le m} \operatorname{rank} S_i$.

Conversely any section of preimages of x of rank greater than zero (that is, any $S \neq \{x\}$) naturally splits into a disjoint union of sections of preimages of $x_1, \dots, x_m$.

Let $\rho_n(x)$ be the supremum of $\Phi(S)$ over all sections of preimages of x with depth at least n and

$$\rho(x) := \lim_{n \to \infty} \rho_n(x) = \inf_{n \to \infty} \rho_n(x).$$

Note that in (5.1.4) $\operatorname{depth}(S_i) \ge \operatorname{depth}(S) - 1$, so

$$\rho(x) = \sum \frac{\rho(x_i)}{Jf(x_i)}.$$

Thus if ρ is finite everywhere (or almost everywhere) then it is a fixed point of the Perron–Frobenius operator (cf. Definition 5.1.7) and hence gives the density of an absolutely continuous f-invariant measure. In order to guarantee that ρ is an L^1 function and hence that $\rho\Omega$ is a finite measure it is sufficient to see that ρ is uniformly bounded. Furthermore if ρ is also bounded from below by a positive number then the measure $\rho\Omega$ is, in fact, equivalent to Ω. Thus we have proved the following counterpart of Theorem 5.1.13 for noninvertible maps:

Theorem 5.1.15. *Let $f: M \to M$ be a differentiable map such that any $x \in M$ has finitely many preimages. If there exist positive constants C_1, C_2 such that for any $x \in M$ and for any section S of preimages of x*

$$C_1 < \Phi(S) < C_2 \tag{5.1.5}$$

then f has an invariant measure $\rho\Omega$, where ρ is a positive bounded function bounded away from zero.

d. Absolutely continuous invariant measures for expanding maps. We will now use Theorem 5.1.15 to show that expanding maps of the circle have an absolutely continuous invariant measure.[2]

Theorem 5.1.16. *Let $f: S^1 \to S^1$ be a C^2 expanding map, that is, $|f'| > 1$. Then f preserves a measure given by a continuous positive density.*

Proof. By passing to f^2 if necessary we may assume that the degree k of f is positive. This is sufficient since for any f^2-invariant measure μ the measure $\mu + f_*\mu$ is f-invariant. Remember from Theorem 2.4.6 that f is topologically conjugate to a linear map E_k. The images Γ_n^m under the conjugacy h of the standard k-nary intervals $\Delta_n^m = [m/k^n, (m+1)/k^n]$, $m = 0, \dots, k^n - 1$, form a partition of S^1. Furthermore, since f is conjugate to E_k, sections over different points can be canonically identified. Namely, given any section $\sigma^0 = \{y_1^0, \dots, y_m^0\}$ of preimages of x^0 there is a unique section $\sigma^1 = \{y_1^1, \dots, y_m^1\}$ of preimages of any other point x^1 that is homotopic to σ^0 in the following sense: If $c: [0,1] \to S^1$ is a curve with $c(0) = x^0$, $c(1) = x^1$ and $f^{n_i}(y_i^0) = x^0$ then σ^1 consists exactly of the endpoints of the curves c_i defined by $c_i(0) = y_i^0$, $f^{n_i} \circ c_i = c$. We denote this situation by $\sigma^0 \sim \sigma^1$. In other words, a section can be defined combinatorially independently of a particular point:

Definition 5.1.17. A *combinatorial section* is a map σ from S^1 to the collection of sections of preimages such that for all $x \in S^1$ $\sigma(x)$ is a section of preimages of x and $\sigma(x) \sim \sigma(y)$ for all $x, y \in S^1$. We define the *total rank n section* σ_n by $\sigma_n(x) = f^{-n}(\{x\})$.

Lemma 5.1.18. *There exist constants $C_1, C_1' \in \mathbb{R}$ such that for $n, m \in \mathbb{N}$ and $x, y \in \Gamma_n^m$ we have*

$$\exp(C_1'|f^n(x) - f^n(y)|) < \frac{(f^n)'(x)}{(f^n)'(y)} < \exp(C_1|f^n(x) - f^n(y)|).$$

Proof. If $x, y \in \Gamma_n^m$ and $x_i := f^i(x)$, $y_i := f^i(y)$ then $|x_i - y_i| < \lambda^{n-i}|x_n - y_n|$, where $\lambda^{-1} > 1$ is a lower bound for f'. Using the Mean Value Theorem we thus have for some point z_i between x_i and y_i:

$$\frac{(f^n)'(x)}{(f^n)'(y)} = \frac{\prod_{i=0}^{n-1} f'(f^i(x))}{\prod_{i=0}^{n-1} f'(f^i(y))} = \prod_{i=0}^{n-1}\left(1 + \frac{f'(x_i) - f'(y_i)}{f'(y_i)}\right)$$

$$= \prod_{i=0}^{n-1}\left(1 + \frac{x_i - y_i}{f'(y_i)} f''(z_i)\right) \le \prod_{i=0}^{n-1}\left(1 + M\lambda^{n-i}|x_n - y_n|\right) \le \exp(C|x_n - y_n|).$$

We used $\log\prod(1+a_n)=\sum\log(1+a_n)\le\sum a_n$. The other inequality follows by symmetry. □

Remark. Inequalities as in this lemma are called *bounded distortion estimates* and will appear several times in the study of one-dimensional systems, both invertible and not.[3] This inequality could be used in place of the C^2 hypothesis of Theorem 5.1.16.

Substituting this inequality into the definition (5.1.3) of Φ, we conclude:

Lemma 5.1.19. *Let σ be a combinatorial section, $x, y \in S^1$. Then*

$$\exp(C_1'|x-y|) < \frac{\Phi(\sigma(x))}{\Phi(\sigma(y))} < \exp(C_1|x-y|),$$

where C_1, C_1' are as before.

Proof. Use the fact that $ca_i \le b_i \le c'a_i$ implies $c\sum a_i \le \sum b_i \le c'\sum a_i$. □

To control the size of Φ in absolute terms we note:

Lemma 5.1.20. *For $n \in \mathbb{N}$ we have $\int_{S^1}\Phi(\sigma_n(x))dx = 1$.*

Proof. $\int_{S^1}\Phi(\sigma_n(x))dx = \int_{S^1}\sum_{f^n(y)=x}\frac{1}{(f^n)'(y)}dx = \int_{S^1}dy = 1$ by a change of variables. □

Corollary 5.1.21. *There exist $C_2 > C_2' > 0$ such that for $n \in \mathbb{N}$ and $x \in S^1$ we have $C_2' < \Phi(\sigma_n(x)) < C_2$.*

Proof. By Lemma 5.1.19 the $\Phi\circ\sigma_n$ are continuous, so by the preceding lemma there exist p_n such that $\Phi(\sigma_n(p_n)) = 1$. Then $\Phi(\sigma_n x) \le \Phi(\sigma_n(p_n))\exp(C_1|x-p_n|) \le e^{C_1}$. Likewise one obtains $C_2' = e^{C_1'}$. □

Next we prove

Lemma 5.1.22. *If S is a section of preimages of x of rank n and $N \ge n$ then*

$$C_2' < \frac{\Phi(S)}{\Phi(\sigma_N(x))} < C_2.$$

Proof. If $y \in S$, $f^l(y) = x$, and we replace y by $f^{l-N}(\{y\}) = \{z_1,\dots,z_s\} \subset \sigma_n(x)$ then the corresponding terms in $\Phi(\sigma_N(x))$ sum to

$$\sum_{i=1}^{s}\frac{1}{(f^N)'(z_i)} = \frac{1}{(f^l)'(y)}\sum_{i=1}^{s}\frac{1}{(f^{N-l})'(z_i)} = \frac{\Phi(\sigma_{l-N}(y))}{(f^l)'(y)},$$

so the total change is within factors C_2, C_2' by Corollary 5.1.21. □

Corollary 5.1.21 and Lemma 5.1.22 yield

Proposition 5.1.23. *There exist $C_3 > C_3' > 0$ such that for all x and all sections S of preimages of x we have*

$$C_3' < \Phi(S) < C_3.$$

Thus by Theorem 5.1.15 we obtain the existence of an invariant measure with positive bounded density bounded away from zero. Continuity (in fact, Lipschitz continuity) of the density follows from Lemma 5.1.19. Namely, notice first that by Lemma 5.1.19 $\log \Phi \circ \sigma$ is Lipschitz continuous (uniformly in σ) and thus by the preceding proposition $\Phi \circ \sigma$ is Lipschitz continuous as well (uniformly in σ). Now $\rho_n(x) = \sup_\sigma \Phi(\sigma(x))$ taken over all combinatorial sections of depth at least n. Thus ρ_n is Lipschitz with the same Lipschitz constant, and the same holds for ρ. □

Proposition 5.1.24. *The invariant measure μ of an expanding map constructed in Theorem 5.1.16 is ergodic.*

Corollary 5.1.25. *Any C^2 expanding map $f\colon S^1 \to S^1$ has a unique absolutely continuous invariant measure.*

Proof. This follows from Proposition 5.1.24 and Proposition 5.1.2. □

Proof of Proposition 5.1.24. We will use the same method as in the second proof of Proposition 4.2.7. The bounded distortion estimate of Lemma 5.1.18 allows us to apply this method in the nonlinear situation.

Suppose $A \subset S^1$ is an f-invariant set of positive Lebesgue measure. Given $\epsilon > 0$ there is an interval $\Gamma_n^m = [a, b]$ such that

$$\lambda(A \cap \Gamma_n^m) > (1-\epsilon)\lambda(\Gamma_n^m) = (1-\epsilon)(b-a), \tag{5.1.6}$$

where λ is Lebesgue measure (because Lebesgue measure is defined via coverings by intervals). By invariance of A we have $f^n(A \cap \Gamma_n^m) \subset A$. By definition of Γ_n^m the nth image of Γ_n^m covers the circle once and hence

$$\int_a^b (f^n)'dx = 1.$$

In particular there exists a $p_n \in \Gamma_n^m$ such that $(f^n)'(p_n) = 1/(b-a)$. Thus by Lemma 5.1.18

$$\max_{\Gamma_n^m} (f^n)' < e^C/(b-a).$$

Integrating $(f^n)'$ over $\Gamma_n^m \setminus A$ and using (5.1.6) we thus obtain

$$\lambda(S^1 \setminus A) \le \max_{\Gamma_n^m} (f^n)' \cdot \lambda(\Gamma_n^m) \le \frac{e^C}{b-a}\epsilon(b-a) = \epsilon e^C.$$

Since ϵ is arbitrary $\lambda(S^1 \setminus A) = 0$ and since the invariant measure μ is absolutely continuous with respect to λ we have $\mu(S^1 \setminus A) = 0$ and μ is ergodic. □

Now we introduce the average expansion rate for the map f:

$$\chi := \int_{S^1} \log |f'(x)| \, d\mu(x).$$

Since $(f^n)'(x) = \prod_{k=0}^{n-1} f'(f^k(x))$ we obtain from Proposition 5.1.24 and Corollary 4.1.9 of the Birkhoff Ergodic Theorem 4.1.2 that for almost every $x \in S^1$

$$\chi = \lim_{n\to\infty} \frac{1}{n} \log(f^n)'(x). \tag{5.1.7}$$

Proposition 5.1.26. $h_\mu(f) = \chi$.[4]

Proof. Again without loss of generality $k = \deg(f) > 0$. Since $\xi_1 := \{\Gamma_1^0, \dots, \Gamma_k^0\}$ is a generator for f it suffices to calculate $h_\mu(f, \xi_1)$. It follows from the construction of the partition ξ_n that $(\xi_1)_{-n}^f = \xi_n$. Thus

$$h_\mu(f, \xi_1) = \lim_{n\to\infty} \Big(- \sum_{i=0}^{k^n-1} \frac{1}{n} \mu(\Gamma_n^i) \log \mu(\Gamma_n^i) \Big). \tag{5.1.8}$$

Since by Theorem 5.1.16 the ratio $\lambda(A)/\mu(A)$ is uniformly bounded from above and below by positive numbers one can replace $\log \mu(\Gamma_n^i)$ on the right-hand side of (5.1.8) by $\log \lambda(\Gamma_n^i)$ without changing the limit. Furthermore, as in the proof of Proposition 5.1.24, by Lemma 5.1.18 $(f^n)'(x)\lambda(\Gamma_n^i)$ is also uniformly bounded on Γ_n^i from above and below by positive numbers, so one can replace $-\log \lambda(\Gamma_n^i)$ in turn by the average of $\log(f^n)'$ on Γ_n^i, that is, by $\int_{\Gamma_n^i} \log(f^n)'(x)dx/\lambda(\Gamma_n^i)$ to get

$$h_\mu(f, \xi_1) = \lim_{n\to\infty} \sum_{i=0}^{k^n-1} \mu(\Gamma_n^i) \int_{\Gamma_n^i} \frac{1}{n} \log(f^n)'(x)dx/\lambda(\Gamma_n^i). \tag{5.1.9}$$

The sequence of functions $\varphi_n(x) = (1/n)\log(f^n)'(x)$ is uniformly bounded and by (5.1.7) converges to χ almost everywhere. This implies that the sequence of averages on the right-hand side of (5.1.9) also converges to χ. □

e. The Moser Theorem. The following result due to Jürgen Moser shows that all smooth positive measures on a compact orientable manifold are equivalent up to a diffeomorphism. This result is useful in various constructions in that it allows us to assume that a diffeomorphism preserving a smooth measure actually preserves any convenient standard volume. Furthermore the proof uses an argument sometimes called a "homotopy trick" which we will use again in the proof of the Darboux Theorem 5.5.9, and in modified form in the local smooth linearization in Section 6.6 and in the smooth orbit classification of area-preserving flows on surfaces in Chapter 14 (see also the proof of the Poincaré Lemma (Theorem A.3.11) in the Appendix).[5]

Theorem 5.1.27. (Moser Theorem) *Let M be a smooth compact orientable manifold and Ω_0 and Ω_1 be two volume forms on M with the same total volume: $\int_M \Omega_0 = \int_M \Omega_1$. Then there exists a diffeomorphism f such that $f^*\Omega_1 = \Omega_0$.*

Proof. Let $\Omega' := \Omega_1 - \Omega_0$. Then $\Omega_t := \Omega_0 + t\Omega'$ is a volume form for $t \in [0,1]$. Furthermore $\int_M \Omega' = 0$ so $\Omega' = d\Theta$ for some $(n-1)$-form Θ by Lemma A.3.13. Since Ω_t is nondegenerate, there is a unique (smooth) vector field X_t such that $\Omega_t \lrcorner X_t = \Omega_t(X_t, \cdot, \dots, \cdot) = -\Theta$. Since M is compact one can integrate X_t to get a 1-parameter family of diffeomorphisms $\{\varphi^t\}_{t\in[0,1]}$ such that $\dot{\varphi}^t = X_t$ and $\varphi^0 = \mathrm{Id}$. Then by (A.3.3)

$$\frac{d}{dt}\varphi^{t*}\Omega_t = \varphi^{t*}(\mathcal{L}_{X_t}\Omega_t) + \varphi^{t*}\frac{d}{dt}\Omega_t = \varphi^{t*}d(\Omega_t \lrcorner X_t) + \varphi^{t*}\Omega' = \varphi^{t*}(-d\Theta + \Omega') = 0,$$

so $\varphi^{1*}\Omega_1 = \varphi^{0*}\Omega_0 = \Omega_0$, that is, φ^1 is the desired coordinate change. □

Exercises

5.1.1. *Let M be an orientable manifold and ω a differential n-form, and define $|\omega|(\xi_1, \dots, \xi_n) := |\omega(\xi_1, \dots, \xi_n)|$ for $x \in M$, $\xi_1, \dots, \xi_n \in T_xM$. Show that $|\omega|$ is an odd n-form.*

5.1.2. *Let M be a smooth manifold, not necessarily compact. Show that there exists an odd n-form on M.*

5.1.3. *Let M be an orientable manifold and α an odd n-form on M. Show that there is an n-form ω on M such that $\alpha = |\omega|$.*

5.1.4. *Prove that if M is a nonorientable manifold, $\pi: M_0 \to M$ is an orientable double cover, and $I: M_0 \to M_0$ is the corresponding involution then there exists a nondegenerate smooth n-form on M_0 such that $I_*\omega = -\omega$ and hence $\pi_*|\omega|$ is an odd n-form on M.*

5.1.5. *Formulate and prove the counterpart of Theorem 5.1.27 for nonorientable manifolds using odd n-forms.*

5.1.6. *Consider a "tentlike" map $f: [0,1] \to [0,1]$ with*
  (1) $f(0) = f(1) = 0$,
  (2) $f(a) = 1$ *for some* $a \in (0,1)$,
  (3) $f'(x) > \lambda > 1$ *when* $0 < x < a$,
  (4) $f'(x) < -\lambda$ *when* $a < x < 1$,
  (5) f *has bounded second derivative away from* a.

Show that f has an invariant measure given by a bounded positive density with respect to Lebesgue measure.

5.1.7. *Prove that if v is a vector field on a smooth manifold M and Ω an n-form then $\rho\Omega$ is invariant under the flow generated by v if and only if $\mathcal{L}_v\rho + \rho \cdot \operatorname{div} v = 0$.*

2. Examples of Newtonian systems

In the rest of this chapter we will discuss several classes of dynamical systems that arise from classical mechanics and differential geometry. Due to the presence of an extra structure all these systems possess a natural smooth invariant volume.

a. The Newton equation. The most basic law of classical mechanics is Newton's Law

$$f = ma$$

describing, for example, the motion in $\mathbb{R}^n$ of a point of mass m under the influence of a force f by giving the acceleration a. This is a description in terms of a second-order ordinary differential equation: If the position of the point is taken to be a point $x \in \mathbb{R}^n$ then the acceleration is $a := \ddot{x} = \frac{d^2x}{dt^2}$. If the force f is supposed to be a function of x only then we get the equation

$$m\frac{d^2x}{dt^2} = f(x).$$

To study such second-order systems of ordinary differential equations it makes sense to reduce the equations to first order by defining the velocity as an extra independent variable by $v := \dot{x} = \frac{dx}{dt} \in \mathbb{R}^n$. Then

$$\begin{aligned} \frac{d}{dt}x &= v \\ \frac{d}{dt}mv &= f(x). \end{aligned}$$

The general solution of this ordinary differential equation defines a dynamical system on $\mathbb{R}^n \times \mathbb{R}^n$ in coordinates (x, v). These equations are divergence free and hence preserve the Lebesgue measure $dxdv$ in the phase space (see Proposition 5.1.10).

The quantity $p := mv$ is called momentum. The *kinetic energy* is given by $\frac{1}{2}m\langle v, v\rangle$.

Whenever the force f is represented by a gradient vector field $f = -\nabla V$, then the function $V\colon \mathbb{R}^n \to \mathbb{R}$ is called *potential energy*. In this case

$$\frac{d}{dt}mv = -\nabla V. \tag{5.2.1}$$

An immediate calculation shows that the *total energy* $H = \frac{1}{2}m\langle v, v\rangle + V$ is preserved:

$$\frac{dH}{dt} = \langle v, m\dot{v}\rangle + \frac{dV}{dt} = \langle v, m\dot{v}\rangle + \langle \dot{x}, \nabla V\rangle = \langle v, m\dot{v} + \nabla V\rangle = 0.$$

Let us discuss several specific examples of this nature.

b. Free particle motion on the torus. Consider the motion of a point mass on the flat torus $\mathbb{T}^n = \mathbb{R}^n/\mathbb{Z}^n$ without external forces. The motion is described by the second-order ordinary differential equation $\ddot{x} = 0$, where x is defined modulo $\mathbb{Z}^n$. Alternatively we can write

$$\dot{x} = v,$$
$$\dot{v} = 0$$

to see that the motion is along straight lines with constant speed, since v is preserved. This means that the n components of v are integrals of motion. For any given v the motion corresponds to the linear flow T_t^v (cf. Section 1.5). Thus we can view the phase space $T\mathbb{T}^n$ as $\mathbb{R}^n \times \mathbb{T}^n$ with dynamics described as follows: The tori $\{v\} \times \mathbb{T}^n$ are invariant and the motion on $\{v\} \times \mathbb{T}^n$ is given by $\{v\} \times T_t^v$. Thus this system is completely integrable and the natural coordinates for this system are action–angle coordinates as discussed in Section 1.5. The Hamiltonian is $H(x, v) = \langle v, v\rangle/2$, kinetic energy, and the nondegenerate 2-form is $\omega = \sum_i dx_i \wedge dv_i$.

c. The mathematical pendulum. Consider a pendulum consisting of a point mass in the plane attached by a rod to a fixed joint. If we take $2\pi x$ to be the angle of deviation from the vertical then the pendulum is described by the differential equation

$$\ddot{x} + \sin 2\pi x = 0,$$

or, equivalently, by the system of first-order ordinary differential equations

$$\begin{aligned} \dot{x} &= v, \\ \dot{v} &= -\sin 2\pi x \end{aligned} \tag{5.2.2}$$

for $x \in S^1$, $v \in \mathbb{R}$. The total energy is given by $H(x, v) = \frac{1}{2}v^2 - \frac{1}{2\pi}\cos 2\pi x$, and is invariant under the flow:

$$\frac{d}{dt}H(x, v) = v\dot{v} + \dot{x}\sin 2\pi x = 0$$

by (5.2.2). The phase space is a cylinder $S^1 \times \mathbb{R}$ and the orbits are on level curves $H = \text{const.}$ For $-1/2\pi < H < 1/2\pi$ each energy level consists of a single closed curve corresponding to oscillations around the stable equilibrium $(x, v) = (0, 0)$. Those orbits are separated from higher-energy orbits corresponding to rotation around the joint by a *homoclinic loop* with $H = 1/2\pi$ containing the unstable equilibrium $(x, v) = (1/2, 0)$. For $H > 1/2\pi$ each energy level consists of two orbits corresponding to rotation in opposite directions.

In the phase space we can write the standard invariant volume $dx \wedge dy$ (that is, area or Lebesgue measure) as $dHdl$, where dH is flow invariant since H is flow invariant, and dl is the length element along the curves $H = \text{const.}$, divided by $\|\nabla H\|$. By (5.2.2), $\|\nabla H\|$ is the speed of motion along $H = \text{const.}$, so dl is also flow invariant.

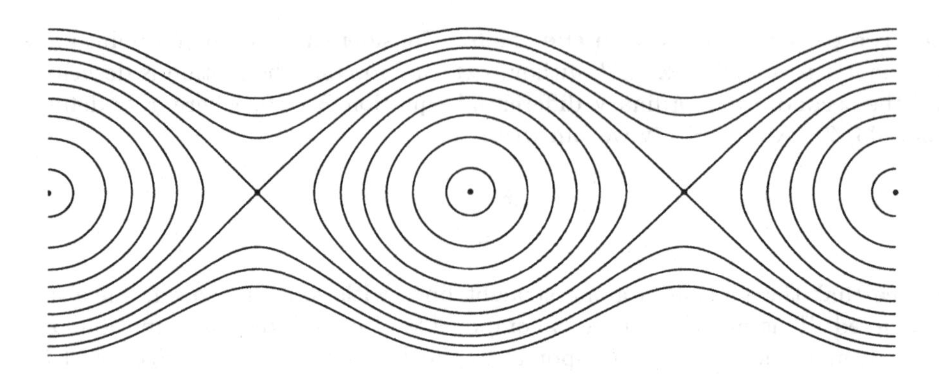

FIGURE 5.2.1. Phase portrait of the mathematical pendulum

d. Central forces.[1] Among the main subjects of classical mechanics is that of celestial mechanics. Let us describe its simplest model. If one considers just two bodies moving freely, but subject to mutual gravitational attraction, then one may either pass to coordinates centered at the center of mass of the system (see Exercise 5.2.2) or assume that one of them (the sun) is much heavier than the other and hence essentially stationary (or rather, moving with constant velocity). Then one can write the position of the second body (the planet) as $x \in \mathbb{R}^3 \setminus \{0\}$ and its velocity as $v \in \mathbb{R}^3$. The potential energy of the gravitational field is given by $V(x) = -1/\|x\|$, so Newton's equation becomes

$$\ddot{x} = \nabla \frac{1}{\|x\|} = -\frac{x}{\|x\|^3}$$

or

$$\begin{aligned} \dot{x} &= v, \\ \dot{v} &= -\frac{x}{\|x\|^3}. \end{aligned}$$

The kinetic energy is $\langle v, v \rangle/2$, as usual. Thus the total energy is $E(x,v) = \langle v, v \rangle/2 - 1/\|x\|$. It is conserved since our equations have the form (5.2.1). Here, as in b, there are other integrals, namely, the components of *angular momentum* given by $x \times v = (x_2v_3 - x_3v_2, x_3v_1 - x_1v_3, x_1v_2 - x_2v_1)$. To check this, note, for example, that

$$\frac{d}{dt}(x_1v_2 - x_2v_1) = \dot{x}_1v_2 + x_1\dot{v}_2 - \dot{x}_2v_1 - x_2\dot{v}_1 = v_1v_2 - \frac{x_1x_2}{\|x\|^3} - v_2v_1 + \frac{x_2x_1}{\|x\|^3} = 0.$$

In fact, this system is completely integrable. We will not introduce the action–angle coordinates as in Section 1.5 directly but we will describe its dynamics by explicitly solving the equations of motion in the spirit of the classical meaning of the notion of "complete integrability". Notice that $v \perp x \times v$, so the motion is in a plane perpendicular to $x \times v$. Thus for any given direction of $x \times v$ the problem reduces to a problem in $\mathbb{R}^2 \setminus \{0\}$, that is, with $x_3 = v_3 = 0$ after a suitable coordinate change.

In passing we note that $x_1 v_2 - x_2 v_1$ is twice the area of the triangle with vertices 0, x, $x + v$. Thus $x_1 v_2 - x_2 v_1$ is twice the derivative of the area swept out by x. The fact of this being constant is known as Kepler's Second Law.

If $A := x_1 v_2 - x_2 v_1 \neq 0$ then we can show that the orbits are on conics. Recall from analytic geometry that in polar coordinates conics are given by $r = ed/(1 + e\cos(\theta - \theta_0))$ with eccentricity $e \in (0, 1)$ for ellipses, $e = 1$ for parabolas, and $e > 1$ for hyperbolas. If we write $r = \|x\|$ then

$$\frac{d}{dt}\left(\frac{x_1}{r}\right) = \frac{v_1 r^2 - x_1 \langle x, v\rangle}{r^3} = -(x_1 v_2 - x_2 v_1)\frac{x_2}{r^3} = A\dot{v}_2,$$

so $Av_2 = x_1/r + C$ for some $C \in \mathbb{R}$. Likewise $Av_1 = -x_2/r - D$. Then $Cx_1 + Dx_2 + r = Ax_1 v_2 - x_1^2/r - Ax_2 v_1 - x_2^2/r + r = A(x_1 v_2 - x_2 v_1) = A^2$, so in polar coordinates $x_1 = r\cos\alpha$, $x_2 = r\sin\alpha$ one has

$$r(\alpha) = \frac{rA^2}{r + Cx_1 + Dx_2} = \frac{A^2}{1 + C\cos\alpha + D\sin\alpha} = \frac{A^2}{1 + \sqrt{C^2 + D^2}\cos(\alpha - \beta)}, \tag{5.2.3}$$

where $\cos\beta = C/\sqrt{C^2 + D^2}$, $\sin\beta = D/\sqrt{C^2 + D^2}$, that is, β is such that $r(\beta)$ is minimal (the *perihelion angle*). Equation (5.2.3) is the equation of a conic with eccentricity $e = \sqrt{C^2 + D^2}$ which is determined by E and A, that is, the values of energy and angular momentum: $e^2 = C^2 + D^2 = (Av_2 - x_1/r)^2 + (Av_1 + x_2/r)^2 = (x_1^2 + x_2^2)/r^2 + 2A^2((v_1^2 + v_2^2)/2) - 2A((x_1 v_2 - x_2 v_1)/r) = 1 + 2EA^2$. Thus the orbit is an ellipse if $E < 0$, a hyperbola if $E > 0$, and a parabola if $E = 0$.

Exercises

5.2.1. *Consider the system of n point masses whose pairwise interaction depends only on their mutual distances, that is, $V(x) = \sum V_{ij}(\|x_i - x_j\|)$. Show that the coordinates of the velocity of the center of gravity and of the angular momentum are first integrals.*

5.2.2. (Two-body problem in the plane) *Show that for the system of two point masses in the plane with interaction as in the previous exercise the four integrals (energy, angular momentum, and coordinates of the velocity of the center of gravity) are independent. Describe the motion relative to the center of gravity.*

5.2.3. *Consider the spherical pendulum, that is, a point mass attached to a point by a rod and subject to gravity. Find the second integral of motion independent of energy and describe the motion for fixed values of both integrals.*

3. Lagrangian mechanics

a. Uniqueness in the configuration space. Lagrangian mechanics, having emerged from Newtonian mechanics, revolves around second-order ordinary differential equations. Therefore before we begin to discuss Lagrangian mechanics we would like to make a comment about second-order ordinary differential equations which will be useful numerous times later. It says that solutions to a second-order ordinary differential equation that start at the same point in different directions will be distinct for some interval of time.

Proposition 5.3.1. *If* $f:\mathbb{R}^n\times\mathbb{R}^n\to\mathbb{R}^n$ *is* C^2, $R>0$, *then there exists* $\epsilon>0$ *such that if for* $i=1,2$ *we have* $v_1\neq v_2\in\mathbb{R}^n$, $\|v_i\|\leq R$, $y_i:\mathbb{R}\to\mathbb{R}^n$, $y_i''=f(y_i,y_i')$, $y_i(0)=0$, $y_i'(0)=v_i$, $t\in(0,\epsilon]$, *then* $y_1(t)\neq y_2(t)$.

Remark. This is not a direct corollary of uniqueness of solutions to an ordinary differential equation, since this proposition asserts uniqueness in the configuration space. Notice that ϵ is independent of the initial conditions.

Proof. Let $g:\mathbb{R}\to\mathbb{R}^n$, $g(t)=y_1''(t)-y_2''(t)$. Since $f\in C^2$ there exist $k,l\in\mathbb{R}$ independent of v_1,v_2 such that $\|g(0)\|=\|f(0,v_1)-f(0,v_2)\|\leq k\|v_1-v_2\|$ and $\|g'(0)\|\leq l\|v_1-v_2\|$. Consequently there exists $\epsilon>0$, independent of v_1,v_2, such that $\|g(t)\|\leq\|v_1-v_2\|/\epsilon$ for $t\in(0,\epsilon)$. Therefore for $t\in(0,\epsilon)$ we have $\|\frac{d}{dt}(y_1-y_2)(t)\|=\|v_1-v_2+\int_0^t g(s)ds\|\geq(1-t/\epsilon)\|v_1-v_2\|>0$. Since $(y_1-y_2)(0)=0$ we have $\|y_1(t)-y_2(t)\|>0$ for $t\in(0,\epsilon]$. □

One way of expressing this result is that the differentiable map

$$\exp:\{v\in\mathbb{R}^n \mid \|v\|\leq R\}\to\mathbb{R}^n,\ v\mapsto y_v(\epsilon),$$

where $y_v(0)=0$, $y_v'(0)=v$, induced by the differential equation is injective, hence a homeomorphism of the R-ball $\{v\in\mathbb{R}^n \mid \|v\|\leq R\}$ onto its image.

b. The Lagrange equation. If we consider the function L given by

$$L(x,v)=\frac{1}{2}m\langle v,v\rangle-V(x) \tag{5.3.1}$$

then the Newton equation (5.2.1) becomes

$$\frac{d}{dt}\frac{\partial L}{\partial v}=\frac{\partial L}{\partial x}\,. \tag{5.3.2}$$

This is called the *Lagrange equation* or *Euler–Lagrange equation.* The reason Lagrange introduced his formalism was that using "$f=ma$" as described earlier can become rather laborious when one considers, for example, constrained systems. For example, a three-dimensional mathematical pendulum consists of a mass attached by a rod to a fixed point and thus constrains the mass point to a sphere (see Exercise 5.2.3). To deal with this, one has to develop notions of constraint forces—forces that are at all times just such that the system will obey the constraint. Here Lagrange's approach greatly simplifies the problem. Constraints are often of the nature that the configurations of the system are restricted to some manifold $M \subset \mathbb{R}^n$. The system is then adequately described by assigning to each point in M a potential energy and to each (tangent) vector a kinetic energy given by a positive definite quadratic form $K(v) = \frac{1}{2}k_x(v, v)$ on TM whose coefficients in local coordinates will depend on the point, that is, a scalar product depending on $x \in M$. Thus fixing a kinetic energy amounts to fixing a Riemannian metric on the configuration space. The Lagrangian now appears in the following form:

$$L(x, v) = \frac{1}{2}k_x(v, v) - V(x). \tag{5.3.3}$$

It is furthermore easy to see that for Lagrangians of the form (5.3.3) the Lagrange equation is invariant under a coordinate change. Let y be a different coordinate such that $x = f(y)$. Then in the tangent space we have $v = \dot{x} = Df\dot{y} = Dfw$. Therefore we can write down the derivatives with respect to y as follows:

$$\begin{aligned}
\frac{\partial L}{\partial y} &= \frac{\partial L}{\partial x}\frac{\partial x}{\partial y} + \frac{\partial L}{\partial v}\frac{\partial v}{\partial y}, \\
\frac{\partial L}{\partial w} &= \frac{\partial L}{\partial v}\frac{\partial v}{\partial w} + \frac{\partial L}{\partial x}\frac{\partial x}{\partial w} = \frac{\partial L}{\partial v}\frac{\partial x}{\partial y}
\end{aligned}$$

since $\frac{\partial x}{\partial w} = 0$. Using the fact that along any curve we have $\frac{d}{dt}x = v$, that is, $\frac{d}{dt}\frac{\partial}{\partial y}x = \frac{\partial v}{\partial y}$, we thus have

$$\begin{aligned}
\frac{d}{dt}\frac{\partial L}{\partial w} - \frac{\partial L}{\partial y} &= \left(\frac{d}{dt}\frac{\partial L}{\partial v}\right)\frac{\partial x}{\partial y} + \frac{\partial L}{\partial v}\left(\frac{d}{dt}\frac{\partial}{\partial y}x\right) - \frac{\partial L}{\partial x}\frac{\partial x}{\partial y} - \frac{\partial L}{\partial v}\frac{\partial v}{\partial y} \\
&= \left(\frac{d}{dt}\frac{\partial L}{\partial v} - \frac{\partial L}{\partial x}\right)\frac{\partial x}{\partial y}
\end{aligned}$$

and hence both sides vanish together since $\frac{\partial x}{\partial y} = Df$ is nonsingular.

c. Lagrangian systems. This allows us to define a global notion of a *Lagrangian dynamical system.* We start with a manifold M which may be called *configuration space.* In general it does not have to be compact, although in most of the following discussion this will be the case. The *phase space* of the dynamical system is the tangent bundle TM. The system itself is determined by a differentiable function $L: TM \to \mathbb{R}$ called the *Lagrangian.* It is given by a second-order ordinary differential equation on M, namely, the first-order ordinary differential equation (5.3.2) on TM. By the previous argument the dynamics determined by this ordinary differential equation is independent of the local coordinate chart chosen to write down (5.3.2). In general this dynamics is only determined for finite time. However, in the case of compact M and L of the form (5.3.3) it is defined for all times and determines a *complete* flow on TM, that is, a flow defined for all t. This can be shown as follows.

Proposition 5.3.2. *For a Lagrangian dynamical system the total energy* $H = \frac{1}{2}k_x(v,v) + V(x)$ *is invariant under the dynamics.*

Proof. Note that the kinetic energy is a homogeneous function of v of degree two, and hence $\langle \partial K/\partial v, v\rangle = 2K$, where $\langle\cdot,\cdot\rangle$ denotes the standard Euclidean inner product. (If $K(v) = \langle A_x v, v\rangle$ then $\langle \partial K/\partial v, w\rangle = \langle A_x v, w\rangle + \langle A_x w, v\rangle$.) Thus $H = \langle \partial L/\partial v, v\rangle - L$ and

$$\frac{d}{dt}H = \langle \frac{d}{dt}\frac{\partial L}{\partial v}, v\rangle + \langle \frac{\partial L}{\partial v}, \dot{v}\rangle - \langle \frac{\partial L}{\partial x}, \dot{x}\rangle - \langle \frac{\partial L}{\partial v}, \dot{v}\rangle = 0$$

since by the Lagrange equation $\frac{d}{dt}\frac{\partial L}{\partial v} = \frac{\partial L}{\partial x}$ and also $\dot{x} = v$ along any curve in the configuration space. □

Corollary 5.3.3. *If the configuration space M is compact then the Lagrange equation defines a global flow in TM.*

Proof. The potential energy is bounded from below by compactness; hence any level set $H =$ const. is compact. By Proposition 5.3.2 any orbit stays in a level set. Therefore for any initial condition on the level set the solution exists for at least a fixed time $\epsilon > 0$ so iteratively any solution can be extended indefinitely. □

The Lagrange formalism provides the basis for the variational description of mechanical systems, which we will present in Section 9.4.

d. Geodesic flows. There is a particular case of Lagrangian system, which we will be interested in later, corresponding to free particle motion in the configuration space.

Definition 5.3.4. Let (M, g) be a Riemannian manifold with Riemannian metric $g_x(\cdot,\cdot)$ and define the Lagrangian

$$L(x, v) = \frac{1}{2}g_x(v, v). \tag{5.3.4}$$

The Lagrangian system on TM corresponding to this Lagrangian as well as its restriction to the unit tangent bundle SM is called the *geodesic flow* of the Riemannian manifold (M, g).

The geodesic flow preserves the length of tangent vectors because the total energy is given by $\frac{1}{2}g_x(v, v)$.

Geodesics are curves on a manifold that are the shortest connection between any two of its points that are sufficiently close. In Section 9.5 we will see that the orbits of the geodesic flow project to geodesics in the configuration space and we will be able to verify the local minimizing property as well as to develop some global analysis of geodesics.

By Proposition 5.3.2 the geodesic flow on any compact manifold is a complete flow. Furthermore, Proposition 5.3.1 shows that geodesics starting at the same point but in different directions do not immediately intersect again. In particular there is a well-defined *exponential map* from a small ball in T_xM to a neighborhood of $x \in M$ defined by sending a tangent vector v to the point $\gamma_v(1)$, where γ_v is the geodesic with $\gamma_v(0) = x$ and $\dot{\gamma}_v(0) = v$. This local diffeomorphism is going to be useful on many occasions. A typical application is that given any basis in T_xM one can find a coordinate chart around x such that the induced basis $\left\{\frac{\partial}{\partial x_1}, \dots, \frac{\partial}{\partial x_n}\right\}$ coincides at x with the given one. Namely, one uses $\exp_x^{-1}$ to get from a neighborhood of the manifold to a ball in the tangent space which can be mapped linearly to $\mathbb{R}^n$.

e. The Legendre transform. Recall that a scalar product in a linear space E defines a natural isomorphism between the space and its dual E^* by assigning to the vector v the form $\langle v, \cdot\rangle$ on E. If we have a Riemannian manifold then via the Riemannian metric this transformation is defined on each tangent space T_xM, so we get a natural identification of the *cotangent bundle* T^*M with TM via a map $\mathcal{L}: TM \to T^*M$. Let us describe this map and the corresponding transformation of the Lagrangian system in local coordinates. Denote the coordinates of a local chart by x_i and recall that the induced coordinates of a vector $v \in T_xM$ are given by the coefficients with respect to the canonical basis $\frac{\partial}{\partial x_i}$ and the coordinates of a form $\omega \in T_x^*M$ are given by the coefficients with respect to the standard basis dx_i dual to $\frac{\partial}{\partial x_i}$. Then

$$\mathcal{L}(x, v) = \left(x, \frac{\partial K}{\partial v}\right), \tag{5.3.5}$$

where, as before, $K(v) = g_x(v, v)$. If the x_i are viewed as coordinates in the configuration space, then the v_i are velocities and the variables $p_i = \partial K/\partial v_i$ are called *momenta*.

Proposition 5.3.5. *The map $\mathcal{L}$ transforms the Lagrange equation into the system of Hamiltonian equations*

$$\dot{p} = -\frac{\partial H}{\partial q}, \qquad \dot{q} = \frac{\partial H}{\partial p}, \tag{5.3.6}$$

where H is the total energy as in Proposition 5.3.2.

Proof. The Lagrange equation can be written as $\dot{p} = \partial L/\partial q$ and $H = \langle p, v\rangle - L$. Thus, on one hand

$$dH = \frac{\partial H}{\partial p} dp + \frac{\partial H}{\partial q} dq$$

while

$$dH = d(p\dot{q} - L) = \dot{q}dp - \frac{\partial L}{\partial q}dq$$

so we get the Hamiltonian equations by comparing coefficients. □

The transformation $\mathcal{L}$ can be defined for a more general class of Lagrangians (notice that we could use a Lagrangian L rather than K in the definition of $\mathcal{L}$), namely for Lagrangians that are C^2 convex functions of v. Thus such a Lagrangian L determines a transformation $\mathcal{L}(x,v) = (x, \partial L/\partial v)$ called the *Legendre transform.* Notice, however, that in this case the Legendre transform is not linear in v any more. A particular case of this situation is described in Exercise 5.3.1.

We showed in Subsection b that the Lagrange equations are invariant under coordinate changes in the configuration space (extended appropriately to the phase space). Hence the Hamiltonian equations are invariant under coordinate changes of the configuration space (extended naturally to the cotangent bundle) as well. More precisely, if $y = f(x)$, then in the tangent space we map

$$(x,v) \mapsto (f(x), Df_x v),$$

and in the cotangent space we map $(x,p) \mapsto (f(x), (Df_x^t)^{-1}p)$, where $A \mapsto A^t$ denotes the transpose. The latter map is a special case of a broader class of transformations that preserve the Hamiltonian equations. The differential two-form $\Omega := dp \wedge dq = \sum dp_i \wedge dq_i$ on T^*M is independent of the choice of coordinates and, as we pointed out in Section 1.5 (equations (1.5.5) and (1.5.6)), the Hamiltonian system is determined by the form Ω and a Hamiltonian H (see Section 5.5c for more details). Thus any transformation preserving Ω preserves the Hamiltonian form of the equations. These are the transformations commonly called *canonical transformations.*

An important qualitative property of Lagrangian and Hamiltonian dynamics is that they preserve a canonically defined volume form. First it immediately follows from the coordinate form (5.3.6) that the Hamiltonian equations are divergence free, so they preserve phase volume in the (x,p)-space, which is, in fact, the nth exterior power of Ω. Going back to the tangent bundle via the inverse of the Legendre transform we see that the invariant volume is given by the product of the volume form on the manifold and the Euclidean volume defined in the tangent space by the Riemannian metric. A Lagrangian system preserves the hypersurfaces $H =$ const., so for every regular value of H it induces an invariant volume form on the hypersurface $H =$ const. This is particularly simple for the case of geodesic flows, where the invariant hypersurfaces are just the sphere bundles $\{\|v\| = \text{const.}\}$ and the invariant volume for the flow is the product of the Riemannian volume and the canonical (angular) volume on the spheres. Thus we have shown the following important fact which will be used on a number of occasions:

Proposition 5.3.6. *The geodesic flow on the unit tangent bundle SM of a complete Riemannian manifold M preserves a smooth measure which is called the Liouville measure. If M is compact then the Liouville measure is finite and can hence be normalized.*

Let us note that geodesic flows are Hamiltonian flows in a natural way. On the one hand the Legendre transform produces from the Lagrangian system given by the geodesic flow a Hamiltonian system in the cotangent bundle. But on the other hand there is a natural identification of tangent bundle and cotangent bundle of a Riemannian manifold. Namely, the correspondence $v \mapsto \langle v, \cdot \rangle$ between vectors and covectors is an isomorphism due to the nondegeneracy of the Riemannian metric. Consequently geodesic flows are Hamiltonian flows on the tangent bundle and hence preserve a smooth volume, and so forth.

Exercises

5.3.1. *Consider a more general class of Lagrangians than (5.3.1), namely,*

$$L(x, v) = K(x, v) - V(x),$$

where K is homogeneous of degree two, twice differentiable away from 0, and convex. Show that the Legendre transform defined by (5.3.5) still defines a diffeomorphism between TM and T^*M away from the zero section.

5.3.2. *The Lagrange transform sends the coordinates (x, v) into the coordinates (q, p). Show that applying it again in these coordinates, with K in (5.3.5) replaced by L, gives the inverse transform.*

4. Examples of geodesic flows

a. Manifolds with many symmetries. In certain cases when a Riemannian manifold possesses a lot of isometries and "symmetries" the geodesic flow (Definition 5.3.4) can be described without explicitly solving the Lagrange equation. We begin with an abstract lemma enabling us to do that, continue with the familiar cases of the round sphere and flat torus, and then proceed to the much more interesting case of the hyperbolic plane and its factors.

Lemma 5.4.1. *Let M be a Riemannian manifold and Γ a group of isometries. Suppose that Γ is transitive on unit vectors, that is, if $v, v' \in SM$ then there exists $\varphi \in \Gamma$ such that $\varphi(v) = v'$. If C is a nonempty family of unit-speed curves $c\colon \mathbb{R} \to M$ with the properties*

(1) *if $c \in C$ and $\varphi \in \Gamma$ then $\varphi \circ c \in C$,*
(2) *if $c, c' \in C$ then there exists $\varphi_{c,c'} \in \Gamma$ such that $\varphi_{c,c'} \circ c = c'$,*
(3) *if $c \in C$ then there exists $\varphi_c \in \Gamma$ such that $\mathrm{Fix}(\varphi_c) = c(\mathbb{R})$,*

then C is the collection of unit-speed geodesics of M.

Proof. To show that C contains all geodesics consider a unit tangent vector v at a point $p \in M$. It determines a unique geodesic γ_v with $\dot{\gamma}_v(0) = v$. We want to show that $\gamma_v \in C$. To that end take some $c \in C$ and let $v' = \dot{c}(0)$. Then there is a $\varphi \in \Gamma$ such that $D\varphi(v') = v$, that is, $c' := \varphi \circ c$ is tangent to γ_v. Now consider $\varphi_{c'} \in \Gamma$. Being an isometry it maps γ_v to a geodesic. It fixes c', so $\varphi \circ \gamma_v$ and γ_v are tangent at p, hence coincide (with parameterization) by uniqueness, that is, $\varphi_c \restriction_{\gamma_v(\mathbb{R})} = \mathrm{Id}$. On the other hand $\mathrm{Fix}(\varphi_{c'}) = c'(\mathbb{R})$, so $\gamma_v = c' \in C$.

Now consider $c \in C$. Since C contains geodesics, property (2) implies that c is the isometric image of a geodesic, hence a geodesic. □

b. The sphere and the torus. Using Lemma 5.4.1 we can now easily describe the geodesics on the sphere and the (flat) torus.

Theorem 5.4.2. *The geodesics on the standard two-sphere are exactly the great circles parameterized with unit speed.*

Proof. The isometry group of the sphere is generated by rotations and reflections in great circles. It is obviously transitive on points: If $x, y \in S^2$ then there is a great circle containing both and an appropriate rotation around the axis perpendicular to the plane of the great circle will send x to y. In addition one can map any unit vector at x to any unit vector at y by subsequently applying a rotation around the axis through y. Rotations and reflections, hence isometries, preserve great circles and can send any great circle to any other with any orientation by the same argument. The theorem follows by Lemma 5.4.1. □

Thus the dynamics of the geodesic flow on the round sphere is extremely simple: All orbits on a fixed energy level are periodic with the same period which is inversely proportional to the square root of the energy.

Theorem 5.4.3. *The geodesics of $\mathbb{R}^2$ are exactly the straight lines.*

Proof. The isometries of $\mathbb{R}^2$ are generated by translations, rotations, and reflections in lines. All of these preserve lines and one can send any tangent vector to any other by translation and rotation, and any parameterized line to any other by translation and rotation. Again Lemma 5.4.1 yields the claim. □

Corollary 5.4.4. *The geodesics on the (flat) torus $\mathbb{T}^2 = \mathbb{R}^2/\mathbb{Z}^2$ are exactly the projections of straight lines.*

We already described the dynamics of the geodesic flow on the flat torus in Section 5.2.b. It is a completely integrable system with frequency vector ω on the invariant torus $\{(x, v) \mid x \in \mathbb{T}^2, v = \omega\}$.

c. Isometries of the hyperbolic plane. Denote by $\mathbb{H}$ the upper half-plane $\mathbb{R} \times \mathbb{R}_+$ in $\mathbb{R}^2$. As an open subset of $\mathbb{R}^2$ this is a smooth manifold. We begin by defining a Riemannian metric on $\mathbb{H}$. If we view $\mathbb{H}$ as

$$\mathbb{H} = \{z \in \mathbb{C} \mid \operatorname{Im} z > 0\}$$

then tangent vectors of $\mathbb{H}$ are naturally written as complex numbers. Thus for $z \in \mathbb{H}$, $u + iv$, $u' + iv' \in T_z\mathbb{H}$ we define

$$\langle u + iv, u' + iv' \rangle_z := \operatorname{Re} \frac{(u + iv)(u' - iv')}{(\operatorname{Im} z)^2}.$$

Since this is clearly symmetric, $\mathbb{R}$-bilinear, and positive definite we have thus defined a Riemannian metric $\langle \cdot, \cdot \rangle$ which is called the *hyperbolic metric*. The half-plane $\mathbb{H}$ endowed with the hyperbolic metric is usually called the *Poincaré upper half-plane*. Viewed as an abstract Riemannian manifold it is sometimes called the *Lobachevsky plane* after the discoverer of the first non-Euclidean geometry. Since the hyperbolic metric differs from the Euclidean metric $\operatorname{Re}(u + iv)(u' - iv')$ by a scalar factor $(\operatorname{Im} z)^2$ only, hyperbolic angles coincide with Euclidean angles.

The principal tool for understanding the geometry of $\mathbb{H}$ will be the isometries of $\mathbb{H}$. We begin with *Möbius transformations*. Denote by $GL_+(2, \mathbb{R})$ the collection of real 2×2 matrices with positive determinant and associate to each $\begin{pmatrix} a & b \\ c & d \end{pmatrix} \in GL_+(2, \mathbb{R})$ the map $T := \psi \begin{pmatrix} a & b \\ c & d \end{pmatrix} : \mathbb{H} \to \mathbb{H}$ defined by

$$T(z) = \frac{az + b}{cz + d}. \tag{5.4.1}$$

Denote the set of maps thus obtained by $\mathcal{M}$. Note that $T'(z) = \dfrac{ad - bc}{(cz + d)^2}$ so

$$\begin{aligned} \operatorname{Im} T(z) = \frac{1}{2i}\left(\frac{az + b}{cz + d} - \frac{a\bar{z} + b}{c\bar{z} + d}\right) &= \frac{(az + b)(c\bar{z} + d) - (a\bar{z} + b)(cz + d)}{2i(cz + d)(c\bar{z} + d)} \\ &= |T'(z)| \operatorname{Im}(z); \end{aligned}$$

and hence T maps the upper half-plane $\mathbb{H} = \{z \in \mathbb{C} \mid \operatorname{Im}(z) > 0\}$ to itself.

Observe that $\mathcal{M}$ is a group under composition and $\psi : GL_+(2, \mathbb{R}) \to \mathcal{M}$ is a homomorphism whose kernel consists of scalar multiples of the identity.

Lemma 5.4.5. *The maps $T \in \mathcal{M}$ are isometries of the hyperbolic metric.*

Proof.

$$\begin{aligned} \langle T'(z)(u + iv), T'(z)(u' + iv') \rangle_{T(z)} &= \operatorname{Re} \frac{T'(z)(u + iv)\overline{T'(z)(u' + iv')}}{(\operatorname{Im} T(z))^2} \\ &= \frac{T'(z)\overline{T'(z)}}{|T'(z)|^2} \operatorname{Re} \frac{(u + iv)(u' - iv')}{(\operatorname{Im}(z))^2} = \langle u + iv, u' + iv' \rangle_z . \end{aligned}$$ □

Note that all $T \in \mathcal{M}$ extend naturally to $\mathbb{H} \cup \mathbb{R} \cup \{\infty\}$ by setting $T(-d/c) = \infty$ and $T(\infty) = a/c$ (or $T(\infty) = \infty$ if $c = 0$). Examples of Möbius transformations are $z \mapsto -1/z$, $z \mapsto z + b$ ($b \in \mathbb{R}$), and $z \mapsto az$ ($a > 0$). They represent correspondingly three types of Möbius transformation from the point of view of the intrinsic geometry of the Lobachevsky plane: *elliptic* (direct counterparts of Euclidean rotations), with a single fixed point inside the plane, *parabolic*, with no fixed points on the plane and no invariant geodesic, and *hyperbolic*, with no fixed points but a unique fixed geodesic (the axis). On $\mathbb{H}$ a parabolic map has a unique fixed point on $\mathbb{R} \cup \{\infty\}$ and a hyperbolic map has two fixed points on $\mathbb{R} \cup \{\infty\}$. Both parabolic and hyperbolic maps are counterparts of parallel translations of the Euclidean plane.

There are also isometries other than Möbius transformations. Clearly $z \mapsto -\bar{z}$ and $z \to 1/\bar{z}$ are examples. Geometrically the former is the reflection in the imaginary axis and the latter is the inversion with respect to the unit circle. We will use Möbius transformations now to study geodesics. We first examine isometric images of the imaginary axis I (parameterized with unit speed by $t \mapsto ie^t$).

Lemma 5.4.6. *If C is any vertical line or a semicircle with center on the real line then there exists a $T \in \mathcal{M}$ with $TI = C$. Furthermore, given any unit tangent vector v at a point of C one can take T such that it maps the upward vertical vector i at $i \in I$ to v.*

Proof. If C is the vertical line $\{z \mid \mathrm{Re}(z) = b\}$ take $T(z) = z + b$. If C is a semicircle with endpoints $x, x+r \in \mathbb{R}$ then note that $T_1 \colon z \mapsto z/(z+1)$ maps I to the semicircle with endpoints 0 and 1 (since $\left|\dfrac{it}{1+it} - \dfrac{1}{2}\right| = \left|\dfrac{2it - (1+it)}{2(1+it)}\right| = \dfrac{1}{2}$) and let $T_2(z) = rz$, $T_3(z) = z + x$, and $T = T_3 \circ T_2 \circ T_1$. In order to map tangent vectors as desired note that there is a Möbius transformation T_0 such that $DT_0(\mathrm{i}) = DT^{-1}(v)$, namely, either $T_0(z) = cz$ or $T_0(z) = -\dfrac{c}{z}$ for some $c \in \mathbb{R}_+$. But then $T \circ T_0$ is as desired. □

Remark. Consider a vertical line or a circle with center on the real axis. Take any unit-speed parameterization of this curve C. We have shown that if I is parameterized by $t \mapsto ie^t$ then there is a Möbius transformation that maps C to I preserving the parameterization.

Corollary 5.4.7. *If $v \in T_z\mathbb{H}$, $w \in T_{z'}\mathbb{H}$, $\|v\| = \|w\| = 1$, then there exists a $T \in \mathcal{M}$ such that $T(z) = z'$ and $T'(z)v = w$, that is, $\mathcal{M}$ acts transitively on the unit tangent bundle to $\mathbb{H}$.*

Let us show that conversely every Möbius transformation maps lines and circles to lines and circles:

Proposition 5.4.8. *If C is a vertical line or a circle with center on the real axis and φ is a Möbius transformation or $\varphi(z) = -\bar{z}$ then $\varphi(C)$ is a vertical line or a circle with center on the real axis.*

Proof. For $\varphi(z) = -\bar{z}$ this is clear. For Möbius transformations note that it suffices to check that the image of a vertical line is a vertical line or a circle with center on the real axis: If C is a vertical line or a circle with center on the real axis and M a Möbius transformation then there is another Möbius transformation N such that N maps the imaginary axis to C. But then it suffices to know that the Möbius transformation $M \circ N$ maps the imaginary axis to a vertical line or a circle with center on the real axis.

Now $\dfrac{az+b}{cz+d} = \dfrac{a}{c} + \dfrac{(bc-ad)/c^2}{z+d/c}$ whenever $c \neq 0$ and $\dfrac{az+b}{d} = \dfrac{a}{d}z + \dfrac{b}{d}$, that is, every Möbius transformation is a composition of maps of the form $z \mapsto z+\alpha$, $z \mapsto \beta z$, and $z \mapsto -1/z$. Thus it suffices to check the assertion for each of these separately. The former two map lines to lines and circles to circles and the latter maps $x + i\mathbb{R}$ to a circle: $\left| -\dfrac{1}{x+it} + \dfrac{1}{2x} \right| = \left| \dfrac{1}{2x} \dfrac{x-it}{x+it} \right| = \left| \dfrac{1}{2x} \right|$. □

d. Geodesics of the hyperbolic plane. Using Lemma 5.4.1 we now have a description of the geodesics of the Lobachevsky plane:

Theorem 5.4.9. *The geodesics of the Poincaré upper half-plane are precisely the vertical lines and the circles with center on the real axis.*

Proof. Consider the group Γ generated by the group $\mathcal{M}$ of Möbius transformations and the transformation $S: x \mapsto -\bar{z}$. It is transitive on unit vectors. Let $\mathcal{C}$ be the family of vertical lines and circles with center on the real axis with all possible unit-speed parameterizations. Then Proposition 5.4.8 and Lemma 5.4.6 show items (1) and (2) of Lemma 5.4.1. Item (3) of Lemma 5.4.1 follows by observing that S is an isometry fixing precisely the imaginary axis I. Thus for $c \in \mathcal{C}$ we can take $\varphi_c = \varphi_{c,I} S \varphi_{I,c}$ to prove (3). □

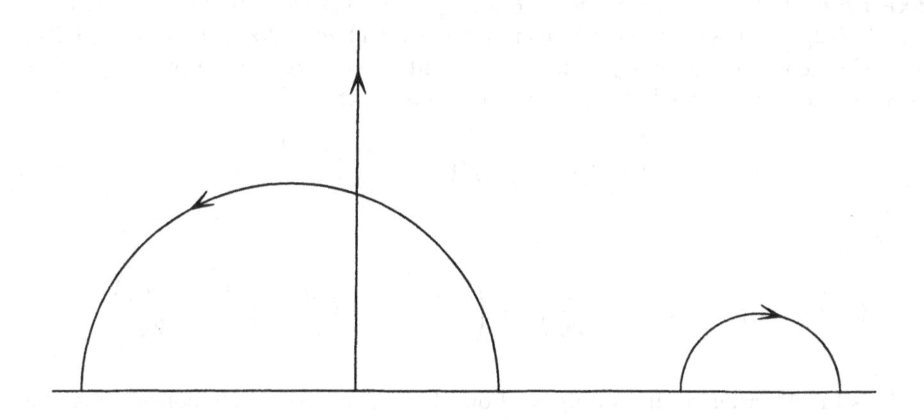

FIGURE 5.4.1. Geodesics on the Lobachevsky plane

Let us point out that the group Γ is, in fact, the entire isometry group:

Proposition 5.4.10. *The group of isometries of* $\mathbb{H}$ *is generated by* $\mathcal{M}$ *and the symmetry* $S: z \mapsto -\bar{z}$.

Proof. Let ϕ be any isometry of $\mathbb{H}$. Note that any isometry that preserves a geodesic and a tangent vector to it is equal to the identity on that geodesic. Since $\phi(I)$ is a geodesic, Theorem 5.4.9 and Lemma 5.4.6 give a $T \in \mathcal{M}$ such that $T^{-1}\phi_{\restriction_I} = \mathrm{Id}_{\restriction_I}$. It suffices to show that $T^{-1}\phi$ is either the identity on $\mathbb{H}$ or coincides with the symmetry S, $S(z) = -\bar{z}$. Consider the geodesic C with endpoints $-r$ and r. It contains the point $ir \in I$ and hence so does $T^{-1}\phi(C)$ (since $T^{-1}\phi_{\restriction_I} = \mathrm{Id}_{\restriction_I}$). Since $T^{-1}\phi$ preserves angles, both these geodesics are orthogonal to I at ir. Hence they coincide up to orientation, that is, we either have $T^{-1}\varphi(z) = z$ for $z \in C$ or $T^{-1}\varphi(z) = -\bar{z}$ for $z \in C$, and hence the derivative of $T^{-1}\varphi$ at ir is either the identity or the reflection in I. Since isometries are smooth, the same case occurs for all points on I; hence the same choice was made for all such geodesics, that is, $T^{-1}\varphi = \mathrm{Id}$ or $T^{-1}\varphi = S$ on $\mathbb{H}$. So $\phi \in \mathcal{M}$ or $\phi \circ S \in \mathcal{M}$. □

We now introduce a convenient distance on the unit tangent bundle of $\mathbb{H}$. For $v, w \in S_z\mathbb{H}$ we take the angle $\measuredangle v, w$ as the distance from v to w. Consider now $z, z' \in \mathbb{H}$ and $v \in S_z\mathbb{H}$, $w \in S_{z'}\mathbb{H}$. There is a unique geodesic $\gamma: [0,1] \to \mathbb{H}$ connecting z and z' and we denote by X the continuous vector field along γ such that $X(0) = v$ and $\measuredangle X(t), \dot{\gamma}(t) = \measuredangle v, \dot{\gamma}(0)$ for all $t \in [0,1]$. Then we define

$$\mathrm{dist}(v, w) = \sqrt{(\measuredangle x(1), w)^2 + (d(z, z'))^2}. \tag{5.4.2}$$

Remark. What we have just described in elementary terms is the procedure of parallel translating v along γ to $z' \in \mathbb{H}$ and measuring angles there. This yields the standard definition of distance. Note in particular that dist indeed defines a distance function (cf. the end of Section A.4).

Example. Let $v \in T_{x+iy}\mathbb{H}$, $w \in T_{x+d+iy}\mathbb{H}$ be vertical unit vectors. If $\alpha = \tan^{-1}(d/2y)$ then with the previous notation $\measuredangle x(1), w = 2\alpha$. If $r^2 = y^2 + (d/2)^2$ then the geodesic segment joining $x + iy$ and $x + d + iy$ can be parameterized as $\gamma(t) = x + d/2 + ire^{it}$, $t \in [-\alpha, \alpha]$, and has length

$$2\log|\sec\alpha + \tan\alpha| = 2\log\left|\left(1 + \left(\frac{d}{2y}\right)^2\right)^{-1/2} + \frac{d}{2y}\right|.$$

Thus

$$(\mathrm{dist}(v,w))^2 = \left(2\tan^{-1}\frac{d}{2y}\right)^2 + \left(2\log\left|\left(1 + \left(\frac{d}{2y}\right)^2\right)^{-1/2} + \frac{d}{2y}\right|\right)^2.$$

It is much easier to find an upper bound in simple form: By considering the horizontal line segment joining $x + iy$ and $x + d + iy$ we find that their distance is less than d/y. Since $\tan^{-1} x < x$ for $x > 0$ we conclude that

$$\mathrm{dist}(v, w) < \sqrt{2}d/y. \tag{5.4.3}$$

Now we can study the geodesic flow $g^t: S\mathbb{H} \to S\mathbb{H}$. We begin by considering the orbit of the upward vertical unit vector $\mathfrak{i}$ at i. Its orbit projects to the geodesic $t \mapsto ie^t$. In fact, for $x \in \mathbb{R}$ let w be the upward vertical vector at $x+i \in \mathbb{H}$. Then the orbit of w projects to the geodesic $t \mapsto x+ie^t$. As we just saw, the distance between the corresponding unit vectors $\mathfrak{i}_t$ at ie^t and w_t at $x+ie^t$ is bounded by $\sqrt{2}xe^{-t}$. Thus the orbits of upward vertical unit vectors at points $x+i \in \mathbb{R}+i$ are positively asymptotic to that of $\mathfrak{i}$.

By using the transformation $z \mapsto -1/z$ one sees then that the orbits of the outward unit normals to the circle of radius $1/2$ centered at $i/2$ are negatively asymptotic to that of $\mathfrak{i}$.

Definition 5.4.11. Horizontal lines $\mathbb{R}+ir = \{t+ir \mid t \in \mathbb{R}\}$ are called *horocycles* centered at ∞. Circles tangent to $\mathbb{R}$ at $x \in \mathbb{R}$ are called *horocycles* centered at x. If $\gamma: \mathbb{R} \to \mathbb{H}$ is a geodesic then $\gamma(-\infty)$ and $\gamma(\infty) \in \mathbb{R} \cup \{\infty\}$ are the limit points of γ as $t \to -\infty$ and $t \to +\infty$, respectively. If $v \in T_z\mathbb{H}$ then let $\pi(v) := z$.

Lemma 5.4.12. *For every horocycle H there exists a Möbius transformation $T \in \mathcal{M}$ such that $T(\mathbb{R}+i) = H$.*

Proof. For horocycles $H = \mathbb{R}+ir$ take $T(z) = rz$. For horocycles centered at $x \in \mathbb{R}$ and of Euclidean diameter r take $T_1(z) = -1/z$, $T_2(z) = rz$, $T_3(z) = z+x$, and $T = T_3 \circ T_2 \circ T_1$. □

To further study the dynamics of the geodesic flow on $\mathbb{H}$ it is useful to parameterize the set $S\mathbb{H}$ of unit vectors on $\mathbb{H}$ by $t, u, v \in \mathbb{R}$ as follows: Given a fixed reference vector $q \in S\mathbb{H}$ and $p \in S\mathbb{H}$ that does not point vertically downwards let H_p be the horocycle with p as inward (or upward) normal vector, γ the geodesic connecting the centers of H_q and H_p (that is, the points of tangency on the real axis), v the oriented hyperbolic length of the arc of H_p between $\gamma \cap H_p$ and the footpoint $\pi(p)$ of p, t the oriented arc length of the segment of γ between H_q and H_p, and u the oriented length of the arc of H_q between $\gamma \cap H_q$ and $\pi(q)$. It is easy to see that *locally* $\phi: (t, u, v) \mapsto p$ is a diffeomorphism between $\mathbb{R}^3$ and $S\mathbb{H}$. Note, however, that this does not parameterize any vertically downward vectors. A second chart starting from $-q$ would cover these.

If $W^s(p)$ denotes the collection of inward (or upward) unit normal vectors to H_p (the *stable manifold* of p), then the orbit of any $p' \in W^s(p)$ is positively asymptotic to that of p by Proposition 5.4.10, since the orbits of upward vertical unit vectors to $\mathbb{R}+i$ have pairwise asymptotic orbits. Note that $W^s(p)$ is a level set of (t, u). Indeed $W^s(q) = \phi(\{0\} \times \{0\} \times \mathbb{R})$. We call $W^{s0}(q) := \phi(\mathbb{R} \times \{0\} \times \mathbb{R})$ the *weak stable manifold* of q. Likewise the points of $W^u(p) := -W^s(-p)$ (the *unstable manifold* of p, outward unit vectors to H_{-p}) have negatively asymptotic orbits and $W^u(q) = \phi(\{0\} \times \mathbb{R} \times \{0\})$. $W^{u0}(q) := \phi(\mathbb{R} \times \mathbb{R} \times \{0\})$ is called the *weak unstable manifold* of of q. For vertically downward vectors we have to use the corresponding chart starting with $-q$ to make these definitions.

The estimate (5.4.3) of the decay of the distance between vertical tangent vectors, Definition 5.4.11, Lemma 5.4.12, and the preceding definitions are summarized as follows:

Proposition 5.4.13. *The stable manifold of $v \in S\mathbb{H}$ with respect to the geodesic flow g^t is the unit normal vector field containing v to the horocycle centered at $\gamma_v(\infty)$. The unstable manifold of $v \in S\mathbb{H}$ is the unit normal vector field containing v to the horocycle centered at $\gamma_v(-\infty)$. In particular all stable and unstable manifolds are one-dimensional and the contraction and expansion rates are e^{-1} and e.*

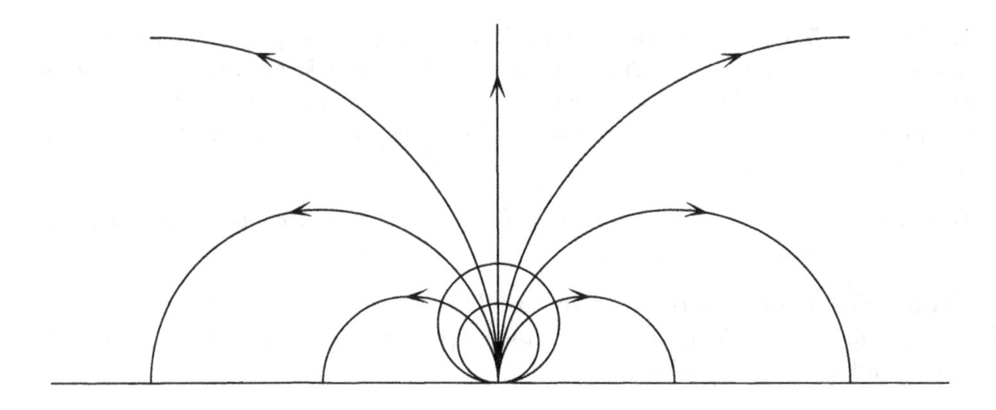

FIGURE 5.4.2. Geodesics and horocycles in the hyperbolic plane

e. Compact factors. For some purposes, in particular for the discussion that follows, it is useful to have an alternative model of the Lobachevsky plane. The map $f\colon \mathbb{H} \to \mathbb{C}, z \mapsto \dfrac{z-i}{z+i}$ maps the Poincaré upper half-plane $\mathbb{H}$ onto the open unit disk $\mathbb{D}$ in $\mathbb{C}$ bounded by the unit circle $S^1 = \{z \in \mathbb{C} \mid |z| = 1\}$ since $|f(z)| = 1$ when $z \in \mathbb{R}$ and $f(i) = 0$. Pushing forward the hyperbolic Riemannian metric $\langle \cdot, \cdot \rangle$ on $\mathbb{H}$ to the metric given by

$$\langle v, w \rangle := \langle Df^{-1}v, Df^{-1}w \rangle$$

on the unit disk makes f an isometry. The unit disk with this metric is called the *Poincaré disk*. Since f maps lines and circles into lines and circles and preserves angles, it immediately follows that the geodesics in the Poincaré disk are diameters of S^1 and arcs of circles perpendicular to S^1. We will use this picture to exhibit an example of a compact manifold that is locally isometric to $\mathbb{H}$. We accomplish this by factoring out by a discrete group of isometries.

Consider the Poincaré disk in $\mathbb{C}$ and draw a regular (hyperbolic) octagon $\mathcal{Q}$ with vertices $v_k = de^{-k\pi i/4}$, $k = 0, \dots, 7$, joined by arcs of circles perpendicular to the unit circle (see Figure 5.4.3). Here $d \in (0,1)$ and as $d \to 1$, the sum

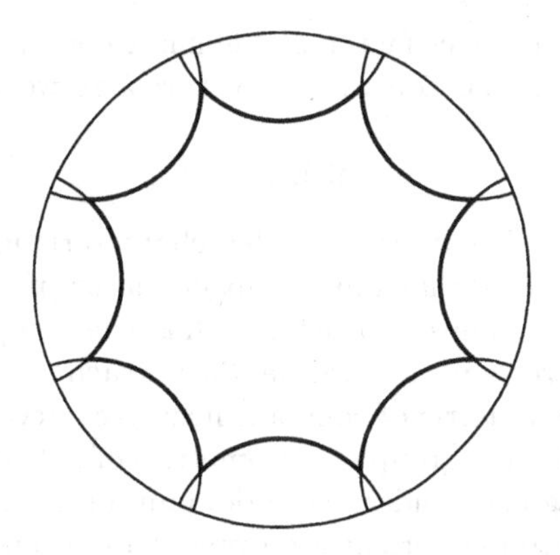

FIGURE 5.4.3. A hyperbolic octagon

of the internal angles converges to zero. On the other hand, the sum of the internal angles goes to 6π, the value for the Euclidean octagon, as $d \to 0$. This becomes clear by keeping d fixed and increasing the size of the Poincaré disk indefinitely so that the arcs of circles approach line segments.

The point of this observation is that since the angles change continuously we can fix d such that the internal angles add up to 2π. To label the edges we name them, in counterclockwise direction, starting from those in the first quadrant, $\alpha_1, \alpha_2, \alpha_3, \alpha_4, \ \alpha_1', \alpha_2', \alpha_3', \alpha_4'$ in this order. Orient the α_i counterclockwise and the α_i' clockwise, and identify α_i with α_i', $i = 1, \dots, 4$, by reflection in the diameter equidistant from the two axes.

The resulting identification space is a surface τ of genus 2. Since the internal angles of $\mathcal{Q}$ add up to 2π, the identification map is smooth at the vertices (which are all identified to one point) and we can therefore push the metric on $\mathcal{Q}$ down to τ. We obtain a compact manifold which is locally isometric to $\mathbb{H}$. Topologically this manifold is homeomorphic to the sphere with two handles, that is, the surface of the simplest "pretzel". One can also show that τ is the space obtained by identifying orbits of the group Γ generated by the isometries A_i, $i = 1, \dots, 4$, mapping α_i to α_i'. In other words, the fundamental group of τ can be identified with a discrete group Γ of hyperbolic Möbius transformations.

Replacing eight arcs in this construction by $4g$ arcs where $g \geq 2$ we obtain a metric locally isomorphic to that of $\mathbb{H}$ on the orientable surface of genus g (sphere with g handles).

If a Möbius transformation γ preserves a geodesic then such geodesic is unique and it is called the axis of γ. In fact, every $\gamma \in \Gamma$ has an axis, but we do not need this fact. The projections of these geodesics to $M := \Gamma \backslash \mathbb{D}$ are precisely the closed geodesics of M. These are, of course, the projections of the closed orbits of the geodesic flow from the tangent bundle to M.

Associated to any C^2 Riemannian metric on a surface is the Gaussian curvature of the metric, an isometry-invariant real-valued function. Since the isometry group of $\mathbb{D}$ is transitive, the curvature of $\mathbb{D}$ is a constant k. Thus the induced metric on the compact factor τ of genus 2 constructed from the octagonal fundamental domain has constant curvature k as well. The Gauss–Bonnet Theorem

$$k \cdot \operatorname{vol} M = 2\pi\chi$$

then shows that $k < 0$ because the Euler characteristic χ of τ is negative. Conversely this then shows that any compact factor of $\mathbb{D}$ has negative Euler characteristic and hence genus at least 2. Thus the compact factors of $\mathbb{D}$ are homeomorphic to spheres with several handles attached. In fact, any compact orientable surface with a metric of constant negative curvature is isometric to a factor $\Gamma\backslash\mathbb{D}$ of $\mathbb{D}$ by a discrete group Γ of isometries of $\mathbb{D}$.[1] To see how the picture developed for the octagonal fundamental domain looks in the general case, let us consider a discrete group of orientation-preserving isometries of the Poincaré disk $\mathbb{D}$ which produces a compact factor. One can choose a fundamental domain for Γ by considering the *Dirichlet domain*

$$D := D_p := \{x \in \mathbb{D} \mid d(x,p) \leq d(x, \gamma p) \text{ for all } \gamma \in \Gamma\}$$

for any given point $p \in \mathbb{D}$. For any $\gamma \in \Gamma$ we evidently have $D_{\gamma p} = \gamma(D_p)$. The interiors of D_p and $D_{\gamma p}$ are disjoint when $\gamma \neq \mathrm{Id}$ and since Γ is discrete, there are only finitely many $\gamma \in \Gamma$ such that $D_p \cap D_{\gamma p} \neq \varnothing$. If $\gamma \in \Gamma$ is one of these elements, then $D_p \cap D_{\gamma p}$ consists of the points equidistant from p and γp, that is, is a geodesic segment (see Exercise 5.4.2). Thus D is a hyperbolic polygon, that is, bounded by finitely many geodesic arcs. Our assumption that $\Gamma\backslash\mathbb{D}$ is compact means that D is compact. By construction we also observe that the sets $D_{\gamma p}$ cover $\mathbb{D}$, so we have, in fact, tessellated $\mathbb{D}$ by the images of D under Γ.[2]

Compact factors of the hyperbolic plane cannot be embedded isometrically in $\mathbb{R}^3$. An illustration of an isometrically embedded surface of constant negative curvature is given by the pseudosphere whose fundamental domain and embedding we show in Figure 5.4.4.

f. The dynamics of the geodesic flow on compact hyperbolic surfaces. Unlike the geodesic flow on the round sphere and the flat torus considered in Subsection b, where the dynamics turned out to be rather simple, compact factors of the hyperbolic plane have geodesic flows of a complicated dynamical nature rather similar to that which appeared for the second group of examples in Chapter 1 (expanding maps of the circle, hyperbolic automorphisms of the torus, and topological Markov chains). Though the whole extent of this similarity will become clear only after developing the theory of hyperbolic dynamical systems (Section 17.5 and Chapter 18), we will now establish for the geodesic flow on compact factors of the hyperbolic plane some of the properties that we tend to consider typical for complicated dynamical behavior, namely, density of closed orbits, topological transitivity, and ergodicity with respect to the Liouville measure (see Proposition 5.3.6). Another such property, positivity of topological entropy, is established in Exercise 5.4.10. The latter

FIGURE 5.4.4. The pseudosphere

property for arbitrary metrics on surfaces of genus at least two will follow from Theorem 9.6.7.

We can now prove density of closed orbits:

Theorem 5.4.14. *Let Γ be a discrete group of fixed-point-free isometries of $\mathbb{D}$ such that $M := \Gamma\backslash\mathbb{D}$ is compact. Then the periodic orbits of the geodesic flow on SM are dense in SM.*

Proof. We will use the model of the Poincaré disk $\mathbb{D}$. Let $v \in SM$ and take a Dirichlet domain D for Γ and let $w \in S\mathbb{D}$ be a lift of v with footpoint in D. Let c be the geodesic with $\dot{c}(0) = w$ in $\mathbb{D}$ and let x and y be the endpoints of c on the boundary of the Poincaré disk. Our strategy is to find a hyperbolic element $\gamma \in \Gamma$ such that the endpoints of its axis lie in given small δ-neighborhoods U and V, respectively, of the points $x =: c(-\infty)$ and $y =: c(\infty)$. Then among the tangent vectors to this axis one can find a vector that is close to w. The projection of the axis to M will be the desired closed geodesic.

To begin let us note that given $\epsilon > 0$ there exists $\delta > 0$ such that when $p \in \mathbb{D}$ is in a δ-neighborhood of $\partial\mathbb{D}$ then any two geodesics through p of Euclidean length greater than ϵ have a mutual angle of at most $\pi/4$. Our aim is to construct a geodesic κ whose endpoints are both very close to x and an element $\gamma \in \Gamma$ of the fundamental group such that the image of κ under γ is a geodesic whose endpoints are very close to y.

Consider the sequence of those images of D that intersect c. In particular there are images D_1 and D_2 of D in U and V, respectively, and a $\gamma \in \Gamma$ such that

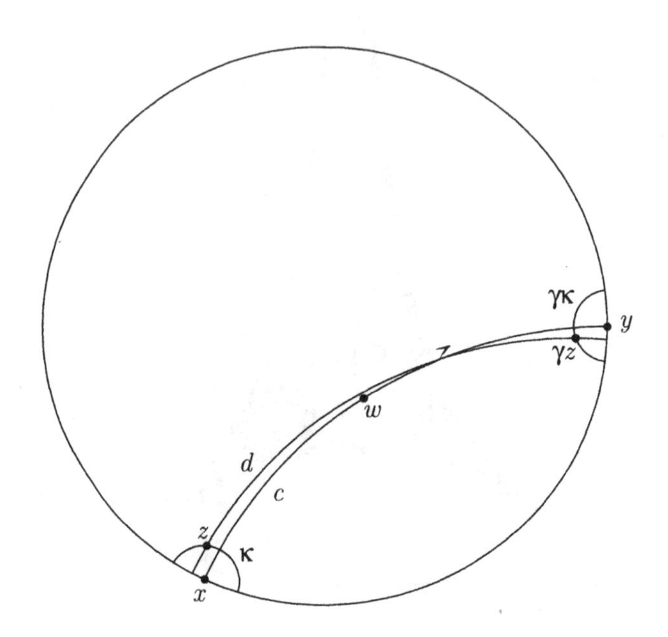

FIGURE 5.4.5. Density of closed geodesics

$\gamma(D_1) = D_2$ and γ preserves the ordering of the sequence of images of D. Choose a point $z \in D_1$. Then $\gamma(z) \in D_2$. But by an earlier observation most geodesics through z are entirely contained in U (after choosing D_1 of sufficiently small Euclidean size). Since γ preserves angles and the same observation applies to geodesics through $\gamma(z)$ we can, in fact, find a geodesic κ through z and contained in U such that $\gamma\kappa$ is contained in V. Now, since γ preserves the ordering of the images of D covering c we conclude that γ maps the region inside U bounded by κ onto a region bounded by $\gamma\kappa$ and containing the complement of V. So γ has two fixed points on $\partial\mathbb{D}$, one each inside U and V. Thus the axis d of γ is uniformly close to c and in particular has a tangent vector as close to the given vector $w = \dot{c}(0)$ as we please, by making δ small enough. □

Theorem 5.4.15. *Let Γ be a discrete group of fixed-point-free isometries of $\mathbb{D}$ such that $M := \Gamma\backslash\mathbb{D}$ is compact. Then the geodesic flow on SM has an orbit which is dense in SM, that is, it is topologically transitive.*[3]

Proof. By Theorem 5.4.14 and Lemma 1.4.2 it is sufficient to show that for any two periodic points $u, v \in SM$ (whose lifts to $\mathbb{D}$ we also denote by u and v) and neighborhoods U, V of u, v, respectively, there is $t \in \mathbb{R}$ such that $g^t(U) \cap V \neq \varnothing$. Take the geodesics c_u and c_v in $\mathbb{D}$ with $\dot{c}_u(0) = u$ and $\dot{c}_v(0) = v$ and denote by c the geodesic with endpoints $c(-\infty) = c_u(-\infty)$ and $c(\infty) = c_v(\infty)$. Replacing, if necessary, u by γu may assume that $c_u(-\infty) \neq c_v(\infty)$. By Proposition 5.4.13 we can find for each $t \in \mathbb{R}$ numbers $f(t), g(t) \in \mathbb{R}$ such that $d(\dot{c}_u(f(t)), c(t))$

converges to zero exponentially as $t \to -\infty$ and $d(\dot{c}_v(g(t)), c(t))$ converges to zero exponentially as $t \to \infty$. Since $\dot{c}_u$ and $\dot{c}_v$ project to closed orbits of the geodesic flow this shows that there exist t_1 and t_2 such that the projection of $\dot{c}(t_1)$ to SM is in U and the projection of $\dot{c}(t_2)$ to SM is in V. This then yields the claim. □

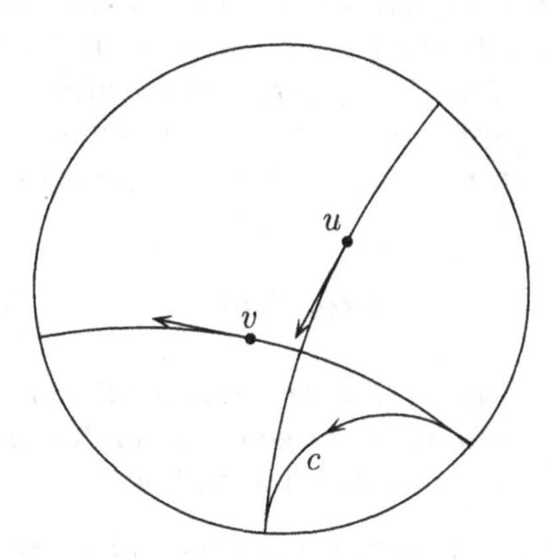

FIGURE 5.4.6. Transitivity of the geodesic flow

We now prove a stronger result, namely, we show that the geodesic flow on a compact factor of $\mathbb{H}$ is ergodic with respect to the natural smooth invariant measure obtained in Proposition 5.3.6.

Theorem 5.4.16. *The Liouville measure for the geodesic flow on a compact connected factor of* $\mathbb{H}$ *is ergodic.*

Proof. The idea of this proof is an important one for hyperbolic dynamics in general and called the *Hopf argument.*[4] Let us note first of all that it suffices to check that the ergodic average $\varphi_{g^t}(x) := \lim_{T\to\infty}(1/T)\int_0^T \varphi(g^t(x))\,dt$ is constant a.e. for every integrable function φ on SM, where M is the compact factor in question and we use the smooth invariant measure. Next observe that it suffices to consider continuous (hence uniformly continuous) functions since these are dense in L^1. Furthermore we can check constancy modulo null sets with respect to *any* smooth measure. Finally it suffices also to check that φ_{g^t} is constant a.e. in any given open set because this constant must then agree on any two overlapping open sets and M is connected.

Consider such φ. As a consequence of the Birkhoff Ergodic Theorem 4.1.2 $\lim_{T\to\infty}(1/T)\int_0^T \varphi(g^t(x))\,dt$ exists a.e. (The extension from discrete time to continuous time is particularly easy for uniformly continuous functions.) We first note that if this limit exists for some $p \in SM$ then it also exists for all $q \in W^s(p)$ and is independent of $q \in W^s(p)$: Given $\epsilon > 0$ there exists $T_0 > 0$ such that $|\varphi(g^t(p)) - \varphi(g^t(q))| < \epsilon$ for $t > T_0$ by uniform continuity. But this means that $|(1/T)\int_0^T (\varphi(g^t(p)) - \varphi(g^t(q)))\,dt| < \epsilon$ for all sufficiently large T, as required. Thus φ_{g^t} is constant on stable leaves. Since existence and the value of the limit are g^t-invariant φ_{g^t} is, in fact, constant on weak stable manifolds.

Consider also the negative time average $\bar{\varphi}_{g^t} := \lim_{T\to\infty} \int_{-T}^{0} \varphi(g^t(x))\,dt$. It exists a.e. as well and is constant on unstable manifolds. Furthermore, however, $\bar{\varphi}_{g^t}$ and φ_{g^t} coincide a.e. by Proposition 4.1.3 of the Birkhoff Ergodic Theorem. In terms of the local C^1 coordinates (t, u, v) introduced at the end of Subsection e on a small open set U the Fubini Theorem then tells us that for t in a set C of full measure $\varphi_{g^t}(t, u, v) = \bar{\varphi}_{g^t}(t, u, v)$ for almost all (u, v). But then for any such t_1, t_2 the corresponding sets of (u, v) intersect (because they both have full measure) and hence $\varphi_{g^t}(t_1, u, v) = \varphi_{g^t}(t_2, u, v)$ for all (u, v), so φ_{g^t} is constant on C, as required. □

Exercises

5.4.1. *Using Lemma 5.4.1 generalize Theorem 5.4.2 and Corollary 5.4.4 to the n-dimensional case, that is, describe the geodesics on the unit sphere $S^n \subset \mathbb{R}^{n+1}$ with the metric induced by the embedding and on the flat torus $\mathbb{T}^n = \mathbb{R}^n/\mathbb{Z}^n$.*

5.4.2. *Given two points p, q on the hyperbolic plane show that the locus of points equidistant from p and q is a geodesic.*

5.4.3. *Given a hyperbolic Möbius transformation $f(z) = \dfrac{az+b}{cz+d}$ of $\mathbb{H}$ and a point z_0 on its axis, calculate the hyperbolic distance between z_0 and $f(z_0)$.*

5.4.4. *Given two hyperbolic Möbius transformations f and g of $\mathbb{H}$ such that there exist z_1 on the axis of f and z_2 on the axis of g for which $d(z_1, f(z_1)) = d(z_2, g(z_2))$ prove that there exists $h \in \mathcal{M}$ such that $f = hgh^{-1}$.*

5.4.5. *Given a geodesic $\gamma \subset \mathbb{H}$ and $p \in \mathbb{H} \setminus \gamma$ prove that the set $E := \{q \in \mathbb{H} \mid d(q, \gamma) = d(p, \gamma)\}$, where $d(q, \gamma) = \inf_{x\in\gamma} d(q, x)$, is a union of two isometric smooth curves one on each side of γ and that those curves are not geodesics. Find how those curves are represented on the upper half-plane.*

5.4.6. *Prove that*

$$d(v, w) = \left| \log \frac{|v - u_1|}{|v - u_2|} \cdot \frac{|w - u_2|}{|w - u_1|} \right|,$$

where u_1 and u_2 are the points of intersection of the real axis with the circle that represents the geodesic passing through v and w; if this geodesic is a line put $u_2 = \infty$.

5.4.7. *Show that if $M = \Gamma\backslash\mathbb{D}$ is compact then the geodesic flow on M has only finitely many closed orbits of period less than any given number, or equivalently, Γ has only finitely many elements whose traces have absolute value less than a given positive number.*

5.4.8*. *Prove that the group of isometries of any compact factor $\Gamma\backslash\mathbb{H}$ is finite.*

5.4.9*. *Let $G = \{g^t\}$ be the geodesic flow on $S(\Gamma\backslash\mathbb{H})$, where $\Gamma\backslash\mathbb{H}$ is a compact surface. Prove that $h_{\text{top}}(G) \geq 1$.*

5. Hamiltonian systems

We have encountered Hamiltonian systems several times now either in examples or as arising from Lagrangian systems. Our aim in this section is to give a brief axiomatic introduction to the modern approach to Hamiltonian dynamics as well as to exhibit some of the structural results of the theory which in certain cases lead to a complete qualitative description of the dynamics.

a. Symplectic geometry. In the examples of Hamiltonian systems we have seen so far, we always obtained a real-valued function H on the phase space and the system was then described by the Hamiltonian equations (5.3.6). In this way the Hamiltonian system was determined by the Hamiltonian and the nondegenerate 2-form $dp \wedge dq$. Now we will elaborate on a more general approach which was first indicated in (1.5.6). To that end we begin with a study of nondegenerate antisymmetric 2-forms on Euclidean spaces.

Definition 5.5.1. Let E be a linear space. A 2-tensor $\alpha: E \times E \to \mathbb{R}$ is said to be *nondegenerate* if $\alpha^\flat: v \mapsto \alpha(v, \cdot)$ is an isomorphism from E to its dual space E^*. It is said to be *antisymmetric* or *skew-symmetric* if $\alpha(v, w) = -\alpha(w, v)$. A nondegenerate antisymmetric 2-form is called a *symplectic* form. A linear space with a symplectic form is called a *symplectic* vector space. If (E, α), (F, β) are symplectic vector spaces then a linear map $T: E \to F$ is said to be *symplectic* if $T^*\beta = \alpha$.

Remark. If a scalar product $\langle \cdot, \cdot \rangle$ on E is fixed it is convenient to write $\alpha(\cdot, \cdot) = \langle \cdot, A \cdot \rangle$, so that we identify the tensor with its matrix representation with respect to a given basis.

Proposition 5.5.2. *Let E be a linear space. If α is a symplectic form on E then $\dim E = 2n$ for some $n \in \mathbb{N}$ and there is a basis $e_1, \dots, e_{2n}$ of E such that $\alpha(e_i, e_{n+i}) = 1$ if $i = 1, \dots, n$ and $\alpha(e_i, e_j) = 0$ if $|i - j| \neq n$. Hence, if one fixes a scalar product with respect to which $e_1, \dots, e_{2n}$ is an orthonormal basis, then $A = \begin{pmatrix} 0 & I \\ -I & 0 \end{pmatrix}$ with respect to this basis, where I is the $n \times n$ identity matrix.*

Proof. Since α is nondegenerate there exist e_1, e_{n+1} such that $\alpha(e_1, e_{n+1}) \neq 0$, and without loss of generality we may assume $\alpha(e_1, e_{n+1}) = 1$. By antisymmetry $\alpha(e_1, e_1) = \alpha(e_{n+1}, e_{n+1}) = 0$ and $\alpha(e_{n+1}, e_1) = -1$, so the matrix of $\alpha_{\restriction E_1}$, where $E_1 = \text{span}\{e_1, e_{n+1}\}$, with respect to (e_1, e_{n+1}) is $\begin{pmatrix} 0 & 1 \\ -1 & 0 \end{pmatrix}$.

Now we use induction: Consider $E_2 := \{v \in E \mid \alpha(v,w) = 0 \text{ for all } w \in E_1\}$. Then $E_1 \cap E_2 = \{0\}$ and $E_1 \oplus E_2 = E$ since for $v \in E$ we have $v - \alpha(v, e_{n+1})e_1 + \alpha(v, e_1)e_{n+1} \in E_2$. Inductively we get the claim. □

Definition 5.5.3. A subspace V of a symplectic linear space (E,α) is said to be *isotropic* if $\alpha_{\restriction V} = 0$. An isotropic subspace of dimension $n = \dim E/2$ is said to be *Lagrangian*.

Remark. Thus the "adapted" basis of the proposition gives a decomposition of E as a direct sum of two Lagrangian subspaces. Notice that by nondegeneracy of α an isotropic subspace has dimension at most $n = \dim E/2$, so Lagrangian subspaces are maximal isotropic subspaces.

An interesting description of nondegeneracy is the following:

Proposition 5.5.4. *Let α be an antisymmetric 2-form on a linear space E. Then α is nondegenerate if and only if E is even-dimensional and the nth exterior power α^n of α is not zero.*

Proof. $\Leftarrow$: If α is degenerate the map $\alpha^\flat$ has nontrivial kernel, that is, there is a vector v such that $\alpha(v,w) = 0$ for all w. But then by definition of exterior powers we have $\alpha^n(v, v_2, \dots, v_n) = 0$ for all $v_2, \dots, v_n$.

$\Rightarrow$: If α is nondegenerate we may assume by Proposition 5.5.2 that $\alpha = \sum_{i=1}^n dx_i \wedge dx_{i+n}$. But then $\alpha^n = \sum_{i_1,\dots,i_n=1}^n dx_{i_1} \wedge dx_{i_1+n} \wedge \cdots \wedge dx_{i_n} \wedge dx_{i_n+n} = n!(-1)^{[n/2]} dx_1 \wedge \cdots \wedge dx_{2n} \neq 0$. □

An immediate observation from the preceding results is

Proposition 5.5.5. *If $T\colon (E,\alpha) \to (F,\beta)$ is a symplectic map then T preserves volume and orientation. In particular T is invertible with Jacobian 1.*

Thus the set of symplectic maps $(E,\alpha) \to (E,\alpha)$ is a group which we call the *symplectic group* of (E,α). Let us now assume that a scalar product $\langle \cdot, \cdot \rangle$ is fixed and that α is in standard form $J = \begin{pmatrix} 0 & I \\ -I & 0 \end{pmatrix}$. Here are some further simple properties of symplectic maps.

Proposition 5.5.6. *Suppose (E,α) is a symplectic vector space and $T\colon (E,\alpha) \to (E,\alpha)$ a symplectic map. If λ is an eigenvalue of T then so are $\bar\lambda$, $1/\lambda$, $1/\bar\lambda$. If T has the form $\begin{pmatrix} A & B \\ C & D \end{pmatrix}$ with respect to a basis for which $\alpha(v,w) = \langle v, Jw \rangle$ then A^tC and B^tD are symmetric and $A^tD - C^tB = I$.*

Proof. If T preserves α and $\alpha(v,w) = \langle v, Jw \rangle$ then symplecticity means $T^tJT = J$. By calculation this implies that A^tC and B^tD are symmetric and $A^tD - C^tB = I$. If λ is an eigenvalue then so is $\bar\lambda$ since the characteristic polynomial $P(\lambda) = \det(T - \lambda I)$ has real coefficients. Furthermore $JTJ^{-1} = (T^{-1})^t$, so

$$\begin{aligned} P(\lambda) = \det(T - \lambda I) &= \det(J(T - \lambda I)J^{-1}) = \det(T^{-1})^t(I - \lambda T^t) \\ &= \det((I - \lambda T)T^{-1}) = \det(\lambda(\lambda^{-1}I - T)) = \lambda^{2n}P(\lambda^{-1}); \end{aligned}$$

hence, since 0 is not an eigenvalue, $P(\lambda) = 0$ if and only if $P(1/\lambda) = 0$. □

Exercise 5.5.3 gives an appropriate version of a converse to this result. Now we are ready to discuss symplectic forms on manifolds.

Definition 5.5.7. Let M be a smooth manifold. A differential 2-form ω is a smooth map from M to the space $\bigwedge^2 T^*M$ of antisymmetric 2-tensor fields, that is, it assigns to each $x \in M$ an antisymmetric 2-tensor on T_xM. A differential 2-form ω is said to be *nondegenerate* if it is nondegenerate at every point. A nondegenerate 2-form ω with $d\omega = 0$ is called a *symplectic form*. A pair (M, ω) of a smooth manifold and a symplectic form is called a *symplectic manifold*. If (M, ω) is a symplectic manifold then a subbundle of the tangent bundle TM of M is said to be *isotropic* if at every point $p \in M$ it defines an isotropic subspace of T_pM, and *Lagrangian* if at every point $p \in M$ it defines a Lagrangian subspace of T_pM. A smooth submanifold of a symplectic manifold is said to be *isotropic* if its tangent bundle is an isotropic subbundle, and *Lagrangian* if its tangent bundle is a *Lagrangian* subbundle of TM. A diffeomorphism $f\colon (M, \omega) \to (N, \eta)$ between symplectic manifolds such that $f^*\eta = \omega$ is called a *symplectic diffeomorphism* or *symplectomorphism*. If $(M, \omega) = (N, \eta)$ it is also called a *canonical transformation*.

Proposition 5.5.4 immediately yields:

Proposition 5.5.8. *If (M, ω) is a symplectic manifold then M is even-dimensional and ω^n is a volume form. In particular M is orientable.*

By Proposition 5.5.2 we can find coordinates around any given point x such that in T_xM the induced coordinates bring the symplectic form into standard form. This can be done by introducing any coordinate system and making an appropriate linear coordinate change in that system. Unlike in the case of a Riemannian metric, it is, however, possible to find a local chart such that the symplectic form is in standard form at *every* point of the chart. We present a proof of this due to Moser, which gives another application of the method we saw in Section 5.1e.

Theorem 5.5.9. (Darboux Theorem) *Let (M, ω) be a symplectic manifold. For every point $x \in M$ there exists a neighborhood U of x and coordinates $\varphi\colon U \to \mathbb{R}^{2n}$ such that at every point $y \in U$ ω is in standard form with respect to the basis* $\left\{\frac{\partial}{\partial x_1}, \dots, \frac{\partial}{\partial x_{2n}}\right\}$.

These coordinates will be referred to as *Darboux* or *symplectic coordinates.*

Proof. As we pointed out we may assume that we already have coordinates such that ω is in standard form at x with respect to the basis $\left\{\frac{\partial}{\partial x_1}, \dots, \frac{\partial}{\partial x_{2n}}\right\}$. Thus we need to find coordinates in which ω is constant. Via the coordinates we already have we may assume that $M = \mathbb{R}^{2n}$ and $x = 0$. Denote by α the form with matrix $J = \begin{pmatrix} 0 & I \\ -I & 0 \end{pmatrix}$. Let $\omega' = \alpha - \omega$ and $\omega_t = \omega + t\omega'$ for $t \in [0, 1]$.

Then there is a ball around 0 on which all ω_t are nondegenerate (since there is such a ball for every t and it depends continuously on t). Thus $\omega' = d\theta$ for some one-form θ by the Poincaré Lemma, and without loss of generality $\theta(0) = 0$.

Since ω_t is nondegenerate, there is a unique (smooth) vector field X_t such that $\omega_t \lrcorner X_t = \omega_t(X_t, \cdot) = -\theta$. Since $X_t(0) = 0$ one can integrate X_t on a small ball around 0 to get a 1-parameter family of diffeomorphisms $\{\varphi^t\}_{t\in[0,1]}$ such that $\dot\varphi^t = X_t$ and $\varphi^0 = \mathrm{Id}$. Then as in Section 5.1e we obtain

$$\frac{d}{dt}\varphi^{t*}\omega_t = \varphi^{t*}(\pounds_{X_t}\omega_t) + \varphi^{t*}\frac{d}{dt}\omega_t = \varphi^{t*}d(\omega_t \lrcorner X_t) + \varphi^{t*}\omega' = \varphi^{t*}(-d\theta + \omega') = 0,$$

so $\varphi^{1*}\omega_1 = \varphi^{0*}\omega_0 = \omega$, that is, φ^1 is the desired coordinate change. $\square$

Remark. As mentioned before, this result is in contrast to the situation for Riemannian metrics, for which such charts exist only for flat metrics. An explanation is that the condition $d\omega = 0$ here may be considered an analog of flatness of a Riemannian metric.

Symplectic C^r diffeomorphisms of a symplectic manifold (M, ω) form a closed subset of $\mathrm{Diff}^r(M)$ with the C^r topology.

b. Cotangent bundles. We now describe a very important class of spaces with a canonical symplectic structure, the cotangent bundle of a smooth manifold which first appeared in Section 5.3e in connection with the Legendre transform. Not only does a cotangent bundle have a canonical symplectic structure, but furthermore the natural coordinates induced by coordinates on the underlying manifold are symplectic coordinates.

Let M be a smooth manifold and consider local coordinates $\{q_1, \dots, q_n\}$. On the cotangent bundle these induce coordinates $\{q_1, \dots, q_n, p_1, \dots, p_n\}$. Define a 1-form θ by setting

$$\theta = -\sum_{i=1}^{n} p_i dq_i. \tag{5.5.1}$$

Then its exterior derivative is

$$\omega = \sum_{i=1}^{n} dq_i \wedge dp_i, \tag{5.5.2}$$

that is, a symplectic form in Darboux coordinates. The next lemma shows that this definition does not depend on the choice of coordinates on the manifold. Alternatively it shows that diffeomorphisms of the manifold induce symplectomorphisms of the cotangent bundle:

Lemma 5.5.10. *Let M be a smooth manifold and $f\colon M \to M$ a diffeomorphism. Then the pullback D^*f on the cotangent bundle is a symplectomorphism.*

Proof. If we write $(Q_1, \dots, Q_n) = f(q_1, \dots, q_n)$ then

$$D^*f(q_1, \dots, q_n, p_1, \dots, p_n) = (Q_1, \dots, Q_n, P_1, \dots, P_n),$$

where $p_j = \sum_{i=1}^n \frac{\partial Q_i}{\partial q_j} P_i$. Thus

$$\sum_{i=1}^n P_i dQ_i = \sum_{i,j=1}^n P_i \frac{\partial Q_i}{\partial q_j} dq_j = \sum_{j=1}^n p_j dq_j$$

and θ, hence ω, is preserved. □

c. Hamiltonian vector fields and flows. Now we can begin to study the Hamiltonian equations in the general setting.

Definition 5.5.11. Let (M, ω) be a symplectic manifold, and $H: M \to \mathbb{R}$ a smooth function. Then the vector field $X_H = dH^{\#}$ defined by $\omega \lrcorner X_H = dH$ is called the *Hamiltonian vector field* associated with H or the *symplectic gradient* of H. The flow φ^t with $\dot{\varphi}^t = X_H$ is called the *Hamiltonian flow* of H.

A Hamiltonian vector field is C^r if and only if the Hamiltonian function is C^{r+1}. Thus one can identify the space of C^r Hamiltonian flows, which is a closed linear subspace of $\Gamma^r(TM)$, with the space $C^{r+1}(M, \mathbb{R})$.

Let us show that this is indeed a formulation of the Hamiltonian equations. The usual Hamiltonian equations are

$$\dot{q}_i = \frac{\partial H}{\partial p_i}, \qquad \dot{p}_i = -\frac{\partial H}{\partial q_i}.$$

Thus to see that $\dot{\varphi}^t = X_H$ gives these, we need to check that $X_H := (\frac{\partial H}{\partial p_i}, -\frac{\partial H}{\partial q_i})$ satisfies $\omega \lrcorner X_H = dH$ in Darboux (symplectic) coordinates. But

$$\begin{aligned}\omega \lrcorner X_H = \sum_{i=1}^n (dq_i \wedge dp_i) \lrcorner X_H &= \sum_{i=1}^n (dq_i \lrcorner X_H) \wedge dp_i - \sum_{i=1}^n dq_i \wedge (dp_i \lrcorner X_H) \\ &= \sum_{i=1}^n \frac{\partial H}{\partial p_i} dp_i + \frac{\partial H}{\partial q_i} dq_i = dH,\end{aligned}$$

since $(dq_i \lrcorner X_H) = \partial H / \partial p_i$ and $(dp_i \lrcorner X_H) = -\partial H / \partial q_i$ by definition of X_H.

It is easy to see that Hamiltonian flows are instances of one-parameter groups of canonical transformations:

Proposition 5.5.12. *Hamiltonian flows are symplectic and hence volume preserving.*

Remark. The conclusion that Hamiltonian flows are volume preserving is known as the *Liouville Theorem.* This generalizes the discussion in Section 5.3e, in particular Proposition 5.3.6.

Proof. Let (M,ω) be a symplectic manifold, $H: M \to \mathbb{R}$ a smooth function, $\omega \lrcorner X_H = dH$, and $\dot{\varphi}^t = X_H$. Then

$$\frac{d}{dt}\varphi^{t*}\omega = \varphi^{t*}(\pounds_{X_H}\omega) = \varphi^{t*}(d(\omega\lrcorner X_H) + (d\omega\lrcorner X_H))$$
$$= \varphi^{t*}(d(\omega\lrcorner X_H)) = \varphi^{t*}(ddH) = 0. \qquad \square$$

We should note that the converse is not true, that is, that there are symplectic flows that are not Hamiltonian. Consider a linear flow on the two-dimensional torus provided with the standard volume 2-form $dx \wedge dy$. Such a flow preserves area and is hence symplectic. Its velocity vector field is constant $\neq 0$. Thus if it were a Hamiltonian flow the Hamiltonian would have to have constant nonzero gradient. On the other hand the Hamiltonian attains its maximum and thus has a critical point, a contradiction. Note, incidentally, that the lift of the linear flow to $\mathbb{R}^2$ is indeed Hamiltonian. If a vector field X generates a symplectic flow the calculation above shows that the 1-form $\omega\lrcorner X$ is closed. Thus the obstruction to being Hamiltonian is, in fact, of a topological nature (namely, vanishing of the cohomology class of the closed 1-form $\omega\lrcorner X$). See Exercise 5.5.4 for a discussion of a related phenomenon.

In our discussion here we will consider only Hamiltonian flows.

In addition we can now show that Hamiltonian flows preserve energy, as we observed in special cases (see Section 5.2a and Proposition 5.3.2 together with Proposition 5.3.5).

Proposition 5.5.13. *Let (M,ω) be a symplectic manifold, $H: M \to \mathbb{R}$ a smooth function, $\omega\lrcorner X_H = dH$, and $\dot{\varphi}^t = X_H$. Then $H(\varphi^t(x))$ does not depend on t.*

Proof.

$$\frac{d}{dt}H\big|_{\varphi^t(x)} = dH(\varphi^t(x))\dot{\varphi}^t(x) = \omega(X_H(\varphi^t(x)), \dot{\varphi}^t(x))$$
$$= \omega(X_H(\varphi^t(x)), X_H(\varphi^t(x))) = 0. \qquad \square$$

d. Poisson brackets. We introduce now a notion which predates the symplectic approach to Hamiltonian mechanics, the Poisson bracket. It was traditionally used in coordinate calculations, but also illuminates the Lie algebraic structure underlying the geometry.

Definition 5.5.14. Let (M,ω) be a symplectic manifold and $f, g: M \to \mathbb{R}$ smooth functions. Then the *Poisson bracket* of f and g is defined by

$$\{f,g\} := \omega(X_f, X_g) = df(X_g),$$

where $X_f = df^\#$ and $X_g = dg^\#$ (cf. Definition 5.5.11), that is, $\omega\lrcorner X_f = df$ and $\omega\lrcorner X_g = dg$. The functions f and g are said to be *in involution* if their Poisson bracket vanishes.

Proposition 5.5.15. *In symplectic coordinates* $\{q_1, \dots, q_n, p_1, \dots, p_n\}$ *we have*

$$\{f, g\} = \sum_{i=1}^{n} \Big(\frac{\partial f}{\partial q_i} \frac{\partial g}{\partial p_i} - \frac{\partial f}{\partial p_i} \frac{\partial g}{\partial q_i} \Big). \tag{5.5.3}$$

The Poisson bracket is antisymmetric and $\{\cdot, f\} = \mathcal{L}_{X_f}$. f *is an integral of the Hamiltonian flow of* H *if and only if* $\{f, H\} = 0$.

Proof. Equation (5.5.3) follows by definition using $X_g = (\partial g/\partial p_i, -\partial g/\partial q_i)$. Antisymmetry follows from antisymmetry of ω. $\{\cdot, f\} = \mathcal{L}_{X_f}$ since $\mathcal{L}_{X_f} g = dg \lrcorner X_f = (\omega \lrcorner X_g) \lrcorner X_f = \omega(X_g, X_f) = \{g, f\}$. If φ^t is the Hamiltonian flow for H then $(d/dt) f \circ \varphi^t = \varphi^{t*} \mathcal{L}_{X_H} f = \varphi^{t*}\{f, H\}$ vanishes if and only if $\{f, H\}$ does. □

Remark. In particular we have reproved invariance of H since $\{H, H\} = 0$.

With this setup it is also easy to prove a well-known result about Hamiltonian systems with symmetries:

Theorem 5.5.16. (Noether Theorem) *Let* (M, ω) *be a symplectic manifold,* $H \colon M \to \mathbb{R}$ *a smooth function,* $\omega \lrcorner X_H = dH$, *and* $\dot{\varphi}^t = X_H$. *If* H *is invariant under a one-parameter family of symplectic transformations generated by a Hamiltonian* f, *then* f *is an integral of* φ^t.

Proof. The hypothesis is that H is an integral for the flow of f, that is, $\{f, H\} = 0$, so conversely f is an integral for the flow of H. □

Remark. An interesting instance may arise when the phase space of the system is a cotangent bundle and the Hamiltonian is invariant under the action on the cotangent bundle of a one-parameter family of diffeomorphisms of the configuration space. Since such symmetries tend to be easy to detect, this result gives an easy way to find integrals of this sort.

Example. Consider the central-force problem of Section 5.2c. The Hamiltonian is clearly invariant under rotations around the origin. In particular it is invariant under rotations in the xy-plane, which are generated by the Hamiltonian $q_1 p_2 - q_2 p_1$, if we choose to label the coordinates (q_1, q_2). Thus $q_1 p_2 - q_2 p_1$ is a first integral. It happens to be the z-component of angular momentum. The other two components are invariant by invariance under rotations in the other planes. Similar arguments apply to the problem considered in the exercises in Section 5.2 (see Exercise 5.5.7).

Proposition 5.5.17. *Let* (M, ω) *be a symplectic manifold. Then the linear space* $C^\infty(M)$ *with the Poisson bracket* $\{\cdot, \cdot\}$ *is a Lie algebra.*

Proof. The Poisson bracket $\{\cdot, \cdot\}$ is bilinear (over $\mathbb{R}$) and antisymmetric by construction. One can prove the *Jacobi identity* $\{\{f, g\}, h\} + \{\{g, h\}, f\} +$

$\{\{h,f\},g\}=0$ by a brute-force coordinate calculation using (5.5.3). Alternatively note that

$$\{\{f,g\},h\}+\{\{g,h\},f\}+\{\{h,f\},g\}=\pounds_{X_h}\{f,g\}+\pounds_{X_f}\{g,h\}+\pounds_{X_g}\{h,f\}$$
$$=\pounds_{X_h}\pounds_{X_g}f+\pounds_{X_f}\pounds_{X_h}g+\pounds_{X_g}\pounds_{X_f}h=\pounds_{X_h}\pounds_{X_g}f+(\pounds_{X_g}\pounds_{X_f}-\pounds_{X_f}\pounds_{X_g})h$$

contains no second-order derivatives of h since $\pounds_{X_g}\pounds_{X_f}-\pounds_{X_f}\pounds_{X_g}$ is a first-order differential operator by a coordinate calculation. Likewise there are no second-order derivatives of f and g. The coordinate calculation then amounts to observing that the first-order terms cancel in pairs. □

This immediately gives a Lie algebra structure of vector fields.

Definition 5.5.18. Let M be a smooth manifold. If X,Y are vector fields on M then the *Lie bracket* $Z=[X,Y]$ is defined to be the unique vector field such that $\pounds_Z=\pounds_Y\pounds_X-\pounds_X\pounds_Y$.

Remark. The Lie bracket measures to which extent the flows of two vector fields fail to commute. Indeed the Lie bracket of two vector fields vanishes identically if and only if the corresponding flows commute.

The Jacobi identity has the

Corollary 5.5.19. $X_{\{f,g\}}=-[X_f,X_g]$.

Remark. Thus Hamiltonian vector fields form a Lie algebra as well.

A further corollary is the following result:

Proposition 5.5.20. (Poisson) *Let (M,ω) be a symplectic manifold, $H\colon M\to\mathbb{R}$ a smooth function, $\omega\lrcorner X_H=dH$, $\dot\varphi^t=X_H$, and f,g integrals of φ^t. Then $\{f,g\}$ is also an integral of φ^t.*

Proof. Since f is an integral of the Hamiltonian flow of H if and only if $\{f,H\}=0$, this follows from the Jacobi identity. □

Remark. This suggests a method for finding new integrals once two are given. This may indeed be successful at times. On the other hand one often obtains "new" integrals that are merely functions of the earlier ones.

e. Integrable systems. Before Poincaré's time the main goal of classical mechanics was to solve the equations of motion explicitly and then to discuss the dynamics. This provided a major motivation for finding integrals of the Hamiltonian system at hand, for if enough integrals are known, the orbits are determined just by the integrals. A priori one should strive to find $2n-1$ independent integrals for a system with $2n$-dimensional phase space, so a joint level set is one-dimensional and thus determines an orbit. It turns out, however, that due to the symplectic structure of the Hamiltonian equations it is indeed enough to have n independent integrals in involution, that is, with pairwise vanishing Poisson brackets, in order to be able to solve the equations of motion. Such systems are thus said to be *completely integrable*, or often just *integrable*. In this case the equations of motion can, in fact, be solved not just in principle, but somewhat explicitly ("by quadrature"). Besides showing how to solve the equations of motion of a completely integrable system, the Liouville Theorem gives a complete description of the orbit structure up to smooth conjugacy, which justifies the dogmatic preview we gave in Section 1.5. Since this part of his theorem is easy to prove and most interesting to us we give a proof here.

Theorem 5.5.21. (Liouville–Arnold Theorem) *Suppose (M,ω) is a $2n$-dimensional symplectic manifold, $H = f_1, f_2, \dots, f_n \in C^\infty(M)$, $\{f_i, f_j\} = 0$ $(i,j = 1,\dots,n)$, and $x \in \mathbb{R}^n$ is such that the differentials Df_i are (pointwise) linearly independent on*

$$M_z := \{x \in M \mid f_i(x) = z_i, i = 1,\dots,n\}.$$

Then:

(1) *M_z is a smooth Lagrangian submanifold invariant under the Hamiltonian flow φ_H^t of H (and, in fact, under any Hamiltonian flow $\varphi_{f_i}^t$).*
(2) *If M_z is compact and connected then M_z is diffeomorphic to the n-torus $\mathbb{T}^n$.*
(3) *Via this diffeomorphism $\varphi_H^t\restriction_{M_z}$ is conjugate to a linear flow.*

Remark. The full force of the Liouville Theorem asserts that in a neighborhood of M_z one can find a symplectic change of coordinates such that in the new coordinates $(y_1,\dots,y_n,\varphi_1,\dots,\varphi_n)$ the Hamiltonian depends only on $(y_1,\dots,y_n)$. These are the so-called *action–angle coordinates*. However, for the purpose of understanding the orbit structure of the Hamiltonian flow this weaker form of the Liouville Theorem is sufficient.[1]

Proof. (1) Since the Df_i are pointwise independent, M_z is a smooth submanifold by the Implicit Function Theorem. Note furthermore that the X_{f_i} are tangent to M_z, since their integral flows preserve all $f_1,\dots,f_n$. Since the Df_i are pointwise independent, so are the X_{f_i} and $\omega(X_{f_i}, X_{f_j}) = \{f_i, f_j\} = 0$, so M_z is indeed Lagrangian.

(2) Define

$$\alpha: \mathbb{R}^n \times M_z \to M_z, \quad \alpha(t,x) := \varphi_{f_1}^{t_1} \circ \varphi_{f_2}^{t_2} \circ \dots \circ \varphi_{f_n}^{t_n}(x).$$

Now $\{f_i, f_j\} = 0$ implies by Corollary 5.5.19 that $[X_{f_i}, X_{f_j}] = 0$ so the flows $\varphi_{f_i}^t$ and $\varphi_{f_j}^t$ commute and hence $\alpha(t+s,x) = \alpha(s,\alpha(t,x))$, that is, α defines an action of the additive group $\mathbb{R}^n$ on M_z.

For $x \in M_z$ the map $g_x := \alpha(\cdot, x): \mathbb{R}^n \to M_z$ has maximal rank since $Dg_x|_0 = (X_{f_1},\dots,X_{f_n})$. Thus g_x is a local diffeomorphism by the Implicit Function Theorem. This shows that the *stabilizer* $S(x_0) := \{t \in \mathbb{R}^n \mid \alpha(t,x_0) = x_0\}$ (which is evidently a subgroup) is *discrete*, that is, that there is a δ-ball $B(0,\delta)$ around $0 \in \mathbb{R}^n$ such that $B(0,\delta) \cap S(x_0) = \{0\}$ and hence $B(t,\delta) \cap S(x_0) = \{t\}$

for every $t \in S(x_0)$. We would like to show furthermore that the action α is *transitive*, that is, that the g_x are surjective. To that end notice first of all that if $y = g_x(t)$ then $g_y(s) = g_x(t+s)$, that is, $g_y(\mathbb{R}^n) \subset g_x(\mathbb{R}^n)$ whenever $y \in g_x(\mathbb{R}^n)$. Now fix $x_0 \in M_z$ and consider $x \in M_z$. M_z is a connected manifold, hence path connected. Thus there is an arc c_x connecting x_0 and x. It has a finite cover by images $g_{x_i}(B(0,\delta_i))$, so by the preceding observation we have $x \in g_{x_0}(\mathbb{R}^n)$ and g_x is surjective for all $x \in M_z$. Thus we have shown that each g_x is a covering of M_z by $\mathbb{R}^n$.

We have thus shown that α defines a transitive action of $\mathbb{R}^n$ with discrete stabilizer subgroup. We now use the fact that a discrete subgroup of $\mathbb{R}^n$ is conjugate to $\mathbb{Z}^k$ for some k, that is, $\Gamma = S(x_0)$ has a linearly independent set $\{\gamma_1, \dots, \gamma_k\}$ of generators such that $t \in \Gamma$ if and only if there exists $a \in \mathbb{Z}^k$ such that $t = \sum_{i=1}^k a_i\gamma_i$. This set of generators can be found by taking as γ_1 an element of Γ with minimal Euclidean norm, as γ_2 an element with minimal nonzero distance to the line $\mathbb{R}\gamma_1$, and so forth. In other words, if $\{e_1 \dots, e_n\}$ is the canonical basis of $\mathbb{R}^n$ and $F: \mathbb{R}^n \to \mathbb{R}^n$ is an isomorphism such that $F(e_i) = \gamma_i$ for $i \in \{1, \dots, k\}$ then $F^{-1}S_{x_0}F = \mathbb{Z}^k$. In particular, if $\pi: \mathbb{R}^n \to \mathbb{T}^k \times \mathbb{R}^{n-k}$ is the standard covering then F induces a unique diffeomorphism $f: \mathbb{T}^k \times \mathbb{R}^{n-k} \to M_z$ such that $g_{x_0} \circ F = f \circ \pi$. Compactness of M_z implies $k = n$, proving (2).

Finally (3) is now obvious by construction. In fact all $\varphi^t_{f_i}$ are translations via f. □

Exercises

5.5.1. *Let L be a Lagrangian subspace of a symplectic vector space E. Prove that L has a Lagrangian complement, that is, a Lagrangian subspace M such that $L \cap M = \{0\}$.*

5.5.2. *Prove that in a Lagrangian subspace $L \subset E$ the basis $e_1, \dots, e_{2n}$ from Proposition 5.5.2 can be chosen in such a way that $e_1, \dots, e_n \in L$.*

5.5.3. *Let $\Lambda = (\lambda_1, \dots, \lambda_{2n})$ be a collection of nonzero complex numbers with the following properties:*

(1) *Λ contains an even number of 1's and an even number of -1's.*
(2) *If $\lambda \in \Lambda$ is real, $\lambda \neq \pm 1$, then $1/\lambda \in \Lambda$ (with the same multiplicity).*
(3) *If $\lambda \in \Lambda$, $|\lambda| = 1$, and $\lambda \neq \pm 1$ then $\bar{\lambda} \in \Lambda$ (with the same multiplicity).*
(4) *If $\lambda \in \Lambda$, $|\lambda| \neq 1$, $\lambda \notin \mathbb{R}$ then $\lambda^{-1}, \bar{\lambda}, \bar{\lambda}^{-1} \in \Lambda$ (with the same multiplicities).*

Prove that there exists a symplectic linear map $T: (\mathbb{R}^{2n}, \omega) \to (\mathbb{R}^{2n}, \omega)$, where ω is the standard symplectic form, such that Λ is the set of eigenvalues of T (with multiplicities).

5.5.4*. *Prove that the 2-cohomology class of any nondegenerate closed 2-form on a $2n$-dimensional compact manifold M is nonzero.*

5.5.5. *Prove that there is no symplectic structure on the* $2n$*-sphere for* $n \geq 2$*, that is, there is no symplectic manifold* (S^{2n}, ω).

5.5.6. *Suppose* $\{\omega_t\}_{0\leq t\leq 1}$ *is a family of nondegenerate closed differential 2-forms on a compact manifold* M*. Prove that there exists a family of diffeomorphisms* $\varphi_t \colon M \to M$ *such that* $\varphi_t^* \omega_t = \omega_0$ *if and only if the cohomology classes of the forms* ω_t *are the same.*

5.5.7. *Deduce the result of Exercise 5.2.1 from the Noether Theorem 5.5.16.*

5.5.8. *Show that the geodesic flow on any surface of revolution has a first integral independent of the total energy. This integral is called the Clairaut integral.*

5.5.9. *Prove that any discrete subgroup of* $\mathbb{R}^n$ *is* $\mathbb{Z}^k$ *for some* $k \leq n$ *using the construction outlined in the proof of Theorem 5.5.21.*

5.5.10*. *Let* (M, ω) *be a symplectic manifold and* $\{\varphi^t\}$ *a Hamiltonian flow all of whose orbits are periodic with the same minimal period* T*. Fix a value* c *of the Hamiltonian and consider the factor space* N *of the level surface* M_c *by the action of the flow. Show that the restriction of* ω *to* M_c *projects to a nondegenerate 2-form on* N.

5.5.11. *Show that the geodesic flow on the standard* n*-dimensional sphere satisfies the conditions of the previous exercise. Apply the procedure from that exercise to obtain a* $(2n-2)$*-dimensional symplectic manifold. Describe that manifold in detail for* $n = 2$.

6. Contact systems

a. Hamiltonian systems preserving a 1-form. From the point of view of classical mechanics the most important (or at least the most traditional) symplectic manifolds are $\mathbb{R}^{2n}$ with the standard symplectic structure and the cotangent bundle of a differentiable manifold M (the *configuration space* of a mechanical system) with the symplectic form ω described in Section 5.5b. Notice that in both cases the symplectic manifold (phase space) itself is not compact, although in the second case the configuration space M may be compact; this is true in many important classical problems such as the motion of a rigid body. Of course $\mathbb{R}^{2n}$ can also be viewed as $T^*\mathbb{R}^n$, so the first case is a particular instance of the second.

In this book we primarily consider dynamical systems with compact phase space. In order to apply the concepts and methods discussed in this book to a Hamiltonian system with Hamiltonian H one considers the restriction of the dynamics to the hypersurfaces $H = c$ which are compact in many situations, for example, for a geodesic flow on a compact Riemannian manifold, where those hypersurfaces are sphere bundles over the configuration space. Sometimes one can make a further reduction using the first integrals other than energy. If c is not a critical value of the Hamiltonian and the hypersurface $H_c{:=}\{x \mid H(x) = c\}$ is compact then the Hamiltonian system preserves a nondegenerate $(2n-1)$-form ω_c which can be described as follows. One can locally decompose the $2n$-dimensional measure generated by ω into $(2n-1)$-dimensional measures on $H_{c+\delta}$ for all sufficiently small $|\delta|$ and consider the conditional measures, each of which is defined up to a multiplicative constant. Thus in this case due to Proposition 5.5.12 one can apply the Poincaré Recurrence Theorem 4.1.19, the Birkhoff Ergodic Theorem 4.1.2, and other facts from ergodic theory to the restriction of the Hamiltonian system to H_c.

There is an important situation when the invariant $(2n-1)$-forms can be described in a particularly natural way. Notice that in the case of both $\mathbb{R}^{2n}$ and T^*M the form ω is not only closed, but also *exact*. The 1-form θ defined by $\sum_{i=1}^n p_i dq_i$—globally in the first case, locally in the second—obviously satisfies $d\theta = \omega$. The calculation in the proof of Lemma 5.5.10 shows that θ is defined on T^*M independently of the choice of local coordinates. Of course in general a Hamiltonian system on T^*M does not preserve θ or any other 1-form whose exterior derivative is equal to ω. Let us see what conditions the invariance of θ imposes on the Hamiltonian. One has

$$\pounds_{X_H}\theta = d\theta \lrcorner X_H + d(\theta \lrcorner X_H) = dH + d(\theta \lrcorner X_H).$$

Thus a 1-form θ is invariant if $\theta \lrcorner X_H = -H$. Local calculation in Darboux coordinates gives $\theta \lrcorner X_H = -\sum p_i \dfrac{\partial H}{\partial p_i}$. Notice that the choice of Hamiltonian for a given vector field X_H is unique up to an additive constant. Thus we have proved the following fact:

Proposition 5.6.1. *The Hamiltonian vector field X_H on T^*M preserves the 1-form θ if and only if the Hamiltonian can be chosen as positively homogeneous in p of degree one, that is, $H(q, \lambda p) = \lambda H(q,p)$ for $\lambda > 0$.*

There is a broader class of Hamiltonians that preserves the form θ *along* the hypersurfaces $H =$ const. In this case the invariance condition becomes

$$d(\theta \lrcorner X_H)(\xi) = 0 \text{ if } dH(\xi) = 0,$$

or in other words the function $\theta \lrcorner X_H$ is constant on every connected component of the hypersurface $H =$ const. This is satisfied if

$$\theta \lrcorner X_H = \varphi(H),$$

that is, using Darboux coordinates,

$$H(q, \lambda p) = \Phi(\lambda) H(q,p),$$

where $\Phi' = \varphi$. If $\varphi(\lambda) \neq 0$ then such Hamiltonians will be called *generalized homogeneous Hamiltonians (in p)*. Away from the zero section every such

Hamiltonian is a function of a homogeneous Hamiltonian of degree one, namely, $H_1(q,p) = \Phi^{-1}(H(q,p))$, where Φ^{-1} is the inverse of Φ. An immediate calculation shows that $X_{\rho(H)} = \rho' X_H$ for any C^1 function ρ, so the flow generated by a generalized homogeneous Hamiltonian is obtained by a time change from the flow generated by a Hamiltonian that preserves θ and the time change is constant on every surface $H = \text{const}$.

In particular, since the Hamiltonian of a geodesic flow is a quadratic function of p, it preserves the restriction of θ to any energy surface.

b. Contact forms. The restriction of the form θ to the surface $H = c$ for a noncritical value of c of H is an example of a 1-form such that $\theta \wedge (d\theta)^{n-1}$ is nondegenerate. This motivates the following definition.

Definition 5.6.2. A 1-form θ on a $(2n-1)$-dimensional orientable manifold M is called a *contact form* if the $(2n-1)$-form $\theta \wedge (d\theta)^{n-1}$ is nondegenerate. Accordingly a pair (M, θ) of a smooth manifold with a contact form is called a *contact manifold.* A *contact flow* is a flow on M that preserves the contact form on M. A diffeomorphism preserving the contact form is called a *contact diffeomorphism.*

Unlike a symplectic manifold, which admits a variety of Hamiltonian vector fields, a contact manifold comes furnished with a canonical vector field v which is defined by $v \lrcorner \theta = 1$ and $v \lrcorner d\theta = 0$. This is unique because the kernel of $d\theta^n$ is one-dimensional and disjoint from that of θ by the nondegeneracy assumption. Note that the Lie derivative $\mathcal{L}_v \theta$ vanishes since $v \lrcorner \theta = \text{const.}$, so the flow of v, which is called the *characteristic flow* of the contact form, preserves θ and hence all structures defined in terms of θ, in particular the volume. Thus the characteristic flow provides a canonical example of a volume-preserving flow.

Suppose now that X is a vector field generating a flow preserving the contact form θ. Then it must preserve $\ker d\theta^n$ as well and hence it must commute with the characteristic flow of θ. Thus contact flows always arise as flows commuting with the characteristic flow of a contact form. The same comment holds for a contact diffeomorphism.

Furthermore if the contact manifold is a level set for a homogeneous Hamiltonian as in the previous section then v is exactly the Hamiltonian vector field. In fact, contact forms always arise in this manner from generalized homogeneous Hamiltonians (see Proposition 5.6.4). Conversely the remark at the end of Section 5.3 and the observations of the preceding subsection are summarized as follows.

Proposition 5.6.3. *Geodesic flows are Hamiltonian flows with a homogeneous Hamiltonian. Hamiltonian flows for homogeneous Hamiltonians, in particular geodesic flows, are characteristic flows, hence contact flows.*

Proposition 5.6.4. *Suppose (M, θ) is a contact manifold. Then M can be embedded into a symplectic manifold (N, ω) in such a way that the restriction of the ambient symplectic form to M is $d\theta$.*

Remark. A contact manifold embedded in this way is called a *submanifold of contact type.*

Proof. If $N = M \times \mathbb{R}$ and $\omega_{x,t} = d(e^t\theta_x)$ then $\omega^n = e^{nt}(ndt \wedge \theta \wedge (d\theta)^{n-1})$ is a volume, so (N, ω) is a symplectic manifold and ω restricted to $M \times \{0\}$ is $d\theta$. □

Proposition 5.6.5. *Let ω be the standard symplectic form on $\mathbb{R}^{2n}$ and $M = f^{-1}(c) \subset \mathbb{R}^{2n}$ a level set of a smooth function $f\colon \mathbb{R}^{2n} \to \mathbb{R}$ with c as a regular value. Then M is a submanifold of contact type if and only if on a neighborhood U of M there is a vector field ξ transverse to M for which $\mathfrak{L}_\xi\omega = \omega$.*

Remark. In the situation of homogeneous Hamiltonians described in the previous subsection ξ is simply $\sum_{i=1}^n p_i\partial/\partial p_i$.

Proof. Let $\theta' = \xi \lrcorner \omega$ and $\theta = \theta'\restriction_M$. Notice that $\mathfrak{L}_\xi\omega = d(\xi\lrcorner\omega)$ since $d\omega = 0$ and consequently $d\theta = \mathfrak{L}_\xi\omega = \omega$ on M. To show that θ is a contact form note first that the transversality assumption means that for the gradient vector field ∇f

$$0 \neq \langle \xi, \nabla f\rangle = Df(\xi) = \omega(X_f, \xi) = (-\xi\lrcorner\omega)X_f = -\theta(X_f)$$

on M. Furthermore $\ker d\theta$ is one-dimensional because θ was obtained from the nondegenerate form ω by contracting with one vector. This shows that $\theta \wedge (d\theta^n)$ defines a volume.

To prove "only if" suppose (M, θ) is of contact type. By the Poincaré Lemma θ can be extended to θ' on a neighborhood U of M in such a way that $d\theta' = \omega$ on U. Then $\xi\lrcorner\omega = \theta'$ uniquely defines ξ and we have $\langle \nabla f, \xi\rangle = \theta(X_f) \neq 0$, so ξ is transverse to M. Finally $\mathfrak{L}_\xi\omega = d(\xi\lrcorner\omega) = d\theta' = \omega$. □

Remark. It is the use of the Poincaré Lemma at the end that necessitates working in $\mathbb{R}^{2n}$.

Locally a contact form, similarly to a symplectic form, can be brought into a standard form. In fact, the following result is a simple consequence of the Darboux Theorem 5.5.9 for symplectic forms.

Theorem 5.6.6. (Darboux Theorem for contact forms) *Let $\theta_0 = x_1dy_1 + \cdots + x_ndy_n + dz$ be the canonical contact form on $\mathbb{R}^{2n+1}$ and (M, θ) a contact $(2n+1)$-manifold. Then for $x \in M$ there exists a neighborhood U of x with coordinates in which $\theta = \theta_0$.*

Proof. For $x \in M$ pick a neighborhood V_0 of 0 in $\ker\theta_x$ and let $V = V_0 \times (-\epsilon, \epsilon)$, $U' = \exp V$, $U'_t = \exp(V_0 \times \{t\}) \subset M$. $d\theta$ restricted to U'_t is a symplectic form so by the Darboux Theorem 5.5.9 each $y \in U'_t$ has a neighborhood $U_t \subset U'_t$ on which there are Darboux coordinates $x_1, \dots, x_n, y_1, \dots, y_n, z$, that is, $d\theta = \sum dx_i \wedge dy_i$. On $U := \bigcup_{-\epsilon<t<\epsilon} U_t$ we thus have $d(\theta - \sum dx_i \wedge dy_i) = 0$ whence $\theta = \sum dx_i \wedge dy_i + dz$ and $x_1, \dots, x_n, y_1, \dots, y_n, z$ are the desired coordinates. □

Exercises

A hypersurface M in $\mathbb{R}^n$ is called *star-shaped* if there exists a point c such that every half-line from c to ∞ intersects M in exactly one point.

5.6.1. *Prove that any star-shaped hypersurface in $\mathbb{R}^{2n}$ provided with the standard symplectic structure is of contact type.*

5.6.2. *Describe the contact form and the characteristic vector field on $S^{2n-1} \subset \mathbb{R}^{2n}$ corresponding to the vector field $\xi = (1/2)\sum_{i=1}^{n}\left(p_i\frac{\partial}{\partial p_i} + q_i\frac{\partial}{\partial q_i}\right)$.*

5.6.3. *In the setting of the previous problem identify $\mathbb{R}^{2n}$ with $\mathbb{C}^n$ by writing $z = p + iq$. Show that all orbits of the vector field v are closed and the factor is complex projective space $\mathbb{CP}^n = S^{2n-1}/S^1$.*

5.6.4. *Consider a hypersurface M in the cotangent bundle T^*N of a smooth manifold that intersects each fiber in a star-shaped hypersurface. Prove that M is of contact type with respect to the standard symplectic structure.*

5.6.5. *Consider the unit tangent bundle $S\mathbb{T}^n$ of the flat torus with the 1-form α that is obtained from the standard 1-form θ given by (5.5.1) via the Legendre transformation (5.3.5). Describe the groups of diffeomorphisms and vector fields on $S\mathbb{T}^n$ preserving α.*

7. Algebraic dynamics: Homogeneous and affine systems

Let G be a locally compact metrizable topological group. This class includes on the one hand compact abelian groups, which we first encountered in Section 1.3 and on the other hand Lie groups such as the group of isometries of the hyperbolic plane (see Sections 5.4 and 17.5). Tori belong to both classes.

Suppose G is *unimodular*, that is, has a locally finite Borel measure invariant with respect to both left and right translations. Let $\Gamma \subset G$ be a lattice in G (see Section A.8). Since right translations on G commute with any left translation, they project to maps of $\Gamma\backslash G$. In fact, since Γ is closed, any right translation defines a homeomorphism of $\Gamma\backslash G$. Since G is unimodular and $\Gamma \subset G$ is a lattice, right translations preserve the finite Haar measure on $\Gamma\backslash G$. Any such translation $T: \Gamma\backslash G \to \Gamma\backslash G$ is called *a homogeneous dynamical system*; a one-parameter group of translations is called *a homogeneous flow*. Thus, similarly to Hamiltonian and contact systems discussed in the previous sections, a homogeneous system is a topological dynamical system and a measure-preserving transformation at the same time. One should point out though, that the phase space is compact only for a uniform lattice.

Translations on compact abelian groups (Sections 1.3 and 1.4) are examples of homogeneous dynamical systems; linear flows on the torus (Section 1.5) are examples of homogeneous flows. We will see in Section 17.5 that the geodesic flow on a compact factor of the hyperbolic plane (Section 5.4e) can be naturally represented as a homogeneous flow on the factor of the group $PSL(2, \mathbb{R})$

of Möbius transformations by a certain cocompact (uniform) lattice Γ. Here $PSL(2,\mathbb{R})$ is the factor of the group $SL(2,\mathbb{R})$ of all 2×2 matrices with determinant one by its center, which consists of $\pm\,\mathrm{Id}$. A reader interested in this interpretation may proceed to Section 17.5 right away. In Section 17.7 we develop this type of construction even further and discuss important homogeneous flows arising from geodesic flows on certain special Riemannian manifolds in higher dimension.

Another class of dynamical systems with a well–defined algebraic structure is represented by linear expanding maps on the circle (Section 1.7) and linear automorphisms and endomorphisms of a torus (Section 1.8). By uniqueness of a normalized Haar measure on a compact abelian group any automorphism of such a group must preserve this measure and any endomorphism multiplies it by a constant; in the latter case the measure is still preserved in the sense of the definition at the end of Section 4.1b. An interesting example of an automorphism of a compact abelian group other than a torus will appear in Exercise 17.1.2.

There are certain situations where automorphisms and endomorphisms of non–compact locally compact groups generate transformations of compact homogeneous spaces. Examples of this kind are discussed in Section 17.3, where G is a *nilpotent* non–abelian Lie group.

Finally, there is a class of algebraic dynamical systems which includes both homogeneous systems and automorphisms; these are *affine* systems which can be described as suitable projections on the finite–measure factors of affine maps on the group G; affine maps are compositions of an endomorphism and a translation. The simplest interesting examples of affine maps, which exhibit properties different from those of both translations and automorphisms, are the maps A_α on the two–torus which have been discussed in Exercises 1.4.4, 3.2.6 and 4.2.3. The subsequent exercises to Section 4.2 demonstrate an intimate connection between the dynamical properties of these maps and similar affine maps in higher dimension on the one hand and the uniform distribution of fractional parts of polynomials on the other hand. This is the first indication of extremely fruitful connections between the dynamics of algebraic systems (homogeneous and affine) and number theory.

The study of asymptotic properties of these classes of algebraic dynamical systems (homogeneous, endomorphisms and affine), both topological and measure–theoretic, is a well–developed area, which combines general ideas and methods from ergodic theory, topological dynamics and differentiable dynamics with more specific methods which take into account the uniform local structure of these systems throughout the phase spase. We saw elementary examples of the appearance of such special methods in the use of Fourier analysis in Sections 1.4 and 4.2. There is a clearly defined invariant structure which makes it reasonable to consider this area as a separate branch of the theory of dynamical systems along with the principal branches discussed in Section 0.1. While no universally accepted name for it exists, the appelation *algebraic dynamics* looks quite appropriate to us.

Part 2

Local analysis and orbit growth

6

Local hyperbolic theory and its applications

In this chapter we carry out a part of the program outlined in Section 0.4. The main paradigm of this analysis is a sort of hyperbolicity of the linearized dynamical system along certain orbits. We show that this translates into a similar behavior for the nonlinear system near a reference orbit (Hadamard–Perron Theorem 6.2.8). The combination of local hyperbolicity coming from the linearized system with nontrivial recurrence, an essentially nonlinear phenomenon, produces an abundance of periodic orbits (Anosov Closing Lemma, Theorem 6.4.15) and an otherwise rich and stable orbit structure which will be further explored in Part 4 of this book.

1. Introduction

In order to avoid notational ambiguity let us consider most of the time a smooth discrete-time dynamical system $f: M \to M$. We fix a "reference" initial condition $p \in M$. Our main objective is to identify those initial conditions x whose evolution under the iterates of f follows that of p sufficiently closely for a sufficiently long time, positive or negative, and to understand the asymptotic behavior of $f^n(x)$ *relative to* $f^n(p)$. We will be particularly interested in those x for which $f^n(x)$ stays close to $f^n(p)$ for *all* positive or negative values of n.

The principal tool in this analysis will be information about the asymptotic behavior of the linear maps $(Df^n)_p$ as n goes to $+\infty$ or $-\infty$, which in a sense reflects the asymptotic behavior of "initial conditions infinitesimally close to p". We will try to show that under certain conditions the behavior of some orbits of the nonlinear system f relative to the reference orbit imitates the behavior of orbits for the linearized system.

The most natural setting for such an analysis appears when the reference orbit is periodic of period m , say, and the differential $(Df^m)_p$ is a hyperbolic linear map in the sense of Definition 1.2.5. Such an orbit is usually called a *hyperbolic periodic orbit*. The local structure of a map near a hyperbolic periodic orbit is studied at the beginning of Section 6.2 and in Section 6.3.

There are several generalizations of this setting to the case of a nonperiodic reference orbit. One of them is discussed in detail in Section 6.2. It involves an invariant splitting for the linearized system along the reference orbit into subspaces of uniformly exponentially contracting and uniformly exponentially expanding vectors (see Definition 6.2.6). We will sometimes call this situation the *uniform hyperbolic splitting.*

Although our interests in this chapter are centered on the hyperbolic case where distinctive qualitative phenomena appear, in view of various applications it is natural to consider a somewhat more general situation of *exponential splitting* (Definition 6.2.6). This is the setting of Theorem 6.2.8, the main technical result of the present chapter.

The local analysis of the situation with uniform hyperbolic splitting along a reference orbit based on that theorem provides the foundation of the global and semilocal theory for a very important class of *hyperbolic* dynamical systems. This class contains all our previous examples of invertible smooth dynamical systems with complicated orbit structure, namely, hyperbolic toral automorphisms (Sections 1.8, 3.2e, 4.2c, 4.4d), their C^1-perturbations (Section 2.6), and various "horseshoes" (Section 2.5c), as well as expanding maps of the circle (Sections 1.7, 2.4, 3.2c, 4.2b, 4.4c, 5.1c), horseshoe-type invariant sets for quadratic maps (Section 2.5b), and in the continuous-time case geodesic flows on compact factors of the hyperbolic plane (Section 5.4f).

The general notion of a hyperbolic dynamical system is introduced in Section 6.4 where, in addition to stating immediate consequences of the preceding local analysis, we make initial steps toward showing that in agreement with the conclusion of our study of examples, hyperbolicity in nonlinear systems tends to produce an abundance of periodic orbits. In Section 6.5 we will see that horseshoe-type hyperbolic sets appear very naturally in the context of semilocal analysis. A more systematic study of hyperbolic dynamical systems is carried out in Part 4 of this book.

As important as the case of uniform hyperbolic splitting is, its assumptions are somewhat too restrictive, especially for the purposes of global (as opposed to semilocal) analysis of dynamical systems. In other words, unlike hyperbolic invariant sets (Definition 6.4.1 and Definition 6.4.3) which appear very often in broad classes of smooth dynamical systems, hyperbolicity of a map or a flow on the whole compact manifold is a very significant but rather special phenomenon. A weaker local condition related to *nonuniform hyperbolicity* or, more generally, "nonuniform exponential splitting" turns out to be a much more suitable and fruitful tool for global analysis of a very broad class of smooth dynamical systems, especially in the study of their stochastic behavior but also in the more topological aspects of the subject. See the Supplement for an introduction to this theory. Let us point out that the setting and methods of Section 6.2 can be adapted to the nonuniform situation with relatively minor adjustments.

So far we have been discussing a local approach based on the assumption that the linearized system serves as a model for the local behavior of a nonlinear system, thus implying that nonlinear terms constitute a rather annoying perturbation which should be kept under control. A natural next step in a local analysis is to try to consider the terms of higher order (than linear) in a more systematic and specific way and try to determine more precisely the extent to which their influence has to be taken into account or whether it can be disregarded altogether. We consider this problem in Section 6.6. Again, a hyperbolic periodic orbit is the best setting of such an analysis. The key phenomena here are certain "resonances" between the eigenvalues of the linearized map. Their presence or absence determines what higher-order terms must be taken into account. In the nonhyperbolic case this kind of analysis is primarily *formal*, that is, it can be carried out only up to terms of (arbitrarily) high order whereas in the hyperbolic case it produces a smooth conjugacy.

Exercise

6.1.1. *Let M be a compact manifold, $f: M \to M$ a C^1 map, p a fixed point of f, and λ an eigenvalue of the differential Df_p. Prove that $|\lambda| \leq l(f)$ where $l(f)$ is as in (3.2.6).*

2. Stable and unstable manifolds

a. Hyperbolic periodic orbits. A natural setting for studying orbits in a neighborhood of an invariant compact set, in particular a periodic orbit, has been described in Section 0.4. Namely, let M be a smooth manifold, $U \subset M$ an open subset, $f: U \to M$ a C^1 diffeomorphism onto its image, and p a periodic point of period n whose orbit lies in U.

Definition 6.2.1. p is a *hyperbolic periodic point* for f if $(Df^n)_p: T_pM \to T_pM$ is a hyperbolic linear map (Definition 1.2.5). Its orbit will be called a *hyperbolic periodic orbit.*

Naturally, a hyperbolic periodic point of period n for f is a hyperbolic fixed point for f^n and vice versa. Hence for the purposes of local analysis it is usually sufficient to consider only hyperbolic fixed points.

For the sake of completeness let us give an analogous definition for continuous-time dynamical systems. In this case we assume that a smooth vector field ξ is defined in U and the orbit of a point $p \in U$ lies in U and closes at time t_0. There are two possible cases: Either $\xi(p) = 0$ or $\xi(p) \neq 0$.

Definition 6.2.2. If $\xi(p) = 0$, the point p is called a *hyperbolic fixed point* of the (local) flow φ^t generated by ξ if $(D\varphi^t)_p: T_pM \to T_pM$ is a hyperbolic linear map for every $t \neq 0$.

If $\xi(p) \neq 0$, the point p is called a *hyperbolic periodic point* of period t for the flow φ^t if $\varphi^t(p) = p$ and the linear operator $(D\varphi_t)_p: T_pM \to T_pM$ has one as a simple eigenvalue and no other eigenvalues of absolute value one.

One can also describe hyperbolicity of a periodic point for a continuous-time system using *Poincaré maps* (Section 0.3). Namely, let N be a small disc of codimension one containing p and transverse to the vector field ξ. Then for an open subset $V \subset N$ with $p \in V$ the Poincaré (first-return) map $F_N: V \to N$ is defined and $F_N(p) = p$. Then p is a hyperbolic periodic point of the flow φ^t if and only if it is a hyperbolic fixed point for the map F_N.

Recall that for a linear map $A: \mathbb{R}^n \to \mathbb{R}^n$ the set of all eigenvalues of A is denoted by $\operatorname{sp}(A)$ (Definition 1.2.1). If A is hyperbolic we define the slowest contraction and expansion rates of A by

$$\lambda(A) := r(A\restriction_{E^-}) = \sup\{|\chi| \mid \chi \in \operatorname{sp}(A), \quad |\chi| < 1\},$$
$$\mu(A) := 1/r(A^{-1}\restriction_{E^+}) = \inf\{|\chi| \mid \chi \in \operatorname{sp}(A), \quad |\chi| > 1\}.$$

(For the definition of the subspaces E^+ and E^- see (1.2.4) and (1.2.5).)

By Proposition 1.2.2 for any $\delta > 0$ one can introduce a norm in $\mathbb{R}^n$ such that $\|A\restriction_{E^-}\| < \lambda(A) + \delta$ and $\|A^{-1}\restriction_{E^+}\| < \mu^{-1}(A) + \delta$.

Theorem 6.2.3. *Let p be a hyperbolic fixed point of a local C^r diffeomorphism $f: U \to M$, $r \geq 1$. Then there exist C^r embedded discs W_p^+, $W_p^- \subset U$ such that $T_pW_p^\pm = E^\pm(Df_p)$, $f(W_p^-) \subset W_p^-$, and $f^{-1}(W_p^+) \subset W_p^+$ and there exists $C(\delta)$ such that for any $y \in W_p^-$, $z \in W_p^+$, $m \geq 0$,*

$$\operatorname{dist}(f^m(y), p) < C(\delta)\,(\lambda(Df_p) + \delta)^m \operatorname{dist}(y, p),$$
$$\operatorname{dist}(f^{-m}(z), p) < C(\delta)\,\left(\mu^{-1}(Df_p) + \delta\right)^m \operatorname{dist}(z, p).$$

Furthermore, there exists $\delta_0 > 0$ such that

$$\text{if } \operatorname{dist}(f^m(y), p) \leq \delta_0 \text{ for } m \geq 0 \quad \text{then} \quad y \in W_p^-,$$
$$\text{if } \operatorname{dist}(f^m(z), p) \leq \delta_0 \text{ for } m \leq 0 \quad \text{then} \quad z \in W_p^+.$$

In fact, there exists a neighborhood $O \subset U$ of p and C^r coordinates $\psi: O \to \mathbb{R}^n$ such that $\psi(W_p^+ \cap O) \subset \mathbb{R}^k \oplus \{0\}$ and $\psi(W_p^- \cap O) \subset \{0\} \oplus \mathbb{R}^{(n-k)}$ (adapted coordinates).

Remark. The discs W_p^+ and W_p^- are not uniquely defined. However, for any two discs satisfying the assertion of this theorem for W_p^+ their intersection contains a neighborhood of p in each of them. In other words, they are open subsets of a common larger submanifold. The same property holds for W_p^-.

Definition 6.2.4. Any disc W_p^+ (correspondingly, W_p^-) satisfying the assertion of Theorem 6.2.3 is called a *local unstable manifold* (correspondingly, a *local stable manifold*) of f at the point p. The manifolds

$$W^u = W_p^u = \bigcup_{m \geq 0} f^m(W_p^+)$$

and

$$W^s = W_p^s = \bigcup_{m \le 0} f^m(W_p^-)$$

are called the (global) *unstable manifold* and the *stable manifold* of f at the point p, respectively.

Unlike the local stable and unstable manifolds the global ones are usually immersed in the phase space in a complicated way. For a typical picture of stable and unstable manifolds see Figure 6.5.2. When it does not cause confusion we will omit the reference to f and will simply talk about local and global stable and unstable manifolds of a point.

Corollary 6.2.5.

$$W_p^u = \{y \in U \mid \operatorname{dist}(f^{-m}(y), p) \xrightarrow[m\to\infty]{} 0\},$$
$$W_p^s = \{y \in U \mid \operatorname{dist}(f^{m}(y), p) \xrightarrow[m\to\infty]{} 0\}.$$

Thus, the stable and unstable manifolds of a hyperbolic fixed point are defined in topological terms. Since Theorem 6.2.3 deals exclusively with dynamics in a neighborhood of the point p it can be restated in local terms. Namely, one can introduce a coordinate chart near p with p as the origin and $E^+(Df_p)$ and $E^-(Df_p)$ tangent to the coordinate planes $\mathbb{R}^k \times \{0\}$ and $\{0\} \times \mathbb{R}^{n-k}$, correspondingly. Theorem 6.2.3 in this local (or Euclidean) form becomes a particular case of the main result in this section (Theorem 6.2.8). Smooth adapted coordinates are obtained from the coordinates $\psi_0 : U \to \mathbb{R}^k \oplus \mathbb{R}^l$ in the proof of Theorem 6.2.8 in which W_{loc}^u is the graph of a function $\varphi^+ : \mathbb{R}^k \to \mathbb{R}^l$ and W_{loc}^s is the graph of $\varphi^- : \mathbb{R}^l \to \mathbb{R}^k$ by passing from $(x, y) \in \mathbb{R}^k \oplus \mathbb{R}^l$ to $(x', y') = (x - \varphi^-(y), y - \varphi^+(x))$.

b. Exponential splitting. Now we proceed to the local analysis near a general (nonperiodic) orbit.

As we discussed in Section 0.4 the differentials of the iterates f^k, $k \in \mathbb{Z}$, along such an orbit can not be reduced to the iterates of a single linear map but should be viewed as products of different linear maps. Thus, we can not talk about eigenvalues any more, but rather should define hyperbolicity in terms of expansion and contraction of tangent vectors. We will also generalize the situation somewhat by allowing a more general kind of exponential splitting for linear maps into "fast-expanding" or "fast-contracting" directions and the rest. As in the case of a single point one can choose appropriate coordinate systems centered at the points of the reference orbit and express both the nonlinear map and its differential in those coordinates. More details on that reduction will be given in Section 6.4 in connection with the proof of Theorem 6.4.9.

Definition 6.2.6. Let $\lambda < \mu$. A sequence of invertible linear maps $L_m: \mathbb{R}^n \to \mathbb{R}^n$, $m \in \mathbb{Z}$, is said to admit a (λ, μ)*-splitting* if there exist decompositions $\mathbb{R}^n = E_m^+ \oplus E_m^-$ such that $L_m E_m^\pm = E_{m+1}^\pm$ and

$$\|L_m \restriction_{E_m^-}\| \leq \lambda, \quad \|L_m^{-1} \restriction_{E_{m+1}^+}\| \leq \mu^{-1}.$$

We will say that $\{L_m\}_{m\in\mathbb{Z}}$ *admits an exponential splitting* if it admits a (λ, μ)-splitting for some λ, μ and $\lambda < 1$, $\dim E_m^- \geq 1$ or $\mu > 1$, $\dim E_m^+ \geq 1$. We will call $\{L_m\}_{m\in\mathbb{Z}}$ *hyperbolic* (or *uniformly hyperbolic*) if it admits a (λ, μ)-splitting for some $\lambda < 1 < \mu$.

By viewing $\mathbb{R}^n$ as a canonical product $\mathbb{R}^k \times \mathbb{R}^{n-k}$ and making a sequence of orthogonal coordinate changes in $\mathbb{R}^n$ one can assume in the previous definition that $E_m^+ = \mathbb{R}^k \times \{0\}$, $E_m^- = \{0\} \times \mathbb{R}^{n-k}$ for some k, $0 \leq k \leq n$, and all m.

Thus we have reduced the problem of the local behavior of the iterates of a diffeomorphism near a reference orbit to the study of a sequence of local diffeomorphisms $f_m: U_m \to \mathbb{R}^n$, where each U_m is a neighborhood of the origin in $\mathbb{R}^n$ containing a ball of some fixed radius, fixing the origin and such that the sequence of linear maps at the origin $(Df_m)_0$, $m \in \mathbb{Z}$, admits an exponential splitting. Although we will be interested only in points whose successive images stay in the neighborhoods, it is convenient to artificially extend our maps from somewhat smaller neighborhoods to the whole space $\mathbb{R}^n$ using the following fact.

Lemma 6.2.7. (Extension Lemma) *Let U be an open bounded neighborhood of $0 \in \mathbb{R}^n$ and $f: U \to \mathbb{R}^n$ a local diffeomorphism with $f(0) = 0$. For $\epsilon > 0$ there exist $\delta > 0$ and a diffeomorphism $\bar{f}: \mathbb{R}^n \to \mathbb{R}^n$ such that $\|\bar{f} - Df_0\|_{C^1} < \epsilon$ and $\bar{f} = f$ on $B(0, \delta)$.*

Proof. For $\eta > 0$ take a C^1 $\rho: \mathbb{R}^n \to [0,1]$ such that $\rho = 1$ on $B(0, \delta)$ and $\rho = 0$ off $B(0, \eta)$, $\|D\rho\| \leq C_0/\eta$, and $\bar{f} = \rho \cdot f + (1 - \rho)Df_0$ (where $\rho \cdot f$ is understood to be zero when $\rho = 0$ even if f is undefined). Then $\bar{f} - Df_0 = \rho \cdot (f - Df_0)$. Since f is C^1 we have $\|f - Df_0\|_{C^0} = o(\eta)$. Furthermore $\|D(\bar{f} - Df_0)\| \leq \|D\rho(f - Df_0)\| + \|\rho(Df - Df_0)\| = o(1)$ in η. The claim follows. □

c. The Hadamard–Perron Theorem. Now we are ready to formulate the central fact of our local analysis, the stable–unstable manifold theorem. We state it in greater generality although our main application will be to the hyperbolic situation. For this reason we mention that in the hyperbolic case the invariant manifolds we obtain are as smooth as the map.

Theorem 6.2.8. (Hadamard–Perron Theorem) *Let $\lambda < \mu$, $r \geq 1$, and for each $m \in \mathbb{Z}$ let $f_m: \mathbb{R}^n \to \mathbb{R}^n$ be a (surjective) C^r diffeomorphism such that for $(x, y) \in \mathbb{R}^k \oplus \mathbb{R}^{n-k}$*

$$f_m(x, y) = (A_m x + \alpha_m(x, y),\ B_m y + \beta_m(x, y))$$

for some linear maps $A_m: \mathbb{R}^k \to \mathbb{R}^k$ *and* $B_m: \mathbb{R}^{n-k} \to \mathbb{R}^{n-k}$ *with* $\|A_m^{-1}\| \leq \mu^{-1}$, $\|B_m\| \leq \lambda$ *and* $\alpha_m(0) = 0$, $\beta_m(0) = 0$.

Then for $0 < \gamma < \min\left(1, \sqrt{\mu/\lambda} - 1\right)$ *and*

$$0 < \delta < \min\left(\frac{\mu - \lambda}{\gamma + 2 + 1/\gamma}, \frac{\mu - (1+\gamma)^2\lambda}{(1+\gamma)(\gamma^2 + 2\gamma + 2)}\right)$$

we have: If $\|\alpha_m\|_{C^1} < \delta$ *and* $\|\beta_m\|_{C^1} < \delta$ *for all* $m \in \mathbb{Z}$ *then there is*

(1) *a unique family* $\{W_m^+\}_{m\in\mathbb{Z}}$ *of* k*-dimensional* C^1 *manifolds*

$$W_m^+ = \{(x, \varphi_m^+(x)) \mid x \in \mathbb{R}^k\} = \operatorname{graph} \varphi_m^+$$

and

(2) *a unique family* $\{W_m^-\}_{m\in\mathbb{Z}}$ *of* $(n-k)$*-dimensional* C^1 *manifolds*

$$W_m^- = \{(\varphi_m^-(y), y) \mid y \in \mathbb{R}^{n-k}\} = \operatorname{graph} \varphi_m^-,$$

where $\varphi_m^+: \mathbb{R}^k \to \mathbb{R}^{n-k}$, $\varphi_m^-: \mathbb{R}^{n-k} \to \mathbb{R}^k$, $\sup_{m\in\mathbb{Z}} \|D\varphi_m^\pm\| < \gamma$, *and the following properties hold:*

(i) $f_m(W_m^-) = W_{m+1}^-$, $f_m(W_m^+) = W_{m+1}^+$.

(ii) $\|f_m(z)\| < \lambda'\|z\|$ *for* $z \in W_m^-$,
$\|f_{m-1}^{-1}(z)\| < (\mu')^{-1}\|z\|$ *for* $z \in W_m^+$,
where $\lambda' := (1+\gamma)(\lambda + \delta(1+\gamma)) < \dfrac{\mu}{1+\gamma} - \delta =: \mu'$.

(iii) *Let* $\lambda' < \nu < \mu'$. *If* $\|f_{m+L-1} \circ \cdots \circ f_m(z)\| < C\nu^L\|z\|$ *for all* $L \geq 0$ *and some* $C > 0$ *then* $z \in W_m^-$.
Similarly, if $\|f_{m-L}^{-1} \circ \cdots \circ f_{m-1}^{-1}(z)\| \leq C\nu^{-L}\|z\|$ *for all* $L \geq 0$ *and some* $C > 0$ *then* $z \in W_m^+$.

Finally, in the hyperbolic case $\lambda < 1 < \mu$ *the families* $\{W_m^+\}_{m\in\mathbb{Z}}$ *and* $\{W_m^-\}_{m\in\mathbb{Z}}$ *consist of* C^r *manifolds.*

Before we proceed to the proof we make several remarks clarifying the statement and the meaning of the theorem. Most estimates do not use the assumption that $\gamma < 1$. For the first reading it may be useful to ignore dependence of our data of m and to think instead of iterates of a single locally defined map f. This gives the local version of Theorem 6.2.3.

Fix $r > 0$ and let

$$D_r = \{(x, y) \in \mathbb{R}^k \oplus \mathbb{R}^{n-k} \mid \|x\| \leq r, \|y\| \leq r\}, \quad W_{m,r}^\pm = W_m^\pm \cap D_r.$$

If $\lambda' < 1$ (this is true if $\lambda < 1$ and γ and δ are sufficiently small), then by (ii) $f_m(W_{m,r}^-) \subset W_{m+1,r}^-$ and W_m^- is contracted under the action of f_m. Thus in this case $W_{m,r}^-$ is determined by the action of f_m on D_r only. Similar comments apply to W_m^+ if $\mu' > 1$. Thus in these situations we obtain meaningful objects from local data.

It is, however, *very important* to be aware that if one tries to apply Theorem 6.2.8 via local charts and the previous extension procedure then one obtains meaningful objects (independent of the extensions and determined by local data) *only* in the two cases of the preceding paragraph ($\lambda' < 1$ for W^- or $\mu' > 1$ for W^+).

In particular in the hyperbolic case for sufficiently small γ, δ we have $\lambda' < 1 < \mu'$ and both W_m^+ and $W_{m,r}^+$ are determined locally. In this case W_m^- and W_m^+ are usually called the *stable manifolds* and the *unstable manifolds* at the origin, correspondingly. Furthermore, we can put $\nu = 1$ in (iii). That shows that stable and unstable manifolds are defined purely topologically, namely,

$$W_m^- = \{z \in \mathbb{R}^n \mid \|f_{m+L-1} \circ \cdots \circ f_m z\| \xrightarrow[L\to\infty]{} 0\},$$
$$W_m^+ = \{z \in \mathbb{R}^n \mid \|f_{m-L}^{-1} \circ \cdots \circ f_{m-1}^{-1} z\| \xrightarrow[L\to\infty]{} 0\}.$$

In the course of the proof we will show that the sequence of differentials $(Df_m)_0$, $m \in \mathbb{Z}$, admits a (λ', μ')-splitting. This will immediately imply that $T_0 W_m^{\pm} = E_m^{\pm}$.

By considering successive images $p_m = f_{m-1} \circ \cdots \circ f_0(p)$ for $m \geq 0$ and $p_m = f_m^{-1} \circ f_{m+1}^{-1} \circ \cdots \circ f_{-1}^{-1}(p)$ for $m < 0$ of any point $p \in \mathbb{R}^n$ and translating the coordinate systems so that they become centered at p_m, we obtain maps

$$f_m^p(z) = f_m(z + p_m) - p_{m+1}$$

satisfying the hypotheses of the theorem. Thus we can construct manifolds $W_{m,p}^+$ and $W_{m,p}^-$ passing through p and satisfying appropriately modified assertions. In particular (ii) and (iii) imply that if $W_{m,p}^+ \cap W_{m,q}^+ \neq \varnothing$ then $W_{m,p}^+ = W_{m,q}^+$ (similarly for $W_{m,p}^-$). Furthermore, $f_m(W_{m,p}^{\pm}) = W_{m+1,f_m p}^{\pm}$. Thus, $\mathbb{R}^n$ splits in two ways into invariant families of manifolds. Naturally, the fields of tangent planes to those manifolds are invariant under the differentials Df_m.

d. Proof of the Hadamard–Perron Theorem. The proof of the Hadamard–Perron Theorem illustrates a method that plays the central role in the theory of hyperbolic dynamical systems.[1] It involves setting the stage for systematic use of the Contraction Mapping Principle in appropriately constructed functional spaces. Our proof consists of five steps:

Step 1. Construction of invariant cone families.

Step 2. Construction of invariant sequences of plane fields inside the invariant cone families. Here we will also explore other implications of the existence of invariant cones which will be used later on a number of occasions.

Step 3. Construction of invariant Lipschitz graphs via an application of the Contraction Mapping Principle to an appropriate operator (graph transform).

Step 4. Verification of differentiability.

Step 5. C^r-smoothness in the hyperbolic case.

In order to distinguish tangent vectors from points in the Euclidean space we will usually denote by $(x, y) \in \mathbb{R}^k \oplus \mathbb{R}^{n-k}$ a point in $\mathbb{R}^n$ and by $(u, v) \in \mathbb{R}^k \oplus \mathbb{R}^{n-k} \cong T_{(x,y)}\mathbb{R}^n$ a tangent vector at (x, y).

Step 1.

Definition 6.2.9. The *standard horizontal γ-cone* at $p \in \mathbb{R}^n$ is defined by

$$H_p^\gamma = \{(u, v) \in T_p\mathbb{R}^n \mid \|v\| \le \gamma\|u\|\}.$$

The *standard vertical γ-cone* at p is

$$V_p^\gamma = \{(u, v) \in T_p\mathbb{R}^n \mid \|u\| \le \gamma\|v\|\}.$$

More generally a *cone* K in $\mathbb{R}^n$ is defined as the image of a standard cone under an invertible linear map.

Let us look at some examples to clarify the picture involved here. In dimension $n = 2$ all cones look alike. A horizontal cone $|x_2| \le \gamma|x_1|$ is shaded in Figure 6.2.1.

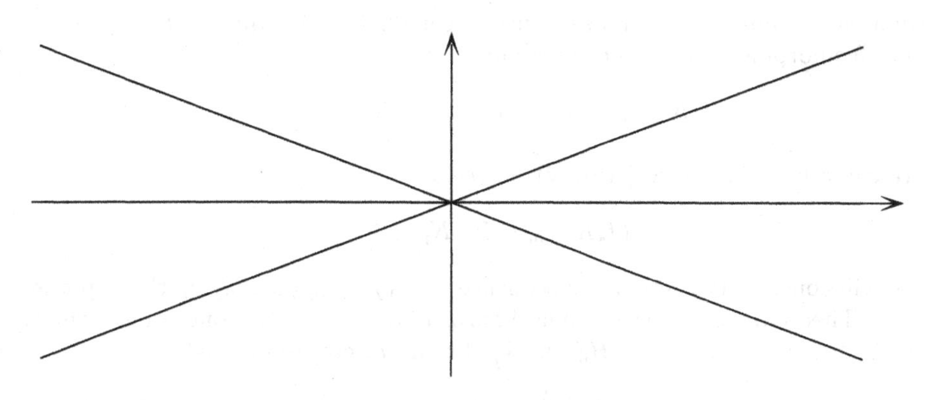

FIGURE 6.2.1. A horizontal cone

The closure of its complement $|x_1| \le |x_2|/\gamma$ is a vertical cone.

In dimension $n = 3$ the following is obviously a cone: Let $u = x_1$, $v = (x_2, x_3)$, $\sqrt{x_2^2 + x_3^2} \le \gamma|x_1|$. But here, too, the closure of the complement of a cone is a cone, so letting $u = (x_2, x_3)$, $v = x_1$, $|x_1| \le \sqrt{x_2^2 + x_3^2}/\gamma$ gives an example of a cone that does not look like those designed to hold ice cream.

By a *cone field* we mean a map that associates to every point $p \in \mathbb{R}^n$ a cone K_p in $T_p\mathbb{R}^n$. A diffeomorphism $f\colon \mathbb{R}^n \to \mathbb{R}^n$ naturally acts on cone fields

$$(f_*K)_p = Df_{f^{-1}(p)}(K_{f^{-1}(p)}).$$

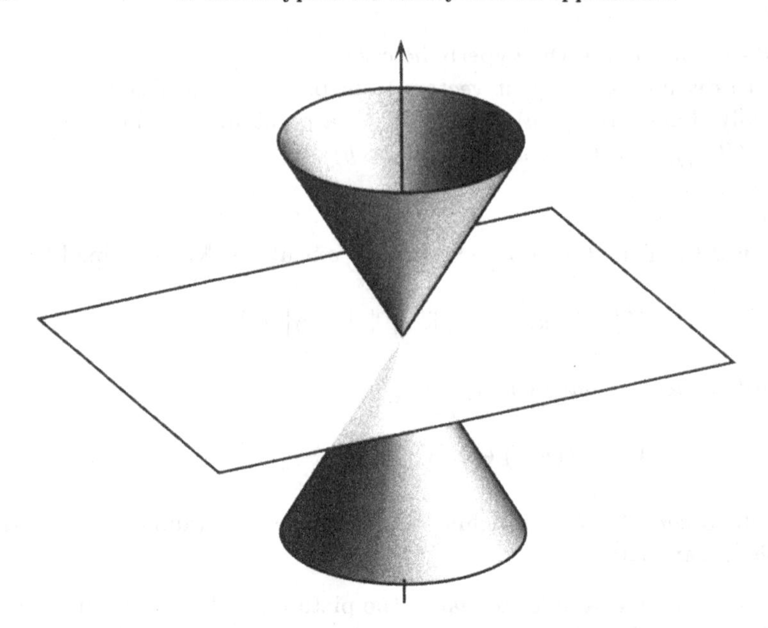

FIGURE 6.2.2. A vertical cone

By a *cone family* we mean a sequence of cone fields. A sequence $f = \{f_m\}_{m\in\mathbb{Z}}$ of diffeomorphisms acts on cone families by

$$(f_*K)_{p,m} = (Df_{m-1})_{f_{m-1}^{-1}(p)}(K_{f_{m-1}^{-1}(p),m-1}).$$

We call a cone family K (strictly) *invariant* if

$$(f_*K)_{p,m} \subset \operatorname{Int} K_{p,m} \cup \{0\}.$$

We will consider the action of a sequence $f = \{f_m\}_{m\in\mathbb{Z}}$ satisfying the hypotheses of Theorem 6.2.8 on the standard horizontal and vertical cone families which assign to $p \in \mathbb{R}^n$ the cones H_p^γ and V_p^γ for all m, correspondingly.

Lemma 6.2.10. *If* $\delta < \dfrac{\mu-\lambda}{2+\gamma+1/\gamma}$ *then*

$$(Df_m)_p(H_p^\gamma) \subset \operatorname{Int} H_{f_m(p)}^\gamma \text{ and } (Df_m)_p^{-1}(V_{f_m(p)}^\gamma) \subset \operatorname{Int} V_p^\gamma.$$

Proof. If $(u,v) \in H_p^\gamma$, that is, $\|v\| \le \gamma\|u\|$, and

$$(u',v') = (Df_m)_p(u,v) = (A_m u + (D\alpha_m)_p(u,v), B_m v + (D\beta_m)_p(u,v))$$

then

$$\|v'\| = \|B_m v + (D\beta_m)_p(u,v)\| \le \|B_m v\| + \|(D\beta_m)_p(u,v)\| < \lambda\|v\| + \delta\|(u,v)\|. \tag{6.2.1}$$

We also have

$$\|u'\| = \|A_m u + (D\alpha_m)_p(u,v)\| \geq \|A_m u\| - \|(D\alpha_m)_p(u,v)\| > \mu\|u\| - \delta\|(u,v)\|$$

and since $\|(u,v)\| \leq \|u\| + \|v\| \leq (1+\gamma)\|u\|$ we get

$$\|u'\| > (\mu - \delta(1+\gamma))\,\|u\|\,. \tag{6.2.2}$$

Now $\delta < \dfrac{\mu-\lambda}{2+\gamma+1/\gamma}$, so $\delta(1+\gamma)^2 < \gamma(\mu-\lambda)$, whence

$$\lambda\gamma + \delta(1+\gamma) < \gamma(\mu - \delta(1+\gamma)) \tag{6.2.3}$$

and

$$\|v'\| < \lambda\|v\| + \delta\|(u,v)\| \leq (\lambda\gamma + \delta(1+\gamma))\|u\| < \gamma(\mu - \delta(1+\gamma))\|u\| < \gamma\|u'\|.$$

To prove invariance of vertical γ-cones note that we need to show that $V^{\gamma}_{f_m(p)} \subset \operatorname{Int}(Df_m)_p V^{\gamma}_p$ or, equivalently, $(Df_m)_p(H^{1/\gamma}_p) \subset \operatorname{Int}(H^{1/\gamma}_{f_m(p)})$. But this fact follows from the observation that the preceding estimates still hold when γ is replaced by $1/\gamma$. □

Let $\tilde{V}^{\gamma} = f_* V^{\gamma}$. We now show that vectors in horizontal cones expand and those in vertical cones contract.

Lemma 6.2.11.

$$\|(Df_m)_p(u,v)\| > \left(\frac{\mu}{1+\gamma} - \delta\right)\|(u,v)\| \;\; \textit{for } (u,v) \in H^{\gamma}_p \textit{ and}$$
$$\|(Df_m)_p(u,v)\| < (1+\gamma)(\lambda+\delta)\|(u,v)\| \;\; \textit{for } (u,v) \in \tilde{V}^{\gamma}_p.$$

Proof. With the above notation (6.2.3) implies

$$\|(u',v')\| \geq \|u'\| > (\mu - \delta(1+\gamma))\,\|u\| \geq \frac{\mu - \delta(1+\gamma)}{1+\gamma}\|(u,v)\|$$

for $(u,v) \in H^{\gamma}_p$. Now (6.2.2) yields

$$\begin{aligned}\|(u',v')\| &\leq \|u'\| + \|v'\| \leq (1+\gamma)\|v'\| \\ &< (1+\gamma)[\lambda\|v\| + \delta\|(u,v)\|] \leq (1+\gamma)(\lambda+\delta)\|(u,v)\|\end{aligned}$$

for $(u,v) \in \tilde{V}^{\gamma}_p$, since $(u',v') \in V^{\gamma}_{f_m(p)}$. □

Note that in these and the subsequent estimates we use the triangle inequality for norms of orthogonal vectors. Instead we could use the Pythagorean Theorem which would result in replacing $1+\gamma$ by $\sqrt{1+\gamma^2}$ and improve the estimates of δ in terms of λ, μ, γ. This may be relevant in applications of the cone method to specific dynamical systems. In our present local setup, however, δ can be chosen arbitrarily small so an improvement in the estimates does not warrant the extra effort resulting from the algebraic complications.

Step 2. Now we are going to explore the relation between the existence of an invariant sequence of cones and exponential splitting for a sequence of linear maps. The conclusions of Lemmas 6.2.10 and 6.2.11 applied along each orbit will then make the results of this step applicable to our setting.

Proposition 6.2.12. *Let $\lambda' < \mu'$ and $\gamma_m, \gamma'_m > 0$ $(m \in \mathbb{Z})$ and let $L_m: \mathbb{R}^k \times \mathbb{R}^{n-k} \to \mathbb{R}^k \times \mathbb{R}^{n-k}$ be a sequence of invertible linear maps such that*

(1) $L_m H^{\gamma_m} \subset \operatorname{Int} H^{\gamma_{m+1}}$;
(2) $L_m^{-1} V^{\gamma'_{m+1}} \subset \operatorname{Int} V^{\gamma'_m}$;
(3) $\|L_m(u,v)\| > \mu' \|(u,v)\|$ *for* $(u,v) \in H^{\gamma_m}$;
(4) $\|L_m(u,v)\| < \lambda' \|(u,v)\|$ *for* $(u,v) \in L_m^{-1} V^{\gamma'_{m+1}}$.

Then

$$E_m^+ := \bigcap_{i=0}^{\infty} L_{m-1} \circ L_{m-2} \circ \cdots \circ L_{m-i} H^{\gamma_{m-i}}$$

is a k-dimensional subspace inside H^{γ_m} and

$$E_m^- := \bigcap_{i=0}^{\infty} L_m^{-1} \circ L_{m+1}^{-1} \circ \cdots \circ L_{m+i}^{-1} V^{\gamma'_{m+i+1}}$$

is an $(n-k)$-dimensional subspace inside $V^{\gamma'_m}$.

Proof. Since $\mathbb{R}^k \times \{0\} \subset H^\gamma$ for all γ, condition (1) implies that

$$\begin{aligned} S_j &:= L_{m-1} \circ L_{m-2} \circ \cdots \circ L_{m-j}(\mathbb{R}^k \times \{0\}) \\ &\subset L_{m-1} \circ L_{m-2} \circ \cdots \circ L_{m-j} H^{\gamma_{m-j}} =: T_j. \end{aligned}$$

For each S_j take an ordered orthonormal basis and consider a subsequence such that the sequences of basis elements all converge. Since the intersection of T_j with the unit sphere is compact it contains the basis consisting of the limits of the basis elements. By the same token any sequence of vectors defined by a fixed set of coefficients converges to a vector in T_j. Hence the span S of the limiting basis belongs to all T_j and thus to the intersection. We need to show that $S = E_m^+$.

If $(u,v) \in E_m^+$ then, since $S \subset H^{\gamma_m}$ is transverse to $\{0\} \times \mathbb{R}^{n-k}$, we can write $(u,v) = (u,v') + (0,v'')$ with $(u,v') \in S$.

If we let

$$\begin{aligned} (u_j, v_j) &= L_{m-j}^{-1} \circ \cdots \circ L_{m-1}^{-1}(u,v), \\ (u'_j, v'_j) &= L_{m-j}^{-1} \circ \cdots \circ L_{m-1}^{-1}(u,v'), \\ (u''_j, v''_j) &= L_{m-j}^{-1} \circ \cdots \circ L_{m-1}^{-1}(0,v'') \end{aligned}$$

then $(u,v) \in E_m^+$ implies that $(u_j, v_j) \in H^{\gamma_{m-j}}$ and by (3) $\|(u_j,v_j)\| \le (\mu')^{-j} \|(u,v)\|$. By the same token $\|(u'_j, v'_j)\| \le (\mu')^{-j} \|(u,v')\|$. Thus since

$(u_j'', v_j'') \in V^{\gamma'_{m-j}}$ we have by (4) that

$$\|v''\| \le (\lambda')^j \|(u_j'', v_j'')\|$$
$$\le (\lambda')^j \left(\|(u_j, v_j)\| + \|(u_j', v_j')\|\right) \le \left(\frac{\lambda'}{\mu'}\right)^j \left(\|(u, v)\| + \|(u, v')\|\right)$$

for all $j \in \mathbb{N}$, whence $v'' = 0$ and $(u, v) \in S$.

The argument for E_m^- is completely similar, using the family $\{L_m^{-1}\}$ instead of L_m. □

Remark. Note that E_m^+ and E_m^- are unique invariant sequences of subspaces inside the cones H_m^γ and $V_m^{\gamma'}$, respectively.

Corollary 6.2.13. *If under the assumptions of Proposition 6.2.12 we have* $\lambda' < 1 < \mu'$ *then* $\{L_m\}$ *is a hyperbolic family of linear maps which admits a* (λ', μ')*-splitting.*

Corollary 6.2.14. *If* $\gamma < \sqrt{(\mu/\lambda) - 1}$ *and*

$$0 < \delta < \min\left(\frac{\mu - \lambda}{\gamma + \frac{1}{\gamma} + 2}, \frac{\mu - (1+\gamma)^2\lambda}{(2+\gamma)(1+\gamma)}\right) \tag{6.2.4}$$

then

$$(E_p^+)_m = \bigcap_{i=0}^{\infty}((f_*)^i H^\gamma)_{p,m} = \bigcap_{i=0}^{\infty}(f_*(f_*(\dots f_*(H^\gamma)\dots)))_{p,m}$$

is a k*-dimensional subspace inside* H_p^γ,

$$(Df_m)_p(E_p^+)_m = \left(E^+_{f_m(p)}\right)_{m+1},$$

and

$$\|(Df_m)_p \xi\| \ge \left(\frac{\mu}{1+\gamma} - \delta\right)\|\xi\|$$

for every $\xi \in (E_p^+)_m$.

Similarly $(E_p^-)_m = \bigcap_{i=0}^{\infty}((f_*^{-1})^i V^\gamma)_{p,m}$ *is an* $(n-k)$*-dimensional subspace inside* V_p^γ,

$$(Df_m)_p(E_p^-)_m = \left(E^-_{f_m(p)}\right)_{m+1},$$

and for every $\xi \in (E_p^-)_m$

$$\|(Df_m)_p \xi\| \le (1+\gamma)(\lambda + \delta)\|\xi\| .$$

Proof. By Lemmas 6.2.10 and 6.2.11 and condition (6.2.4) we can apply Proposition 6.2.12 with $\lambda' = (1+\gamma)(\lambda+\delta)$ and $\mu' = (\mu/(1+\gamma) - \delta)$ since under our assumptions $\lambda' < \mu'$ along each orbit of the sequence $\{f_m\}$. □

Lemma 6.2.15. *For each $m \in \mathbb{Z}$ the subspaces $(E_p^+)_m$ and $(E_p^-)_m$ depend continuously on p.*

Proof. The vectors $v \in (E_p^-)_m$ are characterized by the inequalities

$$\|(Df_{m+j})(Df_{m+j-1})\cdots(Df_m)_p v\| \le (\lambda')^{j+1}\|v\| \qquad (j \in \mathbb{N}) . \tag{6.2.5}$$

For a sequence $p_l \to p$ take orthonormal bases $\xi_1^l, \dots \xi_k^l$ of $(E_{p_l}^-)_m$ and assume without loss of generality that $\lim_{l\to\infty} \xi_i^l = \xi_i$ $(i = 1, \dots, k)$. Since for any fixed i the vectors ξ_i^l satisfy (6.2.5) for all l we conclude by continuity of all Df_m that ξ_i satisfies (6.2.5) and hence $\xi_i \in (E_p^-)_m$. Since $\dim(E_p^-)_m$ does not depend on p this implies that $\lim_{l\to\infty}(E_{p_l}^-)_m = (E_p^-)_m$. □

$(E_p^+)_m$ and $(E_p^-)_m$ $(m \in \mathbb{Z})$ are the invariant sequences of plane fields mentioned in the description of the proof.

Step 3. To obtain invariant graphs, that is, a family $\{\varphi_m^+ : \mathbb{R}^k \to \mathbb{R}^{n-k}\}_{m\in\mathbb{Z}}$ of Lipschitz functions such that $f_m(\operatorname{graph}\varphi_m^+) = \operatorname{graph}\varphi_{m+1}^+$ and $\varphi_m^+(0) = 0$ let $C_\gamma(\mathbb{R}^k)$ be the set of functions $\varphi: \mathbb{R}^k \to \mathbb{R}^{n-k}$ that are Lipschitz continuous with Lipschitz constant γ. Let $C_\gamma^0(\mathbb{R}^k)$ be the space of $\varphi \in C_\gamma(\mathbb{R}^k)$ such that $\varphi(0) = 0$. The following lemma can be viewed as a nonlinear counterpart of Lemma 6.2.10 and shows that the maps f_m act on the spaces $C_\gamma(\mathbb{R}^k)$ and $C_\gamma^0(\mathbb{R}^k)$:

Lemma 6.2.16. *If (6.2.4) holds and $\varphi \in C_\gamma(\mathbb{R}^k)$ then $f_m(\operatorname{graph}\varphi) = \operatorname{graph}\psi$ for some $\psi \in C_\gamma(\mathbb{R}^k)$. The same holds for $C_\gamma^0(\mathbb{R}^k)$.*

Proof. The map $G_\varphi^m: \mathbb{R}^k \to \mathbb{R}^k$ given by

$$G_\varphi^m(x) = A_m x + \alpha_m(x, \varphi(x)) \tag{6.2.6}$$

represents the x-coordinate of f_m acting on graph φ.

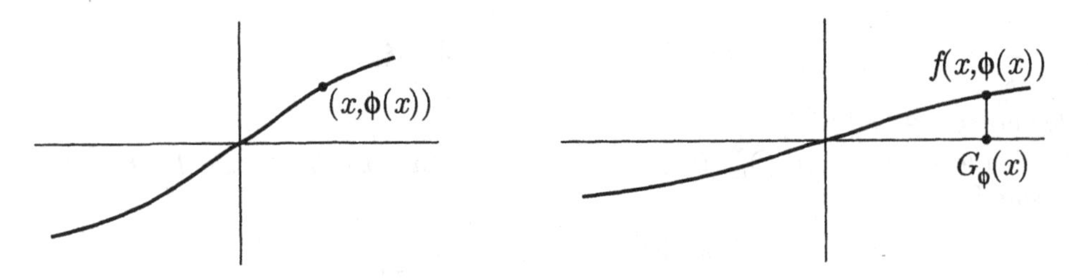

FIGURE 6.2.3. The graph transform

In order to show that $f_m(\operatorname{graph}\varphi)$ is a graph we need to prove that G_φ^m is a bijection. Thus for $x_0 \in \mathbb{R}^k$ we need to find a unique $x \in \mathbb{R}^k$ such that $x_0 = G_\varphi^m(x)$ or equivalently

$$x = F(x), \qquad \text{where} \qquad F(x) = A_m^{-1}x_0 - A_m^{-1}(\alpha_m(x, \varphi(x))) . \tag{6.2.7}$$

$F: \mathbb{R}^k \to \mathbb{R}^k$ is a contracting map since

$$\begin{aligned}\|F(x_1) - F(x_2)\| &= \|A_m^{-1}(\alpha_m(x_1, \varphi(x_1)) - \alpha_m(x_2, \varphi(x_2)))\| \\ &\le \mu^{-1}\|\alpha_m\|_{C^1} \cdot (1+\gamma)\|x_1 - x_2\| < \delta\mu^{-1}(1+\gamma)\|x_1 - x_2\|\end{aligned}$$

and $\delta\mu^{-1}(1+\gamma) < 1$ by the second inequality in (6.2.4). Thus by the Contraction Mapping Principle (Proposition 1.1.2) equation (6.2.7) has a unique solution, that is, $f_m(\operatorname{graph}\varphi) = \operatorname{graph}\psi$.

Next we show that ψ is Lipschitz continuous with Lipschitz constant γ. Suppose $\psi(x_1') = y_1'$ and $\psi(x_2') = y_2'$ and take $(x_1, y_1), (x_2, y_2) \in \operatorname{graph}\varphi$ such that for $i = 1, 2$

$$(x_i', y_i') = f_m(x_i, y_i) = (A_m x_i + \alpha_m(x_i, \varphi(x_i)),\ B_m\varphi(x_i) + \beta_m(x_i, \varphi(x_i))).$$

Then

$$\begin{aligned}\|y_2' - y_1'\| &= \|B_m(\varphi(x_2) - \varphi(x_1)) + \beta_m(x_2,\ \varphi(x_2)) - \beta_m(x_1,\ \varphi(x_1))\| \\ &< \lambda\gamma\|x_2 - x_1\| + \delta(1+\gamma)\|x_2 - x_1\| \\ &= (\lambda\gamma + \delta(1+\gamma))\|x_2 - x_1\|\end{aligned} \tag{6.2.8}$$

and

$$\begin{aligned}\|x_2' - x_1'\| &= \|A_m(x_2 - x_1) + \alpha_m(x_2,\ \varphi(x_2)) - \alpha_m(x_1,\ \varphi(x_1))\| \\ &> \mu\|x_2 - x_1\| - \delta(1+\gamma)\|x_2 - x_1\| \\ &= (\mu - \delta(1+\gamma))\|x_2 - x_1\|.\end{aligned} \tag{6.2.9}$$

Consequently $\|y_2' - y_1'\| \le \dfrac{\lambda\gamma + \delta(1+\gamma)}{\mu - \delta(1+\gamma)}\|x_2 - x_1\| =: \gamma'\|x_2 - x_1\|$. But a straightforward calculation shows that the first condition in (6.2.4) is equivalent to $\gamma' < \gamma$. This shows that f_m acts on $C_\gamma(\mathbb{R}^k)$. The same holds for $C_\gamma^0(\mathbb{R}^k)$ since $f_m(0) = 0$. □

Since we eventually want to apply the Contraction Mapping Principle, we introduce a metric on the space $C_\gamma^0(\mathbb{R}^k)$ and show that the action of f_m is a contraction.

Let us first set

$$d(\varphi, \psi) := \sup_{x \in \mathbb{R}^k \setminus \{0\}} \frac{\|\varphi(x) - \psi(x)\|}{\|x\|}$$

for $\varphi, \psi \in C_\gamma^0(\mathbb{R}^k)$. Since $\varphi(0) = \psi(0) = 0$ and φ and ψ are Lipschitz continuous this is a well-defined metric. It is easy to check that with this metric $C_\gamma^0(\mathbb{R}^k)$ is a complete metric space.

The next lemma shows that the action of f_m on $C_\gamma^0(\mathbb{R}^k)$ given by

$$f_m(\operatorname{graph}\varphi) = \operatorname{graph}((f_m)_*\varphi)$$

is a contracting map.

Lemma 6.2.17. $d((f_m)_*\varphi,\ (f_m)_*\psi) \le \dfrac{\lambda+\delta(1+\gamma)}{\mu-\delta(1+\gamma)} d(\varphi,\psi)$ *for* $\varphi,\psi \in C^0_\gamma(\mathbb{R}^k)$.

Proof. Let $\varphi' = (f_m)_*\varphi$ and $\psi' = (f_m)_*\psi$. Using the map G^m_φ defined by (6.2.6) and the fact that $\psi' \in C^0_\gamma(\mathbb{R}^k)$ we have

$$\begin{aligned}
&\|\varphi'\left(G^m_\varphi(x)\right) - \psi'\left(G^m_\varphi(x)\right)\| \\
&\quad \le \|\varphi'\left(G^m_\varphi(x)\right) - \psi'\left(G^m_\psi(x)\right)\| + \|\psi'\left(G^m_\psi(x)\right) - \psi'\left(G^m_\varphi(x)\right)\| \\
&\quad \le \|\left(B_m(\varphi(x)) + \beta_m(x,\varphi(x))\right) - \left(B_m(\psi(x)) + \beta_m(x,\psi(x))\right)\| \\
&\qquad + \gamma\|G^m_\psi(x) - G^m_\varphi(x)\| \\
&\quad \le \|B_m\left(\varphi(x) - \psi(x)\right)\| + \|\beta_m\left(x,\varphi(x)\right) - \beta_m\left(x,\psi(x)\right)\| \\
&\qquad + \gamma\|\alpha_m\left(x,\psi(x)\right) - \alpha_m\left(x,\varphi(x)\right)\| \\
&\quad < \lambda\|\varphi(x) - \psi(x)\| + \delta\|\varphi(x) - \psi(x)\| + \gamma\delta\|\varphi(x) - \psi(x)\| \\
&\quad = \left(\lambda + \delta(1+\gamma)\right)\|\varphi(x) - \psi(x)\|\,.
\end{aligned}$$

On the other hand

$$\begin{aligned}
\|G^m_\varphi(x)\| &= \|A_m x + \alpha_m\left(x,\varphi(x)\right)\| \ge \|A_m x\| - \|\alpha_m\left(x,\varphi(x)\right)\| \\
&\ge \mu\|x\| - \delta(1+\gamma)\|x\| = \left(\mu - \delta(1+\gamma)\right)\|x\|\,.
\end{aligned}$$

Consequently

$$\begin{aligned}
\frac{\|(f_m)_*\varphi\left(G^m_\varphi(x)\right) - (f_m)_*\psi\left(G^m_\varphi(x)\right)\|}{\|G^m_\varphi(x)\|} &\le \frac{\lambda+\delta(1+\gamma)}{\mu-\delta(1+\gamma)} \cdot \frac{\|\varphi(x)-\psi(x)\|}{\|x\|} \\
&\le \frac{\lambda+\delta(1+\gamma)}{\mu-\delta(1+\gamma)} \cdot d(\varphi,\psi).
\end{aligned}$$

□

Note that $\dfrac{\lambda+\delta(1+\gamma)}{\mu-\delta(1+\gamma)} = \gamma^{-1}\dfrac{\lambda\gamma+\delta\gamma(1+\gamma)}{\mu-\delta(1+\gamma)} \le \gamma^{-1}\dfrac{\lambda\gamma+\delta(1+\gamma)}{\mu-\delta(1+\gamma)} < 1$ by (6.2.4), whenever $\gamma < 1$ (see (6.2.3)).

We now denote by C^0_γ the space of families $\{\varphi_m\}_{m\in\mathbb{Z}}$ of functions in $C^0_\gamma(\mathbb{R}^k)$. The action of $f = \{f_m\}_{m\in\mathbb{Z}}$ on the space C^0_γ given by

$$f_m(\operatorname{graph}\varphi_m) = \operatorname{graph}\left((f_*\varphi)_{m+1}\right)$$

is called the *graph transform*. The preceding lemma shows that the graph transform is a contraction with respect to the metric

$$d\left(\{\varphi_m\}_{m\in\mathbb{Z}},\ \{\psi_m\}_{m\in\mathbb{Z}}\right) := \sup_{m\in\mathbb{Z}} d(\varphi_m,\psi_m).$$

Since C^0_γ is complete with this metric, the Contraction Mapping Principle, Proposition 1.1.2, yields a unique fixed point for this action of f, hence an invariant family $\{\varphi^+_m\}$ of graphs, as claimed.

Remark. If $\lambda < 1$ one can show that $\|\varphi^+_m\|_{C^0} < \delta/(1-\lambda)$ by considering only $\varphi \in C^0_\gamma(\mathbb{R}^k)$ bounded by $\delta/(1-\lambda)$ and showing invariance of this condition under f_*. In this case the first estimate in the proof of Lemma 6.2.17 also shows that the graph transform is a contraction with respect to the C^0 topology.

To construct the functions φ_m^- one argues along the same lines. Using the estimates obtained in this step, with γ replaced by $1/\gamma$, one shows that the maps Df_m^{-1} act on families of γ-Lipschitz functions $\varphi: \mathbb{R}^{n-k} \to \mathbb{R}^k$ vanishing at the origin, and are contracting.

At this point it is natural to prove (ii) since we use the estimates (6.2.8) and (6.2.9). First replace (x_1, y_1) by $(0,0)$ and (x_2, y_2) by $(x, \varphi_m^+ x)$ in (6.2.9). Then

$$\begin{aligned}\|f_m(x, \varphi_m^+(x))\| \geq \|A_m x + \alpha_m(x, \varphi_m^+(x))\| &> (\mu - \delta(1+\gamma))\|x\| \\ &\geq \frac{\mu - \delta(1+\gamma)}{1+\gamma}\|(x, \varphi_m^+(x))\|.\end{aligned}$$

On the other hand, applying (6.2.8) to $(0,0)$ and $(\varphi_m^-(y), y)$ and using the fact that φ_m^- are γ-Lipschitz yields

$$\begin{aligned}&\|f_m(\varphi_m^-(y), y)\| \leq (1+\gamma)\|B_m(y) + \beta_m(\varphi_m^-(y), y)\| \\ &\quad < (1+\gamma)(\lambda\|y\| + \delta(1+\gamma)\|y\|) = (1+\gamma)(\lambda + \delta(1+\gamma))\|(\varphi_m^-(y), y)\|.\end{aligned}$$

Step 4. To prove that the invariant family of functions obtained in the previous step consists of continuously differentiable functions, we introduce the notion of a tangent set for a graph. The results of step 2, the existence of a unique invariant family of continuous plane fields, then imply that the tangent set of each of these graphs is a continuous plane field. But this, by definition, implies that the graphs are graphs of C^1 functions.

Definition 6.2.18. Let $\varphi \in C_\gamma^0(\mathbb{R}^k)$, $x \in \mathbb{R}^k$,

$$\Delta_y \varphi := \frac{(y, \varphi(y)) - (x, \varphi(x))}{\|(y, \varphi(y)) - (x, \varphi(x))\|} \qquad \text{for } y \neq x,$$

$t_x\varphi := \{v \in T_x\mathbb{R}^n \mid \exists \{x_n\}_{n\in\mathbb{N}} \text{ such that } \lim_{n\to\infty} x_n = x \text{ and } \lim_{n\to\infty} \Delta_{x_n}\varphi = v\}$. Then $\tau_x\varphi := \bigcup_{v\in t_x\varphi} \mathbb{R}v$, where $\mathbb{R}v := \{av \mid a \in \mathbb{R}\}$ is the line containing v, is called the *tangent set* of φ at x. The disjoint union $\tau\varphi := \bigcup_{x\in\mathbb{R}^k} \tau_x\varphi$ is called the *tangent set* of φ.

Note that since for every $v \in \mathbb{R}^k$ one can choose $y = x + tv$ in the definition, $\tau_x\varphi$ projects onto $\mathbb{R}^k$.

As an example consider $\varphi(x) = x\sin(1/x) \in C_\gamma^0(\mathbb{R})$ for which $\tau_0\varphi = \{(x,y) \in \mathbb{R}^2 \mid |y| \leq |x|\} = H_0^1$. Indeed, for $\varphi \in C_\gamma^0(\mathbb{R}^k)$ and $x \in \mathbb{R}^k$ we always have $\tau_x\varphi \subset H_x^\gamma$, since φ has Lipschitz constant γ. Another important observation is that $\varphi \in C_\gamma^0(\mathbb{R}^k)$ is differentiable at x if and only if $\tau_x\varphi$ is a k-dimensional plane.

We can now show that the invariant family $\varphi^+ = \{\varphi_m^+\}_{m\in\mathbb{Z}}$ obtained in step 3 consists of C^1 functions. Associated with φ^+ is the family $\tau\varphi^+ := \{\tau\varphi_m^+\}_{m\in\mathbb{Z}}$ of tangent sets for the functions φ_m^+, $m \in \mathbb{Z}$. Since φ^+ is an invariant family of functions for $f = \{f_m\}_{m\in\mathbb{Z}}$, the associated family $\tau\varphi^+$ of tangent sets is invariant under the action of the differentials Df_m. In step 2 we showed that any such invariant family inside the γ-cones is contained in the unique invariant family E^+_m of continuous plane fields obtained there. Since every tangent set $\tau_p\varphi^+_m$ projects onto $\mathbb{R}^k$, we conclude that $\tau_p\varphi^+_m = (E^+_p)_m$, that is, the φ^+_m are C^1 functions.

Smoothness of φ^-_m is proved likewise. This ends the proof of (i).

It remains to prove (iii). We remarked after the formulation of the theorem that we can construct the manifolds $(W^-_m)_p$ and $(W^+_m)_p$ for any point $p = (x, y)$. We still have $(W^+_m)_p = \operatorname{graph}(\varphi^+_m)_p$ and $(W^-_m)_p = \operatorname{graph}(\varphi^-_m)_p$ for some γ-Lipschitz functions $(\varphi^+_m)_p\colon \mathbb{R}^k \to \mathbb{R}^{n-k}$ and $(\varphi^-_m)_p\colon \mathbb{R}^{n-k} \to \mathbb{R}^k$ and properties analogous to (i) and (ii).

Lemma 6.2.19. *For $p, q \in \mathbb{R}^n$ the intersection $(W^+_m)_p \cap (W^-_m)_q$ consists of exactly one point.*

Proof. If $z = (x, y) \in (W^+_m)_p \cap (W^-_m)_q$ then $x = (\varphi^-_m)_q(y)$ and $y = (\varphi^+_m)_p(x)$ and hence $x = (\varphi^-_m)_q \circ (\varphi^+_m)_p(x)$. This in turn implies again that $(x, (\varphi^+_m)_p(x)) \in (W^+_m)_p \cap (W^-_m)_q$. But since we can assume $\gamma < 1$ the map $(\varphi^-_m)_q \circ (\varphi^+_m)_p\colon \mathbb{R}^k \to \mathbb{R}^k$ is a contraction and hence has a unique fixed point. $\square$

Now assume $p \notin (W^-_m)_0$. By Lemma 6.2.19 there is a unique $q \in (W^-_m)_0 \cap (W^+_m)_p$. Using (ii) for $(W^-_m)_0$ and $(W^+_m)_p$ we see that

$$
\begin{aligned}
&\|f_{m+L-1} \circ \cdots \circ f_m(p)\| \\
&\quad \geq \|f_{m+L-1} \circ \cdots \circ f_m(p) - f_{m+L-1} \circ \cdots \circ f_m(q)\| - \|f_{m+L-1} \circ \cdots \circ f_m(q)\| \\
&\quad \geq (\mu')^L \|p - q\| - (\lambda')^L \|q\| = (\mu')^L \Big(\|p - q\| - \Big(\frac{\lambda'}{\mu'}\Big)^L \|q\| \Big).
\end{aligned}
$$

Whenever $\lambda' < \nu < \mu'$ and $C \in \mathbb{R}$ this quantity will exceed $C \cdot \nu^L \|p\|$ for sufficiently large $L \in \mathbb{N}$.

Together with a parallel argument for $(W^+_m)_0$ this proves (iii) and thus also the uniqueness of W^+_m and W^-_m.

This finishes the proof of the general part of the Hadamard–Perron Theorem.

Step 5. To complete the proof of Theorem 6.2.8 we now prove that in the hyperbolic case the leaves are as smooth as the diffeomorphism. We will, in fact, prove the stronger statement that if $\mu \geq 1$ in Theorem 6.2.8 then $\{W^+_m\}_{m\in\mathbb{Z}}$ consists of manifolds as smooth as the diffeomorphism. Df_m has block form $\begin{pmatrix} A^{uu}_m & A^{su}_m \\ A^{us}_m & A^{ss}_m \end{pmatrix}$ with A^{uu}_m a $k \times k$-matrix with $\|(A^{uu}_m)^{-1}\| \leq 1/(\mu - \delta)$, A^{ss}_m an $(n-k) \times (n-k)$-matrix with $\|A^{ss}_m\| \leq \lambda + \delta$, and $\|A^{su}_m\| < \delta$, $\|A^{us}_m\| < \delta$. By the preceding steps, notably Lemma 6.2.16, we can obtain W^+_m by taking smooth functions $\varphi^0_m \in C^0_\gamma(\mathbb{R}^k)$ (such as $\varphi^0_m = 0$), applying the graph transform repeatedly to obtain families $\{\varphi^i_m\}$ for $i \in \mathbb{N}$, and taking the limit as $i \to \infty$. We plan to show inductively that the $r + 1$st derivative of φ^i_m converges as $i \to \infty$, so long as f is C^{r+1}. To that end we note that $D\varphi^i_m$ is the graph of a linear map E^i_m from $\mathbb{R}^k$ to $\mathbb{R}^{n-k}$, or, equivalently, the image of the map

$\begin{pmatrix} I \\ E_m^i \end{pmatrix} : \mathbb{R}^k \to \mathbb{R}^n$. Notice that the image of $D\varphi_m^i$ under Df_m is the image of the linear map

$$\begin{pmatrix} A_m^{uu} & A_m^{su} \\ A_m^{us} & A_m^{ss} \end{pmatrix} \begin{pmatrix} I \\ E_m^i \end{pmatrix} = \begin{pmatrix} A_m^{uu} + A_m^{su} E_m^i \\ A_m^{us} + A_m^{ss} E_m^i \end{pmatrix}.$$

If, referring to (6.2.6), we let $g_m^i := (G_{\varphi_{m-1}^i}^{m-1})^{-1}$ then this has to coincide with the image of $\begin{pmatrix} I \\ E_{m+1}^{i+1} \circ (g_{m+1}^i)^{-1} \end{pmatrix}$ which is the same as that of

$$\begin{pmatrix} A_m^{uu} + A_m^{su} E_m^i \\ (E_{m+1}^{i+1} \circ (g_{m+1}^i)^{-1})(A_m^{uu} + A_m^{su} E_m^i) \end{pmatrix},$$

so

$$(E_{m+1}^{i+1} \circ (g_{m+1}^i)^{-1})(A_m^{uu} + A_m^{su} E_m^i) = A_m^{us} + A_m^{ss} E_m^i.$$

Composing with g_{m+1}^i and differentiating r times we get

$$\begin{aligned} D^r E_{m+1}^{i+1}(\alpha_{m+1,i+1}^u)^{-1} &+ E_{m+1}^{i+1}(A_m^{su} \circ g_{m+1}^i)(D^r E_m^i \circ g_{m+1}^i)(Dg_{m+1}^i)^{\otimes r} \\ &= (A_m^{ss} \circ g_{m+1}^i)(D^r E_m^i \circ g_{m+1}^i)(Dg_{m+1}^i)^{\otimes r} + \zeta_{m+1,i+1}(\alpha_{m+1,i+1}^u)^{-1}, \end{aligned}$$

where $\zeta_{m+1,i+1}$ is a polynomial in lower derivatives of E_{m+1}^{i+1} and E_m^i and $\alpha_{m+1,i+1}^u := [(A_m^{uu} \circ g_{m+1}^i) + (A_m^{su} \circ g_{m+1}^i)(E_m^i \circ g_{m+1}^i)]^{-1}$. Letting $\alpha_{m,i}^s := (A_{m-1}^{ss} \circ g_m^{i-1}) - E_m^i(A_{m-1}^{su} \circ g_m^{i-1})$ this yields

$$\begin{aligned} D^r E_m^i &= \alpha_{m,i}^s (D^r E_{m-1}^{i-1} \circ g_m^{i-1})(Dg_m^{i-1})^{\otimes r} \alpha_{m,i}^u + \zeta_{m,i} \\ &= \alpha_{m,i}^s (\alpha_{m-1,i-1}^s \circ g_m^{i-1}) \times \\ &\quad \times (D^r E_{m-2}^{i-2} \circ g_{m-1}^{i-2} \circ g_m^{i-1})(Dg_{m-1}^{i-2})^{\otimes r}(\alpha_{m-1,i-1}^u \circ g_m^{i-1})(Dg_m^{i-1})^{\otimes r}\alpha_{m,i}^u \\ &\quad + \alpha_{m,i}^s(\zeta_{m-1,i-1} \circ g_m^{i-1})(Dg_m^{i-1})^{\otimes r}\alpha_{m,i}^u + \zeta_{m,i} \\ &= \dots \end{aligned}$$

Applying this inductively we obtain an expression for $D^r E_m^i$ with a leading term involving $D^r E_{m-i}^0$ between i-fold products

$$\alpha_{m,i}^s(\alpha_{m-1,i-1}^s \circ g_m^{i-1})(\alpha_{m-2,i-2}^s \circ g_{m-1}^{i-2} \circ g_m^{i-1}) \dots$$

of terms $\alpha_{m-l,i-l}^s$ and

$$\dots (Dg_{m-2}^{i-3})^{\otimes r}(\alpha_{m-2,i-2}^u \circ g_{m-1}^{i-2} \circ g_m^{i-1})(Dg_{m-1}^{i-2})^{\otimes r}(\alpha_{m-1,i-1}^u \circ g_m^{i-1})(Dg_m^{i-1})^{\otimes r}\alpha_{m,i}^u$$

of $\alpha_{m-l,i-l}^u$ and i occurrences of $(Dg_{m-l}^{i-l-1})^{\otimes r}$. This term converges to 0 uniformly as $i \to \infty$: $\|D^r E_{m-i}^0\|$ is uniformly bounded by choice of φ_{m-i}^0 and $\|\alpha_{m-l,i-l}^s\|\|\alpha_{m-l,i-l}^u\| < 1$ uniformly by taking small δ. Finally, the assumption

$\mu \geq 1$ of this step ensures that the factors $(Dg_{m-l}^{i-l-1})^{\otimes r}$ cause no exponential growth.

The jth of the remaining i summands in the expression for $D^r E_m^i$ similarly consists of $\zeta_{m-j-1,i-j-1}$ between j-fold products of terms $\alpha_{m-l,i-l}^s$ and $\alpha_{m-l,i-l}^u$ as well as j occurrences of $(Dg_{m-l}^{i-l-1})^{\otimes r}$. As before, these terms will tend to 0 uniformly as $j \to \infty$ given uniform control of $\zeta_{m-j-1,i-j-1}$. These, however, involve only lower derivatives of E_l^k's which are uniformly bounded by induction assumption, as well as derivatives of order up to order r of coefficients of Df, which are bounded because $f \in C^{r+1}$. Consequently these remaining terms give partial sums of an exponentially convergent series. We already know that lower-order derivatives of E_m^i converge as $i \to \infty$ and thus conclude that the limit of E_m^i is C^r, as desired. □

Note that $(W_m^+)_p$ and $(W_m^-)_p$ for $p \in \mathbb{R}^n$ depend continuously on p: The characterization (iii) of Theorem 6.2.8 yields

Proposition 6.2.20. *If $p_l \to p \in \mathbb{R}^n$ as $l \to \infty$ and $y_l \in (W_m^+)_{p_l}$ for all $l \in \mathbb{N}$ and $y_l \to y \in \mathbb{R}^n$ as $l \to \infty$ then $y \in (W_m^+)_p$.*

Proof. Fix $L \in \mathbb{N}$. Then (ii) of Theorem 6.2.8 implies for $\nu < \mu'$ that

$$\|f_{m-L}^{-1} \circ \cdots \circ f_{m-1}^{-1}(y_l) - f_{m-L}^{-1} \circ \cdots \circ f_{m-1}^{-1}(p_l)\| \leq \nu^{-L}\|y_l - p_l\|$$

for all $l \in \mathbb{N}$. By continuity of the f_m this implies

$$\|f_{m-L}^{-1} \circ \cdots \circ f_{m-1}^{-1}(y) - f_{m-L}^{-1} \circ \cdots \circ f_{m-1}^{-1}(p)\| \leq \nu^{-L}\|y - p\|$$

and since L was arbitrary the claim follows by (iii). □

Since on any fixed compact set the assumption that y_l converges is redundant (by passing to a subsequence) this means that $(W_m^+)_{p_l} \to (W_m^+)_p$ when $p_l \to p$. Convergence here is in the pointwise sense of the proposition. Since we know that E_m^+ is continuous, we have continuity of W_m^+ together with its tangent spaces. A similar statement holds for W_m^-.

Another pertinent remark is that we obtain in fact continuous dependence of W^+ and W^- on the family f_m of maps we consider. Since the main ingredient of the proof of the Hadamard–Perron Theorem 6.2.8 was obtaining the invariant manifolds and their tangent distributions as fixed points of a contraction operator associated with the family f_m, we may use Proposition 1.1.5 to infer that the invariant manifolds depend continuously on the diffeomorphisms with respect to the C^1 topology.

Proposition 6.2.21. *The invariant manifolds (with the C^1 topology) obtained in the Hadamard–Perron Theorem 6.2.8 depend continuously on the family f_m if we use the C^1 topology defined by calling $\{f_m\}_{m\in\mathbb{N}}, \{g_m\}_{m\in\mathbb{N}}$ C^1-close when $\sup_m d_{C^1}(f_m, g_m)$ is small.*

For the next section we note:

Corollary 6.2.22. *If $p_l \to p \in \mathbb{R}^n$ as $l \to \infty$ and $q \in \mathbb{R}^n$ then the sequence y_l given by $(W_m^+)_{p_l} \cap (W_m^-)_q = \{y_l\}$ converges to y given by $\{y\} = (W_m^+)_p \cap (W_m^-)_q$.*

This follows from Proposition 6.2.21 and Lemma 6.2.19 since the y_l are contained in a compact set because the $(W_m^+)_{p_l}$ are Lipschitz graphs.

A note of caution: The construction and characterization of $(W_m^+)_p$ and $(W_m^+)_p$ other than $(W_m^+)_0$ and $(W_m^-)_0$ depend on the behavior of points whose orbits do not stay in a neighborhood of the origin. Consequently they depend on the extension chosen in the Extension Lemma 6.2.7 and do not represent meaningful objects associated with neighborhoods of a reference orbit on a manifold.

An application of the Hadamard–Perron Theorem 6.2.8 that is more straightforward to use is given by Theorem 6.4.9.

e. The Inclination Lemma. The setting and estimates in the proof of the Hadamard–Perron Theorem 6.2.8 immediately yield a result which will be used in Section 6.5, the Inclination Lemma. It says that successive images of a disk transverse to the stable manifold of a hyperbolic fixed point accumulate (in the C^1 topology) on the unstable manifold of the point. To state this result it is convenient to use adapted coordinates as in Theorem 6.2.3 and to let $\pi_1 \colon \mathbb{R}^k \oplus \mathbb{R}^{(n-k)} \to \mathbb{R}^k$ be the projection to the first coordinate.

Proposition 6.2.23. (Inclination Lemma) *Under the hypotheses of Theorem 6.2.3 consider C^r adapted coordinates on a neighborhood O of a hyperbolic fixed point p of $f \colon U \to M$. Given $\epsilon, K, \eta > 0$ there exists an $N \in \mathbb{N}$ such that if $\mathcal{D}$ is a C^1 disk containing $q \in W_p^- \cap O$ with all tangent spaces in horizontal K-cones and such that $\pi_1(\mathcal{D})$ contains an η-ball around $0 \in \mathbb{R}^k \oplus \{0\}$ and $n \geq N$ then $\pi_1(f^n(\mathcal{D})) = W_p^+ \cap O$ and $T_z f^n(\mathcal{D})$ is contained in a horizontal ϵ-cone for every $z \in f^n(\mathcal{D})$.*

Proof. Since $\mathbb{R}^k \oplus \{0\}$ and $\{0\} \oplus \mathbb{R}^{n-k}$ are f-invariant and f is C^1 the differential of f at points $(x, y) \in \mathbb{R}^k \oplus \mathbb{R}^{n-k}$ takes the form

$$Df_{(x,y)} = \begin{pmatrix} A_z^{uu} & A_z^{su} \\ A_z^{us} & A_z^{ss} \end{pmatrix},$$

where

$$\begin{aligned} A_z^{uu} &\in M_{k,k}, & \|(A_z^{uu})^{-1}\| &\leq \frac{1}{\mu - \delta}, \\ A_z^{ss} &\in M_{n-k,n-k}, & \|A_z^{ss}\| &< \lambda + \delta, \\ A_z^{us} &\in M_{n-k,k}, & \|A_z^{us}\| &= o(\|y\|), \\ A_z^{su} &\in M_{k,n-k}, & A_z^{su} &= o(\|x\|). \end{aligned}$$

Here $\lambda < 1 < \mu$ are the contraction and expansion rates as before. δ can be taken arbitrarily small by possibly shrinking the size of the neighborhood O (and replacing $\mathcal{D}$ by its image under an iterate f^n, so that $\mathcal{D}$ intersects the local stable leaf of p in a point in O). Similarly to the proof of smoothness of stable and unstable manifolds it is convenient now to consider planes in horizontal γ-cones as graphs of linear maps whose operator norm (denoted by $\|\cdot\|$) is bounded by γ. After possibly shrinking $\mathcal{D}$ we assume that $\mathcal{D}\cap(\{0\}\oplus\mathbb{R}^{n-k})=\{z\}$ is a single point. Then our first step consists of showing that $T_{z_n}f^n(\mathcal{D})$ is contained in a horizontal $\epsilon/2$-cone for some $n\in\mathbb{N}$, where $z_i=f^i(z)$. To that end consider a linear map $E_z\colon\mathbb{R}^k\to\mathbb{R}^{n-k}$ with $\|E\|\leq K$. Its graph is parameterized as the image of the linear map

$$\begin{pmatrix} I \\ E_z \end{pmatrix}:\mathbb{R}^k\to\mathbb{R}^k\oplus\mathbb{R}^{n-k},$$

where $I\colon\mathbb{R}^k\to\mathbb{R}^k$ is the identity. It is clear that the image of this graph under Df_z is represented as the image of the linear map

$$Df_z\circ\begin{pmatrix} I \\ E_z \end{pmatrix}:\mathbb{R}^k\to\mathbb{R}^k\oplus\mathbb{R}^{n-k}.$$

In our coordinates this composition is obtained via the matrix product

$$\begin{pmatrix} A_z^{uu} & A_z^{su} \\ A_z^{us} & A_z^{ss} \end{pmatrix}\begin{pmatrix} I \\ E_z \end{pmatrix}=\begin{pmatrix} A_z^{uu}+A_z^{su}E_z \\ A_z^{us}+A_z^{ss}E_z \end{pmatrix} \tag{6.2.10}$$

with $A_z^{su}=0$ in this case. Since $A_z^{uu}\colon\mathbb{R}^k\to\mathbb{R}^k$ is nonsingular, the image of $\begin{pmatrix} A_z^{uu} \\ A_z^{us}+A_z^{ss}E_z \end{pmatrix}$ coincides with the image of $\begin{pmatrix} A_z^{uu} \\ A_z^{us}+A_z^{ss}E_z \end{pmatrix}\circ(A_z^{uu})^{-1}=$ $\begin{pmatrix} I \\ A_z^{us}A_z^{uu-1}+A_z^{ss}E_zA_z^{uu-1} \end{pmatrix}$. In other words, $Df_z(T_z\mathcal{D})$ is the graph of the linear map

$$E_{z_1}=A_z^{us}(A_z^{uu})^{-1}+A_z^{ss}E_z(A_z^{uu})^{-1}.$$

Note that $\|E_{z_1}\|\leq\dfrac{\|A_{z_0}^{us}\|}{\mu-\delta}+\dfrac{\lambda+\delta}{\mu-\delta}\|E_{z_0}\|$ and inductively

$$\|E_{z_n}\|\leq\sum_{i=0}^{n-1}\frac{(\lambda+\delta)^{n-i-1}}{(\mu-\delta)^{n-i}}\|A_{z_i}^{us}\|+\frac{(\lambda+\delta)^n}{(\mu-\delta)^n}\|E_{z_0}\|.$$

Since $\|A_{z_i}^{us}\|=o(\|y_i\|)$, where $z_i=(x_i,y_i)$, there exists $N\in\mathbb{N}$ such that for $n>N$ both $\sum_{i=N}^{n-1}\dfrac{(\lambda+\delta)^{n-i-1}}{(\mu-\delta)^{n-i}}\|A_{z_i}^{us}\|<\epsilon/6$ and $\dfrac{(\lambda+\delta)^n}{(\mu-\delta)^n}\|E_{z_0}\|<\epsilon/6$. If furthermore $N'\in\mathbb{N}$ is such that $\sum_{i=0}^{N-1}\dfrac{(\lambda+\delta)^{N'+N-i-1}}{(\mu-\delta)^{N'+N-i}}\|A_{z_i}^{us}\|<\epsilon/6$ then for $n\geq N+N'=:N_0$ we have $\|E_{z_n}\|<\epsilon/2$.

After possibly increasing N_0 we may assume that $\|A_{(x,y)}^{us}\|<(1-\lambda-\delta)\epsilon$ whenever $\|(x,y)\|\leq\|z_{N_0}\|$. Consequently, by possibly shrinking $\mathcal{D}$, we may assume that all tangent planes to $f^{N_0}(\mathcal{D})$ lie in horizontal ϵ-cones and that

$\|A_z^{us}\| < (1-\lambda-\delta)\epsilon$ for $z \in \bigcup_{i\geq N_0} f^i(\mathcal{D})$. If ϵ is sufficiently small then $\|(A_z^{uu} + A_z^{su}E)^{-1}\| < 1$ whenever $\|E\| < \epsilon$. With this choice of parameters the action of f preserves horizontal ϵ-cones because if $\|E_{z_i}\| < \epsilon$ then, by (6.2.10), $\|E_{z_{i+1}}\|$ is at most

$$\begin{aligned}\|(A_{z_i}^{us} + A_{z_i}^{ss}E_{z_i})(A_{z_i}^{uu} + A_{z_i}^{su}E_{z_i})^{-1}\| &\leq \|A_{z_i}^{us} + A_{z_i}^{ss}E_{z_i}\| \, \|(A_{z_i}^{uu} + A_{z_i}^{su}E_{z_i})^{-1}\| \\ &< \|A_{z_i}^{us}\| + \|A_{z_i}^{ss}E_{z_i}\| < (1-\lambda-\delta)\epsilon + (\lambda+\delta)\epsilon = \epsilon.\end{aligned}$$

Finally, note that the proof of Lemma 6.2.16 shows that $f^n(\mathcal{D})$ covers $W^n_{\text{loc}}(p)$ under the projection π_1 whenever n is sufficiently large. □

Exercises

6.2.1. *Formulate a corollary to Theorem 6.2.8 concerning stable and unstable manifolds for the case of a general (nonhyperbolic) fixed point that generalizes Theorem 6.2.3.*

6.2.2. *Show that in a neighborhood of a nonhyperbolic fixed point there exists an invariant C^1 manifold tangent to the space E^0 defined by (1.2.6).*

6.2.3. *Let f be the time-one map of the vector field $v = (-x, -y^3)$. The x-axis is invariant under f. Show that*

(1) *the orbits of v other than the y-axis are graphs of functions all of whose derivatives vanish at 0;*
(2) *f has a C^∞ invariant manifold tangent to and different from the x-axis which intersects the x-axis at infinitely many points accumulating at the origin.*

6.2.4. *Prove that the local stable and unstable manifolds in a neighborhood of a hyperbolic fixed point of a symplectic diffeomorphism are Lagrangian (Definition 5.5.7). Generalize this statement to the case of a nonhyperbolic fixed point of a symplectic diffeomorphism.*

6.2.5. *Prove two counterparts of Theorem 6.2.3 for flows:*

(1) *near a hyperbolic fixed point;*
(2) *near a hyperbolic periodic point.*

3. Local stability of a hyperbolic periodic point

a. The Hartman–Grobman Theorem. The preceding section has shown that in the presence of hyperbolicity local analysis is a very powerful tool. The Hadamard–Perron Theorem shows that the dynamics of a hyperbolic map is very similar to that of its linear part. In the next section this result will be used for purposes of global and semilocal analysis. Here we digress to prove a local result which shows that near a hyperbolic fixed point a map is topologically conjugate to its linear part.[1]

Theorem 6.3.1. (Hartman–Grobman Theorem) *Let $U \subset \mathbb{R}^n$ be open, $f: U \to \mathbb{R}^n$ continuously differentiable, and $O \in U$ a hyperbolic fixed point of f. Then there exist neighborhoods U_1, U_2, V_1, V_2 of O and a homeomorphism $h: U_1 \cup U_2 \to V_1 \cup V_2$ such that $f = h^{-1} \circ Df_0 \circ h$ on U_1, that is, the following diagram commutes:*

$$\begin{array}{ccc} f: U_1 & \longrightarrow & U_2 \\ {\scriptstyle h}\downarrow & & \downarrow{\scriptstyle h} \\ Df_0: V_1 & \longrightarrow & V_2 \end{array}$$

We will see that the topological character of f near O is determined already by the orientation of f on stable and unstable manifolds and by their dimensions.

Proof. We begin by developing the results of the previous section a bit further.

In the discussion of the Hadamard–Perron Theorem we observed that for the maps $f_m: \mathbb{R}^n \to \mathbb{R}^n$ one can construct the manifolds $W_m^+(p)$ and $W_m^-(p)$ for all $p \in \mathbb{R}^m$. If the reference orbit, around which we localize, is a hyperbolic fixed point then we are led to consider a single map $f: \mathbb{R}^n \to \mathbb{R}^n$. In this case the construction of the stable and unstable manifolds yields for each point $p \in \mathbb{R}^n$ exactly one stable and one unstable manifold, instead of a countable family. This follows from the fact that $W_{m+1}^+(f(p))$ and $f\left(W_m^+(p)\right)$ are characterized in the same way by (iii) in Theorem 6.2.8, and likewise for W^-. Consider a point $p \in \mathbb{R}^m$. The intersection $W^-(p) \cap W^+(0)$ consists of exactly one point $(x, \varphi^+(x))$ and $W^+(p) \cap W^-(0)$ consists of exactly one point $(\varphi^-(y), y)$. Conversely $\{p\} = W^+\left(\varphi^-(y), y\right) \cap W^-\left(x, \varphi^+(x)\right)$, so the point p is uniquely determined by x and y in this way. We can therefore introduce new coordinates $(x(p), y(p))$ by this construction. By Corollary 6.2.22 this is a continuous coordinate change. These coordinates are *adapted* to f in the sense that here we can write

$$f(x, y) = (f_1(x), f_2(y))$$

because by construction $f(x_1, y_1)$ and $f(x_1, y_2)$ have the same x-coordinate and $f(x_1, y_1)$ and $f(x_2, y_1)$ have the same y-coordinate for all x_1, x_2, y_1, and y_2. In these coordinates f_1 and f_2 are indeed C^1 since $f_1(x)$ is the x-coordinate of $f\left(x, \varphi^+(x)\right)$ and $f_2(y)$ is the y-coordinate of $f\left(\varphi^-(y), y\right)$.

Note furthermore that f_1^{-1} and f_2 are contracting maps which can, by virtue of the Extension Lemma 6.2.7, be assumed to be C^1-close to their linear parts $(Df_1^{-1})_0$ and $(Df_2)_0$.

Lemma 6.3.2. *A C^1-perturbation f of an invertible linear contraction $L\colon \mathbb{R}^m \to \mathbb{R}^m$ is topologically conjugate to L.*

This lemma implies the theorem since it provides homeomorphisms h_1 and h_2 such that $h_1 \circ f_1^{-1} = (Df_1^{-1})_0 \circ h_1$ and $h_2 \circ f_2 = (Df_2)_0 \circ h_2$ and hence the Cartesian product $h = (h_1, h_2), (x, y) \mapsto (h_1(x), h_2(y))$ conjugates f and Df_0 since $h \circ f = Df_0 \circ h$. Note finally that the map $f\colon \mathbb{R}^n \to \mathbb{R}^n$ obtained from the map $f\colon U \to \mathbb{R}^n$ in the theorem coincides with $f\colon U \to \mathbb{R}^n$ on some neighborhood of the origin by the Extension Lemma 6.2.7. Thus the Hartman–Grobman Theorem 6.3.1 follows. □

Proof of Lemma 6.3.2. To construct the desired homeomorphism $h\colon \mathbb{R}^m \to \mathbb{R}^m$ we will use the fundamental-domain method (Section 2.7a) which first appeared in a one-dimensional situation in Section 2.1c. On the unit sphere S in $\mathbb{R}^m$ we set $h_0 := h_{\restriction S} := \mathrm{Id}$. Note then that in order to have $h \circ f = L \circ h$ we must set $h_1 := h_{\restriction fS} := L \circ f^{-1}$. We next show that if B denotes the closed unit ball then there exists a homeomorphism $h'\colon B \smallsetminus \operatorname{Int} f(B) \to B \smallsetminus \operatorname{Int} LB$ such that $h'_{\restriction S} = h_0$ and $h'_{\restriction fS} = h_1$. From this we obtain a conjugacy as follows: For $x \in \mathbb{R}^m \smallsetminus \{0\}$ there exists a unique $k(x) \in \mathbb{Z}$ such that $x \in f^{k(x)}(B \smallsetminus f(B))$ and we can thus define

$$h(x) = \begin{cases} L^{k(x)} \circ h' \circ f^{-k(x)}(x) & \text{for } x \in \mathbb{R}^m \smallsetminus \{0\}, \\ 0 & \text{for } x = 0. \end{cases}$$

To construct h' we want to identify $A_1 := B \smallsetminus \operatorname{Int} f(B)$ and $A_2 := B \smallsetminus \operatorname{Int} LB$ with $S \times [0,1]$. In the case of A_2 observe first that any ray $\{tv \mid t \geq 0\}$ through $v \in \mathbb{R}^m$ intersects S in exactly one point $v/\|v\|$. By applying L we see that the same ray $\{tv \mid t \geq 0\} = L\{tL^{-1}v \mid t \geq 0\}$ intersects LS in exactly one point $L\left(\dfrac{L^{-1}v}{\|L^{-1}v\|}\right)$. Thus for each $v \in S$ there is exactly one radial segment connecting it to a point on LS. Parameterizing each of these so that it has constant speed by a parameter in $[0,1]$ yields a homeomorphism $k_2\colon S \times [0,1] \to A_2$ sending $v \in S$ and $t \in [0,1]$ to the point on the segment through v with parameter value t.

To apply the same construction to A_1 we need to show that every radial ray intersects $f(S)$ in exactly one point.

Note that the parameterized curve $(L^{-1}tv) = tL^{-1}v$ has norm increasing at the rate $\|L^{-1}v\| > 0$. Since $f^{-1}(L(L^{-1}tv))$ is a small C^1-perturbation of this curve, its norm also increases as a function of t and hence there is a unique intersection point with S. Consequently the curve $tv = f(f^{-1}(L L^{-1}tv))$ intersects $f(S)$ in exactly one point. Thus the given procedure yields a homeomorphism $k_1\colon S \times [0,1] \to A_1$.

If we choose k_1 and k_2 such that $k_i(\cdot, 0) = \mathrm{Id}_S$ and let $\bar{k_i} := k_i\restriction_{S\times\{1\}}$, then in these coordinates h_0 and h_1 are represented by $h_0' = \mathrm{Id}_{S\times\{0\}}$ and $h_1' = \bar{k_2}^{-1} \circ L \circ f^{-1} \circ \bar{k_1}$ on $S \times \{1\}$, respectively.

Clearly these induce homeomorphisms $\bar{h}_0$ and $\bar{h}_1$ of S, where $\bar{h}_0 = \mathrm{Id}_S$ and $\bar{h}_1$ is close to Id_S. Consequently $\bar{h}(x,t) := \left(\frac{t\bar{h}_0(x) + (1-t)\bar{h}_1(x)}{\|t\bar{h}_0(x) + (1-t)\bar{h}_1(x)\|}, t \right)$ defines a homeomorphism of $S \times [0,1]$ coinciding with h_0' and h_1' on the boundary components and hence $h' := k_2 \circ \bar{h} \circ k_1^{-1}$ is as claimed. □

b. Local structural stability.

Lemma 6.3.3. *Two invertible linear contractions with the same orientation are topologically conjugate.*

Proof. As in the proof of Lemma 6.3.2 we can pass to standard coordinates $S \times [0,1]$. The issue is then to show that an orientation-preserving linear map L is homotopic to the identity. In the absence of negative eigenvalues this is accomplished by a straight-line homotopy as before since $tL + (1-t)\,\mathrm{Id}$ is nonsingular for all $t \in [0,1]$. If there are negative eigenvalues then by the orientation assumption there is an even number of them and the associated root space can be written as a product of copies of $\mathbb{R}^2$. If E_+ is the root space for the remaining eigenvalues then for $t \in [0,1]$ the maps

$$R_t = \mathrm{Id} \times \begin{pmatrix} \cos t & \sin t \\ -\sin t & \cos t \end{pmatrix} \times \cdots \times \begin{pmatrix} \cos t & \sin t \\ -\sin t & \cos t \end{pmatrix} \text{ on } E_+ \times \mathbb{R}^2 \times \cdots \times \mathbb{R}^2$$

yield a homotopy $R_t L$ between L and a linear map L' without negative eigenvalues. The argument for the previous case then applies to L'. □

This lemma yields

Corollary 6.3.4. *Suppose $f: U \to \mathbb{R}^n$, $g: V \to \mathbb{R}^n$ have hyperbolic fixed points $p \in U$ and $q \in V$, respectively, and* $\dim E^+(Df_p) = \dim E^+(Dg_q)$, $\dim E^-(Df_p) = \dim E^-(Dg_q)$, $\operatorname{sign}\det Df_p\restriction_{E^+(Df_p)} = \operatorname{sign}\det Dg_q\restriction_{E^+(Dg_q)}$, *and* $\operatorname{sign}\det Df_p\restriction_{E^-(Df_p)} = \operatorname{sign}\det Dg_q\restriction_{E^-(Dg_q)}$.

Then there exist neighborhoods $U_1 \subset U$ and $V_1 \subset V$ and a homeomorphism $h: U_1 \to V_1$ such that $h \circ f = g \circ h$.

This corollary follows since f and g are locally conjugate to their linear parts by the Hartman–Grobman Theorem and the linear parts are conjugate by Lemma 6.3.3. In particular any C^1 diffeomorphism is locally structurally stable in a neighborhood of a hyperbolic fixed point. Proposition 7.3.1 shows that a converse is also true: Hyperbolicity of a fixed point is necessary for local structural stability.

Exercise

6.3.1. *Let $f: \mathbb{R}^n \to \mathbb{R}^n$ be an orientation-preserving contracting C^1 diffeomorphism. Prove that it is topologically conjugate to the linear map $x \mapsto x/2$.*

4. Hyperbolic sets

a. Definition and invariant cones. In Section 6.2 we concentrated on the study of the behavior of orbits relative to a reference orbit taking the linearized map along that orbit as a model. As we already mentioned interesting qualitative phenomena appear when in the context of global or semilocal analysis the hyperbolic local picture is combined with nontrivial recurrence, which is an essentially nonlinear phenomenon.

Let M be a smooth manifold, $U \subset M$ an open subset, $f: U \to M$ a C^1 diffeomorphism onto its image, and $\Lambda \subset U$ a compact f-invariant set.

Definition 6.4.1. The set Λ is called a *hyperbolic set for the map* f if there exists a Riemannian metric called a *Lyapunov metric* in an open neighborhood U of Λ and $\lambda < 1 < \mu$ such that for any point $x \in \Lambda$ the sequence of differentials $(Df)_{f_x^n}: T_{f_x^n}M \to T_{f_x^{n+1}}M$, $n \in \mathbb{Z}$, admits a (λ, μ)-splitting.

We will usually denote the hyperbolic splitting at $x \in \Lambda$ by $E_x^+ \oplus E_x^-$. In Exercises 6.4.1 and 6.4.2 we give an equivalent definition of a hyperbolic set which is independent of the Riemannian metric.

Definition 6.4.2. A C^1 diffeomorphism $f: M \to M$ of a compact manifold M is called an *Anosov diffeomorphism* if M is a hyperbolic set for f.

The simplest example of a hyperbolic set is a hyperbolic periodic orbit. The hyperbolic automorphism F_L of the 2-dimensional torus

$$F_L(x, y) = (2x + y, x + y) \pmod 1$$

($L = \begin{pmatrix} 2 & 1 \\ 1 & 1 \end{pmatrix}$) discussed in Sections 1.8, 2.5d, 2.6, 3.2e, 4.2d, and 4.4d is an Anosov diffeomorphism. In this case the standard Euclidean metric on $\mathbb{T}^2$ is a Lyapunov metric. $\lambda = \dfrac{3-\sqrt{5}}{2}$, $\mu = \dfrac{3+\sqrt{5}}{2}$ are the eigenvalues of the matrix $\begin{pmatrix} 2 & 1 \\ 1 & 1 \end{pmatrix}$ and the hyperbolic splitting at each point is obtained by translating to that point the eigenspaces of that matrix (cf. Section 1.8).

More generally, any hyperbolic automorphism of the n-torus defined at the end of Section 1.8 is an Anosov diffeomorphism. In this case we can use Proposition 1.2.2 to find a Euclidean norm in $\mathbb{R}^n$ that makes the matrix L contracting in the space $E^-(L)$ and expanding in $E^+(L)$ (cf. (1.2.4) and (1.2.5)), project the Riemannian metric generated by this norm to $\mathbb{T}^n$, and consider the invariant splitting at each point into subspaces parallel to $E^+(L)$ and $E^-(L)$. One can take $\lambda = r\left(L\restriction_{E^-(L)}\right) + \delta$, $\mu = r\left(L^{-1}\restriction_{E^+(L)}\right)^{-1} - \delta$ for any small $\delta > 0$.

Anosov diffeomorphisms are relatively special objects. Much more often one encounters hyperbolic sets that are not manifolds. A prototypical example of such a set is the "horseshoe" described in Section 2.5c. In this case one can take the Euclidean metric, $\lambda = 1/2$, $\mu = 2$, and the splitting into the horizontal and the vertical directions as the hyperbolic splitting. More examples of hyperbolic sets and Anosov systems will be discussed in Section 6.5 and Chapter 17.

Extending the notion of a hyperbolic set to noninvertible systems presents a curious problem. Whereas the contracting part E_x^- of the hyperbolic splitting is determined by the behavior along the positive semiorbit of x and can naturally be defined in this case, the expanding part E_x^+ would have to be determined by considering the negative semiorbit of x which is not uniquely defined for noninvertible maps. This makes a general notion of hyperbolicity less convenient and we will therefore not describe it. However, there is a particular case where there is no ambiguity in choosing the expanding part, namely, when there is no contracting part at all and hence the expanding part must be the entire tangent space. We already discussed expanding maps in Section 2.4a (see Definition 2.4.1). Such maps are somewhat special, similarly to their invertible counterpart, the Anosov diffeomorphisms. A more general phenomenon, similar to hyperbolic sets for invertible maps, is described in the following definition.

Definition 6.4.3. Let Λ be a compact invariant set for a C^1 map $f: U \to M$. Λ is called a *hyperbolic repeller* if there exists a Riemannian metric in a neighborhood of Λ such that for all points $x \in \Lambda$ we have $\|Df^{-1}|_x\| < 1$.

Obviously for an expanding map the whole manifold is a hyperbolic repeller. The invariant sets Λ for quadratic maps described in Section 2.5b are examples of Cantor sets which are hyperbolic repellers. We will encounter more examples of hyperbolic repellers in Sections 16.1, 16.2, and 17.8. Now we will return to the invertible situation.

The following fact is a close counterpart of Lemma 6.2.15:

Proposition 6.4.4. *Let Λ be a hyperbolic set for $f: U \to M$. The dimensions of subspaces E_x^+ and E_x^- are locally constant and those subspaces change continuously with x.*

Proof. Let $\dim M = s$. It follows from Definition 6.4.1 that for any $x \in \Lambda$, $\xi \in E_x^-$, and $n \geq 0$

$$\|Df_x^n \xi\| \leq \lambda^n \|\xi\| \tag{6.4.1}$$

and those inequalities characterize the subspace E_x^-. Let $x_m \to x$. By taking subsequences if necessary one can assume that $\dim E_{x_m}^- = \text{const.} =: k$ and that one can choose an orthonormal basis $(\xi_m^{(1)}, \ldots, \xi_m^{(k)})$ in each $E_{x_m}^-$ such that

$\xi_m^{(i)} \to \xi^{(i)} \in T_xM$, $i = 1, \dots, k$. By continuity of the map Df^n, the inequalities (6.4.1) for $\xi = \xi_m^{(i)}$ imply that

$$\|Df_x^n \xi^{(i)}\| \le \lambda^n \|\xi^{(i)}\| = \lambda^n.$$

Thus $\xi^{(i)} \in E_x^-$, $\dim E_x^- \ge k$, and $E_x^- \supset \lim_{m\to\infty} E_{x_m}^-$.

Similarly, vectors $\eta \in E_x^+$ are characterized by the inequalities

$$\|Df_x^{-n}\eta\| \le \mu^{-n}\|\eta\|$$

for all $n \ge 0$, and a repetition of the previous argument gives $E_x^+ \supset \lim_{m\to\infty} E_{x_m}^+$ and $\dim E_x^+ \ge s - k$.

This implies that $E_x^- = \lim_{m\to\infty} E_{x_m}^-$, $E_x^+ = \lim_{m\to\infty} E_{x_m}^+$. □

Corollary 6.4.5. *The subspaces E_x^- and E_x^+ are uniformly transverse, that is, there exists $\alpha_0 > 0$ such that for any $x \in \Lambda$, $\xi \in E_x^-$, $\eta \in E_x^+$, the angle between ξ and η is at least α_0.*

Proof. Let $\alpha(x)$ be the minimal angle between $\xi \in E_x^+$ and $\eta \in E_x^-$. Since $E_x^+ \cap E_x^- = \{0\}$, $\alpha(x) > 0$. By Proposition 6.4.4 $\alpha(x)$ is a continuous function on Λ. Hence it reaches a positive minimum α_0. □

Obviously, every closed invariant subset of a hyperbolic set for f is also a hyperbolic set. More interestingly, one can sometimes envelop a given hyperbolic set by a larger one.

Proposition 6.4.6. *Let Λ be a hyperbolic set for $f\colon U \to M$. There exists an open neighborhood $V \supset \Lambda$ such that for any g sufficiently close to f in C^1 topology the invariant set*

$$\Lambda_V^g = \bigcap_{n\in\mathbb{Z}} g^n \bar{V}$$

is hyperbolic.

Remark. Obviously $\Lambda_V^f \supset \Lambda$. In a number of cases (for example, for a hyperbolic periodic orbit or for the "horseshoe") $\Lambda_V^f = \Lambda$ if V is a sufficiently small neighborhood of Λ. For $g \ne f$ we have not proved yet that Λ_V^g is nonempty. We will discuss this question later (cf. Theorem 18.2.1).

Proof. Let us extend the invariant hyperbolic splitting $T_xM = E_x^+ \oplus E_x^-$ defined for $x \in \Lambda$ to a continuous (in general, no longer invariant) splitting defined for all x in an open neighborhood $V_1 \supset \Lambda$. Fix a sufficiently small $\gamma > 0$ and consider the family of standard horizontal cones $H_x^\gamma = \{\xi + \eta \mid \xi \in E_x^+,\ \eta \in E_x^-,\ \|\eta\| \le \gamma\|\xi\|\}$ and the family of vertical cones V_x^γ, defined similarly.

For $x \in \Lambda$, having a (λ, μ)-splitting implies that

$$\begin{aligned} &Df_x H_x^\gamma \subset H_{fx}^{\lambda\mu^{-1}\gamma}, \qquad Df_x^{-1} V_x^\gamma \subset V_{f^{-1}\gamma}^{\lambda\mu^{-1}\gamma}, \\ &\|Df_x\xi\| \ge \frac{\mu}{1+\gamma}\|\xi\| \qquad \text{for } \xi \in H_x^\gamma, \\ &\|Df_x\eta\| \ge (1+\gamma)\lambda^{-1}\|\eta\| \text{ for } \eta \in V_x^\gamma. \end{aligned} \tag{6.4.2}$$

Thus if one fixes any $\delta > 0$, by continuity of Df, Df^{-1}, and the extended splitting one can find an open neighborhood $V \supset \Lambda$ such that for any $x \in \bar{V}$ conditions similar to (6.4.2) with λ replaced by $\lambda+\delta$ and μ replaced by $\mu-\delta$ hold not only for f but for any g close enough to f in the C^1 topology. This implies that for any point $x \in \Lambda_V^g$ the assumptions of Proposition 6.2.12 are satisfied with $\lambda' = (1+\gamma)(\lambda+\delta)$, $\mu' = (\mu-\delta)/(1+\gamma)$, $L_m = Dg_{g_x^m}$, $\gamma_m = \gamma_m' = \gamma$. Hence the sequence of differentials $Dg_{g_x^m}$ admits a (λ', μ')-splitting. If γ and δ are sufficiently small then $\lambda' < 1 < \mu'$ so Λ_V^g is a hyperbolic set. □

Corollary 6.4.7. *Any sufficiently small C^1-perturbation of an Anosov diffeomorphism is an Anosov diffeomorphism.*

In addition to being useful in the proof of Proposition 6.4.6, the invariant-cones criterion (Proposition 6.2.12) can be applied more directly to provide a convenient description of hyperbolicity. The following statement is a reformulation of the cone criterion Corollary 6.2.13 in our setting.

Corollary 6.4.8. *A compact f-invariant set Λ is hyperbolic if there exist $\lambda < 1 < \mu$ such that for every $x \in \Lambda$ there is a decomposition $T_xM = S_x \oplus T_x$ (in general, not Df invariant), a family of horizontal cones $H_x \supset S_x$, and a family of vertical cones $V_x \supset T_x$ associated with that decomposition such that*

$$Df_x H_x \subset \operatorname{Int} H_{f(x)}, \quad Df_x{}^{-1} V_{f(x)} \subset \operatorname{Int} V_x,$$

$\|Df_x\xi\| \geq \mu\|\xi\|$ *for* $\xi \in H_x$, *and* $\|Df_x{}^{-1}\xi\| \geq \lambda^{-1}\|\xi\|$ *for* $\xi \in V_{f(x)}$.

b. Stable and unstable manifolds. Next we will show how to apply the Hadamard–Perron Theorem 6.2.8 in the context of hyperbolic sets.

For that we need to describe a certain localization procedure, that is, a nice way of choosing a coordinate system near each point of the manifold.

For each $x \in \Lambda$ let us fix a coordinate system in the tangent space T_xM such that the decomposition $E_x^+ \oplus E_x^-$ is identified with the standard decomposition $\mathbb{R}^n = \mathbb{R}^k \oplus \mathbb{R}^{n-k}$ and the Riemannian metric in T_xM becomes the standard Euclidean metric in $\mathbb{R}^n$. Fix a small enough $\epsilon > 0$ and denote by D_ϵ the product of the ϵ-balls around the origin in $\mathbb{R}^k$ and $\mathbb{R}^{n-k}$. Then use the exponential map (cf. (9.5.1)) $\exp_x\colon D_\epsilon \to M$. Due to the compactness of Λ we can take ϵ such that the map $\exp_x$ is injective for every $x \in \Lambda$ and hence is a diffeomorphism between D_ϵ and a certain neighborhood $P_\epsilon(x)$ of x in M. Accordingly we can define a family of maps $f_{x,\epsilon} = \exp_{fx}^{-1} \circ f \circ \exp_x\colon D_\epsilon \to \mathbb{R}^n$.

The properties of the exponential map, in particular the fact that its differential at the origin is the identity map and the smooth dependence of $\exp_x$ on x, ensure that for sufficiently small $\epsilon > 0$, each map $f_{x,\epsilon}$ is C^1 close to its differential at the origin which is simply $(Df)_x$ expressed in our coordinates. Then one uses the Extension Lemma 6.2.7 to extend the maps f_{x,ϵ_1} for some $\epsilon_1 < \epsilon$ to the whole space $\mathbb{R}^n$. Let us denote those extended maps simply by f_x. Thus, along each orbit $\{f^m(x)\}_{m\in\mathbb{Z}}$ for $x \in \Lambda$ we obtain a sequence of maps

$f_m = f_{f^m(x)}: \mathbb{R}^n \to \mathbb{R}^n$ satisfying the conditions of Theorem 6.2.8. In particular, by decreasing ϵ even further to ϵ_2 we can make the number δ that appears in the formulation of this theorem arbitrarily small. A reformulation of the statement of that theorem in terms of the original map f yields the following result which is usually called the *Stable and Unstable Manifolds Theorem for hyperbolic sets.*

Theorem 6.4.9. *Let Λ be a hyperbolic set for a C^1 diffeomorphism $f: V \to M$ such that Df on Λ admits a (λ, μ)-splitting with $\lambda < 1 < \mu$. Then for each $x \in \Lambda$ there is a pair of embedded C^1 discs $W^s(x)$, $W^u(x)$, called the local stable manifold and the local unstable manifolds of x, respectively, such that*

(1) $T_x W^s(x) = E_x^-$, $T_x W^u(x) = E_x^+$;
(2) $f(W^s(x)) \subset W^s(f(x))$, $f^{-1}(W^u(x)) \subset W^u(f^{-1}(x))$;
(3) *for every $\delta > 0$ there exists $C(\delta)$ such that for $n \in \mathbb{N}$*

$$\operatorname{dist}(f^n(x), f^n(y)) < C(\delta)(\lambda+\delta)^n \operatorname{dist}(x,y) \text{ for } y \in W^s(x),$$
$$\operatorname{dist}(f^{-n}(x), f^{-n}(y)) < C(\delta)(\mu-\delta)^{-n} \operatorname{dist}(x,y) \text{ for } y \in W^u(x);$$

(4) *there exists $\beta > 0$ and a family of neighborhoods O_x containing the ball around $x \in \Lambda$ of radius β such that*

$$W^s(x) = \{y \mid f^n(y) \in O_{f^n(x)}, \quad n = 0, 1, 2, \dots\},$$
$$W^u(x) = \{y \mid f^{-n}(y) \in O_{f^{-n}(x)}, \quad n = 0, 1, 2, \dots\}.$$

Remark. The local stable and unstable manifolds are not unique and in general there is no particularly canonical way to fix them. However, one immediately deduces from (3) and (4) that for any two local stable manifolds, say $W_1^s(x)$ and $W_2^s(x)$, satisfying the assertions of the theorem their intersection contains an open neighborhood of x on each of them. Equivalently, one can say that for some $n \geq 0$ one has $f^n(W_1^s(f^{-n}(x))) \subset W_2^s$ and $f^n(W_2^s(f^{-n}(x))) \subset W_1^s$. In fact such a number n can be chosen uniformly for all $x \in \Lambda$. The same is naturally true for local unstable manifolds with n replaced by $-n$.

This also implies that global stable and unstable manifolds

$$\widetilde{W}^s(x) = \bigcup_{n=0}^{\infty} f^{-n}(W^s(f^n(x)))$$
$$\widetilde{W}^u(x) = \bigcup_{n=0}^{\infty} f^{n}(W^u(f^{-n}(x))) \tag{6.4.3}$$

are defined independently of a particular choice of local stable and unstable manifolds and can be characterized topologically:

$$\widetilde{W}^s(x) = \{y \in U \mid \operatorname{dist}(f^n(x), f^n(y)) \to 0, \quad n \to \infty\},$$
$$\widetilde{W}^u(x) = \{y \in U \mid \operatorname{dist}(f^{-n}(x), f^{-n}(y)) \to 0, \quad n \to \infty\}.$$

Corollary 6.4.10. *The restriction of a diffeomorphism to a hyperbolic set is expansive (see Definition 3.2.11).*

Proof. If $\operatorname{dist}(f^n(x), f^n(y)) < \beta$, for $n \in \mathbb{Z}$, then $f^n(y) \in O_{f^n(x)}$ and $y \in W^s(x) \cap W^u(x) = \{x\}$ by Theorem 6.4.9(4). □

The next statement establishes a sort of uniqueness for local stable and unstable manifolds.

Proposition 6.4.11. *If $x, y \in \Lambda$, $z \in W^s(x) \cap W^s(y)$, then the intersection $W^s(x) \cap W^s(y)$ contains an open neighborhood of z in both manifolds. A similar statement holds for local unstable manifolds.*

Proof. $\operatorname{dist}(f^n(x), f^n(y)) \le \operatorname{dist}(f^n(x), f^n(z)) + \operatorname{dist}(f^n(z), f^n(y)) \to 0$ as $n \to \infty$ by Theorem 6.4.9(3). Thus by Theorem 6.4.9(4), $f^n(y) \in W^s(f^n(x))$ for any sufficiently large n. By the same token we deduce that for any sufficiently large n

$$f^n(W^s(y)) \subset W^s(f^n(x)).$$

Since $f^n(W^s(y))$ and $f^n(W^s(x)) \subset W^s(f^n(x))$ are open discs, their intersection is open in $W^s(f^n(x))$ and hence in $f^n(W^s(x))$ and $f^n(W^s(y))$. The proposition follows since $z \in f^{-n}(f^n(W^s(y)) \cap f^n(W^s(x)))$. □

Corollary 6.4.12. *If for $x, y \in \Lambda$, the global stable manifolds $\widetilde{W}^s(x)$ and $\widetilde{W}^s(y)$ defined in (6.4.3) have nonempty intersection, then $\widetilde{W}^s(x) = \widetilde{W}^s(y)$.*

On the other hand, complementary leaves intersect in exactly one point: Since E^+ and E^- are continuous on Λ and uniformly transverse (Corollary 6.4.5) the smoothness of $W^s(x)$ and $W^u(x)$ implies

Proposition 6.4.13. *Denote by $W^s_\epsilon(x)$ and $W^u_\epsilon(x)$ the ϵ-balls in $\widetilde{W}^s(x)$ and $\widetilde{W}^u(x)$. Then there exists an $\epsilon > 0$ such that then for any $x, y \in \Lambda$ the intersection $W^s_\epsilon(x) \cap W^u_\epsilon(y)$ consists of at most one point $[x, y]$ and there exists a $\delta > 0$ such that whenever $d(x, y) < \delta$ for some $x, y \in \Lambda$ then $W^s_\epsilon(x) \cap W^u_\epsilon(y) \neq \varnothing$.*

Remark. Corollary 6.2.22 shows that $[\cdot, \cdot]$ is continuous.

Proposition 6.4.14. *Let Λ be a compact hyperbolic set for $f: U \to M$. Then there exists an open neighborhood V_0 of Λ and $\alpha_0 > 0$ such that whenever $x, y \in \Lambda$ and $\{z\} = W^s(x) \cap W^u(y) \subset V_0$ then for any $\xi \in T_z W^s(x)$ and $\eta \in T_z W^u(y)$ the angle between ξ and η is greater than α_0.*

Proof. Choose a neighborhood V of Λ as in Proposition 6.4.6. Since Λ is compact the distance between Λ and $M \smallsetminus V$ is positive. Call it δ. If V_0 is the δ-neighborhood of Λ then it satisfies the assertion of Proposition 6.4.6. Choose $W^s(x)$ and $W^u(x)$ to be of size less than δ for all $x \in \Lambda$. Then Theorem 6.4.9(3) shows that if $\{z\} = W^s(x) \cap W^u(y) \subset V_0$ for some $x, y \in \Lambda$ then $f^n(z) \in V_0$ for all $n \in \mathbb{Z}$. Consequently $z \in \Lambda^f_{V_0}$ which is hyperbolic by Proposition 6.4.6 and the claim follows by Corollary 6.4.5. □

c. Closing Lemma and periodic orbits. Next we show that hyperbolicity provides a mechanism for finding many periodic orbits. This is the first indication in a general setting how local behavior modeled on the linear part combined with recurrence caused by nonlinearity necessarily produces a complicated and rich orbit structure. Of course we already saw specific manifestations of this phenomenon in concrete situations (Proposition 1.7.2, Proposition 1.8.1, Corollary 2.5.1, Theorem 5.4.14). Further implications of this will be explored in Chapter 18.

We call a sequence $x_0, x_1, \dots, x_{m-1}$, $x_m = x_0$ of points a *periodic ϵ-orbit* or periodic *pseudo-orbit* if $\operatorname{dist}(f(x_k), x_{k+1}) < \epsilon$ for $k = 0, \dots, m-1$ (see also Definition 18.1.1).

Theorem 6.4.15. (Anosov Closing Lemma) *Let Λ be a hyperbolic set for $f: U \to M$. Then there exists an open neighborhood $V \supset \Lambda$ and C, $\epsilon_0 > 0$ such that for $\epsilon < \epsilon_0$ and any periodic ϵ-orbit $(x_0, \dots, x_m) \subset V$ there is a point $y \in U$ such that $f^m(y) = y$ and $\operatorname{dist}(f^k(y), x_k) < C\epsilon$ for $k = 0, \dots, m-1$.*

Remark. A particular case of an ϵ-periodic orbit is provided by an orbit segment $x_0, f(x_0), \dots, f^{m-1}(x_0)$ such that $\operatorname{dist}(f^m(x_0), x_0) < \epsilon$. Thus the Anosov Closing Lemma implies in particular that near any point in a hyperbolic set whose orbit nearly returns to the point there is a periodic orbit that closely follows the almost-returning segment.

In the latter case, however, the conclusion of the Closing Lemma can be considerably strengthened. This follows from a general property of exponential instability of orbits on and near a hyperbolic set.

Proposition 6.4.16. *Let Λ be a hyperbolic set with a (λ, μ)-splitting for $f: U \to M$. Then for any $\alpha \geq \max(\lambda, \mu^{-1})$ there exists $\delta > 0$ and $C > 0$ such that if $x \in \Lambda$, $y \in U$ and $\operatorname{dist}(f^k(y), f^k(x)) < \delta$ for $k = 0, \dots, n$ then in fact*

$$\operatorname{dist}(f^k(y), f^k(x)) < C\, \alpha^{\min(k, n-k)} \cdot (\operatorname{dist}(x, y) + \operatorname{dist}(f^n(x), f^n(y))).$$

Proof. This statement follows immediately from the localization procedure described at the beginning of the previous subsection with the orbit of x as the reference orbit. Once we are in the setting of Theorem 6.2.8 the assertion is evident from the form of the maps f_m and the estimates on linear and nonlinear parts. □

Corollary 6.4.17. *Let Λ be a hyperbolic set for $f: U \to M$ with a (λ, μ)-splitting. Then for any $\alpha > \max(\lambda, \mu^{-1})$ there exists a neighborhood $U \supset \Lambda$, and C_1, $\epsilon_0 > 0$ such that if $f^k(x) \in U$ for $k = 0, \dots, n$ and $\operatorname{dist}(f^k(x), x) < \epsilon_0$ then there exists a periodic point y such that $f^n(y) = y$ and*

$$\operatorname{dist}(f^k(y), f^k(x)) < C_1 \alpha^{\min(k, n-k)} \operatorname{dist}(f^n(x), x).$$

Proof. Apply Theorem 6.4.15 first to obtain the periodic point y. By Proposition 6.4.6 one can assume that $y \in \Lambda$. Then Proposition 6.4.16 gives the statement. □

Proof of Theorem 6.4.15.[1] The same localization procedure yields for each $x \in \Lambda$ a neighborhood V_x that allows us to reformulate the problem in terms of sequences of small perturbations of hyperbolic linear maps, given by $f_k(u,v) = (A_k u + \alpha_k(u,v),\ B_k v + \beta_k(u,v))$ with $\|\alpha_k\|_{C^1} < C_1\epsilon$ and $\|\beta_k\|_{C^1} < C_1\epsilon$ for all k and some $C_1 > 0$. Unlike the proof of Theorem 6.2.8 we do *not* assume here that the maps f_k fix the origin.

Note that a sequence $(u_k, v_k) \in V_{x_k}$, $k = 0, \dots, m-1$, is a periodic orbit if and only if

$$\begin{aligned}(u,v) &:= ((u_0,v_0),\ (u_1,v_1), \dots, (u_{m-1},v_{m-1}))\\ &= (f_{m-1}(u_{m-1},v_{m-1}),\ f_0(u_0,v_0), \dots, f_{m-2}(u_{m-2},v_{m-2})) =: F(u,v).\end{aligned}$$

Therefore we need to find a fixed point of this map $F\colon \mathbb{R}^N \to \mathbb{R}^N$ ($N = m \cdot \dim M$), where $\mathbb{R}^N$ is given the norm $\|(x_0, x_1, \dots, x_{m-1})\| := \max_{0 \le i \le m-1} \|x_i\|$. To this end we represent F as

$$F(u,v) = L(u,v) + S(u,v),$$

where

$$\begin{aligned}S((u_0,v_0),(u_1,v_1),\dots,(u_{m-1},v_{m-1}))&\\ := \big((\alpha_{m-1}(u_{m-1},v_{m-1}),\beta_{m-1}(u_{m-1},v_{m-1})),&\\ \dots,(\alpha_{m-2}(u_{m-2},v_{m-2}),\beta_{m-2}(u_{m-2},v_{m-2}))\big),&\end{aligned}$$

$$\begin{aligned}&L((u_0,v_0),(u_1,v_1),\dots,(u_{m-1},v_{m-1}))\\ &\quad := ((A_{m-1}u_{m-1}, B_{m-1}v_{m-1}),\ (A_0u_0, B_0v_0), \dots, (A_{m-2}u_{m-2}, B_{m-2}v_{m-2})).\end{aligned}$$

L is hyperbolic: It expands the subspace $((u_0,0),(u_1,0),\dots,(u_{m-1},0))$ and contracts the subspace $((0,v_0),(0,v_1),\dots,(0,v_{m-1}))$. Then $(L-\mathrm{Id})$ is invertible and $\|(L-\mathrm{Id})^{-1}\| \le C_2$ for some $C_2 = C_2(f,\Lambda) > 0$. Note also that $\|S(u,v) - S(u',v')\| \le C_3 \cdot \epsilon \cdot \|(u,v) - (u',v')\|$ for some $C_3 = C_3(f,\Lambda) > 0$.

Consequently the solutions of $Fz = z$ or $Lz + Sz = z$ can be obtained as the solutions of

$$z = -(L-\mathrm{Id})^{-1}S(z) =: \mathcal{F}(z).$$

If we take $\epsilon < 1/C_2C_3$ then $\|\mathcal{F}(z) - \mathcal{F}(z')\| \le C_2 \cdot C_3 \cdot \epsilon\|z - z'\|$ and hence the map $\mathcal{F}\colon \mathbb{R}^N \to \mathbb{R}^N$ is a contraction. By the Contraction Mapping Principle (Proposition 1.1.2) there is a unique fixed point $z_0 = \mathcal{F}(z_0)$ of $\mathcal{F}$ in $\mathbb{R}^N$ and hence a unique solution of $Fz_0 = z_0$.

To complete the proof we need to show that the corresponding periodic orbit stays inside the neighborhoods V_{x_k} and thus is indeed a periodic orbit for f, and that its distance to the x_k does not exceed $C\epsilon$ for some $C > 0$. Indeed, by choosing ϵ sufficiently small, the latter claim implies the former.

Note that $z_0 = \lim_{i\to\infty} \mathcal{F}^i(x)$, where x is the sequence $(x_0, x_1, \dots, x_{m-1})$, and hence

$$\|z_0 - x\| \le \sum_{i=1}^{\infty} \|\mathcal{F}^i(x) - \mathcal{F}^{i-1}(x)\|.$$

Inductively we find that $\|\mathcal{F}^k(x) - \mathcal{F}^{k-1}(x)\| \le C_2C_3\epsilon\|\mathcal{F}^{k-1}(x) - \mathcal{F}^{k-2}(x)\| \le (C_2C_3\epsilon)^{k-1}\|\mathcal{F}(x) - x\|$, and hence

$$\|z_0 - x\| \le \left(\sum_{k=0}^{\infty}(C_2C_3\epsilon)^k\right)\|\mathcal{F}(x) - x\|.$$

Since $Fx = x + v$ for some v with $\|v\| < \epsilon$ we have $Sx = -(L - \mathrm{Id})x + v$ or $\mathcal{F}(x) = x - (L - \mathrm{Id})^{-1}v$ and therefore $\|\mathcal{F}(x) - x\| \le C_2 \cdot \epsilon$. But this shows that $\|z_0 - x\| \le C_2\left(\sum_{k=0}^{\infty}(C_2C_3\epsilon)^k\right) \cdot \epsilon$ and ends the proof. □

d. Locally maximal hyperbolic sets. The Anosov Closing Lemma (Theorem 6.4.15) does not assert that the periodic orbit it provides lies in the hyperbolic set Λ. Although this is indeed not always the case (Exercise 6.4.7), it is true in all the examples we have encountered so far. This is due to the fact that in those examples the hyperbolic sets are maximal sets in an open neighborhood (as remarked after Proposition 6.4.6) in the following sense. There exists an open neighborhood V of Λ such that

$$\Lambda = \Lambda_V^f := \bigcap_{n\in\mathbb{Z}} f^n(\bar{V}).$$

This property will be systematically used in Part 4 and thus we give it a name.

Definition 6.4.18. Let Λ be a hyperbolic set for $f\colon U \to M$. If there is an open neighborhood V of Λ such that $\Lambda = \Lambda_V^f$ then Λ is said to be *locally maximal* or *basic.*[2]

Note that if V is an open neighborhood of Λ then any periodic point in V is contained in Λ_V^f. If V is sufficiently small and Λ is locally maximal then these orbits are therefore in Λ. In this case Λ has many periodic orbits by the following

Corollary 6.4.19. *Let Λ be a hyperbolic set for $f\colon U \to M$ and V such that Λ_V^f is hyperbolic. Then periodic points are dense in $NW(f\restriction_{\Lambda_V^f})$ (see Definition 3.3.3 and the subsequent discussion).*

Proof. For $x \in NW(f\restriction_{\Lambda_V^f})$ and $\epsilon > 0$ denote by U_ϵ the $\epsilon/(2C+1)$-neighborhood of x in Λ_V^f, where C is as in the Closing Lemma. Then there exists $N \in \mathbb{N}$ such that $f^N(U_\epsilon) \cap U_\epsilon \ne \varnothing$. For $y \in f^N(U_\epsilon) \cap U_\epsilon$ we have $\mathrm{dist}(f^N(y), y) < 2\epsilon/(2C+1)$ and hence by the Closing Lemma there is a periodic point z such that $\mathrm{dist}(f^n(z), f^n(y)) < 2C\epsilon/(2C+1)$ for $n \in \{0, \dots, N-1\}$, if ϵ is sufficiently small. If ϵ is small enough we also have $z \in V$ and hence $z \in \Lambda_V^f$. Finally

$$\mathrm{dist}(x, z) \le \mathrm{dist}(x, y) + \mathrm{dist}(y, z) \le \frac{(2C+1)\epsilon}{2C+1} = \epsilon.$$ □

Local maximality makes this corollary much more convenient.

Corollary 6.4.20. *Let Λ be a locally maximal hyperbolic set for $f: U \to M$. Then periodic points are dense in $NW(f_{\restriction \Lambda})$.*

An important technical property that follows from local maximality is the presence of a *local product structure*: We say that a hyperbolic set Λ has local product structure if for sufficiently small $\epsilon > 0$ the intersection points provided by Proposition 6.4.13 are always contained in Λ.

Proposition 6.4.21. *A compact locally maximal hyperbolic set has local product structure.*

Proof. Take ϵ such that the ϵ-neighborhood V of Λ satisfies $\Lambda = \Lambda_V^f$. Then all points $[x, y]$ obtained in Proposition 6.4.13 are in V, hence in Λ. $\square$

We will see in Section 18.4 that the presence of a local product structure is equivalent to local maximality.

A natural example of a closed invariant hyperbolic set that is not locally maximal is given by a hyperbolic periodic orbit together with the orbit of a transverse homoclinic point (see Section 0.4, Definition 6.5.4). This situation appears in the horseshoe (cf. Section 2.5c), for example, coded by the set Λ_0 of sequences of 0's and 1's that have no more than one 1. This set is not locally maximal since for every $N \in \mathbb{N}$ it is contained in the closed set Λ_0^N consisting of all sequences such that any two 1's are separated by at least N 0's and for any open neighborhood V of Λ_0 we have $\Lambda_0^N \subset V$ for sufficiently large N.

It is, in fact, not hard to see that Λ_0^N is indeed locally maximal, so for any neighborhood V of Λ_0 there is an invariant locally maximal hyperbolic set $\widetilde{\Lambda}$ such that $\Lambda_0 \subset \widetilde{\Lambda} \subset V$.

Indeed, although any closed invariant subset of the horseshoe is hyperbolic and may have an extremely complicated structure, it can always be enveloped (by Λ_V^f for an appropriate open neighborhood V as in Proposition 6.4.6) by a locally maximal one (cf. Exercises 6.4.8 and 6.4.9).

In general, however, it is not known whether this is true.

Open problem. *Let Λ be a hyperbolic set for $f: U \to M$ and V an open neighborhood of Λ. Does there exist a locally maximal hyperbolic invariant set $\widetilde{\Lambda}$ such that $\Lambda \subset \widetilde{\Lambda} \subset V$?*

Exercises

6.4.1. *Suppose Λ is a hyperbolic set for a C^1 map $f: U \to M$. Show that for any Riemannian metric on U there exist $C > 0$ and $\lambda \in (0,1)$ such that for $x \in \Lambda$, $v \in E_x^{\pm}$, and $n \in \mathbb{N}$*

$$\|Df_x^{\mp n} v\| \leq C\lambda^n \|v\|, \tag{6.4.4}$$

where the norm is generated by the Riemannian metric.

6.4.2. *Suppose Λ is a compact invariant set for $f: U \to M$ and there is a Df-invariant decomposition $T_\Lambda M = E^+ \oplus E^-$ such that (6.4.4) holds for some Riemannian metric. Show that Λ is a hyperbolic set.*

6.4.3. *Let (M, ω) be a symplectic manifold (see Definition 5.5.7), $U \subset M$ open, and $f: U \to M$ a symplectic diffeomorphism. Let $\Lambda \subset U$ be a hyperbolic set for f. Prove that $\dim E_x^+ = \dim E_x^-$ for all $x \in \Lambda$, $E_x^\pm$ are Lagrangian subspaces of $T_x M$ and that $W^s(x)$ and $W^u(x)$ are Lagrangian submanifolds of M.*

6.4.4. *Formulate and prove the counterpart of the Anosov Closing Lemma (Theorem 6.4.15) for hyperbolic repellers (Definition 6.4.3).*

6.4.5. *Construct an example of a locally maximal hyperbolic set Λ such that periodic points are not dense in Λ.*

6.4.6. *Prove that the number of preimages of any point on a hyperbolic repeller is uniformly bounded. Prove that it is constant if the map is topologically transitive on the repeller.*

6.4.7*. *Show that the invariant set Λ in the horseshoe in Section 2.5c contains a perfect subset on which the map is minimal. Deduce that the periodic points guaranteed by the Anosov Closing Lemma (Theorem 6.4.15) may not belong to the hyperbolic set.*

6.4.8. *Show that for any closed invariant subset S of the invariant set Λ in the horseshoe and any open neighborhood U of S there exists a locally maximal compact invariant set $\tilde{S}$ such that $S \subset \tilde{S} \subset U$.*

6.4.9. *Let $F_L: \mathbb{T}^2 \to \mathbb{T}^2$ be a linear hyperbolic automorphism. Show that for any closed F_L-invariant set S and any open neighborhood U of S there exists a locally maximal compact invariant set $\tilde{S}$ such that $S \subset \tilde{S} \subset U$.*

5. Homoclinic points and horseshoes

In this section we elaborate upon the discussion of Smale's horseshoe in Section 2.5c. It is a nontrivial hyperbolic set with a natural and very convenient way of coding the dynamics. It turns out that the behavior represented by this example can be found in a broad class of dynamical systems.

The treatment in Section 2.5c was limited to 2-dimensional "linear" horseshoes: It was obtained by considering the intersection of the image $f(\Delta)$ of a rectangle Δ under a diffeomorphism f with Δ itself under the assumption that f is an affine hyperbolic map on each component of $\Delta \cap f^{-1}(\Delta)$. Now we give a definition of horseshoes in higher dimensions and for nonlinear maps. From this definition it will be clear that the construction of a coding in Section 2.5c can be carried out verbatim.

Next we pick up the discussion of transverse homoclinic points in Section 0.4. These arose as an example to motivate some aspects of semilocal analysis. Here we study some of these aspects in detail. The description of the dynamics thus obtained leads us to a proof of the fact that in the presence of a transverse homoclinic point for a map f there exists a horseshoe for some iterate of f in an arbitrary small neighborhood of the homoclinic orbit.

a. General horseshoes. By a rectangle in $\mathbb{R}^n$ we mean a set of the form $\Delta = D_1 \times D_2 \subset \mathbb{R}^k \oplus \mathbb{R}^l = \mathbb{R}^n$, where D_1 and D_2 are disks. We denote by $\pi_1: \mathbb{R}^n \to \mathbb{R}^k$ and $\pi_2: \mathbb{R}^n \to \mathbb{R}^l$ the canonical projections. As in Section 6.2 we will refer to the $\mathbb{R}^k$-direction as "horizontal" and the $\mathbb{R}^l$-direction as "vertical".

Definition 6.5.1. Suppose $\Delta \subset U \subset \mathbb{R}^n$ is a rectangle and $f: U \to \mathbb{R}^n$ a diffeomorphism. A connected component $C' = fC$ of $\Delta \cap f(\Delta)$ is called *full* (for f) if

(1) $\pi_2(C) = D_2$, and
(2) for any $z \in C$, $\pi_1 \restriction_{f(C \cap (D_1 \times \pi_2(z)))}$ is a bijection onto D_1.

Geometrically, condition (2) means that the image of every horizontal fiber in C meets Δ and "traverses" Δ completely.

Definition 6.5.2. If $U \subset \mathbb{R}^n$ is open then a rectangle $\Delta = D_1 \times D_2 \subset U \subset \mathbb{R}^k \oplus \mathbb{R}^l = \mathbb{R}^n$ is called a *horseshoe* for a diffeomorphism $f: U \to \mathbb{R}^n$ if $\Delta \cap f(\Delta)$ contains at least two full components Δ_0 and Δ_1 such that for $\Delta' = \Delta_0 \cup \Delta_1$

(1) $\pi_2(\Delta') \subset \text{int } D_2, \quad \pi_1(f^{-1}(\Delta')) \subset \text{int } D_1$,
(2) $D(f\restriction_{f^{-1}(\Delta')})$ preserves and expands a horizontal cone family on $f^{-1}(\Delta')$,
(3) $D(f^{-1}\restriction_{\Delta'})$ preserves and expands a vertical cone family on Δ'.

Conditions (2) and (3) imply by Corollary 6.4.8 that $\Lambda := \bigcap_{n \in \mathbb{Z}} f^{-n}(\Delta')$ is a hyperbolic set for f with "almost horizontal" expanding and "almost vertical" contracting directions. Note also that, as in Section 2.5c, $\Lambda \subset \text{Int}\, \Delta'$ is the maximal f-invariant subset of Δ'.

Thus, by restricting attention to $\Delta' \subset \Delta$ the comments in Section 2.5c up to Corollary 2.5.1 now apply virtually verbatim to the set Λ obtained here. In particular we obtain a coding of the dynamics on Λ via a topological conjugacy with the full 2-shift σ_2.

Since every rectangle that is a horseshoe for f is also a horseshoe for any diffeomorphism f' sufficiently close to f in the C^1 topology we obtain the following semilocal structural-stability result.

Proposition 6.5.3. *Let $\Lambda = \bigcap_{k \in \mathbb{Z}} f^{-k}(\Delta')$ be the maximal invariant set inside a horseshoe for a C^1 diffeomorphism $f: U \to \mathbb{R}^n$. Then for any f' sufficiently C^1-close to f there are an f'-invariant set Λ' and a homeomorphism $h: \Lambda \to \Lambda'$ such that $h \circ f\restriction_\Lambda = f'\restriction_{\Lambda'} \circ h$.*[1]

b. Homoclinic points. We now recall the discussion of the dynamics in the presence of a transverse homoclinic point begun in Section 0.4.

Definition 6.5.4. Let $f: X \to X$ be a homeomorphism of a metric space (X, d). The point $x \in X$ is said to be *homoclinic* to the point $y \in X$ if

$$\lim_{|n|\to\infty} d(f^n(x), f^n(y)) = 0.$$

It is said to be *heteroclinic* to the points $y_1, y_2 \in X$ if $\lim_{n\to\infty} d(f^n(x), f^n(y_1)) = \lim_{n\to-\infty} d(f^n(x), f^n(y_2)) = 0$. If M is a differentiable manifold and $x \in M$ is a hyperbolic fixed point for $f \in \mathrm{Diff}^1(M)$ then we say that $q \in M$ is a *transverse homoclinic point* if q is a point of transverse intersection of the stable and unstable manifolds of x.

We begin by sketching the picture for a transverse homoclinic point of an area-preserving diffeomorphism $f: \mathbb{R}^2 \to \mathbb{R}^2$. Let us assume that the origin is a fixed point for f and that near the origin $f(x, y) = (2x, y/2)$. The local stable and unstable manifolds of the origin are segments of the y-axis and x-axis, correspondingly. We assume that their extensions intersect transversely in a point q.

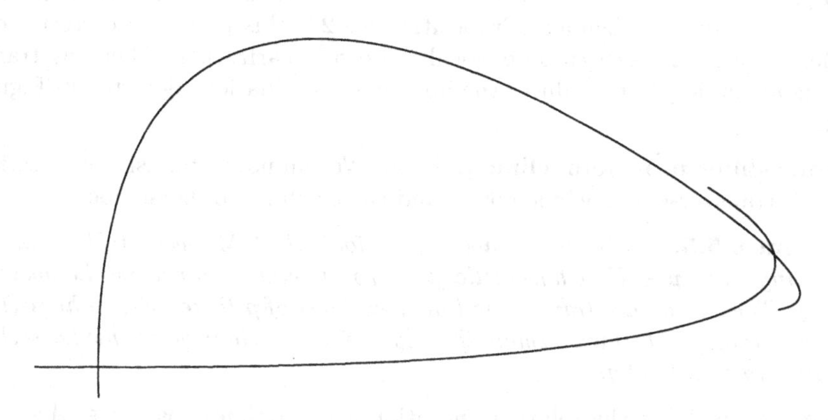

FIGURE 6.5.1. A homoclinic point

As noted in Section 0.4 this clearly implies that $f^n(q) \to 0$ as $|n| \to \infty$. Since the unstable and stable manifolds of 0 are invariant under f, the images $f^n(q)$ are also homoclinic points, that is, intersection points of the unstable and stable manifolds of the origin. Since the intersection at q is transverse and f is a diffeomorphism, the same is true at $f^n(q)$ for any n. We thus immediately obtain a countable number of transverse homoclinic points.

Between any two of these we have homoclinic loops. Since f maps these loops to each other, for example, the loop between q and r to that between $f(q)$ and $f(r)$, these loops have the same area and hence they become large as the proximity of $f^n(q)$ and $f^n(r)$ increases (see Figure 6.5.2).

Since the unstable manifold has no self-intersections, we thus get increasingly thin loops accumulating on the unstable manifold. Since the same arguments

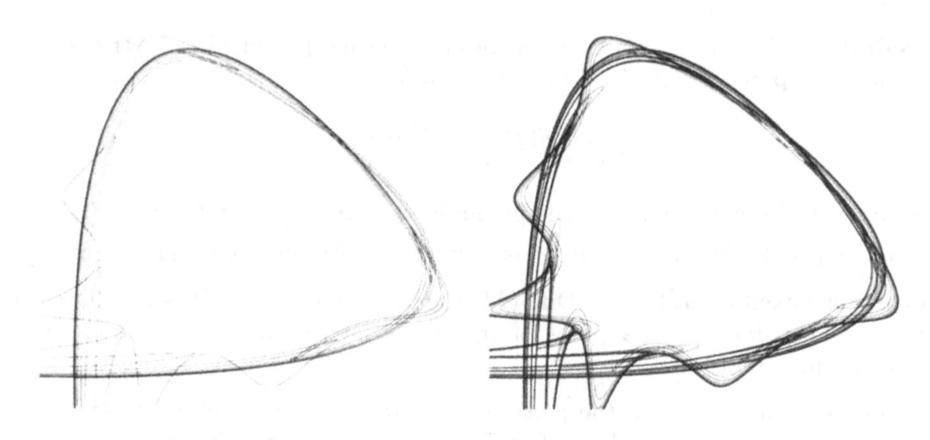

FIGURE 6.5.2. The homoclinic web

work for the stable manifold we obtain similar oscillations for it and thus the complete picture is as in Figure 6.5.2.

In particular we obtain a whole mesh of "new" transverse homoclinic points.

By the Inclination Lemma (Proposition 6.2.23) this picture is correct independently of area preservation or local smooth linearization. Thus any transverse homoclinic point produces the homoclinic oscillations depicted in Figure 6.5.2.

c. Horseshoes near homoclinic points. We can now establish a connection between transverse homoclinic points and the existence of horseshoes.

Theorem 6.5.5. *Let M be a smooth manifold, $U \subset M$ open, $f: U \to M$ an embedding, and $p \in U$ a hyperbolic fixed point with a transverse homoclinic point q. Then in an arbitrarily small neighborhood of p there exists a horseshoe for some iterate of f. Furthermore the hyperbolic invariant set in this horseshoe contains an iterate of q.*

Proof. We will use the following notation several times. For $x \in A \subset \mathbb{R}^n$ denote by $\mathbb{CC}(A, x)$ the connected component of A containing x. Via adapted coordinates on a neighborhood $\mathcal{O}$ we may assume that the hyperbolic fixed point is at the origin and that $W^u_{\text{loc}}(0) := \mathbb{CC}(W^u(0) \cap \mathcal{O}, 0) \subset \mathbb{R}^k \oplus \{0\}$ and $W^s_{\text{loc}}(0) := \mathbb{CC}(W^s(0) \cap \mathcal{O}, 0) \subset \{0\} \oplus \mathbb{R}^l$ where $\mathbb{R}^n = \mathbb{R}^k \oplus \mathbb{R}^l$.

Since $q' := f^{-N_0}(q) \in \operatorname{Int} D_1$ is transverse homoclinic we can take $\delta > 0$ sufficiently small so that if $x \in \delta D_2 := \{\delta z \mid z \in D_2\}$ then $D_1 \times \{x\}$ is transverse to $W^s_{\text{loc}}(q') := \mathbb{CC}(W^s(p) \cap \Delta, q')$ where $\Delta := D_1 \times \delta D_2$. By the Inclination Lemma, Proposition 6.2.23, we can choose $\delta > 0$ and $N_1 \in \mathbb{N}$ such that if $z \in \delta D_2$ and $\mathcal{D}_z := \mathbb{CC}(f^{N_1}(D_1 \times \{z\}) \cap B, f^{N_1}(D_1 \times \{z\}) \cap W^s_{\text{loc}}(q'))$ then $T_x\mathcal{D}_z$ is in a horizontal ϵ-cone for $x \in \mathcal{D}_z$, and $\pi_1 \mathcal{D}_z = D_1$.

This shows that $\Delta_1 := \bigcup_{z \in \delta D_2} \mathcal{D}_z$ is a full component of $\Delta \cap f^{N_1}(\Delta)$. We have in fact shown that in a natural sense this component can be taken arbitrarily close to horizontal. Together with $\Delta_0 := \mathbb{CC}(\Delta \cap f^{N_1}(\Delta), 0)$ which is obviously a full component, we thus have verified (1) of Definition 6.5.2. It remains to

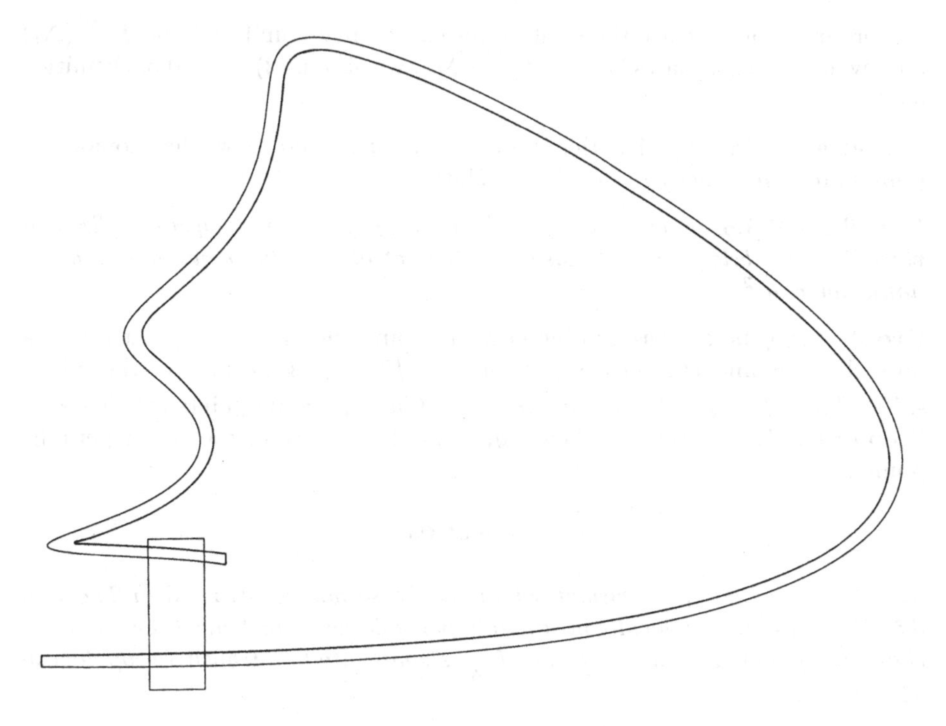

FIGURE 6.5.3. Obtaining a horseshoe

prove the required hyperbolicity. Conditions (2) and (3) of Definition 6.5.2 are easy to check for points $x \in f^{-N_1}(\Delta_0)$ since $f^i(x) \in \Delta$ for $i = 1, 2, \dots, N_1$. Consider $f^{-N_1}(\Delta_1)$. Since $f^{N_1}(q')$ is a transverse homoclinic point we can use the decomposition $\mathbb{R}^n = \mathbb{R}^k \oplus \mathbb{R}^l$ to write $Df^{N_1}(q') = \begin{pmatrix} E & F \\ G & H \end{pmatrix}$ with E nonsingular. The same holds for all $x \in f^{-N_1}(\Delta_1)$ by our choice of δ. If these differentials do not satisfy (2) and (3), replace q' by $q'' = f^{-m}(q')$ and N_1 by $N_2 = N_1 + m + n$ for some $n, m \in \mathbb{N}$ to be specified later. Then

$$Df^{N_2}(q'') = \begin{pmatrix} A_n & B_n \\ C_n & D_n \end{pmatrix} \begin{pmatrix} E & F \\ G & H \end{pmatrix} \begin{pmatrix} A'_m & B'_m \\ C'_m & D'_m \end{pmatrix}.$$

Since E is nonsingular there exists a $\gamma_0 \in \mathbb{R}$ such that the horizontal γ_0-cone is mapped into the horizontal γ_1-cone with $\gamma_1 < \infty$ by $\begin{pmatrix} E & F \\ G & H \end{pmatrix}$. For $\gamma \in \mathbb{R}_+$ take $m \in \mathbb{N}$ such that the horizontal γ-cone is mapped into the horizontal γ_0-cone by $\begin{pmatrix} A'_m & B'_m \\ C'_m & D'_m \end{pmatrix}$ and $n \in \mathbb{N}$ such that the horizontal γ_1-cone is mapped into the horizontal γ-cone by $\begin{pmatrix} A_n & B_n \\ C_n & D_n \end{pmatrix}$. Thus $Df^{N_2}(q'')$ preserves horizontal γ-cones. Enlarging n, m further, if necessary, shows that $Df^{N_2}(q'')$ expands vectors in γ-cones. Since these estimates can be made uniformly on $f^{-N_2}(\Delta_1)$ and even better estimates hold on $f^{-N_2}(\Delta_0)$, we obtain (2) and (3) of Definition 6.5.2. □

Notice in particular that the arguments we used here show that homoclinic points are nonwandering (Definition 3.3.3).

Corollary 6.5.6. *Every transverse homoclinic point to a hyperbolic fixed or periodic point belongs to the closure of the set of periodic points and is hence nonwandering.*[2]

Proof. Let q be a transverse homoclinic point and Λ the invariant set in a horseshoe for an iterate f^n of f such that $f^n\restriction_\Lambda$ is isomorphic to the full 2-shift. Then $f^m(q) \in \Lambda$ for some $m \in \mathbb{N}$. On Λ periodic points are dense by Proposition 1.9.1, so $f^m(q)$, hence q, is in the closure of the set of periodic points.

Exercises

6.5.1. *Prove that the invariant set in the horseshoe constructed in Theorem 6.5.5 is contained in a locally maximal hyperbolic set Λ for f itself that contains the point q. Furthermore show that $f\restriction_\Lambda$ is conjugate to a transitive topological Markov chain.*

6.5.2. *Let $f\colon U \to M$ be a diffeomorphic embedding with hyperbolic fixed points $p_1,\dots,p_k$. We say that $p_1,\dots,p_k = p_0$ form a heteroclinic loop if the dimensions of their stable manifolds are equal and $W^s(p_i)$ intersects $W^u(p_{i+1})$ transversely for $i = 0,\dots,k$. Prove that in this case each P_i has a transverse homoclinic point and conclude that there is a hyperbolic set Λ for f that contains $p_1,\dots,p_k$ and has a dense orbit.*

6.5.3. *Construct a C^∞ diffeomorphism of the torus $f\colon \mathbb{T}^2 \to \mathbb{T}^2$ with two hyperbolic fixed points p and q such that $W^s(p)$ intersects $W^u(q)$ transversely but $h_{\text{top}}(f) = 0$.*

6. Local smooth linearization and normal forms

a. Jets, formal power series, and smooth equivalence. When we introduced C^m equivalence in Definition 2.1.1 we saw that the spectrum of eigenvalues at a fixed point is preserved under smooth conjugacy (since the differential of the conjugacy conjugates the linear parts of the maps). Proposition 2.1.3 shows that in dimension 1 the eigenvalue is the only modulus of local analytic equivalence near a hyperbolic fixed point. Thus we would like to see whether the conjugacy given by the Hartman–Grobman Theorem 6.3.1 can always be taken smooth. In other words, we would like to know whether the local behavior of diffeomorphisms at a hyperbolic fixed point is smoothly classified by the linear parts. As simple examples show, this is not so (cf. Exercise 6.6.1), but we will be able to nevertheless indicate how a reasonable smooth classification at fixed points can be undertaken.

First we should find out whether there are other *infinitesimal* invariants of local smooth conjugacy besides those coming from the linear part of the map at the fixed point. For that purpose fix local coordinates near a fixed point p of a map f and consider the coefficients of the kth Taylor polynomial of f for $k = 2, 3, \dots$. This set of coefficients is called the kth *jet* $J_p^k(f)$ of f at p. Thus two C^k maps f and g have the same k-jet at a (not necessarily fixed) point p if $\|f(x) - g(x)\| = o(\|x - p\|^k)$. Obviously the first jet of a map is determined by the value of the map and its linear part. A real-analytic map is the limit of its kth Taylor polynomials as $k \to \infty$. For C^∞ maps one can write down a Taylor series, but it may not converge at more than one point. Thus we are led to consider *formal* power series, that is, a formal expression consisting of an infinite sum of monomials. The algebraic operations and substitutions are performed by applying the rules familiar from convergent power series.

Obviously the (formal) Taylor series of a C^∞ map at a point determines all jets at that point. k-jets of a map at a point can be identified with polynomial maps in a local coordinate system and their composition is defined by taking the usual composition and discarding higher-order terms. Furthermore near a fixed point a local C^k conjugacy between maps produces a conjugacy between the k-jets at the fixed point. Thus the conjugacy classes of these jets are the infinitesimal invariants we set out to find. For C^∞ maps C^∞ local conjugacy implies that the formal Taylor series at the reference point are conjugate, where the composition of the formal power series is obtained by substitution.

Now we can outline our strategy for solving the smooth conjugacy problem as follows. First we will look for invariants of conjugacy of k-jets for any $k \in \mathbb{N}$. It turns out that all those invariants are completely determined by the linear part of the map, that is, its first jet. Let us call two C^∞ maps f and g C^∞ *tangent at* p if $J_p^k(f) = J_p^k(g)$ for all $k \in \mathbb{N}$, or $\|f(x) - g(x)\| = o(\|x - p\|^k)$ for all $k \in \mathbb{N}$, or equivalently if the formal Taylor series for f and g at p coincide. The second step is to show that if all jets of f and g are conjugate then there exists a C^∞ map h such that f and $f' := h^{-1} \circ g \circ h$ are C^∞ tangent. Finally, if the linear part of f at p is hyperbolic and f' is C^∞ tangent to f at p then f and f' are locally C^∞ conjugate via a local diffeomorphism C^∞ tangent to the identity. This last step can be carried out in various ways. If the linear part of the map f is contracting one can consider the conjugacy equation as the fixed-point equation $h = (f')^{-1} \circ h \circ f$ and use the Contraction Mapping Principle to show that the solution h is C^∞. In the properly hyperbolic case one can use a refined version of the fundamental-domain method called the Sternberg wedge method. We will use a version of the homotopy trick, which was used to prove Theorem 5.1.27 and Theorem 5.5.9, to accomplish this final step. This reduces the problem to analyzing the solution of a twisted cohomological equation (see Section 2.9).

b. General formal analysis. Since we want to conveniently manipulate power series in n variables, we will use multi-indices systematically. Thus for $k = (k_1, \ldots, k_n) \in \mathbb{N}_0^n$ we define the *size* of k to be $|k| := \sum_{i=1}^n k_i$ and if $x = (x_1, \ldots, x_n) \in \mathbb{R}^n$ we let $x^k := \prod_{i=1}^n x_i^{k_i}$.

Proposition 6.6.1. *Suppose $\lambda = (\lambda_1, \ldots, \lambda_n) \in \mathbb{R}^n$ is such that $\lambda_i \neq \lambda^k$ for all i and all $k \in \mathbb{N}_0^n$. Consider a formal power series f given by*

$$f_i(x) = \sum_{k \in \mathbb{N}^n} f_{i,k} x^k$$

with vanishing constant term and whose linear part is $\operatorname{diag} \lambda$ *(that is, the diagonal matrix with eigenvalues $\lambda_1, \ldots, \lambda_n$). Let g be the linear part of f. Then there exists a formal power series h solving the conjugacy equation $h \circ f = g \circ h$.*

A relation of the form $\lambda_i = \lambda^k$ is called a *resonance* and correspondingly the condition "$\lambda_i \neq \lambda^k$ for all i and all $k \in \mathbb{N}_0^n$" is called a *nonresonance assumption.*

Proof. Note first that we can write

$$f_i(x) = \sum_{k \in \mathbb{N}^n} f_{i,k} x^k = \lambda_i x_i + \sum_{|k|>1} f_{i,k} x^k$$

as the linear part plus terms of higher order.

The nonresonance assumption for $|k| = 1$ implies in particular that the λ_i are pairwise distinct. Since the linear part of h commutes with that of f (because the latter coincides with that of g), the linear part of h must be diagonal, with eigenvalues α_i, say. Thus we write a candidate h as

$$h_i(x) = \sum_{k \in \mathbb{N}^n} h_{i,k} x^k = \alpha_i x_i + \sum_{|k|>1} h_{i,k} x^k.$$

The ith coordinate of the conjugacy equation $h \circ f(x) = g \circ h(x)$ now becomes

$$\sum_{k \in \mathbb{N}^n} h_{i,k} (f(x))^k = \lambda_i h_i(x),$$

or, by splitting off the linear parts,

$$\alpha_i f_i(x) + \sum_{|k|>1} h_{i,k} (\lambda_1 x_1 + \sum_{|j_1|>1} f_{1,j_1} x^{j_1})^{k_1} \cdots (\lambda_n x_n + \sum_{|j_n|>1} f_{n,j_n} x^{j_n})^{k_n} = \lambda_i h_i(x).$$

We want to solve this equation inductively in $m := |k|$ for the coefficients $h_{i,k}$ of h in terms of the coefficients $f_{i,k}$ and the λ_i. In other words, we want to construct a conjugacy for m-jets given a conjugacy for $(m-1)$-jets. For $m = 1$ the choices are arbitrary, for example, one can take $Dh = \mathrm{Id}$. Suppose $m \in \mathbb{N}$ and we have determined all $h_{i,k}$ as desired for all $|k| < m$. Then for any k such that $|k| = m$ consider the coefficients of the terms involving x^k. Comparing them yields

$$\alpha_i f_{i,k} + \lambda^k h_{i,k} = \lambda_i h_{i,k} + C_{i,k}, \tag{6.6.1}$$

where $C_{i,k}$ involves only coefficients with indices j of size $|j| < m$, which are thus entirely determined by the previous steps. Thus we solve for $h_{i,k}$ by taking

$$h_{i,k} = \frac{\alpha_i f_{i,k} - C_{i,k}}{\lambda_i - \lambda^k}, \tag{6.6.2}$$

which is possible by the nonresonance assumption. □

Now suppose we are in the situation of the previous proposition but there are $k \in \mathbb{N}^n$ such that $\lambda_i = \lambda^k$ for some i. Then terms with $h_{i,k}$ in (6.6.1) disappear, that is, one cannot remove the term involving x^k from the ith coordinate function. This observation leads to a study of *normal forms*, which are the natural generalizations of the linear part of a map.

Definition 6.6.2. A nonzero term $c \cdot x^k$ in the ith coordinate function that arises in this way from the *resonance* $\lambda_i = \lambda^k$ is called a *resonance term*. A *normal form* of a map is the map in the smooth equivalence class of f whose power series contains only linear and resonance terms.

The previous process can be generalized to the situation where g has nonlinear terms, that is, $g_i(x) = \sum_{k\in\mathbb{N}^n} g_{i,k}x^k = \lambda_i x_i + \sum_{|k|>1} g_{i,k}x^k$. Then (6.6.1) becomes

$$\alpha_i f_{i,k} + \lambda^k h_{i,k} = \lambda_i h_{i,k} + \alpha^k g_{i,k} + C_{i,k}, \tag{6.6.3}$$

where the $C_{i,k}$ are determined by the previous steps and may involve lower-order terms of g. Suppose now that g has *only* resonance terms and the linear part of h is fixed, say, to be the identity. Then (6.6.2) still holds for the nonresonance terms and for resonance terms (6.6.3) implies $g_{i,k} = f_{i,k} - C_{i,k}$, that is, the resonance terms of g are uniquely defined if there is a formal conjugacy. Thus within a formal conjugacy class (with diagonal linear part) the normal form is indeed uniquely defined up to choice of linear part. Interest in these nonlinear normal forms is motivated by the fact that in several natural settings one has "built-in" resonances, that is, in certain natural classes of maps there is a natural collection of resonances exhibited by every one of these maps. The main examples are area-preserving and symplectic maps.

If p is a fixed point of a map preserving a positive absolutely continuous measure then the determinant of the differential Df_p is ± 1 (cf. Proposition 5.1.6 which treats the orientation-preserving case). Thus, if $\lambda_1, \dots, \lambda_n$ are the eigenvalues of Df_p counted with multiplicities then

$$\lambda_1 \cdots \lambda_n = \pm 1$$

and there are resonance relations

$$\lambda_i = \lambda_1 \cdots \lambda_{i-1} \lambda_i^2 \lambda_{i+1} \cdots \lambda_n \tag{6.6.4}$$

or

$$\lambda_i = \lambda_1^2 \cdots \lambda_{i-1}^2 \lambda_i^3 \lambda_{i+1}^2 \cdots \lambda_n^2. \tag{6.6.5}$$

Similarly, if p is a fixed point of a symplectic diffeomorphism then according to Proposition 5.5.6 the eigenvalues can be split into pairs of mutually reciprocal numbers, that is, the vector of eigenvalues can be arranged to look like

$$(\lambda_1, \lambda_1^{-1}, \lambda_2, \lambda_2^{-1}, \dots, \lambda_n, \lambda_n^{-1})$$

and hence there are n resonances

$$\lambda_i = \lambda_i^2 \lambda_i^{-1}, \qquad i = 1, \dots, n. \tag{6.6.6}$$

In the two-dimensional situation where the notion of area-preservation and symplecticity coincide the normal form (6.6.12) is described by Exercise 6.6.4. Note that it is more special than the normal form (6.6.7) for $p = q = 1$.

Now we describe the possible resonances in the two-dimensional hyperbolic case. There are two eigenvalues which we denote by λ_- and λ_+ such that $|\lambda_-| < 1 < |\lambda_+|$. A resonance has the form $\lambda_- = \lambda_-^k \lambda_+^l$ or $\lambda_+ = \lambda_+^k \lambda_-^l$, that is, $\lambda_-^{k-1} = \lambda_+^{-l}$ or $\lambda_+^{k-1} = \lambda_-^{-l}$, where $k, l \in \mathbb{N}_0$ are nonnegative integers. Thus if $\varphi_\pm = |\log|\lambda_\pm||$ then the resonance implies $\varphi_-/\varphi_+ \in \mathbb{Q}$. Conversely it is easy to see that this implies that there is a resonance. If $\varphi_-/\varphi_+ = p/q$ with $p, q \in \mathbb{N}$ relatively prime then we say that there is a (p,q)-resonance. To describe the normal forms in this case let us assume for simplicity that both eigenvalues are positive. Then $\lambda_-^q \lambda_+^p = 1$ and if $\lambda_- = \lambda_-^k \lambda_+^l$ then $k = mq + 1$, $l = mp$ for some $m \in \mathbb{N}$. Similarly $\lambda_+ = \lambda_+^k \lambda_-^l$ implies $k = mp + 1$, $l = mq$. Hence if f is a normal form then

$$f(x,y) = \Big(\lambda_- x\Big(1 + \sum_{m=1}^{\infty} a_m (x^q y^p)^m\Big), \lambda_+ y\Big(1 + \sum_{m=1}^{\infty} b_m (x^q y^p)^m\Big)\Big). \tag{6.6.7}$$

Notice that in the area-preserving case there is a $(1,1)$-resonance. Exercise 6.6.4 provides a specialization of (6.6.7) for that case.

c. The hyperbolic smooth case. One way to justify the formal manipulations with power series is by showing that all power series involved have positive radius of convergence. This approach was exhibited in Section 2.1 for one-dimensional maps. Although the absence of resonances may not be sufficient for analytic linearization even in the hyperbolic case it is so in the C^∞ category.[1] The basic idea is that for any formal power series there is a C^∞ function whose Taylor series coincides with the given power series.

Proposition 6.6.3. *For any sequence $\{a_k\}_{k \in \mathbb{N}_0^n} \subset \mathbb{R}$ there exists a C^∞ function $f \colon \mathbb{R}^n \to \mathbb{R}^n$ such that the a_k are the Taylor coefficients of f.*

Proof. First we introduce a notion that is useful in many places. By a *bump function* we mean a smooth nonzero function with compact support in an open set. An example on the real line is given by the function

$$b_1(x) := \begin{cases} e^{2-(x+1)^{-2}-(x-1)^{-2}} & \text{when } |x| \le 1, \\ 0 & \text{when } |x| > 1, \end{cases}$$

since b_1 vanishes to all orders at ± 1. Note that $b_2(x) := \int_{|x|}^{\infty} b_1(t-2)\,dt / \int_{-1}^{1} b_1(t)\,dt$ defines a bump function which is 1 on $[-1,1]$. Typically one uses bump functions of the second type.

Now we prove the proposition: Set

$$f(x) = \sum_{k \in \mathbb{N}_0^n} a_k x^k b_2(|k|! C_{|k|} \|x\|^2),$$

with $C_N := \sum_{l=0}^{N} \sum_{|i|=l} |a_i|$. Notice that this series converges since for each $x \neq 0$ there are only finitely many nonzero terms. Note that

$$|a_k x^k b_2(|k|!\, C_{|k|} \|x\|^2)| \leq |a_k| \|x\|^{|k|} b_2(|k|!\, C_{|k|} \|x\|^2)| \leq |a_k| \Big(\frac{2C_{|k|}}{|k|!}\Big)^{\frac{|k|}{2}} \leq \Big(\frac{2}{|k|!}\Big)^{\frac{|k|}{2}}, \tag{6.6.8}$$

since $|k|! C_{|k|} \|x\|^2 \leq 2$ for all nonzero terms. Thus the sum converges uniformly and very rapidly. To evaluate the derivatives of order N consider points x such that $|k|! C_{|k|} \|x\|^2 < 1$. For these points we have

$$f(x) = \sum_{|k| \leq N} a_k x^k + \sum_{|k| > N} a_k x^k b_2(|k|! C_{|k|} \|x\|^2).$$

By (6.6.8) the second sum is a bounded multiple of the sum of its lowest-order terms (by factoring them out), so the remainder is of order higher than N and the derivatives up to order N yield the required coefficients. □

Corollary 6.6.4. *Suppose f is a C^∞ map with a fixed point p such that the linear part is* $\operatorname{diag}\lambda$ *(that is, the diagonal matrix with eigenvalues $\lambda_1, \dots, \lambda_n$) satisfying the nonresonance condition $\lambda_i \neq \lambda^k$ for all i and all $k \in \mathbb{N}_0^n$. Then there exists a local C^∞ map h such that $h \circ f \circ h^{-1}$ is C^∞ tangent to the linear part of f.*

Proof. Take the formal power series from Proposition 6.6.1 and construct from it a C^∞ map h using Proposition 6.6.3. □

Therefore local smooth linearization for hyperbolic fixed points with no resonances follows from

Theorem 6.6.5. *Let f be a C^∞ map with a hyperbolic fixed point p and g any C^∞ map C^∞ tangent to f. Then there is a neighborhood U of p and a C^∞ diffeomorphism h which is C^∞ tangent to the identity such that $h \circ f = g \circ h$.*[2]

Proof. First, using Theorem 6.2.3 introduce adapted local coordinates with p at the origin such that the stable and unstable manifolds of p are the coordinate spaces $\mathbb{R}^k$ and $\mathbb{R}^{n-k}$, respectively. Since the stable and unstable manifolds for g are C^∞ tangent to those for f one can conjugate g by a diffeomorphism C^∞ tangent to the identity such that the resulting stable and unstable manifolds coincide with those for f. Next, by the Extension Lemma 6.2.7 we can construct

C^∞ diffeomorphisms of $\mathbb{R}^n$ fixing the origin that coincide with the coordinate representations of f and g, respectively, in a smaller neighborhood of the origin and with the linear part of f and g outside a larger neighborhood, preserve $\mathbb{R}^k$ and $\mathbb{R}^{n-k}$, and are C^1-close to the linear part. We will still denote these maps by f and g. Then $\alpha := f - g$ has zero jets of all orders at the origin and vanishes outside some neighborhood of 0. Next we show that α can be decomposed as

$$\alpha = \alpha^+ + \alpha^-, \tag{6.6.9}$$

where α^+ and all its jets vanish on $\mathbb{R}^k$ and α^- and its jets vanish on $\mathbb{R}^{n-k}$. We will construct conjugacies C^∞ tangent to the identity between f and $w := f + \alpha^-$ and between w and g. To construct the decomposition (6.6.9) take a C^∞ function ρ on the unit sphere S such that $\rho \equiv 1$ on the intersection of S with the horizontal cone $H_{1/2}$ and $\rho = 0$ on the intersection $S \cap V_{1/2}$ and set

$$\alpha^-(x) = \alpha(x)\rho\Big(\frac{x}{\|x\|}\Big) \text{ for } x \neq 0$$

and $\alpha^-(0) = 0$. Then set $\alpha^+ = \alpha - \alpha^-$. Clearly these are as desired, except that we need to verify that both are C^∞ at the origin. Notice that for $k \in \mathbb{N}_0^n$ and $m \in \mathbb{N}$ $\|D^k\alpha(x)\| = o(\|c\|^m)$ and that the derivatives of ρ are bounded. Using the chain rule one sees that the expression for $D^k\alpha^-$ outside the origin is a polynomial in the derivatives of α, ρ, and $\|x\|^{-1}$, and that each monomial contains α or some of its derivatives. This implies that $\|D^k\alpha^-(x)\| = o(\|x\|^m)$ and hence α^- is a C^∞ function.

Now let $f_t := f + t\alpha^-$ for $t \in [0,1]$. We will look for a family of C^∞ diffeomorphisms h_t such that

$$f_0 = h_t^{-1} \circ f_t \circ h_t. \tag{6.6.10}$$

This family is generated by the family of vector fields

$$v_t = \frac{d(h_s h_t^{-1})}{ds}\Big|_{s=t}.$$

Differentiating the relation $h_s \circ h_t^{-1} \circ f_t = f_s \circ h_s \circ h_t^{-1}$ with respect to s we obtain

$$v_t \circ f_t - Df_t(v_t) = \alpha^- \text{ or } v_t - (f_t)_* v_t = \alpha^- \circ f_t^{-1},$$

where $f_* v = Df(v \circ F^{-1})$. Inverting the operator $\operatorname{Id} - (f_t)_*$ formally using the geometric series we obtain

$$v_t = \sum_{m=0}^{\infty} (f_t)_*^m \alpha^- \circ f_t^{-1} = \sum_{m=0}^{\infty} Df_t^m \alpha^- \circ f_t^{-m-1}. \tag{6.6.11}$$

To show that v_t is a C^∞ vector field in a neighborhood of the origin we need to show that this sum converges in the C^∞ topology, that is, that the sum of kth derivatives converges for every $k \in \mathbb{N}_0^n$. Such observations were first made in step 5 in the proof of the Hadamard–Perron Theorem 6.2.8. Note first that by the chain rule and product rule the kth derivative of an m-fold composition grows at a rate of at most $C^m m^{|k|}$, where C is an upper bound for the derivatives up to order $|k|$ of the individual terms. Thus the kth derivative of f^{-m-1} grows at most exponentially with m. Next consider the kth derivative of $\alpha^- \circ f_t^{-m-1}$. By the chain rule this is a polynomial in derivatives of α^- and f_t^{-n-1} and each term contains a derivative of α^- or α^- itself, evaluated at f^{-m-1}, that is, exponentially close to $\mathbb{R}^k$. Thus these factors are superexponentially small by construction of α^- and hence the kth derivative of $\alpha^- \circ f_t^{-m-1}$ converges to zero superexponentially as $m \to \infty$. Again, the kth derivative of the entire summand is a polynomial whose terms each contain a derivative of $\alpha^- \circ f_t^{-m-1}$. Thus each term is superexponentially small (because it consists of a superexponentially small factor and m bounded factors), so in fact the kth derivatives of the summands go to zero superexponentially.

Thus we obtain the desired family h_t and hence the conjugacy between f and $f + \alpha^-$. The second conjugacy between $f + \alpha^-$ and g is constructed similarly using positive iterates of f and α^+ instead of α^-. □

Thus we have completed the proof of the Sternberg Linearization Theorem:

Theorem 6.6.6. *Suppose f is a C^∞ diffeomorphism with a hyperbolic fixed point p such that the linear part of f at p has no resonances. Then near p f is C^∞ conjugate to its linear part.*

In fact, the previous arguments give results about C^∞ conjugacy even in the presence of resonances:

Theorem 6.6.7. *Suppose that f is a C^∞ diffeomorphism with a hyperbolic fixed point p such that the linear part of f at p is diagonal and the normal form of f near p is a convergent power series. Then f is locally C^∞ conjugate to its normal form.*

Proof. First, the discussion after Definition 6.6.2 gives a formal conjugacy between f and its normal form, so by Proposition 6.6.3 we obtain a conjugacy to a map C^∞ tangent to the normal form, which then by Theorem 6.6.5 gives the result. □

Even for analytic maps the normal form may not be convergent. Nevertheless there are some cases where this can be guaranteed. The first is that of contracting maps. In this case there are only finitely many resonances (namely, no more than $-\log r(Df^{-1})/\log r(Df)$, where r denotes spectral radius) and hence the normal form is a polynomial. In particular we can rule out all resonances by a *bunching assumption* (cf. also (19.1.1)):

Corollary 6.6.8. *Suppose that f is a C^∞ diffeomorphism with a hyperbolic fixed point p such that $Df_{|p}$ is diagonal and $-\log r(Df^{-1})/\log r(Df) < 2$. Then f is smoothly linearizable at p.*

Exercise 6.6.1 shows that this condition is sharp. In particular equality cannot be allowed in the bunching condition.

The standing assumption of diagonalizability of the linear part was only used in the formal part of our arguments and these can, in fact, be modified to work in the case of nontrivial Jordan normal forms in the linear part. In particular Corollary 6.6.8 holds without the diagonalizability assumption.

Let us remark in closing that the arguments of this section can be carried out for maps of finite differentiability as well, but with a loss of several degrees of differentiability. This in itself is not a surprising observation, but finding optimal results where this loss is minimal requires much more careful estimates.

Exercises

6.6.1. *Show that the map $(x, y) \mapsto (\lambda x, \lambda^2 y + a x^2)$ cannot be C^2 linearized for any $\lambda \in (0, 1)$ and $a \neq 0$.*

6.6.2. *Describe the formal centralizer of the map of the previous exercise, that is, the set of formal power series that formally commute with the map.*

6.6.3. *Suppose $f: \mathbb{R}^2 \to \mathbb{R}^2$ is C^∞ with $f(0) = 0$ and the linear part at 0 is $\begin{pmatrix} \lambda & 1 \\ 0 & \lambda \end{pmatrix}$ for some $\lambda \in (0, 1)$. Show that f is smoothly linearizable.*

6.6.4*. *Let $f: U \to \mathbb{R}^2$ be a local C^∞ area-preserving diffeomorphism such that $f(0) = 0$ and Df_0 is hyperbolic. Prove that f is formally conjugate to a formal diffeomorphism*

$$g(x, y) = (\lambda x \omega(xy), \lambda^{-1} y (\omega(xy))^{-1}), \tag{6.6.12}$$

where $\omega(t) = 1 + \sum_{k=1}^{\infty} \omega_k t^k$ is a formal power series and $\omega(t)^{-1}$ is its formal inverse.

6.6.5. *Under the assumptions of the previous exercise prove that f is C^∞ conjugate to a map g given by (6.6.12), where ω is a C^∞ function of the product xy.*

7

Transversality and genericity

In this chapter we present several results concerning properties that are exhibited by "most" dynamical systems in various natural classes. This requires first looking into natural notions of "large" or "small" sets in infinite-dimensional spaces. Our main conclusion, the Kupka–Smale Theorem 7.2.6, shows that in a certain rather weak sense hyperbolic behavior is typical. In Section 7.3 we give a very brief glimpse of the important notion of bifurcation which represents an attempt to understand how typical behavior breaks down and how different types of typical behavior are transformed into each other.

1. Generic properties of dynamical systems

a. Residual sets and sets of first category. When we study how various invariants (properties) of dynamical systems change with changes in the system we naturally assume some topology in the space of systems under consideration and it is most desirable to have a property not change at all, at least locally, that is, when the original system is only perturbed slightly.

This is quite natural from the viewpoint of applications: If a mathematical model describes a physical system its parameters are usually only known with limited precision. If a particular property is observed for an open set of dynamical systems and hence an open set of parameters, then it is likely to reflect the true properties of the physical system.

Structural stability represents a good example of such a situation, at least from the topological point of view: Any *topological* property of a structurally stable system is unchanged under small perturbation. This has to be qualified for flows since in that case structurally stable systems may allow nontrivial time changes, so only properties of the *orbit* structure that are not sensitive to time changes are fixed.

For systems that are not structurally stable various elements of the orbit structure may still be stable. For example, hyperbolic periodic points persist under C^1 perturbations (Proposition 1.1.4). However, the size of the perturbation allowed for different elements of the orbit structure, for example, for various hyperbolic periodic points, is not uniform and various pathologies may appear when the stability of a particular property breaks down. This shows that in the nonstructurally stable case there may be no satisfactory description of the orbit structure for an open set of systems. Instead one might try to say something about "most" systems, that is, to describe a "typical" system.

Already for a family of systems depending smoothly on finitely many parameters there are two competing notions of "typical". One uses the natural measure class (Lebesgue measure) on the space of parameters: A property is typical in a certain domain D of parameters if it holds for all parameter values in D except a set of measure zero; it is essential if it holds for a set of parameters of positive measure.

On the other hand one may consider open dense sets as large and then call a property typical or *generic* if it holds for a set of parameters that is a countable intersection of open dense subsets of D. The reason for this is the Baire Theorem A.1.22: In a complete metric space a countable intersection of open dense sets is dense. A countable intersection of open sets is called a G_δ set. A set is called *generic* or *residual* if it contains a dense G_δ. A set is called nowhere dense if its closure has empty interior. (Examples are complements of open dense sets.) Countable unions of nowhere dense sets are said to be of *first category*. A corollary of the Baire Theorem is that the collection of generic sets is closed under countable intersection similarly to the collection of sets of full measure. The complement of a generic set is clearly of first category. Thus the collection of sets of first category which is closed under countable union by definition can be viewed as a topological analog of the collection of sets of measure zero. It follows from the Baire Theorem that these are nonessential in the following sense: Consider a set F of first category and a nonempty open set U. Then $(X \setminus U) \cup F$ is never generic.

Exercises 7.1.4 and 7.1.5 show that there are generic sets of Lebesgue measure zero, so these two notions of "typical" are indeed quite distinct.

Passing to the infinite-dimensional situation we lose the measure-theoretic notion of "typical": There are no natural measure classes in the natural infinite-dimensional spaces of dynamical systems. The topological notion, however, is available. We will study certain generic properties of diffeomorphisms and vector fields in the C^r topology. Let us point out that the C^0 topology is very special from this point of view.

The principal topologies in spaces of dynamical systems are reviewed in the Appendix. Let us point out here that while the C^0 topology is very convenient for using compactness arguments (due to the Arzelá–Ascoli Theorem A.1.24), C^0 generic properties of maps and homeomorphisms may be rather pathological (see Exercise 7.1.9 and 7.1.10). The reason is again the Arzelá–Ascoli Theorem which asserts that any set of "reasonable" maps (differentiable, Lipschitz, Hölder) is σ-compact in $C(X, X)$ and hence rather "thin", in particular of first category (see Exercises 7.1.2 and 7.1.8).[1]

b. Hyperbolicity and genericity. We will see that the key notion in dealing with generic properties in C^r topology for $r \geq 1$ is *transversality.* In the case of a fixed or periodic point transversality means that the differential does not have one as an eigenvalue (see Definition 7.2.1 and Proposition 7.2.2). Proposition 1.1.4 shows that such points are indeed stable as isolated fixed points. In fact, the fixed point of the perturbed map is also transverse. The easiest way to see this is to introduce local coordinates near the original fixed point p of the map f and write Df_p as a matrix. Let q be the fixed point of the perturbed map g. Since g is C^1-close to f the matrix of Dg_q is close to that of Df_p. Since eigenvalues of a matrix are roots of the characteristic polynomial and hence depend continuously on the coefficients (Exercise 1.2.2), Dg_q does not have an eigenvalue 1 if g is sufficiently C^1-close to f and hence q is a transverse fixed point for g. This argument works for periodic points as well and it also shows that hyperbolic periodic points remain hyperbolic under small perturbations.

The following simple example shows that transversality is essential for stability of a fixed point. With any of the natural topologies, namely, the C^r topology for $0 \leq r \leq \infty$, consider the question of whether having a fixed point is a property that persists under perturbation; that is, if M is a smooth manifold, $f \in \mathrm{Diff}(M)$, $x \in M$ is an isolated fixed point of f, and $g \in \mathrm{Diff}(M)$ is "sufficiently close" to f, then does g have a fixed point near x? The answer in general is negative: The diffeomorphism $f\colon \mathbb{R} \to \mathbb{R}$, $f\colon x \mapsto x + (x^2/1 + x^2)$ which first appeared in Exercise 2.1.3 has 0 as the only fixed point. But for no $\epsilon > 0$ does $f + \epsilon$ have any fixed point. This example does not depend on noncompactness of $\mathbb{R}$: If $M = S^1 = \mathbb{R}/\mathbb{Z}$ and $f(x) = x + (1/2\pi)\sin^2 \pi x \pmod 1$ (see Figure 3.3.1) then 0 is the only fixed point and $f+\epsilon$ is fixed-point free when $0 < \epsilon < 1/2$. Thus Proposition 1.1.4 does not hold without the transversality assumption. We will see, however, in the next chapter (Theorem 8.4.4) that a certain purely topological property of an isolated fixed point, namely, nonzero index (Definition 8.4.2), is sufficient for persistence of a fixed point under C^0 perturbations.

We will show in the next section that a certain feature of hyperbolic behavior, namely, hyperbolicity of *all* periodic points, is generic, that is, true for a set of C^r diffeomorphisms that is a dense G_δ set in the C^1 topology (the Kupka–Smale Theorem 7.2.6). Since we have a good understanding of the local topological behavior near hyperbolic periodic points (the Hartman–Grobman Theorem 6.3.1), this property gives certain insight into the orbit structure of "most" systems. In the next chapter we will develop tools for counting periodic points with some algebraic multiplicities (indices) attached (see the Lefschetz Fixed-Point Formula (8.6.1) and its corollaries). Since for hyperbolic periodic points those multiplicities are always ± 1 one obtains effective lower bounds for the growth of the number of periodic points for these generic systems.

Exercises

7.1.1. *Prove that a closed set in a metric space is a G_δ.*

7.1.2. *Prove that a σ-compact subset of an infinite-dimensional Banach space is of first category.*

7.1.3. *Given $c \in (0,1)$ construct an open dense set in $[0,1]$ of measure c.*

7.1.4. *Use the previous exercise to construct a dense G_δ null set and a first category set of full measure.*

7.1.5. *Prove that the set of Diophantine real numbers (see Definition 2.8.1) has full measure (that is, its complement has measure zero) and is of first category.*

7.1.6. *Show that given a positive function $f\colon \mathbb{N} \to \mathbb{R}$ the set of $\alpha \in \mathbb{R}$ for which there exists $c > 0$ such that $\left|\alpha - \frac{p}{q}\right| > cf(q)$ for all relatively prime $p, q \in \mathbb{Z}$ is of first category.*

7.1.7. *Given a C^∞ function f on S^1 that is not a trigonometric polynomial show that for a generic set of numbers α the cohomological equation*

$$\varphi(x) - \varphi(x - \alpha) = f - \int f$$

does not have an L^2 solution φ and that for a set of full measure of numbers it has a C^∞ solution.

7.1.8. *Let $\omega\colon [0,1] \to \mathbb{R}$ be a modulus of continuity, that is, a continuous increasing function with $\omega(0) = 0$. Prove that in the space $C([0,1])$ of continuous functions on the unit interval the functions with modulus of continuity ω, that is, such that there exists $\epsilon > 0$ for which $|f(x) - f(y)| \le \omega(|x - y|)$ when $|x - y| \le \epsilon$, form a set of first category.*

7.1.9. *Prove that functions f for which there exists an interval $(a,b) \subset [0,1]$ on which f is monotone form a set of first category in $C([0,1])$.*

7.1.10. *Prove that in the space* Hom$([0,1])$ *of homeomorphisms of $[0,1]$ there is a residual set of homeomorphisms with uncountably many fixed points.*

2. Genericity of systems with hyperbolic periodic points

a. Transverse fixed points. Fixed points of a diffeomorphism $f\colon M \to M$ correspond to intersections of graph $f = \{(x, f(x)) \in M \times M \mid x \in M\}$ with the diagonal $\Delta = \{(x,x) \in M \times M\}$. This justifies the following definition.

Definition 7.2.1. A fixed point $p = f(p)$ of a smooth map $f\colon M \to M$ is called a *transverse* fixed point if graph $f \pitchfork_{(p,p)} \Delta$ in $M \times M$.

We mentioned earlier that the persistence of such fixed points is related to Proposition 1.1.4. Here is how:

Proposition 7.2.2. *A fixed point p of $f\colon M \to M$ is transverse if and only if 1 is not an eigenvalue of Df_p.*

Proof. If p is not transverse then $T_{(p,p)}\operatorname{graph} f + T_{(p,p)}\Delta \neq T_{(p,p)}(M \times M)$ and hence there is a nonzero vector $v \in T_{(p,p)}\operatorname{graph} f \cap T_{(p,p)}\Delta$, since $\dim T_{(p,p)}\operatorname{graph} f = \dim T_{(p,p)}\Delta = \dim T_pM = \dim T_{(p,p)}(M \times M)/2$. But then there is a $w \in T_pM$ such that $(w,(Df_p)w) = v = (w,w)$, so $w = Df_p w$ is an eigenvector with eigenvalue 1.

The argument runs the other way as well. □

This proposition together with persistence of transverse intersections (Proposition A.3.16) provides an alternative proof of Proposition 1.1.4.

One can introduce a notion corresponding to transversality for critical points of functions as a particular case of transversality of fixed points for maps. Namely, let $f\colon M \to \mathbb{R}$ be a C^2 function. Then the time-one map of its gradient flow with respect to any Riemannian metric on M is a C^1 diffeomorphism whose fixed points are exactly the critical points of f. Thus we call a critical point p of f *nondegenerate* if it is a transverse fixed point of the time-one map of the gradient flow for f. In order to see that this is a well-defined notion we need to show that it is independent of the Riemannian metric chosen to obtain the gradient flow. To that end choose an orthonormal basis in T_pM and local coordinates near p such that this basis is given by $\left\{\frac{\partial}{\partial x_1},\dots,\frac{\partial}{\partial x_n}\right\}$. Then the differential of the time-one map of the gradient flow is $\exp H$ if H is the Hessian (that is, the matrix of second partial derivatives) of f at p with respect to the given coordinates. In particular p is nondegenerate if and only if $\det H \neq 0$. As is well known from calculus, the numbers of positive, negative, and zero eigenvalues of H are independent of the choice of local coordinates and hence of the choice of Riemannian metric in our setting.

Nondegenerate critical points are isolated (see Exercise 7.2.3). The local structure of functions near such points is described by Proposition 9.1.1.

Definition 7.2.3. A C^2 function $f\colon M \to \mathbb{R}$ on a differentiable manifold M is called a *Morse function* if all its critical points are nondegenerate.

b. The Kupka–Smale Theorem. Applying the Transversality Theorem A.3.20 to graphs of diffeomorphisms, that is, when $N = \Delta$ is the diagonal in a Cartesian product, and recalling C^1-openness from the previous section we obtain:

Theorem 7.2.4. *Let $0 \le r \le \infty$ and M a C^r manifold. For any $n \in \mathbb{N}$ those $f \in \operatorname{Diff}^k(M)$ having only transverse periodic points of period at most n are a C^1-open and C^r-dense set.*

Next we will show that diffeomorphisms generically only have hyperbolic periodic points. This is (part of) the content of the Kupka–Smale Theorem 7.2.6.

The next simplest class of orbits after periodic ones are points that are asymptotic to periodic ones. Recall from Definition 6.5.4 that a point q is called *homoclinic* to a periodic point p if $\lim_{|n|\to\infty} d(f^n(p), f^n(q)) = 0$ and *heteroclinic* to a pair p_+ and p_- of periodic points if $\lim_{n\to\pm\infty} d(f^n(p_\pm), f^n(q)) = 0$. If the periodic points in question are hyperbolic, then homoclinic and heteroclinic points are the intersections of the corresponding stable and unstable manifolds (see Sections 0.4, 6.2, and 6.5). It is easy to see that if such an intersection is nontransverse then the homoclinic (heteroclinic) orbit cannot be hyperbolic in the sense that its closure (which contains the orbit plus the points p and q) is not a hyperbolic set (Exercise 7.2.8). The Kupka–Smale theorem asserts that generically all periodic orbits are hyperbolic and all homoclinic and heteroclinic orbits are transverse. As we saw in Section 6.5 the presence of a transverse homoclinic point implies considerable complexity of the orbit structure.

Definition 7.2.5. Suppose M is a C^k manifold, $f \in \mathrm{Diff}^k(M)$. f is said to be *Kupka–Smale to order* n (with respect to a given Riemannian metric) if all periodic points of f of period at most n are hyperbolic and the ball of radius n in the stable manifold of any $x \in \mathrm{Fix}(f^n)$ is transverse to the ball of radius n in the unstable manifold of any $y \in \mathrm{Fix}(f^n)$. f is called a *Kupka–Smale diffeomorphism* if it is Kupka–Smale to all orders.

Theorem 7.2.6. (Kupka–Smale Theorem) *Let $0 < r \le k \le \infty$ and M a compact C^k manifold. Then for any $n \in \mathbb{N}$ Kupka–Smale diffeomorphisms of order n are a C^r-dense C^1-open set in* $\mathrm{Diff}^k(M)$ *and hence (by the Baire Theorem) Kupka–Smale diffeomorphisms are a C^r-dense C^1-G_δ set in* $\mathrm{Diff}(M)$.

Proof. Let us show that being Kupka–Smale of given order is C^1-open. This follows from the facts that hyperbolicity of a periodic point is C^1-open and by compactness there are only finitely many periodic points of any period (if all are hyperbolic), and secondly that stable and unstable manifolds (with the C^1 topology) depend continuously on the map, so by openness of transversality (Corollary A.3.18) the transversality of closed n-balls in stable and unstable manifolds is an open condition. Next we show C^r-density of those diffeomorphisms with only hyperbolic periodic points. By the Transversality Theorem A.3.20 it is sufficient to show that any $f \in \mathrm{Diff}^k(M)$ having only transverse periodic points can be C^r approximated by $g \in \mathrm{Diff}^k(M)$ having only hyperbolic periodic points.

Lemma 7.2.7. *If $f \in \mathrm{Diff}^k(M)$ has only transverse periodic points of period n and $\epsilon > 0$ then there exists $g \in \mathrm{Diff}^k(M)$ having only hyperbolic periodic points of period n and ϵ-close to f in the C^r topology.*

Proof. Let us first consider fixed points. By Proposition 1.1.4 transverse fixed points are isolated, so we can choose pairwise disjoint open sets O_i covering $\mathrm{Fix}(f)$ such that each O_i lies in a coordinate chart for M. Let p_i be the fixed point in O_i. Take $0 < \delta_1 < \delta_2$ small enough and a C^∞ bump function ρ such that $\rho = 1$ on $[0, \delta_1]$ and $\rho = 0$ outside $[0, \delta_2]$. Let $\rho_s = 1 + s\rho$ and let $\psi\colon O_i \to \mathbb{R}^m$

be a coordinate chart such that $\psi(p_i) = 0$. Consider the family f_s of maps that coincide with f outside O_i and are given by $f_s(x) = \psi^{-1}(\rho_s(\|\psi(x)\|)\psi(f(x)))$. Then $f_s \to f$ in the C^∞ topology as $s \to 0$ and thus for small enough s f_s is a C^∞ diffeomorphism C^r-close to f with p_i as the only fixed point in O_i. Furthermore p_i is a hyperbolic fixed point for f_s since all eigenvalues of $D_{p_i}f$ not on the unit circle are a fixed distance away from the unit circle, so $\mathrm{spec}(D_{p_i}f_s) = (1+s)\,\mathrm{spec}(D_{p_i}f)$ does not intersect the unit circle. Fixing any such s and doing this successively for all $p_i \in \mathrm{Fix}(f)$ finishes the argument for fixed points.

Now let p be a periodic point of primer period $n > 1$. Take a neighborhood U of p such that $f^i(U)$, $i = 0, 1, \dots, n-1$, are pairwise disjoint and none of these contains any other periodic point of period n or less. Fix a coordinate system in U, translate it into $f^i(U)$, $i = 1, \dots, n-1$, and consider $f\restriction_{f^{n-1}(U)\cap f^{-1}(U)}$. Now take a small disk $\mathcal{D} \subset f^{n-1}(U) \cap f^{-1}(U)$ and proceed exactly as in the proof for fixed points. The resulting perturbation coincides with f outside $\mathcal{D}$ and has a hyperbolic periodic point of period n in $\mathcal{D}$, and no other periodic points of period n or less in $\mathcal{D}$. □

The lemma yields C^r-density of diffeomorphisms with only hyperbolic periodic points of a given period: By C^r-density of $f \in \mathrm{Diff}^k(M)$ that have only transverse periodic points of period n (Theorem 7.2.4) we obtain C^r-density of $f \in \mathrm{Diff}^k(M)$ that have only hyperbolic periodic points of period n. We observed that the latter set is C^1-open, hence C^r-open, so it is open and dense in the C^r topology. Taking intersections over n yields a dense set by the Baire Theorem A.1.22.

Finally we show that a diffeomorphism with only hyperbolic periodic points can be C^r perturbed such that it is Kupka–Smale of order n. Since transversality of closed n-balls in the stable and unstable manifolds of any two periodic points is an open condition and there are only countably many periodic points, it suffices to show that a map with two hyperbolic periodic points p and q can be perturbed so that $W^s(p) \pitchfork W^u(q)$. Since for any disk $W^s_{\mathrm{loc}}(p) \subset W^s(p)$ we have $W^s(p) \subset \bigcup_{i=1}^\infty f^{-in}(W^s_{\mathrm{loc}}(p))$, where n is the least common multiple of the periods of p and q, it suffices to show that there exists a perturbation g of f for which p and q are periodic points and $W^s_{\mathrm{loc}}(p) \pitchfork W^u(q)$. Let $W^u_{\mathrm{loc}}(q) \subset W^u(q)$ be a disk such that $W^u_{\mathrm{loc}}(q) \subset f^n(W^u_{\mathrm{loc}}(q))$ and $W^s_{\mathrm{loc}}(p) \cap f^n(W^u_{\mathrm{loc}}(q)) = \varnothing$. If $W^s_{\mathrm{loc}}(p) \pitchfork f^{kn}(W^u_{\mathrm{loc}}(q))$ for all $k \in \mathbb{N}$ we are done. Otherwise let m be the maximum of those k. Then $W^s_{\mathrm{loc}}(p) \pitchfork f^{mn}(W^u_{\mathrm{loc}}(q))$ whereas $W^s_{\mathrm{loc}}(p)$ and $f^{(m+1)n}(W^u_{\mathrm{loc}}(q))$ are not transverse.

Let V be a neighborhood of $f^{mn}(W^u_{\mathrm{loc}}(q)) \smallsetminus f^{(m-1)n}(W^u_{\mathrm{loc}}(q))$. Using a construction as in the second part of the proof of the Transversality Theorem A.3.20 we can obtain a diffeomorphism g arbitrarily close to the identity in the C^r topology for a given r such that g is the identity outside V and $g(f^{mn}(W^u_{\mathrm{loc}}(q)) \smallsetminus f^{(m-1)n}(W^u_{\mathrm{loc}}(q)))$ is transverse to $W^s_{\mathrm{loc}}(p)$. Taking $\tilde{f} = g \circ f$ gives the desired perturbation. □

Remark. Notice that, in fact, the compactness assumption can be relaxed to σ-compactness, since it suffices to ensure the presence of at most countably many periodic points of any given period.

By applying the Kupka–Smale Theorem to the gradient flow of a function and noticing that the C^{r+1} topology for functions then coincides with the C^r topology for the gradient flows we obtain

Corollary 7.2.8. *Let $2 \le r \le \infty$ and consider a smooth manifold M. Then Morse functions (see Definition 7.2.3) form a C^r-dense C^2-open set in the space of C^r functions on M.*

Another application of the Kupka–Smale Theorem includes a description of structurally stable diffeomorphisms on the interval.

Proposition 7.2.9. *An orientation-preserving C^r diffeomorphism of $[0,1]$ is C^1 structurally stable if and only if it is Kupka–Smale of order 1.*

Proof. Suppose $f \in \mathrm{Diff}^r([0,1])$ is orientation preserving and has only hyperbolic fixed points. Let $0 = x_0 < x_1 < \cdots < x_k = 1$ be the fixed points. Then those points are alternately repelling and attracting. Thus for any small C^1-perturbation g there are exactly $k+1$ fixed points $0 = y_0, \ldots, y_k = 1$ such that y_i is close to x_i and they are both attracting or both repelling. An immediate application of Proposition 2.1.7 shows that $f\restriction_{[x_i,x_{i+1}]}$ is topologically conjugate to $g\restriction_{[y_i,y_{i+1}]}$. Gluing those conjugacies together gives a conjugacy between f and g.

Conversely suppose f is structurally stable and has a nontransverse fixed point, that is, a point at which $f'(x_0) = 1$. By the Kupka–Smale Theorem f has only isolated fixed points because there are Kupka–Smale diffeomorphisms in any neighborhood of f, hence in the topological conjugacy class of f. Since $f'(x_0) = 1$ there is a small C^1-perturbation that has an interval of fixed points containing x_0, whence f is not structurally stable. Such a map can be constructed by the method used in the Extension Lemma 6.2.7, namely, by constructing an approximation that coincides with the linear part in a neighborhood of the fixed point. □

Together with the Kupka–Smale Theorem this yields

Corollary 7.2.10. *The set of C^1-structurally stable orientation-preserving C^r diffeomorphisms of $[0,1]$ is C^1-dense.*

The notion of transversality also plays a significant role in the context of the Kupka–Smale Theorem for flows. There are two different elements of the orbit structure for flows, namely, fixed and periodic points, which correspond to periodic orbits of maps, and they have to be treated separately. Hyperbolicity of these was defined in Definition 6.2.2.

Definition 7.2.11. A fixed point p of a local flow is said to be *transverse* if the differential at p of any time-t map for $t \neq 0$ does not have 1 as an eigenvalue. Equivalently, the linear part of the vector field at p does not have 0 as an eigenvalue.

A periodic point p of period $t > 0$ for a flow is called *transverse* if 1 is a simple eigenvalue of the differential at p of the time-t map of the flow. Equivalently, p is a transverse fixed point for the Poincaré map on a transverse to the flow near p.

In the case of flows the stable and unstable manifolds of hyperbolic fixed points and periodic orbits can be defined using an appropriate extension of the Hadamard–Perron Theorem 6.2.8 as indicated in Exercise 6.2.5. Accordingly one can speak about transversality of these manifolds. Notice that such manifolds consist of flow orbits, so a transverse intersection can only occur when the sum of the dimensions is strictly greater than the dimension of the manifold.

Definition 7.2.12. A smooth flow is said to be a *Kupka–Smale flow to order* t if all fixed points and all periodic orbits of period less than t are hyperbolic and the t-balls in their stable and unstable manifolds are pairwise transverse. It is called a *Kupka–Smale flow* if it is a Kupka–Smale flow to order t for all $t > 0$.

We formulate the Kupka–Smale Theorem for flows without a proof. The changes in the proof compared to the discrete-time case are fairly routine and can be worked out by a persistent reader.

Theorem 7.2.13. (Kupka, Smale) *Let $0 < r \leq k \leq \infty$ and M a compact C^k manifold. Then for any $t > 0$ Kupka–Smale flows of order t are a C^r-dense C^1-open set and hence Kupka–Smale flows are a C^r-dense C^1-G_δ set in the space of C^r flows.*

There are also counterparts to the Kupka–Smale Theorem for the case of diffeomorphisms and flows preserving an extra structure. The most important and most frequently discussed such structures are smooth positive measures (Definition 5.1.1) and symplectic forms (Definition 5.5.7). Due to the resonances (6.6.4), (6.6.5), and (6.6.6) the relationship between hyperbolicity and genericity changes.

In dimension 2 a symplectic form is simply a volume element. Hence the orientation-preserving area-preserving and symplectic cases coincide. In this case there are two eigenvalues, λ and λ^{-1}, and if a point is transverse then $\lambda \neq 1$. Hence either $\lambda \neq 1$ is real or λ is complex, $|\lambda| = 1$, and $\lambda^{-1} = \bar{\lambda}$. Excluding the case $\lambda = -1$ we obtain two cases, one hyperbolic (λ real) and the other *elliptic* ($|\lambda| = 1$, $\lambda \neq \pm 1$), which are both open. In the orientation-reversing case the eigenvalues are real and any transverse point is hyperbolic. For volume-preserving maps in dimension at least three the resonance conditions (6.6.4) or (6.6.5) do not force nonhyperbolicity. In this case any transverse fixed point can be perturbed to become hyperbolic.

For symplectic maps the phenomenon of persistent nonhyperbolic transverse points exists in any dimension. According to Exercise 5.5.3 the set of eigenvalues of a linear symplectic map in $\mathbb{R}^{2n}$ may contain any number $m \leq n$ of pairs of complex conjugate eigenvalues of absolute value one. Assuming all those eigenvalues are simple one immediately sees that the existence of m distinct pairs of complex eigenvalues of absolute value one persists under small perturbations of a linear symplectic map and hence the same is true for the eigenvalues of the differential of a small C^1-perturbation of a symplectic map at a transverse fixed point. If $m = n$, such a point is called *elliptic*.

Thus, while the Kupka–Smale Theorem is true verbatim for volume-preserving maps in dimension at least three, its counterpart for area-preserving maps in dimension two and symplectic maps only asserts genericity of maps whose periodic points are transverse and have simple eigenvalues. The key point in the proofs is an appropriate modification of the construction in the proof of Lemma 7.2.7.

Although one might hope that Kupka–Smale diffeomorphisms have a simpler orbit structure than nearby "nontransverse" ones, this is not always true. As an illustration let us consider the following important example.

Example. As in Section 5.2b consider the mathematical pendulum described by the differential equation $\ddot{x} + \sin 2\pi x = 0$, that is, the ordinary differential equation

$$\begin{aligned} \dot{x} &= v, \\ \dot{v} &= -\sin 2\pi x \end{aligned}$$

for $x \in S^1$, $v \in \mathbb{R}$. It is a completely integrable system (see Sections 1.5 and 5.5e) with Hamiltonian $H(x, v) = v^2 + (1/2\pi)\cos 2\pi x$, which is invariant under the flow. Recall that the phase space is a cylinder and the orbits are on level curves $H =$ const., that is, some closed curves corresponding to oscillations around the stable equilibrium $(x, v) = (0, 0)$ which are separated from higher-energy orbits corresponding to rotation by a *homoclinic loop* $H = 1/2\pi$ containing the unstable equilibrium $(x, v) = (1/2, 0)$; see Figure 5.2.1. This is clearly a fairly simple orbit structure both for the flow and its time-one map; for example, the topological entropy is zero. But for the time-one-map of this flow we can manufacture a perturbation that has a *transverse* homoclinic point. The Kupka–Smale Theorem (extended to the σ-compact case) guarantees that this perturbation can be arranged so that the resulting map is Kupka–Smale. As we saw in Section 6.5, the presence of a transverse homoclinic point guarantees the presence of a complicated invariant set for an iterate of the map, on which it is topologically conjugate to a full shift σ_2 (Section 2.5) and hence has positive topological entropy. Since the homoclinic orbit of the perturbation stays near the curve $H = 1/2\pi$, this phenomenon has nothing to do with noncompactness of the phase space and can, in fact, be reproduced for diffeomorphisms of compact manifolds, for example, for perturbations of the billiard in an ellipse (see Section 9.2).

Another important remark is that the transversality statement of the Kupka–Smale Theorem cannot be extended to stable and unstable manifolds of non-periodic points. For example, if Λ is a hyperbolic set then every point of Λ has stable and unstable manifolds (cf. Section 6.4), so there may be an uncountable set of points to consider. It is not difficult to construct an example with two disjoint locally maximal hyperbolic sets such that tangencies between the stable manifolds of points in one set and unstable manifolds of points in the other set appear for an open set of perturbations.[1]

Exercises

7.2.1. *Show that transverse fixed points are isolated without using the method or statement of Proposition 1.1.4.*

7.2.2. *Suppose a differentiable manifold M is σ-compact, that is, it is a countable union of compact subsets. Show that a G_δ set of f in* Diff(M) *with respect to the C^1 topology has only transverse fixed points.*

7.2.3. *Prove that if M is a σ-compact differentiable manifold then a G_δ set of $f \in$ (*Diff$(M), C^1$ *topology) has only hyperbolic fixed points.*

7.2.4. *Prove that under the assumption of Exercise 7.2.2 a G_δ set of diffeomorphisms has only transverse periodic points.*

7.2.5. *Prove that under the assumption of Exercise 7.2.3 a G_δ set of diffeomorphisms has only hyperbolic periodic points.*

7.2.6. *Show that for any diffeomorphism with a nontransverse fixed point p there exists an arbitrarily small C^1 perturbation for which p is a nonisolated fixed point. Does this statement hold for C^r perturbations when $r > 1$?*

7.2.7. *Prove that on a compact manifold Morse functions are open in $C^2(M)$.*

7.2.8. *Show that a nontransverse homoclinic or heteroclinic orbit cannot be part of a hyperbolic set.*

7.2.9. *Prove the counterparts of Proposition 7.2.9 and Corollary 7.2.10 for orientation-reversing diffeomorphisms of an interval.*

7.2.10. *Formulate and prove a counterpart of Proposition 7.2.9 for diffeomorphisms of the circle with a fixed point.*

7.2.11. *Consider the negative of the height function on the "vertical" torus (see Figure 1.6.1). Show that the Riemannian metric on the torus may be perturbed in such a way that the time-one map of the gradient flow is Kupka–Smale.*

3. Nontransversality and bifurcations

Transversality holds for "most" individual dynamical systems, but often essential changes in the orbit structure must occur when a system changes, say, as a smooth function of one or several parameters. Nontransverse behavior is responsible for instabilities of the orbit behavior in two ways: inside a given dynamical system and as a qualitative change of the orbit structure under small perturbations. The second aspect is particularly important because usually dynamical systems arising from scientific problems contain parameters and it is crucial to understand how the qualitative behavior changes as the parameters change. Thus even if for a "typical" value of a parameter the system does not exhibit nontransverse behavior, for example, is structurally stable or is Kupka–Smale, there may be values of the parameter where a transition between different kinds of orbit structure occurs. Such changes are usually referred to as *bifurcations.* They are essential for the understanding of properties of typical systems because they show how different types of transverse or typical behavior may appear. The theory of bifurcations is a particular branch of the theory of dynamical systems. It studies finite-parameter families of dynamical systems and follows how various properties of systems in such families vary in a typical way. Naturally the notion of transversality plays a key role in defining what a typical family is. As in the case of the study of individual dynamical systems there are local, semilocal, and global problems in the theory of bifurcations. Without pretending to present any introduction or overview of the subject we give a glimpse of it by considering the simplest and in a way most fundamental local bifurcations. The word "bifurcation" is widely and often carelessly used and there is no agreement on any precise meaning of the term. Here we will restrict ourselves to discussing bifurcations in the following limited sense. Suppose $\{f_\tau\}_{\tau\in I}$ is a one-parameter family of dynamical systems defined locally, semilocally, or globally and suppose that for τ in an open interval $J \subset I$ a certain property holds, but that this property does not hold on any larger interval. Consider an endpoint a of J. We will call it a *bifurcation value.* The property in question may be structural stability, local stability in a neighborhood of a periodic point, the nature of a periodic point, the Kupka–Smale property, and so forth.

a. Structurally stable bifurcations. The most elementary and most thoroughly studied bifurcations are those related to the local behavior near a fixed or periodic point of a discrete-time dynamical system or a periodic orbit of a flow. By Corollary 6.3.4 hyperbolic fixed and periodic points are locally structurally stable. In fact, the appearance of an eigenvalue of absolute value 1 of the differential causes the loss of such structural stability. Thus parameter values at which nonhyperbolic periodic points occur are prime candidates for bifurcation values.

Proposition 7.3.1. *If f is a local diffeomorphism with a nonhyperbolic fixed point p then f is not locally structurally stable near p.*

Proof. First assume f has an eigenvalue 1. We construct two local diffeomorphisms f' and g arbitrarily near f that have distinct orbit structures. First, as in the proof of Proposition 7.2.9 take $f' = Df$, which on a sufficiently small neighborhood is arbitrarily close to f. Since Df has 1 as an eigenvalue, there is a 1-parameter family of fixed points containing p. On the other hand, arbitrarily close to Df there is a hyperbolic linear map L. By putting $g = h$ locally and using the Extension Lemma 6.2.7 we obtain a diffeomorphism with an isolated fixed point. The same argument applies when Df has a root of unity as an eigenvalue, since we can consider an iterate instead. Finally if Df has an eigenvalue on the unit circle with irrational argument, then by an arbitrarily small perturbation the eigenvalue can be made into a root of unity. □

Definition 7.3.2. A family $\{f_\tau\}$ of locally defined C^∞ diffeomorphisms has a *structurally stable bifurcation* at a parameter value τ_0 if f_{τ_0} is not locally structurally stable and for any one-parameter family $\{g_\tau\}$ of locally defined C^∞ diffeomorphisms that is sufficiently C^2-close to $\{f_\tau\}$ there exists an $\epsilon > 0$, an increasing reparameterization $\varphi(\tau)$ of $\{g_\tau\}$, and a continuous one-parameter family $\{h_\tau\}$ of locally defined homeomorphisms such that

$$g_{\varphi(\tau)} = h_\tau^{-1} \circ f_\tau \circ h_\tau, \tag{7.3.1}$$

wherever it is defined.

Structurally stable bifurcations in one-dimensional situations can be described without much difficulty. In this case the only eigenvalue of the differential at the fixed point is real and hence the only eigenvalues on the unit circle are ± 1. Thus f_{τ_0} must have eigenvalue 1 or -1.

We first discuss the case of eigenvalue 1. The simplest bifurcation appears when the graph of the map has a nondegenerate tangency to the diagonal at the bifurcation value and is locally disjoint from the diagonal for any larger nearby value of the parameter, while for smaller values it crosses the diagonal transversely in two nearby points.

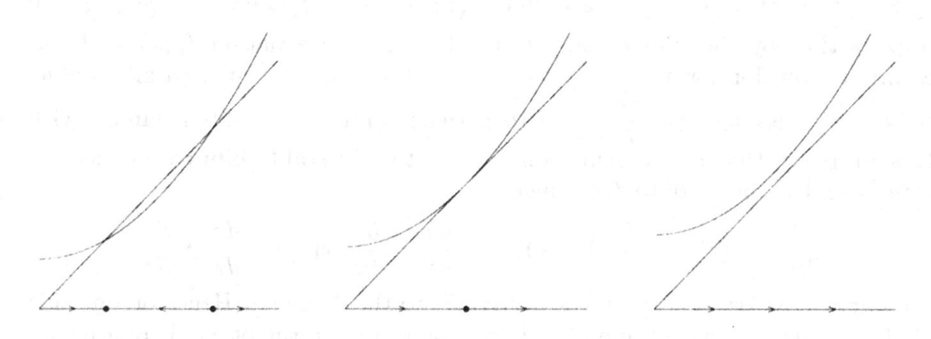

FIGURE 7.3.1. Simplest one-dimensional bifurcation

Dynamically this means that the contracting and expanding fixed points present for smaller values of the parameter collide at the bifurcation value, producing a semistable point (a point that attracts on one side and repels on the other). For larger values of the parameter there are no fixed points nearby. A specific example of this phenomenon is given by the family

$$f_\tau : x \mapsto x + x^2 + \tau \tag{7.3.2}$$

defined near $x = 0$ for τ near 0.

Proposition 7.3.3. *The bifurcation in the family (7.3.2) is structurally stable and any local structurally stable bifurcation in dimension one occurring at a fixed point with derivative* 1 *is (topologically) equivalent (up to reparameterization) to this bifurcation.*

Proof. Note first that in the family (7.3.2) the locus of points p_τ where the derivative is 1 is the origin. Consider a perturbation $g_\tau : x \mapsto x + x^2 + \tau + \eta_\tau(x)$ of (7.3.2) and note that solving for $g'_\tau = 1$ gives the equation $1 + 2x + \eta'_\tau(x) = 1$ or $\eta'_\tau(x) = -2x$. Since η''_τ is small we can use the Implicit Function Theorem to find a solution $x(\tau)$ for this equation, which is C^1-small. Thus the curve $\tau \mapsto (x(\tau), g_\tau(x(\tau)))$ is transverse to the diagonal and crosses it at a single point, that is, for a single value τ_0 of τ. Note that by convexity of g_τ (which follows from convexity of f) this means that for $\tau > \tau_0$ the graph of g_τ does not intersect the diagonal, whereas for $\tau < \tau_0$ the intersection consists of exactly two points, one attracting and one repelling. Using the fundamental-domain method as in the proof of Proposition 2.1.7 onc immediately concludes topological conjugacy with (7.3.2) for $\tau > 0$.

To see that this is the only possible structurally stable bifurcation for eigenvalue 1 let us assume without loss of generality that the point at which the bifurcation occurs is 0 and that the bifurcation value is 0. We first show that the tangency at the bifurcation point is nondegenerate, that is, there is a nontrivial quadratic term. If the tangency is of higher order we have $f_0(x) = x + o(x^2)$. Then we can consider the perturbation $g_{\tau,\epsilon} = f_\tau + \epsilon x^2$. But for any $\epsilon > 0$ and small enough τ_1 and τ_2 the maps $g_{\tau_1,\epsilon}$ and $g_{\tau_2,-\epsilon}$ are not conjugate via a homeomorphism close to the identity.

For each τ near 0 there is a unique $x(\tau)$ at which $f'_\tau(x) = 1$ since $\dfrac{\partial^2 f}{\partial x^2} \neq 0$ implies that by the Implicit Function Theorem the equation $f'_\tau(x) = 1$ has a unique solution for x in terms of τ. The bifurcation is structurally stable, so we may assume that $\dfrac{\partial f}{\partial \tau} \neq 0$ (otherwise adding $\epsilon\tau$ yields a family with this property that is, by definition of structurally stable bifurcation, locally topologically conjugate to f_τ). Since

$$\frac{\partial}{\partial \tau} f_\tau(x(\tau)) = \Big(\frac{\partial f}{\partial \tau}\Big)(x(\tau)) + f'_\tau \frac{dx}{d\tau} = \frac{\partial f}{\partial \tau}(x(\tau)) + \frac{dx}{d\tau} \neq \frac{dx}{d\tau}$$

the curve $\tau \mapsto (x(\tau), f_\tau(x(\tau)))$ is transverse to the diagonal. Hence on one side of the bifurcation value there is no fixed point, and on the other there are two, one repelling and one attracting, by convexity of f_τ near $x(\tau)$. □

Let us point out that in general a perturbed family is not differentiably conjugate to (7.3.2), even if a reparameterization is made. The reason for this is the presence of two fixed points on one side of the bifurcation value. As we showed in Section 2.1c (Corollary 2.1.6) the maps of such type have infinitely many moduli of smooth conjugacy, so in general two 1-parameter families of C^1 maps will consist of maps that are pairwise C^1 inequivalent regardless of the parameterization.

The preceding proof shows in particular that there are no structurally stable bifurcations with derivative 1 where the fixed point persists. The simplest bifurcation of this type occurs in the family $x \mapsto x + \tau x + x^3$ near $\tau = 0$. For $\tau < 0$ there are three fixed points, the stable one at $x = 0$ and two unstable ones, one on each side. For $\tau = 0$ these collide and for $\tau > 0$ the origin is an isolated repeller. To see that this bifurcation is not structurally stable, one perturbs the family to the family $x \mapsto x + \tau x + \epsilon x^2 + x^3$. To find the bifurcation values note that the graph $y = x + \tau x + \epsilon x^2 + x^3$ is tangent to the diagonal $y = x$ at exactly those values of (x, τ) for which the graph $y = \epsilon x^2 + x^3$ is tangent to the line $y = -\tau x$. As one sees by drawing the graph of $y = \epsilon x^2 + x^3$, this happens for two values of τ and at each of these one gets a structurally stable bifurcation of the kind described earlier.

Let us now consider structurally stable bifurcations where the derivative has eigenvalue -1. In this case the fixed point is transverse and hence persistent. It is natural to expect that for a structurally stable bifurcation the eigenvalue changes from less than -1 to greater than -1 or vice versa while the fixed point remains isolated. This is accompanied by the creation of a period-two orbit on one side of the bifurcation value. Thus this type of bifurcation is referred to as *period-doubling bifurcation*. A typical example of this kind is given by

$$f_\tau(x) = -\tau x + x^2$$

near $\tau = 1$. The second iterate f_τ^2 is

$$f_\tau^2(x) = x\left[1 + (\tau - 1)[\tau + 1 + x\tau - 2x^2] + x^2(x - 2)\right].$$

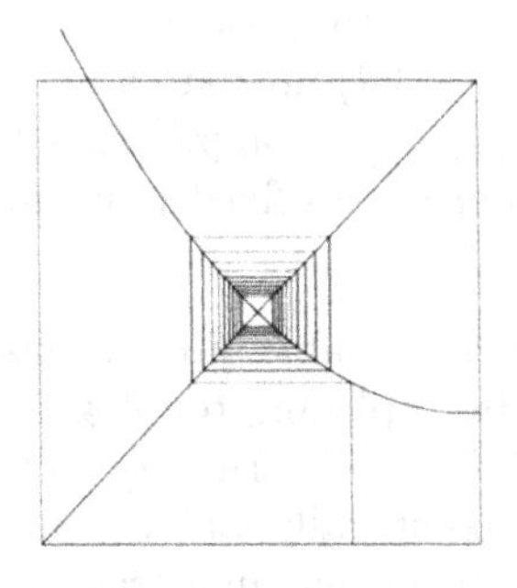
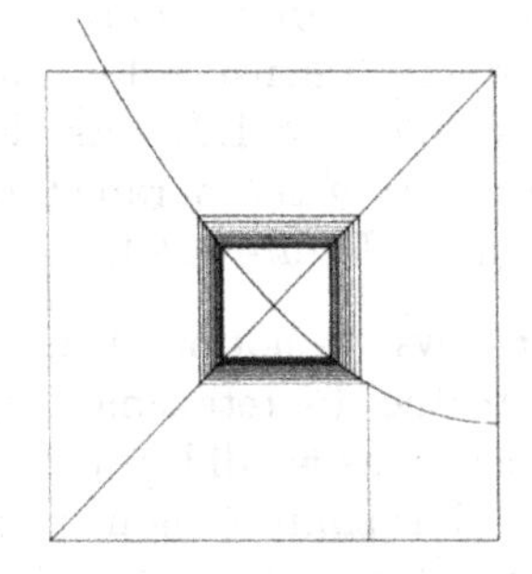
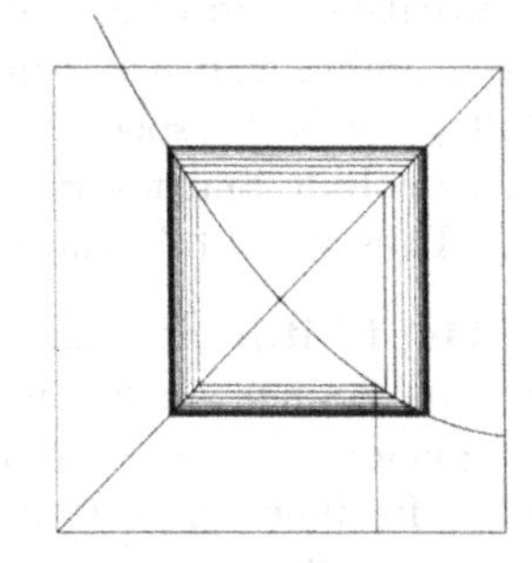

FIGURE 7.3.2. Period-doubling bifurcation

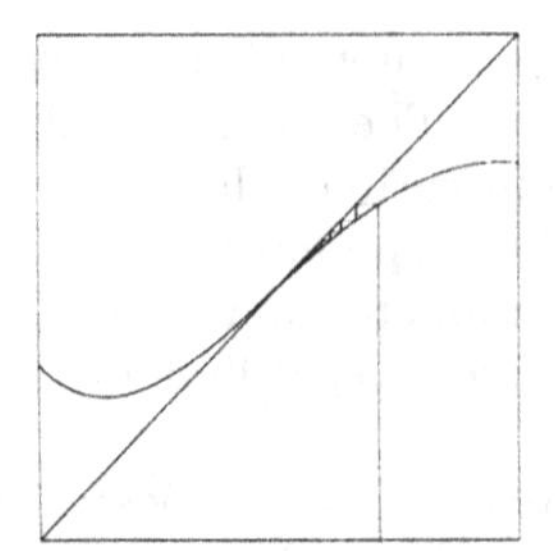
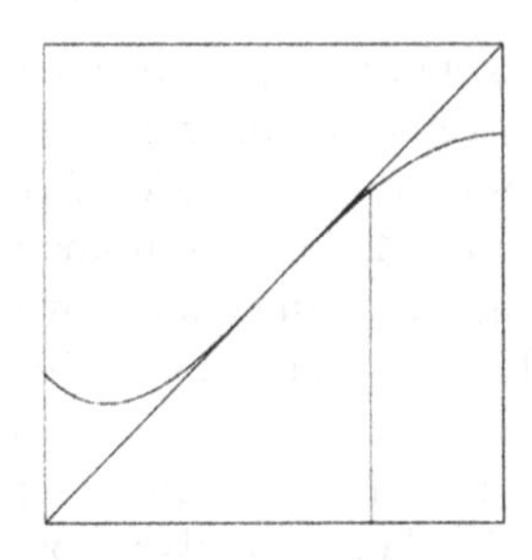
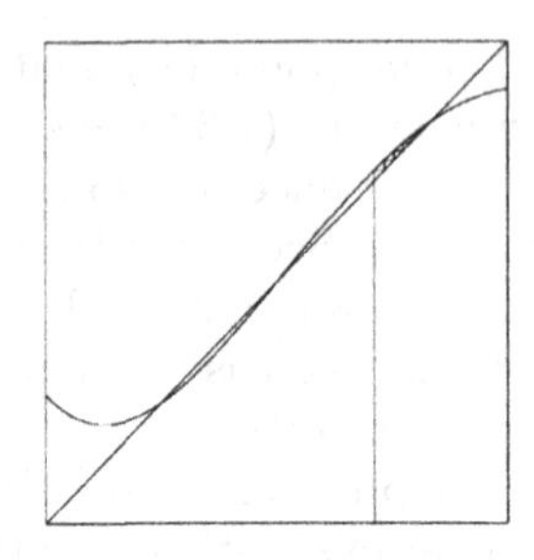

FIGURE 7.3.3. The second iterate

Thus we find periodic points other than 0 by solving

$$(\tau - 1)[\tau + 1 + x\tau - 2x^2] + x^2(x - 2) = 0. \tag{7.3.3}$$

Note that the term $x^2(x-2)$ has a double root at 0 and a simple root at 2. The term we add is $(\tau-1)[\tau+1+x\tau-2x^2]$ and for small values of x it is dominated by $(\tau-1)(\tau+1)$, so for $\tau > 1$ (7.3.3) has two distinct solutions near zero and for $\tau < 1$ there are none. Thus we can describe the dynamics of f_τ^2 as follows. For $\tau < 1$ zero is an isolated fixed point, which is, in fact, contracting for f_τ^2. For $\tau = 1$ the origin is a degenerate fixed point, still contracting topologically. For $\tau > 1$ the origin is an expanding fixed point and there are two new contracting fixed points. Since 0 is an isolated fixed point for f_τ these are not fixed points of f_τ and hence they are points of period two. Notice that this bifurcation looks like the nonstructurally stable bifurcation described when we considered the family f_τ^2 (Figure 7.3.3).

Similarly to the proof of Proposition 7.3.3 one can show that this is a structurally stable bifurcation and that it is the only structurally stable bifurcation that appears with eigenvalue -1 (Exercise 7.3.3).[1]

Structurally stable local bifurcations in higher dimensions appear when one of the eigenvalues of the differential of a diffeomorphism is 1 or -1 and the remaining ones are off the unit circle. As an easy example we describe the example given by the product of (7.3.2) with a linear contraction. The resulting bifurcations are called *saddle–node bifurcations.* The attracting and repelling fixed points that occur in the one-dimensional example (7.3.2) now are a node and a saddle, respectively (see Section 1.2). As the parameter approaches 0 these collide and for values $\tau > 0$ of the parameter there is no fixed point at all. Thus we obtain the picture in Figure 7.3.4.

b. Hopf bifurcations. For flows the notion of structurally stable bifurcations can be defined analogously to the discrete-time case by replacing topological conjugacy with orbit equivalence, as we did in the definition of structural stability for flows in Section 2.3. The simplest nontrivial structurally stable bifurcation for flows near a fixed point appears in dimension two when an attracting focus becomes repelling. This corresponds to having a complex conjugate pair

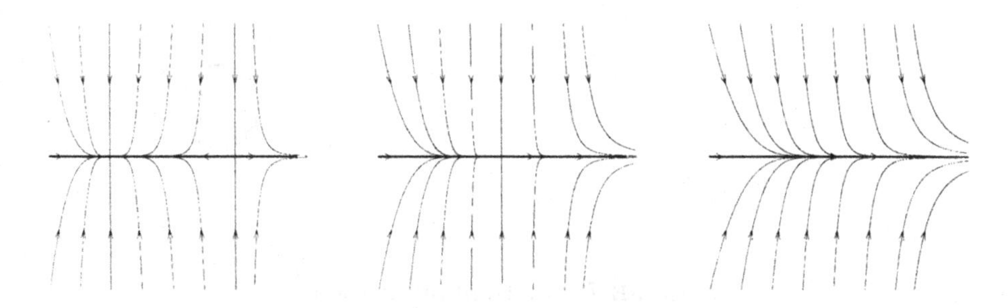

FIGURE 7.3.4. Saddle–node bifurcation

of eigenvalues of the time-t map that passes through the unit circle. An example is given by the flow generated by the ordinary differential equation

$$\begin{aligned}\frac{dx}{dt} &= (\tau - x^2 - y^2)x + \theta y,\\ \frac{dy}{dt} &= -\theta x + (\tau - x^2 - y^2)y.\end{aligned} \tag{7.3.4}$$

For $\tau < 0$ the origin is a globally exponentially attracting focus since

$$\frac{d\log(x^2 + y^2)}{dt} = 2(\tau - x^2 - y^2) \tag{7.3.5}$$

and this is negative when $\tau < 0$. For $\tau = 0$ the origin is still globally attracting since $\dfrac{d\log(x^2 + y^2)}{dt} < 0$ away from 0 by (7.3.5), but not exponentially attracting. Once $\tau > 0$, however, the origin is a repelling focus. A new feature appears here. Namely, note that $\dfrac{d\log(x^2 + y^2)}{dt}$ changes sign on the circle $x^2 + y^2 = \tau$, which is hence an invariant circle, in fact, an attracting periodic orbit (a *limit cycle*).

According to our definition (using orbit equivalence) this bifurcation is structurally stable (cf. Exercise 7.3.4).

There is a discrete-time counterpart of this situation, which appears when the eigenvalues of a map are a complex conjugate pair and cross the unit circle. A simple example of this situation is given by the family of maps

$$(x, y) \mapsto (1+\tau-x^2-y^2)(x\cos\alpha(\tau)+y\sin\alpha(\tau), -x\sin\alpha(\tau)+y\cos\alpha(\tau)) \tag{7.3.6}$$

near $\tau = 0$, where α is a given function. This can, in fact, be embedded in a flow, so that one obtains the same qualitative behavior, namely, the appearance of an invariant circle after the focus changes from attracting to repelling. Both these bifurcations for continuous- and discrete-time systems are usually called *Hopf bifurcations*. This latter bifurcation is not structurally stable. For example, the limit $\alpha(0)$ of the rotation angle of the invariant circle is an invariant. Nevertheless the feature of having an invariant circle appear persists under perturbations of the family. This can be proved by doing a formal analysis of a small perturbation of the family (7.3.6) and showing that the dominant terms can be brought into the form (7.3.6), possibly with a different function α.

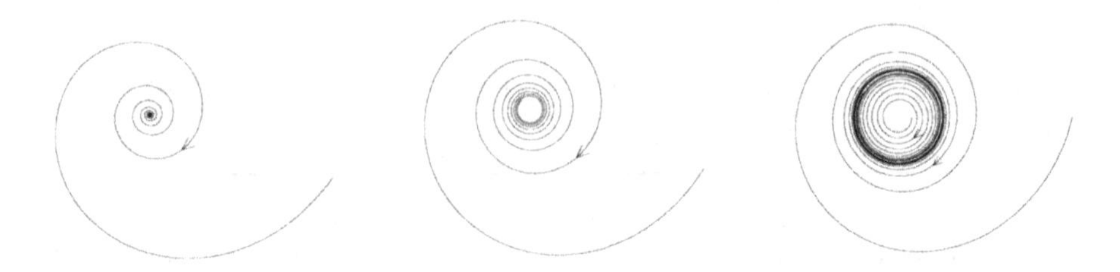

FIGURE 7.3.5. Hopf bifurcation

Exercises

7.3.1. *Consider the quadratic family $f_\lambda(x) = \lambda x(1-x)$ on $[0,1]$. Show that at $\lambda = 3$ a period-doubling bifurcation appears near the fixed point $x = 2/3$.*

7.3.2. *Show that the period-two orbit that appears at $\lambda = 3$ in turn undergoes a period-doubling bifurcation for some larger value of λ.*

7.3.3. *Show that the period-doubling bifurcation is the only structurally stable bifurcation that appears with eigenvalue -1.*

7.3.4. *Show that the bifurcation in (7.3.5) is structurally stable.*

4. The theorem of Artin and Mazur

Though the Kupka–Smale Theorem guarantees finiteness of the number of periodic points of any given period and gives control over the types of periodic points, it says nothing about the growth of their number. Thus it is useful for lower estimates (as we pointed out in Section 7.1b) but says nothing about upper estimates. We now present a general result that gives an estimate of that kind for a *dense* (although not generic) set of dynamical systems. It implies that those diffeomorphisms whose isolated periodic orbits grow at most exponentially are dense. Unfortunately we cannot guarantee that the diffeomorphisms we constructed have only isolated periodic points; however, those diffeomorphisms are of an algebraic nature and their periodic points of a certain period come as finite unions of algebraic varieties, that is, as zero sets of polynomials in appropriate coordinate systems. For such maps the number of connected components of the set of periodic points is a reasonable substitute for the number of isolated periodic points.

In fact, some of the algebraic data of the index type (Section 8.4) can be generalized from isolated fixed points to connected components. Thus global topological information can be connected with the Artin–Mazur diffeomorphisms along the lines of the Lefschetz Fixed-Point Formula. Similarly one can modify the definition (3.1.3) of the zeta function associated to the growth of periodic orbits (see Section 4.1a) to include entire connected components of periodic points. For Artin–Mazur diffeomorphisms this modified zeta function has positive radius of convergence.

Theorem 7.4.1. (Artin–Mazur Theorem) *Let M be a compact manifold. For $f \in \mathrm{Diff}^\infty(M)$ denote by $\mathcal{P}_n(f)$ the number of connected components of* $\mathrm{Fix}(f^n)$. *Then*

$$\{f \in \mathrm{Diff}^\infty(M) \mid \overline{\lim_{n\to\infty}} \frac{1}{n} \log \mathcal{P}_n(f) < \infty\}$$

is dense in $\mathrm{Diff}^\infty(M)$.

We obtain this result as a consequence of an embedding theorem due to Nash which we do not prove.

Proof. Our aim is to arrange for all periodic orbits to arise as solutions of polynomial equations and thus obtain an exponential estimate of the number of components of periodic points. To this end we will realize the manifold as a joint level set of several polynomials and approximate a given diffeomorphism by a polynomial one which can then be controlled. The first object is achieved by the Nash Embedding Theorem:

Theorem 7.4.2. (Nash Embedding Theorem)[1] *Any compact C^∞ manifold can be embedded in Euclidean space as the intersection of the set of zeros for some collection of real-valued polynomials, that is, if M is a compact C^∞ manifold then there exist $N, K \in \mathbb{N}$ and polynomials $P_i: \mathbb{R}^N \to \mathbb{R}$ $(i = 1, \dots, K)$ such that M is diffeomorphic to $\bigcap_{i=1}^K P_i^{-1}(\{0\})$.*

From now on we identify M with this intersection.

Note that although we may have $K > m := \operatorname{codim} M = N - \dim M$, every point $p \in M$ has a neighborhood characterized by the vanishing of m polynomials. By compactness this means that there is a finite open cover $M \subset \bigcup_{i=1}^Z U_i$ such that each U_i is open in $\bigcap_{j=1}^m (P_j^i)^{-1}(\{0\})$ for some subcollection $\{P_j^i\}_{j=1}^m \subset \{P_j\}_{j=1}^K$.

Now we approximate a given diffeomorphism $f \in \mathrm{Diff}^\infty(M)$ by an essentially polynomial one. Consider an extension $\tilde{f}: U \to \mathbb{R}^N$ of f to a diffeomorphism between neighborhoods U and $\widetilde{V}$ of M. We can approximate $\tilde{f}$ arbitrarily well by polynomial diffeomorphisms $\tilde{g}: U \to \mathbb{R}^N$. There exists a neighborhood V of M such that the orthogonal projection $\pi: V \to M$ is well defined on V. If $\tilde{g}$ is sufficiently close to $\tilde{f}$ then $\tilde{g}(M) \subset V$ and $g := \pi \circ \tilde{g}\restriction_M \in \mathrm{Diff}^\infty(M)$ approximates f.

With these choices we can write down equations describing periodicity. Suppose $x \in M$ and $y = g(x) \in U_i$. Then $y = \pi(\tilde{y})$, where $\tilde{y} = \tilde{g}(x)$. So there exists $\lambda = (\lambda_1, \dots, \lambda_{N-m}) \in \mathbb{R}^{N-m}$ such that $\tilde{y} - y = \sum_{j=1}^{N-m} \lambda_j \nabla_y P_j^i$, where ∇_y denotes the gradient (vector) at y. Since $y \in M$ we also have $P_j^i(y) = 0$ for $j = 1, \dots, N-m$. So a fixed point $x \in U_i$ of g satisfies $x = \tilde{g}(x) - \sum_{j=1}^{N-m} \lambda_j \nabla_x P_j^i$ and $P_j^i(x) = 0$. Likewise a periodic point $x \in M$ with $x_k = g_k(x) \in U_{i_k}$ and $x_n = x_0$ satisfies

$$x_{k+1} = \tilde{g}(x_k) - \sum_{j=1}^{N-m} \lambda_j^k \nabla_{x_{k+1}} P_j^{i_{k+1}}, \quad P_j^{i_{k+1}}(x_{k+1}) = 0.$$

These are $nN + (N - m)n$ equations for the $nN + (N - m)n$ variables $\{x_k\}_{k=0}^{n-1}$, $\{\lambda_k\}_{k=0}^{n-1}$ and the system is of the form $P(x, \lambda) = 0$ for a polynomial P. Consequently the number of connected sets of solutions is bounded by the product of the maximal degrees occurring in the equations. This gives an upper bound of the form K^n for some K independent of n.

Finally, in order to account for all periodic points, there are Z^n possible n-itineraries (given by $x_k \in U_{i_k}$) to consider, so $\mathcal{P}_n(g) \leq (ZK)^n$. □

8

Orbit growth arising from topology

In Section 3.1 we described several invariants related to the asymptotic growth of the complexity of the orbit structure. The most direct information of that kind is provided by the growth of periodic orbits (3.1.1) and topological entropy (Definition 3.1.3), which reflects the growth of the number of orbits distinguishable with limited precision. On the other hand we defined the fundamental-group entropy (3.1.23) and the spectral radii of the action on homology (Section 3.1e) which less directly reflect the growth in topological complexity of families of orbits from the homotopical and homological point of view. An obvious advantage of the latter invariants is that in general they can be calculated more easily since they are invariant under homotopy equivalence. For example, since every map of the torus is homotopically equivalent to a linear map (see Sections 2.6 and 8.7 for details), it suffices to consider linear maps when calculating the fundamental-group entropy and the spectral radii of the action on homology. In this chapter we will show how these homotopical and homological invariants provide information related to growth of orbits, that is, we will find quantitative relationships between growth (and in particular existence) of periodic orbits and topological entropy on the one hand, and these topological data on the other.

Let us point out the difference between the one-dimensional topological data (that is, coming from the fundamental group and first homology group) and those of higher dimension: In the first case the connection with orbit growth works for arbitrary continuous maps (see the Manning Theorem 8.1.1 and Proposition 8.2.4 concerning circle maps), whereas in the second smoothness of the map is often essential (Misiurewicz–Przytycki Theorem 8.3.1 and Corollary 8.6.11).

Although all the above notions are global, some relation among these is established through a key local notion of the *index* of a fixed (or periodic) point of a map or a fixed point of a flow which reflects the topological behavior of the map or time-t map of the flow near the fixed point. In particular points of nonzero index are essential in several ways; for example, they do not disappear under C^0-perturbations. The central link in this relation is the Lefschetz Fixed-Point Formula which expresses the sum of the indices of fixed points through homological data. The notion of index is also the basis for Nielsen theory which allows us to estimate from below the number of periodic points through homotopical data. In the next chapter we will show how the concepts related to the index and the structure of critical points of a gradient flow provide information about the orbit structure of some volume-preserving (for example, Lagrangian) dynamical systems via consideration of the gradient of an appropriate function in an auxiliary space of "potential orbits".

1. Topological and fundamental-group entropies

Theorem 8.1.1. *If f is a continuous map of a compact manifold M then $h_{\text{top}}(f) \geq h_*(f)$, that is, the topological entropy of f is at least as large as the fundamental-group entropy.*

Proof. Fix any metric on M. Notice that there is a number $\lambda > 0$ such that any λ-ball lies in a set homeomorphic to a Euclidean ball. Then any two curves with the same endpoints that lie completely inside a λ-ball on M are homotopic via a homotopy fixing the endpoints, namely, the pullback of the obvious straight-line homotopy in a Euclidean ball containing the image of the ball.

Lemma 8.1.2. *Let λ be as before and take a $\lambda/4$-dense set $\{w_1, \dots, w_K\} \subset M$. Fix $p \in M$ and arcs c_i joining p and w_i as well as arcs $c_{ij} \subset B(w_i, \lambda/4) \cup B(w_j, \lambda/4)$ joining w_i and w_j for those (i,j) for which $B(w_i, \lambda/4) \cap B(w_j, \lambda/4) \neq \varnothing$. Then the arcs $c_i c_{ij} c_j^{-1}$ form a set Γ of generators for the fundamental group $\pi_1(M, p)$.*

Proof. Consider a loop γ at p and subdivide it into pieces σ that are contained in the union of two balls $B(w_i, \lambda/4)$ and $B(w_j, \lambda/4)$. Take a path $\eta_0 \subset B(w_i, \lambda/4)$ that connects $\sigma(0)$ and w_i and $\eta_1 \subset B(w_j, \lambda/4)$ that connects $\sigma(1)$ and w_j. Then σ and $\eta_0 c_{ij} \eta_1^{-1}$ have the same endpoints and lie in $B(w_i, \lambda/4) \cup B(w_j, \lambda/4)$ which is contained in a λ-ball, so σ and $\eta_0 c_{ij} \eta_1^{-1}$ are homotopic with fixed endpoints. But $\eta_0 c_{ij} \eta_1^{-1}$ is obviously homotopic to $\eta_0 c_i^{-1} c_i c_{ij} c_j^{-1} c_j \eta_1^{-1}$, so γ is homotopic to the composition of a sequence of paths of the form $c_i c_{ij} c_j^{-1}$ as claimed. □

Since f is uniformly continuous there exists a $\mu \in (0, \lambda/4)$ such that for each $x \in M$ there is a $y \in M$ such that $f(B(x, \mu)) \subset B(y, \lambda)$. Now fix $\epsilon \in (0, \mu/4)$.

Let γ be one of the above generators and fix an arc α connecting p and $f(p)$. Divide γ and α into no more than N segments entirely contained in a $\lambda/2$-ball. Choose an (n, ϵ)-dense set S_n with $S_d(f, \epsilon, n)$ elements and let σ be one of those segments. There are points $x_0, \dots, x_m \in S_n$ and points $\sigma(0) = z_0, \dots, z_m = \sigma(1)$ on σ such that $z_i \in B_f(x_i, \epsilon, n)$ and the subarc $\{z_i, z_{i+1}\}$ of σ connecting z_i and z_{i+1} is contained in $B_f(x_i, \epsilon, n) \cup B_f(x_{i+1}, \epsilon, n)$. Here $B_f(x, \epsilon, n)$ denotes the d_n^f-ball of radius ϵ around x and we shall write

$\{x,y\}$ for $\{y,x\}^{-1}$. Connecting z_i and x_i by an arc $\{z_i,x_i\} \subset B_f(x_i,\epsilon,n)$ we obtain a path σ' given by

$$\{z_0,x_0\}\{x_0,z_0\}\{z_0,z_1\}\{z_1,x_1\}\{x_1,z_1\}\cdots\{z_{m-1},z_m\}\{z_m,x_m\}\{x_m,z_m\}$$

which is clearly homotopic to σ via a homotopy with fixed ends. Being contained in a λ-ball, σ' is homotopic in turn to the arc σ'' obtained from σ' by removing any loops arising from the coincidence of two points x_i and x_j. Thus σ is homotopic to an arc σ'' consisting of at most $3S_d(f,\epsilon,n)$ pieces of the form $\{z_i,x_i\}$, $\{x_i,z_i\}$, and $\{z_i,z_{i+1}\}$. Consider such a segment $\{x,y\}$. Since $\{x,y\} \subset B(x,4\epsilon) \subset B(x,\mu)$ (by choice of x_i and z_i if $\{x,y\} = \{x_i,z_i\}$, by the triangle inequality otherwise) its image $f(\{x,y\})$ is contained in a λ-ball. Since we also have $f(y) \in B(f(x),4\epsilon)$ by the same reason, $f(\{x,y\})$ is homotopic with fixed endpoints to an arc $\eta \subset B(f(x,4\epsilon))$ connecting $f(x)$ and $f(y)$. Inductively we find that for $i \in \{0,\dots,n-1\}$ the image $f^i(\{x,y\})$ is homotopic to an arc in $B(f(x,4\epsilon))$ connecting $f^i(x)$ and $f^i(y)$.

Thus the loop $\alpha f(\alpha) f^2(\alpha)\cdots f^{n-1}(\alpha) f^n(\gamma) f^{n-1}(\alpha)^{-1}\cdots f(\alpha)^{-1}\alpha^{-1}$ is homotopic to the concatenation of at most $(2n+1)NS_d(f,\epsilon,n)$ such arcs σ_k and this is in turn homotopic to a concatenation of no more than $(2n+1)NS_d(f,\epsilon,n)$ arcs of the form $\{\sigma_k(0),w_i\}c_{ij}\{w_j,\sigma_k(1)\}$, where we have used the notation of the lemma. Consequently

$$\alpha f(\alpha) f^2(\alpha)\cdots f^{n-1}(\alpha) f^n(\gamma) f^{n-1}(\alpha)^{-1}\cdots f(\alpha)^{-1}\alpha^{-1}$$

is homotopic to the concatenation of at most $(2n+1)NS_d(f,\epsilon,n)$ generators $\gamma_i \in \Gamma$ and the result follows. □

Combining this result with the inequality (3.1.28) we obtain a lower bound for topological entropy in homological terms:

Corollary 8.1.3. $h_{\text{top}}(f) \geq \log|r(f_{*1})|$.

Exercises

8.1.1. *Let M be a two-dimensional compact manifold, $p_1,\dots,p_n \in M$, and $f\colon M \to M$ a homeomorphism such that $f(p_i) = p_i$, $i = 1,\dots,n$. Prove that $h_{\text{top}}(f) \geq h_*(f\restriction_{M\smallsetminus\{p_1,\dots,p_n\}})$.*

8.1.2. *Using the previous exercise construct a class of homeomorphisms of the two-dimensional disc $\mathbb{D}^2$ that fixes two points p and q and is pairwise homotopic via a homotopy that fixes p and q, such that any map from that class has positive topological entropy.*

2. A survey of degree theory

a. Motivation. The degree of a map $f: N \to M$ between compact orientable manifolds of the same dimension, the notion we explore in this section, is a number that describes the algebraic multiplicity with which the image of N covers M under the map. In connection with the study of complexity of the orbit structure it is interesting for two different reasons. First, the degree of a smooth map of a manifold to itself is directly related to the topological entropy (Theorem 8.3.1) and in a more elementary way to the number of periodic orbits in the case of circle maps. Secondly, the degree of a map to a sphere is the main ingredient in the definition of the index of a fixed point (Definition 8.4.2).

After a brief discussion of the degree for circle maps we give two definitions of degree for smooth maps and give elementary proofs of the main properties invoking only one fundamental result from differential topology (Sard's Theorem A.3.14). The definitions extend to continuous maps by smooth approximation. The most general definition of degree uses homology theory, which is outlined in Section 7 of the Appendix.

b. The degree of circle maps. We defined the degree of a circle map in Definition 2.4.4. We now discuss this elementary notion a bit further. It provides both motivation and a good model to keep in mind when exploring the various definitions of degree in higher dimensions that we give later.

Proposition 8.2.1. *The degree* $\deg(f)$ *is a complete homotopy invariant, that is,* $\deg(f) = \deg(g)$ *if and only if* f *and* g *are homotopic.*

Proof. Homotopy invariance follows from Lemma 2.4.5. Thus it remains to show that if $\deg(f) = k$ then f is homotopic to the linear expanding map E_k. So suppose that $\deg(f) = k$ and let F be a lift of f. Then $F_t := (1-t)F(x) + tkx$ defines a homotopy between F and $k \cdot \mathrm{Id}$ and $F_t(x+1) = (1-t)F(x+1) + tk(x+1) = (1-t)F(x) + (1-t)k + tk + tkx = F_t(x) + k$ so $\pi(x) = \pi(y)$ implies $\pi(F_t(x)) = \pi(F_t(y))$, that is, F_t projects to a homotopy of f and E_k on S^1. □

Notice that a similar straight-line homotopy in higher dimensions was used in the proof of Theorem 8.1.1.

Proposition 8.2.2. $\deg(f \circ g) = \deg(f) \cdot \deg(g)$.

Proof. If $k = \deg(f)$, $l = \deg(g)$ and F and G are lifts fo f and g, respectively, then $F \circ G$ is a lift of $f \circ g$ and $F \circ G(x+1) = F(G(x) + l) = F(G(x)) + kl$. □

Corollary 8.2.3. $\deg(f^n) = (\deg(f))^n$ *for* $n \in \mathbb{N}$.

The main point of this subsection is that for circle maps we are able to fully explore the possibilities of obtaining periodic orbits from the degree. If as in Definition 1.7.1 we let $P_n(f) := \operatorname{card} \operatorname{Fix}(f^n)$ then we have

Proposition 8.2.4. *$P_n(f) \geq |(\deg(f))^n - 1|$ for any continuous map $f: S^1 \to S^1$.*

Proof. Evidently it suffices to consider $f: S^1 \to S^1$ of degree k and to find $|k-1|$ fixed points. Note that by the Intermediate Value Theorem the equation $F(y) - y = 0 \pmod 1$ has at least $|k-1|$ solutions $y \in [0,1)$ since $F(1) - 1 = F(0) - 0 + k - 1$. But any y thus obtained projects to a fixed point of f. □

This estimate will often give us a large number of periodic points, but in some cases there are many more periodic points than this predicts:

Example. The map $f_4: [0,1] \to [0,1], x \mapsto 4x(1-x)$ from the quadratic family projects to a map f on S^1 of degree zero. On the other hand one easily checks (Exercise 1.7.2) that $P_n(f) = 2^n$ by drawing inductively the graph of f_4^n and calculating the number of its points of intersection with the diagonal.

This phenomenon of producing many periodic points for one-dimensional maps due to "folding" will be discussed in full generality in Chapter 15 (Corollary 15.2.2 and Proposition 15.2.13).

c. Two definitions of degree for smooth maps. While the degree of a map is a topological notion and will be defined as such at the end of this section, there are two distinct direct definitions for the case of differentiable maps. We discuss them now and make the standing assumption that N and M are compact oriented smooth n-dimensional manifolds and $f: N \to M$ a C^1 map. Recall a particular case of the notion of a regular value for a map between manifolds (cf. Section 7 of the Appendix).

Definition 8.2.5. A point $x \in N$ is called a *regular point* (for f) if Df is invertible at x. $x \in M$ is called a *regular value* (of f) if $f^{-1}(\{x\})$ consists of regular points, a singular value otherwise.

The set of regular values is obviously open. By Sard's Theorem A.3.14 it has full measure and is hence also dense.

Lemma 8.2.6. *If x is a regular value then $f^{-1}(\{x\})$ is finite.*

Proof. Suppose $x \in M$ and $f^{-1}(\{x\})$ is infinite. Then by compactness of N we can select a sequence $x_i \to x_0$ with $f(x_i) = x$ and hence $f(x_0) = x$ by continuity. But this shows that f is not injective near x_0 so by the Implicit Function Theorem Df is not invertible at x_0 and x is a singular value. □

Definition 8.2.7. Suppose $x \in M$ is a regular value and for each $y \in f^{-1}(\{x\})$ let $\epsilon_y := \pm 1$ according to whether $D_y f$ preserves or reverses orientation. Then the *degree* of f *at* x is defined by

$$\deg_x(f) := \sum_{y \in f^{-1}(\{x\})} \epsilon_y.$$

Note that the sum makes sense and $\deg(f)$ is always finite since $f^{-1}(\{x\})$ is finite by Lemma 8.2.6. We will show that in fact $\deg_x(f)$ is independent of the choice of the regular point $x \in M$.

This definition of degree at x measures how many times f covers M near x, counted with appropriate positive and negative multiplicities. In order to show that it is independent of x we introduce a definition based on integration. We begin with a slightly modified version of a notion introduced in Section 5.1.

Definition 8.2.8. A *positive volume element* on M is a continuous n-form ω that is positive on positively oriented frames. It is said to be normalized if $\int_M \omega = 1$. The *pullback* $f^*\omega$ of ω is a form given by

$$(f^*\omega)(v_1, \dots, v_n) = \omega(Df(v_1), \dots, Df(v_n)).$$

(Compare with Definition 5.1.1 and the definition of push-forward at the beginning of Section 5.1b, where Df is assumed to be invertible.)

Definition 8.2.9. If ω is a positive normalized volume element on M then the *degree* of $f\colon N \to M$ with respect to ω is defined by

$$\deg_\omega(f) := \int_N f^*\omega.$$

We now show that Definition 8.2.7 gives the same number as Definition 8.2.9, so that in particular Definition 8.2.7 is independent of the choice of regular point and Definition 8.2.9 gives an integer that is independent of the choice of positive normalized volume element on M.

Lemma 8.2.10. *Let $p \in M$ be a regular value of $f\colon N \to M$ and ω a positive volume element. Then* $\deg_p(f) = \deg_\omega(f)$.

Proof. Since p is a regular value there are disjoint open neighborhoods $U_1, \dots, U_k \subset N$ of the points $x_1, \dots, x_k$ of $f^{-1}(p)$ such that $\bigcup_{i=1}^k U_i$ is the preimage of a neighborhood V of p and $f\restriction_{U_i}$ is a diffeomorphism for all i. If ν is an n-form supported in V such that $\int_V \nu = \int_M \nu = 1$ then by Lemma A.3.13 $\omega = \nu + d\alpha$ for some $(n-1)$-form α, so $\int_M f^*\omega = \int_M (f^*\nu + f^*d\alpha) = \int_M f^*\nu = \sum_{i=1}^k \int (f\restriction_{U_i})^*\nu = \sum_{i=1}^k \int_{U_i} f^*\nu$. By the transformation rule each of the latter integrals is ± 1 according to whether $f\restriction_{U_i}$, or equivalently Df_{x_i}, preserves or reverses orientation. Consequently $\deg_\omega f = \int_M f^*\omega = \deg_p f$. □

So we can now make the following definition:

Definition 8.2.11. Let $f\colon N \to M$ be a C^1 map. Then the *degree* of f is defined by $\deg(f) := \deg_x(f)$ for any regular value $x \in M$.

Since the degree clearly is continuous in the C^1 topology, hence locally constant, we have in particular:

Lemma 8.2.12. *The degree is invariant under homotopies consisting of C^1 maps.*

Lemma 8.2.13. *Any two C^1 maps $f_0, f_1\colon N \to S^n$ that are never antipodal are smoothly homotopic.*

Proof.

$$f_t(x) := \frac{tf_0(x) + (1-t)f_1(x)}{\|tf_0(x) + (1-t)f_1(x)\|} \tag{8.2.1}$$

gives a smooth homotopy. □

More generally we have

Lemma 8.2.14. *Any two C^1 maps $f_0, f_1\colon N \to M$ that are sufficiently C^0-close are homotopic via C^1 maps.*

Proof. To replace (8.2.1) notice that the arc of a great circle is the unique shortest curve connecting any two non-antipodal points on the sphere. Such shortest curves turn out to be geodesics as discussed after Definition 5.3.4. In the next chapter we show that such curves are uniquely defined for any pair of sufficiently close points on any compact manifold provided with a Riemannian metric (Theorem 9.5.5). Using homotopies along such arcs with the length parameter we obtain the claim. □

Corollary 8.2.15. *The degree of a C^1 map is a homotopy invariant.*

This, in fact, allows us to define the degree for continuous maps by smooth approximation:

Definition 8.2.16. The degree of a continuous map $f\colon N \to M$ is defined by $\deg(f) := \deg(g)$, where $f\colon N \xrightarrow{C^1} M$ is sufficiently close to g.

Evidently the degree of a continuous map is also homotopy invariant. In fact we can use *any* n-form with nonzero integral to define the degree:

Proposition 8.2.17. *Let η be a continuous n-form with $\int_M \eta \neq 0$ and $f\colon N \to M$. Then $\deg(f) = \int_N f^*\eta / \int_M \eta$.*

Proof. We can write $\eta = \omega_1 - \omega_2$ where ω_i are positive volume elements since by compactness if ω is any positive volume element then $\eta + \alpha\omega$ is positive for sufficiently large $\alpha \in \mathbb{R}$. But then by linearity of the integral $\int_M f^*\eta = \int_M f^*\omega_1 - \int_M f^*\omega_2 = \deg(f)(\int_M \omega_1 - \int_M \omega_2)$. □

In our discussion of the index of a fixed point in the next section we will use the following observation.

Proposition 8.2.18. *Let W be an $(n+1)$-dimensional oriented manifold with boundary $N := \partial W$ and $F\colon W \to M$ a continuous map. Let $f := F\restriction_N$ be the restriction of F to N. Then $\deg(f) = 0$.*

Remark. The orientation on N is the one induced from W. If N is connected then any orientation will do.

Proof. If ω is a volume on M then by the Stokes Theorem $\deg(f) = \int_N f_*\omega = \int_W d(f_*\omega) = \int_W f_* d\omega = 0$. □

Let us mention several properties of the degree relevant for dynamics.

Corollary 8.2.19. *If $N \xrightarrow{f} N \xrightarrow{g} K$ are continuous maps then* $\deg(g \circ f) = \deg(g)\deg(f)$.

Proof. We may assume that f and g are C^1. Then Proposition 8.2.17 shows that $\deg(g \circ f) = \int_N f^* g^* \omega / \int_K \omega = \deg(f) \int_M g^*\omega / \int_K \omega = \deg(f)\deg(g)$. □

Since the degree counts the algebraic multiplicity it clearly gives a lower bound on the number of preimages of a regular point:

Corollary 8.2.20. *Let $f\colon M \to M$ be C^1 on an open set with full measure. For almost every point $x \in M$ we have $|\deg(f^m)| \le \operatorname{card}(f^{-m}(\{x\})) < \infty$. In particular if $|\deg(f)| \ge 2$ the cardinality of $f^{-m}(\{x\})$ grows at least exponentially with m.*

Proof. By Sard's Theorem (Theorem A.3.14) almost every point x is regular for all iterates of f, so the claim follows from the definition of the degree. □

In the discussion of the degree of maps of S^1 we showed that the degree is a complete homotopy invariant. This is, in fact, true for S^n as well. This implies in particular that there are only two homotopy types of homeomorphisms of S^n, namely, orientation preserving and reversing.

Let us look at two examples to assess to what extent the topological complexity measured by the degree implies dynamical complexity.

Example. Let $\hat{\mathbb{C}} := \mathbb{C} \cup \{\infty\} \sim S^2$ be the Riemann sphere, that is, the one point compactification of the complex plane. The smooth structure in a neighborhood of $\{\infty\}$ is given by the coordinate $w = 1/z$. Consider the map $f\colon S^2 \to S^2$, $z \mapsto z^2$, extended to ∞ by $f(\infty) := \infty$. Then for z near ∞ we have $f(w) = w^2$ near $w = 0$. Note that $\deg f = 2$ since f covers the sphere twice and preserves orientation. The dynamics of f is as follows: The poles 0 and ∞ are contracting fixed points and the equator $\{z \in \hat{\mathbb{C}} \mid |z| = 1\}$ is an invariant set such that the restriction of f is simply the expanding circle map E_2 defined in (1.7.1). All other points converge to one of the poles under iteration of $z \mapsto z^2$. All periodic points other than the poles are on the equator. They are the solutions of $z = f^n(z) = z^{2^n}$, that is,

$$\{z \in \mathbb{C} \setminus 0 \mid z^{2^n - 1} = 1\}.$$

These are the $2^n - 1$ distinct $(2^n - 1)$th roots of unity. Thus $P_n(f) = 2 + 2^n - 1 = 2^n + 1 = 1 + (\deg(f))^n$ grows approximately as $(\deg(f))^n$.

Example. Now consider $g: S^2 \to S^2$, $z \mapsto z^2/(2|z|)$, extended to ∞ by $g(\infty) := \infty$, so that $g(w) = 2w^2/|w|$ near $w = 0$. Thus ∞ is a (nonsmooth) expanding point whereas 0 is a contracting fixed point. Note that under g all points other than ∞ converge to 0 since $|g(z)| = |z|/2$. Therefore in striking contrast to the earlier example the only periodic points are 0 and ∞. On the other hand g doubly covers S^2 and in fact $g = h \circ f$, where h is the orientation-preserving homeomorphism $z \mapsto z/(2\sqrt{|z|})$. Thus $\deg(g) = \deg(h \circ f) = \deg(h)\deg(f) = \deg(f) = 2$. Since the only invariant measures for g are atomic measures concentrated at 0 and ∞ we have by the Variational Principle, Theorem 4.5.3, $h_{\text{top}}(g) = 0$.

The latter example shows that the degree does not always give information about the dynamical complexity of a map. In particular the exponentially many preimages provided by Corollary 8.2.20 accumulate on a single fixed point at ∞. As we shall see later (via the Shub–Sullivan Theorem 8.5.1), the lack of dynamical complexity in this example can be explained by an essential absence of smoothness at ∞. In the next section we will show that for a C^1 map dynamical complexity is indeed guaranteed by having degree of absolute value at least two (the Misiurewicz–Przytycki Theorem 8.3.1).

d. The topological definition of degree. We close this discussion of the degree by directly giving a topological definition. In order to define the degree of a map $f: N \to M$ between compact manifolds of the same dimension in topological terms recall that f induces an action on homology, so in particular in the case of orientable manifolds N and M, when $H_n(N, \mathbb{Z}) = H_n(M, \mathbb{Z}) = \mathbb{Z}$, f induces a homomorphism $f_{*n}: \mathbb{Z} \to \mathbb{Z}$.

Definition 8.2.21. Let N, M be n-dimensional oriented manifolds and $f: N \to M$ continuous and denote by $f_{*n}: H_n(N, \mathbb{Z}) \to H_n(M, \mathbb{Z})$ the induced homomorphism on $\mathbb{Z}$. Fix an orientation on $H_n(N, \mathbb{Z}) \simeq \mathbb{Z}$ and $H_n(M, \mathbb{Z}) \simeq \mathbb{Z}$. Then the degree of f is defined by

$$\deg(f) := f_{*n}(1).$$

Using the duality between homology and cohomology and the description of de Rham cohomology via differential forms (see Section 3 of the Appendix) one easily establishes the equivalence of this definition with the volume form definition of degree given earlier.

Exercises

8.2.1. *Consider $f: S^2 \to S^2$ given by $z \mapsto z^2$ as discussed before. Show that $h_{\text{top}}(f) = \log 2$.*

8.2.2. *Consider $g: S^2 \to S^2$ given by $z \mapsto z^2/2|z|$ as discussed before. Show that $h_{\text{top}}(f) = 0$ without using the Variational Principle (Theorem 4.5.3).*

3. Degree and topological entropy

We saw in the previous section that the "topological complexity" measured by the degree does not always engender much dynamical complexity. The failure to do so is, however, due only to the absence of smoothness in the example we considered. For C^1 maps we now show that one can obtain positive topological entropy from the degree. We will use the definition of the degree through the preimages of a regular point.

Theorem 8.3.1. (Misiurewicz–Przytycki Theorem)[1] *If M is a smooth compact orientable manifold and $f\colon M \xrightarrow{C^1} M$, then $h_{\mathrm{top}}(f) \geq \log|\deg(f)|$.*

Proof. Fix a volume element ω on M, $\alpha \in (0,1)$. Let $L := \sup_{x\in M} |Jf(x)|$, where Jf is as in Definition 5.1.4, $\epsilon = L^{-\alpha/(\alpha-1)}$, and $B := \{x \mid |Jf(x)| \geq \epsilon\}$. Pick a covering of B by open sets on which f is injective and let δ be the Lebesgue number of the covering. Thus, if $x, y \in B$ and $d(x,y) \leq \delta$ then $f(x) \neq f(y)$.

Fix $n \in \mathbf{N}$ and let

$$A := \{x \in M \mid \operatorname{card}(B \cap \{x, f(x), \dots, f^{n-1}(x)\}) \leq \alpha n\}.$$

If $x \in A$ then

$$|Jf^n(x)| = \prod_{j=0}^{n-1} |Jf(f^j(x))| < \epsilon^{(1-\alpha)n} L^{\alpha n} = (\epsilon^{(1-\alpha)} L^{\alpha})^n = 1$$

so the volume of $f^n(A)$ is less than that of M and by Sard's Theorem A.3.14 there exists a regular value $x \in M \smallsetminus f^n(A)$ of f^n.

We now choose a large (n,δ)-separated set in $f^{-n}(\{x\})$. Since x is regular for f it has at least $N := \deg(f)$ preimages. If N of them are in B (a "good transition") then take Q_1 to consist of N such preimages. Otherwise (a "bad transition") take Q_1 to be a single preimage outside B. Either way, $Q_1 \subset f^{-1}(\{x\})$ consists of regular values of f since x is regular for f^n. Thus we can apply the same procedure to every $y \in Q_1$ and by collecting all of the points chosen that way obtain $Q_2 \subset f^{-2}(\{x\})$, and so on. The set $Q_n \subset f^{-n}(\{x\})$ we thus obtain is (n,δ)-separated: If $y_1, y_2 \in Q_n$ and $d(f^k(y_1), f^k(y_2)) \leq \delta$ for $k \in \{0, \dots, n-1\}$ then $f^{n-1}(y_1) = f^{n-1}(y_2)$ since otherwise by construction $f^{n-1}(y_1) \in B$ and $f^{n-1}(y_2) \in B$ and by choice of δ we have $x = f(f^{n-1}(y_1)) \neq f(f^{n-1}(y_2)) = x$. Likewise $f^{n-2}(y_1) = f^{n-2}(y_2)$, and so forth, so $y_1 = y_2$.

Now $Q_n \subset f^{-n}(\{x\}) \subset f^{-n}(M \smallsetminus f^n(A)) \subset M \smallsetminus A$, that is, $Q_n \cap A = \varnothing$. Thus for any $y \in Q_n$ there are (by definition of A) more than αn numbers $k \in \{0, \dots, n-1\}$ for which $f^k(y) \in B$. So in passing from x to any $y \in Q_n$ there are at least $m := [\alpha n] + 1$ "good transitions" and hence $\operatorname{card} Q_n \geq N^m \geq N^{\alpha n}$. Therefore the maximal cardinality $N_d(f, \delta, n)$ of an (n,δ)-separated set is at least $N^{\alpha n}$ and thus $h_{\mathrm{top}}(f) \geq \alpha \log N$ for every $\alpha \in (0,1)$, which yields the theorem. □

The two properties of smoothness that make the preceding proof work are boundedness of the Jacobian together with the fact that a smooth map is a local homeomorphism near any point where the Jacobian is nonzero. The second property fails at ∞ for the map $z \mapsto z^2/2|z|$.

There are certain cases where the inequality given by Theorem 8.3.1 becomes equality. Expanding maps of the circle provide one example (Corollary 3.2.4). This can be generalized to arbitrary expanding maps of a compact manifold (see Exercise 8.3.2). The map $f(z) = z^2$ discussed in Section 8.2c provides another example, as does the map $f(z) = z^n$ for any n. A much greater generalization involves any rational map of the Riemann sphere, that is, maps of the form $f(z) = \frac{P(z)}{Q(z)}$, where P and Q are relatively prime polynomials, although we only prove one inequality here. Such maps can be characterized as holomorphic maps of the Riemann sphere into itself. Since these maps are orientation preserving, their degree is equal to the number of preimages of a regular point, that is, the number of solutions of

$$\frac{P(z)}{Q(z)} = w \text{ or } R_w(z) := P(z) - wQ(z) = 0.$$

By choosing $w \neq 0$ and so as to avoid cancellation of the leading coefficients of P and Q if $\deg P = \deg Q$, we obtain $\deg R_w = \max(\deg P, \deg Q)$. We may also choose w in such a way that R_w does not have multiple roots since these appear only if w is the image of a critical point. Thus the degree of f is equal to the maximum of the algebraic degrees of P and Q. By Theorem 8.3.1

$$h_{\text{top}}(f) \geq \log \max(\deg P, \deg Q).$$

In fact the reverse inequality also holds, so we have

$$h_{\text{top}}(f) = \log \max(\deg P, \deg Q).$$

At this point we do not have adequate machinery to prove the upper estimate for entropy.[2]

Exercises

8.3.1. *Prove that the topological entropy of any linear fractional transformation* $f(z) = \frac{az+b}{cz+d}$, $a, b, c, d \in \mathbb{C}$, *of the Riemann sphere is zero.*

8.3.2. *Let* M *be a compact Riemannian manifold and* $f\colon M \to M$ *an expanding map (see Definition 2.4.1). Prove that* $h_{\text{top}}(f) = \log|\deg f|$.

8.3.3. *Let* $f\colon S^1 \to S^1$ *be a continuous monotone map (that is, any lift to* $\mathbb{R}$ *is a monotone function). Prove that* $h_{\text{top}}(f) = \log|\deg f|$.

4. Index theory for an isolated fixed point

The degree gives insight into the behavior of isolated fixed points by providing a definition of an index whose most important property is that an isolated fixed point with nonzero index, which in particular includes any transverse fixed point for a smooth map (cf. Definition 7.2.1), persists under C^0 perturbations, whereas our previous results along these lines worked only for C^1 perturbations (Proposition 1.1.4).

Proposition 8.4.1. *Let $x_0 \in U \subset \mathbb{R}^n$ be open and $f\colon U \to \mathbb{R}^n$ continuous such that $f(x) \neq x$ for all $x \in U \smallsetminus \{x_0\}$. Let V be a homeomorphic image of the n-ball with the natural orientation and $x_0 \in V \subset \bar{V} \subset U$ and define the map $v_{f,V}\colon \partial V \to S^{n-1}$, $x \mapsto \dfrac{x - f(x)}{\|x - f(x)\|}$. Then the degree of $v_{f,V}$ is independent of V.*

Proof. Suppose V and V' are as stated and take a ball $B \subset V \cap V'$ containing x_0. It suffices to show that $\deg(v_{f,B}) = \deg(v_{f,V})$ since by the same token we then have $\deg(v_{f,B}) = \deg(v_{f,V'})$ and the claim follows. Given such B note that the map $w\colon A := \bar{V} \smallsetminus B \to S^{n-1}$, $x \mapsto \dfrac{x - f(x)}{\|x - f(x)\|}$ is well defined and continuous and that by Proposition 8.2.18 $\deg(w_{\restriction \partial A}) = 0$. Now ∂A is the union of ∂V and ∂B, but the latter with negative orientation. Thus $\deg(v_{f,V}) - \deg(v_{f,B}) = \deg(w_{\restriction \partial A}) = 0$. $\square$

Remark. We shall usually take ∂V to be the sphere S_ϵ of radius ϵ and then write $v_{f,\epsilon} := v_{f,V}$.

Definition 8.4.2. Let $x_0 \in U \subset \mathbb{R}^n$ be open and $f\colon U \to \mathbb{R}^n$ continuous such that $f(x) \neq x$ for all $x \in U \smallsetminus \{x_0\}$. Then the index of x_0 for f is defined by

$$\operatorname{ind}_f x_0 := \deg(v_{f,V})$$

for any V as in Proposition 8.4.1.

Evidently the index is always an integer. If x_0 is not a fixed point, then the map $x \mapsto \dfrac{x - f(x)}{\|x - f(x)\|}$ is well defined on a ball around x_0 and by Proposition 8.2.18 has degree zero on the boundary, whence $\operatorname{ind}_f(x_0) = 0$.

The index of an isolated fixed point of a map on a manifold is defined via a coordinate chart around the point. Proposition 8.4.1 shows that the definition is independent of the chart chosen.

Example. For maps of an interval the index is the degree of a map of the 0-dimensional sphere $S_\epsilon = \{-\epsilon, \epsilon\}$. There are four maps of $\{-\epsilon, \epsilon\}$. The two constant ones evidently have degree 0, the identity Id has degree 1, and $-\mathrm{Id}$ has degree -1. Consequently for an isolated fixed point of a map f we have

$$\operatorname{ind}_f x = \begin{cases} 0 & \text{if } \mathrm{Id} - f \text{ is not monotone at } x, \\ 1 & \text{if } \mathrm{Id} - f \text{ is increasing at } x, \\ -1 & \text{if } \mathrm{Id} - f \text{ is decreasing at } x. \end{cases}$$

For example, if $f(x) = x^3$ then $\operatorname{ind}_f 0 = 1$ and $\operatorname{ind}_f \pm 1 = -1$.

Homotopy invariance of degree (Corollary 8.2.15) immediately implies a similar statement for the index:

Proposition 8.4.3. *Let $f_t: U \to \mathbb{R}^n$ be a continuous family of maps and $p \in U$ a common isolated fixed point for all f_t. Then the index $\operatorname{ind}_{f_t}(p)$ of p does not depend on t.*

Theorem 8.4.4. *Let $x_0 \in U \subset \mathbb{R}^n$ be an isolated fixed point of $f: U \to \mathbb{R}^n$ with $\operatorname{ind}_f x_0 \neq 0$. Then there exists an $\epsilon > 0$ such that all $g: U \to \mathbb{R}^n$ with $\|g(x) - f(x)\| < \epsilon$ for all $x \in U$ have a fixed point near x_0.*

Remark. Thus fixed points of nonzero index and in particular transverse fixed points persist under C^0 perturbations. Note that the fixed point of the perturbation may not be unique nor isolated. The assumption of nonzero index cannot be dropped, as the example of the isolated fixed point 0 of $f: x \mapsto x + x^2/(1 + x^2)$ on $\mathbb{R}$ shows: $f + \epsilon$ has no fixed point at all (this example already appeared in Exercise 2.1.3(1) and Section 7.1c).

Proof. Let $\epsilon > 0$ be such that the sphere $\partial V = S_\epsilon$ of radius ϵ is as in Proposition 8.4.1, $\delta := \inf\{\|x - f(x)\| \mid x \in S_\epsilon(x_0)\}$, and g a $\delta/3$-perturbation of f. Then $\|(x - f(x)) - (x - g(x))\| = \|f(x) - g(x)\| < \delta/2$ while $\|x - f(x)\| \geq \delta$ for all $x \in U$, that is, $v_{f,\epsilon}$ and $v_{g,\epsilon}$ are never antipodal, hence straight-line homotopic via $\dfrac{tv_{f,\epsilon} + (1-t)v_{g,\epsilon}}{\|tv_{f,\epsilon} + (1-t)v_{g,\epsilon}\|}$. Thus $\deg v_{g,\epsilon} = \deg v_{f,\epsilon} = \operatorname{ind}_f x_0 \neq 0$ so g must have a fixed point in the ϵ-ball around x_0—otherwise we would have $\operatorname{ind}_g(x_0) = 0$. □

To apply this result one has to compute the index of a fixed point. For transverse fixed points of differentiable maps this is quite easy, since one can obtain it from the linear part:

Proposition 8.4.5. *Suppose $f: U \to \mathbb{R}^n$ is differentiable at $0 \in U \subset \mathbb{R}^n$ and 0 is a transverse fixed point for f. Then $\operatorname{ind}_f 0 = \operatorname{ind}_{Df(0)} 0$.*

Proof. If $A := Df(0)$ then the transversality assumption means that 1 is not an eigenvalue of A, that is, $\operatorname{Id} - A$ is invertible, whence there exists $\delta > 0$ such that $\|x - Ax\| > \delta\|x\|$ for all x. On the other hand there exists $\epsilon > 0$ such that $\|x - f(x) - (x - Ax)\| = \|Ax - f(x)\| < \delta\|x\|$ on S_ϵ, since f is differentiable at 0. Consequently $v_{f,\epsilon}$ and $v_{A,\epsilon}$ are never antipodal, hence homotopic. □

For linear maps the index is easy to compute:

Proposition 8.4.6. *If $1 \notin \operatorname{sp}(A)$ then $\operatorname{ind}_A(0) = \operatorname{sign}\det(\operatorname{Id} - A)$.*

Proof. Since A is linear we may consider $v_A := v_{A,1}: x \mapsto \dfrac{x - Ax}{\|x - Ax\|}$. This map is invertible: If $v_A(x) = v_A(y)$ then $x - Ax = \lambda(y - Ay)$ and $x - \lambda y = A(x - \lambda y)$, so $x = \lambda y$ since 0 is the only fixed point for A. But $\|x\| = \|y\| = 1$, so $|\lambda| = 1$ and $x = \pm y$. Since $v_A(-x) = -v_A(x)$ we have $x = y$. So v_A is a homeomorphism of the sphere and hence $\deg(v_A) = 1$ if $\operatorname{Id} - A$ preserves orientation, that is, $\operatorname{sign}\det(\operatorname{Id} - A) = 1$; $\deg(v_A) = -1$ otherwise. □

$\operatorname{sign}\det(\operatorname{Id}-A)$ can be calculated in terms of eigenvalues:

Corollary 8.4.7. $\operatorname{ind}_A(0) = (-1)^{\operatorname{card}\{i \mid \lambda_i > 1\}}$, *where* $\{\lambda_i\}_{i=1}^n$ *are the eigenvalues of* A.

Proof.

$$\begin{aligned}\operatorname{sign}\det(\operatorname{Id}-A) &= \operatorname{sign}\prod_{i=1}^{n}(1-\lambda_i)\\ &= \operatorname{sign}\prod_{\lambda_i>1}(1-\lambda_i)\cdot\operatorname{sign}\prod_{\lambda_i<1}(1-\lambda_i)\cdot\operatorname{sign}\prod_{\lambda_i\in\mathbb{C}\setminus\mathbb{R}}(1-\lambda_i)\\ &= (-1)^{\operatorname{card}\{i\mid\lambda_i>1\}}\cdot 1\cdot\operatorname{sign}\prod_{\lambda_i\in\mathbb{C}\setminus\mathbb{R}}(1-\lambda_i) = (-1)^{\operatorname{card}\{i\mid\lambda_i>1\}}\end{aligned}$$

since pairing complex conjugate eigenvalues shows $\prod_{\lambda_i\in\mathbb{C}\setminus\mathbb{R}}(1-\lambda_i) > 0$. □

A simple example is $A = \begin{pmatrix} 2 & 0 \\ 0 & 1/2 \end{pmatrix}$, where 0 is a hyperbolic fixed point. Here $\operatorname{ind}_A(0) = -1$ and $\operatorname{ind}_{-A}(0) = 1$. It is in fact not hard to make a list of all cases occurring for linear maps $A\colon \mathbb{R}^2 \to \mathbb{R}^2$. Denote by λ_1, λ_2 the eigenvalues of A and assume that A is orientation preserving (that is, $\lambda_1\lambda_2 > 0$) and 1 is not an eigenvalue (that is, 0 is an isolated fixed point). Then there are five possible types of transverse fixed points:

	Eigenvalues	Index of 0	Description
1	$\lvert\lambda_1\rvert < 1,\ \lvert\lambda_2\rvert < 1$	$\operatorname{ind}_A(0) = 1$	contracting
2	$\lvert\lambda_1\rvert > 1,\ \lvert\lambda_2\rvert > 1$	$\operatorname{ind}_A(0) = 1$	expanding
3	$\lvert\lambda_1\rvert = \lvert\lambda_2\rvert = 1 \neq \lambda_i$	$\operatorname{ind}_A(0) = 1$	elliptic
4	$0 < \lambda_1 < 1 < \lambda_2$	$\operatorname{ind}_A(0) = -1$	hyperbolic (saddle)
5	$\lambda_1 < -1 < \lambda_2 < 0$	$\operatorname{ind}_A(0) = 1$	hyperbolic with rotation (inverted saddle)

The notion of index for an isolated zero of a vector field is parallel to that for maps and in some sense even simpler. Namely, let $U \subset \mathbb{R}^n$ be an open set and $v\colon U \to \mathbb{R}^n$ a vector field with an isolated zero $x_0 \in U$. Let $V \subset U$ be a homeomorphic image of the n-ball and $x_0 \in V \subset \bar{V} \subset U$. Then we define the index $\operatorname{ind}_v(x_0)$ as the degree of the map $\xi_{v,V}\colon \partial V \to S^{n-1}$, $\xi(x) = \dfrac{v(x)}{\|v(x)\|}$ which turns out to be independent of the choice of V (see Exercise 8.4.3).

As a simple example let us consider area-preserving flows in dimension 2 that arise from a linear system $\dot{x} = Ax$ of ordinary differential equations with constant coefficients, that is, flows of the form $\{e^{At}\}_{t\in\mathbb{R}}$ for some matrix A with $e^{\operatorname{tr} A} = \det e^A = 1$, or $\operatorname{tr} A = 0$. In this case transverse fixed points can be only elliptic or hyperbolic of the saddle type: The eigenvalues are λ and $-\lambda$ and we may assume A is in Jordan form. If $\lambda = 0$, the origin is not an isolated fixed point. If $\lambda \neq 0$ then λ is either $is \in i\mathbb{R}$ or real. In the first case e^{tA} has eigenvalues $e^{\pm ist}$ for some $s \in \mathbb{R}$ and 0 is an elliptic fixed point. Otherwise e^{tA} has eigenvalues $e^{\pm st}$ for some $s \in \mathbb{R}$ and 0 is a hyperbolic fixed point. In other words here there are only two cases:

	Eigenvalues of A	Index of 0	Description
1	$\pm is \in i\mathbb{R}$	$\text{ind}_{e^A}(0) = 1$	elliptic
2	$\lambda, -\lambda \in \mathbb{R}$	$\text{ind}_{e^A}(0) = -1$	hyperbolic

Since the definition of the index involves the degree of a map of the sphere it is not always easy to determine in those cases where the linear part does not provide this information, that is, for nontransverse fixed points. In dimension 2, however, one can read off the index relatively easily from a local picture.

Example. Consider the time-one map for a flow that near zero looks as in the left half of Figure 8.4.1.

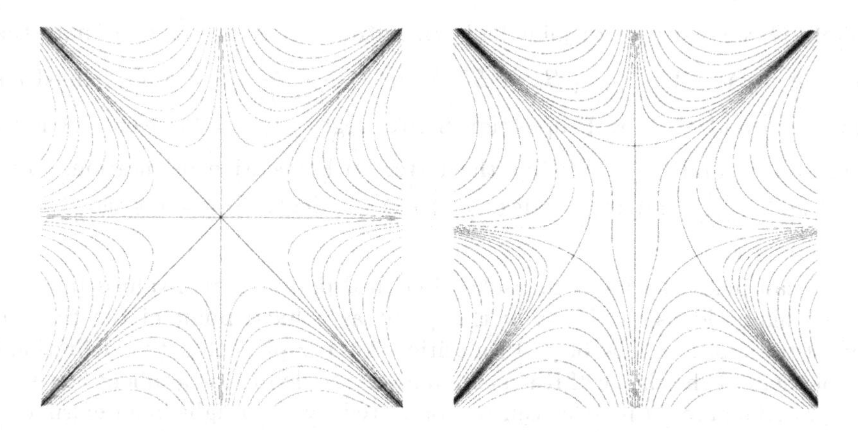

FIGURE 8.4.1. A threefold saddle and a perturbation

A fixed point of this sort is called a *multiple* (here *threefold*) *saddle.* A flow like this can be obtained as the Hamiltonian flow φ_0 induced by the Hamiltonian $H(x, y) = xy(x + y)(x - y)$. To determine the index of 0 from the picture, traverse the unit circle in the positive direction (↺) while noting the direction of the unit tangent vector to the flow. To find the degree of this map $v\colon S^1 \to S^1$ note that v is monotone and list some values, denoting the angle on S^1 by θ:

θ	0	$\frac{\pi}{4}$	$\frac{\pi}{2}$	$\frac{3\pi}{4}$	π	$\dots$	2π
v	π	$\frac{\pi}{4}$	$-\frac{\pi}{2}$	$-\frac{5\pi}{4}$	-2π	$\dots$	-5π

Since v covers S^1 three times and is decreasing, its degree and thus the index of 0 is -3.

Remark. In some sense this saddle can be obtained by "gluing" together three simple saddles of index -1. The index can thus be thought of as additive, that is, if the flow φ_0 occurs in a one-parameter family φ_ϵ of flows such that φ_ϵ has three simple saddles for $\epsilon > 0$, then the origin is in a natural way a multiple saddle obtained from merging three simple saddles, that is, $\epsilon = 0$ is a bifurcation value as defined in Section 7.4. An example is given by the flows for the Hamiltonians $H_\epsilon(x,y) := \epsilon x + xy(x+y)(x-y)$ shown in Figure 8.4.1 for $\epsilon = 0$ and $\epsilon = 1/10$. The sum of the indices is thus preserved in this setting. If we embed this local picture in a compact manifold then this is a consequence of the Lefschetz Fixed-Point Formula (Theorem 8.6.2).

These pictures provide an example of a general phenomenon:

Proposition 8.4.8. *If $f: M \to M$ is continuous and V is an open submanifold inside a coordinate chart for M containing only isolated fixed points $(x_1, \dots, x_k)$ and having no fixed points on ∂V, then $\deg(v_{f,V}) = \sum_{i=1}^k \operatorname{ind}_f(x_i)$, where $v_{f,V} = \dfrac{x - f(x)}{\|x - f(x)\|}$ as before.*

Proof. Given pairwise disjoint balls B_i with $x_i \in B_i \subset \overline{B_i} \subset V$ note that the map $w: A := \bar{V} \setminus \bigcup_{i=1}^k B_i \to S^{n-1}$, $x \mapsto \dfrac{x - f(x)}{\|x - f(x)\|}$ is well defined and continuous and that by Proposition 8.2.18 $\deg(w_{\restriction_{\partial A}}) = 0$. Now ∂A is the union of ∂V and $\bigcup_{i=1}^k \partial B$, but the latter with negative orientation. Thus $\deg(v_{f,V}) - \sum_{i=1}^k \operatorname{ind}_f(x_i) = \deg(v_{f,V}) - \sum_{i=1}^k \deg(v_{f,B_i}) = \deg(w_{\restriction_{\partial A}}) = 0$. □

One can define an index for isolated critical points of real-valued functions f on a manifold as the index of this point as a fixed point of the time-one map of the gradient flow of f. Proposition 8.4.3 implies that this definition is independent of the choice of Riemannian metric to define the gradient flow since any two Riemannian metrics can be connected by a straight-line deformation and hence the gradient flows are homotopic via a homotopy fixing the set of critical points.

For the index of nondegenerate critical points of real-valued functions Corollary 8.4.7 has the following application:

Corollary 8.4.9. *Let f be a smooth real-valued function on a manifold and p a nondegenerate critical point. Then the index of p is* $\operatorname{sign}\det(\mathrm{Id} - \exp H)$, *which is -1 if the Hessian H of f at p has an odd number of positive eigenvalues, and 1 otherwise.*

Exercises

8.4.1. *Suppose 0 is an isolated fixed point of a map $f: \mathbb{R}^2 \to \mathbb{R}^2$, where $f(x,y) = (P(x,y), Q(x,y))$ and P and Q are polynomials. Prove that $\operatorname{ind}_f 0 \le \max(\deg P, \deg Q)$.*

8.4.2*. *Suppose 0 is an isolated fixed point of a map $f: \mathbb{R}^n \to \mathbb{R}^n$ whose coordinates are polynomials. Show that $\operatorname{ind}_f(0)$ is bounded by a function that depends only on the degree of those polynomials.*

8.4.3. *Prove that $\deg \xi_{v,V}$ is independent of the choice of sufficiently small V around an isolated zero of the vector field v.*

8.4.4. *Suppose that v is a vector field with an isolated zero at x_0 and that there exist $\epsilon > 0$ and a neighborhood U of x_0 such that the flow φ^t generated by v does not have periodic orbits of period $\leq \epsilon$ in U. Show that $\operatorname{ind}_v x_0 = \operatorname{ind}_{\varphi^\epsilon} x_0$.*

8.4.5. *Let $P(x,y) = \prod_{i=1}^n (\alpha_i x + \beta_i y)$, where $\alpha_i, \beta_i \in \mathbb{R}$ and α_i/β_i are pairwise distinct. Show that $\operatorname{ind}_{v_P} 0 = 1 - n$ for the Hamiltonian vector field v_P of P.*

8.4.6. *Let $P(x,y) = xy(x+y) + Q(x,y)$, where Q is a homogeneous polynomial of degree at least four. Show that $\operatorname{ind}_{v_P} 0 = -2$.*

5. The role of smoothness: The Shub–Sullivan Theorem

A fixed point of a map f is of course also a fixed point for any iterate of f. A natural question then arises as to whether the indices of the fixed points under the iterates of f are related to that of f or whether they can even be controlled. There are two answers to this question. One is that exponential growth can indeed occur:

Example. Consider the familiar example $g: S^2 \to S^2$, $z \mapsto z^2/(2|z|)$ on the Riemann sphere. Since for $|z| = 1$ the vector $z - g^n(z)$ always points "inside" the unit circle, we have $\operatorname{ind}_{g^n}(\infty) = 2^n$: Around ∞ g is given by $g(w) = 2w^2/w$ in the coordinate $w = 1/z$. But then on the unit circle $|w| = 1$ we have $\dfrac{w - g^n(w)}{|w - g^n(w)|} = \dfrac{w - 2^n w^{2^n}}{|w - 2^n w^{2^n}|}$ which is homotopic to the leading term w^{2^n} via $\dfrac{tw - 2^n w^{2^n}}{|tw - 2^n w^{2^n}|}$, $t \in [0,1]$, when $n \geq 1$. But the leading term $w \mapsto -w^{2^n}$ has degree 2^n.

The other answer is that for smooth maps this does not happen and that indeed the index under the iterates is even bounded.

Theorem 8.5.1. (Shub–Sullivan Theorem)[1] *Let $U \subset \mathbb{R}^m$ be open, $0 \in U$, and $f: U \to \mathbb{R}^m$ continuously differentiable. If 0 is an isolated fixed point for all f^n $(n \in \mathbb{N})$ then $\{\operatorname{ind}_{f^n}(0) \mid n \in \mathbb{N}\}$ is bounded.*

Proof. We begin with two easy observations. First of all $\sum_{i=0}^{n-1} Df^i$ is singular if and only if $n = jk$, where $k > 1$ and Df has a primitive kth root of unity as an eigenvalue:

Lemma 8.5.2. *If* $A := D_0 f$ *and*

$$\mathcal{R}_i^f := \{k \in \mathbb{N} \mid k > 1, \exists \text{ primitive } k\text{th root } \mu \text{ of } 1 \text{ with } \det(\mu \operatorname{Id} - A^i) = 0\}$$

denotes the orders higher than one of primitive roots of unity in the spectrum of A^i *for* $i \in \mathbb{N}$ *and* $\mathbb{N}\mathcal{R}_i^f := \{jk \mid j \in \mathbb{N}, k \in \mathcal{R}_i^f\}$ *then* $\sum_{i=0}^{n-1} A^i$ *is invertible for all* $n \notin \mathbb{N}\mathcal{R}_1^f$.

Proof. If $\sum_{i=0}^{n-1} A^i$ is singular then so is $\operatorname{Id} - A^n = (\sum_{i=0}^{n-1} A^i)(\operatorname{Id} - A)$, and A^n has one as an eigenvalue, that is, $n = jk$ such that there is a kth primitive root of unity in the spectrum of A. We need to check that one can take $k > 1$. If not then 1 is an eigenvalue of A and $\mathbb{R}^m = E_1 \oplus E'$ where $E_1 := \{v \in \mathbb{R}^m \mid (A^n - \operatorname{Id})v = 0 \text{ for some } n\}$ is the root space of 1 and E' is the sum of the remaining root spaces, so $\sum_{i=0}^{n-1} A^i$ decomposes as a product over $\mathbb{R}^m = E_1 \oplus E'$. By construction the action on E_1 is invertible and the previous argument applies on E', now necessarily yielding $k > 1$. □

These k are the divisors of n in $\mathcal{R}_1^f$. We now get "good" values of n from these:

Lemma 8.5.3. *Fix* $n \in \mathbb{N}$ *and let* $l_n := \operatorname{lcm}\{k \in \mathcal{R}_1^f \mid k \mid n\}$ *be the least common multiple of the divisors of* n *in* $\mathcal{R}_1^f$. *Then* $n/l_n \notin \mathbb{N}\mathcal{R}_{l_n}^f$.

Proof. If λ is primitive kth root of unity in $\operatorname{spec}(A)$ then λ^l is a primitive $(\frac{k}{\gcd(k,l)})$th root of unity in $\operatorname{spec}(A^l)$, and vice versa; hence if $\mathcal{R}_1^f = \{k_1, k_2, \dots\}$ then $\mathcal{R}_{l_n}^f = \left\{\frac{k_1}{\gcd(k_1, l_n)}, \frac{k_2}{\gcd(k_2, l_n)}, \dots\right\} \smallsetminus \{1\}$. So if $\frac{n}{l_n} = j\frac{k_i}{\gcd(k_i, l_n)}$ then $n = \frac{jl_n}{\gcd(k_i, l_n)} k_i$, that is, $k_i \mid n$, hence $k_i \mid l_n$ by definition of l_n, and $\gcd(k_i, l_n) = k_i$ so $\frac{k_i}{\gcd(k_i, l_n)} = 1 \notin \mathcal{R}_{l_n}^f$. □

These observations reduce the proof of the theorem to the following key step:

Lemma 8.5.4. *If* $\sum_{i=0}^{n-1} (D_0 f)^i$ *is invertible then* $|\operatorname{ind}_{f^n} 0| = |\operatorname{ind}_f 0|$.

This implies the theorem: Fix $n \in \mathbb{N}$ and let $m = n/l_n \in \mathbb{N}$, $g = f^{l_n}$. Then $\mathcal{R}_1^g = \mathcal{R}_{l_n}^f$ and by Lemma 8.5.3 we have $m \notin \mathbb{N}\mathcal{R}_{l_n}^f = \mathbb{N}\mathcal{R}_1^g$, so $\sum_{i=0}^{m-1} (D_0 g)^i$ is invertible by Lemma 8.5.2 and $|\operatorname{ind}_{f^n} 0| = |\operatorname{ind}_{g^m} 0| = |\operatorname{ind}_g 0| = |\operatorname{ind}_{f^{l_n}} 0|$ by Lemma 8.5.4. Notice, however, that by construction $l_n \le \operatorname{lcm} \mathcal{R}_1^f$ for all n. Thus $|\operatorname{ind}_{f^n} 0| \le \max\{|\operatorname{ind}_{f^l} 0| \mid l \le \operatorname{lcm} \mathcal{R}_1^f\}$ for all n. □

Proof of Lemma 8.5.4. Let $A := D_0 f$, $\Delta_i := f^i - A^i$, $\mathcal{A}_n := \sum_{i=0}^{n-1} A^i$, $\mathcal{F} := \operatorname{Id} - f$, $\mathcal{E}_n := \mathcal{A}_n \Delta_1 - \Delta_n$. Then

$$\begin{aligned} \operatorname{Id} - f^n = \operatorname{Id} - A^n - \Delta_n &= \Big(\sum_{i=0}^{n-1} A^i\Big)(\operatorname{Id} - A) - \Delta_n \\ &= \mathcal{A}_n(\operatorname{Id} - f) + \mathcal{A}_n \Delta_1 - \Delta_n = \mathcal{A}_n \mathcal{F} - \mathcal{E}_n. \end{aligned}$$

Let $\gamma_n = \|\mathcal{A}_n^{-1}\|^{-1}$.

Lemma 8.5.5. *There exists $\epsilon > 0$ such that $\|\mathcal{E}_n(x)\| < \gamma_n\|\mathcal{F}(x)\|$ for $\|x\| < \epsilon$.*

This implies

$$\|(\mathrm{Id} - f^n)(x) - \mathcal{A}_n\mathcal{F}(x)\| = \|\mathcal{E}_n(x)\| < \gamma_n\|\mathcal{F}(x)\| < \|\mathcal{A}_n\mathcal{F}(x)\|,$$

that is, $\mathrm{Id} - f^n$ and $\mathcal{A}_n\mathcal{F}$ are straight-line homotopic in $B(0,\epsilon) \setminus \{0\}$. Thus, after decreasing ϵ a little, the maps of S_ϵ given by $v_{f^n,\epsilon}(x) := \epsilon\dfrac{x - f^n(x)}{\|x - f^n(x)\|}$ and $v_{(\mathrm{Id} - \mathcal{A}_n\mathcal{F}),\epsilon}(x) := \epsilon\dfrac{\mathcal{A}_n\mathcal{F}(x)}{\|\mathcal{A}_n\mathcal{F}(x)\|}$ are homotopic and hence have the same degree, which by definition is also the index of f^n. On the other hand, $\mathcal{A}_n$ is invertible so $|\operatorname{ind}_{f^n} 0| = |\deg(v_{(\mathrm{Id} - \mathcal{A}_n\mathcal{F}),\epsilon})| = |\deg(v_{(\mathrm{Id} - \mathcal{F}),\epsilon})| = |\operatorname{ind}_f 0|$, once ϵ is small enough. Lemma 8.5.4 then follows. □

Proof of Lemma 8.5.5. We prove the following statement by induction on n:

For all $n \in \mathbb{N}$ there exist $\delta_n > 0$ such that

$$\|\mathcal{E}_n(x)\| < \gamma_n\|\mathcal{F}(x)\| \quad \text{when } \|x\| < \delta_n$$

and

$$\|(\mathrm{Id} - f^n)(x)\| \le \Big(\gamma_n + \sum_{j=0}^{n-1}\|A^j\|\Big)\|\mathcal{F}(x)\|.$$

For $n = 0$ and 1 this holds since $\mathcal{E}_0 = \mathcal{E}_1 = 0$. Suppose n is such that both of these hold for all $i < n$. Writing

$$\begin{aligned}\Delta_{i+1} = f^{i+1} - D_0 f^{i+1} &= f \circ f^i - D_0 f \circ f^i + D_0 f \circ f^i - D_0 f \circ D_0 f^i \\ &= \Delta_1 \circ f^i + A \circ \Delta_i\end{aligned}$$

yields $\Delta_n = \sum_{i=0}^{n-1} A^{n-i-1} \circ \Delta_1 \circ f^i$ and hence

$$\mathcal{E}_n = \sum_{i=0}^{n-1} A^i \circ \Delta_1 - \Delta_n = \sum_{i=1}^{n-1} A^{n-i-1}(\Delta_1 - \Delta_1 \circ f^i).$$

Now take δ_n such that

$$\|D\Delta_1\|_{\delta_n} := \sup_{\|x\|<\delta_n} \|D_x\Delta_1\| < \gamma_n\Big(\big(\max_{0\le j\le n-2}\|A^j\|\big)\Big(\sum_{i=1}^{n-1} K_i\Big)\Big)^{-1}.$$

Then by the Mean Value Theorem

$$\begin{aligned}\|\mathcal{E}_n(x)\| &\le \|D\Delta_1\|_{\delta_n}\sum_{i=1}^{n-1}\big(\|A^{n-i-1}\| \cdot \|(\mathrm{Id} - f^i)(x)\|\big) \\ &\le \|D\Delta_1\|_{\delta_n}\big(\max_{0\le j\le n-2}\|A^j\|\big)\left(\sum_{i=1}^{n-1} K_i\|\mathcal{F}(x)\|\right) < \gamma_n\|\mathcal{F}(x)\|.\end{aligned}$$

We also have

$$\begin{aligned}\|(\mathrm{Id}-f^n)(x)\| &= \|\mathcal{A}_n\mathcal{F}(x)+\mathcal{E}(x)\|\\ &\le \|\mathcal{A}_n\|\|\mathcal{F}(x)\|+\gamma_n\|\mathcal{F}(x)\| \le \Big(\gamma_n+\sum_{j=0}^{n-1}\|A^j\|\Big)\|\mathcal{F}(x)\|.\end{aligned}$$

□

6. The Lefschetz Fixed-Point Formula and applications

In our discussion of the degree of circle maps we found that the degree can be used to infer the existence of (many) periodic points. Essentially we counted the number of intersection points of the graph with the shifted diagonals in the product $\mathbb{R}\times\mathbb{R}$ of the universal cover with itself. A more sophisticated version of the same idea is applicable in greater generality and uses the index of a fixed point as a principal tool. The means here is the Lefschetz Fixed-Point Formula which relates the action of a map f on homology groups to the sum of the indices of the fixed points. This establishes a deep connection between the global behavior of a map, as seen on the homology groups on the one hand, and the local behavior at fixed points, as described by their index. In particular if one knows that the indices of fixed points cannot be large (for example, by the Shub–Sullivan theorem), this will yield a lower bound on the number of fixed points, hence periodic points via the iterates of f.

The relevant information about the action of f on homology groups is encoded in the Lefschetz number of f:

Definition 8.6.1. Let M be a compact manifold and $H_k(M,\mathbb{Q})\simeq\mathbb{Q}^{\beta_k}$ the free part of the kth homology group of M, $0\le k\le\dim M$. Let $f_{*k}\colon H_k(M,\mathbb{Q})\to H_k(M,\mathbb{Q})$ be the action of f on $H_k(M,\mathbb{Q})$ and note that f_{*k} has a matrix representation. Then

$$L(f):=\sum_{k=0}^{\dim M}(-1)^k\operatorname{tr} f_{*k}$$

is called the Lefschetz number of f.

Theorem 8.6.2. (Lefschetz Fixed-Point Formula) *Let M be a compact manifold, possibly with boundary, and $f\colon M\to M$ a continuous map all of whose fixed points are isolated. Then*

$$L(f)=\sum_{x\in\mathrm{Fix}(f)}\mathrm{ind}_f\, x. \tag{8.6.1}$$

The most obvious use of the Lefschetz Fixed-Point Formula in dynamics is the following criterion:

Corollary 8.6.3. *A continuous map $f: M \to M$ of a compact manifold M (possibly with boundary) for which $L(f^n) \neq 0$ has a periodic point of period n.*

Here is a simple example of how this works:

Corollary 8.6.4. *Every null-homotopic map of a connected compact manifold (possibly with boundary) has a fixed point.*

Proof. For a null-homotopic map f the maps f_{*k} for $k \geq 1$ are zero. For a connected manifold $H_0(M, \mathbb{Q}) = \mathbb{Q}$ and $f_{*0} = \mathrm{Id}$. Thus $L(f) = 1$. □

Since every map of a closed ball is null-homotopic we obtain

Theorem 8.6.5. (Brouwer Fixed-Point Theorem) *Any map of a closed ball has a fixed point.*

Another important special case is that of maps homotopic to the identity. If $f: M \to M$ is homotopic to the identity then $f_{*k} = \mathrm{Id}$ on $H_k(M, \mathbb{Q}) \sim \mathbb{Q}^{\beta_k}$ and hence $\operatorname{tr} f_{*k} = \beta_k$. In particular $L(f) = L(\mathrm{Id}) = \sum_{k=0}^{\dim M} (-1)^k \beta_k$ and, moreover, since f^n is also homotopic to the identity,

$$L(f^n) = \sum_{k=0}^{\dim M} (-1)^k \beta_k$$

independently of n. The Lefschetz Fixed-Point Formula thus yields

$$\sum_{x \in \mathrm{Fix} f^n} \operatorname{ind}_{f^n} x = \sum_{k=0}^{\dim M} (-1)^k \beta_k$$

independently of n. Recall that by the Euler–Poincaré Formula (Proposition A.7.7) $\sum_{k=0}^{\dim M} (-1)^k \beta_k = \chi(M)$ is the Euler characteristic of M which can be calculated directly by counting simplices in any triangulation of M. Consequently for maps homotopic to the identity with only isolated fixed points

$$\sum_{x \in \mathrm{Fix} f^n} \operatorname{ind}_{f^n} x = \chi(M).$$

Since time-t maps for flows are by construction homotopic to the identity and the indices for isolated fixed points for the flow are equal to the indices of the time-t map by Exercise 8.4.4, we have in particular the following:

Theorem 8.6.6. (Poincaré–Hopf Index Theorem) *For a smooth flow $\varphi^t: M \to M$ with isolated fixed points we have*

$$\sum_{x \in Fix(\varphi^t)} \operatorname{ind}_{\varphi^t} x = \chi(M).$$

Corollary 8.6.7. *Let M be a compact differentiable manifold without boundary and $f: M \to \mathbb{R}$ a C^2 function with isolated critical points. Then the sum of the indices of the critical points is equal to the Euler characteristic $\chi(M)$ of M.*

Proof. Apply Theorem 8.6.6 to the gradient flow for f. □

For manifolds with a particularly simple structure of the homology groups the Lefschetz number can be easily calculated. For example, for maps of the sphere we obtain the following result which should be compared with the discussion of degree for maps of S^1, in particular the number of periodic points obtained there.

Proposition 8.6.8. *If $f: S^m \to S^m$ then $L(f^n) = 1 + (-1)^m (\deg f)^n$.*

Proof. For S^m the Betti numbers are $\beta_0 = \beta_m = 1$ and $\beta_k = 0$ for $0 < k < m$. Thus f_{*0} and f_{*m} are the only nontrivial actions of f and they are given by the 1×1 matrices 1 and $\deg f$, respectively, so $L(f^n) = \sum_{k=0}^{\dim M} (-1)^k \operatorname{tr}(f_{*k})^n = 1 + (-1)^m \deg f^n$. □

Corollary 8.6.9. *If m is even and $f: S^m \to S^m$ then $P_2(f) > 0$.*

Corollary 8.6.10. *Maps of S^m with $|\deg f| \geq 2$ have exponential growth of $L(f^n)$ with n.*

Thus one can try to use the Lefschetz Fixed-Point Formula and its corollaries to establish growth rather than just existence of periodic orbits. Here is a typical specific example (which we discussed in Section 8.2c) where the Lefschetz Fixed-Point Formula gives periodic points:

Example. Consider the map $f: S^2 \to S^2$, $z \mapsto z^2$ on the Riemann sphere $\hat{\mathbb{C}} = \mathbb{C} \cup \{\infty\}$, the one-point compactification of the complex plane. $\deg f = 2$ since f covers the sphere twice. Consequently $L(f^n) = 1 + 2^n$ and as we showed in Section 8.2, $P_n(f) = 2 + 2^n - 1 = 2^n + 1 = L(f^n)$. This implies that all periodic points have the same index, as explained earlier. On the other hand our familiar counterexample does not have as many periodic points as the Lefschetz number would suggest: $g: S^2 \to S^2$, $z \mapsto z^2/(2|z|)$ has degree 2. By the Lefschetz Fixed-Point Formula

$$1 + 2^n = L(g^n) = \sum_{x \in \operatorname{Fix} g^n} \operatorname{ind}_{g^n} x = \operatorname{ind}_{g^n} 0 + \operatorname{ind}_{g^n} \infty.$$

What is happening is that the index $\operatorname{ind}_{g^n}(\infty) = 2^n$ of ∞ grows exponentially.

This example shows a difficulty in using the Lefschetz Fixed-Point Formula to obtain many periodic points: A single periodic point may be able to absorb the growth of the Lefschetz number by having unbounded index. But this phenomenon of having one fixed point absorb most of the Lefschetz number depends on the essential nonsmoothness of g at ∞. The Shub–Sullivan Theorem 8.5.1 shows that for differentiable maps $\operatorname{ind}_{f^n} x$ is bounded independently of n for each x. This, then, often implies that there are infinitely many periodic points. In particular now we are able to state corollaries of the previous results and the Shub–Sullivan Theorem 8.5.1.

Corollary 8.6.11. *Let M be a compact manifold and $f: M \to M$ continuously differentiable such that $\overline{\lim}_{n\to\infty} L(f^n) = \infty$. Then f has infinitely many periodic points.*

Combining Corollary 8.6.10 and Corollary 8.6.11 we obtain

Corollary 8.6.12. *Any C^1 map $f: S^m \to S^m$ with $|\deg(f)| \geq 2$ has infinitely many periodic orbits.*

Note that the Shub–Sullivan theorem asserts boundedness of the index of a given periodic point, but not independently of the point. In other words it leaves open the question of whether $\{\mathrm{ind}_{f^n}(p) \mid n \in \mathbb{N}, p \in M$ an isolated fixed point of $f^n\}$ is bounded. Together with Corollary 8.6.14 this motivates asking the following

Question. *Does the number of periodic points of a smooth map always grow as rapidly as $L(f^n)$, that is, is $\lim_{n\to\infty}(1/n)\log|L(f^n)| \leq p(f)$?*

Coming back to maps of the sphere we notice that in dimension at least two the Misiurewicz–Przytycki Theorem shows that for smooth maps with $|\deg(f)| \geq 2$ the dynamical behavior is complicated, whereas for continuous maps this may not be so. That theorem, however, does not give any information about the growth of periodic orbits. In particular the previous question can be specialized to the case of the sphere:

Question. *Let $f: S^n \to S^n$ be a C^1 map. Is $p(f) \geq \log|\deg(f)|$?*

On the other hand under certain genericity assumptions the Lefschetz Fixed-Point Formula can be used to estimate the growth of the number of periodic orbits.

Corollary 8.6.13. *If $|\mathrm{ind}_{f^n} x| = 1$ for all $x \in \mathrm{Fix}(f^n)$ then $P_n(f) \geq |L(f^n)|$.*

If f is differentiable and $D_x f^n$ has no eigenvalue 1 then, by Proposition 8.4.6, $|\mathrm{ind}_{f^n} x| = 1$. The former condition is transversality of the periodic point. Since generically for differentiable maps all periodic points are transverse (Theorem 7.2.4), Corollary 8.6.13 answers the preceding questions for generic maps:

Corollary 8.6.14. *Generically $L(f^n) \leq P_n(f)$ and $\lim\limits_{n\to\infty} \log|L(f^n)|/n \leq p(f)$.*

Corollary 8.6.15. *C^r maps $f: S^n \to S^n$ generically have $p(f) \geq \log|\deg(f)|$.*

A more specific result is

Corollary 8.6.16. *If $\epsilon = \pm 1$ and $\mathrm{ind}_{f^n} x = \epsilon$ for all $x \in \mathrm{Fix}(f^n)$ then $P_n(f) = \epsilon L(f)$.*

Since $(f^n)_{*k} = {f_{*k}}^n$ one can expect $L(f^n)$ to grow exponentially with n in many cases (see Exercise 8.6.6). Therefore the preceding corollaries often imply at least exponential growth of the number of periodic points. On the other hand in general it is difficult to obtain upper estimates on the number of periodic points even for Kupka–Smale maps due to the difficulties of controlling cancellations of indices for periodic points of growing period (see Section 7.4). This reflects an essential difference between differential topology which operates with algebraic quantities and differentiable dynamics which focuses on geometric information which may or may not be related to the algebraic data. One case where the geometric picture can be described in some algebraic terms is that of locally maximal hyperbolic sets defined in Section 6.4d and studied in detail in Chapter 18.

Exercises

8.6.1. *Compute the Lefschetz number for circle maps.*

8.6.2. (Fuller) *Let f be a homeomorphism of a compact manifold having nonzero Euler characteric. Show that f has a periodic point of period at most* $\max(\sum \beta_{2k}, \sum \beta_{2k-1})$.

8.6.3. *Show that the only compact orientable surface S that admits an Anosov diffeomorphism is the torus.*

8.6.4. *Suppose $f: M \to M$ is an Anosov diffeomorphism and the unstable bundle E^+ is orientable in the following sense: One can assign a sign to each frame in an unstable subspace E_x^+ in such a way that it is constant when moving the frame continuously.*

Show that the ζ-function (3.1.3) of f is rational.

8.6.5*. *Show that no sphere admits a volume-preserving Anosov diffeomorphism.*

8.6.6. *Formulate and prove a version of Theorem 8.6.6 for manifolds with boundary.*

8.6.7. *Show that* $\overline{\lim}_{n\to\infty}(1/n)\log L(f^n) = \max_{0\le i\le \dim M} r(f_{*i})$.

7. Nielsen theory and periodic points for toral maps

In certain cases index-type considerations force exponential growth of the number of periodic points. In the most general setting this kind of question is the subject of Nielsen theory which combines homology and homotopy by considering indices of fixed points for various lifts of a given map to the universal cover. Within the limited collection of manifolds that we specifically discuss in this book this theory yields nontrivial results for maps of tori of arbitrary dimension and surfaces of higher genus. Here we will focus on maps of tori, where the main ideas of Nielsen theory can be presented in a very visual way with little topological complications.

The most elementary example of the statement provided by Nielsen theory is Proposition 8.2.4 which gives an optimal estimate for the number of fixed points of a circle map through purely topological data. The principal purpose of this section is to generalize this Proposition to maps of tori in higher dimensions.

Every toral map $f: \mathbb{T}^n \to \mathbb{T}^n$ is determined up to homotopy by its action f_* on the fundamental group $\mathbb{Z}^n$ which in this case coincides with the first homology group. This action is given by an integer $n \times n$ matrix A which also determines a unique linear map F_A in the homotopy class of f. Our standing assumption will be that F_A has isolated fixed points, which is equivalent to A having no eigenvalue one, or

$$\det(A - \mathrm{Id}) \neq 0.$$

Without this assumption no fixed points can be guaranteed (Exercise 8.7.2). First we observe that the indices of all fixed points of F_A are the same, namely, $\operatorname{sign}\det(\mathrm{Id}-A)$ by Proposition 8.4.6. Thus by the Lefschetz Fixed-Point Formula $L(F_A) = \operatorname{sign}\det(\mathrm{Id}-A)\operatorname{card}\operatorname{Fix}(F_A)$.

Another way to calculate the number of fixed points of F_A is to use the argument from the proof of Proposition 1.8.1, which contains the calculation for the specific case of the matrix $\begin{pmatrix} 2 & 1 \\ 1 & 1 \end{pmatrix}$: If $F_A(x) = x$ for some $x \in \mathbb{T}^n$, then for any $v \in \mathbb{R}^n$ that projects to x we have $Av - v \in \mathbb{Z}^n$. Thus $x = \pi v$, where π is the canonical projection, is a preimage of 0 under the (noninvertible) map $F_{\mathrm{Id}-A}$. But the number of such preimages is exactly equal to the absolute value of the degree of that map, which equals $\det(\mathrm{Id}-A)$ by the volume-form definition. Thus

$$L(F_A) = \det(\mathrm{Id}-A). \tag{8.7.1}$$

Different fixed points of F_A can be identified as follows: Consider different lifts of the map F_A to the universal cover. These lifts differ by deck transformations, that is, integer translations. Thus each lift can be uniquely represented as $v \to Av + m$, where $m \in \mathbb{Z}^n$. If v is a fixed point for any lift then its projection is a fixed point for F_A. Of course, v's that differ by an integer vector represent the same fixed point x. If $Av + m = v$ then $m = (\mathrm{Id}-A)v$. Now suppose m_1 and m_2 are such that the corresponding fixed points v_i for the lifts $v \mapsto Av + m_i$, $i = 1, 2$, project to the same $x \in \mathbb{T}^n$. Thus means that $m_1 - m_2 = (\mathrm{Id}-A)k$ for some $k \in \mathbb{Z}^n$. Thus all integer vectors m, and hence all lifts, are divided into equivalence classes such that the fixed points of two equivalent lifts project to the same fixed point of F_A on $\mathbb{T}^n$.

Before proceeding to the general discussion about bounding the number of periodic points by the Lefschetz number, let us see what happens for perturbations of linear maps. By Proposition 1.1.4 any periodic point of F_A persists under sufficiently small C^1-perturbations and its index remains the same as well. It is remarkable that if A is hyperbolic then in fact sufficiently small perturbations preserve the indices of *all* periodic points. This can be seen as follows: By structural stability (Theorem 2.6.3 for the two-dimensional case which generalizes verbatim to arbitrary dimension) the number $P_n(f)$ of points of period n is constant for any f close enough to F_A and by Corollary 6.4.7 they are all hyperbolic and hence all have the same index since $L(f) = L(F_A)$.

Now let us consider an arbitrary map f homotopic to F_A. Take any lift F of f. It has the form $v \mapsto Av + m + g$ for some $\mathbb{Z}^n$-periodic $g: \mathbb{R}^n \to \mathbb{R}^n$. As before lifts of fixed points are classified by the integer vector m, although the fixed point of F may not be unique any more, and the different m_i give the same class if $m_1 - m_2 = (\mathrm{Id}-A)k$. We will show that for each equivalence class of m's there is a lift that has at least one fixed point. Take a ball B in $\mathbb{R}^n$ around the fixed point of the affine map $v \mapsto Av + m$ such that the map $v_{F,B}: v \mapsto \dfrac{v - F(v)}{|v - F(v)|}$ to the unit sphere S^{n-1} is never antipodal to the corresponding map $v_{A+m,B}$. Since F is a bounded perturbation of $A + m$ this is always true if B is large enough. The degree of $v_{F,B}$ is then the same as that of $v_{A+m,V}$, which in turn is the index of the unique fixed point of $A+m$ on $\mathbb{R}^n$ which equals $\det(\mathrm{Id}-A)$. Consequently F has a fixed point in B. Thus we have proved the following result which can be viewed as a prototype of the results of Nielsen theory.

Theorem 8.7.1. *Let $f\colon \mathbb{T}^n \to \mathbb{T}^n$ be a continuous map such that $f_*\colon \mathbb{Z}^n \to \mathbb{Z}^n$ does not have one as an eigenvalue. Then f has at least $|\det(\mathrm{Id}-f_*)|$ different fixed points.*

More generally let $f\colon M \to M$ be a continuous map of a connected manifold M and $\tilde{M}$ the universal cover of M. Then the fundamental group acts by deck transformations γ. Two lifts F_1, F_2 can then be called equivalent if there exists a deck transformation γ such that $F_2 = \gamma^{-1}F_1\gamma$. Let us denote this relation by $\sim$. The projections of the fixed points of F_1 and F_2 then coincide and give a (possibly empty) equivalence class of fixed points of f. The union of these classes gives all fixed points of f. A class of fixed points is called *essential* if the sum of the indices of their preimages under the projection is nonzero for any lift on the cover. Notice that for a toral map f all classes are essential if $\det(\mathrm{Id}-f_*) \neq 0$ and nonessential otherwise. The central property of these classes is the homotopy invariance of the sum of the indices. In the case of a compact universal cover (which is not a particularly interesting setting for Nielsen theory) this is a consequence of the Lefschetz Fixed-Point Formula. The number of essential fixed-point classes is called the *Nielsen number* of f and denoted $N(f)$. It is a homotopy invariant which can often be calculated by considering particularly simple models in a homotopy class. However, this is much less algorithmic than calculating the Lefschetz number. Thus the Nielsen number gives a lower estimate for the number of fixed points of a map. In fact, in most cases (in particular for the torus) this estimate is optimal as an estimate for a homotopy class of maps.

The equivalence classes of fixed points can be defined directly on the manifold M as follows: Two fixed points p and q of f are said to be equivalent ($p \sim' q$) if there exists an arc c connecting p and q such that c and $f(c)$ are homotopic via a homotopy fixing the endpoints. In Exercises 8.7.4–8.7.6 we see that this definition coincides with the previous one.

Exercises

8.7.1*. *Calculate the Lefschetz number of F_A and find $p(F_A)$ for any integer matrix A using Definition 8.6.1.*

8.7.2. *Show that if 1 is an eigenvalue of an integer matrix A then there exists a map g homotopic to F_A without fixed points.*

8.7.3*. *Consider a toral homeomorphism $f\colon \mathbb{T}^n \to \mathbb{T}^n$ such that $f_*\colon \mathbb{Z}^n \to \mathbb{Z}^n$ does not have one as an eigenvalue and has an eigenvalue that is not a root of unity. Show that $p(f) > 0$.*

8.7.4. *Let $f: M \to M$ and $p \sim' q$ be fixed points. Show that there exists a lift F of f with fixed points p' and q' projecting to p and q, respectively.*

8.7.5. *Let $f: M \to M$ and $p \sim q$ be fixed points. Show that there exists a lift F of f with fixed points p' and q' projecting to p and q, respectively.*

8.7.6. *Let $f: M \to M$ and p, q be fixed points such that there exists a lift F of f with fixed points p' and q' projecting to p and q, respectively. Show that $p \sim' q$.*

8.7.4. Let f [illegible] M and [illegible] have zero points. Show that there exists a [illegible] F of f with fixed points u and [illegible] corresponding to p and q, respectively.

8.7.5. Let $f\colon M$ [illegible] fixed points. Show that there exists a lift [illegible] of f with fixed points $\bar{p}$ and $\bar{q}$ projecting to p and q, respectively.

8.7.6. Let $f\colon M$ [illegible] and p, q fixed points and [illegible] there exists a lift F of f with fixed points [illegible] and $\bar{q}$ projecting to p and [illegible]. Show that [illegible]

9

Variational aspects of dynamics

The purpose of this chapter is to introduce the variational approach to dynamics, that is, to show how interesting orbits in some dynamical systems can be found as special critical points of functionals defined on appropriate auxiliary spaces of potential orbits. This idea goes back to the variational principles in classical mechanics (Maupertuis, d'Alembert, Lagrange, etc.). The classical continuous-time case presents certain difficulties related to infinite-dimensionality of the spaces of potential orbits. In order to demonstrate the essential features of this approach and to avoid those difficulties we start in Section 2 with a model geometric problem describing the motion of a point mass inside a convex domain. Then we consider in Section 3 a more general class of area-preserving two-dimensional dynamical systems, twist maps, which possesses the essential features of that example, but covers many other interesting situations. The main result there is Theorem 9.3.7, which guarantees existence of *infinitely many* periodic orbits with a special behavior for any twist map. At least as important as that result itself is the machinery involving the action functional (9.3.7) for the periodic problem, which will be extended in Chapter 13 to give results about nonperiodic orbits. Furthermore, after developing the necessary local theory, the approach can then be refined to study continuous-time systems as well, although we only carry out the program for geodesic flows, where the action functional has a particularly clear geometric interpretation. An important ingredient here is to reduce the global problem to a finite-dimensional one by considering "broken geodesics" (cf. the proof of Theorem 9.5.8). We concentrate our attention in Sections 6 and 7 on describing the invariant set consisting of globally minimal geodesics, that is, geodesics which on the universal cover are length-minimizing segments between any two of their points. There are two principal conclusions: Theorem 9.6.7 connects the geometrical complexity of the manifold measured by the growth of the volume of balls on the universal cover with the dynamical complexity of the geodesic flow measured by the topological entropy. The second result, Theorem 9.7.2, allows us to produce infinitely many closed geodesics for an arbitrary metric on a surface of genus greater than one. Those geodesics are, in fact, closely analogous to the "minimal" Birkhoff periodic orbits of Theorem 9.3.7.

1. Critical points of functions, Morse theory, and dynamics

In Chapter 5 we outlined two different formalisms to describe the dynamical systems arising in classical mechanics. The Hamiltonian formalism leads to a dynamical system given by a system of first-order ordinary differential equations in an even-dimensional space. In this approach all coordinates appear symmetrically. On the other hand, the Lagrangian formalism involves only a set of coordinates describing the configuration space and gives the dynamics via a set of second-order ordinary differential equations. It turns out that Lagrangian systems can be described by considering "potential" trajectories of a system and obtaining from these the actual trajectories as critical points of a certain functional on the collection of curves in the configuration space. Such a description is usually called *variational* since potential trajectories have to be varied to find actual ones. The Euler–Lagrange equations (5.3.2) are precisely the equations describing the critical curves for the action functional discussed in Section 4.

The study of critical points of functions in finite- or infinite-dimensional spaces has two aspects: the local one dealing with the structure and stability of isolated critical points, and the global one, sometimes called *Morse theory*, which deals with the relation between the global topological properties of the space and the structure of critical points of functions on that space.

The prototypical finite-dimensional local result is the Morse Lemma:

Proposition 9.1.1. *Let p be a nondegenerate critical point of a C^r function, $r \geq 2$, on a smooth manifold M. Then there exist $0 \leq k \leq n$ and a local C^{r-2} coordinate system $(x_1, \dots, x_n)$ with p as the origin such that in these coordinates f is given by*

$$f(x) = f(0) + \sum_{i=1}^{k} x_i^2 - \sum_{i=k+1}^{n} x_i^2$$

Remark. The number k is called the *Morse index* of the point p.

Proof. First, without loss of generality assume that $f(0) = 0$. Second, by a linear change of coordinates we can bring the quadratic part of f into the desired form, that is, we may consider

$$f(x) = \sum_{i=1}^{n} \epsilon_i x_i^2 + g(x) \text{ with } g(x) = o\Big(\sum_{i=1}^{n} x_i^2\Big),$$

where $\epsilon_i = \pm 1$. Now we proceed by induction on the dimension by finding the desired coordinate change and showing that its linear part is the identity. For $n = 1$ we have $f(x) = \epsilon x^2 + g(x)$. Letting $h(x) = g(x)/\epsilon x^2$ and $y = x(1 + h(x))^{1/2}$ we have $f(x) = \epsilon x^2(1 + h(x)) = \epsilon y^2$ and the new coordinate y

is as desired. Now h is C^{r-2} at 0 and C^r away from 0 since $g(x) = o(x^2)$ and g is C^r. But then $(1 + h(x))^{1/2}$ is C^{r-2} since $h(0) = 0$ and therefore y is C^{r-1} (with linear part the identity).

Suppose now that the result holds for n and let $f(x_1, \dots, x_{n+1}) = \sum_{i=1}^{n+1} \epsilon_i x_i^2 + g(x)$ with $\epsilon_i = \pm 1$ and $g(x) = o(\sum_{i=1}^{n+1} x_i^2)$. By the Implicit Function Theorem the solution of $\frac{\partial f}{\partial x_{n+1}} = 0$ near 0 is the graph of a function $x_{n+1} = \varphi(x_1, \dots, x_n)$ and $D\varphi(0) = 0$. Changing coordinates to $y_i = x_i$ for $i \le n$, $y_{n+1} = x_{n+1} - \varphi(x_1, \dots, x_n)$, we may assume $\frac{\partial f}{\partial x_{n+1}} = 0$ for $x_{n+1} = 0$. Now by the induction hypothesis we can bring the restriction of f to the hyperplane $x_{n+1} = 0$ into the desired form by a C^{r-2} coordinate change of the form $y_i = \psi_i(x_1, \dots, x_n)$ for $i \le n$ such that the differential of this transformation at 0 is the identity. Thus this transformation extends to a diffeomorphism on a neighborhood of 0 (with linear part the identity) by letting $y_{n+1} = x_{n+1}$. This brings f into the form

$$f(y_1, \dots, y_{n+1}) = \sum_{i=1}^{n} \epsilon_i y_i^2 + \epsilon_{n+1} y_{n+1}^2 (1 + h(y_1, \dots, y_{n+1})),$$

where, as in the case $n = 1$, h is C^{r-2} and $h(y_1, \dots, y_n, 0) = 0$. This allows us to make the final coordinate change $z_i = y_i$, $i \le n$, $z_{n+1} = y_{n+1}(1 + h(y))^{1/2}$ which is thus C^{r-2} and brings f into the desired form. □

Another important local fact is persistence of a critical point with nonzero index (in particular, any nondegenerate point) under a small perturbation of the function in the C^1 *topology*, which follows from Theorem 8.4.4.

The simplest global fact is that a smooth function on a compact manifold has at least two critical points, one global maximum and one global minimum. A useful variation of this basic result is the following: Let f be a smooth function on a not necessarily compact manifold such that for some t the set $f^{-1}((-\infty, t])$ is compact and nonempty. Then f has at least one critical point, namely a global minimum. The proof of the existence of the first Birkhoff periodic orbit of type (p, q) in Theorem 9.3.7 is based on the last statement.

In some cases there are ways to produce further critical points which are not extrema. The most common arguments of this type are known as "mountain-pass" arguments. Let us describe a simple version.

Proposition 9.1.2. *Let M be a compact manifold, $D \subset M$ a open set with compact closure, and $f: M \to \mathbb{R}$ a C^1 function with the following properties:*

(1) *f has two local minima p and q in D.*
(2) *There is a path γ in D connecting p and q such that*

$$\sup\{f(x) \mid x \in \gamma\} \le \inf\{f(x) \mid x \in \partial D\}.$$

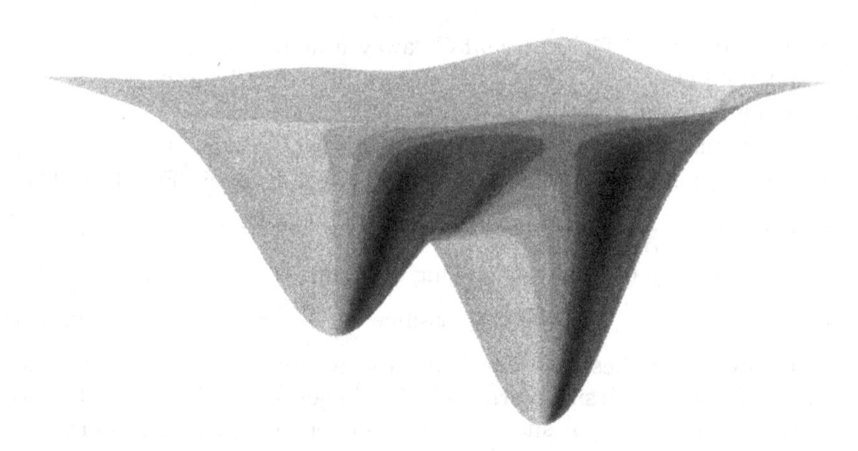

FIGURE 9.1.1. A mountain pass

Then f has another critical point in D which is not a local minimum.

Let us sketch a proof of this fact. Without loss of generality we may assume that γ is piecewise C^1. Consider the space $\mathcal{L}$ of all Lipschitz paths in $\bar{D}$ connecting p and q with sufficiently large Lipschitz constant. $\mathcal{L}$ is compact in the uniform topology by the Arzelá–Ascoli Theorem A.1.24. Thus the function

$$\mathcal{F}(c) := \max\{f(x) \mid x \in c\}$$

on $\mathcal{L}$, being continuous in the uniform topology, attains a global minimum. By (2) at least one path on which the minimum is attained lies in D. Let c_0 be such a path. The maximum along c_0 is attained on a closed subset C of c_0. If C does not contain critical points then move c_0 with the gradient flow for $-f$ reparameterized with a nonnegative function that is positive on C and whose support is disjoint from the set of critical points of f. The new path has a smaller maximum than c_0, a contradiction.

The argument in the proof of Theorem 9.3.7 establishing the existence of the second (minimax) Birkhoff periodic orbit of type (p, q) is, in fact, an example of a mountain-pass–type argument, as one sees from Proposition 9.3.9.

Other basic global results of the finite-dimensional Morse theory include Corollary 8.6.7 as well as the fact that any smooth function on any compact manifold other than the sphere has at least three critical points.

In the case of a continuous-time dynamical system the collection of all potential orbits will, in any reasonable setting, be an infinite-dimensional space. Already the local analysis of critical points of functionals in infinite-dimensional spaces presents technical problems, which we will defer to Section 5 by first considering discrete-time systems that can be described variationally and where the space of potential orbit segments of finite length is finite-dimensional. In some cases infinite orbits can then be described either directly (in the periodic case) or indirectly, by considering appropriate limits of finite segments with properly chosen boundary conditions.

Exercises

9.1.1. *Give a detailed proof of Proposition 9.1.2.*

9.1.2. *Prove that if the dimension of M is greater than 1 then Proposition 9.1.2 produces a critical point that is not a local maximum.*

9.1.3. *Suppose M is a compact smooth surface and f is a smooth function (not necessarily a Morse function) with exactly two critical points. Prove that M is homeomorphic to the sphere.*

9.1.4. *What is the minimum number of critical points for a Morse function on $\mathbb{T}^2$?*

9.1.5*. *Construct a C^∞ function on $\mathbb{T}^n$ with $n+1$ critical points.*

2. The billiard problem

Let us consider the motion of a point mass (or a light ray) inside a convex bounded region D in the plane with smooth boundary B. The orbits of such motion consist of straight line segments inside D joined at boundary points according to the rule that the angle of incidence equals the angle of reflection. The speed of the motion is constant. Since D is bounded, the time between successive collisions with the boundary is bounded, as well. The phase space of this system is conveniently described as the set of all tangent vectors of fixed length (for example, unit length) supported at points of the interior of D together with vectors at boundary points pointing inward. Natural coordinates are given by Euclidean coordinates (x_1, x_2) of the base point and the cyclic angular coordinate α.

The reflections produce discontinuities at the boundary. However, this continuous-time system has a natural global section C in the sense of Section 0.3 on which the return map is continuous if B does not contain line segments. It consists of all inward-pointing vectors on B. Topologically C is a cylinder parameterized by the cyclic length parameter s along B and the angle $\theta \in [0, \pi]$ with the positive tangent direction.

The Poincaré map $f: C \to C$ on the section is usually called the *billiard-ball map* or simply *billiard map* and can be described as follows: A vector $v \in C$ supported at $p \in B$ determines an oriented line l which intersects B in two points p and p'. Then $f(v)$ is a vector at p' pointing inward in the direction of the line obtained by reflecting l in the tangent line to B at p'. The natural coordinates in the phase space are the cyclic length parameter $s \in [0, L)$ on B, where L is the total length of B, and the angle $\theta \in (0, \pi)$ with the direction of the positively oriented tangent direction. In other words, we identify the boundary with the circle $\mathbb{R}/L\mathbb{Z}$.

We will see later that if B is a C^k curve then in these coordinates the map f is C^{k-1} for $1 \leq k \leq \infty$. If the curve B is strictly convex, that is, it contains no

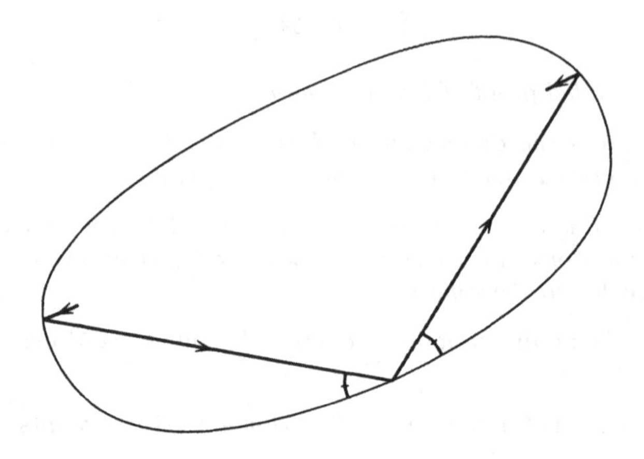

FIGURE 9.2.1. A convex billiard

straight segments then the map f extends continuously to Id on the boundary components of the cylinder. Let $f(s, \theta) = (S(s, \theta), \Theta(s, \theta))$. A calculation of S and Θ is rather unpleasant and, in fact, not necessary to understand the dynamics. We point to two important features of f.

First $S(s_0, \cdot)$ is a monotone function of θ which increases from s_0 to $s_0 + L$ (mod L) when θ changes from 0 to π. In fact, $\dfrac{\partial S}{\partial \theta} = \dfrac{h}{\sin \Theta}$, where h is the length of the chord connecting the boundary points p and P with coordinates (s, θ) and (S, Θ), respectively. Thus

$$\frac{\partial S}{\partial \theta} > 0 \tag{9.2.1}$$

for $0 < \theta < \pi$. (In addition, as shown in Exercise 9.2.4, the limit of $\dfrac{\partial S}{\partial \theta}$ as $\theta \to 0$ or π equals the radius of curvature at p.) This property is called the *twist property* and will play an important role in the subsequent discussion.

Second, f preserves the volume element $\sin \theta ds\, d\theta$. This can be seen as follows. The free motion with unit speed in the plane can be written as

$$\begin{aligned} \dot{\alpha} &= 0, \\ \dot{x}_1 &= \cos \alpha, \\ \dot{x}_2 &= \sin \alpha \end{aligned}$$

and hence is divergence free and preserves the volume form $dx_1 dx_2 d\theta$. At a boundary point p the coordinate α coincides up to an additive constant with the coordinate θ on the section C. Thus this form can be written as $dx_1\, dx_2\, d\theta = (dt\, ds\, \sin \theta)\, d\theta$, where t is the time parameter along an orbit segment. Hence it is of the form $dt(ds \sin \theta d\theta)$, that is, dt multiplied by the volume form $ds \sin \theta d\theta$

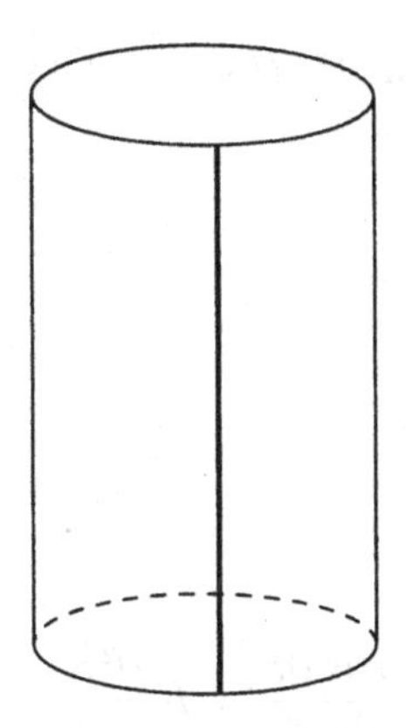
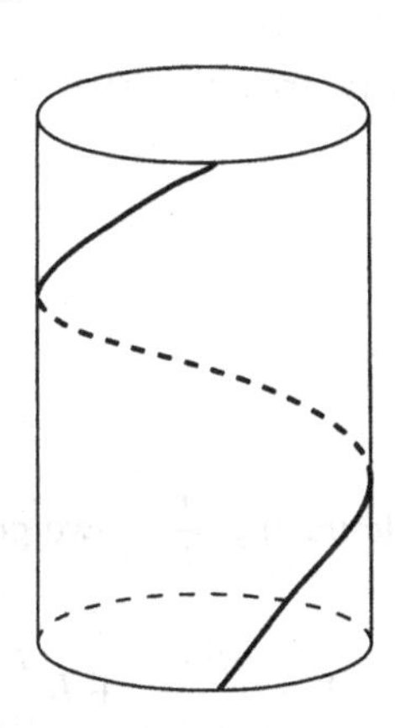

FIGURE 9.2.2. The twist property

on the section C, which in turn is invariant under the reflection in the tangent to B at p. Thus this latter form is invariant under f. (This argument is just a specific case of the argument used in the proof of Proposition 5.1.11.) In particular f preserves orientation as well. It is sometimes convenient to use the coordinate $r = -\cos\theta$ instead of θ because in these coordinates (s, r) the billiard map will preserve area and orientation. The price one pays, however, is that in these coordinates the billiard map fails to be differentiable at the boundary of the phase space, whereas in the standard coordinates it is differentiable as long as the curvature of B does not vanish.

Each orbit of the billiard map f is completely described by the sequence of points of impact on the boundary B. Let us see under which conditions a sequence of points on B corresponds to an orbit of f. Consider first two points p and p' with corresponding coordinates s and s' on B and let $H(s, s')$ be the negative of the Euclidean distance between p and p'. H is called the *generating function* for the billiard. Direct calculation (Exercise 9.2.1) shows that

$$\begin{aligned} \frac{\partial}{\partial s'} H(s, s') &= -\cos\theta', \\ \frac{\partial}{\partial s} H(s, s') &= \cos\theta, \end{aligned} \tag{9.2.2}$$

where θ' is the angle of the segment joining p and p' with the negative tangent at p' and θ is the angle of the segment joining p and p' with the positive tangent at p. If instead we use the coordinate $r = -\cos\theta$ then these equations simplify to

$$\begin{aligned} \frac{\partial}{\partial s'} H(s, s') &= r', \\ \frac{\partial}{\partial s} H(s, s') &= -r, \end{aligned} \tag{9.2.3}$$

where $r' = -\cos\theta'$. These equations can be viewed as a discrete analog of the usual Hamiltonian equations (see (1.5.5) and Section 5.5c). They also permit us to verify area preservation by a simple calculation: If $\tilde{H}(s,r) := H(s, S(s,r))$ then

$$\frac{\partial \tilde{H}}{\partial s} = -r + R\frac{\partial S}{\partial s}$$

and

$$\frac{\partial \tilde{H}}{\partial r} = R\frac{\partial S}{\partial r}$$

so by calculating $\dfrac{\partial^2 \tilde{H}}{\partial s \partial r}$ we get

$$-1 + \frac{\partial R}{\partial r}\frac{\partial S}{\partial s} + R\frac{\partial^2 S}{\partial s \partial r} = \frac{\partial^2 \tilde{H}}{\partial s \partial r} = \frac{\partial^2 \tilde{H}}{\partial r \partial s} = \frac{\partial R}{\partial s}\frac{\partial S}{\partial r} + R\frac{\partial^2 S}{\partial r \partial s},$$

and $\dfrac{\partial R}{\partial r}\dfrac{\partial S}{\partial s} - \dfrac{\partial R}{\partial s}\dfrac{\partial S}{\partial r} = 1.$

The equations (9.2.3) code the dynamics as follows. If the dynamics are given by $(s', r') = f(s,r) = (S(s,r), R(s,r))$ then we can show that (9.2.3) locally determines the functions S and R by the implicit function theorem. In order to apply the implicit function theorem to

$$0 = F(s, s', r, r') := \begin{pmatrix} \dfrac{\partial}{\partial s'} H(s,s') - r' \\ \dfrac{\partial}{\partial s} H(s,s') + r \end{pmatrix}$$

we need to check that the total derivative

$$\begin{pmatrix} \dfrac{\partial^2}{\partial s'^2} H(s,s') & -1 \\ \dfrac{\partial^2}{\partial s \partial s'} H(s,s') & 0 \end{pmatrix}$$

of F with respect to (s', r') is nonsingular, that is, that $0 \neq \dfrac{\partial^2}{\partial s \partial s'} H(s,s') = \dfrac{\partial r'}{\partial s}$, which is indeed a consequence of the twist property. To differentiate $\dfrac{\partial H}{\partial s'} = r'$ with respect to s when s' is fixed, take a unit-speed curve $c(t) = (s(t), r(t))$ such that $S(c(t)) \equiv s'$. Note that $0 = \dfrac{\partial S}{\partial s}\dfrac{ds}{dt} + \dfrac{\partial S}{\partial r}\dfrac{dr}{dt}$ so by the twist property $\dfrac{dr}{dt} = -\left(\dfrac{\partial S}{\partial s}\Big/\dfrac{\partial S}{\partial r}\right)\dfrac{ds}{dt}$, whence $\dfrac{ds}{dt}$ is never zero. Then

$$\begin{aligned} \frac{ds}{dt}\frac{\partial^2}{\partial s \partial s'} H(s,s') &= \frac{d}{dt} R \circ c = \frac{\partial R}{\partial s}\frac{ds}{dt} + \frac{\partial R}{\partial r}\frac{dr}{dt} \\ &= \frac{\partial R}{\partial s}\frac{ds}{dt} - \frac{\partial R}{\partial r}\left(\frac{\partial S}{\partial s}\Big/\frac{\partial S}{\partial r}\right)\frac{ds}{dt} = -\left(\frac{\partial S}{\partial s}\frac{\partial R}{\partial r} - \frac{\partial S}{\partial r}\frac{\partial R}{\partial s}\right)\frac{ds}{dt}\Big/\frac{\partial S}{\partial r} = -\frac{ds}{dt}\Big/\frac{\partial S}{\partial r} \end{aligned}$$

since f preserves area. Consequently by (9.2.1)

$$\frac{\partial^2}{\partial s \partial s'} H(s,s') = -1 \bigg/ \frac{\partial S}{\partial r} < 0. \tag{9.2.4}$$

This calculation should become an obvious geometric argument after some exposure to twist maps in the next section.

Let us assume that the curve B is C^k, that is, the Euclidean coordinates are C^k functions of the length parameter. Then the generating function is also C^k and by the implicit function theorem the functions S and R are C^{k-1} for $0 < r < 1$.

Now consider three points on the boundary. Unlike a pair of points, a triple does not always lie on part of an orbit. Those triples that do can be described as critical points of a certain functional. Consider three points p_{-1}, p_0, and p_1 with corresponding coordinates s_i on B. If they are part of a billiard orbit then by definition the segments joining p_{-1} with p_0 and p_0 with p_1 make the same angles with the tangent at p_0. Consequently

$$\frac{d}{ds} H(s_{-1}, s) + \frac{d}{ds} H(s, s_1) = 0 \text{ at } s = s_0, \tag{9.2.5}$$

that is, p_0 is obtained as a *critical point* of the functional $s \mapsto H(s_{-1}, s) + H(s, s_1)$. This is the first instance where we encounter the description of an orbit segment of a dynamical system as a critical point of a functional defined on a space of "potential" orbit segments of the dynamical system. This procedure can be iterated to produce orbit segments as critical point of functionals depending on several variables. This naturally requires passing to the universal cover of the phase space and will be discussed in the next section.

Now we consider some examples of billiards and discuss some elementary results based on direct geometric intuition.

Example. (The circle) Let D be the unit disk with boundary$B = \{(x,y) \mid x^2 + y^2 = 1\}$. In this case the billiard-ball map can be written explicitly as $(s', \theta') = (s + 2\theta, \theta)$, so the angle θ is an integral of motion and the phase cylinder splits into invariant circles $\theta = \theta_0$. If θ_0 is commensurable with 2π then the billiard map on the circle is periodic and the orbits correspond to inscribed star-shaped polygons. If θ_0 is incommensurable with 2π then by Proposition 1.3.3 all orbits are dense on the circle and each orbit of the billiard map in the circle is dense in the annulus $\cos^2\theta \le x^2 + y^2 \le 1$. The generating function in this case is $H(s, s') = -2 \sin \frac{1}{2}(s' - s)$.

Example. (The ellipse) Consider an elliptic region D with boundary $B = \{(x,y) \mid \frac{x^2}{a^2} + \frac{y^2}{b^2} = 1\}$. We immediately notice two special orbits of period two for the billiard map, namely, the ones corresponding to the symmetry axes of the ellipse. In the circle case we have instead a whole one-parameter family of period-two orbits represented by the diameters. The endpoints of the longer symmetry axis can be characterized by being the only pair of boundary points with maximum mutual distance, that is, the minimum of the generating function H. Likewise the endpoints of the shorter symmetry axis can be characterized by being saddle points for the generating function H. The length of the longer axis is equal to the *diameter* of the ellipse, that is, the maximum distance between any two points in the domain. The length of the shorter axis is equal to the *width*, which is defined to be the minimal width of a strip (between two parallel lines) that contains the ellipse. We will see soon that any billiard has at least two orbits of period two that can be similarly described.

Proceeding to the phase portrait of the elliptic billiard, let us note that there is an integral of motion in this situation as well. It can be described as follows. The line which contains a particular orbit segment is tangential to precisely one quadric confocal to the ellipse. It turns out that the billiard map preserves this property, that is, all lines thus obtained from a single orbit are tangent to the same quadric. Thus any parameter that characterizes the quadric in the family, for example, the eccentricity of the quadric, serves as a first integral. Those quadrics split into two families and two singular cases. Positive eccentricity corresponds to confocal ellipses. Each ellipse corresponds to an invariant curve in the phase space which goes around the cylinder. Negative eccentricity corresponds to confocal hyperbolas. Each such hyperbola determines an invariant curve in the phase space, consisting of two disjoint closed components corresponding to the branches of the hyperbola. The billiard map interchanges those components. The two families are separated by a curve in the phase space which corresponds to the orbits passing through a focus. Each such orbit (except for the long symmetry axis) alternates between these foci. (This is, after all, why these points are called foci!) When the eccentricity goes to minus infinity, the confocal hyperbolas converge to a line and the corresponding orbits shrink to the orbit on the short symmetry axis.

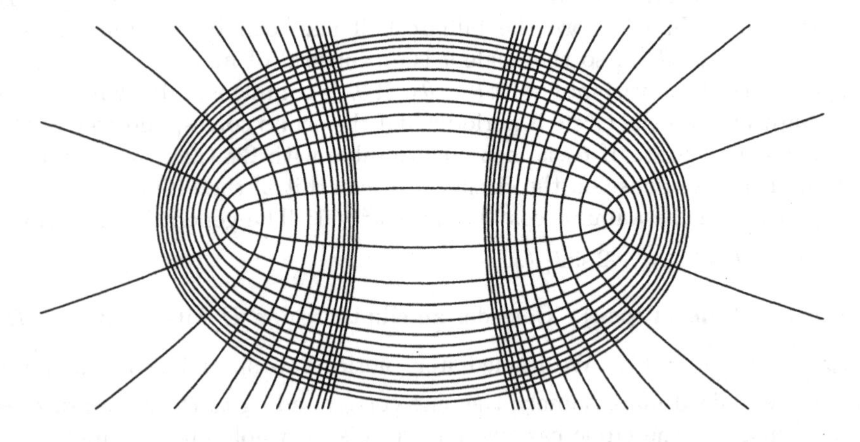

FIGURE 9.2.3. Elliptic billiard with confocal ellipses and hyperbolas

The longer symmetry axis corresponds to a hyperbolic point of period two and the orbits passing through each of the foci form two branches of the singular invariant curve that contains this orbit (Exercise 9.2.5). Those branches are interchanged by the map and each of them comprises a branch of a stable manifold of one point of the periodic orbit and a branch of the unstable manifold of the other. Thus all orbits on this curve are nontransversal heteroclinic orbits, and perturbations of the billiard map according to the Kupka–Smale theorem produce examples of complicated behavior (compare with the example at the end of Section 7.2). The shorter symmetry axis corresponds to an elliptic orbit. Notice that the index of the hyperbolic orbit as a fixed point of the second iterate of the Poincaré map is -1, and the index of the elliptic orbit is $+1$ (see the table in Section 8.4).

The invariance of the family of confocal quadrics under the billiard map can be proved purely geometrically.[1] Instead we will outline an analytic argument that gives this result. The free motion inside the billiard possesses three independent integrals of motion that are linear in the velocities: the two coordinates $v_1 := \dot{x}_1$, $v_2 := \dot{x}_2$ of the velocity, and the angular momentum $A := x_1\dot{x}_2 - x_2\dot{x}_1$ (cf. Section 5.2d). The billiard-ball map does not change the footpoint and amounts to an instantaneous change of the velocity vector, which, in fact, can be described as a reflection with respect to the tangent line of the boundary at the point of reflection. Thus any function of the three integrals that is invariant under reflection at any point of the boundary is an integral of motion for the billiard-ball map. The simplest class of functions that possess this symmetry are functions that are quadratic with respect to the coordinates of the velocity. One such function is obviously the kinetic energy $E := v_1^2 + v_2^2$ which is invariant under *any* reflection. One can try to find billiard tables that generate maps with a second integral of motion by picking another quadratic function of v_1, v_2, and A, for example, $I := v_2^2 - A^2$, constructing a vector field of symmetry axes for that function (in fact there are two such vector fields) and considering integral curves of this vector field as boundaries of billiard tables. In fact, one of the vector fields defined by the function I has closed confocal elliptical orbits and the other produces the family of confocal hyperbolas (cf. Exercises 9.2.8–9.2.9). Now we want to generalize our previous observation that two orbits of period two can be obtained by a geometric description via the diameter and width of the region.

Proposition 9.2.1. *Let D be a strictly differentiably convex bounded region with boundary B, that is, the boundary is C^2 with nonvanishing curvature. Then the associated billiard map has at least two distinct period-two orbits which are described as follows: For one of them the distance between the corresponding boundary points is the diameter of D, for the other it is the width of D.*

Proof. Consider the generating function $H(s, s')$ defined earlier. It is defined and continuous on the torus $B \times B$ and differentiable away from the diagonal. Since it vanishes on the diagonal and is negative elsewhere it attains its minimum d away from the diagonal. Let (s, s') be such that $H(s, s') = d$. Since it is a critical point, (9.2.2) implies that $\theta = \theta' = \pi/2$, so we obtain the first of these period-two orbits. This argument only depends on convexity and can easily be made to work for C^1 curves. Now consider the curve $(s, g(s))$ on the torus, where $s' = g(s)$ is the coordinate of the boundary point other than s on the line through s and s' for which $\theta = -\theta'$. (This line is the one connecting two points with parallel tangents, so the minimal length of such lines should be the width.) On this curve H is bounded from above by a negative number and thus attains a negative maximum w. The nonvanishing of the curvature makes the parameterization of the curve by an absolute angle α differentiable. Note that $\frac{\partial H(s(\alpha), s'(\alpha))}{\partial \alpha} = \frac{\partial H}{\partial s}\frac{\partial s}{\partial \alpha} + \frac{\partial H}{\partial s'}\frac{\partial s'}{\partial \alpha} = \cos\theta(\frac{\partial s}{\partial \alpha} + \frac{\partial s'}{\partial \alpha})$, so at a critical point we have $\cos\theta = 0$, since the terms in parentheses are given by the radii of curvature of B at the respective boundary points and are hence both positive. Consequently we have obtained the second desired point of period two. □

A period-three counterpart of the first orbit from Proposition 9.2.1 is constructed by considering the inscribed triangle with the largest perimeter. A similar construction works for orbits of period four. For higher periods there are different types of orbits, for example, inscribed pentagons versus pentagrams. There are also counterparts of the second type of orbit. The construction of those orbits in the more general setting of area-preserving twist maps will be our main task in the next section.

In the case of an ellipse (and, of course, also of a circle) all orbits of period higher than two come in continuous families. This follows from integrability. Those orbits are parabolic, that is, the matrix of the differential of the billiard map for the period in appropriate coordinates has the form $\begin{pmatrix} 1 & a \\ 0 & 1 \end{pmatrix}$. Before proceeding to the general discussion we will consider another example which demonstrates quite different behavior of periodic orbits.

Example. (The stadium)[2] Let D be the union of the unit square and two half-disks built on two opposite sides of it. The boundary $B = \partial D$ is a C^1 curve, but not C^2, and its shape is reminiscent of an athletic field.

Since there are four points a, b, c, d at which the curvature is discontinuous, the derivative of the Poincaré map on B is discontinuous along four segments. This fact turns out to be crucial for the analysis of the behavior of orbits typical with respect to the absolutely continuous measure $\sin\theta d\theta$. However, when we discuss any finite collection of periodic orbits that do not pass through any of the four points a, b, c, d the lack of smoothness is irrelevant since one can replace the boundary of the stadium by a convex C^∞ curve that coincides with B in a neighborhood of the footpoints of the periodic orbits in question.

There is an isolated periodic orbit of period two which is represented by the horizontal symmetry axis. Similarly to the ellipse, it is hyperbolic (Exercise 9.2.6). It obviously corresponds to the first of the two orbits (the "diameter") constructed in Proposition 9.2.1. The second period-two orbit is, in fact, not unique, unlike in the case of the ellipse; there is a whole family of vertical segments $V_x = \{(x, y) \mid |y| < 1\}$ for $x \in [-1, 1]$. So this is somewhat degenerate.

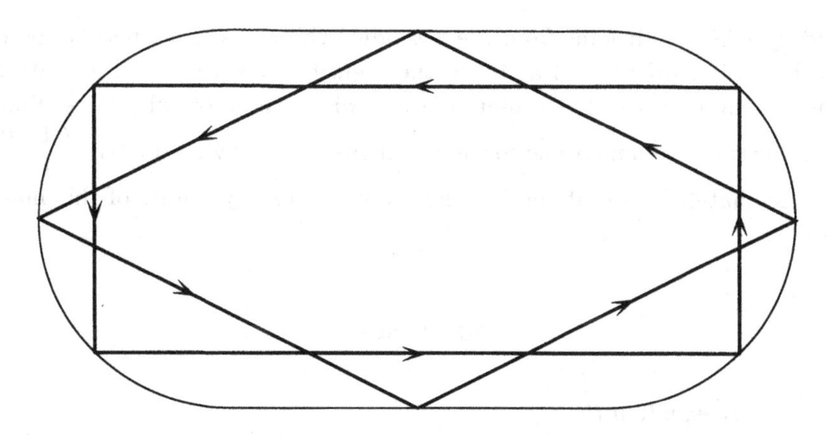

FIGURE 9.2.4. The stadium with two orbits of period four

The situation with period four orbits is much more interesting and more representative. There are two such orbits, both symmetric with respect to the horizontal and vertical axis. The first, γ_1, has the shape of a rectangle, the other, γ_2, of a diamond. An immediate calculation of lengths shows that $l(\gamma_1) = 2\sqrt{2} + 2 > 2\sqrt{5} = l(\gamma_2)$.

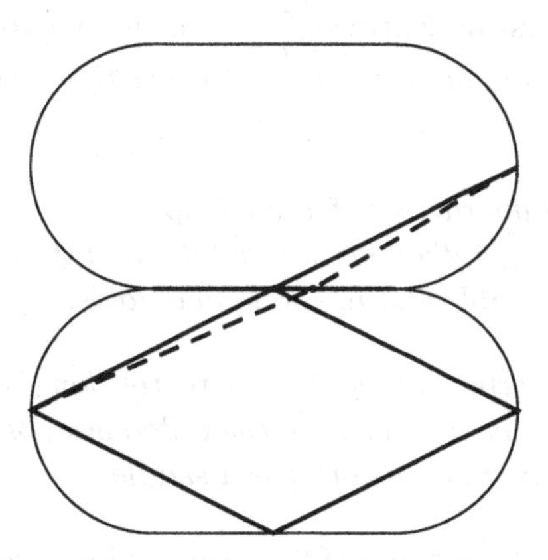

FIGURE 9.2.5. The diamond is not a maximum

A slightly more involved calculation of the derivative (still easy because all curves in which we reflect are segments and circle arcs; see Exercise 9.2.7) shows that γ_1 corresponds to a hyperbolic (saddle) orbit with index -1 and γ_2 not to an elliptic orbit, but to an inverted saddle (line 5 in the table in Section 8.4) which has index 1. Then the sum of the indices is still 0 as for the isolated period-two orbits in the ellipse, but the structure of the second orbit is different. This agrees with the conclusion of the Lefschetz Fixed-Point Formula (Theorem 8.6.2) in the following manner. In a neighborhood of the boundary which is disjoint from the periodic orbits of a given period one can perturb the billiard map so as to be the identity. Then one can identify the boundary components of the annulus to obtain a torus on which the billiard map induces a map homotopic to the linear map given by the matrix $\begin{pmatrix} 1 & 0 \\ 1 & 1 \end{pmatrix}$. One immediately sees that the Lefschetz number of any iterate of this map is indeed 0.

Exercises

9.2.1. *Verify equation (9.2.2).*

9.2.2. *Prove the assertion of Proposition 9.2.1 for any convex D bounded by a C^1 curve.*

9.2.3. *Calculate the total derivative of the billiard map $f(s,\theta)$ as a function of θ, θ', H and the curvatures κ_s and $\kappa_{s'}$ at the points corresponding to s and s'.*

9.2.4. *Calculate the first derivative $\frac{\partial S}{\partial \theta}|_{\theta=0}$ of the billiard map at the value of s corresponding to a boundary point p at which the curvature of B does not vanish.*

9.2.5. *Consider the billiard map for an ellipse. Show that the two points corresponding to the longer diameter are saddles and the orbits passing through the foci constitute the stable and unstable manifolds.*

9.2.6. *Show that the orbit corresponding to the long symmetry axis of the stadium is hyperbolic, that is, calculate the differential of f^2 for the horizontal period-two orbit to show that the orbit is a saddle.*

9.2.7. *Calculate the differential of f^4 for the orbits γ_1 and γ_2 of the stadium billiard. Show that γ_1 is a saddle and γ_2 is an inverted saddle.*

9.2.8. *Consider the quadrics with foci at $(\pm 1, 0)$. Show that reflections in their tangent lines are symmetries of the function $I := v_2^2 - A^2$.*

9.2.9. *Show that the collection of unit tangent vectors to a quadric with foci $(\pm 1, 0)$ is a level set of the function $I := v_2^2 - A^2$.*

3. Twist maps

a. Definition and examples. In this section we consider certain maps of the open cylinder $C := S^1 \times (0,1)$. Its universal cover is the strip $\mathbb{R} \times (0,1)$. Recall that a lift of such a map $f: C \to C$ is a map $F = (F_1, F_2): \mathbb{R}\times(0,1) \to \mathbb{R}\times(0,1)$ such that if $\pi: \mathbb{R} \times (0,1) \to S^1 \times (0,1)$, $(x,y) \mapsto ([x], y)$ denotes the projection to the quotient C, then $\pi \circ F = f \circ \pi$. Thus F_1 commutes with integer shifts in the x-direction, while F_2 is periodic in the first variable. On a number of occasions it has already turned out to be convenient to work on the universal cover (see Sections 2.4, 2.6, 8.2, 8.7, etc.). This will be the case here as well.

Definition 9.3.1. A (surjective) diffeomorphism $f: C \to C$ of the open cylinder $C = S^1 \times (0,1)$ is called an *area-preserving twist map* if

(1) f preserves area,
(2) f preserves orientation,
(3) f preserves boundary components in the sense that there exists an $\epsilon > 0$ such that if $(x,y) \in S^1 \times (0,\epsilon)$ then $f(x,y) \in S^1 \times (0, 1/2)$, and
(4) if $F = (F_1, F_2)$ is a lift of f to the universal cover $S = \mathbb{R} \times (0,1)$ of C then $\dfrac{\partial}{\partial y} F_1(x,y) > 0$.

Remark. We will also use obvious modifications of this definition for $S^1 \times I$, where I is any finite or infinite interval of the real line.

We will use the fact that when we restrict f to a compact subset, for example, $C_\epsilon := S^1 \times [\epsilon, 1-\epsilon]$, then there exists a $\delta > 0$ such that we have $\dfrac{\partial}{\partial y} F_1(x,y) > \delta$ on C_ϵ.

In this definition S^1 corresponds to the configuration space of the billiard system and we will use this analogy systematically. Condition (4) is the twist condition (cf. (9.2.1)). Together with (3) it means that the image of a segment $\{x\} \times (0,1) \in \mathbb{R} \times (0,1)$ is the graph of a function $x' \mapsto h_1(x,x')$ (which need not be monotone) that connects the boundary components, that is, a curve $y \mapsto F(x,y)$ with $\lim_{y\to 0} F_2(x,y) = 0$ and $\lim_{y \to 1} F_2(x,y) = 1$ and such that $y \mapsto F_1(x,y)$ is strictly increasing. Notice that $h_1(\cdot, x')$ is decreasing.

Note that such a map may not extend continuously to the closed cylinder. Natural examples are provided by billiard systems for regions whose boundary has straight pieces such as the stadium billiard.

Let us also point out that the composition of two twist maps is not always a twist map. Even iterates of a twist map may not be twist maps. On the other hand we are interested in studying the asymptotic behavior of these maps, and are thus forced to consider iterates. That does not present technical problems, as we will see, but conceptually it is somewhat unsatisfactory. In fact the principal results about twist maps can be extended to compositions of twist maps (see Definition 9.3.18 and Exercise 9.3.4). This class can be described intrinsically as "positively tilted" maps (see Exercise 9.3.1). The main technical disadvantage in working directly with these maps (rather than as compositions of twist maps) is that they are not determined by globally defined generating functions (see Subsection b).

Definition 9.3.2. The *twist interval* of f is the set of numbers $\alpha \in \mathbb{R}$ for which there exists an $\epsilon > 0$ such that if $(x, y) \in \mathbb{R} \times (0, \epsilon)$ then $F_1(x, y) - x \leq \alpha$ and if $(x, y) \in \mathbb{R} \times (1 - \epsilon, 1)$ then $F_1(x, y) - x \geq \alpha$. This is well defined up to integer translation.

Remark. For example, the twist interval of a billiard system is $(0, 1)$.

If a twist map is defined on the closed cylinder $\bar{C} = S^1 \times [0, 1]$ one may define the full twist interval as (τ_0, τ_1), where τ_0 and τ_1 are the rotation numbers (see Definition 11.1.2) of the restriction of f to $S^1 \times \{0\}$ and $S^1 \times \{1\}$ with respect to the same lift. This may be bigger than the twist interval defined above. However, when the boundary maps are rotations, there is no difference. We will discuss this in detail in Exercise 13.2.6 when we have the notion of rotation number available.

Although we defined twist maps with billiard systems in mind, this concept covers a number of other interesting examples arising from various motivations.

Example. (Integrable twist maps and perturbations) A twist map is called *integrable* if it is of the form

$$f(x, y) = (x + g(y), y).$$

Integrable twist maps leave all circles $S^1 \times \{y\}$ invariant and rotate them by the monotone function g. Thus for each rational value of g we have an invariant circle with rotation number $g(y)$ and hence we have infinitely many families of periodic orbits separated by circles with irrational rotation number. The twist interval is $(\lim_{y \to 0} g(y), \lim_{y \to 1} g(y))$.

An example is the billiard map of a circle discussed in the previous section. On the other hand the billiard map of an ellipse is not an integrable twist map, although it can be viewed as a completely integrable mechanical system.

Another example comes from the free motion of a particle on the flat torus $\mathbb{R}^2/\mathbb{Z}^2$ (Section 5.2b). It is given by $\ddot{x} = 0$. Consider the cylinder $C := \{(x_1, x_2, v_1, v_2) \mid x_1 = 0, \|v\| = 1, v_1 > 0\}$ in the phase space. The natural coordinates on C are $x = x_2 \in S^1$ and $y = v_2/v_1 \in \mathbb{R}$. The induced map is then $(x, y) \mapsto (x + y, y)$. Alternatively one can choose $x = x_2$ and $y = v_2 \in (-1, 1)$ so that the induced map becomes $(x, y) \to (x + \dfrac{y}{\sqrt{1 - y^2}}, y)$. In either coordinate system the twist interval is the whole real line.

If f is a twist map with lift F and $\dfrac{\partial F_1}{\partial y}$ is bounded away from 0 then, in fact, sufficiently small C^1-perturbations of F are twist maps as well. The endpoints of the twist interval depend continuously on the perturbation.

Example. (Forced oscillator) Consider a time-dependent second-order ordinary differential equation for a cyclic coordinate $x \in S^1$ of the form $\ddot{x} = h(x,t)$, or equivalently $\dot{x} = v$, $\dot{v} = h(x,t)$, for a bounded continuously differentiable $h: S^1 \times \mathbb{R} \to \mathbb{R}$. Then for $t'-t$ small enough the map $f_{t,t'}: S^1 \times \mathbb{R} \to S^1 \times \mathbb{R}$ determined by the solutions of the ordinary differential equation from time t to t' is a twist map. Hence $f_{t,t'}$ is a product of twist maps for any t, t'.

Consider in particular the periodically forced mathematical pendulum described by $\ddot{x} + \sin 2\pi x = g(t)$. Let T be the period of g. If T is small enough then the time T map for the pendulum $\ddot{x} + \sin 2\pi x = 0$ (see Section 5.2c) is a twist map and hence this is also true if g is small enough. In general the period map is a product of twist maps.

Example. (The standard map) The map f on the cylinder $S^1 \times \mathbb{R}$ induced by the map

$$F: \mathbb{R} \times \mathbb{R} \to \mathbb{R} \times \mathbb{R},\ (x,y) \mapsto (x+y, y+V(x+y)),$$

with V periodic, is called the standard map. It is immediate that f is a twist map: $x+y$ is increasing in y. The twist interval is $\mathbb{R}$. For $V = 0$ we get a very simple integrable twist. This example is attractive from the analytic and geometric point of view since it is determined by a single function of one variable and since the family of images of vertical lines, which plays a very important role in the analysis of twist maps, is a family of parallel curves. In fact, one can write $F = T \circ F_0$, where $F_0(x,y) := (x+y, y)$ and $T(x,y) := (x, y+V(x))$. On the other hand, even for $V_\lambda(x) = \lambda \sin 2\pi x$ the standard maps exhibit all the complexity of the asymptotic behavior present in the general case. This particular family $(x,y) \mapsto (x+y, y+V_\lambda(x+y))$ of twist maps has been studied extensively both numerically and analytically.[1]

Example. (Neighborhood of an elliptic fixed point) Consider an area-preserving map f of the plane with an elliptic fixed point at 0. This means that Df has eigenvalues $e^{\pm 2\pi i\alpha}$ for some $\alpha \in \mathbb{R}$. We will explain subsequently that for irrational α we can formally bring f into the *Birkhoff normal form* $(\theta, r) \mapsto (\theta+\omega(r), r)$, where ω is a formal power series in r. We will presently see that this implies that for some $n \in \mathbb{N}$ there exist coordinates $(\theta, r) \in S^1 \times (0, \epsilon)$ such that f preserves the area $d\theta dr$ and has the form

$$f(\theta, r) = (\theta + \alpha + c_n r^n + g_n(r,\theta), r + h_n(r,\theta)) \tag{9.3.1}$$

with $g_n, h_n = o(r^n)$ and $c_n \neq 0$. If $c_n > 0$ then

$$\frac{\partial}{\partial r}(\theta + \alpha + c_n r^n + g_n(r,\theta)) > 0$$

for sufficiently small r. Thus there exists $\delta > 0$ such that the twist interval contains $(\alpha, \alpha + c_n\epsilon^n - \delta)$. If $c_n < 0$ then the inverse of f yields a twist map.

The existence of such a reduction is closely related to the analysis given in Section 6.6b. For in this case the eigenvalues are $\lambda_1 = e^{2\pi i\alpha}$ and $\lambda_2 = \lambda_1^{-1}$ and hence there are resonances of the form $\lambda_1 = \lambda_1^{k+1}\lambda_2^k$ and $\lambda_2 = \lambda_1^k\lambda_2^{k+1}$ $(k \in \mathbb{N})$. Any other resonance of the form $\lambda_1 = \lambda_1^k\lambda_2^l$ would imply $\lambda_1 = \lambda_2^{l-k} = \lambda_1^{k-l}$, that is, α would be rational and similarly for resonances of the form $\lambda_2 = \lambda_1^k\lambda_2^l$. Using the complex coordinate that diagonalizes the linear part of f at the origin, applying the procedure from the proof of Proposition 6.6.1, and going back to polar coordinates one can immediately see that *formally* f can be written near the origin as

$$f(\theta, r) = \Big(\theta + \alpha + \sum_{n=1}^{\infty} c_n r^n, r + \sum_{n=2}^{\infty} d_n r^n\Big).$$

The fact that f is area preserving implies then that $d_n = 0$ for $n \geq 2$. Once the formal conjugacy is established one can find a C^∞ coordinate change whose derivatives at 0 coincide with those of the formal power series (see Proposition 6.6.3). Applying this C^∞ coordinate change we obtain coordinates for which (9.3.1) is satisfied. Here n is the smallest natural number such that $c_n \neq 0$. Notice that this situation is formally very similar to the situation of an area-preserving map near a hyperbolic fixed point discussed in Exercises 6.6.4 and 6.6.5. However, in this case the presence of a formal conjugacy does not imply that there is a C^∞ or even a C^0 conjugacy.

Example. (The outer billiard) Consider a strictly differentiably convex bounded region D with oriented boundary B, and the following dynamical system defined on the complement $\mathbb{R}^2 \smallsetminus \operatorname{Int} D$, where $\operatorname{Int} D$ denotes the interior of D: For any point $p \in \mathbb{R}^2 \smallsetminus \operatorname{Int} D$ there exists a unique line l tangent to B at a point q that passes through p and such that the direction from p to q is positive with respect to the orientation on B. Let p' be the point on l such that q is the midpoint between p and p'. The map $p \mapsto p'$ that associates p' to p is called the *outer billiard map.* $\mathbb{R}^2 \smallsetminus \operatorname{Int} D$ can be parameterized by assigning to p the angular coordinate α of the line l with respect to a fixed frame of reference in $\mathbb{R}^2$ and the distance r between p and q. The twist property with respect to these coordinates is clear from the geometry. (See also Exercise 9.3.5.)

b. The generating function.

Definition 9.3.3. Let F be a lift of an area-preserving twist map f. If $(x, x') \in \mathbb{R}\times\mathbb{R}$ is such that there is a point $(x', h_1(x, x')) \in F(\{x\}\times(0,1))\cap(\{x'\}\times(0,1))$ then we denote by $H(x, x')$ the area of the region in S to the right of (or "under") $F(\{x\} \times (0,1))$ and to the left of $\{x'\} \times (0,1)$. $(x, x') \to H(x, x')$ is called the *generating function* of f.

Note that $H(x + k, x' + k) = H(x, x')$ for all $k \in \mathbb{Z}$. By the twist condition the intersection $F(\{x\}\times(0,1))\cap\{x'\}\times(0,1)$ consists of at most one point, that is, h_1 is uniquely defined. When the intersection is nonempty we define h_2 by $F^{-1}(x', h_1(x, x')) = (x, h_2(x, x'))$. In other words, $F(\{x\} \times (0,1))$ is the graph of $x' \mapsto h_1(x, x')$ and $F^{-1}(\{x'\} \times (0,1))$ is the graph of $x \mapsto h_2(x, x')$. H, h_1,

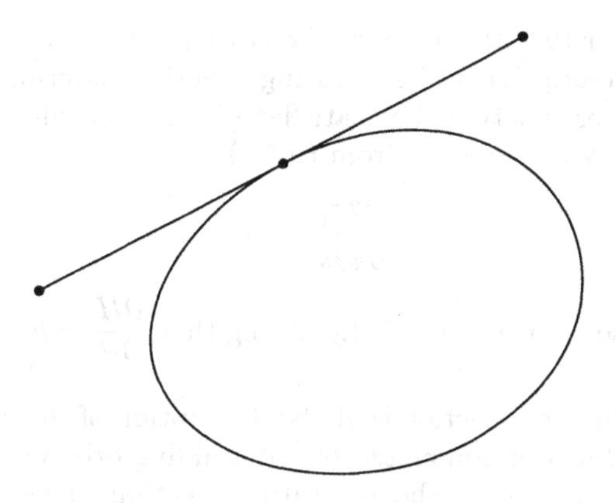

FIGURE 9.3.1. The outer billiard

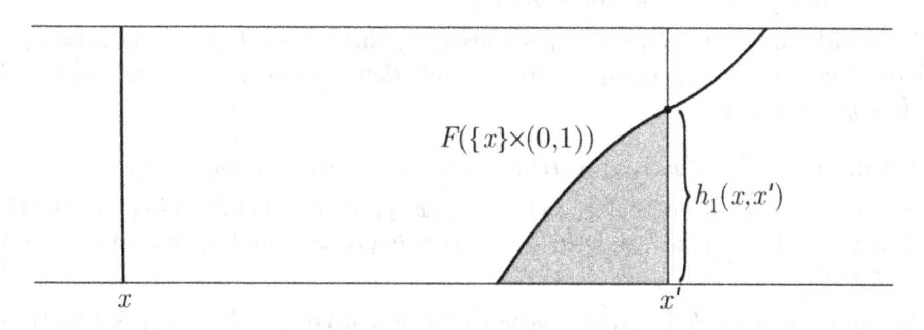

FIGURE 9.3.2. The generating function

and h_2 depend on the lift F used to define them but are invariant under the covering integer shifts $(x, x') \mapsto (x + k, x' + k)$ for $k \in \mathbb{Z}$. It is immediate from the definition of H as the area that

$$\frac{\partial H}{\partial x'} = h_1(x, x').$$

But f preserves area, so $H(x, x')$ is also the area of the preimage of the region considered before, that is, the region to the right of $\{x\} \times (0,1)$ and to the left of $F^{-1}(\{x'\} \times (0,1))$. This shows that we also have

$$\frac{\partial H}{\partial x} = -h_2(x, x').$$

If $(x', y') = F(x, y)$ then $h_1(x, x') = y'$ and $h_2(x, x') = y$. Thus for an area preserving twist map we have a description identical to (9.2.3) for billiards:

$$\begin{aligned} \frac{\partial H}{\partial x'} &= y', \\ \frac{\partial H}{\partial x} &= -y. \end{aligned} \tag{9.3.2}$$

The discussion after (9.2.3) used only the twist property, hence applies here as well so via this description the generating function determines the dynamics uniquely. Any other function that satisfies (9.3.2) coincides with H up to an additive constant. We also recall from (9.2.4) that

$$\frac{\partial^2 H}{\partial s \partial s'} < 0. \tag{9.3.3}$$

In fact, we can also see this directly by noting that $\dfrac{\partial H}{\partial s'} = h_1(s, s')$ and $h_1(\cdot, s')$ is decreasing.

We saw in the previous section that the description of the dynamics in terms of the generating function amounted to determining orbits as critical points of certain functionals built from the generating function. This is going to be the central theme of our subsequent discussion. We start with a simple proposition which reiterates (9.2.5) in this context:

Proposition 9.3.4. *Suppose x_0 is a critical point of $x \mapsto H(x_{-1}, x) + H(x, x_1)$. Then there exist $y_{-1}, y_0, y_1 \in (0,1)$ such that $F(x_{-1}, y_{-1}) = (x_0, y_0)$ and $F(x_0, y_0) = (x_1, y_1)$.*

Proof. $0 = \dfrac{d}{dx}(H(x_{-1}, x) + H(x, x_1))_{|x=x_0} = h_1(x_{-1}, x_0) - h_2(x_0, x_1)$, so $(x_0, h_1(x_{-1}, x_0)) = (x_0, h_2(x_0, x_1)) \in F(\{x_{-1}\} \times (0,1)) \cap F^{-1}(\{x_1\} \times (0,1))$. Thus $y_{-1} = h_2(x_{-1}, x_0)$, $y_0 = h_1(x_{-1}, x_0) = h_2(x_0, x_1)$, and $y_1 = h_1(x_0, x_1)$ are as desired. □

Remark. If $x \mapsto h_1(x_{-1}, x)$ has positive derivative (so $F(\{x_{-1}\} \times (0,1))$ is the graph of an increasing function) and $x \mapsto h_2(x, x_1)$ has negative derivative (so $F^{-1}(\{x_1\} \times (0,1))$ is the graph of a decreasing function) then the critical point x_0 is, in fact, unique (since it can be obtained as a zero of a decreasing function), but moreover, it is a global minimum of $x \mapsto H(x_{-1}, x) + H(x, x_1)$ since $\dfrac{d^2}{dx^2}(H(x_{-1}, x) + H(x, x_1)) = \dfrac{\partial}{\partial x} h_1(x_{-1}, x) - \dfrac{\partial}{\partial x} h_2(x, x_1) < 0$ and there is no other critical point.

c. Extensions. The setup of twist maps presented so far (namely, on an open cylinder) is essentially semilocal. That presents certain difficulties in using variational methods because the spaces where the natural functionals are defined are not compact and attempts to make them compact by imposing a priori boundary conditions lead to topological complications such as nonconnectedness and a complicated structure of the boundary. A convenient method to bypass these difficulties is to replace the given map by one defined globally, that is, on $S^1 \times \mathbb{R}$, that coincides with the given map off a neighborhood of the boundary. In particular we can choose the global map in such a way that the orbits we construct are indeed orbits of the original map. A similar passage from a local to a global problem was useful in Section 6.2b for the same purpose of making the functional spaces involved in the proof of the Hadamard–Perron Theorem 6.2.8 easier to handle.

Proposition 9.3.5. *Let $f: S^1 \times (0,1) \to S^1 \times (0,1)$ be a smooth area-preserving twist map. Then for all $\epsilon > 0$ there exists an area-preserving twist map $\tilde{f}: S^1 \times \mathbb{R}$ such that $\tilde{f} = f$ on $S^1 \times (\epsilon, 1-\epsilon)$ and for any lift $\tilde{F}$ of $\tilde{f}$ we have* $\lim_{y \to \pm\infty} \tilde{F}_1(x,y) - x = \pm\infty$.

Proof. Fix $\epsilon \in (0, 1/2)$ and consider the vector fields V_1 and V_2 defined by $V_1(f(x,y)) = Df_{|_{(x,y)}} e_2$ for $(x,y) \in S^1 \times (0,1)$ and $V_2(x,y) = e_1 + e_2$ for all (x,y), where $\{e_1, e_2\} = \{(1,0),(0,1)\}$ is the standard basis. Take a smooth function $\rho: S^1 \times \mathbb{R} \to [0,1]$ such that $\rho = 1$ on $f(S^1 \times (\epsilon, 1-\epsilon))$ and $\rho = 0$ whenever $y \notin (0,1)$. Let $V = \rho V_1 + (1-\rho)V_2$. Note that the first component of V is positive since the first component of V_1 is positive by the twist condition. Each integral curve γ_t of V contains the image of a segment $\{t\} \times (\epsilon, 1-\epsilon)$ under f. Parameterize γ_t by the parameter $s = x$, say. We now obtain an extension of $f_{\restriction_{S^1 \times (\epsilon, 1-\epsilon)}}$. Parameterize vertical lines by a parameter $y(s)$ that coincides with the standard y-coordinate on $(\epsilon, 1-\epsilon)$ and is defined otherwise as follows. Write $f = (f_1, f_2)$. Given $t \in S^1$ let $s_t = f_1(t, 1-\epsilon)$ and set $y(s) := (1-\epsilon) + \int_{s_t}^{s} \omega(x,t)dx$, where $\omega(s,t)$ is such that $dx\,dy = \omega(s,t)ds\,dt$ at $\gamma_t(s)$, that is, it is the density at $\gamma_t(s)$ of the conditional measure on y_t induced from area by the foliation γ_t. Since ω is bounded away from zero, $y(\cdot)$ is surjective. Set $\tilde{f}(t, y(s)) := \gamma_t(s)$. Then $\tilde{f}$ maps vertical lines to curves y_t and is hence also a twist map (since the first component of V is positive). $\tilde{f}$ preserves area since

$$\int_{y(s_0)}^{y(s_1)} \int_{t_0}^{t_1} 1\,dx\,dy = \int_{t_0}^{t_1} (y(s_1) - y(s_0))dx = \int_{t_0}^{t_1} \int_{s_0}^{s_1} \omega(s,t)\,ds\,dt$$
$$= \iint_{\{\tilde{f}(x,y) \mid y(s_0) \le y \le y(s_1), t_0 \le t \le t_1\}} 1\,dx\,dy.$$

Since outside $\mathbb{R} \times [0,1]$ the images of vertical lines under $\tilde{F}$ are lines of slope 1, we obtain the asymptotic condition in the statement. □

d. Birkhoff periodic orbits. The simple argument that proves Proposition 9.3.4 contains the germ of a powerful approach for finding periodic and other orbits with very specific properties. In this section we find, for any rational number in the twist interval, orbits whose x-coordinates behave like orbits of the rational rotation by that number. In Chapter 13 we will get analogous results for irrational numbers as well.

Definition 9.3.6. A point $w \in C$ is called a *Birkhoff periodic point of type* (p,q) and its orbit a *Birkhoff periodic orbit of type* (p,q) if for a lift $z \in S$ of p there exists a sequence $\{(x_n, y_n)\}_{n \in \mathbb{Z}}$ in S such that

(1) $(x_0, y_0) = z$,
(2) $x_{n+1} > x_n$ $(n \in \mathbb{N})$,
(3) $(x_{n+q}, y_{n+q}) = (x_n + 1, y_n)$,
(4) $(x_{n+p}, y_{n+p}) = F(x_n, y_n)$.

Remark. The sequence (x_n, y_n) does not parameterize the orbit according to the "dynamical ordering" induced by passing from (x, y) to $F(x, y)$, but rather in the "geometric ordering" of its projection to S^1. In fact, this order coincides with the ordering of iterates of the rational rotation $R_{p/q}$ on the circle and moreover, if one considers the projection of a Birkhoff periodic orbit of type (p, q) to the circle, which is a finite set, with the map induced by projecting F, then this map can be extended piecewise linearly to a homeomorphism of the circle.

Theorem 9.3.7. *Let $f: S \to S$ be an area-preserving twist map and p/q be a rational number in the twist interval with p, q relatively prime. Then there exist two Birkhoff periodic orbits of type (p, q) for f.*

This theorem is the first example where variational methods allow us to produce *infinitely many* periodic points. Previously we encountered situations where infinitely many periodic orbits were produced by way of hyperbolicity (Corollary 6.4.19) or topological data (cf. Corollary 8.6.11, Corollary 8.6.12, and Theorem 8.7.1). Later we will be able to use variational methods to produce infinitely many periodic orbits in other instances, namely, for geodesic flows, where we find infinitely many closed geodesics (Theorem 9.5.10) as well as large sets of minimal geodesics (Theorem 9.6.7). The proof of Theorem 9.3.7 is interesting as well in that the issue of finding critical points by variational means is not entirely trivial and requires some topological argument. Whereas the first periodic orbit is obtained as the minimum of some action in a relatively crude (although nontrivial) way, obtaining the second one brings about some interplay of the variational method with differential topology via simple Morse theory or a mountain-pass argument along the lines of Proposition 9.1.2. An interesting aspect of the argument is that it uses the dynamics of a gradient flow in an auxiliary space (Section 1.6). This is somewhat analogous to our use of the contraction mapping principle in function spaces constructed from dynamical systems to obtain periodic points, conjugacies, and so forth (Proposition 1.1.4, the second proof of Theorem 2.4.6 and Proposition 2.4.9, Theorem 2.6.1, the Hadamard–Perron Theorem 6.2.8, and the Anosov Closing Lemma (Theorem 6.4.15), as well as many instances later in this book).

In Proposition 9.2.1 we obtained two kinds of billiard orbits of period two. That is a particular example of the situation we have here. There, too, the orbit corresponding to the minimum of the action was easier to find than the second periodic orbit, which was a "minimax". Just like in the billiard case we use the "configuration space" heavily.

Proof of Theorem 9.3.7.[2] We will obtain one Birkhoff periodic orbit of type (p, q) by obtaining the sequence of its x-coordinates as a global minimum of an appropriate action defined on a certain space of sequences of points in the universal cover $\mathbb{R}$ of S^1. In order to carry out this construction it is, in fact, not necessary to use the extension of Proposition 9.3.5 since the topological complications that arise are very minor. As reasonable candidates for x-coordinates of an orbit consider the following space Σ. First, let $\tilde{\Sigma}$ be the set of nondecreasing sequences $\{s_n\}_{n\in\mathbb{Z}}$ of real numbers such that

$$s_{n+q} = s_n + 1 \tag{9.3.4}$$

and

$$F(s_n \times [\epsilon, 1-\epsilon]) \cap (s_{n+p} \times [\epsilon, 1-\epsilon]) \neq \varnothing, \tag{9.3.5}$$

where $\epsilon > 0$ is as follows: Since p/q is in the twist interval there exists $\delta \in (0, 1/2)$ such that $F_1(x, \delta) - x < p/q$ and $F_1(x, 1-\delta) - x > p/q$ for all x. Take $\epsilon > 0$ such that

$$\bigcup_{i=0}^{q-1} F^i(\mathbb{R} \times ((0,\epsilon] \cup [1-\epsilon, 1))) \subset \mathbb{R} \times ((0,\delta] \cup [1-\delta, 1)). \tag{9.3.6}$$

We call these sequences *ordered states of type* (p, q).

Thus any orbit on the universal cover whose x-coordinates satisfy (9.3.4) and (9.3.5) has y-coordinates in $(\epsilon, 1-\epsilon)$. Define an equivalence relation $\sim$ on $\tilde{\Sigma}$ by $s \sim s'$ if $s_i - s'_i = k$ for all i and some fixed $k \in \mathbb{Z}$. Let $\Sigma := \tilde{\Sigma}/\sim$.

Condition (9.3.4) is periodicity, and condition (9.3.5) ensures that there is a point (s_n, y_n) with $F(s_n, y_n) = (s_{n+p}, y_{n+p})$ for some y_{n+p}. A sequence satisfying (9.3.4) and (9.3.5) is usually not the x-projection of an orbit, but we will find a sequence that is, and the corresponding orbit is the desired Birkhoff periodic orbit of type (p, q).

Note that each sequence only has q "independent variables" $s_0, \dots, s_{q-1}$, say, by (9.3.4), that is, $\tilde{\Sigma}$ is naturally embedded in $\mathbb{R}^q$. Condition (9.3.5) applied inductively shows that $\{s_n - s_0\}_{n=0}^{q-1}$ is bounded for any $s \in \tilde{\Sigma}$, so Σ is a closed and bounded, hence compact, subset of $\mathbb{R}^q/\mathbb{Z} \sim \mathbb{R}^{q-1} \times S^1$. It follows immediately from the definition of the twist interval that $\tilde{\Sigma} \neq \varnothing$; furthermore, for any $x \in \mathbb{R}$ the sequence $\{s_n\}$ defined by $s_n = x + (n/q)$ lies in $\tilde{\Sigma}$.

Define the *action functional*

$$L(s) := \sum_{n=0}^{q-1} H(s_n, s_{n+p}) \tag{9.3.7}$$

on Σ, where H is the generating function from Definition 9.3.3. Since p and q are relatively prime it follows from (9.3.4) that $L(s) = \sum_{n=0}^{q-1} H(s_j, s_{j+np})$ for any $j \in \mathbb{Z}$. Since L is invariant under the integer shift it is defined on the compact set Σ and hence attains its maximum and minimum, but it could be on the boundary. We will show that the minimum corresponds to a Birkhoff periodic orbit of type (p, q) and deduce that it also is not on the boundary.

Consider any sequence $s \in \Sigma$. It is not constant by (9.3.4). So for any $m \in \mathbb{Z}$ there are $n \in \mathbb{Z}$ and $k \geq 0$ such that $n \leq m \leq n+k$ and $s_{n-1} < s_n =$

$\dots = s_{n+k} < s_{n+k+1}$. (If $k > 0$ then s is a boundary point of $\tilde{\Sigma}$.) Since s is nondecreasing, the twist condition implies

$$\epsilon \le h_1(s_{n+k-p}, s_{n+k}) \le \dots \le h_1(s_{n-p}, s_n) \le 1 - \epsilon,$$
$$\epsilon \le h_2(s_n, s_{n+p}) \le \dots \le h_2(s_{n+k}, s_{n+k+p}) \le 1 - \epsilon,$$

so either

$$h_2(s_n, s_{n+p}) < h_1(s_{n-p}, s_n) \tag{9.3.8}$$

or

$$h_1(s_{n+k-p}, s_{n+k}) < h_2(s_{n+k}, s_{n+k+p}) \tag{9.3.9}$$

or

$$h_1(s_{n+l-p}, s_{n+l}) = h_2(s_{n+l}, s_{n+l+p}) \text{ for } l \in \{0, \dots, k\}. \tag{9.3.10}$$

For case (9.3.8) note that considering $x = s_n$ as an independent variable and keeping all other s_i fixed we have

$$\frac{d}{dx}\Big|_{x=s_n} L(s) = \frac{d}{dx}\Big|_{x=s_n} \sum_{i=0}^{q-1} H(s_i, s_{i+p}) = \frac{d}{dx}\Big|_{x=s_n} \big(H(s_{n-p}, x) + H(x, s_{n+p})\big)$$
$$= h_1(s_{n-p}, s_n) - h_2(s_n, s_{n+p}) > 0$$

and that by (9.3.8) we can decrease s_n—and hence $L(s)$—slightly, without leaving Σ, so s is not a minimum. For case (9.3.9) we find similarly that setting $y = s_{n+k}$ we get $\frac{d}{dy}\big|_{y=s_{n+k}} L(s) < 0$ so by (9.3.9) we can increase s_{n+k} slightly—and hence decrease $L(s)$ slightly—without leaving Σ, so s is not a minimum. Thus if s is a minimum then (9.3.10) holds for all $m \in \mathbb{Z}$, hence

$$h_1(s_{m-p}, s_m) = h_2(s_m, s_{m+p}) \text{ for all } m \in \mathbb{Z}. \tag{9.3.11}$$

Setting $(x_n, y_n) = (s_n, h_1(s_{n-p}, s_n))$ now yields a periodic orbit. Notice that we must have $y_n \in (\epsilon, 1-\epsilon)$ for *all* $n \in \mathbb{Z}$ since having $y_n \le \epsilon$ for any $n \in \mathbb{Z}$ implies $y_n < \delta$ for all $n \in \mathbb{Z}$, which is incompatible with (9.3.4) and (9.3.5) by the choice of δ. Thus to show that (x_n, y_n) is, in fact, a Birkhoff periodic orbit of type (p,q) and s is not on the boundary of Σ, it suffices to show that $s_n = x_n$ is strictly increasing. Suppose $s_n = s_{n+1}$. By choosing a different n, if necessary, we may assume that either $s_{n-1} < s_n$ or $s_{n+1} < s_{n+2}$ (since s is not constant). Then, since s is nondecreasing, the twist condition and (9.3.11) yield $y_{n+1} = h_1(s_{n-p+1}, s_{n+1}) \le h_1(s_{n-p}, s_{n+1}) \le h_1(s_{n-p}, s_n) = y_n = h_2(s_n, s_{n+p}) \le h_2(s_n, s_{n+p+1}) = y_{n+1}$ with at least one strict inequality, which is absurd.

Thus we have found a Birkhoff periodic orbit of type (p,q) such that the sequence of its x-coordinates is a global minimum of L in the interior of Σ.

Now we look for the second Birkhoff periodic orbit of type (p,q), the "minimax" orbit. For this construction we will assume that the map has been globalized as in Proposition 9.3.5, so that the space of candidates for solutions is much simpler. We globalize f as follows: Since p/q is in the twist interval there exists $\delta \in (0, 1/2)$ such that $F_1(x, \delta) - x < p/q$ and $F_1(x, 1-\delta) - x > p/q$ for all x. Take $\epsilon > 0$ such that

$$\bigcup_{i=0}^{q-1} F^i(\mathbb{R} \times ((0, \epsilon] \cup [1-\epsilon, 1))) \subset \mathbb{R} \times ((0, \delta] \cup [1-\delta, 1)).$$

So any orbit whose x-coordinates are a sequence in $\tilde{\Sigma}$ has y-coordinates in $(\epsilon, 1-\epsilon)$. We apply Proposition 9.3.5 with this choice of ϵ. Then any Birkhoff periodic orbit of type (p, q) of the extension is actually a Birkhoff periodic orbit of type (p, q) for the original map.

We will, in fact, show, that the second Birkhoff periodic orbit of type (p, q) is "intertwined" with the first one just as for a rotation, that is, the projections of these orbits to S^1 can be mapped by a common homeomorphism to two orbits of a rational rotation. This is a natural byproduct of our method, for we will look for the second orbit by considering sequences thus intertwined with the sequence s of the first Birkhoff periodic orbit of type (p, q). To this end fix once and for all the sequence S representing the first Birkhoff periodic orbit of type (p, q) and let

$$\mathfrak{S} := \{\{s_n\}_{n\in\mathbb{Z}} \mid S_n \le s_n \le S_{n+1} \text{ for } n \in \mathbb{Z}\}.$$

Note that this space is a compact convex polyhedron. Our strategy is now to find a critical point other than S of L on $\mathfrak{S}$. It turns out to be a saddle point. This will yield the second Birkhoff periodic orbit of type (p, q). Let us first make a useful observation:

Lemma 9.3.8. *Suppose $s \in \mathfrak{S}$ is such that $s_k = S_{k+1}$. Then $h_2(s_k, s_{k+p}) \le h_1(s_{k-p}, s_k)$ with equality if and only if $s_{k-p} = S_{k-p+1}$ and $s_{k+p} = S_{k+p+1}$. If $s \in \mathfrak{S}$ is such that $s_k = S_k$ then $h_1(s_{k-p}, s_k) \le h_2(s_k, s_{k+p})$ with equality if and only if $s_{k-p} = S_{k-p}$ and $s_{k+p} = S_{k+p}$.*

Proof. Since $s_n \le S_{n+1}$ for $n \in \mathbb{Z}$, the twist condition yields

$$h_1(S_{k-p+1}, S_{k+1}) = h_1(S_{k-p+1}, s_k) \le h_1(s_{k-p}, s_k)$$

with equality if and only if $S_{k-p+1} = s_{k-p}$. On the other hand

$$F_1(s_k, h_2(s_k, s_{k+p})) = s_{k+p} \le S_{k+p+1} = F_1(S_{k+1}, h_1(S_{k-p+1}, S_{k+1}))$$

with equality if and only if $s_{k+p} = S_{k+p+1}$, so by the twist condition we have

$$h_2(s_k, s_{k+p}) \le h_1(S_{k-p+1}, S_{k+1}) \le h_1(s_{k-p}, s_k)$$

with equality if and only if $s_{k-p} = S_{k-p+1}$ and $s_{k+p} = S_{k+p+1}$. The second part is similar. □

Lemma 9.3.8 shows that there are no critical points of L on the boundary of $\mathfrak{S}$ other than S and S', where $S'_n = S_{n+1}$, which both represent the minimal Birkhoff periodic orbit of type (p, q). Now we show that there is a critical point of L in the interior of $\mathfrak{S}$. For a boundary point of $\mathfrak{S}$ some of the inequalities $S_n \le s_n \le S_{n+1}$ are equalities. Let us restrict attention to the left-hand side and consider a face of $\mathfrak{S}$ of highest dimension. This means that $s_{n-p} > S_{n-p}$ whenever $s_n = S_n$. But in that case Lemma 9.3.8 shows that L decreases with s_n and similarly in the case of reversed inequalities. Consider the gradient flow $\mathcal{L}^t$ of $-L$ on $\mathfrak{S}$ (cf. Section 1.6). The reason we use the gradient of $-L$ is to conform with the principle of "steepest descent". We have observed that the gradient of $-L$ on $\mathfrak{S}$ points inside $\mathfrak{S}$ on every face of full dimension. This makes $\mathcal{L}^t$ defined in $\mathfrak{S}$ for all positive values of t: If the orbit of any point x leaves $\mathfrak{S}$ at time t then the orbits of all points in a neighborhood U of x also leave it, maybe at a slightly later time. Thus $\bigcup_{0<s<2t} \mathcal{L}^s(U) \cap \partial\mathfrak{S}$ is open in $\partial\mathfrak{S}$ and hence intersects a highest-dimensional face in an open set. But this is impossible since the negative gradient points inward on those faces. Thus $\mathcal{L}^t(\mathfrak{S}) \subset \mathfrak{S}$ for any $t > 0$.

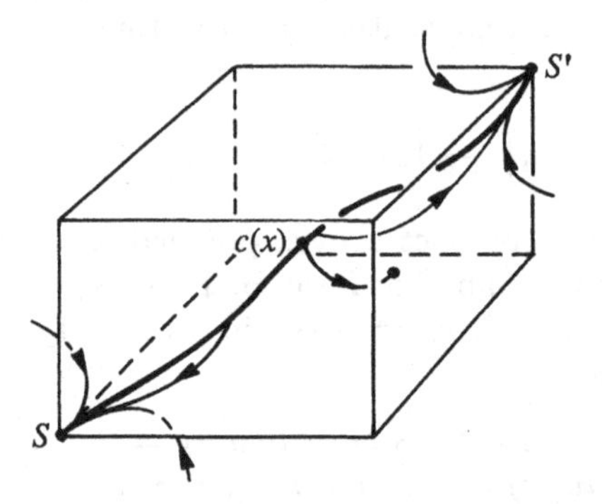

FIGURE 9.3.3. The gradient flow

This gradient flow has two attracting fixed points, namely, the minima S and S'. Consider a continuous curve $c\colon [0, 1] \to \mathfrak{S}$ with $c(0) = S$ and $c(1) = S'$. Then there is a smallest $x > 0$ such that $\mathcal{L}^t(c(x))$ does not converge to S. But by compactness the ω-limit set $\omega(c(x))$ is nonempty and by Proposition 1.6.3 consists of critical points. It cannot be equal to S' because the set of points attracted to S' is open. Since it is disjoint from $\{S, S'\}$, we have found a new critical point. □

The critical point just constructed, corresponding to the second Birkhoff periodic orbit of type (p, q), cannot be a local minimum of the functional L since otherwise the points attracted to it would form an open set containing $c(x)$, which contradicts the definition of x. There may be more than one such point, and at least one of these can be described by a "mountain-pass" argument.

Proposition 9.3.9. *Let $L_{\min\max} := \inf_c \max_{t\in[0,1]} L(c(t))$ taken over all continuous $c\colon [0,1] \to \mathfrak{S}$ with $c(0) = S$ and $c(1) = S'$. Then $L_{\min\max}$ is a critical value of the functional L on $\mathfrak{S}$.*

Proof. Suppose not. Then, since the set of critical values is compact, there exists $\epsilon > 0$ such that there are no critical values in $(L_{\min\max} - \epsilon, L_{\min\max} + \epsilon)$ and moreover $\|\nabla L\| \geq \delta$ on $L^{-1}(L_{\min\max} - \epsilon, L_{\min\max} + \epsilon)$. Thus there exists $T > 0$ such that $\mathcal{L}^t(L^{-1}(-\infty, L_{\min\max}+\epsilon)) \subset L^{-1}(-\infty, L_{\min\max}-\epsilon)$. Therefore if we consider a curve c as before such that $\max_{t\in[0,1]} L(c(t)) < L_{\min\max} + \epsilon$, then the curve $c' := \mathcal{L}^t \circ c$ is a curve with $\max_{t\in[0,1]} L(c'(t)) < L_{\min\max} - \epsilon$, contradicting the definition of $L_{\min\max}$. □

Let us give yet another description of the critical value $L_{\min\max}$. Consider the increasing family $L^{-1}((-\infty, t])$ of subsets of $\mathfrak{S}$. $L_{\min\max}$ is the minimal t for which S and S' belong to a common connected component of $L^{-1}((-\infty, t])$. In some degenerate cases this component of $L^{-1}((-\infty, L_{\min\max}])$ might not be path connected. But for any regular value $t > L_{\min\max}$ the corresponding component of the set $L^{-1}((-\infty, t])$ is path connected.

e. Global minimality of Birkhoff periodic orbits. We constructed the first Birkhoff periodic orbit of type (p,q) as a global minimum of the action functional on the space of ordered states (see (9.3.4) and (9.3.5)). We can instead consider a larger collection of states without requiring monotonicity. For simplicity we consider from the outset globally defined maps as obtained from the extension procedure of Section 9.2c. A state or a *state of type* (p,q) is simply a periodic sequence satisfying (9.3.4) and the action functional on the space of states of type (p,q) is defined by (9.3.7). Although the set of states is not compact, the functional reaches its minimum because for the extended map the area $H(s,s')$ is large for large values of $s' - s$.

Theorem 9.3.10. *The first Birkhoff periodic orbit of type (p,q) corresponds to the minimum of the functional L in (9.3.7) on the entire space of states.*

Proof. The idea of the proof is to show that for any state that is not ordered the functional L is not minimal.[3]

Lemma 9.3.11. *If H is the generating function for the twist map and Δs, $\Delta s' \geq 0$ then*

$$H(s,s') + H(s+\Delta s, s'+\Delta s') \leq H(s, s'+\Delta s') + H(s+\Delta s, s')$$

with equality if and only if $\Delta s \Delta s' = 0$.

Proof. $H(s+\Delta s, s'+\Delta s') - H(s, s'+\Delta s') - H(s+\Delta s, s') + H(s,s') = \int_0^{\Delta s'}\int_0^{\Delta s} \frac{\partial^2 H}{\partial s \partial s'}\, ds\, ds'$ and $\frac{\partial^2 H}{\partial s\partial s'} < 0$ by (9.3.3). □

Proposition 9.3.12. *Consider two states x, x' such that*

$$x_i \leq x'_i, x_{i+p} > x'_{i+p}, \dots, x_{i+(k-1)p} > x'_{i+(k-1)p}, x_{i+kp} \leq x'_{i+kp}$$

for some $i, k \in \mathbb{Z}$. Set $y_j := \min(x_j, x'_j)$ for $j = i + lp$, $0 \leq l \leq k$, and $y_j = x_j$ otherwise. Define y' by $y_j + y'_j = x_j + x'_j$. Then

$$L(y) + L(y') \leq L(x) + L(x').$$

Furthermore, if $x_i < x'_i$ or $x_{i+kp} < x'_{i+kp}$ then $L(y) + L(y') < L(x) + L(x')$.

Proof.

$$\begin{aligned} L(y)+L(y')-L(x)-L(x') &= H(y_i, y_{i+p})+H(y'_i, y'_{i+p})-H(x_i, x_{i+p})-H(x'_i, x'_{i+p}) \\ &\quad +H(y_{i+(k-1)p}, y_{i+kp})+H(y'_{i+(k-1)p}, y'_{i+kp}) \\ &\quad -H(x_{i+(k-1)p}, x_{i+kp})-H(x'_{i+(k-1)p}, x'_{i+kp}) \\ &= H(y_i, y_{i+p})+H(y'_i, y'_{i+p})-H(y_i, y'_{i+p})-H(y'_i, y_{i+p}) \\ &\quad +H(y_{i+(k-1)p}, y_{i+kp})+H(y'_{i+(k-1)p}, y'_{i+kp}) \\ &\quad -H(y'_{i+(k-1)p}, y_{i+kp})-H(y_{i+(k-1)p}, y'_{i+kp}), \end{aligned}$$

so the lemma applied to the first four and last four terms yields the claim. □

If a state s is not ordered and $s'_n := s_{n+1}$ then the sequence $s'_n - s_n$ changes signs between $n = 0$ and $n = q$. Since $s_q > s_0$ there is an n such that $s'_n - s_n > 0$. Hence there are also m, k such that $s'_m - s_m \leq 0$, $s'_{m+lp} - s_{m+lp} < 0$ for $l \in \{0, \dots, k\}$ and $s'_{m+(k+1)p} - s_{m+(k+1)p} \geq 0$. Construct two states v, w as follows. $v_{m+lp+nq} := s'_{m+lp+nq}$, where $l \in \{1, \dots, k-1\}$, $n \in \mathbb{Z}$, and $v_i = s_i$ for all other $i \in \mathbb{Z}$. Take w such that $v_i + w_i = s_i + s'_i$. Applying the proposition yields $2L(s) = L(s) + L(s') \geq L(v) + L(w)$. If the inequality is strict then $L(v) < L(s)$ or $L(w) < L(s)$, so s does not minimize L.

According to the proposition, equality can take place only if there exist m, k such that $s'_m - s_m < 0$, $s'_{m+lp} - s_{m+lp} = 0$ for $l \in \{0, \dots, k\}$ and $s'_{m+(k+1)p} - s_{m+(k+1)p} > 0$. In that case the new states v and w cannot both represent orbits (by the twist property; see the argument following (9.3.11)), and hence they cannot both be critical points of L. Thus they cannot both represent the minimum, so s is not the minimal state either. □

Arguing as in the proof of Theorem 9.3.10 one sees that Proposition 9.3.12 has the following corollary:

Corollary 9.3.13. *For two states s and s' let $x_n := \min(s_n, s'_n)$ and $y_n := \max(s_n, s'_n)$. Then we have $L(s) + L(s') \geq L(x) + L(y)$. The inequality is strict unless there exists $i \in \mathbb{Z}$ such that $s_i = s'_i$.*

Proposition 9.3.14. *Let $f: S \to S$ be an area-preserving twist map, p/q a rational number in the twist interval, and $m \in \mathbb{N}$. Then the minimal Birkhoff periodic orbit of type (p,q) for f realizes the minimum of L on the states of type (mp, mq).*

Proof. The minimum occurs at an orbit whose x-coordinates give an ordered state $\{s_n\}_{n\in\mathbb{Z}}$ of type (mp, mq), that is,

$$s_{n+mq} = s_n + m \text{ and } s_{n+1} > s_n.$$

Consider the state s' defined by $s'_n = s_{n+q} - 1$. Note that $s_n \le s'_n$ for all $n \in \mathbb{Z}$ or vice versa, for otherwise the states x and y with $x_n := \min(s_n, s'_n)$ and $y_n := \max(s_n, s'_n)$ yield $L(x) + L(y) \le L(s) + L(s') = 2L(s)$ by Corollary 9.3.13. If the inequality is strict then s is not minimal. Otherwise arguing as in the proof of Theorem 9.3.10 we see that x and y do not represent orbits, hence are not minimal, so s is not minimal.

Thus $s \le s'$ and hence $s_i \le s'_i = s_{i+q} - 1 \le \cdots \le s_{i+mq} - m = s_i$ for all $i \in \mathbb{Z}$ and we have $s = s'$, whence s is q-periodic and thus gives the minimal Birkhoff periodic orbit of type (p,q). □

Definition 9.3.15. For $a, b \in \mathbb{R}$, $n \in \mathbb{N}$ we call a finite sequence $\{x_i\}_{i=0}^n$ an *(a,b,n)-state* if $x_0 = a$ and $x_n = b$. The set of (a,b,n)-states is denoted by $\Sigma(a,b,n)$ or just Σ when no confusion can arise. For a periodic state s of type (p,q) and $m \in \mathbb{Z}$, $n \in \mathbb{N}$ we obtain an (s_m, s_{m+np}, n)-state x defined by $x_i := s_{m+ip}$. We define an action functional L on the space $\Sigma(a,b,n)$ of (a,b,n)-states by setting

$$L(s) := \sum_{i=0}^{n-1} H(s_i, s_{i+1}), \tag{9.3.12}$$

where H is the generating function from Definition 9.3.3.

Notice that we have reverted to the dynamical ordering, unlike as in (9.3.7). Since $\dfrac{\partial L}{\partial x_i} = h_1(x_{i-1}, x_i) - h_2(x_i, x_{i+1})$ we see that a critical point of L corresponds to an orbit segment. If we say that a global minimum of L on $\Sigma(a,b,n)$ "minimizes between its endpoints", then a segment that minimizes between its endpoints corresponds to an orbit segment. If a segment minimizes between its endpoints, then so does any subsegment.

Definition 9.3.16. Let us call a state of type (p,q) *globally minimal* if for any $m \in \mathbb{Z}$, $k \in \mathbb{N}$ the sum $\sum_{i=0}^{k-1} H(s_{n+ip}, s_{n+(i+1)p})$ minimizes the functional L on the space of (s_m, s_{m+kp}, k)-states.

Proposition 9.3.17. *Any globally minimal state of type (p,q) represents a minimal Birkhoff periodic orbit of type (p,q) and, conversely, any minimal Birkhoff periodic orbit of type (p,q) is globally minimal.*

Proof. Consider a minimal Birkhoff periodic orbit of type (p,q). It determines a state s. Pick $m \in \mathbb{Z}$ and $k \in \mathbb{N}$. Take $N \in \mathbb{N}$ such that $k \le qN$. By

Proposition 9.3.14 the sum $\sum_{i=0}^{Nq-1} H(s_{n+ip}, s_{n+(i+1)p})$ is minimal on the space of all states of type (Np, Nq). In particular this segment $\{s_{n+ip}\}_{i=0}^{N}q$ minimizes between its endpoints, and hence so do all its subsegments. This proves the second statement.

Let s be a globally minimal state. Consider the sum $\sum_{i=0}^{q-1} H(s_{ip}, s_{(i+1)p})$ and assume it is greater than the minimum $L_{\min}$ on the space of states of type (p, q). Let s' be a state such that $\sum_{i=0}^{q-1} H(s'_{ip}, s'_{(i+1)p}) = L_{\min}$ and $|s'_0 - s_0| < 1$. Given $N \in \mathbb{N}$ the segment $\{s_{ip}\}_{i=0}^{Nq-1}$ of the state s minimizes between its endpoints s_0 and $s_{Nq} = s_0 + n$ with action $NL(s)$. Replacing this segment by $\{s_0, s'_p, s'_{2p}, \dots, s'_{(Nq-1)p}, s_0 + N\}$ we obtain an $(s_0, s_0 + N, Nq)$-state whose action differs from $NL_{\min}$ by a constant independent of N. If N is large enough this contradicts global minimality of s. □

In Section 9.7 we will encounter similar arguments when studying minimal geodesics on surfaces. We will return to twist maps in Chapter 13.

Exercises

Definition 9.3.18. A *positive-tilt* map is a C^1 map $f: S^1 \times (0,1) \to S^1 \times (0,1)$ such that the following condition holds. Denote by $\theta(x,y) \in \mathbb{R}$ the angle between $Df|_{(x,y)} e_2$ and e_2, where e_2 is the vertical unit vector, defined in such a way that $\theta(x,0) \in (0,\pi)$ and θ is continuous. Then we require that $\theta > 0$.[4]

9.3.1. *Show that a composition of twist maps is a positive-tilt map.*

Definition 9.3.19. A homeomorphism $f: S^1 \times (0,1) \to S^1 \times (0,1)$ is called a twist homeomorphism if for any lift $F = (F_1, F_2)$ the function F_1 is strictly monotone in y.

9.3.2. *Show that the first Birkhoff periodic orbit of type (p,q) obtained in the proof of Theorem 9.3.7 can be found also for continuous twist homeomorphisms preserving a measure that is positive on open sets.*

9.3.3. *Show that Theorem 9.3.7 holds for the composition $f_2 \circ f_1$ of two twist maps.*

9.3.4. *Show that Theorem 9.3.7 holds for the composition of finitely many twist maps.*

9.3.5. *Show that with respect to the coordinates described in the text outer billiard maps are twist maps with twist interval $[0,\pi]$ preserving Lebesgue measure.*

9.3.6. *Show that the extension $\tilde{f}$ obtained in Proposition 9.3.5 can be made to coincide with the standard integrable twist $(x,y) \mapsto (x+y, y)$ outside a (large) compact set.*

4. Variational description of Lagrangian systems

Now we proceed from the special discrete-time situation of twist maps to the general setting of Lagrangian mechanics as described in Section 5.3. We want to show that solving the Lagrange equation (5.3.2), which is restated as (9.4.2)—and hence describing Newtonian dynamics—in fact amounts to solving a variational problem, that is, finding critical points of a certain functional. Unlike the discrete-time case studied before, the natural action functional is defined on a certain infinite-dimensional space. That leads to considerable technical complications and necessitates development of a local theory. We will later be able to find minima for the action functional defined below, as we just did in the discrete-time case. First we discuss the relation of the Lagrange equation with critical points of a functional.

Suppose L is a smooth function of $(x,v)\in\mathbb{R}^n\times\mathbb{R}^n$. Given $x,y\in\mathbb{R}^n$ and $T>0$ we may consider smooth curves $c\colon[0,T]\to\mathbb{R}^n$ with $c(0)=x$, $c(T)=y$. Then

$$F(c):=\int_0^T L(c(t),\dot c(t))\,dt \tag{9.4.1}$$

is well defined. We would like to minimize it, that is, to find a curve c so that $F(c)$ is minimal. To this end consider curves $c_s\colon[0,T]\to\mathbb{R}^n$ depending smoothly on $s\in(-\epsilon,\epsilon)$ such that $c_0=c$ and $c_s(0)=x$, $c_s(T)=y$. Then $F(c_s)$ is a real-valued function of s and if we want to determine whether $F(c_0)$ is minimal we should first check that $\frac{d}{ds}F(c_s)_{|s=0}=0$. But this is easy:

$$\begin{aligned}\frac{d}{ds}_{|s=0}\int_0^T L(c_s(t),\dot c_s(t))dt &= \int_0^T\left(\frac{\partial L}{\partial x}\frac{dc_s}{ds}_{|s=0}+\frac{\partial L}{\partial v}\frac{d}{ds}_{|s=0}\dot c_s(t)\right)dt\\ &= -\int_0^T\left(\frac{d}{dt}\frac{\partial L}{\partial v}-\frac{\partial L}{\partial x}\right)\frac{dc_s}{ds}_{|s=0}dt\end{aligned}$$

using integration by parts and the fact that $\frac{dc_s}{ds}_{|s=0}$ vanishes for $t=0,T$. Now c_s can be rather arbitrary, save for the values at the endpoints, so $\frac{dc_s}{ds}_{|s=0}$ is arbitrary. Thus the latter integral can only vanish for *all* such variations if in fact

$$\frac{d}{dt}\frac{\partial L}{\partial v}-\frac{\partial L}{\partial x}=0 \tag{9.4.2}$$

(see Exercise 9.4.1). Thus the Lagrange equation arises from minimizing integrals along curves and the critical points are exactly the solutions of Newton's equation.

For later use we would like to restate the preceding calculation for the case of possibly varying endpoints:

Proposition 9.4.1. (First Variation Formula)

$$\frac{d}{ds}_{|s=0}\int_0^T L(c_s(t),\dot c_s(t))dt=\left[\frac{\partial L}{\partial v}\frac{dc_s}{ds}_{|s=0}\right]_0^T-\int_0^T\left(\frac{d}{dt}\frac{\partial L}{\partial v}-\frac{\partial L}{\partial x}\right)\frac{dc_s}{ds}_{|s=0}dt.$$

This calculation shows that for Lagrangians on a manifold the dynamics is described by minimizing the action functional since we showed in Section 5.3b that the Euler–Lagrange equation is independent of the local chart chosen, that is, for Lagrangians

$$L: TM \to \mathbb{R}$$

the orbits are curves c that are critical for

$$F(c) := \int_0^T L(\dot{c})dt.$$

In particular the geodesic flow (Definition 5.3.4) is described in variational terms.

Let us briefly outline the basic formulation of the variational approach in the setting of arbitrary Lagrangians on manifolds. The main point is to get a differentiable structure on the (infinite-dimensional) manifold of smooth curves. Let $\mathcal{C}$ be the collection of smooth curves $c: [0,T] \to M$ with $c(0) = x$, $c(T) = y$. For any $c \in \mathcal{C}$ consider smooth variations $\gamma: [-\epsilon, \epsilon] \times [0,T] \to M$, $(s,t) \mapsto c_s(t)$ such that $c_0 = c$ and $c_s \in \mathcal{C}$ for $-\epsilon < s < \epsilon$. Associated to γ is the vector field Y_γ along c given by $Y_\gamma(t) = \frac{d}{ds}c_s(t)_{|s=0}$.

Proposition 9.4.2. *$T_c\mathcal{C}$ is naturally represented as the space of C^∞ vector fields Y along c such that $Y(0) = Y(T) = 0$.*

This follows from the following easy lemmas whose proofs we leave to the reader:

Lemma 9.4.3. *If $F: \mathcal{C} \to \mathbb{R}$ is such that $\delta_c F(Y_\gamma) := \frac{dF(c_s)}{ds}_{|s=0}$ exists and $Y_\gamma = Y_{\gamma'}$ then $\delta_c F(Y_\gamma) = \delta_c F(Y_{\gamma'})$.*

Lemma 9.4.4. *If Y is a C^∞ vector field along c such that $Y(0) = Y(T) = 0$ then there exists a variation γ of c such that $Y = Y_\gamma$.*

In more modern coordinate-free notation one would describe Lagrangians of the form (5.3.3) as follows: If M is a Riemannian manifold with Riemannian metric g we take kinetic energy to be $g(v,v)/2$ for $v \in TM$. The natural Lagrangian then is of the form

$$L(v) = \frac{1}{2}g(v,v) - V(\pi(v)),$$

where $v \in TM$, $\pi: TM \to M$ is the projection to the footpoint, and $V: M \to \mathbb{R}$ is the potential energy.

The main point of these remarks is to reiterate that when one makes calculations in local coordinates one obtains results whose meaning is independent of the choice of local coordinates. This is the major thrust of Lagrange's approach, and this is what we shall do. In particular in local coordinates vanishing of $\delta_c F: T_c C \to \mathbb{R}$ is equivalent to the Euler–Lagrange equation, which is hence necessary for minimizing L along c.

Let us point out that the traditional notation for work in local coordinates is as follows: A point in a local chart is denoted by q, a vector by $\dot{q}$. Thus the Euler–Lagrange equation in the context of classical mechanics usually looks as follows:

$$\frac{d}{dt}\frac{\partial L}{\partial \dot{q}} = \frac{\partial L}{\partial q}.$$

The Euler–Lagrange equation is in fact equivalent to its solution being a critical point of the functional F defined on the set of all C^1 curves $[0,T] \to M$ with given boundary condition. A priori such a critical point may not even be a local minimum, much less a global minimum. The point of the subsequent discussion is to show that if two boundary conditions are chosen sufficiently close then there is a unique "short" solution of the Euler–Lagrange equation which is in fact the global minimum. This situation changes when the endpoints are far apart. Sufficiently long segments of solutions of the Euler–Lagrange equation that are trajectories of the dynamical system cease to be the minimum. In Section 6 we will discuss examples of orbits that are globally minimizing in the sense that their lifts to the universal cover are minimizing between any two of their points.

Exercises

9.4.1. *Show in detail that if a curve c is a critical point of the functional F given by (9.4.1) on the space of smooth curves with fixed endpoints then the Euler–Lagrange equation (9.4.2) holds.*

9.4.2. *Prove Proposition 9.4.1.*

5. Local theory and the exponential map

From now on we will discuss only geodesic flows, that is, Lagrangian systems with zero potential (cf. Definition 5.3.4).

Definition 9.5.1. For curves $c: [0,T] \to M$ we call

$$A(c) := \int_0^T \frac{1}{2} g_{c(t)}(\dot{c}(t), \dot{c}(t))dt$$

the *action* and $l(c) := \int_0^T \sqrt{g_{c(t)}(\dot{c}(t), \dot{c}(t))}dt$ the *length* of c. We say that c is parameterized with *constant speed* if $\frac{d}{dt}\sqrt{g_{c(t)}(\dot{c}(t), \dot{c}(t))} = 0$. A critical C^2 curve for A is called a *geodesic*. A continuous piecewise-C^2 curve is called a *broken geodesic* if its restriction to any subinterval where it is C^2 is critical for A.

Proposition 5.3.2 immediately implies

Corollary 9.5.2. *Geodesics are parameterized with constant speed.*

Now we can point out a relation between action and length that is particularly useful for curves with constant speed. To simplify notation let us write $\|\dot{c}(t)\| := g_{c(t)}(\dot{c}(t), \dot{c}(t))$.

Proposition 9.5.3. *If $c\colon [0,T] \to M$ is a curve then $A(c)T \geq l^2(c)/2$ with equality if and only if c has constant speed.*

Proof. The Cauchy–Schwarz inequality gives

$$l^2(c) = \Big(\int_0^T \|\dot{c}(t)\| dt\Big)^2 \leq \int_0^T \|\dot{c}(t)\|^2 dt \int_0^T 1^2 dt = 2A(c)T$$

with equality if and only if $\|\dot{c}(t)\| = \text{const}$. □

Since Corollary 9.5.2 shows that we may confine our attention to curves of constant speed when looking for critical curves of the action, this last proposition shows that we may equivalently look for curves of constant speed which are critical for length. Note that this is a simplification: If $c\colon [0,T] \to M$ has constant speed $\|\dot{c}(0)\|$, then $l(c) = T\|\dot{c}(0)\|$.

Let us consider the exponential map now, in the setting of the geodesic flow. Homogeneity of the Lagrangian of the geodesic flow implies:

Lemma 9.5.4. *Let $c\colon [0,T] \to M$ be a geodesic and $\tau\colon [0,T/a] \to [0,T]$, $t \mapsto \tau(t) = at$. Then $\tilde{c} := c \circ \tau$ is a geodesic.*

For $v \in T_xM$ we now denote by c_v the geodesic with $c(0) = x$, $\dot{c}(0) = v$. We found in Section 5.3a that the exponential map $\exp_x\colon v \mapsto \exp_x v := c_v(\epsilon)$ is an embedding of $\{v \in T_xM \mid \|v\| \leq R\}$ into M, where ϵ is a possibly very small constant depending on R. By Lemma 9.5.4 we immediately see, however, that for $\delta = \epsilon R$ we obtain a smooth embedding

$$\exp_x\colon \overline{B(0,\delta)} = \{v \in T_xM \mid \|v\| \leq \delta\} \to M,\ v \mapsto c_v(1). \tag{9.5.1}$$

We define $r_x > 0$ to be the supremum of those δ for which $\exp_x$ is injective on the δ-ball $B(0,\delta)$. r_x is called the injectivity radius of $\exp_x$ or the *injectivity radius* of (M, g) at $x \in M$. We can now show that locally geodesics are unique minima of the action functional.

Theorem 9.5.5. *Let M be a Riemannian manifold, $x \in M$, r_x the injectivity radius of $\exp_x$, $v \in T_xM$, $d := \|v\| < r_x$, $y = \exp_x v$, and $\Gamma_{x,y}$ the space of continuous piecewise-C^1 curves $c\colon [0,1] \to M$ such that $c(0) = x$ and $c(1) = y$. Then the action functional on $\Gamma_{x,y}$ has a unique minimum at the geodesic c_v and $A(c_v) = d^2/2$.*

Remark. Since r_x is (Lipschitz) continuous in x it has a positive minimum on any compact subset of M. In particular for compact manifolds we call $r(M) := \min_{x\in M} r_x > 0$ the *injectivity radius* of M.

Corollary 9.5.6. *Let M be a compact Riemannian manifold and $x, y \in M$ such that $d(x,y) < r(M)$. Then the action functional A on $\Gamma_{x,y}$ has a unique minimum at the geodesic $c_{\exp_x^{-1} y}$.*

Proof. We first prove the

Lemma 9.5.7. (Gauss Lemma) *If $x \in M$, $\sigma: [-1,1] \xrightarrow{C^1} T_xM$, and $\|\sigma(t)\| = r < r_x$ for all $t \in [-1,1]$, then $\frac{d}{ds}\big|_{s=0} \exp_x \sigma(s) \perp \frac{d}{dt}\big|_{t=1} \exp_x t\sigma(0)$.*

Proof. Since $c: t \mapsto \exp_x t\sigma(0)$ is a geodesic, the First Variation Formula (Proposition 9.4.1) yields

$$\frac{d}{ds}\Big|_{s=0} A(c_s) = \sum_{ik} g_{ik}(c_0(1)) \frac{dc_k}{dt}(1) \frac{d(c_s)_i}{ds}\Big|_{s=0}(1),$$

where $c_s(t) = \exp_x t\sigma(s)$. But $A(c_s) = r^2/2$ for all s, so the sum vanishes, as claimed. □

The Gauss Lemma says that if we introduce polar coordinates $B(0, r_x) \setminus \{0\} \to (0, r_x) \times S^{n-1}$ then via $\exp_x^{-1}$ we get polar coordinates on a neighborhood of x with the property that the radial vector $\frac{\partial}{\partial r}$ is orthogonal to ∂S^{n-1}. We use this fact by noting that in particular for $z \in \exp_x B(0, r_x)$ and $w \in T_zM$ we have $\|w\| \geq |g(w, \frac{\partial}{\partial r})|$ with strict inequality unless w is a multiple of $\frac{\partial}{\partial r}$.

Suppose then that $y = \exp_x v$ with $d := \|v\| < r_x$ and $c: [0,1] \to M$ is a curve connecting x and y contained in $\exp_x B(0,d)$. Then we have

$$\int_0^1 g(\dot{c}(t), \dot{c}(t))dt \geq \int_0^1 (g(\dot{c}(t), \frac{\partial}{\partial r}))^2 dt \geq \Big(\int_0^1 g(\dot{c}(t), \frac{\partial}{\partial r})dt\Big)^2 \geq d^2 = 2A(c_v).$$

The first inequality is strict unless c is a reparameterization of c_v. But if c is a reparameterization of c_v then the second inequality (the Cauchy–Schwarz inequality) is strict unless $c = c_v$. Thus if c is as before and $c \neq c_v$ then $A(c) > d^2/2$.

This argument also shows that $l(c) > d$ unless c is a reparameterization of c_v. Varying y we see that any curve c from x that is not contained in $\exp_x B(0,d)$ but with $c(1) \in \exp_x B(0,d)$ has length greater than d and hence action greater than $d^2/2$. □

Our next result deals with global minimization of the action (or length) functional. The proof uses a very fruitful method of "broken geodesics" introduced by Morse, which allows us to use the local theory developed here in order to reduce the problem of global minimization to a finite-dimensional problem.

As before let $\Gamma_{x,y}$ be the space of piecewise-C^1 continuous maps $c: [0,1] \to M$ such that $c(0) = x$ and $c(1) = y$. The number of discontinuity points for each curve $c \in \Gamma_{x,y}$ is finite, but no bound for that number is assumed. Obviously the length and action functionals are defined on $\Gamma_{x,y}$.

Theorem 9.5.8. *Let M be a complete connected Riemannian manifold and $x, y \in M$. Then the action functional A on $\Gamma_{x,y}$ reaches a (not necessarily unique) minimum on a smooth Euler–Lagrange geodesic.*

Proof. By connectedness there exists a smooth curve $\sigma \in \Gamma_{x,y}$. Without loss of generality we may assume that σ is parameterized with constant speed. Let $L := l(\sigma)$. Then

$$A(\sigma) = \frac{L^2}{2} \tag{9.5.2}$$

by Proposition 9.5.3. Consider the ball $B = B(x, 2L) \subset M$. Its closure is compact. Any point $z \in \partial B$ has the property that $d(x, z) = 2L$. Let $c \in \Gamma_{x,y}$. If $c([0,1]) \not\subset B$ then this set contains a point $z \in \partial B$ and hence $l(c) \geq 2L$ and by Proposition 9.5.3 and (9.5.2)

$$A(c) > 2L^2 > A(\sigma). \tag{9.5.3}$$

Thus

$$\inf_{c \in \Gamma_{x,y}} A(c) = \inf\{A(c) \mid c \in \Gamma_{x,y}, c([0,1]) \subset B\}. \tag{9.5.4}$$

Since $\bar{B}$ is compact, the injectivity radius r_x for $x \in \bar{B}$ is bounded from below by some $R > 0$. Let $\tilde{\Gamma} := \{c \in \Gamma_{x,y} \mid l(c) \leq l(\sigma) = L\}$. If $c \in \tilde{\Gamma}$ then $c([0,1]) \subset B = B(x, 2L)$. For each curve $c \in \tilde{\Gamma}$ we can find numbers $0 = t_0 < t_1 < \cdots < t_k = 1$ such that $L(c_{\restriction [t_i, t_{i+1}]}) < R$ and $k \leq [l(\sigma)/R] + 1$. Now we apply Morse's trick of introducing "broken geodesics". We replace σ by the piecewise geodesic $\tilde{\sigma}$ such that

$$\tilde{\sigma}(s) = c_{\exp^{-1}_{\sigma(t_i)} \sigma(t_{i+1})}\left(\frac{s - t_i}{t_{i+1} - t_i}\right) \quad \text{for } s \in [t_i, t_{i+1}].$$

In other words, we replace the segment $\sigma([t_i, t_{i+1}])$ by the appropriate reparameterization of the geodesic giving the unique minimum of the action functional A on $\Gamma_{\sigma(t_i), \sigma(t_{i+1})}$. By Theorem 9.5.5 we have $A(\tilde{\sigma}) \leq A(\sigma)$. Reparameterizing $\tilde{\sigma}$ with constant speed gives a curve $\tilde{\tilde{\sigma}}$ with $A(\tilde{\tilde{\sigma}}) \leq A(\tilde{\sigma})$. The closure Γ^0 of the set of all curves $\tilde{\tilde{\sigma}}$ obtained from $\sigma \in \tilde{\Gamma}$ is compact in the C^0 topology. This follows from the fact that any curve in Γ^0 is determined by an ordered sequence of at most $[l(\sigma)/R] + 1$ points of B and from the observation that as long as Theorem 9.5.5 applies, the unique minimum in $\Gamma_{x,y}$ depends continuously on x, y in the C^0 topology. Thus the functional A attains its minimum on Γ^0 and hence on $\tilde{\Gamma}$. By definition of $\tilde{\Gamma}$ this minimum is also the minimum on $\Gamma_{x,y}$.

It remains to show that any global minimum of A on $\Gamma_{x,y}$ is a C^2 curve so that, since it is a critical point of the action functional A, it is a solution of the Euler–Lagrange equation. Pick $t \in [0,1]$ and ϵ so small that $l(c(t-\epsilon, t+\epsilon)) < R$. Then by Theorem 9.5.5 $c_{\restriction (t-\epsilon, t+\epsilon)}$ is a constant-speed reparameterization of the geodesic $c_{\exp^{-1}_{c(t-\epsilon)} c(t+\epsilon)}$, hence C^2. □

Now let us consider instead of the whole space $\Gamma_{x,y}$ any connected component Σ, that is, the space of all continuous piecewise-C^1 curves homotopic rel endpoints to a given curve. The proof of Theorem 9.5.8 remains valid if we replace $\Gamma_{x,y}$ in the formulation by Σ. The curve c should be chosen from Σ and the set $\tilde{\Gamma}$ should be defined as $\tilde{\Gamma} := \{c \in \Sigma \mid l(c) \leq l(\sigma)\}$. Approximation by "broken geodesics" is local, whence $\tilde{\sigma} \in \Sigma$ and $\tilde{\tilde{\sigma}} \in \Sigma$. Thus we have established the following result:

Theorem 9.5.9. *The action functional A attains its minimum on Σ on a smooth curve, which is a solution of the Euler–Lagrange equation.*

A slight modification of the same argument allows us to find closed geodesics.

Theorem 9.5.10. *Let M be a compact Riemannian manifold and Π a nontrivial free homotopy class of closed piecewise-C^1 curves on M. The action functional A attains its minimum on Π and every curve where the minimum is reached is a closed geodesic.*

Remark. Compactness of M is essential for this theorem, as the example of the pseudosphere shows (see Figure 5.4.4). The curves γ_c represented by horizontal segments $\operatorname{Im} z = c$ are not null-homotopic and $l(\gamma_c) \to 0$ as $c \to \infty$. This homotopy class does not contain any closed geodesics.

Proof. Note first that any closed curve of length less than twice the injectivity radius R is contractible. Thus $\inf_{c \in \Pi} A(c) \geq 2R^2 > 0$. Let σ be a closed curve representing Π. As in the proof of Theorem 9.5.8 every curve $c \in \Pi$ such that $l(c) \leq l(\sigma)$ can be approximated by a shorter broken geodesic $\tilde{\sigma}$ from Π with no more than $l(\sigma)/R$ arcs parameterized with constant speed. The space of such curves is compact in the C^0 topology; hence A attains its (positive) minimum on Π. The argument for smoothness does not change. □

Exercises

9.5.1. *Show that for the unit disk with the Poincaré metric $g(v, w) = (1 - (x^2 + y^2))^{-2}\langle v, w\rangle$, where $\langle \cdot, \cdot \rangle$ denotes the Euclidean metric, the exponential map at every point is a diffeomorphism from the tangent space onto the disk.*

9.5.2. *Show that for any Riemannian manifold the injectivity radius is at most half the length of the shortest closed geodesic. Give an example where equality occurs.*

9.5.3. *Let M be a compact manifold with nontrivial fundamental group. Show that for any Riemannian metric on M there is a nontrivial closed geodesic.*

9.5.4. *Show that no solution of the Euler–Lagrange equation is a local maximum of the action on the space of smooth curves with fixed endpoints.*

9.5.5. *Show that a geodesic satisfies an equation*

$$\frac{d^2c_k}{dt^2} + \sum_{ij=1}^{n} \Gamma_{ij}^k(c(t))\frac{dc_i}{dt}\frac{dc_j}{dt} = 0 \tag{9.5.5}$$

and determine the Γ_{ij}^k. *This equation is called the geodesic equation*

6. Minimal geodesics

We will demonstrate now how the variational approach allows us to identify some special orbits of the geodesic flow.

Definition 9.6.1. Let M be a complete Riemannian manifold. $c\colon \mathbb{R} \to M$ is called *minimal* if for some (hence any) lift $\tilde{c}\colon \mathbb{R} \to \widetilde{M}$ of c to the universal cover $\widetilde{M}$ of M and any two points $t_1, t_2 \in \mathbb{R}$ the renormalized segment $\tilde{c}\restriction_{[t_1,t_2]}$ realizes the minimum of the action functional on the space $\Gamma_{\tilde{c}(t_1),\tilde{c}(t_2)}$. In other words, for any two points on the lift of c the segment of the lift between these points is the shortest curve between its ends.

Since every geodesic parameterized by length is uniquely determined by its tangent vector at any point we can speak about *minimal* vectors in the unit tangent bundle SM. The set $\mathcal{M}$ of all minimal vectors is by definition invariant under the geodesic flow. It is also closed since

$$\mathcal{M} = \bigcap_{n\in\mathbb{N}} \mathcal{M}_n,$$

where

$$\mathcal{M}_T := \{v \in SM \mid \exists s \in [-T + \sqrt{T}, -\sqrt{T}] \text{ so that } c_v \text{ is minimal on } [s, s+T]\},$$

and each $\mathcal{M}_n$ is closed.

Proposition 9.6.2. *If M is compact and connected and the fundamental group $\pi_1(M)$ is infinite, then $\mathcal{M} \neq \varnothing$.*

Proof. Since $\pi_1(M)$ is infinite the universal cover $\widetilde{M}$ is noncompact. In particular for any $n \in \mathbb{N}$ one can find points $x_n, y_n \in \widetilde{M}$ such that $d(x_n, y_n) \geq 3n$. By Theorem 9.5.8 there exists a shortest curve γ_n connecting x_n and y_n and parameterized by arc length. Let $v_n \in S\widetilde{M}$ be the tangent vector to γ_n at its midpoint. Then the projection of v_n to SM is a vector $u_n \in \mathcal{M}_n$. Thus $\mathcal{M}_n$ is nonempty. Since it is compact and $\mathcal{M}_{n+1} \subset \mathcal{M}_n$ we conclude that $\mathcal{M} \neq \varnothing$. □

Remark. In general closed geodesics that realize the minimum of the action functional in a free homotopy class (Theorem 9.5.10) may not be minimal. This is true, however, for orientable surfaces. We will return to this question later.

We have shown that a purely topological property of a compact manifold guarantees existence of minimal geodesics for every Riemannian metric on that manifold. Next we will extend this connection in a quantitative way by showing that a stronger topological property, namely, exponential growth of the fundamental group $\pi_1(M)$, guarantees dynamical complexity of the set of minimal geodesics for any metric on M, namely, positivity of topological entropy of the geodesic flow restricted to that set.

First we define a geometric characteristic of a Riemannian manifold M. Let x be a point on the universal cover $\widetilde{M}$ and $B(x,r)$ the ball around x of radius r.

Proposition 9.6.3. $\lim_{r\to\infty}\dfrac{1}{r}\log\operatorname{vol}(B(x,r))$ *exists and is independent of* x.

Proof. First, since $\operatorname{vol}(B(x,r))$ is invariant under deck transformations (which are isometries of $\widetilde{M}$) and since one can find a compact fundamental domain $\mathcal{D}\subset\widetilde{M}$ for M of diameter less than d, say, we immediately have

$$\operatorname{vol}(B(x,r))\le\operatorname{vol}(B(y,r+d)) \tag{9.6.1}$$

for any $x,y\in\widetilde{M}$. This follows from the fact that

$$B(x,r)\subset B(y,r+d) \tag{9.6.2}$$

for $x,y\in\mathcal{D}$. Inequality (9.6.1) immediately implies independence of the limit if it exists.

In order to prove existence of the limit we will establish the following inequality:

$$\operatorname{vol}(B(x,R_1+R_2))<\frac{1}{a}\operatorname{vol}(B(x,R_1))\operatorname{vol}(B(x,R_2+d+1)),$$

where a is the minimal volume of a 1/2-ball in M (which may be less than the corresponding minimal volume on $\widetilde{M}$). If N is a maximal 1-separated subset of $B(x,R_1)$ then the balls of radius 1/2 around points of N are pairwise disjoint, so

$$a\operatorname{card}N\le\operatorname{vol}(B(x,R_1)). \tag{9.6.3}$$

Now take balls of radius R_2+1 around points of N. Their union covers $B(x,R_1+R_2)$ so

$$\operatorname{vol}(B(x,R_1+R_2))\le\operatorname{card}N\sup_{z\in N}\operatorname{vol}(B(x,R_2+1)).$$

Using (9.6.3) and (9.6.1) we obtain

$$\operatorname{vol}(B(x,R_1+R_2))\le\frac{1}{a}\operatorname{vol}(B(x,R_1))\operatorname{vol}(B(x,R_2+1+d)).$$

This finishes the proof because of the following elementary fact which refines the subadditivity argument used on numerous occasions beginning with Lemma 3.1.5. ☐

Proposition 9.6.4. *If $a_{m+n} \le a_n + a_{m+k} + L$ for all $m, n \in \mathbb{N}$ and some k and L then $\lim_{n\to\infty} a_n/n \in \mathbb{R} \cup \{-\infty\}$ exists.*

Proof. Note first that $a_{m+k} \le a_m + a_{2k} + L$ and hence $a_{m+n} \le a_m + a_n + a_{2k} + 2L = a_m + a_n + L'$, so we may assume $k = 0$. If $\underline{\lim}_{n\to\infty} a_n/n < b < c$, $n > 2L/(c-b)$, $a_n/n < b$ and $l \ge n$, $l(c-b) > 2\max_{r<n} a_r$ write $l = nk + r$ with $r < n$ to get $a_l/l \le (ka_n + a_r + kL)/l \le a_n/n + a_r/l + (L/n) < c$. □

We will also use this fact several times in Chapters 11, 13, 18, and 20.

Definition 9.6.5. $v(M) := \lim_{r\to\infty} \dfrac{1}{r} \log \operatorname{vol}(B(x,r))$ is called the *volume growth* of M.

Proposition 9.6.6. *For $x \in \widetilde{M}$*

$$v(M) = \lim_{r\to\infty} \frac{1}{r} \log \operatorname{card}\{\gamma \in \pi_1(M) \mid \gamma(x) \in B(x,r)\}.$$

Proof. Fix a fundamental domain $\mathcal{D}$ such that x is in the interior of $\mathcal{D}$ and take $d > 0$ such that $B(x,d) \subset \mathcal{D}$ and $R > 0$ such that $\mathcal{D} \subset B(x,R)$. Then $\{B(\gamma(x),d) \mid \gamma \in \pi_1(M)\}$ is a collection of pairwise disjoint sets, so $\operatorname{card}\{\gamma \in \pi_1(M) \mid \gamma(x) \in B(x,r)\} \le \dfrac{\operatorname{vol}(B(x,r))}{\operatorname{vol}(B(x,d))}$. $\{B(\gamma(x),R) \mid \gamma(x) \in B(x,r)\}$ is a covering of $B(x,r)$, so $\operatorname{card}\{\gamma \in \pi_1(M) \mid \gamma(x) \in B(x,r)\} \ge \dfrac{\operatorname{vol}(B(x,r))}{\operatorname{vol}(B(x,R))}$. □

Remark. In particular we thus note that $v(M) > 0$ if and only if the fundamental group of M has exponential growth for any set of generators.

We can now show that if the fundamental group has exponential growth then the set of minimal geodesics is large enough to carry positive topological entropy for the geodesic flow.

Theorem 9.6.7.

$$h_{\text{top}}(g_t\restriction_{\mathcal{M}}) \ge v(M)$$

for the geodesic flow g_t on SM.

Proof. Fix $x \in \widetilde{M}$, $T, \delta > 0$, and a maximal $3\delta T$-separated set N in the annulus $B(x,(1+\delta)T) \smallsetminus B(x,T)$. If $K_T := \sup_{y\in M} \operatorname{vol}(B(y,3\delta T))$ then

$$\operatorname{card} N \ge \frac{1}{K_T}(\operatorname{vol}(B(x,T)) - \operatorname{vol}(B(x,(1-\delta)T))) \ge e^{v(M)(1-3\delta)T}$$

for sufficiently large T.

If $y \in N$ then by Theorem 9.5.8 there exists a geodesic c_y of length $l_y = d(x,y)$ parameterized by arc length joining x and y. If $y_1 \ne y_2 \in N$ and $p_i = c_{y_i}(T)$ then $d(p_1,p_2) \ge 3\delta T - 2\delta T = \delta T$. Thus the orbit segments

$$\{c_y\restriction_{[0,T]} \mid y \in N\}$$

are a $(T, \delta T)$-separated set in $\widetilde{M}$. Now we want to show that the projections of these segments to M are separated as well. Let us write $\gamma_y := \pi \circ c_y$. Suppose that $\delta T > r(M)$, the injectivity radius. Whenever $x, y \in \widetilde{M}$ and $d(x, y) < r(M)$ we have $d(\pi(x), \pi(y)) = d(x, y)$. Now $d(c_{y_1}(t), c_{y_2}(t))$ varies from 0 to at least $2\delta T$ when $t \in [0, T]$ and $y_1 \neq y_2 \in N$, and in particular it attains any value between $r(M)/2$ and $r(M)$.

$$d(\gamma_{y_1}(t), \gamma_{y_2}(t)) > \frac{1}{2} r(M)$$

for some $t \in [0, T]$. Hence

$$\{\gamma_y \restriction_{[0,t]} \mid y \in N\}$$

is $(T, r(M)/2)$-separated. Note that by construction we have

$$V(x, N, T) := \{\dot{c}_y(t) \mid y \in N, t \in [\sqrt{T}, T - \sqrt{T}]\} \subset \mathcal{M}_T,$$

and that $S := \{\dot{\gamma}_y(\sqrt{T}) \mid y \in N\}$ is $(T - 2\sqrt{T}, r(M)/2)$-separated.

If we take $N(T, t, d)$ to be the maximal cardinality of a (t, d)-separated subset of $\mathcal{M}_T$ then we see that $N(T, T, r(M)/2) \geq e^{v(M)(1-3\delta)T}$. In fact, we have found a set of orbits in $\mathcal{M}_T$ that is $(T, r(M)/2)$-separated and has cardinality at least $e^{v(M)(1-3\delta)T}$.

Let $m \in \mathbb{N}$ be such that $T - 2\sqrt{T} = mt + k$ and $0 \leq k \leq t$. For $j \in \{0, \dots, m-1\}$ consider the set $P_j := \{\dot{\gamma}_y(\sqrt{T} + jt) \mid \dot{\gamma}_y(\sqrt{T}) \in S\}$ and let Q_j be the maximal cardinality of a $(t, r(M)/2)$-separated subset of P_j. Since S is $(T - 2\sqrt{T}, r(M)/2)$-separated we get that $Q_1 \cdots Q_{m-1} \geq \operatorname{card} S = \operatorname{card} N \geq e^{(1-3\delta)v(M)t}$, so for some $i \in \{0, \dots, m-1\}$ we must have

$$Q_i \geq \sqrt[m]{\operatorname{card} S} = e^{v(M)(1-3\delta)T/m} = e^{v(M)(1-3\delta)(1+(k+2\sqrt{T})/m)} \geq e^{v(M)(1-4\delta)t}$$

for large T. Thus $\mathcal{M}_T$ and hence $\mathcal{M}$ contains a $(t, r(M)/2)$-separated set of cardinality at least $e^{v(M)(1-4\delta)t}$. Since $\delta > 0$ can be taken arbitrarily small this proves the theorem. □

Exercises

9.6.1. *Calculate the volume growth $v(M)$, where M is a compact factor of the unit disk with the Poincaré metric $g(v, w) = (1 - (x^2 + y^2))^{-2}\langle v, w\rangle$, with $\langle \cdot, \cdot \rangle$ the Euclidean metric.*

9.6.2. *Construct an example of a Riemannian metric on the torus $\mathbb{T}^2$ such that not all geodesics are minimal.*

7. Minimal geodesics on compact surfaces

We now consider compact orientable surfaces with infinite fundamental group, that is, the sphere with $g \geq 1$ handles. See Section 5 of the Appendix for the description of the fundamental group and universal cover for such surfaces. We will consider free homotopy classes of closed geodesics. It is useful to note that powers of a homotopy class are well defined. The following result is an analog of Proposition 9.3.14 for geodesic flows on surfaces.

Theorem 9.7.1. *Let (M, g) be a compact orientable surface with infinite fundamental group and a Riemannian metric g. Let σ be a free homotopy class of closed curves on M and $m \in \mathbb{N}$. Then the shortest curves in σ^m are of the form* $c^m := \underbrace{c \cdot c \cdots c}_{m \text{ times}}$, *with $c \in \sigma$.*

Proof. Let c be such a geodesic. Consider a lift $\tilde{c}$ to the universal cover $\widetilde{M}$ of M. $\widetilde{M}$ is homeomorphic to the Euclidean plane $\mathbb{R}^2$. Since c is a closed geodesic the curve $\tilde{c}$ is invariant under the action of a certain element $\gamma^m \in \pi_1(M)$ as a deck transformation (that is, $\gamma^m \circ \tilde{c}$ is a reparameterization of $\tilde{c}$ with the same orientation). If it is invariant under γ we are done.

If not, then $\gamma^m \circ \tilde{c}$ and $\tilde{c}$ are distinct curves. By the Jordan curve theorem $\tilde{c}$ divides $\mathbb{R}^2$ into two (open) components C_+ and C_-. Thus the image $\gamma \circ \tilde{c}$ either intersects $\tilde{c}$ or is contained in one component, in C_+, say. Since M is an orientable surface the second alternative means that $\gamma C_+ \subset C_+$ and hence $\gamma^m(\tilde{c}) \subset C_+$, a contradiction. Thus $\gamma \circ \tilde{c}$ intersects $\tilde{c}$. Take a point x of intersection. The intersection is transverse since $\gamma \circ \tilde{c}$ and $\tilde{c}$ are geodesics. The point $x' := \gamma^m x$ is also a transverse point of intersection. Construct two new curves connecting x and x' as follows. c_+ is the curve obtained from c by replacing each segment of c between two successive intersection points by the corresponding segment of $\gamma^m \tilde{c}$ if the latter is contained in C_+. c_- is the curve consisting of the remaining segments of the union of $\tilde{c}$ and $\gamma^m \tilde{c}$. c_+ and c_- are γ^m-invariant, hence represent lifts of closed curves in the same free homotopy class as c, and the sum of the lengths of the projections is equal to twice the length of c. But both projections are properly broken geodesics and thus neither of them is minimal in its free homotopy class. Thus s is not minimal either, contradicting the hypothesis. □

Remark. Notice the similarity between this argument and the use of Corollary 9.3.13 in the proof of Proposition 9.3.14.

Theorem 9.7.2. *The shortest geodesic in any nontrivial free homotopy class Σ on M is minimal.*

Proof. As before let c be such a geodesic. Consider a lift $\tilde{c}$ to the universal cover $\widetilde{M}$ of M. Let T be its period. If $t' - t < T$ then $\tilde{c}\restriction_{[t,t']}$ is the shortest curve connecting $\tilde{c}(t)$ and $\tilde{c}(t)$ since otherwise we could shorten c by inserting the projection of a shorter curve connecting these points. By Theorem 9.7.1 c^n

is the shortest geodesic in its homotopy class, so we can replace T by nT in the previous argument. Thus $\tilde{c}$ minimizes length between any of its points. □

It is naturally interesting to inquire about the growth rate of minimal geodesics as a function of the length. The situation here is considerably different for the torus and surfaces of higher genus. For the torus a lower bound is given by the rate for a flat metric, where the classes of minimal geodesics are in one-to-one correspondence with the nonzero elements of the integer lattice Z^2, and hence grow quadratically with length (with a multiplicative constant depending on the flat metric). An arbitrary metric can be viewed as the projection of a periodic metric on $\mathbb{R}^2$. By compactness of the torus this metric is related to the Euclidean one by a bounded factor, so the induced distance is uniformly equivalent to the Euclidean one (that is, the ratio of the distances is bounded between positive numbers) . Since the length of a shortest closed geodesic in the homotopy class corresponding to $k \in \mathbb{Z}^2$ is $\min_{p\in\mathbb{R}^2} d(p, p+k)$, this shows that for any metric the growth rate of the numbers of minimal geodesics is at least quadratic.

For surfaces of higher genus the role of flat metrics as a reference metric for toral metrics is played by metrics of constant negative curvature. Such metrics were discussed in Section 5.4. Later we will show that for any such metric the number of closed geodesics (which are all minimal in that case) grows exponentially with a very precise asymptotic (see Theorem 18.5.7 and Theorem 20.6.9).[1] The universal cover of M can be viewed as the Poincaré disk; the deck transformations are the linear fractional transformations. The metric on M lifts to a metric on $\widetilde{M}$ which is invariant under deck transformations. Since M is compact, such a metric is determined by its restriction to a compact fundamental domain. Since deck transformations preserve both the Poincaré metric and the given metric, the two are uniformly equivalent, so the induced distances have ratio bounded by some constants C and $1/C$. This implies that the number $N(T)$ of minimal geodesics of length at most T satisfies $N(T) \geq N_0(T/C)$, where N_0 is the corresponding number for the metric of constant curvature. Thus $N(T)$ has an exponential lower bound. Notice the difference in the conclusions. Since for the function $f(T) = T^2$ we have $f(cT) = c^2 f(T)$, having a quadratic lower estimate is an invariant in the toral case. For the exponential function $f(T) = e^{aT}$ we have $f(cT) = e^{acT}$, so only the exponential character of the estimate is preserved, but the exponent is not.[2]

Exercises

9.7.1. *Describe a metric on the Möbius strip with the property that the shortest geodesic in the homotopy class Π for the generator γ_1 of the fundamental group and the shortest geodesic in Π^2 are distinct.*

9.7.2. *Let M be a compact nonorientable surface and Π a free homotopy class. Show that the shortest closed geodesic in Π^2 is globally minimal.*

9.7.3*. *Describe a metric on the three-dimensional torus $\mathbb{T}^3$ such that for some free homotopy class some shortest geodesic is not globally minimal.*

9.7.4. *Suppose M is a compact factor of the unit disk with the Poincaré metric $g(v,w) = (1-(x^2+y^2))^{-2}\langle v,w\rangle$, where $\langle\cdot,\cdot\rangle$ denotes the Euclidean metric. Show that $\mathcal{M} = SM$, that is, all geodesics are minimal.*

Part 3

Low-dimensional phenomena

10

Introduction: What is low-dimensional dynamics?

a. Motivation. In Chapters 2, 3, and 4 we described two main thrusts in the theory of dynamical systems. The first one aims at the classification of various classes of systems up to smooth conjugacy, topological conjugacy, semiconjugacy with good properties, and, in the case of flows, orbit equivalence. The main tools for that approach are finding models and describing moduli. The second thrust is to describe (or calculate) the principal asymptotic invariants for various classes of systems as well as for concrete examples. These invariants include growth of periodic orbits, topological entropy, homotopical and homological properties, recurrence properties, and statistical properties reflected in various properties of invariant measures.

For general dynamical systems only partial results in both directions can be obtained. In a very general sense the unifying theme of Part 2 was the identification and exploration of elements in the orbit structure that provide information in both directions, primarily the second one. More specifically we found out that such paradigms as hyperbolicity (Chapter 6), transversality (Chapter 7), the global topological structure (Chapter 8), and variational character (Chapter 9) ensure under certain conditions the presence of sufficiently rich and "interesting" asymptotic behavior.

However, there are two major areas where the program outlined and illustrated in Chapters 2, 3, and 4 can be advanced considerably further. Those are dynamical systems with low-dimensional phase spaces studied in the present part of the book and dynamical systems with hyperbolic structure which we began to discuss in Chapter 6 and will study in depth in Part 4.

When we talk about low-dimensional dynamical systems or, more precisely, dynamical systems with low-dimensional phase space, we have in mind the straightforward old-fashioned concept of dimension of a Euclidean space applied to a connected manifold. In other words we assume that the phase space is a metrizable space where a neighborhood of each point is homeomorphic to a Euclidean space of a particular dimension n which is called the dimension of the space in question. We do not touch upon either more general topological notions of dimension according to which, for example, the Cantor set has dimension zero, or various concepts of dimension depending on a metric such as Hausdorff dimension which, for example, takes arbitrary values between zero and infinity (including both) on various metric spaces homeomorphic to the same Cantor set.

Before proceeding to discuss the precise meaning of "low" dimension in the context of our study we list additional reasons for singling out dynamical systems acting on low-dimensional manifolds as a separate topic of investigation.

First, and least importantly, low-dimensional dynamical systems are expressed analytically via functions of a small number of variables and can be represented visually and graphically with the help of such conventional geometric objects as curves and surfaces. Formulas, equations, and graphics involved in the description of such systems certainly make this area of dynamics, at least superficially, the easiest in terms of immediate perception.

Second, quite often low-dimensional situations serve as excellent test cases and demonstration tools for important dynamical phenomena of general significance. Those phenomena sometimes appear in low-dimensional situations in a clear form stripped of often cumbersome complications. This makes it possible to grasp and demonstrate the essence of the phenomena in question in the most efficient way. As an example, we can point out the use of expanding circle maps, plane horseshoes, and hyperbolic automorphisms of the two-torus to introduce various aspects of hyperbolic behavior in dynamics which we then started to study systematically in Chapter 6 and which constitutes the subject of Part 4 of this book. Another example is given by the concepts of degree and index discussed in Chapter 8. The degree of a circle map introduced in Section 2.4 and the index of a fixed point in a two-dimensional manifold are nice easily visualizable models for the more sophisticated general case.

Finally there is a more speculative reason. Since the physical world is, or at least looks, three-dimensional, some of the geometric intuition that goes into the concept of a higher-dimensional space is based on low-dimensional analogies. This means that in rigorously defined higher-dimensional objects certain phenomena not anticipated by this intuition might (and actually do) appear. Since differentiable and to a lesser extent topological dynamics are based on the aforementioned intuition it makes sense to look more thoroughly into the situation where this intuition is more directly applicable and try to separate phenomena and paradigms that are specific to the low-dimensional cases from those of more universal significance.

b. The intermediate value property and conformality. Passing to a more technical level of discussion we should point out two fundamental mathematical facts, one topological and one differentiable, that distinguish one-dimensional manifolds from higher-dimensional ones.

The topological fact is that in a one-dimensional space a small neighborhood of a point is separated by the point into two connected components, whereas in a manifold of higher dimension a neighborhood remains connected after removing a point. This property very closely corresponds to the well-known *Intermediate Value Theorem:* Every continuous function of one variable defined on a closed interval takes all values between the values at the ends. Instances of the use of the Intermediate Value Theorem for dynamical purposes appeared in Proposition 1.1.6, Lemma 2.4.7, and Proposition 8.2.4.

The differentiable fact is *conformality* of any nonconstant linear function of one variable: If $Ax = ax + b$, where a, b, x are real numbers and $a \neq 0$, then for every x, y, z

$$\frac{\|Az - Ax\|}{\|Ay - Ax\|} = \frac{\|z - x\|}{\|y - x\|}. \tag{10.0.1}$$

For linear maps $A: \mathbb{R}^n \to \mathbb{R}^n$ for $n \geq 2$ (10.0.1) is only true if $A = \lambda U$ where $\lambda \neq 0$ is a scalar and U is an orthogonal map. Let us call such linear maps *conformal.* Accordingly we will call a differentiable map $f: M \to M$ of a Riemannian manifold into itself *conformal* if its differential at every point $p \in M$ is a conformal linear map between the Euclidean spaces T_pM and $T_{f(p)}M$. A prototypical argument based on conformality for one-dimensional smooth maps appeared in the bounded distortion estimate in the proof of Lemma 5.1.18.

Thus every nonsingular differentiable map of a one-dimensional manifold is conformal. In dimension two one can view the sphere S^2 as the Riemann sphere, that is, the complex plane $\mathbb{C}$ combined with the single point at infinity. Then any holomorphic function $f: S^2 \to S^2$, that is, any rational map of the complex variable z, is a conformal map, albeit with critical points. In this particular case, however, the notion of conformality can be extended to the critical points. Of course, conformality of holomorphic functions is the central tenet of complex analysis; it implies much greater rigidity than in the one-dimensional real case. The availability of sophisticated tools from analysis of one complex variable makes complex dynamics such an interesting subject. In dimensions higher than two there are very few conformal maps, which reflects an even greater rigidity of the conformal stucture. While this has far-reaching applications in geometry (Mostow rigidity, etc.), higher-dimensional conformal structures play only a limited role in the traditional theory of dynamical systems.

In the course of the succeeding chapters we will show how the numerous implications of the two fundamental one-dimensional facts, the Intermediate Value Theorem and conformality of one-dimensional differentiable maps, appear in the study of low-dimensional dynamical systems. One should note that the influence of these phenomena is not limited to systems with one-dimensional phase space and holomorphic maps. Sometimes when the phase space has dimension two or three, important invariant structures associated with the system (for example, stable and unstable manifolds of hyperbolic periodic points; see Section 6.2a) are one-dimensional, so the fundamental facts apply to those objects. Furthermore, some orbits in higher-dimensional systems behave as if they came from one-dimensional situations (see Chapter 13, Aubry–Mather sets for twist maps).

In a continuous-time system locally the behavior along the flow or "time" direction is trivial at least away from the fixed points of the flow (compare with Section 0.3). Thus we may expect the asymptotic behavior of flows in an $(n+1)$-dimensional phase space to be only slightly more complicated than that of discrete-time dynamical systems in n-dimensional space. On the other hand, noninvertibility creates extra complexity of the orbit behavior because the number of preimages of a given point may grow with the number of iterates. Furthermore, conformality breaks down for general noninvertible systems even in dimension one. So we also expect that noninvertible dynamical systems exhibit certain phenomena similar to those of invertible systems in higher dimension; in fact the rule of thumb for discrete-time systems is that the asymptotic behavior of noninvertible systems in dimension n is similar to, but less complicated than, that of invertible systems in one dimension higher.

c. Very low-dimensional and low-dimensional systems. We will distinguish between two types of low-dimensional phenomena. For want of better terms we will call them "very low-dimensional" versus simply "low-dimensional". In the very low-dimensional setting only relatively simple types of recurrence and asymptotic orbit structure may occur, for example, the topological entropy of such systems is always equal to zero. The examples discussed in Sections 1.1–1.6 are typical for that class of phenomena even though the phase space in those examples is not always low-dimensional. On the other hand, some phenomena exhibited by those examples, such as the fact that topological transitivity implies minimality and unique ergodicity (cf. Propositions1.4.1, 4.2.2, 4.2.3), do not hold for all very low-dimensional systems where some sort of intermediate complexity may appear such as finiteness of the number of different orbit closures or ergodic measures not supported on periodic orbits (see Theorems 14.6.3 and 14.7.6).

In simply "low-dimensional" situations phenomena characteristic for general dynamical systems, for example, exponential growth rate for periodic points, positive topological entropy (Definition 3.1.3), complicated hyperbolic sets (Definition 6.4.2), and the presence of many invariant measures may appear. Our second group of basic smooth examples, that is, expanding maps from Section 1.7, quadratic maps and two-dimensional horseshoes from Section 2.5, and hyperbolic automorphisms of the two-torus (Section 1.8) are representative for that category. There are, however, two differences between the low-dimensional and the general situation. In the former case, some complicated dynamical phenomena appear in a simplified form; compare, for example, the Markov partition into parallelograms for a hyperbolic automorphism of the two-dimensional torus described in Section 2.5 with the general construction of Markov partitions for hyperbolic systems in Theorem 18.7.3. Furthermore, certain phenomena in low-dimensional situations, unlike in the general one, allow exhaustive classification (for example, the Sharkovsky Theorem 15.3.2 describing possible sets of periods for maps of the interval) or at least a good description. See the relation between topological entropy and periodic orbits for one-dimensional maps (Corollary 15.2.2) and for surface diffeomorphisms (Corollary S.5.11) and the description of invariant measures for the interval maps with zero topological entropy (Theorem 15.4.2).

d. Areas of low-dimensional dynamics. We start our discussion in the next two chapters by describing the prototypical very low-dimensional situation, namely, invertible discrete-time dynamical systems on the circle. Later in Chapter 14 we discuss a somewhat more complicated but still very low-dimensional situation, namely, flows on compact surfaces, and some invertible piecewise-continuous and piecewise-smooth maps of the interval and the circle that appear as section maps for such flows.

In Chapter 13 we continue the study, which began in Section 9.3, of a special class of systems with two-dimensional phase space, the twist maps (Definition 9.3.1), and find their orbits displaying essentially one-dimensional behavior. The guiding paradigm will be the structural theory of circle homeomorphisms from Chapter 11 which was not available in Chapter 9.

To summarize, according to our approach, very low-dimensional phenomena embrace the orbits of invertible continuous and piecewise-continuous maps of one-dimensional manifolds, flows on two-dimensional manifolds, and those special orbits in other situations that display a similar behavior.

In the very low-dimensional situation, a complete topological classification or a very detailed description of the orbit structure is often possible (see Theorems 11.2.7, 12.1.1, 14.1.1, 14.3.1, 14.6.3, 14.7.4, and Section 14.2). One should also emphasize the difference between the topological and smooth situations. The latter often excludes complications that may be present in the topological case.

The low-dimensional cases include:

(1) Noninvertible maps in dimension one (Chapters 15, 16).
(2) Noninvertible conformal maps in complex dimension one, that is, polynomial and rational maps of the Riemann sphere.
(3) Diffeomorphisms of compact surfaces.
(4) Smooth flows on compact 3-dimensional manifolds.

There are also semilocal counterparts of those situations. For example, the counterpart of (2) would include the study of invariant sets for nonrational maps. Only the first class of these systems is discussed in detail in this part of the book. We will touch upon the dynamics of rational maps of S^2 in Section 17.8 and will see there that they provide new topological types of hyperbolic sets. The last two cases have appeared and will continue to appear in the course of our study as suggestive examples and illustrations of particular phenomena (Sections 17.2, 17.5). The principal results in those areas involve the use of nonuniform hyperbolicity, Nielsen–Thurston theory, and other techniques that are not considered in the main part of this book. Some of those results appear in the Supplement; see in particular Theorem S.5.9, Corollary S.5.11, and Corollary S.5.13.

Exercises

Exercises 10.0.1–10.0.4 illustrate the use of the Intermediate Value Theorem in the dynamics of one-dimensional maps. In them we preview some of the ideas that will be developed in Chapter 15.

10.0.1. *Let I be an interval in $\mathbb{R}$ and let $f: I \to \mathbb{R}$ be a continuous map such that $\bar{I} \subset f(I)$. Then f has a fixed point.*

10.0.2. *Suppose $f: I \to \mathbb{R}$ is continuous and $f^2(x)$ and $f^3(x)$ are defined for some point x in the interval I and $f^3(x) < x < f(x) < f^2(x)$. Prove that there is a point $y \in I$ such that $f(y) \neq y$ and $f^3(y) = y$.*

10.0.3. *Suppose $f: [0,1] \to \mathbb{R}$ is continuous and such that $f(0) \leq 0$, $f(1) \leq 0$, and $f(t) > 1$ for some t. Prove that $P_n(f) \geq 2^n$. Prove furthermore that for every $n \geq 1$ there is a periodic point whose minimal positive period is n.*

10.0.4. *Under the same assumptions prove that $h_{\text{top}}(f \restriction_\Lambda) \geq \log 2$, where $\Lambda := \bigcap_{n=0}^{\infty} f^n(I)$ is the set of points for which all iterates are defined.*

The next problem contains another instance of a bounded distortion estimate similar to Lemma 5.1.18.

10.0.5. *Let $f: I \to \mathbb{R}$ or $f: S^1 \to S^1$ be a C^2 map and J an interval such that $f^n(x)$ is defined and $(f')^n(x) \neq 0$ for every $x \in J$. Consider the intervals $f^i(J)$, $i = 0, \dots, n-1$, and let N be the maximal multiplicity with which any point is covered by this system of intervals. Prove that for any $x, y \in J$*

$$\frac{(f^n)'(x)}{(f^n)'(y)} < \exp(KN),$$

where K is a constant depending on f but independent of x, y, and n.

Finally the last two exercises show that invertible conformal maps of the Riemann sphere have a very simple structure.

10.0.6. *Represent the sphere S^2 as the Riemann sphere, that is, the complex plane $\mathbb{C}$ with the single point ∞ added, and a coordinate system near infinity given by $w = 1/z$, where $z \in \mathbb{C}$. The standard Riemannian metric on S^2 is given by $ds^2 = (dx^2 + dy^2)/(x^2 + y^2)^2$, where $z = x + iy$. Prove that every C^1 conformal map $f: S^2 \to S^2$ without critical points, that is, such that the differential Df does not vanish at any point, is*

$$z \mapsto \frac{az+b}{cz+d} \text{ or } z \mapsto \frac{a\bar{z}+b}{c\bar{z}+d},$$

where $a, b, c, d \in \mathbb{C}$ and $ad - bc \neq 0$.

10.0.7. *Prove the assertion of the previous exercise after replacing the assumption that $Df \neq 0$ at every point by the assumption that in a neighborhood of any point f is an injective map.*

11

Homeomorphisms of the circle

Circle homeomorphisms have appeared in earlier chapters. Rotations (Section 1.3) provided examples that are easy to understand thoroughly. Poincaré posed the question of under which conditions a given homeomorphism or diffeomorphism is equivalent to a rotation. It turns out that, at least for sufficiently smooth maps, a single modulus, the rotation number, describes a topological class completely when it takes an irrational value and the complications appearing for rational values can be easily described. Even in the topological case irrationality of the rotation number guarantees a semiconjugacy with the corresponding rotation.

1. Rotation number

We used the lift to the universal cover on several occasions, for example, in Sections 2.4, 2.6, 8.2, 9.3. We have a natural projection $\pi: \mathbb{R} \to S^1 = \mathbb{R}/\mathbb{Z}, x \mapsto x+\mathbb{Z}$. This provides a lift of a homeomorphism $f: S^1 \to S^1$ to a homeomorphism $F: \mathbb{R} \to \mathbb{R}$ with the property

$$f \circ \pi = \pi \circ F. \tag{11.1.1}$$

Such a lift F is unique up to an additive integer constant. We will use the notation $[x] := \max\{k \in \mathbb{Z} \mid k \leq x\}$ for the *integer part* of a real number.

Proposition 11.1.1. *Let $f: S^1 \to S^1$ be an orientation-preserving homeomorphism and $F: \mathbb{R} \to \mathbb{R}$ a lift of f. Then*

$$\tau(F) := \lim_{|n| \to \infty} \frac{1}{n}(F^n(x) - x)$$

exists for all $x \in \mathbb{R}$. $\tau(F)$ is independent of x and well defined up to an integer, that is, if F_1, F_2 are lifts of f then $\tau(F_1) - \tau(F_2) = F_1 - F_2 \in \mathbb{Z}$. If f has a periodic point then $\tau(F)$ is rational.

This proposition justifies the following terminology:

Definition 11.1.2. $\tau(f) := \pi(\tau(F))$ is called the *rotation number* of f.

Proof of Proposition 11.1.1. (i) Independence of x: Since f is an orientation-preserving homeomorphism $\deg(f) = 1$, that is, $F(x+1) = F(x) + 1$, and for $x, y \in [0,1)$ we have $|F(y) - F(x)| < 1$. Consequently

$$\left|\frac{1}{n}|F^n(x) - x| - \frac{1}{n}|F^n(y) - y|\right| \leq \frac{1}{n}(|F^n(x) - F^n(y)| + |x - y|) \leq \frac{2}{n}$$

and the rotation numbers of x and y coincide.

(ii) Existence: Take $x \in \mathbb{R}$ and let $x_n = F^n(x)$, $a_n := x_n - x$, $k := [a_n]$. Then

$$\begin{aligned}a_{m+n} &= F^{m+n}(x) - x = F^m(x_n) - x_n + x_n - x \\ &= (F^m(x+k) - (x+k)) + (x_n - x) + (F^m(x_n) - F^m(x+k)) - (x_n - x - k) \\ &\leq a_m + a_n + 1\end{aligned}$$

since $F^m(y) - F^m(z) \leq 1$ when $y - z \leq 1$ and $x_n - x - k = a_n - [a_n] \geq 0$. Since $a_n/n = (1/n)\sum_{i=0}^{n-1}(F^{i+1}(x) - F^i(x)) = (1/n)\sum_{i=0}^{n-1}(F(x_i) - x_i) \geq \min_{0 \leq y \leq 1} F(y) - y$, a_n/n is bounded from below and thus Proposition 9.6.4 shows that the limit of a_n/n exists. Note also that $\tau(F + k) = \tau(F) + k$ for $k \in \mathbb{Z}$.

If f has a q-periodic point then for a lift x of it and some $p \in \mathbb{Z}$ we have $F^q(x) = x + p$. Thus for $m \in \mathbb{N}$ we have

$$\frac{F^{mq}(x) - x}{mq} = \frac{1}{mq}\sum_{i=0}^{m-1} F^q(F^{iq}(x)) - F^{iq}(x) = \frac{mp}{mq} = \frac{p}{q},$$

so $\tau(F) = p/q$. □

Next we show that the rotation number is invariant under orientation-preserving topological conjugacies:

Proposition 11.1.3. *If $h: S^1 \to S^1$ is an orientation-preserving homeomorphism then $\tau(h^{-1}fh) = \tau(f)$.*

Proof. Let F and H be lifts of f and h, respectively, that is, $\pi F = f\pi$ and $\pi H = h\pi$. Then $\pi H^{-1} = h^{-1}h\pi H^{-1} = h^{-1}\pi H H^{-1} = h^{-1}\pi$, so H^{-1} is a lift of h^{-1}. Also, $H^{-1}FH$ is a lift of $h^{-1}fh$ since $\pi H^{-1}FH = h^{-1}\pi FH = h^{-1}f\pi H = h^{-1}fh\pi$.

Suppose H is such that $H(0) \in [0,1)$. We need to estimate

$$|H^{-1}F^nH(x) - F^n(x)| = |(H^{-1}FH)^n(x) - F^n(x)|.$$

(1) For $x \in [0,1)$ we have $0 - 1 < H(x) - x < H(x) < H(1) < 2$ and by periodicity

$$|H(x) - x| < 2 \qquad \text{for } x \in \mathbb{R}.$$

Similarly we get

$$|H^{-1}(x) - x| < 2 \qquad \text{for } x \in \mathbb{R}.$$

(2) If $|y - x| < 2$ then $|F^n(y) - F^n(x)| < 3$ since $|[y] - [x]| \leq 2$ and thus

$$\begin{aligned} -3 \leq [y] - [x] - 1 = F^n([y]) - F^n([x]+1) &< F^n(y) - F^n(x) \\ &< F^n([y]+1) - F^n([x]) = [y] + 1 - [x] \leq 3. \end{aligned}$$

Those two estimates yield

$$|H^{-1}F^nH(x) - F^n(x)| \leq |H^{-1}F^nH(x) - F^nH(x)| + |F^nH(x) - F^n(x)| < 2+3$$

so $|(H^{-1}FH)^n(x) - F^n(x)|/n < 5/n$ whence $\tau(H^{-1}FH) = \tau(F)$ by (11.1.1). □

We showed that if there is a periodic point then the rotation number is rational. Now we will prove the converse.

Proposition 11.1.4. *Let f be an orientation-preserving homeomorphism of S^1. Then $\tau(f) \in \mathbb{Q}$ if and only if f has a periodic point.*

Proof. Suppose $\tau(f) = p/q \in \mathbb{Q}$. The definition of τ yields

$$\tau(f^m) = \lim_{n\to\infty} \frac{1}{n}((F^m)^n(x) - x) = m \lim_{n\to\infty} \frac{1}{mn}(F^{mn}(x) - x) = m\tau(f) \pmod 1,$$

so $\tau(f^q) = 0$ since the rotation number is defined up to an integer. It thus suffices to show that if $\tau(f) = 0$ then f has a fixed point.

Suppose f has no fixed point and let F be a lift such that $F(0) \in [0,1)$. Then $F(x) - x \in \mathbb{R} \setminus \mathbb{Z}$ for all $x \in \mathbb{R}$ since $F(x) - x \in \mathbb{Z}$ implies that $\pi(x)$ is a fixed point for f. Therefore $0 < F(x) - x < 1$ by the Intermediate Value Theorem. Since $F - \text{Id}$ is continuous on [0,1] it attains its minimum and maximum and therefore there exists a $\delta > 0$ such that

$$0 < \delta \leq F(x) - x \leq 1 - \delta < 1.$$

By periodicity of $F - \text{Id}$ this estimate holds for all $x \in \mathbb{R}$. In particular we can take $x = F^i(0)$ and sum from $i = 0$ to $n - 1$ to get

$$n\delta \leq F^n(0) \leq (1 - \delta)n$$

or

$$\delta \leq \frac{F^n(0)}{n} \leq 1 - \delta.$$

As $n \to \infty$ this gives $\tau(f) \neq 0$, proving the claim by contraposition. □

Proposition 11.1.5. *Let $f: S^1 \to S^1$ be an orientation-preserving homeomorphism with rational rotation number. Then all periodic orbits have the same period.*

Proof. If $\tau(f) = p/q$ with $p, q \in \mathbb{Z}$ relatively prime then we need to show that for any periodic point $\pi(x)$ there is a lift F of f for which $F^q(x) = x + p$. If $\pi(x)$ is periodic and F is a lift then $F^r(x) = x + s$ for some $r, s \in \mathbb{Z}$ and

$$k + \frac{p}{q} = \tau(f) = \lim_{n\to\infty} \frac{F^{nr}(x) - x}{nr} = \lim_{n\to\infty} \frac{ns}{nr} = \frac{s}{r}.$$

We may take F such that $k = 0$ so that $s = mp$ and $r = mq$. If $F^q(x) - p > x$ then by monotonicity

$$F^{2q}(x) - 2p = F^q(F^q(x) - p) - p \geq F^q(x) - p > x$$

and inductively $F^r(x) - s = F^{mq}(x) - mp > x$ contrary to the assumption. Thus $F^{mq}(x) - mp \leq x$ and similarly $F^{mq}(x) - mp \geq x$ as claimed. □

The next few results examine the dependence of the rotation number on the map as the map is varied. Combined with Proposition 11.1.3 the following fact shows that the rotation number is a C^0 modulus of circle homeomorphisms (see Definition 2.1.2).

Proposition 11.1.6. *$\tau(\cdot)$ is continuous in the C^0 topology.*

Proof. Let $\tau(f) = \tau$ and $p'/q', p/q \in \mathbb{Q}$ such that $p'/q' < \tau < p/q$. Pick the lift F of f for which $-1 < F^q(x) - x - p \leq 0$ for some $x \in \mathbb{R}$. Then $F^q(x) < x + p$ for all $x \in \mathbb{R}$ since otherwise $F^q(x) = x + p$ for some $x \in \mathbb{R}$ and $\tau = p/q$. Since the function $F^q - \mathrm{Id}$ is periodic and continuous it attains its maximum. Thus there exists $\delta > 0$ such that $F^q(x) < x + p - \delta$ for all $x \in \mathbb{R}$. This implies that every sufficiently small perturbation $\bar{F}$ of F in the uniform topology satisfies $\bar{F}^q(x) < x + p$ for all $x \in \mathbb{R}$. Therefore $\tau(\bar{f}) < p/q$, where $\bar{f}$ is the circle homeomorphism lifting to $\bar{F}$. A similar argument involving p'/q' completes the proof. □

The definition of the rotation number further suggests that it is monotone: If $F_1 > F_2$ then $\tau(F_1) \geq \tau(F_2)$ follows from the definition. This leads to the following terminology:

Definition 11.1.7. Define an ordering $\prec$ on the collection of orientation-preserving circle homeomorphisms as follows: If $f_0: S^1 \to S^1$ and $f_1: S^1 \to S^1$ are never antipodal construct a straight-line homotopy between f_0 and f_2. In other words, if we think of S^1 as the unit circle in $\mathbb{R}^2$ let $f_t = \dfrac{tf_0 + (1-t)f_1}{\|tf_0 + (1-t)f_1\|}$. Lifting this homotopy to $\mathbb{R}$ yields "compatible" lifts F_0 and F_1 of f_0 and f_1, respectively. We then say $f_0 \prec f_1$ if $F_0(x) < F_1(x)$ for all $x \in \mathbb{R}$.

Notice that the ordering "$\prec$" is not transitive. The definition of rotation number immediately implies:

Proposition 11.1.8. *$\tau(\cdot)$ is monotone: If $f_1 \prec f_2$ then $\tau(f_1) \le \tau(f_2)$ if the representatives of τ are chosen via compatible lifts.*

Remark. In particular if $\{f_t\}$ is a family of orientation-preserving circle homeomorphisms such that $f_t(x)$ is increasing in t for every $x \in \mathbb{R}$ then $\tau(f_t)$ is nondecreasing in t.

At irrational values the rotation number is strictly increasing:

Proposition 11.1.9. *If $f_1 \prec f_2$ and $\tau(f_1) \in \mathbb{R} \smallsetminus Q$ then $\tau(f_1) < \tau(f_2)$.*

Proof. If F_1 and F_2 are lifts as in the definition of "$\prec$" then $F_2(x) - F_1(x) > 0$ for all $x \in \mathbb{R}$ and by continuity and periodicity $F_2(x) - F_1(x) > \delta$ for some $\delta > 0$ and all $x \in \mathbb{R}$. Take $p/q \in \mathbb{Q}$ such that $p/q - \delta/q < \tau(F_1) < p/q$. Then there exists $x_0 \in \mathbb{R}$ such that $F_1^q(x_0) - x_0 > p - \delta$ (otherwise $\tau(F_1) = \lim_{n\to\infty} \dfrac{F_1^{nq}(x) - x}{nq} \le \lim_{n\to\infty} \dfrac{n(p-\delta)}{nq} = p/q - \delta/q$). Since

$$\begin{aligned} F_2^q(x_0) = F_2(F_2^{q-1}(x_0)) &> F_1(F_2^{q-1}(x_0)) + \delta \\ &> F_1(F_1^{q-1}(x_0)) + \delta = F_1^q(x_0) + \delta > x_0 + p \end{aligned}$$

we either have $F_2^q(x) > x + p$ for all $x \in \mathbb{R}$ or $F_2^q(x_1) = x_1 + p$ for some $x_1 \in \mathbb{R}$. In either case $\tau(F_1) < p/q \le \tau(F_2)$. □

While Proposition 11.1.9 shows that having irrational rotation number is not stable, the situation is different for rational rotation numbers:

Proposition 11.1.10. *Let $f: S^1 \to S^1$ be an orientation-preserving homeomorphism with rational rotation number $\tau(f) = p/q$ and some nonperiodic points. Then all sufficiently nearby perturbations $\bar{f}$ with $\bar{f} \prec f$ or all sufficiently nearby perturbations $\bar{f}$ with $f \prec \bar{f}$ have rotation number p/q.*

Proof. Since f has nonperiodic points, $F^q - \operatorname{Id} - p$ does not vanish identically for any lift F of f. (It does have zeros by assumption.) If there exists $x \in \mathbb{R}$ with $F^q(x) - x - p > 0$ then for any sufficiently small perturbation $\bar{f} \prec f$ the corresponding lift $\bar{F}$ of $\bar{f}$ is such that $\bar{F}^q(x) - x - p > 0$ and hence $\tau(\bar{f}) \ge p/q$ and thus $\tau(\bar{f}) \ge p/q$ by Proposition 11.1.8. Otherwise the same holds for perturbations with $f \prec \bar{f}$. □

Remark. The proof shows that circle maps with rational rotation number that have an attracting or repelling periodic orbit (an orbit that lifts to a point where $F^q - \operatorname{Id} - p$ changes sign) can be perturbed (in either direction) without changing the rotation number.

On the other hand, if $F^q - \operatorname{Id} - p$ does not change sign, for example, $F^q - \operatorname{Id} - p \ge 0$, then any perturbation $\bar{f}$ with $f \prec \bar{f}$ will have rotation number $\tau(\bar{f}) > p/q$ since $\bar{F}^q - \operatorname{Id} - p \ge \delta > 0$. In this case the zeros of $F^q - \operatorname{Id} - p$ project to "parabolic" or *semistable* periodic orbits. These are orbits p that attract on one side and repel on the other side, that is, there is some open neighborhood U of p such that for all x in one component of $U \smallsetminus \{p\}$ we have $\lim_{n\to\infty} d(f^n(x), f^n(p)) = 0$ and for all x in the other component $\lim_{n\to-\infty} d(f^n(x), f^n(p)) = 0$ (see Figure 3.3.1).

To see that we have indeed shown that the rotation number depends on f in a very nonsmooth way we reformulate these conclusions as follows:

Recall first that a monotone continuous function $\phi: [0,1] \to \mathbb{R}$ is called a *devil's staircase* if there exists a family $\{I_\alpha\}_{\alpha \in A}$ of disjoint open subintervals of [0,1] with dense union such that ϕ takes distinct constant values on these subintervals.

Proposition 11.1.11. *Let $\{f_t\}_{t\in[0,1]}$ be a monotone continuous family of orientation-preserving circle homeomorphisms such that $\tau: t \mapsto \tau(f_t)$ is nonconstant. If there exists a dense set $S \subset \mathbb{Q}$ such that no map f_t is topologically conjugate to a rotation R_α with $\alpha \in S$, then τ is a devil's staircase.*[1]

Proof. By Proposition 11.1.8 and Proposition 11.1.6 τ is monotone and continuous. Together with Proposition 11.1.10 this also implies that $\tau^{-1}(S)$ is a disjoint union of closed intervals of positive length. We need to show that $\tau^{-1}(S)$ is dense. Assume, by enlarging S if necessary, that whenever $\tau(f_t) = p/q \in \mathbb{Q} \smallsetminus S$ then f_t is topologically conjugate to $R_{p/q}$. Then Proposition 11.1.9 implies that τ is strictly monotone at points $t \in \tau^{-1}([0,1] \smallsetminus S)$: The rotation number is strictly increasing when it takes irrational values and at circle maps conjugate to a rational rotation. Thus for $t \in [0,1) \smallsetminus \tau^{-1}(S)$ and $\epsilon > 0$ we have $\tau(t) \neq \tau(t+\epsilon)$ and hence by density of S, continuity of τ, and the Intermediate Value Theorem there exists a $t_1 \in \tau^{-1}(S) \cap [t, t+\epsilon]$. Together with a similar argument for $t = 1$ this completes the proof. □

In closing we remark that the results of this section depend on monotonicity and continuity of f, but not on invertibility. Thus it suffices to assume that $f: S^1 \to S^1$ is a continuous order-preserving map of degree one, that is, its lift F is nondecreasing. Such a map may take constant values on a finite or countable set of intervals (see Exercise 11.1.1).

Exercises

11.1.1. *Let $f: S^1 \to S^1$ be a monotone (but not necessarily invertible) map of degree one, that is, its lift is a monotone function $F: \mathbb{R} \to \mathbb{R}$ such that $F(x+1) = F(x) + 1$. Prove that the assertions of Propositions 11.1.1, 11.1.3, and 11.1.4 hold for f.*

11.1.2. *Let $f: S^1 \to S^1$ be a continuous map of degree one and $F: \mathbb{R} \to \mathbb{R}$ its lift. Prove that*

$$\tau^+(F) := \lim_{n\to\infty} \max_x \frac{F^n(x) - x}{n} \quad \text{and} \quad \tau^-(F) := \lim_{n\to\infty} \min_x \frac{F^n(x) - x}{n}$$

both exist.

11.1.3. *Under the assumptions of the previous exercise call*

$$R(F) := \left\{ \tau \in \mathbb{R} \mid \exists x \in \mathbb{R} \ \lim_{n\to\infty} \frac{F^n(x) - x}{n} = \tau \right\}$$

the rotation set of F. Prove that $R(F) \neq \varnothing$.

11.1.4. *For $a, b \in [0,1]$ let $f_{a,b}: S^1 \to S^1$, $x \mapsto x + a + b\sin 2\pi x \pmod 1$. Show that for $p/q \in \mathbb{Q} \cap [0,1]$ the regions $A_{p/q} := \{(a,b) \in [0,1] \times [0,1] \mid \tau(f_{a,b}) = p/q\}$ are closed.*

These regions intersect $[0,1] \times \{0\}$ in the point p/q.

11.1.5. *Show that $A_{p/q}$ intersects every line $b =$ const. in a nonempty closed interval and that this interval has nonzero length except for $b = 0$.*

11.1.6. *Are the $A_{p/q}$ connected?*

11.1.7. *Show that the union of the $A_{p/q}$ is dense in $[0,1] \times [0,1]$.*[2]

2. The Poincaré classification

In this section we give a complete description of the possible behavior of orbits for circle homeomorphisms. This allows a description of circle homeomorphisms up to monotone semiconjugacy.

a. Rational rotation number.

Proposition 11.2.1. *Let $f: S^1 \to S^1$ be an orientation-preserving homeomorphism with rational rotation number $\tau(f) = p/q$. Suppose that p and q are relatively prime and let $\bar{x} \in S^1$ be such that $f^q(\bar{x}) = \bar{x}$. Then the ordering of $\{x, f(\bar{x}), f^2(\bar{x}), \dots, f^{q-1}(\bar{x})\}$ on S^1 is the same as that of $\left\{0, \frac{p}{q}, \frac{2p}{q}, \dots, \frac{(q-1)p}{q}\right\}$ on S^1. In particular the periodic orbit induces intervals on S^1 which are permuted by f.*

Remark. This proposition says that the periodic orbits of an orientation-preserving circle homeomorphism behave like those of the circle rotation with the same rotation number.

Proof. Let F be the lift of f such that $F^q(x) = x + p$ for $x \in \pi^{-1}(\{\bar{x}\})$ (see Proposition 11.1.5). The set $A := \pi^{-1}\{\bar{x}, f(\bar{x}), f^2(\bar{x}), \dots, f^{q-1}(\bar{x})\}$ yields a partition of $[x, x+p]$ into $p \cdot q$ intervals. At the same time $[x, x+p]$ is partitioned into the intervals $[x, F(x)], [F(x), F^2(x)], \dots, [F^{q-1}(x), F^q(x)]$ whose interiors are disjoint. Since F is a bijection between any two adjacent intervals in this partition and preserves A, each interval $[F^i(x), F^{i+1}(x)]$ contains exactly $p+1$ points of A. Take $k, r \in \mathbb{Z}$ such that the right neighbor of x in A is $x_1 = F^k(x) - r$. Since $\bar{F} = F^k - r$ is increasing on $\mathbb{R}$ and preserves A, the facts that $x_1 = \bar{F}(x)$ is the nearest right neighbor of x in A and that $[x, F(x)]$ is divided into p subintervals by A show that $\bar{F}^p(x) = F(x)$. Consequently $f^{kp}(\bar{x}) = f(\bar{x})$, where k is the unique number between 0 and $q-1$ such that $kp \equiv 1 \pmod q$. The orbit is therefore ordered as $(\bar{x}, f^k(\bar{x}), f^{2k}(\bar{x}), \dots, f^{(q-1)k}(\bar{x}))$.

Now let $f = R_{p/q}$ be the rotation by p/q. Then $\pi\left\{0, \frac{p}{q}, \dots, \frac{p(q-1)}{q}\right\}$ is a periodic orbit and thus by the preceding argument it is ordered as $\left(0, \frac{kp}{q}, \frac{2kp}{q}, \dots, \frac{(q-1)kp}{q}\right)$ with $kp \equiv 1 \pmod q$ as before. □

The next proposition asserts that for circle homeomorphisms with rational rotation number all nonperiodic orbits are asymptotic to periodic orbits. This yields a complete classification of possible orbits with rational rotation numbers. (Recall Definition 6.5.4 concerning homoclinic and heteroclinic points.)

Proposition 11.2.2. *Let $f\colon S^1 \to S^1$ be an orientation-preserving homeomorphism with rational rotation number $\tau(f) = p/q \in \mathbb{Q}$. Then there are two possible types of nonperiodic orbits for f:*

(1) *If f has exactly one periodic orbit then every other point is heteroclinic under f^q to two points on the periodic orbit. These points are different if the period is greater than one.*
(2) *If f has more than one periodic orbit then each nonperiodic point is heteroclinic under f^q to two points on different periodic orbits.*

Proof. Notice first that f^q can be identified with a homeomorphism of an interval by cutting the circle at a fixed point of f^q, or equivalently by taking a lift z of a fixed point of f^q and restricting a lift $F^q(\cdot) - p$ of f to $[z, z+1]$. But then the statement follows from Proposition 1.1.6 applied to this interval map, except for the last part of (2), that the two periodic orbits in question are different. But if there is an interval $I = [a, b] \subset \mathbb{R}$ such that a and b are adjacent zeros of $F^q - \operatorname{Id} - p$ and a, b project to the same periodic orbit, then f has only one periodic orbit because if $\pi(a) = x \in S^1$, $\pi(b) = f^k(x) \in S^1$ then $\bigcup_{n=0}^{q-1} f^{nk}\pi(a, b)$ covers the complement of $\{f^n(x)\}_{n=0}^{q-1}$ in S^1 and contains no periodic points. By invariance $f^{nk}(\pi(a, b))$ does not either. □

Remark. The above arguments show that if there is only one periodic orbit then it is semistable. It "repels on one side and attracts on the other", as shown in Figure 3.3.1.

Let us note that it is not just nonperiodic points that individually are asymptotic to periodic points, but that this behavior is coherent for iterates of points under f, so for a nonperiodic point x the points $x, f(x), \dots, f^{q-1}(x)$ are all forward asymptotic to the corresponding iterate $y, f(y), \dots, f^{q-1}(y)$ of a periodic point and they are moving in the same direction. This follows immediately from monotonicity:

Lemma 11.2.3. *If $I \subset \mathbb{R}$ is an interval whose endpoints are adjacent zeros of $F^q - \operatorname{Id} - p$ then $F^q - \operatorname{Id} - p$ has the same sign on the interiors of I and $F(I)$.*

Proof. If $(F^q - \operatorname{Id} - p)x > 0$ for $x \in I$ then $F^q - \operatorname{Id} - p$ is increasing at the left endpoint of I. Since F is monotone $(F^q - \operatorname{Id} - p) \circ F$ is increasing at the left endpoint of $F(I)$ and therefore $(F^q - \operatorname{Id} - p)x > 0$ for $x \in F(I)$.

The case of negative sign is similar. □

b. Irrational rotation number. The first step toward a classification in this case is to show that the orbits of a circle map f with rotation number $\tau(f) \in \mathbb{R} \setminus \mathbb{Q}$ are ordered as those for the rotation $R_{\tau(f)}$ by $\tau(f)$.

Proposition 11.2.4. *Let $F: \mathbb{R} \to \mathbb{R}$ be a lift of an orientation-preserving homeomorphism $f: S^1 \to S^1$ with rotation number $\tau = \tau(F) \in \mathbb{R} \setminus \mathbb{Q}$. Then for $n_1, n_2, m_1, m_2 \in \mathbb{Z}$ and $x \in \mathbb{R}$*

$$n_1\tau + m_1 < n_2\tau + m_2 \quad \textit{if and only if} \quad F^{n_1}(x) + m_1 < F^{n_2}(x) + m_2.$$

Proof. First observe that for given $n_1, n_2, m_1, m_2 \in \mathbb{Z}$ the expression $p(x) := F^{n_1}(x) + m_1 - F^{n_2}(x) - m_2$ never changes sign (and thus the second inequality is independent of x): Indeed, by continuity of $p(\cdot)$ a sign change implies the existence of $z \in \mathbb{R}$ with $F^{n_1}(z) - F^{n_2}(z) + m_1 - m_2 = 0$. But then z projects to a periodic point since $F^{n_1}(z) - F^{n_2}(z) \in \mathbb{Z}$, which is impossible.

Now assume $F^{n_1}(0) + m_1 < F^{n_2}(0) + m_2$. With $y := F^{n_2}(0)$ this is equivalent to $F^{n_1 - n_2}(y) - y < m_2 - m_1$. As before this inequality holds for all $y \in \mathbb{R}$, in particular for $y = 0$ and $y = F^{n_1 - n_2}(0)$, whence

$$F^{n_1 - n_2}(0) < m_2 - m_1$$

and

$$F^{2(n_1 - n_2)}(0) < (m_2 - m_1) + F^{n_1 - n_2}(0) < 2(m_2 - m_1).$$

Inductively $F^{n(n_1 - n_2)}(0) < n(m_2 - m_1)$ and

$$\tau = \lim_{n \to \infty} \frac{F^{n(n_1 - n_2)}(0)}{n(n_1 - n_2)} < \lim_{n \to \infty} \frac{n(m_2 - m_1)}{n(n_1 - n_2)} = \frac{m_2 - m_1}{n_1 - n_2}$$

(with strict inequality since $\tau \in \mathbb{R} \setminus \mathbb{Q}$). Consequently

$$n_1\tau + m_1 < n_2\tau + m_2.$$

Thus we have proved the "if". Similarly $F^{n_1}(0) + m_1 > F^{n_2}(0) + m_2$ implies that $n_1\tau + m_1 > n_2\tau + m_2$ and equality never occurs on either side (since $\tau \in \mathbb{R} \setminus \mathbb{Q}$ and f has no periodic orbits). Thus "only if" follows as well. □

The preceding proposition bears some resemblance to the earlier result that in the case of rational rotation number *periodic* orbits are ordered like those for the corresponding rotation. For rational rotation numbers we then determined that nonperiodic orbits are asymptotic to periodic ones. This motivates studying the asymptotic behavior of orbits for homeomorphisms with irrational rotation number.

Proposition 11.2.5. *Let $f: S^1 \to S^1$ be an orientation-preserving homeomorphism of S^1 and assume $\tau(f) \in \mathbb{R} \setminus \mathbb{Q}$. Then the ω-limit set $\omega(x)$ (see Definition 1.6.2) is independent of x and $E := \omega(x)$ is either S^1 or perfect and nowhere dense.*

Lemma 11.2.6. *Let $f: S^1 \to S^1$ be an orientation-preserving homeomorphism, $\tau(f) \in \mathbb{R} \setminus \mathbb{Q}$, $m, n \in \mathbb{Z}$, $m \neq n$, $x \in S^1$, and $I \subset S^1$ a closed interval with endpoints $f^m(x)$ and $f^n(x)$. Then every semiorbit meets I.*

Remark. For $x \neq y \in S^1$ there are exactly two intervals in S^1 with endpoints x and y. The lemma holds for either choice. Since $\tau(f)$ is irrational, x is not periodic and I is not a point.

Proof. Consider positive semiorbits $\{f^n(y)\}_{n \in \mathbb{N}}$. The proof for negative semiorbits is exactly the same. To prove the lemma it suffices to show that the backward iterates of I cover S^1, that is, $S^1 \subset \bigcup_{k \in \mathbb{N}} f^{-k}(I)$.

Let $I_k := f^{-k(n-m)}(I)$ and note that these are all contiguous: If $k \in \mathbb{N}$ then I_k and I_{k-1} have a common endpoint. Consequently if $S^1 \neq \bigcup_{k \in \mathbb{N}} I_k$ then the sequence of endpoints converges to some $z \in S^1$. But then

$$\begin{aligned} z &= \lim_{k \to \infty} f^{-k(n-m)} f^m(x) = \lim_{k \to \infty} f^{(-k+1)(n-m)} f^m(x) \\ &= \lim_{k \to \infty} f^{(n-m)} f^{-k(n-m)} f^m(x) = f^{(n-m)} \lim_{k \to \infty} f^{-k(n-m)} f^m(x) = f^{(n-m)}(z) \end{aligned}$$

is periodic. This contradicts the assumption $\tau(f) \in \mathbb{R} \setminus \mathbb{Q}$. □

Proof of the proposition. Independence of x: We need to show that $\omega(x) = \omega(y)$ for $x, y \in S^1$. Let $z \in \omega(x)$. Then there is a sequence l_n in $\mathbb{N}$ such that $f^{l_n}(x) \to z$. If $y \in S^1$ then by Lemma 11.2.6 there exist $k_m \in \mathbb{N}$ such that $f^{k_m}(y) \in I_m := [f^{l_m}(x), f^{l_{m+1}}(x)]$. But then $\lim_{m \to \infty} f^{k_m}(y) = z$ and thus $z \in \omega(y)$.

Therefore $\omega(x) \subset \omega(y)$ for all $x, y \in S^1$ and by symmetry $\omega(x) = \omega(y)$ for all $x, y \in S^1$.

$E := \omega(x)$ is either S^1 or nowhere dense: Note first that E is the only minimal closed nonempty f-invariant set: If $A \subset S^1$ is a nonempty closed f-invariant set and $x \in A$ then $\{f^k(x)\}_{k \in \mathbb{Z}} \subset A$ since A is invariant and $E = \omega(x) \subset A$ since A is closed. Therefore $\varnothing$ and E are the only closed invariant subsets of E and since the boundary ∂E of E is a closed invariant subset of E we conclude that $\partial E = \varnothing$ or $\partial E = E$.

If $\partial E = \varnothing$ then $E = S^1$. If $\partial E = E$ then E is nowhere dense.

It remains to show that E is perfect. E is closed. Let $x \in E$. Since $E = \omega(x)$ there is a sequence k_n such that $\lim_{n \to \infty} f^{k_n}(x) = x$. Since $\tau(f) \in \mathbb{R} \setminus \mathbb{Q}$ there are no periodic orbits and thus $f^{k_n}(x) \neq x$ for all n. Consequently x is an accumulation point of E since $f^{k_n}(x) \in E$ by invariance. □

The preceding proposition shows that the orbit structure of maps with irrational rotation number is quite different from that of maps with rational rotation number. Whereas in the case of rational rotation number all orbits are either periodic or asymptotic to a periodic orbit, for maps with irrational rotation number either all orbits are dense or all orbits are asymptotic to or in a Cantor set. As our understanding deepens we will see this divergence widen as we return to the conjugacy problem: When is a circle map equivalent to a rotation? In the case of rational rotation number this is clearly hardly ever the case: Since all orbits of a rational rotation are periodic with the same period, any coordinate change will again produce a map with only periodic orbits (a map f, in fact, such that $f^q = \mathrm{Id}$).

This constraint is absent in the case of irrational rotation number and there is an intimate connection with irrational rotations.

Theorem 11.2.7. (Poincaré Classification Theorem) *Let $f: S^1 \to S^1$ be an orientation-preserving homeomorphism with irrational rotation number.*

1. *If f is transitive then f is conjugate to the rotation $R_{\tau(f)}$.*
2. *If f is not transitive then f has the rotation $R_{\tau(f)}$ as a topological factor (cf. Definition 2.3.2) via a noninvertible continuous monotone map $h: S^1 \to S^1$.*

Remark. It is worthwhile to compare this theorem with Proposition 2.4.9 which gives a similar description for maps of degree k with $|k| \geq 2$. However, both proofs of Proposition 2.4.9 (via coding in Section 2.4b and via fixed-point methods in Section 2.4c) are essentially different from the proof of the Poincaré Classification Theorem. They rely on the hyperbolicity of the model map E_k whereas the key idea in the present proof is preservation of order.

Proof. Pick a lift $F: \mathbb{R} \to \mathbb{R}$ of f and let $\tau := \tau(F)$, $x \in \mathbb{R}$, and $B := \{F^n(x) + m\}_{n,m\in\mathbb{Z}}$ the total lift of the orbit of $\pi(x)$. Define $H: B \to \mathbb{R}$, $F^n(x) + m \mapsto n\tau + m$. Proposition 11.2.4 implies that this map is monotone. Note also that $H(B)$ is dense in $\mathbb{R}$ (Proposition 1.3.3). If we abuse notation and write R_τ for the map $\mathbb{R} \to \mathbb{R}$, $x \mapsto x + \tau$ then $H \circ F = R_\tau \circ H$ on B since

$$H \circ F(F^n(x) + m) = H(F^{n+1}(x) + m) = (n+1)\tau + m$$

and

$$R_\tau \circ H(F^n(x) + m) = R_\tau(n\tau + m) = (n+1)\tau + m.$$

Lemma 11.2.8. *H has a continuous extension to the closure $\bar{B}$ of B.*

Proof. If $y \in \bar{B}$ then there is a sequence $\{x_n\}_{n\in\mathbb{N}} \subset B$ such that $y = \lim_{n\to\infty} x_n$. We thus want to define $H(y) := \lim_{n\to\infty} H(x_n)$. To show that $\lim_{n\to\infty} H(x_n)$ exists and is independent of the choice of a sequence approximating y, observe first that the left and right limits exist and are independent of the sequence since H is monotone. If the left and right limits disagree, then $\mathbb{R} \smallsetminus H(B)$ contains an interval, which contradicts the density of $H(B)$. □

H can now easily be extended to $\mathbb{R}$: Since $H: \bar{B} \to \mathbb{R}$ is monotone and surjective (since H is monotone and continuous on B, $\bar{B}$ is closed, and $H(B)$ is dense in $\mathbb{R}$) there is no choice in defining H on the intervals complementary to $\bar{B}$: set $H =$ const. on those intervals, choosing the constant equal to the values at the endpoints. This gives a map $H: \mathbb{R} \to \mathbb{R}$ such that $H \circ F = R_\tau \circ H$ and thus a semiconjugacy $h: S^1 \to S^1$ since for $z \in B$ we have

$$H(z+1) = H(F^n(x) + m + 1) = n\tau + m + 1 = H(z) + 1$$

and this property persists under continuous extension.

The theorem now follows from the observation that in the transitive case we start from a dense orbit and so $\bar{B} = \mathbb{R}$ and h is a bijection. □

Remark. Since $\bar{B}$ projects to the closure of the orbit of $\pi(x)$ it contains the ω-limit set $E = \omega(\pi(x))$ of $\pi(x)$ and by choosing $x \in \pi^{-1}(E)$ we obtain $\pi(\bar{B}) = E$, where E is the universal ω-limit set discussed previously. In the transitive case $\bar{B} = \mathbb{R}$ and $E = S^1$, but in the nontransitive case we find that if $x \in \pi^{-1}(E)$ then $\pi(\bar{B}) = E$ is a Cantor set. Consequently in the case of a semiconjugacy the dynamics on the Cantor set is almost conjugate to the irrational rotation $R_{\tau(f)}$: If the two endpoints of every complementary interval are identified then h becomes a bijection of the identification space $E/\sim$ to S^1 and conjugates $f_{\restriction E/\sim}$ to $R_{\tau(f)}$. All orbits of f in E are dense in E (by the definition of E). On the other hand the construction of $E = \omega(x)$ yields that all points outside E are attracted to E in both positive and negative time because iterates of such a point have to stay inside disjoint complementary intervals of E and the length of these goes to zero.

Conversely one can think of the nontransitive map as being obtained from an irrational rotation by "blowing up" some orbits to intervals whose union then makes up the complement of E. These complementary intervals are thus permuted like the points on an orbit for an irrational rotation. All interior points in these intervals are wandering since they stay within these intervals whose images are all disjoint.

We will later show an explicit construction of such an example. The complete topological classification of circle homeomorphisms with a given irrational rotation number τ is given by a finite or countable collection of orbits of the rotation R_τ modulo a simultaneous translation of all these orbits. Naturally those orbits are the ones blown up under the semiconjugacy in Theorem 11.2.7. Since orbits of an irrational rotation are dense those invariants are not moduli.

c. Orbit types and measurable classification. We end this section by extracting from the preceding discussion a classification of orbits for circle homeomorphisms: On S^1 a point can have six different kinds of orbits under transformations, three each for the cases of rational and irrational rotation number. Let us call an orbit $\mathcal{O}$ of a map *homoclinic* to an invariant set $S \subset S^1 \setminus \mathcal{O}$ if $\alpha(x) = \omega(x) = S$ for any $x \in \mathcal{O}$. Similarly $\mathcal{O}$ is *heteroclinic* to two disjoint invariant sets S_1, S_2 if $\mathcal{O}$ is disjoint from each of them and $\alpha(x) = S_1$, $\omega(x) = S_2$ for $x \in \mathcal{O}$. Then we obtain the list of possible orbits shown in the following table. Note that our description in the case of rational rotation number, as well as our notation, is not standard in the literature.

Finally, the Poincaré Classification Theorem 11.2.7 completely answers the question about invariant measures of circle homeomorphisms. In the case of rational rotation number every ergodic invariant measure is atomic; in fact it is a uniform δ-measure on a periodic orbit. For an irrational rotation number we need to consider the transitive and nontransitive cases separately. Since the

Rotation number $p/q \in \mathbb{Q}$	Rotation number $\alpha \in \mathbb{R} \setminus \mathbb{Q}$
$\mathrm{I}_{\frac{p}{q}}$: Periodic orbit with the same period as $R_{\frac{p}{q}}$ and ordered in the same way as an orbit of $R_{\frac{p}{q}}$.	I_α: An orbit dense in S^1 that is ordered in the same way as an orbit of R_α (as are the two following cases).
$\mathrm{II}_{\frac{p}{q}}$: A homoclinic orbit: It approaches a given periodic orbit as $n \to +\infty$ and as $n \to -\infty$.	II_α: An orbit dense in a Cantor set.
$\mathrm{III}_{\frac{p}{q}}$: A heteroclinic orbit: It approaches two different periodic orbits as $n \to +\infty$ and as $n \to -\infty$ (this happens whenever there is more than one periodic orbit).	III_α: An orbit homoclinic to a Cantor set.

TABLE. The Poincaré classification

irrational rotation is uniquely ergodic (Kronecker–Weyl Equidistribution Theorem, Proposition 4.2.1) and since unique ergodicity is obviously an invariant of topological conjugacy, any transitive homeomorphism of S^1 with irrational rotation number is uniquely ergodic.

Now let $f\colon S^1 \to S^1$ be a nontransitive homeomorphism with irrational rotation number. Since the images of any interval complementary to the unique minimal set E of f are disjoint, such an interval has to have measure zero with respect to any f-invariant probability measure μ, that is, $\mu(S^1 \setminus E) = 0$. But a semiconjugacy between f and the irrational rotation $R_{\tau(f)}$ is one-to-one on $E \setminus \mathcal{D}$, where $\mathcal{D}$ is a countable set. Any f-invariant invariant probability measure is nonatomic since f has no periodic points; hence $\mu(\mathcal{D}) = 0$. If f had two invariant probability measures they would have to differ on a set $A \subset E \setminus \mathcal{D}$ and hence the semiconjugacy would carry them into two different invariant measures for the rotation $R_{\tau(f)}$, which is impossible. Recalling Definition 4.1.20 we thus have

Theorem 11.2.9. *A homeomorphism of the circle with irrational rotation number is uniquely ergodic and as a measure-preserving transformation it is metrically isomorphic to an irrational rotation.*

Corollary 11.2.10. *Any homeomorphism of S^1 has zero topological entropy.*

Proof. This follows from Proposition 11.2.2, Theorem 11.2.9, and the Variational Principle (Theorem 4.5.3). □

Exercises

11.2.1. *Prove that all orientation-preserving homeomorphisms with rational rotation number p/q, without semistable periodic points, and with a given number n of periodic orbits are topologically conjugate. Prove that in this case n is always even.*

11.2.2. *Prove that if $p/q \notin \{0, 1/2\} \pmod 1$ then there are exactly two classes of orientation-preserving homeomorphisms of S^1 with rotation number p/q and with n periodic orbits, all of which are semistable. If $p/q = 0$ or $1/2 \pmod 1$ then there is only one class.*

11.2.3. *Describe all conjugacy classes of orientation-preserving homeomorphisms of S^1 with rotation number p/q and n periodic orbits; prove that the number of such classes is no more than 2^n. Calculate the number for $n = 1, 2, 3$. Consider separately the cases $p/q = 0$ and $p/q = 1/2$.*

11.2.4. *Let $f: S^1 \to S^1$ be an orientation-reversing homeomorphism. Prove that f has exactly two fixed points and that f has rotation number 0.*

11.2.5. *Assume that an orientation-reversing homeomorphism $f: S^1 \to S^1$ has $n \geq 0$ periodic orbits of period two and that none of its periodic orbits is semistable. Prove:*

(1) *If n is even then all such homeomorphisms are topologically conjugate.*
(2) *If n is odd then there are exactly two classes of topological conjugacy.*

11.2.6. *Prove that an orientation-preserving homeomorphism of the circle with rational rotation number has a rotation as a factor if and only if its set of periodic points is uncountable.*

11.2.7. *Show that if $f: S^1 \to S^1$ is topologically conjugate to an irrational rotation, then the conjugating homeomorphism is unique up to a rotation (that is, if $h_i \circ f = R_\tau \circ h_i$ for $i = 1, 2$, then $h_1 \circ h_2^{-1}$ is a rotation).*

11.2.8*. *Show that the restriction of a nontransitive homeomorphism $f: S^1 \to S^1$ with irrational rotation number to its minimal set is topologically conjugate to a symbolic dynamical system, more specifically to the restriction of the full 2-shift σ_2 (cf. (1.9.3)) to a closed invariant set Λ, if and only if it is obtained from a rotation by "blowing up" finitely many orbits.*

11.2.9. *Prove that the irrational rotations R_α and R_β are isomorphic as measure-preserving transformations of the space (S^1, λ) if and only if $\alpha = \pm\beta \pmod 1$.*

11.2.10. *Prove that a circle homeomorphism has zero topological entropy without using the Variational Principle for entropy or the Poincaré Classification.*

12

Circle diffeomorphisms

It turns out that adding sufficient differentiability to the ingredients of the previous chapter produces several new facets in the theory of circle maps. At the end of Section 11.2b we outlined a complete topological classification of circle homeomorphisms with *irrational* rotation number. When we restrict attention to sufficiently smooth diffeomorphisms (Theorem 12.1.1) the situation changes dramatically. The example of Proposition 12.2.1 shows that the smoothness required is almost sharp. The rotation number becomes a complete invariant of *topological* conjugacy. This is not dissimilar to the situation with hyperbolic dynamical systems (cf., for example, Theorems 2.6.1 and 2.6.3). On the other hand, the classification of circle diffeomorphism up to *differentiable* conjugacy is possible only for rotation numbers satisfying extra arithmetic conditions. In Section 12.3 we prove a local result of this kind in the analytic setting, while in Sections 12.5 and 12.6 we show that without an arithmetic condition a variety of pathological behaviors of the conjugacy may be produced at will. Finally we show in Section 12.7 that a certain aspect of the behavior of an irrational rotation, namely, ergodicity with respect to Lebesgue measure, is preserved for all sufficiently smooth circle diffeomorphisms.

1. The Denjoy Theorem

Recall that $g\colon S^1 \to \mathbb{R}$ is of *bounded variation* if its total variation $\mathrm{Var}(g) := \sup \sum_{k=1}^{n} |g(x_k) - g(x'_k)|$ is finite. Here the sup is taken over all finite collections $\{x_k, x'_k\}_{k=1}^{n}$ such that x_k, x'_k are endpoints of an interval I_k and $I_k \cap I_j = \varnothing$ for $k \neq j$.

Remark. Every Lipschitz function and hence every continuously differentiable function has bounded variation.

Theorem 12.1.1. (Denjoy Theorem)[1] *A C^1 diffeomorphism $f\colon S^1 \to S^1$ with irrational rotation number $\tau(f) \in \mathbb{R}\smallsetminus\mathbb{Q}$ and derivative of bounded variation is transitive and hence topologically conjugate to $R_{\tau(f)}$.*

Remark. By the previous remark this theorem holds for any C^2 diffeomorphism with irrational rotation number.

Lemma 12.1.2. *If $f: S^1 \to S^1$ is a homeomorphism with irrational rotation number, then for $x_0 \in S^1$ there are infinitely many $n \in \mathbb{N}$ such that the intervals $f^k((x_0, f^{-n}(x_0)))$ for $0 \le k < n$ are disjoint.*

Proof. Write $x_k = f^k(x_0)$, $I = (x_0, x_{-n})$. The claim only involves the ordering on the orbit of x_0. Since f is either conjugate or semiconjugate to an irrational rotation we can thus assume that f itself is an irrational rotation. In this case it is clear that the claim holds for all $n \in \mathbb{N}$ with the property that for $0 < |k| < n$ no x_k lies in I. But since the orbit of x_0 is dense and thus has a subsequence converging to x_0, there are infinitely many such n. □

Lemma 12.1.3. *Let $X = S^1$ or $X = [0,1]$ and $Y \subset X$. Suppose $f: X \to X$ is such that $f\restriction_Y$ is C^1 with f' of bounded variation and that $|f'|$ is bounded away from 0 on Y. Let $V < \infty$ be the variation of $\varphi: x \mapsto \log|f'(x)|$. If $I \subset Y$ is an interval such that $I, f(I), \dots, f^n(I)$ are pairwise disjoint intervals in Y and $x, y \in I$ then*

$$\exp(-V) \le \frac{|(f^n)'(x)|}{|(f^n)'(y)|} \le \exp(V).$$

Proof.

$$V \ge \sum_{k=0}^{n-1} |\varphi(f^k(x)) - \varphi(f^k(y))| \ge \left|\sum_{k=0}^{n-1} \varphi(f^k(x)) - \sum_{k=0}^{n-1} \varphi(f^k(y))\right|$$
$$= \left|\log\left(\prod_{k=0}^{n-1} |f'(f^k(x))| \cdot \prod_{k=0}^{n-1} |f'^{-1}(f^k(y))|\right)\right| = \left|\log \frac{|(f^n)'(x)|}{|(f^n)'(y)|}\right|$$

The claim follows upon exponentiation. □

Remark. This argument is another example of a bounded distortion estimate. It is similar to the argument used in the proof of Lemma 5.1.18 and the one needed to solve Exercise 10.0.6. The main difference here is the direct use of boundedness of the variation of the derivative instead of an estimate of this variation via the second derivative. The next time it will come up is in the proof of Lemma 12.7.3.

Lemma 12.1.4. *If f is not conjugate to a rotation, I is an interval in the complement of $E = \omega(x)$, $x_0 \in I$, and n is as in Lemma 12.1.2 then $\exp(-V) \le (f^n)'(x) \cdot (f^{-n})'(x) \le \exp(V)$ for every $x \in I$.*

Proof. Since all $x \in I$ have the same image under the semiconjugacy the collection of $n \in \mathbb{N}$ obtained in Lemma 12.1.2 does not depend on $x \in I$. Thus we can apply Lemma 12.1.3 with $y = f^{-n}(x)$. □

Proof of Theorem 12.1.1. Suppose f is not conjugate to a rotation and take an interval $I \subset S^1 \setminus E$, where E is as in Proposition 11.2.5. Note that all its images and preimages are pairwise disjoint. On the other hand the definition of length and Lemma 12.1.4 together with the obvious estimate $a+b \geq \max(a,b) \geq \sqrt{a \cdot b}$ for $a, b \geq 0$ then imply

$$\begin{aligned} l(f^n(I)) + l(f^{-n}(I)) &= \int_I (f^n)'(x)\,dx + \int_I (f^{-n})'(x)\,dx \\ &= \int_I [(f^n)'(x) + (f^{-n})'(x)]\,dx \geq \int_I \sqrt{(f^n)'(x) \cdot (f^{-n})'(x)}\,dx \\ &\geq \int_I \sqrt{\exp(-V)}\,dx = l(I) \cdot e^{-V/2} \end{aligned}$$

for infinitely many $n \in \mathbb{N}$. Since consequently $\sum_{i=-\infty}^{\infty} l(f^i(I)) = \infty$, the intervals $\{f^i(I)\}_{i \in \mathbb{Z}}$ cannot be disjoint, a contradiction. □

The main idea behind the preceding argument is that one can get control of the derivatives of iterates of f by controlling the way in which images of intervals form patterns on S^1. Here the "pattern" was controlled by considering those n for which the first n images of a certain interval are disjoint and this allowed us to use boundedness of the variation to give a strong uniform estimate for the derivatives of those iterates.

This exploits the low-dimensionality of the dynamics to obtain control of the images of intervals using boundedness of the variation. In Section 7 we will use the same idea to prove ergodicity of circle diffeomorphisms of the type discussed here.

2. The Denjoy example

The Denjoy theorem is rather close to being optimal. An example constructed here shows that the regularity hypothesis we used can not be relaxed too much. It gives nontransitive circle diffeomorphisms whose first derivatives have Hölder exponent arbitrarily close to 1. The idea of the construction is to start with an irrational rotation and to replace the points of one orbit by suitably chosen intervals. The resulting map is not transitive. The Denjoy example proves:

Proposition 12.2.1. *For $\tau \in \mathbb{R} \setminus \mathbb{Q}$, $\alpha \in (0,1)$ there exists a nontransitive C^1 diffeomorphism $f\colon S^1 \to S^1$ with α-Hölder derivative and rotation number $\tau(f) = \tau$.*

Proof. Let $k \in \mathbb{N}$, $\alpha = k/k+1$, $l_n = (|n| + (2k)^k + 1)^{-(1+1/k)}$, and $c_n = 2\,((l_{n+1}/l_n) - 1) \geq -1$ and note first that

$$\sum_{n\in\mathbb{Z}} l_n < 2\sum_{n=0}^{\infty} l_n = 2 \sum_{n=(2k)^k+1}^{\infty} \frac{1}{n^{1+1/k}} < 2\int_{(2k)^k}^{\infty} \frac{1}{x^{1+1/k}}\,dx = 1.$$

To "blow up" the orbit $x_n = (R_\tau)^n x$ of the irrational rotation R_τ to intervals I_n of length l_n insert the intervals I_n into S^1 so that they are ordered in the same way as the points x_n and so that the space between any two such intervals I_m and I_n is exactly

$$\left(1 - \sum_{n\in\mathbb{Z}} l_n\right) d(x_m, x_n) + \sum_{x_k\in(x_m,x_n)} l_k .$$

(This is the sum of the lengths of the intervals I_k inserted in between and the length of the arc of the circle between x_m and x_n, appropriately scaled to reflect the fact that the total length of $S^1 \setminus \bigcup_{n\in\mathbb{Z}} I_n$ is $1-\sum_{n\in\mathbb{Z}} l_n$.) To define a circle homeomorphism f such that $f(I_n) = I_{n+1}$ and $f\restriction_{S^1\setminus\bigcup_{n\in\mathbb{Z}} I_n}$ is semiconjugate to a rotation it suffices to specify the derivative $f'(x)$ since f is then obtained by integration.

On the interval $[a, a+l]$ define the tent function

$$h(a,l,x) := 1 - \frac{1}{l}|2(x-a) - l|$$

and note that $h(a,l,a+l/2) = 1$ and $\int_a^{a+l} h(a,l,x)\,dx = l/2$. Denote the left endpoint of I_n by a_n and let

$$f'(x) = \begin{cases} 1 & \text{for } x \in S^1 \setminus \bigcup_{n\in\mathbb{Z}} I_n, \\ 1 + c_n h(a_n, l_n, x) & \text{for } x \in I_n. \end{cases}$$

Since $c_n = 2\left((l_{n+1}/l_n) - 1\right) = 2\left(l_{n+1} - l_n\right)/l_n$ we have

$$\int_{I_n} f'(x)\,dx = \int_{I_n} (1 + c_n h(a_n, l_n, x))\,dx = l_n + \frac{l_n}{2} c_n = l_{n+1}$$

so indeed $f(I_n) = I_{n+1}$.

Now we show that f' has Hölder exponent α. It suffices to find M_α such that $|c_n| \le M_\alpha \cdot (l_n/2)^\alpha$ for all $n \in \mathbb{Z}$, since this implies $|f'(x) - f'(y)| \le M_\alpha d(x,y)^\alpha$ on $I_n (n \in \mathbb{Z})$ and hence on S^1. To prove this claim, consider first the case $n \ge 0$ and let $m = 1 + n + (2k)^k$. Note that since $(1+x)^{-\beta} = 1 - \beta\cdot x + O(x^2)$ (by Taylor expansion) we have

$$\left|\left(1 + \frac{1}{m}\right)^{-\beta} - 1\right| < 2\beta \cdot \frac{1}{m}$$

for all but finitely many $m \in \mathbb{N}$. Therefore

$$\begin{aligned}\frac{1}{2}|c_n| l_n^{-\alpha} = l_n^{-k/(k+1)}\left|\frac{l_{n+1}}{l_n} - 1\right| &= m \cdot \left|\left(\frac{m+1}{m}\right)^{-(k+1)/k} - 1\right| \\ &= m\left|\left(1 + \frac{1}{m}\right)^{-(k+1)/k} - 1\right| < 2\frac{k+1}{k} = \frac{2}{\alpha}\end{aligned}$$

and thus $|c_n|(l_n/2)^{-\alpha} < 2^{\alpha+2}/\alpha$ for all but finitely many $n \in \mathbb{N}$. The case $n < 0$ is treated similarly and we conclude

$$|c_n| \le M_\alpha \left(\frac{l_n}{2}\right)^\alpha$$

so f' is α-Hölder continuous. □

Note that the semiconjugacy to R_τ is established by a map whose derivative vanishes on all intervals I_n and is equal to $(1 - \sum_{n\in\mathbb{Z}} l_n)^{-1}$ elsewhere. A slight modification of this construction shows

Proposition 12.2.2. *For $a \in [0,1)$, $\alpha \in (0,1)$ there is a nontransitive C^1 diffeomorphism $f\colon S^1 \to S^1$ with α-Hölder derivative and such that the Lebesgue measure of the f-invariant minimal set is a.*

Proof. We need to show that the Lebesgue measure of the wandering set $\bigcup_{n\in\mathbb{Z}} I_n$ can be made equal to $1-a$. To that end take $l_n(a):=(1-a)l_n/\sum_{m\in\mathbb{Z}} l_m$ so that $\sum_{m\in\mathbb{Z}} l_m(a) = 1-a$. Since the c_n remain the same the above estimates are unchanged. □

Exercises

12.2.1. *Prove that the unique invariant measure for a diffeomorphism constructed in this section is absolutely continuous if the Lebesgue measure of the Denjoy set $S^1 \setminus \bigcup_{n\in\mathbb{Z}} I_n$ is positive, and singular otherwise.*

12.2.2*. *Modify the construction to obtain for a given $\beta > 0$ a nontransitive C^1 diffeomorphism with a given irrational rotation number, whose derivative satisfies the following condition which is stronger than α-Hölder continuity for any $\alpha < 1$:*

$$|f'(x) - f'(y)| < |x-y| \big|\log|x-y|\big|^{1+\beta}.$$

Explain why this construction does not work for $\beta = 0$.

3. Local analytic conjugacies for Diophantine rotation number

We have seen earlier that the speed of approximation of an irrational rotation number by rationals affects the dynamical properties of a map, namely, in the Siegel Theorem 2.8.2 which gives an analytic linearization of a complex map with linear part λ whose argument is Diophantine. The rational approximation properties of the rotation number are essential in deciding whether the conjugating map given by the Denjoy Theorem 12.1.1 is smooth. There are global results guaranteeing that if a sufficiently smooth diffeomorphism of a circle has rotation number not too well approximable by rationals, then the conjugacy from the Denjoy theory is indeed smooth (analytic in the case of analytic circle diffeomorphisms). In general the interplay between the arithmetic properties of the rotation number and the required smoothness of the map is rather complicated, although by now it is well understood. Those results make full use of the low-dimensionality of the situation and utilize various types of powerful analytic estimates. In this section we prove a theorem of Arnold which was the first result in this direction. It says that an analytic circle map with Diophantine rotation number is analytically conjugate to a rotation if it is sufficiently close to a rotation. Unlike later global results this theorem generalizes to higher dimensions where instead of fixing the rotation number one has to find appropriate parameter values in typical families of maps. The method we use is the same as in the Siegel Theorem 2.8.2 with the only difference that we work on an annulus instead of a disk in the complex plane.

Theorem 12.3.1. (Arnold Theorem) *Given $c > 0$, $d, r > 1$, there exists $\epsilon > 0$ such that if α is of Diophantine type (c, d) (see Definition 2.8.1), u is an analytic function on the annulus $A_r := \{z \in \mathbb{C} \mid 1/r < |z| < r\}$ with $u(z) < \epsilon$ on A_r, and $f(z) := e^{2\pi i\alpha}z + u(z)$ preserves the unit circle S^1 and has rotation number α on S^1, then f is analytically conjugate to the circle rotation by α.*

Proof. We will use the very same setup as in the proof of the Siegel Theorem 2.8.2 adapted to the annulus A_r and we will refer to that proof freely. In particular we assume throughout that $\lambda := e^{2\pi i\alpha}$ is as in (2.8.2). To that end we first adapt several lemmas from complex variables from the setting of the Siegel Theorem to the annulus. The analog of Lemma 2.8.3 is

Lemma 12.3.2.

(1) *Suppose φ is analytic on A_r and continuous on $\overline{A_r}$, with $|\varphi| < \epsilon$ on A_r. Then $\varphi(z) = \sum_{k\in\mathbb{Z}} \varphi_k z^k$ and $|\varphi_k| \le \epsilon r^{-|k|}$.*

(2) *If $|\varphi_k| \le K r^{-|k|}$ and $\varphi(z) = \sum_{k\in\mathbb{Z}} \varphi_k z^k$ then φ is analytic on A_r and $|\varphi| \le 2Kr/\delta$ on $A_{r-\delta}$.*

Proof. (1) Let

$$\varphi^+(z) := \frac{1}{2\pi i}\int_{|\zeta|=r} \frac{\varphi(\zeta)}{\zeta - z}\, d\zeta, \qquad |z| < r,$$

$$\varphi^-(z) := -\frac{1}{2\pi i}\int_{|\zeta|=1/r} \frac{\varphi(\zeta)}{\zeta - z}\, d\zeta, \qquad |z| > 1/r.$$

By the Cauchy Integral Formula φ^+ is analytic on B_r and φ^- is analytic for $1/z \in B_r$. By the Cauchy Theorem $\varphi = \varphi^+ + \varphi^-$. Furthermore

$$\varphi^+(z) = \sum_{k=0}^{\infty} \varphi_k z^k \text{ with } \varphi_k = \frac{1}{2\pi i}\int_{|z|=r} \frac{\varphi(z)}{z^{k+1}} dz, \text{ so } |\varphi_k| \le \epsilon r^{-k}.$$

After some calculation one also finds

$$\varphi^-\left(\frac{1}{z}\right) = \sum_{k=1}^{\infty} \varphi_{-k} z^k \text{ with } \varphi_{-k} = \frac{1}{2\pi i}\int_{|z|=1/r} \frac{\varphi(z)}{z^{k-1}} dz, \text{ so } |\varphi_{-k}| \le \epsilon r^{-k}.$$

(2) This follows by applying Lemma 2.8.3 to $\varphi^\pm$. □

Applying Lemma 2.8.3 to $\varphi^{\pm}$ also yields

Lemma 12.3.3. *Let $\varphi = \sum_{k\in\mathbb{Z}} \varphi_k z^k$ be analytic on A_ρ, $|\varphi| < \delta$ on A_ρ, and λ as in (2.8.2). Then*

$$\psi(z) := \sum_{k\in\mathbb{Z}} \frac{\varphi_k}{\lambda^k - \lambda} z^k$$

is analytic on A_ρ and there exists a $c(d) > 0$ such that

$$|\psi| < 2\delta c_0 c(d) \Delta^{-(d+1)}$$

on $\overline{A_{\rho(1-\Delta)}}$.

Let us now recall the setup for the Newton method from Section 2.7. Consider the operator

$$\mathcal{F}(f,h) = h^{-1} \circ f \circ h$$

and write the conjugacy equation as

$$\mathcal{F}(f,h) = g. \tag{12.3.1}$$

Next we look for an approximate solution h of the conjugacy equation linearized at (g, Id). Thus we write

$$\mathcal{F}(f,h) = \mathcal{F}(g,\mathrm{Id}) + D_1\mathcal{F}(g,\mathrm{Id})(f-g) + D_2\mathcal{F}(g,\mathrm{Id})(h-\mathrm{Id}) + \mathcal{R}(f,h),$$

where $\mathcal{R}(f,h)$ is of second order in $(f-g, h-\mathrm{Id}) =: (u,w)$. Linearizing (12.3.1) by dropping $\mathcal{R}$ leads to

$$\mathcal{F}(g,\mathrm{Id}) + D_1\mathcal{F}(g,\mathrm{Id})u + D_2\mathcal{F}(g,\mathrm{Id})w = g,$$

which simplifies to

$$u + D_2\mathcal{F}(g,\mathrm{Id})w = 0. \tag{12.3.2}$$

Substituting $h = \mathrm{Id} + w$ into $\mathcal{F}(f,h)$ we obtain a function $f_1 = h^{-1} \circ f \circ h = \mathcal{F}(f,h) = g + \mathcal{R}(f,h)$, so the size of $u_1 := f_1 - g$ should be of second order in the size of $u = f - g$. In order to justify this, we will need to estimate the difference between $\mathcal{F}$ and its linearization near (g, Id). In the case $g = \Lambda$ where $\Lambda(z) = \lambda z$, we calculated $D_2\mathcal{F}(g,\mathrm{Id})$ in Section 2.7 and found that (12.3.2) becomes

$$u = w \circ \Lambda - \Lambda \circ w. \tag{12.3.3}$$

Given $u = \sum_{k=1}^{\infty} u_k z^k$ we wrote $w = \sum_{k=1}^{\infty} w_k z^k$ to get

$$w_k = \frac{u_k}{\lambda^k - \lambda}. \tag{12.3.4}$$

Lemma 12.3.3 will be used because we will solve (12.3.3) by obtaining the coefficients of w from (12.3.4). But in (12.3.4) it is necessary that the coefficient

$\eta := u_1$ is zero. Unlike in the Siegel Theorem we do not know this, however, so we first replace u by $\tilde{u}(z) := u(z) - \eta z$ in (12.3.3). We also do not know whether the candidate conjugacy h preserves the unit circle, but it turns out that this is not needed. Nevertheless the conjugacy we obtain in the end conjugates $f\restriction_{S^1}$ to a circle rotation since it conjugates f to a rotation of the annulus.

Then we consider an iterative process as follows. Assuming that $f_1, \dots, f_n$ have been constructed we solve the equation

$$f_n - g + D_2\mathcal{F}(g, \mathrm{Id})w_{n+1} = 0 \tag{12.3.5}$$

and set

$$h_{n+1} = h_n \circ (\mathrm{Id} + w_{n+1}) \text{ and } f_{n+1} = h_{n+1}^{-1} \circ f_n \circ h_{n+1}. \tag{12.3.6}$$

The last step of the construction is the proof of convergence of the sequence h_n in an appropriate topology. It follows from the same estimates that provide the fast decrease of the size of the $f_n - g$.

To keep the notations as similar as possible to the Siegel situation we assume that $|\tilde{u}| < \epsilon$ on A_r. Using Lemma 12.3.2 we may assume the same bound on u', so that we have

$$|u| < \epsilon \text{ and } |u'| < \epsilon \text{ on } A_r. \tag{12.3.7}$$

Replacing u by $\tilde{u}$ leaves (12.3.4) unchanged: We still obtain $w_k = \dfrac{u_k}{\lambda^k - \lambda}$ from (12.3.3). Thus we can apply Lemma 12.3.3 to $\tilde{u}$ with $\rho = r$ and $\delta = \epsilon$ to get

$$|w| < 2\epsilon c_0 c(d)\Delta^{-(d+1)}. \tag{12.3.8}$$

Note that differentiating (12.3.3) and multiplying by z yields $z\tilde{u}(z) = \lambda z w'(\lambda z) - \lambda z w'(z)$, so we can apply Lemma 12.3.3 to conclude

$$|zw'(z)| < 2r\epsilon c_0 c(d)\Delta^{-(d+1)}, \tag{12.3.9}$$

since $z\tilde{u}' < r\epsilon$ on A_r.

Having obtained a candidate conjugacy $h(z) = z + w(z)$ we next want to show that the new map $f_1 = h^{-1} \circ f \circ h$ is defined on $A_{r(1-\Delta)}$ if ϵ is sufficiently small.

Lemma 12.3.4. *If*

$$2\epsilon c_0 c(d) < \Delta^{d+2}(1-\Delta)/r \text{ and } 0 < \Delta < \frac{1}{4} \tag{12.3.10}$$

then

$$h(A_{r(1-4\Delta)}) \subset A_{r(1-3\Delta)} \text{ and } A_{r(1-2\Delta)} \subset h(A_{r(1-\Delta)}). \tag{12.3.11}$$

Proof. To prove the first inclusion in (12.3.11) take $|z| < r(1-4\Delta)$ and use (12.3.8) to get

$$|h(z)| \le |z| + |w(z)| < r(1-4\Delta) + 2\epsilon c_0 c(d)\Delta^{-(d+1)} < r(1-4\Delta+\Delta).$$

A slightly messier calculation shows that $|1/z| < r(1-4\Delta)$ implies $|h(z)| \ge 1/r(1-3\Delta)$, proving the first inclusion.

To show the second inclusion of (12.3.11) note that $|h(z)-z| = |w(z)| < \Delta$ by (12.3.8) and (12.3.10), while for $|z| = r(1-\Delta)$ we clearly have $|z| - r(1-2\Delta) = r\Delta > |w(z)|$. Thus $|h(z)| = |z-(h(z)-z)| \ge |z| - |w(z)| > r(1-2\Delta)$.

Again, a slightly more involved calculation shows that $|1/z| = r(1-\Delta)$ implies $|h(z)| < 1/r(1-2\Delta)$. (This is the only place where the r in (12.3.10) is used.) □

To show that f_1 is defined on $A_{r(1-4\Delta)}$ we write $f_1(z) = \lambda z + u_1(z)$. The next lemma also shows that the error decreases quadratically.

Lemma 12.3.5. *If*

$$2\epsilon c_0 c(d) < \Delta^{d+2}(1-\Delta)/r \text{ and } 0 < \epsilon < \Delta < \frac{1}{5} \tag{12.3.12}$$

then f_1 is defined on $A_{r(1-4\Delta)}$ and

$$|u_1'| \le \epsilon^2 \frac{c' c_0 c(d)}{(1-\Delta)\Delta^{d+2}} \text{ on } A_{r(1-5\Delta)}$$

for some absolute constant c'.

Proof. By Lemma 12.3.4 we have $h(A_{r(1-4\Delta)}) \subset A_{r(1-3\Delta)}$. Equation (12.3.7) yields $|f(z)| \le r(1-3\Delta)+\epsilon < r(1-2\Delta)$ and $|f(z)| \ge 1/(r(1-3\Delta)) - \epsilon > 1/(r(1-2\Delta))$ on $A_{r(1-3\Delta)}$. By Lemma 12.3.4 h^{-1} is defined on $A_{r(1-2\Delta)}$, and hence $f_1 = h^{-1}\circ f \circ h$ is defined on $A_{r(1-4\Delta)}$.

Now rewrite $h \circ f_1 = f \circ h$ as

$$\lambda z + u_1(z) + w(\lambda z + u_1(z)) = \lambda(z + w(z)) + u(h(z)).$$

According to our version of (12.3.3) we have $u(z) - \eta z = w(\lambda z) - \lambda w(z)$. Using this to replace $\lambda w(z)$ we obtain

$$u_1(z) = w(\lambda z) - w(\lambda z + u_1(z)) + u(h(z)) - u(z) + \eta z. \tag{12.3.13}$$

Now by the Mean Value Theorem and (12.3.9)

$$|w(\lambda z) - w(\lambda z + u_1(z))| \le \sup|w'|\sup|u_1| \le \frac{2r\epsilon c_0 c(d)}{\Delta^{d+1}}\sup|u_1| < \frac{1}{5}\sup|u_1|$$

using (12.3.12). Next (12.3.8) yields

$$|u(h(z)) - u(z)| \le \sup|u'||w| < \epsilon^2 2rc_0c(d)\Delta^{-(d+1)} \text{ on } A_{r(1-4\Delta)}.$$

Now comes an important point that did not arise in the proof of the Siegel Theorem. In order to estimate ηz we use the fact that the rotation number of f is α, where $\lambda = e^{2\pi i\alpha}$. This implies that the rotation number of f_1 on the invariant circle $h(S^1)$ is α as well. (Recall that we do not know whether $h(S^1) = S^1$. Here we see that we do not need to know.) We have $f_1(z) = (\lambda+\eta)z + (u_1(z) - \eta z)$. From the above argument we see that

$$|u_1(z) - \eta z| < 3\epsilon^2 r c_0 c(d)\Delta^{(d+1)} =: \kappa. \tag{12.3.14}$$

In order to show that $|\eta|$ is small we will show that both the absolute value and the argument of $(\lambda+\eta)$ are close to those of λ.

First recall that since f_1 is conjugate to f via a diffeomorphism h close to the identity it has an invariant circle $h(S^1)$ which is C^1 close to S^1. We can assume that this circle lies in the annulus $2 \geq |z| > 1/2$. Assume $|\lambda+\eta| \geq 1$ (the case $|\lambda+\eta| < 1$ is completely similar). Let $z_0 \in h(S^1)$ be a point with maximal absolute value. Then $|z_0| \geq |f_1(z_0)| \geq |\lambda+\eta||z_0| - \kappa$ by (12.3.14). Thus

$$(|\lambda+\eta| - 1) \leq \frac{\kappa}{|z_0|} \leq 2\kappa. \tag{12.3.15}$$

Similarly, since the rotation number of f_1 restricted to the invariant circle $h(S^1)$ is equal to $\alpha = (\arg\lambda)/2\pi$, we conclude that $\arg f_1(z) - \arg z - \arg\lambda$ has to vanish at some point $z_1 \in h(S^1)$. But by (12.3.14)

$$\arg f_1(z_1) - \arg(z_1) - \arg(\lambda+\eta) \leq \arcsin\frac{\kappa}{|(\lambda+\eta)z_1|} \leq \frac{4\kappa}{1+2\kappa} \leq 8\kappa.$$

The last two inequalities follow from (12.3.15) and from the fact that κ is small enough. Thus, $\arg(\lambda+\eta) - \arg\lambda \leq 8\kappa$. Combining this with (12.3.15) we obtain $|\eta| \leq 10\kappa$. From here we proceed exactly as in the proof of Lemma 2.8.6. □

Now we have to show convergence of the sequence h_n in (12.3.6). The estimates needed to do this are in fact very much like those in the proof of the Siegel Theorem 2.8.2 and the remaining argument needs only minor and routine adjustments. Thus we have proved Theorem 12.3.1. □

4. Invariant measures and regularity of conjugacies

Without restrictions on the speed of approximation of the rotation number no regularity of the conjugacy to a rotation (beyond the continuity given by the Denjoy Theorem 12.1.1) can be guaranteed. In the next two sections we will show that for C^∞ maps virtually all conceivable possibilities for the regularity of the conjugacy can be realized. A remarkable exception is a simple result that guarantees that a Lipschitz conjugacy must be C^1. The worst, but in a sense most typical, case of pathological conjugacy is the one that is singular, that is, carries some set of Lebesgue measure zero to a set of full measure and vice versa. We will show that this happens for an appropriate parameter value in most 1-parameter families.

As we showed at the end of the previous chapter (Theorem 11.2.9) a transitive homeomorphism $f: S^1 \to S^1$ is uniquely ergodic and its unique invariant probability measure μ is obtained from Lebesgue measure λ, the only invariant probability measure of the rotation $R_{\tau(f)}$, via a conjugacy $h: S^1 \to S^1$ by the formula $\mu = h_*\lambda$, that is, $\mu(A) = \lambda(h^{-1}(A))$.

Proposition 12.4.1. *The invariant measure μ of a C^1 circle diffeomorphism is either absolutely continuous or singular.*

Proof. Suppose $\mu = \mu_1 + \mu_1$ with μ_1 absolutely continuous and μ_2 singular. Since every diffeomorphism preserves sets of Lebesgue measure zero, $\mu_1 + \mu_2 = \mu = f_*\mu = f_*\mu_1 + f_*\mu_2$, so both μ_1 and μ_2 are invariant and by unique ergodicity one of them is zero. □

Using the connection between unique ergodicity and equidistribution of orbits (Corollary 4.1.14) we obtain the following result.

Theorem 12.4.2. *Let $f: S^1 \to S^1$ be a transitive orientation-preserving homeomorphism, μ its unique invariant Borel probability measure, $E \subset S^1$ a finite union of open (or closed) intervals, χ_E its characteristic function, and $x \in S^1$. Then*

$$\lim_{n\to\infty} \frac{1}{n} \sum_{k=0}^{n-1} \chi_E(f^k(x)) = \mu(E)\,.$$

Suppose now that we are given μ and want to obtain information about h. Denote an interval on S^1 by $[\cdot,\cdot]$. If we let $y_0 = h(0)$ then $y = h(x)$ is uniquely determined by the condition

$$\mu([y_0, y]) = h_*\lambda([y_0, y]) = \lambda([0, x]) = x, \tag{12.4.1}$$

since μ is positive on nonempty open sets. Thus the invariant measure μ uniquely determines h up to a rotation. To utilize this we note a straightforward consequence of the Gottschalk–Hedlund Theorem 2.9.4:

Proposition 12.4.3. *Suppose $f: S^1 \to S^1$ is Lipschitz conjugate to an irrational rotation. Then the conjugacy is C^1.*

Proof. The assumption implies that $K := \sup_{n\in\mathbb{Z}} \sup_{x\in S^1} |(f^n)'| < \infty$ which in turn shows that $1/K \le (f^n)' \le K$ for all $n \in \mathbb{Z}$, so $\infty > \sup_{n\in\mathbb{Z}} |\log(f^n)'| = \left|\sum_{i=0}^{n-1} (\log f') \circ f^i\right|$ and by Theorem 2.9.4 there is a continuous $\varphi: S^1 \to \mathbb{R}$ such that $\log f' = \varphi - \varphi \circ f$. Adjusting φ by a constant we have $\int_{S^1} e^{\varphi} = 1$. Now we set $h(x) := \int_0^x e^{\varphi(t)} dt$. Then $h: S^1 \to S^1$ is C^1 with nonvanishing derivative and hence h^{-1} is also C^1 (quotient rule). Furthermore $(h' \circ f)f' = h'$; hence $h \circ f = R_{\tau(f)} \circ h$. □

Now we can summarize the relationship between the properties of the invariant measure and the regularity of the conjugacy to a rotation.

Proposition 12.4.4. *Let f be a transitive orientation-preserving circle diffeomorphism. The invariant measure of f is absolutely continuous with respect to Lebesgue measure with a C^r density if and only if f is conjugate to a rotation R_τ via a C^{r+1} diffeomorphism h. Having density bounded from above and bounded away from zero is equivalent to the fact that the conjugacy is a C^1 diffeomorphism.*

Proof. Suppose the invariant measure is absolutely continuous with density ρ. Then by (12.4.1)

$$\int_{x_0}^{h(x)} \rho(t)dt = x,$$

that is, $\rho(h(x)) = 1/h'(x)$. If $g = h^{-1}$ then $\rho(x) = g'(x)$. Thus g is obtained by integrating ρ. If ρ is continuous it has to be strictly positive since the set A of points where $\rho = 0$ is f-invariant and closed, so $A = \varnothing$ or $A = S^1$ by transitivity of f. But the latter case would imply $\rho \equiv 0$. Thus if ρ is C^r, $r \geq 0$, then g is C^{r+1} and $g' \neq 0$, so $h = g^{-1}$ is also C^{r+1}.

If $0 < a < \rho < b$ we obtain

$$a < \frac{|g(x) - g(y)|}{|x - y|} < b$$

and hence

$$\frac{1}{b} < \frac{|h(x) - h(y)|}{|x - y|} < \frac{1}{a}.$$

Thus the conjugacy and its inverse are Lipschitz continuous and hence the last claim follows from Proposition 12.4.3. □

Therefore in order to find a circle diffeomorphism that is conjugate to a rotation by a nondifferentiable map only, it suffices to check that the invariant measure is singular with respect to Lebesgue measure.

5. An example with singular conjugacy

The idea behind the construction of an example with singular conjugacy is to use an inductive process in order to extract from a one-parameter family of maps a sequence of maps whose invariant measures cluster increasingly around a set of Lebesgue measure zero. The map with the limit parameter then has a singular invariant measure. As we saw in the previous proposition, this shows that there is no differentiable conjugacy to a rotation.

Theorem 12.5.1. *There exists an analytic orientation-preserving diffeomorphism $f\colon S^1 \to S^1$ such that $f = h \circ R_{\tau(f)} \circ h^{-1}$ with a nondifferentiable homeomorphism h.*

Proof. Consider the family f_α of circle diffeomorphisms induced by

$$F_\alpha(x) = x + \alpha + \mu \sin 2\pi x$$

for some fixed μ, $|\mu| < 1/2\pi$. This family appeared as a one-parameter subfamily of the family considered in Exercises 11.1.4–11.1.7. Let us first make a technically useful observation:

Lemma 12.5.2. *f_α never has infinitely many periodic points.*

Proof. Note that F_α extends to an entire function on $\mathbb{C}$ with an essential singularity at ∞. By Picard's theorem F_α is therefore not injective and hence no iterate F_α^q is injective. On the other hand the existence of infinitely many periodic orbits implies that $\tau(f_\alpha) = p/q$ for some $p, q \in \mathbb{Z}$ and the zeros of $F_\alpha^q(x) - x - p$ on $[0,1]$ have an accumulation point. Thus $F_\alpha^q = \mathrm{Id} + p$ on $\mathbb{C}$ by analyticity. But this is injective, a contradiction. □

Since this family is monotone in α, the rotation number $\rho(\alpha) := \tau(f_\alpha)$ is increasing in α (Proposition 11.1.8). The following observation is crucial to the construction:

Suppose α is the right endpoint of some interval I for which $\tau(f_\beta) = p/q$ ($\beta \in I$). The proof of Proposition 11.1.10 shows that in this case $F_\alpha^q(x) - x \geq p$ for all $x \in \mathbb{R}$ (and $F_\alpha^q(x) - x = p$ for some $x \in \mathbb{R}$). Since there are only finitely many periodic orbits, they are all semistable and all other points move in the same direction under iterations of f_α^q. Thus any neighborhood E of the set of periodic points of f_α has the property that for $\epsilon > 0$ there exists an $N \in \mathbb{N}$ such that any orbit segment of length N contains at most ϵN points outside E, or $A_N \chi_E(f_\alpha, x) := (1/N) \sum_{k=0}^{N-1} \chi_E(f_\alpha^k(x)) > 1 - \epsilon$ for all $x \in S^1$. Hence for any sufficiently small perturbation $\bar{f}$ of f_α and the same E and N we have

$$A_N \chi_E(\bar{f}, x) > 1 - 2\epsilon \quad (x \in S^1).$$

From now on let $A_N \chi_E(f) := \inf\{A_N \chi_E(f, x) \mid x \in S^1\}$.

The construction begins by taking $p_1, q_1 \in \mathbb{Z}$, $\alpha_1 = \max \tau^{-1}(p_1/q_1)$, and a neighborhood E_1 of the set of periodic points such that $\lambda(E_1) < 1$ (where λ is Lebesgue measure on S^1). Choose $N_1 \in \mathbb{N}$ such that $A_{N_1} \chi_{E_1}(f_{\alpha_1}) > 1/2$.

For all subsequent α_n we require

$$A_{N_1} \chi_{E_1}(f_{\alpha_n}) > 0 \quad \text{and} \quad \alpha_n \leq \alpha_1 + \frac{1}{10}.$$

Inductively one can choose α_n such that

(1) α_n satisfies the conditions imposed in all previous steps,
(2) $\alpha_n > \alpha_{n-1}$,
(3) $\alpha_n = \max\left(\tau^{-1}(p_n/q_n)\right)$,
(4) $(p_n/q_n) - (p_{n-1}/q_{n-1}) < \left(2(n-1)^2 \max_{1 \leq k \leq n-1} q_k^2\right)^{-1}$.

This can be done by Proposition 11.1.11.

Let E_n be a neighborhood of the set of periodic points for f_{α_n} of Lebesgue measure $\lambda(E_n) < 1/n!$ and consisting of finitely many open intervals. Choose $N_n \in \mathbb{N}$ such that

$$A_{N_n} \chi_{E_n}(f_{\alpha_n}) > 1 - \frac{1}{2n!}$$

and require that for all subsequent α_m

$$A_{N_n}\chi_{E_n}(f_{\alpha_m}) > 1 - \frac{1}{n!}.$$

Let $\alpha = \lim_{n\to\infty}\alpha_n$ and denote by μ the invariant measure for f_α. We will show that μ is not absolutely continuous with respect to Lebesgue measure.

Note that $\lim_{n\to\infty}\alpha_n$ exists since $\{\alpha_n\}$ is a monotone sequence bounded by $\alpha_1 + \frac{1}{10}$. Furthermore the rotation number $\tau(f_\alpha)$ is indeed irrational: By continuity $\tau(f_\alpha) = \lim_{n\to\infty} p_n/q_n$ and therefore

$$\left|\tau(f_\alpha) - \frac{p_n}{q_n}\right| \le \sum_{k=n}^{\infty}\left|\frac{p_{k+1}}{q_{k+1}} - \frac{p_k}{q_k}\right| < \sum_{k=n}^{\infty}\frac{1}{2k^2 \max\limits_{1\le i\le k} q_i^2} \le \sum_{k=n}^{\infty}\frac{1}{2k^2 q_n^2} \le \frac{\pi^2}{12 q_n^2} < \frac{1}{q_n^2},$$

while for $p/q \in \mathbb{Q}$ and $n \in \mathbb{N}$ such that $q_n > q$ we have

$$\left|\frac{p}{q} - \frac{p_n}{q_n}\right| = \left|\frac{pq_n - qp_n}{qq_n}\right| \ge \frac{1}{q_n^2}.$$

In particular the invariant measure for f_α is indeed well defined.

Now $A_{N_n}\chi_{E_n}(f_\alpha) > 1 - 1/n!$ for $n \in \mathbb{N}$. By construction of $A_{N_n}\chi_{E_n}$ this implies that

$$A_m\chi_{E_n}(f_\alpha) > 1 - \frac{1}{n!} \qquad \text{for all } m > N_n.$$

Since E_n is a finite union of open intervals we can use equidistribution (Theorem 11.3.1) to conclude

$$\mu(E_n) > 1 - \frac{1}{n!} \quad (n \in \mathbb{N}).$$

But $E_n \subset F_m := \bigcup\limits_{k\ge m} E_k$ whenever $n \ge m$, so $\mu(F_m) > 1 - 1/n!$ for all $n \in \mathbb{N}$. Consequently $\mu(F_m) = 1$ and $\mu(F) = 1$ for $F := \bigcap\limits_{m\in\mathbb{N}} F_m$.

But $\lambda(F_m) \le \sum_{n=m}^{\infty}\lambda(E_n) \xrightarrow[m\to\infty]{} 0$ and therefore $\lambda(F) = 0$, so μ is singular. Proposition 12.4.4 then shows that f_α is not smoothly conjugate to a rotation. □

Remark. Exercise 12.5.1 demonstrates the scope of the above argument.

Since the conjugating homeomorphism is a monotone function it is almost everywhere differentiable. However, since it is singular its derivative is zero almost everywhere.

Exercise

12.5.1. *Let $g_t: S^1 \to S^1$, $t \in [a,b]$ be a family of diffeomorphisms such that the lift G_t of g_t to $\mathbb{R}$ is a family of entire functions depending continuously on t. Suppose that g_t is not a rotation for any t and that the rotation number $\tau(g_t)$ is not constant. Prove that there are uncountably many t such that g_t is topologically conjugate to an irrational rotation and has a singular invariant measure.*

6. Fast-approximation methods

a. Conjugacies of intermediate regularity. In this section we describe another inductive construction that allows us to produce C^∞ diffeomorphisms with irrational (but very well approximable) rotation number with almost every possible degree of regularity of the conjugacy to the rotation. Specifically we can make the conjugacy singular, absolutely continuous without being Lipschitz, and C^r without being C^{r+1} for any $r \in \mathbb{N}$. Similarly to the case of the Newton method which has numerous applications beyond the two cases discussed in this book our construction is only the simplest application of another inductive procedure which has been used to construct a variety of dynamical systems with unusual properties whose existence was otherwise unknown. Let us point out that the phenomenon of having a singularconjugacy to a rotation is generic in the following sense: In the C^∞-closure of the set of C^∞ circle diffeomorphisms C^∞ conjugate to a rotation there is a residual set of diffeomorphisms that are conjugate to a rotation by a singular homeomorphism. This, of course, implies that diffeomorphisms that are conjugate to a circle rotation via an absolutely continuous homeomorphism, in particular a Lipschitz continuous or smooth one, form a set of first category. We also note that this method is restricted to the C^∞ category, and its application to the analytic category, although not entirely impossible, is problematic.

Theorem 12.6.1. *Given any neighborhood U of 0 in $C^\infty(S^1)$ there exist α and $f: S^1 \xrightarrow{C^\infty} S^1$ such that $f - R_\alpha \in U$ and f is conjugate to R_α via a conjugacy h that has any one of the following three properties:*

1. *h is singular.*
2. *h is absolutely continuous but not Lipschitz continuous.*
3. *h is C^r but not C^{r+1}, where $r \in \mathbb{N}$ is arbitrary.*

Proof. We obtain α as the limit of $\alpha_n = p_n/q_n \in \mathbb{Q}$ defined inductively by $\alpha_{n+1} = \alpha_n + \beta_n$, where $\beta_n = 1/K_n q_n$. The choice of K_n, explained below, is used to ensure smoothness of f. For each n we construct a map $A_n = \operatorname{Id} + a_n: S^1 \to S^1$ with $a_n(x + (1/q_{n-1})) = a_n(x)$ and set $h_n = A_1 \circ \cdots \circ A_n$ and $f_n := h_n \circ R_{p_n/q_n} \circ h_n^{-1}$. Let us explain the choice of K_n and how it ensures that $f_n \to f$ in the C^∞ topology. To that end set

$$\begin{aligned} f_{n+1,K} &= h_{n+1} \circ R_{p_n/q_n} \circ R_{1/Kq_n} \circ h_{n+1}^{-1} \\ &= h_n \circ A_{n+1} \circ R_{p_n/q_n} \circ R_{1/Kq_n} \circ A_{n+1}^{-1} \circ h_n^{-1} \\ &= h_n \circ R_{p_n/q_n} \circ A_{n+1} \circ R_{1/Kq_n} \circ A_{n+1}^{-1} \circ h_n^{-1} \end{aligned}$$

since A_{n+1} commutes with R_{1/q_n} by design. Notice now that for $1/Kq_n = 0$ we would have $f_{n+1,K} = f_n$, so for sufficiently large K the functions $f_{n+1,K}$ and f_n are as C^∞-close as we please. Thus we take K_n large enough so that the sequence defined by $f_{n+1} = f_{n+1,K_n}$ converges in the C^∞ topology to a function f such that $f - R_\alpha \in U$, as desired.

We will see that the conjugacies h_n are C^0 convergent, so that the limit h is the desired conjugacy. We will obtain the desired properties of h by controlling the corresponding properties of the h_n accordingly. Note first that we can always choose the C^0 norm of a_n sufficiently small so that h_{n+1} is as C^0-close to h_n as we wish (and the same holds for the inverses) and hence we obtain C^0 convergence of h_n to some h. Since $f_n \circ h_n = h_n \circ R_{p_n/q_n}$ we obtain $f \circ h = h \circ R_\alpha$ by taking the C^0-limit, where $h: S^1 \to S^1$ is a monotone surjective map. But if h maps an interval to a point then it maps its image under R_α to a point as well and since finitely many such images cover S^1 this is impossible by surjectivity, so that h is a homeomorphism and $f = h \circ R_\alpha \circ h^{-1}$.

We begin by treating case (2) of an absolutely continuous conjugacy that is not C^1. We choose the support of a_n so small that its image under h_{n-1} has measure $\mu_n \xrightarrow[n\to\infty]{} 0$. This has the effect that for almost every point x the sequence $h_n(x)$ stabilizes. This gives absolute continuity of h. Now we need to show that this can be done in such a way that we do not get C^1 convergence. To that end we construct a_n in such a way that h_n has derivative of absolute value at least $M_n \to \infty$ somewhere, and such that the measure of $h_{n-1}(\mathrm{supp}(a_n))$ is less than half the measure of the set on which the derivative of h_{n-1} exceeds M_{n-1}. Thus the limit function h coincides on a set of positive measure with a function whose derivative on this set is of absolute value greater than M_n. Since sets of positive measure have accumulation points, h has unbounded derivative, if it is differentiable. Thus h cannot be C^1 and hence not Lipschitz by Proposition 12.4.3. This finishes case (2).

We now begin our discussion of case (3). We will use that for any given r one can construct a function with arbitrarily small support that is C^r-small and C^{r+1}-large. Such functions are constructed by taking an appropriate C^0-small function that has large derivatives such as locally $\epsilon \sin(x/\delta)$ and integrating it r times.

Let us first consider the case $r = 1$ since the notation in this case is easier. Thus we need to make sure that the h_n have small derivatives and large second derivatives. To that end we examine the relation between the derivatives of h_n and h_{n+1}. Note that $h'_{n+1} = (h_n \circ A_{n+1})' = h'_n \circ A_{n+1} \cdot (1 + a'_{n+1})$. Since we take the a_n C^1-small, we find that we have h_{n+1} as C^1-close to h_n as needed. On the other hand the second derivatives are related via $h''_{n+1} = h''_n \circ A_{n+1} \cdot (1 + a'_{n+1}) + h'_n \circ A_{n+1} \cdot a''_{n+1}$. The first term on the right is close to h''_n. In the second term $h'_n \circ A_{n+1}$ is close to 1 and hence bounded away from 0 (uniformly in x), so we can choose a_{n+1} in such a way as to obtain a second derivative for h_{n+1} that is much larger than that of h_n on some set. Notice that in order to prove that h^{-1} is C^1 we do not have to obtain explicit expressions for derivatives of inverses because the h_n are C^1-close to the identity and hence the derivatives of their inverses are small as long as those of the h_n are small. Making our choices in this manner we obtain the desired example.

Consider next arbitrary r. Calculating the kth derivative of h_{n+1} in terms of the kth derivative of h_n is not very pleasant. It is, however, entirely sufficient to understand how the highest-order terms interact. Notice first that the rth derivative of h_{n+1} involves only derivatives up to order r of a_{n+1} and h_n, all of which are well controlled by construction. In particular the term involving $h_n^{(r)}$ is $h_n^{(r)} \circ A_{n+1} \cdot (1 + a'_{n+1})$ and is hence close to $h_n^{(r)}$, and all remaining terms are small. When considering the $(r+1)$th derivative we should observe that again the term involving $h_n^{(r+1)}$ is $h_n^{(r+1)} \circ A_{n+1} \cdot (1 + a'_{n+1})$ and hence is close to $h_n^{(r+1)}$. Furthermore $a_{n+1}^{(r+1)}$ again appears with a factor $h'_n \circ A_{n+1}$ which, as we explained in the case $r = 1$, is uniformly bounded away from 0 and hence allows us to choose a_{n+1} in such a way as to have the desired effect on $h_{n+1}^{(r+1)}$, because all remaining terms involve only lower-order derivatives and are hence small. This finishes case (3).

Finally we finish case (1) by constructing a singular conjucacy. This time it is not possible to make the sequence of conjugacies stabilize pointwise, but this is also not needed at all. At stage n we consider q_n fundamental domains I of length $1/q_n$ for the rotation R_{p_n/q_n}. We divide each into N equal pieces for some $N \in \mathbb{N}$ and take the "core" C to be the interval containing all but the first and last of these pieces of I. We take A_n smooth such that A_n coincides with the identity in a neighborhood of the endpoints of I and such that its derivative is so small on each C that $h'_n < \delta_n$ on C, where $\delta_n \to 0$. Next we take K_n to be a multiple of N (and large enough as described in the beginning) so that A_{n+1} preserves the partition into the intervals I as well as the cores C from stage n. This has the effect that the measure $\lambda(h_k(\bigcup C)) < \delta_n$ for all $k \geq n$. Therefore $\lambda(h(\bigcup C)) = 0$ regardless of the stage at which the C were picked. Since $\lambda(\bigcup C) \to 1$ as $n \to \infty$ this shows that h is singular. □

b. Smooth cocycles with wild coboundaries. We next show a similar but linear construction which allows us to produce examples of minimal nonergodic transformations via Propositions 4.2.5 and 4.2.6. We will construct an analytic function φ such that $\varphi(x) = \Phi(x+\alpha) - \Phi(x)$ for a very discontinuous Φ. A strong notion of discontinuity is the following:

Definition 12.6.2. Let X, Y be topological spaces and μ a measure on X. A measurable map $f: X \to Y$ is called *metrically dense* with respect to μ if for all nonempty open $U \subset X$, $V \subset Y$ we have $\mu(U \cap f^{-1}(V)) > 0$.

Proposition 12.6.3. *There exists $\alpha \in \mathbb{R}$ and an analytic function $\varphi: S^1 \to \mathbb{R}$ such that $\varphi(x) = \Phi(x+\alpha) - \Phi(x)$ with $\Phi: S^1 \to \mathbb{R}$ measurable and metrically dense with respect to Lebesgue measure.*

We will use Propositions 4.2.5 and 4.2.6 to prove

Corollary 12.6.4. *There exist analytic minimal nonergodic diffeomorphisms of $\mathbb{T}^2$.*

Proof. Consider the map $f: \mathbb{T}^2 \to \mathbb{T}^2$ given by $f(x,y) = (x+\alpha, y+\varphi(x))$, where φ is as in Proposition 12.6.3. By Proposition 4.2.5 f has uncountably many distinct ergodic invariant measures; in particular it has nonergodic invariant measures. If f were not minimal then by Proposition 4.2.6 we would have

$\varphi(x) = \psi(x+\alpha) - \psi(x) + r$ for some continuous $\psi: S^1 \to \mathbb{R}$ and $r \in \mathbb{Q}$. But letting $F = \psi - \Phi$ yields $F(x+\alpha) - F(x) = r$, so $r \neq 0$ since otherwise $F = \text{const.}$ by ergodicity of R_α, which is impossible since Φ is not continuous. Thus we may suppose that $r > 0$ (the case $r < 0$ is similar). Then $F(x+\alpha) = F(x) + r > F(x)$ for all $x \in S^1$. But there is a set $A = F^{-1}(-\infty, c)$ of positive measure, contradicting the Poincaré Recurrence Theorem 4.1.19. □

Proof of Proposition 12.6.3. It is convenient to switch to multiplicative notation on the circle by considering it as the unit circle in $\mathbb{C}$. Then $x \mapsto x+\alpha$ becomes $z \mapsto \lambda z$ with $\lambda = e^{2\pi i\alpha}$. We inductively define Φ_n by taking the *Dirichlet kernel* $D_{q,n}: S^1 \to \mathbb{R}$ given by

$$D_{q,m}(z) = \sum_{j=1}^{m-1} (z^{jq} + z^{-jq}),$$

a function whose density is concentrated around the qth roots of unity, and setting $\Phi_n = \sum_{k=1}^{n} C_k D_{q_k,m_k}(z)$. Our construction rests on making appropriate choices for C_n, q_n, and m_n. The choice of C_n turns out to be quite free. We set $\alpha_n = p_n/q_n$ and finally let $\alpha = \lim_{n\to\infty} \alpha_n$ and $\Phi = \lim_{n\to\infty} \Phi_n$. The latter limit will be a limit *in measure*, that is, we will show that $\mu(\{z \mid |\Phi(z) - \Phi_n(z)| \geq \epsilon\}) \to 0$ as $n \to \infty$ for any $\epsilon > 0$. To that end it suffices to show, of course, that $\mu_{n,\epsilon} := \mu(\{z \mid |\Phi_{n+1}(z) - \Phi_n(z)| \geq \epsilon\})$ is a summable sequence for every ϵ.

Suppose now that we have chosen C_k, q_k, and m_k for all $k \leq n$ in such a way that $\varphi_n(z) := \Phi_n(\lambda_n z) - \Phi(z)$ is analytic in a neighborhood of S^1. Then we can take $\Phi_n(\lambda_{n+1}z) - \Phi(z)$ as close to φ_n as we please by taking a sufficiently large L_n and letting $\alpha_{n+1} = \alpha_n + 1/L_n q_n$. Taking L_n (and m_{n+1}) large enough renders Φ_{n+1} sufficiently close to being metrically dense (since the q_{n+1}th roots of unity will be $2\pi/q_{n+1}$-dense), while choosing m_{n+1} sufficiently large ensures that Φ_{n+1} is sufficiently close to zero in measure to ensure convergence. The latter choice also ensures that the degree of metric density is increased as desired (that is, cancellations are not strong enough).

Notice now that since $\Phi_{n+1} - \Phi_n = C_{n+1} D_{L_n q_n, m_{n+1}}$ is $2\pi/q_n$ periodic we get $\varphi_{n+1}(z) := \Phi_{n+1}(\lambda_{n+1}z) - \Phi_{n+1}(z) = \Phi_n(\lambda_{n+1}z) - \Phi_n(z)$, which by construction is as close to φ_n as desired, ending the proof.

Notice that the constants C_n were not specified and could be chosen in any particular fashion. Accordingly one can choose them to decrease more or less rapidly and obtain either integrable or nonintegrable Φ as a result. □

Exercises

12.6.1. *Show that in the construction for case (2) the derivatives of h and h^{-1} are unbounded and hence discontinuous on every interval.*

12.6.2*. *Make $h': A \to [0, \infty)$ in the construction for case (2) metrically dense.*

12.6.3. *Show that the set of rotation numbers α for which the conclusion of Theorem 12.6.1 can be achieved is residual (that is, contains a dense G_δ).*

7. Ergodicity with respect to Lebesgue measure

We conclude this chapter with a result which indicates that any circle diffeomorphism behaves with respect to Lebesgue measure in a way similar in a certain sense to an irrational rotation. Let us first extend the notion of ergodicity introduced for invariant measures in Definition 4.1.6 to noninvariant measures.

Definition 12.7.1. Let $f: X \to X$ be a bijection and μ a probability measure on X such that $f_*\mu$ is equivalent to μ. Then μ is said to be *quasi-invariant*. If every μ-measurable f-invariant subset of X has μ-measure zero or one, then f is said to be *ergodic* with respect to μ.

Note that Lebesgue measure is quasi-invariant under any circle diffeomorphism. The following theorem is again based on a bounded distortion estimate.

Theorem 12.7.2. *Let $f: S^1 \to S^1$ be an orientation-preserving diffeomorphism whose derivative has bounded variation and such that the rotation number is irrational. Then f is ergodic with respect to Lebesgue measure.*

Remark. By Denjoy's theorem f is topologically conjugate to a rotation and hence uniquely ergodic, as shown earlier. As we saw in the two previous sections the f-invariant measure can be singular, so the Lebesgue measure may not be equivalent to any invariant measure (Proposition 5.1.2).

Lemma 12.7.3. *Let F be a lift of f, $a < b < c < d < a+1 \in \mathbb{R}$, $n \in \mathbb{N}$, V the total variation of $\log f'$, and N the covering multiplicity, that is, the maximal multiplicity with which $\{\pi(F^i([a,d]))\}_{i=0}^{n-1}$ covers points of S^1 (cf. Exercise 10.0.5). Then*

$$\frac{F^n(c) - F^n(b)}{F^n(d) - F^n(a)} \leq \frac{c-b}{d-a} \exp NV.$$

Proof. By the Mean Value Theorem there exist points $x_i \in F^i([b,c])$ and $y_i \in F^i([a,d])$, where $0 \leq i \leq n-1$, such that

$$F^n(c) - F^n(b) = (c-b) \cdot \prod_{i=0}^{n-1} \frac{F^{i+1}(c) - F^{i+1}(b)}{F^i(c) - F^i(b)} = (c-b) \cdot \prod_{i=0}^{n-1} F'(x_i)$$

and

$$F^n(d) - F^n(a) = (d-a) \cdot \prod_{i=0}^{n-1} F'(y_i).$$

Thus

$$\frac{F^n(c)-F^n(b)}{F^n(d)-F^n(a)} = \frac{c-b}{d-a}\frac{\prod_{i=0}^{n-1}F'(x_i)}{\prod_{i=0}^{n-1}F'(y_i)} = \frac{c-b}{d-a}\exp\Big(\sum_{n=0}^{\infty}(\log F'(x_i)-\log F'(y_i))\Big).$$

But if Δ_i is the interval with endpoints x_i, y_i then $\pi(\Delta_i) \subset \pi(F^i([a,d]))$ and $\{\pi(\Delta_i)\}_{i=1}^{n-1}$ therefore covers points of S^1 with multiplicity at most N. Consequently $|\sum_{n=0}^{\infty}(\log F'(x_i) - \log F'(y_i))| \le NV$ and the claim follows. □

If λ denotes Lebesgue measure we have the

Corollary 12.7.4. *Let $a < d < a+1 \in \mathbb{R}$, $B \subset \mathbb{R}$ be a Lebesgue measurable set, $n \in \mathbb{N}$, and F, N, V as before, then*

$$\frac{\lambda(F^n(B\cap[a,d]))}{\lambda(F^n([a,d]))} \le \frac{\lambda(B\cap[a,d])}{\lambda([a,d])}\cdot\exp(NV).$$

Proof. Note that the denominators on both sides are as before and that Lebesgue measure is defined via coverings by intervals. □

Lemma 12.7.5. *For $\epsilon > 0$, $x \in S^1$ there exist $r \in \mathbb{N}$ and an arc $\Delta = \Delta(x,\epsilon) \subset S^1$ such that $x \in \Delta$, $\lambda(f^i(\Delta)) = \epsilon \quad (i = 0,\dots,r), \bigcup_{i=0}^{r} f^i(\Delta) = S^1$, and the multiplicity of this covering does not exceed* 3.

Proof. By the Denjoy Theorem 12.1.1 we may assume that f is an irrational rotation and $x = 0$. Take $k \in \mathbb{N}$ such that $u_k := |f^k(0)| < |f^i(0)|$ for $0 < i < k$ and $u_k < \epsilon$. Then it is easy to see that $\Delta(0,\epsilon) := (-u_k, u_k)$ and $r = k - 1 + \min\{i > 0 \mid |f^i(0)| < |f^k(0)|\}$ are as required. □

Proof of Theorem 12.7.2. Suppose $A \subset S^1$ is Lebesgue measurable, $\lambda(A) > 0$, and $f(A) \subset A$. Let $x_0 \in A$ be a Lebesgue density point for A, that is, such that for any $\delta > 0$ there is an $\epsilon > 0$ such that for all $\epsilon_1, \epsilon_2 \in (0, \epsilon_0)$ we have

$$\lambda(A \cap (x_0 - \epsilon_1, x_0 + \epsilon_2)) > (1-\delta)(\epsilon_1 + \epsilon_2).$$

For x_0, ϵ we can thus find an arc $\Delta = (x_0 - \epsilon_1, x_0 + \epsilon_2)$ with $\epsilon_1, \epsilon_2 < \epsilon$ as in Lemma 12.7.5 and apply Corollary 12.7.4 to get

$$\begin{aligned}\frac{\lambda((S^1\setminus A)\cap f^i(\Delta))}{\lambda(f^i(\Delta))} &\le \frac{\lambda(f^i(S^1\setminus A)\cap f^i(\Delta))}{\lambda(f^i(\Delta))} = \frac{\lambda(f^i((S^1\setminus A)\cap\Delta))}{\lambda(f^i(\Delta))}\\ &\le \frac{\lambda((S^1\setminus A)\cap\Delta)}{\lambda(\Delta)}\cdot\exp 3V \le \delta\exp 3V\,.\end{aligned}$$

Thus $\lambda(S^1 \setminus A) \le \sum_{i=0}^{r}\lambda((S^1\setminus A)\cap f^i(\Delta)) \le \delta\cdot\exp(3V)\cdot\sum_{i=0}^{r}\lambda(f^i(\Delta)) \le 3\delta\exp 3V$. But δ was arbitrary, so $\lambda(S^1 \setminus A) = 0$ and f is ergodic. □

Exercises

12.7.1. *Show that no diffeomorphism with rational rotation number is ergodic with respect to Lebesgue measure and, more generally, any nonatomic quasi-invariant measure.*

12.7.2. *Show that there exists an irrational* α *such that the rotation* R_α *possesses a singular quasi-invariant ergodic measure.*

13

Twist maps

We now return to the study of twist maps which we began in Sections 9.2 and 9.3. The main result of those sections was the existence of at least two special periodic orbits for any rational rotation number from the twist interval (Theorem 9.3.7). Those orbits (Birkhoff periodic orbits of type (p,q)) can be regarded from two different viewpoints. On the one hand they are special minimal and mountain-pass–type minimax critical points of the action functional (9.3.7) on the space of periodic states. The minimal Birkhoff periodic orbits are, in fact, characterized by the property that each of their segments minimizes the action functional (9.3.12) defined on the space of states with the same endpoints. On the other hand those orbits are order preserving, that is, their angular coordinates are in one-to-one order-preserving correspondence to the orbits of the rotation by the angle $2\pi p/q$ (cf. the remark after Definition 9.3.6).

In this chapter we will extend both aspects of this study to include orbits with *irrational* rotation number. We will use extensively the structural theory of circle homeomorphisms developed in Chapter 11. In Section 13.2 we concentrate on the preservation of order and in Sections 13.3–13.4 on the variational description. The most striking conclusion is that whereas for circle homeomorphisms Denjoy-type orbits whose closures are minimal nowhere-dense sets occur only for maps of low regularity (Theorem 12.1.1), for twist maps similar orbits, whose closures (Aubry–Mather sets) project to nowhere-dense Cantor sets in the circle, appear in arbitrarily smooth systems as a rule rather than the exception. (See Section 13.5 for an elaboration of this remark.)

1. The Regularity Lemma

In this section we prove that any order-preserving orbit of a twist map forms a part of the graph of a Lipschitz function with Lipschitz constant which can be taken to be uniform on any closed annulus in $S^1 \times (0,1)$. As in Section 9.3 we will frequently work with lifts.

Lemma 13.1.1. *Let $F: \mathbb{R} \times (0,1) \to \mathbb{R} \times (0,1)$ be the lift of a twist diffeomorphism $f: C \to C$ (not necessarily area preserving). If $(x_i, y_i) = F^i(x_0, y_0)$ and $(x'_i, y'_i) = F^i(x'_0, y'_0)$ and $x'_i > x_i$ for $i = -1, 0, 1$ then there exists $M \in \mathbb{R}$ such that $|y'_0 - y_0| < M|x'_0 - x_0|$. M can be chosen uniformly on any closed annulus in C.*

Proof. Suppose first that $y'_0 < y_0$. If $(\tilde{x}, \tilde{y}) = F(x'_0, y_0)$ then the twist condition Definition 9.3.1(4) yields

$$\tilde{x} > x'_1 + c(y_0 - y'_0),$$

where c is bounded away from zero on any closed annulus in C. On the other hand differentiability of f implies that there is a constant L (bounded on compact annuli in C) such that

$$x'_1 > x_1 > \tilde{x} - L(x'_0 - x_0).$$

Taking $M = Lc^{-1}$ we obtain the claim. If $y'_0 > y_0$ repeat the same argument with f^{-1} replacing f. □

Definition 13.1.2. (Cf. Definition 9.3.6.) Consider a twist diffeomorphism $f: C \to C$. An orbit segment (or orbit) $\{(x_m, y_m), \dots, (x_n, y_n)\}$ with $-\infty \le m < n \le \infty$ of f, which may be infinite in one or both directions, is called *ordered* or *order preserving* if $x_i \ne x_j$ when $i \ne j$ and $(i, j) \ne (n, m)$ and f preserves the cyclic ordering of the x-coordinates, that is, if x_i, x_j, x_k, where $i, j, k < n$, are positively ordered (with respect to a chosen orientation on S^1) then $x_{i+1}, x_{j+1}, x_{k+1}$ are ordered in the same way.

Corollary 13.1.3. *Consider an area-preserving twist map $f: C \to C$ and an order-preserving orbit segment $\{(x_m, y_m), \dots, (x_n, y_n)\}$ with $-\infty \le m < n \le \infty$ of f that is contained in a closed annulus in C. Then $|y_i - y_j| < M|x_i - x_j|$ for all i, j such that $m < i, j < n$.*

Proof. Apply Lemma 13.1.1 to the triples $(i-1, i, i+1)$ and $(j-1, j, j+1)$. □

This corollary shows that the closure E of an ordered orbit is contained in the graph of a Lipschitz function $\varphi: S^1 \to (0,1)$. Note that $f_{\restriction E}$ projects to a homeomorphism of the projection of E to S^1 which we can also extend linearly into the gaps of that set to obtain a circle homeomorphism. We can thus define the *rotation number* of an ordered orbit to be the rotation number of this induced circle homeomorphism. So we also see that the intrinsic dynamics of ordered orbits of a twist map of the two-dimensional annulus is essentially one-dimensional. In the next section we will see that, in fact, the dynamics of the one-dimensional maps corresponding to all rotation numbers from the twist interval is represented in a single twist map.

Exercises

13.1.1. *Let $f: S^1 \times [0,1] \to S^1 \times [0,1]$ be a twist homeomorphism of the closed annulus (see Definition 9.3.19). Prove the following version of the Regularity Lemma 13.1.1: There exists a strictly monotone nonnegative function $\omega: [0,1] \to \mathbb{R}$ (modulus of continuity) such that if $(x_i, y_i) = F^i(x_0, y_0)$ and $(x_i', y_i') = F^i(x_0', y_0')$ and $x_i' > x_i$ for $i = -1, 0, 1$ then $|y_0' - y_0| < \omega(|x_0' - x_0|)$.*

13.1.2. *Calculate the function ω in the case when f is a C^2 diffeomorphism such that $\frac{\partial F_1}{\partial y}\big|_{y=0} \equiv \frac{\partial F_1}{\partial y}\big|_{y=1} \equiv 0$ but the second derivative $\frac{\partial^2 F_1}{\partial y^2}$ is positive for $y = 0$ and negative for $y = 1$.*

2. Existence of Aubry–Mather sets and homoclinic orbits

a. Aubry–Mather sets. We introduce the following metric in the space of closed subsets of a metric space X:

Definition 13.2.1. The *Hausdorff metric* is defined by setting

$$d(A, B) := \sup\{d(x, B) \mid x \in A\} + \sup\{d(A, y) \mid y \in B\}$$

for any two closed sets A, B. We refer to a limit with respect to the topology induced by the Hausdorff metric as a Hausdorff limit.

We make two observations:

Lemma 13.2.2. *The Hausdorff metric on the closed subsets of a compact metric space defines a compact topology.*

Proof. We need to verify total boundedness and completeness. Pick a finite $\epsilon/2$-net N. Any closed set $A \subset X$ is covered by a union of ϵ-balls centered at points of N and the closure of the union of these has Hausdorff distance at most ϵ from A. Since there are only finitely many such sets, we have shown that this metric is totally bounded. To show that it is complete consider a Cauchy sequence (with respect to the Hausdorff metric) of closed sets $A_n \subset X$. If we let $A := \bigcap_{k \in \mathbb{N}} \overline{\bigcup_{n \geq k} A_n}$ then one can easily check that $d(A_n, A) \to 0$. □

Notice that any homeomorphism f of a compact metric space X induces a natural homeomorphism of collection of closed subsets of X with the Hausdorff metric, so we may conclude the following:

Lemma 13.2.3. *The set of closed invariant sets of a homeomorphism f of a metric space is a closed set with respect to the Hausdorff metric.*

Proof. This is just the set of fixed points of the induced homeomorphism, hence is closed. □

Definition 13.2.4. Let $f: C \to C$ be a twist map. A closed invariant set $E \subset C$ is called an *ordered set* if it projects one-to-one to a subset of the circle and f preserves the cyclic order on E. An *Aubry–Mather set* is a minimal ordered invariant set projecting one-to-one on a nowhere-dense Cantor set of S^1.

Note that any orbit in an ordered set is an ordered orbit. The complement of the projection of an Aubry–Mather set is the union of countably many intervals on the circle. We will call those intervals the *gaps* of the Aubry–Mather set. The endpoints of each interval are the projections of points on the Aubry–Mather set which we shall also call *endpoints*. Corollary 13.1.3 immediately yields:

Corollary 13.2.5. *Let $f: C \to C$ be a twist diffeomorphism and A an Aubry–Mather set for f. Then there is a Lipschitz continuous function $\varphi: S^1 \to (0,1)$ whose graph contains A.*

Proof. Corollary 13.1.3 gives us such a function defined in the projection of A to S^1. Extending it linearly through the gaps of that Cantor set gives a function with the same Lipschitz constant. □

We define the *rotation number* of an Aubry–Mather set or an invariant circle to be the rotation number of any of its orbits as defined at the end of Section 13.1. We can now prove one of the central results in the theory of twist maps.

Theorem 13.2.6. *Let $f: C \to C$ be an area-preserving twist diffeomorphism. For any irrational number α from the twist interval of f there exists an Aubry–Mather set A with rotation number α or an invariant circle graph(φ), where φ is a Lipschitz function, with rotation number α.*

Proof. Let p_n/q_n be a sequence of rationals in lowest terms that approximates α. Apply Theorem 9.3.7 and take any sequence w_n of Birkhoff periodic orbits of type (p_n, q_n). According to Corollary 13.1.3 we can construct a Lipschitz function $\varphi_n: S^1 \to (0,1)$ whose graph contains w_n. By an argument similar to the one that yielded (9.3.5) we observe that all these orbits are contained in a closed annulus in C, so the Lipschitz constant can be chosen independently of n. Using precompactness of this equicontinuous family of functions (the Arzelá–Ascoli Theorem A.1.24) we may without loss of generality assume that these functions converge to a Lipschitz function φ. The graph of φ may not be f-invariant, but it always contains a closed f-invariant set A, which is obtained as follows. The domain of φ_n contains the projection of the Birkhoff periodic orbit of type (p_n, q_n) to S^1. These Birkhoff periodic orbits of type (p_n, q_n) are closed f-invariant subsets of C and thus in the topology of the Hausdorff metric they have an accumulation point $A \subset C$. The set A obviously belongs to the graph of φ and is f-invariant by Lemma 13.2.3. Furthermore f preserves the cyclic ordering of A (since this is true for Birkhoff periodic orbit w_n and is a closed property). If we denote by f_n the extensions to S^1 of the projections of f from the Birkhoff periodic orbits of type (p_n, q_n) to S^1, and by f_α the extension of the projection of $f_{\restriction A}$ to S^1, then $f_n \to f_\alpha$ uniformly. Thus by continuity of the rotation number in the C^0 topology (Proposition 11.1.6) the rotation number of A is α. Consider now the minimal set of f_α. By the dichotomy of Proposition 11.2.5 it is either the whole circle or it is an invariant Cantor set. In the latter case the image of this Cantor set under $\mathrm{Id}\times\varphi$ is then an Aubry–Mather set with rotation number α. □

Remark. Note that the Aubry–Mather set obtained in Theorem 13.2.6 may, in fact, be a subset of an invariant circle for f. In this case the restriction of f to this invariant circle is a Denjoy-type homeomorphism similar to the examples constructed in Section 12.2. It is known that this can happen for $C^{3-\epsilon}$ diffeomorphisms,[1] although it is not known whether this is possible for maps with higher smoothness. Of course in the case of a smooth invariant circle there can only be an Aubry–Mather set if f itself is not very smooth (cf. the Denjoy Theorem 12.1.1).

The Hausdorff limit of the Birkhoff periodic orbits of type (p_n, q_n) may be larger than an Aubry–Mather set, although it is always an order-preserving set. If it is not a minimal set then it contains a set of orbits that are homoclinic to the Aubry–Mather set. At this stage we do not know whether such orbits actually exist, but later we will see that this is always the case by taking Hausdorff limits of the minimax Birkhoff periodic orbits and using some careful variational estimates (see Section 13.4).

Let us point out that replacing in the preceding arguments the Birkhoff periodic orbits w_n by arbitrary invariant ordered sets converging in the Hausdorff metric we obtain the following:

Proposition 13.2.7. *The rotation number of an ordered invariant set is continuous in the topology of the Hausdorff metric.*

This, in turn, implies

Corollary 13.2.8. *The rotation number of ordered orbits is a continuous function of the initial condition.*

Proof. Let $x_n \to x$ be a convergent sequence of points with ordered orbits. Without loss of generality we may assume that the rotation numbers α_n of the orbits of the x_n converge. Consider the collection of orbits of the x_n. By compactness of the topology of the Hausdorff metric it contains a subsequence that converges to an ordered set that contains the orbit closure of x. Thus by Proposition 13.2.7 the limit of the rotation numbers of the orbit of x_n is the rotation number of the orbit of x. □

We can now show that for any irrational number there is at most one invariant circle with that rotation number.

Theorem 13.2.9. *Let $f\colon C \to C$ be an area-preserving twist map and α an irrational number in the twist interval. Then f has at most one invariant circle of the form graph(φ) with rotation number α. If there is such an invariant circle* *then f has no Aubry–Mather sets with rotation number α outside this circle, and hence has at most one such Aubry–Mather set.*

Remark. It is, in fact, possible for a twist map to have several invariant circles with the same rational rotation number. This is the case in the elliptic billiard (cf. Figure 9.2.3), where two branches of heteroclinic loops form a pair of invariant circles with rotation number 1/2. A time-t map (for small t) of the mathematical pendulum (Section 5.2c) exhibits a similar phenomenon for rotation number zero.

Proof. We begin with a lemma.

Lemma 13.2.10. *Suppose that f has an invariant circle R (of the form graph(φ)) with rotation number α. Then every order-preserving orbit whose closure is disjoint from R has rotation number different from α.*

Proof. The circle R divides the annulus C into two connected components which we will naturally call the upper and lower components, respectively. Suppose that x is a point in the upper component of $C \setminus R$ whose orbit is order preserving and bounded away from R. Then f restricted to the orbit of x projects to a map of a subset E of the circle S^1. We want to extend it to a map f_2 of S^1, which is strictly ahead (in the sense of Definition 11.1.7) of the map f_1 induced by $f_{\restriction_R}$, that is, $f_1 \prec f_2$. This relation holds on E already, so we need only take care to extend carefully from E. Extending to the closure of E does not change the strict inequalities since we have the twist condition and the assumption that the orbit of x is bounded away from R. Thus we only have to define g on the intervals complementary to $\bar{E}$. This we can do as follows: Denote the endpoints of such an interval by x_1 and x_2. Then denote by δ the smaller of the differences $f_2(x_1) - f_1(x_1)$ and $f_2(x_2) - f_1(x_2)$ and let $f_2(tx_1+(1-t)x_2) = \max(tf_2(x_1)+(1-t)f_2(x_2), \delta+f_1(tx_1+(1-t)x_2))$. Observe that f_2 is a monotone function on S^1 and we still have $f_1 \prec f_2$. Consequently Proposition 11.1.9 implies that the rotation number of f_2 is greater than α. Likewise there cannot be an order-preserving orbit of rotation number α in the lower component of $C \setminus R$. □

Suppose there are two invariant circles with rotation number α. Their intersection is invariant, so if at least one of them is transitive then they are disjoint, which is impossible by the lemma. Otherwise the intersection contains a common Aubry–Mather set A and the two circles form the graphs of two distinct functions φ_1 and φ_2 which coincide on the projection of A. The graphs of both $\max(\varphi_1, \varphi_2)$ and $\min(\varphi_1, \varphi_2)$ are invariant, and hence so is the area between these graphs. But the latter area has to have infinitely many connected components, since it projects into the nonrecurrent complementary intervals to the projection of the Aubry–Mather set. Thus we obtain an open disk with pairwise disjoint images, which is impossible by area preservation (cf. the Poincaré Recurrence Theorem, Theorem 4.1.19). Here we use irrationality of the rotation number, without which there could be finitely many components which are permuted by f.

The lemma also yields the impossibility of having an Aubry–Mather set of rotation number α outside an invariant circle of rotation number α. □

Remark. In the absence of an invariant circle with rotation number α there may be many Aubry–Mather sets with that rotation number. In fact, often there are multiparameter families of such sets.[2]

We now turn the process around and approximate a rational number by irrational ones and consider the limits of the corresponding Aubry–Mather sets in order to construct nonperiodic orbits with rational rotation number.

Proposition 13.2.11. *Let $f: C \to C$ be an area-preserving twist map and p/q a rational number in the twist interval. Then there exists an order-preserving closed f-invariant set with rational rotation number which is either an invariant circle consisting of periodic orbits or contains nonperiodic points. Moreover in the latter case the two endpoints of each complementary interval are nonperiodic.*

Proof. Let $\{\alpha_n\}_{n\in\mathbb{N}}$ be a sequence of irrational numbers in the twist interval approximating p/q. Consider the corresponding invariant minimal order-preserving sets A_n with rotation number α_n. Without loss of generality we may assume that the A_n converge to a set A in the topology of the Hausdorff metric as $n \to \infty$. A is clearly f-invariant and ordered. If infinitely many of the A_n are circles then A is also a circle and by continuity of the rotation number the restriction of f to this circle has rotation number p/q. By the classification of circle maps with rational rotation number (cf. Proposition 11.2.2 or the table in Section 11.2c) we are done in this case. So we may assume that all A_n are Aubry–Mather sets. To understand the dynamics of A we will consider the gaps, that is, the intervals in S^1 complementary to the projection of A to S^1. Each of these gaps $G \subset S^1$ has a well-defined length $l(G)$ and we want to show that the two endpoints of such a gap are not periodic.

To this end let us first note that a gap G of A is the limit of corresponding gaps G_n of A_n in the Hausdorff metric. Let us denote by f_n an extension to a circle homeomorphism of the projection of $f\restriction_{A_n}$ to S^1 and by f_0 the same extension corresponding to $f\restriction_A$. Then, since f_n has irrational rotation number, the images of the gap G_n under the iterates of f_n are pairwise disjoint, so $\sum_{m\in\mathbb{N}} l(f_n^m(G_n)) \le 1$. If both endpoints of G are periodic then the gap G is periodic, that is, $\sum_{n\in\mathbb{N}} l(f_0(G))$ diverges. But $l(f_n^m G_n) \to l(f_0^m G)$ for all $m \in \mathbb{N}$, which gives a contradiction. Thus one of the endpoints of G is nonperiodic.

Note that the other endpoint of the gap G must then also be nonperiodic, since otherwise $f_0^q(G)$ is a gap that intersects G nontrivially without coinciding with G. □

We can thus describe the structure of such an invariant set in the generic case when it contains only finitely many periodic orbits:

Corollary 13.2.12. *If a closed order-preserving f-invariant set A with rational rotation number p/q contains only finitely many periodic orbits then there is a complete set of heteroclinic connections in the following way: If $\gamma_1, \dots, \gamma_s$ denote the periodic orbits in A, ordered according to the induced cyclic ordering of the circle, then there are heteroclinic orbits $\sigma_1, \dots, \sigma_n$ such that either*

$$\begin{aligned}\gamma_1 &= \omega(\sigma_s) = \alpha(\sigma_1),\\ \gamma_2 &= \omega(\sigma_1) = \alpha(\sigma_2),\\ &\vdots\\ \gamma_s &= \omega(\sigma_{s-1}) = \alpha(\sigma_s)\end{aligned}$$

or that the same situation holds with α and ω interchanged. Here α and ω denote the α- and ω-limits sets of an orbit (cf. Definition 1.6.2). If $s = 1$ the orbit σ_1 is, of course, a homoclinic one.

Up to now for any given area-preserving twist map we have constructed ordered (Birkhoff) periodic orbits, which are orbits of type $\mathrm{I}_{p/q}$; nonperiodic orbits of type $\mathrm{II}_{p/q}$ or $\mathrm{III}_{p/q}$ as described in the table in Section 11.2c; dense orbits on an invariant circle, which are orbits of type I_α; and orbits in Aubry–Mather sets, which are of type II_α. Thus we have exhibited in twist maps all types of orbits that appear for circle homeomorphisms, save for those of type III_α, which we will only be able to produce in Section 4 as a result of a more careful variational approach.

b. Invariant circles and regions of instability. The following result by Birkhoff shows that invariant circles divide the annulus into regions inside which orbits freely travel between top and bottom. This follows from a slightly more general semilocal result which does not use differentiability and the assumptions on the invariant set are most conveniently expressed on the half-open annulus $A = S^1 \times [0, \infty)$.

Theorem 13.2.13. (Birkhoff Theorem)[3] *Let $f: A \to A$ be an orientation-preserving twist homeomorphism and $U \subset NW(f)$ an f-invariant open relatively compact set containing $S^1 \times \{0\}$ with connected boundary. Then ∂U is the graph of a continuous $\psi: S^1 \to (0, \infty)$.*

Remark. If f preserves area, or any measure that is positive on open sets and finite on compact sets, then the assumption $U \subset NW(f)$ is automatically satisfied (Proposition 4.1.18).

Using Corollary 13.1.3 we obtain

Corollary 13.2.14. *If under the assumptions of Theorem 13.2.13 f is a C^1 twist diffeomorphism then ψ is Lipschitz.*

Proof of Theorem 13.2.13. Let $V := \{(x, y) \mid (x, y') \subset U \text{ for } y' \in [0, y]\}$. V is open and connected, hence path-connected. We plan to show $V = U$.

Lemma 13.2.15. *If $I = [x_1, x_2] \subset S^1$, $(x_1, y), (x_2, y) \in V$, and $I \times \{y\} \subset U$ then $I \times [0, y] \subset U$.*

Proof. $C := \partial(I \times [0, y]) \subset U$ is a Jordan curve and $A \setminus U$ is unbounded and connected, hence contained in the exterior of C. □

Let ∂_U, Cl_U denote boundary and closure with respect to the topology of U.

Lemma 13.2.16. *$\partial_U V$ is the disjoint union of segments $S_i = \{x_i\} \times (y_{i,1}, y_{i,2})$ ($i \in I \subset \mathbb{N}$) with $0 < y_{i,1} < y_{i,2}$, $(x_i, y_{i,1}), (x_i, y_{i,2}) \in \partial U$. Furthermore S_i divides every sufficiently small neighborhood of $p \in S_i$ into two open sets $V' \subset V$ and $U' \subset U \setminus V$.*

Proof. The connected component of $U \cap (\{x\} \times [0, \infty))$ containing $(x, y) \in \partial_U V$ is $S := \{x\} \times (y_1, y_2)$ with $0 < y_1 < y_2$ (since $(x, 0) \in V$ and U is bounded) and $(x, y_1), (x, y_2) \in \partial U$. Also $V \cap S = \varnothing$. Now we show $S \subset \partial_U V$.

For a sufficiently small open interval $I \subset S^1$ containing x there exists $\epsilon > 0$ such that $W := I \times (y - \epsilon, y + \epsilon) \subset U$. If $(x', y') \in W \cap V$ then $\{x'\} \times (y - \epsilon, y + \epsilon) \subset V$, hence $V \cap W = \bigcup \{\{x'\} \times (y - \epsilon, y + \epsilon) \mid (x', y') \in W \cap V\}$ and $\{x\} \times (y - \epsilon, y + \epsilon) \subset \partial_U (V \cap W) \subset \partial_U V$ since $\{x\} \times (y - \epsilon, y + \epsilon) \subset S \subset U \setminus V$ and $(x, y) \in \partial_U V$. Thus $\partial_U V \cap S$ is open in S and it is closed in S since $S \subset U$, so $S \subset \partial_U V$.

By the preceding lemma $[a, b] \times (y - \epsilon, y + \epsilon) \subset V$ whenever $(a, y), (b, y) \in W \cap V$. Thus for sufficiently small δ and ϵ one component each of $\big((x - \delta, x + \delta) \times (y - \epsilon, y + \epsilon)\big) \setminus \big(\{x\} \times (y - \epsilon, y + \epsilon)\big)$ is in V and $U \setminus V$. In particular there are at most countably many such S. □

Lemma 13.2.17. *For each i the set $U \setminus S_i$ consists of two connected components, one of which is disjoint from V and whose boundary intersects no S_j for $j \neq i$.*

Proof. Denote the connected component of $U \setminus S_i$ containing the open set U' from Lemma 13.2.16 by U_i. Suppose there is a path c in U_i starting in U' and ending either on some S_j for $j \neq i$ or in V', where V' is also as in Lemma 13.2.16. In the latter case c has to cross $\partial_U V$ at least once and by construction this means that c crosses S_j for some $j \neq i$. Thus in either case we obtain a path $\gamma \colon [0, 1] \to U$ such that $\gamma(0, 1) \subset U \setminus \bigcup_{k \in I} S_k$, $\gamma(0) \in \mathrm{Int}\, S_i$, and $\gamma(1) \in \mathrm{Int}\, S_j$ for some $j \neq i$. Now consider the path obtained from γ as follows. Extend γ to $[-\epsilon, 0)$ by a short horizontal segment in V' ending at (x_0, y_0) and then add to it the vertical segment $\{x_0\} \times [0, y_0]$. Next note that by construction the original path γ is disjoint from V, so we can extend it beyond 1 by a horizontal segment into V, to which in turn we append the vertical segment connecting to $S^1 \times \{0\}$. Thus we obtain a simple (since $\gamma \cap V = \varnothing$) path Γ in U with endpoints on $S^1 \times \{0\}$. It follows immediately from the Jordan Curve Theorem A.5.2 that this path separates the annulus into two components. Consider the endpoints of S_i. They belong to ∂U, which is connected by assumption. On the other hand, by construction, they are in different components of $A \setminus \Gamma$, contrary to connectedness of ∂U. □

Notice that $\mathrm{Cl}_U U_i = U_i \cup S_i$ and $V \cap \mathrm{Cl}_U U_i = \varnothing$. Clearly we need to show that such U_i cannot occur.

Lemma 13.2.18.

(1) $U = V \cup (\bigcup_{i \in I} \mathrm{Cl}_U U_i)$.
(2) *The U_i are the connected components of $U \setminus \mathrm{Cl}_U V$.*
(3) *The $\mathrm{Cl}_U U_i$ are the connected components of $U \setminus V$.*

Proof. (1) If $p \in U$ and $p \notin \mathrm{Cl}_U V$ then by path connectedness of U there is an arc γ with $\gamma(0) = p$, $\gamma(1) = q \in \mathrm{Cl}_U V$, and $\gamma([0,1)) \subset U \setminus \mathrm{Cl}_U V$, hence $q \in \partial_U V$, that is, $q \in S_i$. But since $U_i \cup S_i \cup V$ contains a neighborhood of q we have $p \in \gamma([0,1)) \subset U_i$. Thus $U = \mathrm{Cl}_U V \cup \bigcup_{i \in I} U_i = V \cup \bigcup_{i \in I} \mathrm{Cl}_U U_i$.

(2) U_i is open, hence open in $U \setminus \mathrm{Cl}_U V$, and closed in $U \setminus \mathrm{Cl}_U V$ since $\partial_U U_i = S_i \subset \mathrm{Cl}_U V$.

(3) $\mathrm{Cl}_U U_i$ is closed in U, hence closed in $U \setminus V$. $\mathrm{Cl}_U U_i$ is open in $U \setminus V$ since $V \cup \mathrm{Cl}_U U_i$ contains a neighborhood of $\partial_U U_i$ by the previous lemma. □

Now we distinguish those U_i "hanging over" to the right from those that "hang over" on the left: We set $\{R_j\}_{j \in J \subset I} := \{U_i \mid U_i \text{ is to the right of } S_i\}$ and $\{L_k\}_{k \in K \subset I} := \{U_i \mid U_i \text{ is to the left of } S_i\}$. Also $R := \bigcup_{j \in J} R_j$, $L := \bigcup_{k \in K} L_k$. Note that by Lemma 13.2.16 $V \cup (\bigcup_{j \in J} \mathrm{Cl}_U R_j)$ is open in U and hence its complement $\bigcup_{j \in J} \mathrm{Cl}_U R_j$ is closed, so

$$\mathrm{Cl}_U R = \bigcup_{j \in J} \mathrm{Cl}_U R_j \text{ and likewise } \mathrm{Cl}_U L = \bigcup_{k \in K} \mathrm{Cl}_U L_k \tag{13.2.1}$$

and in particular $U = V \cup \mathrm{Cl}_U R \cup \mathrm{Cl}_U L$.

Lemma 13.2.19.

(1) $f(V) \cap \mathrm{Cl}_U L_k = \varnothing$.
(2) $f^{-1}(V) \cap \mathrm{Cl}_U R_j = \varnothing$.
(3) $f^{-1}(\mathrm{Cl}_U L) \subset \mathrm{Cl}_U R \cup \mathrm{Cl}_U L$.
(4) $f^{-1}(\mathrm{Cl}_U L) \cap \mathrm{Cl}_U R = \varnothing$.
(5) $f(\mathrm{Cl}_U R) \cap \mathrm{Cl}_U L = \varnothing$.
(6) $f^{-1}(\mathrm{Cl}_U L) \subset L$.
(7) $f(\mathrm{Cl}_U R) \subset R$.

Proof. (1) If $(x, y) \in V$ the curve $\gamma(s) = f(x, s)$, $s \in [0, y]$, starts on $S^1 \times \{0\}$ and is contained in U. By the twist condition it traverses vertical lines from left to right. But to enter L_k it would have to traverse S_k from right to left. Since $f(V)$ is open in U the claim follows. Claim (2) follows in the same way and (3) is an evident consequence of (1).

(4) If $f^{-1}(\mathrm{Cl}_U L_k) \cap \mathrm{Cl}_U R_j \neq \varnothing$. Then $f^{-1}(\mathrm{Cl}_U L_k) \subset \mathrm{Cl}_U R_j$ since the $\mathrm{Cl}_U U_i$ are pairwise disjoint and connected and by (3). Since $f^{-1}(S_k)$ is transverse to vertical segments and S_j is vertical, this implies $f^{-1}(S_k) \cap R_j \neq \varnothing$ and hence $f^{-1}(V) \cap R_j \neq \varnothing$, contrary to (2). Claim (5) follows likewise.

(6) Claims (3) and (4) imply $f^{-1}(\mathrm{Cl}_U L) \subset \mathrm{Cl}_U L$. Again, since $f^{-1}(\partial_U L)$ is transverse to $\partial_U L$, we must have (6). Claim (7) follows likewise. □

Now we can show that $U = V$. If not then $R \neq \varnothing$ or $L \neq \varnothing$. Suppose the former and note that then $f(\mathrm{Cl}_U R) \subsetneq R$ by (7) and since f is a homeomorphism, whereas $\mathrm{Cl}_U R$ and R are not homeomorphic. But then $f^{-1}(R \smallsetminus f(\mathrm{Cl}_U R))$ is a nonempty subset of U consisting of wandering points, which we ruled out by assumption. By the same token $L = \varnothing$.

To prove Theorem 13.2.13 note that by assumption ∂U projects onto S^1, so it remains to show that the projection restricted to ∂U is injective, which, since $U = V$, amounts to showing that ∂U contains no vertical segments. This follows from the twist condition: If p is an interior point of a segment $S \subset \partial U$ then any q close to p and to the left of S is mapped above $f(S) \subset \partial U$ and hence is not contained in $U = V$. Similarly any point q near p to the right of S has preimage above $f^{-1}(S) \subset \partial U$ and is hence not contained in U. But then $p \notin \partial U$, a contradiction. □

There is an easy consequence of Theorem 13.2.13 that shows that between two invariant circles with irrational rotation number of an area-preserving twist map there is a fair amount of dynamical complexity.

Definition 13.2.20. Let f be a twist map such that there are invariant circles C_1 and C_2 with rotation numbers $\rho_1 < \rho_2$, respectively, and no invariant circle with rotation number in (ρ_1, ρ_2). Then the annulus bounded by C_1 and C_2 is called a *region of instability.*

Note that C_1 and C_2 are necessarily disjoint and that a region of instability cannot contain any invariant circle.

Proposition 13.2.21. *Suppose C_1, C_2 bound a region of instability of an area-preserving twist map f and let $\epsilon > 0$. Then there exists a point p in the ϵ-neighborhood $W_{1,\epsilon}$ of C_1 with an iterate in the ϵ-neighborhood $W_{2,\epsilon}$ of C_2.*

Proof. Otherwise there exists an $\epsilon > 0$ such that $\bigcup_{n\in\mathbb{N}} f^n(W_{1,\epsilon})$ is disjoint from $W_{2,\epsilon}$. Let W be the connected component containing C_2 of the region of instability minus $V_0 = \bigcup_{n\in\mathbb{N}} f^n(W_{1,\epsilon})$. To see that ∂W is connected identify the open cylinder bounded by C_1 and C_2 with the plane $\mathbb{R}^2$ by identifying C_2 with a point and apply Lemma A.7.8. Thus we can apply the Birkhoff Theorem 13.2.13 to the complement V of $\overline{W}$ to obtain an invariant circle strictly between C_1 and C_2, contrary to the hypotheses. □

This result shows that there are orbits that travel from arbitrarily near one boundary component of the region of instability to arbitrarily near the other.

Corollary 13.2.22. *Suppose C_1, C_2 bound a region of instability of an area-preserving twist map f and $\epsilon > 0$. If $f\restriction_{C_1}$ is topologically transitive then for every $q \in C_1$ there exists a point p in the ϵ-neighborhood $B(q, \epsilon)$ of q with an iterate in the ϵ-neighborhood $W_{2,\epsilon}$ of C_2.*

Proof. By assumption a finite union of iterates of $B(q, \epsilon)$ contains a neighborhood of C_1. □

Exercises

13.2.1. *Let f be a twist diffeomorphism of the closed annulus $A = S^1 \times [0,1]$. Let* $\mathrm{Or}(f)$ *be the set of all points whose orbits are ordered (order preserving). Prove that* $\mathrm{Or}(f)$ *is closed and f-invariant and that the topological entropy* $h_{\mathrm{top}}(f\restriction_{\mathrm{Or}(f)})$ *is zero.*

13.2.2. *Describe the set* $\mathrm{Or}(f)$ *for the elliptic billiard (Section 9.2).*

13.2.3. *Let A_1, A_2 be ordered sets for a twist map f with different rotation numbers. Prove that there exists $\epsilon > 0$ such that any ordered orbit is disjoint from an ϵ-neighborhood of A_1 or A_2.*

13.2.4. *Formulate the generalization of Corollary 13.2.12 without the assumption that A contains only finitely many periodic orbits.*

13.2.5. *Generalize Theorem 13.2.6 to twist homeomorphisms (see Definition 9.3.19) preserving a measure that is positive on open sets.*

13.2.6*. *Let f be an area-preserving twist diffeomorphism of the closed annulus, F a lift, and $\tau_0 < \tau_1$ the rotation numbers of the restriction of the lift to the boundary components. We call $[\tau_0, \tau_1]$ the* full twist interval. *Prove that the assertion of Theorem 13.2.6 is true for any $\alpha \in [\tau_0, \tau_1]$.*

13.2.7. *Suppose f is a twist diffeomorphism of $S^1 \times [0,1]$ for which* $\mathrm{Or}(f) = S^1 \times [0,1]$. *Show that f has invariant circles for all rotation numbers in the full twist interval.*

3. Action functionals, minimal and ordered orbits

a. Minimal action. We now return to the variational approach and in particular to the discussion of the action functional defined by (9.3.7) for periodic states and by (9.3.12) for nonperiodic ones. From now on it will be convenient to work on the universal cover most of the time. Since we are going to work with states that are not assumed to be ordered, we will label them using the dynamical ordering. Finally we shall assume that the twist map has been extended as described in Section 9.2c. Let us repeat and extend some of the definitions from Section 9.2e. We use the terminology of Definition 9.3.15.

Definition 13.3.1. An (a,b,n)-state $\{s_0, \dots, s_n\}$ is called *ordered* if $s_i \le s_l - k \le s_j$ implies $s_{i+1} \le s_{l+1} - k \le s_{j+1}$ whenever $0 \le i, j, l < n$ and $k \in \mathbb{Z}$. The minimum of the *action functional* L of (9.3.12) on the space $\Sigma(a,b,n)$ (which exists and is attained on some orbit segment) is denoted by $M(a,b,n)$. A sequence $\{s_i\}_{i\in\mathbb{Z}}$ is called a *globally minimizing state* if every finite segment $\{s_i\}_{i=n}^{m}$ is a minimal $(s_n, s_m, m-n)$-state for L.

We should note again that critical points for L, in particular minimal states, represent orbit segments of f. Since our extension of f ensures that for large

$s - s'$ the values of $H(s, s')$ are large, the action functional is a proper map which goes to (positive) infinity outside compact domains, and hence does have a minimum.

Remark. Note that the definition of globally minimizing states is parallel to that of minimal geodesics on a Riemannian manifold (Definition 9.6.1) and that this notion of an ordered state coincides with that of Definition 13.1.2 for states that represent orbit segments.

First we study the properties of the minima $M(x, y, n)$. By periodicity we evidently have $M(x, y, n) = M(x + k, y + k, n)$ for any $k \in \mathbb{Z}$. Consider now $s_1 \in \Sigma(x, y, n)$ and $s_2 \in \Sigma(y, z, m)$. Then one can naturally concatenate s_1 and s_2 to get a state $s \in \Sigma(x, z, n + m)$ and has $L(s) = L(s_1) + L(s_2)$, so

$$M(x, z, n + m) \leq M(x, y, n) + M(y, z, m) \tag{13.3.1}$$

for all $x, y, z \in \mathbb{R}$ and $n, m \in \mathbb{N}$.

Now we study how $M(x, y, n)$ changes with (x, y).

Lemma 13.3.2. *Suppose that $|y - x|/n < K$ and $|x' - x| \leq 1$, $|y' - y| \leq 1$. Then there exists a constant C depending on K such that*

$$|M(x, y, n) - M(x', y', n)| \leq C. \tag{13.3.2}$$

Proof. Let s be a minimal (x, y, n)-state and take s' to be the $(x', y', n + 2)$-state obtained from s by adding x' to the beginning of s and appending y' to the end. Then $L(s') = H(x', x) + M(x, y, n) + H(y, y')$ and by assumption $H(x', x) + H(y, y') \leq$ const. By compactness there is an a priori bound $|M| < C_1 n$ depending on K, namely, by taking as a test state the uniformly spaced one. Since s is minimal and, as is easy to see from the extension construction in Proposition 9.3.5, H grows quadratically with the size $s_{i+1} - s_i$ of a jump, there cannot be too many long jumps. In particular one can find a constant D depending on K such that $|s_{i+1} - s_i|, |s_{i+2} - s_{i+1}|, |s_{i+3} - s_{i+2}| < D$ for some i. Now define an (x', y', n)-state t by omitting s_{i+1} and s_{i+2} from s'. Then

$$L(t) = L(s') + H(s_i, s_{i+3}) - H(s_i, s_{i+1}) - H(s_{i+1}, s_{i+2}) - H(s_{i+2}, s_{i+3})$$

and

$$\begin{aligned} &|L(t) - M(x, y, n)| \\ &\qquad \leq |H(s_i, s_{i+3}) - H(s_i, s_{i+1}) - H(s_{i+1}, s_{i+2}) - H(s_{i+2}, s_{i+3})| + \text{const.} \leq C, \end{aligned}$$

with C depending on K. □

As a result we note

Corollary 13.3.3. *Whenever* $|x-y|/n < K$ *and* $|x'-y'|/n < K$ *then*

$$|M(x,y,n) - M(x',y',n)| \le C(|x'-x| + |y'-y|).$$

Proposition 13.3.4. *If* $\lim_{n\to\infty}(y_n - x_n)/n = \alpha$ *then*

$$A(\alpha) := \lim_{n\to\infty} \frac{1}{n} M(x_n, y_n, n)$$

exists and depends only on α. *We call* $A(\alpha)$ *the minimal action.*

Proof. To see that the limit exists consider the sequences $x_n = 0$ and $y_n = n\alpha$ and write $a_n := M(0, n\alpha, n)$. Using (13.3.1) and Lemma 13.3.2 we get

$$a_{n+m} = M(0,(n+m)\alpha, n+m) \le M(0,n\alpha,n) + M(n\alpha,(m+n)\alpha,m)$$
$$= a_n + M(n\alpha - [n\alpha], (m+n)\alpha - [n\alpha], m) \le a_n + M(0, m\alpha, m) + C = a_n + a_m + C,$$

where C is as in Lemma 13.3.2, and hence depends only on α. Thus $\{a_n/n\}$ converges by Proposition 9.6.4.

Now the proposition follows once we show that the limit is independent of the sequences $\{x_n\}$ and $\{y_n\}$. Suppose we have two pairs of sequences, $\{x_n\}$, $\{y_n\}$ and $\{x'_n\}$, $\{y'_n\}$. Since $\lim_{n\to\infty}(y_n - x_n)/n = \alpha$ and $\lim_{n\to\infty}(y'_n - x'_n)/n = \alpha$ we can translate the sequences $\{x'_n\}$, $\{y'_n\}$ by integers so as to have $|x'_n - x_n| < 1$ and thus $\lim_{n\to\infty}(y'_n - y_n)/n = \lim_{n\to\infty}((x'_n - x_n) - (y'_n - y_n))/n = 0$. But then Corollary 13.3.3 shows that $\lim_{n\to\infty} |M(x_n,y_n,n) - M(x'_n,y'_n,n)|/n \le C\lim_{n\to\infty}(|x'_n - x_n| + |y'_n - y_n|)/n = 0$ and the limit exists for all such sequences and is independent of the sequence. □

Proposition 13.3.5. *The function* $\alpha \mapsto A(\alpha)$ *is convex.*

Proof.

$$A(t\alpha + (1-t)\beta) = \lim_{n\to\infty} \frac{1}{n} M(0, (t\alpha + (1-t)\beta)n, n)$$
$$\le \lim_{n\to\infty} \frac{1}{n}(M(0, nt\alpha, n) + M(nt\alpha, n(1-t)\beta, n)) = tA(\alpha) + (1-t)A(\alpha). \ \square$$

b. Minimal orbits. We would like to understand the structure of the states on which the functional L attains its minimum. Using Lemma 9.3.11 and adapting the notation of the proof of Proposition 9.3.12 gives the following conclusion analogous to Corollary 9.3.13.

Proposition 13.3.6. *For an* (x,y,n)*-state* s *and an* (x',y',n)*-state* s' *with* $x < x'$ *and* $y < y'$ *let* $x_i := \min(s_i, s'_i)$ *and* $x'_i = \max(s_i, s'_i)$. *Then we have* $L(s) + L(s') \ge L(x) + L(x')$. *The inequality is strict if* $s'_i - s_i$ *attains negative values and is never zero away from the endpoints.*

Remark. Note that the sequence $\{x_i\}$ is an (x,y,n)-state and $\{x'_i\}$ is an (x',y',n)-state. Evidently the minimum of L on $(x+k, y+k, n)$-states coincides with the minimum of L on (x,y,n)-states.

Now consider minimal states of two different kinds. They cannot cross more than once:

Proposition 13.3.7. *Suppose* $s = \{s_i\}_{i=0}^{n}$ *is a minimal* (s_0, s_n, n)*-state and* $s' = \{s'_i\}_{i=0}^{n'}$ *is a minimal* $(s'_0, s'_{n'}, n')$*-state. Then for any* $k \in \mathbb{Z}$ *the sequence* $s'_i - s_i - k$ *changes sign at most once.*

Proof. Suppose that for some $k \in \mathbb{Z}$ the sequence $s'_i - s_i - k$ changes sign more than once. By truncating we may assume that s and s' have the same length m and $s'_i - s_i - k$ changes sign an even number of times. To be definite, assume $s'_0 - s_0 - k > 0$. Define t by $t_i := s'_i - k$ and let $x_i := \min(s_i, t_i)$ and $x'_i = \max(s_i, t_i)$. Then $x_0 = s_0$, $y_0 = t_0$, $x_m = s_m$, and $y_m = t_m$ and $\{x_i\}$ and $\{y_i\}$ are each distinct from s and t. Thus $L(s) + L(s') = L(t) + L(s) \geq L(x) + L(y)$ by Proposition 13.3.6. By minimality of $L(s)$ and $L(s')$ we must have equality. As we argued in the proof of Theorem 9.3.10 x and x' cannot both represent orbits and hence not both be minimal, a contradiction. □

Now we can generalize Proposition 9.3.17 to the case of nonperiodic recurrent (cf. Definition 3.3.2) orbits.

Corollary 13.3.8. *A globally minimal recurrent orbit is (totally) ordered in the sense that every finite segment is an ordered state (cf. Definition 13.3.1).*

Proof. Suppose that (x, y) is a point with globally minimal recurrent orbit and let $(s_n, y_n) := F^n(x, y)$. Suppose the sequence $s_i - s_{i+m} - k$ changes sign, without loss of generality between $i = 0$ and $i = 1$. By recurrence there exists $N \in \mathbb{N}$ such that $|s_N - s_0| + |s_m - s_{m+N}| < |s_0 - s_m - k|$, so $s_N - s_{N+m} - k$ has the same sign as $s_0 - s_m - k$, and the sequence $s_i - s_{i+m} - k$ changes sign at least twice, which is impossible by Proposition 13.3.7. □

In analogy with Section 9.6 we observe that the set of globally minimal orbits is closed. Theorem 9.3.7 shows that for every rational rotation number from the twist interval there exists a minimal Birkhoff periodic orbit of type (p, q), which is then globally minimal by Theorem 9.3.10. Hence orbits on Aubry–Mather sets obtained by the approximation construction of Theorem 13.2.6 from minimal Birkhoff periodic orbits are globally minimal. Thus we have shown:

Corollary 13.3.9. *For every rotation number in the twist interval there exists a globally minimal recurrent orbit with that rotation number.*

Now we describe the structure of the set of all recurrent minimal orbits with a given rotation number.

Theorem 13.3.10. *Let* $f: C \to C$ *be an area-preserving twist map and* α *a number in the twist interval. Then the union* M_α^r *of all recurrent globally minimal orbits with rotation number* α *is an ordered set.*

Proof. The closure of a globally minimal orbit is an ordered set, so we need only show that any two such orbits are properly intertwined, that is, their union is an ordered set, too. Let x and y be recurrent points with globally minimal orbits and rational rotation number α. Then they are both periodic with the same period, so if the orbits cross at all, they do so infinitely often. But by Proposition 13.3.7 they can cross at most once. Thus the orbits are properly intertwined.

Let x and y be recurrent points with globally minimal orbits and irrational rotation number α. Assume first that none of them is an endpoint of the corresponding Aubry–Mather set (if any of these points lies on a transitive invariant circle then this is no restriction). Then for $\epsilon > 0$ and $N \in \mathbb{N}$ one can find $n \in \mathbb{Z}$ such that the cyclic coordinates of $f^{-n}(y)$ and $f^n(y)$ are within ϵ of each other *and* the cyclic coordinates of $f^{-n}(x)$ and $f^n(x)$ are within ϵ of each other. This follows from the facts that both orbit closures are semiconjugate to a circle rotation with rotation number α and that, since neither x nor y is an endpoint of the Aubry–Mather set, the semiconjugacies are injective at x and y, respectively (cf. the Poincaré Classification Theorem 11.2.7). If ϵ is less than the distance between the cyclic coordinates of x and y we have shown that between $-n$ and n the orbits cross an even number of times, hence not at all by Proposition 13.3.7.

If one of the orbits is an endpoint, observe that if the orbit of x is properly intertwined with the orbit of y, then it is also properly intertwined with any orbit in the orbit closure of y. But endpoints are in the orbit closure of non-endpoints, so we are done if one of the orbits is an endpoint. If both x and y are endpoints then, since y is in the orbit closure of a nonendpoint, we get the same conclusion again. □

We immediately obtain uniqueness of minimal Aubry–Mather sets as well as a strong approximation result.

Corollary 13.3.11. *Let $f: C \to C$ be an area-preserving twist map and α an irrational number in the twist interval. Then there is either a transitive invariant circle consisting of globally minimal orbits with rotation number α or a unique globally minimal Aubry–Mather set $\mathcal{A}_\alpha$ with rotation number α. If $p_n/q_n \to \alpha$ and w_n is a sequence of Birkhoff periodic orbits of type (p_n, q_n) then the limit in the topology of the Hausdorff metric is an ordered set with rotation number α which is either a transitive invariant circle or contains the minimal Aubry–Mather set with rotation number α.*

Proof. Existence of a globally minimal circle or Aubry–Mather set follows from the closedness of the set of globally minimal orbits; hence any ordered set in the Hausdorff closure of the Birkhoff periodic orbits of type (p_n, q_n) consists of minimal orbits. Uniqueness follows from the fact that by Theorem 13.3.10 any recurrent minimal orbit with rotation number α is intertwined with the Aubry–Mather set and hence belongs to the Aubry–Mather set. □

For rational rotation numbers p/q Proposition 9.3.17 establishes a counterpart of this result in the following sense: It asserts uniqueness of the value of the action functional on recurrent (that is, periodic) globally minimal orbits. There may, however, be more than one such orbit, as is the case, for example, for an integrable twist map or for a carefully constructed perturbation that destroys the invariant circle with rotation number p/q but preserves more than one Birkhoff periodic orbit with rotation number p/q with the same minimal value of the action functional.

c. Average action and minimal measures. Now we can interpret the minimal action in terms of invariant measures.

Proposition 13.3.12. *Consider an ordered orbit given by* $(x_n, y_n) = f^n(x, y)$. *Then the limits*

$$\mathcal{L}^+(x) := \lim_{n\to\infty} \frac{1}{n}\sum_{i=0}^{n-1} H(x_i, x_{i+1}),$$

$$\mathcal{L}^-(x) := \lim_{n\to\infty} \frac{1}{n}\sum_{i=0}^{n-1} H(x_{i-n}, x_{i+1-n})$$

both exist.

Proof. Suppose the rotation number α of (x, y) is irrational. Then the orbit closure of (x, y) contains a unique minimal set E. By Theorem 11.2.9 this minimal set is uniquely ergodic, that is, carries a unique f-invariant measure μ. Consider the continuous function h defined by

$$h(x, y) = H(x, x') \text{ whenever } f(x, y) = (x', y'). \tag{13.3.3}$$

Notice that $1/n\sum_{i=0}^{n-1} H(x_i, x_{i+1})$ is the time average of this function at (x, y). By Proposition 4.1.13 the time averages of h for orbits in E converge uniformly to the integral of h over E with respect to the measure μ. This proves the proposition for recurrent orbits. For a nonrecurrent point (x, y) the cyclic coordinate of the orbit follows a sequence of gaps in the set E. The orbit is doubly asymptotic to the orbits of the endpoints of the gap containing x and hence by continuity of h the average is the same.

If the rotation number is rational and the orbit of (x, y) is periodic the claim is clear. Otherwise the orbit of (x, y) is positively asymptotic to one periodic orbit and negatively asymptotic to a periodic point. In either case both limits exist, and if the two periodic orbits coincide then so do $\mathcal{L}^+$ and $\mathcal{L}^-$. □

Definition 13.3.13. Consider an ordered orbit given by $(x_n, y_n) = f^n(x, y)$ that is not heteroclinic to two different periodic orbits. Then we define the *average action* $\mathcal{L}$ on the orbit by setting

$$\mathcal{L}(x) := \lim_{n\to\infty} \frac{1}{n} L(\{x_0, \dots, x_{n-1}\}) = \lim_{n\to\infty} \frac{1}{n}\sum_{i=0}^{n-1} H(x_i, x_{i+1}).$$

Proposition 13.3.14. *The minimal action* $A(\alpha)$ *is the infimum over the average actions of all ordered orbits with rotation number* α.

Proof. For such an orbit $L(\{x_0, \dots, x_{n-1}\}) \geq M(x_0, x_{n-1}, n)/n \to \mathcal{A}(\alpha)$ by Proposition 13.3.4. For a minimal orbit we have equality of the limits by Proposition 13.3.12. □

Thus for $\alpha = p/q$ where p and q are relatively prime, $A(p/q)$ is the minimum of the action functional L on the space of states of type (p, q) divided by q, or equivalently, $1/q$ times the value of L on any Birkhoff periodic orbit of type (p, q). This can be interpreted as an integral of the function h defined by (13.3.3) over the uniform δ-measure on any minimal Birkhoff periodic orbit. Similarly for irrational α let us denote by μ_α the unique f-invariant Borel probability measure on the minimal Aubry–Mather set $\mathcal{A}_\alpha$ (see Theorem 11.2.9) with rotation number α. Then $A(\alpha) = \int h \, d\mu_\alpha$.

d. Stable sets for Aubry–Mather sets. We now construct globally minimal semiorbits that follow the gaps of a minimal Aubry–Mather set. This construction is also a first step toward finding orbits homoclinic to a minimal Aubry–Mather set, that is, an orbit whose cyclic coordinates behave like type-III_α orbits for a circle homeomorphism.

Proposition 13.3.15. *Let G be a gap of the minimal Aubry–Mather set $\mathcal{A}_\alpha$, $x \in G$. Then there exist globally minimal positive and negative semiorbits with initial values (x, y^+) and (x, y^-), respectively, that are properly intertwined with every orbit of $\mathcal{A}_\alpha$.*

Proof. Let (p_n, q_n) be a sequence such that $p_n/q_n \to \alpha$ and consider minimal Birkhoff periodic orbits w_n of type (p_n, q_n). According to Corollary 13.3.11 the Hausdorff limit of w_n contains the Aubry–Mather set $\mathcal{A}_\alpha$. If the limit of the w_n contains a point (x, y) then the orbit of (x, y) is a globally minimal orbit that is properly intertwined with all orbits of $\mathcal{A}_\alpha$ and we have an even stronger conclusion than claimed.

Otherwise consider for each n a globally minimal $(x, x+p_n, q_n)$-state. It represents an orbit segment, which is, by Proposition 13.3.7, properly intertwined with w_n (the boundary condition necessitates an even number of crossings, while Proposition 13.3.7 allows at most one). Hence it represents an ordered orbit segment v_n connecting two points with cyclic coordinate x which we denote by (x, y_n^+) and (x, y_n^-). x is contained in a gap G_n of the orbit w_n. Without loss of generality assume that $G_n \to G$. By Corollary 13.1.3 the union W_n of w_n and the orbit segment v_n minus the endpoints (x, y_n^+) and (x, y_n^-) lies on the graph of a Lipschitz function with Lipschitz constant uniform in n. Thus the limit W in the Hausdorff metric of the W_n belongs to the graph of a Lipschitz function and contains $\mathcal{A}_\alpha$ as well as all images of $(x, \lim_n y_n^+) =: (x, y^+)$ and preimages of $(x, \lim_n y_n^-) =: (x, y^-)$. W is almost invariant and ordered in the following sense: $W \setminus f(W) = \{(x, y^-)\}$, $W \setminus f^{-1}(W) = \{(x, y^+)\}$, and f (correspondingly, f^{-1}) preserves order on $W \cap f^{-1}(W)$ (correspondingly, $W \cap f(W)$). □

Exercises

13.3.1. *Prove that for an integrable twist map (see Section 9.3a) all orbits are globally minimal.*

13.3.2. *Find globally minimal orbits for the elliptic billiard (Section 9.2).*

13.3.3*. *Suppose all orbits of an area-preserving twist diffeomorphism of the closed annulus $A = S^1 \times [0,1]$ are globally minimal. Prove that A splits into a disjoint union of f-invariant Lipschitz graphs G_α, where α runs through the full twist interval (see Exercise 13.2.6) and all orbits on G_α are ordered with rotation number α.*

4. Orbits homoclinic to Aubry–Mather sets

In this section we establish the existence for area-preserving twist maps of the last type of orbit for circle maps, by showing that there are nonrecurrent points asymptotic to a minimal Aubry–Mather set whenever there is no invariant circle with the rotation number in question. While our proof of existence of such orbits is totally based on rational approximation, those orbits could, in fact, be constructed as minimax solutions of an infinite-dimensional minimax variational problem which considers all states intertwined with a particular sequence of gaps in the Aubry–Mather set. This method is a direct generalization of our construction of the second (minimax) Birkhoff periodic orbit of type (p,q) in the proof of Theorem 9.3.7.

Theorem 13.4.1. *Suppose there is a minimal Aubry–Mather set $\mathcal{A}_\alpha$ for an area-preserving twist map f and G is any gap of $\mathcal{A}_\alpha$. Then there exists a point projecting into G whose iterates project into the corresponding images of G.*

Proof. Consider a sequence of Birkhoff periodic orbits of type (p_n, q_n) with $p_n/q_n \to \alpha \in \mathbb{R} \setminus \mathbb{Q}$. The Hausdorff limit contains the minimal Aubry–Mather set with rotation number α. If it contains a point in the gap G of the minimal Aubry–Mather set, then the orbit of this point is as desired. Otherwise we consider the minimax Birkhoff periodic orbit w_n' intertwined with the minimal Birkhoff periodic orbit w_n of type (p_n, q_n). We will study the limiting behavior of the quantity

$$\Delta L(p_n, q_n) := L(w_n') - L(w_n),$$

that is, the difference in action between minimax and minimal Birkhoff periodic orbits of type (p_n, q_n). We will prove that one of the following occurs:

(1) The Hausdorff limit of w_n' contains a point whose projection lies in G. The orbit of this point is properly intertwined with the minimal Aubry–Mather set and hence is as desired.

(2) $\Delta L(p_n, q_n) \xrightarrow[n\to\infty]{} 0$.

In the second case we consider the globally minimal positive and negative semiorbits with initial values (x, y^+) and (x, y^-), respectively, constructed in Proposition 13.3.15 and show that $y^+ = y^-$, so that the pair is part of a single orbit which is hence as desired. Thus, in this case, we get a much stronger conclusion, by showing that every point in a gap is the projection of an orbit of the desired type. We prove that $y^+ = y^-$ by showing that the differences $y_n^+ - y_n^-$, where $y_n^{\pm}$ are defined in the proof of Proposition 13.3.15, can be bounded in terms of $\Delta L(p,q)$.

So we need to show two things: First, if the Hausdorff limit of w'_n does not contain a point whose projection lies in G, then $\Delta L(p_n, q_n) \xrightarrow[n\to\infty]{} 0$. Second, we then want to give a bound for the differences y_n^+ and y_n^- in terms of $\Delta L(p,q)$.

As before in this chapter we will label states and orbits by their dynamical ordering rather than the geometric ordering we used earlier in Chapter 9.

We now prove that if the Hausdorff limit of w'_n does not contain a point whose projection lies in G then $\Delta L(p_n, q_n) \xrightarrow[n\to\infty]{} 0$. Consider periodic states $s = \{s_0, \dots, s_q\}$ such that $s_q = s_0 + p$. We call these states (p,q)-states. Recall the functions h_1 and h_2 defined in and after Definition 9.3.3. We say that a state s *matches* at k if $h_1(s_{k-1}, s_k) = h_2(s_k, s_{k+1})$ (cf. Proposition 9.3.4). In this case we write $y_k := h_1(s_{k-1}, s_k)$ and if a state s' matches at k we let $y'_k := h_1(s'_{k-1}, s'_k)$. If s and s' are two periodic states we denote by $O(s,s')$ the set of $k \in \mathbb{Z}$ such that s, s' match at $k-1$, k, and $k+1$ and $s'_{k-1} - s_{k-1}$, $s'_k - s_k$, and $s'_{k+1} - s_{k+1}$ have the same sign. The proof of the Regularity Lemma 13.1.1 shows that there exists a constant C_1 depending on f such that for $k \in O(s,s')$ we have

$$|h_1(s'_{k-1}, s'_k) - h_1(s_{k-1}, s_k)| = |y'_k - y_k| < C_1 |s'_k - s_k|. \tag{13.4.1}$$

Numbers in $O(s,s')$ are going to be "good" points for our estimates. Consider the states s^n and s'^n obtained from w_n and w'_n, respectively. Let ρ_n be such that $s^n_{\rho_n}$ projects to the right neighbor of s^n_0 on the circle and $\sigma_n \in \mathbb{Z}$ such that $s^n_{\rho_n} - \sigma_n - s^n_0$ is the fractional part of $s^n_{i+\rho} - s^n_i$. We now consider the case that the Hausdorff limit of w'_n does not contain a point whose projection lies in a gap G of the minimal Aubry–Mather set. We need to show that then $\Delta L(p_n, q_n) \xrightarrow[n\to\infty]{} 0$.

By our hypotheses there is a gap $G = (l_0, r_0)$ containing no point of the Hausdorff limit of the w'_n. Thus without loss of generality

$$s^n_0 \to l_0$$

and

$$s'^n_0 \to l_0 \text{ or } s'^n_0 \to r_0.$$

For any $\epsilon > 0$ there exists $m(\epsilon) \in \mathbb{N}$ such that the minimal Aubry–Mather set has at most $m(\epsilon)$ gaps of length $\epsilon/2$ or more. Setting

$$A_{n,\epsilon} := \{k \mid 0 \le k < q_n,\ s^n_{k+\rho_n} - \sigma_n - s^n_k \ge \epsilon\}$$

we conclude that for all $\epsilon > 0$ and sufficiently large n we have

$$\operatorname{card} A_{n,\epsilon} \le m(\epsilon).$$

Note that if $k \notin A_{n,\epsilon}$ then

$$s'^n_k - s^n_k < \epsilon \text{ and } s^n_{k+\rho_n} - \sigma_n - s'^n_k < \epsilon. \qquad (13.4.2)$$

By assumption the Hausdorff limits of the orbits w'_n and w_n coincide, so that

$$\lim_{n\to\infty} \sup_{k\in\mathbb{Z}} \max(s'^n_k - s^n_k, s^n_{k+\rho_n} - \sigma_n - s'^n_k) = 0,$$

so for any $\delta \in (0, \frac{\epsilon}{2})$ and sufficiently large n we have

$$s'^n_k - s^n_k < \delta \text{ or } s^n_{k+\rho_n} - \sigma_n - s'^n_k < \delta$$

for all k. Let

$$\begin{aligned} L_{n,\epsilon} &:= \{k \in A_{n,\epsilon} \mid s'^n_k - s^n_k < \delta\}, \\ R_{n,\epsilon} &:= \{k \in A_{n,\epsilon} \mid s^n_{k+\rho_n} - \sigma_n - s'^n_k < \delta\}. \end{aligned} \qquad (13.4.3)$$

Lemma 13.4.2. $\min\{|k_1 - k_2| \mid k_1 - \rho_n \in A_{n,\epsilon}, k_2 \in A_{n,\epsilon}\} \xrightarrow[n\to\infty]{} \infty.$

Proof. If not then there exist gaps $G = (l_0, r_0)$, $G' = (l'_0, r'_0)$ of $\mathcal{A}_\alpha$, $i_n, k_n \in \mathbb{N}$, such that

$$x_n := s^n_{k_n} \xrightarrow[n\to\infty]{} l_0, \quad s^n_{k_n+i_n} \xrightarrow[n\to\infty]{} r'_0 \bmod 1,$$

and $\{i_n\}_{n\in\mathbb{N}}$ is bounded. But if $(s^n_{k_n+i_n}, \tilde{y}_n) = F^{i_n}(x_n, y_n)$ converges and $\{i_n\}_{n\in\mathbb{N}}$ is bounded then $i_n = i \in \mathbb{N}$ is constant. Since A is a perfect set by minimality we conclude that the image of G under f^i (defined in the natural way) is G'. Thus $F^i(x_n, y_n) \to r'_0$, hence $s^n_{k_n+i} \to r'_0 \bmod 1$ but by monotonicity $F^i(x_n, y_n) \to l'_0$, that is, $s^n_{k_n+i} \to l'_0 \bmod 1$, a contradiction. □

The same argument also yields

Lemma 13.4.3. $\min\{|k_1 - k_2| \mid k_1 \in L_{n,\epsilon}, k_2 \in R_{n,\epsilon}\} \xrightarrow[n\to\infty]{} \infty.$

Let us fix an interval $\Delta \subset [0,1)$ such that the projection of Δ to the circle intersects no gap of length $l(\Delta)/2$ or greater and write

$$C_{n,\Delta} := \{k \in [0, q_n - 1] \mid s^n_k - \sigma_n \in \Delta\}.$$

An interval $S = [k, k+s-1]$ is called an A_n-segment if

$$S \cap C_{n,\Delta} = \{k\}, \; S \cap A_{n,\epsilon} \neq \varnothing, \text{ and } k+s \in C_{n,\Delta}.$$

The collection of A_n-segments $S = [k, k+s-1]$ with $k \in [1, q_n - 1]$ is denoted by $\mathfrak{A}_n$. Using unique ergodicity of f on $\mathcal{A}_\alpha$ and the fact that $w_n \to \mathcal{A}_\alpha$ we note that there is an $N(\Delta) \in \mathbb{N}$ such that for any sufficiently large n any interval of length $N(\Delta)$ intersects $C_{n,\Delta}$. In particular for any A_n-segment $S = [k, k+s-1]$ we have $s - 1 \leq N(\Delta)$. If S and S' are A_n-segments, then $S \cap (S' + \rho_n) = \varnothing$ by Lemma 13.4.2. For $S = [k, k+s-1] \in \mathfrak{A}_n$ let $\mathcal{D}_S := \sum_{i=k}^{k+s-1} \left(H(s^n_i, s^n_{i+1}) - H(s^n_{i+\rho_n}, s^n_{i+\rho_n+1})\right)$. We call $S \in \mathfrak{A}_n$ a left A_n-segment if $\mathcal{D}_S < 0$ and denote the collection of left A_n-segments by $\mathfrak{A}_{n,\mathrm{l}}$. $S \in \mathfrak{A}_n$ is a right A_n-segment if $\mathcal{D}_S > 0$ and the collection of right A_n-segments is denoted by $\mathfrak{A}_{n,\mathrm{r}}$.

We now show that one can change s^n along A_n-segments from left to right endpoints without changing the action too much.

Lemma 13.4.4. *There exists $c \in \mathbb{R}$ such that $\overline{\lim}_{n\to\infty} \sum_{S\in\mathfrak{A}} |\mathcal{D}_S| < c \cdot l(\Delta)$.*

Proof. Define

$$t_k^n := \begin{cases} s_{k-\rho_n}^n + \sigma_n & \text{if } k-\rho_n \text{ is in a left } A_n\text{-segment},\\ s_{k+\rho_n}^n - \sigma_n & \text{if } k \text{ is in a right } A_n\text{-segment},\\ s_k^n & \text{otherwise.}\end{cases}$$

This is periodic and $t_{k+\rho_n}^n \geq t_k^n$. t^n is also properly intertwined with s^n and, since s^n is a minimal state, $L(t^n) \geq L(s^n)$. If B_{l} (B_{r}) and E_{l} (E_{r}) denote the sets of first elements and last elements, respectively, of intervals in $\mathfrak{A}_{n,\mathrm{l}}$ ($\mathfrak{A}_{n,\mathrm{r}}$) then we obtain

$$\begin{aligned} 0 \leq L(t^n) - L(s^n) &= \sum_{S\in\mathfrak{A}_{n,\mathrm{l}}} \mathcal{D}_S - \sum_{S\in\mathfrak{A}_{n,\mathrm{r}}} \mathcal{D}_S \\ &+ \sum_{k-\rho_n\in B_{\mathrm{l}}} \left[H(s_{k-1}^n, s_{k-\rho_n}^n) - H(s_{k-1}^n, s_k^n)\right] + \sum_{k-\rho_n\in E_{\mathrm{l}}} \left[H(s_{k-\rho_n}^n, s_{k+1}^n) - H(s_k^n, s_{k+1}^n)\right] \\ &+ \sum_{k\in B_{\mathrm{r}}} \left[H(s_{k-1}^n, s_{k+\rho_n}^n) - H(s_{k-1}^n, s_k^n)\right] + \sum_{k\in E_{\mathrm{r}}} \left[H(s_{k+\rho_n}^n, s_{k+1}^n) - H(s_k^n, s_{k+1}^n)\right] \\ &\leq - \sum_{S\in\mathfrak{A}_{n,\mathrm{l}}} |\mathcal{D}_S| + \text{const.}\Big(\sum_{k-\rho_n\in B_{\mathrm{l}}\cup E_{\mathrm{l}}} (s_k^n - s_{k-\rho_n}^n) + \sum_{k-\rho_n\in B_{\mathrm{r}}\cup E_{\mathrm{r}}} (s_{k+\rho_n}^n - s_k^n)\Big) \end{aligned}$$

by differentiability of H. Since A_n-segments are pairwise disjoint the last two sums are bounded by $c \cdot l(\Delta)$ for sufficiently large n, as required. □

Now we can show that if the Hausdorff limit of w_n' does not contain a point whose projection lies in G then $\Delta L(p_n, q_n) \xrightarrow[n\to\infty]{} 0$.

Lemma 13.4.2 implies that if the interval Δ is sufficiently small and n is sufficiently large then every A_n-segment S intersects only one of the sets $L_{n,\epsilon}$ and $R_{n,\epsilon}$ defined by (13.4.3). S is then called an L-segment or R-segment accordingly. Suppose S is an R-segment. Since its length is at most $N(\Delta)$ we have by definition

$$\sum_{k\in S} (s_{k+\rho_n}^n - \sigma_n - s'^n_k) < N(\Delta)\delta. \tag{13.4.4}$$

Let us write

$$\mu_k := H(s'^n_k, s'^n_{k+1}) - H(s_k^n, s_{k+1}^n).$$

Then

$$\Delta L(p_n, q_n) = L(s'^n) - L(s^n) = \sum_{k=0}^{q_n-1} \mu_k = \sum_{\substack{S \text{ is an}\\ R\text{-segment}}} \sum_{k\in S} \mu_k + \sum_{\substack{k \text{ not in any}\\ R\text{-segment}}} \mu_k.$$

To estimate the first sum note that if S is an R-segment then

$$\begin{aligned} \sum_{n\in S} \mu_k &= -\mathcal{D}_S + \sum_{n\in S} [H(s'^n_k, s'^n_{k+1}) - H(s_{k+\rho_n}^n, s_{k+\rho_n+1}^n)] \\ &< |\mathcal{D}_S| + c \sum_{n\in S} (s_{k+\rho_n}^n - \sigma_n - s'^n_k) < |\mathcal{D}_S| + cN(\Delta)\delta, \end{aligned}$$

by differentiability of H and (13.4.4), so, since there are at most $m(\epsilon)$ R-segments,

$$\sum_{\substack{S \text{ is an} \\ R\text{-segment}}} \sum_{k\in S} \mu_k < \sum_{S\in\mathfrak{A}} |\mathcal{D}_S| + c\,m(\epsilon)N(\Delta)\delta \le c(l(\Delta) + m(\epsilon)N(\Delta)\delta),$$

which is arbitrarily small if one fixes ϵ first, then takes Δ sufficiently small, and finally chooses δ small enough.

To estimate the second sum we need a lemma.

Lemma 13.4.5. *There exists $C \in \mathbb{R}$ such that for any (p,q) and any (p,q)-states s and s' and every interval $I = [l,m] \subset O(s,s')$ we have*

$$\Big|\sum_{k=l}^{m}[H(s'_k, s'_{k+1}) - H(s_k, s_{k+1})]\Big| \le C\sum_{k=l}^{m}(s'_k - s_k)^2 + |s_{m+1} - s'_{m+1}| + |s_{l-1} - s'_{l-1}|.$$

Proof. Let $\mu_k := H(s'{}_k, s'{}_{k+1}) - H(s_k, s_{k+1})$. Then $L(s') - L(s) = \sum_{k=0}^{q-1}\mu_k$. But

$$\begin{aligned}
&\Big|\sum_{k=l}^{m}[H(s'_k, s'_{k+1}) - H(s_k, s_{k+1})]\Big| \\
&\quad= \Big|\sum_{k=l}^{m}([H(s'_k, s'_{k+1}) - H(s_k, s'_{k+1})] - [H(s_k, s_{k+1}) - H(s_k, s'_{k+1})])\Big| \\
&\quad= \Big|\sum_{k=l}^{m}\Big([H(s'_k, s'_{k+1}) - H(s_k, s'_{k+1})] - [H(s_{k-1}, s_k) - H(s_{k-1}, s'_k)] \\
&\qquad\qquad - [H(s_{k-1}, s_k) - H(s_{k-1}, s'_k) - H(s_k, s_{k+1}) - H(s_k, s'_{k+1})]\Big)\Big| \\
&\quad\le \Big|\sum_{k=l}^{m}[H(s_{k-1}, s_k) - H(s_{k-1}, s'_k) - H(s_k, s_{k+1}) - H(s_k, s'_{k+1})]\Big| \\
&\qquad + \sum_{k=l}^{m}|[H(s'_k, s'_{k+1}) - H(s_k, s'_{k+1})] - [H(s_{k-1}, s_k) - H(s_{k-1}, s'_k)]| \\
&\quad= |H(s_{l-1}, s_l) - H(s_{l-1}, s'_l) - H(s_m, s_{m+1}) - H(s_m, s'_{m+1})| \\
&\qquad + \sum_{k=l}^{m}|[H(s'_k, s'_{k+1}) - H(s_k, s'_{k+1})] - [H(s_{k-1}, s_k) - H(s_{k-1}, s'_k)]|.
\end{aligned}$$

The first term is bounded by $C_4(|s_{m+1} - s'_{m+1}| + |s_l - s'_l|) < C_5(|s_{m+1} - s'_{m+1}| + |s_{l-1} - s'_{l-1}|)$ using differentiability of f. Next

$$\begin{aligned}
&[H(s'_k, s'_{k+1}) - H(s_k, s'_{k+1})] - [H(s_{k-1}, s_k) - H(s_{k-1}, s'_k)] \\
&\qquad\qquad = (s'_k - s_k)\frac{\partial H}{\partial x}(\xi, s'_{k+1}) - (s_k - s'_k)\frac{\partial H}{\partial x'}(s_{k-1}, \eta) \\
&\qquad\qquad = (s'_k - s_k)h_2(\xi, s'_{k+1}) - (s'_k - s_k)h_1(s_{k-1}, \eta)
\end{aligned}$$

with both ξ and η between s'_k and s_k by the Mean Value Theorem. But if $k \in I \subset O(s, s')$ then $h_1(s'_{k-1}, s'_k) = h_2(s'_k, s'_{k+1})$ and hence $h_2(\xi, s'_{k+1}) - h_1(s_{k-1}, \eta)$ is at most linear in $(s'_k - s_k)$. Therefore

$$[H(s'_k, s'_{k+1}) - H(s_k, s'_{k+1})] - [H(s_{k-1}, s_k) - H(s_{k-1}, s'_k)] \le C_3(s'_k - s_k)^2. \square$$

To estimate $\sum_{\substack{k \text{ not in any} \\ R\text{-segment}}} \mu_k$, note that every k that is not in any R-segment is contained in an interval $[i, j]$ such that

$$s_i^n - \sigma_n \in \Delta \text{ and } s_{j+1}^n - \sigma_n \in \Delta. \tag{13.4.5}$$

Since $O(s, s') = [1, q_n - 1]$, Lemma 13.4.5 yields

$$\mu_i + \cdots + \mu_j \le c\Big(s'^n_i - s^n_i + s'^n_{j+1} - s^n_{j+1} + \sum_{k=i+1}^{j} (s'^n_k - s^n_k)^2\Big).$$

Adding all these expressions yields a sum of linear terms which is bounded by $c \cdot l(\Delta)$ by (13.4.5) as well as a sum of quadratic terms. The terms being squared are each less than ϵ by (13.4.2) and (13.4.3) and their sum is at most 1. Thus the sum of the quadratic terms is less than ϵ and hence $\Delta L(p_n, q_n)$ is arbitrarily small as $n \to \infty$.

We have now shown that if the Hausdorff limit of w'_n does not contain a point whose projection lies in G then $\Delta L(p_n, q_n) \xrightarrow[n\to\infty]{} 0$. The following lemma finishes the proof of Theorem 13.4.1:

Lemma 13.4.6. *Using the notation from the statement and proof of Proposition 13.3.15 let s be the state represented by the Birkhoff periodic orbit w_n and s^x the globally minimal $(x, x + p_n, q_n)$-state. Then $L(s^x) - L(s) \ge c|y_n^+ - y_n^-|$.*

Proof. If $s^x = \{s'_0, \dots, s'_{q_n}\}$ then let $s(t) = \{s'_0 + t, s'_1, \dots, s'_{q_n-1}, s'_{q_n} + t\}$. Then

$$\frac{dL(s(t))}{dt}\Big|_{t=0} = \frac{dH(x, s'_1)}{dx} + \frac{dH(s'_{q_n-1}, x)}{dx} = -h_2(x, s'_1) + h_2(s'_{q_n-1}, x) = y_n^+ - y_n^-.$$

For $\epsilon > 0$ let $t = \epsilon \operatorname{sgn}(y_n^+ - y_n^-)$. If ϵ is sufficiently small and such that $s_0 \le x + t \le s_{p_n} - \sigma_n$ then

$$L(s) \le L(s(t)) \le L(s^x) + \frac{\epsilon}{2}|y_n^+ - y_n^-|.$$

This implies the claim. $\square$

Notice that the lemma implies that the semiorbits obtained in Proposition 13.3.15 are part of a single orbit with initial value $(x, y^+) = (x, y^-)$. Thus for every x in a gap G of $\mathcal{A}_\alpha$ we have an orbit of the desired kind whose projection contains x. $\square$

It is interesting to note that if the Hausdorff limit of w_n contains an entire circle then $\Delta L(p_n, q_n) \xrightarrow[n\to\infty]{} 0$. This involves a relatively easy application of Lemma 13.4.5.

$$\max_i(s^n_{i+p_n} - \sigma_n - s^n_i) \xrightarrow[n\to\infty]{} 0$$

by assumption, so

$$\max_i(s'^n_i - s^n_i) \xrightarrow[n\to\infty]{} 0 \tag{13.4.6}$$

and hence, since $O(s^n, s'^n) = \{1, \dots, q_n - 1\}$, Lemma 13.4.5 yields

$$\Delta L(p_n, q_n) = L(s'^n) - L(s^n) \le \text{const.} \sum_{i=0}^{q-1} (s'^n_i - s^n_i)^2.$$

Since $\sum_{i=0}^{q-1}(s'^n_i - s^n_i) < 1$, this and (13.4.6) imply that $\Delta L(p_n, q_n) \xrightarrow[n\to\infty]{} 0$.

The converse is given in Exercise 13.4.1.

Exercise

13.4.1*. *Prove that if under the assumptions of Theorem 13.4.1 $\Delta L(p_n, q_n) \to 0$ then f has an invariant circle.*

5. Nonexistence of invariant circles and localization of Aubry–Mather sets

In this section we shall use an idea related to the Regularity Lemma 13.1.1 to give a criterion which can be used to show that a point is not an accumulation point of any ordered set. If this criterion is satisfied along a curve connecting the two boundary components of the annulus then we can conclude that the map has no invariant circles.

Proposition 13.5.1. *Let $f: C \to C$ be a twist map, $F = (F_1, F_2)$ a lift to the strip S, $\hat{F} = (\hat{F}_1, \hat{F}_2)$ its inverse, and $(x, y) \in S$ a point that projects to an accumulation point of an ordered set in C. Then*

$$\frac{\partial F_2/\partial y}{\partial F_1/\partial y}(F^{-1}(x,y)) \ge -\frac{\partial \hat{F}_2/\partial y}{\partial \hat{F}_1/\partial y}(F(x,y)). \tag{13.5.1}$$

Proof. Write $(x_{-1}, y_{-1}) := F^{-1}(x_0, y_0)$, $(x_1, y_1) = F(x_0, y_0)$. If (x'_0, y'_0) is a nearby point from an ordered set then the preimage (x'_{-1}, y'_{-1}) and image (x'_1, y'_1) are intertwined with those of (x_0, y_0), that is, the differences $x'_i - x_i$ all have the same sign, no matter how close (x', y') is. This implies that the slope of $F(x_{-1} \times (0,1))$ at (x_0, y_0) exceeds that of $F^{-1}(x_1 \times (0,1))$ at (x_0, y_0), that is,

$$\frac{\partial F_2/\partial y}{\partial F_1/\partial y}(F^{-1}(x,y)) \ge -\frac{\partial \hat{F}_2/\partial y}{\partial \hat{F}_1/\partial y}(F(x,y)).$$

□

This condition expresses the fact that it is necessary to have a vector pointing "right" at x whose image and preimage still point rightward. It can alternatively be expressed via the generating function by using the representation of $F(x \times (0,1))$ as the graph of the function $h(x,\cdot)$.

In the case of an area-preserving twist map (13.5.1) has a particularly convenient interpretation via the generating function. Namely, by (9.3.2) the slope of the curve $F(x_{-1} \times (0,1))$ at (x_0, y_0) equals $\frac{\partial^2 H(x,x')}{\partial x'^2}(x_{-1}, x_0)$ and the slope of $F^{-1}(x_1 \times (0,1))$ at (x_0, y_0) is $\frac{\partial^2 H(x,x')}{\partial x^2}(x_0, x_1)$.

This immediately gives the following corollary.

Corollary 13.5.2. *Suppose $y \in \mathbb{R}$ is such that*

$$\frac{\partial^2 H(x,x')}{\partial x'^2}(x,y) < -\frac{\partial^2 H(x,x')}{\partial x^2}(y,z)$$

for all y in the domain of $h_2(\cdot, x)$ and z in the domain of $h_1(x,\cdot)$ (cf. Definition 9.3.3 and the paragraph following it). Then there is no invariant circle in C and, moreover, there is a neighborhood of $\{x\} \times (0,1)$ that is disjoint from all Aubry–Mather sets of f.

Let us give two simple applications. First consider the standard map of Section 9.2a:

$$F: \mathbb{R} \times \mathbb{R} \to \mathbb{R} \times \mathbb{R},\ (x,y) \mapsto (x+y, y+V(x+y))$$

with V periodic. Let U be an antiderivative of V. Then a direct calculation shows that this map has generating function $H(x,x') = \frac{1}{2}(x-x')^2 + U(x')$. Then we get

$$\frac{\partial^2 H}{\partial x^2} = 1 \text{ and } \frac{\partial^2 H}{\partial x'^2} = 1 + V'(x').$$

Suppose $V'(x_0) < -2$ for some x_0. Then the condition of the corollary is satisfied and hence there are no invariant circles and, moreover, there is a neighborhood of $\{x_0\} \times (0,1)$ that is disjoint from all Aubry–Mather sets of f. In particular for the family f_λ, where $V_\lambda = (\lambda/2\pi)\sin 2\pi x$, we conclude that there are no invariant circles when $|\lambda| > 2$.[1]

As a second application let us consider the billiard-ball problem discussed in Section 9.2. In this case invariant circles correspond to *caustics* or singularities of the wave fronts. Let R be an invariant circle in the phase space. Consider all the oriented segments l arising from orbits in R. Their *evolute* is a curve which is sometimes convex and sometimes contains concave arcs and singular points. In the convex case it can be described as follows. The orientation of the segments l allows us to distinguish a "left" and a "right" region of $D \setminus l$. The boundary c of the intersection of the left regions thus obtained is referred to as a convex caustic. In fact, every segment arising from an orbit in R is tangent to c. If one thinks of the billiard as a convex mirrored room illuminated with light from a tangent of the caustic, then one will see the caustic outlined as a particularly well lit curve surrounding a dark region. As we showed in (9.2.2) and (9.2.3), the generating function with respect to the coordinates (s,r), where s is the length parameter along the boundary curve, is simply the distance between the boundary points corresponding to the values s and s'. An example of a nonconvex caustic in the shape of an astroid is shown in Figure 13.5.1.[2]

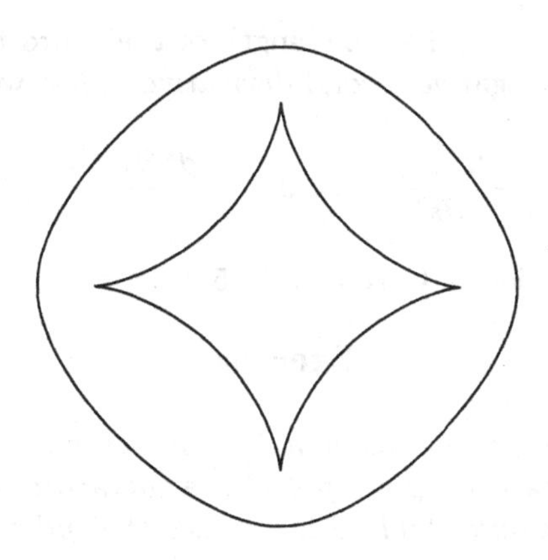

FIGURE 13.5.1. A nonconvex caustic

Proposition 13.5.3. (Mather)[3] *Let D be a convex bounded region in the plane with C^2 boundary B that has a point where the curvature vanishes. Then there are no caustics for the billiard map.*

Proof. Suppose the boundary is flat at the point with parameter s_0.

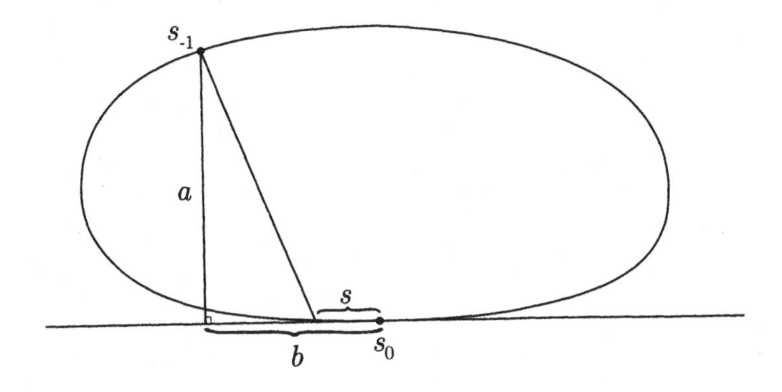

FIGURE 13.5.2. Billiard with a flat boundary point

Given a point s_{-1} consider the length $f(s) := -H(s_{-1}, s_0 - s)$ of the chord from s_{-1} to $s_0 - s$. Up to second order it is given by $f(s) = \sqrt{a^2 + (b-s)^2}$, where a and b are indicated in Figure 13.5.2, namely, the distance from the corresponding point on the tangent line at s_0. Then

$$f''(0) = (a^2 + b^2)^{-\frac{1}{2}} - b^2(a^2 + b^2)^{-\frac{3}{2}} > 0.$$

A similar calculation shows that the length of the chord from s_1 to $s_0 + s$ as a function of s also has negative second derivative. Thus we have

$$\frac{\partial^2 H(s_{-1}, s)}{\partial s^2} < 0 < -\frac{\partial^2 H(s, s_1)}{\partial s^2},$$

which yields the condition of Corollary 13.5.2. □

Exercise

13.5.1. (Gutkin, Katok) *Consider a C^1 piecewise-C^2 convex curve B and assume that near at least one point $p \in B$ the curvature at x goes to infinity as $x \to p$. Prove that the outer billiard for B (see the end of Section 9.3a) has no caustics.*

14

Flows on surfaces and related dynamical systems

In this chapter we study a class of continuous-time dynamical systems with very low-dimensional behavior according to the description given in Chapter 10, namely, smooth flows on closed compact surfaces. We will also pay attention to flows on surfaces with boundary such as the closed disc or the cylinder and on open surfaces such as the plane. This, in particular, will allow us to treat semilocal problems. Another natural object associated with such flows are Poincaré maps induced on transversals to the flow. If the flow preserves a nonatomic measure positive on open sets (for example, an area) then such Poincaré return maps are topologically conjugate to a locally isometric map with finitely many discontinuities. The term "interval exchange transformation" gives a visual description of such a map.

In general, the asymptotic behavior of flows on surfaces is characterized by slow orbit growth but they have less uniform types of recurrence and of statistical behavior than invertible one-dimensional maps studied in Chapters 11 and 12. The former aspect is closely related to the fact that both orbits and one-dimensional transversals to a flow locally separate the surface; the latter is due primarily to the more complicated topology of surfaces of genus greater than one compared to the circle (and the torus) and to a lesser extent to the effects of time change. Typical manifestations of this type of complexity, intermediate between the simple behavior of our first group of examples (Sections 1.3–1.6) and the circle diffeomorphisms on the one hand, and that of the examples with positive topological entropy (Sections 1.7–1.9, 5.4, 9.6) on the other, are finiteness results for the number of nontrivial orbit closures (Theorem 14.6.3) and nonatomic ergodic invariant measures (Theorem 14.7.6) for flows on surfaces of genus greater than one, which replace uniqueness of a minimal set (Proposition 11.2.5) and unique ergodicity (Theorem 11.2.9) for circle homeomorphisms.

Even though most of these results hold in full generality we prove them under simplifying assumptions, such as finiteness and not too degenerate structure of the fixed points and preservation of area or a measure positive on open sets. This is justified by the facts that first unlike in the case of circle maps these assumptions do not restrict us to essentially trivial situations and indeed convey the entire possible complexity of flows on surfaces, and second that this case appears in several interesting situations such as billiards in polygons with rational angles.

1. Poincaré–Bendixson theory

a. The Poincaré–Bendixson Theorem. We begin with the study of flows on those surfaces whose topology allows only simple types of recurrent orbits, that is, fixed points and periodic orbits. This does not mean that the global orbit structure of any flow on such a surface is trivial, but only that the complexity of that structure would have be due to the combinatorial picture of fixed points, periodic orbits, and saddle connections rather than to recurrent behavior. The sphere, the plane, and the disc are prime examples of surfaces with that property, but this class also includes the cylinder, the Möbius strip, and the projective plane. The arguments are based on the Jordan Curve Theorem A.5.2. We obtain the following:

Theorem 14.1.1. (Poincaré–Bendixson Theorem) *Let M be a surface that is an open subset of the sphere S^2 or the projective plane. Let X be a C^1 vector field on M. Then all positively or negatively recurrent orbits are periodic. Furthermore, if the ω-limit set of a point contains no fixed points, then it consists of a single periodic orbit.*

Proof. Suppose first that M is a subset of the sphere. Denote by φ^t the flow generated by X and suppose p is positively recurrent and nonperiodic. Take a short transversal γ at p and let t be the smallest positive number for which $\varphi^t(p) \in \gamma$. Then the union of the orbit segment $\{\varphi^s(p)\}_{0 \le s \le t}$ and the piece of γ between p and $\varphi^t(p)$ is a simple closed curve $\mathcal{C}$ called a *pretransversal* (because we shall later use such curves to construct transversals). By the Jordan Curve Theorem A.5.2 the complement of $\mathcal{C}$ consists of two disjoint open sets A and B. We may label them such that near γ the flow goes from A to B. This implies that the positive semiorbit of $\varphi^t(p)$, hence the ω-limit set $\omega(p)$ of p, is in B. Since p is recurrent we have $A \ni \varphi^{-\epsilon}(p) \in \mathcal{O}(p) \subset \omega(p) \subset B$, a contradiction.

If M is a subset of the projective plane then there is an orientable double cover of M. If $p \in M$ is a positively recurrent nonperiodic point, then consider the two points p_1 and p_2 that cover it. The orbit of p_1 under the flow generated by the lift of the vector field X accumulates on $\{p_1, p_2\}$. If it accumulates on p_1 then we are done by the previous argument. Otherwise it accumulates on p_2 and we can construct a pretransversal near p_2, which again leads to a contradiction.

Now consider the ω-limit set W of a point p and assume that it contains no fixed points. By Corollary 3.3.7 there are recurrent points in W. By the above these are periodic. Thus let $q \in W$ be a periodic point. Note that in the case of the projective plane the lift of q to the orientable double cover is still periodic, so we may assume that M is orientable. Consider a small transverse segment

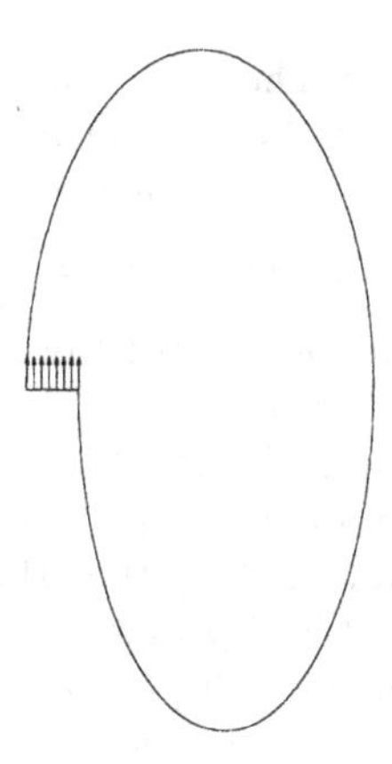

FIGURE 14.1.1. A pretransversal

γ containing q. By continuity the return map to this segment is defined on a neighborhood of q in γ. Take a one-sided neighborhood I of q small enough so that the first point $\varphi^t(p)$ in γ is not in I, but infinitely many of these returns are. Parameterizing this neighborhood by $[0,\delta)$ gives a continuous map f from an interval $[0,\delta)$ to an interval $[0,\delta')$ that fixes 0. The orbit of p provides infinitely many $x \in (0,\delta)$ for which $f(x) < x$, so either $f(x) < x$ for all $x \in [0,\delta)$ or $[0,\delta)$ contains a fixed point y. The latter case is impossible, since the interval $[0,y]$ would be invariant under f and hence there would be an invariant annulus for the flow that separates the orbit of q from that of p, so $q \notin \omega(p)$. But if $f(x) < x$ then all $x \in (0,\delta)$ are positively and monotonically asymptotic to 0. Since the return times to I are bounded this means that the orbit segments of p between successive returns converge to the orbit of q, so $\omega(p)$ coincides with the orbit of q. □

b. Existence of transversals.

Definition 14.1.2. A *transversal* to a vector field on a surface is a simple closed curve such that the vector field is nowhere tangent to the curve.

Let τ be a transversal to a vector field X and fix an orientation of τ. Then at each point of τ we can define the angle between τ and X. This angle is either in $(0,\pi)$ or in $(0,-\pi)$ for all points.

Proposition 14.1.3. *Let M be a surface with a Riemannian metric and X a C^1 vector field on M with a positively recurrent nonperiodic orbit. Then for every $\epsilon > 0$ there exists an oriented transversal τ such that the angle between X and τ is in $(0,\epsilon)$ at every point.*

We note that the proof is easier in the case of orientable surfaces, but that nonorientability is only a minor complication.

Proof. We begin with a construction which is a more careful version of our construction of pretransversals. Naturally, since we are on an arbitrary surface we cannot use the Jordan Curve Theorem A.5.2 and we do not get a contradiction as in the proof of Theorem 14.1.1. Let p be a positively recurrent point and consider a *flow box*, that is, a neighborhood $\mathcal{U}$ of p on which there are C^1 coordinates (x, y), $-\epsilon \le y \le \epsilon$, in which $X = \frac{\partial}{\partial y}$ and $p = 0$. The boundary curve given by $y = -\epsilon$ is called the *base* of the flow box, and the curve $y = \epsilon$ the *roof*.

Denote the flow generated by X by φ^t. Since p is positively recurrent, but not periodic, there are infinitely many values of t for which $\varphi^t(p)$ is on the line $y = 0$ in $\mathcal{U}$. Since X does not vanish along the orbit of p, we can choose a vector field Y along the positive semiorbit of p such that (X, Y) is a frame, that is, X and Y are linearly independent at each point. This defines an orientation for all points of the orbit of p. For any two points of $\varphi^t(p)$ in $\mathcal{U}$ we can compare these orientations. Given $\delta > 0$ let us call a time t_0 a *closest-return* time if the point $\varphi^{t_0}(p)$ is on the line $y = 0, |x| < \delta$ in $\mathcal{U}$ and closer to 0 than any point $\varphi^t(p)$ on the same line for any $t \in (0, t_0)$. We would like to have infinitely many closest returns whose orientation agrees with that at p. If that is not the case then the orientations at p and $\varphi^t(p)$ differ for infinitely many closest returns, so we can consider two successive closest-return times t_0 and t_1 such that the orientations at $\varphi^{t_0}(p)$ and $\varphi^{t_1}(p)$ agree and the distance between $\varphi^{t_0}(p)$ and $\varphi^{t_1}(p)$ is as small as we please. Replacing p by $\varphi^{t_0}(p)$ then puts us in the first case. Thus we may assume that the point p has a closest return $\varphi^{t_0}(p)$ at which the orientation coincides with that at p. Notice that the preceding argument is only needed to take care of the case of nonorientable surfaces.

Consider now a narrow strip around the orbit segment of p for $t \in [0, t_0]$. This strip may be assumed to be orientable since the frame (X, Y) along the orbit of p can be extended to a frame on a small neighborhood of the orbit segment. Using the Riemannian structure we can thus define a rotated vector field $Z = R_\theta X$ as the vector field that has angle θ with X. If the angle is small enough and has the right sign then the orbit of p under the flow ψ_θ generated by Z stays within the strip and returns to a point $\psi_\theta^{t'}(p)$ between p and $\varphi^{t_0}(p)$ on the line $y = 0$. Consider the curve c defined by connecting the points $\psi_\theta^{t_1}$ and $\psi_\theta^{t_2}(p)$ with a straight line in $\mathcal{U}$, where t_1 and t_2 are chosen such that $\psi_\theta^t(p) \in \mathcal{U}$ for $t \in [0, t_1]$ and $t \in [t_2, t']$. If t_1 is not too small and t_2 is not too close, then for sufficiently small δ the angle between $\dot{c}$ and X is less than ϵ where defined. Thus we can take a smooth curve τ sufficiently close to c such that the angle between $\dot{\tau}$ and X is between 0 and ϵ along τ. This is the desired transversal. □

We would like to see where the return map to such a transversal is defined. This construction clearly yields a transversal τ such that there is at least one point $q \in \tau$ that returns to τ in positive time. Thus the return map to τ is well defined and continuous on a nonempty open subset of τ. The following result gives a way to show that this is a large set in some cases. It applies not only to closed transversals, but to transverse segments as well. Notice that the set of points on the transversal that return to it is open and hence a union of disjoint intervals.

Proposition 14.1.4. *Let M be a closed surface with a C^1 vector field X. Suppose τ is a transversal, not necessarily closed. If τ_0 is the set of points returning to τ and $p \in \partial\tau_0$ an endpoint of an interval in τ_0 that is not an endpoint of τ and does not return to an endpoint of τ, then the ω-limit set $\omega(p)$ consists of fixed points of X.*

Proof. Denote by φ^t the flow generated by X and let us suppose that there is a point $y \in \omega(p) \smallsetminus \mathrm{Fix}(\varphi^t)$. Observe that in this case the orbit of p has infinite length since there is an $\epsilon > 0$ and a neighborhood of y (for example, a flow box) to which the orbit returns infinitely many times and such that upon each return the orbit segment in the neighborhood has length ϵ. Since M is compact, the orbit of p thus has a nonperiodic recurrent limit point. This means that for $\epsilon > 0$ there are times $0 < t_1 < t_2$ such that $d(\varphi^{t_1}(p), \varphi^{t_2}(p)) < \epsilon$ and the orientations at $\varphi^{t_1}(p)$ and $\varphi^{t_2}(p)$ (as defined in the previous proof) coincide. By closing $\{\varphi^t(p) \mid t_1 \leq t \leq t_2\}$ by a short transverse curve we obtain a pretransversal. So from the orbit of p we have just produced pretransversals with arbitrarily short transverse pieces.

Note that given two segments τ_1 and τ_2 transverse to X for which there is an interval $I \subset \tau_1$ of points whose positive orbits intersect τ_2, there is a flow box consisting of orbit segments beginning on I and ending on τ_2. It is useful to note that any closed transversal that intersects an orbit segment of a flow box must intersect the base or roof or else traverse the entire flow box, that is, intersect every orbit segment in the flow box. The same holds for the transverse segment of a pretransversal as long as its ends are not in the flow box.

Given $q \in I$ let r be the midpoint (in I) between p and q and denote by J the interval with endpoints q and r. Let d be the minimum width of the flow box with base J and roof in τ (that is, the infimum of lengths of curves intersecting all orbit segments in the flow box). As we have observed, we can construct a pretransversal from the orbit of p whose transverse part γ has length less than $d/2$ and is disjoint from τ. Note that the positive semiorbit of every point p' of I sufficiently close to p must intersect γ (either at time approximately t_1 or at time approximately t_2). The interval with endpoints p and p' is the base of a flow box consisting of orbit segments ending on γ. Except for the base, τ is disjoint from this flow box, since otherwise τ would intersect all orbit segments in the flow box, in particular that of p. But by assumption p does not return to τ.

Consider now the flow box whose base is the interval in τ with endpoints q and p' and whose roof is in τ. We may assume that this interval contains the midpoint r between p and q. The previous argument shows that γ intersects this flow box and hence traverses it completely (since it is disjoint from τ).

Hence the length of γ exceeds d by choice of d. On the other hand we chose γ to have length less than $d/2$. This contradiction ends the proof. □

Corollary 14.1.5. *If X is a fixed-point-free vector field on a closed surface and τ a closed transversal, then the return map to τ is either defined on all of τ or not defined at all.*

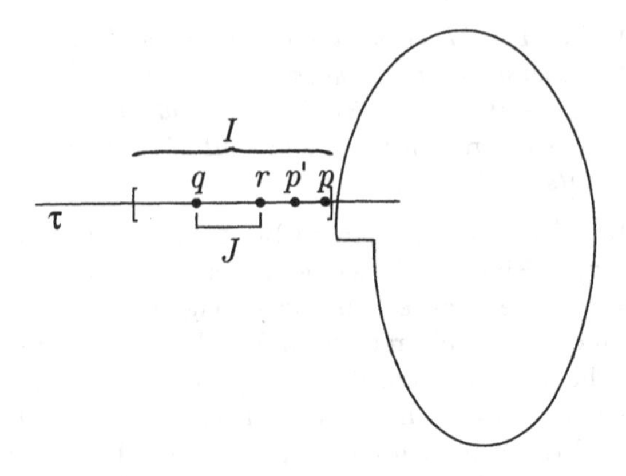

FIGURE 14.1.2. Arrangement of the transversals

Proposition 14.1.6. *Let X be an area-preserving vector field on a closed compact surface M all of whose fixed points are centers and (possibly multiple) saddles. Let τ be a not necessarily closed transversal to X. Then the return map on τ is defined and continuous at all but finitely many points and at those points both one-sided limits exist.*

Proof. First, the set of fixed points of X is finite since saddles are isolated and M is compact. There are finitely many positive semiorbits whose ω-limit sets consist of fixed points. Those are incoming (stable) separatrices of the saddles. In fact, the ω-limit set of any separatrix consists of a single saddle.

By Proposition 4.1.18(1) applied to a time-1 map the set of recurrent points is dense in M; hence its intersection with any flow box $\mathcal{U}$ around $p \in \tau$ is dense in $\mathcal{U}$. Since the set of recurrent points is flow-invariant, the set τ_0 of points in τ that return to τ is dense in τ. Recall that τ_0 is open and hence consists of intervals. Hence by Proposition 14.1.4 the positive semiorbit of any endpoint of any component of τ_0 must be an incoming separatrix of one of the fixed points. Thus τ_0 consists of finitely many intervals, and since it is dense in τ, its complement is finite.

Note that on each interval of τ_0 the return map is injective, hence monotone, so the one-sided limits exist at the endpoints. □

Remark. The only way invariance of the area was used was through Proposition 4.1.18, which can be applied if the flow has an invariant measure whose support is the whole surface, that is, which is positive on open sets. However, in the area-preserving case extra information can be obtained about the return map.

Corollary 14.1.7. *Under the assumptions of the previous proposition there exists a smooth parameter on τ such that the return map to τ is an isometry on each component of τ_0.*

Proof. Consider the smooth invariant measure on τ induced by the (invariant) area (cf. Proposition 5.1.11). Pick an initial point $p \in \tau$ and an orientation on τ. Then parameterize points $x \in \tau$ by the (signed) measure of the interval between p and x. □

In Section 14.5 we will study the return maps that appear in this corollary.

Exercises

14.1.1*. *Show that there is a C^r-open dense set of C^r vector fields on a compact closed surface that have only finitely many fixed and periodic orbits, all of them hyperbolic.*

14.1.2. *Show that for any area-preserving vector field on the sphere there is a dense invariant set that consists of fixed points and periodic orbits.*

14.1.3. *Show that the Poincaré–Bendixson Theorem 14.1.1 holds for any flow generated by a C^0 vector field so long as the latter is uniquely integrable.*

2. Fixed-point-free flows on the torus

a. Global transversals. According to the Poincaré–Hopf Index Theorem 8.6.6 the only compact surfaces (without boundary) that admit fixed-point-free flows are the torus and the Klein bottle. It turns out that on the Klein bottle nontrivial recurrence is impossible (see Exercise 14.2.3) and that on the torus any flow with nontrivial recurrence is equivalent to a flow under a function built from a circle diffeomorphism.

We will use the following observation. Consider any compact orientable surface M with a Riemannian metric. Then given a vector field X on M there is a one-parameter family of vector fields $R_\theta X$ obtained by rotating $X(p)$ by an angle θ at every point p. By orientability this is well defined. Obviously the set of fixed points of $R_\theta X$ coincides with that of X.

Proposition 14.2.1. *A fixed-point-free C^1 flow on the torus $\mathbb{T}^2$ admits a closed transversal.*

Proof. Fix a Riemannian metric on $\mathbb{T}^2$ and consider the vector field $Z = R_{\pi/2}X$ perpendicular to X. If Z has a periodic orbit then this gives a closed transversal to X. Otherwise Z has a positively recurrent nonperiodic orbit and we use Proposition 14.1.3 to obtain an oriented transversal to Z that makes an angle less than $\pi/4$ with Z. But this then is also a transversal to X since the angle with X cannot be zero. □

Proposition 14.2.2. *If a fixed-point-free flow with a nonperiodic recurrent orbit on a closed surface admits a closed transversal then every orbit intersects the transversal.*

Proof. We will show that the set of points whose orbit intersects a transversal τ is open and closed in $\mathbb{T}^2$. Note first that by Corollary 14.1.5 the return map is defined on the entire transversal. Note that by compactness of τ the return times are bounded. Thus the set of points whose orbit intersects τ is a finite union of images of the closed set $\bigcup_{t\in[-\epsilon,\epsilon]} \varphi^t(\tau)$, hence is closed. Likewise, however, this set is also a finite union of open sets $\bigcup_{t\in(-\epsilon,\epsilon)} \varphi^t(\tau)$, hence is open. □

These two results now yield the advertised equivalence

Corollary 14.2.3. *A fixed-point-free C^1 flow on $\mathbb{T}^2$ with a nonperiodic recurrent orbit is smoothly conjugate to the flow under a function over an orientation-preserving circle diffeomorphism.*

Proof. By Proposition 14.2.1 the flow admits a transversal τ and by Proposition 14.2.2 every orbit intersects τ. If we parameterize τ by an angle $\theta \in S^1$, then the return map to τ defines a circle diffeomorphism f. If the return time of θ is $h(\theta)$, then we can coordinatize $\mathbb{T}^2$ by (θ, y) with $0 \leq y < h(\theta)$ and in these coordinates the vector field is $\frac{\partial}{\partial y}$, that is, the flow is a special flow. □

This close relation to circle maps will allow us to apply results from Chapters 11 and 12. A first instance is the following classification.

Proposition 14.2.4. *A C^2 fixed-point-free flow on $\mathbb{T}^2$ with a nonperiodic recurrent orbit is topologically conjugate to a time change of a linear flow.*

Proof. By Proposition 14.2.1 there is a transversal, which we may assume to be smooth, so the flow is conjugate to a special flow over a C^2 circle diffeomorphism. By Theorem 12.1.1 this diffeomorphism is topologically conjugate to a rotation, so via Corollary 14.2.3 the flow is topologically conjugate to a special flow over a rotation, that is, a time change of a linear flow on $\mathbb{T}^2$. □

The complications we found in the theory of circle maps of regularity less than C^2, namely, the Denjoy example (cf. Proposition 12.2.1), arise in this situation as well. Namely, the special flow over a Denjoy map from Section 12.2 is a flow on the torus which exhibits Denjoy-type behavior and in particular is not topologically conjugate to a special flow over a rotation.

b. Area-preserving flows. In certain cases the topological conjugacy obtained in Proposition 14.2.4 is actually a smooth conjugacy. We first consider area-preserving flows, where the Poincaré Recurrence Theorem 4.1.19 provides the recurrence needed in the last two statements. .

Proposition 14.2.5. *Let φ^t be a C^k flow of $\mathbb{T}^2$ preserving a C^r area element and let $n = \min(k, r+1)$. Then φ^t is C^n conjugate to a special flow over a circle rotation.*

Proof. Note first that the C^∞ transversal τ obtained in Proposition 14.2.1 has a $C^{\min(k,r)}$ length parameter invariant under the return map by Proposition 5.1.11. Thus the return map is C^n conjugate to a circle rotation by Proposition 12.4.4. By extending this conjugacy as in the proof of Corollary 14.2.3 we obtain a C^n conjugacy to a special flow. □

Note that a given linear flow can be represented as a suspension of different rotations by choosing different transversals. It is easy to see, however, that the rotation number depends only on the homotopy type of the transversal. In particular consider the flow T_t^ω and the transversal $\tau_{k,l}$ that lifts to the straight line with slope l/k. Then consider a linear transformation given by $A = \begin{pmatrix} k & m \\ l & n \end{pmatrix}$ with integers m, n such that $\det A = 1$. The rotation number with respect to the transversal $\tau_{k,l}$ is the rotation number obtained from $T_t^{A^{-1}\omega}$ using the horizontal transversal $\tau_{1,0}$, that is, the reciprocal of the slope of $A^{-1}\omega$, which is given by $\frac{n\omega_1 - m\omega_2}{-l\omega_1 + k\omega_2}$. Notice that (m,n) is unique up to integer multiples of (k,l), so this expression for the rotation number is well defined modulo 1. In particular the rotation numbers ρ and ρ' obtained from two different transversals are related by a linear fractional transformation

$$\rho' = \frac{a\rho + b}{c\rho + d},$$

where $\begin{pmatrix} a & b \\ c & d \end{pmatrix}$ is an integer matrix with unit determinant.

Recall (Definition 2.8.1) that an irrational number α is called *Diophantine* if there exist $k, r > 0$ such that for any nonzero $p, q \in \mathbb{Z}$ we have $|q\alpha - p| > kq^{-r}$. This property is invariant under linear fractional transformations.

Lemma 14.2.6. *If α is a Diophantine irrational and $\begin{pmatrix} a & b \\ c & d \end{pmatrix}$ a nonsingular integer matrix then $\frac{a\alpha + b}{c\alpha + d}$ is Diophantine as well.*

Proof. Note that by assumption a and c cannot both vanish. Thus we find

$$\left|q\left(\frac{a\alpha + b}{c\alpha + d}\right) - p\right| = \left|\frac{(qa - pc)\alpha + (qb - pd)}{c\alpha + d}\right| > \frac{k|qa - pc|^{-r}}{|c\alpha + d|} = \frac{k'}{|qa - pc|^r} > \frac{k''}{q^r}$$

since $|qa - pc| < \text{const.}\, q$ for those p/q that approximate $\frac{a\alpha + b}{c\alpha + d}$ well. □

Returning to general area-preserving flows it is natural to ask, how to see whether the smoothly conjugate special flow obtained in Proposition 14.2.5 is smoothly conjugate to a suspension. According to the Proposition 2.9.5 a sufficient condition for that is that the base is rotated by a Diophantine angle. Thus we have

Corollary 14.2.7. *Let φ^t be a C^∞ flow of $\mathbb{T}^2$ preserving a C^∞ area element and that possesses a transversal such that the induced map has Diophantine rotation number. Then φ^t is C^∞ conjugate to a linear flow.*

We will see in Section 14.7a how this question can be decided without referring to a section.

Exercises

14.2.1. *Show that for a generic (dense G_δ) set of numbers ρ there exists a real-analytic function φ such that the special flow under φ built over the rotation R_ρ is not C^0 flow equivalent to a linear flow.*

14.2.2. *Prove that every fixed-point-free flow on the Klein bottle has a periodic orbit.*

14.2.3. *Show that any recurrent orbit for a fixed-point-free flow on the Klein bottle is periodic.*

14.2.4. *Prove that for $r \geq 0$ there is a C^∞ flow on the torus that is C^r, but not C^{r+1}, orbit equivalent to a linear flow.*

3. Minimal sets

The examples of minimal sets for flows on surfaces we have seen so far are very limited. Of course fixed points and periodic orbits clearly occur for flows on any compact surface. Furthermore the irrational linear flow on $\mathbb{T}^2$ is minimal. Finally we have mentioned that minimal Cantor sets occur when one builds a special flow over a Denjoy example. The latter is possible for C^1 but impossible for fixed-point-free C^2 flows on the torus. In fact, we will now show that for C^2 flows on any surface the first three examples are the only possible kinds of compact minimal sets. This is a generalization of the Denjoy theorem and the proof again uses a bounded distortion estimate.

Theorem 14.3.1. (Schwartz)[1] *Let M be a smooth surface, φ^t a C^2 flow, and A a nonempty compact minimal set. Then A is either a fixed point or a periodic orbit or $A = M$.*

Remark. Note that if $A = M$ we can conclude that M is compact and φ^t is fixed-point-free, so by the classification of compact surfaces and the Poincaré–Hopf Theorem 8.6.6 M is either the torus or the Klein bottle. Exercise 14.2.3 excludes the latter possibility and thus $M = \mathbb{T}^2$.

As the Denjoy-type flows on $\mathbb{T}^2$ show, the C^2 assumption is really needed. We employ it as follows. A minimal set A has no proper closed invariant subsets, so $\partial A = A$ or $\partial A = \varnothing$. Thus A is nowhere dense unless $A = M$. Thus we need only rule out the possibility of Cantor-type sets, which we accomplish by an argument similar to the proof of the Denjoy Theorem 12.1.1 using the C^2 hypothesis.

Proof. We assume for purposes of contradiction that A is a nowhere dense compact invariant minimal set without fixed or periodic points. Let us take a point in A and consider a C^∞ transverse segment τ through this point whose endpoints are not in A. By taking a flow box $\bigcup_{|t|<\epsilon} \varphi^t(\tau)$ we identify τ with $I = (-\eta, \eta) \subset \mathbb{R}$ and $\tau \cap A$ with a compact nonempty nowhere dense subset C of I. Thus $W := I \smallsetminus C$ is dense and a countable disjoint union of open intervals (a_l, b_l), $l \in \mathbb{N}$. If U is the set of $x \in I$ whose orbit returns to τ at some time $t > 0$ but does not hit an endpoint of τ for time $0 < s < t$, then U is a neighborhood of C in I and the return map f is well defined, injective, and continuous, in fact C^2 by transversality, on U. Since τ is transverse to the vector field $\dot{\varphi}^t$ we also have $f' \neq 0$ on U. If V is an open neighborhood of C whose closure is in U then by compactness of $\bar{V}$ there exists $K > 1$ such that

$$\frac{1}{K} < |f'| < K, \qquad |f''| < K \tag{14.3.1}$$

on V. Of course C is a nonempty compact invariant minimal set for f without periodic points. Furthermore the set $\tilde{C} := \bigcup_{l\in\mathbb{N}}\{a_l, b_l\} \smallsetminus \{-\eta, \eta\} \subset C$ of endpoints is f-invariant as well. Thus if (a_l, b_l) is a component of W, $(a_l, b_l) \subset U$, $a_l \neq -\eta$, and $b_l \neq \eta$ then $f((a_l, b_l))$ is also a component of W.

Now consider the distance

$$\epsilon := d(C, \partial V) \in (0, \eta)$$

and note that the set Q of points $a_l, b_l \in \tilde{C}$ such that $b_l - a_l \geq \epsilon$ is finite. Thus there exists an $N \in \mathbb{N}$ such that $f^n(a_1) \notin Q$ for $n \geq N$. Since $\tilde{C}$ is invariant there exists $l \in \mathbb{N}$ such that $f^n(a_1) = a_l$ or $f^n(a_1) = b_l$. Thus $(a, b) := (a_l, b_l)$ has the property that $f^n((a, b))$ is a component of W of length less than ϵ for all $n \in \mathbb{N}$. This implies that

$$f^n([a, b]) \subset V \text{ for } n \in \mathbb{N}$$

since $\{a, b\} \subset C$. These intervals are pairwise disjoint since C contains no periodic points and thus their lengths are summable. We will use this fact now in a Denjoy-type argument which ultimately shows that there must be a periodic point in C after all.

Proposition 14.3.2. *Suppose $f\colon [0,1] \to [0,1]$ is C^2, $K > 0$. Then there exists $C \in \mathbb{R}$ such that if $I \subset [0,1]$ and (14.3.1) holds on $\bigcup_{i=0}^{k} f^i(I)$ then*

$$\left|\log \frac{(f^{k+1})'(x)}{(f^{k+1})'(y)}\right| \leq C \sum_{i=0}^{k} |f^i(x) - f^i(y)|$$

for all $x, y \in I$.

Proof. Since by the chain rule $f^{k+1'}(x) = \prod_{i=1}^{k} f'(f^i(x))$, the Mean Value Theorem yields $\xi_i \subset f^i((p,q))$ such that

$$\begin{aligned}\left|\log\left|\frac{f^{k+1'}(p)}{f^{k+1'}(q)}\right|\right| &\le \sum_{i=0}^{k} |\log|f'(f^i(p))| - \log|f'(f^i(q))|| \\ &\le \sum_{i=0}^{k} |D\log|f'(\xi_i)|| \cdot |f^i(p) - f^i(q)| \\ &= \sum_{i=0}^{k} |f'(\xi_i)|^{-1}|f''(\xi_i)| \cdot |f^i(p) - f^i(q)| \\ &\le K^2 \sum_{i=0}^{k} |f^i(p) - f^i(q)|,\end{aligned}$$

by (14.3.1). □

For later use we note the following immediate consequence:

Lemma 14.3.3. *If $f\colon[0,1] \to [0,1]$ is a C^2 map and I has pairwise disjoint images then the ω-limit set of I contains a critical point.*

Suppose $n \in \mathbb{N}$ and $[p,q] \subset [s,t] \subset V$ such that $f^k([s,t]) \subset V$ for $0 \le k \le n$. Then by Lemma 14.3.3

$$\left|\frac{f^{k+1'}(p)}{f^{k+1'}(q)}\right| \le \exp\Big(K^2 \sum_{i=0}^{k} |f^i(p) - f^i(q)|\Big). \tag{14.3.2}$$

We first use this observation to show that $|f^{i'}(a)|$ is summable. To that end note that by the Mean Value Theorem there exist $\xi_i \in (a,b)$ such that $f^i(b) - f^i(a) = f^{i'}(\xi_i) \cdot (b-a)$ and hence

$$(b-a)\sum_{i=0}^{\infty} |f^{i'}(\xi_i)| = \sum_{i=0}^{\infty} |f^i(b) - f^i(a)| \le 2\eta$$

since the intervals $f^i((a,b))$ are pairwise disjoint. Since by (14.3.2) we thus have $|f^{i'}(a)| \le |f^{i'}(\xi_i)|e^{2K^2\eta}$, we have shown

$$1 \le d := \sum_{i=0}^{\infty} |f^{i'}(a)| \le \frac{2\eta}{b-a} e^{2K^2\eta}. \tag{14.3.3}$$

We want to see that (14.3.3) implies

$$\lim_{i\to\infty} f^{i'}(x) = 0 \text{ uniformly for } |x-a| \le \delta := \frac{\epsilon}{3K^2 d(1+\eta)} < \frac{\epsilon}{3d} < \epsilon. \tag{14.3.4}$$

To this end we show

Lemma 14.3.4. *For all $n \in \mathbb{N}$*

(1) $f^n([a-\delta, a+\delta]) \subset V$,
(2) $|f^n(x) - f^n(a)| < \epsilon$ *when* $|x-a| \le \delta$,
(3) $|f^{n\prime}(x)| \le e|f^{n\prime}(a)|$ *when* $|x-a| \le \delta$.

Note that (3) indeed implies (14.3.4); (1) and (2) are only used for purposes of induction in the proof.

Proof. We proceed by induction. For $n = 0$ the claim is trivial. Suppose it holds for $k \le n$. We first show (3) for $n+1$. Equation (14.3.2) yields

$$|f^{n+1\prime}(x)| \le |f^{n+1\prime}(a)| \exp\left(K^2 \sum_{k=0}^{n} |f^k(x) - f^k(a)|\right)$$

and the Mean Value Theorem together with (3) yields

$$\sum_{k=0}^{n} |f^k(x) - f^k(a)| \le |x-a| \sum_{k=0}^{n} |f^{k\prime}(\xi_k)| \le 3|x-a| \sum_{k=0}^{n} |f^{k\prime}(a)|.$$

By (14.3.3) we get

$$|f^{n+1\prime}(x)| \le |f^{n+1\prime}(a)| e^{3K^2 d\delta} < |f^{n+1\prime}(a)| e^{\frac{\epsilon}{\eta}} < e|f^{n+1\prime}(a)|,$$

proving (3) for $n+1$.

To prove (2) note that by the Mean Value Theorem and (3) we have

$$|f^{n+1}(x) - f^{n+1}(a)| \le |x-a| \cdot |f^{n+1\prime}(\xi)| \le 3\delta |f^{n+1\prime}(a)| \le 3d\delta < \epsilon,$$

proving (2).

Finally, since $f^{n+1}(a) \in C$, (2) and the choice of ϵ imply (1). □

Now we can finish the proof of the theorem. Since C is minimal there exist infinitely many $k \in \mathbb{N}$ such that $|f^k(a) - a| \le \delta/2$. For large enough such k we have $|f^{k\prime}(x)| < 1/2$ for $|x-a| \le \delta$ by (14.3.4) and hence $|\frac{d}{dx}(f^k(x) - x)| > 1/2$, so $f^k(x) - x$ changes sign on $[a-\delta, a+\delta]$ and hence has a zero $z \in [a-\delta, a+\delta]$. We could of course use the Contraction Mapping Principle, Proposition 1.1.2, but in the one-dimensional case the Intermediate Value Theorem is sufficient. Thus $f^k(z) = z$ and furthermore $f^{kn}(a) \xrightarrow[n\to\infty]{} z$ by (14.3.4). Since $a \in C$ and C is closed and invariant we conclude that $z \in C$ contrary to the assumption that C contains no periodic points. This proves the theorem. □

Exercises

14.3.1. *Given an orientable surface of genus g and $1 \le k \le g$ show that there exists a C^1 flow with exactly k nowhere-dense minimal sets that are not fixed points or circles.*

14.3.2. *Show that any C^1 flow on the orientable surface of genus g has no more than g different minimal sets that are not fixed points or periodic orbits.*

4. New phenomena

New types of dynamical behavior appear for smooth flows on the torus with fixed points and for flows on surfaces of higher genus.

a. The Cherry flow.[1] Now we show that in the presence of fixed points behavior that produces Denjoy-type minimal sets on transversals to the flow appears for arbitrarily smooth flows on the torus. The idea is to modify a linear flow in such a way that there is a transversal containing a dense set of points which return to the transversal only finitely may times and are then attracted to a fixed point. The remaining points form a Cantor set which then inevitably exhibits Denjoy-type behavior and consists of points that have a saddle in their orbit closure. This example is then easily modified to give a flow on a surface of genus 2 that has no attracting fixed points and two saddles and exhibits similar phenomena.

Here is a description of this construction. Consider a local vector field on a disk $D_1 \subset \mathbb{R}^2$ with a saddle–node phase portrait as in Figure 7.3.4 and interpolate it via bump functions to the constant vector field $X = (0,1)$ outside a neighborhood D_2 of D_1. Take D_2 of diameter ϵ and rotate the vector field by $\tan^{-1}\alpha$. Translate D_2 to the center of the unit square $[0,1]^2$ and project the restriction to $[0,1]^2$ of this vector field to $\mathbb{T}^2 = [0,1]^2/\mathbb{Z}^2$. This gives a vector field X_0 with a node p and a saddle s in a disk $D \subset \mathbb{T}^2$ which is constant outside D. If ϵ and α are not too big then there will be an interval $[a,b] \subset \{0\} \times S^1$ such that $a, b \in W^s(s)$, $\omega(y) = p$ for $y \in (a,b) \times S^1$ and the return map on $\{0\} \times S^1$ is well defined outside $[a,b]$. Note that this map extends (by constant interpolation across $[a,b]$) to a continuous monotone circle map f_0 of degree one, hence has rotation number $\tau(f_0)$ depending continuously on this construction. Modifying X_0 on $[1-\delta, 1] \times S^1$ we can make $Df(x) > 1$ outside $[a,b]$.

Let us show that we can make $\tau(f_0)$ irrational. Let $Y_\lambda = (0, h(x))$ be a C^∞ vector field on $\mathbb{R}^2$ with $\operatorname{supp}(h) \subset [1-\delta, 1]$ such that the map induced between $\{0\} \times \mathbb{R}$ and $\{1\} \times \mathbb{R}$ by the flow of $(\cos\alpha, 0) + Y_\lambda$ is a translation by λ. The vector fields $X_\lambda = X_0 + Y_\lambda$ generate flows on $\mathbb{T}^2$ with induced circle maps f_λ which lift to F_λ on $\mathbb{R}$ such that $\tau(F_1) - \tau(F_0) = 1$, and hence there is a λ_0 for which $\tau(f_{\lambda_0}) \notin \mathbb{Q}$. Let $f := f_{\lambda_0}$.

We call the basin $T = \{q \mid \omega(q) = p\}$ of the sink the *tail* and its complement $\Lambda = \mathbb{T}^2 \setminus T$ the *Cherry set*. To see that $K := \Lambda \cap (\{0\} \times S^1)$ is a Cantor set let $K_0 \subset K$ be a maximal closed interval and $K_n := f^n(K_0)$. Then $l(K_{n+1}) \geq l(K_n)$ and $K_i \cap K_j = \varnothing$ since otherwise we have inclusion by maximality and thus a periodic point of f by Lemma 15.1.2, which is impossible since $\tau(f) \notin \mathbb{Q}$. But then we must have $l(K_0) = 0$ and K has empty interior. Next the pairwise disjoint intervals $I_n := f^{1-n}((a,b))$ have dense union in $\{0\} \times S^1$ and the endpoints belong to different components of $W := W^s(s) \setminus \{s\}$, so each of these components is dense in Λ and $\alpha(x) = \Lambda$ for $x \in W$. This shows that K is perfect, hence a Cantor set.

For the reversed flow the Cherry set is an attractor of a type we have not encountered in the Poincaré–Bendixson setting or in fixed-point-free toral flows.

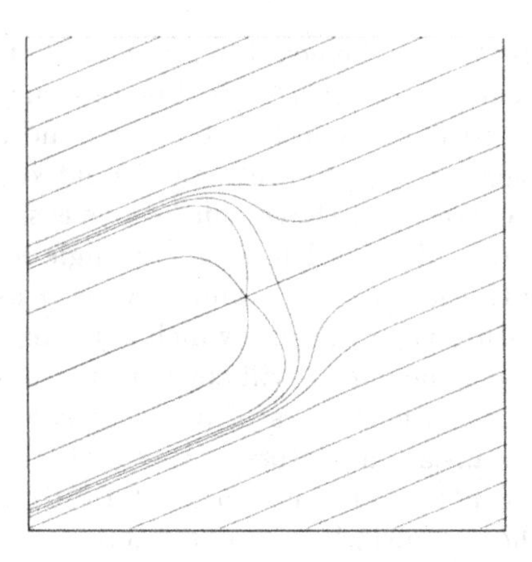

FIGURE 14.4.1. The Cherry flow

It is useful to define a *quasi-minimal set* to be a set containing finitely many fixed points and such that every semiorbit that is not attracted to a fixed point is dense in the set. We can summarize our discussion by saying that the nonwandering set of the Cherry flow consists of an attracting fixed point and a quasi-minimal set containing a hyperbolic saddle. We will see later that quasi-minimal sets are a typical phenomenon for flows on surfaces of higher genus, including area-preserving ones. They produce minimal sets for the Poincaré return map on closed transversals with an appropriately modified topology (see Section 14.5b and Exercise 14.5.2).

The Cherry flow can be modified to produce a flow on the double torus, that is, the sphere with two handles attached: Remove a small neighborhood of the attracting fixed point and consider a second copy of the torus with the same disk removed and the flow reversed. Then these flows can be glued together along the boundaries of the two disks to give a flow on the double torus which has no attracting or repelling fixed points, two saddles, and two disjoint closed invariant nowhere-dense quasi-minimal sets C^+ and C^- containing a saddle each. For every point x outside these two sets the α-limit set is C^- and the ω-limit set is C^+. This flow is obviously not area preserving. Next we will consider an interesting example of an area-preserving flow on the same surface of genus 2.

b. Linear flow on the octagon. If one views the linear flow on the torus as a flow on the unit square whose orbits are parallel, one can naturally try to generalize this construction by replacing the square with another centrally symmetric polygon with opposite sides identified by *translations* and considering the linear flow on the interior extended to the closed surface obtained from the identification. In order to obtain a smooth, or even only continuous flow, certain care has to be taken in defining the flow near the vertices. However, from the point of view of the global orbit structure and the recurrence behavior of the nonfixed points this is not particularly important.

For this kind of construction the next obvious candidate after the square is a regular hexagon. One can see, however, that this construction produces nothing new: The translations of the hexagon tile the plane and the three translations identifying opposite pairs of sides are rationally related, namely, their sum is zero. Thus the group generated by these translations is simply the lattice whose generators are two vectors of equal length at an angle $\pi/3$. This allows us to extend the linear flow to the tiled plane and to view it as a linear flow on the factor by the lattice with two generators, which is again a torus.

To produce a new phenomenon we will consider the next candidate, namely, the regular octagon. It has four pairs of opposite sides which are identified by translations. The translation vectors have equal length and their mutual angles are multiples of $\pi/4$. It is easy to see (cf. Exercise 14.4.3) that the group generated by these translations is not discrete, that is, when we apply the translations to the octagon we will return and cover every point infinitely many times.

When opposite sides of the octagon are identified, all eight vertices are glued together. Notice that from the topological point of view this construction is equivalent to the construction of the genus-two surface from the hyperbolic octagon in Section 5.4e, so the surface thus obtained is homeomorphic to the sphere with two handles. We will give another proof of this fact by constructing a vector field on the surface and calculating its Euler characteristic. This will be the vector field we will later study from the dynamical point of view. Pick a direction in the plane not parallel to any side and take a family of oriented line segments inside the octagon parallel to this direction. The identification of parallel sides allows us to glue those line segments together, except for the ones beginning and ending in a vertex. There are exactly three segments beginning in a vertex and three segments ending in a vertex. A neighborhood of the vertex in the identification space consists of eight sectors glued together in such a way that incoming and outgoing segments alternate and divide the neighborhood into six sectors. No other orbit crosses any of the separating lines. Thus the picture of the orbits near this special point looks similar to the example showed in Figure 8.4.1, except for having six separatrices rather than eight. By making an appropriate time change the flow can be made into a smooth flow with a double saddle.

Now we will give a description of a natural differentiable structure in a neighborhood of the identified vertices which will also provide a natural time change for the flow. The problem is that the Euclidean differentiable structure does not behave well under the projection to the identification space because the total angle is 6π rather than 2π. In the construction on the hyperbolic plane in Section 5.4e the total angle was indeed 2π. In our case the natural way to fix this is to introduce a complex coordinate w on the neighborhoods of the vertex in the octagon such that the standard Euclidean complex coordinate $z = x + iy$ is given by $z - z_0 = w^3$, where z_0 is the coordinate of a vertex. Gluing the edges together with these local coordinates gives a total angle of 2π in the factor. Note that on any open set not containing a vertex the coordi-

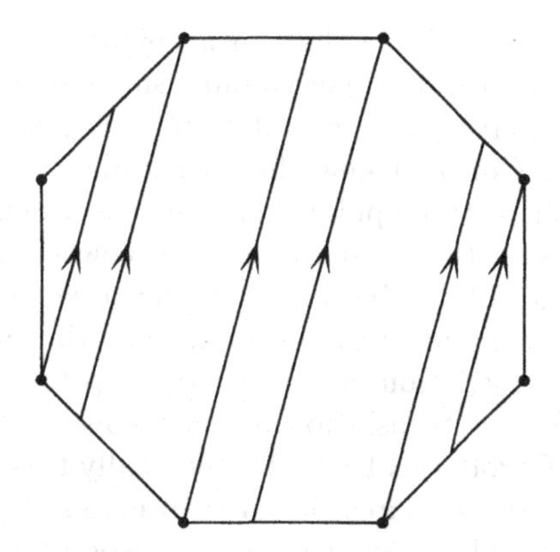

FIGURE 14.4.2. Linear flow with separatrices

nate w is locally differentiably compatible with z, so we have indeed defined a differentiable structure on the surface. We can, in fact, describe the Euclidean area in the coordinate w explicitly. Namely, if $z = x + iy$ then the Euclidean area is given by $dx \wedge dy$. If $w = u + iv$ then $z = w^3 = u^3 - 3uv^2 + i(3u^2v - v^3)$ and thus

$$dx \wedge dy = [(3u^2 - 3v^2)du - 6uvdv] \wedge [6uvdu + (3u^2 - 3v^2)dv] = (3u^2 + 3v^2)^2 dudv.$$

We will now use the fact (which follows from Proposition 5.1.9) that if the flow of a vector field v preserves a volume $\rho\Omega$ then the flow of ρv preserves Ω. Thus in a neighborhood of the vertex we can multiply the vector field by $(3u^2 + 3v^2)^2 = 9|w|^4$ to get a vector field that preserves the standard Euclidean area element of the w-coordinate in the neighborhood. To obtain a vector field on the surface that preserves a smooth area element we multiply by a function ρ that has the following properties: On a small neighborhood of the vertex ρ is equal to the scalar factor we just described. Outside a slightly larger neighborhood we set $\rho = 1$ and interpolate smoothly. The resulting flow preserves Euclidean area outside a neighborhood of the vertex and the w-standard area on a small neighborhood of the vertex. In a small collar around the vertex the invariant area is a smooth multiple of Euclidean area.

Let us now show that the vector field we have thus defined is, in fact, a smooth vector field. Note that the only problem is at the vertex, so we need to check smoothness there. In z coordinates the original vector field is given by a constant vector field $Y = (a, b)$. The coordinate change is given by $z = w^3$, whose derivative is given by $Y = 3w^2X$ or $X = Y/(3w^2)$. Thus the scaled vector field is given by $9|w|^4X = 9w^2\bar{w}^2Y/(3w^2) = 3\bar{w}^2Y$, which is indeed a smooth vector field with a saddle having six separatrices.

Let us now consider the return map to a convenient transversal. As a transversal we take a line connecting the midpoints of two opposite segments of the boundary and such that the angle α between the transversal and the vector field is in $(\pi/4, \pi/2)$. Note that on a neighborhood of such a line the flow-invariant volume is the Euclidean volume, since we are away from the vertex. Consequently the naturally induced volume is just (a constant multiple of) the length element. Consider now the return map to this transversal. It is continuous except at those three points that lie on segments of the flow ending on the saddle. Otherwise the return map is piecewise orientation preserving and, since the invariant volume induces the length element, the image of any interval not containing a point of discontinuity has the same length as the interval itself. Thus the restriction of the return map to every interval without points of discontinuity is a translation and we have a particular case of the situation described in Corollary 14.1.7. Topologically this transversal is a circle and it is the union of the closures of three intervals Δ_1, Δ_2, and Δ_3 without discontinuity points whose lengths are correspondingly $l_1 = 1/(1+\sqrt{2})$, $l_2 = (1+\cot\alpha)/(2+\sqrt{2})$, and $1 - l_1 - l_2$, if we take the transversal to have unit length.

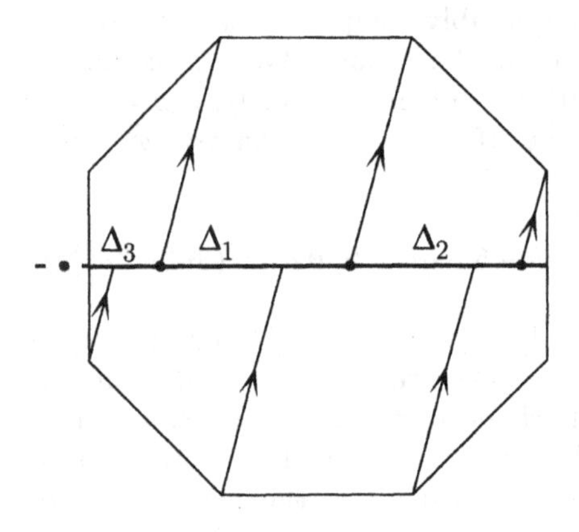

FIGURE 14.4.3. The induced map

The one-parameter family of linear flows we discussed here can also be obtained by considering the billiard problem on a triangular table T with angles $\pi/2$, $\pi/8$, and $3\pi/8$. Denote the vertex with angle $\pi/8$ by v. Note that by reflecting the triangle repeatedly in a side adjacent to the vertex v we obtain an octagon centered at v as a union of 16 copies of T.

Consider an orbit of the billiard in T. Reflecting it in the edges of T, as required, amounts to continuing along a straight line into a reflected copy of T. When the orbit hits the edge of the octagon, reflection in the boundary of T amounts to jumping to the opposite side and continuing in the same direction, that is, the continuation of the orbit corresponds exactly to the translation of the orbit to another copy of the triangle in the opposite orientation. Thus the orbit lifts exactly to an orbit of the linear flow on the octagon of the slope of the billiard orbit. We have thus established a one-to-one correspondence between the collection of orbits of the billiard in the triangle on the one hand and the

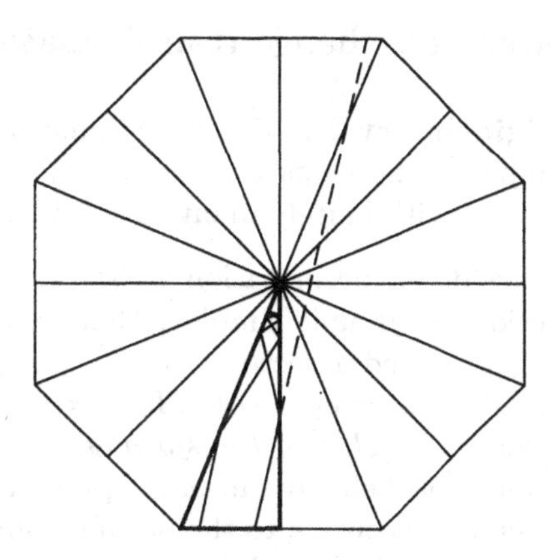

FIGURE 14.4.4. The octagon decomposed

collection of all orbits of the various linear flows on the octagon. It is good to note that the orbits for a fixed linear flow on the octagon correspond to billiard orbits whose angles with respect to a fixed side take at most 16 values, namely, for a given angle α we obtain the angles $\pm\alpha + k\pi/4$, that is, the angles obtained by reflecting α with the various reflections we used to generate the octagon.

We will soon prove a general result (Corollary 14.6.6) which implies that for all but countably many directions the linear flow on the octagon is topologically transitive and quasi-minimal. This is similar to the situation for linear flows on the torus (see Proposition 1.5.1).

Exercises

14.4.1. *Given an orientable surface of genus g and $1 \le k \le g$ show that there exists a C^∞ flow with exactly k nowhere-dense quasi-minimal sets that are not circles.*

14.4.2. *Show that any C^∞ flow on the orientable surface of genus g has no more than g different quasi-minimal sets that are not fixed points or circles.*

14.4.3. *Prove that the orbit of any point with respect to the group generated by translations identifying opposite sides of the regular octagon is dense.*

14.4.4. *For the linear flow on the octagon consider the transversal given by a diagonal (a diameter) of the octagon. Discuss the map on it induced by the flow by describing the topology of the transversal, giving the lengths of the maximal intervals without discontinuities and the way in which they are permuted.*

14.4.5. *Generalize the construction of Subsection b to the regular $2n$-gon ($n \ge 4$). Calculate the genus of the resulting surface and the number and indices of the fixed points for the flow, and describe the differentiable structure that makes a time change smooth.*

5. Interval exchange transformations

a. Definitions and rigid intervals. The return map for the linear flow on the octagon is an example of the maps that appear as section maps of area-preserving flows on surfaces with finitely many saddles (see Corollary 14.1.7).

Definition 14.5.1. Consider a permutation π of $\{1,\dots,n\}$, a vector $v = (v_1,\dots,v_n)$ in the interior of the unit simplex, that is, such that $v_i > 0$ for $i = 1,\dots,n$ and $\sum_{i=1}^n v_i = 1$, and a vector $\epsilon = (\epsilon_1,\dots,\epsilon_n)$ whose coordinates are either 1 or -1. Let $u_0 = 0$, $u_i = v_1+\dots+v_i$ for $i = 1,\dots,n$, $\Delta_i = (u_{i-1}, u_i)$ for $i = 1,\dots,n$. The *interval exchange transformation* $I_{v,\pi,\epsilon}\colon [0,1] \to [0,1]$ is the map that is continuous and Lebesgue-measure preserving on every interval Δ_i, rearranges those intervals according to the permutation π, and preserves or reverses orientation on Δ_i according to the sign of ϵ_i $(i = 1,\dots,n)$. If $\epsilon_i = 1$ for all i we write $I_{v,\pi}$ instead of $I_{v,\pi,\epsilon}$. Such a map will be called an *oriented interval exchange transformation.*

Remark. One can similarly define an arc exchange transformation of the circle. Obviously every arc exchange transformation of n arcs is also an interval exchange transformation of $n+1$ intervals.

Thus the map $I_{v,\pi,\epsilon}$ restricted to each interval Δ_i is either a translation of Δ_i (if $\epsilon_i = 1$) or a reflection with respect to a point (if $\epsilon_i = -1$).

The return map to the midpoint transversal for the linear flow on the octagon with slope α with respect to the horizontal is an interval exchange of four intervals (obtained from an exchange transformation of three arcs of a circle).

While talking about interval exchange transformations it is convenient to use the term "partition" in an extended sense as a decomposition of $[0,1]$ or an interval $\Delta \subset [0,1]$ into intervals with piecewise disjoint interiors. If two successive interiors Δ_i and Δ_{i+1} are mapped to $\Delta_k \cup \Delta_{k+1}$ preserving orientation or to $\Delta_k \cup \Delta_{k-1}$ reversing orientation, then we can lump $\Delta_i \cup \Delta_{i+1}$ together and consider $I_{v,\pi,\epsilon}$ as an exchange of a smaller number of intervals. Thus, without loss of generality we may always assume that $u_1,\dots,u_{n-1}$ are discontinuity points of I and that the partition $\xi := \xi(I)$ is the partition into intervals of continuity of I.

There is an ambiguity in the definition of the map $I_{v,\pi,\epsilon}$ at the points of discontinuity, that is, at $u_1, u_2,\dots,u_{n-1}$. Sometimes there is a natural way to extend the definition to some of those points and obtain a one-to-one map. For example, for $n = 2$, $\pi = (2,1)$, there is only one discontinuity point v_1 inside the interval and if we set $I_{v,\pi}(v_1) = 0$ then by identifying 0 and 1 we obtain the rotation of the circle by the angle $2\pi v_2$. More often, however, such a natural extension is not possible as in the octagon example in Section 4b. A more useful approach is the following. At each point u_i of discontinuity the map $I_{v,\pi,\epsilon}$ has left and right limits, which we will denote by w_i^- and w_i^+, correspondingly. It makes sense to think of the point u_i as having two "ends" and of w_i^- and w_i^+ as the images of those "ends".

Definition 14.5.2. An interval exchange transformation $I = I_{v,\pi,\epsilon}$ is said to have a *saddle connection* if for some $i, j \in \{0, \ldots, n\}$ and for some $k \in \mathbb{N}$ we have $I^k(w_i^+) = u_j$ (correspondingly, $I^k(w_i^-) = u_j$) but $I^l(w_i^+)$ (correspondingly, $I^l(w_i^-)$) are points of continuity for $0 < l < k$. The orbit segment $u_i, w_i^+, I(w_i^+), \ldots, I^{k-1}(w_i^+), u_j$ (correspondingly, $u_i, w_i^-, \ldots, I^{k-1}(w_i^-), u_j$) is called a *connecting segment*. Let us call a permutation π of $\{1, \ldots, n\}$ *irreducible* if it does not preserve any subset of the form $\{1, \ldots, k\}$ for $k = 1, \ldots, n-1$. An open interval $\Delta \subset [0, 1]$ is called *rigid* under I if all positive iterates of I are defined and continuous on Δ. A rigid interval Δ is a *maximal* rigid interval if any other rigid interval is either disjoint from it or contained in it. An interval exchange transformation is called *generic* if it has no rigid intervals. A point $x \in [0, 1]$ is called a *generic point* for I if all positive and negative iterates of x are defined, that is, no image or preimage of x is a discontinuity point.

Evidently we have

Lemma 14.5.3. *The number of different connecting segments for any interval exchange transformation $I_{v,\pi,\epsilon}$ does not exceed $2n - 2$, where n is the number of intervals of continuity.*

Lemma 14.5.4. *Any rigid interval Δ for an interval exchange transformation I consists of periodic points. Any maximal rigid interval either consists of points of the same period or all of its points except for the midpoint have even period $2k$ while the midpoint has period k. Any endpoint of a maximal rigid interval belongs to a connecting segment.*

Proof. It suffices to consider maximal rigid intervals since the union of rigid intervals intersecting a given rigid interval Δ is again a rigid interval and clearly maximal.

For any rigid interval Δ any image $I^k(\Delta)$ is an interval of the same length and if Δ is maximal then $I^k(\Delta) \cap \Delta$ is either empty or equal to Δ. If $I^k(\Delta) \cap \Delta = \varnothing$ for all k then $I^k(\Delta) \cap I^l(\Delta) = \varnothing$ for all $k, l \geq 0$ by invertibility and continuity of I on every image of Δ. But this is impossible since the sum of the lengths of these intervals is infinite. Thus there exists $k \in \mathbb{N}$ such that $I^k(\Delta) = \Delta$. Since I^k preserves Lebesgue measure and is continuous on Δ, it is either the identity transformation or the reflection in the midpoint, which is thus fixed, and I^{2k} is the identity on Δ.

Let x be the left endpoint of Δ. I^k (correspondingly, I^{2k}) must be discontinuous at x because otherwise it would be the identity on a neighborhood of x, contradicting maximality of Δ. The same argument applies to I^{-k} (correspondingly, I^{-2k}). Since discontinuity appears only when the image of a point is one of the points $u_1, \ldots, u_{n-1}$, we see that x belongs to a connecting segment. The same argument applies to the right endpoint of Δ. □

Thus we can associate with any maximal rigid interval its *orbit*, that is, the union of its images. Since for an exchange transformation of n intervals the number of different connecting segments does not exceed $2n - 2$ and the same connecting segment can only be an endpoint for one maximal rigid interval, there are at most $2n - 2$ maximal rigid intervals whose orbits are different.

Corollary 14.5.5. *If an interval exchange transformation has no saddle connections then it is generic.*

Corollary 14.5.6. *On any rigid interval Δ all (positive and negative) iterates of I are defined and continuous.*

The joint partition $\xi^I_{-m} := \xi \vee I^{-1}(\xi) \vee \cdots \vee I^{1-m}(\xi)$ into intervals of continuity of I^k, $k = 1, \dots, m$, consists of at most $m(n-1)+1$ intervals and exactly this many if there are no saddle connections. For the nested sequence ξ^I_{-m}, $m \in \mathbb{N}$, there is an obvious dichotomy:

(1) $\max_{c \in \xi^I_{-m}} l(c) \to 0$ as $m \to \infty$, where $l(\cdot)$ is the length. This happens for generic interval exchange transformations.
(2) There is a nested sequence of elements $c_m \in \xi^I_{-m}$ such that $c_\infty := \bigcap_{m=1}^{\infty} c_m$ is an interval of positive length. Then c_∞ is the closure of a maximal rigid interval and obviously any maximal rigid interval appears in this way.

b. Coding. Let L be the union of the closures of all maximal rigid intervals. The partition ξ offers a natural way of coding the interval exchange transformation on the invariant set $[0,1] \setminus L$. Namely, let

$$\Omega_I := \Big\{ \omega \in \Omega_n \mid \bigcap_{m \in \mathbb{Z}} I^{-m}(\Delta_{\omega_m + 1}) \neq \varnothing \Big\}. \tag{14.5.1}$$

Ω_I is closed (Exercise 14.5.1) and obviously shift invariant. If I is generic then the map $h: \Omega_I \to [0,1]$, $\omega \mapsto \bigcap_{m \in \mathbb{Z}} I^{-m}(\Delta_{\omega_m+1})$ is a finite-to-one continuous surjective map and injective on preimages of generic points. If I has no saddle connection then any nongeneric point has exactly two preimages. Otherwise 2^n is obviously an upper bound for the number of preimages of a point.

If I has rigid intervals then h is not defined on Ω_I. However, it is still defined and continuous on the complement of finitely many periodic orbits corresponding to the orbits of maximal rigid intervals, and its image is $[0,1] \setminus L$. Even though this map cannot be extended to a semiconjugacy between a symbolic system and I, it is a measure-theoretic isomorphism for any nonatomic ergodic shift-invariant measure on Ω_I since both the set of discontinuity and the set of nonuniqueness can only carry atomic ergodic measures.

Notice the analogy to the measure-theoretic classification of circle maps (Theorem 11.2.9) in the nontransitive case.

c. Structure of orbit closures. We will use notions from topological dynamics, such as recurrence, freely in the sequel. Although these were defined for continuous maps this is justified because on one hand these definitions do not require continuity and we do not use properties that do, and on the other hand because we have a symbolic model at hand, where all these notions appear in the standard way.

The construction of the first-return map plays a very important role in the study of interval exchange transformations. This is due to the following lemma which shows that the class of interval exchange transformations is closed under the operation of inducing (taking the return map) on subintervals and the number of intervals increases by at most 2.

Lemma 14.5.7. *Let I be an exchange transformation of n intervals or arcs of the circle. Then for any interval (or arc) Δ the first-return map I_Δ is defined and continuous everywhere except for at most $n+1$ points and is an exchange transformation of $k \le n+2$ intervals.*

Proof. By the Poincaré Recurrence Theorem 4.1.19 the set of points that return to Δ has full Lebesgue measure and is hence dense. Suppose $x \in \Delta$ is such that $y = I_\Delta(x) = I^k(x) \in \operatorname{Int}(\Delta)$ and $I^l(x)$ $(0 \le l < k)$ are points of continuity of I. Then I^k is a local isometry near x and hence maps a neighborhood of x onto a neighborhood of y in Δ. On the other hand, $\min_{1 \le l < k} \operatorname{dist}(I^l(x), \Delta) = \epsilon > 0$, so $I_\Delta = I^k$ in a neighborhood of x. Thus I_Δ is continuous at x, hence I_Δ is defined and continuous on an open set.

Let z be the left endpoint of a maximal interval where I_Δ is defined and continuous. This means that $I_\Delta = I^k$ in a one-sided neighborhood of z for some k. But if $I^l(z)$, $0 \le l \le k$ are points of continuity of I and not endpoints of Δ then $I^k(z) \in \operatorname{Int}\Delta$ and by the previous argument $I_\Delta = I^k$ in a two-sided neighborhood of z, a contradiction. Thus an iterate $I^l(z)$ for $0 \le l \le k$ is either a point of discontinuity of I or an endpoint of Δ. Consider the smallest such l. Each of the $n-1$ points of discontinuity of I and each of the two endpoints of Δ can appear in this way as an iterate of at most one left endpoint of an interval of continuity of I_Δ. Thus I_Δ is defined away from at most $n+1$ points in $\operatorname{Int}\Delta$ and is an isometry on the complementary intervals. □

Remark. The above argument essentially reproves Proposition 14.1.6 in the setting of interval exchange transformations.

Corollary 14.5.8. *Every generic point for an interval exchange transformation is recurrent.*

Now we can prove our first important finiteness result for interval exchange transformations.

Proposition 14.5.9. *Let x be a nonperiodic recurrent point for an interval exchange transformation I. Then the complement of the orbit closure of x consists of finitely many intervals whose endpoints belong to connecting segments.*

Proof. Let $\Delta = (a, b)$ be such a complementary interval. By Lemma 14.5.7 I_Δ is defined at all but finitely many points of Δ, so the right-hand limit $I_\Delta(a^+)$ at a of $I_\Delta(x)$ is well defined. There are two possibilities:

(1) $I_\Delta(a^+) \in \operatorname{Int}\Delta$. Then $I_\Delta(a^+) = I^k(a^+)$ for some minimal $k > 0$. If I^k is continuous at a then I^k is an isometry near a and hence points from the orbit of x that accumulate at a are mapped to points accumulating on $I^k(a) \in \operatorname{Int}\Delta$, a contradiction. Thus $I^l(a)$ is a discontinuity point of I for some $l < k$.

(2) $I_\Delta(a) \in \partial\Delta$, that is, $I_\Delta(a^+) = a$ or $I_\Delta(a^+) = b$. In the first case $I_\Delta = I^k = \mathrm{Id}$ in a right-hand neighborhood of a. If a is a continuity point of I^k then $I^k = \mathrm{Id}$ on a neighborhood of a contrary to the fact that nonperiodic recurrent orbits accumulate on a; hence an iterate of a is a discontinuity point of I. The case $I_\Delta = I(a^+) = b$ reduces to the previous one or (1) by considering I_Δ^2.

Thus we find in all cases that the positive semiorbit of a contains a discontinuity point of I. The same argument and conclusion apply for I^{-1}, so a belongs to a connecting segment. By symmetry the same holds for b. □

Corollary 14.5.10. *Every generic orbit is either periodic or its orbit closure is a finite union of intervals.*

Corollary 14.5.11. *For a generic interval exchange transformation the closure of all but finitely many orbits is a union U of finitely many intervals. The orbit of any generic point in U is dense in U.*

Corollary 14.5.12. *If an interval exchange transformation does not have saddle connections then every generic orbit as well as every semiorbit that does not contain a discontinuity point is dense.*

Note that the latter corollary describes a situation as close to topologically minimal as it could be for an interval exchange. In fact, the symbolic model Ω_I is a minimal set in this case (Exercise 14.5.2).

Let us call the orbit closures consisting of finitely many intervals (of nonzero length) the *transitive components* of the interval exchange transformation I. The interiors of different transitive components are disjoint. Similarly let us call the orbit of a maximal rigid interval a *periodic component.* We thus have found that all points, except maybe those lying on connecting segments, have to belong either to a transitive or to a periodic component. It follows from Lemma 14.5.4 and Proposition 14.5.9 that the boundary of each (transitive or periodic) component consists of complete connecting segments. The number of connecting segments does not exceed $2n - 2$. Each connecting segment may belong to the boundary of at most 2 components. Thus the total number of components does not exceed $4n - 4$. Furthermore in the oriented case each connecting segment comes with two orientations and the boundary of each component must contain at least one positively oriented and one negatively oriented segment, which decreases the possible number of components to $2n-2$. Thus the topological structure of orbits of interval exchange transformations can be summarized as follows:

Theorem 14.5.13. *Let I be an exchange map of n intervals. Then $[0,1]$ splits into a finite union of connecting segments and $k \leq 4n-4$ disjoint open invariant sets each of which is either a transitive or a periodic component and is a finite union of open intervals. If in addition I is oriented then $k \leq 2n-2$.*

d. Invariant measures. The last theorem gives a topological finiteness result for interval exchange transformations. In fact, every transitive component is *quasi-minimal*: Any semiorbit that does not begin or end in a discontinuity point is dense in it. A corresponding measure-theoretic property would be uniqueness of the nonatomic invariant measure on a transitive component. We shall see in the next subsection that this is not the case even under the stronger assumption of no saddle connections. Nevertheless there is the following general finiteness result for invariant measures:

Theorem 14.5.14. *Let I be an exchange transformation of n intervals. Then there are at most n mutually singular invariant nonatomic Borel probability measures for I supported on the union of the transitive components.*[1]

Corollary 14.5.15. *There are at most n distinct invariant nonatomic ergodic Borel probability measures for I.*

Corollary 14.5.16. *For a generic interval exchange there are no more than n disjoint invariant sets of positive Lebesgue measure.*

Proof. Since such a measure is supported on the transitive components, the joint partitions constructed from ξ are dense with respect to the metric $\mathcal{D}$ of (4.3.9), so ξ is a one-sided generator. (See Section 4.3 for a general discussion.) Thus an invariant measure is determined by its values on the elements of the joint partitions ξ_m^I. The key observation is that it is indeed determined by its values on the intervals in ξ. Namely, these determine the measures of the elements of the joint partition $\xi \vee I(\xi)$ as follows: Start from the left endpoint and notice that the first interval of $\xi \vee I(\xi)$ is the shorter of the leftmost interval of ξ and the leftmost image of an interval in ξ, so its measure is determined. The next interval is again the shorter of the remainder of the other interval on the left and the next image, and so forth. Similarly we can proceed by induction to determine from the values of the measure on the intervals of $\xi \vee \cdots \vee I^n(\xi)$ and ξ those on the intervals of $\xi \vee \cdots \vee I^{n+1}$ by superimposing $I(\xi \vee \cdots \vee I^n(\xi))$ on ξ. This defines a map h from the I-invariant measures to the $(n-1)$-dimensional simplex σ in $\mathbb{R}^n$ which is evidently affine, continuous (in the weak* topology), and, as we saw, injective. Let us note that mutually singular measures correspond to linearly independent elements of σ. Namely, if $a_1 h(\mu_1) + \cdots + a_l h(\mu_l) = 0$ then on a set A with $\mu_i(A) > 0$ and $\mu_j(A) = 0$ for $i \neq j$ we have $0 = a_1\mu_1(A) + \cdots + a_l\mu_l(A) = a_i\mu_i(A)$, whence $a_i = 0$. $\square$

Remark. The image of this set of measures is always a simplex and its vertices correspond to ergodic measures.

Let us notice that different invariant measures for an interval exchange generate conjugacies between this and other interval exchanges. To facilitate this discussion assume the interval exchange is topologically transitive, which is weaker than absence of saddle connections but stronger than being generic. In this case every nonatomic invariant measure is positive on open sets, and hence the map taking an interval $[0,t]$ to its measure is a homeomorphism and takes this interval exchange to another one such that the image of the given invariant measure is Lebesgue measure. Thus, if a topologically transitive interval exchange has k ergodic invariant measures then we have a $(k-1)$-simplex of topologically conjugate interval exchanges.

e. Minimal nonuniquely ergodic interval exchanges. We will soon see that the absence of saddle connections which (by Corollary 14.5.12) implies essential minimality, is a typical property in many families of interval exchange transformations. We now give an example of an interval exchange transformation that has no saddle connections and is not uniquely ergodic. Both the method of construction and the result are very similar to those that appear in Corollary 12.6.4. Suppose $s\colon [0,1] \to \{0,1\} = \mathbb{Z}/2$ has three points of discontinuity. Then the extension

$$I_s\colon [0,1]\times\{0,1\} \to [0,1]\times\{0,1\}, \quad I_s(x,i) := (I(x), i+s(x))$$

of an exchange I of m intervals can be viewed as an exchange of the at most $2(m+3)$ intervals obtained from the two copies of the (at most $m+3$) intervals defined by subdividing the intervals of continuity of I, if necessary, at the discontinuity points of s. Our example then becomes a corollary of the following coboundary construction:

Theorem 14.5.17. *Let I be an oriented interval exchange transformation without saddle connections. Then there is a function $s\colon [0,1] \to \{0,1\}$ with three discontinuity points such that*

(i) *$s(x) = h(I(x)) - h(x)$ for some measurable $h\colon [0,1] \to \{0,1\}$, and*
(ii) *for $O \subset [0,1]$ open $\lambda(h^{-1}(\{1\}) \cap O) > 0$, $\lambda(h^{-1}(\{0\}) \cap O) > 0$.*[2]

Thus h is metrically dense (Definition 12.6.2) and s is exactly a wild coboundary, similarly to Proposition 12.6.3. Thus, similarly to Corollary 12.6.4 we obtain our desired example:

Corollary 14.5.18. *The interval exchange given by the extension I_s corresponding to s has no saddle connections and is not ergodic.*

Proof. By (i) of Theorem 14.5.17, s is a measurable coboundary and hence I_s is metrically isomorphic to $I \times \mathrm{Id}$ via $(x,i) \mapsto (x, i+h(x))$, so I_s preserves graph h and graph$(1-h)$, both of positive (product) measure. On the other hand we can see that there are no saddle connections by considering a point w_i^- (as before Definition 14.5.2) and noting that its positive I-semiorbit is dense by Corollary 14.5.12, so by (ii) of Theorem 14.5.17 it is dense for I_s and hence not part of a connecting segment. Points w_i^+ are taken care of similarly and evidently no points outside the positive semiorbit of the $w_i^\pm$ could be part of a connecting segment. □

Proof of Theorem 14.5.17. We will use addition modulo 1, that is, we are allowed to switch + and − signs. Replacing I by I^{-1} we may replace (i) by

(i′) $s(x) = h(I^{-1}(x)) - h(x)$.

We call an interval $\Delta \subset [0,1]$ *k-clear* if I^i is continuous on $\operatorname{Int}\Delta$ for $|i| \leq k$. For $n \in \mathbb{N}_0$ let $a_n := I^n(0)$. Due to the absence of saddle connections this sequence is well defined and by Corollary 14.5.12 dense in $[0,1]$. For $a_n < a_m$ let $\chi_{m,n} = \chi_{[a_n,a_m)}$ be the characteristic function.

We now begin an inductive construction similar to that of Proposition 12.6.3, although more explicit. First let $k_0 = 0$ and $k_1 > 0$ be such that $[a_0, a_{k_1}] \cap [a_1, a_{k_1+1}] = \varnothing$. Inductively we will determine an increasing sequence $\{k_m\}_{m\in\mathbb{N}_0}$ and a related sequence l_m defined by

$$l_0 = 1, \quad l_1 = k_1+1, \quad l_{m+1} = k_{m+1}-k_m+l_{m-1}, \text{ that is, } l_m = 1+\sum_{i=0}^{m-1}(-1)^i k_{m-i}$$

so that

(1) $a_{k_0} < \cdots < a_{k_m} < a_{l_0} < a_{l_2} < \cdots < a_{l_{2[m/2]}} < a_{l_1} < \cdots < a_{l_{2[(m+1)/2]-1}}$,
(2) $[a_{k_m}, a_{k_{m+1}}]$ is k_m-clear,
(3) $a_{k_{m+1}} - a_{k_m} < (a_{k_m} - a_{k_{m-1}})/3k_m$.

To see that (1)–(3) can be satisfied inductively assume they hold up to m. By (1) there is a c such that $a_{k_m} < c < a_{l_0}$ and $[a_{k_m}, c)$ is k_m-clear. By density of $\{a_n\}$ there exists $k_{m+1} > k_m$ such that $a_{k_{m+1}}$ is in the left half of $[a_{k_m}, c)$, showing (2) for $m+1$. Taking $a_{k_{m+1}}$ still closer to a_{k_m} also yields (3). To verify (1) note that $a < \cdots < a_{k_{m+1}} < a_{l_0}$ is already known. Next note (by induction) that $l_m = k_m - l_{m-1} + 2 \leq k_m + 1 \leq k_{m+1}$ and hence $-k_m < l_{m-1} - k_m \leq 0$, so (2) implies that $[a_{l_{m-1}}, a_{l_{m+1}}] = I^{l_{m-1}-k_m}([a_{k_m}, a_{k_{m+1}}])$ is a translate of $[a_{k_m}, a_{k_{m+1}}]$, which yields the remaining inequalities in (1).

To construct s let $s_1(x) := \chi_{0,k_1}(x) + \chi_{1,k_1+1}(x) = \chi_{0,k_1}(x) - \chi_{0,k_1}(I^{-1}(x))$ and

$$\begin{aligned} s_{m+1}(x) :=& s_m(x) + \chi_{k_m,k_{m+1}}(x) - \chi_{l_{m-1},l_{m+1}}(x) \\ =& s_m(x) + \chi_{k_m,k_{m+1}}(x) - \chi_{k_m,k_{m+1}}(I^{k_{m+1}-l_{m+1}}(x)) \\ =& s_m(x) + g_{m+1}(I^{-1}(x)) - g_{m+1}(x), \end{aligned}$$

where $g_{m+1}(x) = \sum_{i=1}^{k_m-l_{m-1}} \chi_{k_m-i,k_{m+1}-i}(x)$. With (1) this yields

$$\begin{aligned} s_m(x) &= \sum_{i=1}^{m} \chi_{k_{i-1},k_i}(x) + \chi_{l_0,l_1}(x) - \sum_{i=1}^{m-1} \chi_{l_{i-1},l_{i+1}}(x) \\ &= \chi_{0,k_m}(x) + \chi_{l_{2[m/2]},l_{2[(m+1)/2]-1}}(x) \to \chi_{[0,b)}(x) + \chi_{[c,d)}(x) =: s(x) \end{aligned}$$

(note that we add mod 2), where $b = \lim_{m\to\infty} a_{k_m}$, $c = \lim_{m\to\infty} a_{l_{2m}}$, $d = \lim_{m\to\infty} a_{l_{2m+1}}$. Thus s has three points of discontinuity. Let us show that s is a coboundary. First, $c_m(x) = h_m(I^{-1}(x)) - h_m(x)$, where

$$h_m(x) := \chi_{0,k_1}(x) + \sum_{i=2}^{m} g_i(x) = \chi_{0,k_1}(x) + \sum_{j=2}^{m} \sum_{i=2}^{k_{j-1}-l_{j-2}} \chi_{k_{j-1}-i,k_j-i}(x).$$

Since $\lambda(g_{m+1}^{-1}(\{1\})) \le (k_m - l_{m-1})(a_{k_{m+1}} - a_{k_m}) < (a_{k_m} - a_{k_{m-1}})/3$ by (3), h_m converges in L^1 to a function h which clearly satisfies (i′).

Next we prove (ii). We call $[a_{k_{m-1}+i}, a_{k_m+i})$ an interval of rank m if $i \in \{0, \dots, k_{m-1} - l_{m-2} - 1\}$. Such intervals are either disjoint or there is an inclusion: Suppose $n \le m$, $i \in \{0, \dots, k_{m-1} - l_{m-2} - 1\}$, $j \in \{0, \dots, k_{n-1} - l_{n-2} - 1\}$, and $a_{k_{m-1}+i} \le a_{k_{n-1}+j} < a_{k_m+i}$. Since $k_{n-1} + j < 2k_{n-1} \le 2k_{m-1}$, applying $I^{-(k_{n-1}+j)}$ gives $a_{k_{m-1}+(j-k_{n-1}-j)} \le 0 < a_{k_m+(j-k_{n-1}-j)}$, so we have equality on the left and hence $a_{k_{m-1}+i} = a_{k_{n-1}+j}$. In particular intervals of a given rank are pairwise disjoint and h_m is constant on every interval of rank m.

For any interval $\Delta \subset [0,1]$ there is an interval $\Delta' \subset \Delta$ of rank m for some m by density of $\{a_m\}$. h_m is constant on Δ' and for $n > m$ (3) yields

$$\begin{aligned}\lambda(\{x \in \Delta' \mid h_n(x) = h_m(x)\}) &\ge \lambda(\Delta') - \sum_{i=1}^{n-m} \lambda(g_{n+i}^{-1}(\{1\})) \\ &\ge \lambda(\Delta')\Big(1 - \frac{1}{3} - \frac{1}{9} - \dots - \frac{1}{3^{n-m}}\Big) \ge \lambda(\Delta')/3\end{aligned}$$

and $\lambda(\{x \in \Delta' \mid h(x) = h_m(x)\}) \ge \lambda(\Delta')/3$. On the other hand for the smallest $m' > m$ such that Δ' contains an interval Δ'' of rank m' the constant value of $h_{m'}$ on Δ'' differs from that of h_m on Δ' while the same argument as before shows that $\lambda(\{x \in \Delta'' \mid h(x) = h_{m'}(x)\}) \ge \lambda(\Delta'')/3$. □

Exercises

14.5.1. *Show that the set Ω_I defined in (14.5.1) is closed in Ω_n.*

14.5.2. *Show that if I has no saddle connections then the shift on Ω_I is minimal.*

14.5.3. *Prove that every interval exchange has zero entropy with respect to any invariant measure.*

14.5.4. *Consider the permutation $\pi = (3,2,1)$. Show that the oriented interval exchange transformation $I_{v,\pi}$ for any vector v can be obtained by inducing a circle rotation on an interval. Prove that in this case minimality implies unique ergodicity and find a necessary and sufficient condition for minimality.*

14.5.5. *Consider an exchange of two arcs on the circle that changes orientation on one arc. Prove that all orbits are periodic.*

14.5.6. *Consider the map $I \times \mathbb{Z}_2$, $(x, j) \mapsto (x + \alpha, j\chi_\beta(x))$, where χ_β is the characteristic function of $[0, \beta]$. Show that there exist α, β such that this map has no saddle connection but is not uniquely ergodic.*

6. Application to flows and billiards

a. Classification of orbits. Corollary 14.1.7 shows that interval exchange transformations appear as (or, more precisely, are smoothly conjugate to) Poincaré maps induced by area-preserving flows on compact surfaces with fixed points of the saddle type. There is also a more general case where the theory of interval exchange transformations applies.

Let $f: [0, 1] \to [0, 1]$ be a piecewise-monotone map that is one-to-one and continuous away from finitely many points. A convenient way to represent such a transformation is as $f = I \circ h$, where h is a homeomorphism and I an interval exchange transformation. Let μ be a nonatomic f-invariant Borel probability measure. Then the map $g: [0, 1] \to [0, 1]$, $g(x) := \mu([0, x])$ is monotone and defines a semiconjugacy between f and an interval exchange transformation, for the factor map preserves Lebesgue measure and has only finitely many discontinuity points. There is an obvious but important case where the semiconjugacy in fact turns out to be a conjugacy.

Proposition 14.6.1. *Any $f = I \circ h$ as above that preserves a measure positive on open intervals is topologically conjugate to an interval exchange transformation.*

Using Poincaré maps for flows yields the following result.

Proposition 14.6.2. *Let φ be a C^0 flow on a closed compact surface defined by a uniquely integrable C^0 vector field X, and τ a transversal to X. Suppose that φ has a finite number of fixed points, which are orbit equivalent to (multiple) saddles or centers. Furthermore suppose that φ preserves a Borel probability measure that is positive on open sets. Then the first-return map induced on τ is topologically conjugate to an interval exchange transformation.*

Proof. By the remark after Proposition 14.1.6 the return map is defined away from finitely many points, which are the last points of intersection of the stable separatrices of a saddle with τ. Assume τ does not pass through a center. The return map preserves the measure ν defined by $\nu(A) := \mu(\bigcup_{t=0}^{\epsilon} \varphi^t(A))/\epsilon$ for any $\epsilon > 0$ that is smaller than the minimal return time for τ. Since $\nu(A) > 0$ for any open $A \subset \tau$, Proposition 14.6.1 applies. $\square$

This result allows us to apply Theorem 14.5.13 to measure-preserving flows on surfaces.

Theorem 14.6.3. *Under the assumptions of Proposition 14.6.2 the surface M splits in a φ-invariant way as $M = \bigcup_{i=1}^{k} P_i \cup \bigcup_{j=1}^{l} T_j \cup C$, where l is at most equal to the genus of M, P_i are open sets consisting of periodic orbits, each T_j is open, every semiorbit in T_j that is not an incoming separatrix of a fixed point is dense in T_j, and C is a finite union of fixed points and saddle connections.*[1]

Corollary 14.6.4. *If in addition φ has no saddle connections it is quasi-minimal, that is, every semiorbit other than the fixed points and separatrices is dense in M.*

Remark. Similarly to the previous section we will call the P_i *periodic components* and the T_j *transitive components* of the flow.

Proof. One can construct a finite family of closed transversals that intersects every semiorbit of φ except for fixed points. Now apply Theorem 14.5.13 to the return map on each transversal and take the images of each periodic and transitive component under the flow. They produce the sets P_i and T_j. Any orbit not included in these sets must be a saddle connection for the flow. It remains to show that $l \leq \text{genus}(M)$. First pick an orbit segment in each of the T_j that almost returns back and close it up to obtain a pretransversal γ_j. Preservation of measure implies that $M \setminus \gamma_j$ is connected. Since the γ_j are disjoint the same argument applies to $M \setminus \bigcup_{j=1}^{l} \gamma_j$, so $l \leq \text{genus}(M)$ (see the remarks after Theorem A.5.2). □

Remark. In the absence of centers one can give estimates on $k+l$. An easy one involves counting the number of incoming separatrices using the Poincaré–Hopf Index Formula (Theorem 8.6.6); a more subtle argument gives $k+l \leq \text{genus}(M)$.

b. Parallel flows and billiards in polygons. The case of best recurrence properties appears when the decomposition of Theorem 14.6.3 contains a single transitive component and no periodic component, that is, when M is a quasi-minimal set for the flow as defined at the end of Section 14.4a. By Corollary 14.6.4 the absence of saddle connections is sufficient for quasi-minimality. In the next section we will show how to parameterize the set of smooth orbit equivalence classes of area-preserving flows in such a way that in the absence of homologically trivial closed orbits most flows have no saddle connection. Right now we will show that in natural one-parameter families of area-preserving flows similar to linear flows on the octagon with slope as a parameter, all but countably many flows have no saddle connection.

First, let us generalize the octagon construction from Section 14.4b. Let $P \subset \mathbb{R}^2$ be a polygon whose angles may be less than, equal to, or greater than π. In other words, P is an ordinary (not necessarily convex) polygon and some of its sides may be artificially subdivided into several pieces. Furthermore assume that the sides of P are divided into pairs of parallel arcs of equal length.

Denote the translations identifying the sides in each pair by $T_1^{\pm 1}, \ldots, T_m^{\pm 1}$. Any centrally symmetric polygon, convex or not, with pairs of opposite sides identified is an example of such an arrangement. Another example is given by the L-shaped polygon with two vertices added on the longer (outer) sides. Identifying the pairs of sides by the translations makes P into a compact closed surface $\tilde{P}$. As in Section 14.4b there is an obvious smooth structure at all points except for those that come from vertices. At these points the smooth structure can also be defined, but it is not essential for describing the orbit structure of the parallel flow. For each value of the angle $\alpha \in [0, 2\pi)$ one can define the parallel flow on $\tilde{P}$ as the motion with unit speed along the oriented lines that form an angle α with the fixed direction (and taking into account the identification). Such a flow is discontinuous; it is defined for all values of time only at points whose orbits never hit a vertex. Since the set of such points has full Lebesgue measure λ it is defined as a measure-preserving flow for all values of time. In order to make Theorem 14.6.3 applicable, multiply the vector field X_α defining the flow by a nonnegative function ρ vanishing precisely at the vertices and such that ρ^{-1} is Lebesgue integrable. The vector field ρX_α is C^0 and integrates uniquely to a C^0 flow preserving the measure $\rho^{-1}\lambda$ which is positive on open sets. The vertices are saddle-type fixed points and the first-return map on any transversal coincides with that for the original discontinuous flow. Denote by $\mathcal{T}$ the group of parallel translations generated by the translations $T_1, \ldots, T_m$. Let $p_1, \ldots, p_{2m}$ be the vertices of P.

Proposition 14.6.5. *If a parallel flow on $\tilde{P}$ has a saddle connection then its direction is parallel to a vector of the form $g + p_j - p_i$ for some $g \in \mathcal{T}$, $i, j = 1, \ldots, 2m$.*

Proof. Suppose there is a saddle connection, that is, an orbit that starts at p_i and ends at p_j. Consider the following process of "unfolding". Start at p_i and each time the orbit reaches a side, instead of applying the appropriate translation to the point, apply its inverse to the entire polygon. This way we obtain a correspondence between the orbit and a segment of the straight line beginning at p_i. Since there is a saddle connection, after finitely many crossings the segment will reach a vertex of a shifted polygon $T\tilde{P}$. Obviously the shift T is a linear combination of basic translations T_k, $k = 1, \ldots, m$, with integer coefficients and the vertex is the translate of p_j. Thus the orbits are parallel to $T + p_j - p_i$. □

Corollary 14.6.6. *For all but countably many values of α the linear flow on $\tilde{P}$ has no saddle connection and is hence quasi-minimal. The same applies to the interval exchange transformation induced by the flow on any straight line segment.*

At the end of Section 14.4 we described a correspondence between the family of parallel flows in the regular octagon and the billiard flow inside the right triangle with an angle $\pi/8$. This construction allows a generalization which we briefly sketch here.

Consider the billiard inside a polygon P whose angles are commensurable with π. We will call such polygons *rational.* Let G be the group of plane motions generated by the reflections in the sides of P. It has a finite-index normal subgroup G_0 of parallel translations and the factor group G/G_0 is isomorphic to a dihedral group D_r; it corresponds to the action of G on the set of directions. In other words, the direction of any billiard orbit after reflection belongs to the same orbit of G/G_0. Now one can pick elements $g_0, \dots, g_{2N-1}$ of G in each coset of G_0. They can be ordered in such a way that $g_0 = \mathrm{Id}$, $g_{m+1} = R_m g_m$, where R_m is one of the reflections in the sides of P generating G. Now we take the $2N$ copies $P, R_1P, R_2R_1P, \dots, R_{2N-1}\cdots R_1P$ of P and identify them using the corresponding reflections. The resulting figure may have overlaps, unlike the case of the 16 triangles in Figure 14.4.4. Nevertheless, for any "free" side of one of the polygons there is a free side of another one which can be identified by a translation from G_0. Thus one can construct a closed surface S and for any orbit of G/G_0 a flow on P lifts to a parallel flow on S which is defined unambiguously despite the presence of several "sheets". Let us consider the billiard-ball map on the set of unit tangent vectors with footpoint on ∂P. A convenient way to study it is to pick a side l and a direction α and consider the return map to $l \times (G/G_0)\{\alpha\}$. The latter is a union of $2N$ intervals and after putting them side by side and appropriately normalizing Lebesgue measure on each of them we obtain a family $I^{(\alpha)}$ of interval exchange transformations. Arguing exactly as in the proof of Proposition 14.6.5 and passing to the section we obtain

Proposition 14.6.7. *For any side l of a rational polygon the interval exchange transformation $I^{(\alpha)}$ has no saddle connection for all but countably many values of α.*[2]

We have seen that Poincaré maps on transversals to measure-preserving flows are isomorphic to interval exchange transformations. There is, in fact, a construction showing that at least in the oriented case any interval exchange transformation appears in this way, that is, for any oriented interval exchange transformation $I_{v,\pi}$ (see Definition 14.5.1) there exists a compact orientable surface M, a smooth area-preserving flow φ with finitely many fixed points of saddle type, and a transversal τ such that the first-return map of φ on τ is smoothly conjugate to $I_{v,\pi}$. Thus in this sense these theories are equivalent.

Exercises

14.6.1. *Construct an example of an area-preserving flow whose fixed points are simple saddles on the sphere with n handles that has n transitive components and no periodic components.*

14.6.2. *Consider the parallel flow in the L-shaped polygon with identifications as described near the beginning of Subsection b. Show that the resulting surface is homeomorphic to the sphere with 2 handles and after an appropriate time change the flow has one topological double saddle.*

14.6.3. *Consider the parallel flow in the L-shaped polygon with identifications as described above. Reduce this to a parallel flow in a polygon and calculate the genus of the resulting surface.*

7. Generalizations of rotation number

a. Rotation vectors for flows on the torus. In the discussion of circle diffeomorphisms in Chapter 12 we saw that the possibility of a smooth conjugacy to a linear map is related to the arithmetic properties of the rotation number of the circle diffeomorphism. As is to be expected, the situation is similar for flows on the torus. This is one of many motivations for developing a notion corresponding to rotation number. As in the definition of the rotation number of circle diffeomorphisms we need to pass to the universal cover of the space. However, there is an interesting distinction. For the circle the choice of generator in the first homology group is unique up to an orientation, but the lift of a map to the universal cover is defined up to a deck transformation and the latter factor is responsible for the rotation number being defined only modulo 1. For a vector field on the torus (or any manifold) the lift to the universal cover is unique, but the choice of generators in $H_1(\mathbb{T}^2, \mathbb{Z})$ is not. Accordingly we will be able to define a rotation *vector* for any given choice of basis in $H_1(\mathbb{T}^2, \mathbb{Z})$ and thus it is determined up to an action of $SL(2, \mathbb{Z})$.

Proposition 14.7.1. *Let φ^t be a fixed-point-free C^1 flow on $\mathbb{T}^2$ and denote the lift of φ^t to the universal cover $\mathbb{R}^2$ by Φ^t. Then for every $x \in \mathbb{R}^2$ the limit*

$$\rho(\varphi) := \lim_{t\to\infty} \frac{1}{t}\Phi^t(x) \in \mathbb{R}^2 \tag{14.7.1}$$

exists and is independent of x. We will call it the rotation vector of φ.

Proof. First notice that existence and independence of x of $\rho(\varphi)$ is an invariant of flow conjugacy. Let $h: \mathbb{T}^2 \to \mathbb{T}^2$ be a homeomorphism and $H = L + G$ its lift to the universal cover $\mathbb{R}^2$, where L is a linear map and G is periodic. Let $x \in \mathbb{R}^2$ and $y = H^{-1}(x)$. Then

$$\frac{1}{t}H(\Phi^t(H^{-1}(x))) = L\Big(\frac{1}{t}\Phi^t(y)\Big) + \frac{1}{t}G(\Phi^t(y)).$$

$G(\Phi^t(y))$ is bounded, so $\lim_{t\to\infty} H(\Phi^t(H^{-1}(x)))/t = L\lim_{t\to\infty}\Phi^t(y)/t$.

Using Proposition 14.2.1 we construct a closed transversal τ to the flow φ^t. By Proposition 14.2.2 we obtain a C^1 diffeomorphism $h: \mathbb{T}^2 \to \mathbb{T}^2$ which maps τ into the standard "horizontal" circle $\tau_0 := S^1 \times \{0\} = \{(s, 0) \mid s \in \mathbb{R}/\mathbb{Z}\}$ (we use additive notation). We will show existence of the limit (14.7.1) for the flow $h \circ \varphi^t \circ h^{-1}$. Since every point returns to τ_0 and the return time to τ_0 is bounded it is sufficient to show existence of the limit for points on τ_0. Furthermore by the same reason it is sufficient to consider only the sequence of moments $t_n(s)$ of returns to τ_0. Denote the return map to τ_0 by f and its lift to $\mathbb{R} \times \{0\}$ by F. Notice that on the universal cover a return to τ_0 corresponds to a change of the second coordinate by 1 or -1; without loss of generality we consider the first case. Then $\Phi^{t_n(s)}(s,0) = (F^n(s), n)$. Notice that $t_n(s) = t(s) + t(f(s)) + \cdots + t(f^{n-1}(s))$, where $t(s)$ is the return time to τ_0. We use existence of the rotation number for s, that is, $\lim_{n\to\infty} F^n(s)/n = \tau(f)$, and unique ergodicity of F, which implies $\lim_{n\to\infty} t_n(s)/n = \int t(s)\, d\mu =: t_0$, where μ is the unique invariant Borel probability measure for F. Then

$$\lim_{n\to\infty} \frac{\Phi^{t_n(s)}(s,0)}{t_n(s)} = \lim_{n\to\infty} \Big(\frac{n}{t_n(s)} \cdot \frac{F^n(s)}{n}, \frac{n}{t_n(s)}\Big) = \Big(\frac{\tau(f)}{t_0}, \frac{1}{t_0}\Big).$$

□

Now we can reformulate Corollary 14.2.7 without referring to a section.

Corollary 14.7.2. *Let φ^t be a C^∞ flow preserving an area element. If the coordinates of the rotation vector $\rho(\varphi)$ are Diophantine numbers then φ^t is C^∞ conjugate to a linear flow.*[1]

b. Asymptotic cycles.[2] Now consider a more general situation. Let M be a compact differentiable manifold and φ^t a C^1 flow generated by the vector field X and preserving a measure μ. Let ω be a closed differential 1-form. Then the integral

$$\int X \lrcorner \omega \, d\mu$$

in fact depends only on the cohomology class of ω, for if $\omega_2 - \omega_1 = dF$ then $X \lrcorner (\omega_2 - \omega_1) = \pounds_X F$ and

$$\int X \lrcorner \omega_2 \, d\mu - \int X \lrcorner \omega_1 \, d\mu = \int \pounds_X F \, d\mu = \Big(\frac{d}{dt} \int F \circ \varphi^t \, d\mu\Big)_{|t=0} = 0$$

due to the preservation of μ. Thus the map $\omega \mapsto \int X \lrcorner \omega \, d\mu$ defines a linear functional on the first de Rham cohomology group of M which by duality can be identified with an element $\rho_\mu \in H_1(M, \mathbb{R})$, which is called the *asymptotic cycle* of the flow with respect to the measure μ.

Suppose now that μ is ergodic. In this case we can give a geometric interpretation of the asymptotic cycle. By the Birkhoff Ergodic Theorem 4.1.2

$$\lim_{t\to\infty} \frac{1}{t} \int_0^t \omega(X(\varphi^s(x)))\, ds = \int X \lrcorner \omega \, d\mu \tag{14.7.2}$$

for μ-a.e. $x \in M$. Let $\gamma_t(x)$ be the oriented orbit segment from x to $\varphi^t(x)$. By definition of integration of differential forms $\int_0^t \omega(X(\varphi^s(x)))\, ds = \int_{\gamma_t(x)} \omega$. Now we proceed similarly to the construction of homotopical entropy for a flow at the end of Section 3.1. Namely, pick a family of arcs $\gamma_{x,y}$ of bounded length

Replace $\gamma_t(x)$ by the closed loop $\tilde{\gamma}_t(x) := \gamma_t(x) \cdot \gamma_{y,x}$. From (14.7.2) and (14.7.3) we obtain

$$\lim_{t\to\infty} \frac{1}{t} \int_{\tilde{\gamma}_t(x)} \omega = \int X \lrcorner \omega \, d\mu.$$

Since any homology class in $H_1(M, \mathbb{R})$ is uniquely determined by its values on a basis of closed 1-forms we can deduce that for μ-a.e. $x \in M$

$$\rho_\mu = \lim_{t\to\infty} \frac{1}{t} [\tilde{\gamma}_t(x)],$$

where $[\cdot]$ denotes the homology class.

Note that for a uniquely ergodic flow all the above a.e. convergences are uniform. It is not difficult to see (Exercise 14.7.1) that the rotation vector for a fixed-point-free flow on the two-torus is simply a coordinate representation of the asymptotic cycle with respect to the standard basis in the first cohomology group.

In the two-dimensional orientable case which is currently our prime concern, we can also interpret the asymptotic cycles as elements of the first *cohomology group*. In general, for flows on n-dimensional oriented manifolds the corresponding elements belong to the $(n-1)$st cohomology group. Namely, a vector field X and an invariant measure μ define a flux *current*, an object similar to closed $(n-1)$-forms, which can be integrated over $(n-1)$-submanifolds. If τ is an $(n-1)$-dimensional oriented transversal to X and $A \subset \tau$ a Borel subset then the flux $\mathcal{F}(A)$ is defined as $\pm\mu(\bigcup_{t=0}^{\epsilon} \varphi^t(A))/\epsilon$, where φ^t is the flow generated by X, $\epsilon > 0$ is any small number, and the sign is determined according to the agreement of the orientation of M and the orientation obtained from the orientation of X and that on τ. To simplify the discussion we consider the special case when the measure μ is given by the volume Ω and that the vector field is C^1. Then the flux current is defined by integrating the $(n-1)$-form $X \lrcorner \Omega$ called the *flux form*.

Lemma 14.7.3. *The flux form is closed if and only if the vector field X preserves Ω.*

Proof. By (A.3.3) and $d\Omega = 0$ we have $\mathcal{L}_X \Omega = d(X \lrcorner \Omega)$. □

Any closed $(n-1)$-form ω determines a linear functional l_ω in $H^1(M, \mathbb{R})$ (that is, on $H_1(M, \mathbb{R})$) via $l_\omega(\alpha) = \int_M \omega \lrcorner \alpha$. Applying this to our case $\omega = X \lrcorner \Omega$ we obtain

$$l_{X \lrcorner \Omega}(\alpha) = \int_M (X \lrcorner \Omega) \wedge \alpha = \int_M (X \lrcorner \alpha) \Omega = \int_M (X \lrcorner \alpha) \, d\mu = \rho_\mu(\alpha).$$

Next we will show how in the two-dimensional area-preserving case the asymptotic cycle can be extended to an invariant giving a complete local (in the space of vector fields) classification up to smooth orbit equivalence.

c. Fundamental class and smooth classification of area-preserving flows.[3] We say that a zero p of an area-preserving vector field on a surface is a *generic saddle of index* $-n$ (or a *generic* n*-fold saddle*) if in local coordinates near p the vector field is Hamiltonian with Hamiltonian function $H(x,y) = \prod_{i=1}^{n+1}(\alpha_i x - \beta_i y) + R(x,y)$, where all ratios β_i/α_i are different and R has zero $(n+1)$-jet at 0. Thus any standard linear saddle is a generic 1-fold saddle and the saddle shown at the left of Figure 8.4.1 is a generic threefold saddle.

Consider a closed compact orientable surface M of genus g, a smooth 2-form Ω on M, $p_1, \dots, p_r \in M$, and $n_1, \dots, n_r \in \mathbb{N}$ such that $\sum_{i=1}^{r} n_i = 2g-2$. Let $\mathcal{X}$ be the space of C^∞ vector fields on M preserving the form Ω such that the point p_i is a generic saddle of index $-n_i$ for $i \in \{1, \dots, r\}$ and there are no other zeros. This description of critical points agrees with the Poincaré–Hopf Index Formula (Theorem 8.6.6). We call $\Delta := \{p_1, \dots, p_r\}$ the *critical set* and consider the space of 1-cycles on M with real coefficients *relative to* Δ. Such a cycle can be represented as a linear combination of oriented arcs in M whose boundaries belong to Δ. Relative boundaries are the same as ordinary boundaries. Thus the space of relative cycles factored by relative boundaries has dimension $2g+r-1$; one needs to add to $2g$ independent cycles a collection of arcs connecting the points in Δ and forming a tree.

The restriction of the flux form for any vector field $X \in \mathcal{X}$ to the space of relative cycles is called the *fundamental class* of X and will be denoted by $FC(X)$. The first classification result for area-preserving vector fields on orientable surfaces of genus $g \geq 2$ can be summarized as follows.

Theorem 14.7.4. *Suppose $X_t \in \mathcal{X}$, $0 \leq t \leq 1$, is a smooth family such that $FC(X_t) = \lambda_t FC(X_0)$, where λ_t is a positive scalar. Then there exists a family $h_t \colon M \to M$ of Lipschitz homeomorphisms that are C^∞ diffeomorphisms away from the critical set, and a positive function μ_t such that $h_t(p_i) = p_i$ for $i = 1, \dots, r$ and $(h_t)_* X_0 = \mu_t X_t$. In other words, h_t effects a Lipschitz orbit equivalence between the flows generated by X_0 and X_t that is C^∞ away from the critical set.*

Proof. Not surprisingly we will use a version of the "homotopy trick" that was first used in the proof of the Moser Theorem 5.1.27 and then appeared several more times. Consider the one-form $\omega_t = X_t \lrcorner \Omega$ and let $\alpha_t := \dfrac{d\omega_t}{dt}$. We will look for the infinitesimal generator $H_t := \dfrac{dh_t}{dt}$ of the family h_t. Suppose we found h_t such that $h_t^* \omega_t = \omega_0$. Then if $h_t^* \Omega = \lambda_t \Omega$ we have

$$X_0 \lrcorner \Omega = h_t^*(X_t \lrcorner \Omega) = (h_t)_* X_t \lrcorner h_t^* \Omega = \lambda_t^{-1} (h_t)_* X_t \lrcorner \Omega,$$

that is, $(h_t)_* X_t = \lambda_t X_0$. By (A.3.3) we have

$$\frac{d}{dt}(h_t^* \omega_t) = h_t^* \mathcal{L}_{H_t} \omega_t + h_t^* \alpha_t = h_t^*(d(H_t \lrcorner \omega_t) + \alpha_t),$$

so we want to find a vector field H_t for which the right-hand side of this identity vanishes.

By assumption α_t is an exact form, that is, $\alpha_t = dP_t$, where the function P_t is defined up to a constant. Furthermore, since α_t vanishes on relative cycles $P_t(p_i) = P_t(p_j)$ for $i, j = 1, \ldots, r$, we can assume that $P_t(p_i) = 0$ for all i. Thus it suffices to solve

$$H_t \lrcorner \omega_t = -P_t. \tag{14.7.4}$$

It follows from the definition that $\ker \omega_t = X_t$. The solution of (14.7.4) is defined up to a term from $\ker \omega_t$. Fixing a Riemannian metric we define the solution uniquely by making it orthogonal to X_t and such that (X_t, H_t) is a positively oriented pair. Naturally $H_t(p_i) = 0$. Thus H_t is defined and continuous and C^∞ away from the zeros of X_t. At the point p_i the form w_t has zeros of order n_i and the same is true for α_t since the Taylor coefficients are differentiable in t. Hence P_t has zeros of order $n_i + 1$ and H_t chosen along the gradient lines of p_i decreases near p_i in proportion to the distance to p_i. Hence the H_t are Lipschitz vector fields and H_t is uniquely integrable to a one-parameter family of Lipschitz homeomorphisms that are smooth away from the critical set and define orbit equivalences between X_t and X_0. □

Remark. The source of nonsmoothness at the critical set is the presence of local invariants of smooth orbit equivalence near multiple saddles. Namely, an n-fold saddle has $2n+2$ separatrices and if two such saddles are smoothly orbit equivalent then the tangent directions at the separatrices must be carried to each other by the derivative of the conjugacy. Thus we have to consider the action of $GL(n, \mathbb{R})$ on $(n+1)$-tuples of lines. This action is transitive for $n \leq 2$, but there are invariants (cross-ratios) for $n \geq 3$.

In fact, if all saddles are no more than double then in Theorem 14.7.4 smooth orbit equivalence can be achieved. First notice that if the vector fields are identical near the critical set then the resulting conjugacy will be the identity nearby as well. Thus one can first find a local coordinate change near each critical point that brings the saddle into a standard form. For the case of a simple saddle this is a continuous-time counterpart of Exercises 6.6.4–6.6.5. Making a time change and carefully applying the Moser Theorem 5.1.27 we reduce to the situation of flows that are identical near the critical set.

Proposition 14.7.5. *Consider an area-preserving vector field on a surface with finitely many fixed points of the saddle type. Then invariant measures supported on transitive components are determined uniquely by their asymptotic cycles.*

Remark. This is an analog of Theorem 14.5.14 for flows.

Proof. By Theorem 14.6.3 it suffices to show that the flux through a small transversal inside a transitive component is determined by the asymptotic cycle of the measure, that is, by the fluxes through closed curves. To that end we take, using Theorem 14.6.3, an orbit segment starting very close to one end of the transversal and ending very close to the other end, and close it using the piece of transversal. The fluxes through these closed curves coincide with those through their transverse portion and on the other hand each of them and hence their limit, which is also the flux through the transversal, is indeed determined by the asymptotic cycle. □

Theorem 14.7.6. *There are at most* genus(M) *nontrivial ergodic invariant measures for any area-preserving vector field on a surface* M.[4]

Proof. We begin by showing that the asymptotic intersection number (see the remarks after Theorem A.5.2) of any two dense orbits for such a vector field is zero. To that end take a transversal and for any two orbits consider segments beginning and ending on the transversal, and intersecting the transversal n times. We refer to the construction of the asymptotic cycle by closing orbits by transverse pieces and close these two orbit segments by the pieces of the transversal connecting their ends. The lengths of the resulting curves then are of order n each. They can intersect each other no more than $2n$ times, namely, on the transversal, so their intersection number is of order $2n/n^2$ after normalizing by length, whence the limit is indeed zero. Thus all dense orbits lie in a g-dimensional Lagrangian subspace for the (symplectic) intersection form (see the remarks after Theorem A.5.2).

Now nontrivial flow-invariant measures are determined by their asymptotic cycle, that is, we have an injective map from ergodic measures to asymptotic cycles. It is affine and hence mutually singular measures correspond to linearly independent points as in the proof of Theorem 14.7.6; hence the image lies in this g-dimensional subspace, so there are at most g distinct ergodic measures. □

Exercises

14.7.1. *Show that the rotation vector for a fixed-point-free flow on the two-torus is the coordinate representation of the asymptotic cycle with respect to the standard basis in the first cohomology group.*

14.7.2*. *Consider the double torus and pick a standard basis* $(\gamma_1, \gamma_2, \gamma_3, \gamma_4)$ *in the first homology group that consists of generators of the homology of one of the joined tori. Pick two points* p, q *and join them by a short curve* γ_5. *Show that for any* (x_1, x_2, x_3, x_4) *with positive coordinates there exists an* $\epsilon > 0$ *such that for* $|x_5| < \epsilon$ *there exists a* C^∞ *area-preserving flow on the double torus with simple saddles at* p *and* q, *and the flux through* γ_i *is* x_i.

14.7.3. *Under the assumptions of the previous exercise show that if* $(x_1, \dots, x_5)$ *are rationally independent then the resulting flow is quasi-minimal.*

14.7.4*. *Show that among the flows constructed in the previous exercises there are quasi-minimal nonergodic ones.*

15

Continuous maps of the interval

Continuous maps of the interval provide an ideal situation in which one can develop a structural theory based on the idea of coding and semiconjugacies with topological Markov chains. The intermediate value theorem allows us to deduce all needed information from data on how subintervals of I are contained in images of other subintervals. This approach yields remarkably precise results concerning topological entropy, the growth of periodic orbits, the presence of orbits of various periods, and the structure of maps with zero topological entropy. Later we introduce a coding-type machinery which goes beyond topological Markov chains and provides a sufficient collection of models up to an almost invertible semiconjugacy for piecewise-monotone maps.

1. Markov covers and partitions

Consider a continuous map $f: I \to I$. We say that $J \subset I$ *covers* $K \subset I$ (under f) if $K \subset f(J)$ and denote this situation by $J \to K$. If $J \to K$ then the covering takes place in the obvious way:

Lemma 15.1.1. *If J, K are intervals, K is closed, and $J \to K$ then there exists a closed interval $L \subset J$ such that $f(L) = K$.*

Proof. Let us write $K = [a, b]$. Then we let $c := \max f^{-1}(\{a\})$ and take $L = [c, d]$ with $d := \min((c, \infty) \cap f^{-1}(\{b\}))$, if this is defined. Otherwise $L = [c', d']$ with $c' := \max((-\infty, c) \cap f^{-1}(\{b\}))$ and $d' := \min((c', \infty) \cap f^{-1}(\{a\}))$ is as desired. □

Thus if $J \to K$ then there are several intervals $L_1, \dots, L_k \subset J$ with pairwise disjoint interiors such that $f(L_i) = K$. Sometimes we compactly write this as $J \rightrightarrows K$ with k arrows, if k is the maximal number of such subintervals L_i. These L_i are called *full components* associated to $J \to K$. Note that the preimage of K in J may contain infinitely many intervals, even though there are only finitely many full components by compactness.

The next two lemmas provide a connection between this covering relation and periodic points.

Lemma 15.1.2. *If $J \to J$ then f has a fixed point $x \in J$.*

Proof. If $J = [a, b]$ then $J \to J$ implies that there are $c, d \in J$ such that $f(c) = a \leq c$ and $f(d) = b \geq d$, so by the Intermediate Value Theorem $f(x) - x$ has a zero in J. □

Lemma 15.1.3. *If $I_0 \to I_1 \to I_2 \to \cdots \to I_n$ then $\bigcap_{i=0}^{n} f^{-i}(I_i)$ contains an interval Δ_n such that $f^n(\Delta_n) = I_n$.*

Proof. Take a full component Δ_1 associated to $I_0 \to I_1$ and then a full component $\delta_2 \subset I_1$ associated to $I_1 \to I_2$, and let Δ_2 be the full component associated to $\Delta_1 \to \delta_2$. Continue inductively to obtain Δ_n. □

By virtue of Lemma 15.1.2 this has the

Corollary 15.1.4. *If $I_0 \to I_1 \to I_2 \to \cdots \to I_{n-1} \to I_0$ then there exists $x \in \text{Fix}(f^n)$ such that $f^i(x) \in I_i$ for $0 \leq i < n$.*

Note that n is not necessarily the prime (minimal) period.

Let us consider a collection $\mathcal{C} = \{I_1, \ldots, I_n\}$ of closed subintervals of I with pairwise disjoint interiors. The relation "$\to$" then yields the edges of a directed graph, the *Markov graph*, associated to $\mathcal{C}$, whose vertices are the intervals in $\mathcal{C}$. Let A be the 0-1 matrix determined by the graph (see Section 1.9c) and σ_A^R the one-sided topological Markov shift defined by A. We will say that A is *associated* to the collection $\mathcal{C}$. The following fact establishes the basic coding on "horseshoe-type" components for one-dimensional maps.

Theorem 15.1.5. *Let $\mathcal{C} = \{I_1, \ldots, I_n\}$ be a collection of pairwise disjoint closed subintervals of I, $J := \bigcup I_i$, and A the matrix associated to $\mathcal{C}$. Then there exists a closed f-invariant subset $S \subset J$ such that σ_A^R is a factor of $f\restriction_S$ via a semiconjugacy $h \colon S \to \Omega_A^R$. There are at most countably many points with more than one preimage under h and the preimages of these points are intervals.*

Proof. We first obtain S. Denote by S_1 a union of full components, one each for every arrow in the Markov graph. This is a pairwise disjoint collection of intervals each of which maps onto an interval of $\mathcal{C}$. We now argue similarly to the proof of Lemma 15.1.3, but make sure choices are made coherently for successive n. For $n \in \mathbb{N}$ we inductively define a set S_n as follows: By our induction assumption S_{n-1} consists of intervals that under f^{n-1} cover intervals of $\mathcal{C}$. Consider now the intervals of S_1. Being contained in intervals of $\mathcal{C}$, we can define full components associated with the coverings of intervals of S_1 by intervals of S_{n-1} under f^{n-1}. Taking one full component associated to each arrow of the Markov graph we obtain a set S_n which is again as in the induction assumption. Note that by construction $S_n \subset S_{n-1}$, so the intersection S of the compact sets S_n is nonempty if all S_n are. The connected components of S are closed intervals or possibly isolated points. By construction there is an obvious map to Ω_A^R that is constant on each component of S and defines the desired semiconjugacy. □

Corollary 15.1.6. *For each periodic orbit ω of σ_A^R in Ω_A^R there is a periodic orbit of the same period in the preimage $h^{-1}(\{\omega\})$ of ω and hence $P_n(f) \geq P_n(f\restriction_S) \geq P_n(\sigma_A^R)$.*

Proof. This is obvious if $h^{-1}(\{\omega\})$ is a point. If it is an interval use Lemma 15.1.2. □

Remark. The construction of the invariant Cantor set for the quadratic map f_λ in Section 2.5b is an example of the above procedure. In that case, with $\mathcal{C} = \{\Delta^0, \Delta^1\}$, each arrow corresponds to a single full component due to the monotonicity of f_λ on Δ^0 and Δ^1, and all components of S are points due to the hyperbolicity of f_λ on Δ^0 and Δ^1. Hence S is uniquely defined and the semiconjugacy is a conjugacy.

The uniqueness of S in the quadratic case is an instance of the following situation:

Corollary 15.1.7. *If under the assumptions of Theorem 15.1.5 f is monotone on each interval of $\mathcal{C}$ then S is uniquely defined and $S = \bigcap_{n\in\mathbb{N}} f^{-n}(J)$.*

Proof. There is only one full component per arrow at each stage. □

In the next chapter we will give a condition for C^1 maps generalizing the situation of the quadratic examples which ensures that every component of S is a point (Corollary 16.1.2).

Corollary 15.1.8. *Under the assumptions of Theorem 15.1.5 the topological entropy of $f\restriction_{\bigcap_{n\in\mathbb{N}} f^{-n}(J)}$ is at least $h_{\text{top}}(\sigma_A^R)$, where A is the matrix associated to $\mathcal{C}$. In particular $h_{\text{top}}(f) \geq h_{\text{top}}(\sigma_A^R)$.*

Proof. This follows from Proposition 3.1.7(1) and Proposition 3.1.6. □

We now turn to the situation of a collection $\mathcal{C}$ of intervals that are not entirely disjoint, namely, by possibly having endpoints in common. In this case there is a difficulty in constructing a semiconjugacy as before since itineraries of points are not uniquely defined. The example of tent maps (Exercise 2.4.1) shows that the map itself may indeed be a (proper) factor of the Markov chain. In general, there may be no semiconjugacy in either direction at all (see Section 2.5a). However, the identifications that occur are sufficiently benign to yield the same conclusions about entropy and growth of periodic points.

Theorem 15.1.9. *Let $f\colon [0,1] \to [0,1]$ be continuous, $\mathcal{C} = \{I_1, \dots, I_m\}$ a collection of closed subintervals of I with pairwise disjoint interiors, $J := \bigcup I_i$, and A the 0-1 matrix associated to $\mathcal{C}$. Then there exists a closed f-invariant subset $S \subset J$ such that $(f\restriction_S, S)$ and (σ_A^R, Ω_A^R) have a common topological factor (σ, X) with $h_{\text{top}}(\sigma) = h_{\text{top}}(\sigma_A^R)$ and the exponential growth rate $p(\sigma)$ of periodic points of σ coincides with $p(\sigma_A^R)$.*

Proof. Construct S as in the proof of Theorem 15.1.5. Since the intervals in $\mathcal{C}$ may have boundary points in common, there is no canonical map to Ω_A^R.

Let X be the space obtained from σ_A^R by identifying two sequences if they are itineraries of the same point in S. Then the shift σ_A^R and $f\restriction_S$ both naturally project to (σ, X). Notice that the semiconjugacy $g: \Omega_A^R \to X$ is injective outside a countable set, namely, the itineraries of points in the backward orbit of boundary points. Notice further that in the Variational Principle (Theorem 4.5.3) it suffices to consider nonatomic measures, since purely atomic measures have zero entropy. But g establishes a bijective correspondence between nonatomic invariant measures for σ_A^R and σ, so $h_{\text{top}}(\sigma) = h_{\text{top}}(\sigma_A^R)$ by the Variational Principle, Theorem 4.5.3.

We trivially have $p(\sigma) \le p(\sigma_A^R)$. To get the reverse inequality note that g only identifies periodic orbits of σ_A^R if they are itineraries of common boundary points (which must then be periodic) of intervals in $\mathcal{C}$ and that there are at most two possible *admissible* itineraries corresponding to an endpoint. □

A particularly useful simple application of this entropy estimate concerns horseshoes.

Definition 15.1.10. If $I \subset \mathbb{R}$ is a closed interval, $f: I \to \mathbb{R}$ continuous, and $a < c < b \in I$, then we say that $[a, b]$ is a *horseshoe* for f if $[a, b] \subset f([a, c]) \cap f([c, b])$.

The presence of a horseshoe clearly produces a full two-shift σ_2^R as a factor of the restriction of f to an invariant set, so we have

Corollary 15.1.11. *If $f: I \to R$ has a horseshoe then $h_{\text{top}}(f) \ge \log 2$.*

Another immediate application of these techniques is an easy proof of the following special case of the Sharkovsky Theorem 15.3.2.[1]

Proposition 15.1.12. *Suppose $f: [0, 1] \to [0, 1]$ has a periodic point of period 3. Then f has periodic points of all periods.*

Proof. Consider the period-3 orbit and label its points by $\{x_1 < x_2 < x_3\}$. Consider the intervals $I_1 = [x_1, x_2]$ and $I_2 = [x_2, x_3]$. Suppose $f(x_2) = x_3$. Then $f^2(x_2) = x_1$ and hence I_2 f-covers both I_1 and I_2, and I_1 covers I_2. If $f(x_2) = x_1$ we relabel I_1 and I_2 to get the same conclusion. Thus the Markov graph associated to I_1, I_2 contains the graph

$$I_1 \rightleftarrows \underset{\circlearrowleft}{I_2} \tag{15.1.1}$$

and for any $n \in \mathbb{N}$ we have a loop $I_1 \to I_2 \to I_2 \to \cdots \to I_2 \to I_1$ (with $n - 1$ occurrences of I_2) which, by Corollary 15.1.4, gives a periodic point of period n which evidently cannot have any smaller period. □

In the exercises we will consider a generalization of the tent map that provides a standard model for a given Markov graph.

Definition 15.1.13. A map $f: [0,1] \to [0,1]$ is called a *standard map* (sometimes also *horseshoe map*) if there exists $m \in \mathbb{N}$ such that:

(1) The set $\{0, 1/m, 2/m, \dots, 1\}$ is forward f-invariant.
(2) On each interval $I_k := [k/m, k+1/m]$ the map is nonconstant and linear.
(3) The Markov graph of such an f with respect to the collection $\mathcal{C} = \{I_k \mid 0 \le k < m\}$ contains no "periodic traps", that is, loops with no arrow that leaves the loop.

We let A be the Markov matrix associated to the Markov graph of such an f with respect to the collection $\mathcal{C} = \{I_k \mid 0 \le k < m\}$. Note that by (1) the number of arrows leaving I_k is also the absolute value of f' on I_k.

Exercises

15.1.1. *Show that any standard map f is a factor of the topological Markov chain obtained from the Markov graph of f. The number of preimages of every point exceeds one for at most countably many points.*

15.1.2. *Suppose f is a standard map with Markov matrix A and with variation $V(f)$. Show that $\lim_{n\to\infty}(1/n)\log V(f^n) = h_{\text{top}}(\sigma_A^R)$.*

15.1.3. *Suppose $\mathcal{C} = \{I_1, \dots, I_n\}$ is a collection of closed intervals with pairwise disjoint interiors and that $J := \bigcup_{i=1}^n I_i$ is an interval. Show that there exists a closed f-invariant set $S \subset J$ such that $f\restriction_S$ has a standard map as a factor.*

2. Entropy, periodic orbits, and horseshoes

We noticed in Section 3.2e that in many cases the exponential growth rate $p(f)$ of periodic orbits is equal to the topological entropy $h_{\text{top}}(f)$. In Section 18.5 we will show that the equality of $p(f)$ and $h_{\text{top}}(f)$ is a general feature of dynamical systems with hyperbolic behavior. For more general classes of systems periodic orbits may not be isolated so one can only expect an inequality $p(f) \ge h_{\text{top}}(f)$. Our principal aim now is to show that for interval maps topological entropy is indeed a lower bound for the growth of periodic orbits. Moreover we show that entropy is approximated by the entropy of invariant sets like those in Theorem 15.1.5. Let us emphasize that in the sequel the word "interval" may mean a single point or an open, half-open, or closed interval.

Theorem 15.2.1. *Suppose f is a continuous interval map. Then for $n \in \mathbb{N}$ there exist*

(1) *an interval J_n,*
(2) *a collection $\mathcal{D}_n$ of pairwise disjoint subintervals of J_n, and*
(3) $k_n \in \mathbb{N}$

such that

$$\lim_{n\to\infty} \frac{1}{k_n} \log \operatorname{card} \mathcal{D}_n = h_{\text{top}}(f) \text{ and } J_n \subset f^{k_n}(I) \text{ for all } I \in \mathcal{D}_n.$$

This result may be a bit technical, but it immediately gives the advertised relation between periodic points and topological entropy:

Corollary 15.2.2. *For every continuous map f of an interval $p(f) \geq h_{\text{top}}(f)$.*

Proof. By Lemma 15.1.2 each $I \in \mathcal{D}_n$ contains a point of period k_n. □

Furthermore we can infer that a map with positive topological entropy must have an iterate with a horseshoe (see Definition 15.1.10):

Corollary 15.2.3. *Suppose f is a continuous interval map and $h_{\text{top}}(f) > 0$. Then there is a $k \in \mathbb{N}$ such that f^k has a horseshoe.*

Proof. Take n such that $\operatorname{card} \mathcal{D}_n > 1$ and let $I_1 \neq I_2 \in \mathcal{D}_n$. Then $I_i \to J_n$ under f^{k_n} for $i = 1, 2$. This means that J_n is a horseshoe. □

This in turn yields:

Corollary 15.2.4. *Suppose f is a continuous interval map and $h_{\text{top}}(f) > 0$. Then there is a periodic point whose period is not a power of 2.*

Proof. Take n as in the previous proof and recall that in a horseshoe all periods occur. □

Thus maps whose periodic points all have powers of 2 as periods must have zero topological entropy.

Proof of Theorem 15.2.1. Without loss of generality take the interval to be $[0, 1]$. We begin with an elementary lemma.

Lemma 15.2.5. *If $a_n, b_n \geq 0$ then*

$$\overline{\lim_{n\to\infty}} \frac{1}{n} \log\Big(\sum_{k=0}^{n} e^{a_k + b_{n-k}}\Big) \leq M := \max\Big(\overline{\lim_{n\to\infty}} \frac{a_n}{n}, \overline{\lim_{n\to\infty}} \frac{b_n}{n}\Big).$$

Proof. If $M < \infty$ let $S := \max_{n\in\mathbb{N}}(1/n)\max(a_n, b_n)$ and for $U > M$ take $N \in \mathbb{N}$ such that $\max(a_n, b_n) \leq nU$ for $n \geq N$. If $n \geq 2N$ and $k \leq n$ we thus have $k \geq N$ or $n - k \geq N$, so $a_k + b_{n-k} \leq NS + (n - N)U \leq NS + nU$ and

$$\overline{\lim_{n\to\infty}} \frac{1}{n} \log\Big(\sum_{k=0}^{n} e^{a_k + b_{n-k}}\Big) \leq \lim_{n\to\infty} \Big(\frac{1}{n}\log(n+1) + \frac{1}{n}(NS + nU)\Big) = U.$$

But $U > M$ was arbitrary. □

Now we adapt the Adler–Konheim–McAndrew definition of topological entropy (Exercises 3.1.7–3.1.9) for maps of the interval where one can use partitions by intervals instead of covers. The easy Lemma 15.2.8 will then establish coincidence with our standard definition.

Let $\mathcal{C}$ be a finite partition of $[0,1]$ by intervals and $\mathcal{C}^n := \bigvee_{i=0}^{n-1} f^{-i}(\mathcal{C})$. Furthermore let $N(J,\mathcal{C}) := \min\{\operatorname{card} C \mid C \subset \mathcal{C},\ J \subset \bigcup_{I\in C} I\}$ for any $J \subset [0,1]$ and define $h(f,\mathcal{C}) := \overline{\lim}_{n\to\infty}(1/n)\log\operatorname{card}(\mathcal{C}^n)$. Take

$$\varnothing \neq E_1 := E := \{J \in \mathcal{C} \mid \overline{\lim_{n\to\infty}} \frac{1}{n}\log N(J,\mathcal{C}^n) = h(f,\mathcal{C})\} \subset \mathcal{C}.$$

Define E_n inductively: If $I \in E_{n-1}$, $J \in E$ then $I \to f^{n-1}(I) \cap J$ under f^{n-1}, whenever the latter is nonempty, so in this case by Lemma 15.1.1 there is an interval $K = K(I,J) \subset I$ such that $f^{n-1}(K) = f^{n-1}(I) \cap J$. Let $E_n := \{K(I,J) \mid I \in E_{n-1},\ J \in E,\ f^{n-1}(I) \cap J \neq \varnothing\}$.

In the sequel it will be convenient to use the following notation: If $\mathcal{A}$ is a collection of subsets of a space X and $Y \subset X$ then $\mathcal{A}_{\restriction Y} := \{I \cap Y \mid I \in \mathcal{A}\}$.

Lemma 15.2.6. $\overline{\lim}_{n\to\infty}\log\operatorname{card}(E_{n\restriction J}) = h(f,\mathcal{C})$ *for all* $J \in E$.

Proof. Clearly $\operatorname{card}(E_{n\restriction J}) \leq \operatorname{card}(\mathcal{C}^n{}_{\restriction J}) = N(J,\mathcal{C}^n) \leq \operatorname{card}(\mathcal{C}^n)$. To show the reverse inequality let $a_n := \log\operatorname{card}(E_{n\restriction J})$, $b_n := \log\left(\sum_{I\in\mathcal{C}\setminus E}\operatorname{card}(\mathcal{C}^n{}_{\restriction I})\right)$ and $a_0 = b_0 = 1$. We use a modified submultiplicativity argument. Observe that

$$\operatorname{card}(\mathcal{C}^n{}_{\restriction J}) \leq \sum_{k=0}^{n}\operatorname{card}(E_{k\restriction J})\cdot\Big(\sum_{I\in\mathcal{C}\setminus E}\operatorname{card}(\mathcal{C}^{n-k}{}_{\restriction I})\Big) = \sum_{k=0}^{n} e^{a_k+b_{n-k}}. \quad (15.2.1)$$

Namely, if $L \in \mathcal{C}^n{}_{\restriction J}$ then $f^i(L) \subset K \in \mathcal{C}$ for $0 \leq i \leq n$. Let k_0 be the smallest integer such that $f^{k_0}(L) \subset I \in \mathcal{C}\setminus E$. One can associate to such an L a unique element of $E_{k_0\restriction J}$ and a unique element of $\mathcal{C}^{n-k_0}{}_{\restriction I}$. Thus the total number of such L is bounded by the summand for $k = k_0$ on the right-hand side of (15.2.1).

By definition of E we thus have

$$h(f,\mathcal{C}) \leq \overline{\lim_{n\to\infty}}\frac{1}{n}\log\operatorname{card}(\mathcal{C}^n{}_{\restriction J}) \leq \overline{\lim_{n\to\infty}}\frac{1}{n}\log\sum_{k=0}^{n} e^{a_k+b_{n-k}} \text{ if } J \in E \quad (15.2.2)$$

and also $\overline{\lim}_{n\to\infty}(1/n)\log\operatorname{card}(\mathcal{C}^n{}_{\restriction J}) < h(f,\mathcal{C})$ for $J \in \mathcal{C}\setminus E$, so

$$\overline{\lim_{n\to\infty}}\frac{1}{n}b_n < h(f,\mathcal{C}). \quad (15.2.3)$$

Relations (15.2.2) and (15.2.3) imply $\overline{\lim}_{n\to\infty} a_n/n \geq h(f,\mathcal{C})$ by Lemma 15.2.5. □

The following main lemma shows that at least one $J \in E$ is covered by many $I \in E_n\restriction_J$ under f^n or, more specifically, that the Markov matrices of f^n on J have a substantial proportion of blocks of 1's around the diagonal.

Lemma 15.2.7. *If* $\gamma(J,I,n):=\operatorname{card}\{L \in E_n\restriction_J \mid I \subset f^n(L)\}$ *and* $h(f,\mathcal{C}) > \log 3$ *then there exists* $K \in E$ *such that*

$$\overline{\lim_{n\to\infty}} \frac{1}{n} \log \gamma(K,K,n) = h(f,\mathcal{C}).$$

Proof. Clearly we always have "$\leq$".

Fix $J \in E$, $u \in (\log 3, h(f,\mathcal{C}))$. Then for all $N \in \mathbb{N}$ there exists $n \geq N$ such that

$$\begin{aligned} \frac{1}{n} \log \operatorname{card}(E_n\restriction_J) &> u, \\ \operatorname{card}(E_{n+1}\restriction_J) &\geq 3 \operatorname{card}(E_n\restriction_J), \end{aligned} \tag{15.2.4}$$

since otherwise there exists $N \in \mathbb{N}$ such that $(1/n)\log\operatorname{card}(E_n\restriction_J) > u$ implies $\operatorname{card}(E_{n+1}\restriction_J) < 3\operatorname{card}(E_n\restriction_J)$ for $n \geq N$ and therefore we have $\overline{\lim}_{n\to\infty}(1/n)\log\operatorname{card}(E_n\restriction_J) \leq u$, contrary to Lemma 15.2.6.

If $L \in E_n\restriction_J$ then $f^n(L)$ is an interval, so there are at most 2 intervals of $\mathcal{C}$ that intersect $f^n(L)$ without being covered by it. Hence

$$\operatorname{card}\{I \in E \mid I \subset f^n(L)\} + 2 \geq \operatorname{card}\{I \in E \mid I \cap f^n(L) \neq \varnothing\} \geq \operatorname{card}(E_{n+1}\restriction_L),$$

where the last inequality follows from the definition of E_{n+1}. This is the key point in the proof where one-dimensionality is used. The point is that there are enough complete intersections to produce elements of E_{n+1}. Adding these inequalities for all $L \in E_n\restriction_J$ gives

$$\begin{aligned} \sum_{I\in E} \gamma(J,I,n) &= \sum_{I \in E} \operatorname{card}\{L \in E_n\restriction_J \mid I \subset f^n(L)\} \\ &= \sum_{L \in E_n\restriction_J} \operatorname{card}\{I \in E \mid I \subset f^n(L)\} \geq \operatorname{card}(E_{n+1}\restriction_J) - 2\operatorname{card}(E_n\restriction_J), \end{aligned}$$

so by (15.2.4) $\overline{\lim}_{n\to\infty}(1/n)\log(\sum_{I\in E}\gamma(J,I,n)) \geq u$ for all $u < h(f,\mathcal{C})$, whence $\overline{\lim}_{n\to\infty}(1/n)\log(\sum_{I\in E}\gamma(J,I,n)) \geq h(f,\mathcal{C})$. Since E is finite there is a $\varphi(J) \in E$ such that $\overline{\lim}_{n\to\infty}(1/n)\log(\sum_{I\in E}\gamma(J,\varphi(J),n)) \geq h(f,\mathcal{C})$. Again, since E is finite the transformation $\varphi\colon E \to E$ has a periodic point K of period m, say. Since

$$\gamma\Big(K,K,\sum_{i=0}^{m-1} n_i\Big) \geq \prod_{i=0}^{m-1} \gamma(\varphi^i(K),\varphi^{i+1}(K),n_i)$$

for any $n_i \in \mathbb{N}$, we find that $\overline{\lim}_{n\to\infty}(1/n)\log\gamma(K,K,n)) \geq h(f,\mathcal{C})$. □

Lemma 15.2.8. *If $\mathcal{C}$ is a partition by intervals of length at most ϵ, then $h(f,\mathcal{C}) \geq h_d(f, 2\epsilon)$ (see (3.1.10)).*

Proof. A maximal 2ϵ-separated set has at most one point in each element of $\mathcal{C}^n$, so the claim follows by (3.1.15). □

Now we complete the proof of the theorem. If $h_{\text{top}}(f) = 0$ set $J_n = \bigcap_{j=0}^{\infty} f^j([0,1])$, $\mathcal{D}_n = \{J_n\}$, $k_n = n$ for all $n \in \mathbb{N}$.

Otherwise take a finite partition $\mathcal{C}_n$ of $[0,1]$ into intervals of length at most $1/2n$. Fix $n \in \mathbb{N}$ and take $r \in \mathbb{N}$ such that $\log 3 < rh(f, \mathcal{C}_n)$. By Lemma 15.2.7 applied to f^r there exist an interval J_n and $m_n > m_{n-1}$ such that

$$\frac{1}{m_n} \log \gamma(J_n, J_n, m_n) \geq h(f^r, \mathcal{C}_n) - \frac{1}{n}.$$

Let $k_n = rm_n$ and $\mathcal{D}_n = \{I \in E_{m_n, f^r \restriction_{J_n}} \mid J_n \subset f^{k_n}(I)\}$. Then $J_n \subset f^{k_n}(I)$ for all $I \in \mathcal{D}_n$ and $\operatorname{card} \mathcal{D}_n = \gamma(J_n, J_n, m_n)$. Now

$$\begin{aligned}\varliminf_{n\to\infty} \frac{1}{k_n} \operatorname{card} \mathcal{D}_n &= \frac{1}{r} \varliminf_{n\to\infty} \frac{1}{m_n} \log \gamma(J_n, J_n, m_n) \\ &\geq \frac{1}{r} \varliminf_{n\to\infty} h(f^r, \mathcal{C}_n) - \frac{1}{n} \geq \frac{1}{r} \varliminf_{n\to\infty} h_d\Big(f^r, \frac{1}{n}\Big) - \frac{1}{n} = h_{\text{top}}(f).\end{aligned}$$

The reverse inequality is clear. □

We note that the Markov graphs of f^{k_n} associated to $\mathcal{D}_n$ have entropy approximating $h_{\text{top}}(f)$. Thus we have, using Theorem 15.1.5,

Corollary 15.2.9. *If $f\colon [0,1] \to [0,1]$ is continuous then the topological entropy of f is approximated arbitrarily well by the topological entropy of Markov graphs for iterates of f associated to collections of subintervals.*

A closer reformulation of Theorem 15.2.1 is the following

Corollary 15.2.10. *If $f\colon [0,1] \to [0,1]$ is continuous and $\epsilon > 0$ then there exists $n \in \mathbb{N}$ such that f^n has an invariant set Λ that is a factor of a full shift and such that $h_{\text{top}}(f^n \restriction_{\Lambda}) \geq n(h_{\text{top}}(f) - \epsilon)$.*

It is interesting to point out that a similar fact holds for $C^{1+\alpha}$ diffeomorphisms of two-dimensional manifolds; namely, by Corollary S.5.10 any such diffeomorphism has an invariant hyperbolic horseshoe-type set whose entropy approximates the topological entropy arbitrarily well. Unlike the one-dimensional case this is not a topological fact. For example, Mary Rees constructed a minimal homeomorphism of the two-torus with positive topological entropy.[1] The role played by the Intermediate Value Theorem is taken by hyperbolicity in the two-dimensional setting. Hyperbolicity is due to the Ruelle inequality (Theorem S.2.13) which asserts that positive topological entropy implies some exponential expansion in the linearized system. A similar fact holds for holomorphic maps of the Riemann sphere and for holomorphic diffeomorphisms of complex two-dimensional surfaces. In both cases hyperbolicity is also used. In the first case it works due to conformality of the map itself, in the second due to conformality of the restrictions to local stable and unstable manifolds.

We now use Theorem 15.2.1 to see how topological entropy depends on the map:

Theorem 15.2.11. *$h_{\text{top}}: C^0([0,1],[0,1]) \to \mathbb{R} \cup \{\infty\}$ is lower semicontinuous.*

Proof. Using the notations of Theorem 15.2.1 fix $f \in C^0([0,1],[0,1])$ and $\epsilon > 0$. Pick $n \in \mathbb{N}$ such that $(1/k_n)\log(\operatorname{card}\mathcal{D}_n - 2) \geq h_{\text{top}}(f) - \epsilon$. Take an interval $K_n \subset J_n$ such that $\bar{K}_n \subset \operatorname{Int} J_n$, no $I \in \mathcal{D}_n$ intersects both K_n and $J_n \setminus K_n$, and at most two $I \in \mathcal{D}_n$ are disjoint from K_n. Then $\bar{K}_n \subset \operatorname{Int}(f^{k_n}(I))$ for all $I \in \mathcal{D}_n$, so for any $g \in C^0([0,1],[0,1])$ sufficiently C^0-close to f we have $\bar{K}_n \subset \operatorname{Int}(g^{k_n}(I))$ for all $I \in \mathcal{D}_n$. Modify the $I \in \mathcal{D}_n$ slightly to get a disjoint collection $\mathcal{F}_n$ of intervals such that

(1) if x is an endpoint of $I \in \mathcal{F}_n$ but not of K_n then $g^{k_n}(x) \notin \bar{K}_n$;
(2) $\operatorname{card}\mathcal{F}_n \geq \operatorname{card}\mathcal{D}_n - 2$;
(3) $\bar{K}_n \subset \operatorname{Int}(g^{k_n}(I))$ for all $I \in \mathcal{F}_n$.

Then $Y := \bigcap_{i=0}^{\infty} g^{-ik_n}(\bar{K}_n) \subset g^{k_n}(I)$ for every $I \in \mathcal{F}_n$. By considering δ-separated sets one sees that this implies $h_{\text{top}}(g^{k_n}\restriction_Y) \geq \log\operatorname{card}\mathcal{F}_n$. Thus

$$h(g) = \frac{h(g^{k_n})}{k_n} \geq \frac{h(g^{k_n}\restriction_Y)}{k_n} \geq \frac{\log\operatorname{card}\mathcal{F}_n}{k_n} \geq \frac{\log(\operatorname{card}\mathcal{D}_n - 2)}{k_n} \geq h(f) - \epsilon.$$

□

Notice that the conclusion of Theorem 15.2.11 holds also for the the aforementioned higher-dimensional smooth and holomorphic situations (see Corollary S.5.13).

We now turn to relations between entropy and the variation $V(f)$ of a map f. A special case of such a relation appeared in Exercise 15.1.2. An obvious connection in the general case is as follows.

Proposition 15.2.12. *Let $f: [0,1] \to [0,1]$ be continuous. Then $h_{\text{top}}(f) \leq \underline{\lim}_{n\to\infty}(1/n)\log V(f^n)$. Suppose that f is piecewise-monotone, that is, the number $c(f)$ of maximal intervals of monotonicity is finite. Then $h_{\text{top}}(f) \leq \lim_{n\to\infty} \frac{1}{n}\log c_n \leq \frac{1}{n}\log c_n$, where $c_n := c(f^n)$.*

Proof. If $\{x_1, \dots, x_N\} \subset [0,1]$ is a maximal (n,ϵ)-separated set, that is, $N = N_d(f,\epsilon,n)$ (see (3.1.16)), then for $1 \leq i \leq N$ there exists $0 \leq k_i < n$ such that $|f^{k_i}(x_i) - f^{k_i}(x_{i+1})| > \epsilon$ and hence $\sum_{k=0}^{n-1} V(f^k) \geq \epsilon(N_d(f,\epsilon,n) - 1)$. Letting $n \to \infty$ and then $\epsilon \to 0$ yields the first claim by (3.1.15). To prove the second note that the limit exists and is the infimum of $(\log c_n)/n$ since $\{c_n\}_{n\in\mathbb{N}}$ is clearly submultiplicative. But every interval of monotonicity contributes at most 1 to the variation, whence $c_n \geq V(f^n)$. □

We now show that for piecewise-monotone maps these estimates are equalities.

Proposition 15.2.13. *If* $f: [0,1] \to [0,1]$ *is a piecewise-monotone continuous map then*

$$\lim_{n\to\infty} \frac{1}{n} \log c_n = h_{\mathrm{top}}(f).$$

Proof. Take 2ϵ less than the minimal distance between turning points. Let $\mathcal{C}$ be a partition of $[0,1]$ by intervals of length between ϵ and 2ϵ on which f is monotone. Then $\operatorname{card} \mathcal{C}^n \geq c_n$. By choice of ϵ every d_n^f-ball of radius ϵ covers at most 3^n elements of $\mathcal{C}^n$. Thus a minimal cover of $[0,1]$ by d_n^f-ϵ-balls has cardinality $S = S_d(f,\epsilon,n) \geq 3^{-n} \operatorname{card} \mathcal{C}^n \geq 3^{-n} c_n$. Thus $\lim_{n\to\infty} \frac{1}{n} \log c_n \leq h_{\mathrm{top}}(f) + \log 3$. Applying this result to f^k yields $\lim_{n\to\infty} \frac{1}{n} \log c_n \leq (1/k) h_{\mathrm{top}}(f^k) + (1/k)\log 3 = h_{\mathrm{top}}(f) + (1/k)\log 3$. Letting $k \to \infty$ yields the claim. □

Corollary 15.2.14. *If* $f: [0,1] \to [0,1]$ *is a piecewise-monotone continuous map then*

$$\lim_{n\to\infty} \frac{1}{n} \log V(f^n) = h_{\mathrm{top}}(f).$$

Proof. $h_{\mathrm{top}}(f) \leq \underline{\lim} \dfrac{\log V(f^n)}{n} \leq \overline{\lim} \dfrac{\log V(f^n)}{n} \leq \lim \dfrac{\log c_n}{n} = h_{\mathrm{top}}(f).$ □

In certain cases this result allows us to calculate topological entropy explicitly. Consider, for example, the tent map

$$\tau_s: x \mapsto s(1 - |1 - 2x|)/2 \tag{15.2.5}$$

with slope $s \in [1,2]$. It, together with all its iterates, has constant absolute value of the slope which therefore gives the variation, that is, $V(\tau_s^n) = s^n$. Thus Corollary 15.2.14 shows that

$$h_{\mathrm{top}}(\tau_s) = \log s. \tag{15.2.6}$$

Of course this argument applies to any map whose slope is of constant absolute value. This result is useful in connection with the Milnor–Thurston Classification, Theorem 15.6.1.

Since there are only countably many 0-1 matrices the topological entropy of topological Markov chains can only take countably many values, whereas tent maps can by (15.2.6) have any value of topological entropy. Thus we immediately see that Markov chains are not sufficient to model all interval maps. We will return to this theme in Section 15.5. But first we show that Markov models are sufficient to understand how periodic points of different periods appear.

Exercises

15.2.1. *Show that for every $s > 1$ there exist infinitely many maps of constant slope with absolute value s no two of which are topologically conjugate.*

15.2.2. *Given $\epsilon > 0$ and $0 \le h \le \infty$ construct a map of $[0,1]$ that is C^0-close to the identity and has topological entropy h.*

3. The Sharkovsky Theorem

In the previous section we showed that entropy gives a lower bound for the growth rate of the number of periodic orbits. Now we will show that the mere existence of periodic points of certain periods forces lower bounds on the entropy. We begin with a remarkable theorem due to Sharkovsky which describes in a precise way the order in which periodic orbits of different periods appear. A very special case was given in Proposition 15.1.12.

Definition 15.3.1. The Sharkovsky ordering of the natural numbers $\mathbb{N}$ is defined by

$$1 \lhd 2 \lhd 2^2 \lhd 2^3 \lhd \cdots \lhd 2^m \lhd \cdots \lhd 2^k(2n-1) \lhd \cdots \lhd 2^k \cdot 3 \lhd \cdots \lhd 2 \cdot 3 \lhd \cdots \lhd 2n-1 \lhd \cdots \lhd 9 \lhd 7 \lhd 5 \lhd 3.$$

Recall that for $n \in \mathbb{N}$ the periodic points of *prime period* n are the points in $\operatorname{Fix}(f^n) \smallsetminus (\bigcup_{i=1}^{n-1} \operatorname{Fix}(f^i))$.

Theorem 15.3.2. (Sharkovsky Theorem)[1] *Let $I \subset \mathbb{R}$ be a closed interval and $f \colon I \to I$ a continuous map. If f has a periodic point of prime period p and $q \lhd p$, then f has a periodic point of prime period q.*

Proof. The following main lemma describes a "minimal Markov model" which is present in any interval map with periodic points of odd periods. It also serves as the base of the lower estimate of entropy (Theorem 15.3.5).

Lemma 15.3.3. *Let $I \subset \mathbb{R}$ be a closed interval and $f \colon I \to I$ a continuous map. Let $x \in I$ be a periodic orbit of odd period $p > 1$ such that there is no periodic orbit of odd period $q < p$. If $x_{\min}$ and $x_{\max}$ are the minimum and maximum of the orbit of x then the Markov graph of the partition $\mathcal{C}$ of $J := [x_{\min}, x_{\max}]$ induced by the orbit $\mathcal{O}(x)$ of x contains the following subgraph:*

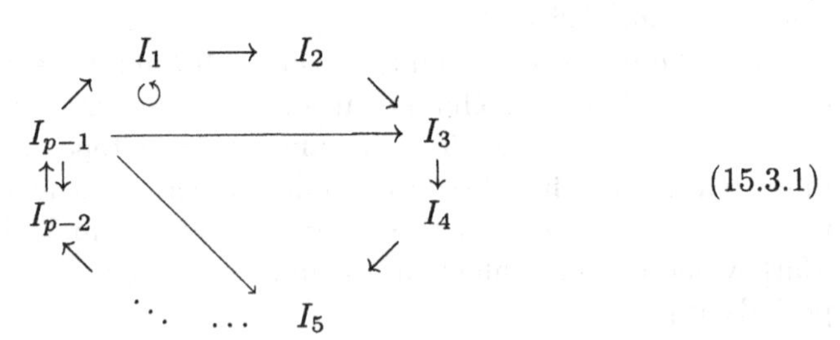

That is, one can label the intervals of the partition as $\{I_1, \dots, I_{p-1}\}$ in such a way that $I_1 \to I_1 \to I_2 \to \dots \to I_{p-1}$ and $I_{p-1} \to I_k$ for every odd k.

Proof. Take $a := \max\{y \in \mathcal{O}(x) \mid f(y) > y\}$ and $I_1 = [a, b]$, where b is the closest point of the orbit $\mathcal{O}(x)$ to the right of a. Then $f(a) \geq b$ and $f(b) \leq a$, hence $I_1 \subset f(I_1)$, that is, $I_1 \to I_1$. In fact, the inclusion is proper since a has odd period. Now $f(I_1) \subset f(f(I_1))$ so inductively $I_1 \subset f(I_1) \subset \dots \subset f^p(I_1)$; thus $f^p(I_1)$ contains the orbit of a and hence J.

Let us now see that there is an element $I \in \mathcal{C} \setminus \{I_1\}$ such that $I \to I_1$. Let $l := \text{card}\{y \in \mathcal{O}(x) \mid y < a\}$ and $r := \text{card}\{y \in \mathcal{O}(x) \mid y > b\}$. Then $l + r = p - 2$ so $l \neq r$. Thus on at least one side of I_1 there is a point of $\mathcal{O}(x)$ whose image is on the same side. On the other hand both sides have points of $\mathcal{O}(x)$ that change sides under f, since x is periodic. Thus there is an adjacent pair of points c, d such that exactly one of them changes sides, that is, $I = [c, d]$ is as desired.

Now let us label the intervals in such a way that $I_1 \to \dots \to I_k \to I_1$ is the shortest nontrivial loop containing I_1. We would like to show that $k = p - 1$. Since $k \leq p - 1$ by minimality of the loop it suffices to show $k \geq p - 1$. To that end let $q \in \{k, k+1\}$ be odd. By Corollary 15.1.4 the existence of the loop $I_1 \to \dots \to I_k \to I_1$ or of the loop $I_1 \to \dots \to I_k \to I_1 \to I_1$ yields a point $y \in \text{Fix}(f^q)$. y is not a fixed point since otherwise $y \in I_1 \cap \dots \cap I_k \subset I_1 \cap I_2$ is in the orbit of x, a contradiction. By assumption on p we thus have $q \geq p$, which yields $k \geq p - 1$.

It remains to show that $I_{p-1} \to I_k$ for odd k. To that end let us argue that the I_i are, up to orientation, ordered like this:

$$I_{p-1}, I_{p-3}, \dots, I_2, I_1, I_3, \dots, I_{p-2}. \tag{15.3.2}$$

Note that since $I_1 \to \dots \to I_{p-1} \to I_1$ is the shortest nontrivial loop $I_1 \to \dots \to I_1$ we have that $I_k \to I_j$ implies $j \leq k + 1$. Thus I_1 covers I_1 and I_2, hence by connectedness I_2 is adjacent to $I_1 = [a, b]$, so up to orientation we have $I_2 = [c, a]$ and $f(a) = b$ as well as $c = f(b)$. Let us determine $f(I_2)$. Since $f(a) = b$ and I_2 does not cover I_1 (since $I_1 \to \dots \to I_{p-1} \to I_1$ is the shortest nontrivial loop $I_1 \to \dots \to I_1$), $f(I_2)$ lies entirely to the right of a. It covers I_3, so $I_3 = [b, d]$. Since I_2 covers no other intervals we have, in fact, $d = f^2(b)$. Similarly one sees that $I_4 = [e, c]$ and obtains the claimed ordering inductively.

Writing $a_i := f^i(a)$ we have shown $x_{\min} = a_{p-1} < a_{p-3} < \dots < a_2 < a < a_1 < a_3 < \dots < a_{p-2} = x_{\max}$, hence evidently $I_{p-1} = [a_{p-1}, a_{p-3}] \to I_k$ for odd k. □

Lemma 15.3.4. *If f has a periodic point of even period then it has a point of period 2.*

Proof. Let p be the smallest even period and $x \in \text{Fix}(f^p)$. If $p = 2$ we are done. Otherwise the period-2 point may arise in two different ways. Suppose first that there exists a pair c, d of adjacent points of $\mathcal{O}(x)$ such that the interval $I_k = [c, d] \neq I_1$ covers I_1 (with notation as before). Label the intervals generated by

$\mathcal{O}(x)$ such that $I_1 \to \cdots \to I_k \to I_1$ is the shortest nontrivial loop $I_1 \to \cdots \to I_1$. Then clearly $k \le p-1$.

Let $q \in \{k, k+1\}$ be even. Then $q \le p$. By Corollary 15.1.4 the existence of the loop $I_1 \to \cdots \to I_k \to I_1$ or of the loop $I_1 \to \cdots \to I_k \to I_1 \to I_1$ yields a point $y \in \text{Fix}(f^q)$. y is not a fixed point since otherwise $y \in I_1 \cap \cdots \cap I_k \subset I_1 \cap I_2$ is in the orbit of x, a contradiction. By assumption on p we thus have $q = p$ and hence $k \ge p-1$.

Furthermore, we can argue as before that the I_i are, up to orientation, ordered like this:

$$I_{p-2}, \dots, I_2, I_1, I_3, \dots, I_{p-1},$$

and that $I_{p-1} \to I_k$ for even k. Thus by Corollary 15.1.4 the loop $I_{p-1} \to I_{p-2} \to I_{p-1}$ gives a point of prime period two.

In the second case, when there is no interval $[c,d]$ covering I_1, we would like to show that $[x_{\min}, a] \to [b, x_{\max}] \to [x_{\min}, a]$. To this end note that $f(a) \ge b$ so $f([x_{\min}, a])$ contains points to the right of I_1. Our assumption implies that $[x_{\min}, a]$ does not cover I_1, so $f([x_{\min}, a])$ is to the right of a. Likewise $f([b, x_{\max}])$ lies to the left of b. But then, since f permutes $\mathcal{O}(x)$, we must have $[x_{\min}, a] \to [b, x_{\max}] \to [x_{\min}, a]$. Now Corollary 15.1.4 again yields a period-2 point. □

Now we can prove the theorem.

(0) If $p = 2^k$ and $q \lhd p$ then $q = 2^l$ for some $l < k$.

It is easy to see that there is a fixed point, so let us consider the case $l > 0$. If x is a periodic point with prime period p then it is a periodic point of $f^{q/2} = f^{2^{l-1}}$ with period 2^{k-l+1}, so $f^{2^{l-1}}$ has a periodic point of prime period 2 by Lemma 15.3.4. This point has prime period $2^{l-1+1} = q$ for f.

Otherwise, if $p = r \cdot 2^k$ with $r > 1$ odd and $q \lhd p$ then there are three cases:

(1) $q = 2^l$, $l \le k$,

(2) $q = s \cdot 2^k$, s even,

(3) $q = s \cdot 2^k$, $s > r$ odd.

To prove the theorem in case (1) note that by applying it in case (2) with $s = 2$ yields a periodic point of period 2^{k+1}, so we have periodic points of periods 2^l, $l \le k$, by the theorem in case (0).

To prove the theorem in case (2) we may assume without loss of generality that r is minimal, that is, that f has no points of period $t \cdot 2^k$ for $t < r$ odd. Then r is the minimal odd period for f^{2^k}. By Lemma 15.3.3 we have a nontrivial loop of length s, namely, $I_{r-1} \to I_{r-s} \to \cdots \to I_{r-2} \to I_{r-1}$ if $s < r$, otherwise $I_1 \to I_2 \to \cdots \to I_{r-1} \to I_1 \to I_1 \to \cdots \to I_1$. By Corollary 15.1.4 f^{2^k} has a point of prime period s, which is hence a point of prime period $s \cdot 2^k = q$ for f (prime since otherwise $s/2$ would be a period of this point for f^{2^k}).

Finally, in case (3), the loop $I_1 \to I_2 \to \cdots \to I_{r-1} \to I_1 \to \cdots \to I_1$ gives a point of prime period s for f^{2^k}. If the prime period under f is $s \cdot 2^k$ we are done. Otherwise it is $s \cdot 2^t$ for some $t < k$. But then we can take $p' = s \cdot 2^t$ and

$s' = s \cdot 2^{k-t}$ to get a periodic orbit of prime period $s' \cdot 2^t = s \cdot 2^k = q$ via case (2).

Thus we have proved the Sharkovsky Theorem. □

The method of the proof of the Sharkovsky Theorem allows us to develop information concerning types of periodic orbits (defined by the permutation induced on the periodic orbit by the map). One says that one periodic orbit *forces* another if any map having the first kind of periodic orbit also has the second kind. Thus period p forces period $q \triangleleft p$. In Exercise 15.3.1 we will see an example of a type of orbit of period four that forces the existence of all other periods.

The combinatorial insight gained in the preceding arguments enables us to obtain lower bounds on the entropy of an interval map when the presence of certain periodic points is known. We will use Proposition 3.2.5 in conjunction with the information about the Markov graph given by Lemma 15.3.3. The connection is given by Theorem 15.1.9.

Theorem 15.3.5. *Let $I \subset \mathbb{R}$ be a closed interval and $f: I \to I$ a continuous map with a periodic point of period $p2^m$ with $p > 1$ odd. Then $h_{\text{top}}(f) \geq 2^{-m} \log \lambda_p$, where λ_p is the largest root of $x^p - 2x^{p-2} - 1$.*

Proof. Assume that p is minimal and consider $\tilde{f} := f^{2^m}$. Note that by Lemma 15.3.3 the Markov graph of $\tilde{f}$ with respect to the partition induced by the periodic orbit contains a subgraph (15.3.1). By Theorem 15.1.9 and the Perron–Frobenius Theorem 1.9.11 it suffices to show that the entropy of (15.3.1) is the largest root of $x^p - 2x^{p-2} - 1$. Thus we need to calculate the characteristic polynomial of the Markov matrix associated to (15.3.1), that is, we need a formula for the largest eigenvalue of an $n \times n$ matrix

$$A_n := \begin{pmatrix} 1 & 0 & 0 & & & & 1 \\ 1 & 0 & 0 & & & & 0 \\ 0 & 1 & 0 & & & & 1 \\ & 0 & 1 & \ddots & & & 0 \\ & & 0 & \ddots & 0 & 0 & \vdots \\ & & & \ddots & 1 & 0 & 1 \\ & & & & 0 & 1 & 0 \end{pmatrix},$$

where n is even. If J is the $n \times n$ matrix with $J_{11} = 1$ and all other entries zero then $P_n(\lambda) := \det(A_n - \lambda \operatorname{Id})$ and $Q_n(\lambda) := \det(A_n - \lambda \operatorname{Id} - J)$ are related by $P_n(\lambda) = (1 - \lambda)(-\lambda)Q_{n-2}(\lambda) - 1$. On the other hand $Q_n(\lambda) = \lambda^2 Q_{n-2}(\lambda) - 1$ and $Q_2(\lambda) = \lambda^2 - 1$ so inductively $Q_n(\lambda) = \lambda^n - \lambda^{n-2} - \cdots - 1$ and hence $Q_n(\lambda) - Q_{n-2}(\lambda) = \lambda^n - 2\lambda^{n-2}$. Since $P_n(\lambda) = (\lambda^2 - \lambda)Q_{n-2}(\lambda) - 1 = Q_n(\lambda) - \lambda Q_{n-2}(\lambda)$ we have

$$\begin{aligned} (1+\lambda)P_n(\lambda) &= (1+\lambda)(Q_n(\lambda) - \lambda Q_{n-2}(\lambda)) \\ &= Q_n(\lambda) - \lambda Q_{n-2}(\lambda) + \lambda Q_n(\lambda) - (\lambda^2 Q_{n-2}(\lambda) - 1) - 1 \\ &= \lambda(Q_n(\lambda) - Q_{n-2}(\lambda)) - 1 = \lambda^{n+1} - 2\lambda^{n-1} - 1. \end{aligned}$$

The largest root of this polynomial is positive and hence coincides with the largest root of $P_n(\lambda)$. Applying this to $n = p - 1$ yields the claim. □

Notice in particular that we have shown that a map with zero entropy cannot have periodic points of periods other than powers of two, a complement to Corollary 15.2.4. Combining these we have

Corollary 15.3.6. *Suppose $f: [0,1] \to [0,1]$ is continuous. Then $h_{\text{top}}(f) = 0$ if and only if the period of every periodic point of f is a power of two.*

Finally we show that there are no restrictions on the appearance of periodic orbits other than those of the Sharkovsky Theorem and that the entropy estimates of Theorem 15.3.5 are sharp. To that end let $S_p := \{p\} \cup \{q \in \mathbb{N} \mid q \lhd p\}$ for $p \in \mathbb{N}$ and $S_\infty = \{2^n \mid n \in \mathbb{N}\}$. Also let $\mathcal{P}(f) := \{n \in \mathbb{N} \mid \operatorname{Fix}(f^n) \setminus \bigcup_{k<n} \operatorname{Fix}(f^k) \neq \varnothing\}$. Then we have

Theorem 15.3.7.[2] *For $p \in \mathbb{N} \cup \{\infty\}$ there exists a continuous $f: [0,1] \to [0,1]$ such that $\mathcal{P}(f) = S_p$ and if we write $p = n2^k$ with n odd then $h_{\text{top}}(f) = 2^{-k} \log \lambda_n$, where λ_n is as in Theorem 15.3.5.*

Proof. We first consider the case of p a power of 2. To that end we show a period-doubling construction, the "square-root trick". Consider any map $f: [0,1] \to [0,1]$ and define $\tilde{f}: [0,1] \to [0,1]$ by $\tilde{f}(x) = (2 + f(3x))/3$ for $x \in [0, 1/3]$, $\tilde{f}(x) = x - 2/3$ for $x \in [2/3, 1]$, and extending linearly into $[1/3, 2/3]$. Note that $\tilde{f}^2(x) = f(3x)/3$ on $[0, 1/3]$, so for every period-n point of f there is a period-$2n$ point of $\tilde{f}$. Moreover, the restriction to $[1/3, 2/3]$ is a linear expanding homeomorphism and thus has a unique fixed point and no other periodic points. Thus the periods for $\tilde{f}$ are precisely twice those of f and there is an additional fixed point.

If $p = 1$ then the map $f_1(x) = x(1-x)$ has zero as a fixed point and no periodic points since all points are attracted to 0. If $p = 2$ take $f_2 = \tilde{f}_1$. By the foregoing observations it has a period-2 point and it clearly also has a fixed point in $[1/3, 2/3]$. Inductively let $f_n = \tilde{f}_{n-1}$ and note that f_n has points of period 2^k for $k < n$ and no others. At this point it is also worthwhile to note that $f_n = f_m$ on $[2 \cdot 3^{-\min(n,m)}, 1]$ and it follows easily that $f_n \to f_\infty$ uniformly and $f_\infty(0) = 1$. f_∞ is an example of a map having all powers of 2 as periods, and no more. All these maps also have zero entropy.[3]

Suppose next that p is odd and consider the standard map f_p (Definition 15.1.13) with Markov graph (15.3.1) (constructed using the prescription (15.3.2)). Then by construction f_p has a period-p point. There are no points of smaller period (other than 1) because all loops of odd length less than p in (15.3.1) only involve I_1 and hence arise from the fixed point of f_p. By Theorem 15.1.9, Exercise 15.1.1, and the Variational Principle, Theorem 4.5.3, $h_{\text{top}}(f) = \log \lambda_p$.

Finally suppose $p = 2^k n$ for some odd n. Then for the map $f_{n,k}$ obtained from f_n by applying the square-root trick k times we have $\mathcal{P}(f_{n,k}) = S_p$. Furthermore the period-doubling construction has the effect that from a map f

it produces a map $\tilde{f}$ that interchanges $[0,1/3]$ and $[2/3,1]$ and the only non-wandering point of $\tilde{f}$ in $(1/3,2/3)$ is a fixed point. Thus $[0,3^{-k}]$ is invariant under $f_{n,k}^{2^k}$ and the union of the images of this interval contains all nonperiodic nonwandering points of $f_{n,k}$. But then Proposition 3.1.7(3) implies that $h_{\text{top}}(f_{n,k}) = 2^{-k}\log\lambda_n$. □

Exercises

15.3.1. *Suppose $f: [0,1] \to [0,1]$ has a periodic orbit $\{x_1 < x_2 < x_3 < x_4\}$ such that $f(x_i) = x_{i+1}$ for $i < 4$ and $f(x_4) = x_1$. Show that f has periodic points of all periods.*

15.3.2. *Show that for any $n \geq 3$ there exists a permutation of n elements such that a periodic orbit $\{x_1 < \cdots < x_n\}$ of prime period n that realizes this permutation forces the existence of periodic points of all other periods.*

4. Maps with zero topological entropy

Our next result describes interval maps with zero topological entropy in terms of invariant measures. To an extent it may be viewed as an analog of the measurable classification of circle homeomorphisms with irrational rotation number (Theorem 11.2.9). There topological entropy was also zero and rotations gave the only models in the measurable category or under semiconjugacy. Here we show that for noninvertible maps of an interval there is exactly one model with nonatomic measure with zero entropy. An important ingredient of the proof is the observation that the presence of a horseshoe produces positive topological entropy (Corollary 15.1.11). This is repeatedly used to rule out combinatorial complexity. We begin by describing the standard model for these maps which first appeared in Exercise 1.3.3.

Definition 15.4.1. The map $\alpha_2: \Omega_2^R \to \Omega_2^R$ given by

$$(\alpha_2\omega)_i = \begin{cases} 1-\omega_i & \text{if } \omega_i = 1 \text{ for all } j < i, \\ \omega_i & \text{otherwise} \end{cases}$$

of the space Ω_2^R with the Bernoulli measure $\nu = \mu_{(1/2,1/2)}$ is called the *dyadic adding machine.*

Note that the inverse of α is

$$(\alpha_2^{-1}\omega)_i = \begin{cases} 1-\omega_i & \text{if } \omega_i = 0 \text{ for all } j < i, \\ \omega_i & \text{otherwise.} \end{cases}$$

Furthermore α maps cylinders to cylinders of the same length and is thus a homeomorphism preserving ν.

Theorem 15.4.2. *Let $f\colon [0,1] \to [0,1]$ be continuous with $h_{\text{top}}(f) = 0$ and μ a nonatomic ergodic f-invariant Borel probability measure. Then (f,μ) is metrically isomorphic (Definition 4.1.20) to the dyadic adding machine and $f\restriction_{\operatorname{supp}\mu}$ has the dyadic adding machine as a topological factor.*

Proof. Let $S = \operatorname{supp}\mu$ and set

$$S_0 := \{x \in S \mid f(x) = x\}, S_1 := \{x \in S \mid f(x) > x\}, S_2 := \{x \in S \mid f(x) < x\}.$$

Then

(1) S is a closed invariant set,
(2) $f\restriction_S$ is topologically transitive (by Proposition 4.1.18(2)),
(3) S is infinite,
(4) S_1 and S_2 are nonempty.

The strategy of our proof is to show that $S_0 = \varnothing$, $f(S_1) = S_2$, and $f(S_2) = S_1$. This follows from several lemmas which use the absence of horseshoes in f and its iterates (Definition 15.1.10 and Corollary 15.1.11). Another major device is topological transitivity of f on S which follows from ergodicity by Proposition 4.1.18(2).

Lemma 15.4.3. $\sup S_1 \le \inf(S_0 \cup S_2)$ *and* $\sup(S_0 \cup S_1) \le \inf S_2$.

Proof. If we let $\bar{f}(x):=1-f(1-x)$ then the second claim follows from the first by considering $\bar{f}$. To prove the first claim let $p:=\inf(S_0 \cup S_2)$ and assume there is a $q \in S_1 \cap (p,1]$. $f\restriction_{(S_0 \cup S_1)\cap[p,1]}$ attains a maximum at c, say. By definition $f(c) \ge c \in S_0 \cup S_1$ and since $f(p) \le p$ by continuity and $f(q) \ge q > p$ we must have $c > p$.

(1)In order to get a contradiction suppose that $c \in S_1$ and let $a := \sup\{x < c \mid f(x) = x\}$. Notice that $S \cap [a,c] \subset S_1 \cup S_0$ since $f(x) < x \le c < f(c)$ implies by the Intermediate Value Theorem that there exists $y \in [x,c] \cap \operatorname{Fix}(f)$; hence $x \le y \le a$ by choice of a and clearly $x \ne a$.

To get a horseshoe we want to show that there exists $b \in S \cap [c, f(c)]$ such that $f(b) \le a$. To that end suppose $f(x) > a$ for all $x \in S \cap [c, f(c)]$. In this case $A := S \cap [a, f(c)]$ is a closed invariant set with nonempty interior: By choice of c we have $f(x) \le f(c)$ for all $x \in A$; furthermore $f(x) \ge x \ge a$ for $x \in S \cap [a,c]$ as noted before, and $f(x) \ge a$ for $x \in S \cap [c, f(c)]$ by assumption. But $f\restriction_S$ is topologically transitive and hence has no such invariant sets. Consequently we have $a < c < b$ and $f(b) \le a = f(a)$ and $f(c) \ge b$ so $[a,b] \subset f([a,c]) \cap f([c,b])$, a horseshoe. This is impossible because $h_{\text{top}}(f) = 0$ (Corollary 15.1.11).

(2)Thus $c \in S_0$, that is, $f(c) = c$. Suppose the minimum of f on $(S_0 \cup S_2) \cap [0,c]$ is attained at u. If $u \in S_2$ we can apply the argument of the preceding paragraph to $\bar{f}$ instead of f, with u instead of c and c instead of a, to get a horseshoe. Thus $u \in S_0$ and hence $u = p$ and $f(S \cap [p,c]) \subset [p,c]$. By topological transitivity this implies $S \subset [p,c]$.

Notice that when $x < c$ then $[x,c] \cap S \not\subset S_1$ because otherwise $[x,c] \cap S$ would be an invariant set with nonempty interior, contradicting topological transitivity of f on S. Let us show that we also have $[x,c] \cap S \not\subset S_2$ for $x < c$. If $p \le x < c$ and $[x,c] \cap S \in S_2$ then $M := \max\{f(y) \mid p \le y \le x\} = c$ since otherwise $[p, M] \cap S$ is a proper closed f-invariant set, contradicting topological transitivity. Thus take $v \in [p,x] \cap S$ such that $f(v) = c = f(c)$. Then $v \in S_1$ and we could have chosen $c = v$ to begin with. But above we showed that $c \notin S_1$, a contradiction.

Thus for $x < c$ there are points of both S_1 and S_2 in $[x,c]$, whence $(x,c) \cap \operatorname{Fix} f \neq \varnothing$. We use this to construct a horseshoe for an iterate, leading to the contradiction proving Lemma 15.4.3. Take $x \in (p,c) \cap S$, $\beta \in (x,c) \cap \operatorname{Fix}(f)$. By topological transitivity there is some $y \in (p,x)$ and $N \in \mathbb{N}$ such that $f^N(y) \in [\beta, c]$. Since $f(p) \le p$ and by continuity there is some $\alpha \in (p,x)$ such that $f^N(\alpha) = \beta$, we have $f^n(\alpha) = \beta$ for $n \ge N$. By topological transitivity there also exist $\gamma \in (\alpha, \beta)$ and $n \ge N$ such that $f^n(\gamma) < \alpha$. Thus $[\alpha, \beta] \subset f^n([\alpha, \gamma]) \cap f^n([\gamma, \beta])$, a horseshoe for f^n. □

Our aim is to show that $S_0 = \varnothing$. Here is a step in that direction:

Lemma 15.4.4. *$S_0 = \varnothing$ or $S_0 = \{\sup S_1\} = \{\inf S_2\}$.*

Proof. Lemma 15.4.3 implies that $S_0 \subset [p,q] := [\sup S_1, \inf S_2]$, so by topological transitivity $S_0 \subset \{p,q\}$. If $S_0 \neq \varnothing$ then, after possibly replacing f by $\bar{f}$, we have $p \in S_0$. Assume $p \neq q$ to get a contradiction. If U is a sufficiently small neighborhood of p and $x \in U \cap S \setminus \{p\}$ then $x \in S_1$, $x < p = f(p)$, and $f(x) < q$ by continuity, so $x < f(x) \le p$ by invariance of S, and hence $\omega(x) = \{p\}$, contradicting topological transitivity. □

Next we show that f interchanges S_1 and S_2. It then easily follows that S_0 is, in fact, empty, and thus yields a complete description of the measurable dynamics on S.

Lemma 15.4.5. *$f(S_1) = S_2$ and $f(S_2) = S_1$.*

Proof. We will show that $f(S_1) \subset S_2$. Applying this to $\bar{f}$ yields $f(S_2) \subset S_1$, which then implies the lemma by topological transitivity. In order to get a contradiction assume that there exists a $p \in S_1$ with $f(p) \in S_0 \cup S_1$ and let $a := \sup\{x \in S_1 \mid f(x) \in S_0 \cup S_1\} \in S_0 \cup S_1$.

(1) Suppose that $a \in S_1$. We will show that this produces a horseshoe and is thus impossible. First note that there exists $r \in S_0 \cup S_1$ such that $f(r) \le a$: Otherwise $s = \inf\{f(x) \mid x \in S_0 \cup S_1\} > a$ and $[s, 1] \cap (S_2 \cup S_0)$ is a proper f-invariant subset of S (if $x \in [s,1] \cap (S_2 \cup S_0)$ then $x \in S_0 \cup S_2$ whence $f(x) \ge s$) and f is not topologically transitive.

We have $r \notin S_0$ since otherwise $S_0 \ni r = f(r) \le a \in S_1$, a contradiction. Thus $a < t := \inf\{x \in S_2 \mid f(x) \le a\} \in S_2$. We want to conclude that there is a $c \in [a,t] \cap S$ such that $f(c) \ge t$. If this is not the case then $M := \max\{f(x) \mid x \in [a,t] \cap S\} < t$ and $(a,t) \cap S$ is invariant: If $x \in (a,t) \cap S$ then $f(x) \le M < t$ by definition, while either $x \in (a,t) \cap S_2$, so $f(x) > a$, or $x \in (a,t) \cap (S_0 \cup S_1)$, so $f(x) \ge x > a$. Again this is incompatible with topological transitivity.

By continuity there is a fixed point $b = f(b)$ in $[\sup S_1, \inf S_2]$. By definition $c \in S_1$, so $c \leq b$. By choice $f(a) \in S_1 \cup S_0$ and hence $f(a) \leq b$ as well. Consequently $[b,t] \subset f([a,c]) \cap f([c,b])$. Since $f(t) \leq a$ by choice we have $[a,b] \subset f([b,t])$, so we get $[a,b] \subset f^2([a,c]) \cap f^2([c,b])$, a horseshoe for f^2, which is impossible.

(2) Thus we have shown that $a \in S_0$. We want to conclude that this, too, produces a horseshoe for an iterate of f. By definition of a there is an increasing sequence $w_n \to a$ with $w_n \in S_1$ and $w_n < f(w_n) \in S_0 \cup S_1$. $a \in S_0$ implies that $a \geq \sup S_1$ and hence $f(w_n) \leq a$ for all n. Since $[w_n, a] \cap S$ cannot be invariant there are $z_n \in (w_n, a) \cap S_1$ such that $f(z_n) > a$. By continuity $f(z_n) \to f(a) = a$, so there is a $k \in \mathbb{N}$ such that $\gamma := z_2 < w_k =: \beta < a$. If $m \in \mathbb{N}$ is such that $f(z_m) < f(\gamma)$ then we may assume without loss of generality that z_m is on a dense orbit of $f\restriction_S$ (z_m is defined by an open condition). Thus there exists $N \in \mathbb{N}$ such that $f^N(z_m) < w_2 =: \alpha$ and $[\alpha, f(z_m)] \subset f([\alpha,\gamma]) \cap f([\gamma,\beta])$. But then $[\alpha,\beta] \subset [f^N(z_m), \alpha] \subset f^N([\alpha,\gamma]) \cap f^N([\gamma,\beta])$, a horseshoe for f^N. □

Lemma 15.4.6. $S_0 = \varnothing$.

Proof. If $S_0 = \{p\} = [\sup S_1, \inf S_2]$ then consider f^2 instead of f and the f^2-invariant set $S_0 \cup S_1$ instead of S. f^2 is topologically transitive on this set. Thus we can apply Lemma 15.4.3, using the fact that $f^2(p) = p$. We conclude that the set $\{x \in S_0 \cup S_1 \mid f^2(x) < x\}$ is empty, contrary to the observation (4) from the beginning of the proof applied to f^2. Thus Lemma 15.4.4 shows Lemma 15.4.6. □

Now we construct the isomorphism to the adding machine. Define inductively $S^0 := S$, $S_i^0 = S_i$ $(i = 1,2)$, $S^n := S_1^{n-1}$, $S_1^n := \{x \in S^n \mid f^{2^n}(x) > x\}$, $S_2^n := \{x \in S^n \mid f^{2^n}(x) < x\}$. Using Lemma 15.4.5 and Lemma 15.4.6 inductively shows that S_i^n are closed, $S_1^n \cup S_2^n = S^n = f^{2^n}(S^n)$, and $f^{2^n}\restriction_{S^n}$ is topologically transitive. For any $n \in \mathbb{N}$ the collection $\{f^j(S^n) \mid 0 \leq j < 2^n\}$ is a partition of S by sets of measure 2^{-n}. Define $h: S \to \Omega_2^R$ by $\pi(S^n) = \{\omega \in \Omega_2^R \mid \omega_1 = \cdots = \omega_n = 1\}$ and $h \circ f = \alpha_2 \circ h$. Then h is well defined, continuous, surjective, and measure preserving. To complete the proof of Theorem 15.4.2 we need to show that h is injective μ-almost everywhere. We will, in fact show that there is an at most countable set where h is two-to-one and outside of which h is injective. Namely, the preimage under h of a point $\omega \in \Omega_2^R$ is the intersection of S and a collection of intervals, hence either a point or the intersection of S with an interval and of measure zero since ν is nonatomic (hence this intersection consists of at most two points). These intervals are pairwise disjoint, so there are at most countably many. □

As a byproduct of the proof we observe that a map with zero topological entropy and an ergodic invariant measure corresponds to the double limit point in the Sharkovsky order.

Proposition 15.4.7. *Under the assumptions of Theorem 15.4.2 the map f has periodic orbits with prime periods 2^k, $k = 0, 1, \ldots$, and no orbits of other periods.*

Proof. By Theorem 15.3.5 f can only have periods that are powers of 2. With the same notation as before let $\Delta_0 = [\sup S_1, \inf S_2]$. Then $\Delta_0 \to \Delta_0$ by Lemma 15.4.5 and hence Δ_0 contains a fixed point for f by Lemma 15.1.2. We proceed by induction. Define $\Delta_n = [\sup S_1^n, \inf S_2^n]$. Applying Lemma 15.4.5 inductively shows that $\Delta_n \to \Delta_n$ under f^{2^n} and hence Δ_n contains a point of prime period 2^n for f by Lemma 15.1.2. □

Notice from this construction that the periodic orbits are *properly intertwined* in the following sense: In the set of periodic points of period up to 2^{n+1} the immediate neighbors of any point of period 2^n have period 2^{n+1}.

Note also that S is contained in the closure of the set of periodic points, although it is disjoint from all periodic points. Namely, the set S is obtained by the following construction: For given n let $\mathcal{S}_n$ be the union of intervals $[\inf A, \sup A]$, where A is an image of S^{n+1} under an iterate of f. Any complementary interval contains a periodic point and S is the boundary of $\bigcap_{n\in\mathbb{N}} \mathcal{S}_n$.

We can reverse this statement as follows:

Theorem 15.4.8. *Let f be an interval map that has a properly intertwined system of periodic points with periods 2^n for all $n \in \mathbb{N}$ and no other periods. Then it has an ergodic invariant measure μ that is an adding machine and whose support is in the closure of the periodic points.*

Remark. This result fails without the assumption that the orbits are properly intertwined. By Theorem 15.3.5 one can replace the assumption of absence of other periods by the assumption of zero entropy. The map f_∞ constructed in the proof of Theorem 15.3.7 satisfies the conditions of Theorem 15.4.8.

Proof. We begin by studying the structure of the intertwined set of periodic points. Regarding the fixed point as the middle we have one period-two point on either side and these are interchanged by f. There are two period-four points on either side. We will use the absence of periods other than powers of two to see first that the two left period-four points are mapped to the two right period-four points. The period-four points are adjacent to the period-two points and the six points of period two and four define a collection of five intervals. We will study its Markov graph assuming that a period-four point on the left is mapped to the other period-four point on the left. Label these periodic points by $x_0, \dots, x_5$ (with periods 4, 2, 4, 4, 2, 4) and the intervals $[x_{i-1}, x_i]$ by I_i. If the left and right period-four points are not interchanged then f induces one of the following permutations on the period-four points: (1234), (1243), (2134), (2143). The first forces the presence of all periods by Exercise 15.3.1 and is hence forbidden. The same argument rules out the last (which can be rewritten as (4321)). The two remaining permutations have in common the fact that $\{x_2, x_3\}$ is mapped to $\{x_1, x_4\}$, that is, the interval I_3 f-covers all other intervals. Since I_3 is f-covered by (at least two) other intervals in either case, we obtain a subgraph isomorphic to (15.1.1) and hence have a period-three point, contrary to our assumption.

Refining this argument slightly shows that the left and right halves of the period-2^n orbit of our intertwined system are exchanged by f for any $n \in \mathbb{N}$. Furthermore we can study the dynamics of the period-eight orbit in more detail by considering the action of f^2 on either half of it. Since the halves are interchanged by f, this is well defined and we find ourselves in the same situation as in the preceding argument, so we have a description of how f^2 acts on the left half. Let us show that the left half $\{x_1, \dots, x_4\}$ is mapped to the right half $\{x_5, \dots, x_8\}$ by $f(\{x_1, x_2\}) = \{x_i, x_{i+1}\}$ for $i = 5$ or $i = 7$ (that is, in "packets") by assuming the contrary to conclude that there must be a period-six orbit. The period-eight orbit defines six intervals not containing the fixed point of the intertwined system. Labeling them as I_1 to I_6 we need to show (as a representative case) that $I_1 \to I_5$ is not allowed. But in this case we must have $I_5 \to I_3$, since we know f^2 on the left half of the orbit, and $I_3 \to I_j$ for $j = 4, 5, 6$, since the endpoints of I_3 are necessarily mapped to the extreme points of the right half. Since we have $I_j \to I_1$ for at least one $j = 4, 5, 6$, we obtain a Markov subgraph $I_1 \to I_6 \rightleftarrows I_3 \to I_j \to I_1$ that contains a loop of length six, forcing a period-six orbit by Corollary 15.1.4. An equivalent description of this conclusion is that none of the intervals containing a period-four point can f-cover an interval containing a period-two point. In general the same arguments show that of the intervals defined by a period-2^{n+1} orbit none containing a period-2^n point of the intertwined system can f-cover one containing a period-2^{n-1} point of the system.

We will use the preceding description of the dynamics to construct a set S that will be the support of the desired ergodic invariant measure. First consider the interval I whose endpoints are the period-two points and let $\mathcal{S}^0$ be a full component of $I \to I$. There are two intervals whose endpoints are on the period-four orbit and that contain a period-two point, I_1 on the left and I_2 on the right. We let $\mathcal{S}^1$ be the union of one full component each of $I_1 \to I_2 \to I_1$ and $I_2 \to I_1 \to I_2$. Notice that $\mathcal{S}^1$ has positive distance from the fixed point. $\mathcal{S}^2$ is likewise obtained as a union of full components corresponding to the length-four loops associated with the four intervals obtained similarly from the period-eight orbit. Thus $\mathcal{S}^2$ is separated from the points with period one and two. Likewise we obtain sets $\mathcal{S}^n$ consisting of 2^{n-1} intervals separated from the points of period up to 2^{n-1}. Next we let S be the boundary of $\bigcup_{m=1}^{\infty} \bigcap_{n=m}^{\infty} \mathcal{S}^n$, that is, S is the collection of nonperiodic points in the closure of the set of periodic points. Define the measure μ as the weak limit of the uniform measures δ_n on the orbits of period 2^n. Finally denote by y_n the right endpoint of the leftmost interval in $\mathcal{S}^n$ and let $S^n = \{x \in S \mid x \leq y_n\}$. To these S^n the final argument in the proof of Theorem 15.4.2 applies verbatim. □

Exercise

15.4.1*. *Suppose $f: [0,1] \to [0,1]$ is piecewise monotone with finitely many fixed points and $h_{\text{top}}(f) < (\log 2)/2$. Let μ be an f-invariant Borel probability measure such that f and f^2 are ergodic. Show that $\mu = \delta_x$ is concentrated on a point.*

5. The kneading theory

As we noticed at the end of Section 2 topological Markov chains do not provide a sufficiently rich class of models even for piecewise-monotone interval maps. Nevertheless a more general class of symbolic dynamical systems may be used to describe the essential features of such maps. The idea, due to Milnor and Thurston, is to consider the coding of critical points with respect to the partition into intervals of monotonicity. This approach can be carried out for arbitrary piecewise-monotone maps. In order to avoid rather heavy notation we choose to simplify the discussion by considering the simplest such maps, namely, unimodal maps.

In the remainder of this chapter we denote by Ω the space Ω_3^R of one-sided sequences of symbols 0, 1, 2 and by σ the one-sided shift σ^R on Ω.

Definition 15.5.1. Let $I = [0,1]$ denote the unit interval. Then a continuous map $f: I \to I$ is called *unimodal* if $f(0) = f(1) = 0$ and there exists a $c \in (0,1)$ such that f is increasing on $I_0 := [0,c)$ and decreasing on $I_2 := (c,1]$. c is then called the *turning point* of f and we let $I_1 := \{c\}$. Let $i^f := i: I \to \Omega$, $x \mapsto i(x)$ such that $f^n(x) \in I_{i_n}$. Then $i(x)$ is called the *itinerary* of x. Let $I(x) := \{y \mid i(y) = i(x)\}$. $\nu := \nu(f) := i(c)$ is called the *kneading sequence* of f.

The maps $x \mapsto \lambda x(1-x)$ from the quadratic family give useful standard examples. Clearly we have $\sigma \circ i = i \circ f$. Note also that the one-sided limits

$$i(x^-) := \lim_{y \to x^-} i(y) \text{ and } i(x^+) := \lim_{y \to x^+} i(y)$$

are well defined and coincide with $i(x)$ if and only if $f^i(x) \neq c$ for all $i \in \mathbb{N}_0$. Conversely, if $f^i(x) = c$ for some $i \in \mathbb{N}_0$ then $i(x) \neq i(x^+) \neq i(x^-) \neq i(x)$. Note that $\nu_0 = 1$ always.

An example of how one obtains dynamical information from itineraries is the following lemma.

Lemma 15.5.2. *Let $f: I \to I$ be unimodal, $x \in I$, and $n, p \in \mathbb{N}$ such that $\sigma^{p+n}(i(x)) = \sigma^n(i(x))$. Then the ω-limit set $\omega(x)$ is a periodic orbit of period p or $2p$.*

Proof. Replacing x with $f^n(x)$, assume $n = 0$. If $f^i(x) = c$ for some i, then c and hence x is p-periodic and we are done. Otherwise $\sigma^p(i(x)) = i(x)$ implies that $\{f^{kp+j}(x) \mid k \in \mathbb{N}\} \subset I_{l_j}$, where $l_j \in \{0,2\}$. Let $K_j \subset I_{l_j}$ be the smallest closed interval containing $\{f^{kp+j}(x) \mid k \in \mathbb{N}\} \subset I_{l_j}$. Then $f_{\restriction K_j}$ is monotone, $f(K_j) \subset K_{j+1}$ when $0 \le j < p-1$, and $f(K_{p-1}) \subset K_0$, so $f^p{}_{\restriction K_0}: K_0 \to K_0$ is monotone. If $f^p{}_{\restriction K_0}$ is increasing, then by Proposition 1.1.6 every point of K_0, hence x, is asymptotic to some $y \in \operatorname{Fix}(f^p)$. If $f^p{}_{\restriction K_0}$ is decreasing, then $f^{2p}{}_{\restriction K_0}$ is increasing and x is asymptotic to some $y \in \operatorname{Fix}(f^{2p})$ by Proposition 1.1.6. □

In order to understand the relation between the itinerary of a point and its dynamics better we define now an ordering on Ω which for itineraries corresponds to that of the points.

Definition 15.5.3. If $\alpha \in \Omega$ let $\epsilon_n(\alpha) := (1-\alpha_0)(1-\alpha_1)(1-\alpha_2)\cdots(1-\alpha_{n-1})$. If $\omega \in \Omega$ then we say $\alpha \prec \omega$ if and only if there exists an $n \in \mathbb{N}$ such that $\alpha_i = \omega_i$ for $i < n$ and $\epsilon_n(\alpha)(\alpha_n+1) < \epsilon_n(\omega)(\omega_n+1)$. Furthermore let $\alpha \preceq \omega$ mean that $\alpha \prec \omega$ or $\alpha = \omega$.

Remark. Note that this defines a linear ordering.

Lemma 15.5.4. *For $x, y \in I$ we have $i(x) \prec i(y) \Rightarrow x < y \Rightarrow i(x) \preceq i(y)$.*

Proof. If $i(x) \prec i(y)$ and n is as in the definition of "$\prec$" then $i_j(x) \neq 1$ for $0 \leq j \leq n-1$ and moreover f^n is monotone on the interval with endpoints x and y and increasing or decreasing according to whether $\epsilon_n(i(x))$ is positive or negative. Together with the observation that in this situation $i_n(x) < i_n(y) \Leftrightarrow f^n(x) < f^n(y)$ and $i_n(x) > i_n(y) \Leftrightarrow f^n(x) > f^n(y)$ this shows that $x < y$.

By contraposition this also implies that $x < y \Rightarrow i(x) \preceq i(y)$. □

Remark. The preceding lemma shows that $I(x) = \{y \mid i(y) = i(x)\}$ is an interval (possibly of zero length) on which f^n is monotone for all $n \in \mathbb{N}$ and such that this holds for no open interval containing $I(x)$. Note further that unless the intervals $f^j(I(x))$ $(j \in \mathbb{N})$ are pairwise disjoint, there is a $y \in f^{n+p}(I(x)) \cap f^n(I(x))$ for some $n, p \in \mathbb{N}$, so $\sigma^{n+p}(i(x)) = i(y) = \sigma^n(i(x))$, that is, $i(x)$ is eventually periodic, and thus by Lemma 15.5.2 every $y \in I(x)$ is asymptotic to a periodic point of f.

The point of these observations is that we can now decide whether a sequence in Ω is the itinerary of a point. We will see that the following obvious necessary condition is essentially sufficient:

Corollary 15.5.5. *If $\alpha = i(x) \in \Omega$ then*

$$\alpha_j = 1 \Rightarrow \sigma^j(\alpha) = \nu \text{ and } \sigma^{j+1}(\alpha) \preceq \sigma(\nu) \text{ for all } j. \tag{15.5.1}$$

Proof. If $\alpha_j = 1$ then $f^j(x) = c$, hence $\sigma^j(\alpha) = \sigma^j(i(x)) = i(f^j(x)) = \nu$. Since $f^{j+1}(x) < f(c)$ we have $\sigma(\nu) = i(f(c)) \succeq i(f^{j+1}(x)) = \sigma^{j+1}(\alpha)$. □

If we assume strict inequality in (15.5.1) then the condition becomes sufficient.

Proposition 15.5.6. *If $f: I \to I$ is unimodal with kneading sequence ν and $\alpha \in \Omega$ satisfies*

$$\alpha_j = 1 \Rightarrow \sigma^j(\alpha) = i(c), \qquad \sigma^{j+1}(\alpha) \prec \sigma(\nu) \text{ for all } j \tag{15.5.2}$$

then $\alpha = i(x)$ for some $x \in I$.

Proof. If not then $L := \{x \in I \mid i(x) \prec \alpha\}$ and $R := \{x \in I \mid \alpha \prec i(x)\}$ are intervals by Lemma 15.5.4 and $I = L \cup R$ and $0 \in L$, $1 \in R$. Thus, since I is connected, we have $\sup L = \inf R =: a$ and $a \in L$ or $a \in R$. Suppose $a \in L$. Then

$$i(a) \prec \alpha \preceq i(a^+), \tag{15.5.3}$$

so there exists an $n \in \mathbb{N}$ with $i_j(a) = \alpha_j$ for $j < n$ and

$$\epsilon_n(\alpha)(i_n(a) + 1) < \epsilon_n(\alpha)(\alpha_n + 1). \tag{15.5.4}$$

Note that $i_j(a) \neq 1$ for $j < n$ since otherwise $\sigma^j(\alpha) = i(c)$, so $\alpha = i(a)$. Thus f^n is monotone on some open neighborhood of a. In particular $i_j(a^+) = i_j(a)$ for $0 \leq j \leq n$. On the other hand

$$f^n(a) = c \tag{15.5.5}$$

since otherwise there is an open interval J containing a and $f^j(J) \subset I_{i_j(a)}$ for $j \leq n$, and hence $i(y) \prec \alpha$ for all $y \in J$, that is, $a \in J \subset L$, contradicting the definition of a. This shows that $i_{n+1}(a^+) \neq i_{n+1}(a)$ and hence

$$\epsilon_n(\alpha)(i_{n+1}(a^+) + 1) < \epsilon_n(\alpha)(i_{n+1}(a) + 1)$$

since $i(a) \preceq i(a^+)$. Furthermore (15.5.4) and (15.5.5) yield

$$\epsilon_n(\alpha) \cdot 2 < \epsilon_n(\alpha)(\alpha_n + 1)$$

and hence $\alpha_n = 2$ if $\epsilon_n(\alpha) = 1$ and $\alpha_n = 0$ if $\epsilon_n(\alpha) = -1$. In the first case $\nu_0 = 1 < 2 = \alpha_n$, so $\nu \prec \sigma^n(\alpha)$. Since $\epsilon_n(\alpha) = 1$ this implies $i(a^+) \prec i(a) \prec \alpha$, contradicting (15.5.5). In the second case $\nu_0 = 1 > 0 = \alpha_n$ so $\sigma^n(\alpha) \prec \nu$. Since $\epsilon_n(\alpha) = -1$ this implies $i(a) \prec i(a^+) \prec \alpha$, contradicting (15.5.5). Thus $a \notin L$ and by replacing $f(x)$ with $1 - f(1 - x)$ we also get $a \notin R$. □

There is one interesting case in which all itineraries satisfy the sufficient condition (15.5.2) rather than just (15.5.1). Namely, when c is not periodic and $I(f(c)) = \{f(c)\}$, that is, the full preimage $C_f := \{x \in [0,1] \mid f^n(x) = c \text{ for some } n \in \mathbb{N}\}$ of c is dense. In this case we can get a lot of structural information.

Proposition 15.5.7. *Suppose $f, g\colon I \to I$ are unimodal with the same kneading sequence. Then there is a strictly monotone $h\colon C_f \to C_g$ such that $h \circ f = g \circ h$ on C_f. If $\bar{C}_g = I$ then h extends to a semiconjugacy $h\colon I \to I$ between f and g. In particular if we consider only maps with C_f dense, then the kneading sequence is a complete invariant of topological conjugacy.*

Proof. If $x \in C_f$ then $i^f(x)$ satisfies (15.5.2), so $i^f(x) = i^g(y)$ for some $y \in I$ and clearly $y \in C_g$; hence y is unique (since $I(y)$ is mapped to the critical point by some iterate of g it is a point by piecewise monotonicity). Thus $h(x) := y$ is well defined and also monotone by Lemma 15.5.4. Clearly $i^f(h \circ f(x)) = i^g(g \circ h(x))$, so $h \circ f = g \circ h$ on C_f. By monotonicity f extends to $\bar{C}_f \subset I$. If (a, b) is a connected component of $I \smallsetminus C_f$ and $\bar{C}_g = I$ then $h(a) = h(b)$ and we set $h(x) = h(a)$ for $x \in (a, b)$. Then $h\colon I \to I$ is monotone and $h \circ f = g \circ h$. □

6. The tent model

Theorem 15.6.1. (Milnor, Thurston) *Let $f: I \to I$ be a unimodal map with positive topological entropy. Then f is semiconjugate to the tent map $\tau_s: x \mapsto s(1-|1-2x|)/2$ with slope $s = e^{h_{\text{top}}(f)}$.*

The proof is rather indirect and demonstrates the use of complex analysis to establish existence of certain asymptotics (in this case $\Lambda(J)$; see Definition 15.6.7), which is used to construct the semiconjugacy. More specifically, essential asymptotic information about the map is organized in convergent power series similarly to the way it was done for the zeta function. Then analytic properties of the resulting holomorphic or meromorphic functions are used to produce the desired asymptotic. We will use certain estimates to establish convergence and functional equations to locate the zeros and poles of the functions. Similar methods are extensively used in the theory of zeta functions for hyperbolic maps.[1]

Proof. Let $B(0,1) \subset \mathbb{C}$ be the open unit disk and denote by $C^\omega(B(0,1))$ the space of holomorphic functions endowed with the topology of uniform convergence on compact sets. It will be convenient to set

$$E_n(J) := \{x \in J \mid f^n(x) = c, f^k(x) \neq c \text{ for } k < n\},$$

for any interval $J \subset I$.

Definition 15.6.2. Let $\tilde{k} = (\tilde{k}^0, \tilde{k}^2): \Omega \to C^\omega(B(0,1)) \times C^\omega(B(0,1))$,

$$\tilde{k}^i(\alpha) = \sum_{n=0}^{\infty} \tilde{k}^i_n(\alpha) t^n,$$

where

$$\tilde{k}^i_n(\alpha) = \begin{cases} \epsilon_n(\alpha) & \text{if } \alpha_n = i, \\ \frac{1}{2}\epsilon_n(\alpha) & \text{if } \alpha_n = 1, \\ 0 & \text{otherwise,} \end{cases}$$

and define $k: I \to C^\omega(B(0,1)) \times C^\omega(B(0,1))$ by $k := \tilde{k} \circ i$, where i is the itinerary mapping. We write $k(x,t) = \sum_{n=0}^{\infty} k_n(x) t^n$, that is, $k^i_n(x) = \tilde{k}^i_n(i(x))$. We define the *kneading invariant* by

$$K(t) := k(c^+, t) - k(c^-, t).$$

Note that $\tilde{k}(\alpha)$ is holomorphic on $B(0,1)$ and a polynomial if $\alpha_n = 1$ for some $n \in \mathbb{N}_0$. Note also that $K(0) = (-1,1)$.

Lemma 15.6.3. *$\tilde{k}$ is continuous in α.*

Proof. If $\alpha \in \Omega$, $A \subset B(0,1)$ is compact, and $\epsilon > 0$ then let $r < 1$ such that $A \subset B(0,r)$, $N \in \mathbb{N}$ such that $2r^N/(1-r) < \epsilon$, and V a neighborhood of α in Ω such that $\beta_n = \alpha_n$ for all $\beta \in V$ and $n < N$. Then

$$|\tilde{k}^i(\alpha)(t) - \tilde{k}^i(\beta)(t)| \leq \sum_{n=N}^{\infty} |\tilde{k}^i_n(\alpha) - \tilde{k}^i_n(\beta)||t|^n \leq \sum_{n=N}^{\infty} 2|t|^n \leq \frac{2r^N}{1-r} < \epsilon$$

on A. □

Lemma 15.6.4.

(1) $(1-t)k^0(x,t)+(1+t)k^2(x,t)=1$.
(2) $(1+t)K_2(t)=-(1-t)K_0(t)$.
(3) $k(x^+,t)=k(x,t)+\frac{1}{2}t^nK(t)$ *for* $x\in E_n(I)$.
(4) $k(x^-,t)=k(x,t)-\frac{1}{2}t^nK(t)$ *for* $x\in E_n(I)$.
(5) $k(x,t)$ *is an nth degree polynomial for* $x\in E_n(I)$.

Proof. Note that

$$k_0^0(x)+k_0^2(x)=1. \tag{15.6.1}$$

We also have

$$k_n^0(x)-k_n^2(x)=k_{n+1}^0(x)+k_{n+1}^2(x). \tag{15.6.2}$$

Recall that $I_0=[0,c)$ and $I_2=(c,1]$. To see (15.6.2) note that if $\epsilon_n(i(x))=0$ then both sides vanish. Otherwise consider three cases:

I. $f^n(x)\in I_i$, $f^{n+1}(x)\in I_j$. Then $k_{n+1}^j(x)=(1-i)k_n^i(x)$, $k_{n+1}^{2-j}(x)=0$, and $k_n^{2-i}(x)=0$, proving (15.6.2).
II. $f^n(x)=c$. Then $k_n^0(x)-k_n^2(x)=\frac{1}{2}\epsilon_n(i(x))-\frac{1}{2}\epsilon_n(i(x))=k_{n+1}^0(x)+k_{n+1}^2(x)$.
III. $f^n(x)\in I_i$, $f^{n+1}(x)=c$. Then $k_{n+1}^0(x)+k_{n+1}^2(x)=\frac{1}{2}(1-i)k_n^i(x)+\frac{1}{2}(1-i)k_n^i(x)=(1-i)k_n^i(x)=k_n^0(x)-k_n^2(x)$.

Using (15.6.2) we now prove (1) of Lemma 15.6.4:

$$\begin{aligned}(1-t)\sum_{n=0}^{N}k_n^0(x)t^n+(1+t)\sum_{n=0}^{N}k_n^2(x)t^n\\ &=\sum_{n=0}^{N}[(k_n^0(x)+k_n^2(x))t^n-(k_n^0(x)-k_n^2(x))t^{n+1}]\\ &=\sum_{n=0}^{N}[(k_n^0(x)+k_n^2(x))t^n-(k_{n+1}^0(x)+k_{n+1}^2(x))t^{n+1}]\\ &=1-(k_{N+1}^0(x)+k_{N+1}^2(x))t^{N+1}\xrightarrow[N\to\infty]{}1\end{aligned}$$

uniformly on compact sets. This shows (1). Claim (2) follows immediately.

Statement (5) is evident. To prove (3) and (4) one compares the Taylor coefficients on both sides. This is a bit tedious, but uses only the following easy observations:

$$\begin{gathered}i(x)_j=i(x^-)_j=i(x^+)_j \text{ for } j<n,\\ \sigma^{n+1}(i(x^-))=\sigma^{n+1}(i(x^+))=\sigma(i(c^-))=\sigma(i(c^+)),\\ \epsilon_{n+k}(i(x^-))=-\epsilon_{n+k}(i(x^+)),\\ \epsilon_n(i(x^-))\cdot\epsilon_k(i(c^-))=\epsilon_{n+k}(i(x^-)),\\ \epsilon_n(i(x^+))\cdot\epsilon_k(i(c^+))=\epsilon_{n+k}(i(x^+)),\end{gathered}$$

for $k\geq 1$. To compare coefficients of t^n note also that $K(0)=(-1,1)$. □

By Lemma 15.6.4(2) we may define

$$K_f(t) := \frac{K_2(t)}{1-t} = -\frac{K_0(t)}{1+t}.$$

Note that K_f is holomorphic on $B(0,1)$.

Definition 15.6.5. Recall that in Proposition 15.2.12 we defined $c(f^n) = c_n$ to be the number of turning points of f^n and let $s := \lim_{n\to\infty} \sqrt[n]{c(f^n)} = e^{h_{\mathrm{top}}(f)}$ (see Proposition 15.2.13) and for an interval $J \subset I$ set $C_n(J) := \operatorname{card} E_n(J)$ and

$$C(J)(t) := \sum_{n=0}^{\infty} C_n(J)t^n.$$

We define the *lap function* by

$$L_f(J,t) := \sum_{n=0}^{\infty} c(f^n\restriction_J)t^n.$$

We assume from now on that $s > 1$.

Lemma 15.6.6. *If* $[a,b] = J \subset I$ *then* $C(J)$ *is holomorphic on* $B(0,1/s)$,

$$C(J)(t)K(t) = k(b^-,t) - k(a^+,t) \text{ for } |t| < \frac{1}{s}, \tag{15.6.3}$$

and $C(J)$ *is meromorphic on* $B(0,1)$ *with poles at most at the zeros of* K_f. *Furthermore* $L_f(J,\cdot)$ *is meromorphic on* $B(0,1)$ *with poles at most at the zeros of* K_f, *and* $L_f(I,\cdot)$ *has a pole at* $1/s$.

Proof. $C_n(J) \le c(f^n\restriction_J) \le c(f^n)$, so $C(J)$ is holomorphic on $B(0,1/s)$. Now $F_n(J) := \bigcup_{k\le n} E_k(J)$ is the collection of turning points of $f^n\restriction_J$. If x and y are adjacent points of $F_n(J)$, $x<y$, then $k(y^-,t) = k(z,t) = k(x^+,t)$ up to nth order for any $z \in (x,y)$. Thus

$$\begin{aligned} k(b^-,t) - k(a^+,t) + o(t^n) = \sum_{x\in F_n(J)} (k(x^+,t) - k(x^-,t)) &= \sum_{k\le n}\sum_{x\in E_k(J)} t^k K(t) \\ &= \sum_{k\le n} C_k(J)K(t)t^k, \end{aligned}$$

using (3) and (4) from the previous lemma. This proves (15.6.3). To see that $C(J)$ is meromorphic on $B(0,1)$ with poles at most at the zeros of K_f note that for $|t| < 1/s$

$$C(J)(t)K_2(t) = k^2(b^-,t) - k^2(a^+,t),$$

so $C(J)K_f(t) = (1-t)\big(k^2(b^-,t) - k^2(a^+,t)\big)$.

To study L_f note the identity

$$\Big(\sum_{n=0}^{\infty} t^{n+1}\Big)\Big(\sum_{m=0}^{\infty} a_m t^m\Big) = \sum_{n=0}^{\infty}\Big(\sum_{m=0}^{n-1} a_m\Big)t^n.$$

Since $c(f^n\restriction_J) = 1 + \sum_{m=0}^{n-1} C_m(J)$ this yields

$$L_f(J,t) = \sum_{n=0}^{\infty} c(f^n\restriction_J)t^n = \frac{1}{1-t} + \sum_{n=0}^{\infty}\sum_{m=0}^{n-1} C_m(J)t^n = \frac{1}{1-t}(1+tC(J)(t)),$$

so $L_f(J,\cdot)$ is indeed meromorphic on $B(0,1)$ with poles at most at the poles of $C(J)$, hence at most at the zeros of K_f. To see that $L_f(I,\cdot)$ has a pole at $1/s$ note that since the coefficients $c(f^n)$ of $L_f(I,\cdot)$ are positive, hence $|L_f(I,t)| \le \sum_{n=0}^{\infty} c(f^n)|t|^n$, we have $\lim_{t\to 1/s}\sum_{n=0}^{\infty} c(f^n)|t|^n = \infty$, because $c(f^n) \ge s^n$ by submultiplicativity. □

Lemma 15.6.6 implies in particular that $\dfrac{L_f(J,t)}{L_f(I,t)}$ is meromorphic, and for $0 < t < 1/s$ we clearly have $0 \le L_f(J,t) \le L_f(I,t)$. This is the key point where complex analysis guarantees the existence of certain limits to produce the semiconjugacy.

Definition 15.6.7. For intervals $J \subset I$ we set $\Lambda(J) := \lim_{t\to 1/s} \dfrac{L_f(J,t)}{L_f(I,t)} \in [0,1]$ and $h\colon I \to I, x \mapsto \Lambda([0,x])$.

Clearly h is nondecreasing. We will finish the proof of the theorem by showing that h is a semiconjugacy between f and the tent map with slope s.

Lemma 15.6.8.

(1) *If $J_1 \cap J_2$ is a point then $\Lambda(J_1 \cup J_2) = \Lambda(J_1) + \Lambda(J_2)$.*
(2) *If f is monotone on J then $\Lambda(f(J)) = s\Lambda(J)$.*
(3) *h is continuous.*
(4) *$h(f(x)) = sh(x)$ on $[0,c]$.*
(5) *$h(f(x)) = s(1-h(x))$ on $[c,1]$.*

Proof. (1) $|c(f^n\restriction_{J_1}) + c(f^n\restriction_{J_2}) - c(f^n\restriction_{J_1\cup J_2})| \le 1$, so

$$|L_f(J_1,t) + L_f(J_2,t) - L_f(J_1\cup J_2,t)| \le \sum_{n=0}^{\infty} |t|^n = \frac{1}{1-|t|}.$$

Dividing by $L_f(I,t)$ and letting $t \to 1/s$ yields (1) since $L_f(I,\cdot)$ has a pole at $1/s$.

(2) Note that $c(f^{n+1}\restriction_J) = c(f^n\restriction_{f(J)})$ implies $L_f(J,t) = 1 + tL_f(f(J),t)$ and hence $\Lambda(J) = \lim_{t\to 1/s} \dfrac{1+tL_f(f(J),t)}{L_f(I,t)} = \dfrac{1}{s}\Lambda(f(J))$.

(3) If $x < y$ such that $f^n \restriction_{[x,y]}$ is monotone then $h(y) = \Lambda([0,y]) = \Lambda([0,x]) + \Lambda([x,y]) = h(x) + s^{-n}\Lambda(f^n([x,y])) \leq h(x) + s^{-n}$.
(4) Use (2) together with $f(0) = 0$ and $h(0) = \Lambda([0,0]) = 0$.
(5) For $x \in [c,1]$ we have

$$\begin{aligned} h(f(x)) = \Lambda([0,f(x)]) = \Lambda(f([x,1])) &= s\Lambda([x,1]) \\ &= s(\Lambda([0,1]) - \Lambda([0,x])) = s(1-h(x)). \end{aligned} \square$$

This proves Theorem 15.6.1 □

Remark. Similarly to the kneading theory the result of this section generalizes to arbitrary piecewise-monotone maps. A very attractive conclusion is that *up to semiconjugacy* such maps with positive entropy admit a finite-dimensional family of models.

Exercise

15.6.1. *Show that if f is unimodal, $h_{\text{top}}(f) > 0$ then* $\lim_{n\to\infty} \dfrac{V(f^n)}{e^{nh_{\text{top}}(f)}} = 1$.

16

Smooth maps of the interval

Similarly to the case of invertible circle maps studied in Chapters 11 and 12 for interval maps differentiability allows us to sharpen many results concerning the orbit structure. Unlike Chapter 12 where the main theme was conjugacy to a rotation, a prototypical nonhyperbolic model, here the main new feature which distinguishes the smooth case is the presence of hyperbolic sets as defined in Section 6.4.

1. The structure of hyperbolic repellers

As we saw in the previous chapter, noninvertible interval maps may have periodic points of different periods. For an f-periodic point p the ***basin*** of attraction B of p is the collection of all points positively asymptotic to p (p can be attracting or semistable). We call the union of the connected components that contain a point of the orbit $\mathcal{O}(p)$ of p the ***immediate basin*** of attraction of p. Basins and immediate basins of attracting points are clearly open. Consider the union R of the semistable points and the complement of the union of all basins of periodic points of f. This set is called the ***universal repelling set*** of f. By construction it is closed and f-invariant. It is also "f^{-1}-invariant" in the sense that $f^{-1}(R) = R$. Any complicated dynamics clearly takes place in R. For example, every nonatomic f-invariant measure is supported on R, so by the Variational Principle (Theorem 4.5.3) $h_{\text{top}}(f) = h_{\text{top}}(f\restriction_R)$. If there are only finitely many attracting periodic points then R is a repeller in the traditional sense, that is, for a sufficiently small neighborhood U of R and $x \in U \smallsetminus R$ there exists $n \in \mathbb{N}$ such that $f^n(x) \notin U$. This motivates the study of hyperbolic repellers. A repelling hyperbolic set (Definition 6.4.3) is said to be *locally maximal* if it has an open neighborhood that contains no larger invariant set.

Theorem 16.1.1. *If $f\colon [0,1] \xrightarrow{C^1} [0,1]$ then any locally maximal hyperbolic repeller Λ is topologically conjugate to a topological Markov chain.*

Proof. Fix a neighborhood U of Λ containing no critical points and no invariant sets properly containing Λ and consider a pairwise disjoint open cover $\mathcal{C}$ of Λ by intervals in U on which f is monotone. Since $f\restriction_\Lambda$ is hyperbolic, there exists $n \in \mathbb{N}$ such that the nth preimage of each of these intervals is in $\bigcup_{I\in\mathcal{C}} I$. Then this collection $\mathcal{C}_n = \{J_i \mid 0 \le i \le N\}$ is clearly a Markov cover of Λ for f^n. We may assume that f is monotone on $f^k(I_i)$ for $0 \le k \le n$, $0 \le i \le N$, so that the collection of intervals Δ_n obtained from Lemma 15.1.3 is canonically defined and by construction a Markov cover of Λ for f. By our choice of U, Λ is the intersection of the preimages of the unions of these Δ_n. □

Remark. Later (Proposition 18.7.8) we will prove an analogous result for totally disconnected hyperbolic sets.

Corollary 16.1.2. *Let $f\colon [0,1] \xrightarrow{C^1} [0,1]$ and suppose the universal repelling set R is a hyperbolic repeller. Then $f\restriction_R$ is topologically conjugate to a topological Markov chain.*

Exercise

16.1.1. *Given $n \in \mathbb{N}$ construct a C^∞ interval map whose universal repelling set is topologically conjugate to the full shift on n symbols.*

2. Hyperbolic sets for smooth maps

In this section we prove that away from the critical points any smooth interval maps is to a large extent hyperbolic.

Theorem 16.2.1. (Mañé)[1] *If $f \in C^2([0,1],[0,1])$ and U is a neighborhood of the critical points of f then:*

(1) *All periodic orbits of f in $[0,1] \smallsetminus U$ of sufficiently large period are hyperbolic repelling.*
(2) *If B_0 is the union of the immediate basins of attraction of the attracting periodic orbits in $[0,1]\smallsetminus U$ and if all periodic orbits of f in $[0,1]\smallsetminus U$ are hyperbolic then there exist $C > 0$ and $\lambda > 1$ such that for every orbit segment $\{x, \dots, f^n(x)\} \subset [0,1] \smallsetminus (U \cup B_0)$ we have $|Df^n(x)| \ge C\lambda^n$.*

Corollary 16.2.2. *If $f \in C^2([0,1],[0,1])$ and Λ is a closed invariant set that contains neither critical points nor attracting or nonhyperbolic periodic points then Λ is hyperbolic.*

Proof. If $U \subset [0,1] \smallsetminus \Lambda$ is a neighborhood of the critical points then Theorem 16.2.1(1) gives $N \in \mathbb{N}$ such that all periodic points in $[0,1] \smallsetminus U$ of period at least N are hyperbolic repelling. The remaining periodic points are a compact set and hence there is a neighborhood V of Λ that contains no critical points or nonhyperbolic periodic points. If $W \subset \overline{W} \subset V$ is a neighborhood of Λ then Theorem 16.2.1(2) together with a variant of Exercise 6.4.2 implies that f is hyperbolic on $\Lambda \subset W$. □

Corollary 16.2.3. *If $f \in C^2([0,1],[0,1])$, all critical points are in the basin of a hyperbolic attracting periodic orbit, and all periodic points are hyperbolic then there are only finitely many hyperbolic attracting periodic orbits and the universal repelling set is hyperbolic.*

Proof of Theorem 16.2.1.

Lemma 16.2.4. *If p is a period-n point and I an interval containing p such that $f^n(I) \cap \mathcal{O}(p) = \{p\}$ then the covering multiplicity of $\{I, f(I), \dots, f^{n-1}(I)\}$ is at most 3.*

Proof. $f^n(I)$ is contained in the maximal interval J for which $J \cap \mathcal{O}(p) = \{p\}$. Clearly $f^i(I) \cap \mathcal{O}(p) = \{f^i(p)\}$ and $A := \{i < n \mid f^i(p) \in \bar{J}\}$ has at most three elements. But if $f^i(I) \cap J \neq \varnothing$ then $i \in A$.

Suppose there are numbers $0 \le i_1, i_2, i_3, i_4 < n$ such that $f^{i_1}(I) \cap f^{i_2}(I) \cap f^{i_3}(I) \cap f^{i_4}(I) \neq \varnothing$, that is, $\varnothing \neq f^{n-i_4+i_1}(I) \cap f^{n-i_4+i_2}(I) \cap f^{n-i_4+i_3}(I) \cap f^n(I) \subset f^{n-i_4+i_1}(I) \cap f^{n-i_4+i_2}(I) \cap f^{n-i_4+i_3}(I) \cap J$. Since card $A \le 3$, two of these numbers agree, as claimed. □

Lemma 16.2.5. *If I is an interval such that $f^n\restriction_I : I \to f^n(I)$ is a homeomorphism for all $n \in \mathbb{N}$ then either I has pairwise disjoint images or there are an interval J and $i, k \in \mathbb{N}$ such that $f^i(I) \subset J$, $f^k(J) \subset J$, and $f^k\restriction_J$ is monotone.*

Remark. In the latter case every point in I is positively asymptotic to a period-$2k$ orbit by Proposition 1.1.6 applied to f^{2k}.

Proof. If the images are not pairwise disjoint, that is, there exist $i, k \in \mathbb{N}$ such that $f^i(I) \cap f^{i+k}(I) \neq \varnothing$ and hence $f^{i+nk}(I) \cap f^{i+(n+1)k}(I) \neq \varnothing$, then $J := \bigcup_{n=0}^{\infty} f^{i+nk}(I)$ is as desired. □

The next lemma completes the proof of (1) of Theorem 16.2.1.

Lemma 16.2.6. *If U is a neighborhood of the critical points then there exist $M_n \to \infty$ such that if p has period n with $\mathcal{O}(p) \cap U = \varnothing$ then $|(f^n)'(p)| > M_n$.*

Proof. We first modify f conveniently in U. Let K be an interval containing a neighborhood of $[0,1]$ and let $g \in C^2(K,K)$ be such that

(1) $g\restriction_{[0,1]\smallsetminus U} = f\restriction_{[0,1]\smallsetminus U}$,
(2) ∂K is an attracting periodic orbit of g with immediate basin in $K \smallsetminus [0,1]$,
(3) for every critical point c of f there are open intervals $c \in W_c \subset V_c \subset U$ such that for either component J of $V_c \smallsetminus W_c$ there is a repelling periodic point in $g(J) \cup g^2(J)$.

Clearly it suffices to prove Lemma 16.2.6 for g with $V := \bigcup_c V_c$ instead of U. We also let $W := \bigcup_c W_c$.

The intent of (3) is to provide some amount of expansion. Indeed, it implies that there exists $\delta > 0$ such that for any component J of $V \smallsetminus W$ and any $i \in \mathbb{N}$ we have $l(g^i(J)) > \delta$.

Consider now a period-n orbit in $K \setminus V$ and pick a point p on it and a critical point c such that there are no points of $\mathcal{O}(p)$ between p and c. Let I be the maximal interval containing p for which $g^i(I) \cap W = \varnothing$ $(i \leq n)$ and $g^n(I) \cap \mathcal{O}(p) = \{p\}$.

Let us now show that $l(I) \to 0$, that is, for $\epsilon > 0$ there is an $N \in \mathbb{N}$ such that $n \geq N$ implies $l(I) < \epsilon$. To that end suppose to the contrary that there is a sequence p_n of period-m_n points with $m_n \to \infty$ for which the corresponding intervals I_n have length at least ϵ and converge (without loss of generality) to an interval I. But then $g^k(I_n) \cap W = \varnothing$ for $k \leq m_n$, hence $g^k(I) \cap W = \varnothing$ for $k \in \mathbb{N}$. Thus by Lemma 14.3.3 I does not have pairwise disjoint images and by Lemma 16.2.5 there exist $i, k \in \mathbb{N}$ and an interval J such that $g^i(I) \subset J$ and every $x \in J$ is asymptotic to a period-$2k$ point. But this is absurd since $g^k(I_n)$, and hence p_n, are in J for sufficiently large k. This proves that $l(I) \to 0$.

To complete the proof let I_c be the component of $I \setminus \{p\}$ whose image under g^n is between p and c. Since I was chosen maximal there is an $i \leq n$ and a component J of $V \setminus W$ such that $J \subset g^i(I_c)$, hence $l(g^n(I_c)) > \delta$ by choice of δ. As $n \to \infty$ we have $l(I) \to 0$ and hence $l(g^n(I_c))/l(I_c) \to \infty$. On the other hand Lemma 16.2.4 implies that $\sum_{i \leq n} l(g^i(I)) \leq 3l(K)$. This fact, together with Proposition 14.3.2, shows that $|(g^n)'(p)| \geq \text{const.}\, l(g^n(I_c))/l(I_c) \to \infty$ as claimed. □

Since (1) of Theorem 16.2.1 is proved we pick a neighborhood $V \subset \bar{V} \subset U$ of the critical points and assume from now on that all periodic orbits in $[0,1] \setminus V$ are hyperbolic. We furthermore denote by B the union of the immediate basins of the attracting periodic points whose orbits are in $[0,1] \setminus V$. For $X \subset [0,1]$ let

$$\Lambda_n(X) := \{x \in [0,1] \mid f^i(x) \in [0,1] \setminus (X \cup B) \text{ for } i \leq n\}.$$

Note that the assertion of Theorem 16.2.1 involves $\Lambda_n(U)$. Let d_n be the maximal length of a connected component of $\Lambda_n(V)$.

Next we obtain some hyperbolicity for points in $\Lambda_n(V)$.

Lemma 16.2.7. *There exist $\delta > 0$ and $\lambda > 1$ such that for all intervals $I \subset J \subset \Lambda_{n-1}(V)$ with $l(J) < \delta$ and such that $J, f(J), \dots, f^{n-1}(J)$ are pairwise disjoint and $J \subset f^n(J)$ we have $l(f^n(I)) > \lambda l(I)$.*

Proof. Since $J \subset f^n(J)$ there is a period-n point in J (Corollary 15.1.4) to which we can apply Lemma 16.2.6. Since $J \subset \Lambda_{n-1}(V)$ and $J, f(J), \dots, f^{n-1}(J)$ are pairwise disjoint, Proposition 14.3.2 gives $C > 0$ such that $l(f^n(I)) \geq CM_n l(I)$. Thus there exists $N \in \mathbb{N}$ such that $l(f^n(I)) \geq 2l(I)$ for $n \geq N$, and we are done for $n \geq N$.

For $n < N$ there are only finitely many period-n orbits; hence there exist $\delta > 0$, $\lambda > 1$ such that for any repelling period-n point x and $|y - x| < \delta$ we have $|(f^n)'(y)| > \lambda$. Thus if J is as before then $l(J) < \delta$ implies $|(f^n)'(x)| > \lambda$ on J. □

Lemma 16.2.8. *For $\epsilon > 0$ there exists $n_\epsilon \in \mathbb{N}$ such that $d_n < \epsilon$ for all $n > n_\epsilon$.*

Proof. Otherwise there exist $\epsilon > 0$ and intervals $I_n \subset \Lambda_n(V)$ with $l(I_n) \geq \epsilon$ and without loss of generality converging to an interval $I \subset \bigcap_{n \in \mathbb{N}} \Lambda_n(V)$. By construction the images of I do not accumulate to a critical point, and hence are not pairwise disjoint by Lemma 14.3.3. Thus Lemma 16.2.5 yields an interval J and $i, k \in \mathbb{N}$ such that $f^i(I) \subset J$ and $f^k\restriction_J$ is monotone. Thus every point of J is asymptotic to a periodic orbit by Proposition 1.1.6 and hence $f^i(I) \cap B \neq \varnothing$, so $f^i(I_n) \cap B \neq \varnothing$ for large n, which is impossible. □

The next lemmas are aimed at showing that for any interval $I \subset \Lambda_n(V)$ the sum of lengths of $f^i(I)$ up to $i = n$ is bounded by a constant independent of n. (This is then used by applying Proposition 14.3.2.) To that end we use some technical notions:

Definition 16.2.9. An interval I is said to be *n-Markov* if $f^i(I) \cap f^j(I) \neq \varnothing$ implies $f^i(I) \subset f^j(I)$ whenever $i \leq j \leq n$. If in addition $f^i(I), f^j(I) \subset f^k(I)$ implies $l(f^j(I)) > \lambda l(f^i(I))$ when $i < j < k \leq n$ then we say that I is *(λ, n)-hyperbolic.*

Hyperbolic intervals have bounded sums of lengths:

Lemma 16.2.10. *If $\lambda > 1$ and I is (λ, n)-hyperbolic then $\sum_{i=0}^{n} l(f^i(I)) < \lambda/(\lambda - 1)$.*

Proof. Suppose $\{i_1, \dots, i_k\} \subset \{0, \dots, n\}$ is such that $f^{i_1}(I), \dots, f^{i_k}(I)$ are pairwise disjoint and if $i \leq n$ then $f^i(I) \subset f^{i_t}(I)$ for some $t \leq k$. This is possible by the Markov property. For given t let $\{j_1, \dots, j_r\} = \{i \leq n \mid f^i(I) \subset f^{i_t}(I)\}$. The hyperbolicity assumption implies that $l(f^{j_{s+1}}(I)) > l(f^{j_s}(I))$ for $s < r$ and $j_r = i_t$, so $\sum_{s=1}^{r} l(f^{j_s}(I)) \leq (\lambda/(\lambda-1)) l(f^{i_t}(I))$. Thus $\sum_{i=0}^{n} l(f^i(I)) = \sum_{t=1}^{k} \sum_{f^i(I) \subset f^{i_t}(I)} l(f^i(I)) \leq \sum_{t=1}^{k} (\lambda/(\lambda-1)) l(f^{i_t}(I)) \leq \lambda/(\lambda - 1)$. □

This lemma together with the following three provides the crucial estimate of Lemma 16.2.13. Let δ be as in Lemma 16.2.7 and let $\epsilon < \delta$ be a lower bound for the length of connected components of $U \setminus V$. Take n_ϵ as in Lemma 16.2.8.

Lemma 16.2.11. *If I is a connected component of $\Lambda_n(V)$ and $I \cap \Lambda_n(U) \neq \varnothing$ then I is $(n - n_\epsilon)$-Markov.*

Proof. If not then there exist $i < j \leq n - n_\epsilon$ such that $f^i(I) \cap f^j(I) \neq \varnothing$ and $f^i(I) \not\subset f^j(I)$, that is, there is an endpoint a of I such that $f^j(a) \subset \operatorname{Int} f^i(I)$. We will show that I contains a neighborhood of a, which is absurd. Note first that $f^t(a) \notin \partial B$ for $t < j$ since otherwise $\partial B \ni f^{j-t}(f^t(A)) = f^j(a) \in \operatorname{Int} f^i(I)$, a contradiction. Furthermore $f^t(a) \notin \partial V$ for $t \leq n - n_\epsilon$ since otherwise $f^t(I)$ would contain a component of $U \setminus V$ (since it contains a point of U) and hence be of length at least ϵ, so $f^t(I) \not\subset \Lambda_{n-t}(V)$ by Lemma 16.2.8. This shows the existence of a neighborhood W of a such that $f^j(W) \subset f^i(I)$ and $f^t(W) \cap (B \cap V) = \varnothing$ for $t \leq j$. Thus $W \subset \Lambda_n(V) \subset I$, a contradiction. □

Lemma 16.2.12. *If I is as in Lemma 16.2.11, then I is (λ, n)-hyperbolic, where λ is as in Lemma 16.2.7.*

Proof. Lemma 16.2.11 already yields the required Markov property. Given $i < k \le n - n_\epsilon$ suppose j is the minimal integer for which there exist $i < j < k$ with $f^i(I) \cup f^j(I) \subset f^k(I)$.

We first show that $J := f^{k-j+i}(I) \subset f^{j-i}(J) = f^k(I)$ and that $J, \dots, f^{j-i-1}(J)$ are pairwise disjoint. It is useful to keep in mind that f^n is a diffeomorphism on I. Note that $f^{j-i}(f^i(I)) = f^j(I) \subset f^k(I)$ and that $J \subset \Lambda_{j-i}(V)$ is the maximal interval such that $f^{j-i}(J) \subset f^k(I)$. Thus $f^i(I) \subset J$ and by assumption $f^i(I) \subset f^k(I) = f^{j-i}(J)$, so $J \cap f^{j-i}(J) \ne \varnothing$. Since I is k-Markov, J is $(j-i)$-Markov, and hence $J \subset f^{j-i}(J) = f^k(I)$. To see that $J, \dots, f^{j-i-1}(J)$ are pairwise disjoint suppose first that $f^t(J) \cap f^{j-i}(J) \ne \varnothing$ for some $t < j - i$. Then using the Markov property $f^{i+t}(I) \subset f^t(J) \subset f^{j-i}(J) = f^k(I)$, contradicting minimality of j. Thus this is impossible. But if $f^n(J) \cap f^m(J) \ne \varnothing$ for any $n < m < j - i$ then also $f^{n+(j-i-m)}(J) \cap f^{j-i}(J) \ne \varnothing$, which we just ruled out.

The previous paragraph shows that we can apply Lemma 16.2.7 to conclude that $l(f^j(I)) > \lambda l(f^i(I))$.

In the general case of $i < j < k \le n - n_\epsilon$ and $f^i(I) \cup f^t(I) \subset f^k(I)$ we first pick a minimal $j \le t$ as before to obtain $l(f^j(I)) > \lambda l(f^i(I))$. Then we replace i by j and repeat this until we find $l(f^t(I)) > \lambda^m l(f^i(I)) > \lambda l(f^i(I))$ as required. □

Lemma 16.2.13. *There exists $C \in \mathbb{R}$ such that if $I \subset \Lambda_n(V)$ is an interval then $\sum_{i=0}^n l(f^i(I)) \le C$.*

Proof. If $n \le n_\epsilon$ then n_ϵ is an obvious upper bound. If $n > n_\epsilon$ Lemmas 16.2.10 and 16.2.12 show that

$$\sum_{i=0}^{n} l(f^i(I)) = \sum_{i=0}^{n-n_\epsilon} l(f^i(I)) + \sum_{i=n-n_\epsilon+1}^{n} l(f^i(I)) \le \lambda/(\lambda - 1) + n_\epsilon =: C.$$

□

We now use Lemma 16.2.13 and Proposition 14.3.2 to prove (2) of Theorem 16.2.1. We argue by contradiction and thus assume that there is a sequence $n_i \to \infty$ and $x_i \in \Lambda_{n_i}(V)$ such that $|(f^{n_i})'(x_i)| \to 1$. Suppose δ is smaller than the length of any connected component of $U \smallsetminus V$ and smaller than the distance between any $x, y \in \partial B$. Then we can show that if I_i is the connected component of x_i in $\Lambda_{n_i}(V)$ then there exists $m_i < n_i$ such that $l(f^{m_i}(I_i)) > \delta$.

To see this note that there are two possibilities:

(1) There exists $t < n_i$ such that $f^t(I_i) \cap \partial V \ne \varnothing$. Since $f^t(x_i) \notin U$ this means that $f^t(I_i)$ contains a component of $U \smallsetminus V$ and hence has length larger than δ.
(2) There exist $m < t < n_i$ such that $f^m(a) \cup f^t(b) \subset \partial B$, where a, b are the endpoints of I_i. Then $f^t(a) \in \partial B$ and hence $l(f^t(I_i)) > \delta$.

Note also that $l(I_i) \to 0$ by Lemma 16.2.8, so $m_i \to \infty$ and we may, after passing to a subsequence, assume that $f^{m_i}(I_i)$ converges to an interval I. Clearly $l(I) \geq \delta$.

Meanwhile note that $|(f^{n_i})'(x_i)| \to 1$ and Lemma 16.2.13 together with Proposition 14.3.2 imply that $l(f^{n_i}(I_i)) \leq 2l(I_i)$ for large i. Since $l(I_i) \to 0$ we find that $l(f^{n_i}(I_i)) \to 0$ and hence $n_i - m_i \to \infty$, so $I \subset \bigcap_{n\in\mathbb{N}} \Lambda_n(V)$, which is impossible. This completes the proof of Theorem 16.2.1. □

3. Continuity of entropy

Theorem 16.3.1. *Consider the space $C_0^2([0,1],[0,1])$ of C^2 maps of $[0,1]$ with nondegenerate critical points (that is, $|f'| + |f''| > 0$) with the C^2 topology. Then $h_{\text{top}}\colon C_0^2([0,1],[0,1]) \to \mathbb{R}$ is continuous.*

Proof. Lower semicontinuity is provided by Theorem 15.2.11. To prove upper semicontinuity we use Proposition 15.2.13 together with the fact that for small C^2-perturbations $c(f)$ can be controlled. Fix $f \in C_0^2([0,1],[0,1])$ and $\epsilon > 0$. Note that f is piecewise monotone. Suppose $g \in C_0^2([0,1],[0,1])$ is sufficiently C^2-close to f, so that it, too, is piecewise monotone. By Proposition 15.2.13 we can fix $n \geq (2(c_1+3)\log 2)/\epsilon$ such that $(1/n)\log c_n \leq h_{\text{top}}(f) + (\epsilon/2)$. The turning points for f^n are $0 = x_0 < x_1 < \cdots < x_{c_n} = 1$. Consider pairwise disjoint open intervals $\{I_i \mid 0 \leq i \leq c_n\}$ with $x_i \in I_i$. Since all critical points of f are turning points, the same holds for f^n and thus we may assume that g^n has no turning points in $[0,1] \smallsetminus \bigcup_{i=0}^{c_n} I_i$.

Now fix i. Let $r := \operatorname{card}\{0 \leq k < n \mid f^k(x_i) \text{ is a turning point for } f\}$. Note that if g is sufficiently close to f then g^n has at most $2^r + 1$ turning points in I_i. (Consider the number of maximal intervals in I_i on which g^n is monotone. For any k such that $g'(x) = 0$ at some $x \in g^k(I_i)$ this number is at most doubled, so it is at most 2^r.)

Consider the simple (and generic) case when there is no $k < n$ such that $f^k(x_i)$ is a periodic turning point. Then $r \leq c_1 + 1$ and g^n has at most $2^{c_1+1}+1$ turning points in I_i.

Now we take care of the complications that arise from the presence of periodic critical points. Denote the set of such points by C and let $X = \bigcup_{i\in\mathbb{N}} f^i(C)$. For g sufficiently close to f there is an open neighborhood U of X such that $g(U) \subset U$ and U contains only finitely many nonwandering points (periodic critical points). Thus $Y := [0,1] \smallsetminus \bigcup_{i\in\mathbb{N}} g^{-i}(U)$ is closed and g-invariant and every nonatomic invariant measure is supported on Y, so $h_{\text{top}}(g{\restriction}_Y) = h_{\text{top}}(g)$ by the Variational Principle (Theorem 4.5.3). If it happens that $f^k(x_i) \in X$ for some $k < n$ then take I_i so small that for g close enough to f we have $I_i \cap Y = \varnothing$. Thus $g^m{\restriction}_Y$ has at most $(c_n+1)(2^{c_1+1}+1)$ turning points. Extending linearly to $[0,1] \smallsetminus Y$ allows us to apply Proposition 15.2.12 to conclude

$$\begin{aligned} h_{\text{top}}(g) = \frac{1}{n} h_{\text{top}}(g^n) &\leq \frac{1}{n}\log\left((c_n+1)(2^{c_1+1}+1)\right) \\ &\leq \frac{1}{n}\log(c_n 2^{c_1+3}) \leq h_{\text{top}}(f) + \frac{\epsilon}{2} + \frac{(c_1+3)\log 2}{n} \leq \epsilon. \end{aligned}$$ □

Exercises

16.3.1. *Prove that in the quadratic family $f_\lambda(x) = \lambda x(1-x)$, $0 < \lambda < 4$, there is an uncountable set of parameters λ for which the universal repelling set is not hyperbolic.*

16.3.2. *Assume $r \geq 2$, f is a C^r map and the rth jets at all critical points of f are nonzero, that is, for each such point x_0 there is a $k \leq r$ such that the kth derivative of f is nonzero. Prove that topological entropy is continuous at f in the C^r topology.*

4. Full families of unimodal maps

We now give a sufficient condition for all possible kneading sequences to occur in a family of smooth unimodal maps. This is interesting because we saw in Proposition 15.5.7 that in some cases two maps having the same kneading sequence must be conjugate. We will use the notation from Section 15.5, notably the notions related to the kneading sequence $\nu(f)$ of a unimodal map (Definition 15.5.1).

Definition 16.4.1. A family $f_\lambda\colon [0,1] \to [0,1]$ $(\lambda \in [a,b])$ of unimodal maps is called a *full family* or *transition family* if for every unimodal $g\colon [0,1] \to [0,1]$ there exists some $\lambda \in [a,b]$ such that $\nu(g) = \nu(f_\lambda)$.

Theorem 16.4.2. *A family $f_\lambda\colon [0,1] \to [0,1]$ $(\lambda \in [a,b])$ of C^1 unimodal maps is a full family if $\nu(f_a) = i^{f_a}(c_{f_a}) = (1,0,0,0,\dots)$ and $\nu(f_b) = i(c_{f_b}) = (1,2,0,0,\dots)$, that is, the itinerary of the turning point of f_a is $(1,0,0,0,\dots)$ and the itinerary of the turning point of f_b is $(1,2,0,0,\dots)$.*

Proof. Let $\alpha = \sigma(\nu)$ for some unimodal map and suppose $\alpha \neq \sigma(\nu(f_t))$ for all $t \in [a,b]$. Then $a \in L := \{t \in [a,b] \mid \sigma(\nu(f_t)) \prec \alpha\}$, $b \in R := \{t \in [a,b] \mid \sigma(\nu(f_t)) \succ \alpha\}$, and $[a,b] = L \cup R$, so L, R cannot both be open. To show that this is impossible we show that L is open. Considering $1 - f(1-x)$ instead of $f(x)$ then implies that R is also open and contradicts connectedness of $[a,b]$.

It will be convenient to write $\nu^t := \nu(f_t)$. If $t \in L$ and the turning point c_t of f_t is not periodic for f_t then there is a minimal $n \in \mathbb{N}_0$ such that $\nu^t_{n+1} = i^{f_t}_n(f_t(c_t)) \neq \alpha_n$. For $j \leq n$ we have $\nu^t_{j+1} \neq 1$ and $s \mapsto f^{j+1}_s(c_s)$ is continuous, so there is a neighborhood $U \subset [a,b]$ such that for $s \in U$ we have $i^{f_s}_j(f_s(c_s)) = i^{f_t}_j(f_t(c_t))$ when $j \leq n$, that is, $s \in L$. This part of the argument did not use differentiability.

Thus it remains to consider the case when c_t is f_t-periodic with prime period n, say. Since $(f^n_t)'(c_t) = 0$ there is an $\epsilon > 0$ such that $|(f^n_t)'(x)| < \dfrac{1}{2}$ on $[c_t - \epsilon, c_t + \epsilon]$ and hence $f^n_t([c_t - \epsilon, c_t + \epsilon]) \subset \left[c_t - \dfrac{\epsilon}{2}, c_t + \dfrac{\epsilon}{2}\right]$. Furthermore we may assume that $f_t^j([c_t-\epsilon,c_t+\epsilon])\cap[c_t-\epsilon,c_t+\epsilon]=\varnothing$ for $0<j<n$. Therefore there is a neighborhood U of t in $[a,b]$ such that for $s\in U$

(1) $|(f_t^n)'(x)|<\frac{1}{2}$ on $[c_t-\epsilon,c_t+\epsilon]$;
(2) $f_t^n([c_t-\epsilon,c_t+\epsilon])\subset[c_t-\epsilon,c_t+\epsilon]$;
(3) $\nu_{j+1}^s=\nu_{j+1}^t$, that is, $i_j^{f_s}(f_s(c_s))=i_j^{f_t}(f_t(c_t))$, when $j<n$;
(4) $c_s\in[c_t-\epsilon,c_t+\epsilon]$;
(5) $f_s^j([c_t-\epsilon,c_t+\epsilon])\cap[c_t-\epsilon,c_t+\epsilon]=\varnothing$ for $0<j<n$.

By (1) and (2), $[c_t-\epsilon,c_t+\epsilon]$ is in the basin of attraction of a unique attracting fixed point $x_s\in[c_t-\epsilon,c_t+\epsilon]$ of f_s^n.

If $x_s=c_s$ then $\sigma(\nu^s)=\sigma(i(x_s))=\sigma(i(c_t))\prec\alpha$ and $s\in L$. Thus it remains to consider the cases $x_s>c_s$ and $x_2<c_s$. Both are handled similarly, so let us suppose $x_s>c_s$.

In this case $f_s^{kn}(c_s)\in(c_s,x_s)$ for all $k\in\mathbb{N}$ since by (4) $c_s\in[c_t-\epsilon,c_t+\epsilon]$ is in the basin of x_s. Thus $\nu_{kn}^s=i_{kn}^{f_s}(c_s)=2$ and by (5) $\nu_{j+n}^s=i_{j+n+1}^{f_s}(c_s)=i_{j+1}^{f_s}(c_s)=\nu_j^s$ for $j\in\mathbb{N}_0$. This shows that unless $\nu^s\prec\nu^t$, in which case $s\in L$ as desired, we must have $\epsilon_n(\sigma(\nu^s))\cdot 3>\epsilon_n(\sigma(\nu^s))\cdot 2$, where ϵ_n is as in Definition 15.5.3, and hence $\epsilon_n(\sigma(\nu^s))=1$.

If $s\notin L$ then

$$\sigma(\nu^t)\prec\alpha\prec\sigma(\nu^s),$$

since $\alpha\neq\sigma(\nu^s)$ by assumption. This means (by Definition 15.5.3) that there exist $k\geq 1$ and $l<n$ such that $\alpha_j=\nu_{j+1}^s$ (hence $\epsilon_j(\alpha)=\epsilon_j(\sigma(\nu^s))$) for $j<kn+l$ and

$$\epsilon_{kn+l}(\sigma(\nu^s))\alpha_{kn+l}<\epsilon_{kn+l}(\sigma(\nu^s))\nu_{kn+l+1}^s.$$

Since $\nu_n^s=2$ we have $\epsilon_{n+1}(\sigma(\nu^s))=-\epsilon_n(\sigma(\nu^s))=-1$; hence

$$\epsilon_{kn+l}(\sigma(\nu^s))=(\epsilon_{n+1}(\sigma(\nu^s)))^k\epsilon_l(\sigma(\nu^s))=(-1)^k\epsilon_l(\sigma(\nu^s))$$

since ν^s is n-periodic. The same holds for α in place of $\sigma\nu^s$. Thus

$$\begin{aligned}(-1)^k\epsilon_l(\alpha)\alpha_{kn+l}=\epsilon_{kn+l}(\alpha)\alpha_{kn+l}&<\epsilon_{kn+l}(\sigma(\nu^s))\nu_{kn+l+1}^s\\&=(-1)^k\epsilon_l(\sigma(\nu^s))\nu_{l+1}^s=(-1)^k\epsilon_l(\alpha)\alpha_l.\end{aligned}$$

For k odd this yields $\epsilon_l(\sigma^{kn}(\alpha))\alpha_{kn+l}=\epsilon_l(\alpha)\alpha_{kn+l}>\epsilon_l(\alpha)\alpha_l$, that is, $\alpha\prec\sigma^{kn}(\alpha)$, contradicting (15.5.1). For k even we get

$$\epsilon_l(\sigma^{kn-n}(\alpha))(\sigma^{kn-n}(\alpha))_{l+n}=\epsilon_l(\alpha)\alpha_{kn+l}<\epsilon_l(\alpha)\alpha_l=\epsilon_l(\alpha)\alpha_{l+n}.$$

Multiplying by $-1=\epsilon_n(\alpha)=\epsilon_n(\sigma^{kn-n}(\alpha))$ yields

$$\epsilon_{l+n}(\alpha)\alpha_{l+n}<\epsilon_{l+n}(\sigma^{kn-n}(\alpha))(\sigma^{kn-n}(\alpha))_{l+n},$$

so $\alpha\prec\sigma^{kn-n}(\alpha)$, again contradicting (15.5.1). Thus L is open and the proof is complete. $\square$

Corollary 16.4.3. *The quadratic maps $f_\lambda\colon[0,1]\to[0,1]$, $x\mapsto\lambda x(1-x)$ for $\lambda\in[1,4]$ form a full family.*

Exercises

16.4.1. *Show that the family of tent maps (15.2.5) is not a full family.*

16.4.2. *Let* $\varphi\colon [0,1] \to [0,1]$ *be a* C^1 *unimodal map with* $\max \varphi = 1$. *Show that the family* $\varphi_\lambda := \lambda\varphi$, $\lambda \in [0,1]$, *is a full family.*

Part 4

Hyperbolic dynamical systems

17

Survey of examples

In this part of the book we develop a systematic theory of locally maximal (basic) hyperbolic sets for smooth dynamical systems (see Definitions 6.4.1 and 6.4.18). We will study both topological and metric properties of a dynamical system restricted to such a set or to a neighborhood as well as stochastic properties of various important invariant measures supported by locally maximal hyperbolic sets.

This general structural theory occupies the next three chapters. Before proceeding to it we will expand in the present chapter our collection of examples.

So far our examples of hyperbolic sets include the following basic items as well as their C^1 perturbations, invariant closed subsets, unions, and Cartesian products:

(1) Isolated hyperbolic periodic orbits (see Definition 6.2.1).
(2) Markov-type repellers for interval maps (see Sections 2.5b and 16.1).
(3) The Smale horseshoe (Section 2.5c) and its modifications and generalizations including invariant sets appearing near hyperbolic periodic points (Section 6.5).
(4) Hyperbolic automorphisms of the two-dimensional torus and, more generally, of the n-torus (Section 1.8).
(5) Geodesic flows on compact factors of the hyperbolic plane (Section 5.4f); the notion of hyperbolicity for flows will be defined in Section 17.4.

Some of those examples have already been studied in a fairly comprehensive way. For instance, for the basic hyperbolic automorphism F_L of the two-dimensional torus determined by the matrix $L = \begin{pmatrix} 2 & 1 \\ 1 & 1 \end{pmatrix}$ we proved topological transitivity and density of periodic points, calculated their numbers (Proposition 1.8.1), constructed a Markov partition and consequently a semiconjugacy with a topological Markov chain in Section 2.5d, proved topological stability and structural stability (Theorems 2.6.1 and 2.6.3, respectively), calculated topological entropy (Proposition 3.2.6), established ergodicity and mixing (Proposition 4.2.12), and calculated the measure-theoretic entropy with respect to Lebesgue measure (4.4.7) and used this to show that Lebesgue measure represents the limit distribution of periodic orbits (Exercise 4.4.8). All those results indicate directions of study for more general systems which will be pursued in the subsequent chapters.

In the present chapter we proceed in several directions.

First, we will look for hyperbolic sets that are also attractors (See Definition 3.3.1). So far the only examples of that kind at our disposal are rather trivial, from the geometric point of view, namely, contracting periodic orbits, hyperbolic automorphisms of a torus, where the whole torus is an attractor, and the product of these two situations where an invariant torus carrying a hyperbolic automorphism attracts all points in a neighborhood. In the first two sections we will describe much more intricate examples of hyperbolic attractors.

Then we describe the essentially only known (topological conjugacy) classes of Anosov diffeomorphisms other than those of toral automorphisms (Section 17.3).

Next, we extend the notions of hyperbolic set and Anosov system to the continuous-time case (Section 17.4) and discuss a very important class of Anosov flows, namely, geodesic flows on compact Riemannian manifolds with negative sectional curvature. First, in Section 17.5, we consider the basic 2-dimensional examples which already appeared in Section 5.4f, then we introduce the general setting (Section 17.6), and finally we describe the general class of algebraic examples in Section 17.7.

Finally we discuss a class of hyperbolic repellers that appear for rational maps of the Riemann sphere (Section 17.8). Those sets are called Julia sets and carry most of the dynamical complexity of the map. They are similar to the universal repelling set for interval maps (Section 16.1). However, the topological structure of these sets might be quite different from the simple Markov structure in the interval case. In particular this provides examples that are topologically distinct from the previous examples which were all modeled on Cantor sets or manifolds or products of these.

1. The Smale attractor

The following construction can be visualized as somewhat similar to doubling over a rubber band. However it is useful to note that the resulting transformation cannot be produced by a continuous deformation in $\mathbb{R}^3$.

Consider the solid torus $M = S^1 \times D^2$, where D^2 is the unit disk in $\mathbb{R}^2$. On it we define coordinates (φ, x, y) such that $\varphi \in S^1$ and $(x, y) \in D^2$, that is, $x^2 + y^2 \leq 1$. Using these coordinates we define the map

$$f\colon M \to M, \quad f(\varphi, x, y) = \Big(2\varphi, \frac{1}{10}x + \frac{1}{2}\cos\varphi, \frac{1}{10}y + \frac{1}{2}\sin\varphi\Big).$$

To verify that this map is well defined, that is, $f(M) \subset M$, we check

$$\left(\frac{1}{10}x + \frac{1}{2}\cos\varphi\right)^2 + \left(\frac{1}{10}y + \frac{1}{2}\sin\varphi\right)^2 =$$

$$\frac{1}{100}(x^2+y^2) + \frac{1}{10}(x\cos\varphi + y\sin\varphi) + \frac{1}{4}(\cos^2\varphi + \sin^2\varphi) \leq \frac{1}{100} + \frac{2}{10} + \frac{1}{4} < 1.$$

Thus, in fact, $f(M)$ is contained in the interior of M.

Next let us show that f is injective: Suppose $f(\varphi_1, x_1, y_1) = f(\varphi_2, x_2, y_2)$. Then

$$\begin{aligned} 2\varphi_1 &= 2\varphi_2 \pmod{2\pi}, \\ \frac{1}{10}x_1 + \frac{1}{2}\cos\varphi_1 &= \frac{1}{10}x_2 + \frac{1}{2}\cos\varphi_2, \\ \frac{1}{10}y_1 + \frac{1}{2}\sin\varphi_1 &= \frac{1}{10}y_2 + \frac{1}{2}\sin\varphi_2 . \end{aligned}$$

If $\varphi_1 = \varphi_2$ we immediately see that $x_1 = x_2$ and $y_1 = y_2$. If $\varphi_1 = \varphi_2 + \pi$ then

$$\begin{aligned} \frac{1}{10}x_1 + \frac{1}{2}\cos\varphi_1 &= \frac{1}{10}x_2 - \frac{1}{2}\cos\varphi_1, \\ \frac{1}{10}y_1 + \frac{1}{2}\sin\varphi_1 &= \frac{1}{10}y_2 - \frac{1}{2}\sin\varphi_1 \end{aligned}$$

or

$$\frac{1}{10}(x_2 - x_1) = \cos\varphi_1 \text{ and } \frac{1}{10}(y_2 - y_1) = \sin\varphi_1$$

which implies

$$(x_2 - x_1)^2 + (y_2 - y_1)^2 = 100.$$

Since the left-hand side is bounded by 8, this is impossible.

Thus if we consider any cross section $C = \{\theta\} \times D^2$ of M, the image $f(M)$ will intersect C in two disjoint disks of radius $1/10$ as shown in Figure 17.1.1.

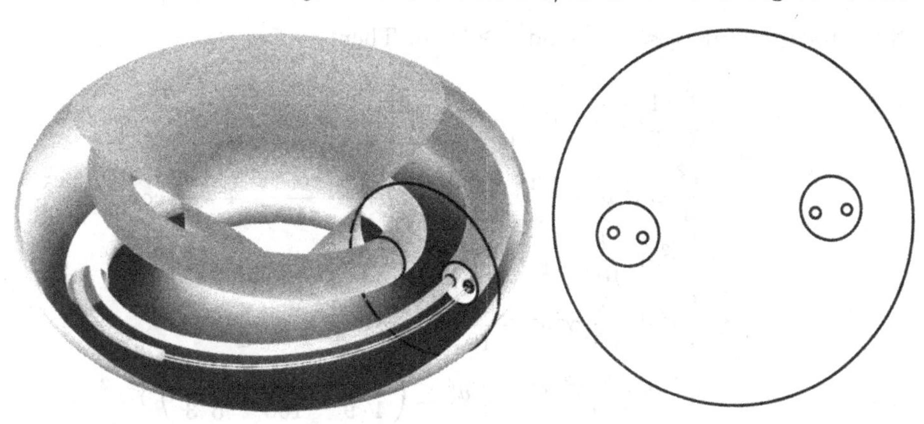

FIGURE 17.1.1. The Smale attractor and a cross section

Note that $C \cap f(M)$ can be written as $f(C_1) \cup f(C_2)$, where C_1 and C_2 are two cross sections.

Now consider $f^2(M)$. Clearly $f^2(M) \subset f(M)$, but moreover $C \cap f^2(M) = f(C_1 \cap f(M)) \cup f(C_2 \cap f(M))$, where C_1 and C_2 are as before. Thus $C \cap f^2(M)$ consists of four little disks, two each inside the disks $f(C_1)$ and $f(C_2)$ shown in Figure 17.1.1.

$f^2(M)$ thus winds around M four times. Pictorially we have doubled up the rubber band a second time.

As we continue to consider successive images $f^l(M)$ we thus find that $C \cap f^{l+1}(M)$ consists of 2^{l+1} disks, two each in the disks of $C \cap f^l(M)$.

Since $f(M) \subset M$ the maximal invariant set inside M is $\Lambda := M \cap f(M) \cap f^2(M) \cap \cdots = \bigcap_{l \in \mathbb{N}_0} f^l(M)$. Obviously Λ is an attractor in the sense of Definition 3.3.1.

To understand the topology of Λ we consider $C \cap \Lambda$ first. $C \cap \Lambda$ is obtained by a Cantor process. Indeed, $C \cap \Lambda$ is clearly closed (being the intersection of closed sets) and perfect. Therefore Λ is locally homeomorphic to the Cartesian product of an interval and a Cantor set. The global structure of Λ is somewhat complicated, however, since Λ is connected by construction (Exercise 17.1.1). Thus Λ appears like a *solenoid* wrapped in a complicated fashion.

Proposition 17.1.1. *Λ is a hyperbolic set for the map f.*

Proof. First we show that cones of the form

$$\{(u, v_1, v_2) \mid v_1^2 + v_2^2 \leq \gamma^2 u^2\}$$

for appropriate γ are invariant under $Df_{(\varphi,x,y)}: T_{(\varphi,x,y)}M \to T_{(\varphi,x,y)}M$. In our coordinates

$$Df(u, v_2, v_2) = \left(2u, -\frac{1}{2}u \sin\varphi + \frac{1}{10}v_1, \frac{1}{2}u\cos\varphi + \frac{1}{10}v_2\right) =: (u', v_1', v_2').$$

Now suppose $v_1^2 + v_2^2 < \gamma^2 u^2$ and $\gamma \geq 3/10$. Then

$$\begin{aligned}
(v_1')^2 + (v_2')^2 &= \left(-\frac{1}{2}u\sin\varphi + \frac{1}{10}v_1\right)^2 + \left(-\frac{1}{2}u\cos\varphi + \frac{1}{10}v_2\right)^2 \\
&= \frac{u^2}{4}\sin^2\varphi - \frac{uv_1}{10}\sin\varphi + \frac{v_1^2}{100} + \frac{u^2\cos^2\varphi}{4} + \frac{uv_2}{10}\cos\varphi + \frac{v_2^2}{100} \\
&= \frac{1}{4}u^2 + \frac{1}{100}(v_1^2 + v_2^2) + \frac{1}{10}u(v_2\cos\varphi - v_1\sin\varphi) \\
&\leq \frac{1}{4}u^2 + \frac{1}{100}\gamma^2u^2 + \frac{1}{10}u(v_2\cos\varphi - v_1\sin\varphi) \\
&\leq \left(\frac{1}{4} + \frac{1}{100}\gamma^2 + \frac{1}{5}\gamma\right)u^2 \leq \left(\frac{100}{4\cdot 9} + \frac{1}{100} + \frac{1}{5}\frac{10}{3}\right)\gamma^2u^2 \\
&< \left(\frac{27}{9} + \frac{1}{100} + \frac{2}{3}\right)\gamma^2u^2 < 4\gamma^2u^2 = \gamma^2(u')^2.
\end{aligned}$$

Thus horizontal cones with $\gamma \geq 3/10$ are Df-invariant. This calculation shows simultaneously that vertical cones with $\gamma \leq 10/3$ are invariant under Df^{-1}. (Note that f^{-1} is well defined on $f(M)$, whose interior is an open neighborhood of Λ.)

In order to obtain hyperbolicity by the cone condition (Corollary 6.4.8), we need to show that Df is expanding in horizontal cones and contracting in vertical cones.

Suppose $v_1^2 + v_2^2 \leq \gamma^2 u^2$. Then

$$\|Df(u, v_1, v_2)\|^2 > 4u^2 \geq \frac{4}{1+\gamma^2}\|(u, v_1, v_2)\|^2 > \|(u, v_1, v_2)\|^2$$

whenever $\gamma^2 < 3$. So Df preserves and expands horizontal γ-cones for $3/10 \leq \gamma < \sqrt{3}$.

On the other hand, if $v_1^2 + v_2^2 \geq u^2/\gamma^2$ then

$$\begin{aligned}\|Df(u, v_1, v_2)\|^2 &= 4u^2 + \left(-\frac{1}{2}u\sin\varphi + \frac{1}{10}v_1\right)^2 + \left(\frac{1}{2}u\cos\varphi + \frac{1}{10}v_2\right)^2 \\ &\leq 4\gamma^2\left(v_1^2 + v_2^2\right) + \frac{u^2}{4} + \frac{1}{100}\left(v_1^2 + v_2^2\right) + \frac{u}{10}(v_2\cos\varphi - v_1\sin\varphi) \\ &\leq \frac{1}{4}u^2 + \left(\frac{1}{100} + \frac{1}{5}\gamma + 4\gamma^2\right)\left(v_1^2 + v_2^2\right) \\ &\leq \frac{1}{4}u^2 + \left(\frac{1}{100} + \frac{2}{25} + \frac{16}{25}\right)\left(v_1^2 + v_2^2\right) < \frac{3}{4}\|(u, v_1, v_2)\|^2\end{aligned}$$

whenever $\gamma \geq 2/5$. Thus Df^{-1} preserves and is expanding on vertical γ-cones for $2/5 \leq \gamma \leq 10/3$.

Consequently Λ is a hyperbolic set by Corollary 6.4.8. □

In fact, it is rather easy to determine the stable manifolds: For each point (φ_0, x_0, y_0) the cross section $C_{\varphi_0} = \{(\varphi, x, y) \in M \mid \varphi = \varphi_0\}$ is contracted by f and mapped into another cross section. Thus cross sections form an invariant family of submanifolds of M that are contracted by f, hence are the stable manifolds.

The unstable manifolds cannot be given as explicitly. The unstable manifold of any point in Λ is contained in Λ, since Λ is a maximal invariant subset of M for f^{-1} by construction, that is, no subset of M containing Λ properly is f^{-1}-invariant. But the union of the unstable manifolds for all $p \in \Lambda$ is f^{-1}-invariant and contains Λ, hence is equal to Λ.

Finally we describe a natural coding procedure between the Smale attractor and the 2-shift $\sigma_2 \colon \Omega_2 \to \Omega_2$ (see (1.9.3)).

It is convenient to replace the angular coordinate φ determined mod 2π by the normalized coordinate $w = \varphi/2\pi$ determined mod 1.

Let

$$\Lambda_0 = \{(w, x, y) \in \Lambda \mid 0 \leq w \leq 1/2\},$$
$$\Lambda_1 = \{(w, x, y) \in \Lambda \mid 1/2 \leq w \leq 1\}.$$

Obviously $\Lambda_0 \cup \Lambda_1 = \Lambda$ and the interiors of Λ_0 and Λ_1 in the topology of Λ are disjoint.

Let $\pi: \Lambda \to S^1$ be the natural projection, $\pi(w, x, y) = w$, and $E_2: S^1 \to S^1$ the linear expanding map, $E_2(w) = 2w \pmod 1$. Then obviously $E_2 \circ \pi = \pi \circ f$, that is, the linear expanding map E_2 is a factor of our map f restricted to Λ.

We now describe a coding $h: \Omega_2 \to \Lambda$ by setting

$$\{h(\omega)\} = \bigcap_{n \in \mathbb{Z}} \overline{\operatorname{Int}(\bigcap_{|i| \le n} f^{-i}(\Lambda_{\omega_i}))}$$

for all $\omega \in \Omega_2$, analogously to (2.5.2).

We have to show that this is well defined, that is, the intersection consists of exactly one point. It is not hard to see that $\operatorname{Int}(\bigcap_{|i| \le n} f^{-i}(\Lambda_{\omega_i})) \neq \varnothing$ and

$$\pi(\operatorname{Int}(\bigcap_{|i| \le n} f^{-i}(\Lambda_{\omega_i}))) = \operatorname{Int}(\bigcap_{|i| \le n} \pi(f^{-i}(\Lambda_{\omega_i}))) = \operatorname{Int}(\bigcap_{|i| \le n} E_2^{-i}(\Delta_1^{\omega_i}))$$

has diameter 2^{-n-1}, so if $p_1, p_2 \in \bigcap_{n \in \mathbb{Z}} \overline{\operatorname{Int}(\bigcap_{|i| \le n} f^{-i}(\Lambda_{\omega_i}))}$ then $\pi(p_1) = \pi(p_2)$. Since

$$f^k(p_j) \in \bigcap_{n \in \mathbb{Z}} \overline{\operatorname{Int}(\bigcap_{|i| \le n} f^{-i}(\Lambda_{(\sigma_2^k \omega)_i}))} \text{ for } j = 1, 2 \text{ and } k \in \mathbb{Z}$$

we have $\pi(f^k(p_1)) = \pi(f^k(p_2))$ for $k \in \mathbb{Z}$. Write $f^{-k}(p_j) = (w^{(k)}, x_j^{(k)}, y_j^{(k)})$ to get

$$100^k[(x_1^{(0)} - x_2^{(0)})^2 + (y_1^{(0)} - y_2^{(0)})^2] < (x_1^{(k)} - x_2^{(k)})^2 + (y_1^{(k)} - y_2^{(k)})^2 \le 8$$

for all $k \ge 0$, and hence $p_1 = p_2$. Thus h is well defined.

It is easy to see that h is surjective: If $p \in \Lambda$ then it is the image of any sequence $\omega(p)$ such that $f^i(p) \in \Lambda_{\omega(p)_i}$ for $i \in \mathbb{Z}$. We can not expect h to be injective, as should be anticipated from the fact that our coding is related to binary expansion. Since two sequences $(\dots, \omega_{-2}, \omega_{-1}, 0, 1, 1, 1, \dots)$ and $(\dots, \omega_{-2}, \omega_{-1}, 1, 0, 0, 0, \dots)$ have the same image, h is 2–1 on

$$A := \bigcup_{k \in \mathbb{Z}} \sigma_2^k(\{\omega \in \Omega_2 \mid \omega_i = \omega_j \quad \forall i, j \in \mathbb{N}\})$$

and injective only on the complement $\Omega_2 \setminus A$.

Since unstable manifolds of points in Ω_2 with respect to σ_2 are dense and the coding map h is continuous we obtain

Proposition 17.1.2. *Every unstable manifold for F is dense in Λ.*

Exercise 17.1.2 demonstrates an interesting similarity between the Smale attractor and toral automorphisms and shifts.

Exercises

17.1.1. *Show that the Smale attractor is connected.*

17.1.2. *Show that the map $f: \Lambda \to \Lambda$ of the Smale attractor is topologically conjugate to an automorphism of a compact abelian group G (cf. the discussion of the group structure on Ω_N in Section 4.2f).*

17.1.3. *Describe a construction of a hyperbolic attractor $f: \Lambda \to \Lambda$ such that f is topologically conjugate to an automorphism of the dual group to the discrete group of k-ary rationals $\{m \cdot k^n \mid m, n \in \mathbb{Z}\}$.*

17.1.4. *Describe a diffeomorphic embedding of the solid torus into itself that can be effected by a continuous deformation in $\mathbb{R}^3$ and has a hyperbolic attractor.*

2. The DA (derived from Anosov) map and the Plykin attractor

As noted before, the map considered in the example of the Smale attractor cannot be effected by a continuous deformation of the solid torus in $\mathbb{R}^3$.

One way to modify that map would be to describe a transformation of the solid torus in $\mathbb{R}^3$ corresponding to the actual doubling over of a rubber band when tightening it around a cylindrical object (see Exercise 17.1.4). However, it is even more interesting to see that hyperbolic attractors can be obtained by a continuous deformation performed on the 2-sphere where there is seemingly very little space for complicated stretching and bending. We will obtain such an attractor as a byproduct of a certain surgery performed on a hyperbolic automorphism of the two-torus. The surgery is similar to the procedure of constructing the Cherry flow from the linear flow in $\mathbb{T}^2$.

a. The DA map.[1] Let $F = F_L$ be the Anosov diffeomorphism of $\mathbb{T}^2$ induced by $L = \begin{pmatrix} 2 & 1 \\ 1 & 1 \end{pmatrix}$. Denote by v^u and v^s the normalized eigenvectors corresponding to the eigenvalues $\lambda_1 = (3+\sqrt{5})/2$ and $\lambda_2 = \lambda_1^{-1} = (3-\sqrt{5})/2$, respectively, and let e^u and e^s be the stable and unstable vector fields obtained from v^u and v^s by parallel translation. Then $E^u(p) = \operatorname{span}\{e^u(p)\}$ and $E^s(p) = \operatorname{span}\{e^s(p)\}$ and $DF_p e^u(p) = \lambda_1 e^u(F(p))$ and $DF_p e^s(p) = \lambda_2 e^s(F(p))$ for all $p \in \mathbb{T}^2$.

On a disk U centered at 0 introduce coordinates (x_1, x_2) diagonalizing A, that is, such that points with coordinates $(x_1, 0)$ are in the unstable manifold of 0 and points with coordinates $(0, x_2)$ are in the stable manifold of 0. Then $F(x_1, x_2) = (\lambda_1 x_1, \lambda_2 x_2)$ on U.

To define a new nonlinear diffeomorphism f on $\mathbb{T}^2$ let $\phi: \mathbb{R} \to [0, 1]$ be a C^∞ function such that

$$\phi(-t) = \phi(t) \qquad (t \in \mathbb{R}),$$

$$\phi(t) = \begin{cases} 1 & \text{if } |t| \le 1/8, \\ 0 & \text{if } |t| \ge 1/4, \end{cases}$$

$$\phi'(t) < 0 \quad \text{if } 1/8 < t < 1/4.$$

Now take $k \in \mathbb{R}$ sufficiently large (to be specified later) and define $f\colon \mathbb{T}^2 \to \mathbb{T}^2$ by

$$f\restriction_{\mathbb{T}^2 \smallsetminus U} = F\restriction_{\mathbb{T}^2 \smallsetminus U}$$

and

$$f(x_1, x_2) = F(x_1, x_2) + (0, (2 - \lambda_2)\phi(x_1)\phi(kx_2)x_2) \text{ on } U.$$

Let us find the fixed points of f. There are none outside U and on U the equation

$$(x_1, x_2) = f(x_1, x_2)$$

is equivalent to

$$x_1 = \lambda_1 x_1,$$
$$x_2 = \lambda_2 x_2 + (2 - \lambda_2)\phi(x_1)\phi(kx_2)x_2.$$

The first equation implies $x_1 = 0$ and thus the second equation reduces to

$$\left(1 - \frac{2 - \lambda_2}{1 - \lambda_2}\phi(kx_2)\right) x_2 = 0$$

with solutions $x_2 = 0, \bar{x}, -\bar{x}$, where $\bar{x}$ is such that $\phi(k\bar{x}) = \dfrac{1 - \lambda_2}{2 - \lambda_2}$.

In order to determine the type of these fixed points we note that

$$Df_{(x_1,x_2)} = \begin{pmatrix} \lambda_1 & 0 \\ (2 - \lambda_2)\phi'(x_1)\phi(kx_2)x_2 & h(x_1, kx_2) \end{pmatrix}$$

with $h(x_1, kx_2) := \lambda_2 + (2 - \lambda_2)\phi(x_1)\left(\phi'(kx_2)kx_2 + \phi(kx_2)\right)$. In particular $Df_0 = \begin{pmatrix} \lambda_1 & 0 \\ 0 & 2 \end{pmatrix}$ and 0 is a repelling fixed point. Also

$$Df_{(0,x_2)} = \begin{pmatrix} \lambda_1 & 0 \\ 0 & \lambda_2 + (2 - \lambda_2)(\phi'(kx_2)kx_2 + \phi(kx_2)) \end{pmatrix}.$$

Since $\phi'(t) \cdot t < 0$ unless $\phi(t) \in \{0, 1\}$ we see that

$$\lambda_2 + (2 - \lambda_2)(\phi'(k\bar{x})k\bar{x} + \phi(k\bar{x})) = \lambda_2 + (2 - \lambda_2)\left(\phi'(k\bar{x})k\bar{x} + \frac{1 - \lambda_2}{2 - \lambda_2}\right) < \lambda_2 + (1 - \lambda_2)$$

and hence that the points $(0, \bar{x})$ and $(0, -\bar{x})$ are hyperbolic fixed points.

Note also that f preserves the stable manifold of 0 and that Df preserves the stable distribution E^s for F although it may not contract vectors in E^s everywhere, and in fact permutes the stable manifolds for F in the same way as F does. Consider now the set $W = W^u(0) = \{p \in \mathbb{T}^2 \mid \alpha(p) = \{0\}\} = \bigcup_{n \in \mathbb{N}} f^n(U_0)$ for a sufficiently small open neighborhood U_0 of 0. It is open and we will show later that it is dense. $\Lambda := \mathbb{T}^2 \smallsetminus W$ is an attractor by definition.

Proposition 17.2.1. *For any sufficiently large k, Λ is a hyperbolic set.*

Lemma 17.2.2. *There exists $\lambda' < 1$ such that $h(x_1, kx_2) < \lambda'$ on Λ.*

Proof. Note that it suffices to show that $h(x_1, kx_2) < 1$ on the complement of a compact subset of $\mathbb{T}^2 \setminus \Lambda$ since Λ is compact.

Let $V := \{(x_1, x_2) \in U \mid h(x_1, kx_2) \geq 1\} = \bigcup_{(x_1,x_2)\in U} (\{x_1\} \times V_{x_1})$ where $V_{x_1} = \{x_2 \mid h(x_1, kx_2) \geq 1\}$. Note that $h(x_1, t) = \lambda_2 + (2 - \lambda_2)\phi(x_1)\frac{d}{dt}t\phi(t) \geq 1$ if and only if $\frac{d}{dt}t\phi(t) \geq \frac{1-\lambda_2}{2-\lambda_2} \cdot \frac{1}{\phi(x_1)}$. Since $\frac{d}{dt}t\phi(t)$ is an even function it follows that V_{x_1} is a symmetric interval for all x_1 and furthermore if $x > y \geq 0$ then $\phi(x) \leq \phi(y)$ and $V_x \subset V_y$. On the other hand if we write $f(x_1, x_2) = (f_1(x_1, x_2), f_2(x_1, x_2)) = (x_1', x_2')$ then $h(x_1, kx_2) = \frac{\partial f_2}{\partial x_2}(x_1, x_2)$ and $f^{-1}(\{x_1'\} \times V_{x_1'}) = \{x_1\} \times V'_{x_1}$ (since f permutes unstable leaves of F) with V'_{x_1} symmetric and of length no more than that of V_{x_1} (since $h(x_1', kx_2') \geq 1$). Thus $V'_{x_1} \subset V_{x_1}$ and since $f(x_1, 0) = F(x_1, 0)$ we conclude that $f^{-1}(V) \subset V$. Moreover, since $\{0\} \times V_0 \subset \mathbb{T}^2 \setminus \Lambda$ we find $\{x_1\} \times V_{x_1} \subset \mathbb{T}^2 \setminus \Lambda$ for all x_1 sufficiently close to 0. Consequently $f^{-n}(v) \subset \mathbb{T}^2 \setminus \Lambda$ for some $n \in \mathbb{N}$ and hence $V \subset \mathbb{T}^2 \setminus \Lambda$. □

Proof of Proposition 17.2.1. The lemma shows that on Λ the diagonal elements of Df are λ_1 and a function bounded from above by $\lambda' < 1$. For large k the off-diagonal element is close to zero since $\phi'(x_1)$ is bounded and $\phi(kx_2)x_2 \neq 0$ only for $|x_2| \leq k/4$ and $\phi \leq 1$, so $\phi(kx_2)x_2 \leq 1/4k$. One obtains hyperbolicity of Λ as follows: If we write $Df_x = \begin{pmatrix} \lambda_1 & 0 \\ G(x) & H(x) \end{pmatrix}$ and take $\epsilon \in (0, \sqrt{\lambda_1^2 - 2})$ such that $\frac{1}{\lambda_1 - \lambda'} < \sqrt{1 + 1/\epsilon^2} - 1$, then $|G(x)| < \epsilon$ for all $x \in \Lambda$ if k is sufficiently large. Consider now horizontal cones of the form $|v| < \gamma|u|$ with $\frac{\epsilon}{\lambda_1 - \lambda'} < \gamma < \sqrt{\epsilon^2 + 1} - \epsilon < 1$. Note that these are invariant under Df since if we let $(u', v') := Df_x(u, v)$ then

$$|v'| = |G(x)u + H(x)v| < \epsilon|u| + \lambda'|v| < (\lambda'\gamma + \epsilon)|u| < \gamma\lambda_1|u| = \gamma|u'|$$

since $\epsilon < (\lambda_1 - \lambda')\gamma$. To see that vectors in γ-cones expand, note that

$$\begin{aligned}
|(u', v')|^2 &= u'^2 + v'^2 = \lambda_1^2 u^2 + (G(x)u + H(x)v)^2 \\
&\geq \lambda_1^2 u^2 + G^2(x)u^2 - 2|G(x)|\, H(x)\, |u||v| + H^2(x)v^2 \\
&\geq [\lambda_1^2 + G^2(x) - 2\gamma|G(x)|\, H(x)]u^2 - [1 - H^2(x)]v^2 + v^2 \\
&\geq [\lambda_1^2 + G^2(x) - 2\gamma|G(x)|\, H(x) - \gamma^2(1 - H^2(x))]u^2 + v^2 \\
&> [\lambda_1^2 - \epsilon^2 - 2\gamma\epsilon H(x) - \gamma^2 + \gamma^2 H^2(x)]u^2 + v^2 \\
&> (\lambda_1^2 - \epsilon^2 - 2\gamma\epsilon - \gamma^2)u^2 + v^2 = (\lambda_1^2 - (\gamma + \epsilon)^2)u^2 + v^2 \\
&> (\lambda_1^2 - 1 - \epsilon^2)u^2 + v^2 > u^2 + v^2.
\end{aligned}$$

The last two inequalities used $\gamma + \epsilon < \sqrt{\epsilon^2+1}$ and $\epsilon^2 < \lambda_1^2 - 2$. Similarly one can show that vertical cones are invariant under Df_x^{-1} and that Df_x^{-1} expands vectors in vertical cones so that hyperbolicity follows from Corollary 6.4.8. But since here, as in the previous section, the stable manifolds are already given, hyperbolicity follows directly without any discussion of vertical cones. □

Let us show that $W = \mathbb{T}^2 \setminus \Lambda$ is dense in $\mathbb{T}^2$. To this end consider $p \in \Lambda$ and any open neighborhood U_p of p. Then there is a point $q \in U_p$ that is periodic for F, with period n, say. The stable manifold L of q (under F) is thus f^n-invariant and dense. Density of W follows if we can find $N \in \mathbb{N}$ such that $f^{-Nn}(L^1) \cap W \neq \varnothing$, where $L^1 = L \cap U_p$. But this is necessarily the case, since otherwise $L_f := \bigcup_{n\in\mathbb{N}} f^{-Nn}(L^1) \subset \Lambda$ and by hyperbolicity f^{-n} expands L_f so $L_f = L$. But L is dense in $\mathbb{T}^2$, so we would have $\Lambda = \mathbb{T}^2$, a contradiction. Thus Λ is the complement of an open dense set.

We have thus produced a hyperbolic attractor on $\mathbb{T}^2$. To a certain extent the relation between this attractor and the hyperbolic automorphism F is similar to the relation between the Denjoy minimal set for a nontransitive homeomorphism of the circle (see Section 11.2) and the corresponding irrational rotation.

b. The Plykin attractor. To obtain a hyperbolic attractor on S^2 let $J: \mathbb{T}^2 \to \mathbb{T}^2$, $J(x) = -x \pmod 1$ and note that the construction of the DA map is J-invariant, that is, $f \circ J = J \circ f$. Furthermore note that $(1/2, 1/2)$ is a periodic point for f since $f(1/2,1/2) = F(1/2,1/2) = (1/2,0)$, $f(1/2,0) = (0,1/2)$, and $f(0,1/2) = (1/2,1/2)$. Now we replace F by F^3 and note that F^3 fixes these four fixed points of J and perform the construction described in Subsection a simultaneously around the four fixed points of F^3. We will thus obtain a map $f: \mathbb{T}^2 \to \mathbb{T}^2$ which commutes with J, has four fixed points as repelling fixed points, and has a hyperbolic attractor Λ.

Note that on $\mathbb{T}^2$ we have

$$-\left(\frac{1}{2},\frac{1}{2}\right) = \left(\frac{1}{2},\frac{1}{2}\right), -\left(\frac{1}{2},0\right) = \left(0,\frac{1}{2}\right), -\left(0,\frac{1}{2}\right) = \left(0,\frac{1}{2}\right).$$

Thus if v_i, $i = 1, \dots, 4$, are disks around $(0,0)$, $(1/2,1/2)$, $(1/2,0)$, $(0,1/2)$, respectively, contained in $\mathbb{T}^2 \setminus \Lambda$, then $M = (\mathbb{T}^2 \setminus \bigcup_{i=1}^4 V_i)_{/(x\sim -x)}$ is a smooth manifold. It is not hard to see that M is a 2-sphere with four holes (Exercise 17.2.2). Since $f(-x) = -f(x)$ we obtain an induced map $f': M \to M$ which is smooth and injective. Filling $S^2 \setminus M$ with four repellers (one fixed and one period-3 cycle) gives a diffeomorphism $\tilde{f}: S^2 \to S^2$ with a hyperbolic attractor (obtained by projecting Λ to M). This is the *Plykin attractor.*[2]

Remark. The map $\mathbb{T}^2 \to S^2$ given by the identification of the J-orbits is an example of a *branched covering*, that is, a map that looks like a covering map away from a finite collection of points and like $z \mapsto z^k$ at those points.

Exercises

17.2.1. *Justify the simultaneous performance of the "surgery" in Subsection a at several fixed points of the linear map. Generalize this construction to periodic orbits.*

17.2.2. *Verify that if* V_i, $i = 1, 2, 3, 4$, *are as before, then*

$$M := \left(\mathbb{T}^2 \setminus \bigcup_{i=1}^{4} V\right)_{/(x\sim -x)}$$

is a 2-sphere with four holes.

17.2.3. *Use Exercise 17.2.1 to obtain a map of* S^2 *with a hyperbolic attractor starting from* F *instead of* F^3.

17.2.4*. *Using the fact that every surface of higher genus is a branched cover of the torus describe, for any such surface, the construction of a diffeomorphism* f *such that* $NW(f)$ *consists of a hyperbolic attractor and finitely many fixed points.*

3. Expanding maps and Anosov automorphisms of nilmanifolds

So far the only examples of Anosov diffeomorphisms (Definition 6.4.2) at our disposal have been hyperbolic automorphisms of the n-torus (Section 1.8) and their perturbations, which by Theorem 2.6.3 and Exercise 2.6.1 are topologically conjugate to the linear models. As we already mentioned in Section 6.4a Anosov diffeomorphisms are rather special objects. For example Exercises 8.6.2 and 8.6.4 establish topological restrictions on manifolds that can carry an Anosov diffeomorphism. Right now we proceed to the description of essentially the only known construction of Anosov diffeomorphisms that are not topologically conjugate to a toral automorphism. The idea of this construction due to Smale is to think of the toral automorphisms as factors of automorphisms of the simply connected abelian Lie group $\mathbb{R}^n$ and look for other (nonabelian) Lie groups for which a similar construction can be carried out. It turns out that the only nonabelian Lie groups admitting a hyperbolic automorphism are *nilpotent* Lie groups. (See Section A.8.) Throughout the discussion it is useful to keep in mind the familiar special case of toral automorphisms.

Suppose that G is a simply connected Lie group and that Γ is a discrete subgroup such that $\Gamma\backslash G$ is compact. Such a subgroup is called a *(uniform) lattice.* Equivalently one can find a compact fundamental domain for the action of Γ on G by left translations. Suppose $F\colon G \to G$ is an automorphism such that $F(\Gamma) = \Gamma$ (hence F projects to $\Gamma\backslash G$) and $DF_{|\mathrm{Id}}$ is hyperbolic. Corollary 1.2.6 shows that there is a splitting of the Lie algebra $\mathfrak{L}(G) = T_{\mathrm{Id}}G = E^+ \oplus E^-$ and a norm on $\mathfrak{L}(G)$ such that $DF^{-1}{\restriction}_{E^+}$ and $DF{\restriction}_{E^-}$ are contractions. It is the existence of such a hyperbolic automorphism of the Lie algebra that makes it necessary for G to be nilpotent. For $g \in G$ we obtain a corresponding splitting and norm by applying the differential of the left translation $x \mapsto gx$. Thus we obtain a splitting which is a hyperbolic splitting for F. By construction this splitting and this norm are invariant under left translations, so they induce a splitting and a norm on the compact quotient $\Gamma\backslash G$. The factor $f: \Gamma\backslash G \to \Gamma\backslash G$ of F is then an Anosov diffeomorphism.

In the more general situation when $F(\Gamma) \subset \Gamma$ we still have a map on the factor $\Gamma\backslash G$ which is, however, not one-to-one. Let us point out that unlike in the abelian case Γ is not a normal subgroup, so the factor, which is called a *nilmanifold*, does not have a group structure.

Let us consider two examples of this situation. Consider the *Heisenberg group* defined by

$$H := \left\{ \begin{pmatrix} 1 & x & z \\ 0 & 1 & y \\ 0 & 0 & 1 \end{pmatrix} \middle| \, x, y, z \in \mathbb{R} \right\}$$

with the usual matrix multiplication, that is, in coordinates x, y, z multiplication is given by the formula

$$(x_1, y_1, z_1) \times (x_2, y_2, z_2) = (x_1 + x_2, y_1 + y_2, z_1 + z_2 + x_1 y_2).$$

The one-parameter subgroup $\begin{pmatrix} 1 & 0 & z \\ 0 & 1 & 0 \\ 0 & 0 & 1 \end{pmatrix}$ is the *center* of H. Thus H is 3-dimensional, simply connected, nonabelian, and nilpotent. An obvious lattice is the integer lattice of matrices for which $x, y, z \in \mathbb{Z}$.

The Lie algebra of H is given by

$$\mathfrak{L}(H) = \left\{ \begin{pmatrix} 0 & x & z \\ 0 & 0 & y \\ 0 & 0 & 0 \end{pmatrix} \middle| \, x, y, z \in \mathbb{R} \right\}$$

with generators $X = \begin{pmatrix} 0 & 1 & 0 \\ 0 & 0 & 0 \\ 0 & 0 & 0 \end{pmatrix}$, $Y = \begin{pmatrix} 0 & 0 & 0 \\ 0 & 0 & 1 \\ 0 & 0 & 0 \end{pmatrix}$ and $Z = \begin{pmatrix} 0 & 0 & 1 \\ 0 & 0 & 0 \\ 0 & 0 & 0 \end{pmatrix}$. The Lie brackets are given by $[X, Y] = Z$ with all other brackets of generators being zero.

To construct an expanding map on H consider the map given by

$$F\left(\begin{pmatrix} 1 & x & z \\ 0 & 1 & y \\ 0 & 0 & 1 \end{pmatrix} \right) = \begin{pmatrix} 1 & 2x & 4z \\ 0 & 1 & 2y \\ 0 & 0 & 1 \end{pmatrix}.$$

It is easy to see that F is a group automorphism and F clearly maps the integer lattice $H_{\mathbb{Z}}$ into $H_{\mathbb{Z}}$. Thus it induces a map f on the factor $H_{\mathbb{Z}}\backslash H$. This map f, however, is not invertible, much like the expanding circle maps. Furthermore no automorphism of H that is one-to-one on $H_{\mathbb{Z}}$ is hyperbolic because such an automorphism A must preserve the center $Z(H)$ of H and hence the intersection $Z(H) \cap H_{\mathbb{Z}}$ which implies that $A\begin{pmatrix} 1 & 0 & 1 \\ 0 & 1 & 0 \\ 0 & 0 & 1 \end{pmatrix} = \begin{pmatrix} 1 & 0 & \pm 1 \\ 0 & 1 & 0 \\ 0 & 0 & 1 \end{pmatrix}$ and thus $DA(Z) = \pm Z$. In fact, one can show that no hyperbolic automorphism of H preserves *any* uniform lattice (Exercise 17.3.4). Thus similarly to the abelian case, where one has to pass to the torus to obtain an invertible hyperbolic map, one needs to pass to higher-dimensional nilpotent Lie groups, for example, products of Heisenberg groups, in order to construct invertible hyperbolic maps on their compact factors.

In order to motivate the following construction let us present an interpretation of the standard hyperbolic automorphism of $\mathbb{T}^2$ which is somewhat unusual from the geometric viewpoint, but perfectly natural from the point of view of algebraic number theory.

The eigenvalues $\lambda_1 = \dfrac{3+\sqrt{5}}{2}$ and $\lambda_2 = \dfrac{3-\sqrt{5}}{2} = \lambda_1^{-1}$ are units in the algebraic number field $\mathbb{K} := \mathbb{Q}(\sqrt{5}) = \{a + b\sqrt{5} \mid a, b \in \mathbb{Q}\}$, which is a quadratic extension of the field $\mathbb{Q}$ of rationals. This field possesses a unique nontrivial automorphism σ, which in fact interchanges the eigenvalues, namely, $\sigma(a + b\sqrt{5}) = a - b\sqrt{5}$ for $a, b \in \mathbb{Q}$. The matrix $L := \begin{pmatrix} 2 & 1 \\ 1 & 1 \end{pmatrix}$ acts on the space $\mathbb{K}^2$ and can be diagonalized over $\mathbb{K}$. In the coordinates with respect to this eigenbasis the map has the form

$$L(x_1, x_2) = (\lambda_1 x_1, \lambda_2 x_2)$$

and the vectors of the form

$$(\mu, \sigma(\mu)), \tag{17.3.1}$$

where μ are algebraic integers in $\mathbb{K}$, form a lattice which is invariant under L. This is easy to see because there are two linearly independent vectors of the form (17.3.1) corresponding to $\mu = 1$ and $\mu = \sqrt{5}$, say, and because there are only finitely many vectors of the form (17.3.1) whose Euclidean norm is bounded by a given constant. This lattice is different from the standard integer lattice which is also preserved by L. For example, this lattice can be generated by the vectors (17.3.1) corresponding to $\mu_1 = 1$ and $\mu_2 = (1+\sqrt{5})/2$. In that basis the automorphism of $\mathbb{K}^2$ is represented by the same matrix $\begin{pmatrix} 2 & 1 \\ 1 & 1 \end{pmatrix}$.

Now let $G = H \times H$ with generators $X_1, Y_1, Z_1, X_2, Y_2, Z_2$ such that $[X_i, Y_i] = Z_i$ and all other brackets of generators are zero. The Lie algebra of G is

$$\mathfrak{L}(G) = \left\{ \begin{pmatrix} A & 0 \\ 0 & B \end{pmatrix} \mid A, B \in \mathfrak{L}(H) \right\}.$$

Let Γ be the subgroup of G given by $\exp_{\mathrm{Id}} \gamma$, where $\exp_{\mathrm{Id}}: \mathcal{L}(G) \to G$ and $\gamma \subset \mathcal{L}(G)$ is given by

$$\gamma = \left\{ \begin{pmatrix} A & 0 \\ 0 & \sigma(A) \end{pmatrix} \mid A \in \mathcal{L}(H) \text{ with entries in algebraic integers in } \mathbb{K} \right\},$$

with $\sigma(A)_{ij} = \sigma(A_{ij})$. Similarly to the abelian case it can be shown that Γ is a lattice. Now define two Lie algebra automorphisms f_1', f_2' on $\mathcal{L}(G)$ by

$$\begin{aligned} f_1'(X_1) &= \lambda_1 X_1, & f_1'(Y_1) &= \lambda_1^2 Y_1, & f_1'(Z_1) &= \lambda_1^3 Z_1, \\ f_1'(X_2) &= \lambda_1^{-1} X_2, & f_1'(Y_2) &= \lambda_1^{-2} Y_2, & f_1'(Z_2) &= \lambda_1^{-3} Z_2, \\ f_2'(X_1) &= \lambda_1 X_1, & f_2'(Y_1) &= \lambda_1^{-3} Y_1, & f_2'(Z_1) &= \lambda_1^{-2} Z_1, \\ f_2'(X_2) &= \lambda_1^{-1} X_2, & f_2'(Y_2) &= \lambda_1^3 Y_2, & f_2'(Z_2) &= \lambda_1^2 Z_2. \end{aligned}$$

Then there exist unique automorphisms $F_i: G \to G$ with $DF_i\big|_{\mathrm{Id}} = f_i'$. Since λ_1 and λ_2 are units in $\mathbb{K}$, that is, integers whose inverses are also integers, and $\sigma(\lambda_1) = \lambda_2$ we also have $F_i(\Gamma) = \Gamma$ for $i = 1, 2$. Thus the F_i project to Anosov diffeomorphisms of $\Gamma \backslash G$.

Exercises

17.3.1. *Describe all hyperbolic automorphisms of the Heisenberg group H such that $F(H_{\mathbb{Z}}) \subset H_{\mathbb{Z}}$.*

17.3.2. *Let G be a Lie group, χ a Haar measure, and Γ a uniform lattice. Show that if $F: G \to G$ is an automorphism such that $F(\Gamma) = \Gamma$ then $F_* \chi = \pm \chi$.*

17.3.3. *Show that any Haar measure on the Heisenberg group is proportional to the volume element $dx\, dy\, dz$.*

17.3.4. *Suppose that F is an automorphism of the Heisenberg group H such that $F(\Gamma) = \Gamma$ for some uniform lattice Γ. Show that $DF\big|_{\mathrm{Id}}$ has 1 or -1 as an eigenvalue.*

4. Definitions and basic properties of hyperbolic sets for flows

In this section the basic hyperbolic theory for discrete-time dynamical systems developed in Sections 6.2 and 6.4 is reformulated for flows. The theory looks much the same and most basic results are obtained by straightforward application of the results from Chapter 6. On the other hand we will encounter at the end of this section a phenomenon peculiar to flows, a time change (cf. Section 2.2). We show there, however, that time changes do not affect hyperbolicity.

Definition 17.4.1. Let M be a smooth manifold, $\varphi: \mathbb{R} \times M \to M$ a smooth flow, and $\Lambda \subset M$ a compact φ^t-invariant set. The set Λ is called a *hyperbolic set for the flow* φ^t if there exist a Riemannian metric on an open neighborhood U of Λ and $\lambda < 1 < \mu$ such that for all $x \in \Lambda$ there is a decomposition $T_xM = E_x^0 \oplus E_x^+ \oplus E_x^-$ such that $\frac{d}{dt}\big|_{t=0}\varphi^t(x) \in E_x^0 \setminus \{0\}$, $\dim E_x^0 = 1$, $D\varphi^t E_x^\pm = E_x^\pm$, and

$$\|D\varphi^t\restriction_{E_x^-}\| \leq \lambda^t, \quad \|D\varphi^{-t}\restriction_{E_x^+}\| \leq \mu^{-t}.$$

Definition 17.4.2. A C^1 flow $\varphi^t: M \to M$ on a compact manifold M is called an *Anosov flow* if M is a hyperbolic set for φ^t.

The following theorem is the Stable and Unstable Manifolds Theorem for flows, an analog of Theorem 6.4.9.

Theorem 17.4.3. *Let Λ be a hyperbolic set for a C^r flow $\varphi^t: M \to M$, $r \in \mathbb{N}$, λ, μ as in Definition 17.4.1, and $t_0 > 0$. Then for each $x \in \Lambda$ there is a pair of embedded C^r-discs $W^s(x)$, $W^u(x)$, called the local strong stable manifold and the local strong unstable manifold of x, respectively, such that*

(1) $T_xW^s(x) = E_x^-$, $T_xW^u(x) = E_x^+$;
(2) $\varphi^t(W^s(x)) \subset W^s(\varphi^t(x))$ *and* $\varphi^{-t}(W^u(x)) \subset W^u(\varphi^{-t}(x))$ *for* $t \geq t_0$;
(3) *for every* $\delta > 0$ *there exists* $C(\delta)$ *such that*

$$\begin{aligned} \operatorname{dist}(\varphi^t(x), \varphi^t(y)) &< C(\delta)(\lambda + \delta)^t \operatorname{dist}(x, y) \quad \textit{for } y \in W^s(x),\ t > 0, \\ \operatorname{dist}(\varphi^{-t}(x), \varphi^{-t}(y)) &< C(\delta)(\mu - \delta)^{-t} \operatorname{dist}(x, y) \quad \textit{for } y \in W^u(x),\ t > 0; \end{aligned}$$

(4) *there exists a continuous family U_x of neighborhoods of $x \in \Lambda$ such that*

$$\begin{aligned} W^s(x) &= \{y \mid \varphi^t(y) \in U_{\varphi^t(x)},\ \ t > 0,\ \operatorname{dist}(\varphi^t(x), \varphi^t(y)) \xrightarrow[t\to\infty]{} 0\}, \\ W^u(x) &= \{y \mid \varphi^{-t}(y) \in U_{\varphi^{-t}(x)},\ t > 0,\ \operatorname{dist}(\varphi^{-t}(x), \varphi^{-t}(y)) \xrightarrow[t\to\infty]{} 0\}. \end{aligned}$$

Proof. As remarked before Theorem 6.4.9 we can pass to local coordinates and use the Hadamard–Perron Theorem 6.2.8. Consider the time-t_0 map φ^{t_0}. Note that although φ^{t_0} is not hyperbolic ($D\varphi^{t_0}$ has an eigenvalue 1 with eigenspace E^0), we may write $T_xM = (E_x^0 \oplus E_x^+) \oplus E_x^-$ and, via the localization in coordinates, apply Theorem 6.2.8 with $\mu = 1$. This yields the existence of $W^s(x)$ satisfying (1)–(4) for $t \in \mathbb{N}t_0$. Since $T_xM = (E_x^0 \oplus E_x^-) \oplus E_x^+$, an application of Theorem 6.2.8 with $\lambda = 1$ yields $W^u(x)$ satisfying (1)–(4) with $-t \in \mathbb{N}t_0$.

Observe now that (4) holds for positive multiples of t_0 if and only if it holds for real t. Once (3) holds for $t \in \mathbb{N}t_0$ it trivially holds for $t > 0$ by adjusting the constant $C(\delta)$ since $\{\varphi^t\}_{t\in[0,t_0]}$ is equicontinuous and M is compact.

Finally, to obtain C^r manifolds note that the proof of C^r smoothness in Theorem 6.2.8 works in this situation as well, as remarked there. The theorem follows. □

Remark. With a certain amount of care one can replace the condition $t \geq t_0$ in (2) by $t > 0$.

The sets

$$\widetilde{W}^s(x) := \bigcup_{t>0} \varphi^{-t}(W^s(\varphi^t(x))) \text{ and } \widetilde{W}^u(x) := \bigcup_{t>0} \varphi^t(W^u(\varphi^{-t}(x))) \quad (17.4.1)$$

are defined independently of a particular choice of local stable and unstable manifolds, are smooth injectively immersed manifolds called the global *strong stable* and *strong unstable* manifolds, and can be characterized by

$$\widetilde{W}^s(x) = \{y \in M \mid \operatorname{dist}(\varphi^t(x), \varphi^t(y)) \xrightarrow[t\to\infty]{} 0\},$$
$$\widetilde{W}^u(x) = \{y \in M \mid \operatorname{dist}(\varphi^{-t}(x), \varphi^{-t}(y)) \xrightarrow[t\to\infty]{} 0\}.$$

The manifolds

$$\widetilde{W}^{0s}(x) := \bigcup_{t\in\mathbb{R}} \varphi^t(\widetilde{W}^s(x)) \text{ and } \widetilde{W}^{0u}(x) := \bigcup_{t\in\mathbb{R}} \varphi^t(\widetilde{W}^u(x)) \quad (17.4.2)$$

are called the *weak stable* and *weak unstable* manifolds of x. Note that $T_x\widetilde{W}^{0s} = E_x^0 \oplus E_x^-$, $T_x\widetilde{W}^{0u} = E_x^0 \oplus E_x^+$.

To prove hyperbolicity in the examples of Sections 17.1 and 17.2 we used the cone criterion of Corollary 6.4.8. The analogous criterion for flows is given by the following statement.

Proposition 17.4.4. *A compact φ^t-invariant set $\Lambda \subset M$ is hyperbolic if there exist constants $\lambda < 1 < \mu$ such that for all $x \in \Lambda$ there is a decomposition $T_xM = E_x^0 \oplus S_x \oplus T_x$ (in general not $D\varphi^t$-invariant), a family of horizontal cones $H_x \supset S_x$ associated with the decomposition $S_x \oplus (E_x^0 \oplus T_x)$, and a family of vertical cones $V_x \supset T_x$ associated with the decomposition $(S_x \oplus E_x^0) \oplus T_x$ such that for $t > 0$*

$$D\varphi^t H_x \subset \operatorname{Int} H_{\varphi^t(x)}, \qquad D\varphi^{-t} V_x \subset \operatorname{Int} V_{\varphi^{-t}(x)},$$
$$\frac{d}{dt}\|D\varphi^t\xi\| \geq \|\xi\| \log \mu \quad \textit{for} \quad \xi \in H_x,$$
$$\frac{d}{dt}\|D\varphi^{-t}\xi\| \geq \|\xi\| \log \lambda \quad \textit{for} \quad \xi \in V_x\,.$$

Remark. Requiring existence of the distributions S and T is used simply as a convenient way of expressing the fact that the cones are "complementary". Hence in applications we will simply exhibit the invariant cones.

Proof. Apply Proposition 6.2.12 once with $\lambda' = 1$ and once with $\mu' = 1$ to $D\varphi^1$ and obtain distributions E^+ and E^-. The assumptions then guarantee that these are as required in the definition of hyperbolicity of Λ for φ^t. □

One can prove a counterpart of the Anosov Closing Lemma for flows using Poincaré maps between successive transversals to a pseudo-orbit (see Exercise 17.4.2). Alternatively it follows from the Shadowing Theorem 18.1.7 for flows.

We close this section with some remarks about time changes. Time changes in a general setting were discussed in Section 2.2.

Proposition 17.4.5. *Let Λ be a hyperbolic set for φ^t. If ψ^t is a time change of φ^t then Λ is hyperbolic for ψ^t.*

Proof. Write $\psi^t(x) = \varphi^{\alpha(t,x)}(x)$ as in Section 2.2 and note that $\alpha(0,\cdot) = 0$. As we did when deducing Theorem 6.4.9 from the Hadamard–Perron Theorem 6.2.8 we obtain for each $x \in \Lambda$ local coordinates $x = (x^0, x^u, x^s)$ centered at x and adapted to the splitting $T_xM = E_x^0 \oplus E_x^+ \oplus E_x^-$ so that with respect to these coordinates

$$D\varphi^t(0) = \begin{pmatrix} 1 & 0 & 0 \\ 0 & A_t & 0 \\ 0 & 0 & B_t \end{pmatrix}.$$

We assume $\|B_t\| \le \lambda^t < 1$ and $\|A_t^{-1}\| \le \mu^{-t} < 1$. We immediately note that in these coordinates

$$D\psi^t(0) = \begin{pmatrix} 1 & \alpha_{x^u}(t,x) & \alpha_{x^s}(t,x) \\ 0 & A_{\alpha(t,x)} & 0 \\ 0 & 0 & B_{\alpha(t,x)} \end{pmatrix},$$

where $\alpha_{x^u}(t,x)$ and $\alpha_{x^s}(t,x)$ are the partial derivatives of α with respect to x^u and x^s, respectively. By compactness of Λ we may take $Kt > 0$ as an upper bound for their size when $t > 0$. In order to prove hyperbolicity of ψ^t we now use Proposition 17.4.4. We begin by writing vectors in $T_x\Lambda = E_x^0 \oplus E_x^+ \oplus E_x^-$ as (u, v, w) with $u \in E_x^0$, $v \in E_x^+$, $w \in E_x^-$ and letting

$$\|u, v, w\|^2 := \epsilon^2\|u\|^2 + \|v\|^2 + \|w\|^2,$$

where a sufficiently small $\epsilon > 0$ will be specified later. For $\gamma < \sqrt{\mu^2 - 1}$ we now check whether the γ-cone given by

$$\epsilon^2\|u\|^2 + \|w\|^2 \le \gamma^2\|v\|^2$$

is $D\psi^t$-invariant for $t \in [0, 1]$. By taking ϵ small enough we may assume that

$$K^2t^2\epsilon^2 + \lambda^{2\alpha(t,x)} \le 1 \text{ for } t \in [0, 1].$$

If $(u', v', w') = D\psi^t(u, v, w)$ then

$$\begin{aligned}
\epsilon^2\|u'\|^2 + \|w'\|^2 &= \epsilon^2\|u + \alpha_{x^u} v + \alpha_{x^s} w\|^2 + \|B_{\alpha(t,x)} w\|^2 \\
&\le \epsilon^2 \left(\|u\| + Kt\|v\| + Kt\|w\|\right)^2 + \lambda^{2\alpha(t,x)}\|w\|^2 \\
&= \epsilon^2\|u\|^2 + (K^2t^2\epsilon^2 + \lambda^{2\alpha(t,x)})\|w\|^2 \\
&\quad + \epsilon^2 Kt(Kt\|v\|^2 + 2\|u\|\|v\| + 2\|u\|\|w\| + 2Kt\|v\|\|w\|) \\
&\le \gamma^2\|v\|^2 + \epsilon^2 Kt\Big(Kt\|v\|^2 + \frac{2\gamma}{\epsilon}\|v\|^2 + \frac{2\gamma^2}{\epsilon}\|v\|^2 + 2\gamma Kt\|v\|^2\Big) \\
&= \gamma^2\Big(1 + \frac{\epsilon Kt}{\gamma^2}(\epsilon Kt(1 + 2\gamma) + 2\gamma(1 + \gamma))\Big)\|v\|^2 \\
&< \gamma^2 \mu^{2\alpha(t,x)}\|v\|^2 \le \gamma^2\|v'\|
\end{aligned}$$

for sufficiently small $\epsilon > 0$ and $t \in (0, 1]$. Thus γ-cones are ψ^t-invariant. To check that vectors in γ-cones are expanded note that $\epsilon^2\|u'\|^2 + \|w'\|^2 \ge \delta^{\alpha(t,x)}(\epsilon^2\|u\|^2 + \|w\|^2)$ for some $\delta > 0$ and take $\gamma > 0$ small enough so that

$$\frac{\mu^{2\beta} + \delta^\beta\gamma^2}{1 + \gamma^2} \ge \eta^\beta \tag{17.4.3}$$

for some $\eta > 1$ and all $\beta > 0$. Then if $\epsilon^2\|u\|^2 + \|w\|^2 \le \gamma^2\|v\|^2$ we have

$$\begin{aligned}
\epsilon^2\|u'\|^2 + \|v'\|^2 + \|w'\|^2 &\ge \delta^{\alpha(t,x)}(\epsilon^2\|u\|^2 + \|w\|^2) + \|A_{\alpha(t,x)} v\|^2 \\
&\ge \eta^{\alpha(t,x)}(\epsilon^2\|u\|^2 + \|w\|^2) \\
&\quad + (\delta^{\alpha(t,x)} - \eta^{\alpha(t,x)})(\epsilon^2\|u\|^2 + \|w\|^2) \\
&\quad + \mu^{2\alpha(t,x)}\|v\|^2 \\
&\ge \eta^{\alpha(t,x)}(\epsilon^2\|u\|^2 + \|w\|^2) \\
&\quad + [(\delta^{\alpha(t,x)} - \eta^{\alpha(t,x)})\gamma^2 + \mu^{2\alpha(t,x)}]\|v\|^2 \\
&\ge \eta^{\alpha(t,x)}(\epsilon^2\|u\|^2 + \|v\|^2 + \|w\|^2),
\end{aligned}$$

where the last inequality follows from (17.4.3).

Since ψ^{-t} is a time change of φ^{-t} there is a corresponding cone family for ψ^{-t} and thus the result follows from Proposition 17.4.4. □

Exercises

17.4.1. *Deduce that the stable and unstable manifolds in Theorem 17.4.3 are C^r directly from the Hadamard–Perron Theorem 6.2.8 by considering smooth transversals to the orbit and the family of maps between transversals.*

17.4.2. *Use Poincaré maps to formulate and prove a version of the Anosov Closing Lemma for flows.*

5. Geodesic flows on surfaces of constant negative curvature

In Section 5.4f we established some properties of the geodesic flow on compact factors of the hyperbolic plane which are characteristic for systems with hyperbolic behavior, namely, density of periodic orbits, topological transitivity, and ergodicity with respect to the smooth invariant measure. We now want to show that the geodesic flow on a compact factor of the hyperbolic plane is an Anosov flow. We will adopt the notation of Section 5.4. Consider the geodesic flow on a compact quotient τ of $\mathbb{H}$, that is, the geodesic flow on a surface τ obtained by factorizing $\mathbb{H}$ by any discrete group Γ of isometries acting without fixed points such that the factor $\Gamma\backslash\mathbb{H}$ is compact: Since τ is locally isometric to $\mathbb{H}$ we can use Proposition 5.4.13 and compactness of τ to obtain

Theorem 17.5.1. *The geodesic flow on τ is an Anosov flow.*

A rather different viewpoint on the geodesic flow on the unit tangent bundle $S\mathbb{H}$ is given by the following algebraic interpretation: If $PSL(2,\mathbb{R}) := GL_+(2,\mathbb{R})/_{\ker\psi}$, where $\psi\begin{pmatrix} a & b \\ c & d \end{pmatrix}$ is the Möbius transformation $z \mapsto \dfrac{az+b}{cz+d}$, then $PSL(2,\mathbb{R})$ is isomorphic to $\mathcal{M}$ via the natural map $PSL(2,\mathbb{R}) \to \mathcal{M}$ induced by ψ.

Since for any $v, w \in S\mathbb{H}$ there is a unique $T \in \mathcal{M}$ such that $(DT)(v) = w$ we can identify $S\mathbb{H}$ with $PSL(2,\mathbb{R})$ by taking $v_0 = i \in S_i\mathbb{H}$ and defining

$$\phi\colon S\mathbb{H} \to PSL(2,\mathbb{R}) \quad \text{by} \quad D\left(\psi(\phi(v))\right)(v_0) = v. \tag{17.5.1}$$

To be explicit we view $PSL(2,\mathbb{R})$ as $SL(2,\mathbb{R})/_{\pm\,\mathrm{Id}}$, where $SL(2,\mathbb{R})$ is the group of 2×2 matrices with unit determinant. Then we can write transformations in $\mathcal{M}$ as matrices in $SL(2,\mathbb{R})$. Möbius transformations lifted to $S\mathbb{H}$ correspond to left multiplications on $PSL(2,\mathbb{R})$. The classification into elliptic, parabolic, and hyperbolic Möbius transformations mentioned in Section 5.4c corresponds to the classification of matrices according to the absolute value T of the trace: $T < 2$ for elliptic, $T = 2$ for parabolic, and $T > 2$ for hyperbolic transformations.

To describe $\mathbb{H}$ itself in this algebraic fashion notice that a given point p is the image of i under many isometries, but that any two of these differ by an isometry fixing i. Thus $\mathbb{H}$ is naturally identified as the quotient $PSL(2,\mathbb{R})/K$, where K is the *isotropy subgroup* of $I = i$, that is, the compact group of $A \in PSL(2,\mathbb{R})$ such that $\psi(A)$ fixes i. The Riemannian metric on $\mathbb{H}$ thus corresponds to the left-invariant metric on $PSL(S,\mathbb{R})$ that is right-invariant under K.

It is helpful later to point out that K is a *maximal* compact subgroup, namely, if L is a subgroup containing K and some $\gamma \notin K$ then either γ is nonelliptic and hence its iterates are not contained in any compact subgroup, or else γ is elliptic with fixed point other than i. Then consider the point 0 in the boundary of $\mathbb{H}$. Since K acts transitively on the boundary there is a $g \in K$ such that $g\gamma(0) = 0$. Since $g\gamma \neq \mathrm{Id}$, $g\gamma$ is not elliptic and its iterates are not contained in any compact subgroup.

Now we consider two interesting one-parameter subgroups acting on the right.

Examples. (1) If $v_t = ie^t \in S_{ie^t}\mathbb{H}$ then $\phi(v_t) = G_t = \begin{pmatrix} e^{t/2} & 0 \\ 0 & e^{-t/2} \end{pmatrix}$.

(2) If $v_t = i \in S_{t+i}\mathbb{H}$ then $\phi(v_t) = H_t = \begin{pmatrix} 1 & t \\ 0 & 1 \end{pmatrix}$.

In regard to example (1) note that $v_t = g^t v_0$ so $\phi(g^t v_0) = \phi(v_t) = G_t$. In fact, if we write $\bar{\phi}(v) := D\left(\psi(\phi(v))\right)$ and recall that g^t commutes with differentials of isometries then we find

$$\begin{aligned}\phi(g^t v) &= \phi(g^t \bar{\phi}(v) v_0) = \phi(\bar{\phi}(v) g^t v_0) = \phi(\bar{\phi}(v)\bar{\phi}(g^t v_0) v_0) \\ &= \phi\left(D(\psi(\phi(v)\phi(g^t v_0)))(v_0)\right) = \phi(v)\phi(g^t v_0) = \phi(v) G_t.\end{aligned}$$

This shows that the action of $\{G_t\}_{t\in\mathbb{R}}$ by right multiplication on $PSL(2,\mathbb{R})$ represents the geodesic flow of $\mathbb{H}$ under the identification (17.5.1).

As for example (2) note that $H_t = \phi(v_t)$ parameterizes the stable manifold of Id. Aside from the geometry of v_t one can observe this by noting that

$$\begin{pmatrix} 1 & t \\ 0 & 1 \end{pmatrix}\begin{pmatrix} e^{t/2} & 0 \\ 0 & e^{-t/2} \end{pmatrix} = \begin{pmatrix} e^{t/2} & te^{-t/2} \\ 0 & e^{-t/2} \end{pmatrix}$$

and

$$\mathrm{Id}\begin{pmatrix} e^{t/2} & 0 \\ 0 & e^{-t/2} \end{pmatrix} = \begin{pmatrix} e^{t/2} & 0 \\ 0 & e^{-t/2} \end{pmatrix}$$

are positively asymptotic.

The right action of $\{H_t\}_{t\in\mathbb{R}}$ is referred to as the *horocycle flow*. Geometrically the tangent vector v to the geodesic γ moves with unit speed along the horocycle determined by $\gamma(-\infty)$ parameterized by length, remaining orthogonal to it.

We will see in Section 7 how both the geometric and algebraic interpretation of the geodesic flow on surfaces of constant negative curvature can be extended to produce interesting examples of Anosov geodesic flows on manifolds of higher dimension.

Exercises

17.5.1. *Give a geometric interpretation in terms of the unit tangent bundle of the action on $PSL(2,\mathbb{R})$ by right multiplication of the following one-parameter groups:*

(1) $\begin{pmatrix} 1 & 0 \\ -t & 1 \end{pmatrix}$, (2) $\begin{pmatrix} \cos t & \sin t \\ -\sin t & \cos t \end{pmatrix}$, (3) $\begin{pmatrix} \cosh t & \sinh t \\ \sinh t & \cosh t \end{pmatrix}$.

17.5.2. *Consider an action of a one-parameter subgroup on $\Gamma\backslash PSL(2,\mathbb{R})$ by right multiplication. Prove that it is smoothly conjugate up to a constant reparameterization to a geodesic flow, horocycle flow, or the action of (2) of the previous exercise on $\Gamma'\backslash PSL(2,\mathbb{R})$ for some other lattice Γ'.*

6. Geodesic flows on compact Riemannian manifolds with negative sectional curvature

From the detailed and very explicit description of the geometry and dynamics of the geodesic flow on the hyperbolic plane and its compact factors we now turn to a general discussion of geodesic flows on Riemannian manifolds with negative sectional curvature. The principal aim of this section is to show that these provide a class of examples of Anosov flows.

Recall that geodesic flows in the general setting were discussed in Sections 5.3 (Definition 5.3.4) and 9.5.

We begin by stating some facts about curvature in a form suitable for our purpose. Let M be a compact Riemannian manifold. Denote by TM the tangent bundle, by $SM := \{v \in TM \mid \|v\| = 1\}$ the unit tangent bundle, and by R the curvature tensor (see Section A.4). Then for $u, v, w, x \in T_pM$ we have

$$R(u,u) = 0 \text{ and } \langle R(u,v)w, x\rangle = \langle R(w,x)u, v\rangle.$$

If u, v are independent then the expression

$$\frac{\langle R(u,v)u, v\rangle}{\langle u,u\rangle\langle v,v\rangle - \langle u,v\rangle^2} \tag{17.6.1}$$

depends only on the linear span S of the vectors u and v and is called the *sectional curvature* of the 2-plane $S \subset T_pM$. It is equal to the Gaussian curvature at p of the 2-manifold $\exp_p S$ with respect to the Riemannian metric induced from M. We assume from now on that this quantity is always negative and hence, by compactness, bounded from above by $-k < 0$.

Jacobi fields $Y: t \mapsto Y(t) \in T_{\gamma(t)}M$ along a geodesic $\gamma: \mathbb{R} \to M$ are obtained as solutions of the Jacobi equation

$$\ddot{Y}(t) + K(t)Y(t) = 0,$$

where dots denote differentiation with respect to t and $K(t) := R(\dot\gamma(t), \cdot)\dot\gamma(t)$.

A tangential Jacobi field is of the form $Y(t) = f(t)\dot\gamma(t)$ with $\ddot{f}(t) = 0$ (since $\ddot\gamma(t) = 0$ and $K(t)\dot\gamma(t) = 0$) and hence linear in time. On the other hand the projection Y_T onto $\dot\gamma$ of any Jacobi field Y is of the same form with $f(t) = \langle Y(t), \dot\gamma(t)\rangle$. But $\ddot{f} = \langle \ddot{Y}, \dot\gamma\rangle = -\langle K\dot\gamma, Y\rangle = 0$ and thus the tangential projection Y^T of Y is a Jacobi field. By linearity of the Jacobi equation the same holds for $Y^\perp := Y - Y^T$ which is orthogonal to $\dot\gamma$.

The interest in Jacobi fields stems from the fact that they arise from variations of geodesics which causes their behavior to reflect the dynamics of the geodesic flow g^t in a way that is made precise as follows.

For $p \in M$, $v \in T_pM$ denote by γ_v the geodesic with $\gamma_v(0) = p$, $\dot\gamma_v(0) = v$. Then there exist isomorphisms

$$\psi_v: T_vTM \to T_pM \oplus T_pM, \quad \xi \mapsto (x, x')$$

such that

$$\psi_{g^t v}(Dg^t\xi) = \Big(Y(t), \dot{Y}(t)\Big),$$

where Y is the Jacobi field along γ_v with $Y(0) = x$ and $\dot{Y}(0) = x'$.

This allows us to describe the dynamics of the geodesic flow in terms of the evolution of Jacobi fields and to speak of the action g^t (or Dg^t, rather) on Jacobi fields.

The two linearly independent tangential Jacobi fields with linear growth correspond to affine reparameterizations of the geodesic, that is, shifts of the initial point and uniform changes of speed. The first variation corresponds to the flow direction for the geodesic flow in the unit tangent bundle SM; the second is transversal to SM. Thus in order to establish that the geodesic flow in SM is an Anosov flow it is sufficient to show that the space of *orthogonal* Jacobi fields admits a splitting into exponentially contracting and exponentially expanding invariant subspaces.

To study orthogonal Jacobi fields it suffices to know that they are solutions of the Jacobi equation

$$\ddot{Y} + KY = 0$$

with a negative-definite symmetric operator K and that our curvature assumption together with compactness implies the existence of $k, \kappa > 0$ such that

$$\langle KY, Y\rangle \leq -k\langle Y, Y\rangle$$

whenever $Y \perp \dot{\gamma}$, and such that

$$\langle KY, KY\rangle < \frac{1}{\kappa^2}$$

for all $Y \in SM$. Any upper bound for the sectional curvature can be used as $-k^2$.

To show hyperbolicity of the geodesic flow we introduce a norm on $T_pM \oplus T_pM$ by

$$\|u, v\| := \sqrt{\langle u, u\rangle + \epsilon\langle v, v\rangle}$$

for $u, v \in T_pM$ and some $\epsilon < 1/\kappa$ and note that $\langle Y, \dot{Y}\rangle / \|Y, \dot{Y}\|^2 \geq \delta$ defines a cone in the sense of Definition 6.2.9.

Lemma 17.6.1. *For* $\delta < \dfrac{k}{1+\kappa^{-3/2}}$ *the family of cones* C_δ *given by*

$$\frac{\langle Y, \dot{Y}\rangle}{\|Y, \dot{Y}\|^2} \geq \delta$$

is strictly invariant.

It would be easier to prove this for $\delta = 0$ only, but we need some positive δ later to get expansion inside these cones.

Proof. It suffices to show that $\frac{d}{dt}\frac{\langle Y,\dot{Y}\rangle}{\|Y,\dot{Y}\|^2}$ is positive when $\frac{\langle Y,\dot{Y}\rangle}{\|Y,\dot{Y}\|^2}=\delta$. To do this note that for $a,b\in\mathbb{R}$ we have $\frac{\sqrt{ab}}{a+\epsilon b}\leq\frac{1}{2\sqrt{\epsilon}}$ since $a^2+2\epsilon ab+\epsilon^2b^2\geq 4\epsilon ab$ and hence $a+\epsilon b\geq 2\sqrt{\epsilon}\sqrt{ab}$. Also $\frac{b+ka}{a+\epsilon b}=k\left(1+\frac{\frac{1}{k}-\epsilon}{\frac{a}{b}+\epsilon}\right)\geq\min(k,\frac{1}{\epsilon})=k$. Thus

$$\frac{\sqrt{\langle Y,Y\rangle\langle\dot{Y},\dot{Y}\rangle}}{\|Y,\dot{Y}\|^2}\leq\frac{1}{2\sqrt{\epsilon}}\quad\text{and}\quad\frac{\langle\dot{Y},\dot{Y}\rangle+k\langle Y,Y\rangle}{\|Y,\dot{Y}\|^2}\geq k.$$

Now, using the Cauchy–Schwarz inequality and taking $\frac{\langle Y,\dot{Y}\rangle}{\|Y,\dot{Y}\|^2}=\delta<1/2$ gives

$$\begin{aligned}
\frac{d}{dt}\frac{\langle Y,\dot{Y}\rangle}{\|Y,\dot{Y}\|^2}&=\frac{(\langle\dot{Y},\dot{Y}\rangle+\langle\ddot{Y},Y\rangle)\|Y,\dot{Y}\|^2-2\langle Y,\dot{Y}\rangle(\langle Y,\dot{Y}\rangle+\epsilon\langle\ddot{Y},\dot{Y}\rangle)}{\|Y,\dot{Y}\|^4}\\
&=\frac{\langle\dot{Y},\dot{Y}\rangle-\langle KY,Y\rangle}{\|Y,\dot{Y}\|^2}-2\frac{\langle Y,\dot{Y}\rangle}{\|Y,\dot{Y}\|^2}\frac{\langle Y-\epsilon KY,\dot{Y}\rangle}{\|Y,\dot{Y}\|^2}\\
&\geq\frac{\langle\dot{Y},\dot{Y}\rangle+k\langle Y,Y\rangle}{\|Y,\dot{Y}\|^2}-2\delta\left(\frac{\langle Y,\dot{Y}\rangle}{\|Y,\dot{Y}\|^2}-\epsilon\frac{\langle KY,\dot{Y}\rangle}{\|Y,\dot{Y}\|^2}\right)\\
&\geq k-2\delta\left(\delta+\epsilon\frac{\sqrt{\langle KY,KY\rangle\langle\dot{Y},\dot{Y}\rangle}}{\|Y,\dot{Y}\|^2}\right)\\
&\geq k-2\delta\left(\delta+\frac{\epsilon}{\kappa}\frac{\sqrt{\langle Y,Y\rangle\langle\dot{Y},\dot{Y}\rangle}}{\|Y,\dot{Y}\|^2}\right)\\
&\geq k-2\delta\left(\delta+\frac{\sqrt{\epsilon}}{2\kappa}\right)\geq k-\delta\left(1+\frac{1}{\kappa^{3/2}}\right)>0.
\end{aligned}$$

□

Having picked δ in Lemma 17.6.1 independently of $\epsilon<1/\kappa$ we next take $\epsilon<4\kappa^2\delta^2$ and $(Y,\dot{Y})\in C_\delta$. Then

$$\begin{aligned}
\frac{\frac{d}{dt}\|Y,\dot{Y}\|}{\|Y,\dot{Y}\|}&=\frac{\langle Y,\dot{Y}\rangle+\epsilon\langle\ddot{Y},\dot{Y}\rangle}{\|Y,\dot{Y}\|^2}=\frac{\langle Y,\dot{Y}\rangle}{\|Y,\dot{Y}\|^2}-\epsilon\frac{\langle KY,\dot{Y}\rangle}{\|Y,\dot{Y}\|^2}\\
&\geq\delta-\epsilon\frac{\sqrt{\langle KY,KY\rangle\langle\dot{Y},\dot{Y}\rangle}}{\|Y,\dot{Y}\|^2}\geq\delta-\frac{\epsilon}{\kappa}\frac{\sqrt{\langle Y,Y\rangle\langle\dot{Y},\dot{Y}\rangle}}{\|Y,\dot{Y}\|^2}\geq\delta-\frac{\sqrt{\epsilon}}{2\kappa}>0.
\end{aligned}$$

This proves that Jacobi fields in C_δ expand exponentially.

It is now an easy exercise to construct cones that are strictly invariant and expanding in negative time, but it is much easier to note that their existence follows from the fact that by definition

$$g^{-t}v=-g^t(-v).$$

We can thus invoke Proposition 17.4.4 to conclude:

Theorem 17.6.2. *The geodesic flow on a compact Riemannian manifold with negative sectional curvature is an Anosov flow.*

By the Hadamard–Perron Theorem 17.4.3 there are stable and unstable manifolds through each point $v \in SM$. We close this section by describing these in geometric terms. As we do this it may be very instructive to compare the argument with the discussion for surfaces of constant negative curvature at the end of Section 5.4d and in the previous section. Let us pass to the universal cover $\widetilde{M}$ of M which is diffeomorphic to $\mathbb{R}^n$ (Exercise 17.6.3). We begin with unstable manifolds. Fix $v \in S\widetilde{M}$ and let

$$B_T := \{\gamma(0) \mid \gamma \text{ geodesic}, \ \gamma(-T) = \gamma_v(-T)\}$$

and

$$W_T := \{\dot{\gamma}(0) \mid \gamma \text{ geodesic}, \ \gamma(-T) = \gamma_v(-T)\},$$

the outside unit normal vectors to B_T. W_T is a smooth submanifold of $S\widetilde{M}$ of dimension $n-1$, where $n = \dim M$.

Consider any curve in W_T. Associated with the corresponding geodesic variation is a Jacobi field Y with $Y(-T) = 0$. Unless $Y = 0$ we have $\langle Y(t), \dot{Y}(t)\rangle > 0$ for $t > -T$ since $\dot{Y}(-T) \neq 0$ and $Y(t-T) = t\dot{Y}(-T) + o(t)$ whence $\langle Y(t-T), \dot{Y}(t-T)\rangle > 0$ for small positive values of t. But we showed that this must then hold for all $t > 0$.

This shows that every tangent vector to W_T is contained in a cone from the invariant family. As we showed in proving the Hadamard–Perron theorem this implies that as $T \to \infty$ the W_T converge to $W^u(v)$, a smooth $(n-1)$-dimensional submanifold of $S\widetilde{M}$. Since the projection $\pi\colon S\widetilde{M} \to \widetilde{M}$ is smooth the spheres B_T converge to a smooth submanifold B_∞ called a *horosphere* (which means limit sphere).

$W^u(v)$ in fact consists of the outward unit normals to B_∞, which itself can be described as $\{\gamma(0) \mid \gamma \text{ geodesic}, d(\gamma(t), \gamma_v(t)) \xrightarrow[t\to-\infty]{} 0\}$.

Exercises

17.6.1. *Let M be an m-dimensional Riemannian manifold with negative sectional curvature bounded from below by $-K^2$ and above by $-k^2$. Prove that for the volume growth $v(M)$ defined by Definition 9.6.5 we have $k \leq v(M)/(m-1) \leq K$.*

17.6.2. *Prove that the fundamental group $\pi_1(M)$ of a compact manifold that admits a metric of negative sectional curvature has exponential growth, that is, for any given system Γ of generators of $\pi_1(M)$ the number of elements $\gamma \in \pi_1(M)$ that can be represented by words of length at most n grows exponentially with n.*

17.6.3. *Prove that the universal cover of a manifold of negative sectional curvature is diffeomorphic to Euclidean space.*

17.6.4. *Prove that all geodesics on a manifold of negative sectional curvature are minimal (see Definition 9.6.1).*

7. Geodesic flows on rank-one symmetric spaces

An important class of manifolds of negative curvature is obtained by an algebraic construction which generalizes the algebraic description of surfaces of constant negative curvature in Section 5. The geometric property that enabled us to describe the geodesic flow on the sphere, the torus, and the hyperbolic plane was the presence of an isometry group that is transitive on unit tangent vectors (Lemma 5.4.1). In general such spaces are called (globally) symmetric spaces. We begin with the traditional definition and then prove transitivity of the isometry group in the case of nonvanishing curvature.

Definition 17.7.1. A *Riemannian locally symmetric space* is a connected Riemannian manifold M such that for all $p \in M$ there is a neighborhood U such that $\exp_p \circ(-\mathrm{Id}) \circ \exp_p^{-1}: U \to M$ is an isometry. M is called a *globally symmetric space* if this local isometry extends to an isometry of M, that is, for every $p \in M$ there is an isometry σ_p of M with $\sigma_p(p) = p$ and $D\sigma_{p}\big|_p = -\mathrm{Id}$. σ_p is called the (global) symmetry at p. The space is said to have *rank one* if there is no totally geodesic isometrically embedded Euclidean plane.

Remarks.

(1) An alternative definition is that the curvature tensor is parallel, that is, $\nabla R = 0$.
(2) Since the endpoints of any geodesic segment are exchanged by the symmetry at the midpoint and any two points are connected by a broken geodesic, the isometry group of a globally symmetric space or compact locally symmetric space is clearly transitive on points.
(3) Having rank 1 implies that all sectional curvatures are nonzero.
(4) S^n, $\mathbb{R}^n$, and $\mathbb{H} = \mathbb{RH}^2$ are globally symmetric spaces; $\mathbb{T}^n$ is locally symmetric.
(5) A complete simply connected locally symmetric space is globally symmetric.
(6) Thus the universal cover of a complete locally symmetric space is a globally symmetric space.

Proposition 17.7.2. *If M is a rank-one symmetric space then the isometry group is transitive on SM.*

Proof. Since transitivity on points is known we only need to show that the isometry group is transitive on any particular unit sphere S_pM. To that end it suffices to show that for every 2-plane $\Pi \subset T_pM$ the isometry group is transitive on $\Pi \cap S_pM$, which in turn follows once we see that there exists an $\epsilon > 0$ such that for $v \in \Pi \cap S_pM$ there exists a family of isometries such that the images of v under their differentials cover an arc of length ϵ in $\Pi \cap S_pM$.

To that end consider a disk $D = \exp_p B(0, \delta)$ and a triangle in D with p as one vertex and interior angles α, β, γ. Consider the isometry I obtained by composing the three symmetries about the midpoints of the edges (in cyclic order). Since isometries preserve angles one easily sees by a picture that the angle between v and $DI(v)$ is $\alpha + \beta + \gamma$. Since Π has nonzero curvature the sum $\alpha + \beta + \gamma$ converges to π as the diameter of the triangle tends to 0 but it never equals π. Thus we obtain an arc of images whose size is independent of v. □

All symmetric spaces arise from an algebraic construction which generalizes the construction of Section 17.5. To give an indication of how this comes about we begin with a direct generalization of a *geometric* construction of the hyperbolic space.

The Poincaré disk with the group of Möbius transformations can be obtained as follows. Consider the upper sheet $\mathcal{H}$ of the hyperboloid in $\mathbb{R}^3$ given by $Q(x) := x_1^2 + x_2^2 - x_3^2 = -1$, $x_3 > 0$. The group $SO(2,1)$ of real 3×3 matrices preserving the indefinite quadratic form Q acts on the hyperboloid, and the index-two subgroup preserving $x_3 > 0$ therefore acts on $\mathcal{H}$. Since the action is linear in $\mathbb{R}^3$ it sends planes through 0 (that is, planes given by $ax_1 + bx_2 - cx_3 = 0$) to planes through 0 and hence the family $\mathcal{C}$ of curves given by the intersection of such planes with $\mathcal{H}$ is preserved.

If we change variables to $\eta_1 = x_1/x_3$, $\eta_2 = x_2/x_3$, $\eta_3 = 1/x_3$ the hyperboloid becomes the hemisphere $\eta_1^2 + \eta_2^2 + \eta_3^2 = 1$, $\eta_3 > 0$ and a plane $ax_1 + bx_2 - cx_3 = 0$ is mapped to the plane $a\eta_1 + b\eta_2 = c$ perpendicular to the $\eta_1\eta_2$-plane. Thus curves from $\mathcal{C}$ are mapped to circles orthogonal to the equator $\eta_3 = 0$. Finally apply the stereographic projection centered at $(0,0,-1)$ from the upper hemisphere to the disk $\eta_1^2 + \eta_2^2 < 1$. It is known to be conformal, so the curves from $\mathcal{C}$ now are (lines and) circles perpendicular to the boundary, that is, the geodesics of the Poincaré disk. One can show that the transformations that arise from $SO(2,1)$ in this process are exactly the Möbius transformations. In fact, the hyperboloid is an isometric embedding of the Poincaré disk into Minkowski space $(\mathbb{R}^3, q)$ with the pseudometric q induced by the form Q.

This geometric construction generalizes to give n-dimensional real hyperbolic spaces $\mathbb{RH}^n$. Consider the upper sheet of the hyperboloid $\mathcal{H}$ in $\mathbb{R}^{n+1}$ given by $Q(x) := x_1^2 + \cdots + x_n^2 - x_{n+1}^2 = -1$, $x_{n+1} > 0$. Again let $\mathcal{C}$ be the family of curves that lie on planes through 0, that is, on planes given by n simultaneous equations of the form $a_1x_1 + \cdots + a_nx_n - a_{n+1}x_{n+1} = 0$. The group $SO(n,1)$ of matrices preserving Q acts on $\mathcal{H}$. Change variables to $\eta_1 = x_1/x_{n+1}, \dots,$ $\eta_n = x_n/x_{n+1}$, $\eta_{n+1} = 1/x_{n+1}$ and then apply the stereographic projection centered at $(0,\dots,0,-1)$ to map the resulting hemisphere to the open unit ball in $\mathbb{R}^n$. As before curves in $\mathcal{C}$ map to (lines and) circles perpendicular to the boundary of the unit ball $\mathbb{RH}^n$.

These spaces $\mathbb{RH}^n$ have (sectional) curvature -1 as well. This is clear for all tangent planes Π at $(0,\dots,0,1)$ since in the three-dimensional subspace of $\mathbb{R}^{n+1}$ containing Π the entire picture looks like the description of $\mathbb{RH}^2$.

For purposes of generalization it is more convenient to view $\mathbb{RH}^n$ as a subset of the n-dimensional real projective space $\mathbb{RP}^n$ of lines through 0 in $\mathbb{R}^{n+1}$ by identifying a point p on the upper hyperboloid with the line through 0 containing p. The Riemannian metric is, of course, not the induced one, but the tangent vectors to $\mathbb{RH}^n$ are tangent vectors of $\mathbb{RP}^n$. Hyperbolic distances are given as follows. Two points in this space correspond to two lines in $\mathbb{R}^{n+1}$. The plane defined by these intersects the cone $Q=0$ in two more lines. The hyperbolic distance is given by the logarithm of the cross ratio of the four points in projective space determined by these four lines.

This latter description works over the complex field $\mathbb{C}$ as well. Namely, we obtain the n-dimensional complex hyperbolic space $\mathbb{CH}^n$ as a subset of complex projective space $\mathbb{CP}^n$, that is, the space of complex lines through the origin of $\mathbb{C}^{n+1}$, with a distance similarly defined by cross ratios. There is an important new phenomenon, however. Any tangent space can be viewed simultaneously as an n-dimensional complex linear space or a $2n$-dimensional real linear space. Thus a real vector v in a tangent space can be multiplied by $i=\sqrt{-1}$ to give a unique direction that is perpendicular to v with respect to the real structure but collinear to v with respect to the complex structure. One can check that this real 2-dimensional subspace has (sectional) curvature -4 and that multiplication by i is an isometry of the unit tangent bundle. Thus one has a natural real 1-dimensional distribution on the unit tangent bundle $S\mathbb{CH}^n$ given by these directions. There is naturally a complementary distribution defined by the vectors that are complex orthogonal to v and iv. Inside this distribution all sectional curvatures are -1. This distribution turns out to be nonintegrable.

For the geodesic flow these distributions correspond to distributions of vectors with expansion rates e^{2t} and e^t, respectively, and corresponding contraction rates.

For the quaternions $\mathbb{Q}$ one obtains hyperbolic spaces $\mathbb{QH}^n$ with a similar structure, but here one obtains a (real) 3-dimensional distribution corresponding to planes of curvature -4. Even for the octonians $\mathbb{O}$ (Cayley numbers) one obtains a hyperbolic plane $\mathbb{OH}^2$, here with a corresponding 7-dimensional distribution. The last construction, however, does not extend to higher dimension due to nonassociativity of the Cayley numbers. These examples in fact exhaust the list of Riemannian globally symmetric spaces of negative curvature.[1] All of these spaces admit compact Riemannian factors obtained by the left action of a uniform lattice in the isometry group, so the geodesic flows on such factors provide examples of Anosov geodesic flows.

We now give, also without proof, an indication of the general *algebraic* description of globally symmetric spaces.

Proposition 17.7.3. *If M is a globally symmetric space then the identity component G of the isometry group of M acts transitively on M and the isotropy group K of any point is compact.*

Definition 17.7.4. A globally symmetric space M is said to be of noncompact type if G is semisimple with no compact factors and K is a maximal compact subgroup of G.

Remark. Unlike in the case of $\mathbb{RH}^2$ the group G for other globally symmetric spaces of rank 1 will be substantially larger than the unit tangent bundle of the manifold we are considering.

Conversely for every connected semisimple Lie group with no compact factors and a maximal compact subgroup K (which is unique up to conjugacy by an inner automorphism of G) there is a natural globally symmetric structure on $M := G/K$, namely, every left-invariant Riemannian metric on G that is right-invariant under K then makes M a Riemannian manifold and the quotient of M under the left action of a lattice Γ in G will then be a compact Riemannian factor of M. This is the analog of the torus and compact factors of the hyperbolic plane $\mathbb{RH}^2$ in Section 5.4.

In this model geodesics through Id are given by one-parameter subgroups of G/K.

The general algebraic description of the geodesic flow on rank-one Riemannian symmetric spaces of noncompact type is as follows. Let G be a simple noncompact Lie group of real rank one. Such groups are $SO(n,1)$, $SU(n,1)$, $Sp(n,1)$, and F_4. Let K be a maximal compact subgroup of G. Then G/K is a globally symmetric space and its unit tangent bundle is of the form G/T, where T is a compact subgroup of K (namely, the isotropy subgroup of a tangent vector). The symmetric spaces are, correspondingly, n-dimensional real, complex, and quaternionic hyperbolic spaces and the 2-dimensional hyperbolic Cayley plane. The geodesic flow corresponds to the right action of a one-parameter subgroup that commutes with T. (Note that in the two-dimensional case $T = \{\mathrm{Id}\}$.)

Let us point out that unlike Anosov diffeomorphisms, which seem to be rather rigid objects from the point of view of topological conjugacy, Anosov flows appear more often. On the one hand there are other constructions of Riemannian manifolds of negative curvature besides perturbations of symmetric spaces. In particular there are Riemannian manifolds with negative curvature in dimension greater than three with fundamental group different from that of any locally symmetric space. Thus the phase space of the geodesic flow is also topologically distinct from the unit tangent bundle of any locally symmetric space and hence the geodesic flows are not orbit equivalent. Furthermore, already in dimension three there are compact manifolds that admit Anosov flows without being homeomorphic to the unit tangent bundle of a surface. Among those examples there are even Anosov flows with nowhere dense nonwandering sets which are therefore dynamically very different from any volume-preserving flow.[2]

Exercise

17.7.1*. *Prove that for a compact n-dimensional manifold of constant negative sectional curvature $-k^2$ $(k>0)$, that is, a compact factor of the real hyperbolic space $\mathbb{RH}^m$, the topological entropy of the geodesic flow is given by $(m-1)k$.*

8. Hyperbolic Julia sets in the complex plane

a. Rational maps of the Riemann sphere. These were first mentioned in Section 8.3, where we obtained a lower estimate for the topological entropy. For the simplest polynomial map $z \mapsto z^n$ with $n \geq 2$ the nonwandering set consists of two attracting fixed points and the equator. In this case the critical points are *superattracting* (that is, have zero derivative) and their basins are the components of the complement of the equator. By the *basin* of an attracting periodic point p we mean the (obviously open) set of points that are positively asymptotic to p. We will call the union of the connected components containing points of the orbit of p the *immediate basin* of p. (Compare with the discussion in Section 16.1 for interval maps.) The equator is a hyperbolic set and the restriction of the map to the equator is the expanding circle map E_n. This is a particular case of a general phenomenon. To avoid the complications of the most general case we prove a relevant result under simplifying assumptions.

Theorem 17.8.1. *Let $f: \hat{\mathbb{C}} \to \hat{\mathbb{C}}$, $z \mapsto \dfrac{P(z)}{Q(z)}$ be a rational map of the Riemann sphere $\hat{\mathbb{C}}$, where P and Q are relatively prime polynomials. Assume that*

1. *every critical point is attracted to (or itself) a periodic point, and*
2. *for every n-periodic point p the derivative of f^n at p is not of absolute value 1 (that is, every periodic point is hyperbolic or superattracting).*

Then the closure of the periodic orbits of f is hyperbolic.

Remark. For a small perturbation of $z \mapsto z^n$ to a polynomial map of algebraic degree n the attracting points 0 and ∞ persist and their basins are still topological disks. The complement of their union is still a topological circle, but may not be smooth.

Proof. The first step in the proof is the following classical result of Julia.

Proposition 17.8.2. (Julia) *If p is an attracting periodic point of a rational map f then the immediate basin of p contains a critical point.*

Proof. Assume the immediate basin of p contains no critical point of f and denote by n the period of p. Let B_p be the connected component of the basin of p that contains p. Then the immediate basin of p is $\bigcup_{i=0}^{n-1} f^i(B_p)$. Thus B_p contains no critical point of f^n. Since B_p is the immediate basin of p under f^n this means we may assume without loss of generality that $n=1$, that is, p is a fixed point. Notice that $B_p = \bigcup_{i\in\mathbb{N}} f_0^{-i}(U)$ for some small disk U, where f_0^{-1}

denotes the appropriate branch of the inverse. Since B_p contains no critical points of f this shows that B_p is simply connected, that is, a topological disk. By the Riemann mapping theorem B_p is biholomorphically equivalent to the unit disk D in $\mathbb{C}$, hence possesses a hyperbolic metric induced by the hyperbolic metric of the Poincaré disk. Since biholomorphic bijective mappings of the unit disk are Möbius transformations and hence isometries, this hyperbolic metric on B_p is unique. Thus the following result applies:

Lemma 17.8.3. *Let $f: D_1 \to D_2$ be a holomorphic map between simply connected domains with a hyperbolic metric. Then either f is a holomorphic isometry of the hyperbolic metric, or f strictly contracts the hyperbolic metric.*

Proof. The Riemann mapping theorem gives us biholomorphic maps $D_i \to D$, so it suffices to consider the case $f: D \to D$. For any $z \in D$ pick Möbius transformations φ, ψ such that $\varphi(z) = 0$ and $\psi(f(z)) = 0$. Then $h := \psi \circ f \circ \varphi^{-1}$ is a biholomorphic map of D fixing 0 and hence by the Schwarz lemma either h is a Möbius transformation or $|Dh(0)| < 1$. But $Dh = D\psi Df D\varphi^{-1}$ and $|D\psi| = |D\varphi^{-1}| = 1$. □

In particular we see that f is an isometry whenever it is surjective. In this case $f: B_p \to B_p$ is clearly surjective by construction, hence an isometry. In particular the derivative at any fixed point in B_p has absolute value 1. But this contradicts the presence of the attracting fixed point p in B_p. □

Corollary 17.8.4. *The number of attracting periodic points is bounded by the number of critical points of $f = P/Q$, which in turn is bounded by* $\deg P + \deg Q$.

Now we return to the proof of Theorem 17.8.1. Consider the complement S in $\hat{\mathbb{C}}$ of the closure C of the orbits of the critical points. Since the critical points are attracted to the finitely many attracting periodic points, S is connected. Thus the closure J of the repelling periodic points is a compact invariant set. We will show that J is hyperbolic. Then the closure of the periodic points of f is the union of J with the finitely many attracting periodic points and hence is hyperbolic. Suppose first that $f^{-1}(C) \neq C$. If $\operatorname{card}(C) < 3$ note that either $C = \varnothing$ and f is a Möbius transformation and hence, since any Möbius transformation has at most 2 periodic points, Theorem 17.8.1 is trivial in this case, or there is a critical point c_0 with infinite negative semiorbit, that is, there is a point p such that $f^2(p) \neq c_0$ and $f^3(p) = c_0$, so we can replace S by $S \smallsetminus \{f(p), f^2(p)\} \supset J$ and henceforth assume $\operatorname{card}(\hat{\mathbb{C}} \smallsetminus S) \geq 3$. Then S can be considered as a Riemann surface whose (holomorphic) universal cover is the unit disk (Koebe uniformization theorem). Thus S has a unique hyperbolic metric whose norm we denote by $\|\cdot\|^S$. Given $x \in S$ such that $f(x) \in S$ let R be the connected component containing x of $f^{-1}(S)$. The inclusion $R \to S$ is not surjective, hence a contraction from $\|\cdot\|^R$ to $\|\cdot\|^S$, so $\|\cdot\|^S < \|\cdot\|^R$. Since $f_{\upharpoonright R}: (R, \|\cdot\|^R) \to (S, \|\cdot\|^S)$ is a surjective holomorphic map without critical points, it is a local isometry, hence expanding with respect to $\|\cdot\|^S$. Since $J \subset S$ is compact it is hyperbolic.

Now consider the exceptional cases, that is, where $f^{-1}(C) = C$. If $\operatorname{card} C = 1$ then up to a coordinate change $C = \{\infty\}$ and $f(\infty) = f^{-1}(\infty) = \infty$, so f is a polynomial without critical points in $\mathbb{C}$, hence linear, and so ∞ is not a critical point, a contradiction. If $\operatorname{card} C = 2$ then there are two cases. Either f has two fixed critical points (and no others), without loss of generality $0 = f(0) = f^{-1}(0)$ and $\infty = f(\infty) = f^{-1}(\infty)$, and is thus conjugate to $z \mapsto z^k$ by a Möbius transformation, in which case the conclusion of Theorem 17.8.1 is not new, or, up to Möbius transformation, we have $f(\infty) = f^{-1}(\infty) = 0$ and f is conjugate to $z \mapsto z^{-k}$ by a Möbius transformation, again yielding Theorem 17.8.1.

If $\operatorname{card} C \geq 3$ then $S\hat{\mathbb{C}} \setminus C$ admits a hyperbolic metric as before and f is an isometry with respect to this metric (because it is surjective) hence $\{f^n\}_{n\in\mathbb{N}}$ is equicontinuous on S. Since it is trivially equicontinuous on C it is equicontinuous on $\hat{\mathbb{C}}$ and thus by Theorem A.1.24 f accumulates on a meromorphic map of $\hat{\mathbb{C}}$. But then considering the topological degree, we find $\deg g = \lim \deg f^{n_k}$, so $\deg f = 1$ and f is a Möbius transformation, contradicting $\operatorname{card} C \geq 3$. □

Assumption (2) of Theorem 17.8.1 was made as a matter of convenience. In the general case one studies the dynamics of complex maps by starting from a classical dichotomy for the behavior of orbits. Orbits with simple behavior form the *Fatou set* of a map. In the case of holomorphic maps a natural notion of simplicity is provided by the concept of a *normal family* of functions, that is, an equicontinuous family. The Fatou set is the set of points having a neighborhood on which the iterates of the map f form a normal family. Thus it is open. This notion is very natural since a normal family is not only compact in the C^0 topology (Arzelá–Ascoli Theorem A.1.24), but, in fact, compact in the holomorphic topology. The *Julia set* is defined as the complement of this set and is hence closed. Let us show that in the interesting cases it is nonempty:

Proposition 17.8.5. *If $f:\hat{\mathbb{C}} \to \hat{\mathbb{C}}$ is a rational map of degree greater than one then the Julia set is nonempty.*

Proof. Otherwise the family $\{f^n\}_{n\in\mathbb{N}}$ is normal on $\hat{\mathbb{C}}$ and there is a subsequence converging uniformly to some meromorphic $g:\hat{\mathbb{C}} \to \hat{\mathbb{C}}$. Considering the topological degree, we find $\deg g = \lim \deg f^{n_k}$, so $\deg f = 1$. □

In this case the Julia set can, in fact, be shown to be the closure of the repelling periodic points. Thus the set J considered in the proof of Theorem 17.8.1 is actually the Julia set. We shall, however, not need this fact. Under the assumptions of Theorem 17.8.1 the Fatou set happens to be the union of the basins of attracting periodic points. In general, there may be other phenomena, however. For example, the presence of a period-n point p where the differential of f^n has an eigenvalue $e^{2\pi i\alpha}$ with real Diophantine α (Definition 2.8.1) allows us to apply Theorem 2.8.2 to f^n in order to obtain a neighborhood on which f^n is analytically conjugate to a rotation of a disk. Such a neighborhood is called a *Siegel disk*. Obviously every point in a Siegel disk belongs to the Fatou set.

Let us discuss the structure of the Julia set for quadratic polynomials. We first show that the quadratic family $z \mapsto z^2 + c$ exhibits all possibilities for quadratic polynomials. Namely, consider a quadratic polynomial $P(z) = az^2 + 2bz+d$ and let $h(z) = az+b$, $c = ad+b-b^2$. Then one checks that $h \circ P \circ h^{-1}(z) = z^2 + c$, that is, P is analytically conjugate to $z \mapsto z^2 + c$. Thus we shall consider examples of hyperbolic Julia sets arising from different values of c in $z \mapsto z^2 + c$.

Notice first that for a quadratic map there are two critical points, one at zero and one at infinity, the latter being a superattracting fixed point. Thus there are three possibilities:

(1) Zero belongs to the basin of infinity. In this case there is only one basin by Proposition 17.8.2 and the Julia set is totally disconnected. Similarly to the one-dimensional case (Corollary 16.1.2) the restriction of the quadratic map to such a set is topologically conjugate to a topological Markov chain (see Exercise 17.8.1). This happens, for example, for $|c| > 2$.

(2) There is an attracting periodic point different from infinity, which by Proposition 17.8.2 attracts zero. In this case the Julia set is the common boundary of the basins of attraction of the two points. The basin of attraction of infinity is always connected. The other one may or may not be connected.

  (a) Both basins of attraction are connected. Then both attracting points are fixed. This happens, for example, for values of c near 0.

  (b) The period of the finite attracting point is greater than one. Then its basin of attraction has infinitely many connected components and the Julia set has a natural Markov structure. However, the topological structure of the Julia set is different from all our previous examples: It is not locally a product of a manifold with a Cantor set, but rather a perfect connected set of topological dimension one whose complement has infinitely many connected components. This occurs when $c = -1$, that is, when $z \mapsto z^2 - 1$.

(3) Zero belongs to the Julia set. In this case the Julia set is not hyperbolic.

These three cases determine sets I_1, I_2, and I_3, correspondingly, of parameters in the complex plane. The sets I_1 and I_2 are obviously open and within each connected component of one of these the Julia set and the restriction of the map to the Julia set remain topologically the same. I_3 divides the complex plane. It is conjectured that I_3 is nowhere dense. The union of I_2 and I_3 is usually called the Mandelbrot set or sometimes the Brooks–Matelski set. It has been the subject of extremely intense numerical studies which are used among other things to produce many intricate colorful pictures intended to show the beauty of fractals and similar notions.

b. Holomorphic dynamics. The results proved and outlined in the previous subsection offer a glimpse into the world of *holomorphic dynamics*. The underlying structure of this branch of the theory of dynamical systems is a complex manifold (not necessarily compact; an open set in $\mathbb{C}^n$ is an example) and a holomorphic map defined in a neighborhood of a compact invariant set (the semi–local setting). The corresponding global setting is a holomorphic map of a compact complex manifold (for example, complex projective space $\mathbb{C}P(n)$) into itself. Holomorphic maps, both in one and several complex variables, possess a certain rigidity, manifested both locally (Taylor coefficients at a point define the map in an open set) and globally (Liouville Theorem, maximum modulus principle etc.). This sets holomorphic dynamics apart from general differentiable dynamics (where different locally defined maps can be easily glued together) and to a lesser extent Hamiltonian dynamics (where there are no local restrictions either, but there are some global ones). In this respect holomorphic dynamics is closer to the algebraic dynamics of translations and affine maps on homogeneous spaces (Section 5.7) although the dynamical paradigms for the two areas tend to be quite different, for example, no nice invariant measure is usually present in the holomorphic case and dissipative behavior is quite common. One of the characteristic features of holomorphic dynamics is the important role played by singularities of holomorphic maps. Since the singular set has positive complex codimension and hence real codimension at least two the singularities tend to be more manageable than in the real case, even the real–analytic case.

Holomorphic dynamics in one variable is very well developed. In fact, the classical works of Fatou, Julia and Montel appeared at a time when real differentiable dynamics, not to mention ergodic theory, was in its infancy. It rests on two pillars: conformality and uniformization. The former is an *infinitesimal* property; it was discussed in Chapter 10b. as the characteristic property in low–dimensional differentiable dynamics. From this point of view one can define the area of conformal dynamics which essentially includes differentiable dynamics in real dimension one (see Chapters 12 and 16) and holomorphic dynamics in complex dimension one. That this short list is exhaustive follows from the fact that any conformal map in real dimension two is holomorphic and that in higher dimension there are too few conformal maps (only higher–dimensional counterparts of Möbius transformations, see Section 5.4) to have any interesting dynamics. The main technical corollaries of conformality which are crucial for the analysis of a growing number of iterates of a map are various kinds of bounded distortion estimates (see Sections 5.1, 12.1, 12.7., 14.3). Thus the emphasis on conformality brings together one–dimensional real dynamics and one–dimensional complex holomorphic dynamics.

On the other hand, uniformization, whose most elementary manifestation is the Riemann mapping theorem and a more advanced one the Koebe uniformization theorem, (both used in a crucial way in the proof of Theorem 17.8.1) is an essentially one–dimensional *complex* phenomenon.

Multi–dimensional complex dynamics is a much newer and less developed field. While neither conformality nor uniformization are available as a tool, there are other powerful tools from complex analysis which make it possible to understand the structure of certain classes of holomorphic maps (for example, polynomials) to a considerably greater degree than in the case of real differentiable dynamics. At the root of those tools are certain extremal properties of holomorphic maps which allow, for example, to prove a proper formula for the topological entropy. A useful observation is that in complex dimension two hyperbolic behavior of invertible maps forces both stable and unstable manifoalds to be one–dimensional complex submanifolds. Thus the dynamics on these families of manifolds is conformal and some tools from one–dimensional complex analysis can be adapted to this situation.

Exercises

17.8.1. *Show that if the Julia set J for a rational map f is hyperbolic and totally disconnected then $f\restriction_J$ is topologically conjugate to a topological Markov chain.*

17.8.2. *Describe an almost bijective semiconjugacy between the 2-shift σ_2^R and the map $z \mapsto z^2 - 1$ on its Julia set.*

18

Topological properties of hyperbolic sets

We return to the general structure theory of hyperbolic sets for smooth dynamical systems starting from the point reached at the end of Section 6.4. We first concentrate on the robustness of the global orbit structure of such systems. Then we show how the central idea of approximating "almost orbits" by real orbits enables us to give a good description of recurrence, fine asymptotics of the growth of periodic orbits, and an almost invertible semiconjugacy with topological Markov chains.

1. Shadowing of pseudo-orbits

An outstanding feature of hyperbolicity is a very sensitive dependence of an orbit on its initial point. This poses the problem of extracting meaningful information from approximate knowledge of an orbit segment. We already saw that an approximate periodic orbit is always "shadowed" by a real one (Anosov Closing Lemma, Theorem 6.4.15). Now we consider the nonperiodic situation.

Definition 18.1.1. Let (X, d) be a metric space, $U \subset M$ open and $f: U \to X$. For $a \in \mathbb{Z} \cup \{-\infty\}$ and $b \in \mathbb{Z} \cup \{\infty\}$ a sequence $\{x_n\}_{a<n<b} \subset U$ is called an *ϵ-orbit* or ϵ-pseudo-orbit for f if $d(x_{n+1}, f(x_n)) < \epsilon$ for all $a < n < b$.
It is said to be *δ-shadowed* by the orbit $\mathcal{O}(x)$ of $x \in U$ if $d(x_n, f^n(x)) < \delta$ for all $a < n < b$.

Theorem 18.1.2. (Shadowing Lemma) *Let M be a Riemannian manifold, $U \subset M$ open, $f: U \to M$ a diffeomorphism, and $\Lambda \subset U$ a compact hyperbolic set for f. Then there exists a neighborhood $U(\Lambda) \supset \Lambda$ such that whenever $\delta > 0$ there is an $\epsilon > 0$ so that every ϵ-orbit in $U(\Lambda)$ is δ-shadowed by an orbit of f.*

The Shadowing Lemma can be proved in a similar way to the Anosov Closing Lemma (Exercise 18.1.1) by considering local coordinates and a sequence of maps of $\mathbb{R}^n$ close to hyperbolic linear maps. We derive this result as a particular case of the more general Theorem 18.1.3. Shadowing as such does not guarantee that the shadowing orbit is in any sense *typical.* If one considers, for example, the map $f: \mathbb{R}/\mathbb{Z} \to \mathbb{R}/\mathbb{Z}, x \mapsto 2x \pmod 1$ then any computer-generated orbit will eventually become zero since the initial condition is internally represented by a binary fraction and at each step the number of nonzero binary digits after the point decreases. Thus the computer will always compute an actual orbit—but always one attracted to zero. But this is not typical behavior for the system (see Section 1.7). Likewise for toral automorphisms all points with only rational coordinates, and hence all digitally representable points, are periodic (see Proposition 1.8.1).

There is an intimate connection between shadowing and structural stability: The orbits of a perturbation of a dynamical system are ϵ-orbits for the original system. Since they are shadowed by unperturbed orbits, the correspondence that sends perturbed orbits to the unperturbed orbits shadowing them may give a candidate for an orbit equivalence. Conversely, structural stability (with continuous dependence of the conjugating homeomorphism on the perturbation) certainly implies that orbits of a perturbation are shadowed by unperturbed orbits. More generally shadowing is related to the richness and robustness of the orbit structure of a dynamical system as opposed to the instability of individual orbits. This motivates asking for coherent shadowing for entire continuous families of ϵ-orbits.

Theorem 18.1.3. (Shadowing Theorem) *Let M be a Riemannian manifold, d the natural distance function, $U \subset M$ open, $f: U \to M$ a diffeomorphism, and $\Lambda \subset U$ a compact hyperbolic set for f.*

Then there exist a neighborhood $U(\Lambda) \supset \Lambda$ and $\epsilon_0, \delta_0 > 0$ such that for all $\delta > 0$ there is an $\epsilon > 0$ with the following property:

If $f': U(\Lambda) \to M$ is a C^2 diffeomorphism ϵ_0-close to f in the C^1 topology, Y a topological space, $g: Y \to Y$ a homeomorphism, $\alpha \in C^0(Y, U(\Lambda))$, and $d_{C^0}(\alpha g, f'\alpha) := \sup_{y \in Y} d(\alpha g(y), f'\alpha(y)) < \epsilon$ then there is a $\beta \in C^0(Y, U(\Lambda))$ such that $\beta g = f'\beta$ and $d_{C^0}(\alpha, \beta) < \delta$.

Furthermore β is locally unique: If $\bar{\beta} g = f'\bar{\beta}$ and $d_{C^0}(\alpha, \bar{\beta}) < \delta_0$, then $\bar{\beta} = \beta$.

Remarks. (1) Note that local maximality of Λ is not required.

(2) To get the Shadowing Lemma take $Y = (\mathbb{Z}, \text{discrete topology})$, $f' = f$, $\epsilon_0 = 0$, and $g(n) = n + 1$ and replace $\alpha \in C^0(Y, U(\Lambda))$ by $\{x_n\}_{n \in \mathbb{Z}} \subset U(\Lambda)$ and "$\beta \in C^0(Y, U(\Lambda))$ such that $\beta g = f'\beta$" by $\{f^n(x)\}_{n \in \mathbb{Z}} \subset U(\Lambda)$. Then $d(x_n, f^n(x)) < \delta$ for all $n \in \mathbb{Z}$. It may be useful to keep these simplifications in mind while reading the proof.

(3) It might also help to read again the proof of the Anosov Closing Lemma (Theorem 6.4.15), since the method of proof there is similar to the one employed here. In fact, the Closing Lemma is another particular case of the Shadowing Theorem corresponding to $f' = f$, $Y = \mathbb{Z}/n\mathbb{Z}$, $g(k) = k + 1 \pmod n$.

(4) There is an even closer similarity to the proof of topological stability of the linear toral automorphism (Theorem 2.6.1).

Proof. Similarly to the proof of Theorem 2.6.1 the essence of the proof is setting up an appropriate (in the present case rather elaborate) system of notation which allows a direct application of the Contraction Mapping Principle (Proposition 1.1.2). The difference from the aforementioned case is that the appropriate functional space does not have a natural linear structure and such structure needs to be imposed. In fact the proof is a variation of the contraction mapping method for finding conjugacies discussed in Section 2.7a.

An additional difference from the proof of the Anosov Closing Lemma (Theorem 6.4.15) is that the Contraction Mapping Principle is applied in an infinite-dimensional space.

The desired map β is a fixed point of the operator

$$F: C^0(Y, U(\Lambda)) \to C^0(Y, M), \beta \mapsto f' \circ \beta \circ g^{-1}.$$

In order to find it we want to decompose F into linear and nonlinear parts. This requires a linear structure which we introduce similarly to that for the space $\mathrm{Diff}^r(M)$ described in Section 3 of the Appendix. Namely, we define the "difference" between $\beta \in C^0(Y, U(\Lambda))$ and the given map $\alpha \in C^0(Y, U(\Lambda))$ by considering the space

$$C^0_\alpha(Y, TM) := \{v \in C^0(Y, TM) \mid v(y) \in T_{\alpha(y)}M \quad (y \in Y)\}$$

of vector fields along α with the norm $\|\cdot\|$ of uniform convergence, where, for sufficiently small $\theta > 0$ independent of Y, g, and α, the following map $\mathcal{A}$ of the θ-ball $B_\theta(\alpha)$ around α in $C^0(Y, U(\Lambda))$ is well defined:

$$\mathcal{A}: B_\theta(\alpha) \to C^0_\alpha(Y, TM) \text{ given by } \mathcal{A}\beta(y) = \exp^{-1}_{\alpha(y)} \beta(y)$$

for $\beta \in B_\theta(\alpha)$ and $y \in Y$. Furthermore, $\mathcal{A}$ is a homeomorphism onto $B^\alpha_\theta(0) \subset C^0_\alpha(Y, TM)$.

If v is a fixed point of

$$F^\alpha := \mathcal{A}F\mathcal{A}^{-1}: B^\alpha_\theta(0) \to C^0_\alpha(Y, TM),$$
$$F^\alpha(v)(y) = \exp^{-1}_{\alpha(y)}(f'(\exp_{\alpha(g^{-1}(y))} v(g^{-1}(y))))$$

then $\mathcal{A}^{-1}v$ is a fixed point of F. Note that F^α is smooth in v and that by the chain rule its derivative

$$((DF^\alpha)_v \xi)(y) = (D\exp^{-1}_{\alpha(y)})\big|_{f' \exp_{\alpha(g^{-1}(y))} v(g^{-1}(y))}$$
$$\times Df'\big|_{\exp_{\alpha(g^{-1}(y))} v(g^{-1}(y))} (D\exp_{\alpha(g^{-1}(y))})\big|_{v(g^{-1}(y))} \xi(g^{-1}(y))$$

is Lipschitz in v.

Remark. This becomes a bit clearer when $M = \mathbb{R}^m$, where $\exp_{\alpha(y)} v(y) = v(y) + \alpha(y)$ and hence $(F^\alpha(v))(y) = f'(v(g^{-1}(y)) + \alpha(g^{-1}(y))) - \alpha(y)$. (On the other hand $\mathbb{R}^m$ already has a linear structure so there is no need for this procedure—compare with the proof of Theorem 2.6.1.)

Lemma 18.1.4. *There exist a neighborhood $U(\Lambda) \supset \Lambda$, $\epsilon_0, \epsilon > 0$, and $R > 0$ independent of Y, g, and α such that $\|((DF^\alpha)_0 - \mathrm{Id})^{-1}\| < R$ whenever $d_{C^1}(f, f') < \epsilon_0$, $d_{C^0}(\alpha g, f'\alpha) < \epsilon$.*

We first show how this completes the proof of the theorem. As announced we write $F^\alpha(v) = (DF^\alpha)_0 v + H(v)$. Thus a fixed point v of F^α satisfies

$$((DF^\alpha)_0 - \mathrm{Id})v = -H(v)$$

or

$$v = -((DF^\alpha)_0 - \mathrm{Id})^{-1} H(v) =: T(v).$$

Since DF^α and hence DH is Lipschitz in v with Lipschitz constant K, say, (independent of Y, g, and α) we observe that

$$\|T(v_1) - T(v_2)\| < RK \max(\|v_1\| \|v_2\|) \|v_1 - v_2\|$$

so T is a contraction near 0. Since

$$H(0)(y) = F^\alpha(0)(y) = \exp^{-1}_{\alpha(y)} f'\alpha(g^{-1}(y))$$

we have $\|H(0)\| = d_{C^0}(\alpha, f'\alpha g^{-1}) = d_{C^0}(\alpha g, f'\alpha)$ and

$$\|T(0)\| < R\|H(0)\| = R d_{C^0}(\alpha g, f'\alpha).$$

Take now $\delta_0 = 1/2RK$, $\theta = \min(\delta, \delta_0)$, and $\epsilon < \theta/2R$ as in the lemma. Then

$$\|T(v_1) - T(v_2)\| < \frac{1}{2}\|v_1 - v_2\|$$

for $v_1, v_2 \in B^\alpha_{\delta_0}(0) \subset C^0_\alpha(Y, TM)$ and $\|T(0)\| < \theta/2$ whenever α is such that $d_{C^0}(\alpha g, f'\alpha) < \epsilon$.

Thus $T(B^\alpha_{\delta_0}(0)) \subset B^\alpha_{\delta_0}(0)$. By the Contraction Mapping Principle (Proposition 1.1.2), T has a unique fixed point $v \in B^\alpha_{\delta_0}(0)$ and F has a unique fixed point $\beta = \mathcal{A}^{-1} v \in B_{\delta_0}(\alpha)$ which is in fact in $B_\delta(\alpha)$ since $T(B^\alpha_\theta(0)) \subset B^\alpha_\theta(0) \subset B^\alpha_\delta(0)$. Thus β is as required and unique in $B_{\delta_0}(\alpha)$. □

Proof of Lemma 18.1.4. For $\delta > 0$ there exist $\epsilon_0 > 0$ and $\mu < 1$ and a neighborhood $U(\Lambda) \supset \Lambda$ such that the hyperbolic splitting $T_\Lambda M = E^u \oplus E^s$ extends to a (possibly not invariant) splitting on $U(\Lambda)$ which we will also denote $E^u \oplus E^s$ and for all $f' \in C^1(U(\Lambda), M)$ with $d_{C^1}(f, f') < \epsilon_0$ we have, with respect to $E^u \oplus E^s$,

$$Df' = \begin{pmatrix} a_{uu} & a_{su} \\ a_{us} & a_{ss} \end{pmatrix}$$

with $\|a_{uu}\|^{-1} < \mu$, $\|a_{su}\| < \delta^2\mu$, $\|a_{us}\| < \delta^2\mu$, $\|a_{ss}\| < \mu$. To see that $(DF^\alpha)_0$ is hyperbolic note that

$$((DF^\alpha)_0 \xi)(y) = D(\exp^{-1}_{\alpha(y)})\big|_{f'\alpha(g^{-1}(y))} Df'\big|_{\alpha(g^{-1}(y))} \xi(g^{-1}(y))$$

and $D(\exp_p^{-1})_{|_0} = \mathrm{Id}$. This means that for any $\epsilon_2 > 0$ there is a $\delta_2 > 0$ such that $\|(D\exp_{p_1}^{-1})_{|_{p_2}} - \mathrm{Id}\,\| < \epsilon_2$ when $d(p_1, p_2) < \delta_2$ and that with respect to the decomposition $C^0_\alpha(Y, TM) = \mathcal{F}^u \oplus \mathcal{F}^s$ into

$$\mathcal{F}^u = \{v \in C^0_\alpha(Y, TM) | v(y) \in E^u_{\alpha(y)} \quad (y \in Y)\},$$
$$\mathcal{F}^s = \{v \in C^0_\alpha(Y, TM) | v(y) \in E^s_{\alpha(y)} \quad (y \in Y)\}$$

we have

$$(DF^\alpha)_0 = \begin{pmatrix} A_{uu} & A_{su} \\ A_{us} & A_{ss} \end{pmatrix},$$

where, if $d_{C^0}(\alpha, f'\alpha g^{-1}) < \epsilon$ and $d_{C^1}(f, f') < \epsilon_0$,

$$\|A_{uu}\|^{-1} < \frac{1+\mu}{2}, \ \|A_{su}\| < \delta\mu, \ \|A_{us}\| < \delta\mu, \ \|A_{ss}\| < \frac{1+\mu}{2}.$$

This implies the lemma. □

We now formulate the Shadowing Lemma and Shadowing Theorem for flows. We first define ϵ-orbits and shadowing:

Definition 18.1.5. Let M be a Riemannian manifold and $\varphi: \mathbb{R} \times M \to M$ a smooth flow. A differentiable curve $c: \mathbb{R} \to M$ is called an *ϵ-orbit* for φ^t if $\|\dot{c}(t) - \dot{\varphi}(c(t))\| < \epsilon$ for all $t \in \mathbb{R}$. If c is periodic then it is called a *closed* ϵ-orbit. A differentiable curve $c: \mathbb{R} \to M$ is said to be *δ-shadowed* by the orbit of $x \in M$ if there exists a function $s: \mathbb{R} \to \mathbb{R}$ with $\left|\frac{d}{dt}s - 1\right| < \delta$ such that $d(c(s(t)), \varphi^t(x)) < \delta$ for all $t \in \mathbb{R}$.

Notice that unlike in the discrete-time case where by expansiveness the shadowing orbit is unique for small δ, here the choice of the function s is *not* unique, although the *orbit* is still unique.

Now we can formulate the shadowing results:

Theorem 18.1.6. (Shadowing Lemma for flows) *Let M be a Riemannian manifold, φ^t a smooth flow, and Λ a hyperbolic set for φ^t. Then there exists a neighborhood $U(\Lambda) \supset \Lambda$ of Λ so that for all $\delta > 0$ there is an $\epsilon > 0$ such that every ϵ-orbit is δ-shadowed by an orbit of φ^t.*

Theorem 18.1.7. (Shadowing Theorem for flows) *Let M be a Riemannian manifold, d the natural distance function, φ^t a smooth flow, and Λ a compact hyperbolic set for φ^t. Then there exist a neighborhood $U(\Lambda) \supset \Lambda$ of Λ and $\epsilon_0, \delta_0 > 0$ such that for all $\delta > 0$ there is an $\epsilon > 0$ with the following property:*

If $\psi^t: U(\Lambda) \to M$ is a flow ϵ-close to φ^t in the C^1 topology, Y a topological space, $\gamma^t: Y \to Y$ a continuous flow, $\alpha \in C^0(Y, U(\Lambda))$ such that $\alpha(\gamma^t(y))$ is a C^1 curve for each $y \in Y$ whose tangent vector $(\alpha\gamma^t)\dot{}_{|_0}(y)$ at $\alpha(y)$ depends continuously on y, and

$$\sup_{y \in Y}((\alpha\gamma^t)\dot{}_{|_0}(y), (\psi\alpha)\dot{}_{|_0}(y)) < \epsilon$$

then there is a map $s: Y \times \mathbb{R} \to \mathbb{R}$ *with* $\left|\frac{d}{dt}s_y - 1\right| < \delta$ *and* $\beta \in C^0(Y, U(\Lambda))$ *such that*

$$\beta\gamma^{s(t)} = \psi^t\beta \tag{18.1.1}$$

and $\sup_{y\in Y} d(\alpha, \beta) < \delta$.

Furthermore β *is unique up to time change: If* $\bar{\beta}\gamma^{\sigma_y(t)} = \psi^t\bar{\beta}$ *for some* $\sigma_y: \mathbb{R} \to \mathbb{R}$, $\left|\frac{d}{dt}\sigma_y - 1\right| < \delta$, *and* $\sup_{y\in Y} d(\alpha, \bar{\beta}) < \delta_0$, *then* $\bar{\beta}(y) = \beta\gamma^{s_y(t+\tau_y)-\sigma_y(t)}(y)$ *for some small* $\tau_y: \mathbb{R} \to \mathbb{R}$.

Proof. The only difference between the proofs of Theorem 18.1.7 and Theorem 18.1.3 is the reduction to a "transverse" problem. The idea is to restrict the considerations to a subset of perturbations of the map $\alpha: Y \to U(\Lambda)$ in which the solution β of the equation (18.1.1) is unique. To that end let $E \subset TM$ be the distribution of codimension-one subspaces orthogonal to the vector field $v = \dot{\varphi} = (d/dt)\varphi^t|_{t=0}$ that generates the flow φ^t, D_x a small ball in E_x, and $S_x = \exp_x D_x$. In a small neighborhood U_x of $x \in M$ there is a canonically defined projection $\pi_x: U_x \to S_x$ along the orbits of φ^t. Consider the space $\mathcal{S}$ of maps $\beta: Y \to U(\Lambda)$ such that $\beta(y) \in S_{\alpha(y)}$. For any flow ψ sufficiently close to φ we define an operator F on $\mathcal{S}$ by

$$(F\beta)(y) := \pi_{\alpha(y)}\psi^1(\beta(\gamma^{-1}(y))).$$

As before $\mathcal{S}$ can be identified via $\exp_{\alpha(y)}^{-1}$ with a ball in the space Γ of vector fields $u: Y \to TM$ such that $u(y) \in E_{\alpha(y)}$. Accordingly the operator F can be transformed into an operator F^α on Γ and the solution of the fixed-point equation $F^\alpha(u) = u$ produces via $\exp_{\alpha(x)}$ a map β that takes γ-orbits into ψ-orbits. The hyperbolicity estimates for this "transverse" version of the operator F which allow us to apply the Contraction Mapping Principle are completely parallel to the estimates in the proof of Theorem 18.1.3. Finally, since the fixed point β is a map in $\mathcal{S}$, that is, $\beta(y) \in D_{\alpha(y)}$, and since D_x depends differentiably on x one sees that β is differentiable along orbits of γ and that the derivative ds_y/dt of the time change is close to 1. □

Obviously we obtain an Anosov Closing Lemma for flows:

Corollary 18.1.8. *Let* M *be a Riemannian manifold,* φ^t *a smooth flow, and* Λ *a compact hyperbolic set for* φ^t. *Then there exist a neighborhood* $U(\Lambda) \supset \Lambda$ *of* Λ *and* $\epsilon_0, \delta_0 > 0$ *such that for all* $\delta > 0$ *there is an* $\epsilon > 0$ *such that every closed* ϵ*-orbit is* δ*-shadowed by a periodic orbit.*

Exercises

18.1.1. *Prove the Shadowing Lemma, Theorem 18.1.2, using the method of proof of the Anosov Closing Lemma, Theorem 6.4.15.*

18.1.2. *Prove the Shadowing Lemma for the time-one map of the gradient flow for the negative of the height function on the round sphere (cf. (1.6.1)). Show that in this case the shadowing orbit is not always unique.*

18.1.3. *Prove the Shadowing Lemma for the time-one map of the gradient flow for the negative of the height function on the vertical torus and the tilted torus (cf. Section 1.6).*

18.1.4. *Show that the Shadowing Theorem does not hold for the time-one map of the gradient flow for the negative of the height function on the vertical torus.*

18.1.5. *Show that if Λ is a compact locally maximal hyperbolic set for a flow φ^t then periodic orbits are dense in $NW(\varphi_{\restriction_\Lambda})$.*

2. Stability of hyperbolic sets and Markov approximation

The Shadowing Theorem is directly used to establish a number of properties showing that the orbit structure on a hyperbolic set is stable, structured, and generally rich. A prototype of this kind of result is Corollary 6.4.19 to the Anosov Closing Lemma (Theorem 6.4.15). In this section we prove two more basic results of that kind.

Structural stability (Definition 2.3.3) of hyperbolic sets has been alluded to before several times. We proved structural stability for linear expanding maps of the circle, for hyperbolic linear maps of the two-torus (see Theorems 2.4.6, 2.6.3), and for general horseshoes (Proposition 6.5.3).

Then we obtained partial results of a general nature in Chapter 6, notably Proposition 6.4.6, which asserted that for a perturbation f' of a diffeomorphism f with hyperbolic set Λ there exists a neighborhood U of Λ such that any f'-invariant subset of U is hyperbolic for f'. At that point it was not clear, however, whether there is any nontrivial invariant set. The shadowing theorem of the previous section provides the means to assert the existence of such an f'-invariant set homeomorphic to Λ. At the same time it produces a topological conjugacy. Thus we obtain structural stability, in fact even strong structural stability (Definition 2.3.4):

Theorem 18.2.1. (Strong structural stability of hyperbolic sets) *Let $\Lambda \subset M$ be a hyperbolic set of the diffeomorphism $f: U \to M$. Then for any open neighborhood $V \subset U$ of Λ and every $\delta > 0$ there exists $\epsilon > 0$ such that if $f': U \to M$ and $d_{C^1}(f_{\restriction_V}, f') < \epsilon$ there is a hyperbolic set $\Lambda' = f'(\Lambda') \subset V$ for*

f' and a homeomorphism $h\colon \Lambda' \to \Lambda$ with $d_{C^0}(\mathrm{Id}, h) + d_{C^0}(\mathrm{Id}, h^{-1}) < \delta$ such that $h \circ f'\restriction_{\Lambda'} = f\restriction_{\Lambda} \circ h$. Moreover h is unique when δ is small enough.

Proof. In the proof we use the Shadowing Theorem 18.1.3 three times. First take $\delta_0 < \delta$ as in the Shadowing Theorem and apply the Shadowing Theorem with $\epsilon < \delta_0/2$, $y = \Lambda$, $\alpha = \mathrm{Id}\restriction_{\Lambda}$ the inclusion, and $g = f$ to obtain a unique $\beta\colon \Lambda \to U(\Lambda)$ such that $\beta \circ f = f' \circ \beta$. By Proposition 6.4.6 $\Lambda' := \beta(\Lambda)$ is hyperbolic.

To show that β is injective apply the Shadowing Theorem the other way around: Take ϵ as before, $y = \Lambda'$, $\alpha' = \mathrm{Id}\restriction_{\Lambda'}$ the inclusion, and $g = f'$ to obtain a map h such that $h \circ f' = f \circ h$. It is important to keep in mind that we are allowed to use f' instead of f in the Shadowing Theorem if ϵ here is chosen small enough. We claim that $h \circ \beta = \mathrm{Id}$ and hence $h = \beta^{-1}$ is a homeomorphism.

Apply the uniqueness part of the Shadowing Theorem now in the "$f = f'$" case, when trivially $\alpha \circ f = f \circ \alpha$ and at the same time by the above $\bar\beta \circ f = f \circ \bar\beta$, where $\bar\beta := h \circ \beta$.

Since $d_{C^0}(\alpha, \bar\beta) = d_{C^0}(\mathrm{Id}, h \circ \beta) \le d_{C^0}(\mathrm{Id}, \mathrm{Id} \circ \beta) + d_{C^0}(\mathrm{Id} \circ \beta, h \circ \beta) = d_{C^0}(\mathrm{Id}, \beta) + d(\mathrm{Id}, h) < \delta_0$, the uniqueness assertion of the Shadowing Theorem implies $\bar\beta = \alpha = \mathrm{Id}\restriction_{\Lambda}$ as claimed. □

Remark. The uniqueness part of the Shadowing Theorem takes the place of the use of expansiveness in the proof of Theorem 2.6.3.

Corollary 18.2.2. *Anosov diffeomorphisms are structurally stable. The conjugacy is unique when chosen near the identity.*

Using the Shadowing Theorem 18.1.7 for flows a very similar argument yields strong structural stability of hyperbolic sets for flows:

Theorem 18.2.3. *Let $\Lambda \subset M$ be a hyperbolic set of the smooth flow φ^t on M. Then for any open neighborhood V of Λ and every $\delta > 0$ there exists $\epsilon > 0$ such that if ψ^t is another smooth flow and $d_{C^1}(\varphi, \psi) < \epsilon$ then there is an invariant hyperbolic set Λ' for ψ and a homeomorphism $h\colon \Lambda \to \Lambda'$ with $d_{C^0}(\mathrm{Id}, h) + d_{C^0}(\mathrm{Id}, h^{-1}) < \delta$ that is smooth along the orbits of φ and establishes an orbit equivalence of φ and ψ. Furthermore the vector field $h_*\dot\varphi$ is C^0 close to $\dot\psi$ and if h_1, h_2 are two such homeomorphisms then $h_2^{-1} \circ h_1$ is a time change of φ (close to the identity).*

Corollary 18.2.4. *Anosov flows are C^1 strongly structurally stable in the sense of Definition 2.3.6.*

Another immediate application of the Shadowing Theorem is the construction of Markov approximations according to which a compact locally maximal hyperbolic set is a factor of a topological Markov chain.

Theorem 18.2.5. *Any compact locally maximal hyperbolic set Λ for a diffeomorphism f is a factor of a topological Markov chain σ_A. Furthermore for any $\epsilon > 0$ one can choose A such that the images of the basic cylinders*

$\Omega_A^i := \Omega_A \cap C_i^0$ *under the semiconjugacy* $h: \Omega_A \to M$ *have diameter less than* ϵ *and* $h_{\text{top}}(\sigma_A) < h_{\text{top}}(f_{\restriction_\Lambda}) + \epsilon$.

Proof. For $\delta > \epsilon > 0$ as in the Shadowing Theorem let $\mathcal{A} = \{X_0, \dots, X_{N-1}\}$ be an open cover of Λ with $\text{diam}(X_i) < \epsilon/2$ and $\text{diam}(f(X_i)) < \epsilon/2$ for all i. Define A_{ij} for $i, j \in \{0, \dots, N-1\}$ by $A_{ij} = 1$ if $f(X_i) \cap X_j \neq \varnothing$ and $A_{ij} = 0$ otherwise. Pick points $p_i \in X_i$ and define $\alpha: \Omega_A \to \Lambda$ by $\alpha(\omega) = p_{\omega_0}$.

Note that α is continuous: If $\omega, \omega' \in \Omega_A$ are close then $\omega_0 = \omega_0'$ and hence $\alpha(\omega) = \alpha(\omega')$. By the choice of p_i and X_i there exists an $x \in f(X_{\omega_0}) \cap X_{\omega_1}$ and hence $d(\alpha(\sigma_A(\omega)), f(\alpha(\omega))) = d(p_{\omega_1}, f(p_{\omega_0})) \leq d(p_{\omega_1}, x) + d(x, f(p_{\omega_0})) < \epsilon/2 + \epsilon/2$ whence $d_{C^0}(\alpha\sigma_A, f\alpha) < \epsilon$. By the Shadowing Theorem the ϵ-orbits $\alpha(\sigma_A^i(\omega)) = p_{\omega_i}$ are δ-shadowed by $\beta(\omega)$ where $\beta \in C^0(\Omega_A, \Lambda)$ and $\beta\sigma_A = f\beta$. β is, in fact, surjective: If $x \in \Lambda$ take $\omega \in \Omega_A$ such that $f^i(x) \in X_{\omega_i}$. Then x and $\beta(\omega)$ both δ-shadow $(\alpha(\sigma_A^i(\omega)))_{i \in \mathbb{Z}}$ and hence coincide by uniqueness. Since $d(\beta, \alpha) < \delta$, the images of the basic cylinders $\Omega_A^i = \alpha^{-1}(p_i)$ under the semiconjugacy β have diameter less than 2δ.

In order to obtain the inequality between topological entropies we refine this construction and use the description of topological entropy through coverings given in Exercises 3.1.7–3.1.9.

Let $\mathcal{A}$ be the cover of X by open sets X_i as before. Fix $n \in \mathbb{N}$ and consider the cover $\mathcal{A} \vee f^{-1}(\mathcal{A}) \vee \dots \vee f^{1-n}(\mathcal{A}) =: \mathcal{A}_n$ by sets of the form

$$\bigcap_{k=0}^{n-1} f^{-k}(X_{i_k}). \tag{18.2.1}$$

Take a subcover $\mathcal{B}$ of $\mathcal{A}_n$ with minimal possible number of elements. This number was denoted by $N(\mathcal{A}_n)$ in Exercise 3.1.7.

The states of our topological Markov chain will be coded by those multi-indices $I = (i_0, \dots, i_{n-1})$ for which the intersection (18.2.1) belongs to $\mathcal{B}$ (see Definition 1.9.10 of an n-step topological Markov chain). For two such multi-indices $I = (i_0, \dots, i_{n-1})$, $I' = (i_0', \dots, i_{n-1}')$ we define the entry $A_{I,I'}$ of the transition matrix as follows. $A_{I,I'} = 1$ if $i_k' = i_{k+1}$ for $0 \leq k \leq n-2$ and

$$\left(\bigcap_{k=0}^{n-1} f^k(X_{i_k})\right) \cap \left(\bigcap_{k=0}^{n-1} f^{k+1}(X_{i_k'})\right) \neq \varnothing.$$

A surjective continuous map from Ω_A to Λ is constructed exactly as before, namely, by picking a point p_I in every element $\bigcap_{k=0}^{n-1} f^k(X_{i_k})$ of the cover $\mathcal{B}$, sending the cylinder Ω_A^I into p_I, and using the Shadowing Theorem. By Proposition 3.1.7(3)

$$h_{\text{top}}(\sigma_A) = \frac{1}{n} h_{\text{top}}(\sigma_A^n).$$

Obviously $h_{\text{top}}(\sigma_A^n) \leq \log N(\mathcal{A}_n)$ since $(\sigma_A)^n$ is a subsystem of the full shift with $N(\mathcal{A}_n)$ symbols. Thus for any $\epsilon > 0$ we can take n such that $h_{\text{top}}(\sigma_A) \leq \frac{1}{n}\log(N(\mathcal{A}_n)) \leq h_{\text{top}}(f_{\restriction_\Lambda}, \mathcal{A}) + \epsilon \leq h_{\text{top}}(f_{\restriction_\Lambda}) + \epsilon$. □

Some of the coding maps considered before, for example, in Sections 2.5c and 16.1, were homeomorphisms. Others, including the map between a topological Markov chain and the hyperbolic automorphism of the 2-torus constructed in Section 2.5d, and the coding for the Smale attractor in Section 17.1, were noninvertible semiconjugacies. However, in all of those cases every point has only finitely many preimages and the set where the semiconjugacy is not injective is dynamically not very significant. In particular the topological entropy of the topological Markov chain and that of the system it codes agree precisely, not just up to ϵ as above. We will show in Section 18.7 that these properties can be achieved for any locally maximal hyperbolic set via the construction of *Markov partitions.*

Exercises

18.2.1. *Formulate and prove counterparts of Theorems 18.2.1 and 18.2.5 for hyperbolic repellers (see Definition 6.4.3).*

18.2.2*. *Let $F\colon \mathbb{T}^2 \to \mathbb{T}^2$ be a noninvertible and nonexpanding hyperbolic toral endomorphism, that is, a map given by an integer matrix with determinant of absolute value greater than one and one eigenvalue of absolute value less than one. Show that F is not structurally stable.*

3. Spectral decomposition and specification

a. Spectral decomposition for maps. We proceed now to the study of recurrence of orbits in locally maximal hyperbolic sets. The principal construction (spectral decomposition) shows that the nonwandering set of a compact locally maximal hyperbolic set for a diffeomorphism f breaks up into finitely many components permuted by f such that on each of these components the appropriate iterate of f is topologically mixing.

Theorem 18.3.1. (Spectral decomposition) *Let M be a Riemannian manifold, $U \subset M$ open, $f\colon U \to M$ a diffeomorphism, and $\Lambda \subset U$ a compact locally maximal hyperbolic set for f. Then there exist disjoint closed sets $\Lambda_1, \dots, \Lambda_m$ and a permutation σ of $\{1, \dots, m\}$ such that $NW(f\restriction_\Lambda) = \bigcup_{i=1}^m \Lambda_i$, $f(\Lambda_i) = \Lambda_{\sigma(i)}$, and when $\sigma^k(i) = i$ then $f^k\restriction_{\Lambda_i}$ is topologically mixing.*

Proof. We define a relation on $\operatorname{Per}(f\restriction_\Lambda)$ (which is dense in $NW(f\restriction_\Lambda)$ by Corollary 6.4.19) by $x \sim y$ if and only if $W^u(x) \cap W^s(y) \neq \varnothing$ and $W^s(x) \cap W^u(y) \neq \varnothing$ with both intersections transverse at at least one point. We want to show that this is an equivalence relation and obtain each Λ_i as the closure of an equivalence class.

Note that $\sim$ is trivially reflexive and symmetric. To check transitivity suppose that $x, y, z \in \operatorname{Fix}(f^k\restriction_\Lambda)$ and $p \in W^u(x) \cap W^s(y), q \in W^u(y) \cap W^s(z)$ are transverse intersection points. By the Inclination Lemma (Proposition 6.2.23) the images of a ball around p in $W^u(p) = W^u(x) = f^k(W^u(x))$ accumulate on $W^u(y)$ so $W^u(x)$ and $W^s(z)$ have a transverse intersection.

By Proposition 6.4.21 any two sufficiently near points are equivalent, so by compactness we have finitely many equivalence classes whose (pairwise disjoint) closures we denote by $\Lambda_1, \dots, \Lambda_m$. They are permuted by f with permutation σ, that is, $f(\Lambda_i) = \Lambda_{\sigma(i)}$. Let k be the order of σ. By Corollary 6.4.19 $NW(f\restriction_\Lambda) \subset \overline{\operatorname{Per}(f\restriction_\Lambda)}$ since Λ is locally maximal, so $\bigcup_{i=1}^m \Lambda_i = NW(f\restriction_\Lambda)$.

It remains to show that $f^k\restriction_{\Lambda_i}$ is topologically mixing. To that end notice first that if $p \in \Lambda_i$ is periodic and $p \sim q$ with q periodic, then there is by definition a heteroclinic point $z \in W^u(p) \cap W^s(q)$. If N is the common period then an application of the Inclination Lemma shows that $W^u(p)$ accumulates on q, so $W^u(p)$ is dense in $\Lambda_i \cap \operatorname{Per}(f\restriction_\Lambda)$, hence in Λ_i. To simplify notations assume $k = 1$, $\Lambda = \Lambda_i$.

Now recall (Definition 1.8.2) that we need to show that for any two open sets U and V in Λ there exists an $M \in \mathbb{N}$ such that $f^m(U) \cap V \neq \varnothing$ for all $m \geq M$. For open $U, V \subset \Lambda$ density of periodic points implies the existence of a $p \in U$ such that $f^n(p) = p$. Since U is open it contains in fact a neighborhood $W^u_\delta(p)$ of p in the unstable manifold of p. Since $W^u(p) = \bigcup_{i=0}^\infty f^{in}(W^u_\delta(p))$ is dense there exists $m_0 \in \mathbb{N}$ such that $V \cap \bigcup_{i=0}^{m_0} f^{in}(W^u_\delta(p)) \neq \varnothing$. Since $f^k(W^u_\delta(p))$ is a neighborhood of $f^k(p)$ in $W^u(f^k(p))$ one also obtains $m_1, \dots, m_{n-1} \in \mathbb{N}$ such that $V \cap \bigcup_{i=0}^{m_k} f^{k+in}(W^u_\delta(p)) \neq \varnothing$. Take $M = \max_k (n+1)m_k$. Then for $m \geq M$ we have $V \cap \bigcup_{i=0}^m f^i(W^u_\delta(p)) \neq \varnothing$ and hence $V \cap f^m(U) \neq \varnothing$. □

Corollary 18.3.2. *If a compact locally maximal hyperbolic set Λ is topologically mixing then periodic points are dense in Λ and the unstable manifold of every periodic point is dense in Λ.*

Proof. The spectral decomposition must be trivial. □

Corollary 18.3.3. *A diffeomorphism f restricted to a compact locally maximal hyperbolic set is topologically transitive if and only if the permutation σ from Theorem 18.3.1 is cyclic.*

Corollary 18.3.4. *Let Λ be a connected compact locally maximal hyperbolic set for a diffeomorphism f such that $\Lambda = NW(f\restriction_\Lambda)$ (or equivalently periodic points are dense in Λ). Then $f\restriction_\Lambda$ is topologically mixing.*

Proof. The spectral decomposition must be trivial. □

Corollary 18.3.5. *Let $f: M \to M$ be an Anosov diffeomorphism of a compact connected manifold such that $NW(f) = M$. Then f is topologically mixing.*

Remarks. (1) Since the suspension of an Anosov diffeomorphism is an Anosov flow but no suspension is topologically mixing Corollary 18.3.5 fails for flows.
(2) Whether $NW(f) = M$ for every Anosov diffeomorphism $f: M \to M$ is not known, although it is highly probable. It is true when $M = \mathbb{T}^n$ by Proposition 18.6.5. However, as we noted at the end of Section 17.7, for Anosov flows the nonwandering set is not always the entire manifold.

b. Spectral decomposition for flows. A decomposition into topologically transitive components for flows can be easily obtained by using weak stable and unstable manifolds (Exercise 18.3.7). As we remarked, however, Corollary 18.3.5 fails for flows because suspensions are not topologically mixing. Thus the possibility of a suspension has to be ruled out by additional assumptions in order to obtain topological mixing.

We now show that contact Anosov flows are topologically mixing. This class in particular includes geodesic flows on negatively curved Riemannian manifolds (see Section 5.6b and Theorem 17.6.2). The next result plays an important role in deriving a multiplicative asymptotic for the growth of the number of closed geodesics on a manifold of negative sectional curvature (Theorem 20.6.10).

Theorem 18.3.6. *Contact Anosov flows on connected manifolds are topologically mixing.*

Proof. Denote the contact Anosov flow by $\varphi^t\colon M \to M$. We will show that the strong unstable manifold $W^u(p)$ is dense in M for every periodic point p of φ^t. This implies topological mixing in much the same way as it does for diffeomorphisms. Let $\dim M = 2m-1$. The contact form θ provides a smooth invariant measure via the volume element $\theta \wedge (d\theta)^{m-1}$, so $NW(\varphi^t) = M$ by the Poincaré Recurrence Theorem 4.1.19. Thus topological transitivity follows from connectedness and the spectral decomposition. It suffices to show that $W^u(p)$ is dense in a neighborhood U of p, because this shows that the equivalence classes defined by intersections of manifolds are open, so there is only one and $W^u(p)$ is dense.

To that end note first that if q is a periodic point such that there exists a $z \in W^s_{\text{loc}}(q) \cap W^u_{\text{loc}}(p)$ then we can take integers $a_n, b_n \to \infty$ such that if τ, σ are the periods of p, q, respectively, then $a_n/b_n \to \sigma/\tau$. But then $\varphi^{a_n\tau}(z) \to q$ and $q \in \overline{W^u(p)}$. Furthermore, if U is sufficiently small then for every periodic point $q \in U$ the local weak stable manifold of q and the local strong unstable manifold of p intersect and thus there is a number t near 0 such that $W^s_{\text{loc}}(\varphi^t(q)) \cap W^u_{\text{loc}}(p) \neq \varnothing$. Thus we have shown that $W^u(p)$ is "transversely dense" in U, that is, if T is a disk in U transverse to the flow then the projection of $W^u(p) \cap U$ to T along the flow is dense in T.

Let us note also that for such q the closure of $W^u(q)$ is also contained in $\overline{W^u(p)}$ because $W^u(p)$ accumulates on $W^u_{\text{loc}}(q)$ by continuity of the strong unstable foliation.

Now we have to address the problem of ruling out a suspension and it is here that contactness is used in an essential way.

Lemma 18.3.7. $\ker\theta = E^+ \oplus E^-$.

Proof. We begin by showing that the contact form θ vanishes on (strong) stable vectors. To that end note first of all that the unit tangent bundle of M is compact, so by continuity θ is bounded, that is, if v is any vector then $|\theta(v)| \le \text{const.}\,\|v\|$. If $v \in E^-(x)$ is a strong stable vector then φ^t-invariance of θ yields $|\theta(v)| = |\varphi^t_*\theta(v)| = |\theta(\varphi^t(v))| \le \text{const.}\,\|\varphi^t(v)\|$. But the latter term goes to zero, so $\theta(v) = 0$. Likewise one sees of course that θ vanishes on unstable vectors as well. □

This has the consequence that if $\dot\varphi$ denotes the vector field generating φ then $d\theta(\dot\varphi, v) = 0$ for all vectors v. Thus if we decompose vectors v, w into their components v^t, w^t in the flow direction and $v^\perp, w^\perp$ in the direction of $E^+ \oplus E^-$ then $d\theta(v, w) = d\theta(v^\perp, w^\perp)$.

We now use this to show that $\varphi^t(p) \in \overline{W^u(p)}$ for a dense set of t near 0. To that end take two vectors $v \in E^+(p)$, $w \in E^-(p)$ such that $d\theta(v, w) \neq 0$, which is possible because θ is nondegenerate. Now consider short arcs $c_v\colon [0, \epsilon] \to W^u_{\text{loc}}(p)$, $c_w\colon [0, \epsilon] \to W^s_{\text{loc}}(p)$ that are geodesic segments in these submanifolds. For sufficiently small ϵ there is a point $z \in W^{0u}_{\text{loc}}(c_w(\epsilon)) \cap W^s_{\text{loc}}(c_v(\epsilon))$. Then take t_ϵ such that $z' = \varphi^{-t_\epsilon}(z) \in W^u_{\text{loc}}(c_w(\epsilon))$. There are smooth arcs $\gamma_w \subset W^s_{\text{loc}}(c_v(\epsilon))$ and $\gamma_v \subset W^u_{\text{loc}}(c_w(\epsilon))$ connecting to z and z', respectively. Since the strong foliations are continuous in the C^1 topology these arcs can be taken to be "almost parallel" to c_w and c_v, respectively. For instance, one can parallel translate the tangent vectors of c_w along geodesics to corresponding points of γ_w and ensure that this vector field along γ_w is as close to the tangent vector field as desired as long as ϵ is sufficiently small. Notice also that up to an arbitrarily small smooth perturbation we may take z to be periodic. Translating the arcs c_w and γ_w by φ^{t_ϵ} gives a quadrilateral connecting p with $\varphi^{t_\epsilon}(p)$ by arcs along the strong stable and unstable foliations. Together with a small orbit segment of p we thus obtain a piecewise-smooth closed curve c. This curve projects to a simple curve in the transversal T, so it is the boundary of a surface A which projects injectively to a surface $\pi(A)$ in T. Now note that up to a rescaling of θ by a constant factor we have by the Stokes Theorem

$$t_\epsilon = \int_c \theta = \int_A d\theta = \int_{\pi(A)} d\theta.$$

The latter integral is $\epsilon^2 d\theta(v, w)(1 + o(\epsilon))$ and therefore can be arranged to assume a dense set of values in a neighborhood of 0.

As we noted before this argument, any such point z as we have encountered here is in $\overline{W^u(p)}$ and furthermore so are the points $\varphi^t(p)$ which we just obtained. Consequently $\overline{W^u(p)}$ contains an orbit segment of p and projects onto a local transversal along the flow direction. This implies that $W^u(p)$ is dense in a sufficiently small neighborhood U of p. As we noted at the beginning this implies topological mixing. □

In Exercise 18.3.4 the reader is encouraged to try a variation of this argument to give another criterion for topological mixing.

c. Specification. Next we give the strongest result yet about the abundance of periodic orbits in a hyperbolic set. It says that one can prescribe the evolution of a periodic orbit to the extent of specifying a finite collection of arbitrarily long orbit segments and any fixed precision: As long as one allows for enough time between the specified segments one can find a periodic orbit approximating this itinerary. Let us emphasize that the time between the segments depends only on the quality of the approximation and not on the length of the specified segments. As we will see in Section 18.5 and in Chapter 20, Bowen's Specification Theorem is a tool of great utility and importance in studying both the topological structure of hyperbolic sets and statistical properties of orbits within such sets.

We will prove two slightly different specification theorems. The stronger form of the specification property follows from the assumption that the hyperbolic set under consideration is topologically mixing. As a corollary we obtain a weaker conclusion assuming only topological transitivity. The connection arises from the spectral decomposition into mixing components (for an iterate), Theorem 18.3.1.

Definition 18.3.8. Let $f: X \to X$ be a bijection of a set X. A *specification* $S = (\tau, P)$ consists of a finite collection $\tau = \{I_1, \dots, I_m\}$ of finite intervals $I_i = [a_i, b_i] \subset \mathbb{Z}$ and a map $P: T(\tau) := \bigcup_{i=1}^m I_i \to X$ such that for $t_1, t_2 \in I \in \tau$ we have $f^{t_2 - t_1}(P(t_1)) = P(t_2)$. S is said to be *n-spaced* if $a_{i+1} > b_i + n$ for all $i \in \{1, \dots, m\}$ and the minimal such n is called the *spacing* of S. We say that S *parameterizes* the collection $\{P_I \mid I \in \tau\}$ of orbit segments of f.

We let $T(S) := T(\tau)$ and $L(S) := L(\tau) := b_m - a_1$. If (X, d) is a metric space we say that S is *ϵ-shadowed* by $x \in X$ if $d(f^n(x), P(n)) < \epsilon$ for all $n \in T(S)$.

Thus a specification is a parameterized union of orbit segments $P_{\restriction I_i}$ of f.

If (X, d) is a metric space and $f: X \to X$ a homeomorphism then f is said to have the *specification property* if for any $\epsilon > 0$ there exists an $M = M_\epsilon \in \mathbb{N}$ such that any M-spaced specification S is ϵ-shadowed by some $x \in X$ and such that moreover for any $q \geq M + L(S)$ there is a period-q orbit ϵ-shadowing S.

Theorem 18.3.9. (Specification Theorem) *Let Λ be a topologically mixing compact locally maximal hyperbolic set for a diffeomorphism f. Then $f_{\restriction \Lambda}$ has the specification property.*

Remark. It is easy to show that the specification property implies that $f_{\restriction \Lambda}$ is topologically mixing (Exercise 18.3.8).

We first show that in a topologically mixing hyperbolic set all unstable manifolds are uniformly dense (compare with the earlier discussion in Section 1.8 for the special case of the hyperbolic toral automorphisms):

Proposition 18.3.10. *If Λ is a compact locally maximal hyperbolic set for f and $f: \Lambda \to \Lambda$ is topologically mixing then if $\alpha > 0$ there is an $N \in \mathbb{N}$ such that for $x, y \in \Lambda$ and $n \geq N$ we have $f^n(W^u_\alpha(x)) \cap W^s_\alpha(y) \neq \varnothing$.*

Proof. Recall that a set Y in a metric space X is called ϵ-dense if X is an ϵ-neighborhood of Y. By Proposition 6.4.21 we have a local product structure, that is, there is a $\delta^* > 0$ such that $W^s_\delta(x) \cap W^u_\delta(y)$ consists of at most one point whenever $0 < \delta \leq \delta^*$ and a function $\epsilon(\delta)$ such that $d(x, y) < \epsilon(\delta) \Rightarrow W^s_\delta(x) \cap W^u_\delta(y) \neq \varnothing$. Let $\delta := \min\{\delta^*, \alpha/2, \epsilon(\alpha/2)/4\}$. To choose N take a $\epsilon(\alpha/2)/2$-dense set $\{p_k \mid k = 1, \dots, r\}$ of periodic points (with periods t_k). By

Corollary 18.3.2 the unstable manifold of every periodic point in Λ is dense in Λ, so there exist m_k such that $f^{mt_k}(W^u_\delta(p_k))$ is $\epsilon(\delta)$-dense for all $m \geq m_k$ and all k. Let $N = \prod_{k=1}^r m_k t_k$ and note that $f^N(W^u_\delta(p_k))$ is $\epsilon(\delta)$-dense for all k.

Now we show that N is as desired: For $x, y \in \Lambda$ take j such that $d(x, p_j) < \epsilon(\alpha/2)/2$, $z \in f^N(W^u_\delta(p_j))$ such that $d(y,z) \leq \epsilon(\delta)$, and $w \in W^u_\delta(z) \cap W^s_\delta(y)$. Then $f^{-N}(w) \in W^u_\delta(f^{-N}(z)) \subset W^u_{2\delta}(p_j) \subset W^u_{\epsilon(\alpha/2)/2}(p_j)$, so $d(f^{-N}(w), x) \leq \epsilon(\alpha/2)$ by the triangle inequality. So there exists $v \in W^s_{\alpha/2}(f^{-N}(w)) \cap W^u_{\alpha/2}(x)$ and $f^N(v) \in f^N(W^u_{\alpha/2}(x)) \cap W^s_{\alpha/2}(w) \subset f^N(W^u_\alpha(x)) \cap W^s_\alpha(y) \neq \varnothing$ since $\delta < \alpha/2$. For $x, y \in \Lambda$, $n \geq N$ note that $f^n(W^u_\alpha(x)) \cap W^s_\alpha(y) \supset f^N(W^u_\alpha(f^{n-N}(x))) \cap W^s_\alpha(y) \neq \varnothing$. □

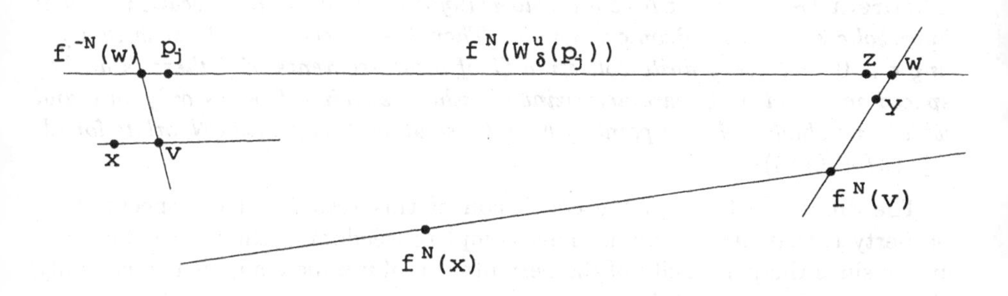

FIGURE 18.3.1. Density of stable and unstable manifolds

Proof of Theorem 18.3.9. For $\beta \leq \min\{\epsilon, \delta^*\}$ and $\alpha = \beta/3$ we take the corresponding N obtained in Proposition 18.3.10 and let $M \geq N$ be such that $\lambda^M < 1/2$, where λ is the contraction rate in the Definition 6.4.1 of hyperbolicity. We let $x_1 = f^{-a_1}(P(a_1))$ and define $x_2, x_3, \dots, x_m$ as follows: Given x_k there exists, by the previous Proposition, an x_{k+1} such that

$$f^{a_{k+1}}(x_{k+1}) \in f^{a_{k+1}-b_k}\left(W^u_\alpha(f^{b_k}(x_k))\right) \cap W^s_\alpha(P(a_{k+1}))$$

since $a_{k+1} - b_k \geq N$ by assumption.

To show that $x := x_m$ is the desired orbit note that, since $f^{a_k}(x_k) \in W^s_\alpha(P(a_k))$ by construction, $d(f^n(x_k), P(n)) = d(f^n(x_k), f^{n-a_k}(P(a_k))) \leq \alpha = \beta/3$ and therefore we are done by the triangle inequality once we prove

Lemma 18.3.11. *$d(f^n(x), f^n(x_k)) \leq 2\beta/3$ for all $n \in I_k$, $k \in [1,m]$.*

Proof. We show that $f^{b_k}(x) \in W^u_{2\beta/3}(f^{b_k}(x_k))$. Since $\sum \lambda^{Mj} < 2$ and $\alpha = \beta/3$ this follows once we have $f^{b_k}(x_{k+r}) \in W^u_{\alpha+\alpha\lambda^M+\dots+\alpha\lambda^{M(r-1)}}(f^{b_k}(x_k))$. For $r = 1$ this holds by construction. Now $f^{b_k}(x_{k+r+1}) \in W^u_{\alpha\lambda^{Mr}}(f^{b_k}(x_{k+r}))$ since $f^{b_{k+r}}(x_{k+r+1}) \in W^u_\alpha(f^{b_{k+r}}(x_{k+r}))$ and $b_{k+r} - b_k \geq rM$, so we obtain the claim by induction. □

To show that the shadowing orbit can be taken periodic we assume that $\beta \leq \min\{\epsilon/2C, \epsilon, 2\delta^*\}/2$, where C is as in the Anosov Closing Lemma (Theorem 6.4.15) and take M correspondingly as before. If $q \geq M + L(S)$ is the desired period, we "close" S by augmenting it to a specification $S' = (\tau', P')$ with $\tau' = \tau \cup \{\{a_1 + q\}\}$ and $P' \restriction_{T(\tau)} = P$, $P'(a_1 + q) = P(a_1)$, which is clearly M-spaced. We thus obtain a point $x' := f^{a_1}(x) \in \Lambda$ such that $d(x', f^q(x')) \leq d(x', P(a_1)) + d(f^q(x'), P(a_1)) \leq \epsilon/2C$ and hence, by the Anosov Closing Lemma, a period-q point $z \in \Lambda$ such that $d(f^{n+a_1}(z), f^n(x')) \leq \epsilon/2$ for all $n \in [0, q]$. The theorem follows by the triangle inequality. □

By the spectral decomposition into mixing components and in particular Corollary 18.3.3 we immediately have

Theorem 18.3.12. *Let Λ be a topologically transitive compact locally maximal hyperbolic set for a diffeomorphism f. Then there exists $N \in \mathbb{N}$ such that for any $\epsilon > 0$ and every finite collection C of orbit segments of f there is an M-spaced specification S parameterizing C, whose spacing depends only on ϵ and which is ϵ-shadowed by a point of Λ and ϵ-shadowed by period-qN orbits for all $q \geq (M + L(S))/N$.*

The difference between the conclusion of this result and the specification property is that here we do not have complete freedom in the choice of specification since the periodicity of the permutation of mixing components may only allow transitions at certain times.

Let us describe the analog of the specification property for flows.

Definition 18.3.13. Suppose φ is a flow on a set X. A *specification* $S = (\tau, P)$ consists of a finite collection $\tau = \{I_1, \dots, I_m\}$ of bounded intervals $I_i = [a_i, b_i] \subset \mathbb{R}$ and a map $P: T(\tau) := \bigcup_{I \in \tau} I \to X$ such that for $t_1, t_2 \in I \in \tau$ we have $f^{t_2 - t_1}(P(t_1)) = P(t_2)$. S is said to be *θ-spaced* if $a_{i+1} > b_i + \theta$ for all $i \in \{1, \dots, m\}$ and the minimal such θ is called the *spacing* of S. We say that S *parameterizes* the collection $\{P_I \mid I \in \tau\}$ of orbit segments of f.

We let $T(S) := T(\tau)$ and $L(S) := L(\tau) := \max T(\tau) - \min T(\tau)$. If (X, d) is a metric space we say that S is *ϵ-shadowed* by $x \in X$ if $d(\varphi^t(x), P(t)) < \epsilon$ for all $t \in T(S)$.

Thus a specification is a parameterized union of orbit segments $P\restriction_{I_i}$ of f.

If (X, d) is a metric space and φ a flow then φ is said to have the *specification property* if for any $\epsilon > 0$ there exists an $M = M_\epsilon \in \mathbb{R}$ such that any M-spaced specification S is ϵ-shadowed by a point of Λ and moreover such that for any $s \geq M + L(S)$ there is a period-s' orbit ϵ-shadowing S with $|s - s'| < \epsilon$.

With this notion of specification we have the Specification Theorem for flows:

Theorem 18.3.14. *Let Λ be a topologically mixing compact locally maximal hyperbolic set for a smooth flow φ. Then $\varphi\restriction_\Lambda$ has the specification property.*

This theorem can be proved by combining the Anosov Closing Lemma for flows (Corollary 18.1.8) with density of unstable manifolds of periodic points similarly to the proof of Theorem 18.3.9.

Exercises

18.3.1. *Use the results of Exercise 1.9.9 to prove that the nonwandering set $NW(\sigma_A)$ of any topological Markov chain has a decomposition similar to that of Theorem 18.3.1.*

18.3.2. *Construct a locally maximal hyperbolic set Λ for a diffeomorphism $f: M \to M$ of a compact manifold M such that $NW(f) \cap \Lambda \neq NW(f\restriction_\Lambda)$.*

18.3.3. *Extend the map $f: \Delta \to \mathbb{R}^2$ from Section 2.5c to a diffeomorphism of the Riemann sphere $S^2 = \mathbb{R}^2 \cup \{\infty\}$ such that $NW(f) = \{p\} \cup \{q\} \cup \Lambda$, where p is an attracting fixed point, q a repelling fixed point, and Λ the invariant hyperbolic set of Section 2.5c.*

18.3.4. *Show that a transitive Anosov flow is topologically mixing if it has two periodic orbits with incommensurate periods.*

18.3.5. *Prove that any transitive topological Markov chain has the specification property.*

18.3.6. *Prove counterparts of Theorems 18.3.1 and 18.3.9 for hyperbolic repellers.*

18.3.7. *Consider a compact locally maximal hyperbolic set Λ for a flow φ^t. Show that there is a decomposition of $\Lambda' := NW(\varphi^t\restriction_\Lambda)$ into finitely many disjoint topologically transitive sets Λ_i.*

18.3.8. *Show that any homeomorphism of a compact metric space with the specification property is topologically mixing.*

18.3.9. *Prove that any factor of a system with the specification property also has the specification property.*

4. Local product structure

In Proposition 6.4.21 we obtained the presence of a local product structure as a consequence of local maximality. We now want to show that it is, in fact, equivalent to local maximality.

Theorem 18.4.1. *A compact hyperbolic set Λ with local product structure is locally maximal.*

Proof. We want to show that every orbit staying sufficiently close to Λ for all time is indeed in Λ. We begin with a special case:

Lemma 18.4.2. *Let Λ be a compact hyperbolic set with local product structure. Then there exist $\delta_1, \delta_2 > 0$ such that if $x \in \Lambda$, $y \in W^u_{\delta_1}(x)$, and $d(f^n(y), \Lambda) < \delta_2$ for all $n \in \mathbb{N}$, then $y \in \Lambda$.*

Proof. Let $K := \sup \|Df\|$, taken over a neighborhood of Λ, and $\epsilon > 0$ an admissible size of local stable and unstable manifolds in the definition of local product structure. Denote by d^u the distance inside unstable leaves W^u. Then there exists $M > 0$ such that $d^u(p, W^s_\epsilon(q) \cap W^u_\epsilon(p)) < Md(p,q)$ whenever $d(p,q)$ is sufficiently small and $W^s_\epsilon(q) \cap W^u_\epsilon(p) \neq \varnothing$. For any $\delta_1 \leq \min\{1/MK,\ 1/2,\ \epsilon/2\}$ take $\delta_2 \leq \delta_1/MK$ such that $d(x,y) < \delta_2 \Rightarrow W^s_{\delta_1}(x) \cap W^u_{\delta_1}(y) \neq \varnothing$. To finish the proof we show the following: If $x \in \Lambda$, $y \in W^u_{\delta_1}(x)$, and $d(f^n(y), \Lambda) < \delta_2$ for all $n \in \mathbb{N}$, then $a_n := \min_{z \in \Lambda \cap W^u(f^n(x))} d^u(f^n(y), z) = 0$ for some $n \in \mathbb{N}$.

To that end we first show inductively that $a_n < \delta_1$ for all $n \in \mathbb{N}_0$. For a_0 this is true by assumption. If $a_n < \delta_1$ take $z_n \in \Lambda$ such that $d(f^n(y), z_n) < \delta_2$ and note that $\{p_n\} := \{[z_n, f^n(y)]\} = W^s_{\delta_1}(z_n) \cap W^u_{\delta_1}(f^n(y)) = W^s_{\delta_1}(z_n) \cap W^u_{2\delta_1}(f^n(x)) = \{[z_n, f^n(x)]\} \subset \Lambda$ by local product structure, so $a_n \leq d^u(f^n(y), p_n) < M\delta_2 \leq \delta_1/K$. By choice of K we thus have $a_{n+1} \leq Ka_n < \delta_1$ as well.

On the other hand by Theorem 6.4.9(3) for sufficiently small $\delta_1 > 0$ there exists $\mu > 1$ for which $a_n < \delta_1 \Rightarrow a_{n+1} \geq \mu a_n$. Consequently $a_n = 0$ for all $n \in \mathbb{N}_0$. □

To reduce the proof of the theorem to the lemma we use heteroclinic points. Let $\epsilon > 0$ be such that $d(p,q) < \epsilon$ implies $W^s_{\delta_2/2}(p) \cap W^u_{\delta_2/2}(q) \neq \varnothing$ and take $\delta_3 < \min\{\delta_2/2, \epsilon\}$. If y is such that $d(f^n(y), \Lambda) < \delta_3$ for all $n \in \mathbb{Z}$, then there exists $x \in \Lambda$ such that $\varnothing \neq W^s_{\delta_2/2}(y) \cap W^u_{\delta_2/2}(x) =: \{p\}$. But $d(f^n(p), \Lambda) \leq d(f^n(y), \Lambda) + d(f^n(p), f^n(y)) < \delta_2/2 + \delta_2/2 < \delta_2$ for $n > 0$ and $d(f^n(p), \Lambda) \leq d(f^n(p), f^n(y)) \leq d(p,y) < \delta_2$ for $n \leq 0$, so $p \in \Lambda$ by Lemma 18.4.2. Likewise one can show that $\{q\} := W^u_{\delta_2/2}(y) \cap W^s_{\delta_2/2}(x) \subset \Lambda$ so by the local product structure $\{y\} = W^s_{\delta_2/2}(p) \cap W^u_{\delta_2/2}(q) \subset \Lambda$. □

Exercises

18.4.1. *Show that the following property of a hyperbolic set Λ for $f: U \to M$ is equivalent to local maximality:*

There exists an open neighborhood V of Λ such that given any point y with $f^n(y) \in V$ for all $n \in \mathbb{N}$ there is a point $x \in \Lambda$ such that $y \in W^s(x)$.

18.4.2. *Prove that the image of any n-step topological Markov chain (Definition 1.9.10) under the conjugacy between the full shift σ_2 and the set Λ in the horseshoe map of Section 2.5c is a locally maximal hyperbolic set for the horseshoe map.*

18.4.3. *With Λ as in the previous exercise prove that any closed subset of Λ that is locally maximal is the image of some n-step topological Markov chain in the full shift σ_2. (Compare with Theorem 16.1.1.)*

18.4.4. *Construct an example of a perfect topologically transitive hyperbolic set that does not have a local product structure.*

5. Density and growth of periodic orbits

In earlier sections we have seen several aspects of the relationship between the abundance of closed orbits and hyperbolicity, such as density of periodic points in the nonwandering set of a hyperbolic set by the Anosov Closing Lemma (Corollary 6.4.19). In several earlier examples the exponential growth rate $p(f) := \overline{\lim}_{n\to\infty}(1/n)\log^+ P_n(f)$ of periodic orbits was calculated and turned out to be equal to the topological entropy of the map f (Proposition 3.2.3, Corollary 3.2.4, Proposition 3.2.6, and Proposition 3.2.5). Corollary 15.2.2 and later Corollary S.5.11 show that the inequality $h_{\text{top}}(f) \leq p(f)$ is a general feature of various classes of low-dimensional systems. At this point we have the means to see that, as announced at the end of Section 3.2e, hyperbolicity implies coincidence of topological entropy and the growth rate of periodic orbits.

Theorem 18.5.1. *Let M be a compact Riemannian manifold, $U \subset M$ open, $f: U \to M$ a diffeomorphism, and $\Lambda \subset U$ a compact locally maximal hyperbolic set for f. Then $p(f\restriction_\Lambda) = h_{\text{top}}(f\restriction_\Lambda)$.*

Proof. Since $f_1 := f\restriction_\Lambda$ is expansive we have $p(f_1) \leq h_{\text{top}}(f_1)$ by Proposition 3.2.14. To prove the reverse inequality we reduce to the case where f_1 is topologically mixing: First of all we may consider $f_2 := f_1\restriction_{NW(f_1)}$ since $h_{\text{top}}(f_1) = h_{\text{top}}(f_2)$ by (3.3.1) which is in turn a consequence of the variational principle (Theorem 4.5.3). The Spectral Decomposition Theorem 18.3.1 tells us that there is some $n \in \mathbb{N}$ such that $NW(f_1)$ decomposes into finitely many components on which f_2^n is topologically mixing. By Proposition 3.1.7(2) we may restrict $f_3 = f_2^n$ to the topologically mixing component X of highest topological entropy, that is, $p(f_3) = np(f_2) = np(f_1)$ and $h_{\text{top}}(f_3) = nh_{\text{top}}(f_2) = nh_{\text{top}}(f_1)$.

This reduction to the topologically mixing case allows us to apply the Specification Theorem 18.3.9. Any element of an (n, ϵ)-separated set E_n can be $\epsilon/2$-shadowed by a periodic point of period $m + M_{\epsilon/2}$. These points are distinct by the triangle inequality in the d_n^f metric. Thus there are at least $\text{card}(E_n)$ distinct periodic points of period $n + M_{\epsilon/2}$ and consequently $P_{n+M_{\epsilon/2}}(f_3) \geq N(f_3, \epsilon, n)$, implying $p(f_3) \geq h_{\text{top}}(f_3)$. □

Note that after the reduction to the topologically mixing case we used specification only. Thus we have shown that for expansive maps with the specification property (Definition 18.3.8) topological entropy is equal to the growth rate of periodic orbits. It is worthwhile to point this out because although locally maximal hyperbolic sets of diffeomorphisms provide the prime examples of expansive maps with specification, other important classes of transformations are covered, too. Notably transitive topological Markov chains and some more general classes of symbolic systems such as sofic systems (see Exercise 20.1.2) are expansive with specification.

The remainder of this section is dedicated to improving our understanding of the growth rates of separated sets and periodic orbits for such maps, leading ultimately to a refined version of Theorem 18.5.1. As before we only use expansiveness and the specification property.

Lemma 18.5.2. *Let $f: X \to X$ be an expansive homeomorphism of a compact metric space with expansivitiy constant δ_0 (see Definition 3.2.11). Then for $0 < \epsilon < \delta_0/2$ and $\delta > 0$ there exists $C_{\delta,\epsilon}$ such that $N(f,\delta,n) \le C_{\delta,\epsilon} N(f,\epsilon,n)$ for all $n \in \mathbb{N}$.*

Proof. For $\epsilon < \delta_0/2$ let $N \in \mathbb{N}$ and $\alpha > 0$ be such that

$$d^f_{2N+1}(f^{-N}(x), f^{-N}(y)) \le 2\epsilon \Rightarrow d(x,y) < \delta$$

and

$$d(x,y) \le \alpha \Rightarrow d^f_{2N+1}(f^{-N}(x), f^{-N}(y)) < \delta.$$

If E is a maximal (n,δ)-separated set and F a maximal (n,ϵ)-separated set then for $x \in E$ one can choose a point $z(x) \in F$ such that $d^f_n(x, z(x)) \le \epsilon$. Thus $\operatorname{card}(E) = \sum_{z \in F} \operatorname{card}(E_z)$, where $E_z := \{x \in E \mid z(x) = z\}$, and the claim follows once we bound $\operatorname{card}(E_z)$ independently of n.

But if $x, y \in E_z$ then $d^f_n(x,y) \le 2\epsilon$ by definition of E_z; hence $d(f^i(x), f^i(y)) \le \delta \quad (i \in [N, n-N))$ by choice of N and thus, by choice of α and since the points x and y are (n,δ)-separated, $d(x,y) > \alpha$ or $d(f^n(x), f^n(y)) > \alpha$.

Therefore $\operatorname{card}(E_z) = \operatorname{card}\{(x, f^n(x)) \mid x \in E_z\} \le \max\{\operatorname{card} A \mid A \subset X \times X, d(a,b) > \alpha\ \forall (a,b) \in A\}$ since the $(x, f^n(x))$ form just such an α-separated set. □

Remark. Obviously we can take $C_{\delta,\epsilon} = 1$ for $\delta > \epsilon$.

Lemma 18.5.3. *Let $f: X \to X$ be an expansive homeomorphism of a compact metric space with the specification property. Then for $0 < \epsilon < \delta_0/3$ there exist $k_\epsilon, K_\epsilon > 0$ such that*

$$\prod_{j=1}^{m} k_\epsilon N(f,\epsilon,n_j) \le N\Big(f,\epsilon,\sum_{j=1}^{m} n_j\Big) \le \prod_{j=1}^{m} K_\epsilon\, N(f,\epsilon,n_j)$$

for $n_1, \dots, n_m \in \mathbb{N}$.

Proof. (i) The upper bound does not use the assumptions. If E is $(\sum_{j=1}^m n_j, \epsilon)$-separated and F_j a maximal $(n_j, \epsilon/2)$-separated set, then for $x \in E$ there exists $z(x) := (z_1(x), \dots, z_m(x)) \in F_1 \times \cdots \times F_m$ such that $d^f_{n_j}(f^{n_1+\cdots+n_{j-1}}(x), z_j(x)) \le \epsilon/2$ and $z(\cdot)$ is injective. Thus $\operatorname{card}(E) \le \prod_{j=1}^m \operatorname{card}(F_j) = \prod_{j=1}^m N(f,\epsilon/2,n_j)$ and we can take $K_\epsilon = C_{\epsilon/2,\epsilon}$.

(ii) If E_j is $(n_j, 3\epsilon)$-separated, $a_j = n_1 + \cdots + n_{j-1} + (j-1)M_\epsilon$, with M_ϵ as given by the specification property, and $I_j = [a_j, a_j + n_j - 1]$, then specification implies that for $x := (x_1, \dots, x_m) \in E_1 \times \cdots \times E_m$ there is a $z = z(x)$ such that $d^f_{n_j}(f^{a_j}(z), x_j) < \epsilon$. Since by construction $E := \{z(x) \mid x \in E_1 \times \cdots \times E_n\}$ is $(a_m + n_m, \epsilon)$-separated, we have shown that

$$N(f,\epsilon,a_m+n_m) \ge \prod_{j=1}^{m} N(f,3\epsilon,n_j).$$

Since $a_m + n_m = (m-1)M_\epsilon + \sum_{j=1}^m n_j$, part (i) yields

$$N(f,\epsilon,a_m+n_m) \le K_\epsilon^m\, N(f,\epsilon,\sum_{j=1}^m n_j)\, N(f,\epsilon,M_\epsilon)^{m-1}$$

and

$$N\Big(f,\epsilon,\sum_{j=1}^m n_j\Big) \ge K_\epsilon^{-m}\, N(f,\epsilon,M_\epsilon)^{1-m} \prod_{j=1}^m N(f,3\epsilon,n_j)$$
$$\ge \prod_{j=1}^m (C_{\epsilon,3\epsilon}\, K_\epsilon\, N(f,\epsilon,M_\epsilon))^{-1}\, N(f,\epsilon,n_j).$$

□

Proposition 18.5.4. *Let X be a compact metric space and $f: X \to X$ an expansive homeomorphism with expansivity constant δ_0 and with the specification property. If $0 < \epsilon < \delta_0/3$, $n \in \mathbb{N}$, and k_ϵ and K_ϵ are as in Lemma 18.5.3, then*

$$\frac{1}{K_\epsilon}\, e^{nh_{\text{top}}(f)} \le N(f,\epsilon,n) \le \frac{1}{k_\epsilon}\, e^{nh_{\text{top}}(f)}.$$

Proof. By Corollary 3.2.13 $h(f) = \lim_{n\to\infty} \frac{1}{n} \log N(f,\epsilon,n)$ for $\epsilon < \delta_0$. Lemma 18.5.3 thus implies

$$\frac{1}{n}\log k_\epsilon N(f,\epsilon,n) \le \lim_{k\to\infty} \frac{1}{kn} \log N(f,\epsilon,kn) = h_{\text{top}}(f) \le \frac{1}{n}\log K_\epsilon N(f,\epsilon,n).$$ □

With these prerequisites it is now very easy to refine the asymptotics of Theorem 18.5.1. In the proofs of Proposition 3.2.14 and Theorem 18.5.1 we showed that

$$N(f,2\epsilon,n-M_\epsilon) \le P_n(f) \le N(f,\epsilon,n)$$

by expansiveness and specification. This and Proposition 18.5.4 immediately yield

Theorem 18.5.5. *Let X be a compact metric space and $f: X \to X$ an expansive homeomorphism with the specification property. Then there exist $c_1, c_2 > 0$ such that for $n \in \mathbb{N}$*

$$c_1 e^{nh_{\text{top}}(f)} \le P_n(f) \le c_2 e^{nh_{\text{top}}(f)}. \tag{18.5.1}$$

The Spectral Decomposition Theorem 18.3.1, Specification Theorem 18.3.9, Corollary 6.4.10, and Proposition 3.1.7(2) allow us to apply Theorem 18.5.5 and thus obtain (18.5.1) (for all $n \in \mathbb{N}$) for $f = F{\restriction}_\Lambda$, where Λ is a locally maximal topologically mixing hyperbolic set of a diffeomorphism F or, more generally, a locally maximal hyperbolic set all of whose topologically transitive parts are topologically mixing. For an arbitrary locally maximal set one can find an iterate N such that F^N has the latter property (Theorem 18.3.1). Then taking into account Proposition 3.1.7(3) one sees that (18.5.1) holds for such a set Λ for all $n \in \mathbb{N}$ that are multiples of N. Thus we have shown

Theorem 18.5.6. *Let Λ be a compact locally maximal hyperbolic set for a diffeomorphism f all of whose topologically transitive parts are topologically mixing. Then there exist $c_1, c_2 > 0$ such that (18.5.1) holds for $n \in \mathbb{N}$. If Λ is a compact locally maximal hyperbolic set for a diffeomorphism f then there exists $N \in \mathbb{N}$ such that (18.5.1) holds for all $n = kN \in \mathbb{N}$.*

The two preceding theorems have analogs for flows. There are some modifications, however. First it is more convenient to consider the number $P_{T,\epsilon}(\varphi^t) = P_{T+\epsilon}(\varphi^t) - P_{T-\epsilon}(\varphi^t)$ of periodic orbits with periods in $(T-\epsilon, T+\epsilon]$ rather than all orbits of up to a certain period. Second the arguments for the discrete-time case rely on the fact that periodic points up to a certain period are an appropriately separated set, whereas in the setting of flows we have to count instead the size of a separated set of points sitting on orbits of period approximately t. This will cause factors of t to appear in the asymptotics. Thus we obtain the following result:

Theorem 18.5.7. *Let X be a compact metric space and φ an expansive flow with the specification property. Then there exist $c_1, c_2 > 0$ such that for $t > 0$*

$$c_1 e^{t h_{\text{top}}(\varphi^t)} \le t P_{t,\epsilon}(\varphi^t) \le c_2 e^{t h_{\text{top}}(\varphi^t)}. \tag{18.5.2}$$

This immediately also yields

Theorem 18.5.8. *Let Λ be a compact locally maximal topologically mixing hyperbolic set for a flow φ. Then there exist $c_1, c_2 > 0$ so that (18.5.2) holds for $t > 0$.*

Exercises

18.5.1. *Show that for a transitive topological Markov chain σ_A the limit*

$$\lim_{n\to\infty} \frac{P_n(\sigma_A)}{e^{n h_{\text{top}}(\sigma_A)}}$$

exists and calculate it. This is a stronger conclusion than (18.5.2).

18.5.2. *Let L be an $m \times m$ integer matrix with determinant one that has no roots of unity as eigenvalues but has a pair of eigenvalues of absolute value one. Let $F_L\colon \mathbb{T}^m \to \mathbb{T}^m$ be the induced automorphism of the torus. Prove that $\lim_{n\to\infty} (\log P_n(F_L))/n = h_{\text{top}}(F_L)$ but $P_n(F_L) e^{-n h_{\text{top}}(F_L)}$ is not bounded away from 0.*

18.5.3. *Show that $h_{\text{top}}(F_L) = \sum_{|\lambda_i|>1} \log \lambda_i$ for a hyperbolic toral automorphism F_L, where λ_i are the eigenvalues of L.*

6. Global classification of Anosov diffeomorphisms on tori

Now we will show that in the case of Anosov diffeomorphisms of the torus structural stability can be extended to a global classification. We already saw in Section 2.6 that within the homotopy class of a hyperbolic automorphism any map f (and hence any Anosov map) has the linear model as a factor. We first show that any Anosov diffeomorphism is homotopic to a hyperbolic linear one. In this Theorem 18.5.6 and the Lefschetz Fixed-Point Formula (8.6.1) will play a key role. Next we prove that the semiconjugacy to the linear model is indeed injective, hence a homeomorphism. As an intermediate step of independent interest this involves showing that the nonwandering set is the entire torus.

Theorem 18.6.1. *Every Anosov diffeomorphism of the n-torus is topologically conjugate to a linear hyperbolic automorphism.*

Proof. Suppose $f\colon \mathbb{T}^n \to \mathbb{T}^n$ is Anosov. Its homotopy class contains precisely one linear map $F_L\colon \mathbb{T}^n \to \mathbb{T}^n$.

Lemma 18.6.2. *The map F_L is hyperbolic.*

Proof. This is equivalent to hyperbolicity of the linear map f_* induced by f on the fundamental group $\mathbb{Z}^n$ of $\mathbb{T}^n$ because the associated matrix defines F_L on $\mathbb{T}^n$. By the Spectral Decomposition Theorem 18.3.1 the nonwandering set $NW(f)$ is a union of components on which some power f^N is topologically mixing, so $P_n(f^N)$ has a multiplicative exponential asymptotic by Theorem 18.5.6. After passing to a double cover of $\mathbb{T}^n$ (which is still a torus) we may assume that the unstable bundle E^+ of f is orientable in the sense explained in Exercise 8.6.4. Then on each mixing component f^N consistently either preserves or reverses the orientation of E^+, so passing to f^{2N} we may assume that f preserves the orientation of E^+ (and $P_n(f)$ has a multiplicative exponential asymptotic). Then all periodic points of a given period have the same index, either 1 or -1 (Corollary 8.4.7). Thus by Corollary 8.6.16 of the Lefschetz Fixed-Point Formula and (8.7.1) which gives the Lefschetz number of a toral map we have

$$P_n(f) = |L(f^n)| = |L(f_*^n)| = |\det(\mathrm{Id} - f_*^n)|.$$

If $\lambda_1, \dots, \lambda_k$ are the eigenvalues of f_* with multiplicities then this gives

$$P_n(f) = \Big|\prod_{i=1}^{k}(\lambda_i^n - 1)\Big| = \Big|\prod_{|\lambda_i|>1}(\lambda_i^n - 1)\Big| \cdot \Big|\prod_{|\lambda_i|<1}(\lambda_i^n - 1)\Big| \cdot \Big|\prod_{|\lambda_i|=1}(\lambda_i^n - 1)\Big|,$$

with $\prod_{|\lambda_i|>1}(\lambda_i^n - 1) = \prod_{|\lambda_i|>1}\lambda_i^n(1 + o(n))$ and $|\prod_{|\lambda_i|<1}(\lambda_i^n - 1)| \to 1$. To see that f_* is hyperbolic note that otherwise $\underline{\lim}_{n\to\infty} |\prod_{|\lambda_i|=1}(\lambda_i^n - 1)| = 0$. This is clear if λ_i is a qth root of unity because then $|\lambda_i^{kq} - 1| = 0$ for all $k \in \mathbb{N}$. If $\lambda_i = e^{2\pi i\alpha}$ with irrational α then by Proposition 1.3.3 the fractional part of $n\alpha$ can be arbitrarily small, so λ_i^n can be arbitrarily close to 1 while $|\lambda_j^n - 1| \le 2$. Likewise $\overline{\lim}_{n\to\infty} |\prod_{|\lambda_i|=1}(\lambda_i^n - 1)|$ is always bounded away from 0 (by considering roots of unity and irrational arguments separately). This contradicts existence of an exponential multiplicative asymptotic of $P_n(f)$. □

Given Lemma 18.6.2 the proof of Theorem 2.6.1 shows that F_L is a factor of f via a unique semiconjugacy h homotopic to the identity (and is hence surjective). To apply this proof in the higher-dimensional setting one only needs to adapt (2.6.4) and (2.6.5) by using vector-valued functions and replacing λ_1^{-1} and λ_2^{-1} by the inverses of the restriction of Df to the expanding and contracting subspaces. Using uniform continuity one also sees that h (and likewise its lift H to $\mathbb{R}^n$) sends stable manifolds into stable manifolds and likewise unstable manifolds into unstable manifolds.

Any lift F of f has no more fixed points than L, which in turn has exactly one by hyperbolicity. This follows from the argument that proves Theorem 8.7.1. The $|\det(\mathrm{Id} - f_*)|$ different fixed points of f lift to points which are shifted by different integer vectors. In fact they belong to different essential (as defined in Section 8.7) equivalence classes. If the lift had more than one fixed point then f would have too many, contrary to the index calculation. Thus we have

Lemma 18.6.3. *If $F: \mathbb{R}^n \to \mathbb{R}^n$ is the lift of an Anosov diffeomorphism and F_L the lift of a hyperbolic automorphism such that $d(F(x), F_L(x))$ is bounded then F has at most one fixed point.*

Lemma 18.6.4. *There is a topologically transitive compact locally maximal hyperbolic set Λ of f such that $h(\Lambda) = \mathbb{T}^n$.*

Proof. First we show that $h(NW(f)) = \mathbb{T}^n$. If this is not the case then $h(NW(f))$ is a proper closed F_L-invariant subset of $\mathbb{T}^n$ and thus there is a periodic orbit of F_L in $\mathbb{T}^n \setminus h(NW(f))$, whose preimage under h is a nonempty compact f-invariant subset of $\mathbb{T}^n \setminus NW(f)$, contradicting Corollary 3.3.5.

Now if x is a point with dense F_L-orbit and $y \in NW(f) \cap h^{-1}(\{x\})$ let Λ be the topologically transitive component containing y from the Spectral Decomposition Theorem 18.3.1. Then $\mathbb{T}^n = \overline{\mathcal{O}(x)} \subset h(\Lambda)$. □

Proposition 18.6.5. *If $f: \mathbb{T}^n \to \mathbb{T}^n$ is an Anosov diffeomorphism of $\mathbb{T}^n$ then $NW(f) = \mathbb{T}^n$ (and thus f is topologically mixing by Corollary 18.3.5).*

Proof. If not then f has a topologically transitive component from the spectral decomposition other than the Λ obtained in Lemma 18.6.4. Thus for some l there is a $q \in \mathrm{Fix}(F_L^l) \setminus h(\mathrm{Fix}(f^l \restriction_\Lambda))$ for which l is the minimal positive period. Here F_L is as before. $\Lambda' := \Lambda \cap h^{-1}(\{q\})$ is a locally maximal hyperbolic set because any orbit in a sufficiently small neighborhood maps under h into a small neighborhood of the orbit of q and hence into the orbit of q by expansivity. Since the nonwandering set of $f \restriction_{\Lambda'}$ is nonempty by Corollary 3.3.5 and periodic points are dense in it by Corollary 6.4.19 there is a periodic point $p \in \Lambda'$. By assumption l is not a period of p, so $p' := f^l(p) \neq p$ while $h(p') = q$. Take $k \in \mathbb{N}$ such that $f^k(p) = p$. Denote by $\pi: \mathbb{R}^n \to \mathbb{T}^n$ the projection. If $\pi(a) = p$, $\pi(b) = p'$, and $\pi(c) = q$ then the maps

$$\Phi(x) := F^k(x + a) - a, \quad \Psi(x) := F_L^k(x + c) - c, \quad N(x) := H(x + a) - c$$

of $\mathbb{R}^n$ fix 0 and descend to maps ϕ, ψ, η of $\mathbb{T}^n$. Since η is homotopic to Id we have $N(x+v) = N(x)+v$ for $v \in \mathbb{Z}^n$. Now $\eta(\pi(b-a)) = \pi(N(b-a)) = h(\pi(b)) - \pi(c) = 0$ so there exists $m \in \pi^{-1}(b-a)$ with $N(m) = 0$. Since $\phi(\pi(b-a)) = \pi(F^k(b)-a) = f^k(p')-p = \pi(b-a)$ we have $\Phi(m) = m+v$ for some $v \in \mathbb{Z}^n$. But $0 = \Psi(0) = \Psi(N(m)) = N(\Phi(m)) = N(m+v) = N(m)+v = v$, so $\Phi(m) = m$. If $\Theta\colon \mathbb{R}^n \to \mathbb{R}^n$ is the hyperbolic linear map obtained by extending $\Phi_{\restriction \mathbb{Z}^n}$ linearly then $d(\Phi(x+v), \Theta(x+v)) = d(\Phi(x)+\Phi(v), \Theta(x)+\Phi(v)) = d(\Phi(x), \Theta(x))$ for $x \in \mathbb{R}^n$, $v \in \mathbb{Z}^n$, so this distance is bounded independently of x by periodicity. Lemma 18.6.3 shows that Φ cannot have both m and 0 as fixed points. □

We will need the observation that H is proper:

Lemma 18.6.6. *If $Y \subset \mathbb{R}^n$ is bounded then so is $H^{-1}(Y)$.*

Proof. If $I = [0,1]^n \subset \mathbb{R}^n$ is the canonical fundamental domain for $\mathbb{T}^n = \mathbb{R}^n/\mathbb{Z}^n$ then $K := H(I)$ is a compact fundamental domain as well and $A := \{l \in \mathbb{Z}^n \mid I \cap (K+l) \neq \varnothing\}$ is finite. Since $K + l = H(I + h_*^{-1}(l))$ we have $H^{-1}(I) \subset \bigcup_{l \in A}(I + h_*^{-1}(l))$, and the latter is compact since A is finite. Since a bounded set is covered by finitely many translates of I this proves the claim. □

Next we establish a global product structure on $\mathbb{R}^n$.

Lemma 18.6.7. *If $x \neq y$ then the stable manifold of x and the unstable manifold of y (both for F) intersect in exactly one point.*

Proof. First we prove that the intersection of a stable and an unstable manifold for F consists of at most one point. To that end we will argue by contradiction and suppose that $y \in W^s(x) \cap W^u(x)$ and $y \neq x$. Take a local product neighborhood P of x that does not contain y. Since $W^s(P) = \{z \mid W^s(z) \cap P \neq \varnothing\}$ and $W^u(P) = \{z \mid W^u(z) \cap P \neq \varnothing\}$ are open, $W^s(P) \cap W^u(P)$ is a neighborhood of y. Since by Corollary 6.4.19 periodic points are dense in $NW(f) = \mathbb{T}^n$ there is a lift y' of an f-periodic point in $W^s(P) \cap W^u(P) \setminus \bar{P}$. But $W^u(y') \cap P \neq \varnothing$ and $W^s(y') \cap P \neq \varnothing$, so there is a point $x' \in W^s(y') \cap W^u(y') \cap P$ by the product structure on P. Thus without loss of generality we may assume that $y \in W^s(x) \cap W^u(x)$, $y \neq x$, and x is the lift of a fixed point of f (after possibly passing to an iterate). Changing the lift F of f we may assume that x is a fixed point of F. The F-homoclinic point y is nonwandering for F by Corollary 6.5.6 so there is a periodic point z of F near y by density of periodic points in $NW(F)$. But if n is the period of z then this shows that F^n has two fixed points, contrary to Lemma 18.6.3.

To show that $W^u(x) \cap W^s(y) \neq \varnothing$ for all x, y note that for given y and $W_0^s := W^s(y)$ the set $G := \{x \mid W^u(x) \cap W_0^s \neq \varnothing\}$ is open because there is a neighborhood U of W_0^s such that $G = \{x \mid W^u(x) \cap U \neq \varnothing\}$. To show $G = \mathbb{R}^n$ we prove that G is closed. To that end take $w \in \bar{G}$, $x \in G$ near w, and $\{y\} = W^u(x) \cap W_0^s$. To reduce to two dimensions let $\gamma\colon [0,1] \to W^u(x)$ and

$\eta\colon [0,1] \to W^s(x)$ be arcs such that $\gamma(0) = \eta(0) = x$ and $\gamma(1) = y$, $\eta(1) = w$. We will show that there is a map $\theta\colon [0,1] \times [0,1] \to \mathbb{R}^n$ such that

$$\theta(s,0) = \gamma(s), \quad \theta(0,t) = \eta(t), \quad \{\theta(s,t)\} = W^s(\gamma(s)) \cap W^u(\eta(t)). \quad (18.6.1)$$

This implies $w \in G$ since $\theta(1,1)$ is the desired intersection point.

Equation (18.6.1) defines θ on $[0,1] \times \{0\}$ and $\{0\} \times [0,1]$. Since there is a local product structure the domain $\mathcal{D}$ of θ in $[0,1]^2$ is open. Let

$$t_0 := \sup\{t \mid [0,1]\times[0,t] \subset \mathcal{D}\}, \quad s_0 := \sup\{s \mid [0,s]\times\{t_0\} \subset \mathcal{D}\}, \quad J := [0,1]\times[0,t_0).$$

To see that $\theta(J)$ is bounded note first that the projection of $h(\theta(J))$ to the unstable manifold $V^u(0)$ of 0 for F_L along stable leaves V^s of F_L is the projection of $H(\gamma([0,1]))$, hence is compact, and the projection to $V^s(0)$ along V^u is contained in the projection of $H(\eta([0,1]))$, so $H(\theta(J))$ is bounded and hence $\theta(J)$ is bounded by Lemma 18.6.6. Thus there is a sequence $(s_n, t_n) \to (s_0, t_0)$ with $s_n \le s_{n+1}$ such that $\theta(s_n, t_n) \to p$ for some $p \in \mathbb{R}^n$. Without loss of generality assume that this sequence is contained in a local product neighborhood O of p. Then $\theta(s_1, t_n) \to W^u(p) \cap W^s(\theta(s_1, t_1)) = \theta(s_1, t_0)$ and $\theta(s_n, t_1) \to W^s(p) \cap W^u(\theta(s_1, t_1)) = \theta(s_0, t_1)$, so for any $(S_n, T_n) \to (s_0, t_0)$ we have

$$\lim_{n\to\infty} \theta(S_n, T_n) = \lim_{n\to\infty} W^u(\theta(s_1, T_n)) \cap W^s(\theta(S_n, t_1))$$
$$= W^u(\lim_{n\to\infty} \theta(s_1, T_n)) \cap W^s(\lim_{n\to\infty} \theta(S_n, t_1)) = W^u(\theta(s_1, t_0)) \cap W^s(\theta(s_0, t_1)) = p,$$

hence setting $\theta(s_0, t_0) = p$ gives a continuous extension of θ which implies that $(s_0, t_0) = (1,1)$ and ends the proof. □

Since the foliations (that is, the local product structure) are continuous there is a homeomorphism $\Phi\colon \mathbb{R}^n \to \mathbb{R}^k \times \mathbb{R}^{n-k}$ sending each $W^u(x)$ to a leaf $\mathbb{R}^k \times \{c\}$ and each leaf $W^s(x)$ to $\{c\} \times \mathbb{R}^{n-k}$ by $\Phi(x) := (W^s(x) \cap W^u(0), W^u(x) \cap W^s(0))$. To complete the proof of the global classification we need to prove

Lemma 18.6.8. *The semiconjugacy is injective.*

Proof. We first show that H is injective on unstable leaves of f. If we denote by $B^u(0,r)$ and $B^s(0,r)$ the r-balls in the unstable and stable manifold of 0, respectively (using the intrinsic metrics d^u and d^s), then by Lemma 18.6.6 there exists $r > 0$ such that $H^{-1}(I) \subset D(r) := \{x \mid W^u(x) \cap B^s(0,r) \ne \varnothing$ and $W^s(x) \cap B^u(0,r) \ne \varnothing\}$. Since $D(r)$ has compact closure we find, using the homeomorphism Φ from the global product structure, that $\mu := \sup\{d^u(x,y) \mid x,y \in D(r),\ y \in W^u(x)\} < \infty$. Now if $y \in W^s(x)$ and $H(x) = H(y)$ let $x_m := F^m(x)$, $y_m := F^m(y)$. Then $H(y_m) = H(F^m(y)) = F_L^m(H(y)) = F_L^m(H(x)) = H(F^m(x)) = H(x_m)$. But there exist $l_m \in \mathbb{Z}^n$ such that $H(x_m + l_m) = H(y_m + l_m) \in I$ so $x_m + l_m, y_m + l_m \in D(r)$ and hence $d^u(F^m(x), F^m(y)) = d^u(x_m + l_m, y_m + l_m) \le \mu$ for all $m \in \mathbb{N}$. By hyperbolicity this implies $x = y$, showing that H is injective on unstable manifolds.

By the same token H is injective on stable manifolds. Suppose now that $H(x) = H(y)$ for some $x, y \in \mathbb{R}^n$. If there is a heteroclinic point z for x and y then $H(z)$ is a heteroclinic point for $H(x)$ and $H(y)$ (under F_L), which implies that $H(x) = H(z) = H(y)$ because F_L has no nontrivial heteroclinic points. But by injectivity of H on stable and unstable leaves $x = z = y$. Thus the lift H of the semiconjugacy is injective. The semiconjugacy itself must then also be injective because H is given as the identity plus a periodic map: If $h(\pi(x)) = h(\pi(y))$ then $H(x) = H(y) + k$ for some $k \in \mathbb{Z}^n$ and hence $H(x-k) = H(y)$, so $\pi(x) = \pi(y)$. □

Thus we have obtained a surjective semiconjugacy between the Anosov diffeomorphism f and its linear model F_L and shown that it is injective, hence a conjugacy, completing the proof of Theorem 18.6.1. □

7. Markov partitions

We have encountered the notion of Markov partition on several occasions: The coding for expanding maps (Section 2.4b), the horseshoe-type set for quadratic maps (Section 2.5b), the Smale horseshoe (Section 2.5c), the hyperbolic automorphism of $\mathbb{T}^2$ (Section 2.5d), hyperbolic repellers for general one-dimensional systems (Theorem 16.1.1), and the Smale attractor (Section 17.1). In all those examples Markov partitions either produced a conjugacy with a topological Markov chain or a semiconjugacy where the identifications were described in rather elementary terms. This is, in fact, a low-dimensional phenomenon due to the fact that here the boundary of each of these sets is a finite union of pieces of stable and unstable leaves. Already for a hyperbolic automorphism of $\mathbb{T}^3$ one necessarily has to construct the elements in such a way that the boundary contains uncountably many pieces of stable or unstable leaves. Thus the geometric structure of Markov partitions in higher-dimensional situations turns out to be much more complicated. It is nevertheless possible to produce Markov partitions and we will be able to obtain from them sufficient similarity between the Markov model and the compact locally maximal hyperbolic set Λ.

We will write $\operatorname{Int}_\Lambda$ and ∂_Λ to refer to the interior and boundary with respect to the topology of Λ.

Definition 18.7.1. Let Λ be a compact locally maximal hyperbolic set and take ϵ, δ and $[x, y]$ as in Proposition 6.4.13 and let $\eta = \epsilon$. Then $R \subset \Lambda$ is called a *rectangle* if the diameter of R is smaller than $\eta/10$ and $[x, y] \in R$ whenever $x, y \in R$. A rectangle R is called proper if $R = \overline{\operatorname{Int}_\Lambda R}$. We write $W^i_R(x) := W^i_\eta(x) \cap R$ for $x \in R$, $i = u, s$, and set $\partial^s R := \{x \in R \mid x \notin \operatorname{Int}_{\Lambda \cap W^u_\eta(x)} W^u_R(x)\}$, $\partial^u R := \{x \in R \mid x \notin \operatorname{Int}_{\Lambda \cap W^s_\eta(x)} W^s_R(x)\}$.

A *Markov partition* is a finite cover $\mathcal{R} = \{R_0, \dots, R_{m-1}\}$ of Λ by proper rectangles such that

(1) $\operatorname{Int} R_i \cap \operatorname{Int} R_j = \varnothing$ for $i \neq j$;
(2) whenever $x \in \operatorname{Int} R_i$ and $f(x) \in \operatorname{Int} R_j$ then $W^u_{R_j}(f(x)) \subset f(W^u_{R_i}(x))$ and $f(W^s_{R_i}(x)) \subset W^s_{R_j}(f(x))$.

It is useful to observe the following:

Lemma 18.7.2. *If R is a rectangle then $\partial_\Lambda R = \partial^s R \cup \partial^u R$.*

Proof. $x \in \operatorname{Int}_\Lambda R \Rightarrow x \in \operatorname{Int}_{\Lambda\cap W^u_\eta(x)}(R \cap W^u_\eta(x) \cap \Lambda) = \operatorname{Int}_{\Lambda\cap W^u_\eta(x)} W^u_R(x)$ since R is a neighborhood of x in Λ. Thus $\partial^s R \subset \partial_\Lambda R$. Likewise $\partial^u R \subset \partial_\Lambda R$. If $x \in (\operatorname{Int}_{\Lambda\cap W^s_\eta(x)} W^s_R(x)) \cap (\operatorname{Int}_{\Lambda\cap W^u_\eta(x)} W^u_R(x))$ then by continuity of $[\cdot,\cdot]$ (see Proposition 6.4.13) there is a neighborhood U of x in Λ such that for all $y \in U$ we have $[x,y],[y,x] \in R$ hence $y' := [[y,x],[x,y]] \in R \cap W^s_\eta(x) \cap W^u_\eta(y) \subset W^s_\eta(x) \cap W^u_\eta(y) \subset \{y\}$, so $x \in \operatorname{Int}_\Lambda R$. □

Theorem 18.7.3. *A compact locally maximal hyperbolic set admits Markov partitions of arbitrarily small diameter.*

Proof. First take $\delta > 0$ small, ϵ as in Theorem 18.1.3, $\gamma < \epsilon/2$ such that $d(f(x), f(y)) < \epsilon/2$ when $d(x,y) < \gamma$, and a γ-dense set $P := \{p_0, \dots, p_{N-1}\}$ in the hyperbolic set Λ. Note that $\Omega(P) := \{\omega \in \Omega_N \mid d(f(p_{\omega_i}), p_{\omega_{i+1}}) < \epsilon\}$ is a topological Markov chain. For each ϵ-orbit from $\Omega(P)$ there is a unique $\beta(\omega) \in \Lambda$ that δ-shadows $\alpha(\omega) := \{p_{\omega_i}\}_{i\in\mathbb{Z}}$. As argued in the proof of Theorem 18.2.5, β is surjective and continuous. We extend $[\cdot,\cdot]$ to ϵ-orbits by setting

$$[\omega, \omega']_i = \begin{cases} \omega_i & \text{for } i \geq 0, \\ \omega'_i & \text{for } i \leq 0, \end{cases}$$

for any $\omega, \omega' \in \Omega(P)$ with $\omega_0 = \omega'_0$. Then $[\cdot,\cdot]$ commutes with β, that is, $\beta([\omega,\omega']) \in W^s_{2\delta}(\beta(\omega)) \cap W^u_{2\delta}(\beta(\omega')) = \{[\beta(\omega), \beta(\omega')]\}$.

Let $R'_i := \{\beta(\omega) \mid \omega_0 = i\}$. Then R'_i is a rectangle since for $x = \beta(\omega), y = \beta(\omega') \in R'_i$ we have $[\omega,\omega']_0 = i$ and thus $[x,y] = [\beta(\omega), \beta(\omega')] = \beta([\omega,\omega']) \in R'_i$. Next note that $\mathcal{R}' := \{R'_i \mid 1 \leq i \leq N\}$ satisfies a condition similar to (2) in Definition 18.7.1. Namely, suppose that $x = \beta(\omega)$ with $(\omega_0, \omega_1) = (i,j)$ and that $y = \beta(\omega') \in W^s_{R'_i}(x)$ with $\omega'_0 = i$. Then $y \in W^s_\eta(f(x))$ but also $y = [x,y] = \beta([\omega,\omega'])$ and hence $f(y) \in \beta(\sigma([\omega,\omega'])) \in R'_j$, so $f(y) \in W^s_{R'_j}(f(x))$. This proves half of the analog of (2) and the other half follows similarly, that is,

$$f(W^s_{R'_i}(x)) \subset W^s_{R'_j}(f(x)) \text{ and } W^u_{R'_j}(f(x)) \subset f(W^u_{R'_i}(x)). \tag{18.7.1}$$

Note also that by continuity of β the R'_i are closed. To obtain a Markov partition we need, however, proper rectangles with pairwise disjoint interiors. To that end we modify these rectangles.

For $x \in \Lambda$ let $\mathcal{R}(x)$ be the set of rectangles from $\mathcal{R}'$ that contain x and $\mathcal{R}^*(x)$ be the set of rectangles from $\mathcal{R}'$ that intersect a rectangle from $\mathcal{R}'(x)$. Then $A := \{x \in \Lambda \mid W^s_\eta(x) \cap \partial^s R'_i = \varnothing,\ W^u_\eta(x) \cap \partial^u R'_i = \varnothing \text{ for all } i\}$ is open and dense. If $R'_i \cap R'_j \neq \varnothing$ then we cut R'_j into four rectangles as follows:

$$\begin{aligned}
R(i,j,su) &:= R'_i \cap R'_j, \\
R(i,j,0u) &:= \{x \in R'_j \mid W^s_{R'_i}(x) \cap R'_j = \varnothing,\ W^u_{R'_i}(x) \cap R'_j \neq \varnothing\}, \\
R(i,j,s0) &:= \{x \in R'_j \mid W^s_{R'_i}(x) \cap R'_j \neq \varnothing,\ W^u_{R'_i}(x) \cap R'_j = \varnothing\}, \\
R(i,j,00) &:= \{x \in R'_j \mid W^s_{R'_i}(x) \cap R'_j = \varnothing,\ W^u_{R'_i}(x) \cap R'_j = \varnothing\},
\end{aligned} \tag{18.7.2}$$

and for $x \in A$ let $R(x) := \bigcap\{\mathrm{Int}_\Lambda R(i,j,q) \mid x \in R'_i, R'_i \cap R'_j \neq \varnothing, x \in R(i,j,q), q \in \{su, 0u, s0, 00\}\}$. Then $\overline{R(x)}$ are rectangles covering $R'_i \cap A$ and the $R(x)$ are finitely many pairwise disjoint open rectangles, so

$$\mathcal{R} := \{\overline{R(x)} \mid x \in A\} =: \{R_0, \dots, R_{m-1}\}$$

is a finite cover of Λ by proper rectangles with pairwise disjoint interiors. We will show that this is the desired Markov partition by showing that the Markov condition (2) of Definition 18.7.1 holds. It suffices to show that $f(W^s_{R_i}(x)) \subset W^s_{R_j}(f(x))$ for $x \in R_i \cap f^{-1}(R_j)$ since the second half then follows by considering f^{-1}.

We begin by showing that for $x, y \in A \cap f^{-1}(A)$, $y \in W^s_{R(x)}(x)$ we have $R(f(x)) = R(f(y))$. First notice that if $x = \beta(\omega)$ with $(\omega_0, \omega_1) = (i,j)$ then $f(y) \in f(W^s_{R(x)}(x)) \subset f(W^s_{R'_i}(x)) \subset W^s_{R'_j}(f(x)) \subset R'_j$ by (18.7.2). Thus $f(x), f(y) \in R'_j$. Next suppose $R'_j \cap R'_k \neq \varnothing$. To show that $f(x), f(y) \in R(j,k,q)$ for some $q \in \{su, 0u, s0, 00\}$ note that $W^s_{R'_j}(f(x)) = W^s_{R'_j}(f(y))$, so (by symmetry) it suffices to check that if $f(z) \in W^u_{R'_j}(f(x)) \cap R'_k$ then $\varnothing \neq W^u_{R'_j}(f(y)) \cap R'_k$. We will show $[f(z), f(y)] \in W^u_{R'_j}(f(y)) \cap W^s_{R'_k}(f(z))$. Write $x = \beta(\omega)$, $(\omega_0, \omega_1) = (i,j)$, $z = \beta(\omega')$, $(\omega'_0, \omega'_1) = (l,k)$. Then $f(z) \in W^u_{R'_j}(f(x)) \subset f(W^u_{R'_i}(x))$ and hence $z \in W^u_{R'_i}(x) \cap R'_l$. $R(x) = R(y)$ implies $x, y \in R(i,j,q)$ for some q, so there exist $z' \in W^u_{R'_i}(y) \cap R'_l$ and $z'' := [x,y] = [z,z'] \in W^s_{R'_l}(z) \cap W^u_{R'_i}(y)$. Now $z = \beta(\omega')$, so $f(W^s_{R'_l}(z)) \subset W^s_{R'_k}(f(z))$. Since $f(y), f(z) \in R'_j$, which is a rectangle, we conclude that $[f(z), f(y)] \in W^s_{R'_k}(f(z)) \cap W^u_{R'_j}(f(y))$. Thus $R(f(x)) = R(f(y))$ as claimed.

To obtain the Markov condition (2) let $C^s := \bigcup\{W^s_\zeta(x) \mid x \in \bigcup_i \partial^s R'_i\}$, $C^u := \bigcup\{W^u_\zeta(x) \mid x \in \bigcup_i \partial^u R'_i\}$, and $B := \Lambda \smallsetminus ((C^s \cup C^u) \cap f^{-1}(C^s \cup C^u))$. Choose ζ such that $B \subset A' := A \cap f^{-1}(A)$. If $x \in B$ then $W^s_{R(x)}(x) \cap A'$ is open and dense in $W^s_{\overline{R(x)}}(x)$, so $R(f(y)) = R(f(x))$ by the preceding result and therefore $f(W^s_{\overline{R(x)}}(x)) \subset \overline{R(f(x))}$, so $f(W^s_{\overline{R(x)}}(x)) \subset W^s_{\overline{R(f(x))}}(f(x))$. It only remains to verify this condition for arbitrary x. But if $x \in \mathrm{Int}\, R_i \cap f^{-1}(\mathrm{Int}\, R_j)$ then there exists $x' \in B \cap \mathrm{Int}\, R_i \cap f^{-1}(\mathrm{Int}\, R_j)$ and

$$\begin{aligned} f(W^s_{R_i}(x)) = f(\{[x,y] \mid y \in W^s_{R_i}(x')\}) &= \{[f(x), f(y)] \mid y \in W^s_{R_i}(x')\} \\ &\subset \{[f(x), z] \mid z \in W^s_{R_j}(f(x'))\} \subset W^s_{R_j}(f(x)). \end{aligned}$$

□

The most immediate consequence is the existence of a semiconjugacy between a compact locally maximal hyperbolic set and a topological Markov chain:

Theorem 18.7.4. *Suppose that Λ is a compact locally maximal hyperbolic set, $\mathcal{R} = \{R_1, \dots, R_m\}$ a Markov partition of sufficiently small diameter, and $A_{ij} := \begin{cases} 1 & \text{if } R_i \cap f^{-1}(R_j) \neq \varnothing, \\ 0 & \text{otherwise} \end{cases}$. Then $f_{\restriction \Lambda}$ is a topological factor of the topological*

Markov chain (Ω_A, σ_A). The semiconjugacy $h: \Omega_A \to \Lambda$ is injective on $h^{-1}(\Lambda')$, where $\Lambda' := \Lambda \setminus \bigcup_{i\in\mathbb{Z}} f^i(\partial^s\mathcal{R} \cup \partial^u\mathcal{R})$ and $\partial^s\mathcal{R} := \bigcup_{R\in\mathcal{R}} \partial^s R$, $\partial^s\mathcal{R} := \bigcup_{R\in\mathcal{R}} \partial^u R$.

Proof. We begin by studying the intersections that arise in the Markov partition. Let us call a proper subrectangle $S \neq \varnothing$ of a rectangle R an unstable subrectangle if $W^u_S(x) = W^u_R(x)$ for all $x \in S$. Then we have

Lemma 18.7.5. *If S is an unstable rectangle of R_i and $A_{ij} = 1$ then $f(S) \cap R_j$ is an unstable subrectangle of R_j.*

Proof. Let $x \in R_i \cap f^{-1}(R_j)$ and $D := W^s_{R_i}(x) \cap S \neq \varnothing$. Then $S = \bigcup_{y\in D} W^u_{R_i}(y)$ and $f(S) \cap R_j = \bigcup_{y\in D} f(W^u_{R_i}(y)) \cap R_j$. But $f(y) \in R_j$ (as shown in the proof of Theorem 18.7.3) and $f(W^u_{R_i}(y)) \cap R_j = W^u_{R_j}(f(y))$, so $f(S) \cap R_j = \bigcup_{z\in f(D)} W^u_{R_j}(z) = [W^u_{R_j}(f(x)), f(D)]$. This is proper because D, and hence $f(D)$, is, and $R_j = [W^u_{R_j}(f(x)), W^s_{R_j}(f(x))]$, hence $W^u_{R_j}(f(x))$ is as well. It is nonempty since $f(D) \neq \varnothing$. Finally, if $w \in f(S) \cap R_j$ then $w \in W^u_{R_j}(z)$ for some $z \in f(D)$, so $W^u_{R_j}(w) = W^u_{R_j}(z) \subset f(S) \cap R_j$. □

For $\omega \in \Omega_A$ we define $h(\omega) = \bigcap_{i\in\mathbb{Z}} f^{-i}(R_{\omega_i})$. The intersection is nonempty since we inductively have the finite intersection property by Lemma 18.7.5 and it cannot contain more than one point by expansivity. h is continuous by the arguments proving Theorem 18.2.5 and surjective because $h(\Omega_A)$ is a compact set containing Λ'. Clearly $h \circ \sigma_A = f \circ h$ and it is clear that every $x \in \Lambda'$ has only one preimage. □

It is also useful to note:

Lemma 18.7.6. *With the above notation $f(\partial^s\mathcal{R}) \subset \partial^s\mathcal{R}$ and $\partial^u\mathcal{R} \subset f(\partial^u\mathcal{R})$.*

Proof. For $x \in R_i$ there exist $j, x_n \in \operatorname{Int} R_i \cap f^{-1}(R_j))$ such that $x_n \to x$. Then $A_{ij} = 1$ and $x \in R_i \cap f^{-1}(R_j)$, hence $W^u_{R_j}(f(x)) \subset f(W^u_{R_i}(x))$. Thus for $x \notin \partial^s\mathcal{R}$, that is, such that $W^u_{R_j}(f(x))$ is a neighborhood of $f(x)$ in $W^s_\eta(f(x)) \cap \Lambda$, $W^u_{R_i}(X)$ is a neighborhood of x in $W^s_\eta(x) \cap \Lambda$, so $x \notin \partial^s\mathcal{R}$. The other inclusion follows by considering f^{-1}. □

Another intuitive and useful consequence is that topological transitivity and topological mixing are equivalent for the hyperbolic set and its Markov model:

Proposition 18.7.7. *If Λ is a compact locally maximal hyperbolic set for f then the topological Markov chain σ_A obtained from the coding in Theorem 18.7.4 is topologically transitive (correspondingly, mixing) if $f\restriction_\Lambda$ is topologically transitive (correspondingly, mixing).*

Proof. If $\varnothing \neq U, V \in \Omega_A$ are open there exist $\omega, \omega' \in \sigma_A$ and $m \in \mathbb{N}$ such that $U' := C^{-m,\dots,m}_{\omega_{-m},\dots,\omega_m} \subset U$, $V' := C^{-m,\dots,m}_{\omega'_{-m},\dots,\omega'_m} \subset V$ (using cylinder sets defined by (1.9.1)), hence $\varnothing \neq U_\Lambda := \operatorname{Int} \bigcap_{-m\leq i\leq m} f^{-i}(\operatorname{Int} R_{\omega_i})$ and $\varnothing \neq V_\Lambda := \operatorname{Int} \bigcap_{-m\leq i\leq m} f^{-i}(\operatorname{Int} R_{\omega'_i})$ and, of course, $h^{-1}(U_\Lambda) \subset U' \subset U$ and $h^{-1}(V_\Lambda) \subset V' \subset V$. Thus topological transitivity or mixing of f imply the corresponding property for σ_A. □

One may ask when the semiconjugacy obtained in Theorem 18.7.4 is a conjugacy. Clearly it is necessary that Λ be totally disconnected, because Ω_A is. By now we have enough machinery to show that this condition is indeed sufficient (cf. Theorem 16.1.1 in the one-dimensional case).

Proposition 18.7.8. *Let Λ be a totally disconnected compact locally maximal hyperbolic set for a diffeomorphism $f: U \to M$. Then $f\restriction_\Lambda$ is topologically conjugate to a topological Markov chain.*

Proof. We outline the proof, only skipping details that are easy to fill in.

Construct a cover by rectangles that are both open and closed. This is accomplished by taking closed open sets on stable and unstable manifolds of a point and using $[\cdot,\cdot]$ to produce such a rectangle. For each pair of such rectangles apply the cutting construction (18.7.2). If the diameters of the original rectangles are small enough then the rectangles thus obtained will also be open and closed. Thus we have a partition of Λ into disjoint closed rectangles and hence there is a $\gamma > 0$ such that if two points are closer than γ then they are in the same rectangle and their bracket is in the same rectangle. Now code according to this partition. The coding map is an injective continuous map. The image is a closed invariant subset of the full shift which has a local product structure by the previous remark. Thus by Theorem 18.4.1 the image is a locally maximal subset of the full shift. (Even though this theorem was proved for hyperbolic sets it applies to shifts because they can be viewed as a horseshoe-type invariant set of a smooth system.) Finally use the conclusion of Exercise 18.4.3 generalized to the case of the shift σ_N. Thus the image of Λ under the coding is an N-step topological Markov chain (Definition 1.9.10). As explained after that definition any n-step topological Markov chain is topologically conjugate to a topological Markov chain. □

Exercises

18.7.1. *Deduce the Specification Theorem 18.3.9 from Theorem 18.7.4.*

18.7.2. *Prove that if under the assumptions of Theorem 18.7.4 the dimension of M is 2 then the semiconjugacy is injective away from the union of at most countably many stable and unstable manifolds.*

18.7.3. *Prove that if under the assumptions of Theorem 18.7.4 the dimension of M is 2 then $h_{\text{top}}(f\restriction_\Lambda) = h_{\text{top}}(\Sigma_A)$, where A comes from Theorem 18.7.4.*

19

Metric structure of hyperbolic sets

In the previous chapter we concentrated on the behavior of purely topological invariants associated with hyperbolic sets. The topological dynamics of a hyperbolic set is closely related to that of a topological Markov chain and since a hyperbolic set appears as an invariant set of a smooth dynamical system it is relevant to study how this topological dynamics is embedded in a smooth manifold. The main conclusion is that all principal structures associated with the dynamics are Hölder continuous and sometimes possess a moderate degree of differentiability (for example, C^1). Higher differentiability is very exceptional. It turns out that Hölder regularity is also natural for treating cohomological equations of the kind discussed in Section 2.9 over hyperbolic dynamical systems. Our main conclusion, the Livschitz Theorem 19.2.1, asserts that periodic obstructions provide complete systems of invariants of Hölder cocycles up to Hölder coboundaries. This result as well as its C^1 version has a number of useful applications.

1. Hölder structures

a. The invariant class of Hölder-continuous functions. Earlier (Section 1.9a and Exercises 1.9.1–1.9.3) we encountered the class of Hölder-continuous functions on the phase space of a dynamical system. It arose there naturally since the space was a sequence space and there was a one-parameter family of naturally defined metrics. We observed that these metrics not only induced the same topology, but had the same class of Hölder-continuous functions.

In this section we will see that the class of Hölder-continuous functions on a hyperbolic set also arises rather naturally. One of the main points of studying such functions is that they will enable us to study the ergodic theory of smooth hyperbolic systems in greater detail. They will enter through the definition of *pressure*, a generalization of entropy, and the study of pressure in turn will give us insights into the behavior of smooth invariant measures—showing their ergodicity, for example.

On the other hand the class of Hölder-continuous functions is a natural class of functions to study, since we will see that the principal structures associated with hyperbolicity are Hölder continuous with respect to the smooth structure although they usually do not possess any higher regularity (such as Lipschitz continuity or C^1). In particular we will find that topological conjugacies between hyperbolic sets are always Hölder continuous and that therefore the class of Hölder-continuous functions is preserved under topological conjugacies, that is, is an invariant of the topological dynamics.

We will observe in addition that the stable and unstable foliations are Hölder continuous. We will use this fact to construct an important Hölder-continuous function, the expansion rate, whose pressure (see Definition 20.2.1) we study in order to learn about smooth invariant measures.

Before proceeding to the general theory we would like to point out that a simple example of invariance of the class of Hölder-continuous functions already occurred earlier. The hyperbolic set in the Smale horseshoe (Section 2.5c) is topologically conjugate to the full 2-shift. With the right choice of contraction and expansion rates one can easily see that it is, in fact, isometric to a metric d_λ on the 2-shift as considered in Section 1.9a. Consequently the class of Hölder-continuous functions for that symbolic system is exactly the class of Hölder-continuous functions on the horseshoe with respect to the Euclidean metric.

b. Hölder continuity of conjugacies. Structural stability (Theorem 18.2.1) asserts that if $\Lambda \subset U \subset M$ is a hyperbolic set for an embedding $f\colon U \to M$ and f' is sufficiently close to f in the C^1 topology, then there is a hyperbolic set $\Lambda' \subset U$ for f' and a homeomorphism $h\colon \Lambda \to \Lambda'$ such that $hf = f'h$.

In this section we want to show that this conjugacy, and, in fact, *any* topological conjugacy between hyperbolic sets of smooth dynamical systems, is effected by a Hölder-continuous homeomorphism. This implies that the class of Hölder-continuous functions is an invariant of topological conjugacy.

It will be convenient to obtain Hölder continuity of conjugacies by showing that a conjugating homeomorphism is Hölder continuous along stable and unstable leaves separately. Let us first show that this is indeed sufficient. Since this follows from the fact alone that the stable and unstable manifolds are uniformly transverse continuously varying Lipschitz submanifolds, we give this result as an abstract lemma about metric spaces, where we think of stable and unstable foliations as defining two ("vertical" and "horizontal") equivalence relations.

Proposition 19.1.1. *Let Λ be a metric space with two equivalence relations $\sim_h$ and $\sim_v$ such that there exist $\epsilon > 0$, $K_1 \in \mathbb{R}$ so that for $x \sim_h y \sim_v z$ and $d(x,z) < \epsilon$ we have*

$$d(x,y)^2 + d(y,z)^2 \le K_1 d(x,z)^2$$

and such that for sufficiently nearby $x, y \in \Lambda$ there exists a w such that $x \sim_h w \sim_v y$. Let $\varphi\colon \Lambda \to X$ be a map to a metric space X such that there are $K_2, \alpha > 0$ so that for $d(x,y) < \epsilon$ and $x \sim_h y$ or $x \sim_v y$ we have

$$d(\varphi(x), \varphi(y)) \le K_2 d(x,y)^\alpha.$$

Then there exists $K_3 > 0$ such that $d(\varphi(x), \varphi(y)) \le K_1K_2K_3 d(x,y)^\alpha$ for all sufficiently close $x, y \in \Lambda$.

Proof. It is a standard exercise in linear algebra to show that there is a $K_3 > 0$ such that for $(x,y) \in \mathbb{R}^2$ one has

$$(|x|^\alpha + |y|^\alpha)^{1/\alpha} \le K_3^{1/\alpha}(x^2 + y^2)^{1/2}.$$

(One sees this by drawing $\{(x,y) \in \mathbb{R}^2 \mid |x|^\alpha + |y|^\alpha = 1\}$.) But then for $x, y \in \Lambda$ take w such that $x \sim_h w \sim_v y$ and note

$$\begin{aligned} d(\varphi(x), \varphi(y)) &\le d(\varphi(x), \varphi(w)) + d(\varphi(w), \varphi(y)) \le K_2(d(x,w)^\alpha + d(w,y)^\alpha) \\ &\le K_2K_3(d(x,w)^2 + d(w,y)^2)^{\alpha/2} \le K_1K_2K_3 d(x,y)^\alpha. \end{aligned}$$

□

Theorem 19.1.2. *Let Λ and Λ' be compact hyperbolic sets for diffeomorphisms f and f', respectively, and $h = f'hf^{-1}\colon \Lambda \to \Lambda'$ a topological conjugacy. Then both h and h^{-1} are Hölder continuous.*

Proof. Since f and f' appear symmetrically in the statement it suffices to check that h itself is Hölder continuous. Furthermore we just showed that it is indeed enough to show Hölder continuity of h along stable and unstable manifolds. Since h also conjugates f^{-1} and ${f'}^{-1}$ (for which stable and unstable manifolds reverse roles) it is, in fact, enough to prove that $h\restriction_{W^u(x)\cap\Lambda}$ is Hölder continuous for every $x \in \Lambda$ (with uniform constant and exponent).

To this end take $c < 1 < C$ such that C is a Lipschitz constant for f and c is a Lipschitz constant for ${f'}^{-1}\restriction_{W^u}$ and let $\alpha > 0$ be such that $cC^\alpha < 1$. Fix $\epsilon_0 > 0$. Since Λ is compact and h is continuous, hence uniformly continuous, there exists $\delta_0 > 0$ such that $d(x,y) < \delta_0$ implies $d(h(x), h(y)) < \epsilon_0$.

Now if $x, y \in \Lambda$, $y \in W^u(x)$, and $\delta := d(x,y)$ is sufficiently small, then there exists $n \in \mathbb{N}$ such that $d(f^n(x), f^n(y)) \le C^n\delta < \delta_0 \le C^{n+1}\delta$. Hence $d(h(f^n(x)), h(f^n(y))) < \epsilon_0$, by choice of δ_0, so using $cC^\alpha < 1$ we have

$$\begin{aligned} d(h(x), h(y)) &= d({f'}^{-n}hf^n(x), {f'}^{-n}hf^n(y)) < c^n\epsilon_0 \\ &= c^n\delta_0^\alpha \cdot \epsilon_0/\delta_0^\alpha \le (cC^\alpha)^n C^\alpha(\epsilon_0/\delta_0^\alpha)\delta^\alpha < C^\alpha(\epsilon_0/\delta_0^\alpha)(d(x,y))^\alpha. \end{aligned}$$

□

This result shows that any Hölder-continuous structure associated with a compact hyperbolic set has a corresponding Hölder-continuous structure in any topologically conjugate compact hyperbolic set. Of particular interest are the local charts defining the local product structure. Equivalently, if $x, y \in \Lambda$ are sufficiently close, one can consider the *holonomy map* $\Lambda \cap W^s(x) \to \Lambda \cap W^s(y), z \mapsto W^u(z) \cap W^s(y)$, where all manifolds are taken to be local ones, or one may consider the bracket $[x,y]$ defined in Proposition 6.4.13. We make Hölder continuity of the holonomy maps precise as follows: If we take all $W^u(x)$,

$W^s(x)$ to be of size δ_0, then there exist $\epsilon, \alpha, K > 0$ such that if $z, z' \in \Lambda$, $x, y \in W^s(z)$, $d(x,y) < \epsilon$, $d(z,z') < \epsilon$ then

$$d(W^u(x) \cap W^s(z'), W^u(y) \cap W^s(z')) < Kd(x,y)^\alpha,$$

and if $z, z' \in \Lambda$, $x, y \in W^u(z)$, $d(x,y) < \epsilon$, $d(z,z') < \epsilon$ then

$$d(W^s(x) \cap W^u(z'), W^s(y) \cap W^u(z')) < Kd(x,y)^\alpha.$$

We refer to this property as Hölder continuity of the dependence of the unstable manifold on a point or as Hölder continuity of the unstable foliation. From Theorem 19.1.2 we then immediately have

Corollary 19.1.3. *Let M_1, M_2 be Riemannian manifolds, $U_i \subset M_i$ open, and $f_i: U \to M$ embeddings with compact invariant hyperbolic sets $\Lambda_i \subset U_i$ $(i = 1, 2)$. If the stable and unstable manifolds of f_1 depend Hölder continuously on the base point and f_1 and f_2 are topologically conjugate then so do those of f_2.*

Since by Theorem 18.6.1 toral Anosov diffeomorphisms are conjugate to a linear model where the holonomy maps are smooth we obtain

Corollary 19.1.4. *Anosov diffeomorphisms on tori have Hölder-continuous holonomy maps.*

This latter fact does not depend on the Anosov condition or on the manifold. Any compact hyperbolic set has Hölder-continuous foliations, with the same Hölder exponent as obtained for the distributions in Theorem 19.1.6. Since we do not need this result and a detailed proof would be a bit too lengthy, we do not provide it.[1]

c. Hölder continuity of orbit equivalence for flows. An analog of the preceding result applies to flows as well by very similar reasoning. There is, however, a new aspect to be taken into account here, namely, the lack of uniqueness in the flow direction. We thus obtain the following result:

Theorem 19.1.5. *Let $\Lambda \subset M$, $\Lambda' \subset M'$ be compact hyperbolic sets for flows φ and ψ, respectively, and suppose that φ and ψ are orbit equivalent via $h: \Lambda \to \Lambda'$. Then there is a Hölder-continuous orbit equivalence arbitrarily C^0-close to h.*

Proof. We begin with a local construction of a Hölder orbit equivalence. Take small smooth transversals $\mathcal{T}$ at $p \in \Lambda$ and $\mathcal{T}'$ at $q = h(p) \in \Lambda'$. Then locally $h(\mathcal{T})$ projects canonically to $\mathcal{T}'$ along the orbits of ψ and the composition of h with this projection is Hölder continuous by the same arguments as in the proof of Theorem 19.1.2 using the intersections of $\mathcal{T}$ with weak unstable and weak stable foliations as the equivalence classes in Proposition 19.1.1.

Now fix some $\delta > 0$ and cover Λ by flow boxes whose floors are small smooth transversals and fix corresponding smooth transversals in Λ'. From the Hölder-continuous map between these transversals construct local conjugacies on the flow boxes by taking them to be time preserving. This gives local homeomorphisms from these flow boxes to Λ'. To assemble these into one global map take a smooth partition of unity on Λ subordinate to the covering by these flow boxes. Now all images of a point $x \in \Lambda$ lie on an orbit segment and thus one can take the average of the corresponding time parameters weighted by the values of the members of the partition of unity at x. This gives a well-defined Hölder-continuous map $\tilde{h}$ which is C^0-close to h and takes orbits of φ to orbits of ψ. $\tilde{h}$ is also differentiable along the orbits of φ. The remaining problem is that $\tilde{h}$ may not be monotone along orbits.

To find a homeomorphism with the desired properties we use the fact that $\tilde{h}$ is as C^0-close to the homeomorphism h as we please, so long as δ is sufficiently small. This implies that there is an $\eta > 0$ such that for any $x \in \Lambda$ and $t > \eta$ we have $\tilde{h}(\varphi^t(x)) = \psi^s(\tilde{h}(x))$ with $s > 0$. This implies that defining $h'(x) := (1/\eta)\int_0^\eta \tilde{h}(\varphi^t(x))dt$ (the integral interpreted as one involving the real parameter along the orbit of x) gives a homeomorphism h' with all desired properties. □

d. Hölder continuity and differentiability of the unstable distribution. Let us now show that the unstable distribution of a hyperbolic set $\Lambda \subset U$ of an embedding $f: U \to M$ is Hölder continuous as well. Our approach is to use the way it is obtained in the proof of the Hadamard–Perron Theorem 6.2.8. There it arises as the fixed point of a contraction, so it is the limit of any "reasonable" distribution nearby under iterates of the differential of the diffeomorphism. We will show that the action of the differential preserves the Hölder property for appropriate Hölder exponent and constant. By compactness of the set of such distributions (by the Arzela–Ascoli Theorem A.1.24) starting from a Hölder-continuous candidate one obtains a Hölder-continuous unstable distribution.

Theorem 19.1.6. *Let M be a Riemannian manifold, $U \subset M$ open, and $f: U \to M$ a $C^{1+\beta}$ embedding with a compact invariant hyperbolic set $\Lambda \subset U$. Then the stable and unstable distributions are Hölder continuous.*

Proof. We choose local coordinates near every point that will make estimates easier. Each tangent plane $T_pM = E^+ \oplus E^-$ for $p \in \Lambda$ can be coordinatized linearly in such a way that E^+ is the x-"plane" $\mathbb{R}^k \times \{0\}$ and E^- is the y-"plane" $\{0\} \times \mathbb{R}^{n-k}$. These coordinates can be chosen to depend continuously on the point p. There exists $\epsilon > 0$ such that the exponential map $\exp_p$ (see (9.5.1)) is an embedding of the closed ϵ-ball $\overline{B(0,\epsilon)} \subset T_pM$ into M for every point $p \in \Lambda$. This gives us closed neighborhoods $V_p = \overline{B(p,\epsilon)}$ of $p \in \Lambda$ on which we have coordinates such that $p \sim 0$ and at p the space E_p^+ is tangent to the x-"plane" $\mathbb{R}^k \times \{0\}$ and E^- is tangent to the y-"plane" $\{0\} \times \mathbb{R}^{n-k}$. These coordinates depend continuously on p in the C^∞ topology, that is, all derivatives depend continuously on p, because this holds for $\exp_p$.

As "candidates" for the unstable distribution let us consider the space $C(\delta)$ of k-dimensional distributions inside the horizontal δ-cones. The action of f on

$E \in C(\delta)$ is given by

$$f_* E(p) = Df(E(f^{-1}(p))).$$

Then $f_* C(\delta) \subset C(\delta)$.

Suppose $E \in C(\delta)$. If $d(p,q) < \delta$ then we can pass to coordinates on V_p in which q corresponds to a point z with $\|z\| = d(p,q) < \delta$ (since $\exp_p$ is a "radial isometry"). The subspace in $T_z\mathbb{R}^n$ defined by E can be viewed as the graph of a linear map $E(z)$ from the x-"plane" $\mathbb{R}^k \times \{0\}$ to the y-"plane" $\{0\} \times \mathbb{R}^{n-k}$. The operator norm of $E(z)$ gives a bound for the distance between $E^+(p)$ and $E(q)$. In particular we want to show that if $E = E^+$ then there are constants $K > 0$ and $\alpha \in (0,1)$ such that we have $\|E(z)\| \le K\|z\|^\alpha$ for all z in a coordinate chart and all coordinate charts. To consider candidate distributions with some degree of Hölder continuity let

$$C(\delta, \epsilon_0, K) := \{E \in C(\delta) \mid \|E(z)\| \le K\|z\|^\alpha \text{ when } \epsilon_0 \le \|z\| \le \epsilon\}.$$

While $C(\delta, \epsilon_0, K) \subset C(\delta)$ by definition, the extra condition is, in fact, vacuous for sufficiently large K:

Lemma 19.1.7. *For $\delta, \epsilon_0 > 0$ there exists $K = K(\delta, \epsilon_0) > 0$ such that $C(\delta) \subset C(\delta, \epsilon_0, K)$.*

Proof. Since E^+ is continuous, hence uniformly continuous, there is a constant $C > 0$ such that $\|E_z\| \le C$ for all $E \in C(\delta)$ and $\|z\| < \epsilon$. But then if $\epsilon_0 > 0$ is fixed, $K = C\epsilon_0^{-\alpha}$ is as desired. □

Thus we know in particular that for this choice of K we have $f_*^n(C(\delta)) \subset C(\delta) \subset C(\delta, \epsilon_0, K)$. Our next step is to show that f_* improves the estimates for distributions in $C(\delta)$. To that end we use local coordinates in much the same way as step 5 of the proof of the Hadamard–Perron Theorem 6.2.8 and the estimates are of a very similar nature.

With respect to the decomposition of coordinate neighborhoods into x-"plane" $\mathbb{R}^k \times \{0\}$ and y-"plane" $\{0\} \times \mathbb{R}^{n-k}$ we can use the obvious block form $Df = \begin{pmatrix} A_z & B_z \\ C_z & D_z \end{pmatrix}$ with

$$\|A_z^{-1}\| < \frac{\mu^{-1}}{1+2\epsilon}, \quad \|D_z\| < \lambda - \epsilon, \quad \|B_z\| \le L\|z\|^\beta < \epsilon, \quad \|C_z\| \le L\|z\|^\beta < \epsilon$$

for some $\lambda_x = \lambda < 1 < \mu = \mu_x$ depending on the point $x \in M$. If $E \in C(\delta)$ then we can write it in coordinates as graphs of linear operators $E_z \colon \mathbb{R}^k \to \mathbb{R}^{n-k}$. More conveniently we write it equivalently as images of linear maps $\begin{pmatrix} I \\ E_z \end{pmatrix} \colon \mathbb{R}^k \to \mathbb{R}^n$, for then $f_* E \circ f = (Df)E$ is simply the map represented by the composition

$$Df \circ \begin{pmatrix} I \\ E_z \end{pmatrix} = \begin{pmatrix} A_z & B_z \\ C_z & D_z \end{pmatrix} \begin{pmatrix} I \\ E_z \end{pmatrix},$$

so $f_*E \circ f$ is represented by

$$(Df)E_z = (C_z + D_zE_z)(A_z + B_zE_z)^{-1}.$$

Note that we are moving from one local coordinate system to another. Now $\|B_zE_zA_z^{-1}\| \le \|B_z\|\|E_z\|\|A_z^{-1}\| < \epsilon$ since $\|E_z\| \le \delta \le 1$, hence

$$\|(A_z + B_zE_z)^{-1}\| = \|A_z^{-1}(I + B_zE_zA_z^{-1})^{-1}\| \le \frac{\mu^{-1}}{1+2\epsilon}(1+2\epsilon) = \mu^{-1}$$

for sufficiently small ϵ. In particular $A_z + B_zE_z$ is indeed invertible. Since

$$\|C_z + D_zE_z\| \le \|C_z\| + \|D_zE_z\| \le (\lambda - \epsilon)\|E_z\| + L\|z\|^\beta$$

we thus have

$$\|(Df)E_z\| \le \mu^{-1}((\lambda - \epsilon)\|E_z\| + L\|z\|^\beta).$$

$(Df)E_z$ represents f_*E at the image z' of the point with coordinate z, that is, $(Df)E_z = f_*E_{z'}$. Notice that if ν_x^{-1} is a Lipschitz constant for f^{-1} at $f(x)$ then $\|z'\| \ge \nu\|z\|$. Thus if $\alpha < \beta$ is such that

$$\lambda_x\mu_x^{-1}\nu_x^{-\alpha} < 1 \tag{19.1.1}$$

at every point and $K \ge L/\epsilon$ then $\mu^{-1}((\lambda - \epsilon)K + L) \le \mu^{-1}\lambda K$ and hence

$$\|E(z)\| \le K\|z\|^\alpha \text{ implies } \|f_*E(z')\| \le \mu^{-1}\lambda K\nu^{-\alpha}\|z'\|^\alpha < K\|z'\|^\alpha, \tag{19.1.2}$$

where z' corresponds to the image of the point with coordinate z. Thus for all $n \in \mathbb{N}$ we have $f_*^n(C(\delta)) \subset C(\delta, \lambda^n, K)$ with $K := K(\delta, \sup_x \lambda_x)$. We will refer to (19.1.1) as the *bunching condition*. Now let

$$R_n := \{(f^n(p), f^n(q)) \in M \times M \mid d_n^f(p,q) \le \epsilon,\ \epsilon_0 \le d(p,q) \le \epsilon\}$$

and set $S_n := \bigcup_{i=0}^n R_i$. Applying (19.1.2) inductively we see that for any $E \in C(\delta)$ and $(x, \exp_x z) \in S_n$ we have $\|(f_*^n)E(z)\| \le K\|z\|^\alpha$. By the Hadamard–Perron Theorem 6.2.8

$$\{(x,y) \in M \times M \mid y \notin W^u(x),\ d(x,y) \le \epsilon\} \subset S := \bigcup_{n=0}^{\infty} S_n.$$

Thus the fixed point F of f_* which is simply the coordinate representation of E^+ satisfies $\|F(z)\| \le K\|z\|^\alpha$ (in coordinates at x) if $(x, \exp_x z) \in \bar{S}$. Since $\bar{S}$ contains a neighborhood of the diagonal in $M \times M$ this proves the theorem. □

Remark. It is, in fact, possible to show by a similar argument that one obtains the same Hölder exponent for derivatives of E in its own direction.

Next we show that if (19.1.1) holds uniformly in x for some $\alpha > 1$ then the unstable distribution is C^1. This will imply in particular that the unstable distribution is C^1 if it has codimension one.

To make sense of the statement it is necessary to note that the property of being C^1 can be defined on subsets of a manifold even if these are not submanifolds. Namely, the differentiability condition merely requires the existence of a linear approximation at every point, and being C^1 requires that one can take these linear approximations to depend continuously on their base points. Both these notions only use the presence of an ambient smooth structure.

Theorem 19.1.8. *Let M be a Riemannian manifold, $U \subset M$ open, and $f: U \to M$ an embedding with a compact invariant hyperbolic set $\Lambda \subset U$. Suppose further that for some $\alpha > 1$ the bunching condition (19.1.1) holds at all points of Λ. Then the stable and unstable distributions are C^1.*

Proof. In order to use a similar strategy as before we first observe that differentiability can be proved via estimates: A map $T: \mathbb{R}^n \to \mathbb{R}^m$ is differentiable at x if for $(v_1, v_2, v_3) \in (\mathbb{R}^n)^3$ with $v_1 + v_2 + v_3 = 0$

$$\Big|\frac{T(x+v_1h_1)}{h_1} + \frac{T(x+v_2h_2)}{h_2} + \frac{T(x+v_3h_3)}{h_3} - \Big(\frac{1}{h_1} + \frac{1}{h_2} + \frac{1}{h_3}\Big)T(x)\Big| \to 0 \tag{19.1.3}$$

uniformly as $(h_1, h_2, h_3) \to 0$. One sees this by setting $v_3 = 0$ to show existence of directional derivatives; then (19.1.3) shows that these depend linearly on the direction. Uniformity guarantees that this is sufficient and yields, in fact, continuity of the derivatives. (Similar estimates as before show that the derivatives are indeed Hölder continuous.) For a function on a manifold replace $(x + v_ih_i)$ by $\exp_x(v_ih_i)$.

To show that (19.1.3) holds for the coordinate representations of E^+ consider

$$\mathcal{B} := \left\{\begin{matrix}\text{triples } (v_1, v_2, v_3) \text{ of vector fields on } M \text{ with} \\ v_1 + v_2 + v_3 = 0 \text{ and } \|v_1\| + \|v_2\| + \|v_3\| = 1\end{matrix}\right\}.$$

Set $v_i'(p) = Dfv_i(f^{-1}(p))/\xi(v_1(f^{-1}(p)), v_2(f^{-1}(p)), v_3(f^{-1}(p)))$ with $\xi \in \mathbb{R}$ such that $(v_1', v_2', v_3') \in \mathcal{B}$. Then f acts (invertibly) on $\mathcal{B}$ by $\mathcal{P}(v_1, v_2, v_3) := (v_1', v_2', v_3')$ and $\mathcal{P}(\mathcal{B}) = \mathcal{B}$. To avoid large subscripts we now write $E(z)$ for E_z. To prove differentiability we begin with a lemma:

Lemma 19.1.9. *There exist $\epsilon, \eta > 0$ such that for all $p \in M, (v_1, v_2, v_3) \in \mathcal{B}, 0 < h_1, h_2, h_3 < \epsilon$*

$$\Big|\frac{E(\exp(v_1(p)h_1))}{h_1} + \frac{E(\exp(v_2(p)h_2))}{h_2} + \frac{E(\exp(v_3(p)h_3))}{h_3}\Big| < K$$

implies for $(v_1', v_2', v_3') = \mathcal{P}(v_1, v_2, v_3)$ and $h_i' := \xi \cdot h_i$ that

$$\Big|\frac{E(\exp(v_1'(p)h_1'))}{h_1'} + \frac{E(\exp(v_2'(p)h_2'))}{h_2'} + \frac{E(\exp(v_3'(p)h_3'))}{h_3'}\Big| < (1-\eta)K.$$

Proof. To avoid clutter let us write $C_i := C_{\exp(v_i(p)h_i)}$, $D_i := D_{\exp(v_i(p)h_i)}$. We also let $A_i := (A_{\exp(v_i(p)h_i)} + B_{\exp(v_i(p)h_i)} E(\exp(v_i(p)h_i)))^{-1}$ and define E_i by $h_i E_i = h_1 h_2 h_3 E(\exp(v_i(p)h_i))$. Then we need to estimate

$$\begin{aligned}
&(h_2' h_3' E(\exp(v_1'(p)h_1')) + h_1' h_3' E(\exp(v_2'(p)h_2')) + h_1' h_2' E(\exp(v_3'(p)h_3')))\\
&= h_2' h_3'\Big(C_1 + \frac{D_1 E_1}{h_2 h_3}\Big)A_1 + h_1' h_3'\Big(C_2 + \frac{D_2 E_2}{h_1 h_3}\Big)A_2 + h_1' h_2'\Big(C_3 + \frac{D_3 E_3}{h_1 h_2}\Big)A_3\\
&= \xi^2 D_1(E_1 + E_2 + E_3)A_1\\
&\quad + (D_2 - D_1)E_2 A_2 + D_1 E_2(A_2 - A_1) + (D_3 - D_1)E_3 A_3 + D_1 E_3(A_3 - A_1)\\
&\quad + h_2' h_3' C_1 A_1 + h_1' h_3' C_2 A_2 + h_1' h_2' C_3 A_3.
\end{aligned}$$

To estimate the first term we note that

$$\|\xi^2 D_1(E_1 + E_2 + E_3)A_1\| < \xi^2(\lambda - \epsilon)\mu^{-1} K h_1 h_2 h_3 = (\lambda - \epsilon)\mu^{-1}\xi^{-1} K h_1' h_2' h_3'.$$

Again, if ν^{-1} is a Lipschitz constant for f^{-1} then $\xi > \nu$ and if (19.1.1) holds for some $\alpha > 1$ then $(\lambda - \epsilon)\mu^{-1}/\xi < 1 - 2\eta$ uniformly for some $\eta > 0$. Thus the first term is bounded by $(1 - 2\eta)K h_1' h_2' h_3'$. It remains to show that the remaining terms are sufficiently smaller than the first. But the second set of terms converges to zero faster than $h_1' h_2' h_3'$ because A and D are Lipschitz continuous and E is continuous and vanishes at 0, so this term contributes at most $\eta K h_1' h_2' h_3'/2$ once the h_i are small enough. Likewise the last set of terms goes to zero faster than $h_1' h_2' h_3'$ since C is Lipschitz continuous and vanishes at 0. Thus we obtain the desired bound. □

To use this lemma restrict attention to triples (v_1, v_2, v_3) in the stable distribution and denote the space of these by $\mathcal{B}^s$. Then $\xi < 1 - \rho$ for some constant $\rho > 0$ depending only on f. Notice now that since the unstable distribution is Lipschitz continuous (using (19.1.1) with $\alpha = 1$ in Theorem 19.1.6) there is a K such that the first equation in Lemma 19.1.9 holds for any $(v_1, v_2, v_3) \in \mathcal{B}$ and h_1, h_2, h_3. Take ϵ as in Lemma 19.1.9 and for any given $\delta \leq \epsilon$ take the smallest $n \in \mathbb{N}$ for which there is a $(v_1, v_2, v_3) \in \mathcal{B}^s$ such that $\prod_{i=0}^{n-1} \xi(\mathcal{P}^i(v_1, v_2, v_3)) < \delta/\epsilon$. Then $n \to \infty$ as $\delta \to 0$ (because $\xi < 1 - \rho$) and for $h_1, h_2, h_3 < \delta$ an $(n-1)$-fold application of Lemma 19.1.9 yields

$$\Big|\frac{E(\exp(v_1(p)h_1))}{h_1} + \frac{E(\exp(v_2(p)h_2))}{h_2} + \frac{E(\exp(v_3(p)h_3))}{h_3}\Big| < (1 - \eta)^{n-1} K$$

for all $(v_1, v_2, v_3) \in \mathcal{B}^s$. This proves (19.1.3) along E^s.

Next we note:

Lemma 19.1.10. *Suppose $\varphi\colon \mathbb{R}^n \to \mathbb{R}$ is C^1 along the leaves of two continuous transverse foliations W^u and W^s in $\mathbb{R}^n$. Then φ is C^1.*

Proof. We repeat the argument proving that continuity of partial derivatives implies that a function is C^1. Given two nearby points x, y we need to show that

$\varphi(y) - \varphi(x) = L(y-x)$ up to higher-order terms in $|y-x|$ for some linear map L. Since φ is C^1 along the leaves of W^u and W^s we have for $z \in W^u(x) \cap W^s(y)$ near x and y up to higher order $\varphi(y) - \varphi(x) = \varphi(y) - \varphi(z) + \varphi(z) - \varphi(x) = L^s_z(y-z) + L^u_x(z-x)$ for two linear maps L^u and L^s depending continuously on the base point. But then $L^s_z \to L^s_x$ as $z \to x$, hence as $y \to x$, that is, up to higher order we have $L^s_z(y-z) = L^s_x(y-z)$ and hence we can take L to be the linear map that restricts to L^u on $TW^u(x)$ and to L^s on $TW^s(x)$. □

Note that we know E^u to be C^1 along W^u by Theorem 6.2.8. Thus Lemma 19.1.10 applied to components in local coordinates proves Theorem 19.1.8. □

Let us show that the hypotheses of the preceding theorem, that is, (19.1.1) with $\alpha > 1$, are fulfilled for some interesting situations.

Corollary 19.1.11. *Let M be a Riemannian manifold, $U \subset M$ open, and $f: U \to M$ an embedding with a compact invariant hyperbolic set $\Lambda \subset U$. If the unstable distribution has codimension one then it is C^1.*

Proof. If the unstable distribution has codimension one then we can take $\lambda = \nu$, which immediately yields (19.1.1) with $\alpha > 1$. □

Remark. Evidently the preceding result applies to the stable distribution as well by considering f^{-1} in place of f. In particular we see, therefore, that the stable and unstable distributions of a hyperbolic set in a two-dimensional manifold are always C^1.

Corollary 19.1.12. *Let M be a Riemannian manifold, $U \subset M$ open, and $f: U \to M$ a volume-preserving embedding with a compact invariant hyperbolic set $\Lambda \subset U$. Suppose the stable distribution has codimension one. Then the stable and unstable distributions are C^1.*

Proof. The stable distribution is C^1 by the previous result. To see that the unstable distribution is also C^1 consider two cases. If the stable distribution also has dimension one then we are done by the preceding result. Otherwise take λ to be $Df\restriction_{E^+}$ and note that volume preservation (that is, having Jacobian identically equal to one in appropriate coordinates) implies that we may assume $\lambda\mu\nu \geq 1$ which implies that $\lambda\mu^{-1}\nu^{-\alpha} \leq \lambda^2\nu^{1-\alpha}$. The latter is less than one whenever $\alpha > 1$ is sufficiently close to 1. □

At times it is useful to extract from the preceding argument the optimal Hölder exponent. This involves looking for the optimal α for which (19.1.1) can be satisfied, that is, we consider the unstable bunching constant $B^u(f) := \inf_x \dfrac{\log \mu_x - \log \lambda_x}{\log \nu_x}$. The stable bunching constant is $B^s(f) := B^u(f^{-1})$ and the bunching constant is $B(f) := \min(B^u(f), B^s(f))$. Note that these bunching constants depend on the metric used, whereas the Hölder exponent clearly does not, so we can in all bunching constants take any metric, or the supremum over all metrics.[2]

When the bunching constant exceeds $k \in \mathbb{N}$ one can similarly show that the distributions are C^k with α-Hölder kth derivatives for any $\alpha < B^u(f) - [B^u(f)]$. However, this does not yield simultaneous high smoothness for both foliations since $B(f) \leq 2$ always (Exercise 19.1.3).

e. Hölder continuity of the Jacobian. When we study smooth invariant measures in Section 20.4, we will be interested in the distortion of volume of unstable manifolds under the diffeomorphism. This is measured by the (unstable) Jacobian of the map. It is a variant of the Jacobian discussed in Section 5.1.

Suppose M is a Riemannian manifold, $U \subset M$ open, and $g: U \to M$ a C^2 embedding. If $x \in U$ and $E \subset T_xM$ is a linear subspace then the Riemannian metric on M induces volume forms ω_E and ω_{DgE} on E and DgE. Note that these depend smoothly on E. On the other hand we can define the pullback of ω under g, that is, a volume form $g^*\omega_{DgE}$ on E, by

$$g^*\omega_{DgE} = \omega_{DgE}(Dg(\cdot), \dots, Dg(\cdot)).$$

This volume form also depends smoothly on E. Now any two volume forms on E are linearly dependent, so there exists a number $J'g(x)$ such that

$$g^*\omega_{DgE} = J'g(x)\omega_E.$$

$J'g(x)$ depends smoothly on E, since ω_E and $g^*\omega_{DgE}$ do. In particular if $N \subset U$ is a smooth submanifold and $O \subset N$, then the volume of $g(O)$ can be computed two ways:

$$\int_{g(O)} \omega_{DgTN} = \int_O J'g\omega_{TN}.$$

We will be interested in the case when $f: U \to M$ has an invariant hyperbolic set $\Lambda \subset U$ and $N \subset M$ is a piece of unstable manifold. We then obtain a real-valued function on Λ defined by

$$J^u f^{-1}(f(x)) := J'f^{-1}(f(x))$$

for $x \in \Lambda$, called the *unstable Jacobian.* As noted already, this function depends smoothly on $E_{f(x)}$. Theorem 19.1.6 thus has the following consequence:

Corollary 19.1.13. *Let M be a Riemannian manifold, $U \subset M$ open, $f: U \to M$ a smooth embedding, and $\Lambda \subset U$ a compact hyperbolic set for f. Then the unstable Jacobian $J^u f^{-1}(f(\cdot))$ is Hölder continuous on Λ.*

In Section 20.4 we will use this corollary by invoking Hölder continuity of $\varphi^u := \log J^u f^{-1}(f(\cdot))$ on Λ.

Exercises

19.1.1. *Show that the weak and strong stable and unstable distributions for a compact hyperbolic set of a flow are Hölder continuous.*

19.1.2. *Prove an analog of Corollary 19.1.11 for the weak unstable distribution on a hyperbolic set for a flow.*

19.1.3. *Show that the bunching constant $B(f)$ is at most 2.*

19.1.4. *Show that the geodesic flow of a compact surface with negative Gaussian curvature is ergodic with respect to the Liouville measure.*

19.1.5. *Construct an example of an Anosov diffeomorphism on $\mathbb{T}^2$ that has analytic unstable foliation but is not C^1 conjugate to a linear Anosov diffeomorphism.*

19.1.6. *Construct an example of an area-preserving Anosov diffeomorphism on $\mathbb{T}^3$ whose stable foliation is 2-dimensional and C^∞ but such that there is no C^1 conjugacy to a linear map.*

19.1.7*. *Suppose f is a sufficiently small analytic perturbation of a linear Anosov diffeomorphism T on $\mathbb{T}^2$ with analytic stable and unstable foliations. Show that f is analytically conjugate to T.* [3]

2. Cohomological equations over hyperbolic dynamical systems

Untwisted cohomological equations (see Section 2.9) have appeared in several contexts: Existence of absolutely continuous measures (Section 5.1), time change for flows and more general relations between orbit equivalence and flow equivalence (Section 2.2), and the classification of S^1 extensions of dynamical systems (Subsection 4.2b). In this section we prove a general result for hyperbolic sets which describes a complete set of invariants of Hölder and C^1 cocycles. Then we show how this yields results concerning these issues in the context of hyperbolic sets. Other applications will appear in Sections 20.3 and 20.4.

a. The Livschitz Theorem. We have seen in various places that in hyperbolic sets there is an abundance of periodic orbits. The Anosov Closing Lemma (Theorem 6.4.15) and the Specification Theorem 18.3.9 give strong density assertions while Theorems 18.5.1 and 18.5.6 show that the growth rate of periodic orbits reflects the full dynamical complexity of a hyperbolic set. In this subsection we show that solutions of cohomological equations with Hölder data over a hyperbolic set are completely determined by the data on periodic orbits. This adds a new method for finding solutions of cohomological equations and establishing their regularity to the two methods for solving untwisted cohomological equations discussed in Section 2.9 (and the pathological construction of wild coboundaries in Section 12.6). The method consists of the simpleminded approach of defining $\Phi(f^n(x)) = \sum_{i=0}^{n-1} \varphi(f^i(x))$ along a dense orbit and showing that if the periodic obstructions vanish then the solution can be extended off the dense orbit by virtue of the Anosov Closing Lemma (in fact, Corollary 6.4.17).

Theorem 19.2.1. (Livschitz Theorem) *Let M be a Riemannian manifold, $U \subset M$ open, $f: U \to M$ a smooth embedding, $\Lambda \subset U$ a compact topologically transitive hyperbolic set, and $\varphi: \Lambda \to \mathbb{R}$ Hölder continuous. Suppose that for every $x \in \Lambda$ such that $f^n(x) = x$ we have $\sum_{i=0}^{n-1} \varphi(f^i(x)) = 0$. Then there exists a continuous $\Phi: \Lambda \to \mathbb{R}$ such that $\varphi = \Phi \circ f - \Phi$. Moreover Φ is unique up to an additive constant and Hölder with the same exponent as φ.*

Proof. Since $f_{\restriction \Lambda}$ is topologically transitive there exists a point $x_0 \in \Lambda$ such that the orbit $\mathcal{O}(x_0) = \{f^n(x_0)\}_{n\in\mathbb{Z}}$ is dense in Λ. Once we choose a value $\Phi(x_0) \in \mathbb{R}$ we must have $\Phi(f^n(x_0)) = \Phi(x_0) + \varphi(n, x_0)$, where $\varphi(n, x)$ is as in (2.9.4).

Lemma 19.2.2. *The function Φ thus defined on $\mathcal{O}(x_0)$ is Hölder continuous with the same exponent as φ.*

Proof. Suppose $n, m \in \mathbb{N}$ are such that $\epsilon := d(f^n(x_0), f^m(x_0))$ is small enough to apply the Anosov Closing Lemma (Theorem 6.4.15) together with the amplifying Proposition 6.4.16. Then we obtain $C > 0$, $\mu \in (0, 1)$, and $y = f^{m-n}(y)$ such that $d(f^{n+i}(x_0), f^i(y)) \le C\epsilon\mu^{\min(i,m-n-i)}$. Since φ is Hölder continuous with exponent $\alpha \in (0, 1]$ there exists $M > 0$ such that whenever $d(x_1, x_2)$ is small enough we have $|\varphi(x_1) - \varphi(x_2)| \le Md(x_1, x_2)^\alpha$. Consequently

$$\begin{aligned}
|\Phi(f^n(x_0)) - \Phi(f^m(x_0))| &= \Big| \sum_{i=0}^{m-n-1} \varphi(f^{n+i}(x_0)) \Big| \\
&= \Big| \sum_{i=0}^{m-n-1} (\varphi(f^{n+i}(x_0)) - \varphi(f^i(y))) + \sum_{i=0}^{m-n-1} \varphi(f^i(y)) \Big| \\
&= \Big| \sum_{i=0}^{m-n-1} (\varphi(f^{n+i}(x_0)) - \varphi(f^i(y))) \Big| \\
&\le \sum_{i=0}^{m-n-1} |\varphi(f^{n+i}(x_0)) - \varphi(f^i(y))| \\
&\le \sum_{i=0}^{m-n-1} MC^\alpha \epsilon^\alpha \mu^{\alpha \min(i,m-n-i)} \\
&\le 2MC^\alpha \epsilon^\alpha \sum_{i=0}^{m-n-1} \mu^{\alpha i} < 2MC^\alpha \epsilon^\alpha \frac{1}{1-\mu^\alpha} \\
&= \frac{2MC^\alpha}{1-\mu^\alpha} d(f^n(x_0), f^m(x_0))^\alpha.
\end{aligned}$$

□

Thus Φ is in particular uniformly continuous on $\mathcal{O}(x_0)$ and hence extends uniquely to a continuous function on Λ, which we also denote by Φ. The uniqueness assertion follows since the choice of $\Phi(x_0)$ determines Φ uniquely. (Alternatively note that if $\Phi \circ f - \Phi = \Psi \circ f - \Psi$ then $\Phi - \Psi$ is a continuous f-invariant function, hence constant by topological transitivity.) Clearly the extension has the same Hölder exponent. Finally note that φ and $\Phi \circ f - \Phi$ are continuous functions on Λ that coincide on a dense set. Therefore they coincide and Φ solves the cohomological equation. □

Remark. It is useful to note that the same argument works verbatim if φ takes values in an abelian group with a translation-invariant analog of absolute value, for example, to tori with the standard distance function or any compact abelian group for that matter. We will, in fact, apply the Livschitz Theorem in a case where φ is a map to a torus.

Using the Spectral Decomposition Theorem 18.3.1 we obtain a version of the Livschitz Theorem that does not use transitivity:

Corollary 19.2.3. *Let Λ be a compact locally maximal hyperbolic set for $f: U \to M$ and $\Lambda' := NW(f_{\restriction \Lambda})$. Suppose $\varphi: \Lambda' \to \mathbb{R}$ is Hölder continuous and that for every $x \in \Lambda'$ such that $f^n(x) = x$ we have $\sum_{i=0}^{n-1} \varphi(f^i(x)) = 0$. Then there exists a continuous $\Phi: \Lambda \to \mathbb{R}$ such that $\varphi = \Phi \circ f - \Phi$ and Φ is Hölder continuous with the same exponent as φ. Moreover Φ is unique up to addition of a function that is constant on each topologically transitive component of Λ'.*

With the obvious modifications for the case of flows the same proof yields the analogous result for continuous time:

Theorem 19.2.4. *Let M be a Riemannian manifold, φ^t a smooth flow, $\Lambda \subset M$ a compact topologically transitive hyperbolic set, and $g: \Lambda \to \mathbb{R}$ Hölder continuous such that for every periodic $x = \varphi^T(x)$ we have $\int_0^T g(\varphi^t(x))dt = 0$. Then there is a continuous $G: \Lambda \to \mathbb{R}$ such that $g = \dot{\varphi}G$, that is, the derivative of G in the flow direction exists and equals g. Furthermore G is unique up to an additive constant and Hölder continuous with the same exponent as g.*

There is a C^1 version of the Livschitz Theorem 19.2.1 as well:[1]

Theorem 19.2.5. *Let M be a Riemannian manifold, $f: U \to M$ a smooth embedding with a compact topologically transitive hyperbolic set, and $\varphi: \Lambda \to \mathbb{R}$ a C^1 function. Suppose that for every $x \in M$ such that $f^n(x) = x$ we have $\sum_{i=0}^{n-1} \varphi(f^i(x)) = 0$. Then there exists a C^1 function $\Phi: \Lambda \to \mathbb{R}$ such that $\varphi = \Phi \circ f - \Phi$ and Φ is unique up to an additive constant.*

Proof. By Theorem 19.2.1 we have a Lipschitz-continuous solution Φ and it remains to show that it is C^1. To that end we show that the derivatives of Φ along stable and unstable leaves exist and are continuous. If x and y are nearby points of a stable leaf then $\Phi(y) - \Phi(x) = \lim_{n\to\infty} \big(-\sum_{i=0}^{n}(\varphi(f^i(y)) - \varphi(f^i(x))) + \Phi(f^n(x)) - \Phi(f^n(y))\big) = -\sum_{i=0}^{\infty}(\varphi(f^i(y)) - \varphi(f^i(x)))$. Keeping x fixed and differentiating with respect to $y = x + tv$ at $t = 0$ gives by the chain rule $D_v\Phi(x) = -\sum_{i=0}^{\infty} D_{v_i}\varphi(f^i(x))D_v(f^i)(x)$, where $v_i = Df^i v$. Note that $D_v(f^i)$ is exponentially small since v is a stable vector. The first factor is exponentially small as well since φ is C^1 and the v_i are exponentially small. Thus the series on the right converges uniformly and hence to a well-defined and continuous function which is thus the left-hand side. Likewise one obtains differentiability of Φ in the unstable direction. We have thus shown that Φ has continuous partial derivatives. By Lemma 19.1.10 this implies that Φ is C^1. □

Corollary 19.2.6. *Let M be a Riemannian manifold, $f: U \to M$ a smooth embedding with a compact topologically transitive hyperbolic set, $\varphi: \Lambda \to \mathbb{R}$ a Hölder-continuous (correspondingly, C^1) function, and $\varphi = \Phi \circ f - \Phi$ for a bounded (everywhere defined) function Φ. Then $\varphi = \Phi' \circ f - \Phi'$ for some Hölder-continuous (correspondingly, C^1) Φ'.*

Remark. Using Theorem 2.9.3 one can reformulate the last corollary by requiring the cocycle determined by φ to be bounded.

The Gottschalk–Hedlund Theorem 2.9.4 has a very similar conclusion in the continuous category for minimal dynamical systems. The methods are, of course, very different, since hyperbolic systems are very far from minimal.

b. Smooth invariant measures for Anosov diffeomorphisms. We now show that the necessary condition for existence of a smooth invariant measure found in Proposition 5.1.6 is indeed sufficient for topologically transitive Anosov diffeomorphisms. Recall that $Jf(\cdot)$ denotes the Jacobian of a differentiable map with respect to an ambient volume form.

Theorem 19.2.7. *Let M be a Riemannian manifold with volume Ω and $f: M \to M$ a C^2 topologically transitive Anosov diffeomorphism. Then the following three conditions are equivalent:*

(1) *There is an f-invariant measure with bounded density that is bounded away from 0.*
(2) *$Jf^n(x) = 1$ for every $x \in \text{Fix}(f^n)$.*
(3) *There is an f-invariant measure with positive C^1 density.*

Proof. The Jacobian Jf is C^1 by assumption. If $\varphi(x) := -\log Jf(x)$ then $e^{\Phi}\Omega$ is f-invariant if and only if

$$\begin{aligned}0 = (f^*e^{\Phi}\Omega)_x - (e^{\Phi}\Omega)_x &= e^{\Phi(f^{-1}(x))}\Omega_{f^{-1}(x)}(Df^{-1}(\cdot), \dots, Df^{-1}(\cdot)) - e^{\Phi(x)}\Omega_x \\ &= e^{\Phi(f^{-1}(x))}(Jf(x))^{-1}\Omega_x - e^{\Phi(x)}\Omega_x = (e^{\Phi(f^{-1}(x))}e^{\varphi(x)} - e^{\Phi(x)})\Omega,\end{aligned}$$

that is, if and only if Φ solves the cohomological equation

$$\Phi(f^{-1}(x)) - \Phi(x) = -\varphi(x).$$

Evidently then (1) implies (2). But assuming (2) Theorem 19.2.5 shows that there is a C^1 solution and hence (3) follows. Clearly (3) implies (1). □

Remark. Note that by Corollary 18.3.5 for Anosov diffeomorphisms $f: M \to M$ topological transitivity is equivalent to $NW(f) = M$.

c. Time change and orbit equivalence for hyperbolic flows. We noted early on that the periods of periodic orbits are moduli of flow equivalence and hence obstructions to trivializing a time change (2.2.5) as well as to orbit equivalence being flow equivalence. We now see that for hyperbolic sets these are a complete set of moduli.

Proposition 19.2.8. *Let φ^t be a flow on a manifold M with a compact topologically transitive hyperbolic set Λ and ψ^t a time change of φ^t. If the periods of all periodic orbits of φ^t and ψ^t agree then φ^t and ψ^t are flow equivalent via a homeomorphism which is Hölder continuous if the time change is and C^1 if the time change is C^1.*

Proof. Recall from Section 2.2 that a time change $\psi^t(x) = \varphi^{\alpha(t,x)}(x)$ arises from a flow equivalence, that is, is trivial, if there is a function $\beta: \Lambda \to \mathbb{R}$, differentiable along orbits, such that $\alpha(t,x) - t = \beta(x) - \beta(\varphi^t(x))$. Notice, however, that by assumption the values of the cocycle on the left-hand side over periodic orbits are zero, so by the Livschitz Theorem 19.2.4 for flows there is a solution β which is Hölder if α is and C^1 when α is C^1 (by the counterpart of Theorem 19.2.5 for flows; see Exercise 19.2.4). □

Theorem 19.2.9. *Suppose $\varphi^t: M \to M$ and $\psi^t: M' \to M'$ are flows that are orbit equivalent on hyperbolic sets Λ, Λ', respectively, and that the periods of corresponding periodic orbits in Λ and Λ' agree. Then φ^t and ψ^t are flow equivalent.*

Proof. By Theorem 19.1.5 the orbit equivalence h can be taken to be Hölder continuous. Thus $h \circ \varphi^t \circ h^{-1}$ is a Hölder-continuous time change of ψ^t with the same periods as ψ^t; hence by Proposition 19.2.8 it is Hölder flow equivalent to ψ^t and hence so is φ. □

d. Equivalence of torus extensions. We now apply the Livschitz Theorem to a situation related to the circle extensions introduced in Section 4.2b. It gives a sufficient criterion for C^1 conjugacy of extensions over a hyperbolic set.

Proposition 19.2.10. *Suppose M is a compact manifold and $f: U \to M$ a diffeomorphism with a compact topologically transitive hyperbolic set Λ. For $\varphi \in C^1(\Lambda, \mathbb{T}^k)$ define $F_\varphi(x,t) := (f(x), t + \varphi(x))$ on $\Lambda \times \mathbb{T}^k$. If $\psi \in C^1(\Lambda, \mathbb{T}^k)$ and $\sum_{i=0}^{l} \varphi(f^i(x)) = \sum_{i=0}^{l} \psi(f^i(x))$ for every $x \in \operatorname{Fix}(f^l) \cap \Lambda$ then F_φ and F_ψ are C^1 conjugate.*

Proof. By the C^1 Livschitz Theorem 19.2.5 (for values in $\mathbb{T}^k$) there is a C^1 function $\Phi: \Lambda \to \mathbb{T}^k$ such that $\varphi - \psi = \Phi \circ f - \Phi$. Then one can easily check that $h(x,t) := (x, t - \Phi(x))$ is the desired conjugacy. □

Remark. The preceding result does not depend on the product structure of the space. In fact the same arguments would yield an analogous result for extensions to a torus bundle over Λ.

Exercises

19.2.1. *Let φ^u be as at the end of Section 19.1 and $\varphi^s := \log Jf\restriction_{E^s}$. Show that φ^u and φ^s are cohomologous if and only if there is a smooth invariant measure for f.*

19.2.2*. *Suppose A is a hyperbolic automorphism of $\mathbb{T}^2$ and φ a C^∞ cocycle as in Theorem 19.2.1. Show that $\varphi = \Phi \circ f - \Phi$ for some $\Phi \in C^\infty$.*

19.2.3*. *Show that the conclusion of the previous exercise holds for hyperbolic diffeomorphisms of $\mathbb{T}^2$ with C^∞ stable and unstable foliations.*

19.2.4. *Formulate and prove the counterpart of Theorem 19.2.5 for flows.*

20

Equilibrium states and smooth invariant measures

In this chapter we describe some of the principal results in the ergodic theory of hyperbolic dynamical systems. There are two distinguished kinds of invariant measures for a smooth dynamical system: Measures with maximal entropy and smooth measures. We will see that for hyperbolic systems these measures are particular cases of equilibrium states, which are counterparts of Gibbs measures in statistical mechanics. The first four sections are dedicated to the study of equilibrium states, with particular emphasis on those distinguished measures, and complete the line of development from Chapters 18 and 19. The two key tools are specification and the Livschitz Theorem.

The final two sections have a somewhat distinct character. Our aim there is to obtain a multiplicative asymptotic of the growth of closed orbits for topologically mixing Anosov flows, and hence a similar asymptotic for the number of closed geodesics on a compact Riemannian manifold of negative sectional curvature, following the original approach of Margulis. It is based on an alternative description of the measure of maximal entropy which we give in Section 5. In Section 6 we derive the multiplicative asymptotic following the doctoral thesis of Toll. Neither the original proof by Margulis nor that of Toll were ever published.

1. Bowen measure

The Variational Principle (Theorem 4.5.3) asserts that topological entropy is the supremum of metric entropies. We observed also that for expansive maps the supremum is indeed attained (Theorem 4.5.4). Thus it is natural to try to investigate those preferred measures whose entropy is maximal. For the linear expanding maps E_m of the circle and the full N-shift the measures of maximal entropy were apparent and for hyperbolic automorphisms of the torus we found that Lebesgue measure has maximal entropy (4.4.7). We saw in Proposition

4.4.2 that a special Markov measure μ_Π, the Parry measure, has maximal entropy for any topological Markov chain. Furthermore, Exercise 4.4.2 asserts that this measure appears as the limit distribution of periodic orbits. The same is obvious for Lebesgue measure in the case of the linear expanding map. Now we will see that in the presence of the specification property (Definition 18.3.8) in addition to expansivity there is exactly one measure with maximal entropy. We will exhibit it as a limit distribution of periodic orbits and show uniqueness. Due to Corollary 6.4.10 and the Specification Theorem 18.3.9 this analysis, which by itself lies in the realm of topological dynamics, is applicable to the restriction of a smooth dynamical system to a topologically mixing locally maximal hyperbolic set. The construction we give in this section is a special case of the construction of equilibrium states in Section 20.3. It is worthwhile to consider this special case because we have several technical facts available already and because the notation is significantly simpler, so the idea is easier to discern.

More specifically, our strategy is as follows: If δ_x denotes the probability measure with support $\{x\}$ then consider the f-invariant measures $\mu_n := \sum_{x \in \mathrm{Fix}(f^n)} \delta_x / P_n(f)$. By weak*-compactness of $\mathfrak{M}(f)$ this sequence has at least one weak*-accumulation point μ. We will show that if $\nu \in \mathfrak{M}(f)$ is such that $h_\nu(f) = h_{\mathrm{top}}(f)$ then $\nu = \mu$, so in particular $h_\mu(f) = h_{\mathrm{top}}(f)$ and there is, in fact, only one accumulation point, that is, $\mu_n \to \mu$.

To fix notations let us pick a subsequence n_k such that

$$\mu = \lim_{k\to\infty} \mu_{n_k}. \tag{20.1.1}$$

Let us begin our study of μ with two lemmas concerning its dynamics.

Lemma 20.1.1. *Let (X,d) be a compact metric space, $f: X \to X$ an expansive homeomorphism with the specification property, μ as in (20.1.1), $\epsilon > 0$, M_ϵ as in Definition 18.3.8, $K_{3\epsilon}$ as in Proposition 18.5.4, c_2 as in Theorem 18.5.5, and $A_\epsilon := (1/c_2 K_{3\epsilon}) e^{-2M_\epsilon h_{\mathrm{top}}(f)} > 0$. For $y \in X$ and $n \in \mathbb{N}$ let $B := B_f(y,\epsilon,n)$ be the ϵ-ball in the d_n^f metric centered at y. Then we have*

$$\mu(\bar{B}) \geq A_\epsilon e^{-nh_{\mathrm{top}}(f)}.$$

Remark. In fact, there is an analogous upper estimate, which we prove in the more general case of equilibrium states. For the purposes of this section a one-sided estimate is, however, sufficient.

Proof. Take $r \geq n + 2M_\epsilon$ and define m by $r = n + m + 2M_\epsilon$. Suppose E_m is a maximal $(m, 3\epsilon)$-separated set, $x \in E_m$. By the specification property there is a $z(x) \in \mathrm{Fix}(f^r) \cap B$ with $d_m^f(f^{n+M_\epsilon}(z(x)), x) < \epsilon$. $z(\cdot)$ is injective on E_m by choice of E_m, and E_m is maximal, so Proposition 18.5.4 and Theorem 18.5.5 yield

$$\mu_r(B) = \frac{\mathrm{card}(\mathrm{Fix}(f^r) \cap B)}{P_r(f)} \geq \frac{\mathrm{card}(E_m)}{P_r(f)} \geq \frac{\mathrm{card}(E_m)}{c_2 e^{rh_{\mathrm{top}}(f)}}$$

$$\geq \frac{N_d(f, 3\epsilon, m)}{c_2 e^{rh_{\mathrm{top}}(f)}} \geq \frac{e^{mh_{\mathrm{top}}(f)}}{c_2 K_{3\epsilon} e^{rh_{\mathrm{top}}(f)}} = \frac{e^{-nh_{\mathrm{top}}(f)}}{c_2 K_{3\epsilon} e^{2M_\epsilon h_{\mathrm{top}}(f)}}.$$

Since $\bar{B}$ is closed,

$$\mu(\bar{B}) = \inf\left\{ \int \varphi \, d\mu \,\middle|\, 0 \le \varphi \in C^0(X), \bar{B} \subset \varphi^{-1}(\{1\}) \right\}.$$

But for any such φ we have $\int \varphi \, d\mu = \lim_{k\to\infty} \int \varphi \, d\mu_{n_k} \ge \overline{\lim}_{k\to\infty} \mu_{n_k}(\bar{B}) \ge A_\epsilon e^{-nh_{\text{top}}(f)}$, since $\mu_r(B) \ge A_\epsilon e^{-nh_{\text{top}}(f)}$. □

Lemma 20.1.2. *μ (as defined in (20.1.1)) is ergodic.*

Proof. We will show that there is a $c > 0$ such that

$$\underline{\lim}_{n\to\infty} \mu(P \cap f^{-n}(Q)) \ge c\mu(P)\mu(Q) \tag{20.1.2}$$

for all measurable sets $P, Q \subset X$. This implies ergodicity since if $f(P) = P = X \smallsetminus Q$ and $\mu(P)\mu(Q) > 0$ the inequality fails. In fact, given $\delta > 0$, $A, B \subset X$ compact with boundary of measure zero, and U, V δ-neighborhoods of A, B, respectively, we will show that

$$\underline{\lim}_{n\to\infty} \mu(\bar{U} \cap f^{-n}(\bar{V})) \ge c\mu(A)\mu(B).$$

For compact sets with boundary of measure zero this implies

$$\underline{\lim}_{r\to\infty} \mu(A \cap f^{-r}(B)) \ge c\mu(A)\mu(B).$$

But by Lemma 4.5.1 compact sets with boundary of measure zero form a sufficient collection in the sense of Definition 4.2.9 and similarly to the proof of Proposition 4.2.10(1) it follows that if (20.1.2) holds for any sufficient collection, then (20.1.2) holds in general.

Fix $\epsilon \in (0, \delta_0/2)$, where δ_0 is the expansivity constant (Definition 3.2.11), and let $N(\delta) \in \mathbb{N}$ be such that $d(x,y) \le \delta$ whenever $d(f^i(x), f^i(y)) \le \epsilon$ for all $|i| < N(\delta)$ (existence of such N follows from expansivity). Let $n \ge 2N(\delta)$ and for $s, t \in \mathbb{N}$ take a maximal $(s, 3\epsilon)$-separated set E_s and a maximal $(t, 3\epsilon)$-separated set E_t. Corresponding to A, E_s, B, E_t define intervals $I_i = [a_i, b_i] \subset \mathbb{Z}$ for $i = 1, 2, 3, 4$ by

$$\begin{array}{llll} a_1 = -\left[\dfrac{n}{2}\right], & a_2 = b_1 + M_\epsilon, & a_3 = b_2 + M_\epsilon, & a_4 = b_3 + M_\epsilon, \\ b_1 = a_1 + n, & b_2 = a_2 + s, & b_3 = a_3 + n, & b_4 = a_4 + t \end{array}$$

and let $W = f^{-[n/2]}(\text{Fix}(f^n) \cap A) \times E_s \times f^{-[n/2]}(\text{Fix}(f^n) \cap B) \times E_t$. Then for each $x = (x_1, x_2, x_3, x_4) \in W$ there is, by specification, a periodic point $z = z(x) \in \text{Fix}(f^{b_4 - a_1 + M_\epsilon})$ with $d^f_{b_i - a_i}(f^{a_i}(z), x_i) < \epsilon$ for $i = 1, 2, 3, 4$. $z(\cdot)$ is injective (any $\{x_i, x_i'\}$ is $(b_i - a_i, 3\epsilon)$-separated) and $z(\cdot) \in U \cap f^{-s-2M_\epsilon}V$ by choice of $N(\delta)$. So if $m := b_4 - a_1 + M_\epsilon = t + s + 2n + 4M_\epsilon$ then

$$\begin{aligned}\operatorname{card}(\operatorname{Fix}(f^m) \cap U \cap f^{-s-M_\epsilon}(V)) &\geq \operatorname{card}(W)\\ &= \operatorname{card}(\operatorname{Fix}(f^n) \cap A) \operatorname{card}(\operatorname{Fix}(f^n) \cap B) N_d(f, 3\epsilon, s) N_d(f, 3\epsilon, t)\\ &= \mu_n(A) P_n(f) \mu_n(B) P_n(f) N_d(f, 3\epsilon, s) N_d(f, 3\epsilon, t)\end{aligned}$$

and

$$\begin{aligned}\mu_m(U \cap f^{-s-2M_\epsilon}(V)) &\geq \frac{1}{P_m(f)} \operatorname{card}(W)\\ &\geq \frac{\mu_n(A)\mu_n(B)}{P_m(f)} P_n(f) P_n(f) N_d(f, 3\epsilon, s) N_d(f, 3\epsilon, t)\\ &\geq \mu_n(A)\mu_n(B) \frac{c_1^2}{c_2 K_{3\epsilon}^2} e^{-4M_\epsilon h_{\text{top}}(f)} =: c\mu_n(A)\mu_n(B)\end{aligned}$$

by Proposition 18.5.4 and Theorem 18.5.5.

Keeping n and s fixed we let $t \to \infty$ in such a manner that $m = n_k$. Then

$$\mu(\bar{U} \cap f^{-s-2M_\epsilon}(\bar{V})) \geq \varlimsup_{k\to\infty} \mu_{n_k}(U \cap f^{-s-2M_\epsilon}(V)) \geq c\mu_n(A)\mu_n(B)$$

and

$$\varliminf_{r\to\infty} \mu(\bar{U} \cap f^{-r}(\bar{V})) \geq c\mu_n(A)\mu_n(B)$$

as $s \to \infty$. Finally, when $n \to \infty$ we get

$$\varliminf_{r\to\infty} \mu(\bar{U} \cap f^{-r}(\bar{V})) \geq c \varlimsup_{k\to\infty} \mu_{n_k}(A)\mu_{n_k}(B).$$

Since $\mu(\partial A) = \mu(\partial B) = 0$ we get

$$\varliminf_{r\to\infty} \mu(\bar{U} \cap f^{-r}(\bar{V})) \geq c\mu(A)\mu(B),$$

as advertised. □

Theorem 20.1.3. (Bowen) *Let (X, d) be a compact metric space and $f: X \to X$ an expansive homeomorphism with the specification property. Then there is exactly one $\mu \in \mathfrak{M}(f)$ with $h_\mu(f) = h_{\text{top}}(f)$. It is called the Bowen measure of f and is given by*

$$\mu = \lim_{n\to\infty} \frac{1}{P_n(f)} \sum_{x \in \operatorname{Fix}(f^n)} \delta_x,$$

where δ_x denotes the probability measure with support $\{x\}$.

Remark. Thus the Bowen measure is the unique measure for which periodic points are equidistributed. It is also called Bowen–Margulis measure since it was discovered in a different guise by Margulis (cf. Theorem 20.5.15).

Proof. We show that if $\nu \in \mathfrak{M}(f)$ is such that $h_\nu(f) = h_{\text{top}}(f)$ then $\nu = \mu$, where μ is as in (20.1.1), so $h_\mu(f) = h_{\text{top}}(f)$ and there is only one accumulation point of the μ_n.

If $\nu \in \mathfrak{M}(f)$ then $\nu = \alpha\nu' + (1-\alpha)\mu'$ for some $\alpha \in [0,1]$ and $\nu', \mu' \in \mathfrak{M}(f)$ such that $\mu' \ll \mu \perp \nu'$. Since μ is ergodic, the (f-invariant) density of μ' with respect to μ is constant μ-a.e. and hence $\mu' = \mu$. Since $\mu \perp \nu'$ we have $h_\nu(f) = \alpha h_{\nu'}(f) + (1-\alpha)h_\mu(f)$ by Corollary 4.3.17. Assume now that $h_\nu(f) = h_{\text{top}}(f)$. Since $h_{\nu'}(f) \le h_{\text{top}}(f)$ and $h_\mu(f) \le h_{\text{top}}(f)$ we have only two possibilities: Either $\alpha = 0$, $\nu = \mu$, and $h_\mu(f) = h_{\text{top}}(f)$ so we are done, or $h_{\nu'}(f) = h_{\text{top}}(f)$. We rule out this latter case by showing that if $\nu \perp \mu$ then $h_\nu(f) < h_{\text{top}}(f)$.

For $n \in \mathbb{N}$ and a maximal $(n, 2\epsilon)$-separated set $E_n = \{x_1, \dots, x_k\}$ we can find a partition $\mathfrak{B}_n := \{\beta_x \mid x \in E_n\}$ by Borel sets β_x such that $B_f(x,\epsilon,n) \subset \beta_x \subset B_f(x, 2\epsilon, n)$. Namely, since $X \subset \bigcup_{x \in E_n} B_f(x, 2\epsilon, n)$, we take

$$\beta_{x_1} = B_f(x_1, 2\epsilon, n) \smallsetminus \bigcup_{i=2}^{k} B_f(x_i, \epsilon, n),$$

$$\beta_{x_{j+1}} = B_f(x_{j+1}, 2\epsilon, n) \smallsetminus \bigcup_{i=j+2}^{k} B_f(x_i, \epsilon, n) \smallsetminus \bigcup_{i=1}^{j} \beta_{x_j}.$$

Note that we may pick an arbitrarily small $\epsilon > 0$ such that $(\mu + \nu)(\partial\mathfrak{B}_n) = 0$ (Lemma 4.5.1).

Since f is expansive, $\operatorname{diam} f^{-[n/2]}(\mathfrak{B}_n) \xrightarrow[n\to\infty]{} 0$. This makes the partition $\mathfrak{B}_n$ sufficiently fine to separate the supports of μ and ν, that is, if $f(B) = B \subset X$ such that $\mu(B) = 0$ and $\nu(B) = 1$ then there exist finite unions C_n of elements of $\mathfrak{B}_n$ such that

$$(\mu+\nu)(C_n \bigtriangleup B) = (\mu+\nu)(f^{-[n/2]}(C_n) \bigtriangleup B) \xrightarrow[n\to\infty]{} 0.$$

Furthermore, if $\epsilon < \delta_0/2$, where δ_0 is the expansivity constant, then $\mathfrak{B}_n$ is a generator for f^n, and hence by Proposition 4.3.16(4) and Corollary 4.3.14 $nh_\nu(f) = h_\nu(f^n) = h_\nu(f^n, \mathfrak{B}_n) \le H_\nu(\mathfrak{B}_n)$. To use this note that by convexity of $\varphi(x) := x \log x$ we have $-\frac{1}{m}\sum_{i=1}^m \varphi(a_i) \le -\varphi\Big(\frac{1}{m}\sum_{i=1}^m a_i\Big)$ or

$$-\sum_{i=1}^{m} \varphi(a_i) \le -m\varphi\Big(\frac{1}{m}\sum_{i=1}^{m} a_i\Big) = -\sum_{i=1}^{m} a_i \log\Big(\frac{1}{m}\sum_{i=1}^{m} a_i\Big)$$
$$= \sum_{i=1}^{m} a_i \log m - \varphi\Big(\sum_{i=1}^{m} a_i\Big) \le \sum_{i=1}^{m} a_i \log m + \frac{1}{e}$$

since $-\varphi(x) \le 1/e$. This yields

$$\begin{aligned} nh_\nu(f) &\le -\sum_{\beta_x\in\mathfrak{B}_n} \varphi(\nu(\beta_x)) \\ &\le -\sum_{\beta_x\subset C_n} \varphi(\nu(\beta_x)) - \sum_{\beta_x\cap C_n=\varnothing} \varphi(\nu(\beta_x)) \\ &\le \nu(C_n)\log\operatorname{card}\{x\in E_n \mid \beta_x\subset C_n\} \\ &\quad + \nu(X\smallsetminus C_n)\log\operatorname{card}\{x\in E_n \mid \beta_x\cap C_n=\varnothing\} + \frac{2}{e}. \end{aligned}$$

Obviously $\operatorname{card}\{x\in E_n \mid \beta_x\subset C_n\} \le \mu(C_n)/\min_{x\in E_n}\mu(\beta_x)$ and since every element β_x contains an ϵ-ball in the d_n^f metric, Lemma 20.1.1 implies that $\operatorname{card}\{x\in E_n \mid \beta_x\subset C_n\} \le \mu(C_n)e^{nh_{\text{top}}(f)}A_\epsilon^{-1}$ and hence

$$\begin{aligned} n(h_\nu(f)-h_{\text{top}}(f)) - \frac{2}{e} &\le \nu(C_n)\log(e^{-nh_{\text{top}}(f)}\operatorname{card}\{x\in E_n \mid \beta_x\subset C_n\}) \\ &\quad + \nu(X\smallsetminus C_n) \\ &\qquad \times \log(e^{-nh_{\text{top}}(f)}\operatorname{card}\{x\in E_n \mid \beta_x\cap C_n=\varnothing\}) \\ &\le \nu(C_n)\log(A_\epsilon^{-1}\mu(C_n)) \\ &\quad + \nu(X\smallsetminus C_n)\log(A_\epsilon^{-1}\mu(X\smallsetminus C_n)). \end{aligned}$$

But since $\nu(C_n)\to 1$ and $\mu(C_n)\to 0$ the right-hand side goes to $-\infty$ as $n\to\infty$. Consequently $h_\nu(f) < h_{\text{top}}(f)$. □

Corollary 20.1.4. *If Λ is a locally maximal hyperbolic set for $f\colon U\to M$ and $f\restriction_\Lambda$ is topologically transitive then $f\restriction_\Lambda$ has a unique measure of maximal entropy.*

Proof. By Corollary 18.3.3 Λ is the disjoint union $\Lambda_1\cup\cdots\cup\Lambda_m$ of closed subsets which are cyclically interchanged by f and f^m is topologically mixing on Λ_i. The Specification Theorem 18.3.9 and Corollary 6.4.10 allow us to apply Theorem 20.1.3 to $f^m\restriction_{\Lambda_i}$ for each i. Obviously the unique measures of maximal entropy for f^m are interchanged by f and their average is f-invariant. Conversely, if μ is f-invariant and has maximal entropy then the conditional measure on any topologically mixing component of f^m is the unique measure with maximal entropy. □

The same argument using the conclusion of Exercise 1.9.9 produces the first statement of the next corollary.

Corollary 20.1.5. *A topologically transitive topological Markov chain has a unique measure of maximal entropy. For a transitive (that is, topologically mixing) topological Markov chain the Parry measure μ_Π (see (4.2.13), (4.4.5), and (4.4.6)) is the unique measure of maximal entropy. In particular in this case the measure of maximal entropy is a Markov measure.*

Combining the characterization of the measure of maximal entropy with the results obtained in the development of Markov partitions we obtain an asymptotic exponential estimate for the growth rate of periodic orbits (see (3.1.1)) of a compact locally maximal hyperbolic set based on Corollary 1.9.12 and Proposition 3.2.5 which is much sharper than the estimate given by Theorem 18.5.6.

Theorem 20.1.6. *Let Λ be a compact locally maximal topologically mixing hyperbolic set for f. Then $|P_n(f_{\restriction\Lambda}) - e^{nh_{\text{top}}(f_{\restriction\Lambda})}| < K\lambda^n$ for some $\lambda < e^{h_{\text{top}}(f_{\restriction\Lambda})}$, $K > 0$.*

Proof. By Corollary 1.9.12 and Proposition 3.2.5 this result holds for a transitive topological Markov chain in place of $f_{\restriction\Lambda}$. To obtain it for $f_{\restriction\Lambda}$ we note that since $f_{\restriction\Lambda}$ is topologically mixing the topological Markov chain obtained from a Markov partition is transitive. Thus we want to show that one may disregard the closed f-invariant set $C := \bigcap_{i\in\mathbb{Z}} f^i(\partial^s\mathcal{R} \cup \partial^u\mathcal{R})$ (using the notation of Theorem 18.7.4). To that end note first that Lemma 18.7.6 implies that all f-periodic points in $\Lambda \setminus \Lambda'$ are in C. Furthermore σ_A is expansive on $h^{-1}(C)$, so $\sigma_{A\restriction_{h^{-1}(C)}}$ has a measure μ of maximal entropy by Theorem 4.5.4. This measure is invariant for σ_A, but since it is not positive on open sets (because $h^{-1}(C)$ is closed and not equal to Ω_A), Lemma 20.1.1 shows that it is not the measure of maximal entropy for σ_A, so $h_{\text{top}}(\sigma_{A\restriction_{h^{-1}(C)}}) < h_{\text{top}}(\sigma_A)$. By Proposition 3.2.14 we have $p(\sigma_{A\restriction_{h^{-1}(C)}}) \le h_{\text{top}}(\sigma_{A\restriction_{h^{-1}(C)}}) < h_{\text{top}}(\sigma_A)$ and therefore $|P_n(\sigma_{A\restriction_{h^{-1}(\Lambda\setminus C)}}) - P_n(\sigma_A)| < K\lambda^n$ for some $\lambda < h_{\text{top}}(\sigma_A)$.

By the same argument we have, of course, that the topological entropy of $f_{\restriction C}$ is less than that of f, so by Proposition 3.2.14 $p(f_{\restriction C}) < p(f)$. Since the semiconjugacy is bijective on the set of periodic points in $h^{-1}(\Lambda \setminus C)$ we obtain the theorem. □

Using Markov partitions and uniqueness of the measure of maximal entropy we can also strengthen the conclusion of Lemma 20.1.2 as follows:

Proposition 20.1.7. *If Λ is a compact locally maximal hyperbolic set for $f: U \to M$ and $f_{\restriction\Lambda}$ is topologically mixing then the Bowen measure of $f_{\restriction\Lambda}$ is mixing.*

Remark. In our study of equilibrium states we will obtain the same conclusion via different means, namely, using Lemma 20.3.5 which is a more general and two-sided version of Lemma 20.1.1 and a criterion of mixing from ergodic theory (Proposition 20.3.6).

Proof. The preceding proof shows that the semiconjugacy obtained from a Markov partition preserves the exponential growth rate of periodic points which is equal to topological entropy. Therefore the measure of maximal entropy for the shift induces a measure of maximal entropy on Λ via the semiconjugacy which is hence the Bowen measure. But the measure of maximal entropy for the shift is the Parry measure and hence mixing by Proposition 4.2.15 and Proposition 4.4.2. □

Let us quickly outline the description of the Bowen measure for flows. The details are left to the reader because they are similar to the proof of Theorem 20.1.3.

Let $\operatorname{Per}(t,\epsilon) := \{\mathcal{O}(x) \mid \varphi^s(x) = x \text{ for some } s \in (t-\epsilon, t+\epsilon)\}$, $l(\mathcal{O})$ be the minimal period of $\mathcal{O} \in \operatorname{Per}(t,\epsilon)$, and $L_{t,\epsilon}(\varphi) = \sum_{\mathcal{O}\in\operatorname{Per}(t,\epsilon)} l(\mathcal{O})$. We denote by $\delta_{\mathcal{O}}$ the φ^t-invariant Borel measure corresponding to arc length on $\mathcal{O} \in \operatorname{Per}(t,\epsilon)$. As in the proof of Theorem 20.1.3 the measures

$$\mu_{t,\epsilon} := \frac{1}{L_{t,\epsilon}} \sum_{\mathcal{O}\in\operatorname{Per}(t,\epsilon)} \delta_{\mathcal{O}}$$

converge (in the weak* topology) to a φ^t-invariant Borel probability measure μ_B, called the *Bowen measure*. If the flow is *topologically mixing* then one obtains similar (two-sided) estimates as in Lemma 20.1.1 and the proof of mixing for equilibrium states (Proposition 20.3.6) goes through almost verbatim. Thus the Bowen measure for topologically mixing flows is mixing.

Exercises

20.1.1. *Prove that the topological entropy of an expansive homeomorphism with the specification property is positive.*

20.1.2. *A symbolic dynamical system (Definition 1.9.2) is called sofic if it is a factor of a topological Markov chain. Prove that every topologically transitive sofic system has a unique measure of maximal entropy.*

20.1.3. *Let $B_k \subset \Omega_2$ be the set defined in Exercise 1.9.10 and $S_k = \sigma_2\restriction_{B_k}$. Prove that S_k has a unique measure of maximal entropy.*

20.1.4. *Formulate and prove counterparts of Theorem 20.1.3 and Corollary 20.1.5 for flows.*

20.1.5. *Find a counterpart of Theorem 20.1.6 assuming only that $f\restriction_\Lambda$ is topologically transitive.*

20.1.6. *Show that if $L \in SL(n,\mathbb{Z})$ is a hyperbolic matrix then the set* $\operatorname{Fix}(F_L)$ *of fixed points of the toral automorphism F_L lifts to a lattice in $\mathbb{R}^n$.*

20.1.7. *Prove that Lebesgue measure λ is the unique measure of maximal entropy for a hyperbolic toral automorphism F_L.*

20.1.8. *Prove that $h_\lambda(F_L) = \sum_{|\lambda_i|>1} \log \lambda_i$ for a hyperbolic toral automorphism, where λ_i are the eigenvalues of L.*

2. Pressure and the variational principle

It is fruitful, as we shall see later, to generalize the notion of entropy in such a way as to have different invariants associated to a dynamical system. One application we will be able to explore is a description of smooth invariant measures for a smooth dynamical system.

The way to extend the notion of entropy is to recall that it is calculated by counting the elements of a maximal (n, ϵ)-separated set. If one thinks of counting a set as summing 1 over the elements of the set, it is natural to more generally allow weighted sums over separating or spanning sets. This leads to the notion of pressure, again a term motivated by statistical mechanics.[1]

Definition 20.2.1. Let X be a compact metric space and $f: X \to X$ a homeomorphism. For $\varphi \in C^0(X)$, $x \in X$, and $n \in \mathbb{N}_0$ define $S_n\varphi(x) := \sum_{i=0}^{n-1} \varphi(f^i(x))$. For $\epsilon > 0$, $n \in \mathbb{N}$ let

$$N_d(f, \varphi, \epsilon, n) := \sup \Big\{ \sum_{x \in E} e^{S_n\varphi(x)} \mid E \subset X \text{ is } (n, \epsilon)\text{-separated} \Big\},$$

$$S_d(f, \varphi, \epsilon, n) := \inf \Big\{ \sum_{x \in E} e^{S_n\varphi(x)} \mid X = \bigcup_{x \in E} B_f(x, \epsilon, n) \Big\}.$$

These expressions are sometimes called *statistical sums.* Then

$$P(\varphi) := P(f, \varphi) := \lim_{\epsilon \to 0} \overline{\lim_{n \to \infty}} \frac{1}{n} \log S_d(f, \varphi, \epsilon, n)$$

is called the *topological pressure* of f with respect to φ and if $\mu \subset \mathfrak{M}(f)$ then $P_\mu(\varphi) := P_\mu(f, \varphi) := h_\mu(f) + \int \varphi \, d\mu$ is called the *pressure* of μ.

This definition as well as the subsequent arguments can be carried out for flows as well (Exercise 20.2.2). If one defines $S_t\varphi(x) := \int_0^t \varphi(\psi^t(x))\, dt$ then the definitions of $N_d(\psi^t, \varphi, \epsilon, t)$, $S_d(\psi^t, \varphi, \epsilon, t)$, $P(\varphi)$, and $P_\mu(\varphi)$ go through verbatim after replacing n with t. Note that the pressure of ψ^t thus obtained coincides with the pressure of the time-one map ψ^1. The proof of Proposition 3.1.2 extends immediately to show that pressure is independent of the metric (inducing a given topology) used to define it, thus justifying some of our notation. This immediately implies that pressure is invariant under topological conjugacy, that is, if $f = h^{-1} \circ g \circ h$ and $\psi = \varphi \circ h$ then $P(f, \varphi) = P(g, \psi)$.

When one takes $\varphi = 0$ in Definition 20.2.1 one obtains topological entropy and metric entropy. Note that one always has

$$S_d(f, \varphi, \epsilon, n) \le \|e^{S_n\varphi}\|_{C^0} \cdot S_d(f, \epsilon, n)$$

and thus

$$\overline{\lim_{n \to \infty}} \frac{1}{n} \log S_d(f, \varphi, \epsilon, n) \le \|\varphi\|_{C^0} + \overline{\lim_{n \to \infty}} \frac{1}{n} \log S_d(f, \epsilon, n) < \infty.$$

Analogously to (3.1.13) and (3.1.14) we have

$$N_d(f,\varphi,2\epsilon,n) \le S_d(f,\varphi,\epsilon,n) \le N_d(f,\varphi,\epsilon,n), \qquad (20.2.1)$$

which shows that we can use N_d instead of S_d in the definition of pressure. The quantity

$$D_d(f,\varphi,\epsilon,n){:=}\inf\Big\{\sum_{C\in E}\inf_{x\in C} e^{S_n\varphi(x)}\;\Big|\; X\subset\bigcup_{C\in E} C \text{ and } \operatorname{diam}_{d_n^f}(C)\le\epsilon \text{ for } C\in E\Big\}$$

is related to S_d by inequalities analogous to (3.1.11) and (3.1.12). Similarly to Lemma 3.1.5 D_d is submultiplicative and hence

$$P(\varphi) = \lim_{\epsilon\to 0}\varliminf_{n\to\infty}\frac{1}{n}\log S_d(f,\varphi,\epsilon,n)$$

by an argument similar to that following Lemma 3.1.5.

Our purpose is to generalize Theorem 4.5.3 to a variational principle for pressure. Our proof closely follows that of Theorem 4.5.3. We begin with another consequence of convexity of the function $\phi(x) := x\log x$.

Lemma 20.2.2. *If $\sum_{i=1}^n p_i = 1$, $p_i \ge 0$, $a_i\in\mathbb{R}$, and $A = \sum_{i=1}^k e^{a_i}$ then $\sum_{i=1}^n p_i(a_i - \log p_i) \le \log A$ with equality if $p_i = e^{a_i}/A$.*

Proof. Set $\alpha_i = e^{a_i}/A$, $x_i = p_i/\alpha_i$. Then $0 = 1\cdot\log(1) \le \sum_{i=1}^k \alpha_i x_i\log(x_i) = \sum_{i=1}^k p_i(\log(p_i) + \log A - a_i)$ with equality if and only if $p_i = \alpha_i$. □

Next we prove a counterpart of Lemma 4.5.2.

Lemma 20.2.3. *Let (X,d) be a compact metric space, $f\colon X\to X$ a homeomorphism, and $\varphi\in C^0(X)$. For $n\in\mathbb{N}$ select an (n,ϵ)-separated set $E_n\subset X$. Let $\nu_n := \big(\sum_{x\in E_n} e^{S_n\varphi(x)}\big)^{-1}\sum_{x\in E_n} e^{S_n\varphi(x)}\delta_x$ and $\mu_n := \frac{1}{n}\sum_{i=0}^{n-1} f_*^i\nu_n$. Then there exists an accumulation point μ of $\{\mu_n\}_{n\in\mathbb{N}}$ (in the weak* topology) that is f-invariant and satisfies*

$$\varlimsup_{n\to\infty}\frac{1}{n}\log\sum_{x\in E_n} e^{S_n\varphi(x)} \le P_\mu(f,\varphi).$$

Proof. Consider a subsequence n_k such that $\lim_{k\to\infty}\log\sum_{x\in E_{n_k}} e^{S_{n_k}\varphi(x)} = \varlimsup_{n\to\infty}\log\sum_{x\in E_n} e^{S_n\varphi(x)}$. Take any accumulation point μ of the sequence μ_{n_k}. Existence of an accumulation point follows from weak*-compactness of $\mathfrak{M}$, and f-invariance of μ is clear. Consider thus a partition ξ with elements of diameter less than ϵ and $\mu(\partial\xi) = 0$. If we write $E_n = \{x_1,\dots,x_m\}$ and $a_i := S_n\varphi(x_i)$ then Lemma 20.2.2 yields

$$\begin{aligned} H_{\nu_n}(\xi_{-n}^f) + n\int\varphi\,d\mu_n &= H_{\nu_n}(\xi_{-n}^f) + \int S_n\varphi\,d\nu_n \\ &= \sum_{x\in E_n}[-\nu_n(\{x\})\log(\nu_n(\{x\})) + \nu_n(\{x\})S_n\varphi(x)] = \log\sum_{x\in E_n} e^{S_n\varphi(x)}. \end{aligned} \qquad (20.2.2)$$

Now we proceed as in the proof of Lemma 4.5.2. Suppose $0 < q < n$ and let $a(k) := [(n-k)/q]$ be the integer part of $(n-k)/q$ whenever $0 \le k < q$. Using (20.2.2) and the inequality (4.5.1) for $H_{\nu_n}(\xi^f_{-n})$ we obtain

$$\begin{aligned}\frac{q}{n}\log\sum_{x\in E_n} e^{S_n\varphi(X)} &= \frac{q}{n}H_{\nu_n}(\xi^f_{-n}) + q\int\varphi\,d\mu_n\\ &= \frac{1}{n}\sum_{k=0}^{q-1} H_{\nu_n}(\xi^f_{-n}) + q\int\varphi\,d\mu_n\\ &\le \sum_{k=0}^{q-1}\Big[\sum_{r=0}^{a(k)-1}\frac{1}{n}H_{f_*^{rq+k}\nu_n}(\xi^f_{-q}) + \frac{2q}{n}\log\operatorname{card}(\xi)\Big] + q\int\varphi\,d\mu_n\\ &\le H_{\mu_n}(\xi^f_{-q}) + \frac{2q^2}{n}\log\operatorname{card}(\xi) + q\int\varphi\,d\mu_n,\end{aligned}$$

where the last inequality follows from Proposition 4.3.3(6). Therefore

$$\lim_{k\to\infty}\frac{1}{n_k}\log\sum_{x\in E_{n_k}} e^{S_{n_k}\varphi(X)} \le \frac{1}{q}\lim_{k\to\infty} H_{\mu_{n_k}}(\xi^f_{-q}) + \int\varphi\,d\mu_{n_k} = \frac{1}{q}H_\mu(\xi^f_{-q}) + \int\varphi\,d\mu$$

and $\overline{\lim}_{n\to\infty}\dfrac{1}{n}\log\sum_{x\in E_n} e^{S_n\varphi(X)} \le h_\mu(f,\xi) + \int\varphi\,d\mu \le P_\mu(f,\varphi)$. □

Theorem 20.2.4. (Variational Principle for pressure) *Let $f\colon X\to X$ be a homeomorphism of a compact metric space X and $\varphi\in C^0(X)$. Then $P(\varphi) = \sup\{P_\mu(\varphi) \mid \mu\in\mathfrak{M}(f)\}$.*

Proof. Note first that if $\{E_n\}_{n\in\mathbb{N}}$ are (n,ϵ)-separated sets in X such that

$$\sum_{x\in E_n} e^{S_n\varphi(x)} \ge N_d(f,\varphi,\epsilon,n) - \delta$$

then applying Lemma 20.2.3 yields

$$\overline{\lim_{n\to\infty}}\frac{1}{n}\log N_d(f,\varphi,\epsilon,n) \le P_\mu(f,\varphi)$$

for a corresponding accumulation point $\mu\in\mathfrak{M}(f)$ of the measures μ_n of the lemma. Taking the supremum over μ and letting $\epsilon\to 0$ and $\delta\to 0$ proves $P(\varphi)\le\sup\{P_\mu(\varphi)\mid\mu\in\mathfrak{M}(f)\}$.

To prove the reverse inequality suppose $\xi = \{C_1,\dots,C_k\}$ is a measurable partition of X. Then $\mu(C_i) = \sup\{\mu(B) \mid B\subset C_i \text{ closed}\}$. Thus we can choose compact sets $B_i\subset C_i$ such that for $\beta = \{B_0, B_1,\dots,B_k\}$ with $B_0 = X\setminus\bigcup_{i=1}^k B_i$ we have $H(\xi|\beta) < 1$. By Proposition 4.3.10(3) we thus have

$$h_\mu(f,\xi)\le h_\mu(f,\beta) + H_\mu(\xi|\beta)\le h_\mu(f,\beta)+1.$$

Now take $d := \min\{d(B_i, B_j) \mid i, j \in \{1, \dots, k\},\ i \neq j\} > 0$ and $\delta \in (0, d/2)$ such that $|\varphi(x) - \varphi(y)| \le 1$ whenever $d(x, y) < \delta$. For $C \in \beta^f_{-n}$ there is an $x_C \in \bar{C}$ such that $(S_n\varphi)(x_C) = \sup\{(S_n\varphi)(x) \mid x \in C\}$. Next suppose that $E \subset X$ is an (n, δ)-spanning set. Then we can take $y_C \in E$ such that $d^f_n(x_C, y_C) \le \delta$ and hence $S_n\varphi(x_C) \le S_n\varphi(y_C) + n$. Note also that $\delta < d/2$ implies $\operatorname{card}\{C \in \beta^f_{-n} \mid y_C = y\} \le 2^n$ for all $y \in E$. Using Lemma 20.2.2,

$$\begin{aligned} H_\mu(\beta^f_{-n}) + \int S_n\varphi\, d\mu &\le \sum_{C \in \beta^f_{-n}} \mu(C)(-\log \mu(C) + S_n\varphi(x_C)) \\ &\le \log \sum_{C \in \beta^f_{-n}} e^{S_n\varphi(y_C) + n} \le n + \log\Big(2^n \sum_{C \in E} e^{S_n\varphi(x)}\Big) \end{aligned}$$

and

$$\frac{1}{n} H_\mu(\beta^f_{-n}) + \int \varphi\, d\mu = \frac{1}{n} H_\mu(\beta^f_{-n}) + \frac{1}{n} \int S_n\varphi\, d\mu \le 1 + \log 2 + \frac{1}{n} \log \sum_{x \in E} e^{S_n\varphi(x)}.$$

Consequently

$$h_\mu(f, \xi) + \int \varphi\, d\mu \le h_\mu(f, \beta) + 1 + \int \varphi\, d\mu \le 2 + \log 2 + P(f, \varphi)$$

and thus $P_\mu(f, \varphi) \le 2 + \log 2 + P(f, \varphi)$ for any f, φ. Applying this to f^n and $S_n\varphi$ yields $P_\mu(f, \varphi) \le (2 + \log 2)/n + P(f, \varphi)$, and hence the desired inequality by letting $n \to \infty$. □

Definition 20.2.5. We denote by $C^f = C^f(X)$ the set

$$\{\varphi \in C^0(X) \mid \exists K, \epsilon > 0 \text{ such that } d^f_n(x, y) \le \epsilon \Rightarrow |S_n\varphi(x) - S_n\varphi(y)| \le K\}.$$

Let us explain the choice of C^f as the class of admissible functions. The functions in C^f are precisely those for which we have good control of the $S_n\varphi$ uniformly for $n \in \mathbb{N}$. This is exactly the property needed to prove existence and uniqueness of the measure of maximal pressure. In order to show that this class is large enough we show that in the case of a hyperbolic set all Hölder-continuous functions are in C^f:

Proposition 20.2.6. *Let M be a Riemannian manifold, $U \subset M$ open, $f\colon U \to M$ a smooth embedding, and $\Lambda \subset U$ a hyperbolic set. Then every Hölder-continuous function on Λ is in $C^f(\Lambda)$.*

Proof. Let $\operatorname{Var}_m(\varphi, f, \epsilon) := \sup\{|\varphi(x) - \varphi(y)| \mid d(f^j(x), f^j(y)) \le \epsilon \text{ for } |j| \le m\}$. By hyperbolicity there are $\epsilon > 0$ and $\mu < 1$ such that if $x, y \in \Lambda$, $d(f^i(x), f^i(y)) < \epsilon$ for $|i| \le n$ then $d(x, y) < \mu^n$. Thus if $|\varphi(x) - \varphi(y)| \le c d(x, y)^\alpha$ then $\operatorname{Var}_m(\varphi, f, \epsilon) \le c\mu^{\alpha m}$ and $K := \sum_{n \in \mathbb{N}} \operatorname{Var}_n(\varphi, f, \epsilon) < \infty$.

Now if $d_n^f(x,y) < \epsilon$ and $0 \le k \le n$ then $d(f^j(f^k(x)), f^j(f^k(y))) < \epsilon$ for $|j| \le m_k := \min(k, n-k)$ and $|\varphi(f^k(x)) - \varphi(f^k(y))| \le \mathrm{Var}_{m_k}(\varphi, f, \epsilon)$. Thus

$$|S_n\varphi(x) - S_n\varphi(y)| \le 2 \sum_{m=0}^{[n/2]+1} \mathrm{Var}_m(\varphi, f, \epsilon) \le \sum_{M\in\mathbb{N}} \mathrm{Var}_m(\varphi, f, \epsilon) = K,$$

since $m_k = m$ for at most two values of k. Thus $\varphi \in C^f$. □

As we have seen in Section 19.1, the class of Hölder-continuous functions is invariant under topological conjugacies between hyperbolic sets and hence a very natural class of functions to consider. Moreover, we have seen a particular example that will be important in this chapter, the unstable Jacobian. It plays a central role in our study of smooth invariant measures.

But while our motivation and primary application for this theory concern hyperbolic sets and Hölder-continuous functions, it is illuminating to take an axiomatic approach to point out precisely what is used in our development of the theory.

Analogously to Corollary 3.2.13 we have

Proposition 20.2.7. *If f is expansive with expansivity constant δ_0, $\epsilon < \delta_0/2$, and $\varphi \in C^f(X)$ then*

$$P(\varphi) = \lim_{n\to\infty} \frac{1}{n} \log S_d(f, \varphi, \epsilon, n).$$

By (20.2.1) this is an immediate corollary of the following analog of Lemma 3.2.12:

Lemma 20.2.8. *Let $f: X \to X$ be an expansive homeomorphism of a compact metric space with expansivity constant δ_0 (cf. Definition 3.2.11). Then for $\varphi \in C^f(X)$, $\epsilon \in (0, \delta_0/2)$, and $\delta > 0$ there exists $C_{\delta,\epsilon}$ such that*

$$N_d(f, \varphi, \delta, n) \le C_{\delta,\epsilon} N_d(f, \varphi, \epsilon, n).$$

Remark. Obviously we can take $C_{\delta,\epsilon} = 1$ for $\delta > \epsilon$.

Proof. First proceed as in the proof of Lemma 3.2.12. With the notation from there take ϵ, K as in Definition 20.2.5 so that $|S_n\varphi(x) - S_n\varphi(z)| \le K$ for $x \in E_z$ and

$$\sum_{x\in E} e^{S_n\varphi(x)} \le \sum_{z\in F} (\operatorname{card} E_z) e^K e^{S_n\varphi(z)} \le M e^K N_d(f, \varphi, \epsilon, n).$$

Thus we can take $C_{\delta,\epsilon} = Me^K$. □

Definition 20.2.9. Let X be a compact metric space, $f: X \to X$ a homeomorphism, and $\varphi \in C^0(X)$. A measure $\mu \in \mathfrak{M}(f)$ is called an *equilibrium state* for φ if $P_\mu(f, \varphi) = P(f, \varphi)$.

Analogously to Theorem 4.5.4 an immediate consequence of Lemma 20.2.3 and Proposition 20.2.7 is the following result:

Theorem 20.2.10. *Let X be a compact metric space, $f: X \to X$ an expansive homeomorphism, and $\varphi \in C^f(X)$. Then there is an equilibrium state for φ.*

Exercises

20.2.1. *Calculate the pressure for the full N-shift σ_N for the function $\varphi(\omega) = \omega_0$.*

20.2.2. *Show that the pressure of a flow is the pressure of the time-one map of the flow.*

20.2.3. *If $h \circ f = g \circ h$, h is surjective, and $\varphi = \psi \circ h$ then $P(f, \varphi) \geq P(g, \psi)$.*

3. Uniqueness and classification of equilibrium states

a. Uniqueness of equilibrium states. In Section 20.1 we saw that expansive maps with specification have a unique measure of maximal entropy. This generalizes to uniqueness of equilibrium states for expansive maps with specification.

The proof of Theorem 20.1.3 about uniqueness of the measure of maximal entropy used several facts from Section 18.5, however, which we first need to adapt to the context of pressure. Thus we are now going to generalize the relevant lemmas about the growth rates of periodic orbits and separated sets to results about statistical sums $\sum e^{S_n\varphi(x)}$ over separated sets or periodic orbits.

For the remainder of this section we assume that $f: X \to X$ is an expansive homeomorphism with expansivity constant δ_0 and with the specification property and that $\varphi \in C^f(X)$ with ϵ, K as in the definition of $C^f(X)$ (Definition 20.2.5). In analogy to Lemma 18.5.3 we then have

Lemma 20.3.1. *For $\epsilon \in (0, \delta_0/3)$ there exist k_ϵ, $K_\epsilon > 0$ such that*

$$\prod_{j=1}^{m} k_\epsilon N_d(f, \varphi, \epsilon, n_j) \leq N_d\Big(f, \varphi, \epsilon, \sum_{j=1}^{m} n_j\Big) \leq \prod_{j=1}^{m} K_\epsilon N_d(f, \varphi, \epsilon, n_j)$$

for $n_1, \dots, n_m \in \mathbb{N}$.

Proof. (i) If E is $(\sum_{j=1}^{m} n_j, \epsilon)$-separated and F_j a maximal $(n_j, \epsilon/2)$-separated set, then for $x \in E$ there exists $z(x) := (z_1(x), \dots, z_m(x)) \in F_1 \times \cdots \times F_m$ such that $d^f_{n_j}(f^{n_1+\cdots+n_{j-1}}(x), z_j(x)) \leq \epsilon/2$ and $z(\cdot)$ is injective. Since furthermore

$$\Big|S_{\sum_{j=1}^{m} n_j}\varphi(x) - \sum_{j=1}^{m} S_{n_j}\varphi(z_j(x))\Big| \leq \sum_{j=1}^{m} |S_{n_j}\varphi(f^{\sum_{i=1}^{j-1} n_i}(x)) - S_{n_j}\varphi(z_j(x))| \leq mK$$

we have

$$\sum_{x \in E} \exp(S_{\sum_{j=1}^{m} n_j}\varphi(x)) \leq \prod_{j=1}^{m} N_d(f, \varphi, \frac{\epsilon}{2}, n_j)$$

so we can take $K_\epsilon = C_{\epsilon/2,\epsilon} e^K$.

(ii) If E_j is $(n_j, 3\epsilon)$-separated, $a_j = n_1 + \cdots + n_{j-1} + (j-1)M_\epsilon$, with M_ϵ given by the specification property, and $I_j = [a_j, a_j + n_j - 1]$, then specification implies that for $x := (x_1, \dots, x_m) \in E_1 \times \cdots \times E_m$ there is a $z = z(x)$ such that $d^f_{n_j}(f^{a_j}(z), x_j) < \epsilon$. Note that by construction $E := \{z(x) \mid x \in E_1 \times \cdots \times E_n\}$ is $(a_m + n_m, \epsilon)$-separated. Since furthermore

$$S_{a_m+n_m}\varphi(z(x)) \geq -mM_\epsilon\|\varphi\|_{C^0} - mK + \sum_{j=1}^{m} S_{n_j}\varphi(x_j)$$

we have shown that

$$N_d(f, \varphi, \epsilon, a_m + n_m) \geq e^{-m(M_\epsilon\|\varphi\|_{C^0}+K)} \prod_{j=1}^{m} N_d(f, \varphi, 3\epsilon, n_j).$$

Since $a_m + n_m = (m-1)M_\epsilon + \sum_{j=1}^m n_j$, part (i) yields

$$N_d(f, \varphi, \epsilon, a_m + n_m) \leq K_\epsilon^m N_d(f, \varphi, \epsilon, \sum_{j=1}^{m} n_j) N_d(f, \varphi, \epsilon, M_\epsilon)^{m-1}$$

so by Lemma 20.2.8

$$\begin{aligned} N_d\Big(f, \varphi, \epsilon, \sum_{j=1}^{m} n_j\Big) &\geq K_\epsilon^{-m} N_d(f, \varphi, \epsilon, M_\epsilon)^{1-m} e^{-m(M_\epsilon\|\varphi\|_{C^0}+K)} \prod_{j=1}^{m} N_d(f, \varphi, 3\epsilon, n_j) \\ &\geq \prod_{j=1}^{m} \frac{e^{-m(M_\epsilon\|\varphi\|_{C^0}+K)}}{C_{\epsilon,3\epsilon}\, K_\epsilon N_d(f, \varphi, \epsilon, M_\epsilon)} N_d(f, \varphi, \epsilon, n_j) \end{aligned}$$

□

Proposition 20.3.2. *Let X be a compact metric space and $f: X \to X$ an expansive homeomorphism with expansivity constant δ_0 and with the specification property. If $0 < \epsilon < \delta_0/3$, $n \in \mathbb{N}$, and k_ϵ and K_ϵ are as in Lemma 20.3.1, then*

$$\frac{1}{K_\epsilon} e^{nP(\varphi)} \leq N_d(f, \varphi, \epsilon, n) \leq \frac{1}{k_\epsilon} e^{nP(\varphi)}$$

Proof. By Proposition 20.2.7

$$P(\varphi) = \lim_{n\to\infty} \frac{1}{n} \log S_d(f, \varphi, \epsilon, n).$$

for $\epsilon < \delta_0/2$. Lemma 20.3.1 thus implies

$$\frac{\log k_\epsilon N_d(f, \varphi, \epsilon, n)}{n} \leq \lim_{k\to\infty} \frac{\log N_d(f, \varphi, \epsilon, kn)}{kn} = P(\varphi) \leq \frac{\log K_\epsilon N_d(f, \varphi, \epsilon, n)}{n}.$$

□

Finally we consider statistical sums over periodic points in an analog of Theorem 18.5.5. To simplify notation let

$$P_Y(f,\varphi,n) := \sum_{x\in \mathrm{Fix}(f^n)\cap Y} e^{S_n\varphi(x)}$$

for $Y \subset X$.

Proposition 20.3.3. *Let X be a compact metric space, $f: X \to X$ an expansive homeomorphism with the specification property, and δ_0 its expansivity constant. Then there exist $c_1, c_2 > 0$ such that for sufficiently large $n \in \mathbb{N}$*

$$c_1 e^{nP(\varphi)} \le P_X(f,\varphi,n) \le c_2 e^{nP(\varphi)}. \tag{20.3.1}$$

Proof. To show Proposition 3.2.14 we proved that $\mathrm{Fix}(f^n)$ is (n,δ_0)-separated. This, together with Proposition 20.3.2, gives the inequality on the right.

On the other hand for $n \ge M_\epsilon$, E an $(n-M_\epsilon, 3\epsilon)$-separated set, and $x \in E$ there exists $z = z(x) \subset \mathrm{Fix}(f^n)$ with $d^f_{n-M_\epsilon}(x,z) \le \epsilon$. $z(\cdot)$ is injective and $S_n\varphi(z) \ge S_{n-M_\epsilon}\varphi(x) - K - M_\epsilon\|\varphi\|_{C^0}$, so

$$P_X(f,\varphi,n) = \sum_{x\in \mathrm{Fix}(f^n)} e^{S_n\varphi(z)} \ge e^{-K-M_\epsilon\|\varphi\|_{C^0}} N_d(f,\varphi,3\epsilon,n-M_\epsilon)$$

$$\ge \frac{e^{-K-M_\epsilon\|\varphi\|_{C^0}}}{K_{3\epsilon}e^{M_\epsilon P(\varphi)}} e^{nP(\varphi)}.$$

□

In analogy to the Bowen measure we define a measure μ as follows: By compactness of $\mathfrak{M}(f)$ there exists an accumulation point

$$\mu = \lim_{k\to\infty} \mu_{n_k} \in \mathfrak{M}(f) \tag{20.3.2}$$

of the sequence $\mu_n := \dfrac{1}{P_X(f,\varphi,n)} \sum_{x\in\mathrm{Fix}(f^n)} e^{S_n\varphi(x)}\delta_x$; in other words $\mu_n(Y) = P_Y(f,\varphi,n)/P_X(f,\varphi,n)$. From now on we fix $\{n_k\}_{k\in\mathbb{N}}$ and μ.

Now we generalize the key Lemma 20.1.1 to the present situation. In order to obtain mixing we will also need an estimate from below.

Lemma 20.3.4. *Let (X,d) be a compact metric space, $f: X \to X$ an expansive homeomorphism with the specification property, μ as in (20.3.2), and $\epsilon > 0$ as in the Definition 20.2.5 of $C^f(X)$. Then there exist $A_\epsilon, B_\epsilon > 0$ such that for $y \in X$ and $n \in \mathbb{N}$ we have $A_\epsilon e^{S_n\varphi(y)-nP(\varphi)} \le \mu(\overline{B_f(y,\epsilon,n)}) \le B_\epsilon e^{S_n\varphi(y)-nP(\varphi)}$.*

Proof. Take $r \ge n + 2M_\epsilon$ and $m = r - n - 2M_\epsilon$. Suppose E_m is a maximal $(m, 3\epsilon)$-separated set, $x \in E_m$. By the specification property there is a $z(x) \in \mathrm{Fix}(f^r) \cap B_f(y,\epsilon,n)$ with $d^f_m(f^{n+M_\epsilon}(z(x)), x) < \epsilon$. $z(\cdot)$ is injective on E_m by choice of E_m, and E_m is maximal. Note that if K is as in the definition of $C^f(X)$ then

$$|S_r\varphi(z(x)) - S_n\varphi(y) - S_m\varphi(x)| \leq 2M_\epsilon\|\varphi\|_{C^0} + 2K$$

so

$$\begin{aligned}\mu_r(B_f(y,\epsilon,n)) &= \frac{P_{B_f(y,\epsilon,n)}(f,\varphi,r)}{P_X(f,\varphi,r)} = \frac{1}{P_X(f,\varphi,r)} \sum_{z\in \mathrm{Fix}(f^r)\cap B_f(y,\epsilon,n)} e^{S_r\varphi(z)} \\ &\geq \frac{1}{P_X(f,\varphi,r)} e^{S_n\varphi(y)-2M_\epsilon\|\varphi\|_{C^0}-2K} \sum_{x\in E_m} e^{S_m\varphi(x)}.\end{aligned}$$

Since $N_d(f,\varphi,3\epsilon,m) = \sup\{\sum_{x\in E_m} e^{S_m\varphi(x)} \mid E_m \quad (m,3\epsilon)\text{-separated}\}$ we get by Proposition 20.3.3 and Proposition 20.3.2

$$\begin{aligned}\mu_r(B_f(y,\epsilon,n)) &\geq \frac{e^{-2(M_\epsilon\|\varphi\|_{C^0}+K)}}{P_X(f,\varphi,r)} N_d(f,\varphi,3\epsilon,m)e^{S_n\varphi(y)} \\ &> \frac{1}{c_2}e^{-rP(\varphi)}e^{-2(M_\epsilon\|\varphi\|_{C^0}+K)}\frac{1}{K_{3\epsilon}}e^{mP(\varphi)}e^{S_n\varphi(y)} \geq A_\epsilon e^{S_n\varphi(y)-nP(\varphi)},\end{aligned}$$

where $A_\epsilon = e^{-2(M_\epsilon\|\varphi\|_{C^0}+K+M_\epsilon P(\varphi))}/c_2K_{3\epsilon}$. Letting $n_k = r \geq n + 2M_\epsilon$ and $k \to \infty$ gives the lower bound.

To obtain the upper bound note that for $x \in \mathrm{Fix}(f^r)\cap B_f(y,\epsilon,n)$ we have $|S_r\varphi(x) - S_n\varphi(y) - S_{r-n}\varphi(f^n(x))| \leq K$ and that $f^n(\mathrm{Fix}(f^r)\cap B_f(y,\epsilon,n))$ is $(r-n,\delta_0)$ separated if $\epsilon < \delta_0$. Thus

$$\begin{aligned}\mu_r(B_f(y,\epsilon,n)) &= \frac{P_{B_f(y,\epsilon,n)}(f,\varphi,r)}{P_X(f,\varphi,r)} = \frac{1}{P_X(f,\varphi,r)} \sum_{x\in \mathrm{Fix}(f^r)\cap B_f(y,\epsilon,n)} e^{S_r\varphi(x)} \\ &\leq \frac{1}{P_X(f,\varphi,r)} \sum_{x\in \mathrm{Fix}(f^r)\cap B_f(y,\epsilon,n)} e^{S_n\varphi(y)+S_{r-n}\varphi(f^n(x))+K} \\ &\leq \frac{e^K}{c_1}e^{-rP(\varphi)}e^{S_n\varphi(y)}N_d(f,\varphi,\delta_0,r-n) \leq \frac{e^K}{c_1k_{\delta_0}}e^{S_n\varphi(y)-nP(\varphi)}\end{aligned}$$

using Proposition 20.3.3 and Proposition 20.3.2. □

Lemma 20.3.5. *For the measure μ in (20.3.2) there exist constants $c, C > 0$ such that*

$$c\mu(P)\mu(Q) \leq \varliminf_{n\to\infty} \mu(P\cap f^{-n}(Q)) \leq \varlimsup_{n\to\infty} \mu(P\cap f^{-n}(Q)) \leq C\mu(P)\mu(Q) \tag{20.3.3}$$

for all measurable sets $P, Q \subset X$.

Proof. We begin with the lower estimate. As in the proof of Lemma 20.1.2 it is sufficient to consider $\delta > 0$, $A, B \subset X$ compact, and U, V δ-neighborhoods of A, B, respectively, and to show that $\underline{\lim}_{n\to\infty}\mu(\bar{U}\cap f^{-n}(\bar{V}))\geq c\mu(A)\mu(B)$. Let $\epsilon>0$ and $N(\delta)\in\mathbb{N}$ such that $d(x,y)\leq\delta$ whenever $d(f^i(x),f^i(y))\leq\epsilon$ for all $|i|<N(\delta)$. Let $n\geq 2N(\delta)$ and for $s,t\in\mathbb{N}$, $\eta>0$ take an $(s,3\epsilon)$-separated set E_s and a $(t,3\epsilon)$-separated set E_t such that $\sum_{x\in E_s}e^{S_s\varphi(x)}\geq(1-\eta)N_d(f,\varphi,3\epsilon,s)$ and $\sum_{x\in E_t}e^{S_t\varphi(x)}\geq(1-\eta)N_d(f,\varphi,3\epsilon,t)$. We assume ϵ is such that $N_d(f,\varphi,3\epsilon,s)>0$ and $N_d(f,\varphi,3\epsilon,t)>0$. Corresponding to A,E_s,B,E_t define intervals $I_i=[a_i,b_i]\subset\mathbb{Z}$ $(i=1,2,3,4)$ by

$$\begin{aligned} a_1&=-[n/2], & a_2&=b_1+M_\epsilon, & a_3&=b_2+M_\epsilon, & a_4&=b_3+M_\epsilon,\\ b_1&=a_1+n, & b_2&=a_2+s, & b_3&=a_3+n, & b_4&=a_4+t \end{aligned}$$

and let $W=f^{-[n/2]}(\mathrm{Fix}(f^n)\cap A)\times E_s\times f^{-[n/2]}(\mathrm{Fix}(f^n)\cap B)\times E_t$ as well as $m:=b_4-a_1+M_\epsilon=t+s+2n+4M_\epsilon$. Then for each $x=(x_1,x_2,x_3,x_4)\subset W$ there is, by specification, a periodic point $z=z(x)\in\mathrm{Fix}(f^m)$ with $d^f_{b_i-a_i}(f^{a_i}(z),x_i)<\epsilon$ for $i=1,2,3,4$. $z(\cdot)$ is injective (any $\{x_i,x_i'\}$ is $(b_i-a_i,3\epsilon)$-separated) and $z(x)\in U\cap f^{-s-2M_\epsilon}(V)$ for all x by choice of $N(\delta)$. So

$$S_m\varphi(z(x))\geq S_n\varphi(x_1)+S_s\varphi(x_2)+S_n\varphi(x_3)+S_t\varphi(x_4)-4K-4M_\epsilon\|\varphi\|_{C^0}.$$

Now $S_n\varphi(x_i)=S_n\varphi(f^{[n/2]}(x_i))$ for $i=1,3$ by periodicity, so with $c_0:=e^{-4K-4M_\epsilon\|\varphi\|_{C^0}}$ we have

$$\begin{aligned}
\sum_{x\in W}e^{S_m\varphi(z(x))}&\geq c_0\sum_{x_1\in f^{-[n/2]}(\mathrm{Fix}(f^n)\cap A)}e^{S_n\varphi(x_1)}\times\sum_{x_2\in E_s}e^{S_s\varphi(x_2)}\\
&\quad\times\sum_{x_3\in f^{-[n/2]}(\mathrm{Fix}(f^n)\cap B)}e^{S_n\varphi(x_3)}\times\sum_{x_4\in E_t}e^{S_t\varphi(x_4)}\\
&\geq c_0(1-\eta)^2\sum_{x\in\mathrm{Fix}(f^n)\cap A}e^{S_n\varphi(x)}\sum_{x\in\mathrm{Fix}(f^n)\cap B}e^{S_n\varphi(x)}\\
&\quad\times N_d(f,\varphi,3\epsilon,s)N_d(f,\varphi,3\epsilon,t)\\
&=c_0(1-\eta)^2P_A(f,\varphi,n)P_B(f,\varphi,n)N_d(f,\varphi,3\epsilon,s)N_d(f,\varphi,3\epsilon,t)\\
&=c_0(1-\eta)^2P_X(f,\varphi,n)\mu_n(A)P_X(f,\varphi,n)\mu_n(B)\\
&\quad\times N_d(f,\varphi,3\epsilon,s)N_d(f,\varphi,3\epsilon,t)\\
&\geq P_X(f,\varphi,m)\frac{c_0(1-\eta)^2}{c_2e^{mP(\varphi)}}c_1e^{nP(\varphi)}c_1e^{nP(\varphi)}\\
&\quad\times\frac{1}{K_{3\epsilon}}e^{sP(\varphi)}\frac{1}{K_{3\epsilon}}e^{tP(\varphi)}\mu_n(A)\mu_n(B)\\
&=P_X(f,\varphi,m)\frac{c_0(1-\eta)^2c_1^2}{c_2K_{3\epsilon}^2e^{4M_\epsilon P(\varphi)}}\mu_n(A)\mu_n(B),
\end{aligned}$$

where the last inequality follows from Proposition 20.3.3 and Proposition 20.3.2. Since $\sum_{x\in\mathrm{Fix}(f^m)\cap U\cap f^{-s-M_\epsilon}(V)}e^{S_m\varphi(x)}\geq\sum_{x\in W}e^{S_m\varphi(z(x))}$ we get

$$\mu_m(U\cap f^{-s-2M_\epsilon}(V))\geq\frac{1}{P_X(f,\varphi,m)}\sum_{x\in W}e^{S_m\varphi(z(x))}\geq c\mu_n(A)\mu_n(B).$$

Keeping n and s fixed we let $t \to \infty$ in such a manner that $m = n_k$. Then

$$\mu(\bar{U} \cap f^{-s-2M_\epsilon}(\bar{V})) \geq \overline{\lim_{k\to\infty}} \mu_{n_k}(U \cap f^{-s-2M_\epsilon}(V)) \geq c\mu_n(A)\mu_n(B)$$

and

$$\underline{\lim_{r\to\infty}} \mu(\bar{U} \cap f^{-r}(\bar{V})) \geq c\mu_n(A)\mu_n(B)$$

as $s \to \infty$. Finally, when $n \to \infty$ we get

$$\underline{\lim_{r\to\infty}} \mu(\bar{U} \cap f^{-r}(\bar{V})) \geq c \overline{\lim_{k\to\infty}} \mu_{n_k}(A)\mu_{n_k}(B).$$

Since $\mu(\partial A) = \mu(\partial B) = 0$ this proves the lower estimate.

To prove the other inequality we reason very similarly. For $m = s + t + 2n$ and $x \in f^{-[n/2]}(\mathrm{Fix}(f^m) \cap A \cap f^{-r}(B))$ there exists a unique $z(x) = (z_1, z_2, z_3, z_4)$ such that $z_1 \in \mathrm{Fix}(f^{n+M_\epsilon})$, $z_2 \in \mathrm{Fix}(f^{s+M_\epsilon})$, $z_3 \in \mathrm{Fix}(f^{n+M_\epsilon})$, $z_4 \in \mathrm{Fix}(f^{t+M_\epsilon})$ and $d^f_{b_i-a_i}(f^{a_i}(x), z_i) < \epsilon$ for $i = 1,2,3,4$, where

$$\begin{aligned} a_1 &= -[n/2], & a_2 &= b_1, & a_3 &= b_2, & a_4 &= b_3, \\ b_1 &= a_1 + n, & b_2 &= a_2 + s, & b_3 &= a_3 + n, & b_4 &= a_4 + t. \end{aligned}$$

Then $z_1 \in U$ and $z_3 \in V$ and

$$S_m(x) \leq S_{n+M_\epsilon}(z_1) + S_{s+M_\epsilon}(z_2) + S_{n+M_\epsilon}(z_3) + S_{t+M_\epsilon}(z_4) + 4K + 4M_\epsilon\|\varphi\|_{C^0},$$

so

$$\begin{aligned} \sum_{x\in\mathrm{Fix}(f^m)\cap A\cap f^{-r}(B)} e^{S_m(X)} &\leq \frac{1}{c_0} \sum_{z_1\in\mathrm{Fix}(f^{n+M_\epsilon})\cap U} e^{S_{n+M_\epsilon}(z_1)} \sum_{z_2\in\mathrm{Fix}(f^{s+M_\epsilon})} e^{S_{s+M_\epsilon}(z_2)} \\ &\quad \times \sum_{z_3\in\mathrm{Fix}(f^{n+M_\epsilon})\cap V} e^{S_{n+M_\epsilon}(z_3)} \sum_{z_4\in\mathrm{Fix}(f^{t+M_\epsilon})} e^{S_{t+M_\epsilon}(z_1)} \\ &\leq \frac{1}{c_0} P_U(f,\varphi,n+M_\epsilon)P_V(f,\varphi,n+M_\epsilon)N_d(f,\varphi,\epsilon,s)N_d(f,\varphi,\epsilon,t) \\ &= \frac{1}{c_0} P_X(f,\varphi,n+M_\epsilon)\mu_{n+M_\epsilon}(U)P_X(f,\varphi,n+M_\epsilon)\mu_{n+M_\epsilon}(V) \\ &\quad \times N_d(f,\varphi,\epsilon,s)N_d(f,\varphi,\epsilon,t) \\ &\leq CP_X(f,\varphi,m)\mu_{n+M_\epsilon}(U)\mu_{n+M_\epsilon}(V). \end{aligned}$$

Continuing as before we obtain the upper bound. □

Proposition 20.3.6. *Let f be a homeomorphism of a compact metric space X and μ an f-invariant Borel probability measure such that for any Borel sets P, Q the inequalities (20.3.3) hold. Then f is mixing.*

Remarks. (1) In particular we have a new proof of Proposition 20.1.7.

(2) In fact the statement is true for measure-preserving transformations of a measure space. In the proof one has to replace the use of the weak* topology by some purely measure-theoretic considerations.

Proof. First we show that the left inequality in (20.3.3) implies that the Cartesian square $f \times f$ is ergodic with respect to the product measure $\mu \times \mu$. Let $A, B, C, D \subset X$ be Borel sets. Then

$$\begin{aligned}\underline{\lim}_{n\to\infty} (\mu \times \mu)((f \times f)^n(A \times C) \cap (B \times D)) &= \underline{\lim}_{n\to\infty} [\mu(f^n(A) \cap B) \cdot \mu(f^n(C) \cap D)] \\ &\geq c^2\mu(A) \cdot \mu(B) \cdot \mu(C) \cdot \mu(D) = c^2(\mu \times \mu)(A \times B) \cdot (\mu \times \mu)(C \times D).\end{aligned}$$

The same inequality holds if we replace $A \times C$ and $B \times D$ by finite disjoint unions of product sets and hence, since such sets approximate every measurable $P, Q \subset X \times X$, we have

$$\underline{\lim}_{n\to\infty} (\mu \times \mu)((f \times f)^n(P) \cap Q) \geq c^2(\mu \times \mu)(P) \cdot (\mu \times \mu)(Q),$$

and $f \times f$ is ergodic with respect to $\mu \times \mu$.

Now let ν be the diagonal measure in $X \times X$ given by $\nu(E) = \mu(\pi_1(E \cap \Delta))$, where $\Delta = \{(x, x) \mid x \in X\}$ and $\pi_1 \colon X \times X \to X$ is the projection to the first coordinate. The measure ν as well as its shift ν_n under the map $f^n \times \mathrm{Id}$ are $(f \times f)$-invariant. Explicitly, $\nu_n(A \times B) = \mu(f^n(A) \cap B)$. By the right inequality in (20.3.3) we have

$$\overline{\lim}_{n\to\infty} \nu_n(A \times B) = \overline{\lim}_{n\to\infty} \mu(f^n(A) \cap B) < C\mu(A) \cdot \mu(B) = C(\mu \times \mu)(A \times B). \tag{20.3.4}$$

Let η be any weak limit point of the sequence ν_n. If $A, B \subset X$ are closed sets then $\eta(A \times B) \leq C(\mu \times \mu)(A \times B)$ by (20.3.4). Taking disjoint unions of products of closed sets and using approximation we deduce that $\eta(P) < C(\mu \times \mu)(P)$ for any Borel set $P \subset X \times X$ and hence η is absolutely continuous with respect to $\mu \times \mu$. Since η is $(f \times f)$-invariant and $\mu \times \mu$ is ergodic we have $\eta = \mu \times \mu$ by Proposition 5.1.2, so for any closed sets A, B with $\mu(\partial A) = \mu(\partial B) = 0$ we have

$$\lim_{n\to\infty} \mu(f^n(A) \cap B) = \lim_{n\to\infty} \nu_n(A \times B) = (\mu \times \mu)(A \times B) = \mu(A) \cdot \mu(B).$$

Since the collection of all such sets is sufficient, f is mixing with respect to μ by Proposition 4.2.10. □

Before we prove uniqueness of equilibrium states let us quickly note a useful consequence of convexity of $g(x) := x \log x$ (and $g(0) = 0$). Namely, if $a_i, x_i \geq 0$ for $i = 1, \dots, n$ and $A = \sum_{i=1}^m a_i$ then

$$\begin{aligned}\frac{1}{A}\sum_{i=1}^m x_i \log\left(\frac{x_i}{a_i}\right) &= \sum_{i=1}^m \frac{a_i}{A} g\left(\frac{x_i}{a_i}\right) \\ &\geq g\left(\sum_{i=1}^m \frac{a_i}{A}\frac{x_i}{a_i}\right) = g\left(\frac{\sum_{i=1}^m x_i}{A}\right) = \frac{\sum_{i=1}^m x_i}{A} \log\left(\frac{\sum_{i=1}^m x_i}{\sum_{i=1}^m a_i}\right),\end{aligned}$$

so in particular for $b_i := \log a_i$ using $g(x) \geq -1/e$ yields

$$\sum_{i=1}^{m} x_i(b_i - \log x_i) \leq \sum_{i=1}^{m} x_i \log\Big(\sum_{j=1}^{m} e^{b_j}\Big) + \frac{1}{e}. \tag{20.3.5}$$

Theorem 20.3.7. (Bowen) *Let (X, d) be a compact metric space, $f\colon X \to X$ an expansive homeomorphism with the specification property, and $\varphi \in C^f(X)$. Then there is exactly one $\mu_\varphi = \mu \in \mathfrak{M}(f)$ with $P_\mu(f, \varphi) = P(f, \varphi)$. It is mixing and*

$$\mu_\varphi = \lim_{n\to\infty} \frac{1}{P_X(f, \varphi, n)} \sum_{x\in \mathrm{Fix}(f^n)} e^{S_n\varphi(x)}\delta_x.$$

Proof. Let us fix a weak*-accumulation point

$$\mu = \lim_{k\to\infty} \mu_{n_k}$$

of the sequence $\mu_n := \dfrac{1}{P_n(f)} \sum_{x\in \mathrm{Fix}(f^n)} \delta_x \subset \mathfrak{M}(f)$. Proposition 20.3.6 implies that f is mixing with respect to μ.

We show that if $P_\nu(f, \varphi) = P(f, \varphi)$ then $\nu = \mu$, so $P_\mu(f, \varphi) = P(f, \varphi)$ and there is only one accumulation point.

As in the proof of Theorem 20.1.3 it suffices to show that if $\nu \perp \mu$ then $P_\nu(f, \varphi) < P(f, \varphi)$.

For $n \in \mathbb{N}$ and a maximal $(n, 2\epsilon)$-separated set E_n take Borel sets β_x such that $B_f(x, \epsilon, n) \subset \beta_x \subset B_f(x, 2\epsilon, n)$, $\mathfrak{B}_n := \{\beta_x \mid x \in E_n\}$ is a partition, and $(\mu + \nu)(\partial\mathfrak{B}_n) = 0$. Since f is expansive, $\operatorname{diam} f^{-[n/2]}(\mathfrak{B}_n) \xrightarrow[n\to\infty]{} 0$ so if $f(B) = B \subset X$ such that $\mu(B) = 0$ and $\nu(B) = 1$ then there exist finite unions C_n of elements of $\mathfrak{B}_n$ such that

$$(\mu + \nu)(C_n \Delta B) = (\mu + \nu)(f^{-[n/2]}(C_n)\Delta B) \xrightarrow[n\to\infty]{} 0.$$

Furthermore, if $\epsilon < \delta_0/2$ then $\mathfrak{B}_n$ is generating for f^n, that is, $nh_\nu(f) = h_\nu(f^n) = h_\nu(f^n, \mathfrak{B}_n) \leq H_\nu(\mathfrak{B}_n)$. Noting that $S_n\varphi \leq K + S_n\varphi(x)$ on $\mathfrak{B}_n$ yields

$$\begin{aligned}
nP_\nu(f, \varphi) &\leq - \sum_{x\in E_m;\beta_x\in\mathfrak{B}_n} \Big(\varphi(\nu(\beta_x)) + \int_{\beta_x} S_n\varphi\, d\nu\Big) \\
&\leq K + \sum_{x\in E_m;\beta_x\subset C_n} \nu(\beta_x)(S_n\varphi(x) - \log \nu(\beta_x)) \\
&\quad + \sum_{x\in E_m;\beta_x\cap C_n=\varnothing} \nu(\beta_x)(S_n\varphi(x) - \log \nu(\beta_x)) \\
&\leq K + \nu(C_n) \log \sum_{x\in E_m;\beta_x\subset C_n} e^{S_n\varphi(x)} \\
&\quad + \nu(X \smallsetminus C_n) \log \sum_{x\in E_m;\beta_x\cap C_n=\varnothing} e^{S_n\varphi(x)} + \frac{2}{e},
\end{aligned}$$

where the last estimate used (20.3.5). We now apply Lemma 20.3.4 to get

$$\begin{aligned} n(P_\nu(f,\varphi) - P(f,\varphi)) - \frac{2}{e} - K \\ &\leq \nu(C_n)\log\Big(\sum_{x\in E_m;\beta_x\subset C_n} e^{S_n\varphi(x)-nP(f,\varphi)}\Big) \\ &\quad + \nu(X\smallsetminus C_n)\log\Big(\sum_{x\in E_m;\beta_x\cap C_n=\varnothing} e^{S_n\varphi(x)-nP(f,\varphi)}\Big) \\ &\leq \nu(C_n)\log(A_\epsilon^{-1}\mu(C_n)) + \nu(X\smallsetminus C_n)\log(A_\epsilon^{-1}\mu(X\smallsetminus C_n)). \end{aligned}$$

But since $\nu(C_n) \to 1$ and $\mu(C_n) \to 0$ the right-hand side goes to $-\infty$ as $n \to \infty$. Consequently $P_\nu(f,\varphi) < P(f,\varphi)$. □

Corollary 20.3.8. *Let Λ be a topologically transitive compact locally maximal hyperbolic set for an embedding $f: U \to M$ and $\varphi: \Lambda \to \mathbb{R}$ a Hölder-continuous function. Then there is a unique equilibrium state for φ which we denote by μ_φ.*

Proof. This is just like the proof of Corollary 20.1.4 from Theorem 20.1.3. □

Analogously to Corollary 20.1.5 there is a parallel statement for topological Markov chains.

b. Classification of equilibrium states. Theorem 20.3.7 establishes uniqueness of the equilibrium state μ_φ for a given function $\varphi \in C^f$. On the other hand, however, it is clear that several functions may have the same equilibrium state. For example, adding a constant to φ adds the same constant to $P(\varphi)$ and $P_\mu(\varphi)$ for any μ, so we obtain the same equilibrium state. Equilibrium states are evidently also unaffected if φ is changed in such a way that the statistical sums $P_Y(f,\varphi,n)$ do not change, for example, by changing φ in such a way that its sums over any periodic orbit remain the same. This is the case, for example, whenever φ and ψ are in the same cohomology class. In other words, we have

Proposition 20.3.9. *Let (X,d) be a compact metric space, $f: X \to X$ an expansive homeomorphism with the specification property, and $\varphi, \psi \in C^f(X)$ such that $S_n\varphi(x) = S_n\psi(x) + nc$ for some constant c independent of $x \in \operatorname{Fix}(f^n)$ and $n \in \mathbb{N}$. Then the equilibrium states μ_φ and μ_ψ for φ and ψ coincide.*

Remark. The hypothesis holds in particular whenever $\varphi - \psi$ is cohomologous to c.

We now have the tools to show that for hyperbolic sets the converse holds as well.

Proposition 20.3.10. *Let Λ be a topologically transitive compact locally maximal hyperbolic set for an embedding $f: U \to M$ and $\varphi, \psi: \Lambda \to \mathbb{R}$ Hölder-continuous functions such that the equilibrium states μ_φ and μ_ψ for φ and ψ*

coincide. Then $\psi(x) = \varphi(x) + c + h(f(x)) - h(x)$ *for some Hölder-continuous* h.

Proof. First we can adjust ψ by an additive constant such that $P(\varphi) = P(\psi)$. By the Livschitz Theorem 19.2.1 it suffices to show then that all sums $S_n\varphi(y)$ and $S_n\psi(y)$ over period-n orbits coincide. To that end pick $y \in \mathrm{Fix}(f^n)$ and note that Lemma 20.3.4 yields $A_\epsilon^\varphi e^{S_n\varphi(y)} \leq B_\epsilon^\psi e^{S_n\psi(y)}$ and hence $S_n\varphi(y) + \log A_\epsilon^\varphi \leq S_n\psi(y) + \log B_\epsilon^\psi$. Since $y \in \mathrm{Fix}(f^{kn})$ for all $k \in \mathbb{N}$ this implies that $S_n\varphi(y) = \lim_{k\to\infty} S_{kn}\varphi(y)/k \leq \lim_{k\to\infty} S_{kn}\psi(y)/k = S_n\psi(y)$. By symmetry one also obtains the reverse inequality and hence the claim. □

Exercises

20.3.1. *Let* $p = (p_0, \dots, p_{m-1})$, $p_i > 0$ *for* $i = 1, \dots, m$, $\sum_{i=0}^{m-1} p_i = 1$. *Prove that the Bernoulli measure* μ_p *is an equilibrium state for the full shift* σ_m *and some Hölder-continuous function* φ.

20.3.2. *Let* $f: X \to X$ *be an expansive homeomorphism of a compact metric space with the specification property and* $\varphi \in C^f(X)$. *Prove that* $\operatorname{supp}\mu_\varphi = X$.

20.3.3. *Prove that furthermore* $h_{\mu_\varphi}(f) > 0$.

20.3.4. *Let* Λ *be a hyperbolic repeller for a map* $f: U \to M$ *(Definition 6.4.3). Let* $\hat{\Lambda} := \{(x_0, x_1, \dots) \mid x_i \in \Lambda,\ f(x_i) = x_{i-1}\}$ *with the distance* $d(x,y) = \sum_{n=0}^{\infty} 2^{-n} d(x_n, y_n)$. *Define* $\hat{F}: \hat{\Lambda} \to \hat{\Lambda}$ *as the shift:* $\hat{f}(x_0, x_1, \dots) = (x_1, x_2, \dots)$. *Prove that* $\hat{f}$ *is an expansive homeomorphism with the specification property. The map* $\hat{f}$ *is called the natural extension of* f.

20.3.5. *Outline the theory of measures of maximal entropy and equilibrium states for hyperbolic repellers.*

4. Smooth invariant measures

In the preceding section we studied invariant measures for expansive homeomorphisms with the specification property in some generality. We now consider a specific class of such maps, namely, transitive Anosov diffeomorphisms. One aim in developing the theory of equilibrium states to the extent that we have done so is to give interesting results in this setting. In the case of diffeomorphisms of a smooth manifold it is natural to be interested in smooth invariant measures, as we were in Chapter 5.

a. Properties of smooth invariant measures. We can now describe the ergodic theory of smooth invariant measures for Anosov diffeomorphisms by using the theory of equilibrium states.

Theorem 20.4.1. *Let M be a compact connected smooth Riemannian manifold and $f\colon M \to M$ an Anosov diffeomorphism. Then f has at most one smooth invariant measure. If f has a smooth invariant measure λ then f is topologically mixing and λ is equal to the equilibrium state μ_φ for $\varphi{:=}\log J^u f^{-1}(f(\cdot))$, where $J^u f^{-1}(f(\cdot))$ is the unstable Jacobian from Section 19.1e, and hence is mixing. Furthermore $h_\lambda(f) = -\int \varphi\, d\lambda$.*[1]

Remark. Compare the last statement of the theorem with Proposition 5.1.26 for expanding maps.

Proof. If f has a smooth invariant measure λ then the Poincaré Recurrence Theorem 4.1.19 implies that λ-a.e. point is nonwandering. Since $NW(f)$ is closed and λ is positive on open sets this means that $NW(f) = M$. By Corollary 18.3.5 this implies that f is topologically mixing.

The remaining statements of the theorem follow from the results of the previous section and the following lemma which gives multiplicative bounds on the volume of balls in the d_n^f metric.

Lemma 20.4.2. *Let $f\colon M \to M$ be an Anosov diffeomorphism and λ a smooth positive measure on M, not necessarily invariant. Then for every $\epsilon > 0$ there exist $C_\epsilon, D_\epsilon > 0$ such that for all $n \in \mathbb{N}$*

$$D_\epsilon J^u f^{-n}(f^n(x)) \le \lambda(B_f(x,\epsilon,n)) \le C_\epsilon J^u f^{-n}(f^n(x)).$$

To deduce the theorem assume that λ is f-invariant. Lemma 20.4.2 implies that

$$\frac{1}{C_\epsilon}\lambda(B_f(x,\epsilon,n) \le e^{S_n\varphi(x)} \le \frac{1}{D_{\epsilon/2}}\lambda(B_f(x,\frac{\epsilon}{2},n))$$

for all $x \in M$, $\epsilon > 0$, $n \in \mathbb{N}$. If E is (n,ϵ)-separated then $\{B_f(x,\epsilon,n)\}_{x\in E}$ cover M and $\{B_f(x,\epsilon/2,n)\}_{x\in E}$ are disjoint. Consequently

$$\begin{aligned}
P(\varphi) &= \lim_{\epsilon\to 0}\lim_{n\to\infty}\frac{1}{n}\log\sup\Big\{\sum_{x\in E} e^{S_n\varphi(x)}\ \Big|\ E \text{ is } (n,\epsilon)\text{-separated}\Big\}\\
&\le \lim_{\epsilon\to 0}\lim_{n\to\infty}\frac{1}{n}\log\sup\Big\{\sum_{x\in E}\frac{1}{D_{\epsilon/2}}\lambda\Big(B_f\Big(x,\frac{\epsilon}{2},n\Big)\Big)\ \Big|\ E \text{ is } (n,\epsilon)\text{-separated}\Big\}\\
&\le \lim_{\epsilon\to 0}\lim_{n\to\infty}\frac{1}{n}\log\frac{1}{D_{\epsilon/2}}\lambda(M) = 0
\end{aligned}$$

and likewise $P(\varphi) \ge 0$. But then (20.3.5) and Lemma 20.3.4 immediately show that $\lambda \ll \mu_\varphi$ and hence $\lambda = \mu_\varphi$ by ergodicity of μ_φ. By Theorem 20.3.7 λ is mixing. Finally note that since the topological pressure of φ, and hence the pressure of λ, is zero, we obtain the expression for entropy by definition of P_λ. ∎

Proof of Lemma 20.4.2. Let $m = \dim M$. We first replace the balls $B_\epsilon(x, \epsilon, n)$ by sets that are easier to handle. For each $x \in M$ introduce adapted coordinates in a neighborhood V_x of x as described in Section 19.1d. Similarly to the case of a periodic point (Theorem 6.2.3) we may take these coordinates to be such that the local unstable manifold $W^u(x)$ is parameterized by a disk in the coordinate "plane" $\mathbb{R}^k \times \{0\}$ and the local stable manifold $W^u(x)$by a disk in $\{0\} \times \mathbb{R}^{m-k}$. If $y \in V_x$ is parameterized by $(y_1, y_2) \in \mathbb{R}^k \times \mathbb{R}^{m-k}$ then there are constants $c_1, c_2 > 0$ independent of x and y such that

$$c_1 \max(\|y_1\|, \|y_2\|) \leq \operatorname{dist}(x, y) \leq c_2 \max(\|y_1\|, \|y_2\|).$$

Let $B_\epsilon(x)$:=

$$\{y \in V_x \mid V_{f^i(x)} \ni f^i(y) = (y_1^{(i)}, y_2^{(i)}),\ \max(\|y_1^{(i)}\|, \|y_2^{(i)}\|) \leq \epsilon \text{ for } 0 \leq i < n\}.$$

Then $B_{\epsilon/c_2}(x) \subset B_f(x, \epsilon, n) \subset B_{\epsilon/c_1}(x)$ and in order to prove the lemma it suffices to get a similar estimate for $\lambda(B_\epsilon(x))$. Furthermore the measure λ in V_x is given by a density which is bounded and bounded away from 0. Thus instead of estimating $\lambda(B_\epsilon(x))$ we can estimate the volume of $B_\epsilon(x)$ in adapted coordinates. Similarly the Jacobian, which is taken with respect to a fixed Riemannian metric, can be replaced by the Jacobian of f^{-n} from adapted local coordinates near $f^n(x)$ to those near x. We assume that ϵ is small enough so that it satisfies several conditions to be specified later on.

To describe $B_\epsilon(x)$ note that first of all it contains a piece S of the stable manifold $W_\epsilon^s(x)$ of size ϵ in adapted coordinates. Hence the $(m-k)$-dimensional volume of S is of order ϵ^{m-k}. Now fix $y \in S$. In local coordinates $y = (0, y_2)$. Let $U_{y_2} = \{z \in B_\epsilon(x) \mid z = (y_1, y_2)\}$, that is, the horizontal "slice" of $B_\epsilon(x)$ through y. Obviously

$$\operatorname{vol}^m(B_\epsilon(x)) = \int_S \operatorname{vol}^k(U_{y_2}) dy_2. \tag{20.4.1}$$

Thus we need to estimate the k-dimensional volume of the sets U_{y_2}. To that end notice that for small enough ϵ the tangent space $T_z U_{y_2}$ is close to the unstable subspace E_z^+ and in particular inside an invariant cone family around E^+. This implies that if in local coordinates around $f^i(x)$ we write $f^i(z) = (z_1^{(i)}, z_2^{(i)})$ then $\|z_1^{(i)}\| \geq \mu \|z_1^{(i-1)}\|$ for some $\mu > 1$ and $1 \leq i \leq n$. Hence $\|z_1^{(i)}\|$ is maximal for $i = n$ and the image $f^n(U_{y_2})$ is the graph of a Lipschitz function in the adapted coordinates around $f^n(x)$ defined over the whole ϵ-ball around the origin. Thus $\operatorname{vol}^k(f^n(U_{y_2}))$ is of order ϵ^k. Now if ω is the k-dimensional volume element on $f^n(U_{y_2})$ then

$$\operatorname{vol}^k(U_{y_2}) = \int_{f^n(U_{y_2})} J(f^{-n})\restriction_{Tf^n(U_{y_2})} d\omega. \tag{20.4.2}$$

Together with (20.4.1) this means that Lemma 20.4.2 follows once we show that for $w = f^n(z) \in f^n(U_{y_2})$ the ratios

$$\frac{J(f^{-n})\restriction_{Tf^n(U_{y_2})}}{J^u f^{-n}(f^n(x))}$$

are bounded from above and below by positive constants. To that end compare the orbits

$$z, f(z), \dots, f^n(z) = w \text{ and } x, f(x), \dots, f^n(x)$$

and the spaces

$$TU_{y_2}, (Df)(TU_{y_2}) = Tf(U_{y_2}), \dots, Df^n(TU_{y_2}) \text{ and } E^+_x, E^+_{f(x)}, \dots, E^+_{f^n(x)}.$$

Proposition 6.4.16 bounds the distance between $f^i(z)$ and $f^i(x)$ by $C\epsilon\alpha^{\min(i,n-i)}$ for some $\alpha < 1$. Hence by Hölder continuity of the unstable distribution E^+ (Theorem 19.1.6)

$$\left| \frac{Jf^{-1}\restriction_{E^+_{f^i(x)}}}{Jf^{-1}\restriction_{E^+_{f^i(z)}}} - 1 \right| < C_1\alpha_1^{\min(i,n-i)}. \tag{20.4.3}$$

Furthermore the distance between $E^+_{f^i(z)}$ and $Df^i(TU_{y_2})$ converges exponentially, so

$$\left| \frac{Jf^{-1}\restriction_{E^+_{f^i(z)}}}{Jf^{-1}\restriction_{Df^i(TU_{y_2})}} - 1 \right| < C_2\alpha_2^i \tag{20.4.4}$$

Combining (20.4.3) and (20.4.4) and multiplying the resulting inequalities for $i = 0, \dots, n-1$ we obtain a uniform bound on the ratios

$$\frac{Jf^{-n}\restriction_{Df^n(TU_{y_2})}}{Jf^{-n}\restriction_{E^+_{f^n(x)}}},$$

which is what we needed to get a two-sided estimate for $\operatorname{vol}^k(U_{y_2})$ by a constant multiple of ϵ^k. Integrating over S ends the proof. □

b. Smooth classification of Anosov diffeomorphisms on the torus. The characterization of smooth invariant measures as equilibrium states has several natural applications. Among them is a classification of a certain class of Anosov maps up to smooth conjugacy via natural moduli. Here we present a basic result of this kind, namely, a C^1 classification.

Theorem 20.4.3. *Suppose $f, g: \mathbb{T}^2 \to \mathbb{T}^2$ are C^2 area-preserving Anosov diffeomorphisms and $f \circ h = h \circ g$ for a homeomorphism h homotopic to the identity. Then f and g are C^1 conjugate if and only if their eigenvalues at corresponding periodic points p and $h(p)$ coincide.*

Proof. The "only if" part was observed in Section 2.1a. To prove the converse consider the function $\psi_f := \log(J^u f) \circ h$. Since both $\log(J^u f)$ and h are Hölder continuous (Corollary 19.1.13 and Theorem 19.1.2), so is ψ_f. The eigenvalue hypothesis implies that ψ_f and $\varphi_g := \log J^u g$ have the same sums over periodic orbits and hence the same equilibrium state by Proposition 20.3.9. By Theorem 20.4.1 the equilibrium state for φ_g is the smooth measure, that is, area. Note next that equilibrium states are equivariant under Hölder homeomorphism by construction, that is, the equilibrium state for ψ_h (which we just identified as area) is equal to the pullback of the equilibrium state for $\varphi_f = \log J^u f$ by h. This means that h is area preserving.

By Corollary 19.1.11 the stable and unstable foliations for f and g are C^1. This means that the local product structure is C^1, that is, there are local C^1 coordinates for f and g in which the foliations are linear. The image of area under these coordinates is a measure with continuous density, which therefore induces continuous densities on every leaf. Each of those conditional densities is defined up to a multiplicative constant. Thus we obtain continuous densities on the leaves of the foliations for f and g. Since h preserves area it preserves these densities. This shows that on a leaf h is locally obtained by integration of a continuous density, hence is itself C^1. Thus h is a C^1 diffeomorphism by Lemma 19.1.10. □

Corollary 20.4.4. *Let $f: \mathbb{T}^2 \to \mathbb{T}^2$ be a C^2 area-preserving Anosov diffeomorphism and suppose there exists $\lambda \in \mathbb{R}$ such that for every $x \in \operatorname{Fix}(f^n)$ the expanding eigenvalue of $Df^n(x)$ is λ^n. Then f is C^1 conjugate to a linear automorphism A.*

Proof. First of all there exists a topological conjugacy to the linear model F_L by Theorem 18.6.1, that is, there is a Hölder homeomorphism $h: \mathbb{T}^2 \to \mathbb{T}^2$ such that $f \circ h = h \circ F_L$. Then h maps area (the equilibrium state for $\varphi_{F_L} = \text{const.}$) to the equilibrium state for $\psi_f = \varphi_f \circ h$. But by Theorem 20.4.1 the equilibrium state for φ_f is area and by assumption $\varphi_f = \text{const.}$, hence $\varphi_f = \varphi_f \circ h = \psi_f$. Consequently h preserves area and the preceding proof shows that h is C^1. □

Corollary 20.4.5. *Suppose $f: \mathbb{T}^2 \to \mathbb{T}^2$ is a C^2 area-preserving Anosov diffeomorphism such that $h_\lambda(f) = h_{\text{top}}(f)$. Then f is C^1 conjugate to a linear automorphism.*

Proof. The assumption is that the equilibrium state for φ_f is equal to the equilibrium state for the zero function. Thus φ_f is cohomologous to zero by Proposition 20.3.10 and the preceding result applies. □

The results of this subsection can be strengthened as follows: If f is assumed to be C^∞ then the smooth conjugacies obtained above are, in fact, C^∞. This follows from two facts. First, although the stable and unstable foliations are only C^1, their leaves are C^∞ manifolds (see the Hadamard–Perron Theorem 6.2.8) and the conditional densities are in fact C^∞. Such a density is defined up to a multiplicative constant and if $y \in W^u(x)$ then one can show that

$$\frac{\rho(y)}{\rho(x)} = \lim_{n\to\infty} \frac{J^u f^{-n}(x)}{J^u f^{-n}(y)} \tag{20.4.5}$$

and the latter expression is in fact C^∞ along the leaves and all derivatives are continuous. A similar argument holds for the stable leaves. The proof of these facts is similar to the proof of Theorem 6.6.5.

Second one can show that any vector function that is continuous together with its partial derivatives of all orders along the stable and unstable foliation is C^∞.[2]

c. Smooth classification of contact Anosov flows on 3-manifolds. We now study flows that are Anosov flows and preserve a contact structure (see Sections 5.6 and 18.3b). We consider the lowest-dimensional situation (flows on 3-manifolds) and the role of the linear model is played by geodesic flows on factors of the hyperbolic plane.

Theorem 20.4.6. *Suppose φ^t and ψ^t are orbit-equivalent contact Anosov flows on a three-dimensional manifold and that the periods of corresponding periodic orbits coincide. Then φ^t and ψ^t are C^1 flow equivalent.*

Proof. Note first that the flows are Hölder flow equivalent by Theorem 19.2.9. The flow equivalence preserves the strong stable and unstable foliations and the orbits. Next the weak stable and unstable foliations are C^1 by the flow analog of Corollary 19.1.11. The strong stable and strong unstable distributions are the intersections of the weak stable and weak unstable distributions with the kernel of the contact form by Lemma 18.3.7. Since the contact form is C^1 this implies that the strong foliations are C^1.

By the counterpart of Theorem 20.4.1 for flows (Exercise 20.4.1) the equilibrium state for the unstable Jacobian is the invariant volume induced by the contact form. Again this implies that the flow equivalence preserves volume, and since it preserves the three one-dimensional foliations (orbits and the strong foliations), which are C^1, it preserves the conditional measures on those foliations and hence it is C^1. ☐

Exercises

20.4.1. *Prove the counterpart of Theorem 20.4.1 for Anosov flows.*

20.4.2*. *Show that for any Anosov diffeomorphim f the conditional measures on the unstable leaves for the equilibrium state μ_φ, where $\varphi = \log J^u f^{-1}(f(\cdot))$, are absolutely continuous and given by (20.4.5).*

20.4.3. *Generalize Theorem 20.4.3 to arbitrary (not only area-preserving) Anosov diffeomorphisms on $\mathbb{T}^2$.*

20.4.4. *Give an example of a C^∞ volume-preserving Anosov diffeomorphism of $\mathbb{T}^3$ that is conjugate to a linear automorphism via a volume-preserving conjugacy, but not C^1 conjugate to a linear model.*

5. Margulis measure

In this section we present an alternative construction due to Margulis of the unique measure of maximal entropy in the case of topologically mixing Anosov flows. Unlike the Bowen construction from Section 20.1 which produces this measure as a limit distribution of periodic orbits, the Margulis construction considers limits of normalized Lebesgue measure on very long pieces of unstable leaves. Naturally this construction also works in the discrete-time case, but there it does not lead to any particularly interesting new results. In the flow case, however, it will allow us in the next section to obtain the most precise asymptotic known of the growth rate for the number of periodic orbits. Thus throughout this section we assume that $\varphi^t: M \to M$ is a topologically mixing Anosov flow. We first establish some notation.

We will write $A \subset W^{0u}$ when $A \subset W^{0u}(p)$ for some $p \in M$. Likewise we use the notions of openness, compactness, continuity, and measurability for sets and functions defined on an unstable leaf; sometimes we will write W^{0u}*-open*, and so on. Thus a W^{0u}*-neighborhood* of a point $p \in M$ is a W^{0u}-open set containing p. We let $C(W^{0u}) := \{f: M \to \mathbb{R} \mid \operatorname{supp}(f) \subset W^{0u} \text{ compact}, f\restriction_{\operatorname{supp}(f)} \text{ continuous }\}$. The distance function on a leaf $W^{0u}(p)$ induced by the Riemannian structure of this leaf is denoted by d^{0u} and λ^{0u} is the (Lebesgue) measure on each unstable leaf $W^{0u}(p)$ induced by the Riemannian volume in the leaf. $B^{0u}(p,r) := \{q \in W^{0u}(p) \mid d^{0u}(p,q) < r\}$ is the r-ball around p in $W^{0u}(p)$. This definition carries over to the other foliations (W^u, W^{0s}, W^s) as well. Similarly to the holonomy maps for the case of diffeomorphisms in Section 19.2b we have that if $x, y \in \Lambda$ are sufficiently close, then there is a well-defined *holonomy map* $\mathcal{H}: B^{0u}(x,\epsilon) \to W^{0u}(y), z \mapsto W^s(z) \cap B^{0u}(y,\delta)$, where ϵ and δ depend on x and y. We call $A, B \subset W^{0u}$ *ϵ-equivalent* if there is a well-defined holonomy $\mathcal{H}$ from A to B and $d^s(x,\mathcal{H}(x)) < \epsilon$ for all $x \in A$. We call $f, g \in C(W^{0u})$ *ϵ-equivalent* if $\operatorname{supp}(f)$ and $\operatorname{supp}(g)$ are ϵ-equivalent via $\mathcal{H}$ and $f = g \circ \mathcal{H}$.

Applying analogs of Corollary 18.3.5 and Proposition 18.3.10 for flows to φ^t shows that for $p_1, p_2 \in M$ and $r > 0$ we have $W^s(p_1) \cap B^{0u}(p_2, r) \neq \varnothing$. Thus a simple compactness argument shows in turn that for any open $A \subset W^{0u}$ there exist $\epsilon(A), r(A) > 0$ such that for all $p \in M$ the ball $B^{0u}(p,r)$ is $\epsilon(A)$-equivalent to a subset of A. This yields

Lemma 20.5.1. *If $A \subset W^{0u}$ is open then there exists $C(A)$ such that for $p \in M$ and $t \geq 0$*

$$\lambda^{0u}(\varphi^t B^{0u}(p, r(A))) < C(A)\lambda^{0u}(\varphi^t(A)).$$

Proof. Suppose $T > 0$. Then the claim holds for all $p \in M$ and $t \in [0,T]$ since the holonomy maps establishing ϵ-equivalence of $\varphi^t(B^{0u}(p, r(A)))$ to a subset C of $\varphi^t(A)$ are a uniformly equicontinuous family of local homeomorphisms with uniformly equicontinuous inverses for $t \in [0,T]$, $s \in [0, \epsilon(A)]$.

If T is a sufficiently large number (depending on A) and $t \geq T$ then the set $\varphi^t(B^{0u}(p,r(A)))$ is ϵ-equivalent to a subset C of $\varphi^t(A)$ with ϵ small enough that $\lambda^{0u}(\varphi^t(B^{0u}(p,r(A)))) <$ const. $\cdot\lambda^{0u}(C)$ with a bounded constant depending on A. (This is possible because the curvature of the boundary of $\varphi^t(B^{0u}(p,r(A)))$ is bounded independently of t.) □

Lemma 20.5.2. *If $0 \leq f \in C(W^{0u})$ and $K \in W^{0u}$ compact then there exists $C(K,f) > 0$ such that for any bounded W^{0u}-measurable g with support in K and for $t \geq 0$*

$$\int g \circ \varphi^{-t}\, d\lambda^{0u} < C(K,f)\|g\|_\infty \int f \circ \varphi^{-t}\, d\lambda^{0u},$$

where $\|\cdot\|_\infty$ is the essential-supremum norm.

Proof. Let $A = \{x \in M \mid f(x) > \epsilon\}$ and cover K by balls $B^{0u}(x_i, r(A))$, $i = 1,\dots,N$. Using Lemma 20.5.1 we then have

$$\begin{aligned}\int g \circ \varphi^{-t}\, d\lambda^{0u} &\leq \lambda^{0u}(\varphi^t(K))\|g\|_\infty < \sum_{i=1}^{N} \lambda^{0u}(B^{0u}(x_i, r(A)))\|g\|_\infty \\ &< NC(A)\lambda^{0u}(\varphi^t(A))\|g\|_\infty < C(K,f)\|g\|_\infty \int f \circ \varphi^{-t}\, d\lambda^{0u},\end{aligned}$$

where $C(K,f) = NC(A)/\epsilon$. □

For $p \in M$ we define a function $f_p\colon M \to \mathbb{R}$ by

$$f_p(x) := \begin{cases} \left(1 + \lambda^{0u}\left(B^{0u}\left(p, d^{0u}\left(p,x\right)\right)\right)\right)^{-2} & \text{if } x \in W^{0u}(p), \\ 0 & \text{otherwise.}\end{cases}$$

Let $r_p(s)$ be such that $\lambda^{0u}(B^{0u}(p, r_p(s))) = s$ and set

$$\begin{aligned} U_p^i :=& \{x \in W^{0u}(p) \mid i \leq \lambda^{0u}(B^{0u}(p, d^{0u}(p,x))) < i+1\} \\ =& B^{0u}(p, r_p(i+1)) \smallsetminus B^{0u}(p, r_p(i)) \end{aligned}$$

for $i \in \mathbb{N}_0$. Then $\lambda^{0u}(U_p^i) \leq 1$ and hence

$$\int f_p(x)\, d\lambda^{0u} < \sum_i \int_{U_p^i} f_p(x)\, d\lambda^{0u} \leq \sum_i 1/(i+1)^2 < 2. \tag{20.5.1}$$

For $A \subset W^{0u}$ open and $p \in M$ we set $\chi_{p,A} = \chi_{\overline{B^{0u}(p,r(A))}}$ (the characteristic function of the closed $r(A)$-ball) and

$$g_{p,A}(x) := \int \chi_{x,A}(y) f_p(y)\, d\lambda^{0u}(y).$$

Then $g_{p,A}$ is W^{0u}-continuous and positive on $W^{0u}(p)$.

Lemma 20.5.3. *For $A \subset W^{0u}$ open, $q \in M$, $t \geq 0$ and with $C(A)$ as in Lemma 20.5.1 we have*

$$\int g_{p,A}(\varphi^{-t}(x))\,d\lambda^{0u}(x) < 2C(A)\lambda^{0u}(\varphi^t(A)).$$

Proof. Using $\chi_{p,A}(x) = \chi_{x,A}(p)$, the Fubini Theorem, Lemma 20.5.1, and (20.5.1) yields

$$\begin{aligned}\int g_{p,A}(\varphi^{-t}(x))\,d\lambda^{0u}(x) &= \iint \chi_{\varphi^{-t}(x),A}(y) f_p(y)\,d\lambda^{0u}(y)\,d\lambda^{0u}(x)\\ &= \iint \chi_{y,A}(\varphi^{-t}(x))\,d\lambda^{0u}(x) f_p(y)\,d\lambda^{0u}(y)\\ &= \int \lambda^{0u}(\varphi^t(B^{0u}(\varphi^{-t}(y), r(A)))) f_p(y)\,d\lambda^{0u}(y)\\ &< 2C(A)\lambda^{0u}(\varphi^t(A)).\end{aligned}$$

□

Lemma 20.5.4. *If $f_1 \in C(W^{0u})$ and $\epsilon > 0$ then there exists $\delta > 0$ (depending continuously on f_1 in the C^0 topology) such that if f_2 is δ-equivalent to f_1 then*

$$\left|\int f_1\,d\lambda^{0u} - \int f_2\,d\lambda^{0u}\right| < \epsilon \int |f_1|\,d\lambda^{0u}.$$

Proof. There are step functions $\underline{\xi}$ and $\overline{\xi}$ representing upper and lower Riemann sums for $\int f_1\,d\lambda^{0u}$ that are accurate to within $\frac{\epsilon}{2}\int |f_1|\,d\lambda^{0u}$. Then it suffices to show the result for $\underline{\xi}$ and $\overline{\xi}$ because if f_2 is δ-equivalent to f_1 then the corresponding δ-equivalent step functions give upper and lower bounds for f_2. In each case it suffices to show that for a given open $O \subset W^{0u}$ and $\alpha > 0$ there exists η such that any η-equivalent set O' has the same volume up to α, that is, $|\operatorname{vol} O - \operatorname{vol} O'| < \alpha$. But this follows from the fact that the holonomy maps converge to isometries as $\eta \to 0$. □

Now we fix once and for all a W^{0u}-open set $\mathcal{K}$ with W^{0u}-compact closure and a function

$$f_{\mathcal{K}} > \chi_{\mathcal{K}} \tag{20.5.2}$$

in $C(W^{0u})$, where $\chi_{\mathcal{K}}$ is the characteristic function of $\mathcal{K}$.

Definition 20.5.5. $f_1, f_2 \in C(W^{0u})$ are said to be *ϵ-close* if there exist $\tilde{f}_1, \tilde{f}_2 \in C(W^{0u})$ and $x_1, x_2 \in M$ such that

(1) $\tilde{f}_1, \tilde{f}_2$ are ϵ-equivalent,
(2) $|f_i(x) - \tilde{f}_i(x)| < \epsilon g_{x_i,\mathcal{K}}(x)$ for all $x \in M$.

Lemma 20.5.6. *If $0 \le f_1 \in C(W^{0u})$ and $\epsilon > 0$ then there exists $\delta(\epsilon, f_1)$ such that for $f_2 \in C(W^{0u})$ δ-close to f_1 we have*

$$\left|\int f_1 \circ \varphi^{-t}\, d\lambda^{0u} - \int f_2 \circ \varphi^{-t}\, d\lambda^{0u}\right| < \epsilon \int f_{\mathcal{K}} \circ \varphi^{-t}\, d\lambda^{0u}.$$

Proof. By definition

$$\begin{aligned}\left|\int f_1 \circ \varphi^{-t}\, d\lambda^{0u} - \int f_2 \circ \varphi^{-t}\, d\lambda^{0u}\right| &\le \left|\int f_1 \circ \varphi^{-t}\, d\lambda^{0u} - \int \tilde{f}_1 \circ \varphi^{-t}\, d\lambda^{0u}\right| \\ &+ \left|\int (\tilde{f}_1 \circ \varphi^{-t} - \tilde{f}_2 \circ \varphi^{-t})\, d\lambda^{0u}\right| \\ &+ \left|\int \tilde{f}_2 \circ \varphi^{-t}\, d\lambda^{0u} - \int f_2 \circ \varphi^{-t}\, d\lambda^{0u}\right| \\ &\le \delta \int g_{x_1,\mathcal{K}} \circ \varphi^{-t}(x)\, d\lambda^{0u} \\ &+ \frac{\epsilon}{2\|\tilde{f}_1\|_\infty C(\mathcal{K}, f_1)} \int |\tilde{f}_1| \circ \varphi^{-t}\, d\lambda^{0u} \\ &+ \delta \int g_{x_2,\mathcal{K}} \circ \varphi^{-t}(x)\, d\lambda^{0u},\end{aligned}$$

where the middle term was estimated using Lemma 20.5.4 for sufficiently small δ (using the fact that $\tilde{f}_1 \circ \varphi^{-t}$ and $\tilde{f}_2 \circ \varphi^{-t}$ are $\delta\lambda^t$-equivalent, where $\lambda < 1$ is as in Definition 17.4.1). The other two terms are by Lemma 20.5.3 each bounded by $2C(A)\delta\lambda^{0u}(\varphi^t\mathcal{K})$. Thus, taking $\delta < \epsilon/8C(A)$, Lemma 20.5.2 shows that $\epsilon \int f_{\mathcal{K}} \circ \varphi^{-t}\, d\lambda^{0u}$ is indeed an upper bound. □

Although $C(W^{0u})$ is not a linear space (it is not closed under addition), we can define a notion of linearity: A function $F\colon C(W^{0u}) \to \mathbb{R}$ is said to be *linear* or to be a *linear functional* if $F(\alpha f) = \alpha F(f)$ and $F(f+g) = F(f) + F(g)$ whenever $f, g, f+g \in C(W^{0u})$, $\alpha \in \mathbb{R}$. The space C^* of functionals on $C(W^{0u})$ has a topology induced by the natural embedding into $\prod_{f\in C(W^{0u})} \mathbb{R}_f$, where $\mathbb{R}_f$ is a copy of $\mathbb{R}$. This product topology is the topology of pointwise convergence (which over a linear space is the weak* topology).

Now define $\{F_t\}_{t\in\mathbb{R}} \subset C^*$ by $F_t(f) := \int f \circ \varphi^{-t}\, d\lambda^{0u}$. Let

$$C_0^* := \left\{F \in C^* \,\middle|\, F = \sum_{i=1}^m c_i F_{t_i} \text{ for some } c_i, t_i \ge 0 \text{ and } F(f_{\mathcal{K}}) = 1\right\}.$$

Lemma 20.5.7. (1) *If $f \in C(W^{0u})$ then there exists $C_1(f)$ such that $|F(f)| \le C_1(f)$ for all $F \in \overline{C_0^*}$.*

(2) *If $0 \le f \in C(W^{0u}) \setminus \{0\}$ then there exists $C_2(f)$ such that $|F(f)| \ge C_2(f)$ for all $F \in \overline{C_0^*}$.*

(3) *If $f \in C(W^{0u})$ and $\epsilon > 0$ then there exists $\delta > 0$ such that if $g \in C(W^{0u})$ is δ-close to f then $|F(f) - F(g)| < \epsilon$ for all $F \in \overline{C_0^*}$.*

Proof. Let $A := \{\hat{F}_t := F_t/F_t(f_\mathcal{K}) \mid t \in \mathbb{R}\}$. Lemma 20.5.2 and Lemma 20.5.6 imply that (1)–(3) hold with A in place of $\overline{C_0^*}$. Thus they remain true for the convex hull $C_0^* = \mathrm{co}(A)$ of A and thus for the closure $\overline{\mathrm{co}}(A) = \overline{C_0^*}$ of C_0^*. □

We set $\varphi^{t*}(F)(f) := F(f \circ \varphi^{-t})$. The action of φ^t on $\overline{C_0^*}$ is then given by $\widehat{\varphi^t}^*(F)(f) := F(f \circ \varphi^{-t})/F(f_\mathcal{K} \circ \varphi^{-t})$.

Lemma 20.5.8. *There exist* $\mathfrak{m} \in \overline{C_0^*}$ *and* $h^u > 0$ *such that*

$$\varphi^{t*}\mathfrak{m} = e^{h^u t}\mathfrak{m}. \tag{20.5.3}$$

Remark. This gives the uniform-expansion property that distinguishes the Margulis measure.

Proof. Lemma 20.5.7(1) implies that $\overline{C_0^*}$ is compact in the topology of pointwise convergence (using the Tychonoff Theorem which shows that the product of compact sets is compact). Then the Tychonoff Fixed-Point Theorem A.2.11 shows that there is an $\mathfrak{m} \in \overline{C_0^*}$ such that $\widehat{\varphi^t}^*\mathfrak{m} = \mathfrak{m}$ for all $t \geq 0$. Thus we obtain (20.5.3) for $t \geq 0$ and hence for all $t \in \mathbb{R}$.

To see that $h^u > 0$ let $0 \leq f \in C(W^{0u}) \setminus \{0\}$ and $t_1, t_2 \geq 0$. Then

$$\begin{aligned}\varphi^{t_1*}F_{t_2}(f) = F_{t_2}(f \circ \varphi^{-t_1}) = \int (f \circ \varphi^{-t_1}) \circ \varphi^{-t_2}\, d\lambda^{0u} &\geq \lambda^{-t_1} \int f \circ \varphi^{-t_2}\, d\lambda^{0u} \\ &= \lambda^{-t_1} F_{t_2}(f),\end{aligned}$$

where $\lambda < 1$ is as in Definition 17.4.1. Thus $\varphi^{t*}F = \lambda^{-t}F$ for all $F \in A$ and hence for all $F \in \overline{\mathrm{co}}(A) = \overline{C_0^*}$. Thus $h^u > 0$. □

Lemma 20.5.9. *If f, g are ϵ-equivalent then* $\mathfrak{m}(f) = \mathfrak{m}(g)$.

Remark. This gives holonomy invariance, another property that characterizes Margulis measure.

Proof. By considering positive parts and using linearity we may assume that f and g are nonnegative. Since $f \circ \varphi^{-t}$ and $g \circ \varphi^{-t}$ are $\lambda^t\epsilon$-equivalent, Lemma 20.5.4 shows that $\lim_{t\to\infty} F_t f/F_t g = 1$ so for $\eta > 0$ there exists $T_\eta > 0$ such that

$$|F_t(f) - F_t(g)| \leq \eta F_t(g)$$

for $t \geq T_\eta$. Thus if $F = \sum c_i F_{t_i}$ with $c_i, t_i \geq 0$ then

$$\begin{aligned}|\varphi^{t*}F(f) - \varphi^{t*}F(g)| &\leq \sum c_i |\varphi^{t*}F_{t_i}(f) - \varphi^{t*}F_{t_i}(g)| \\ &= \sum c_i |F_{t_i+t}(f) - F_{t_i+t}(g)| \leq \eta \sum c_i F_{t_i+t}(g) = \eta\varphi^{t*}F(g).\end{aligned}$$

The same estimate then holds for $F \in A$, hence for all $F \in \overline{\mathrm{co}}(A) = C_0^*$, hence for $\mathfrak{m}$. Thus by Lemma 20.5.8 we indeed have

$$\frac{\mathfrak{m}(f)}{\mathfrak{m}(g)} = \lim_{t\to\infty} \frac{C_{0u}^t \mathfrak{m}(f)}{C_{0u}^t \mathfrak{m}(g)} = \lim_{t\to\infty} \frac{\varphi^{t*}\mathfrak{m}(f)}{\varphi^{t*}\mathfrak{m}(g)} = 1.$$

□

Now we show that $\mathfrak{m}$ corresponds to a family of measures on leaves of W^{0u}. To that end let $OC(W^{0u}(p))$ be the collection of open sets in $W^{0u}(p)$ with compact closure and $OC(W^{0u}) := \bigcup_{p\in M} OC(W^{0u}(p))$. If $U \in OC(W^{0u})$ let $C_U(W^{0u}) := \{f \in C(W^{0u}) \mid \operatorname{supp}(f) \subset \bar{U}\}$ endowed with the supremum norm $\|\cdot\|_\infty$. By Lemma 20.5.7(3) $\mathfrak{m}$ is a continuous linear functional on $C_U(W^{0u})$. By the Hahn–Banach Theorem A.2.4 $\mathfrak{m}$ thus extends to the space $C(\bar{U})$ of continuous functions on $\bar{U}$, hence by the Riesz Representation Theorem A.2.7 there is a measure μ_U on $\bar{U}$ such that for $f \in C_U(W^{0u})$ we have

$$\mathfrak{m}(f) = \int f\, d\mu_U. \tag{20.5.4}$$

If $U_1 \subset U_2$ in $OC(W^{0u})$ then there exist $\{f_j\}_{j\in\mathbb{N}} \subset C_U(W^{0u})$ such that $f_i \nearrow \chi_{U_1}$ and hence $\mu_{U_2}(U_1) = \mathfrak{m}(\chi_{U_1}) = \lim_{j\to\infty} \mathfrak{m}(f_j)$, so we have

Theorem 20.5.10. *There is a map $\mu^{0u}\colon OC(W^{0u}) \to \mathbb{R}$ such that*

(1) $\mu^{0u}\restriction_{OC(W^{0u}(p))}$ *extends to a measure on $W^{0u}(p)$;*
(2) $\mu^{0u}(\varphi^t(U)) = e^{h^u t}\mu^{0u}(U)$ *for $U \in OC(W^{0u})$, $t \in \mathbb{R}$;*
(3) *if $\varnothing \neq U \in OC(W^{0u})$ then $0 < \mu^{0u}(U) < \infty$;*
(4) *if $U_1, U_2 \in OC(W^{0u})$ are ϵ-equivalent then $\mu^{0u}(U_1) = \mu^{0u}(U_2)$.*

We will see in Lemma 20.5.16 that $h^u = h_{\text{top}}(\varphi^t)$.

Replacing φ^t by φ^{-t} we obtain a measure μ^{0s} for which the same results hold, except that in (2) we obtain a constant $h^s < 0$. (In fact, $h^s = -h^u$; see (20.5.9).)

Adapting the notation to W^u, W^{0s}, and W^s we find that $\bigcup_{t_1<t<t_2} \varphi^t(U) \in OC(W^{0u})$ whenever $t_1 < t_2$ and $U \in OC(W^u)$. Furthermore there exist $r_0, t_0 > 0$ such that

$$\varphi^{t_1}(U) \cap \varphi^{t_2}(U) = \varnothing \text{ if } 0 \le t_1 < t_2 \le t_0,\ U \subset B^i(p, r_0) \text{ open},\ i = u, s. \tag{20.5.5}$$

Thus for $U \subset B^i(p, r_0)$ $(i = u, s)$ we let $\mu^i(U) := \mu^{0i}(\bigcup_{0<t<t_0} \varphi^t(U))$, $i = u, s$. This induces a measure on $B^i(p, r_0)$, and indeed on $OC(W^i)$ $(i = u, s)$, for which

$$\mu^i(\varphi^t(U)) = e^{h^i t}\mu^i(U) \text{ and } 0 < \mu^i(U) < \infty \text{ for } \varnothing \neq U \in OC(W^i), \tag{20.5.6}$$

where $i = u, s$. Theorem 20.5.10(2) then yields

$$\mu^{0i}\Big(\bigcup_{t_1<t<t_2} \varphi^t(U)\Big) = \frac{\int_{t_1}^{t_2} e^{h^i t} dt}{\int_0^{t_0} e^{h^i t} dt}\mu^i(U) \text{ if } 0 \le t_1 < t_2 \le t_0,\ U \subset B^i(p, r_0) \text{ open}, \tag{20.5.7}$$

where $i = u, s$ and t_0, r_0 are as in (20.5.5).

With the notion of ϵ-equivalence adapted to sets in the strong foliations W^i $(i = u, s)$, we get

Lemma 20.5.11. *For all $\epsilon > 0$ there is a $\zeta > 0$ such that for all $A_1, A_2 \subset W^i$ measurable ($i = u, s$) and A_1, A_2 ζ-equivalent we have*

$$\left| \frac{\mu^i(A_1)}{\mu^i(A_2)} - 1 \right| < \epsilon.$$

Proof. Note first that if A_1, A_2 are ζ-equivalent then there are partitions $A_l = \bigcup_k A_l^k$ with $A_l^k \subset B^i(p_l^k, r_0)$ ($l = 1, 2$) and A_1^k, A_2^k ζ-equivalent. Thus assume without loss of generality that $A_l \subset B^i(p_l, r_0)$. But for any $\alpha > 0$ there exists a $\zeta > 0$ such that if $A_1, A_2 \subset W^i$ are ζ-equivalent and $A_l \subset B^i(p_l, r_0)$, then $\bigcup_{\alpha<t<t_0-\alpha} \varphi^t(A_2)$ is ζ-equivalent to a subset of $\bigcup_{0<t<t_0} \varphi^t(A_1)$ and vice versa $\bigcup_{\alpha<t<t_0-\alpha} \varphi^t(A_1)$ is ζ-equivalent to a subset of $\bigcup_{0<t<t_0} \varphi^t(A_2)$.

For these sets, (20.5.7) and Theorem 20.5.10(4) yield

$$\begin{aligned} \mu^i(A_1) &= \frac{\int_0^{t_0} e^{h^u t} dt}{\int_\alpha^{t_0-\alpha} e^{h^u t} dt} \mu^{0i}\Big(\bigcup_{\alpha<t<t_0-\alpha} \varphi^t(A_1) \Big) \\ &\le \frac{\int_0^{t_0} e^{h^u t} dt}{\int_\alpha^{t_0-\alpha} e^{h^u t} dt} \mu^{0i}\Big(\bigcup_{0<t<t_0} \varphi^t(A_2) \Big) = \frac{\int_0^{t_0} e^{h^u t} dt}{\int_\alpha^{t_0-\alpha} e^{h^u t} dt} \mu^i(A_2), \end{aligned}$$

and vice versa, which in turn implies the claim. □

Corollary 20.5.12. *For all $r > 0$ there exists $C > 1$ such that if A_1, A_2 are r-equivalent then*

$$\frac{1}{C} < \frac{\mu^i(A_1)}{\mu^i(A_2)} < C.$$

From these measures on leaves we now construct a finite φ^t-invariant measure on M. We do this by locally defining a *weighted product measure* as follows. Every $p \in M$ has a neighborhood $U(p)$ which is a *local product cube*, that is, using the local product structure we can write $U(p)$ as $U^{0u}(p) \times U^s(p)$, where $U^{0u}(p) \subset W^{0u}(p)$ and $U^s(p) \subset W^s(p)$. If $O \subset U(p)$ let

$$f_O(q) := \mu^s((\{q\} \times U^s(p)) \cap O) \quad (q \in U^{0u}(p)).$$

Lemma 20.5.13. *f_O is upper semicontinuous (hence locally integrable).*

Proof. For $x \in U^{0u}$ and $\epsilon > 0$ there exists $\delta > 0$ such that

$$\frac{\mu^s\left((\pi^s(O \cap (\{x\} \times U^s)) \times U^{0u}) \cap (\{y\} \times U^s) \right)}{\mu^s((\{y\} \times U^s) \cap O)} > 1 - \epsilon$$

when $d^{0u}(x, y) < \delta$, where π^s is the projection to W^s. Now apply Lemma 20.5.11. □

For $q \in U^s(p)$, $A \subset U^{0u}(p)$ let $\mu_q(A) := \mu^{0u}(A \times \{q\})$ wherever defined. By Theorem 20.5.10(4) this is independent of $q \in U^s(p)$. Together with Lemma 20.5.13 this shows that

$$\mu(O) := \int f_O(x)\, d\mu_q(x) \tag{20.5.8}$$

is well defined.

Proposition 20.5.14. *The measure on M obtained from (20.5.8) by extending to Borel sets is finite and φ^t-invariant.*

Proof. Finiteness is clear since M is compact and local product cubes have finite measure. Theorem 20.5.10(2), (20.5.6) and (20.5.8) show that

$$\mu(\varphi^t(A)) = e^{(h^u + h^s)t}\mu(A)$$

for all measurable $A \in M$. Setting $A = M$ thus yields

$$h^u = -h^s =: h. \tag{20.5.9}$$

□

Proposition 20.5.14 in particular shows that, by proper choice of $\mathcal{K}$, for example, we can normalize the measure obtained from (20.5.8) to be a φ^t-invariant Borel probability measure μ which is called the *Margulis measure* for φ^t.

Theorem 20.5.15. *The Bowen measure and the Margulis measure coincide.*

Proof. We show equality via volume estimates for d_t^f ϵ-balls.

Lemma 20.5.16. $E_\epsilon e^{-th^u} \le \mu(B_f(x,\epsilon,t)) \le F_\epsilon e^{-th^u}$ *for some constants* E_ϵ, F_ϵ.

Proof. We will use a Lyapunov metric (as in Theorem 20.4.1). Note that it suffices to show that this inequality holds for "boxes" $B_f^1(x,\epsilon,t) := B_f^{0s}(x,\epsilon,t) \times B_f^u(x,\epsilon,t)$ since for $\epsilon > 0$ there exist $\epsilon_1, \epsilon_2 > 0$ such that $B_f^1(x,\epsilon_1,t) \subset B_f(x,\epsilon,t) \subset B_f^1(x,\epsilon_2,t)$. But since we use a Lyapunov metric we have $B_f^1(x,\epsilon,t) = B^{0s}(x,\epsilon) \times \varphi^{-t}(B^u(\varphi^t(x),\epsilon))$, which immediately yields the claim by the uniform-expansion property of μ^u. □

Lemma 20.5.17. $h^u = h_{\text{top}}(\varphi)$.

Proof. If E is a maximal d_t^f-ϵ-separated set then $M = \bigcup_{x\in E} B_f(x,\epsilon,t)$ and hence $1 \le \sum_{x\in E} \mu(B_f(x,\epsilon,t)) \le \text{const.}\, e^{t(h_{\text{top}}(\varphi)-h^u)}$ by the previous lemma and (the flow analog of) Proposition 18.5.4. Thus $h^u \le h_{\text{top}}(\varphi)$. Conversely $B_f(x,\epsilon/2,t)$ are pairwise disjoint, so $1 \ge \sum_{x\in E} \mu(B_f(x,\epsilon/2,t)) \ge \text{const.}\, e^{t(h_{\text{top}}(\varphi)-h^u)}$ and $h^u \ge h_{\text{top}}(\varphi)$. □

The preceding two lemmas imply that Margulis measure is absolutely continuous with respect to Bowen measure by (the flow analog of) Lemma 20.1.1. Since Bowen measure is ergodic this implies the claim. □

Remark. We will simply write h for $h_{\text{top}}(\varphi)$ in the sequel.

6. Multiplicative asymptotic for growth of periodic points

The aim of this section is to establish a *multiplicative* asymptotic of the growth of periodic orbits for flows. This asymptotic is stronger than an exponential one. Let us note that for the discrete-time case Theorem 20.1.6 establishes this fact together with an extra exponential estimate for the error term. First we begin with a description of "local product flow boxes" and their *full components* of intersection analogous to Chapter 15. These provide the setup in which the equality of Bowen measure and Margulis measure as well as the earlier estimates give the multiplicative asymptotic of the growth of periodic points.

Throughout this section we have the same standing assumption as in the preceding section: $\varphi = \{\varphi^t\}_{t\in\mathbb{R}}$ is a topologically mixing Anosov flow on a compact Riemannian manifold M.

a. Local product flow boxes. We now discuss local product flow boxes. They are simple to describe and involve two size parameters ϵ and δ. We first obtain several technically useful properties which depend on δ being sufficiently small with respect to ϵ.

We begin by picking a positive $\epsilon < 1/2\min\{1,\delta_0\}$, where δ_0 is an expansivity constant of φ^t (see Definition 3.2.11), that is, if $d(\varphi^t(x),\varphi^t(y)) < \delta_0$ for all $t \in \mathbb{R}$ then $y \in \mathcal{O}(x)$. Furthermore we will assume that ϵ is less than the least period of orbits of φ^t. We may choose to decrease ϵ further, but those subsequent choices will not depend on δ. Given $p \in M$ let $C := B^{0u}(p) := \bigcup_{0\le t\le\epsilon}\varphi^t(\overline{B^u(p,\delta)})$. If ϵ is sufficiently small then $B := \bigcup_{z\in C} B^s(z)$ is a local product cube, where $B^s(z) \subset \overline{B^s(z,\delta)}$. For $x \in B$ denote by $B^u(x)$ the connected component of $W^u(x)\cap B$ containing x and similarly define $B^{0u}(x)$ and $B^s(x)$. Define π_C by $C \cap B^s(x) = \{\pi_C(x)\}$. If $z \in C$ then clearly B contains an orbit segment of (parameter) length ϵ. This is true for all $x \in B$:

Lemma 20.6.1. *If $x \in B$ then there exists $t_0 \in [0,\epsilon]$ such that $\{\varphi^t(x) \mid t \in [t_0-\epsilon,t_0]\} \subset B$.*

Proof. Let $z = \pi_C(x)$. If $\varphi^t(z) \in C$ then $\varphi^t(B^s(x)) = B^s(\varphi^t(x))$, hence $\{\varphi^t(x)\} = B^s(\varphi^t(z)) \cap B^{0u}(x) \subset B$. □

Analogously to Proposition 6.4.13 we have

Theorem 20.6.2. *There are η, $\gamma > 0$ such that if $d(x,y) < \eta$ then there is a unique $\theta = \theta(x,y) \in (-\gamma,\gamma)$ such that $\varnothing \ne B^s(x,\gamma)\cap B^u(\varphi^\theta(y)) =: \{[x,y]\}$. θ and $[\cdot,\cdot]$ are continuous on $\{(x,y)\in M\times M \mid d(x,y)<\eta\}$.*

Define $\tau\colon B \to [0,\epsilon]$ by $z \in \varphi^{\tau(z)}(B^u(w))$ when $z \in C$ and $\tau(x) := \tau(\pi_C(x))$ for $x \in B$. For sufficiently small δ uniform continuity of the unstable foliation clearly yields

$$\tau(B^u(x)) \subset [t-\epsilon^2, t+\epsilon^2] \text{ whenever } \tau(x) = t \tag{20.6.1}$$

and $x \in B$. We furthermore assume that δ is also small enough that if $y \in B^u(x)$ then $B^s(x)$ and $B^s(y)$ are ζ-equivalent, where ζ is as in Lemma 20.5.11 applied with our choice of ϵ. Thus

$$\left|\frac{\mu^s(B^s(x))}{\mu^s(B^s(y))} - 1\right| < \epsilon.$$

Lemma 20.6.3. *There exists $K > 0$ (independent of ϵ and δ) such that if $x, y \in B$ then*

$$\left|\frac{\mu^s(B^s(x))}{\mu^s(B^s(y))} - 1\right| < K\epsilon.$$

Proof. Suppose $\tau(x) = t$, $\tau(y) = t'$, $\tau(w) = 0$. Then

$$\begin{aligned}\frac{\mu^s(B^s(x))}{\mu^s(B^s(y))} &= \frac{\mu^s(B^s(x))}{\mu^s(B^s(\varphi^t(w)))}\frac{\mu^s(B^s(\varphi^t(w)))}{\mu^s(B^s(\varphi^{t'}(w)))}\frac{\mu^s(B^s(\varphi^{t'}(w)))}{\mu^s(B^s(y))}\\ &= \frac{\mu^s(B^s(x))}{\mu^s(B^s(\varphi^t(w)))}e^{h(t-t')}\frac{\mu^s(B^s(\varphi^{t'}(w)))}{\mu^s(B^s(y))}.\end{aligned}$$

Both fractions differ from 1 by at most ϵ and thus we obtain the claim. □

We write $x \sim y$ if $x, y \in B$ lie on a common orbit segment contained in B and let $[x] := \{y \in B \mid y \sim x\}$. Relation (20.6.1) shows that if $y \in B^{0u}(x)$ and $\tau(x), \tau(y) \in (\epsilon^2, \epsilon - \epsilon^2)$ then $\bigcup_{z \in B^u(x)}[z] = \bigcup_{z \in B^u(y)}[z]$. For $x \in B$ we set $\Delta(x) := \sup\{r > 0 \mid B^u(x, r) \subset B\}$. For $T > 0$ there exists $r(T)$ (exponentially decreasing as $T \to \infty$) such that if $x, \varphi^T(x) \in B$, $\Delta(x) > r(T)$ then $B^u(\varphi^T(x)) \subset \varphi^T(B)$.

Definition 20.6.4. Let $B^\circ(T) := \{x \in B \mid \epsilon^2 < \tau(x) < \epsilon - \epsilon^2,\ \Delta(x) > r(T)\}$. If Δ_0 is a connected component of $B^\circ(T) \cap \varphi^T(B^\circ(T))$ then $\Delta := \bigcup_{x \in \Delta_0}[x] \cap \varphi^T(B)$ is called a *full component of intersection.*

An important observation is that full components of intersection essentially correspond bijectively to periodic orbits. This is the content of the next two lemmas.

Lemma 20.6.5. *If Δ is a full component of intersection of $B \cap \varphi^T(B)$ then Δ intersects a unique orbit of period in $[T - \epsilon, T + \epsilon]$.*

Proof. This is a standard argument: Consider the action of the iterates of φ^T on the projection of Δ to $B/\sim$ to obtain a unique fixed point which corresponds to the desired orbit. □

Conversely it is easy to check

Lemma 20.6.6. *Each orbit segment in $B^\circ(T)$ of length $\epsilon - 2\epsilon^2$ that belongs to a periodic orbit of period in $[T-(\epsilon-2\epsilon^2), T+(\epsilon-2\epsilon^2)]$ intersects a unique full component of intersection of $B \cap \varphi^T(B)$.*

Next we estimate the number of full components of intersection:

Proposition 20.6.7. *Let $\Delta(T)$ be the number of full components of intersection in $B \cap \varphi^t(B)$. Then $\Delta(T) = 2e^{hT}\mu(B)(1+O(\epsilon))(1+o(T^0))$, with $O(\epsilon)$ independent of B.*

Proof. We will calculate $\mu(\varphi^T(B))$ in (20.6.3) and $\mu(B \cap \varphi^T(B))$ in (20.6.6). Since μ is mixing (see the discussion at the end of Section 20.1), this yields the claim.

If $x, y \in C$ then $B^s(x)$ and $B^s(y)$ are ζ-equivalent, hence $\mu^s(B^s(x)) = (1+O(\epsilon))\mu^s(B^s(y))$, and by (20.5.6) $\mu^s(B^s(x)) = (1+O(\epsilon))\mu^s(B^s(y))$ as well and hence

$$\mu^s(B^s(x)) = C(T)(1+O(\epsilon)), \tag{20.6.2}$$

with $C(T)$ independent of $x \in B$. Since $\mu^{0u}(\varphi^T(C)) = e^{hT}\mu^{0u}(C)$ we get

$$\mu(\varphi^T(B)) = e^{hT}\mu^{0u}(C)C(T)(1+O(\epsilon)), \tag{20.6.3}$$

with $O(\epsilon)$ independent of B.

To calculate $\mu(B \cap \varphi^T(B))$ we first show that it suffices to consider full components of intersection. Any point of $B \cap \varphi^T(B)$ that is not contained in a full component of intersection is in the set

$$A_T := \Big(\varphi^T\Big(\bigcup_{0\le t\le \epsilon^2} B_t\Big) \cap \bigcup_{\epsilon-\epsilon^2\le t\le\epsilon} B_t\Big) \cup \Big(\varphi^T\Big(\bigcup_{\epsilon-\epsilon^2\le t\le\epsilon} B_t\Big) \cap \bigcup_{0\le t\le\epsilon^2} B_t\Big)$$
$$\cup \left(\{x \in B \mid \Delta(x) \le r(x)\} \cap \varphi^T(B_t)\right),$$

namely, such a point is too close to the boundary of B either in the time direction or in an unstable leaf. By mixing each of the first two sets has measure $\epsilon^2\mu(B)^2(1+o(T^0))$. The measure of the third set decreases exponentially with T, so by absorbing it into the error we have

$$\mu(A_T) \le 2\epsilon^2\mu(B)^2(1+o(T^0)). \tag{20.6.4}$$

To prove the proposition it is clearly useful to calculate the measure of full components of intersection via $\Delta(T)$ and their average measure. To calculate the average measure note that a full component of intersection Δ is of the form $\Delta = B \cap \varphi^T(\Delta')$, where $\Delta' = \bigcup\{[x] \mid x \in B,\ \varphi^T(x) \in \Delta\}$ and either $\varphi^T(\Delta' \cap B_\epsilon) \subset \Delta$ (a "front intersection component") or $\varphi^T(\Delta' \cap B_0) \subset \Delta$ (a "back intersection component"). In the first case we define the thickness of Δ by

$$\theta(\Delta) := \inf\{\tau(x) \mid x \in \varphi^T(\Delta' \cap B_\epsilon)\},$$

and in the second by

$$\theta(\Delta) := \epsilon - \sup\{\tau(x) \mid x \in \varphi^T(\Delta' \cap B_0)\}.$$

By (20.6.1) every orbit segment in Δ has (parameter) length $\epsilon(1+O(\epsilon))$; hence by (20.6.2) we have

$$\mu(\Delta) = \frac{1}{\epsilon}\theta(\Delta)\mu^{0u}(C)C(T)(1+O(\epsilon)). \tag{20.6.5}$$

Naturally we have

Lemma 20.6.8. *The average thickness of full components of intersection of* $B \cap \varphi^T(B)$ *is* $(\epsilon/2)(1+O(\epsilon))(1+o(T^0))$.

Proof. Partition B into $n := [1/\epsilon]$ sets $S_i := \bigcup\{B_t \mid \epsilon j \leq tn \leq \epsilon(j+1)\}$ $(0 \leq j < n)$ and note that for sufficiently large T the number of components of $\varphi^T(S_{n-1}) \cap S_i$ is independent of j by mixing, hence the average thickness is $(\epsilon/2)(1+O(1/n))(1+o(T^0))$ as claimed. □

Noticing that the error in (20.6.4) may be absorbed into $O(\epsilon)(1+o(T^0))$ we obtain from (20.6.5) and Lemma 20.6.8

$$\mu(B \cap \varphi^T(B)) = \frac{1}{2}\Delta(T)C(T)\mu^{0u}(C)(1+O(\epsilon))(1+o(T^0)). \tag{20.6.6}$$

Since μ is mixing, (20.6.3) yields

$$\mu(B \cap \varphi^T(B)) = \mu(B)\mu^{0u}(C)e^{hT}C(T)(1+O(\epsilon))(1+o(T^0)),$$

which, with (20.6.6), yields the claim. □

b. The multiplicative asymptotic of orbit growth. In this section we give a multiplicative asymptotic for the growth of the number $P_T(\varphi)$ of periodic orbits of φ^t with period at most T.

Theorem 20.6.9. *Suppose* $\varphi = \{\varphi^t\}_{t\in\mathbb{R}}$ *is a topologically mixing Anosov flow on a compact Riemannian manifold* M. *Then*

$$\lim_{t\to\infty} th_{\text{top}}(\varphi)P_t(\varphi)e^{-th_{\text{top}}(\varphi)} = 1, \text{ that is, } P_t(\varphi) \sim \frac{e^{th_{\text{top}}(\varphi)}}{th_{\text{top}}(\varphi)}.$$

Proof. First note that if μ_B denotes the Bowen measure then $\mu_B(B) = \mu_B(B^\circ(T))(1+O(\epsilon))$ for sufficiently large T and by Lemmas 20.6.5 and 20.6.6

$$\begin{aligned}\frac{\epsilon\Delta(T)}{P_{T,\epsilon}} &= \frac{1}{P_{T,\epsilon}} \sum_{\mathcal{O}\in\text{Per}(T,\epsilon)} \delta_{\mathcal{O}}(B) \leq \mu_B(B)(1+o(T^0)) \\ &= \mu_B(B^\circ(T))(1+O(\epsilon))(1+o(T^0)) \\ &= \frac{1}{P_{T,\epsilon-2\epsilon^2}} \sum_{\mathcal{O}\in\text{Per}(T,\epsilon-2\epsilon^2)} \delta_{\mathcal{O}}(B^\circ(T))(1+O(\epsilon)) \leq \frac{\epsilon\Delta(T)}{P_{T,\epsilon-2\epsilon^2}}(1+O(\epsilon)).\end{aligned}$$

By Proposition 20.6.7 this yields

$$P_{t,\epsilon-2\epsilon^2} \le \frac{2\epsilon e^{hT}\mu(B)}{\mu_B(B)}(1+O(\epsilon))(1+o(T^0)) \le P_{t,\epsilon}.$$

Replacing ϵ by ϵ' with $\epsilon' - 2\epsilon'^2 = \epsilon$ introduces another factor of $1+O(\epsilon)$ and by Theorem 20.5.15 we get

$$P_{T,\epsilon} = 2\epsilon e^{hT}(1+O(\epsilon))(1+o(T^0)). \tag{20.6.7}$$

Since $P_{T,\epsilon}$ is the number of periodic orbits with *a* period in $(T-\epsilon, T+\epsilon]$, we have

$$P_{T,\epsilon} = T_1(P_{T+\epsilon}(\varphi) - P_{T-\epsilon}(\varphi)) + T_2(P_{(T+\epsilon)/2}(\varphi) - P_{(T-\epsilon)/2}(\varphi)) + \cdots,$$

where $iT_i \in [T-\epsilon, T+\epsilon]$. By (20.6.7) this simplifies to $P_{T,\epsilon} = T_1(P_{T+\epsilon}(\varphi) - P_{T-\epsilon}(\varphi))(1+o(T^0))$. Using $T_1 = T(1+o(T^0))$ and (20.6.7) we find

$$P_{T+\epsilon}(\varphi) - P_{T-\epsilon}(\varphi) = \frac{2\epsilon}{T}e^{hT}(1+O(\epsilon))(1+o(T^0)). \tag{20.6.8}$$

Now fix $T_0 > 1/h$ such that $|o(T^0)| < \epsilon$ for all $T \ge T_0$ on the right-hand side. Writing

$$\begin{aligned} P_{T+\epsilon}(\varphi) = {} & (P_{T+\epsilon}(\varphi) - P_{T-\epsilon}(\varphi)) + (P_{T-\epsilon}(\varphi) - P_{T-3\epsilon}(\varphi)) \\ & + \cdots + (P_{T-2j\epsilon+\epsilon}(\varphi) - P_{T-2j\epsilon-\epsilon}(\varphi)) + P_{T-2j\epsilon-\epsilon}(\varphi) \end{aligned}$$

for $T-(2j+1)\epsilon \le T_0 < t - 2j\epsilon$ and estimating the differences by (20.6.8) gives

$$P_{T+\epsilon}(\varphi) = \frac{e^{hT}}{T}S(T)(1+O(\epsilon))(1+o(T^0)),$$

where $S(T) := 2\epsilon\Big(1 + \frac{T}{T-2\epsilon}e^{-2\epsilon h} + \cdots + \frac{T}{T-2j\epsilon}e^{-2j\epsilon h}\Big)$ and we have absorbed $P_{T-2j\epsilon-\epsilon}(\varphi)$ into $(1+o(T^0))$. Observe now that $S(T)$ is a Riemann sum for

$$\begin{aligned} \int_0^{T-T_0} \frac{T}{T-x}e^{-hx}\,dx &= Te^{-hT}\int_{T_0}^T \frac{e^{hu}}{u}du \\ &= Te^{-hT}\left(\frac{e^{hT}}{hT} - \frac{e^{hT_0}}{hT_0} - T\int_{T_0}^T \frac{e^{hu}}{hu^2}du\right) = \frac{1}{h}(1+o(T^0)). \end{aligned}$$

The integrand decreases on $[0, T-T_0]$, so $S(T) = \frac{1}{h}(1+O(\epsilon))(1+o(T^0))$ and hence

$$\begin{aligned} P_T(\varphi) = P_{T+\epsilon}(\varphi)(1+O(\epsilon)) &= \frac{e^{hT}}{T}S(T)(1+O(\epsilon))(1+o(T^0)) \\ &= \frac{e^{hT}}{hT}(1+O(\epsilon))(1+o(T^0)), \end{aligned}$$

that is, $\lim_{T\to\infty} \frac{P_T(\varphi)}{e^{ht}/hT} = (1+O(\epsilon))$ for any ϵ, proving the claim. □

Theorem 20.6.10. (Margulis Theorem) *Let M be a compact Riemannian manifold of negative sectional curvature, $G(t)$ the number of different closed geodesics of length at most t, and h the topological entropy of the geodesic flow. Then* $\lim_{t\to\infty} G(t)2the^{-th} = 1$.

Proof. Each closed geodesic of length t defines exactly two period-t orbits of the geodesic flow. The geodesic flow is a contact flow by Proposition 5.6.3 and Anosov by Theorem 17.6.2, hence topologically mixing by Theorem 18.3.6. Thus Theorem 20.6.9 applies. □

Supplement

Supplement

Dynamical systems with nonuniformly hyperbolic behavior

by
Anatole Katok and
Leonardo Mendoza

1. Introduction

The purpose of this work is to present some of the key ideas, methods, and applications of the theory of smooth dynamical systems with nonuniformly hyperbolic behavior, often called the Pesin theory due to the landmark work of Ya. B. Pesin in the mid-seventies [Pes1], [Pes2], [Pes3]. The central idea of the theory is the combination of the hyperbolic behavior for the linearized system, which can be sharpened and amplified by the crucial technical devices of the Lyapunov metric (see Section S.2d) and regular neighborhoods (Definition S.3.3), with the presence of nontrivial recurrence, an essentially nonlinear phenomenon. This combination produces a rich and complicated orbit structure manifested by the growth of the number of periodic orbits at an exponential rate determined by entropy, the presence of large hyperbolic invariant sets of the horseshoe type, the determination of cohomology classes of Hölder cocycles by periodic data, and so forth. By the classical theorem of Luzin the hyperbolic estimates (which are obviously given by Borel functions) are uniform on a set of large measure. Controlling the recurrence on such a set allows us to produce the aforementioned phenomena. Remarkably, a considerable part of the theory can be developed without constructing the families of stable (contracting) and unstable (expanding) manifolds which are usually presented as the cornerstone of the subject. Instead it is sufficient to use their approximations, admissible manifolds (see Definition S.3.4), which can be constructed in a straightforward fashion.

It is exactly this part of the nonuniformly hyperbolic theory that is treated in the present work. In order to make the presentation lighter, we for the most part restrict the argument to the case of invertible maps of a two-dimensional manifold. Some of the principal applications of the theory (see [K5], [K9], and Section S.5) deal with this case anyway.

This work can be viewed as a modest step toward the more ambitious goal of presenting the theory of nonuniformly hyperbolic dynamical systems in its up-to-date form from a unified point of view based on the technical devices of ϵ-reduction (see Section S.2d), regular neighborhoods, and admissible manifolds. A core of that project is realized in our unpublished and unfinished notes "Smooth Ergodic Theory" which were allowed a limited circulation. These notes contain all the material of the present work in the general case including a complete proof of the multiplicative ergodic theorem. Aside from that there is an extensive treatment of several classes of examples, a much more thorough discussion of regular neighborhoods including volume estimates, a proof of local ergodicity, and some material concerning families of stable and unstable manifolds. At the present moment it is hard to predict the future of this project. We hope, however, to come back to it, maybe within the context of an even broader exposition of smooth ergodic theory.

This work is not an expository article in the narrow sense of the word, that is, a presentation with or without proofs of results that have appeared elsewhere. It is closer to an "Uspehi" article, that is, the type of survey article published in the Russian journal "Uspehi Mathematichekih Nauk", translated into English under the name "Russian Mathematical Surveys". In fact, the word "uspehi" is literally translated as "advances" or "achievements". An "Uspehi" article usually presents the author's point of view on its subject and contains some previously unpublished results. Our point of view has been briefly explained above. Among previously unpublished (although known to us for a while) results are the Shadowing Lemma (Theorem S.4.14), the Livschitz-type Theorem S.4.17, the Spectral Decomposition Theorem S.5.8, and most importantly Theorem S.5.9 and its corollaries which where announced in the first author's 1983 ICM address [K9].

2. Lyapunov exponents

a. Cocycles over dynamical systems. We begin by summarizing the theory of Lyapunov exponents in the context of linear cocycles over measure-preserving transformations. Let $f: X \to X$ be an invertible measure-preserving transformation of a Lebesgue space $(X, \mathcal{B}, \mu)$. We will always assume that μ is a probability measure, that is, $\mu(X) = 1$.

Let $GL(n, \mathbb{R})$ denote the group of invertible linear transformations of $\mathbb{R}^n$.

Then for any measurable function $A: X \to GL(n, \mathbb{R})$ and $x \in X$ if we set

$$\begin{aligned} \mathcal{A}(x, m) &= A(f^{m-1}(x)) \cdots A(x) && \text{for } m > 0, \\ \mathcal{A}(x, m) &= A(f^{m}(x))^{-1} \cdots A(f^{-1}(x))^{-1} && \text{for } m < 0, \end{aligned} \tag{S.2.1}$$

and

$$\mathcal{A}(x, 0) = \mathrm{Id},$$

then it follows that

$$\mathcal{A}(x, m+k) = \mathcal{A}(f^k(x), m)\mathcal{A}(x, k). \tag{S.2.2}$$

Definition S.2.1. We call any measurable function $\mathcal{A}: X \times \mathbb{Z} \to GL(n, \mathbb{R})$ satisfying (S.2.2) a *measurable linear cocycle over* f, or simply a cocycle.

Any cocycle $\mathcal{A}$ has the form (S.2.1) with $A(x) = \mathcal{A}(x, 1)$. The map A is called the *generator* of the cocycle $\mathcal{A}$. Sometimes, if it does not cause confusion, we will not make a distinction between a cocycle and its generator and we will refer to the latter as a cocycle.

A cocycle $\mathcal{A}$ over f induces a linear extension F of f to $X \times \mathbb{R}^n$ defined by

$$F(x, v) = (f(x), A(x)v),$$

so for $m \in \mathbb{Z}$ the mth iterate of F is equal to

$$F^m(x, v) = (f^m(x), \mathcal{A}(x, m)v).$$

Definition S.2.2. A measurable function $C: X \to GL(n, \mathbb{R})$ is said to be *tempered with respect to* f, or simply *tempered*, if for almost every $x \in X$

$$\lim_{m \to \infty} \frac{1}{m} \log \|C^{\pm 1}(f^m(x))\| = 0.$$

Definition S.2.3. If $A, B: X \to GL(n, \mathbb{R})$ are measurable maps defining cocycles $\mathcal{A}, \mathcal{B}$ over f, then $\mathcal{A}$ and $\mathcal{B}$ are said to be *equivalent* if there exists a measurable tempered function $C: X \to GL(n, \mathbb{R})$ such that for almost every $x \in X$

$$A(x) = C^{-1}(f(x))B(x)C(x).$$

This is clearly an equivalence relation and if two cocyles $\mathcal{A}, \mathcal{B}$ are equivalent we write $\mathcal{A} \sim \mathcal{B}$. Also we say that a cocycle $\mathcal{A}$ over f is *tempered* if its generator A is tempered.

We say that cocycle $\mathcal{A}$ is *rigid* if it is equivalent to a cocycle independent of x, that is, given by the powers of a single matrix.

Lemma S.2.4. *If $f: X \to X$ is a measure-preserving transformation, $C: X \to GL(n, \mathbb{R})$ is measurable, and*

$$\max\{\log \|C(x)\|, \log \|C^{-1}(x)\|\} \in L^1(X, \mu)$$

then C is tempered with respect to f.

Proof. By the Birkhoff Ergodic Theorem 4.1.2 applied to $\log \|C(x)\|$ for almost every $x \in X$

$$\begin{aligned}\lim_{n\to\infty} \frac{1}{n} \sum_{k=0}^{n-1} \log \|C(f^k(x))\| &= \lim_{n\to\infty} \frac{1}{n} \sum_{k=0}^{n} \log \|C(f^k(x))\| \\ &= \lim_{n\to\infty} \frac{1}{n} \Big(\log \|C(f^n(x))\| + \sum_{k=0}^{n-1} \log \|C(f^k(x))\| \Big),\end{aligned}$$

so $\lim_{n\to\infty} \frac{1}{n} \log \|C(f^n(x))\| = 0$. Similarly for $n \to -\infty$. □

Definition S.2.5. For a cocycle $\mathcal{A}: X \to GL(n, \mathbb{R})$ over a transformation $f: X \to X$ and for $(x, v) \in X \times \mathbb{R}^n$ the (possibly infinite) number

$$\overline{\chi}^+(x, v, \mathcal{A}) := \overline{\chi}^+(x, v) := \overline{\lim_{m\to\infty}} \frac{1}{m} \log \|\mathcal{A}(x, m)v\|$$

is called the *upper Lyapunov exponent of (x, v) with respect to the cocycle $\mathcal{A}$.* If $\lim_{m\to\infty}(1/m) \log \|\mathcal{A}(x, m)v\|$ exists then we denote it by $\chi^+(x, v)$ and call it the *Lyapunov exponent of (x, v) with respect to the cocycle $\mathcal{A}$.*

Lemma S.2.6. *Given a linear cocycle $\mathcal{A}$ over f:*

(1) *For $(x, v) \in X \times \mathbb{R}$ and $\lambda \in \mathbb{R} \setminus \{0\}$ we have $\overline{\chi}^+(x, v) = \overline{\chi}^+(x, \lambda v)$.*
(2) *If $v, w \in \mathbb{R}^n$, then $\overline{\chi}^+(x, v + w) \le \max\{\overline{\chi}^+(x, v), \overline{\chi}^+(x, w)\}$.*
(3) *If $\overline{\chi}^+(x, v) \ne \overline{\chi}^+(x, w)$, then $\overline{\chi}^+(x, v + w) = \max\{\overline{\chi}^+(x, v), \overline{\chi}^+(x, w)\}$.*

Proof. Part (1) is obvious. Part (2) follows from

$$\|\mathcal{A}(x, m)(v + w)\| \le 2 \max\{\|\mathcal{A}(x, m)v\|, \|\mathcal{A}(x, m)w\|\}.$$

For (3) suppose $\overline{\chi}^+(x, v) < \overline{\chi}^+(x, w)$; then $\overline{\chi}^+(x, v+w) \le \overline{\chi}^+(x, w) = \overline{\chi}^+(x, w - v + v) \le \max\{\overline{\chi}^+(x, w + v), \overline{\chi}^+(x, -v)\} = \overline{\chi}^+(x, v + w)$. □

It follows from Lemma S.2.6 that given a cocycle $\mathcal{A}$, then for each real number χ and each $x \in X$ the set $E_\chi(x) = \{v \in \mathbb{R}^n \mid \overline{\chi}^+(x, v) \le \chi\}$ is a linear subspace of $\mathbb{R}^n$ and if $\chi_1 \ge \chi_2$ we have $E_{\chi_2}(x) \subset E_{\chi_1}(x)$. Furthermore for each $x \in X$ there exists an integer $k(x) \le n$ and a collection of numbers and linear subspaces

$$\chi_1(x) < \chi_2(x) < \cdots < \chi_{k(x)}(x),$$
$$\{0\} \subset E_{\chi_1}(x) \subset E_{\chi_2}(x) \subset \cdots \subset E_{\chi_{k(x)}}(x) = \mathbb{R}^n$$

such that if $v \in E_{\chi_{i+1}}(x) \setminus E_{\chi_i}(x)$ then $\chi^+(x,v) = \chi_{i+1}(x)$. These numbers will be called the *upper Lyapunov exponents at x with respect to the cocycle $\mathcal{A}$.* We refer to this collection of linear subspaces as the *filtration at x associated to the cocycle $\mathcal{A}$.* Let us call the number

$$l_i(x) = \dim E_{\chi_i}(x) - \dim E_{\chi_{i-1}}(x)$$

the *multiplicity of the exponent* $\chi_i(x)$; we define the *spectrum of $\mathcal{A}$ at x* as the collection of pairs

$$\mathrm{Sp}_x \mathcal{A} = \{(\chi_i(x), l_i(x)) \mid i = 1, \dots, k(x)\}.$$

Proposition S.2.7. *If $\mathcal{A}$ and $\mathcal{B}$ are two equivalent cocycles over the measure-preserving transformation $f: X \to X$, then for almost every $x \in X$*

$$\mathrm{Sp}_x \mathcal{A} = \mathrm{Sp}_x \mathcal{B}.$$

Proof. Since $\mathcal{A} \sim \mathcal{B}$ there exists a tempered measurable function $C: X \to GL(n,\mathbb{R})$ such that $\mathcal{A}(x,m) = C^{-1}(f^m(x))\mathcal{B}(x,m)C(x)$ for almost every $x \in X$. Therefore for $v \in \mathbb{R}^n$

$$\varlimsup_{m\to\infty} \frac{\log \|\mathcal{B}(x,m)v\|}{m} \le \varlimsup_{m\to\infty} \frac{\log \|C(f^m(x))\|}{m} + \varlimsup_{m\to\infty} \frac{\log \|\mathcal{A}(x,m)C^{-1}(x)v\|}{m},$$

which implies the proposition. □

b. Examples of cocycles.

i. Let $X = \{x\}$ consist of a single element, so if $A \in GL(n,\mathbb{R})$ then $\mathcal{A}(x,m) = A^m$ and the upper Lyapunov exponents are given by the logarithms of the absolute values of the eigenvalues of A. Moreover in this case the limits always exist.

ii. Suppose $f: M \to M$ is a diffeomorphism of a compact n-dimensional Riemannian manifold M preserving a Borel probability measure μ. The derivative Df of f acts on the tangent bundle TM and for $x \in M$

$$D_x f: T_x M \to T_{f(x)} M$$

is a linear map of a vector space isomorphic to $\mathbb{R}^n$.

Now represent M as a finite union of diffeomorphic copies of the n-simplex, say $M = \bigcup \Delta_i$ such that in each Δ_i we can introduce local coordinates so that $T\Delta_i = \Delta_i \times \mathbb{R}^n$ and all the nonempty intersections $\Delta_i \cap \Delta_j$ are $(n-1)$-dimensional manifolds. By slightly perturbing the boundaries $\partial\Delta_i$ of the Δ_i if necessary we can always obtain a decomposition $\{\Delta_i\}_{i=1}^r$ such that $\mu(\partial\Delta_i)=0$ for all $i=1,\dots,r$. Thus we have that

$$M=\bigcup_i \operatorname{Int}\Delta_i \quad (\mathrm{mod}\ 0)$$

and on each Δ_i the tangent bundle is trivial. Therefore the derivative Df can be interpreted as a linear cocycle

$$Df\colon M=\bigcup_i \operatorname{Int}\Delta_i \to \mathbb{R}^n$$

with $Df(x)$ being the matrix representing the derivative at x in local coordinates, so the cocycle Df depends on the choices of decomposition of M and local coordinates. However, for another choice of local coordinates the coordinate change sending one representation into another is uniformly bounded together with its inverse, so the conditions of the lemma hold and therefore the coordinate change is tempered almost everywhere. In particular Proposition S.2.7 implies that the spectrum of the derivative cocycle does not depend on the coordinate representation.

iii. Let $f\colon X\to X$ be a measure-preserving transformation of the Lebesgue space (X,μ). If $\mathcal{A}$ is a measurable linear cocycle over f then for $n\geq 1$ we can consider the transformation $f^n\colon X\to X$ and the linear cocycle $\mathcal{A}^n$ with generator

$$A^n(x)=A(f^{n-1}(x))\cdots A(x).$$

Clearly $\bar{\chi}_1^+(x)(x,v,\mathcal{A}^n)=n\bar{\chi}_1^+(x)(x,v,\mathcal{A})$ for every $v\neq 0$.

iv. Let $f\colon X\to X$ be a measure-preserving transformation of the Lebesgue space (X,μ). For any measurable subset $Y\subset X$ with $\mu(Y)>0$ the Poincaré Recurrence Theorem 4.1.19 produces a transformation $f_Y\colon Y\to Y$ mod 0 defined as follows: For $x\in Y$, let

$$r_Y(x):=\min\{k\geq 1 \mid f^k(x)\in Y\} \text{ and } f_Y(x):=f^{r_Y}(x).$$

We call f_Y the *induced transformation* on Y. If $\mathcal{A}$ is a measurable linear cocycle over f define the *induced cocycle* $\mathcal{A}_Y$ *over* f_Y by

$$A_Y(x)=A^{r_Y(x)}(x).$$

Almost everywhere we can define the ergodic average of the return time r_Y as

$$\tau(x)=\lim_{k\to\infty}\frac{1}{k}\sum_{i=0}^{k-1} r_Y(f_Y^i(x)).$$

Lemma S.2.8. *Let $Y \subset X$ with $\mu(Y) > 0$. Then for $v \neq 0$*

$$\bar{\chi}^+(x, v, \mathcal{A}_Y) = \tau(x) \cdot \bar{\chi}^+(x, v, \mathcal{A}).$$

Proof. Let $\tau_k(x) = \sum_{i=0}^{k-1} r_Y(f^i(x))$; then

$$\begin{aligned}\bar{\chi}^+(x, v, \mathcal{A}_Y) &= \lim_{k\to\infty} \frac{1}{k} \log \|(A_Y)^k v\| \\ &= \lim_{k\to\infty} \frac{1}{k} \log \|A^{\tau_k(x)} v\| = \bar{\chi}^+(x, v, \mathcal{A}) \lim_{k\to\infty} \frac{\tau_k(x)}{k}.\end{aligned}$$ □

c. The Multiplicative Ergodic Theorem. In this section we recall the Osedelec Multiplicative Ergodic Theorem [Os] for measurable cocycles over measure-preserving transformations of Lebesgue spaces. This theorem has been reproved and generalized a number of times; see, for example, [Rue6], [Ma]. In the case of surface diffeomorphisms, Kingman's Subadditive Ergodic Theorem would be sufficient for our purposes. However, the wide scope of Osedelec's result suggested its inclusion in these notes.

Theorem S.2.9. (Osedelec Multiplicative Ergodic Theorem) *Suppose $f: X \to X$ is a measure-preserving transformation of a Lebesgue space (X, μ) and let $A: X \to \mathbb{R}^n$ be a measurable cocycle over X. If*

$$\log^+ \|A^{\pm 1}(x)\| \in L^1(X, \mu),$$

then there exists a set $Y \subset X$ such that $\mu(X \setminus Y) = 0$ and for each $x \in Y$:

(1) *There exists a decomposition of $\mathbb{R}^n$ as*

$$\mathbb{R}^n = \bigoplus_{i=1}^{k(x)} H_i(x)$$

that is invariant under the linear extension of f determined by A. The Lyapunov exponents

$$\chi_1(x) < \cdots < \chi_{k(x)}(x)$$

exist and are f-invariant and

$$\lim_{m\to\pm\infty} \frac{1}{|m|} \log \frac{\|\mathcal{A}(x, m)v\|}{\|v\|} = \pm\chi_i(x) \tag{S.2.3}$$

uniformly in $v \in H_i(x) \setminus \{0\}$.

(2) *For $S \subset N := \{1, \ldots, k(x)\}$ let $H_S(x) := \bigoplus_{i\in S} H_i(x)$. Then*

$$\lim_{m\to\infty} \frac{1}{m} \log \sin |\measuredangle(H_S(f^m(x)), H_{N\setminus S}(f^m(x)))| = 0.$$

d. Osedelec–Pesin ϵ-Reduction Theorem. Let $f: X \to X$ be a measure-preserving transformation of a Lebesgue space (X, μ) and consider a measurable cocycle $A: X \to GL(n, \mathbb{R})$ satisfying the integrability condition $\log^+ \|A^{\pm 1}(x)\| \in L^1(X, \mu)$. The best cocycles to deal with are the rigid ones. There are several situations where one can show that all cocycles from certain classes are rigid: the classical Floquet theory (where the dynamical system in the base is a periodic flow), smooth cocycles over translations on the torus with rotation vector satisfying a Diophantine condition, measurable cocycles over actions of "large" groups [Z], and certain classes of smooth cocycles over hyperbolic actions of higher-rank Abelian groups [KS1], [KS2]. However, in our situation this is usually not the case. Instead there is a weaker property which allows us to reduce every cocycle to a constant one up to arbitrarily small error. Namely, we will show that for a given $\epsilon > 0$ there exists an equivalent cocycle A_ϵ having block form and such that for each block $A_\epsilon^i(x)$ and $v \in H_i(x)$ we have

$$e^{\chi_i(x)-\epsilon} \leq \|A_\epsilon^i(x)v\| \leq e^{\chi_i(x)+\epsilon}$$

for all x in the set Y provided by the Osedelec theorem. We say that A_ϵ is the *ϵ-reduced form.*

Theorem S.2.10. (Osedelec–Pesin ϵ-Reduction Theorem) *Suppose that $A: X \to GL(n, \mathbb{R})$ is a measurable cocycle over the measure-preserving transformation $f: X \to X$ of the Lebesgue space (X, μ). If $\log^+ \|A^{\pm 1}(x)\| \in L^1(X, \mu)$ then there exists a measurable f-invariant function $k: X \to \mathbb{N}$, and numbers $\chi_1(x), \dots, \chi_{k(x)}(x) \in \mathbb{R}$,*

$$l_1(x), \dots, l_{k(x)}(x) \in \mathbb{N}$$

depending measurably on x with $\sum l_i(x) = n$ such that for every $\epsilon > 0$ there exists a tempered map

$$C_\epsilon: X \to GL(n, \mathbb{R})$$

such that for almost every $x \in X$ the cocycle

$$A_\epsilon(x) = C_\epsilon^{-1}(f(x))A(x)C_\epsilon(x)$$

has the following Lyapunov block form

$$A_\epsilon(x) = \begin{pmatrix} A_\epsilon^1(x) & & & \\ & A_\epsilon^2(x) & & \\ & & \ddots & \\ & & & A_\epsilon^{k(x)}(x) \end{pmatrix},$$

where each $A_\epsilon^i(x)$ is an $l_i(x) \times l_i(x)$ matrix and

$$e^{\chi_i(x)-\epsilon} \leq \|A_\epsilon^i(x)^{-1}\|^{-1}, \ \|A_\epsilon^i(x)\| \leq e^{\chi_i(x)+\epsilon}.$$

Furthermore $k(x)$ and $\chi_i(x)$ are as in Theorem S.2.9 and for almost every $x \in X$ we can decompose $\mathbb{R}^n$ as $\bigoplus_{i=1}^{k(x)} H_i(x)$ such that $l_i(x) = \dim H_i(x)$ and $C_\epsilon(x)$ sends the standard decomposition $\bigoplus_{i=1}^{k(x)} \mathbb{R}^{l_i(x)}$ to $\bigoplus_{i=1}^{k(x)} H_i(x)$.

Proof. Let $Y \subset X$ be the set provided by Theorem S.2.9. Thus if $x \in Y$ then $\mathbb{R}^n = \bigoplus H_i(x)$. Define a new scalar product on each $H_i(x)$ and $\epsilon > 0$ as follows: If $u, v \in H_i(x)$ then

$$\langle u, v\rangle'_{x,i} := \sum_{m\in\mathbb{Z}} \langle \mathcal{A}(x,m)u, \mathcal{A}(x,m)v\rangle e^{-2m\chi_i(x)} e^{-2\epsilon|m|},$$

where $\langle \cdot, \cdot\rangle$ denotes the standard scalar product on $\mathbb{R}^n$. We call the sequence $\{e^{-2m\chi_i(x)} e^{-2\epsilon|m|}\}_{m\in\mathbb{Z}}$ the *Pesin tempering kernel.* Now observe that by (S.2.3) for each $x \in Y$ and $\epsilon > 0$ there exists a constant $C_i(x, \epsilon)$ such that

$$\|\mathcal{A}(x,m)v\| \le C_i(x,\epsilon) e^{m\chi_i(x)} e^{\epsilon|m|/2} \|v\|,$$

and therefore

$$\langle u, v\rangle'_{x,i} \le C_i^2(x,\epsilon) \sum_{m\in\mathbb{Z}} e^{-|m|\epsilon},$$

which implies that $\langle u, v\rangle'_{x,i}$ is well defined.

We recall that $\mathcal{A}(x, m+1) = \mathcal{A}(f(x), m)A(x)$, whence

$$\begin{aligned}\langle A(x)v, A(x)v\rangle'_{f(x),i} &= \sum_{m\in\mathbb{Z}} \|\mathcal{A}(f(x),m)A(x)v\|^2 e^{-2\chi_i(f(x))m} e^{-2\epsilon|m|} \\ &= \sum_{m\in\mathbb{Z}} \|\mathcal{A}(x,m+1)v\|^2 e^{-2\chi_i(x)m} e^{-2\epsilon|m|} = \sum_{k\in\mathbb{Z}} \|\mathcal{A}(x,k)v\|^2 e^{-2\chi_i(x)k} e^{-2\epsilon|k|} e^{\psi}\end{aligned}$$

with $2\chi_i(x) - 2\epsilon \le \psi := 2\chi_i(x) - 2\epsilon(|k-1| - |k|) \le 2\chi_i(x) + 2\epsilon$, so we have

$$e^{2(\chi_i(x)-\epsilon)} \le \frac{\langle A(x)v, A(x)v\rangle'_{f(x),i}}{\langle v, v\rangle'_{x,i}} \le e^{2(\chi_i(x)+\epsilon)}. \tag{S.2.4}$$

To extend the scalar product to $\mathbb{R}^n$, consider

$$\langle u, v\rangle'_x = \sum_{i=1}^{k(x)} \langle v_i, u_i\rangle'_{x,i},$$

where v_i is the projection of v to $H_i(x)$ with respect to $\bigoplus_{j=1}^{k(x)} H_j(x)$.

Now let $C_\epsilon(x)$ be the positive symmetric matrix such that if $u, v \in \mathbb{R}^n$ then

$$\langle u, v\rangle'_x = \langle C_\epsilon(x)u, C_\epsilon(x)v\rangle$$

and define

$$A_\epsilon(x) = C_\epsilon^{-1}(f(x))A(x)C_\epsilon(x).$$

Thus if $u, v \in H_i(x)$, then

$$\begin{aligned}\langle A(x)u, A(x)v\rangle &= \langle C_\epsilon^{-1}(f(x))A(x)u, C_\epsilon^{-1}(f(x))A(x)v\rangle'_{f(x),i}\\ &= \langle A_\epsilon(x)C_\epsilon^{-1}(x)u, A_\epsilon(x)C_\epsilon^{-1}(x)v\rangle'_{f(x),i},\end{aligned}$$

so applying (S.2.4) to $v = C_\epsilon^{-1}(x)u$ we obtain

$$e^{2(\chi_i(x)-\epsilon)} \leq \frac{\|A_\epsilon(x)v\|^2}{\|v\|^2} \leq e^{2(\chi_i(x)+\epsilon)}.$$

It remains to prove that $C_\epsilon(x)$ is tempered. Since the angles between the different subspaces satisfy a subexponential lower estimate due to Theorem S.2.9(2), it is enough to consider just block matrices. Notice that the function $\log \|A_\epsilon^{\pm 1}\|$ is bounded and hence Theorem S.2.9 can be applied to A_ϵ. Set $X_N = \{x \in X \mid \|C_\epsilon^{\pm 1}(x)\| < N\}$. For some $N > 0$ large enough, by the Poincaré Recurrence Theorem 4.1.19 there exists a set $Y \subset X_N$ such that $\mu(X_N \setminus Y) = 0$ and the orbit of $y \in Y$ returns infinitely many times to Y. Thus let m_k be a sequence such that $f^{m_k}(y) \in Y$ for all k. Then

$$\|\mathcal{A}_\epsilon(y, m_k)\| \leq \|C_\epsilon^{-1}(f^{m_k}(y))\| \|\mathcal{A}(y, m_k)\| \|C_\epsilon^{-1}(y)\|$$

and therefore for almost every point $y \in Y$ the spectra of $\mathcal{A}_\epsilon$ and A are the same. Since N is chosen arbitrarily this is true for almost every $x \in X$. Observe that

$$C_\epsilon(f^n(x)) = \mathcal{A}(x,n)C_\epsilon(x)\mathcal{A}_\epsilon^{-1}(x,n) \text{ and } \mathcal{A}(x,n) = C_\epsilon(f^n(x))\mathcal{A}_\epsilon(x,n)C_\epsilon^{-1}(x),$$

so by taking growth rates in both equations we find that

$$\overline{\lim_{n\to\infty}} \frac{1}{n} \log \|C_\epsilon(f^n(x))\| = 0$$

for all x for which $\mathcal{A}_\epsilon$ and $\mathcal{A}$ have the same spectrum. □

Definition S.2.11. The new scalar product $\langle\cdot,\cdot\rangle'_x$ is called the *Lyapunov scalar product*, and the norm $\|\cdot\|'_x$ induced by it the *Lyapunov norm* and the *Lyapunov metric*, correspondingly. We refer to $C_\epsilon(x)$ as the Lyapunov change of coordinates.

Lemma S.2.12. (Tempering-Kernel Lemma) *Let $f: X \to X$ be a measure-preserving transformation of a Lebesgue space (X, μ). If $K: X \to \mathbb{R}$ is a measurable function satisfying* (1) $K(x) > 0$ *and*

(2) $\lim_{m\to\infty}(1/m)(\log K(f^m(x)) = 0$

then for any $\epsilon > 0$ there exists a measurable $K_\epsilon: X \to \mathbb{R}$ such that

$$K_\epsilon(x) > K(x) \text{ and } e^{-\epsilon} \leq \frac{K_\epsilon(x)}{K_\epsilon(f(x))} \leq e^\epsilon.$$

Proof. By (2) there exists, for a given $\epsilon > 0$, a measurable function $C(x,\epsilon)$ such that

$$K(f^m(x)) \leq C(x,\epsilon)e^{|m|\epsilon/2}.$$

Then

$$K_\epsilon(x) := \sum_{m\in\mathbb{Z}} K(f^m(x))e^{-|m|\epsilon} \leq C(x,\epsilon)\sum_{m\in\mathbb{Z}} e^{-|m|\epsilon/2}$$

is well defined and

$$\begin{aligned} K_\epsilon(f(x)) &= \sum_{m\in\mathbb{Z}} K(f^{m+1}(x))e^{-|m|\epsilon} \\ &= \sum_{k\in\mathbb{Z}} K(f^k(x))e^{-|k-1|\epsilon}e^{(-|k|+|k|)\epsilon} = \sum_{k\in\mathbb{Z}} K(f^k(x))e^{-|k|\epsilon}e^{(-|k|-|k-1|)\epsilon}, \end{aligned}$$

from which the result follows. □

Applying the Tempering-Kernel Lemma S.2.12 to the function $\|C_\epsilon(x)\|$ which is nonzero almost everywhere we obtain the following inequality between the Euclidean norm and the Lyapunov norm: $\|\cdot\| \leq \|\cdot\|'_x \leq K_\epsilon(x)\|\cdot\|$, and $e^{-\epsilon} \leq K_\epsilon(f(x))/K_\epsilon(x) \leq e^\epsilon$.

e. The Ruelle inequality. In this section we prove a basic result in the ergodic theory of diffeomorphisms. It is known as the Ruelle inequality, although it should be mentioned that Margulis proved it earlier for the volume-preserving case. This result connects metric entropy with the sum of the positive Lyapunov exponents, and it will prove to be a very useful tool in proving the existence of measures with some exponents different from zero. The importance of this inequality is based on the immediate corollary that if the topological entropy of a diffeomorphism is not zero then there exists a measure with some of its exponents positive. In the case of surfaces it implies nonzero exponents. Let us point out that the positivity of topological entropy can sometimes be determined by different methods as we noted on various occasions in Chapter 8. Some of those methods require only topological information (Theorem 8.1.1, Corollary 8.1.3, Theorem 8.3.1).

Theorem S.2.13. (Ruelle inequality) [Rue4] *Let $f \in \mathrm{Diff}^1(M)$ and M a compact Riemannian manifold and suppose μ is an f-invariant Borel probability measure. Then*

$$h_\mu(f) \leq \int \chi^+(x)\,d\mu,$$

where $\chi^+(x) = \sum_{i:\chi_i(x)>0}\chi_i(x)$.

Corollary S.2.14. *Let $f \in \mathrm{Diff}^1(M)$ with $h(f) > 0$; then there exists an ergodic f-invariant measure μ with at least one positive and one negative Lyapunov exponent.*

Proof. By the Variational Principle for entropy (Theorem 4.5.3) there is an f-invariant Borel probability measure μ with $h_\mu(f) > 0$. Let $\{\mu_\alpha\}_{\alpha\in A}$ be an ergodic decomposition of μ (Theorem 4.1.12), so

$$h_\mu(f) = \int_A h_{\mu_\alpha}(f)\, d\alpha \tag{S.2.5}$$

(cf. Corollary 4.3.17 and the subsequent discussion). Hence for some $\alpha \in A$ we have $h_{\mu_\alpha}(f) > 0$. Now apply the Ruelle inequality to see that $\chi^+_{\mu_\alpha} > 0$ and hence μ_α has at least one positive Lyapunov exponent. Applying the Ruelle inequality to f^{-1} and using the fact that $h_{\mu_\alpha}(f^{-1}) = h_{\mu_\alpha}(f) > 0$ shows that f^{-1} must also have a positive exponent for μ_α which is the negative of the negative exponent for f. □

Proof of Theorem S.2.13. By (S.2.5) we may assume that μ is ergodic, whence $\chi_i(x) = \chi_i$ are constant a.e. Fix $m > 0$ and using compactness choose d_m such that for $0 < d \le d_m$ and $x \in B(y,d)$ (the ball around y of radius d)

$$\frac{1}{2} D_x f^m(\exp_x^{-1}(B(y,d))) \subset \exp^{-1}_{f^m(x)}(f^m(B(y,d))) \subset 2D_x f^m(\exp_x^{-1}(B(y,d))), \tag{S.2.6}$$

where $\alpha B = \{\alpha x \mid x \in B\}$ for $\alpha \in \mathbb{R}$.

Now fix $\epsilon > 0$ and choose a partition ξ satisfying the following conditions:

(1) $\operatorname{diam} \xi \le d_m/10$ and $h_\mu(f^m, \xi) \ge h_\mu(f^m) - \epsilon$.

(2) For every $C \in \xi$ there exist balls B, B' with radii r, r', respectively, such that $r < 2r' \le d_m/20$ and $B' \subset C \subset B$.

(3) If $C \in \xi$, there exists $0 < r < d_m/20$ such that $C \subset B(y,r)$ for some $y \in M$, and if $x \in C$ we have

$$\frac{1}{2} D_x f^m(\exp_x^{-1} B(y,r)) \subset \exp^{-1}_{f^m(x)}(f^m(C)) \subset 2D_x f^m(\exp_x^{-1} B(y,r)).$$

The idea here is that at the scale of order d_m the mth iterate of f and its derivative do not differ very much.

To show the existence of such partitions first consider for any $\alpha > 0$ and a maximal α-separated set Γ the *Dirichlet regions*

$$\mathcal{D}_\Gamma(x) = \{y \in M \mid d(y,x) \le d(y,z) \text{ for all } z \in \Gamma, z \ne x\}.$$

Obviously $B(x,\alpha/2) \subset \mathcal{D}_\Gamma(x) \subset B(x,\alpha)$. Since the $\mathcal{D}_\Gamma(x)$ only intersect on the boundaries, that is, submanifolds of codimension one, we may move the boundaries slightly if necessary, so that they have measure 0 and therefore we have a partition.

For each $\alpha < d_m/20$ we can find such a partition, so there exists a partition ξ with

$$h_\mu(f^m, \xi) > h_\mu(f^m) - \epsilon$$

and $\operatorname{diam} \xi < d_m/10$. Hence (1) and (2) are satisfied. Part (3) follows from (S.2.6).

If $H(\xi \mid D)$ is the entropy of ξ with respect to the conditional measure on D then

$$h_\mu(f^m,\xi) = \lim_{k\to\infty} H(\xi \mid f^m(\xi) \vee \cdots \vee f^{km}(\xi)) \le H(\xi \mid f^m(\xi))$$
$$= \sum_{D\in f^m(\xi)} H(\xi \mid D)\mu(D) \le \sum_{D\in f^m(\xi)} \log(\operatorname{card}\{C \in \xi \mid C\cap D \ne \varnothing\})\mu(D).$$

We first give a uniform exponential estimate for $\operatorname{card}\{C \in \xi \mid C \cap D \ne \varnothing\}$ and then we will get a finer exponential bound for those D containing regular points. Let $n = \dim M$.

Lemma S.2.15. *There exists a constant $K_2 > 0$ such that*
$$\operatorname{card}\{C \in \xi \mid D\cap C \ne \varnothing\} \le e^{K_2}\|Df\|_{C^0}^{mn}.$$

Proof. By the Mean Value Theorem
$$\operatorname{diam}(f^m(C)) \le \big(\sup_{x\in M}\|D_xf\|\big)^m \operatorname{diam} C.$$
Thus if $C \cap D \ne \varnothing$ then C is contained in the $4r'$-neighborhood of D whose diameter is at most $((\sup_{x\in M}\|D_xf\|)^m + 2)4r'$ and whose volume is hence bounded by const. $r'^n(\sup_{x\in M}\|D_xf\|)^{mn}$. On the other hand C contains a ball B' of radius r' by (2) and hence the volume of C is at least const.$(r')^n$. Comparing these estimates gives the result. □

We say that x is *(m,ϵ)-regular* if for $k > m$, $1 \le i \le k(x)$, $v \in H_i(x)$, $\|v\| = 1$
$$\left|\frac{1}{k}\log\|D_xf^k(v)\| - \chi^+(x,v)\right| < \epsilon.$$
Write $R_{m,\epsilon}$ for the set of (m,ϵ)-regular points of μ.

Lemma S.2.16. *If $D \in f^m(\xi)$ contains an (n,ϵ)-regular point, then there exists a constant K_3 such that*
$$\operatorname{card}\{C \in \xi \mid D\cap C \ne \varnothing\} \le e^{K_3+\epsilon m}\prod_{\chi_i>0} e^{m(\chi_i+\epsilon)}.$$

Proof. Let $C' \in \xi$ such that $C' \cap R_{m,\epsilon} \ne \varnothing$ and $f^m(C') = D$. Pick some $x \in C' \cap R_{m,\epsilon}$ and let $B = B(x, 2\operatorname{diam} C')$. Then $D \subset B_0 := D_xf^m(\exp_x^{-1}(B))$. If $C \in \xi$, $B_0 \cap C \ne \varnothing$ then $C \subset B_1 := \{y \mid d(y,B_0) < \operatorname{diam}\xi\}$, so
$$\operatorname{card}\{C \in \xi \mid D\cap C \ne \varnothing\} \le \frac{\operatorname{vol}(B_1)}{(\operatorname{diam}\xi)^n}.$$
Up to a bounded factor the volume of B_1 is bounded by the product of the lengths of the axes of the ellipsoid B_0. Those corresponding to nonpositive exponents are small or at most subexponentially large. The remaining ones are of size at most $e^{\chi_i+\epsilon}$, up to a bounded factor. Thus $\operatorname{vol}(B_1)$ is bounded by
$$e^{K_3+\epsilon m}\prod_{i=1}^{n}\operatorname{diam} B\, e^{m(\chi_i+\epsilon)} \le e^{K_3+\epsilon m}(\operatorname{diam} B)^n \prod_{\chi_i>0} e^{m(\chi_i+\epsilon)}$$
$$\le e^{K_3+\epsilon m}(2\operatorname{diam} C')^n \prod_{\chi_i>0} e^{m(\chi_i+\epsilon)} < e^{K_3+\epsilon m}(\operatorname{diam}\xi)^n \prod_{\chi_i>0} e^{m(\chi_i+\epsilon)}.$$
□

Now

$$\begin{aligned} h_\mu(f^m,\xi) &\le \sum_{D\in f^m(\xi)} \mu(D)\log\operatorname{card}\{C\in\xi \mid C\cap D\ne\varnothing\} \\ &= \sum_{D\cap R_{m,\epsilon}\ne\varnothing} \mu(D)\log\operatorname{card}\{C\in\xi \mid C\cap D\ne\varnothing\} \\ &\quad+ \sum_{D\cap R_{m,\epsilon}=\varnothing} \mu(D)\log\operatorname{card}\{C\in\xi \mid C\cap D\ne\varnothing\}. \end{aligned}$$

Applying Lemmas S.2.15 and S.2.16 we obtain

$$mh_\mu(f)-\epsilon = h_\mu(f^m)-\epsilon < h_\mu(f^m,\xi)$$

$$\le \sum_{D\cap R_{m,\epsilon}\ne\varnothing} \mu(D)\Big(K_3+\epsilon m+m\sum_{\chi_i>0}(\chi_i+\epsilon)\Big)+\sum_{D\cap R_{m,\epsilon}=\varnothing}\mu(D)\Big(K_2+nm\log\|Df\|_{C^0}\Big)$$

$$\le K_3+\epsilon m+m\sum_{\chi_i>0}(\chi_i+\epsilon)+\Big(K_2+nm\log\|Df\|_{C^0}\Big)\mu(M\smallsetminus R_{m,\epsilon}).$$

Now $\mu(M\smallsetminus R_{m,\epsilon})\to 0$ as $m\to\infty$ by the Oseledec Theorem S.2.9, so

$$h_\mu(f)\le 2\epsilon+\sum_{\chi_i>0}(\chi_i+\epsilon).$$

Letting $\epsilon\to 0$ we obtain the Ruelle inequality. □

Corollary S.2.17. *Let $f\colon M\to M$ be a C^1 diffeomorphism of a compact manifold M. Then*

$$h_{\text{top}}(f)\le(\dim M)\log\sup_{x\in M}\|D_xf\|.$$

Furthermore

$$h_{\text{top}}(f)\le\dim M\lim_{m\to\infty}\frac{1}{m}\sup_{x\in M}\|D_xf^m\|.$$

3. Regular neighborhoods

Now we begin a systematic study of the dynamics of diffeomorphisms of compact smooth manifolds, based on the theory of linear cocycles reviewed in Section S.2, and in particular the Osedelec–Pesin ϵ-Reduction Theorem S.2.10. We recall that for a diffeomorphism $f\colon M\to M$ of a compact smooth n-dimensional Riemannian manifold M preserving a Borel probability measure μ we can consider (by introducing local coordinates) the derivative Df as a linear cocycle over f; therefore the ϵ-Reduction Theorem says that for a given $\epsilon>0$ and for almost every $x\in M$ there exists a linear transformation $C_\epsilon(x)\colon T_xM\to\mathbb{R}^n$ such that

$$D_\epsilon(x)=C_\epsilon(f(x))D_xf\,C_\epsilon^{-1}(x) \tag{S.3.1}$$

has the Lyapunov block form as in Theorem S.2.10 and C_ϵ is a tempered function.

a. Existence of regular neighborhoods. Our first aim is to construct, for almost every $x \in M$, a neighborhood $N(x)$ such that f acts on $N(x)$ very much like the linear map $D_\epsilon(x)$ in a neighborhood of the origin.

Let us denote by $\mathrm{Diff}^{1+\alpha}(M)$, $\alpha > 0$, the set of $C^{1+\alpha}$ diffeomorphisms of a compact smooth Riemannian manifold M, that is, the set of diffeomorphisms of M for which the derivative Df is *α-Hölder continuous.* From now on we assume that $f \in \mathrm{Diff}^{1+\alpha}(M)$ for some $\alpha > 0$. This assumption seems to be crucial in the nonuniformly hyperbolic theory. Similarly to before $B(0,r)$ be the standard Euclidean r-ball in $\mathbb{R}^n$ centered at the origin.

Theorem S.3.1. [Pes1] *Let $f \in \mathrm{Diff}^{1+\alpha}(M)$, $\alpha > 0$, $\dim M = n$ and suppose that f preserves a Borel probability measure μ. Then there exists a set $\Lambda \subset M$ of full measure such that for $\epsilon > 0$:*

(1) *There exists a tempered function $q: \Lambda \to (0,1]$ and a collection of embeddings $\Psi_x: B(0,q(x)) \to M$ such that $\Psi_x(0) = x$ and $e^{-\epsilon} < q(x)/q(f(x)) < e^{\epsilon}$.*
(2) *If $f_x = \Psi_{f(x)}^{-1} \circ f \circ \Psi_x: B(0,q(x)) \to \mathbb{R}^n$, then $D_0 f_x$ has the Lyapunov block form.*
(3) *The C^1 distance $d_{C^1}(f_x, D_0 f_x) < \epsilon$ in $B(0,q(x))$.*
(4) *There exist a constant $K > 0$ and a measurable function $A: \Lambda \to \mathbb{R}$ such that for $y, z \in B(0,q(x))$*

$$K^{-1} d(\Psi_x(y), \Psi_x(z)) \le \|y - z\| \le A(x) d(\Psi_x(y), \Psi_x(z)),$$

with $e^{-\epsilon} < A(f(x))/A(x) < e^{\epsilon}$.

Definition S.3.2. The points $x \in \Lambda$ are called *completely regular points.*

Proof. Fix $\epsilon > 0$ and let $\Lambda \subset M$ be the set for which the Osedelec–Pesin ϵ-Reduction Theorem S.2.10 works. For each $x \in \Lambda$ consider $C_\epsilon(x)$ as a linear map from $T_x M$ to $\mathbb{R}^n$, where $C_\epsilon(x)$ is the Lyapunov change of coordinates. Thus for $x \in \Lambda$

$$D_\epsilon(x) = C_\epsilon(f(x)) \circ D_x f \circ C_\epsilon^{-1}(x) \tag{S.3.2}$$

has the Lyapunov block form.

For $x \in M$, $r > 0$ let $T_x M(r) := \{w \in T_x M \mid \|w\| \le r\}$; choose $r_0 > 0$ so that for every $x \in M$ the exponential map $\exp_x: T_x M(r_0) \to M$ is an embedding, $\|D_w \exp_x\| \le 2$, and $\exp_{f(x)}$ is injective on $\exp_{f(x)}^{-1} \circ f \circ \exp_x(T_x M(r_0))$. Define

$$f_x := C_\epsilon(f(x)) \circ \exp_{f(x)}^{-1} \circ f \circ \exp_x \circ C_\epsilon^{-1}(x),$$

so f_x is defined in the ellipsoid $P(x) = C_\epsilon(x) T_x M(r_0) \subset \mathbb{R}^n$. Let $p(x) = r_0 \min\{\|C_\epsilon(x)\|, \|C_\epsilon(x)^{-1}\|\}$; thus if $w \in T_x M$ and $\|w\| \le p(x)$ then $w \in P(x)$, that is, the Euclidean ball $B(0,p(x))$ is contained in $P(x)$. Now write $f_x(w) = D_\epsilon(x) w + h_x(w)$, see (S.3.2), and observe that $D_0 f_x = D_\epsilon(x)$, by the chain rule, so $D_0 h_x = 0$. Write $\exp_{f(x)}^{-1} \circ f \circ \exp_x = D_x f + g_x$. Since f is $C^{1+\alpha}$, there exists $L > 0$ such that $\|D_u g_x\| \le L\|u\|^\alpha$, and thus

$$\|D_w h_x\| = \|D_w (C_\epsilon(f(x)) \circ g_x \circ C_\epsilon^{-1}(x))\| \le L\|C_\epsilon(f(x))\|\|C_\epsilon^{-1}(x)\|^{1+\alpha}\|w\|^\alpha.$$

Hence if $\|w\|$ is sufficiently small the contribution of the nonlinear part of f_x is negligible. In particular

$$\|D_w h_x\| < \epsilon \text{ for } \|w\| < \delta(x) := (L\|C_\epsilon(f(x))\|\|C_\epsilon^{-1}(x)\|^{1+\alpha}/\epsilon)^{-1/\alpha}. \quad \text{(S.3.3)}$$

By the Mean Value Theorem also

$$\|h_x w\| < \epsilon \text{ for } \|w\| \le \delta(x). \quad \text{(S.3.4)}$$

From the definition of $\delta(x)$ we have

$$\lim_{m\to\infty} \frac{1}{m} \log \delta(f^m(x)) = \lim_{m\to\infty} -\frac{1}{\alpha m}\|C_\epsilon(f^{m+1}(x))\| = 0.$$

Applying the Tempering-Kernel Lemma S.2.12 to $\|C_\epsilon(x)\|$ we find a measurable $K_\epsilon\colon \Lambda \to \mathbb{R}$ such that $K_\epsilon(x) \ge \|C_\epsilon(x)\|$ and $e^{-\alpha\epsilon} \le K_\epsilon(x)/K_\epsilon(f(x)) \le e^{\alpha\epsilon}$. Define $q(x) := \epsilon L^{-1/\alpha} K_\epsilon(f(x))^{-1/\alpha} \le \delta(x)$. Then

$$e^{-\epsilon} < q(x)/q(f(x)) < e^\epsilon.$$

$\Psi_x\colon B(0, q(x)) \to M$, $x \mapsto \exp_x \circ C_\epsilon^{-1}(x)$ is obviously an embedding. Condition (2) follows from the definition and (3) from (S.3.3) and (S.3.4). It remains to prove (4). From the definition of $C_\epsilon(x)$ it follows that $d(\Psi_x(y), \Psi_x(z)) \ge \|y - z\|/K$ for some constant K depending only on M. On the other hand

$$\|y - z\| = \|\Psi_x^{-1}\Psi_x(y) - \Psi_x^{-1}\Psi_x(z)\| \le 2\|C_\epsilon(x)\| d(\Psi_x(y), \Psi_x(z)),$$

that is, (4) holds with $A(x) = 2K_\epsilon(f(x))^{-\alpha}$. □

Definition S.3.3. The set $N(x) = \Psi_x(B(0, q(x)))$ is called a *regular neighborhood* of $x \in \Lambda$.

Remarks. (1) Let us point out that this is the first of two places where the $C^{1+\alpha}$ condition comes up. The second time, it arises to establish a Hölder condition for the invariant family of plane fields.

(2) For any $x \in \Lambda$ we can extend f_x restricted to a smaller ball $B(0, \eta q(x))$ for some $\eta < 1$ to all of $\mathbb{R}^n$ via the Extension Lemma 6.2.7. So to each $x \in \Lambda$ we can associate a sequence $\{f_{x,m}\}_{m\in\mathbb{Z}}$ of diffeomorphisms of $\mathbb{R}^n$ such that $f_{x,m}\restriction_{B(0,\eta q(x))} = f_x$, $f_{x,m}(0) = 0$, and the C^1 distance $d_{C^1}(f_{x,m}, D_0 f_x) < \epsilon$ in all of $\mathbb{R}^n$.

(3) For most of our technical statements we will be working with sequences of diffeomorphisms of $\mathbb{R}^n$. The conclusions drawn for these sequences are carried over to the manifold via Ψ_x, as long as they are applied only to the points within the regular neighborhoods.

b. Hyperbolic points, admissible manifolds, and the graph transform. Let $f \in \mathrm{Diff}^{1+\alpha}(M)$, $\alpha > 0$, and M be a compact Riemannian manifold. Suppose that f preserves a Borel probability measure μ defined on M. If we consider the derivative Df as a cocycle over f, then by Theorem S.2.9 there exists a Borel set $\Lambda \subset M$ such that

(1) $\mu(\Lambda) = 1$,
(2) $T_xM = H_1(x) \oplus \cdots \oplus H_{k(x)}(x)$, where $H_i(x)$ is the linear subspace for which $v \in H_i(x)$ if and only if $\lim_{n\to\infty}(1/n)\log\|D_xf^n(v)\| = \chi_i(x)$.

Here $\chi_1(x) < \chi_2(x) < \cdots < \chi_{k(x)}(x)$ denote the Lyapunov exponents at x.

In the context of the linear theory of Sections S.2a–S.2d no special value of the exponents has any particular importance. In fact, multiplying any given cocycle by a constant scalar cocycle changes the values of the Lyapunov exponents without changing the Osedelec decomposition and other structures associated to the cocycle. However, when we start using linear theory for the study of nonlinear systems, 0 acquires a particular role as a value for the Lyapunov exponents. Accordingly we pay special attention to the sign of the exponents. This was already apparent in the proof of the Ruelle inequality (Theorem S.2.13).

Now let $s(x)$ be the largest integer s such that $\chi(x) < 0$ for $1 \le i \le s$ and $u(x)$ the smallest integer u such that $\chi(x) > 0$ for $u \le i \le k(x)$. Set

$$E^s(x) = H_1(x) \oplus \cdots \oplus H_{s(x)}(x),$$
$$E^0(x) = H_{s(x)+1}(x) \oplus \cdots \oplus H_{u(x)-1}(x), \text{ and}$$
$$E^u(x) = H_{u(x)}(x) \oplus \cdots \oplus H_{k(x)}(x),$$

so $T_xM = E^s(x) \oplus E^0(x) \oplus E^u(x)$. We call these subspaces *stable*, *neutral*, and *unstable*, respectively. If $E^0(x) = \{0\}$ then we say that x is a *hyperbolic point* for f, that is, all the Lyapunov exponents of x are different from zero. The proof of Corollary S.2.14 shows that there always is a hyperbolic point for surface diffeomorphisms leaving invariant an ergodic measure of positive entropy. From now on we only consider hyperbolic points. Moreover, to simplify our exposition we will most of the time assume that $\dim M = 2$. Therefore the stable and unstable subspaces are one-dimensional. By the Osedelec–Pesin Theorem S.2.10 and a version of Lemma 6.2.7 we can just consider families of maps on $\mathbb{R}^2$, and identify the stable subspace with the x-axis and the unstable one with the y-axis. In this context let $R_\delta = [-\delta, \delta] \times [-\delta, \delta]$.

Definition S.3.4. A 1-dimensional submanifold $V \subset R_\delta$ is called an *admissible (s, γ, δ)-manifold near* 0 if $V = \operatorname{graph}\varphi = \{(\varphi(v), v) \mid v \in [-\delta, \delta]\}$, where $\varphi: [-\delta, \delta] \to [-\delta, \delta]$ is a C^1 map such that $\varphi(0) \le \delta/4$ and $|D\varphi| \le \gamma$.

Similarly, a 1-dimensional submanifold $V \subset R_\delta$ is called an *admissible (u, γ, δ)-manifold near* 0 if $V = \operatorname{graph}\varphi = \{(v, \varphi(v)) \mid v \in [-\delta, \delta]\}$, where $\varphi: [-\delta, \delta] \to [-\delta, \delta]$ is a C^1 map such that $\varphi(0) \le \delta/4$ and $|D\varphi| \le \gamma$. Now suppose that x is a hyperbolic point and let $R(x, \delta) = \Psi_x(R_\delta)$. We say that $W \subset R(x, \delta)$ is an *admissible (s, γ, δ)-manifold near x* if $W = \Psi_x(V)$ with V an *admissible (s, γ, δ)-manifold near* 0. Similarly define admissible (u, γ, δ)-manifolds near x.

Proposition S.3.5. *Let $f: \mathbb{R}^2 \to \mathbb{R}^2$ be a C^1 diffeomorphism satisfying the following conditions:*

(1) $f(0) = 0$.
(2) *There exist $\chi > 0$ and $\epsilon \in (0,1)$ such that $1/(1-2\epsilon) \le 1+4\epsilon < e^\chi < 1/\epsilon$ and for $(u,v) \in \mathbb{R}^2$*

$$f(u,v) = (Au + h_1(u,v), Bv + h_2(u,v)),$$

with $|D_z h_i| < \epsilon$ for $z \in \mathbb{R}^2$ $(i = 1,2)$, $|A| < e^{-\chi}$, and $|B^{-1}| < e^{-\chi}$.

If V is an admissible (u, γ, δ)-manifold near 0 for $\gamma \in (0, \epsilon e^{-\chi}]$ then $f(V)$ is an admissible (u, γ, δ)-manifold near 0. Furthermore there exists $\lambda > 1$ such that if $y, z \in V$ then

$$\|f(y) - f(z)\| > \lambda \|y - z\|.$$

Proof. (Compare with Section 6.2.) Let V be an admissible (u, γ, δ)-manifold near 0 with $V = \operatorname{graph} \varphi$ and $0 < \gamma < \epsilon e^{-\chi}$. We will prove that $V' = f(V)$ can be represented as the graph of a C^1 function which has a restriction $\psi: [-\delta, \delta] \to [-\delta, \delta]$ satisfying the assertion of the theorem. For $w \in [-\delta, \delta]$ consider

$$f(\varphi(w), w) = (A\varphi(w) + h_1(\varphi(w), w), Bw + h_2(\varphi(w), w)).$$

Let us define $\tau: [-\delta, \delta] \to \mathbb{R}$ by

$$\tau(w) := Bw + h_2(\varphi(w), w).$$

τ is an expanding map, that is, for $y, z \in [-\delta, \delta]$

$$\begin{aligned} |\tau(y) - \tau(z)| &= |B(y-z) + h_2(\varphi(y), y) - h_2(\varphi(z), z)| \\ &\ge |B||y-z| - |h_2(\varphi(y), y) - h_2(\varphi(z), z)| \\ &\ge e^\chi |y-z| - \epsilon(1+\gamma)|y-z| = (e^\chi - \epsilon(1+\gamma))|y-z| \end{aligned} \tag{S.3.5}$$

with $e^\chi - \epsilon(1+\gamma) > e^\chi - \epsilon - e^{-\chi} > 1 + 3\epsilon - \epsilon > 1$. So

$$\psi(w) := A(\varphi(\tau^{-1}(w))) + h_1(\varphi(\tau^{-1}(w)), \tau^{-1}(w))$$

is a well-defined map. Notice that for $v = \tau(w)$ we have $f(\varphi(w), w) = (\psi(v), v)$. Now let w_0 be such that $f(\varphi(w_0), w_0) = (\psi(0), 0)$. Then by (S.3.5) we have $|w_0| \le |\tau(0) - \tau(w_0)|/(1+2\epsilon) = |\tau(0)|/(1+2\epsilon) = h_2(\varphi(0), 0)/(1+2\epsilon) \le (\epsilon/(1+2\epsilon)) \cdot (\delta/4)$ and thus

$$\begin{aligned} |\psi(0)| &= \|A(\varphi(w_0)) + h_1(\varphi(w_0), w_0)\| \le e^{-\chi}(1+\gamma)\|w_0\| + \epsilon(1+\gamma)\|w_0\| \\ &\le (e^{-\chi} + \epsilon)(1+\gamma)(\epsilon/(1+2\epsilon)) \cdot (\delta/4) \le (1-\epsilon)(1+\gamma)\delta < (\epsilon/(1+2\epsilon)) \cdot (\delta/4). \end{aligned}$$

Let us look at $D\psi$. Suppose that (u, v) is a tangent vector to V at $(\varphi(w_0), w_0)$; then by definition $|u| \le \gamma |v|$. Consider

$$Df(u,v) = (Au + Dh_1(u,v), Bv + Dh_2(u,v)) =: (u_1, v_1).$$

Then $|u_1| \leq e^{-\chi}|u| + \epsilon|(u,v)|$ and $|v_1| \geq e^{\chi}|v| - \epsilon|(u,v)|$. Since $|(u,v)| \leq (1+\gamma)|v|$ we have

$$|u_1| \leq e^{-\chi}|u| + \epsilon(1+\gamma)|v| \leq \frac{\gamma e^{-\chi} + \epsilon(1+\gamma)}{e^{\chi} - \epsilon(1+\gamma)}|v_1| \leq e^{-2\chi}|v_1|$$

after possibly decreasing ϵ. Thus ψ is differentiable and its derivative is less than $e^{-2\chi}$.

It only remains to prove that f expands distances on admissible manifolds. Let $y, z \in V$, say $y = (\varphi(s), s)$ and $z = (\varphi(t), t)$. Then

$$\begin{aligned}|f(\varphi(s),s) - f(\varphi(t),t)| &\geq |B(s-t)| - |A(\varphi(s) - \varphi(t))| - |h(\varphi(s),s) - h(\varphi(t),t)| \\ &\geq e^{\chi}|s-t| - e^{-\chi}|\varphi(s) - \varphi(t)| - \epsilon(1+\gamma)|s-t| \geq (e^{\chi} - \gamma e^{-\chi} - \epsilon(1+\gamma))|s-t|,\end{aligned}$$

so

$$\|f(y) - f(z)\| \geq \frac{e^{\chi} - \gamma e^{-\chi} - \epsilon(1+\gamma)}{1+\gamma}|y-z|,$$

where $\dfrac{e^{\chi} - \gamma e^{-\chi} - \epsilon(1+\gamma)}{1+\gamma} \geq \dfrac{e^{\chi} - 3\epsilon}{1+\epsilon} =: \lambda > 1$ because $\gamma < \epsilon e^{-\chi}$. □

Now consider a hyperbolic point $x \in M$. Let $\chi(x) := \min\{-\chi_{s(x)}(x), \chi_{u(x)}(x)\}$ and $\epsilon(x) := \sup\{\epsilon > 0 \mid 2\epsilon < e^{-\chi(x)+\epsilon} < (e^{\epsilon} + 3\epsilon)^{-1}\}$.

For $\epsilon > 0$ set $\lambda(x,\epsilon) = e^{\chi(x)-\epsilon}$. Then Proposition S.3.5 and the remarks made during its proof concerning admissible (u,γ,δ)-manifolds only defined on a neighborhood of the origin yield the following corollary.

Corollary S.3.6. *Let $x \in M$ be a hyperbolic point and $0 \leq \epsilon \leq \epsilon(x)$. If V is an admissible (u,γ)-manifold near x for some $\gamma \in (0, \epsilon/\lambda(x,\epsilon)]$ then $f(V) \cap R(f(x))$ is an admissible $(u, \epsilon e^{-\chi}/\lambda(x,\epsilon))$-manifold near $f(x)$. Furthermore there exists $\kappa(x) > 1$ such that if $y, z \in \Psi_x^{-1}(V)$ then*

$$\|f_x(y) - f_x(z)\| \geq \kappa(x)\|y - z\|.$$

The following statements show that the notion of admissible manifolds will allow us to use some sort of local product structure in regular neighborhoods.

Proposition S.3.7. *Let $f: \mathbb{R}^2 \to \mathbb{R}^2$ be a C^1 diffeomorphism satisfying the following conditions:*

1. *$f(0) = 0$.*
2. *There exist a splitting of $\mathbb{R}^2 = \mathbb{R}^s \oplus \mathbb{R}^u$, $\chi > 0$, and $\epsilon > 0$ such that $1 + 4\epsilon < e^{\chi} < 1/\epsilon$ and for $(u,v) \in \mathbb{R}^s \oplus \mathbb{R}^u$*

$$f(u,v) = (Au + h_1(u,v), Bv + h_1(u,v)),$$

with $\|D_z h_i\| < \epsilon$ for $z \in \mathbb{R}^2$ $(i = 1,2)$, $\|A\| < e^{-\chi}$, and $\|B^{-1}\| < e^{-\chi}$.

Then any admissible (s,γ,δ)*-manifold near* 0 *intersects any admissible* (u,γ,δ)*-manifold near* 0 *at exactly one point and the intersection is transverse.*

Proof. Let $\varphi^s, \varphi^u: \mathbb{R} \to \mathbb{R}$ be C^1 maps such that $\operatorname{graph}\varphi^s$ and $\operatorname{graph}\varphi^u$ are admissible (u,γ,δ)- and (s,γ,δ)-manifolds, respectively. $\varphi^u \circ \varphi^s: \mathbb{R} \to \mathbb{R}$ is a contraction because

$$|\varphi^u \circ \varphi^s(v) - \varphi^u \circ \varphi^s(w)| \le \gamma^2 |v - w|,$$

so by Proposition 1.1.2 it has a unique fixed point v. Now

$$(v, \varphi^s(v)) = (\varphi^u \circ \varphi^s(v), \varphi^s(v))$$

is obviously the only point in $\operatorname{graph}\varphi^u \cap \operatorname{graph}\varphi^s$.

For transversality observe that $(\eta,\xi) \in T_{(v,\varphi^s(v))}\operatorname{graph}\varphi^s$ implies $\|\xi\| \le \gamma\|\eta\|$ and $(\eta,\xi) \in T_{(v,\varphi^s(v))}\operatorname{graph}\varphi^u$ implies $|\eta| \le \gamma|\xi|$. Thus if $(\eta,\xi) \in T_{(v,\varphi^s(v))}\operatorname{graph}\varphi^s \cap T_{(v,\varphi^s(v))}\operatorname{graph}\varphi^u$ then $\eta = \xi = 0$, which means that the intersection is transverse. □

Corollary S.3.8. *Let* $x \in M$ *be a hyperbolic point for* $f \in \operatorname{Diff}^{1+\alpha}(M)$. *Then any admissible* (s,γ)*-manifold near* x *intersects any admissible* (u,γ)*-manifold near* x *transversely and in exactly one point.*

Corollary S.3.9. *Let* $x \in M$ *be a hyperbolic point and* $0 \le \epsilon \le \epsilon(x)$. *If* V *is an admissible* (s,γ)*-manifold near* x *for some* $\gamma \in (0, \epsilon/\lambda(x,\epsilon)]$ *then* $f^{-1}(V) \cap R(f^{-1}(x))$ *is an admissible* $(s, \epsilon e^{-\chi}/\lambda(x,\epsilon))$*-manifold near* $f^{-1}(x)$. *Furthermore there exists* $\kappa(x) < 1$ *such that if* $y, z \in \Psi_x^{-1}(V)$ *then*

$$\|f_x(y) - f_x(z)\| \le \kappa(x)\|y - z\|.$$

4. Hyperbolic measures

We now extend some techniques widely used in the theory of locally maximal hyperbolic sets to measures with nonzero exponents. These tools are not only important for applications but they give a certain geometric structure to measures with nonzero exponents. We explain how we can close recurrent orbits, shadow pseudo-orbits, construct almost Markov covers, and determine the cohomology class of a Hölder cocycle by periodic data.

a. Preliminaries.

Definition S.4.1. We say that an f-invariant Borel probability measure μ for $f \in \mathrm{Diff}^{1+\alpha}(M)$, $\alpha > 0$, is an *f-hyperbolic measure* if for almost every $x \in M$ all its Lyapunov exponents are different from zero, that is, $E^0(x) = \{0\}$ a.e.

In particular, any ergodic measure with positive entropy for surface diffeomorphisms is hyperbolic.

In this section we study some of the properties of regular neighborhoods and admissible manifolds for hyperbolic measures. For this we need to define certain sets where our estimates can be made uniform.

Definition S.4.2. For a completely regular point $x \in M$ (Definition S.3.2) let $r(x)$ be the radius of the maximal ball contained in the regular neighborhood $N(x)$; we say that $r(x)$ is the *size* of $N(x)$ (see Definition S.3.3).

Theorem S.4.3. *Let μ be an f-hyperbolic measure for $f \in \mathrm{Diff}^{1+\alpha}(M)$, $a > 0$, and M compact. Then for any $\delta > 0$ there exists a compact set Λ_δ and $\epsilon = \epsilon(\delta) > 0$ such that $\mu(\Lambda_\delta) > 1 - \delta$ and the following conditions hold:*

(1) *The functions*

$$x \mapsto A_\epsilon(x), \quad x \mapsto C_\epsilon(x), \quad x \mapsto q(x), \quad x \mapsto r(x), \quad x \mapsto \chi(x)$$

are continuous on Λ_δ. Here the functions A_ϵ, q, r, and χ are provided by the ϵ-Reduction Theorem S.2.10 for $\epsilon = \epsilon(\delta)$.

(2) *The splitting $T_xM = E^s(x) \oplus E^u(x)$ varies continuously on Λ_δ.*

Proof. Consider the measurable function $x \mapsto \chi(x)$ and choose compact Λ_δ^1 such that $\mu(\Lambda_\delta^1) > 1 - \delta/4$, for $x \in \Lambda_\delta^1$ we have $\chi(x) > c > 0$ for some $c > 0$, and (Luzin Theorem) $z \mapsto \chi(x)$ is continuous on Λ_δ^1. Let $\chi_\delta = \min\{\chi(x) \mid x \in \Lambda_\delta^1\}$ and choose $\epsilon > 0$ such that $0 < \epsilon < \chi_\delta/100$ and $2\epsilon < e^{-\chi_\delta + \epsilon} < (e^\epsilon + 3\epsilon)^{-1}$.

Now for this $\epsilon > 0$ consider the Oseledec–Pesin ϵ-reduced cocycle $D_\epsilon(x)$ from (S.3.2), construct the regular neighborhoods $R(x)$, and define $q(x)$ for this cocycle. Apply Luzin's Theorem once more to get that

$$x \mapsto q(x), \quad x \mapsto r(x), \quad x \mapsto A_\epsilon(x), \quad x \mapsto E^s(x), \quad x \mapsto E^u(x)$$

are continuous on Λ_δ^2 and $\mu(\Lambda_\delta^2) > 1 - \delta/2$.

Then set $\Lambda_\delta = \Lambda_\delta^1 \cap \Lambda_\delta^2$ and the theorem is proved. □

Remark. Here, as the reader may have noticed, we have abused notation, since we referred to $q(x)$, which depends on the choice of ϵ, long before it is explicitly chosen.

Let

$$A_\delta = \max\{A(x) \mid x \in \Lambda_\delta\},$$
$$q_\delta = \min\{q(x) \mid x \in \Lambda_\delta\},$$
$$r_\delta = \min\{r(x) \mid x \in \Lambda_\delta\},$$

and

$$\chi_\delta = \min\{|\chi_i(x)| \mid x \in \Lambda_\delta,\ 1 \le i \le k(x)\}.$$

Then the results of the previous section are true for all $x \in \Lambda_\delta$ with $0 < \gamma = \gamma(\delta) < \epsilon\lambda$,

$$\lambda = \lambda(\delta) = e^{(-\chi_\delta + \epsilon)},$$

and $\epsilon = \epsilon(\delta)$ such that $2\epsilon < \lambda < (e^\epsilon + 3\epsilon)^{-1}$. Set

$$\Lambda_\delta^k = \{x \in \Lambda_\delta \mid \dim E^s(x) = k\}.$$

Definition S.4.4. For $\delta > 0$ we call any set Λ_δ a *δ-Pesin set*, or simply a Pesin set.

Proposition S.4.5. *Let $f \in \mathrm{Diff}^{1+\alpha}(M)$, $\alpha > 0$, and M a compact surface. Suppose μ is an f-hyperbolic measure; then for $\tau \in (0,1)$ there exists $\beta > 0$ such that if $x, y \in \Lambda_\delta$ and $d(x,y) < \beta$ we have*

$$d_{C^1}(\Psi_x^{-1}\Psi_y, \mathrm{Id}) < 1 - \tau$$

on $\Psi_y^{-1}(R(x) \cap R(y))$.

Proof. $\Psi_y = \exp_y \circ C_\epsilon^{-1}(y)$, so $\Psi_x^{-1} \circ \Psi_y = C_\epsilon(x) \circ \exp_x^{-1} \circ \exp_y \circ C_\epsilon^{-1}(y)$. Let $T_{yx}\colon T_yM \to T_xM$ be a linear map from T_yM to T_xM that preserves the hyperbolic splitting and is an isometry on each of these subspaces. Clearly if $x, y \in \Lambda_\delta$ are sufficiently close then $T_{yx}^{-1} \circ \exp_x^{-1} \circ \exp_y$ will be close to Id_{T_yM} in the C^1 topology. Also since $x, y \in \Lambda_\delta$ then by the continuity of the splitting on Λ_δ and of $x \mapsto C_\epsilon(x)$ we have that $C_\epsilon(x) \circ T_{yx} \circ C_\epsilon^{-1}(y)$ is close to $\mathrm{Id}_{\mathbb{R}^2}$ provided that x and y are close enough. □

Corollary S.4.6. *There exists $\rho = \rho(\delta)$ such that if $x, y \in \Lambda_\delta$, $d(x,y) < \rho$, and $W \in U_{\gamma/4}(y)$, then $\Psi_x(\Psi_y^{-1}(W) \cap Q(y, q_\delta/2))$ is part of an admissible (u,γ)-manifold near x.*

For many applications we need to reduce the size of the regular rectangle $R(x)$, so for $0 < h \le 1$ and $x \in \Lambda_\delta$ let us define

$$\tilde{R}(x,h) = \Psi_x(Q(x, hq_\delta)).$$

Definition S.4.7. We say that $W \cap \tilde{R}(x, h/2)$ is an *admissible (u,γ,δ,h)-manifold near $x \in \Lambda_\delta$* if W is an admissible (u,γ)-manifold near x such that if

$W = \Psi_x(\operatorname{graph}\varphi)$ then $\|\varphi(0)\| \leq hq_\delta/4$. Let $U_{\gamma,h}(x)$ be the set of all admissible (u,γ,δ,h)-manifolds near x. Similarly define admissible (s,γ,δ,h)-manifolds near x and denote by $S_{\gamma,h}(x)$ the set of all admissible (s,γ,δ,h)-manifolds near x.

The main difference between admissible (u,γ,δ,h)-manifolds and admissible (u,γ,δ)-manifolds is that admissible (u,γ,δ,h)-manifolds are subsets of the given manifold, whereas admissible (u,γ,δ)-manifolds are in a Euclidean space.

The following proposition will play a similar role as the Inclination Lemma (Proposition 6.2.23) in the uniformly hyperbolic theory. For the rest of this section let us suppose that M is compact, $f \in \operatorname{Diff}^{1+\alpha}(M)$, $\alpha > 0$, and μ is an f-hyperbolic measure on M. In general for $0 < h \leq 1$ let

$$R(x,h) = \Psi_x(Q(x,h\,q(x))).$$

Proposition S.4.8. *For δ, h given, there exists $\beta = \beta(\delta,h) > 0$ such that if $x, y \in \Lambda_\delta$ are such that $d(y, f^m(x)) < \beta$ and $f^m(x) \in \Lambda_\delta$ for some $m \geq 0$, and W_0 is an admissible (u,γ,δ,h)-manifold near x with $\gamma = \gamma(\delta)$, then W_1 defined by*

$$\begin{aligned} W_0^1 &= f(W_0), \\ W_0^i &= f(W_0^{i-1} \cap R(f^{i-1}(x),h)), \quad i = 2,\dots,m-1, \\ W_1 &= f(W_0^{m-1}) \cap \widetilde{R}(y,h/2) \end{aligned} \tag{S.4.1}$$

is an admissible (u,γ,δ,h)-manifold near y.

Proof. Let W_0 be an admissible (u,γ,δ,h)-manifold near x. Proposition S.3.5 implies that W_1 is part of a manifold $\widetilde{W}_i \in U_{\lambda^n\gamma,h}(f^m(x))$ such that $W_1 = \Psi_{f^m(x)}(\operatorname{graph}\tau)$ with $\tau: Q^u(x,h/2) \to Q^s(x,h/2)$ of class C^1. By Corollary S.4.6 W_1 is part of an admissible (u,γ,δ,h)-manifold near y. Let $W_0 = \Psi_x(\operatorname{graph}\varphi)$ where $\varphi \in C^1(Q^u(x,h/2), Q^s(x,h/2))$. Now extend W_0 to an admissible (u,γ,δ,h)-manifold $\widetilde{W}_0$ near x such that

$$\widetilde{W}_0 = \Psi_x(\operatorname{graph}\tilde{\varphi})$$

with $\tilde{\varphi} \in C^1(Q^u(x,h), Q^s(x,h))$, $\|\varphi(0)\| \leq hq/4$, and $\varphi = \tilde{\varphi}\restriction_{Q^u(x,h/2)}$. Define W_0^i as in (S.4.1) and set

$$W_1 = f((W_0^{m-1}) \cap R(f^{m-1}(x),h)) \cap R(f^m(x),h).$$

By Proposition S.3.5 we have that for $i = 1,\dots,m-1$

$$f(W_0^{i-1} \cap R(f^{i-1}(x),h)) = \Psi_{f^i(x)}(\operatorname{graph}\tilde{\varphi}_i),$$

where $\tilde{\varphi}_i \in C^1(Q^u(f^i(x),h), Q^s(f^i(x),h))$, $\|\tilde{\varphi}_i(0)\| \leq hq/4\lambda^i$, $\|D\tilde{\varphi}_i\| \leq \gamma\lambda^i$, and $\lambda = \lambda_\delta = e^{\chi_\delta+\epsilon}$. Similarly $\widetilde{W}_i = \Psi_{f^m(x)}(\operatorname{graph}\tilde{\varphi}_m)$ satisfies the same conditions as $\tilde{\varphi}_i$, $i = 1, \ldots, m-1$. Obviously, if $d(f^m(x), y)$ is small enough $W_1 \subset \widetilde{W}_1$. Set $z_m = \Psi_{f^m(x)}(\tilde{\varphi}(0), 0)$ and $z_i := f^{i-m}(z_m)$, $i = 0, 1, \ldots, m-1$. Since $z_i \in \widetilde{W}_0^i$ then $z_i = \Psi_{f^i(x)}(\tilde{\varphi}_i(v_i), v_i)$ for some $v_i \in Q^u(f^i(x), h)$, and clearly

$$\|\tilde{\varphi}_i(v_i)\| \leq \|\tilde{\varphi}_i(0)\| + \gamma\|v_i\| \leq hq/4\lambda^i + \gamma\|v_i\|.$$

Now let us consider the manifold $S = \Psi_{f^m(x)}(Q^s(f^m(x), hq_\delta)) \times \{0\}$. Clearly S is an admissible (s, γ, δ, h)-manifold. Observe that S contains $f^m(x)$ and z_m; therefore $z_i \in S_i$ and consequently $\|v_i\| \leq \gamma\|\tilde{\varphi}_i(v_i)\|$. It follows that

$$\|\tilde{\varphi}_i(v_i)\| \leq hq/4\lambda^i + \gamma^2\|\tilde{\varphi}_i(v_i)\|$$

and

$$\|v_i\| \leq hq/4\lambda^i + \gamma\|v_i\|.$$

Thus $z_i \in W_0^i \cap R(f^i(x), h)$ for all i.

Represent $W_0^i \cap R(f^i(x), h)$ as $\Psi_{f^i(x)}(\operatorname{graph} \tilde{\varphi}_i\restriction_{D_i})$ with $D_i \subset Q^u(f^i(x), h)$. Then if

$$\pi\tilde{\varphi}_i(v_i) = B_{f^i(x)}v + (h_2)_{f^i(x)}(\varphi_i(v), v),$$

where B and h are as in Proposition S.3.5, the proof of Proposition S.3.5 shows that

$$D_{i+1} = \pi\tilde{\varphi}_i D_i \cap Q^u(f^i(x), h),$$

where $D_0 = Q^u(x, hq/2)$.

Since every map $\pi\tilde{\varphi}_i$ is expanding with coefficient of expansion bigger than $e^\epsilon/(1+\epsilon\lambda)$, if every D_i contains a ball of radius r around v_i then there exists r' such that the ball of radius r' around v_{i+1} is contained in D_{i+1}. For this take

$$r' = \min\left\{\frac{e^\epsilon}{1+\epsilon\lambda}r,\ h\,q(f^{i+1}(x)) - \frac{\gamma\, h\, q_\delta}{4(1+\gamma^2)}\left(\frac{1+\lambda}{2}\right)^{i+1}\right\}.$$

Since $q(f(x))/q(x) > e^\epsilon$, we have

$$q(f^{i+1}(x)) > e^{\epsilon(i+1)}q(x) \geq e^{\epsilon(i+1)}q_\delta$$

and $r' \geq \min\left\{\dfrac{e^\epsilon}{1+\epsilon\lambda}r, h(1-\gamma)e^{-\epsilon(i+1)}q_\delta\right\}$.

Therefore, since D_0 contains a ball around v_0 of radius $(1-\gamma)h\,q_\delta/2$ we have that D_n contains a ball around the origin of radius

$$\min\left\{(1-\gamma)\left(\frac{e^\epsilon}{1+\epsilon\lambda}\right)^n\frac{hq}{2}, h(1-\gamma)q(f^n(x))\right\},$$

which is obviously greater than $hq/2\left(\dfrac{e^\epsilon}{1+\epsilon\lambda}\right)^{1/2}$. Therefore if τ is sufficiently close to 1 then

$$\Psi_y^{-1}\Psi_{f^m(x)}(\operatorname{graph}\tilde{\varphi}_m\restriction_{D_m}) = \operatorname{graph}\varphi_m$$

with Domainφ_m covering $Q^u(f^n(x), qh/2)$ and so

$$W_1 = \Psi_{f^m(x)}(\operatorname{graph}\tilde{\varphi}_m\restriction_{D_m}) \cap \widetilde{R}(y, h/2)$$

is an admissible (u, γ, δ, h)-manifold near y. □

Proposition S.4.9. *For δ, h given, there exists $\beta = \beta(\delta, h) > 0$ such that if $x, y \in \Lambda_\delta$ are such that $d(y, f^{-m}(x)) < \beta$ and $f^{-m}(x) \in \Lambda_\delta$, for some $m \geq 0$, and W_0 is an admissible (s, γ, δ, h)-manifold near y with $\gamma = \gamma(\delta)$, then W_1 defined by*

$$\begin{aligned}
W_0^0 &= W_0 \cap R(f^m(x), h),\\
W_0^1 &= f^{-1}(W_0^0) \cap R(f^{m-1}(x), h),\\
W_0^i &= f^{-i}(W_0^{i-1}) \cap R(f^{m-i}(x), h), \qquad i = 2, \dots, m-1,\\
W_1 &= f^{-1}(W_0^{m-1}) \cap \widetilde{R}(x, h/2)
\end{aligned}$$

is an admissible (s, γ, δ, h)-manifold near x.

The proof of this proposition is similar to the corresponding one for (u, γ, δ, h)-manifolds.

b. The Closing Lemma. In this subsection we address one of the fundamental problems in dynamics: Given a recurrent point x is it possible to find a nearby periodic point y that follows the orbit of x during the period of time that the point needs to return very close to its original position? This sort of problem is called a closing problem.

We will show that for diffeomorphisms that preserve a hyperbolic measure it is possible to close certain orbits, and furthermore the periodic points thus obtained are hyperbolic. This result will be very much the basis of the next section. Similarly to the closing problem there is the shadowing problem or shadowing property. This is that, given a sequence of points on the manifold such that the image of x_i is very close to x_{i+1}, it is possible to find a point x such that $f^i(x)$ is close to x_i. In other words, if we have a sequence of points that resemble an orbit (a pseudo-orbit) then we can find a real orbit that shadows (or closely follows) the pseudo-orbit.

These results are counterparts of the Closing and Shadowing Lemmas (Theorems 6.4.15 and 18.1.2) in the uniformly hyperbolic case.

Lemma S.4.10. *Let $f \in \mathrm{Diff}^{1+\alpha}(M)$, $\alpha > 0$, and M a compact Riemannian manifold. Suppose that μ is an f-invariant hyperbolic measure. For every $h, \delta > 0$ there exists a number $\beta = \beta(\delta, h)$ such that if $x \in \Lambda_\delta$, $f^m(x) \in \Lambda_\delta$ for some $m > 0$, and $d(x, f^m(x)) < \beta$ then there exists a point $z = z(x)$ satisfying* (1) *$f^m(z) = z$ and*

$$(2)\ \ d(f^i(x), f^i(z)) \leq 3hA_\delta \max\left\{\left(\frac{e^\epsilon}{1+\epsilon\lambda}\right)^{i-m}, \left(\frac{e^\epsilon}{1+\epsilon\lambda}\right)^{-i}\right\},\ 1 \leq i \leq m.$$

Here λ and ϵ are as in the previous section.

Proof. First we prove the existence of the periodic point $z = z(x)$. Let $\beta = \beta(\delta, h) > 0$ satisfy Propositions S.4.8 and S.4.9. Consider the reduced regular neighborhood $Q(x, hq_\delta))$ for $x \in \Lambda_\delta$. Then $d_{C^1}(\Psi_x \Psi^{-1}_{f^m(x)}, \mathrm{Id})$ is small since $f^m(x) \in \Lambda_\delta$ and $d(x, f^m(x)) < \beta$, so the rectangles $\widetilde{R}(x, h)$ and $\widetilde{R}(f^m(x), h)$ look like superimposed ones of almost the same size.

Thus if B_0 is an admissible (u,γ,δ,h)-manifold near x for $\gamma=\gamma(\delta)>0$, then B_1 defined as follows

$$\begin{aligned} B_0^1 &= f(B_0),\\ B_0^i &= f(B_0^{i-1}\cap R(f^{i-1}(x),h)), \quad i=2,\dots,m-1,\\ B_1 &= f(B_0^{m-1})\cap \widetilde{R}(x,h/2), \end{aligned}$$

with $y_0\in\Psi_x(Q^s(x,h)\times\{0\})\cap B_0$, is an admissible (u,γ,δ,h)-manifold near x. For instance, let us take $B_0^1=\Psi_x(\{0\}\times Q^u(x,h))$.

Similarly if $A_0=\Psi_{f^m(x)}(Q^s(x,h)\times\{0\})\cap\widetilde{R}(x,h/2)$ define

$$\begin{aligned} A_0^0 &= A_0\cap\widetilde{R}(f^m(x),h),\\ A_0^1 &= f^{-1}(A_0^0\cap R(f^{m-1}(x),h)),\\ A_0^i &= f^{-1}(A_0^{i-1}\cap R(f^{m-i}(x),h)), \quad i=2,\dots,m-1,\\ A_1 &= f^{-1}(A_0^{m-1}\cap\widetilde{R}(x,h/2)), \end{aligned}$$

where $y_0\in\Psi_x(\{0\}\times Q^u(x,h))\cap A_0$. Then by Proposition S.4.9 A_1 is an admissible (s,γ,δ,h)-manifold near x. By Proposition S.3.7 A_1 intersects B_1 at exactly one point, say $z_{1,1}$. Now define inductively $A_k, A_k^1,\dots,A_k^{m-1}$ for $k=1,2,\dots$ as follows

$$\begin{aligned} A_k^1 &= f^{-1}(A_k)\cap R(f^{m-1}(x),h),\\ A_k^i &= f^{-1}(A_k^{i-1})\cap R(f^{m-i}(x),h), \quad i=2,\dots,m-1,\\ A_{k+1} &= f^{-1}(A_k^{m-1})\cap\widetilde{R}(x,h/2). \end{aligned}$$

Similarly

$$\begin{aligned} B_k^1 &= f(B_k),\\ B_k^i &= f(B_k^{i-1}\cap R(f^{i-1}(x),h)), \qquad i=2,\dots,m-1,\\ B_{k+1} &= f(B_k^{m-1}\cap\widetilde{R}(x,h/2)). \end{aligned}$$

Here we have omitted taking connected components to simplify the notation.

By Propositions S.4.8 and S.4.9 A_k is an admissible (s,γ,δ,h)-manifold near x for every $k=1,2,\dots$. Therefore each A_k intersects every B_j at a point $z_{k,j}$ with $x=z_{1,0}$ and $f^m(x)=z_{0,1}$.

We claim that if $k\geq 1$, $j\geq 0$ then

$$f^m(z_{k,j})=z_{k-1,j+1}.$$

To prove this observe that

$$f^m(z_{k,j})\in f^m(A_k)\subset f^{m-1}(A_{k-1}^{m-1})\subset\cdots\subset A_{k-1},$$

so it remains to show that for $i = 1, \dots, m-1$

$$f^i(z_{k,j}) \in B^i_j,$$

because $f^m(z_{k,j}) = f(f^{m-1}(z_{k,j})) \in f(B^{m-1}_j)$ and $f^m(z_{k,j}) \in A_{k-1} \subset \widetilde{R}(x, h/2)$ imply that $f^m(z_{k,j}) \in B_{j+1}$.

Now we proceed by induction on i, so suppose $f^{i-1}(z_{k,j}) \in B^{i-1}_j$. Since $f^i(z_{k,j}) \in f(A^{m-i+1}_{k-1}) = A^{m-i}_{k-1} \cap f(R(f^{i-1}(x), h))$ we have

$$f^i(z_{k,j}) \in f(B^{i-1}_j) \cap f(R(f^{i-1}(x), h)) = B^i_j,$$

since $f^{i-1}(z_{k,j}) \in B^{i-1}_j$ by the inductive hypothesis.

Therefore $f^m(z_{k,k-1}) = z_{k-1,k}$ and if $z_{k-1,k}$ converges to a point z then

$$f^m(z) = \lim_{k\to\infty} f^m(z_{k-1,k}) = \lim_{k\to\infty} z_{k,k+1} = \lim_{k\to\infty} z_{k-1,k} = z.$$

Clearly since f^i is continuous and $f^i(z_{k,k-1}) \in R(f^i(x), h)$ for all $k \geq 1$ we have $f^i(z) = \lim_{k\to\infty} f^i(z_{k,k-1}) \in R(f^i(x), h)$ for $i = 0, \dots, m-1$.

To finish the proof of (1) let us show that $z_{k-1,k}$ is a convergent sequence. For this we first observe that for

$$\tau > \left(\frac{e^\epsilon}{1+\epsilon}\right)^{-m}$$

there exists $0 < \lambda' < 1$ such that for all $k_1, k_2 \geq 0$, $j \geq 0$ we have

$$\|\Psi_x^{-1}(z_{k_1,j}) - \Psi_x^{-1}(z_{k_2 j})\| < \lambda' \|\Psi_x^{-1}(z_{k_1-1,j+1}) - \Psi_x^{-1}(z_{k_2-1,j+1})\|.$$

Since $f^i(z_{k,j}) = f^{-1}(f^{i+1}(z_{k,j})) \in f^{-1}(B^{i+1}_j) = B^i_j \cap R(f^i(x), h)$ and each manifold $B^i \cap R(f^i(x), h)$ is part of an admissible manifold near $f^i(x)$ for all $i = 1, \dots, m$,

$$\begin{aligned}\|\Psi_x^{-1}(z_{k_1-1,j+1}) - \Psi_x^{-1}(z_{k_2-1,j+1})\| &\geq \tau \|\Psi_{f^{m-1}(x)}(z_{k_1-1,j+1})\| \\ &\geq \tau \left(\frac{e^\epsilon}{1+\epsilon\lambda}\right)^m \|\Psi_x^{-1}(z_{k_1,j}) - \Psi_x^{-1}(z_{k_2,j})\|,\end{aligned}$$

so we can set $\lambda' = \left(\frac{e^\epsilon}{1+\epsilon\lambda}\right)^{-m} \tau^{-1} < 1$.

Similarly for $k \geq 0$, $j_1, j_2 \geq 1$

$$\lambda' \|\Psi_x^{-1}(z_{k+1,j_1-1}) - \Psi_x^{-1}(z_{k+1,j_2-1})\| \geq \|\Psi_x^{-1}(z_{k_1,j_1}) - \Psi_x^{-1}(z_{k_2,j_2})\|.$$

Using these estimates we prove that

$$\|\Psi_x^{-1}(z_{k+1,k}) - \Psi_x^{-1}(z_{k,k-1})\| \leq 4q\, h(\lambda')^k,$$

which not only implies that $z_{k,k-1}$ is a convergent sequence but also that

$$\sum_{k=1}^{\infty} \|\Psi_x^{-1}(z_{k+1,k}) - \Psi_x^{-1}(z_{k,k+1})\| < \infty,$$

and therefore $z \in \widetilde{R}(x, h/2)$.

By definition $z_{k,k-1}$ and $z_{k-1,k-1}$ belong to the same manifold B_{k-1}, so

$$f^{m(k-1)} z_{k-1,k-1} = z_{0,2k-2} \in B_{2k-2}$$

and

$$f^{m(k-1)} z_{k,k-1} = z_{1,2k-2} \in B_{2k-2}.$$

For every $i = 0, \dots, k-2$

$$\|\Psi_x^{-1}(z_{k-1+i,k-1+i}) - \Psi_x^{-1}(z_{k-1,k-1+i})\| \le \lambda' \|\Psi_x^{-1}(z_{k-2-i,k+i}) - \Psi_x^{-1}(z_{k-1-i,k+i})\|$$

and therefore

$$\begin{aligned}\|\Psi_x^{-1}(z_{k,k-1}) - \Psi_x^{-1}(z_{k-1,k-1})\| &\le (\lambda')^{k-1} \|\Psi_x^{-1}(z_{2k-2,1}) - \Psi_x^{-1}(z_{2k-2,0})\| \\ &\le (\lambda')^{k-1} 2qh,\end{aligned}$$

which concludes the proof of (1).

To estimate $d(f^i(x), f^i(z))$ for $i = 0, \dots, m-1$ we consider the following manifolds: B_0 defined as before, A_0^s the image of an $\tilde{s}$-dimensional plane parallel to $Q^s(f^m(x), hq_\delta)$ passing through z, and $A_i^s = f^{-1}(A_{i-1}^s) \cap R(f^{m-1}(x), h)$ for $i = 1, \dots, m$. Thus B_0 and A_m^s intersect at a unique point y, by Proposition S.3.5 and its correspondent for f^{-1}, and with $\Xi_m^i := \max\left\{\left(\frac{e^\epsilon}{1+\epsilon\lambda}\right)^{i-m}, \left(\frac{e^\epsilon}{1+\epsilon\lambda}\right)^{-i}\right\}$ we have

$$\begin{aligned}d(f^i(x), f^i(z)) &\le d(f^i(x), f^i(y)) + d(f^i(y), f^i(z)) \\ &\le d(f^m(x), f^m(y)) \left(\frac{e^\epsilon}{1+\epsilon}\right)^{i-m} + d(y,z) \left(\frac{e^\epsilon}{1+\epsilon\lambda}\right)^{-i} \\ &\le \Xi_m^i [d(f^m(x), f^m(y)) + d(y,z)] \\ &\le \Xi_m^i [d(f^m(x), f^m(z)) + d(f^m(z), f^m(y)) + d(y,z)] \le 3hA_\delta \Xi_m^i.\end{aligned}$$

□

Proposition S.4.11. *Let $x \in \Lambda_\delta$ be such that $f^m(x) \in \Lambda_\delta$ and $d(x, f^m(x)) < \beta$ for some $m > 0$. If $z \in R(x,1)$ is a periodic point of period m with $f^i(z) \in R(f^i(x),1)$ for $i = 0, \dots, m-1$, then z is a hyperbolic periodic point.*

Proof. Recall that for $y \in \mathbb{R}^2$ and $\gamma > 0$ we define the cones

$$C^u_\gamma(y) = \{(v,u) \in T_y\mathbb{R}^2 \approx \mathbb{R} \oplus \mathbb{R} \mid \|v\| \le \gamma\|u\|\}$$

and $C^s_\gamma(y) = \{(v,u) \in T_y\mathbb{R}^2 \mid \|u\| \le \gamma\|v\|\}$.

By the same arguments as in the proof of Proposition S.3.5 or in the construction of an invariant cone family if $x \in \Lambda_\delta$, $y \in Q(x,1)$ then

$$Df_x\big|_y C^u_\gamma(y) \subset C^u_{\lambda\gamma}(f_x(y)).$$

for λ and γ chosen as in Proposition S.3.5; furthermore we have that if $w \in C^u_\gamma$ then $\|Df_x\big|_y w\| \ge K\|w\|$ with $K = K(\lambda) > 1$.

Similarly, $Df_x^{-1}\big|_y C^s_\gamma(y) \subset C^s_{\lambda\gamma}(f_x^{-1}(y))$ and if $w \in C^s_\gamma(y)$ then

$$\|Df_x^{-1}\big|_y w\| > K\|w\|.$$

Now let z be a periodic point satisfying the condition of the proposition. Set

$$F^m_{x,z} := Df_{f^{m-1}(x)}\big|_{\tilde{z}_{m-1}} \cdots Df_x\big|_{\tilde{z}_0},$$

where $\tilde{z}_0 = \Psi_x^{-1}(z)$ and $\tilde{z}_i = f_{f^{i-1}(x)}(\tilde{z}_{i-1})$ for $i = 1, \dots, m-1$. So it follows that $F^m_{x,z}(C^u_\gamma(\tilde{z}_0)) \subset C^u_{\lambda^m\gamma}(\tilde{z}_0)$ and if $w \in C^u_\gamma(\tilde{z}_0)$ then $\|F^m_{x,z}w\| \ge K^m\|w\|$. Similarly $(F^m_{x,z})^{-1}(C^s_\gamma(\tilde{z}_0)) \subset C^s_{\lambda^m\gamma}(\tilde{z}_0)$ and for $w \in C^s_\gamma(\tilde{z}_0)$ we have $\|(F^m_{x,z})^{-1}w\| \ge K^m\|w\|$. Now we consider cones in T_zM. For $0 < \beta < 1$ let

$$\tilde{C}^u_\beta(z) = D\Psi_x\big|_{\tilde{z}_0} C^u_\beta(\tilde{z}_0) \text{ and } \tilde{C}^s_\beta(z) = D\Psi_x\big|_{\tilde{z}_0} C^s_\beta(\tilde{z}_0).$$

Observe that

$$D_z f^m = D\Psi_{f^m(x)}\big|_{\tilde{z}_m} \circ F^m_{x,z} \circ D\Psi_x^{-1}\big|_z = D\Psi_{f^m(x)}\big|_{\tilde{z}_m} \circ D\Psi_x^{-1}\big|_z \circ D\Psi_x\big|_{\tilde{z}_0} \circ F^m_{x,z} \circ D\Psi_x^{-1}\big|_z$$

and

$$D\Psi_x\big|_{\tilde{z}_0} \circ F^m_{x,z} \circ D\Psi_x^{-1}\big|_z : T_zM \to T_zM.$$

The tangent space T_zM can be written as $E^s(z) \oplus E^u(z)$ with $E^s(z) = D\Psi_x\big|_{\tilde{z}_0}(\mathbb{R} \times \{0\})$ and $E^u(z) = D\Psi_x\big|_{\tilde{z}_0}(\{0\} \times \mathbb{R})$. Since $d(x, f^m(x)) < \beta$ it follows that

$$D\Psi_{f^m(x)}\big|_{\tilde{z}_m} C^u_{\lambda\gamma}(\tilde{z}_m) \subset \tilde{C}^u_{\lambda^{1/2}\gamma}(z) \text{ and } D\Psi_{f^m(x)}\big|_{\tilde{z}_m} C^s_{\lambda\gamma}(\tilde{z}_m) \subset \tilde{C}^s_{\lambda^{1/2}\gamma}(z).$$

Thus

$$\begin{aligned} D_z f^m(\tilde{C}^u_\gamma(z)) &= D\Psi_{f^m(x)}\big|_{\tilde{z}_m} \circ D\Psi_x^{-1}\big|_z \circ D\Psi_x\big|_{\tilde{z}_0} \circ F^m_{x,z} \circ D\Psi_x^{-1}\big|_z(\tilde{C}^u_\gamma(z)) \\ &= D\Psi_{f^m(x)}\big|_{\tilde{z}_m} \circ D\Psi_x^{-1}\big|_z \circ D\Psi_x\big|_{\tilde{z}_0} \circ F^m_{x,z}(C^u_\gamma(\tilde{z}_0)) \\ &\subset D\Psi_{f^m(x)}\big|_{\tilde{z}_m} \circ D\Psi_x^{-1}\big|_z \circ D\Psi_x\big|_{\tilde{z}_0}(C^u_{\lambda^m\gamma}(\tilde{z}_0)) \\ &= D\Psi_{f^m(x)}\big|_{\tilde{z}_m} \circ D\Psi_x^{-1}\big|_z(\tilde{C}^u_{\lambda^m\gamma}(\tilde{z}_0)) \subset \tilde{C}^u_{\lambda^{m/2}\gamma}(z). \end{aligned}$$

Similarly $D_z f^{-m}(\tilde{C}^s_\gamma(z)) \subset \tilde{C}^s_{\lambda^{m/2}\gamma}$, so if we take $\lambda' = \lambda^{m/2}$ we have that

$$D_z f^m(\tilde{C}^u_\gamma(z)) \subset C^u_{\lambda'\gamma}(z) \text{ and } D_z f^{-m}(\tilde{C}^s_\gamma(z)) \subset C^s_{\lambda'\gamma}(z).$$

Now take $w \in \tilde{C}^u_\gamma(z)$; then

$$\begin{aligned} \|D_z f^m(w)\| &= \|D\Psi_{f^m(x)}\big|_{\tilde{z}_m} \circ F^m_{x,z} \circ D\Psi_x^{-1}(w)\big|_z\| \\ &\geq \|F^m_{x,z} \circ D_z\Psi_x^{-1}(w)\|/K_\epsilon(\delta) \geq K^m\|D_z\Psi_x^{-1}(w)\|/K_\epsilon(\delta) \geq C\,K^m\|w\|, \end{aligned}$$

where $C = C(\Lambda_\delta, \beta)$. These conditions imply hyperbolicity by Corollary 6.4.8. □

Remark. This proof shows that in general if we have a closed set Λ such that $\Lambda \subset R(x,1)$ and $f^i(\Lambda) \subset R(f^i(x),1)$ for $i = 0,\dots,m-1$ and $f^m(\Lambda) = \Lambda$ then $f^i(\Lambda)$ is a hyperbolic set.

Now we prove a statement about hyperbolic periodic points that remain inside the regular neighborhoods of some point with nonzero exponents.

Proposition S.4.12. *If x and $z(x)$ are as in Lemma S.4.10 then the stable [unstable] manifold of $z(x)$ is an admissible $(s,\gamma,1)$- [$(u,\gamma,1)$-] manifold near x.*

Proof. Let us consider the set S_x of all admissible $(s,\gamma,1)$-manifolds near x. If $W_1, W_2 \in S_x$ are such that $W_i = \Psi_x(\operatorname{graph}\varphi_i)$ for $i = 1,2$ define the C^0 distance between W_1 and W_2 by

$$d_{C^0}(W_1,W_2) = \max_{v\in Q^s(x,1)} \|\varphi_1(v) - \varphi_2(v)\|.$$

It follows from the definition that $\bar{S}_x$ is a compact set in this metric. Consequently, the sequence $\{A_k\}_{k\in\mathbb{N}}$ defined in the proof of Lemma S.4.10 contains a subsequence $\{A_{k_l}\}_{l\in\mathbb{N}}$ such that A_{k_l} converges in this metric to some manifold $A \subset \bar{S}_x$.

For $w \in A$, we prove that

(1) $f^{km}(w) \in \widetilde{R}(x,1)$ for $k \in \mathbb{N}$,
(2) $d(f^{km}(w), z) < C\tilde{\lambda}^{km}d(w,z)$ for some constants $C > 0$ and $\tilde{\lambda} < 1$.

Let us fix k and find a sequence of points $w_l \in A_{k_l}$ such that $w = \lim_{l\to\infty} w_l$. Thus for $k_l \geq k$ it follows from the proof of Lemma S.4.10 that

$$f^{km}(w) \in A_{k_l-k} \subset \widetilde{R}(x,1),$$

so we have checked (1).

To estimate $d(f^{km}(w), z)$ observe that

$$\begin{aligned} d(f^{km}(w), z) &= d(\Psi_x^{-1}\Psi_x f^{km}(w), \Psi_x^{-1}\Psi_x(z)) \leq K^{-1}\|\Psi_x f^{km}(w) - \Psi_x(z)\| \\ &\leq K^{-1}\alpha^{-km}\|\Psi_x(w) - \Psi_x(z)\| \leq K^{-1}\alpha^{-km}A_\delta d(w,z), \end{aligned}$$

so if we take $C = K^{-1}A_\delta$ and $\tilde{\lambda} = \alpha^{-1}$ then (2) is verified.

Statements (1) and (2) show that A is contained in the local stable manifold $W^s(z)$ of the point z for f^m: Since $T_z W^s(z) = E^s(z)$ and $D\Psi_x^{-1}\big|_z E^s(x) \subset C_\gamma(\tilde{z}_0)$ we may conclude that locally near z the manifold $W^s(z)$ has the form

$$W^s(z) = \Psi_x(\operatorname{graph}\varphi),$$

where $\varphi\colon \mathbb{R}^{\tilde{s}} \to \mathbb{R}^{n-\tilde{s}}$ is a C^1 function defined on a neighborhood of 0 and $\|D_0\varphi\| < \gamma$.

Since A has the same form, we conclude that, locally, A coincides with $W^s(z)$, and as f is a diffeomorphism their extensions have to be the same and so A has to be the local stable manifold. □

Now we assemble all the previous statements of this chapter in the following theorem, which is called the *Closing Lemma* for nonuniformly hyperbolic systems and was originally proved in [K5].

Theorem S.4.13. (Closing Lemma for nonuniformly hyperbolic systems) *Let $f \in \mathrm{Diff}^{1+\alpha}(M)$, $\alpha > 0$, and M a compact Riemannian surface. Suppose that μ is an f-invariant hyperbolic measure. For every h, $\delta > 0$, and $0 < k \leq \dim M$ there exists $\beta = \beta(h,\delta) > 0$ such that if $x \in \Lambda_\delta$ and $f^m(x) \in \Lambda_\delta$ for some $m > 0$ and $d(x, f^m(x)) < \beta$ then there exists a point $z = z(x)$ satisfying the following conditions:*

(1) $f^m(z) = z$.

(2) $d(f^i(z), f^i(x)) \leq 3h\, A_\delta \max\left\{\left(\frac{e^\epsilon}{1+\epsilon\lambda}\right)^{i-m}, \left(\frac{e^\epsilon}{1+\epsilon\lambda}\right)^{i}\right\}$ *for* $0 \leq i \leq m$.

(3) *z is a hyperbolic periodic point whose stable and unstable manifolds are admissible $(s,\gamma,1)$- and $(u,\gamma,1)$-manifolds, respectively.*

c. The Shadowing Lemma. We now give an analog of the Shadowing Lemma (Theorem 18.1.2) for the case of nonuniformly hyperbolic systems. We use the notions of pseudo-orbit and shadowing from Definition 18.1.1.

Theorem S.4.14. (Shadowing Lemma for nonuniformly hyperbolic systems) *Let μ be an f-invariant hyperbolic measure for $f \in \mathrm{Diff}^{1+\alpha}(M)$, $\alpha > 0$, and M a compact Riemannian surface. For $\delta > 0$ let $\widetilde{\Lambda}_\delta = \bigcup_{x \in \Lambda_\delta} \widetilde{R}(x, 1)$; then for $\alpha > 0$ sufficiently small there exists $\beta = \beta(\alpha, \Lambda_\delta)$ such that for every β-pseudo-orbit $\{x_m\} \subset \widetilde{\Lambda}_\delta$ there exists $y \in M$ such that its orbit $\mathcal{O}(y)$ α-shadows $\{x_m\}$.*

Since we do not use this theorem later and it resembles the one of the Closing Lemma, we just sketch its proof.

Sketch of the proof. Consider a β-pseudo-orbit $\{x_m\} \subset \widetilde{\Lambda}_\delta$ for some small $\beta > 0$. For $m \in \mathbb{Z}$ consider the points x_{m-1}, x_m, x_{m+1}, x_{m+2}, and x_{m+3} and choose z_{m-1}, z_m, z_{m+1}, $z_{m+2}, z_{m+3} \in \Lambda_\delta$ such that $x_{m+i} \in \widetilde{R}(z_{m+i}, 1)$, $i = -1, 0, \dots, 3$. We will produce a pseudo-orbit $\{x'_m\}$ satisfying $d(f(x'_m), x'_{m+1}) < \zeta(f(x_m), x_{m+1})$ with $0 < \zeta < 1$.

We work in the Lyapunov chart. As we said before the idea is basically the same as that of the Closing Lemma, but here we work on different charts and move two steps forward and backward.

Let us choose an admissible (u, γ)-manifold V^u_{m-1} near z_{m-1} with $x_{m-1} \in V^u_{m-1}$, for γ small. Similarly choose an admissible (s, γ)-manifold V^s_m near z_{m+1} with $f(x_m) \in V^s_m$, and consider $f(V^u_{m-1})$ and $f^{-1}(V^s_m)$. If β is sufficiently small then $f(V^u_{m-1})$ is an admissible manifold near z_m, and $f^{-1}(V^s_m)$ as well, so let $y_m \in f(V^u_{m-1}) \cap f^{-1}(V^s_m)$. Now choose V^u_m admissible near z_m with $x_m \in V^u_m$, similarly let V^s_{m+1} be an admissible s-manifold near z_{m+2} containing $f(x_{m+1})$, and pick $x'_m \in f^{-2}(V^s_{m+1}) \cap f(V^u_{m+1})$. To produce x'_{m+1}, define V^s_{m+2} and let $x'_{m+1} \in f^{-2}(V^s_{m+2}) \cap f(V^u_m)$. The same arguments as in the Closing Lemma show the existence of $\xi < 1$ such that $d(f(x'_m), x'_{m+1}) < \zeta d(f(x_m), x_{m+1})$.

Using the above method and induction one can prove that for all $k > 0$ there exists a $\zeta^k\beta$-pseudo-orbit $\{x^k_m\} \subset \widetilde{A}_\delta$. Thus let

$$y = \lim_{k \to \infty} x^k_0.$$

Observe that

$$d(f^i(y), x_i) = \lim_{k \to \infty} d(f^i(y), x^k_i) \le \frac{3K_s\beta}{1 - \alpha},$$

so if $\beta = \dfrac{(1-\zeta)\alpha}{3K_\delta}$ the result follows. □

d. Pseudo-Markov covers. Fix $0 < \gamma < 1$. Given C^1 maps $\varphi_1, \varphi_2 \colon [-1, 1] \to [-1, 1]$ such that $\|\varphi_1(v)\| > \|\varphi_2(v)\|$ for $|v| \le 1$ and $|D\varphi_i| \le \gamma$ $(i = 1, 2)$, we say that the set $H = \{(v, u) \in [-1, 1]^2 \mid u = \theta\varphi_1(v) + (1 - \theta)\varphi_2(v),\ 0 \le \theta \le 1\}$ is an admissible (s, γ)-rectangle in $Q(1)$.

Similarly define an admissible (u, γ)-rectangle in $Q(1)$ as

$$V = \{(v, u) \in [-1, 1]^2 \mid v = \theta\tilde{\varphi}_1(u) + (1 - \theta)\tilde{\varphi}_2(u),\ 0 \le \theta \le 1\},$$

where $\tilde{\varphi}_1, \tilde{\varphi}_2 : [-1,1] \to [-1,1]$, $|D\tilde{\varphi}_i| \le \gamma$, and $|\tilde{\varphi}_1| > |\tilde{\varphi}_2|$. As for regular neighborhoods, if $\Psi_x : [-1,1]^2 \to M$ is a C^1 embedding such that $\Psi_x(0) = x$ we say that $R(x) = \Psi_x([-1,1]^2)$ is a rectangle in M centered at x. A set $\tilde{H}$ is an admissible (s,γ)-rectangle in $R(x)$ if $\tilde{H} = \Psi_x(H)$ for some admissible (s,γ)-rectangle in $[-1,1]^2$. Similarly define an admissible (u,γ)-rectangle in $R(x)$.

Definition S.4.15. Given a diffeomorphism $f: M \to M$ of a compact surface and $\Lambda \subset M$ compact we say that a finite collection of rectangles $\{R(x_1), R(x_2), \dots, R(x_2)\}$ is a (ρ,β,λ)-rectangle cover for $\rho > 0$, $1 > \lambda > 0$ if there exists $\gamma = \gamma(\rho,\beta,\lambda) \in (0,1)$, satisfying:

(1) $\Lambda \subset \bigcup_{i=1}^{t} \beta(x_i,\beta)$, with $\beta(x_i,\beta) \subset \operatorname{int} R(x_i)$ and $x_i \in \Lambda$.
(2) $\operatorname{diam} R(x_i) \le \rho/3$ for $i = 1, \dots, t$.
(3) If $x \in \Lambda$, $f^m(x) \in \Lambda$ for some $m > 0$, $x \in \beta(x_i,\beta)$, and $f^m(x) \in \beta(x_j,\beta)$, then the connected component $\mathbb{CC}(R(x_i) \cap f^{-m}(R(x_j)), x)$ of $R(x_i) \cap f^{-m}(R(x_j))$ containing x is an admissible (s,γ)-rectangle in $R(x_i)$ and $f^m(\mathbb{CC}(R(x_i) \cap f^{-m}(R(x_j)), x))$ is an admissible (u,γ)-rectangle in $R(x_j)$.
(4) $\operatorname{diam} f^k(\mathbb{CC}(R(x_i) \cap f^{-m}(R(x_j)), x)) \le 3 \operatorname{diam} R(x_i) \max\{\lambda^k, \lambda^{m+k}\}$ for $0 \le k \le m$.

Remark. To simplify the definition of (ρ,β,λ)-rectangle cover we have chosen rectangles $R(x) = \Psi_x([-1,1]^2)$. However, as we will soon see, in the main applications we are given Ψ_x, the Lyapunov chart; so to reduce the size of the regular neighborhood we just consider $[-h,h]^2$ for $0 < h \le 1$, and in that case $\Psi_x(Q(h)) =: R(x,h)$. We say that Λ admits a (ρ,β,λ)-rectangle cover if the above definition is satisfied.

If μ is an ergodic hyperbolic measure for $f \in \operatorname{Diff}(M)$ let

$$\chi(\mu) = \min\{|\chi_i| \mid i = 1,2\}. \tag{S.4.2}$$

Theorem S.4.16. *Let $f \in \operatorname{Diff}^{1+\alpha}(M)$, $\alpha > 0$, and M a compact Riemannian manifold. If μ is a hyperbolic measure then for every $\delta > 0$, $\rho > 0$ there exists a compact set Λ_δ with $\mu(\Lambda_\delta) > 1 - \delta$, a constant $\beta = \beta(\rho,\delta) > 0$, and $0 < \lambda < 1$ such that Λ_δ admits a (ρ,β,λ)-rectangle cover $R = \{R(x_1), \dots, R(x_2)\}$ with $x_i \in \Lambda_\delta$, $i = 1, \dots, t$, and $e^{(-\chi_\mu - \delta)} \le \lambda \le e^{(-\chi_\mu + \delta)}$.*

Proof. Fix $\delta > 0$ and choose a Pesin set Λ_δ as in Theorem S.4.3 such that $\mu(\Lambda_\delta) > 1 - \delta$. For $\rho > 0$ choose $0 < h < 1$ such that the regular neighborhood $R(x,h/2)$ has diameter less than ρ. Now choose $\beta = \beta(\delta,\rho) = \beta(\delta,h)$ as in the Closing Lemma and for each $x \in \Lambda_\delta$ assume that $\beta(x,\rho) \subset \operatorname{Int} R(x,h/2)$. Since Λ_δ is compact consider $\bigcup_{x \in \Lambda_\delta} B(x,\rho) \supset \Lambda_\delta$ and taking a finite subcover we find rectangles $R(x_i,h/2), \dots, R(x_t,h/2)$ satisfying (1) and (2) of the definition of a (ρ,β,λ)-rectangle cover. Choose $\gamma = \gamma(\rho,\delta)$.

To check (3) observe that for each x_i the boundaries of $R(x_i, h/2)$ are admissible (γ, h)-manifolds (u or s corresponding to the different sides). By Proposition S.3.5 and the construction of the Closing Lemma we have that if $x \in B(x_i, \rho)$, $f^m(x) \in B(x_j, \rho)$, and x, $f^m(x) \in \Lambda_\delta$ then if $m > 0$ the μ-boundaries of $R(x_i, h/2)$ will be mapped by f^m to admissible (u, γ, δ, h)-manifolds near x_g. Thus the connected component $\mathbb{CC}(R(x_g, h) \cap f^m(R(x_i, h)), x)$ is an admissible (u, γ)-rectangle near x_j. Similarly we prove (3) for s-rectangles.

For (4) we refer to the analogous property in the Closing Lemma. Finally the choice of λ can be made arbitrarily close to $\chi(\mu)$ by an appropriate choice of ϵ in the ϵ-reduction process. □

e. The Livschitz Theorem.

Theorem S.4.17. (Livschitz Theorem for nonuniformly hyperbolic systems) *Let $f \in \mathrm{Diff}^{1+\alpha}(M)$, $\alpha > 0$, and M a compact Riemannian manifold. Suppose that μ is a hyperbolic measure for f and $\varphi: M \to \mathbb{R}$ is a Hölder-continuous function such that for each periodic point p with $f^m(p) = p$ we have $\sum_{i=0}^{m-1} \varphi(f^i(p)) = 0$. Then there exists a measurable Borel function h such that for μ-almost every x*

$$\varphi(x) = h(f(x)) - h(x).$$

Proof. Suppose μ is ergodic and continuous (the modifications for the general case are not significant). Then there exists $x \in \operatorname{supp}\mu$ such that $\operatorname{supp}\mu \subset \bar{\mathcal{O}}(x)$. For $\delta > 0$ let Λ_δ be the δ-Pesin set for μ and assume that there is $x \in \Lambda_\delta$ such that $\operatorname{supp}\mu \subset \bar{\mathcal{O}}(x)$. Moreover we assume that for every $\delta > 0$ the intersection $\mathcal{O}(x) \cap \Lambda_\delta$ is dense in Λ_δ. Define $h(x) = 0$, $h(f(x)) = \varphi(x)$,

$$h(f^2(x)) = \varphi(x) + \varphi(f(x)),$$

and so forth. To extend h to Λ_δ continuously we need to show that h is uniformly continuous on $\mathcal{O}(x) \cap \Lambda_\delta$. Let $\beta = \beta(h, \delta)$ be as in the Closing Lemma for some $h > 0$. If $n_2 > n_1$ are such that $d(f^{n_2}(x), f^{n_1}(x)) < \beta$, then by the Closing Lemma there exists a hyperbolic periodic point z satisfying

$$d(f^i(f^{n_1}(x)), f^i(z)) \le 3h\, A_\delta \max\left\{\left(\frac{e^\epsilon}{1+\epsilon\lambda}\right)^{-i}, \left(\frac{e^\epsilon}{i+\epsilon\lambda}\right)^{-(n_2-n_1-i)}\right\}$$

for $i = 0, \dots, n_2 - n_1 - 1$.

Now since φ is Hölder continuous, that is,

$$|\varphi(x) - \varphi(y)| \le C\, d(x, y)^r$$

for some $C > 0$ and $0 < r \le 1$, we have

$$|\varphi(f^{n_1+i}(x)) - \varphi(f^i(z))| \le C(3h\, A_\delta)^r \max\left\{\left(\frac{e^\epsilon}{1+\epsilon\lambda}\right)^{-ir}, \left(\frac{e^\epsilon}{1+\epsilon\lambda}\right)^{-r(n_2-n_1-i)}\right\},$$

so

$$\begin{aligned}
|h(f^{n_2}(x))-h(f^{n_1}(x))| &= \Big| \sum_{i=0}^{n_2-n_1-1} (\varphi(f^{n_1}(f^i(x))) - \varphi(f^i(x))) + \sum_{i=0}^{n_2-n_1-1} \varphi(f^i(z))\Big| \\
&\le C(3h\,A_\delta)^r \sum_{i=0}^{n_2-n_1-1} \max\left\{ \left(\frac{e^\epsilon}{1+\epsilon\lambda}\right)^{-ir}, \left(\frac{e^\epsilon}{1+\epsilon\lambda}\right)^{-r(n_2-n_1-1)} \right\} \\
&\quad + \Big| \sum_{i=0}^{n_2-n_1-1} \varphi(f^i(z))\Big| \\
&\le C(3h\,A_\delta)^r N
\end{aligned}$$

for some $N > 0$ independent of n_2, n_1.

Thus φ is defined and continuous on Λ_δ. Now we extend φ to $\bigcup_{i=0}^\infty f^i(\Lambda_\delta)$ as follows: If $y \in f(\Lambda_\delta) \setminus \Lambda_\delta$ then $h(y) = \varphi(f^{-1}(y)) + h(f^{-1}(y))$, and so on. Since $\mu\left(\bigcup_{i=0}^\infty f^i(\Lambda_\delta)\right) = 1$, h is defined almost everywhere and clearly satisfies the conditions of the theorem. □

Remarks. (1) In the proof we actually do not use Hölder continuity. The argument goes through if φ is a Borel function whose restriction to Λ_δ is Hölder continuous with respect to a Lyapunov metric, with uniform Hölder exponent and constant for all $\delta > 0$. Notice that the function h obtained in the theorem has the same property.

(2) As in the uniformly hyperbolic case one can, in fact, show that the unstable distribution is Hölder continuous with respect to a Lyapunov metric on Λ_δ, similarly to the previous remark. Hence the restriction of the Jacobian to the unstable distribution is also Hölder continuous with respect to a Lyapunov metric (compare with the arguments in Section 19.1.e). Indeed, the most important application of the Livschitz Theorem concerns the logarithm of the unstable Jacobian.

5. Entropy and dynamics of hyperbolic measures

In this section we describe some of the consequences of the existence of hyperbolic measures for $C^{1+\alpha}$ diffeomorphisms of compact surfaces. The main goal is to relate the existence of nonzero exponents with the presence of hyperbolic periodic orbits, transverse homoclinic points, and hyperbolic horseshoes. As the reader might have guessed the main tools will be the Closing Lemma and the recurrence provided by the invariant measures.

a. Hyperbolic measures and hyperbolic periodic points. Let $f \in \mathrm{Diff}^{1+\alpha}(M)$, $\alpha > 0$, and M a compact Riemannian manifold; moreover let us assume that f preserves a hyperbolic measure μ.

The *support of a measure* μ, $\operatorname{supp}\mu$, is defined as the set of points $x \in M$ such that for any $\epsilon > 0$ we have $\mu(B(x,\epsilon)) > 0$. Obviously $\operatorname{supp}\mu$ is a closed set, and any other closed set of measure 1 contains $\operatorname{supp}\mu$. Now let $\mathrm{Per}(f)$ denote the set of all periodic points for f and $\mathrm{Per}_n(f)$ the set of hyperbolic periodic points for f.

Theorem S.5.1. *If μ is a hyperbolic measure for $f \in \mathrm{Diff}^{1+\alpha}(M)$, $\alpha > 0$, and M compact, then* $\operatorname{supp}\mu \subset \overline{\mathrm{Per}(f)}$.

Proof. Let $x_0 \in \operatorname{supp}\mu$ and fix $\alpha > 0$. Since $\mu(B(x_0, \alpha/4)) > 0$ we can find $\delta > 0$ such that $\mu(B(x_0, \alpha/4)) \cap \Lambda_\delta > 0$ and choose $\beta = \beta(\alpha/(12A_\delta), \delta)$ as in the Closing Lemma.

Now pick a set $B \subset B(x_0, \alpha/4) \cap \Lambda_\delta$ of diameter less than β and of positive measure. By the Poincaré Recurrence Theorem 4.1.19, for almost every $x \in B$ there exists a positive integer $n(x)$ such that $f^{n(x)}(x) \in \beta$, and consequently $d(x, f^{n(x)}(x)) < \beta$. Applying the Closing Lemma then we obtain that there exists a hyperbolic periodic point z of period $n(x)$ such that $d(x, z) < A_\delta 3\alpha/(12A_\delta) = \alpha/4$, and clearly $d(x_0, z) < d(x_0, x) + d(x, z) < \alpha/2$. □

Now we look at the more trivial situation where all the Lyapunov exponents are negative. Recall that a periodic point p is called a *sink* if all its Lyapunov exponents are negative.

Corollary S.5.2. *Let $f \in \mathrm{Diff}^{1+\alpha}(M)$, $\alpha > 0$, and M a compact Riemannian surface. If μ is an ergodic hyperbolic measure with all its exponents negative then it is concentrated on the orbit of a periodic sink p, that is, $\exists m > 0$ such that* $\operatorname{supp}\mu = \{p, f(p), \dots, f^{m-1}(p)\}$.

Proof. This follows from the definition of regular neighborhoods and ergodicity. □

b. Continuous measures and transverse homoclinic points. By Theorem 6.5.5 the existence of a transverse homoclinic point (Definition 6.5.4) implies that there exists a hyperbolic horseshoe Λ for f in any neighborhood of the homoclinic point. This, of course, suggests complicated dynamics for the diffeomorphism, which is preserved under small perturbations. We will see that if μ is a (continuous) *nonatomic measure* then there exist homoclinic phenomena.

Theorem S.5.3. *Let $f \in \mathrm{Diff}^{1+\alpha}(M)$, $\alpha > 0$, and M compact. If μ is an ergodic hyperbolic continuous measure then* $\operatorname{supp}\mu$ *is contained in the closure of the transverse homoclinic points of a hyperbolic periodic point p.*

Proof. We have shown already that if μ is ergodic and all its Lyapunov exponents are negative then it is concentrated on the orbit of a periodic sink. Similarly if all the exponents are positive then μ is concentrated on the orbit of a source (a sink for f^{-1}). Since μ is continuous this shows that its Lyapunov exponents have different signs.

Now, as in the proof of Theorem S.5.1, let $x_0 \in \operatorname{supp}\mu$ and $\alpha > 0$ be a small number. Choose $\delta > 0$ such that $\mu(B(x_0, \alpha/4) \cap \Lambda_\delta) > 0$. Pick $x_1, x_2 \in B(x_0, \alpha/4)$ such that for any $r > 0$ we have $\mu(B(x_i, r) \cap \Lambda_\delta) > 0$, for $i = 1, 2$, and

$$d(x_1, x_2) < 1/2 \min\{\alpha/4, \rho(\delta)\}.$$

The existence of such points is guaranteed by the continuity of μ. Furthermore we can take two subsets

$$B_i \subset \Lambda_\delta \cap B(x_i, d(x_1, x_2)/10) \qquad \text{for } i = 1, 2$$

such that $\mu(B_i) > 0$ and

$$\operatorname{diam} B_i < \beta = \beta(3d(x_1, x_2)/100A_\delta, k, \delta) \quad \text{for } i = 1, 2.$$

Again using the Poincaré Recurrence Theorem 4.1.19 we can find points $y_1 \in B_1$, $y_2 \in B_2$ and positive integers $m(y_1), m(y_2)$ such that $f^{m(y_1)}(y_1) \in B_1$, $f^{m(y_2)}(y_2) \in B_2$. Therefore, we can apply the Closing Lemma and find periodic points z_1, z_2 such that

$$d(z_i, y_i) < d(x_1, x_2)/100 \text{ for } i = 1, 2.$$

Clearly

$$\begin{aligned} d(x_1, x_2) - d(y_1, x_1) - d(z_1, y_1) - d(y_2, x_2) - d(z_2, y_2) \\ \leq d(z_1, z_2) \leq d(x_1, x_2) + d(y_1, x_1) + d(z_1, y_1) + d(y_2, x_2) + d(z_2, y_2) \end{aligned}$$

so

$$\frac{1}{2} d(x_1, x_2) < d(z_1, z_2) < \frac{3}{2} d(x_1, x_2).$$

The inequality on the left shows that $z_1 \neq z_2$, the Closing Lemma also guarantees that locally $W^s(z_i)$ is an admissible $(s, 1)$-manifold rear y_1, and similarly $W^u(z_i)$ is an admissible $(\mu, 1)$-manifold near y_i, $i = 1, 2$. Now $d(z_1, z_2) < \rho = \rho(\delta)$, so $W^s(z_i)$ is also locally an admissible manifold for z_2, and by Proposition S.3.7 they intersect transversely. The rest follows from ergodicity. □

Corollary S.5.4. *Under the assumptions of Theorem S.5.3:*

1. *f has a compact invariant set Λ such that $f\restriction_\Lambda$ is a horseshoe.*
2. $\overline{\lim}_{m\to\infty} 1/m \log \operatorname{card} \operatorname{Fix}(f^m) > 0$.
3. *If μ is just continuous then* $\operatorname{supp}\mu$ *is contained in the closure of the hyperbolic periodic points that have transverse homoclinic points.*

Let $\chi(x) := \min\{|\chi_i(x)| \mid \chi_i(x) \neq 0\}$; this is an f-invariant function and therefore if μ is an ergodic measure it is constant almost everywhere, hence is equal to $\chi(\mu)$ (see (S.4.2)), and it characterizes the minimal rate of typical exponential behavior of the system.

Theorem S.5.5. *Let $f \in \operatorname{Diff}^{1+\alpha}(M)$, $\alpha > 0$, and M compact. If μ is an ergodic hyperbolic measure and $x \in \operatorname{supp}\mu$ then for any $\rho > 0$, any neighborhoods V of x and W of* $\operatorname{supp}\mu$, *and any collection of continuous functions $\phi_1, \dots, \phi_k$ there exists a hyperbolic periodic point $z \in V$ such that the orbit of z is contained in W,*

$$\chi(z) \geq \chi(\mu) - \rho,$$

and $\left|\frac{1}{m(z)}\sum_{k=0}^{m(z)-1}\phi_i(f^k(z)) - \int\phi_i\,d\mu\right| < \rho$ *for* $i = 1,\dots,k$, *where* $m(z)$ *denotes the period of* z.

Proof. The idea of this proof is the same as the one of Theorem S.5.1; we just have to be more careful about the choice of constants. Thus let $x \in \operatorname{supp}\mu$ and δ, V, W, and $\phi_1,\dots,\phi_k$ be as in the hypotheses of the theorem. First observe that if $z_m = z_m(x)$ is a sequence of periodic points obtained through the Closing Lemma such that $z_m \to x \in \operatorname{supp}\mu \cap \Lambda$ then $\chi(z_m) \to \chi(x)$. Second suppose that $B(x,\alpha) \subset V$ for small $\alpha > 0$ and let $B \subset B(x,\alpha)$ be as in the proof of Theorem S.5.1. Choose a point x_0 such that there exists $m_k > 0$ such that $f^{m_k}(x_0) \to x_0$ and

$$\frac{1}{m_k}\sum_{i=0}^{m_k-1}\delta(f^i(x_0)) \to \mu$$

as $m_k \to \infty$. Fix Λ_δ also as in Theorem S.4.3; furthermore we may assume that $f^{m_k}(x_0) \in \Lambda_\delta$ for all $k \geq 0$, similarly to Proposition S.4.11, and $x_0 \in \operatorname{supp}\mu$. Given $\phi_1,\dots,\phi_k$ pick $m_k > 0$ such that

$$\left|\frac{1}{m_k}\sum_{k=0}^{m_k-1}\phi_i(f^k(x_0)) - \int\phi_i\,d\mu\right| < \frac{\rho}{2}.$$

Now choose $r > 0$ such that $B(f^i(x_0), r) \subset W$ for $0 \leq i \leq m_k$. Since we are only dealing with a finite number of continuous functions pick $h' > 0$ such that if $d(w_1,w_2) \leq 3h'A_\delta$ then

$$|\phi_i(w_1) - \phi_i(w_2)| < \rho/2$$

Now apply the Closing Lemma for $h = \min(h', r/3h'A_\delta)$ to obtain a periodic point $z(x_0)$ such that $f^{m_k}(z(x_0)) = z(x_0)$,

$$d(f^i(x_0), f^i(z(x_0))) \leq 3h'A_\delta,$$

and

$$\left|\frac{1}{m_k}\left(\sum_{j=0}^{m_k-1}\phi_i\left(f^j(z(x_0))\right) - \sum_{j=0}^{m_k-1}\phi_i\left(f^j(x_0)\right)\right)\right| < \frac{\rho}{2}.$$

□

Corollary S.5.6. *With f and μ as in Theorem S.5.5 if f_n is a sequence converging to f in the C^1 topology, then f_n has an invariant hyperbolic probability measure μ_n such that μ_n converges to μ weakly. Furthermore μ_n may be assumed to be supported on hyperbolic periodic points.*

This corollary suggests a "weak stability" of hyperbolic measures. It also says that the measures supported on hyperbolic periodic points are dense in the set of hyperbolic measures.

c. The Spectral Decomposition Theorem. By Theorem S.5.3 we know that if μ is an ergodic hyperbolic measure then its support is either a periodic sink or is contained in the closure of the transverse homoclinic points of a hyperbolic periodic orbit. As we will see now this implies that there exists $x \in M$ such that $\operatorname{supp} \mu \subset \overline{\mathcal{O}(x)}$. In this section we give a partial extension of this theorem for hyperbolic measures that are not ergodic.

Proposition S.5.7. *Let $f: M \to M$ be a C^1 diffeomorphism of a compact manifold M. Suppose $x \in M$ is a hyperbolic periodic point for f with transverse homoclinic points. Then the set Λ formed by the closure of the transverse homoclinic points of x has a dense orbit. Furthermore, if m is the period of x there exist $\Lambda_0, \dots, \Lambda_{m-1}$ such that for $i = 0, \dots, m-1$ $f(\Lambda_i) = \Lambda_{i+1} \mod m$, $f^m(\Lambda_i) = \Lambda_i$, and $f^m\restriction_{\Lambda_i}$ is topologically mixing (Definition 1.8.2).*

Proof. To obtain a dense orbit we prove that if U, V are open sets in Λ then there exists $N = N(U, V)$ such that $f^N(U) \cap V \neq \varnothing$ (see Lemma 1.4.2). So let U, V be open sets in Λ. Since the transverse homoclinic points are dense in Λ there exist transverse homoclinic points $q_1 \in U$, $q_2 \in V$ for the orbit of x. By taking a power k of f we can assume that $f^k(q_2)$ is on the same stable and unstable manifolds as q_1, say $q_1, f(q_2) \in W^s(x) \cap W^u(x)$. Now we only consider powers of f^m, so we fix x and $W^s(x)$, $W^u(x)$ are f^m-invariant. Let us take two small disks $D^u(q_1)$ and $D^s(f^k(q_2))$ on $W^u(x)$, $W^s(x)$, respectively, containing q_1, $f^k(q_2)$. The Inclination Lemma implies that large iterates of $f^{im}(D^u(q_1))$ converge locally in the C^1 topology to $W^u(x)$ and similarly $f^{-im}(D^s(f^k(q_2)))$ converges locally to $W^s(x)$. Take a "rectangle" around x of small diameter. For large i the connected components

$$\mathcal{CC}(f^{-im}(D^s(f^k(q_2))), f^{k-im}(q_2)) \text{ and } \mathcal{CC}(f^{im}(D^u(q_1)), f^{im}(q_1))$$

are C^1 submanifolds C^1-close to the local stable and unstable manifolds. Since the latter are transverse

$$f^{-im}(D^s(f^k(q_2))) \cap f^{im}(D^u(q_1)) \neq \varnothing.$$

The rest of the proof is obvious from here. □

Theorem S.5.8. (Spectral Decomposition Theorem) *Let $\alpha > 0$, $f \in \operatorname{Diff}^{1+\alpha}(M)$, and M a compact Riemannian surface. Suppose that μ is a hyperbolic measure for f. Then for $\delta > 0$ the Pesin set $\Lambda_\delta = \Lambda_{1,\delta} \cup \dots \cup \Lambda_{t,\delta}$, where the $\Lambda_{i,\delta}$'s are closed f-invariant sets such that for each i there exists $x \in M$ such that $\Lambda_{i,\delta} \subset \overline{\mathcal{O}(x)}$.*

Proof. Consider Λ_δ for some fixed $\delta > 0$ and let $\rho = \rho(\delta)$ be as in Corollary S.4.6. Thus if $x_1, x_2 \in \Lambda_\delta$ are such that $\mu(B(x_i, \epsilon) \cap \Lambda_\delta) > 0$ for $\epsilon > 0$, $i = 1, 2$, and $d(x_1, x_2) \leq \rho/2$ then by the proof of Theorem S.5.3 there exist hyperbolic periodic points $z(x_1), z(x_2)$ such that $W^s(z(x_1))$ intersects $W^u(z(x_2))$ transversely and $W^u(z(x_1))$ intersects $W^s(z(x_2))$ transversely. This is an equivalence relation for hyperbolic periodic points. Similarly we define an equivalence relation on Λ_δ as follows: $x_1 \sim x_2$ if $x_1, x_2 \in \Lambda_\delta$ and they belong to the closure of transverse homoclinic points of the same periodic point x. Therefore it remains to prove that we can only have a finite number of such classes on Λ_δ. Observe that if $x_1 \in \Lambda_\delta$ then all $y \in B(x, \rho; 2) \cap \Lambda_\delta$ are in the same class, so by compactness we only have a finite number of equivalence classes. Hence by Proposition S.5.7 the theorem follows. □

d. Entropy, horseshoes, and periodic points for hyperbolic measures. Recall that a compact f-invariant set Λ is a horseshoe for $f \in \text{Diff}^1(M)$ if there exist s, k and sets $\Lambda_0, \dots, \Lambda_{k-1}$ such that $\Lambda = \Lambda_0 \cup \dots \cup \Lambda_{k-1}$, $f^k(\Lambda_i) = \Lambda_i$, $f(\Lambda_i) = \Lambda_{i+1}$ mod k, and $f^k\restriction_{\Lambda_0}$ is conjugate to a full shift in s-symbols. For Λ a hyperbolic horseshoe we can define $\chi(\Lambda) = \inf\{\chi(\mu) \mid \mu$ is supported on a periodic orbit$\}$.

The main result of this subsection is the following:

Theorem S.5.9. *Let $f \in \text{Diff}^{1+\alpha}(M)$, $\alpha > 0$, and M a compact Riemannian surface M. Suppose that μ is an ergodic hyperbolic measure for f with $h_\mu(f) > 0$. Then for any $\rho > 0$ and any finite collection of functions $\varphi_1, \dots, \varphi_k \in C(M)$ there exists a hyperbolic horseshoe Λ satisfying the following conditions:*

(1) $h_\mu(f) - \rho < h_{\text{top}}(f\restriction_\Lambda)$.
(2) *Λ is contained in a ρ-neighborhood of* supp μ.
(3) $\chi(\mu) - \rho < \chi(\Lambda)$.
(4) *There exists a measure $\nu = \nu(\Lambda)$ supported on Λ such that for $i = 1, \dots, k$*

$$\left| \int \varphi_i \, d\nu - \int \varphi_i \, d\mu \right| < \rho.$$

Before giving the proof of Theorem S.5.9 we discuss some corollaries of it.

Corollary S.5.10. *For f and μ as in Theorem S.5.9 there exists a sequence of f-invariant measures μ_n supported on hyperbolic horseshoes Λ_n such that* (1) *$\mu_n \to \mu$ in the weak* topology and* (2) $h_{\mu_n}(f) \to h_\mu(f)$.

Corollary S.5.11. *For f and μ as in Theorem S.5.9 and $\epsilon > 0$*

$$h_\mu(f) \le \overline{\lim_{m\to\infty}} \frac{1}{m} \log \operatorname{card}\{x \in M \mid f^m(x) = x,\ \chi(x) \ge \chi(\mu) - \epsilon\}.$$

In particular $h_{\text{top}}(f) \le p(f)$, where $p(f)$ is as in (3.1.1).

Proof. By Theorem 18.5.1, if Λ is a hyperbolic horseshoe for f then

$$h_{\text{top}}(f\restriction_\Lambda) = \overline{\lim_{m\to\infty}} \frac{1}{m} \log \operatorname{card}\{x \in \Lambda \mid f^m(x) = x\}.$$

Then by (1) and (2) of Theorem S.5.9 the corollary follows. □

The following corollary shows that hyperbolic measures are "stable" or persistent under C^1 perturbations. This, of course, is a consequence of the structural stability of the hyperbolic horseshoes.

Corollary S.5.12. *Given f and μ as in Theorem S.5.9 and $f_n \in \mathrm{Diff}^{1+\alpha}(M)$ such that $f_n \to f$ in the C^1 topology there exists a collection of f_n-invariant ergodic measures μ_n satisfying*

(1) *$\mu_n \to \mu$ in the weak topology,*
(2) *$h_{\mu_n}(f_n) \to h_\mu(f)$,*
(3) *$\chi(\mu_n) \to \chi(\mu)$.*

Corollary S.5.13. *The entropy function $h\colon \mathrm{Diff}^{1+\alpha}(M^2) \to \mathbb{R}$ is lower semicontinuous.*

Proof. By Theorem S.5.9 $h(f) = \sup\{h(f_{\restriction\Lambda}) \mid \Lambda \text{ is a hyperbolic horseshoe}\}$. By structural stability of horseshoes lower semicontinuity follows. □

Proof of Theorem S.5.9. Write $C(M)$ for the space of real-valued continuous functions defined on M, $G_\mu(f,m,\epsilon,\delta) = \max \operatorname{card}\{(m,\epsilon)\text{-separated points on a set of measure } 1-\delta\}$ [K5] and let $\varphi_1,\dots,\varphi_k \in C(M)$. For $\varphi > 0$ choose $\delta > 0$, $\epsilon > 0$ such that

(1) $\overline{\lim}_{m\to\infty} 1/m \log G_\mu(f,m,\epsilon,2\delta) > h_\mu(f) - r$,
(2) if $d(x,y) < \epsilon$ then $|\varphi_i(x) - \varphi_i(y)| < r/2$.

By Theorem S.4.16 we can choose a Pesin set with $\mu(\Lambda_\delta) > 1-\delta$ admitting an $(\epsilon/3,\beta,\lambda)$-rectangle cover for $\beta = \beta(\delta,\epsilon)$ and $\lambda = \lambda(x_\mu)$. Now let ζ be a finite measurable partition of M with $\operatorname{diam}\zeta < \beta/2$ and $\zeta > \{\Lambda_\delta, M \smallsetminus \Lambda_\delta\}$. Set

$$\Lambda_{\delta,m} = \Big\{x \in \Lambda_\delta \mid f^q(x) \in \zeta(x) \text{ for some } q \in [m,(1+r)m]$$
$$\text{and } \Big|\frac{1}{s}\sum_{j=0}^{s-1}\varphi_i(f^i(x)) - \int \varphi_i\, d\mu\Big| < r/2 \quad \text{for } s \geq m, i = 1,\dots,k\Big\}.$$

We claim that $\mu(\Lambda_{\delta,m}) \to \mu(\Lambda_\delta)$ as $m \to \infty$. Take an element $C \in \zeta$ such that $C \subset \Lambda_\delta$ and consider the sets C_m given by

$$\Big\{x \in C \Big| \sum_{i=0}^{m-1}\frac{\chi_C(f^i(x))}{m} < \mu(C)\Big(1+\frac{r}{3}\Big),\ \mu(C)\Big(1+\frac{2r}{3}\Big) < \sum_{i=0}^{[m(1+r)]}\frac{\chi_C(f^i(x))}{m}\Big\},$$

where χ_C denotes the characteristic function of C. By the Birkhoff Ergodic Theorem 4.1.2 $\mu(C \smallsetminus C_m) \to 0$. Thus applying the same argument to each element $C \in \zeta$ contained in Λ_δ we obtain that $\mu(\Lambda_{\delta,m}) \to \mu(\Lambda_\delta)$. So the claim is proved.

Now choose m sufficiently large that $\mu(\Lambda_{\delta,m}) > 1-2\delta$. Let $E_m \subset \Lambda_{\delta,m}$ be an (m,ϵ)-separated set of maximal cardinality. Clearly $\Lambda_{\delta,m} \subset \bigcup_{x\in E_m} B_n^f(x,g)$ and therefore there exist infinitely many m such that

$$\operatorname{card} E_m \geq e^{m(h_\mu(f)-2r)}.$$

For each $q \in [m, (1+r)m]$ let $V_q = \{x \in E_m \mid f^q(x) \in \zeta(x)\}$ and n be the value of q that maximizes $\operatorname{card} V_q$. Since $e^{mr} > mr$ we have that $\operatorname{card} V_n \geq e^{m(h_\mu(f)-3r)}$. Let $R = \{R(x_1), R(x_2), \dots, R(x_t)\}$, with $x_j \in \Lambda_{\delta,m}$, be an $(\epsilon/e, \beta, l)$-rectangle cover, and consider $V_m \cap R(x_j)$ for $1 \leq j \leq t$ and choose the value i of j that maximizes $\operatorname{card} V_m \cap R(x_j)$. Thus if we write D_m for $V_m \cap R(x_j)$ then

$$\operatorname{card} D_n \geq \frac{1}{t} \operatorname{card} V_m \geq \frac{1}{t} e^{m(h_\mu(f)-3r)}.$$

So consider $R(x_i)$ and D_n. Each $x \in D_n$ returns to $R(x_i)$ in m iterations, thus $\mathbb{CC}(R(x_i) \cap f^m(R(x_i)), f^m(x))$ is an admissible u-rectangle in $R(x_i)$ and $f^{-n}(\mathbb{CC}(R(x_i) \cap f^m(R(x_i)), f^m(x)))$ is an admissible s-rectangle in $R(x_i)$. This follows from the facts that $d(x_i, x) < \beta$ and $d(f^n(x), x_i) < \beta$, and (2) of the definition of an $(\epsilon/\beta, \beta, \lambda)$-rectangle cover.

If $y \in \mathbb{CC}(R(x_i) \cap f^{-n}(R(x_i)), x)$ then by (4) of the definition of an $(\epsilon/3, \beta, \lambda)$-rectangle cover $d(f^i(x), f^i(y)) \leq 3 \operatorname{diam} R(x_i) \leq \epsilon$ for $l \in [0, n]$, which implies that if $y \in \mathbb{CC}(R(x_i) \cap f^{-n}(R(x_i)), x)$ and $y \neq x$ then $y \notin V_n$; otherwise it would contradict the separability of V_n. Hence there exist $\operatorname{card} V_n$ disjoint admissible s-rectangles mapped by f^n onto $\operatorname{card} V_n$ admissible u-rectangles.

So let

$$\Lambda^* = \Lambda(m) = \bigcap_{n \in \mathbb{Z}} f^{nl}\Big(\bigcup_{\zeta \in D_n} \mathbb{CC}(R(x_i) \cap f^{-n}(R(x_i)), x)\Big).$$

Then $f^n \restriction_{\Lambda^*}$ is conjugate to the full shift in $\operatorname{card} V_n$ symbols. Now observe that for each $y \in \Lambda^*$ its orbit remains in the union of the regular neighborhoods $R(x_i), \dots, R(f^n(x_i))$, so $f^n \restriction_{\Lambda^*}$ is a hyperbolic horseshoe.

The entropy of $f^n \restriction_{\Lambda^*}$ equals $\log \operatorname{card} D_n$, so

$$h_{\text{top}}(f \restriction_{\Lambda^*}) = \frac{1}{n} \log \operatorname{card} D_n \geq \frac{1}{n} \log \frac{1}{t} e^{m(H_\mu(f)-3r)} \geq \frac{1}{n} \log \frac{1}{t} + \frac{m}{n}(h_\mu(f) - 3r).$$

Since $m/n > 1/(1+r)$

$$h_{\text{top}}(f \restriction_{\Lambda^*}) \geq \frac{1}{m} \log 1/t + (h_\mu(f) - \epsilon r)(1+r)^{-1}\Big(-\frac{1}{m} \log t + h_\mu(f) - 3r\Big),$$

which proves the existence of the horseshoes satisfying (1) and obviously (2). For part (3) reduce the size of ϵ in the Osedelec–Pesin ϵ-reduction process and the sizes of the regular neighborhoods if necessary.

Finally, by the choice of $\Lambda_{\delta,m}$ and the sizes of the rectangles we have (4). □

Appendix

Appendix

Background material

1. Basic topology

a. Topological spaces.

Definition A.1.1. A *topological space* $(X, \mathcal{T})$ is a set X endowed with a collection $\mathcal{T} \subset \mathcal{P}(X)$ of subsets of X, called the *topology* of X, such that

(1) $\varnothing, X \in \mathcal{T}$,
(2) if $\{O_\alpha\}_{\alpha \in A} \subset \mathcal{T}$ then $\bigcup_{\alpha \in A} O_\alpha \in \mathcal{T}$ for any set A,
(3) if $\{O_i\}_{i=1}^k \subset \mathcal{T}$ then $\bigcap_{i=1}^k O_i \in \mathcal{T}$,

that is, $\mathcal{T}$ contains X and $\varnothing$ and is closed under union and finite intersection.

The sets $O \in \mathcal{T}$ are called *open* sets, and their complements are called *closed* sets. If $x \in X$ then an open set containing x is called a *neighborhood* of x. The *closure* $\bar{A}$ of a set $A \subset X$ is the smallest closed set containing A, that is, $\bar{A} := \bigcap\{C \mid A \subset C \text{ and } C \text{ closed}\}$. x is said to be an *accumulation point* of $A \subset X$ if every neighborhood of x contains infinitely many points of A.

A *base* for the topology $\mathcal{T}$ is a subcollection $\beta \subset \mathcal{T}$ such that for every $O \in \mathcal{T}$ and $x \in O$ there exists $B \in \beta$ such that $x \in B \subset O$. A topology $\mathcal{S}$ is said to be *finer* than $\mathcal{T}$ if $\mathcal{T} \subset \mathcal{S}$, *coarser* if $\mathcal{S} \subset \mathcal{T}$. If $Y \subset X$ then Y can be made into a topological space in a natural way by taking the *induced topology* $\mathcal{T}_Y := \{O \cap Y \mid O \in \mathcal{T}\}$.

A sequence $\{x_i\}_{i \in \mathbb{N}} \subset X$ is said to *converge* to $x \in X$ if for every open set O containing x there exists $N \in \mathbb{N}$ such that $\{x_i\}_{i>N} \subset O$.

$(X, \mathcal{T})$ is called a *Hausdorff space* if for any two $x_1, x_2 \in X$ there exist $O_1, O_2 \in \mathcal{T}$ such that $x_i \in O_i$ and $O_1 \cap O_2 = \varnothing$. It is called *normal* if it is Hausdorff and for any two closed $X_1, X_2 \subset X$ there exist $O_1, O_2 \in \mathcal{T}$ such that $X_i \subset O_i$ and $O_1 \cap O_2 = \varnothing$.

$\{O_\alpha\}_{\alpha \in A} \subset \mathcal{T}$ is called an *open cover* of X if $X = \bigcup_{\alpha \in A} O_\alpha$, and a finite open cover if A is finite. $(X, \mathcal{T})$ is called *compact* if every open cover has a finite subcover, *locally compact* if every point has a neighborhood with compact closure, and *sequentially compact* if every sequence has a convergent subsequence. X is called σ-compact if it is a countable union of compact sets.

If $(X_\alpha, \mathcal{T}_\alpha)$, $\alpha \in A$ are topological spaces and A is any set, then the *product topology* on $\prod_{\alpha \in A} X$ is the topology generated by the base $\{\prod_\alpha O_\alpha \mid O_\alpha \in \mathcal{T}_\alpha,\ O_\alpha \neq X_\alpha$ for only finitely many $\alpha\}$.

Let $(X, \mathcal{T})$ be a topological space. A set $D \subset X$ is said to be *dense* in X if $\bar{D} = X$. X is said to be *separable* if it has a countable dense subset.

$\mathbb{R}^n$ with the usual open and closed sets is a familiar example. The open balls (open balls with rational radius, open balls with rational center and radius) form a base. Points of a Hausdorff space are closed sets.

Proposition A.1.2. *A closed subset of a compact set is compact.*

Proof. If K is compact, $C \subset K$ is closed, and Γ is an open cover for C then $\Gamma \cup \{K \smallsetminus C\}$ is an open cover for K, hence has a finite subcover $\Gamma' \cup \{K \smallsetminus C\}$, so Γ' is a finite subcover (of Γ) for C. □

Proposition A.1.3. *A compact subset of a Hausdorff space is closed.*

Proof. If X is Hausdorff and $C \subset X$ compact fix $x \in X \smallsetminus C$ and for each $y \in C$ take neighborhoods U_y of y and V_y of x such that $U_y \cap V_y = \varnothing$. The cover $\bigcup_{y \in C} U_y \supset C$ has a finite subcover $\{U_{x_i} \mid 0 \leq i \leq n\}$ and hence $N_x := \bigcap_{i=0}^n V_{y_i}$ is a neighborhood of x disjoint from C. Thus $X \smallsetminus C = \bigcup_{x \in X \smallsetminus C} N_x$ is open and C is closed. □

Proposition A.1.4. *A compact Hausdorff space is normal.*

Proof. First we show that a closed set K and a point $p \notin K$ can be separated by open sets. For $x \in K$ there are open sets O_x, U_x such that $x \in O_x$, $p \in U_x$ and $O_x \cap U_x = \varnothing$. Since K is compact there is a finite subcover $O := \bigcup_{i=1}^n O_{x_i} \supset K$, and $U := \bigcap_{i=1}^n U_{x_i}$ is an open set containing p disjoint from O. Now suppose K, L are closed sets and for $p \in L$ consider open disjoint sets $O_p \supset K$, $U_p \ni p$. By compactness of L there is a finite subcover $U := \bigcup_{j=1}^m U_{p_j} \supset L$ and $O := \bigcap_{j=1}^m O_{p_j} \supset K$ is an open set disjoint from U. □

A useful consequence of normality is the following extension result:

Theorem A.1.5. *If X is a normal topological space, $Y \subset X$ closed, and $f\colon Y \to \mathbb{R}$ continuous, then there is a continuous extension of f to X.*

A collection of sets is said to have the *finite intersection property* if every finite subcollection has nonempty intersection.

Proposition A.1.6. *A collection of compact sets with the finite intersection property has nonempty intersection.*

Proof. It suffices to show that in a compact space every collection of closed sets with the finite intersection property has nonempty intersection. To that end consider a collection of closed sets with empty intersection. Their complements form an open cover. Since it has a finite subcover the finite intersection property does not hold. □

Definition A.1.7. The *one-point compactification* of a noncompact Hausdorff space $(X, \mathcal{T})$ is $\hat{X} := (X \cup \{\infty\}, \mathcal{S})$, where $\mathcal{S} := \mathcal{T} \cup \{(X \cup \{\infty\}) \setminus K \mid K \subset X \text{ compact}\}$.

It is easy to see that $\hat{X}$ is a compact Hausdorff space.

Theorem A.1.8. (Tychonoff Theorem) *The product of compact spaces is compact.*

This result is very useful in many situations because often a useful topology can be viewed as a product topology or is induced by a product topology. An obvious example is the topology of pointwise convergence.

Definition A.1.9. Let $(X, \mathcal{T})$ and $(Y, \mathcal{S})$ be topological spaces. A map $f: X \to Y$ is said to be *continuous* if $O \in \mathcal{S}$ implies $f^{-1}(O) \in \mathcal{T}$, *open* if $O \in \mathcal{T}$ implies $f(O) \in \mathcal{S}$, and a *homeomorphism* if it is continuous and bijective with continuous inverse. If there is a homeomorphism $X \to Y$ then X and Y are said to be *homeomorphic*. We denote by $C^0(X, Y)$ the space of continuous maps from X to Y and write $C^0(X)$ for $C^0(X, \mathbb{R})$. A map f from a topological space to $\mathbb{R}$ is said to be *upper semicontinuous* if $f^{-1}(-\infty, c) \in \mathcal{T}$ for all $c \in \mathbb{R}$, *lower semicontinuous* if $f^{-1}(c, \infty) \in \mathcal{T}$ for $c \in \mathbb{R}$.

A property of a topological space that is the same for any two homeomorphic spaces is said to be a *topological invariant.*

Proposition A.1.10. *The image of a compact set under a continuous map is compact.*

Proof. If C is compact and $f: C \to Y$ continuous and surjective then any open cover Γ of Y induces an open cover $f_*\Gamma := \{f^{-1}(O) \mid O \in \Gamma\}$ of C which by compactness has a finite subcover $\{f^{-1}(O_i) \mid i = 1, \dots, n\}$. By surjectivity $\{O_i\}_{i=1}^n$ is a cover for Y. □

A useful application of the notions of continuity, compactness, and being Hausdorff is the following result, sometimes referred to as *invariance of domain:*

Proposition A.1.11. *A continuous bijection from a compact space to a Hausdorff space is a homeomorphism.*

Proof. Suppose X is compact, Y Hausdorff, $f: X \to Y$ bijective and continuous, and $O \subset X$ open. Then $C := X \setminus O$ is closed, hence compact, and $f(C)$ is compact, hence closed, so $f(O) = Y \setminus f(C)$ (by bijectivity) is open. □

Definition A.1.12. A topological space $(X, \mathcal{T})$ is said to be *connected* if no two disjoint open sets cover X. $(X, \mathcal{T})$ is said to be *path connected* if for any two points $x_0, x_1 \in X$ there exists a continuous curve $c: [0, 1] \to X$ with $c(i) = x_i$. A connected component is a maximal connected subset of X. $(X, \mathcal{T})$ is said to be *totally disconnected* if every point is a connected component.

A Cantor set and $\mathbb{Q} \subset \mathbb{R}$ are totally disconnected. It is not hard to see that connected components are closed. Thus connected components are open if there are only finitely many and, more generally, if every point has a connected neighborhood (that is, the space is locally connected). This is not the case for $\mathbb{Q}$.

Theorem A.1.13. *A continuous image of a connected space is connected.*

Remark. This implies that a path-connected space is connected. The converse is false as is shown by the union of the graph of $\sin 1/x$ and $\{0\} \times [-1, 1]$ in $\mathbb{R}^2$.

Theorem A.1.14. *The product of two connected topological spaces is connected.*

Definition A.1.15. A *topological manifold* is a Hausdorff space X with a countable base for the topology such that every point is contained in an open set homeomorphic to a ball in $\mathbb{R}^n$. A pair (U, h) of such a neighborhood and a homeomorphism $h: U \to B \subset \mathbb{R}^n$ is called a *chart* or a system of *local coordinates.* A topological *manifold with boundary* is a Hausdorff space X with a countable base for the topology such that every point is contained in an open set homeomorphic to an open set in $\mathbb{R}^{n-1} \times [0, \infty)$.

Remark. One easily sees that if X is connected then n is constant. In this case it is called the *dimension* of the topological manifold. Path connectedness and connectedness are equivalent for topological manifolds.

Definition A.1.16. Consider a topological space $(X, \mathcal{T})$ and suppose there is an equivalence relation $\sim$ defined on X. Then there is a natural projection π to the set $\hat{X}$ of equivalence classes. The *identification space* or *factor space* $X/\!\sim\, := (\hat{X}, \mathcal{S})$ is the topological space obtained by calling a set $O \subset \hat{X}$ open if $\pi^{-1}(O)$ is open, that is, taking on $\hat{X}$ the finest topology with which π is continuous.

An important class of factor spaces appears when there is a group G of homeomorphisms acting on X such that the orbits are closed. Then one identifies points on the same orbit and obtains an identification space which in this case is denoted by X/G and called the *quotient* of X by G. For the case where $X = S^1$ and G is the cyclic group of iterates of a rational rotation we get $X/G \simeq X$. If $X = \mathbb{R}^2$ and G is the group of translations parallel to the x-axis then $X/G \simeq \mathbb{R}$. The torus is obtained from $\mathbb{R}^n$ by identifying points modulo $\mathbb{Z}^n$, that is, two points are equivalent if their difference is in $\mathbb{Z}^n$. Equivalently, one obtains it from identifying pairs of opposite sides of the unit square (or any rectangle) with the same orientation. A *cone* over a space X is the space obtained from identifying all points of the form $(x, 1)$ in $(X \times [0, 1], \text{product topology})$. The sphere is obtained by identifying all boundary points of a closed ball.

A priori the topology of an identification space may not be very nice. Unless the equivalence classes are all closed, for example, the identification space will not even be a Hausdorff space. In particular the space of orbits of a dynamical system with some recurrent behavior is not a very good object from the topological point of view. An example is X/G, where $X = S^1$ and G is the group of iterates of an irrational rotation.

b. Homotopy theory.

Definition A.1.17. Two continuous maps $h_0, h_1: X \to Y$ between topological spaces are said to be *homotopic* if there exists a continuous map $h: [0,1] \times X \to Y$ (the *homotopy*) such that $h(i, \cdot) = h_i$ for $i = 1, 2$. If $h_0(x) = h_1(x) = p$ for some $x \in X$ then $h_0, h_1: X \to Y$ are called *homotopic rel* p if h can be chosen such that $h(\cdot, x) = p$. If $X = [0,1]$, $h_0(0) = h_1(0)$, and $h_0(1) = h_1(1)$ then we say that h_0, h_1 are *homotopic rel endpoints* if $h(\cdot, 0)$ and $h(\cdot, 1)$ can be taken constant. h is called *null-homotopic* if it is homotopic to a constant map. If h_1, h_2 are homeomorphisms they are called *isotopic* if h can be taken such that every $h(t, \cdot)$ is a homeomorphism. X and Y are called *homotopically equivalent* if there exist maps $g: X \to Y$ and $h: Y \to X$ such that $g \circ h$ and $h \circ g$ are homotopic to the identity. X is called *contractible* if it is homotopic to a point. A property of a topological space which is the same for any homotopically equivalent space is called a *homotopy invariant.*

Obviously homeomorphic spaces are homotopically equivalent. The circle and the cylinder are homotopically equivalent but not homeomorphic. Balls and cones are contractible. Contractible spaces are connected.

Definition A.1.18. Let M be a topological manifold, $p \in M$, and consider the collection of curves $c: [0,1] \to M$ with $c(0) = c(1) = p$. If c_1 and c_2 are such curves then let $c_1 \cdot c_2$ be the curve given by

$$c_1 \cdot c_2(t) := \begin{cases} c_1(2t) & \text{when } t \leq \frac{1}{2}, \\ c_2(2t-1) & \text{when } t \geq \frac{1}{2}. \end{cases}$$

Upon identifying curves homotopic rel endpoints one obtains a group called the *fundamental group* $\pi_1(M, p)$ of M at p. A space with trivial fundamental group is said to be *simply connected,* 1-connected if it is also connected.

We are mostly interested in connected manifolds where path connectedness ensures that the groups obtained at different p are isomorphic. Thus we simply write $\pi_1(M)$. Since the fundamental group is defined modulo homotopy, it is the same for homotopically equivalent spaces, that is, it is a homotopy invariant. The *free* homotopy classes of curves (that is, with no fixed base point) correspond exactly to the conjugacy classes of curves modulo changing base point, so there is a natural bijection between the classes of freely homotopic closed curves and conjugacy classes in the fundamental group.

Definition A.1.19. If M, M' are topological manifolds and $\pi: M' \to M$ is a continuous map such that $\operatorname{card} \pi^{-1}(y)$ is independent of $y \in M$ and every $x \in \pi^{-1}(y)$ has a neighborhood on which π is a homeomorphism to a neighborhood of $y \in M$ then M' (or (M', π)) is called a *covering (space)* or *cover* of M. If $n = \operatorname{card} \pi^{-1}(y)$ is finite then (M', π) is said to be an n-fold covering. If $f: N \to M$ is continuous and $F: N \to M'$ is such that $f = \pi \circ F$ then F is said to be a *lift* of f. If $f: M \to M$ is continuous and $F: M' \to M'$ is continuous such that $f \circ \pi = \pi \circ F$ then F is said to be a lift of f as well. A simply connected covering is called the *universal cover.* A homeomorphism of a covering M' of M is called a *deck transformation* if it is a lift of the identity on M.

Examples. $(\mathbb{R}, \exp(2\pi i(\cdot)))$ is a covering of the unit circle. Geometrically one can view this as the helix $(e^{2\pi i x}, x)$ covering the unit circle under projection. The map defined by taking the fractional part likewise defines a covering of the circle $\mathbb{R}/\mathbb{Z}$ by $\mathbb{R}$. The torus is covered by the cylinder which is in turn covered by $\mathbb{R}^2$. Notice that the fundamental group $\mathbb{Z}$ of the cylinder is a subgroup of that of the torus ($\mathbb{Z}^2$) and $\mathbb{R}^2$ is a simply connected cover of both. The expanding maps on the circle (1.7.1) define coverings of the circle by itself. Factors of the Poincaré upper half-plane are covered by the upper half-plane (Sections 5.4c,e).

There is a natural bijection between conjugacy classes of subgroups of $\pi_1(M)$ and classes of covering spaces modulo homeomorphisms commuting with deck transformations. In particular the universal cover is unique. This bijection can be described as follows. Suppose (M', π) is a covering of M and $x_0, x_1 \in \pi^{-1}(y)$. Since M' is path connected there are curves $c\colon [0,1] \to M'$ with $c(i) = x_i$ for $i = 1, 2$. Under π these project to loops on M. Any continuous map induces a homomorphism between the fundamental groups. Any continuous map possesses a lift, so a homotopy of the loop $\pi \circ c$ rel y can be lifted to a homotopy of the curve c and since by hypothesis $\pi^{-1}(y)$ is discrete, this homotopy is a homotopy rel endpoints. In particular homotopic curves project to homotopic curves and, by considering the case $x_1 = x_2$, the fundamental group of M' injects into the fundamental group of M as a subgroup. This is the subgroup corresponding to the covering. Furthermore this subgroup is a proper subgroup whenever π is not a homeomorphism, that is, the cover is a nontrivial covering. Thus a simply connected space has no proper coverings. One can also see that any two coverings M_1' and M_2' of M have a common covering M'', so the universal cover is unique. Any topological manifold has a universal cover.

c. Metric spaces. For several quite natural notions a topological structure is not adequate, but one rather needs a *uniform* structure, that is, a topology in which one can compare neighborhoods of different points. This can be defined abstractly and is realized for topological vector spaces (see Definition A.2.1), but it is a little more convenient to introduce these concepts for metric spaces.

Definition A.1.20. If X is a set then $d\colon X \times X \to \mathbb{R}$ is called a *metric* if

(1) $d(x, y) = d(y, x)$,
(2) $d(x, y) = 0 \Leftrightarrow x = y$,
(3) $d(x, y) + d(y, z) \geq d(x, z)$ *(triangle inequality).*

If d is a metric then (X, d) is called a *metric space.* The set $B(x, r) := \{y \in X \mid d(x, y) < r\}$ is called the *(open) r-ball* around x.

$O \subset X$ is called *open* if for every $x \in O$ there exists $r > 0$ such that $B(x, r) \subset O$.

For $A \subset X$ the set $\bar{A} := \{x \in X \mid \forall r > 0 \quad B(x, r) \cap A \neq \varnothing\}$ is called the *closure* of A. A is called *closed* if $\bar{A} = A$.

Let (X, d), (Y, dist) be metric spaces. A map $f: X \to Y$ is said to be *uniformly continuous* if for all $\epsilon > 0$ there is a $\delta > 0$ such that for all $x, y \in X$ with $d(x, y) < \delta$ we have $\mathrm{dist}(f(x), f(y)) < \epsilon$. A uniformly continuous bijection with uniformly continuous inverse is called a *uniform homeomorphism*. A family $\mathcal{F}$ of maps $X \to Y$ is said to be *equicontinuous* if for every $x \in X$ and $\epsilon > 0$ there is a $\delta > 0$ such that $d(x, y) < \delta$ implies $\mathrm{dist}(f(x), f(y)) < \epsilon$ for all $y \in X$ and $f \in \mathcal{F}$. A map $f: X \to Y$ is said to be *Hölder continuous* with exponent α, or *α-Hölder*, if there exist $C, \epsilon > 0$ such that $d(x, y) < \epsilon$ implies $d(f(x), f(y)) \leq C(d(x, y))^{\alpha}$, *Lipschitz continuous* if it is 1-Hölder, and biLipschitz if it is Lipschitz and has a Lipschitz inverse. It is called an *isometry* if $d(f(x), f(y)) = d(x, y)$ for all $x, y \in X$.

A sequence $\{x_i\}_{i \in \mathbb{N}}$ is called a *Cauchy sequence* if for all $\epsilon > 0$ there exists an $N \in \mathbb{N}$ such that $d(x_i, x_j) < \epsilon$ whenever $i, j \geq \mathbb{N}$. X is said to be *complete* if every Cauchy sequence converges.

Remark. The collection of open sets induces a topology with the open balls as a base. Closed sets have open complements. The definitions are consistent with those made for topological spaces. For metric spaces the notions of compactness and sequential compactness are equivalent.

For an open cover of a compact metric space there exists a number δ such that every δ-ball is contained in an element of the cover. The largest such number is called the *Lebesgue number* of the cover.

It is not hard to show that a uniformly continuous map from a subset of a metric space uniquely extends to the closure.

Completeness is a very important property since it allows us to perform limit operations which arise frequently in our constructions. Notice that it is not possible to define a notion of Cauchy sequences in an arbitrary topological space since one lacks the possibility of comparing neighborhoods at different points. A useful observation is that compact sets are complete by sequential compactness. A metric space can be made complete in the following way:

Definition A.1.21. If X is a metric space and there is an isometry from X onto a dense subset of a complete metric space $\hat{X}$ then $\hat{X}$ is called the *completion* of X.

Up to isometry the completion of X is unique: If there are two completions X_1 and X_2 then by construction there is a bijective isometry between dense subsets and therefore this isometry extends (by uniform continuity) to the entire space. On the other hand completions always exist by virtue of the construction used to obtain the real numbers from the rational numbers. This completion is obtained from the space of Cauchy sequences on X by identifying two sequences if the distance between corresponding elements converges to zero. The distance between two (equivalence classes of) sequences is defined as the limit of the distance between corresponding elements. The isometry maps points to constant sequences.

Theorem A.1.22. (Baire Category Theorem) *In a complete metric space a countable intersection of open dense sets is dense. The same holds for a locally compact Hausdorff space.*

Proof. If $\{O_i\}_{i\in\mathbb{N}}$ are open and dense in X and $\varnothing \neq B_0 \subset X$ is open then inductively choose a ball B_{i+1} of radius at most ϵ/i such that $\bar{B}_{i+1} \subset O_{i+1} \cap B_i$. The centers converge by completeness, so $\varnothing \neq \bigcap_i \bar{B}_i \subset B_0 \cap \bigcap_i O_i$. For locally compact Hausdorff spaces take B_i open with compact closure and use the finite intersection property. □

A topological space is said to be *metrizable* if there exists a metric on it that induces the given topology. Any metric space is normal and hence Hausdorff. A metric space has a countable base if and only if it is separable. Conversely (using Proposition A.1.4) we have

Proposition A.1.23. *A normal space with a countable base for the topology, hence any compact Hausdorff space with a countable base, is metrizable.*

If X is a compact metrizable topological space (for example, a compact manifold), then the space $C(X,X)$ of continuous maps of X into itself possesses the C^0 or *uniform* topology. It arises by fixing a metric ρ in X and defining the distance d between $f,g \in C(X,X)$ by

$$d(f,g) := \max_{x\in X} \rho(f(x),g(x)).$$

The subset $\mathrm{Hom}(X)$ of $C(X,X)$ of homeomorphisms of X is neither open nor closed in the C^0 topology. It possesses, however, a natural topology as a complete metric space induced by the metric

$$d_H(f,g) := \max(d(f,g), d(f^{-1},g^{-1})).$$

If X is σ-compact we introduce the compact–open topologies for maps and homeomorphisms, that is, the topologies of uniform convergence on compact sets.

We sometimes use the fact that equicontinuity gives some compactness of a family of continuous functions in the uniform topology.

Theorem A.1.24. (Arzelá–Ascoli Theorem) *Let X, Y be metric spaces, X separable, and $\mathcal{F}$ an equicontinuous family of maps. If $\{f_i\}_{i\in\mathbb{N}} \subset \mathcal{F}$ such that $\{f_i(x)\}_{i\in\mathbb{N}}$ has compact closure for every $x \in X$ then there is a subsequence converging uniformly on compact sets to a function f.*

Thus in particular a closed bounded equicontinuous family of maps on a compact space is compact in the uniform topology (induced by the maximum norm).

Let us sketch the proof. First use the fact that $\{f_i(x)\}_{i\in\mathbb{N}}$ has compact closure for every point x of a countable dense subset S of X. A diagonal argument shows that there is a subsequence f_{i_k} which converges at every point of

S. Now equicontinuity can be used to show that for every point $x \in X$ the sequence $f_{i_k}(x)$ is Cauchy, hence convergent (since $\{f_i(x)\}_{i \in \mathbb{N}}$ has compact, hence complete, closure). Using equicontinuity again yields continuity of the pointwise limit. Finally a pointwise convergent equicontinuous sequence converges uniformly on compact sets.

2. Functional analysis

Many interesting spaces we encounter either directly or as spaces of objects (functions or maps) we are studying have a linear structure. Combining this with appropriate topological structures yields new and useful insights. We will mostly work over the field $\mathbb{R}$ of real numbers but it is useful to allow for the complex field $\mathbb{C}$, for example, when trying to diagonalize a linear map. If we want to allow for both cases we write $\mathbb{F}$ for the field.

Definition A.2.1. A *topological vector space* is a linear space endowed with a Hausdorff topology that is invariant under translation and scalar multiplication (that is, translations and multiplication by a nonzero scalar are homeomorphisms). An *isomorphism* of a topological vector space is a linear homeomorphism.

Often the topology of a linear space is induced by a more convenient metric structure.

Definition A.2.2. A *norm* on a linear space V is a function $\|\cdot\|\colon V \to \mathbb{R}$ such that for every $v, w \in V$ and every $\alpha \in \mathbb{F}$

(1) $\|v\| = 0 \Rightarrow v = 0$,
(2) $\|\alpha v\| = |\alpha|\|v\|$,
(3) $\|v + w\| \le \|v\| + \|w\|$.

A vector v is called a *unit vector* if $\|v\| = 1$. A *normed linear space* is a linear space V with a norm $\|\cdot\|$. A *Banach space* is a normed linear space that is complete with respect to the metric $d(v, w) := \|v - w\|$ induced by the norm. Two norms $\|\cdot\|$ and $\|\cdot\|'$ are said to be *equivalent* if there exists $C > 0$ such that $\frac{1}{C}\|\cdot\|' \le \|\cdot\| \le C\|\cdot\|'$, that is, if the identity map is a uniform homeomorphism with respect to $\|\cdot\|$ and $\|\cdot\|'$.

An *inner product* on a linear space V is a positive-definite symmetric bilinear form, that is, a map $V \times V \to \mathbb{R}$, $(u, v) \mapsto \langle u, v\rangle$ such that

(1) $\langle v, v\rangle \ge 0$, with equality only for $v = 0$,
(2) $\langle u, v\rangle = \overline{\langle v, u\rangle}$,
(3) $\langle au + bv, w\rangle = a\langle u, w\rangle + b\langle v, w\rangle$.

An inner product then induces a *norm* $\|v\| := \sqrt{\langle v, v\rangle}$. A *pre-Hilbert space* is a linear space V with an inner product. A *Hilbert space* is a complete pre-Hilbert space. $u, v \in V$ are said to be *orthogonal* if $\langle u, v\rangle = 0$. An *orthogonal system* in a pre-Hilbert space is a set of pairwise-orthogonal vectors, and an *orthonormal system* is an orthogonal system of unit vectors. An orthonormal system is called *complete* if its span is dense.

On a finite-dimensional normed linear space all norms are equivalent. Thus all finite-dimensional normed linear spaces are isomorphic to Euclidean space via a biLipschitz isomorphism.

An important example of a normed linear space is the linear space $C(X)$ of continuous real-valued functions on a topological space X with the norm $\|f\| := \sup\{|f(x)| \mid x \in X\}$ inducing the C^0 topology. A standard result is that this is a complete norm.

The primary infinite-dimensional example of a Hilbert space is given by the space L^2 of square integrable functions on a measure space (X, μ) with the inner product defined by $\langle f, g\rangle = \int f\bar{g}d\mu$ (which is well defined by the Hölder inequality). If (X, μ) is the unit interval $[0,1]$ with Lebesgue measure then the collection of functions 1, $\sin 2\pi nx$ ($n \in \mathbb{N}$), and $\cos 2\pi nx$ ($n \in \mathbb{N}$) forms a complete orthonormal system. This also gives a complete orthonormal system for the L^2 functions on the unit circle with Lebesgue measure. On the unit circle the space of square integrable functions has the more convenient complete orthonormal system $\{e^{2\pi inx}\}_{n\in\mathbb{Z}}$.

It is sometimes useful to note that if $\|v\| = \sqrt{\langle v, v\rangle}$ then on a real vector space one can recover the inner product from $\|\cdot\|$ via

$$\langle u, v\rangle = \frac{1}{4}(\|u+v\|^2 - \|u-v\|^2). \tag{A.2.1}$$

This is called *polarization.*

Important information is often obtained by considering linear maps to the scalar field such as projections to a given coordinate or, in spaces of integrable functions, the integral.

Definition A.2.3. A *linear map* or *linear operator* from a linear space V to a linear space Y is a map $A: V \to Y$ such that $A(\alpha v + \beta w) = \alpha A(v) + \beta A(w)$ for all $v, w \in V$ and all $\alpha, \beta \in \mathbb{F}$. A linear map $A: V \to Y$ between normed linear spaces is said to be *bounded* if $\|A\| := \sup_{\|v\|=1} \|A(v)\| < \infty$ and in this case $\|A\|$ is called the *operator norm* of A. (Bounded linear operators are clearly continuous.) It is said to be an *isometry* or an *isometric operator* if $\|Av\| = \|v\|$ for all $v \in V$. It is called *unitary* if it is an invertible isometry.

A *linear functional* on a linear space V is a linear map from V to $\mathbb{F}$. The space of bounded linear functionals on a normed linear space V is called the *dual* of V and denoted by V^*. The *weak topology* on a normed linear space V is the weakest topology in which all bounded linear functionals are continuous. In the separable case it can equivalently be defined by saying that $v_i \to 0$ iff $f(v_i) \to 0$ for every $f \in V^*$. Since V^* is a normed linear space itself (using $\|f\|$ as defined above), it too has a weak topology. More useful is the *weak* topology* defined by $f_i \to 0 \Leftrightarrow f_i(v) \to 0$ for all $v \in V$, that is, the topology of pointwise convergence on V^*.

To see that the notions relating to the dual of a normed linear space are not vacuous one needs the following theorem about existence of linear functionals:

Theorem A.2.4. (Hahn–Banach Theorem) *Let V be a normed linear space, $W \subset V$ a linear subspace, and $f: W \to \mathbb{F}$ a bounded linear functional. Then there is an extension $F: V \to \mathbb{F}$ of f to a linear functional on V such that $\|F\| = \|f\|$.*

This result immediately yields that the dual of a normed linear space is nonempty and that the weak topology is a Hausdorff topology.

If $v \in V$ then $\varphi_v: V^* \to \mathbb{F}$, $f \mapsto f(v)$ is a bounded linear functional on V^* (with $\|\varphi_v\| = \|v\|$) and the map $\Phi: V \to V^{**}, v \mapsto \varphi_v$ is an isometric homomorphism (by the Hahn–Banach Theorem). If it is an isomorphism then V is said to be *reflexive*. (This is stronger than requiring V and V^{**} to be isometrically isomorphic.)

The fundamental importance of the weak* topology is related to the following useful compactness result:

Theorem A.2.5. (Alaoglu Theorem) *The (norm-) unit ball in the dual of a normed linear space is weak*-compact.*

This result follows from the Tychonoff Theorem A.1.8 since any topology of pointwise convergence is induced by the product topology of $\prod_{x \in X} Y = \{f: X \to Y\}$. Namely, let X be the unit ball in the normed linear space and $Y = [-1, 1]$. The unit ball in the dual corresponds naturally to the collection of "linear" maps $X \to Y$ (and linearity is a closed condition). The Alaoglu Theorem implies that norm-bounded weak*-closed sets are compact.

An important class of functionals in this book are those corresponding to (Borel probability) measures via integrals. In particular the collection of probability measures is a weakly closed norm-bounded subset of the dual to $C(X)$ and hence compact by Theorem A.2.5.

The dual of a finite-dimensional space is isomorphic to the space itself. In the case of a Hilbert space, there is a natural isomorphism: Every continuous linear functional is given by $\langle v, \cdot \rangle$ for some vector v. Sometimes this happens for other linear spaces as well, but it is not very common. As an example consider the situation of the L^p spaces of measurable functions whose p-norm $\|\varphi\|_p := (\int |\varphi|^p)^{1/p} d\mu$ is finite:

Theorem A.2.6. (Riesz Representation Theorem) *Let (X, μ) be a measure space, $1 \le p < \infty$, and $\dfrac{1}{p} + \dfrac{1}{q} = 1$. Then $(L^p)^*$ is naturally isometrically isomorphic to L^q, that is, if $f \in (L^p)^*$ then there exists $\psi \in L^q$ such that $f(\varphi) = \int \varphi\psi d\mu$ for $\varphi \in L^p$ and $\|\psi\|_q = \|f\|$.*

Thus L^p is reflexive for $1 < p < \infty$, but only L^2 is isometrically isomorphic to its dual. If we view $\psi d\mu$ as a measure on X which represents the functional f then the following result is natural.

Theorem A.2.7. (Riesz Representation Theorem) *Let X be a compact Hausdorff space. Then for each bounded linear functional f on $C^0(X)$ there* *exists a unique mutually singular pair μ, ν of finite Borel measures such that $f(\varphi) = \int \varphi d\mu - \int \varphi d\nu$ for all $\varphi \in C^0(X)$.*

In particular, when f is positive (that is, nonnegative on positive functions) there is a unique finite Borel measure μ such that $f(\varphi) = \int \varphi d\mu$.

Definition A.2.8. A subset C of a linear space is said to be *convex* if $tv + (1-t)w \in C$ whenever $v, w \in C$ and $t \in [0,1]$. If $A \subset V$ then the *convex hull* $\operatorname{co}(A)$ is the smallest convex set containing A, that is, $\operatorname{co}(A) := \bigcap\{C \mid A \subset C,\ C \text{ convex}\}$. In a topological vector space the *closed convex hull* of A is the closure $\overline{\operatorname{co}}(A)$ of $\operatorname{co}(A)$. An *extreme* point of a convex set C is a point v such that $v = ta + (1-t)b$ for $a, b \in C$, $t \in [0,1]$ implies $t \in \{0,1\}$ or $a = b = v$, that is, v is not a proper convex combination of other points. The set of extreme points of C is denoted by $\operatorname{ex}(C)$.

A topological vector space is said to be *locally convex* if every open set contains a convex open set.

A subset A of a topological vector space is said to be *balanced* if $\alpha A \subset A$ for every $\alpha \in \mathbb{F}$, $|\alpha| \leq 1$.

Clearly the set of measures on a space is convex, as is the set of probability measures. Normed linear spaces are locally convex since balls are convex. The following result is clear in the finite-dimensional case and helps the intuition about convex sets.

Theorem A.2.9. (Krein–Milman Theorem) *A compact convex set in a locally convex topological vector space is the closed convex hull of its extreme points, that is, $C = \overline{\operatorname{co}} \operatorname{ex}(C)$.*

More explicitly the connection is given as follows.

Theorem A.2.10. (Choquet Theorem) *Suppose x is a point of a compact metrizable convex set C in a locally convex topological vector space. Then there is a probability measure μ supported on $\operatorname{ex} C$ such that $x = \int_{\operatorname{ex} C} z\, d\mu(z)$.*

Another important property of convex sets is the following:

Theorem A.2.11. (Tychonoff Fixed-Point Theorem) *Let E be a locally convex topological vector space and $K \subset E$ compact and convex. Then every continuous map $f\colon K \to K$ has a fixed point.*

We sketch a proof using the following lemma:

Lemma A.2.12. *A continuous map $f\colon X \to X$ of a Hausdorff space has a fixed point if and only if for every open cover $\{U_\alpha\}_{\alpha \in A}$ there exist an $x \in X$ and $\alpha \in A$ such that $x \in U_\alpha$ and $f(x) \in U_\alpha$.*

Notice that it suffices to check this property for subcovers or refinements of any given cover. Consider now an open cover $\{U_\alpha\}_{\alpha \in A}$ of K. Since in a locally convex topological vector space every open cover has a convex refinement, we may assume that the U_α are convex and by compactness also that A is finite. There is a finite star-refinement $\{V_i\}_{i=1}^k$, that is, for each $i \in \{1, \dots, k\}$ there exists an $\alpha \in A$ such that $\mathrm{star} V_i := \bigcup\{V_j \mid V_i \cap V_j \neq \varnothing\} \subset U_\alpha$. For $i \in \{1, \dots, k\}$ pick $x_i \in K \cap V_i$ and let $S := \mathrm{co}(\{x_1, \dots, x_k\})$ be the convex hull of $\{x_1, \dots, x_k\}$, that is, a simplex of dimension $d \leq k-1$. Triangulate S so that the closure of each d-simplex of the triangulation is in some $f^{-1}(V_i)$. Define $F \colon S \to S$ for vertices y of the triangulation by $F(y) = x_i$, where i is chosen such that $f(y) \in V_i$, and extend it linearly to S.

If $x \in S$ then $x \in \bar{\sigma} \subset f^{-1}(V_i)$ for some simplex σ and $i \in \{1, \dots, k\}$, that is, $f(x) \in f(\bar{\sigma}) \subset V_i$ and in particular there exists an $\alpha \in A$ such that $f(y) \in \bigcup\{V_j \mid V_i \cap V_j \neq \varnothing\} = \mathrm{star} V_i \subset U_\alpha$ for every vertex $y \in \mathrm{ex}(\sigma)$ of σ. By convexity of U_α we thus have $F(x) \in \mathrm{co}(\mathrm{ex}(\sigma)) \subset U_\alpha$. Since $f(x) \in V_i \subset \mathrm{star} V_i \subset U_\alpha$ we have shown that for each $x \in S$ there is an α such that $f(x)$ and $F(x)$ are in U_α.

But by the Brouwer Fixed-Point Theorem 8.6.5 F has a fixed point $x_0 \in S$, so $f(x_0) \in U_\alpha$ and $x_0 = F(x_0) \in U_\alpha$ for some α and f has a fixed point by the lemma.

Exercises

A.2.1. *Show that the one-point compactification of a Hausdorff space X is a compact Hausdorff space with X as a dense subset.*

A.2.2. *Prove Theorem A.1.13.*

A.2.3. *Prove that a path-connected space is connected.*

A.2.4. *Prove Theorem A.1.14.*

A.2.5. *Verify the claims made in the remark after Definition A.1.18.*

A.2.6. *Prove a polarization formula (analogous to (A.2.1)) for complex vector spaces.*

A.2.7. *Show that if V is a Hilbert space and $f \in V^*$ then there exists a unique $v \in V$ such that $f(w) = \langle w, v \rangle$ for all $w \in V$ and that $\|v\| = \|f\|$. Thus V^* is isometrically isomorphic to V and V is reflexive.*

3. Differentiable manifolds

In this section we give some definitions and results about manifolds that also have a differentiable structure.

a. Differentiable manifolds.

Definition A.3.1. A n-dimensional topological manifold M is called a differentiable manifold if it is covered by a family $\mathfrak{A} = \{(U_i, h_i)\}_i$ of charts such that for any two charts (U_1, h_1) and (U_2, h_2) in $\mathfrak{A}$ with $h_i: U_i \to B_i \subset \mathbb{R}^n$ the *coordinate change* $h_2 \circ h_1^{-1}$ is differentiable on $h_1(U_1 \cap U_2) \subset B_1$. Here differentiable can be taken to mean C^r for any $r \in \mathbb{N} \cup \infty$, or analytic. A collection of such charts covering M is called an *atlas* of M. Any atlas defines a unique *maximal atlas* by taking all charts compatible with the present ones. A maximal atlas is called a *differentiable structure.* A *submanifold* V of M (of dimension k) is a differentiable manifold that is a subset of M such that the maximal atlas for M contains charts $\{(U, h)\}$ for which the induced maps $h_{\restriction U \cap V}$ on $U \cap V$ map to $\mathbb{R}^k \times \{0\} \subset \mathbb{R}^n$ and define charts for V compatible with the differentiable structure of V. An important example is an open subset of M with the induced atlas (in which case $k = n$). A function $f: M \to \mathbb{R}$ is said to be *differentiable* if $f \circ h^{-1}$ is differentiable on $B \subset \mathbb{R}^n$ for any chart (U, h). We denote by $C^r(M)$ the functions that are C^r in this sense.

One should note that the differentiable structure is obtained in a very different way from the topological structure: The latter is given a priori, whereas a differentiable structure is obtained from $\mathbb{R}^n$ via a compatibility condition on charts.

$\mathbb{R}^n$ is a smooth manifold with the identity as a chart, as are its open subsets. An interesting example is obtained by viewing the linear space of $n \times n$ matrices as $\mathbb{R}^{n^2}$. The condition $\det A \neq 0$ then defines an open set, hence a manifold, which is familiar as the *general linear group* $GL(n, \mathbb{R})$ of invertible $n \times n$ matrices. Simple smooth curves and surfaces in $\mathbb{R}^n$ are manifolds: Any local piece of the parameterization gives a chart (its inverse). In particular the standard sphere is a manifold (as charts one can take the six parallel projections of hemispheres to coordinate planes or the stereographic projections of the sphere minus a pole). The embedded torus (doughnut) is a manifold via any pieces of the parameterization. (Note that also nonsmooth curves can be viewed as smooth manifolds, for example, a simple curve with a corner (like "$\lrcorner$") is homeomorphic to $\mathbb{R}$, so this single global chart defines a differentiable structure. Of course this structure is incompatible with the ambient one so this example is not a smooth submanifold of $\mathbb{R}^n$.) *Manifolds defined by equations,* namely, level sets of smooth functions into $\mathbb{R}$ or $\mathbb{R}^m$ corresponding to regular values, are an interesting general class of manifolds. Charts are provided by the implicit function theorem. Examples are the sphere in $\mathbb{R}^n$ and the *special linear group* $SL(n, \mathbb{R})$ of $n \times n$ matrices with unit determinant. Viewing the space of $n \times n$ matrices as $\mathbb{R}^{n^2}$ we obtain $SL(n, \mathbb{R})$ as the manifold defined by the equation $\det A = 1$. One can check that 1 is a regular value for the determinant. Thus this is a manifold defined by one equation. Examples of manifolds defined by several equations are the symplectic group of matrices A satisfying $AJA^t = J$, where $J = \begin{pmatrix} 0 & \mathrm{Id} \\ -\mathrm{Id} & 0 \end{pmatrix}$, and the *orthogonal group* of matrices A satisfying $AA^t = \mathrm{Id}$. All the preceding examples are by construction examples of submanifolds. Conversely every smooth manifold can be viewed as a submanifold of R^n in a way described after Definition A.3.3.

Identification spaces can also be smooth manifolds, for example, the unit circle viewed as $\mathbb{R}/\mathbb{Z}$, the torus $\mathbb{R}^n/\mathbb{Z}^n$, or compact factors of the hyperbolic plane. In either case the obvious charts that give these spaces the structure of a topological manifold are smoothly compatible. Conversely a smooth structure on a manifold always lifts to a smooth structure on any cover. An important result for analysis on manifolds is the fact that (using our assumption of second countability, that is, that there is a countable base for the topology) every smooth manifold admits a partition of unity, which is defined as follows and will be used later to define the volume element of a manifold:

Definition A.3.2. A *partition of unity* is a collection $\{(U_i, \psi_i)\}_{i\in I}$, where $\{U_i\}$ is a locally finite open cover and $\psi_i \in C^\infty(M)$ is nonnegative and has compact support in U_i, such that $\sum_i \psi_i = 1$. It is said to be *subordinate* to a cover $\{O_j\}_{j\in J}$ if every U_i is contained in some O_j.

The derivative of functions can be calculated in coordinates, of course. But there is an invariant way of defining it using a local linear structure, that is, tangent vectors. This is obtained exactly by differentiating:

Definition A.3.3. Let M be a C^∞ manifold and $p \in M$. Consider curves $c\colon (a,b) \to M$, where $a < 0 < b$, with the property that $h \circ c$ is differentiable at 0 for one (hence any) chart (U,h) with $p \in U$. Each such curve acts on $C^\infty(M)$ by the *derivation* (that is, an operator satisfying the product rule) $f \mapsto \frac{d}{dt}\big|_0 f\circ c$. Many different curves will induce the same derivation and we identify two such curves. The space of these equivalence classes (at p) obtained in this way, which is also the space of derivations at p, has a linear structure (since each derivation is a real-valued function) and turns out to have dimension n. It is called the *tangent space* at p of M and denoted T_pM. We denote the derivation, that is, tangent vector, induced by $c\colon (a,b) \to M$ by $\dot c(0)$ or $\frac{d}{dt}\big|_0 c$. Given a specific chart (U,h) we define the standard basis of T_pM by taking the canonical basis $\{e_1,\dots,e_n\}$ of $\mathbb{R}^n$ and letting $\frac{\partial}{\partial x^i} := \dot c_i(0)$, where $c_i(t) = h^{-1}(h(p) + te_i)$. We define the *tangent bundle* of M to be the disjoint union $TM := \bigcup_{p\in m} T_pM$ of the tangent spaces with the *canonical projection* $\pi\colon TM \to M$ such that $\pi(T_pM) = \{p\}$. Any chart (U,h) of M then induces a chart $(U \times \bigcup_{p\in U} T_pU, H)$ by taking $H(p,v) := (h(p),(v^1,\dots,v^n)) \in \mathbb{R}^n \times \mathbb{R}^n$, where the v^i are the coefficients of $v \in T_pM$ with respect to the basis $\left\{\frac{\partial}{\partial x^1},\dots,\frac{\partial}{\partial x^n}\right\}$ of T_pM. In this way TM is a differentiable manifold itself. A *vector field* is a map $X\colon M \to TM$ such that $\pi \circ X = \mathrm{Id}_M$, that is, X assigns to each p a tangent vector at p. We denote by $\Gamma(M)$ the space of smooth vector fields on M. Thus vector fields act on functions as derivations. We write this as vf or $v(f)$. We shall see later that

$\pounds_v w := [v, w] := vw - wv$ is a derivation, that is, a vector field, and we call $[v, w]$ the *Lie bracket* of v and w and $\pounds_v$ the *Lie derivative*.

We can now define the morphisms of the differentiable structure:

Definition A.3.4. Let M and N be differentiable manifolds. A map $f: M \to N$ is said to be *differentiable* if for any charts (U, h) of M and (V, g) of N the map $g \circ f \circ h^{-1}$ is differentiable on $h(U \cap f^{-1}(V))$. A differentiable map f acts on derivations by sending curves $c: (a, b) \to M$ to $f \circ c: (a, b) \to N$. Differentiability means that curves inducing the same derivation have images inducing the same derivation. Thus we define the differential of f to be the map $Df: TM \to TN$, $\frac{d}{dt}\big|_0 c \mapsto \frac{d}{dt}\big|_0 f \circ c$ and the differential (or derivative) $D_p f = Df\big|_p$ at p to be the restriction to T_pM. A *diffeomorphism* is a differentiable map with differentiable inverse. Two manifolds M, N are said to be *diffeomorphic* or *diffeomorphically equivalent* if and only if there is a diffeomorphism $M \to N$. An *embedding* of a manifold M in a manifold N is a diffeomorphism $f: M \to V$ of M onto a submanifold V of N. We often abuse terminology and refer to an embedding of an open subset of M into M as a *(local) diffeomorphism* as well. An *immersion* of a manifold M into a manifold N is a differentiable map $f: M \to V$ onto a subset of N whose differential is injective everywhere.

Clearly diffeomorphic manifolds are homeomorphic. The converse is, however, not true. There are "exotic" spheres and other manifolds whose differentiable structure is not diffeomorphic to the usual differentiable structure. Notice that an immersion need not be injective, for example, the unit circle in $\mathbb{R}^2$ is an immersed line. But even injectively immersed manifolds may fail to be topological submanifolds. For example, nontrivial orbits of a flow are immersed lines, but the immersion topology of a nontrivially recurrent orbit (such as an orbit of a linear flow on $\mathbb{T}^2$ with irrational slope) is not the same as the topology induced from the ambient space. Nevertheless we will refer to such objects as immersed submanifolds.

It turns out that any smooth manifold can be smoothly embedded into a Euclidean space, so every smooth manifold is, in fact, diffeomorphic to a submanifold of R^n.

Let us briefly review the commonly used topologies on spaces of differentiable maps. First for differentiable maps there is naturally the C^1 topology. In the case of a compact smooth manifold M it can be described by fixing a Riemannian metric on M (see Section A.4), lifting it to the tangent bundle by using the induced Euclidean metric in the fibers, and considering the uniform metric in the space of differentials of C^1 maps restricted to the unit tangent bundle $\{v \in TM \mid \|v\| = 1\}$. Alternatively one can cover M by a finite set of coordinate neighborhoods and take the supremum of the usual C^1 distance over all charts. This approach extends immediately to the C^r topology for any $r \in \mathbb{N}$. The first definition can also be extended using spaces of jets but we will not go into any details about jets here (see Section 6.6a). For σ-compact manifolds the C^r topology is defined by considering restrictions to compact subsets.

Diffeomorphisms form an open set $\mathrm{Diff}^r(M)$ in the space $C^r(M,M)$ of C^r maps of M to itself due to the Inverse Function Theorem, so unlike in the case of homeomorphisms we have a natural induced topology from $C^r(M,M)$. The C^r topology allows us to locally model $\mathrm{Diff}^r(M)$ on a Banach space. Namely, given an $f \in \mathrm{Diff}^r(M)$, any nearby $g \in \mathrm{Diff}^r(M)$ can be described as

$$g = (\exp v) \circ f,$$

where v is a C^r map that associates to a point $x \in M$ a tangent vector $v(x) \in T_{f(x)}M$ and exp is the exponential map first mentioned in Section 5.3d and rigorously discussed in (9.5.1). Since such v form a Banach space this can be shown to give local charts modeling $\mathrm{Diff}^r(M)$ on a Banach space. We will not pursue this approach.

The C^∞ topology in turn is the topology on $\mathrm{Diff}^\infty(M)$ given by convergence in all C^r topologies. Unlike the C^r topologies this topology is not a norm topology, so $\mathrm{Diff}^\infty(M)$ is not locally modeled on a Banach space. This is a source of some analytic difficulties, such as when trying to apply fixed-point theorems.

Let us also mention real-analytic topologies, which appear in the proof of the Poincaré–Siegel Theorem 2.8.2. In the simplest case, when the manifold in question is a domain $D \subset \mathbb{R}^n$ and we consider real-analytic maps $f \colon D \to \mathbb{R}^n$, we extend f to a holomorphic map of an ρ-neighborhood of D in $\mathbb{C}^n$ and consider the uniform topology on holomorphic maps in D. This construction extends to analytic manifolds, but we shall not need this.

The foregoing discussion translates to the case of flows with one important comment. For $r \geq 1$ there are two distinct C^r topologies for flows. The straightforward one is obtained by considering a C^r flow as a C^r map $M \times \mathbb{R} \to \mathcal{M}$ and using the C^r topology induced from the space of such maps. For $r = 0$ this is the only possibility. For $r \geq 1$ one can alternatively consider the vector field generating the flow. For $r \geq 1$ a C^r flow is generated by a C^{r-1} vector field, but may not be generated by a C^r vector field. Conversely there are C^r vector fields that do not generate a C^{r+1} flow. The space $\Gamma^r(TM)$ of C^r vector fields is a linear space which possesses a natural Banach structure given, for example, in the case of C^1 vector fields on a compact Riemannian manifold by the norm induced by the Riemannian metric.

We now define a "lamination", which is a suitably coherent collection of immersed submanifolds. This notion is relevant in the context of the theory of dynamical systems with a hyperbolic structure.

Definition A.3.5. Suppose M is a differentiable manifold and $X \subset M$. A family $\mathcal{N}$ of smoothly injectively immersed manifolds $\{N_\alpha\}_{\alpha \in A}$ (called the *leaves*) is called a *lamination* on X if $N_\alpha \cap N_\beta = \varnothing$ when $\alpha \neq \beta$, $X \subset \bigcup_\alpha N_\alpha$, and for each $x \in X$ there is a neighborhood U and a homeomorphism $h \colon U \to \mathbb{R}^n$ such that h maps every connected component of $\bigcup_\alpha N_\alpha \cap U$ to $h(U) \cap (\mathbb{R}^k \times \{y\}) \subset \mathbb{R}^n$ for some $y \in \mathbb{R}^{n-k}$.

This is a weakening of the better-known notion of a *foliation*, where $X = M$ and h is assumed smooth. In particular a foliation is a partition of M into smoothly injectively immersed submanifolds. Here we only obtain submanifolds for points of a subset X. Sometimes the collections of stable and unstable leaves of an Anosov diffeomorphism or flow are referred to as a foliation even though h is not usually smooth in this case.

By the theorems of existence, uniqueness, and smooth dependence for solutions of ordinary differential equations a C^1 vector field on M induces a local flow, that is, for every $p \in M$ there is a curve $c_{v,p}\colon (-\epsilon, \epsilon) \to M$ such that $c_{v,p}(0) = p$ and $\dot{c}_{v,p}(t) := \frac{d}{dt}c_{v,p}(t) = v(c_{v,p}(t))$. Here ϵ can be chosen to depend continuously on p. Where defined the map $\varphi_v\colon (p,t) \mapsto \varphi^t(p) := c_{v,p}(t)$ is as smooth as v. By continuity of ϵ it is bounded on any compact manifold and hence by the group property $c_{v,p}(t+s) = c_{v(c_{v,p}(t)),c_{v,p}(t)}(s)$ (which follows from uniqueness) every vector field on a compact manifold induces a *complete* flow, that is, φ_v^t is defined for *all* times. If φ_v^t and φ_w^s are the flows for vector fields v and w, respectively, then usually the diffeomorphisms φ_v^t and φ_w^t do not commute, that is, $\varphi_v^t \circ \varphi_w^s \neq \varphi_w^s \circ \varphi_v^t$. If they do, the vector fields v and w are said to *commute*. The extent to which two vector fields v, w fail to commute is measured by their Lie bracket $[v,w]$ which can be computed as $[v,w](p) = \lim_{t\to 0} (w - d\varphi_v^t w)(\varphi_v^t(p))/t$.

Let us now show briefly how these invariant notions appear in local coordinates. If (U,h) is a chart then we say that we have coordinates $(x^1,\dots,x^n)$ on U. For $p \in U$ the canonical basis of T_pM is the set of *derivations* $\partial/\partial x^i$ induced by the curves $c_i(t) := h^{-1}(h(p) + te_i)$, where e_i is the ith standard basis vector in $\mathbb{R}^n$. A tangent vector $v \in T_pM$ can then be written as $v = \sum_{i=1}^n v^i \partial/\partial x^i$ and if $f\colon M \to \mathbb{R}$ is smooth then $vf = \sum_{i=1}^n v^i \partial(f \circ h^{-1})/\partial x^i$. Thus the induced coordinates of TM are $(x^1,\dots,x^n,v^1,\dots,v^n)$, where the v^i are the components we just defined. Likewise a vector field is locally given by a representation $v(p) = \sum_{i=1}^n v^i(p)\partial/\partial x^i$ and it is smooth if and only if the v^i are. To see that the Lie bracket of two vector fields v, w defines a derivation, that is, a vector field, we calculate in local coordinates. Namely, write $v = \sum_{i=1}^n v^i \partial/\partial x^i$, $w = \sum_{i=1}^n w^i \partial/\partial x^i$ and for convenience write f for $f \circ h$. Then using the theorem of H. A. Schwarz that second partial derivatives commute we obtain

$$
\begin{aligned}
(vw - wv)f &= v\sum_{i=1}^n w^i \frac{\partial f}{\partial x^i} - w\sum_{i=1}^n v^i \frac{\partial f}{\partial x^i} \\
&= \sum_{i,j=1}^n v^j \frac{\partial w^i}{\partial x^j}\frac{\partial f}{\partial x^i} + \sum_{i,j=1}^n v^j w^i \frac{\partial^2 f}{\partial x^i \partial x^j} - \sum_{i,j=1}^n w^j \frac{\partial v^i}{\partial x^j}\frac{\partial f}{\partial x^i} + \sum_{i,j=1}^n v^i w^j \frac{\partial^2 f}{\partial x^j \partial x^i} \\
&= \sum_{i,j=1}^n \left(v^j \frac{\partial w^i}{\partial x^j} - w^j \frac{\partial v^i}{\partial x^j} \right) \frac{\partial f}{\partial x^i},
\end{aligned}
\tag{A.3.1}
$$

that is, $[v, w]$ is indeed a vector field given locally by $v^j \frac{\partial w^i}{\partial x^j} - w^i \frac{\partial v^j}{\partial x^i}$. In particular $[\partial/\partial x^i, \partial/\partial x^j] = 0$. There are several important properties of Lie brackets that are not hard to check in local coordinates. By definition we obviously have $[v, w] = -[w, v]$ and $[\cdot, \cdot]$ is $\mathbb{R}$-bilinear, that is, $[\alpha v + \beta w, z] = \alpha[v, z] + \beta[w, z]$ for $\alpha, \beta \in \mathbb{R}$. Next observe that for functions as coefficients we get $[fv, gw] = fg[v, w] + f(vg)w - g(wf)v$ by a coordinate calculation similar to the preceding one. This means in particular (for $f \equiv 1$) that the Lie derivative is a derivation, that is, satisfies the product rule $\mathcal{L}_v(gw) = g\mathcal{L}_v w + \mathcal{L}_v g\, w$. Furthermore there is the fundamental *Jacobi identity*

$$[v, [w, z]] + [w, [z, v]] + [z, [v, w]] = 0. \tag{A.3.2}$$

This is straightforward in coordinates. Namely, we know from (A.3.1) that only first-order derivatives occur, so we may simplify the calculation by discarding all higher-order derivatives. The symmetry then makes the remaining terms cancel. Alternatively write $[v, w] = vw - wv$ and expand (A.3.2) accordingly to see that all terms cancel.

Differentiating differentiable maps between manifolds is also straightforward calculus on local coordinates: If $f: M \to N$ and (U, h), (V, k) are local charts around $p \in M$ and $f(p) \in N$, respectively, then the differential of f at p is represented by the matrix of partial derivatives of the map $k \circ f \circ h^{-1}$ in Euclidean space with respect to the standard bases.

b. Tensor bundles. The tangent bundle is an example of the following:

Definition A.3.6. A differentiable *vector bundle* with *structure group* G, a subgroup of $GL(m, \mathbb{R})$, over M (the *base space*) is a manifold P, called the *total space* or *bundle space*, such that the projection $\pi: P \to M$ is differentiable and furthermore locally $P = M \times \mathbb{R}^m$, that is, every $x \in M$ has a neighborhood U such that there is a diffeomorphism $h: \pi^{-1}(U) \to U \times \mathbb{R}^m$, $u \mapsto (\pi(u), \varphi(u))$ and such that for any point x in the intersection $U_1 \cap U_2$ of two such neighborhoods the trivialization differs by an element of G. A *subbundle* or *distribution* is a bundle whose fibers are contained in those of P. For two distributions E, F we define the *Whitney sum* $E + F$ to be the distribution with $(E + F)_p = E_p + F_p$. We use "$\oplus$" if the sum is (pointwise) direct, that is, $E_p \cap F_p = \{0\}$ for all $p \in M$. A *section* of P is a map $v: M \to P$ such that $\pi \circ v = \mathrm{Id}_M$.

Example. The tangent bundle TM of M is of this form: Here m is the dimension of M and $G = GL(m, \mathbb{R})$ acts by the linear coordinate changes in the tangent fibers induced by coordinate change in the base. The sections are the vector fields. If there is a nonvanishing vector field on M then the one-dimensional subspaces it spans at every point define a one-dimensional distribution.

Note that the differentiable manifold TM has in turn a tangent bundle TTM. This is an important object. On one hand it allows us to differentiate vector fields. On the other hand classical mechanics involves second-order differential equations and the natural setting for second derivatives is the second (or double) tangent bundle TTM.

The second tangent bundle TTM is obviously a vector bundle over TM, but it is, in fact, a vector bundle over M as well. To that end notice that coordinate changes in M change coordinates in TTM by a coordinate change determined again by the linear part of the coordinate change in M. We will return to this in the setting of Riemannian manifolds.

From the linear structure in the tangent spaces arise linear objects other than vectors and linear maps (for example, differentials). Namely, it is often important to consider *multilinear* maps. The easiest examples, and a building block, are *1-forms.*

Definition A.3.7. We denote by T^*M the *cotangent bundle* consisting of the spaces $T_p^*M = (T_pM)^*$ of linear maps (covectors) $T_pM \to \mathbb{R}$. A section of T_p^*M is called a *1-form*. A multilinear map $\underbrace{T_p^*M \oplus \cdots \oplus T_p^*M}_{k \text{ times}} \oplus \underbrace{T_pM \oplus \cdots \oplus T_pM}_{l \text{ times}} \to \mathbb{R}$ (that is, linear in each entry independently) is called a (k,l)-*tensor.* A section of the bundle $TM \otimes \cdots \otimes TM \otimes T^*M \otimes \cdots \otimes T^*M = (TM)^{\otimes k} \otimes (T^*M)^{\otimes l}$ is a (k,l)-*tensor field* (or tensor). A tensor is called *smooth* if its values on smooth vector and covector fields define a smooth function. (Alternatively, if its coefficients in local coordinates are smooth.)

Thus a vector is a $(1,0)$-tensor, a 1-form is a $(0,1)$-tensor, and the Riemannian metrics defined in Definition A.4.4 are $(0,2)$-tensors. A basis for the space of 1-forms on T_pM is given by the forms dx^i which are given by the derivatives of the coordinate functions x^i, that is, $dx^i(\partial/\partial x^j) = \delta^i_j := \begin{cases} 0 & \text{if } i \neq j, \\ 1 & \text{if } i = j. \end{cases}$ The derivative of a function f is a 1-form $Df(v) := vf = \sum_{i=1}^n \partial f/\partial x_i \, dx^i$. If T is a (k,l)-tensor then $T = T^{j_1,\dots,j_k}_{i_1,\dots,i_l} \partial/\partial x^{j_1} \otimes \cdots \otimes \partial/\partial x^{j_k} \otimes dx^{i_1} \otimes \cdots \otimes dx^{i_l}$ with $T^{j_1,\dots,j_k}_{i_1,\dots,i_l} = T(dx^{j_1},\dots,dx^{j_k},\partial/\partial x^{i_1},\dots,\partial/\partial x^{i_l})$. There is a natural way to extend the Lie derivative to tensors. Namely, note first that for $(1,0)$-tensors (vector fields) it is already defined and that for $(0,0)$-tensors (functions) we can define $\mathcal{L}_v f := vf$. Now extend to $(0,1)$-tensors ξ by setting $\mathcal{L}_v(\xi(w)) = \mathcal{L}_v(\xi)(w) + \xi(\mathcal{L}_v w)$. Likewise one can extend $\mathcal{L}_v$ to any tensor field by postulating the product rule $\mathcal{L}_v(\xi \otimes \eta) = \mathcal{L}_v \xi \otimes \eta + \mathcal{L}_v \eta \otimes \xi$. If ω is a $(0,1)$-tensor on N and $f: M \to N$ differentiable then we can define the *pullback* $f^*\omega$ of ω on M by $f^*\omega(v) := \omega(Dfv)$. This, of course, works for $(0,k)$-tensors just as well. Likewise one can send vectors from M to N via Df, but this can be expected to send vector fields to vector fields only if f is injective (if $f(p) = f(q)$ and v is a vector field such that $Dfv(p) \neq Dfv(q)$ then there is no well-defined vector field "f_*v" on $f(M)$). If f is a diffeomorphism then this is no problem, however. Using pullbacks the Lie derivative of a $(0,k)$-tensor can be computed by using the flow φ^t defined by the vector field v to write

$$\mathcal{L}_v \omega = \lim_{t \to 0} (1/t)((\varphi^t)^*\omega - \omega).$$

The Lie derivative of any (k,l)-tensor can be computed similarly.

An important special class of tensors is that of alternating ones:

Definition A.3.8. A $(0,k)$-tensor ω on a linear space is said to be an *alternating tensor* or an *(exterior) form* if $\omega(v_1,\dots,v_k)=0$ whenever $v_i=v_j$ for some $i\neq j$. A $(0,k)$-tensor field is said to be alternating if it is alternating at every point. Alternating $(0,k)$-tensor fields are called *k-forms*, and the space of k-forms is denoted by $\Gamma(\bigwedge^k T^*M)$. In analogy to the asymmetric part of a matrix the alternating part $\mathcal{A}\lambda$ of a $(0,k)$-tensor is defined by $\mathcal{A}\lambda = 1/k!\sum_{\pi\in S_k}\operatorname{sgn}\pi\,\eta\circ\pi$, where π permutes the entries and $\operatorname{sgn}\pi$ is its sign, that is, -1 if π is odd, 1 otherwise. Thus $\mathcal{A}$ is a projection of $(T^*M)^{\otimes k}$ to $\bigwedge^k T^*M$. We define the *wedge product* or *exterior product* of $\omega\in\bigwedge^k T^*M$ and $\eta\in\bigwedge^l T^*M$ by

$$\omega\wedge\eta := \frac{(k+l)!}{k!\,l!}\mathcal{A}(\omega\otimes\eta)\in\bigwedge^{k+l} T^*M.$$

Nonzero elements of $\Gamma(\bigwedge^n T^*M)$ are called volume elements and two volume elements Ω,Ω' are said to be equivalent if $\Omega'=f\Omega$ for some $f\in C^\infty(M)$, $f>0$. An equivalence class of volume forms is called an *orientation* of M and M is called *orientable* if there exists an orientation on M.

With these definitions one gets the following standard facts: $\omega\wedge\eta$ is $\mathbb{R}$-bilinear in ω and η, $\eta\wedge\omega=(-1)^{kl}\omega\wedge\eta$ (hence $\omega\wedge\omega=0$ for odd k), $f^*(\omega\wedge\eta)=(f^*\omega)\wedge(f^*\eta)$, and $\omega\wedge(\eta\wedge\lambda)=(\omega\wedge\eta)\wedge\lambda=:\omega\wedge\eta\wedge\lambda$. A basis for $\bigwedge^k T_p^*M$ is given by $\{dx^{i_1}\wedge\cdots\wedge dx^{i_k}\mid 1\le i_j\le n\}$, where $\{dx^i\mid 1\le i\le n\}$ is the dual basis for $\{\partial/\partial x^i\mid 1\le i\le n\}$. Thus $\dim\bigwedge^k T_p^*M=\binom{n}{k}$. In fact, $\beta^1\wedge\cdots\wedge\beta^k\neq 0$ if and only if $\{\beta^1,\dots,\beta^k\}\subset T_p^*M$ is linearly independent.

A manifold is orientable if and only if $\Gamma(\bigwedge^n T^*M)$ is one-dimensional over $C^\infty(M)$. (Namely, there exists a volume, hence the dimension is at least one, and for two volumes Ω and Ω' the function $\varphi:=\Omega'/\Omega$ is well defined, since $\Gamma(\bigwedge^n T_p^*M)$ is one-dimensional, and smooth as well.) One can also check that orientability is equivalent to the existence of an oriented atlas, that is, an atlas where $h\circ h'$ preserves the orientation of $\mathbb{R}^n$ for any two charts h,h'. On a compact manifold a volume form can be integrated to give the total volume. This is done via charts as follows. In $\mathbb{R}^n$ we define $\int\Omega := \int\Omega_{1,\dots,n}dx^1\cdots dx^n$ for any volume $\Omega=\Omega_{1,\dots,n}dx^1\wedge\cdots\wedge dx^n$. For orientation-preserving diffeomorphisms f we get $\int f^*\Omega=\int\Omega$. Thus we can define $\int\Omega$ for a manifold M by taking a partition of unity $\{U_i,\psi_i\}$ subordinate to a covering by charts (V_i,h_i) and define $\int\Omega:=\sum_i\int(h_i)_*(\psi_i\Omega)$, and this definition via charts is coordinate independent.

c. Exterior calculus. Next we want to study the calculus of exterior forms, also called exterior calculus.

Definition A.3.9. The *exterior derivative* $d\colon \Gamma(\bigwedge^k T^*M) \to \Gamma(\bigwedge^{k+1} T^*M)$ (for any k) is defined by the following axioms (which uniquely determine d): $df = Df$ for functions, d is $\mathbb{R}$-linear and $d(\omega \wedge \eta) = d\omega \wedge \eta + (-1)^k \omega \wedge d\eta$, $d \circ d = 0$, and d is locally defined, that is, if two forms coincide on an open set O then their derivatives coincide on O as well.

By induction on dimension one sees that this is well defined. Namely, if $\omega = \varphi d\psi^1 \wedge \cdots \wedge d\psi^k$ then necessarily $d\omega = d\varphi \wedge d\psi^1 \wedge \cdots \wedge d\psi^k$. The last property is also satisfied inductively since it holds for functions: $dd\varphi = \sum_{i,j=1}^n (DD\varphi)_{ij} dx^i \wedge dx^j = \frac{\partial^2 \varphi}{\partial x^i \partial x^j} dx^i \wedge dx^j = 0$. Furthermore d commutes with pullback and the Lie derivative: $f^* d\omega = d(f^*\omega)$ (and $f_* d\omega = d(f_*\omega)$ if f is a diffeomorphism) and $\mathcal{L}_v(\omega^1 \wedge \cdots \wedge \omega^k) = \mathcal{L}_v \omega^1 \wedge \cdots \wedge \omega^k + \cdots + \omega^1 \wedge \cdots \wedge \mathcal{L}_v \omega^k$, whence $d\mathcal{L}_v = \mathcal{L}_v d$.

We occasionally use the convenient notation of the *contraction* of ω with a vector v defined by $v \lrcorner \omega := \omega(v, \cdot, \dots, \cdot)$. This is $\mathbb{R}$-linear and $C^\infty(M)$-linear in v. Furthermore $v \lrcorner (\omega \wedge \eta) = (v \lrcorner \omega) \wedge \eta + (-1)^k \omega \wedge (v \lrcorner \eta)$ and $v \lrcorner df = \mathcal{L}_v f$ and

$$\mathcal{L}_v \omega = v \lrcorner d\omega + d(v \lrcorner \omega). \tag{A.3.3}$$

Finally $f^* v \lrcorner f^* \omega = f^*(v \lrcorner \omega)$ and $f_* v \lrcorner f_* \omega = f_*(v \lrcorner \omega)$ for any diffeomorphism f.

Associated with forms is a cohomology theory which is based on the following notion and theorem:

Definition A.3.10. $\omega \in \Gamma(\bigwedge^k T^*M)$ is said to be *closed* if $d\omega = 0$ and *exact* if $\omega = d\eta$ for some $\eta \in \Gamma(\bigwedge^{k-1} T^*M)$.

Since $d^2 = 0$ every exact form is closed. Locally the converse holds:

Theorem A.3.11. (Poincaré Lemma) *If ω is closed then for all $p \in M$ there is a neighborhood U of p on which ω is exact.*

Proof. We use the homotopy trick (see Theorems 5.1.27, 5.5.9): Assume $p = 0 \in \mathbb{R}^n$ and let $v_t(x) = x/t$. v_t generates the flow $\varphi^t(x) = tx$ for $t > 0$, so $d/dt(\varphi^t)^*\omega = (\varphi^t)^* \mathcal{L}_{v_t} \omega = (\varphi^t)^*(d(v_t \lrcorner \omega)) = d((\varphi^t)^*(v_t \lrcorner \omega))$ since $d\omega = 0$, and $\omega - (\varphi^{t_0})^*\omega = d\int_{t_0}^1 (\varphi^t)^*(v_t \lrcorner \omega) dt \xrightarrow[t_0 \to 0]{} d\int_0^1 (\varphi^t)^* v_t \lrcorner \omega dt =: d\eta$. The limit exists since $(\varphi^t)^*(v_t \lrcorner \omega)_x(w_x) = \omega_{tx}(x/t, tw_x) = \omega_{tx}(x, w_x)$. $\square$

The advertised cohomology theory for compact manifolds is the *de Rham cohomology*: The kth cohomology group is the factor of the space of closed k-forms by the space of exact k-forms. This is a finite-dimensional vector space by virtue of the Poincaré Lemma. It is naturally dual to the kth homology group with real coefficients (see Section A.7) of the manifold. It also possesses a natural multiplicative structure induced by the wedge product.

To obtain this duality we need to define integrals of forms. To that end notice first that the integral of a k-form over an immersed k-simplex can be defined as the integral of the pullback by the immersion. By the change of variables formula this depends only on the image of the immersion. A useful result for integration of forms is the theorem of Stokes:

Theorem A.3.12. *If M is an n-manifold, possibly with boundary, and ω an $(n-1)$-form on M then $\int_M d\omega = \int_{\partial M} \omega$.*

In particular an exact form on any boundaryless manifold integrates to zero. Now any immersed k-dimensional submanifold can be partitioned (up to overlapping boundaries) into immersed k-simplices (triangulation), so the integral of a k-form over immersed k-dimensional submanifolds is well defined (again, this is independent of the triangulation). In particular k-forms can be integrated over embedded k-cycles (see Section A.7). It is convenient to think of cycles as embedded boundaryless manifolds, and indeed by the Stokes Theorem these integrals depend only on the cohomology class of the form. On the other hand for a given form these integrals depend only on the homology class (with real coefficients) of the cycle because two homologous manifolds form the boundary of a manifold. This gives the duality between the de Rham cohomology of forms and the simplicial homology of the manifold.

We shall have occasion to invoke a result related to the preceding ones:

Lemma A.3.13. *On an n-manifold an n-form with zero integral is exact.*

d. Transversality. On a smooth manifold we have a natural notion of "measure zero" because if the image of a set in one chart has measure zero then the same holds for any chart. If $f: M \to N$ is a differentiable map then $x \in N$ is called a *regular value* if for all $y \in f^{-1}(\{x\})$ the differential Df_y has maximal rank, that is, its rank is $\min(\dim M, \dim N)$. Otherwise x is called a *singular value.*

Theorem A.3.14. (Sard Theorem) *If $f \in C^\infty(M,N)$ then the set of singular values of f has Lebesgue measure zero.*

One implication is that by the Implicit Function Theorem almost any value of a smooth f has a manifold as a level set. This theorem is local because we can take M to be a neighborhood of a point in some manifold. Sard's Theorem is the central background result for transversality theory.

Definition A.3.15. Let M be a smooth manifold and K, $N \subset M$ smooth submanifolds. K and N are said to be *transverse* at $x \in M$ if $x \notin K \cap N$ or $T_xK + T_xN = T_xM$. We write $K \pitchfork_x N$. In particular, if $\dim K + \dim N = \dim M$ and $x \in K \cap M$ the latter condition is equivalent to $T_xK \cap T_xN = \{0\}$.

We say that K and N are transverse (to each other), written $K \pitchfork N$, if $K \pitchfork_x N$ for all $x \in K \cap N$. If K and M are manifolds with boundary (see Definition A.1.15) then they are said to be transverse if the (boundaryless) manifolds ∂K, $K \setminus \partial K$, ∂M, $M \setminus \partial M$ are pairwise transverse in the previous sense.

Transversality is a C^1-open condition:

Proposition A.3.16. *Transverse intersections are stable in the C^1 topology.*

Let us first observe that transverse intersections can be brought into a "normal form":

Lemma A.3.17. (Adapted coordinates) *If $K \pitchfork_x N$ in M, $x \in K \cap N$, and $k = \dim K$, $n = \dim N$, $m = \dim M$ then there is a neighborhood U of x and coordinates $(x_1, \dots, x_m)$ on U such that in these coordinates*

$$K \cap U = \{(x_1, \dots, x_m) \mid x_{k+1} = \dots = x_m = 0\}$$

and

$$N \cap U = \{(x_1, \dots, x_m) \mid x_1 = \dots = x_{m-n} = 0\}.$$

Proof. By extending a basis for $T_xK \cap T_xN$ to a basis for T_xK and a basis for T_xN one obtains local coordinates $\psi: U \to \mathbb{R}^m$ on a neighborhood U of x in which $D_x\psi T_xK = \{v \in \mathbb{R}^m \mid v_{k+1} = \dots = v_m = 0\}$ and $D_x\psi T_xN = \{v \in \mathbb{R}^m \mid v_1 = \dots = v_{m-n} = 0\}$. Thus, after possibly shrinking U, there exists $\varphi: \mathbb{R}^k \to \mathbb{R}^{m-k}$ such that $\psi(K \cap U) = \operatorname{graph} \varphi$. Then $\phi: (x_1, \dots, x_m) \mapsto (x_1, \dots, x_m) - (0, \dots, 0, \varphi(x_1, \dots, x_k))$ has the effect that $\phi \circ \psi$ represents K as required. "Straightening out" N similarly yields the claim. □

Proof of the proposition. We will use the Implicit Function Theorem. Since the problem is a local one we pass to adapted coordinates and note that C^1-perturbations K', N' of K and N are graphs of maps $\xi: \mathbb{R}^k \times \{0\} \to \mathbb{R}^{m-k}$ and $\eta: \{0\} \times \mathbb{R}^n \to \mathbb{R}^{m-n}$, respectively. The space of pairs of C^1 maps of this kind is a Banach space E. So is the space $F = \mathbb{R}^{m-n} \times \{0\} \times \mathbb{R}^{m-k}$. Consider the map

$$f: E \times F \to F, \quad ((\xi, \eta), (x, 0, y)) \mapsto (x, 0, \xi(x, 0, 0)) - (\eta(0, 0, y), 0, y).$$

It vanishes at 0, and the derivative at 0 in the F-direction is nonsingular by transversality. The Implicit Function Theorem yields $x_{(\xi,\eta)}$ and $y_{(\xi,\eta)}$ such that $f((\xi, \eta), (x_{(\xi,\eta)}, 0, y_{(\xi,\eta)})) = 0$ or $(x_{(\xi,\eta)}, 0, \xi(x_{(\xi,\eta)}, 0, 0)) = (\eta(0, 0, y_{(\xi,\eta)}), 0, y_{(\xi,\eta)})$. Therefore K' and N' do intersect. Transversality of the intersection follows because in local coordinates the tangent spaces to K' and N' are small perturbations of those of K and N and the spanning condition is open. □

Here is an immediate consequence of Lemma A.3.17:

Corollary A.3.18. *If K and M are compact transverse manifolds (possibly with boundary) then any sufficiently small C^1-perturbations $\tilde{K}$ and $\tilde{M}$ are transverse.*

Definition A.3.19. Let $0 \le r \le \infty$ and M a C^r manifold. Two submanifolds K_1 and K_2 of M are said to be C^r-close if there exist a C^r manifold K_0 and C^r embeddings $f_i: K_0 \to K_i$ such that f_1 and f_2 are C^r-close.

Theorem A.3.20. (Transversality Theorem) *Let M be a C^∞ manifold of dimension m, and $N \subset M$ a submanifold of dimension n. Then among the k-dimensional submanifolds $K \subset M$, those transverse to N are C^∞-dense.*

Proof. Consider a coordinate neighborhood U such that

$$N \cap U = \{(x_1, \dots, x_n) \mid x_1 = \dots = x_{m-n} = 0\}$$

and let T be the natural transversal

$$T := \{(x_1, \dots, x_m) \mid x_{m-n+1} = \dots = x_m = 0\}.$$

For $s = (s_1, \dots, s_{m-n})$ let furthermore

$$N^s := \{(x_1, \dots, x_m) \mid x_1 = s_1, \dots, x_{m-n} = s_{m-n}\}$$

and let π be the projection of U onto T along $N^0 = N \cap U$, that is,

$$\pi(x_1, \dots, x_m) = (x_1, \dots, x_{m-n}, 0, \dots, 0).$$

Now let K be represented as the image of the C^∞ embedding $\varphi\colon K_0 \to M$ and let $\tilde{K}_0 = \varphi^{-1}(K \cap U)$. Consider the C^∞ map $\pi \circ \varphi\colon \tilde{K}_0 \to T$. Whenever the point $\{(s_1, \dots, s_{m-n}, 0 \dots, 0)\} = T \cap N^s$ is a regular value of $\pi \circ \varphi$ then $(K \cap U) \pitchfork N^s$. By the Sard Theorem the set of regular values has full Lebesgue measure in T, whence there are regular values arbitrarily close to 0. Thus arbitrarily close to $N \cap U$ there are manifolds transverse to K.

To extract a global result from these considerations we first take a slightly smaller neighborhood $U' \subset U$ and pick a nonnegative C^∞ bump function ρ that is 1 on U' and vanishes outside U. Then for any $s \in \mathbb{R}^{m-n}$ one constructs a C^∞ vector field given by $\rho \sum_{i=1}^{m-n} s_i \dfrac{\partial}{\partial x_i}$ in U and 0 outside U. Let T_s be the time-one map for the flow generated by this vector field. For any $r \in \mathbb{N}$ the diffeomorphism T_s is C^r-close to the identity as long as $\|s\|$ is sufficiently small. For such s

$$\tilde{N}^s := (T_s \circ \varphi)(N)$$

is C^r-close to N. Obviously $\tilde{N}^s \cap U = N^s$, so if s is a regular value of $\pi \circ \varphi$ then $\tilde{N}^s$ is transverse to $K \cap U$. Now cover M by coordinate neighborhoods U_i ($i \in \mathbb{N}$) in a locally finite way, that is, such that every compact set intersects only finitely many U_i. Furthermore take $U_i' \subset U_i$ such that $\bar{U}_i' \subset U_i$ and the U_i' still cover M. Then the foregoing procedure can be used inductively to produce vectors s^i and corresponding diffeomorphisms $T^i_{s^i}$ for $i \in \mathbb{N}$ such that $\tilde{N} := (\dots \circ T^k_{s^k} \circ \dots \circ T^1_{s^1})N$ (locally this is a finite composition) is transverse to K and C^r-close to N. □

4. Differential geometry

We begin by returning briefly to the discussion of bundles in general. Since a vector bundle P is a differentiable manifold, it has a tangent bundle. There is an additional structure in this case:

Definition A.4.1. In T_uP we call

$$V_u := \{v \in T_uP \mid v \text{ is tangent to a fiber of } P\}$$

the *vertical subspace* at u.

This is a useful notion, since we can now differentiate vector fields to get a vector field on TTM and project to the vertical subbundle which is identified with TM, thus obtaining a new vector field in TM which we then call the derivative of the original vector field. But in order to project to V_u we need a complementary subspace, and this will have to be chosen:

Definition A.4.2. A *connection* is a smooth distribution H in TP such that $T_uP = H_u \oplus V_u$ and $H_{ug} = g(H_u)$ for all $u \in P$, $g \in G$ (G-invariance).

While the general notion of a connection is quite versatile and useful we will almost only encounter it on the tangent bundle and, in fact, in the context of Riemannian manifolds where it is given more directly in terms of differentiation. Thus for tangent bundles we give a more operational definition:

Definition A.4.3. Let M be a smooth manifold. A *connection* on M is an $\mathbb{R}$-bilinear map $\nabla\colon \Gamma(M) \times \Gamma(M) \to \Gamma(M)$ such that $\nabla_{\varphi v} w = \varphi \nabla_v w$ and $\nabla_v \varphi w = \varphi \nabla_v w + v(\varphi) w$. $\nabla_v w$ is called the *covariant derivative* of w along v. The *covariant derivative* of a vector field w *along a curve* c is defined as $\nabla_{\dot{c}} w$. w is said to be *parallel* along c if $\nabla_{\dot{c}} w = 0$. A curve c is called a *geodesic* if $\nabla_{\dot{c}} \dot{c} = 0$, that is, if $\dot{c}$ is parallel along c (no acceleration).

The relation to the previous definition is simply that the derivative of a vector field w in the direction v is horizontal if and only if $\nabla_v w = 0$.

For coordinate calculations we let Γ^i_{jk} be the *Christoffel symbols* defined by $\nabla_{\partial/\partial x^j}(\partial/\partial x^k) = \sum_i \Gamma^i_{jk} \partial/\partial x^i$. Coordinate calculations show that the covariant derivative along a curve is well defined regardless of which extension of $\dot{c}$ one chooses and depends only on the values of the vector field along c. In terms of the Christoffel symbols the *geodesic equation* $\nabla_{\dot{c}} \dot{c}$ reads

$$\ddot{c}^i + \sum_{jk} \Gamma^i_{jk} \dot{c}^j \dot{c}^k = 0. \tag{A.4.1}$$

Now we introduce the setting where these notions appear in this book.

Definition A.4.4. A Riemannian manifold is a differentiable manifold with an inner product $g_p(\cdot,\cdot) = \langle \cdot,\cdot \rangle$ on each tangent space, which depends smoothly on the base point. An *isometry* between Riemannian manifolds (M, g) and (N, g') is a differentiable map $f\colon M \to N$ such that $f^* g' = g$.

Remark. Since a Riemannian metric is a $(0,2)$-tensor, smooth dependence is well defined: It can be checked either by checking that for any two smooth vector fields their inner product is a smooth function, or by obtaining a matrix representation of g in local charts and checking smoothness of the coefficients. The Riemannian structure provides us immediately with a notion of length of vectors, hence of smooth curves, as well as angles. Often it suffices to have a notion of length, that is, to define a norm on each tangent space. Such manifolds are called Finsler manifolds. In fact, we usually use a Riemannian metric, as an auxiliary structure. Our definition of hyperbolicity, for example, refers to a particular (Lyapunov) Riemannian metric, but Exercises 6.4.1 and 6.4.2 give a definition that is independent of the choice of Riemannian metric (or Finsler metric). The only occasions on which we take a substantial interest in a Riemannian structure are when we study geodesic flows.

A natural topology on a Riemannian manifold is given by the *length metric* defined as the infimum of lengths of curves connecting two points. The topology induced by this metric coincides with the topology of the Riemannian manifold as a topological manifold.

Every Riemannian manifold has a natural connection described as follows.

Theorem A.4.5. *If M is a Riemannian manifold then there exists a unique connection ∇ on M, called the Levi-Cività connection, such that*

(1) $\nabla_v w - \nabla_w v = [v, w]$,
(2) $u\langle v, w\rangle = \langle \nabla_u v, w\rangle + \langle v, \nabla_u w\rangle$.

Proof. ∇ is given by the *Koszul formula*

$$2\langle v, \nabla_u w\rangle = u\langle v, w\rangle + \langle u, [v, w]\rangle + w\langle v, w\rangle + \langle w, [v, u]\rangle - v\langle w, u\rangle - \langle v, [w, u]\rangle. \quad \square$$

Condition (2) means that g is *parallel*, that is, $\nabla g = 0$. Local coordinate calculations (Exercise 9.5.5) show that our geodesic equation (A.4.1) for the Levi-Cività connection coincides with the geodesic equation (9.5.5) derived by minimizing length. In either form the geodesic equation shows that the curves $\dot{c}$ in TM are the orbits of a flow which is complete if and only if M is complete in the sense of Definition A.1.20, for example, compact (the Hopf–Rinow Theorem). In this case the exponential map (9.5.1) is globally defined. Isometries map geodesics to geodesics and are isometries of the Riemannian manifolds viewed as metric spaces (with the length metrics). If $f_1, f_2\colon M \to N$ are isometries, M is connected and there exists $p \in M$ such that $f_1(p) = f_2(p)$ and $Df_1 = Df_2$ at p then $f_1 = f_2$, that is, an isometry is determined by its derivative at a point. This follows since the set of such p is by assumption nonempty and trivially closed, but, by considering the exponential map (see (9.5.1)) as a local parameterization, also open, hence all of M.

For two unit tangent vectors whose footpoints are joined by a unique geodesic segment one can define the distance between them by squaring the length of this geodesic as well as the angle between one of these vectors and the parallel translate of the other along the geodesic and taking the square root of the result. This is the basis of (5.4.2).

A very important invariant of isometries is the *sectional curvature* of a Riemannian manifold. In general, the *curvature tensor* of a Riemannian manifold is defined by $R(u, v, w) := \nabla_u \nabla_v w - \nabla_v \nabla_u w - \nabla_{[u,v]} w$. The sectional curvature of a 2-dimensional plane $\Pi_{v,w}$ spanned by orthonormal vectors v and w is then defined by $K(\Pi) := \langle v, R(v, w, w)\rangle$. It is useful to describe it geometrically. For a surface the sectional curvature is clearly just a function of the point called the *Gaussian curvature.* It can be explained nicely for surfaces embedded in $\mathbb{R}^3$. Namely, near $p \in M$ consider the normal vector field as a map from M to the unit sphere (the Gauss map). Then the curvature at p is the Jacobian of this map. Thus it is one for the unit sphere, $1/r^2$ for the sphere of radius r, negative for a saddle-shaped surface, and zero for a plane, but also for a cylinder. In general the sectional curvature of a plane Π is the Gaussian curvature of the surface $\exp \Pi$ locally embedded in M.

5. Topology and geometry of surfaces

In this section we review the classification of compact surfaces (2-dimensional manifolds) from various points of view. Every orientable compact surface is homeomorphic to a space obtained from the sphere by attaching handles. Attaching a handle means deleting two disjoint disks and identifying the resulting two boundary circles with the boundary circles of a cylinder. The number g of attached handles is called the *genus* of the surface and is a topological invariant. As differentiable manifolds, surfaces are, in fact, determined up to diffeomorphism by their genus. The genus is related to the *Euler characteristic* χ of a surface via $\chi = 2-2g$. The Euler characteristic can be described in various different ways. First consider a triangulation of the surface (see Definition A.7.1), that is, a representation as a polyhedron with triangular faces, and let f be the number of faces, e the number of edges, and v the number of vertices. Then $\chi = f - e + v$ independently of the triangulation. (In fact $\chi = \beta_2 - \beta_1 + \beta_0$, where the β_i are the Betti numbers of Definition A.7.4. For the surface of genus g we have $\beta_0 = \beta_2 = 1$ and $\beta_1 = 2g$, so we do get $\chi = 2 - 2g$.) Second, we can consider a vector field v with finitely many fixed points. Then by the Poincaré–Hopf Index Formula (Theorem 8.6.6) the sum of the indices of the fixed points of v is χ. Finally let $\langle\cdot,\cdot\rangle$ be a Riemannian metric on the surface and denote by $K(x)$ the Gaussian curvature at the point x. Then we have

Theorem A.5.1. (Gauss–Bonnet Theorem) $\chi = (1/2\pi)\int K(x)\,d\,\mathrm{vol}$, *where* vol *is the volume induced by the Riemannian metric.*

The fundamental group of a surface can be represented in various ways. From the process of attaching handles one obtains generators a_i, b_i for $i = 1,\dots,g$, where each pair corresponds to a handle and $a_1b_1a_1^{-1}b_1^{-1}\cdots a_gb_ga_g^{-1}b_g^{-1} = \mathrm{Id}$. This representation also corresponds to the identifications made on the $4g$-gon to obtain the surface as an identification space. (For genus 1, the torus, this is the description as $\mathbb{R}^2/\mathbb{Z}^2$; for genus 2, the double torus, this is similar to the description via the octagon given in Sections 5.4e and 14.4b, although the identifications are different.)

Recall that a simple closed curve on a manifold M is a homeomorphic image of the circle S^1 in M, or equivalently the image of S^1 under a continuous injection $S^1 \to M$.

Theorem A.5.2. (Jordan Curve Theorem) *A simple closed curve C on the sphere S^2 separates the sphere into two connected components, each homeomorphic to the disk, that is, $S^2 \smallsetminus C = \mathcal{D}_1 \cup \mathcal{D}_2$ with $\mathcal{D}_1$ and $\mathcal{D}_2$ homeomorphic to a disk.*

More generally the genus of a surface can be defined as the maximal number of disjoint closed curves such that the complement of their union is connected. This is easily visualized in terms of the description of genus by adding handles to a sphere.

For any two oriented closed curves on an orientable surface that have only transverse intersections we can define their intersection index by counting ± 1

according to whether their tangent vectors at an intersection point form a positively or negatively oriented pair. This index actually depends only on the homology classes of the curves and defines a nondegenerate skew-symmetric (symplectic) 2-form on $H_1(M,\mathbb{R}) \times H_1(M,\mathbb{R})$.

The universal cover of any orientable surface other than the sphere is R^2. We can also view surfaces as one-dimensional complex manifolds: The sphere is the Riemann sphere $\mathbb{C} \cup \{\infty\}$, the torus is the factor of $\mathbb{C}$ by a lattice, and surfaces of higher genus are obtained from the upper half-plane $\mathbb{H} := \{z \in \mathbb{C} \mid \operatorname{Im} z > 0\}$ or the unit disk $\mathbb{D}$ in $\mathbb{C}$ as the factor of a group of Möbius transformations as described in Section 5.4e. The Riemann sphere, $\mathbb{R}^2$, and the Poincaré disk each admit a metric of constant Gaussian curvature (positive, zero, and negative, correspondingly) and these descend to compact factors, so all compact surfaces admit a Riemannian metric of constant curvature.

Nonorientable surfaces are classified in a similar way. It is useful to begin with the best-known example, the *Möbius strip*, which is the nonorientable surface with boundary obtained by identifying two opposite sides of the unit square $[0,1] \times [0,1]$ via $(0,t) \sim (1,1-t)$. Its boundary is a circle.

Any compact nonorientable surface is obtained from the sphere by attaching several *Möbius caps*, that is, deleting a disk and identifying the resulting boundary circle with the boundary of a Möbius strip. Attaching m Möbius caps yields a surface of genus $2-m$. Alternatively one can replace any pair of Möbius caps by a handle, so long as at least one Möbius cap remains, that is, one may start from a sphere and attach one or two Möbius caps and then any number of handles.

All compact surfaces with boundary are obtained by deleting several disks from a closed surface. In general then a sphere with h handles, m Möbius strips, and d deleted disks has Euler characteristic $\chi = 2 - 2h - m - d$. In particular there is a finite list of surfaces with nonnegative Euler characteristic:

Surface	h	m	d	χ	Orientable?
Sphere	0	0	0	2	yes
Projective plane	0	1	0	1	no
Disk	0	0	1	1	yes
Torus	1	0	0	0	yes
Klein bottle	0	2	0	0	no
Möbius strip	0	1	1	0	no
Cylinder	0	0	2	0	yes

6. Measure theory

a. Basic notions.

Definition A.6.1. Let X be a set. A nonempty collection $\mathcal{S}$ of subsets is called a *σ-algebra* if:

(1) If $A \in \mathcal{S}$ then $X \setminus A \in \mathcal{S}$.

(2) If $\{A_i \mid i \in \mathbb{N}\} \subset \mathcal{S}$ then $\bigcup_i A_i \in \mathcal{S}$.

A *measure* is a function $\mu: \mathcal{S} \to \mathbb{R}_+ \cup \{\infty\}$ that is *σ-additive*, that is, $\mu(\bigcup_i A_i) = \sum_i \mu(A_i)$ if the A_i are pairwise disjoint. A *measured σ-algebra* is a pair $(\mathcal{S}, \mu)$ of a σ-algebra and a measure, and a *measure space* is a triple $(X, \mathcal{S}, \mu)$ of a set, a σ-algebra of subsets, and a measure.

Clearly a σ-algebra is also closed under taking differences and countable intersections and contains $\varnothing$ and X. A set $A \subset X$ is called a *null set* if it is contained in a set $A \in \mathcal{S}$ of measure zero. A property is said to hold *almost everywhere* (a.e.) if it holds on the complement of a null set. A measure is called *finite* if $\mu(X) < \infty$ (hence $\mu(A) < \infty$ for all $A \in \mathcal{S}$), σ-finite if $X = \bigcup_i A_i$ with $\mu(A_i) < \infty$, and a *probability measure* if $\mu(X) = 1$. A measure space with a probability measure is also called a *probability space* We only consider σ-finite measures and, most of the time, finite or probability measures. A map from a measure space to a topological space is said to be *measurable* if the preimage of every open set is measurable. $(\mathcal{S}, \mu)$ is said to be *complete* if every subset of a null set is in $\mathcal{S}$. The minimal σ-algebra $\bar{\mathcal{S}}$ containing $\mathcal{S}$ and all μ-null sets is called the *completion* of $\mathcal{S}$. If $A, B \in \mathcal{S}$ then define the pseudometric $d_\mu(A, B) := \mu(A \vartriangle B)$. Two sets $A, B \in \mathcal{S}$ are said to be *equivalent mod* 0 if $d_\mu(A, B) = 0$. We denote by $\mathcal{S}_0$ the set of equivalence classes in $\mathcal{S}$ of equivalence mod 0. The notions of countable union, intersection, complementation, difference, and inclusion naturally project to this factor and the pseudometric d_μ projects to a metric on $\mathcal{S}_0$. $(\mathcal{S}, \mu)$ and $(\mathcal{T}, \nu)$ are said to be *isomorphic* if there is a bijective map $\mathcal{F}: \mathcal{S}_0 \to \mathcal{T}_0$ such that $\nu(\mathcal{F}(S)) = \mu(S)$ for all $S \in \mathcal{S}_0$. Clearly the measure of the space and the presence and measure of atoms are isomorphism invariants. An element A of $\mathcal{S}$ is called an *atom* if every $B \subset A$ is equivalent to either A or $\varnothing$. A measure is called *nonatomic* if there are no atoms. $(\mathcal{S}, \mu)$ is said to be *separable* if there is a countable d_μ-dense subset $\mathcal{D}$ of $\mathcal{S}$, that is, the σ-algebra generated by $\mathcal{D}$ contains the completion of $\mathcal{S}$. In other words, for every $A \in \mathcal{S}$ and $\epsilon > 0$ there exists $D \in \mathcal{D}$ such that $\mu(A \vartriangle D) < \epsilon$. If $(\mathcal{S}, \mu)$ is separable then the restriction of μ to any sub-σ-algebra is itself separable.

The construction of Lebesgue measure on the real line starts from the definition of length for intervals and then proceeds along the lines of a general extension construction which we now briefly describe. A collection $\mathcal{A}$ of subsets of a set X is called a *semialgebra* if the complement of any $A \in \mathcal{A}$ is a finite disjoint union of elements of $\mathcal{A}$ and an intersection of two elements of $\mathcal{A}$ is in $\mathcal{A}$. (An example are intervals in $\mathbb{R}$.) A monotone nonnegative set function μ on a semialgebra $\mathcal{A}$ that is zero on the empty set (if this is in $\mathcal{A}$), additive, that is, $\mu(A \cup B) = \mu(A) + \mu(B)$ if $A, B, A \cup B \in \mathcal{A}$ and $A \cap B = \varnothing$, and such that $\mu\left(\bigcup_{i \in \mathbb{N}} A_i\right) \le \sum_{i \in \mathbb{N}} \mu(A_i)$ if $A_i, \bigcup_{i \in \mathbb{N}} A_i \in \mathcal{A}$, extends to an additive function μ on the algebra $\mathcal{C}$ consisting of the empty set and all finite disjoint unions of sets in $\mathcal{A}$ by postulating additivity. Next, this function μ induces an *outer measure* on all subsets of X via $\mu^*(E) := \inf\{\sum_i \mu(A_i) \mid E \subset \bigcup_i A_i,\ A_i \in \mathcal{C}\}$. A set E is called measurable then, if for every $A \subset X$ one has

$\mu^*(A) = \mu^*(A \cap E) + \mu^*(A \setminus E)$. The collection of measurable sets then is a σ-algebra and μ^* defines a measure on it.

Up to isomorphism this yields no essentially new benign spaces:

Theorem A.6.2. *Every separable nonatomic σ-algebra with probability measure is isomorphic to the σ-algebra of Borel sets on $[0, 1]$ with Lebesgue measure.*

This theorem is Theorem C of Section 41 in [Hal]. Every measure can be extended canonically to a complete measure on the completion of $\mathcal{S}$. As an example note that the σ-algebra of Lebesgue measurable sets is the completion with respect to Lebesgue measure of the σ-algebra of Borel sets. If $(X, \mathcal{S}, \mu)$ and $(Y, \mathcal{T}, \nu)$ are measure spaces then a map $f: X \to Y$ (defined a.e.) is called an *isomorphism* of the measure spaces X and Y if f induces an isomorphism $\mathcal{F}: \bar{\mathcal{S}} \to \bar{\mathcal{T}}$ of the completions of $\mathcal{S}$ and $\mathcal{T}$. Measure spaces can thus be classified up to isomorphism and they are isomorphic if and only if their measured σ-algebras are isomorphic.

A *basis* $\mathcal{B}$ for a measure space $(X, \mathcal{S}, \mu)$ is a countable collection $\mathcal{B} = \{B_i\}_{i \in \mathbb{N}} \subset \mathcal{S}$ whose union is X and such that there is a null set N such that for $x, y \in X \setminus N$ there exists $B \in \mathcal{B}$ such that $x \in B$, $y \notin B$. Clearly a d_μ-dense set is a basis. A basis induces a one-to-one *coding* of sets and points in the following way: A sequence $\omega = (\omega_0, \dots) \in \Omega_2^R$ with $\omega_i \in \{0, 1\}$ corresponds to the intersection $\bigcap \{B_i \mid \omega_i = 1\}$. Preimages of the B_i under this coding are *cylinder sets* in Ω_2^R and they generate a unique minimal σ-algebra. A basis is said to be *complete* if the collection of sequences coding points is measurable with respect to the completion of the σ-algebra generated by cylinder sets.

Theorem A.6.3. *A measure space $(X, \mathcal{S}, \mu)$ is isomorphic to the standard measure space $([0, 1], \mathcal{M}, \lambda)$, where $\mathcal{M}$ is the σ-algebra of Lebesgue measurable sets on $[0, 1]$, if and only if μ is a nonatomic separable probability measure with a complete basis. In this case every basis is complete.*

The main idea behind this result is the natural correspondence between $[0, 1]$ and Ω_2^R which is given by binary expansion and hence one-to-one away from $\mathbb{Q}$ and thus also one-to-one a.e. with respect to any nonatomic measure. On $[0, 1]$ a basis for $\mathcal{M}$ is given by the sets B_i of irrational numbers with ith coefficient zero in their binary expansion. Conversely any nonatomic separable probability measure induces a nonatomic probability measure on Ω_2^R which, via the preceding remark, induces a nonatomic Borel (cylinder sets are Borel sets) probability measure on $[0, 1]$. This measure is given by a distribution function $f(x) = \mu([0, x])$ and hence isomorphic to Lebesgue measure. The statement about completeness of every basis can be found in [Rok1].

Definition A.6.4. A measure space $(X, \mathcal{S}, \mu)$ with finite μ is called a *Lebesgue space* if it is isomorphic to the union of $[0, a]$ with Lebesgue measure with at most countably many points of positive measure.

By definition the standard results from the theory of Lebesgue measure apply verbatim to any Lebesgue space: Fatou's Lemma, the Monotone Convergence Theorem, the Lebesgue (Dominated) Convergence Theorem, and so forth.

If $(X, \mathcal{S}, \mu)$ and $(X, \mathcal{T}, \nu)$ are measure spaces then ν is said to be *absolutely continuous* with respect to μ, written $\nu \ll \mu$, if every set of μ-measure zero is a null set for ν.

Theorem A.6.5. *If $(X, \mathcal{S}, \mu)$ and $(X, \mathcal{T}, \nu)$ are finite measure spaces and $\nu \ll \mu$ then there exists a function $\rho: X \to \mathbb{R}$ measurable with respect to the completion $\bar{\mathcal{S}}$ of $\mathcal{S}$ such that $\nu(A) = \int_A \rho d\bar{\mu}$, where $\bar{\mu}$ is the completion of μ, for every A in the completion of $\mathcal{T}$.*

This follows from Theorem B in Section 31 of [Hal]. In particular we assert that the completion of $\mathcal{S}$ contains that of $\mathcal{T}$.

b. Measure and topology. Next we consider measures on spaces that have a topological structure compatible with the measurable one.

Definition A.6.6. Let X be a separable locally compact Hausdorff space and $\mathcal{B}$ the σ-algebra of Borel sets, that is, the σ-algebra generated by closed sets. Then a *Borel measure* is a measure μ defined on $\mathcal{B}$ such that $\mu(B) < \infty$ when B is compact.

The main property of Borel measures is that they are *regular*, that is, for every $B \in \mathcal{B}$ we have $\mu(B) = \inf\{\mu(O) \mid B \subset O \text{ open}\} = \sup\{\mu(K) \mid K \subset B \text{ compact}\}$. Furthermore every continuous function $f: X \to \mathbb{R}$ is Borel measurable, that is, preimages of open sets are Borel sets, and for every compact set K there is a decreasing sequence $\{f_n\}_{n \in \mathbb{N}}$ of nonnegative continuous functions with compact support such that $f_n \to \chi_K$ pointwise, where χ_K is the characteristic function of K. By separability of X such a measure is separable: For every point x_i of a countable dense subset consider a countable collection of open neighborhoods with compact closure B_{ij} such that $\bigcap_j B_{ij} = \{x_i\}$. This defines a basis. Furthermore every atom is a point by regularity and the Hausdorff assumption.

Theorem A.6.7. *Any Borel probability measure μ on a separable locally compact Hausdorff space X defines a Lebesgue space.*

Proof. Without loss of generality we assume that X is compact and μ is nonatomic. Then X is metrizable. We use the following result:

Lemma A.6.8. *There exists a finite partition of X into sets of arbitrarily small diameter that are contained in the closures of their interiors and whose boundaries have measure zero.*

Proof. Given the desired diameter d consider a finite set $\{x_i\}_{i=1}^k$ such that the $d/2$-balls centered at the x_i cover X. Given an x_i the boundaries of the r-balls for $d/2 < r < d$ are pairwise disjoint, so there exists an r_i such that the boundary of the r_i-ball around x_i has measure zero. Consider the collection of $B(x_i, r_i)$. Take $C_1 = B(x_1, r_1)$ and inductively take $C_i = B(x_i, r_i) \setminus \bigcup_{j \leq i} C_i$. This is as desired. □

Construct a sequence of partitions inductively as follows: Begin with any partition into nonempty sets as in the lemma. Given the nth partition ξ_n choose a partition $\tilde{\xi}_{n+1}$ into nonempty sets of diameter 2^{-n} as before, *subordinate to* ξ_n in such a way that each element of $\tilde{\xi}_{n+1}$ is contained in an element of ξ_n, and furthermore take $\tilde{\xi}_{n+1}$ such that each $c \in \xi_n$ contains the same number of elements of $\tilde{\xi}_{n+1}$. Finally take ξ_{n+1} to be a partition into sets each of which is the union of elements of $\tilde{\xi}_{n+1}$, exactly one from each $c \in \xi_n$.

Next observe that this construction implies that any intersection of the closures of a sequence $B_i \in \xi_i$ is a point. The set of points that are obtained like this in two different ways is contained in the union of the boundaries of the partitions, hence a null set. Thus on a set of full measure there is a bijective correspondence between X and a sequence space, which therefore carries a Borel measure induced by μ. □

7. Homology theory

This section introduces some notions from homology theory which are referred to when we define the degree of a map and discuss the Lefschetz number.

Definition A.7.1. For $k, N \in \mathbb{N}$, $v_0, \dots, v_k \in \mathbb{R}^N$ such that $\{v_i - v_0\}_{i=1}^k$ is linearly independent, the convex hull σ of $\{v_0, \dots, v_k\}$ is called the (k-dimensional) *simplex* spanned by $\{v_0, \dots, v_k\}$ and the v_i are called the *vertices* of the simplex. The simplices spanned by a subset of $\{v_0, \dots, v_k\}$ are called the *faces* of σ. A *simplicial complex* S is a finite collection of simplices such that any two simplices intersect in a common face. The union of the simplices of S is denoted by $|S|$. A *triangulation* of a manifold M is a pair (S, h) consisting of a simplicial complex S and a homeomorphism $h: S \to M$ from the simplicial complex S to M. The images of simplices, faces, and vertices under h are also called simplices, faces, and vertices. Identifying two orderings of the set $\{v_0, \dots, v_k\}$ of vertices of a simplex in M if they differ by an even permutation, we call a simplex with a choice of ordering of the vertices (modulo even permutations) an *oriented simplex.* The chosen orientation is then called the *positive orientation*, the other one the *negative orientation* If σ is an oriented simplex then denote by $-\sigma$ the same simplex with the negative orientation.

Note that an orientation of a simplex induces an orientation on each face since the vertices of a face form a subset of the vertices of the simplex.

Now fix a triangulation (S, h) of M and consider formal sums $\sum n_i \sigma_i$, where σ_i are oriented k-simplices and $n_i \in \mathbb{Z}$. For $n_i < 0$ we define $n_i \sigma_i := (-n_i)(-\sigma_i)$. The set of such formal sums with the obvious additive structure is the free group generated by the k-simplices and subject to the relations $\sigma + (-\sigma) = 0$ and $\sigma_i + \sigma_j = \sigma_j + \sigma_i$, that is, a finitely generated free abelian group.

Definition A.7.2. A formal sum $\sum n_i \sigma_i$ of oriented k-simplices is called a k-*chain.* Denote by C_k the set of k-chains. The *boundary operator* $\partial: C_k \to C_{k-1}$

is defined by setting

$$\partial\sigma := \sum_{i=0}^{k} \sigma_i$$

for an oriented k-simplex σ, where the σ_i are the $(k-1)$-dimensional faces of σ with induced orientation, and extending linearly to C_k. For a triangulation S let $\chi := \sum_{k=0}^{\dim M}(-1)^k \operatorname{card}\{\sigma \mid \sigma \text{ is a } k\text{-dimensional simplex in S}\}$.

It turns out that the number χ is the same for different triangulations of a given manifold M and thus provides a topological invariant $\chi(M)$ called the *Euler characteristic* of M.

An important combinatorial fact is that the boundary of a boundary is zero:

Theorem A.7.3. (Poincaré Lemma) $\partial^2 = 0$.

Since $\partial: C_k \to C_{k-1}$ is by definition a homomorphism with respect to the additive structure, the Poincaré Lemma shows that the set $B_k := \partial(C_{k+1}) \subset C_k$ of k-dimensional *boundaries* is a subgroup of the group $Z_k := \ker \partial = \{c \in C_k \mid \partial C = 0\}$ of k-dimensional *cycles*. Since C_k is abelian, B_k is normal in C_k.

Definition A.7.4. $H_k(M,\mathbb{Z}) := H_k := Z_k/B_k$ is called the *kth homology group* of M over the integers. H_k, as a finitely generated abelian group, can be written as $\mathbb{Z}^{\beta_k} \oplus F$, where F is a finite abelian group. $\mathbb{Z}^{\beta_k}$ is then called the *free part* of H_k, and β_k is called the *kth Betti number*.

The zeroth homology group of a manifold always is $\mathbb{Z}^n$, where n is the number of connected components.

If we define the commutator subgroup $[\pi_1(M,p), \pi_1(M,p)] := \{aba^{-1}b^{-1} \mid a, b \in \pi_1(M)\}$ then we have

Theorem A.7.5. (Hurewicz) *The first homology group $H_1(M,\mathbb{Z})$ of a manifold M is isomorphic to the abelianization*

$$\pi_1(M,p)/[\pi_1(M,p), \pi_1(M,p)]$$

of the fundamental group $\pi_1(M,p)$ of M at any point p. The isomorphism is called the Hurewicz isomorphism.

Proposition A.7.6. $H_n(M,\mathbb{Z}) \simeq \begin{cases} \mathbb{Z} & \text{if } M \text{ is orientable,} \\ 0 & \text{if } M \text{ is not orientable,} \end{cases}$ *where* $n = \dim M$.

Proof. Note that $B_n = 0$ and thus $H_n = Z_n$. Suppose a simplex σ occurs i times in an n-cycle c. Since the boundary is zero, the $(n-1)$-faces of σ have to be canceled by other terms in ∂c. Thus each neighboring simplex has to appear i times with appropriate orientation. Thus c is the i-fold sum of all n-simplices of the triangulation, all oriented compatibly. In the nonorientable case the simplices cannot be ordered compatibly so that there is no n-cycle and $H_n = Z_n = 0$. In the orientable case there is exactly one such chain for each $i \in \mathbb{Z}$, so $H_n = Z_n = \mathbb{Z}$. □

Remark. All homology groups are homotopy invariants.

Proposition A.7.7. (Euler–Poincaré Formula) $\sum_{k=0}^{\dim M}(-1)^k\beta_k = \chi(M)$, *where* $\chi(M)$ *is the Euler characteristic of* M.

Example. For S^m the Betti numbers are $\beta_0 = \beta_m = 1$ and $\beta_k = 0$ for $0 < k < m$.

Example. Since the fundamental group $\mathbb{Z}^2$ of $\mathbb{T}^2 = \mathbb{R}^2/\mathbb{Z}^2$ is abelian, it coincides with the first homology group, that is, $\beta_1 = 2$. The other Betti numbers are $\beta_0 = 1$ (number of connected components) and $\beta_2 = 1$ (since $\mathbb{T}^2$ is orientable). For $\mathbb{T}^n$ one has $\beta_k = \binom{n}{k}$.

Example. For the ball the only nonzero Betti number is $\beta_0 = 1$ since it is homotopic to a point.

In order to understand the behavior of a map f with respect to the homology we need to adapt f to a given triangulation.

Suppose S and T are two simplicial complexes triangulating M. A map $s: T \to S$ that maps simplices of T linearly onto simplices of S is then called a *simplicial map.* T is called a *refinement* of S if every simplex of S is triangulated by simplices of T. As subsets of $\mathbb{R}^n$, S and T are then naturally identified and we will assume that they triangulate M via a common homeomorphism h.

Fact. *For any map* $f: M \to M$ *and any triangulation* S *of* M *there exists a refinement* S' *of* S *and a simplicial map* $s: S' \to S$ *homotopic to* $h^{-1} \circ f \circ h$.

Such a map S is called a *simplicial approximation* of f. Via it f acts on homology: Note that a k-chain $c_k = \sum a_i\sigma_i$ on S induces a k-chain c'_k on S' by replacing every simplex σ_i by a k-chain in S' triangulating it. The simplicial map s sends c'_k to a k-chain $s_*c'_k$ on S again. It can be shown that this induces an action on the kth homology which is independent of the choice of simplicial approximation. Thus, in particular in the case of an orientable manifold M when $H_n(M,\mathbb{Z}) = \mathbb{Z}$, a map f induces a homomorphism $f_*: \mathbb{Z} \to \mathbb{Z}$.

More generally the construction of the homology groups can be carried out over any commutative ring R instead of $\mathbb{Z}$ to give homology groups $H_k(M, R)$ which are modules over R. One starts with formal sums of simplices with coefficients from R and proceeds as before. In particular, if $\mathbb{K}$ is a field then $H_k(M,\mathbb{K})$ is isomorphic to $\mathbb{K}^{\beta_k}$ and is thus determined by the free part of the homology group over $\mathbb{Z}$. As we mentioned in Section 3 the kth de Rham cohomology group is naturally isomorphic via integration over cycles to the dual of $H_k(M,\mathbb{R})$.

Let us close with an elementary fact about plane topology. It does not use much homology, but the proof is easier to formulate with the basic definitions in mind.

Lemma A.7.8. *Suppose* $O_1, O_2 \subset \mathbb{R}^2$ *are disjoint connected open sets such that* $\partial O_1 \subset \partial O_2$. *Then* ∂O_1 *is connected.*

Proof. We first prove that if $F_1, F_2 \subset \mathbb{R}^2$ are disjoint closed sets such that $\mathbb{R}^2 \smallsetminus F_i$ is connected ($i = 1,2$), then $\mathbb{R}^2 \smallsetminus (F_1 \cup F_2)$ is path connected. To that end triangulate $\mathbb{R}^2$ and take $x, y \in \mathbb{R}^2 \smallsetminus (F_1 \cup F_2)$. Then there are 1-chains κ_1, κ_2 in $\mathbb{R}^2 \smallsetminus F_1$, $\mathbb{R}^2 \smallsetminus F_2$, respectively, such that $\{x, y\} = \partial\kappa_i$. We can orient κ_2 such that $\kappa_1 + \kappa_2$ is homologous to 0 in $\mathbb{R}^2$, that is, $\kappa_1 + \kappa_2$ bounds a compact 2-chain K in $\mathbb{R}^2$. After possibly refining the triangulation we may assume that K has no cell that meets both F_1 and F_2 and we denote by K_1 the complex of cells meeting F_1. Now $\kappa_0 := \kappa_2 + \partial K_1$ has boundary $\{x, y\}$ since ∂K_1 is a cycle and evidently $|\kappa_2| \cap F_2 = \varnothing$ and $|K_1| \cap F_2 = \varnothing$, so $|\kappa_0| \cap F_2 = \varnothing$. Thus it remains to show that $|\kappa_0| \cap F_1 = \varnothing$. Note that $|\kappa_0| = |\kappa_2 - \partial K_1| = |-\kappa_1 + (\kappa_1 + \kappa_2 - \partial K_1)| \subset |\kappa_1| \cup |\kappa_1 + \kappa_2 - \partial K_1|$ and $|\kappa_1| \cap F_1 = \varnothing$. Furthermore F_1 is disjoint from $K - K_1$ and its boundary $|\kappa_1 + \kappa_2 - \partial K_1|$.

To prove the lemma take $F \subsetneq \partial O_1$ closed and note that any $x \in \partial O_1 \smallsetminus F$ has a connected neighborhood disjoint from F, so F does not separate O_1 from O_2. But that means that if we have $\partial O_1 = F_1 \cup F_2$ disjoint and closed then we get a contradiction to the above observation. □

8. Locally compact groups and Lie groups

In this section we introduce groups which carry a topology invariant under the group operations. A *topological group* is a group endowed with a topology with respect to which all *left translations* $L_{g_0}: g \mapsto g_0 g$ and *right translations* $R_{g_0}: g \mapsto g g_0$ as well as $g \mapsto g^{-1}$ are homeomorphisms. Familiar examples are $\mathbb{R}^n$ with the additive structure as well as the circle or, more generally, the n-torus, where translations are clearly diffeomorphisms, as is $x \mapsto -x$. Important other examples are matrix groups, for example, $GL(n, \mathbb{R})$, $SL(n, \mathbb{R})$, and so forth, as described after Definition A.3.1. A topological group is said to be locally compact if every point (or equivalently, the identity) has a neighborhood with compact closure. Such a group possesses a locally finite Borel measure invariant with respect to all right translations, which is unique up to a scalar multiple and called the *right Haar measure*. Similarly, the *left Haar measure* is, up to a scalar multiple, the unique measure invariant with respect to all left translations. These measures are finite if and only if the group is compact. In most interesting cases right invariant Haar measures are also left invariant, for example, when the group is abelian, compact, or, most importantly, a *unimodular linear group*, that is, a closed subgroup of the group $SL(n, \mathbb{R})$ of all $n \times n$ matrices with determinant one. In general, groups for which the left and right Haar measures coincide (and naturally are simply called Haar measures) are called *unimodular* and we will restrict our discussion to such groups.

A subgroup Γ of a group G is called discrete if it is closed and all of its points are isolated in the induced topology. In this case the homogeneous space G/Γ (corr. $\Gamma\backslash G$) of orbits of R_g (corr. L_g), $g \in \Gamma$, is called the right (corr. left) quotient of G by Γ. (Unless Γ is a normal subgroup these quotients are not groups.) If either quotient (hence both) is compact in the quotient–topology then Γ is said to be a *uniform* or *cocompact lattice*. It is not difficult to see that for a uniform lattice Γ any right (corr. left) Haar measure on G projects to a finite Borel measure on the homogeneous space G/Γ (corr. $\Gamma\backslash G$). More generally, a discrete subgroup Γ is called a *lattice* in G if a right Haar measure projects to a finite measure on G/Γ.

A *non–uniform* lattice is a lattice whose homogeneous space is not compact but still has finite Haar measure. The simplest example, but an extremely important one, especially in number theory, is the subgroup $SL(2,\mathbb{Z})$ of all matrices with integer elements in the group $SL(2,\mathbb{R})$ of all 2×2 matrices with determinant one; this generalizes to $SL(n,\mathbb{Z})$ being a non–uniform lattice in $SL(n,\mathbb{R})$.

Definition A.8.1. A *Lie group* is a differentiable manifold with a group structure such that all left and right translations as well as $g \mapsto g^{-1}$ are diffeomorphisms. A *Lie subgroup* is a subgroup that is a submanifold as well.

Examples of Lie subgroups of $\mathbb{R}^n$ are linear subspaces as well as integer or rational multiples of a fixed vector and products of these. $\mathbb{Z}^n$ is a discrete subgroup of $\mathbb{R}^n$ and $SL(n,\mathbb{Z})$ (integer matrices with unit determinant) is a discrete subgroup of $SL(n,\mathbb{R})$. Note that $GL(n,\mathbb{Z})$ is *not* a subgroup of $GL(n,\mathbb{R})$.

Definition A.8.2. A *Lie algebra* is a linear space g with an antisymmetric bilinear operation $[\cdot,\cdot]: g \times g \to g$ satisfying the *Jacobi identity* $[v,[w,z]] + [w,[z,v]] + [z,[v,w]] = 0$. An *ideal* is a subalgebra $a \subset g$ such that $[g,a] \subset a$.

If a, b are ideals of g then so is $[a,b]$ and thus we have the *derived series* $g \supset g' := [g,g] \supset g'' := [g',g'] \supset \cdots$ of ideals. g is said to be *abelian* if $g' = 0$ and *solvable* if the derived series ends at 0 after finitely many steps. A Lie algebra is said to be *semisimple* if 0 is the only solvable ideal, and *simple* if in addition there are no ideals other than g and 0. The *descending central series* of g is the sequence $g^1 := g \supset g^2 := [g,g^1] \supset g^3 := [g,g^2] \supset \cdots$ of ideals. g is said to be *nilpotent* if this series terminates at 0.

The vector fields on any differentiable manifold form an infinite-dimensional Lie algebra with the Lie bracket defined in Definition A.3.3. Since left translations are diffeomorphisms they act on vector fields. The Lie bracket is defined on the space $\mathcal{L}(G)$ of left-invariant vector fields (that is, vector fields invariant under all left translations), so this linear space becomes a Lie algebra called the *Lie algebra of* G. $\mathcal{L}(G)$ is naturally isomorphic to the tangent space of G at the identity. Thus its dimension is finite and it coincides with the dimension of G. Conversely every Lie algebra is the Lie algebra of a unique simply connected Lie group. Important examples of Lie groups are the matrix group $GL(n,\mathbb{R})$ (non-singular matrices) and its closed subgroups, such as $SL(n,\mathbb{R})$ (matrices with unit determinant, the *special linear group*), $O(\mathbb{R})$ (orthogonal matrices, the *orthogonal group*), and $SO(n,\mathbb{R})$ (orthogonal matrices with determinant one, the *special orthogonal group*). Here the Lie bracket is given by the commutator $[A,B] = AB - BA$. The *exponential map* $\exp(A) := e^A = \sum_{n=0}^{\infty} \frac{A^n}{n!}$ defines a map from the tangent space at the identity, that is, the Lie algebra, to the matrix group. This is, in fact, the explicit representation of the abstract exponential map (9.5.1) in differential geometry. (This is how the name "exponential" map came about.) On any Lie group a choice of inner product on $T_{\text{Id}}G$ produces a left-invariant Riemannian metric by left translations. The geodesics through Id for such a Riemannian metric are exactly the one-parameter subgroups of G. Then G is a complete differentiable manifold with respect to this metric structure, and the exponential map is always defined on $T_{\mathrm{Id}}G$. For matrix groups geodesics through Id are of the form $t \mapsto e^{tA}$ because these are exactly the one-parameter subgroups. Note that therefore $\mathcal{L}(G)$ is canonically given as the linear space of matrices obtained from entrywise differentiation of one-parameter subgroups of the matrix group G. This helps to identify the Lie algebras corresponding to these matrix groups. They are $\mathcal{L}(GL(n,\mathbb{R})) = \mathfrak{gl}(n,\mathbb{R})$, the space of all $n \times n$ matrices; $\mathcal{L}(SL(n,\mathbb{R})) = \mathfrak{sl}(n,\mathbb{R})$, the space of traceless matrices (because $\det e^{A} = e^{\operatorname{tr} A}$); $\mathcal{L}(SO(n,\mathbb{R})) = \mathcal{L}(O(n,\mathbb{R})) = \mathfrak{o}(n,\mathbb{R})$, the space of skew-symmetric matrices. The *symplectic group* $Sp(n,\mathbb{R}) \subset GL(2n,\mathbb{R})$ of matrices A such that $A^tJA = J$, where $J = \begin{pmatrix} 0 & \mathrm{Id} \\ -\mathrm{Id} & 0 \end{pmatrix}$, with Id the identity in $GL(n,\mathbb{R})$, has Lie algebra $\mathfrak{sp}(n,\mathbb{R}) = \{A \in \mathfrak{gl}(2n,\mathbb{R}) \mid JA + A^tJ = 0 = A^t + A\}$. The group $SO(p,q,\mathbb{R})$ of matrices A such that $A^tI_{p,q}A = I_{p,q}$, where $I_{p,q} = \begin{pmatrix} \mathrm{Id}_p & 0 \\ 0 & \mathrm{Id}_q \end{pmatrix}$, with Id_k the $k \times k$ identity, has as its Lie algebra the space $\mathfrak{so}(p,q,\mathbb{R})$ of matrices $\begin{pmatrix} A & B \\ B & C \end{pmatrix}$ with A, C skew-symmetric of size p, q, respectively. An automorphism (that is, diffeomorphic group isomorphism) of a connected Lie group G is uniquely determined by its differential at the identity. Since a basis in $T_{\mathrm{Id}}G$ defines a field of bases on G via left translations there is a left-invariant volume defining the Haar measure. Abelian, compact, discrete, semisimple, and connected nilpotent Lie groups (that is, Lie groups whose Lie algebras are semisimple and nilpotent, correspondingly) are unimodular. The existence of lattices is not always clear, but $\mathbb{R}^n$ has plenty, all isomorphic to $\mathbb{Z}^k$ for some $k \leq n$. Lattices in the Heisenberg group came up in Section 17.3. The action of the group G by left translations on a right quotient defines important examples of dynamical systems, where "time", of course, may be multidimensional. For example, the action of $\mathbb{R}$ on $S^1 = \mathbb{R}/\mathbb{Z}$ is the unit-speed flow around the circle.

Notes

Chapter 0

Section 1. In addition to the comments made in the text we present here an incomplete and admittedly subjective survey of principal sources in the main branches of the theory of dynamical systems.

Ergodic theory. The book by Walters [W] comes closest to qualify as a standard textbook in ergodic theory despite the absence of many proofs, especially those of important background results, and, of course, the fact that it was written almost twenty years ago. Since its publication several important developments have taken place including Kakutani (monotone) equivalence theory, combinatorial ergodic theory, and finitary isomorphism theory.

The book by Cornfeld, Fomin, and Sinai [CFSin] is a very valuable although far from comprehensive source in ergodic theory. In particular, it reflects a broad view of the subject in the context of the general theory of dynamical systems somewhat similar to the view of the theory of smooth dynamical systems developed in the present book.

The book by Petersen [Pet] covers many key subjects in ergodic theory very well, including ergodic theorems, but the connections between ergodic theory and other branches of dynamics are almost totally missing. It is well endowed with exercises.

The short book by Rudolph [Ru2] may serve as a useful supplement to any of the three aforementioned sources because it covers several subjects that are missed in all of them.

The brilliant short monograph by Halmos [Ha2] played an important role in the development of ergodic theory; by now it has primarily historical value. It gives a good panorama of the subject on the verge of the entropy breakthrough. For those interested in looking back even further to the origin of the subject we recommend von Neumann's original article [vN] and the 1937 monograph by E. Hopf [Ho1].

Among the major sources for particular branches of ergodic theory are [O2] for the Ornstein isomorphism theory, [K3] and [ORW] for Kakutani equivalence theory, [KeS] and [Ru1] for the finitary isomorphism theory, [Fu3] for combinatorial ergodic theory, [Kr] for ergodic theorems, and [Kri] for the orbit equivalence of transformations with quasi-invariant measures.

Topological and symbolic dynamics. Earlier developments in topological dynamics are summarized in [GH]. An important topic in topological dynamics is the theory of distal extensions developed by Furstenberg [Fu2]. Probably a comprehensive text on topological dynamics as a self-contained subject cannot be written since many important aspects of topological dynamics are closely related with other branches of the theory of dynamical systems. The book [DGS] by Denker, Grillenberger, and Sigmund represents an attractive synthesis of ergodic theory and topological dynamics seen primarily from the point of view of applications to smooth dynamical systems and symbolic dynamics.

The survey article [Boy] by Boyle contains an accessible up-to-date introduction to symbolic dynamics as well as a description of the principal results in the field. A comprehensive introduction to the field is given in the book [LM] by D. Lind and B. Marcus.

Differentiable dynamics and smooth ergodic theory. Most existing sources on the theory of smooth dynamical systems, both textbooks and monographs, concentrate on particular aspects of the subject.

Hyperbolic theory treated primarily from the topological and geometric viewpoints first appeared in modern form in the pioneering article by Smale [Sm5]. This point of view is well presented in different ways in [Nit1], [PMe], [Shu3], and [I]. The book by C. Robinson [Rob2] is the most comprehensive and up-to-date source of that kind.

Another tendency in modern hyperbolic dynamics, associated primarily with the Russian school, emphasizes analytic and probabilistic aspects as well as geometric and topological ones. It originated in the fundamental monograph by Anosov [An3]. Although heavy and somewhat archaic in presentation it is still a first-rate source on the subject. This work together with the survey article by Anosov and Sinai [AnSin] are viewed as the origin of modern smooth ergodic theory.

Among the texts reflecting a synthetic view of differentiable dynamics with strong emphasis on smooth ergodic theory one should mention a relatively elementary and very useful book by Szlenk [Sz] and an excellent advanced text by Mañé [Ma4]. For a presentation of smooth ergodic theory see the original articles by Pesin [Pes1, Pes2, Pes3] and the later work in [KSt], [Pol], [Rue6], [PuShu], and [LedY].

Two excellent but outdated books by Arnold and Avez [ArAv] and Moser [Mos6] played a very important role in forming the modern multifaceted view of the theory of smooth dynamical systems. The authors certainly feel indebted to these sources.

The volumes [AKK] and [GMN] consist of major expository articles covering various aspects of differentiable dynamics and some related fields.

A very important book by Ruelle [Rue5] contains the most comprehensive treatment of the connection between symbolic dynamics, ergodic theory, and differentiable dynamics based on ideas from statistical mechanics. An earlier pioneering article by Sinai [Sin5] ought to be mentioned as a key contribution to this line of development.

There are a number of books on low-dimensional dynamics, including texts (for example, [De], [Be]) as well as monographs ([Me], [MeS], [CE]). From among the rapidly growing literature attempting to use low-dimensional dynamics and connections to applications to a wider audience we particularly recommend that by Strogatz [Str] for its lively and intelligent treatment of applications.

The books [Hen], [HMO], and [Te] provide the framework for the treatment of infinite-dimensional dynamical systems coming from parabolic partial differential equations and similar situations with highly dissipative behavior based on appropriate generalizations of the stable–unstable manifold theory and extensive use of concepts of topological dynamics and ergodic theory. They contain a good collection of specific cases that can be treated with these methods.

Hamiltonian dynamics. The book [Ar6] by Arnold established the theory of Hamiltonian systems as a discipline within the core of mathematics in its own right and placed it in the realm of the study of dynamical systems. The book [AM] by Abraham and Marsden provides a lot of useful background and develops selected topics quite far. A number of textbooks in classical mechanics have been influenced by this point of view and can themselves serve as good sources for the field ([G] is a good representative of these). In the absence of a comprehensive monograph on the Kolmogorov–Arnold–Moser "KAM" theory the book [Mos6] and articles [Mos3] by Moser still provide the best available introduction to this theory. Moser's article in [GMN] serves as a good introduction to the modern theory of finite-dimensional completely integrable Hamiltonian systems. Variational methods are a very important tool in Hamiltonian dynamics. A recent account of these is in [FM].

Finite-dimensional symplectic dynamics is only one of the aspects of an extensive branch of mathematics which encompasses also symplectic topology, symplectic geometry, certain aspects of partial differential equations based on infinite-dimensional symplectic structures, and so forth. Symplectic topology and its relations to Hamiltonian dynamics are presented in the recent book [HZ] by Hofer and Zehnder, which starts at a basic level and explains connections between a certain rigidity of symplectic maps and periodic phenomena in dynamics in terms of symplectic capacities. In infinite-dimensional Hamiltonian dynamics two particular areas have been the subject of extensive study. First is the theory of infinite-dimensional completely integrable systems which turned out to include a number of important nonlinear partial differential equations such as the Korteweg–deVries equation. Although a comprehensive introduction or survey of the field does not exist, the book [BDT] by Beals, Deift, and Tomei is a good source which presents the main methods and representative applications. The second is an infinite-dimensional generalization of the KAM theory [Ku].

Section 2. A detailed discussion of the connection between differential equations, vector fields, and flows can be found in many modern books on the theory of ordinary differential equations such as [Ar4] or the theory of differentiable manifolds [BiC].

Section 3. The suspension construction plays an important role in topology as a way of constructing manifolds and other interesting topological spaces.

Section 4. Transverse homoclinic points and interesting invariant sets found in their neighborhoods are among the most popular and powerful paradigms in dynamics. Finding these is by far the most common method of rigorously establishing complicated orbit behavior in various classes of dynamical systems, both finite- and infinite-dimensional. A good example is M. Levi's work on periodically forced oscillations [Le]. Finding those phenomena has become a veritable industry in applied dynamics. Among more far-reaching developments of the semilocal approach Conley's theory of isolating blocks [Co] and Alekseyev's work on *quasi-random* dynamical systems [A3] stand out.

Chapter 1

Section 1. The Contraction Mapping Principle penetrates analysis beginning with such elementary but basic applications as the proof of the Implicit Function Theorem and existence and uniqueness of solutions of ordinary differential equations.

Section 3. For a detailed discussion of dynamics and ergodic theory of translations on compact abelian groups see [CFSin, Chapter 12] or [W, Chapter 3]. The first proof of topological transitivity of irrational rotations can be traced to an 1828 work of Jacobi. Another early source is a work of Dirichlet in 1841.

Section 6. See the first chapter of the classical book [Mi] by Milnor for the use of gradient flows in the study of the topology of manifolds. In fact, our presentation of the three elementary examples follows Milnor. A useful dynamical generalization of the notion of gradient flow of a "typical" function is the concept of a Morse–Smale system [Sm5].

Section 7. **(1)** The quadratic family f_λ has been the subject of very extensive study, including traditional analytical proofs, computer-assisted proofs, and computer simulation. See, for example, [MeS], [CE].

Section 8. Similarly to rotations of a circle and translations of a torus which are particular cases of translations of compact abelian groups toral automorphisms and endomorphisms are the simplest representatives of the class of automorphisms and endomorphisms of such groups. The full shift which is discussed in the next section and the Smale attractor which is discussed in Section 17.1 can also be viewed as automorphisms of compact abelian groups. The study of the dynamics and ergodic theory of automorphisms of compact abelian groups is related to questions in commutative algebra, algebraic geometry, and especially algebraic number theory. The book [Sc] by Schmidt is a good source on this subject.

Section 9. **(1)** Topological Markov chains first appeared in the work of Parry [P1] under the name "subshifts of finite type" which has been commonly used in the Western literature on the subject. The term "topological Markov chain" which is used throughout the book was introduced by Alekseyev [A1], [A3]. We prefer this term since it better reflects the nature of the object as well as the parallelism between topological and probabilistic concepts (similarly to "topological entropy" versus "measure-theoretic entropy" (metric entropy) in Sections 3.1 and 4.3, correspondingly). Among other things Alekseyev used topological Markov chains to construct some previously unknown types of behavior in such classical problems as the three-body problem [A3].

(2) The name Perron–Frobenius Theorem is traditionally applied to several different but related results. What we state is a result due to Perron [Per1] and later generalized by Frobenius to irreducible nonnegative matrices (see, for example, [Ga]). Some proofs are shorter because they use the Brouwer Fixed-Point Theorem [Rob2]. Infinite-dimensional generalizations of the Perron–Frobenius Theorem play a very important role in hyperbolic dynamics, especially in the theory of Gibbs measures and zeta functions [Rue3], [PP].

(3) The classification of topological Markov chains follows Alekseyev [A4]. Classifying topological Markov chains up to topological conjugacy (see Section 2.3) is the central problem of symbolic dynamics. Williams [Wi2], [Boy] suggested an answer known as the "Williams Conjecture". Attempts to prove it produced numerous fruitful developments in symbolic dynamics concluding, somewhat anticlimactically, in a recent counterexample by Kim and Roush.

Chapter 2

Section 1. Moduli as in Subsection c, and, in particular, results along the lines of Corollary 2.1.6 seem to have been part of the folklore in geometric dynamics for a long time. Several people claimed to have discovered such results independently. Since we are not sure as to the priority, and have not done any research, we abstain from any particular attribution. We certainly do not claim priority ourselves and discuss those moduli simply as an example. Y. Ilyashenko pointed out to us that similar moduli can be used to classify expanding maps on the circle (see Section 2.4) up to smooth conjugacy. In that case those moduli do not seem to shed much light on the essential features of the orbit structure.

(1) This proof was shown to us by Jürgen Moser.

Section 3. A short historical account of structural stability is given in Section 0.1. We should add that Andronov and Pontryagin in their original 1937 work [AP] used a French term ("grossièreté") which literally translates into English as "roughness". Smale started to study this notion in the late fifties and, influenced by Levinson [Lev], discovered the famous "horseshoe" example (See Section 2.5c). Smale's example, presented in a memorable session at the 1961 conference on nonlinear oscillations in Kiev to an audience which contained several leading young Russian mathematicians, gave a great impetus to the Russian school [Sm2]. Originally Smale hoped to show that "most" systems are structurally stable but soon he found a counterexample [Sm4] which was followed by counterexamples to several weakened versions of the conjecture; see, for example, [ASm]. After that the quest to find necessary and sufficient conditions for structural stability became one of the principal driving forces in the development of the theory in the West. Sufficient conditions for C^1 strong structural stability were found by Robbin [Ro] and sharpened by C. Robinson [Rob1]. Mañé's proof of necessity of these conditions [Ma3] is one of the highest points in the development of the subject. Liao [Li] found a proof in the low-dimensional case independently and more or less simultaneously with Mañé. It is still an open and intriguing problem for C^r systems to decide when C^m structural stability for $1 < m \leq r$ is equivalent to C^1 structural stability.

Section 4. The basic theory of expanding diffeomorphisms on manifolds of arbitrary dimension was developed by Shub [Shu1].

Section 5. As we mentioned before, the horseshoe example was the first case where structural stability was established for a system with complicated orbit structure [Sm3]; see Proposition 6.5.3.

The construction of Markov partitions for 2-dimensional toral automorphisms is due to Adler and Weiss [AW]. This short work represents a landmark in establishing direct connections between symbolic dynamics and the theory of smooth dynamical systems. It was shortly followed by works of Sinai [Sin4] and later Bowen [Bo2], where much more complicated Markov partitions were constructed for a broad class of systems with hyperbolic behavior. We discuss the general construction in Section 18.7. Markov partitions quickly became one of the most powerful tools in hyperbolic dynamics and in particular in ergodic theory of hyperbolic systems. This is due to the fact that the thermodynamical formalism, the counting of periodic orbits, and other techniques which are well understood in the symbolic situation, become available for smooth systems with hyperbolic behavior via Markov partitions. See [Sin5], [PP].

(1) The same conclusion can, in fact, be made for any $\lambda > 4$ by much more sophisticated techniques using that the Schwarzian derivative of f_λ is negative. These methods are treated in [MeS].

Section 6. Our proof of structural stability is modeled on Moser's proof for a more general case [Mos5], which appeared after geometric proofs by Smale and Anosov but was the first where a fixed-point method in an appropriate functional space was used. The global character of the argument in the toral case (existence of a semiconjugacy within a homotopy class) was established by Franks [Fr1].

Sections 7 and 8. The formal description of the iteration method as well as the proof of the Siegel theorem are taken almost verbatim from Moser's articles [Mos3]. The method itself is most commonly known as the KAM (Kolmogorov–Arnold–Moser) method. Its history is well known and we only give a brief account here. Around 1953 Kolmogorov discovered that in an analytic "non-degenerate" completely integrable Hamiltonian system after a small analytic perturbation of the Hamiltonian, those invariant tori remain whose frequency vectors are not too well approximable by rationals, that is, Diophantine. That contradicted a widely held belief, which goes back to Poincaré, that the behavior in such systems becomes "chaotic". Kolmogorov published an announcement and a short outline of his proof [Ko2]. A detailed proof of the Kolmogorov theorem was published in 1962 by his student Arnold [Ar2], who also made an advance in the partially degenerate case needed to treat the question of stability of the solar system [Ar3]. Notice that in the completely degenerate case the assertion of the Kolmogorov theorem is not valid [K1]. Independently, Moser developed methods which allowed him to prove a version of Kolmogorov's theorem for systems of finite differentiability, first for two degrees of freedom [Mos1] and then for the general case. He also put the method in a general functional-analytic framework which greatly simplified its use and led to new applications [Mos3]. Among major later works in the area are: Lazutkin [L], Rüssmann [Rü], Nekhoroshev [Ne], Salamon and Zehnder [SZ], and so on.

(1) For the original proof of the Siegel Theorem using an advanced version of the majorization method which we develop in a rudimentary form in our proof of Proposition 2.1.3 see [Si].

Section 9. Proposition 2.9.5, a prototypical result in a problem with "small denominators", was proved by Kolmogorov [Ko1] shortly before his discovery of the KAM method, where "small denominators" play a crucial role.

Chapter 3

Section 1. **(1)** The zeta function for dynamical systems was introduced by M. Artin and Mazur [ArM] and popularized by Smale [Sm5]. See [PP] for a nearly comprehensive treatment of the subject.

(2) Topological entropy was introduced by Adler, Konheim, and McAndrew in 1965 [AKM], who modeled in the topological setting the Kolmogorov definition of entropy with respect to an invariant measure (see Section 4.3, Exercises 3.1.7–3.1.9). The definition using separated sets was introduced independently by Dinaburg [Di] and Bowen [Bo1].

(3) Connections between entropy and volume growth were established by Yomdin [Yo2]. See the work of Newhouse for consequences of this fact for dynamics [New2].

(4) Fundamental-group entropy without the name appears in [K4] and in [Bo8]. The notion certainly has been a part of the mathematical folklore for a while.

(5) A very influential entropy conjecture by Shub asserts that for a C^1 map of a compact smooth manifold $h_{\mathrm{top}}(f) = \sup_i \log r(f_{*i})$ [Shu2]. It has stimulated a number of important works. It is not true for continuous maps or even homeomorphisms. It has been proven in a number of special cases (see Chapter 8 and [K4]). Its validity for C^∞ maps follows from Yomdin's result. The general case of finite smoothness is still open.

Section 3. Our list of important recurrence properties is incomplete. For example, the notion of nonwandering set was important for Smale's program of classification of "typical" smooth dynamical systems. Later it became apparent that for general classes of dynamical systems the weaker notion of *chain recurrence* is crucial [Co], [Fr3].

One way to quantify the recurrence behavior is related with invariant measures (see the summary at the beginning of Section 4.3). In topological dynamics there is another approach exemplified by such notions as *proximal orbits* or *syndetic sets* [E].

Chapter 4

Section 1. For historical remarks on the development of ergodic theory prior to the introduction of entropy see [Rok2], [Ha2].

(1) For the original proof of the Krylov–Bogolubov Theorem see [KB].

(2) For the original proof of the Birkhoff Ergodic Theorem see [Bi4]. Our proof was communicated to us by U. Schmock with attribution to J. Neveu. We incorporated a further shortcut due to A. Fieldsteel and B. Bassler. There are other short proofs by Shields, Katznelson and Weiss, and others. The Birkhoff Ergodic Theorem has been generalized in many directions—see [Kr].

(3) Key contributions in establishing connections between invariant measures and the asymptotic distribution of orbits were made by Oxtoby [Ox].

Section 2. (1) Before the introduction of entropy, mixing was viewed as a prime "stochastic" property of measure-preserving transformations (see, for example, a discussion in [Ha2]). It is, however, one of those notions that is easy and natural to define but very difficult to study. An outstanding example of a natural question that turned out to be notoriously elusive is Rokhlin's "multiple-mixing problem" (see [Ha2]).

(2) The original proof of uniform distribution of the fractional parts of polynomials is due to H. Weyl and uses estimates of trigonometric sums. The argument using ergodic theory presented in the exercises is due to H. Furstenberg.

Section 3. Measure-theoretic entropy was discovered by Kolmogorov [Ko3]. It was closely modeled on the notion of entropy of an information channel, that is, a stationary random process, introduced by Shannon in 1948 [Sh]. In 1956 Khinchin gave Shannon's theory a rigorous mathematical form which essentially coincides with our description of entropy of a transformation with respect to a fixed partition [Kh]. This set the stage for Kolmogorov's crucial observation that entropy can be made an invariant of the transformation itself. The introduction of entropy was a turning point in the development of ergodic theory. Kolmogorov's original application was to distinguish shifts with different Bernoulli measures—see Section 4.4c. He also introduced the important notion of a K-system, that is, a dynamical system whose entropy with respect to any partition is positive. This turned out to be a "right" kind of a mixing-type notion. Among the most important developments in ergodic theory that followed the introduction of entropy were the weak-isomorphism theorem by Sinai [Sin1], Ornstein isomorphism theory [O1], [O2], the theory of finitary isomorphism by Keane and Smorodinsky [KeS], and Kakutani (monotone) equivalence theory [K3], [ORW]. Entropy also plays a central role in smooth ergodic theory (see Chapter 20 and the Supplement). The best standard sources on entropy and K-systems are the article by Rokhlin [Rok3] and the book by Parry [P2].

(1) This follows from an equivalent definition of entropy given in [K5].

Section 4. The invariant measure for topological Markov chains defined by (4.4.5) and (4.4.6) was introduced by Parry [P1]. It was used by Adler and Weiss to show that automorphisms of the 2-torus with equal entropy are metrically isomorphic by means of the Markov partition described in Section 2.5. Although this fact could be derived from the Ornstein isomorphism theory it preceded Ornstein's work on isomorphism of Bernoulli shifts and was one of the early nontrivial examples of isomorphism in dynamics.

Section 5. Goodwyn [Gw] first proved the inequality $h_{\text{top}} \geq h_\mu$. The opposite inequality was first proved by Dinaburg [Di] under the extra assumption that the phase space has finite topological dimension and then by T. Goodman [Goo] for the general case. A number of simplifications and generalizations appeared afterward. See Section 20.2 for one of those. We follow Misiurewicz's proof which is to be particularly recommended for its simplicity and versatility [Mis1].

Chapter 5

Section 1. **(1)** The problem of existence of an invariant (finite or σ-finite) measure equivalent to a given quasi-invariant measure is a central issue in the part of ergodic theory that deals with transformations with quasi-invariant measure. For a review of early results in that direction see [Ha2]. Later it was realized that the problem is intimately related to a classification of von Neumann algebras. See [Kri], [Con].

(2) Apparently the first proof of existence of an absolutely continuous invariant measure for an expanding map (not only on the circle) was given by Krzyzewski and Szlenk in 1969 [KSz]. Our proof seems to be different from the numerous proofs in the literature. Existence of an absolutely continuous invariant measure for more general classes of interval and circle maps is one of the central problems of one-dimensional dynamics. A relatively straightforward but useful generalization is provided by [LY]. For further discussion see the notes to Section 16.2.

(3) Bounded distortion estimates play a crucial role in one-dimensional differentiable dynamics. This was first realized by Denjoy [D]. We will encounter bounded distortion estimates many times.

(4) Proposition 5.1.26 is a very simple manifestation of the phenomenon that plays a central role in smooth ergodic theory: Entropy is closely related to infinitesimal exponential divergence of orbits. This was first established by Sinai [Sin2] in the hyperbolic case (compare with Theorem 20.4.1), and by Ruelle, Pesin, Ledrappier, and Young in the general case [Rue4], [Pes2], [LedY]. The Ruelle inequality is proved in the Supplement (Theorem S.2.13).

(5) For the two original proofs of Theorem 5.1.27 see [Mos2]. We reproduce the second of these. The homotopy trick was introduced by R. Thom. It is used extensively in the theory of singularities of differentiable mappings.

Section 2. **(1)** Our presentation of the central-force problem follows [AM].

Section 4. **(1)** This follows from the Koebe regularization theorem which can be found in advanced books on the theory of Riemann surfaces.

(2) The fundamental groups of compact factors of the hyperbolic plane are a particular case of Fuchsian groups of the first kind. A good and accessible standard reference on this subject, which also covers the needed results from hyperbolic geometry, is [Ka].

(3) This was originally shown by E. Artin [Art].

(4) The original proof by Hedlund [He2] uses methods of function theory in a way somewhat similar to the proof in Section 4.2 based on Fourier series. E. Hopf [Ho2] proved ergodicity of geodesic flows on surfaces of variable negative curvature. Our proof presents Hopf's method in a somewhat different language. For further developments of the method see [AnSin] for the case of Anosov systems and [Pes2] for nonuniformly hyperbolic systems.

Section 5. Hamiltonian systems and related subjects are covered by [AM], [Ar6].

(1) The full force of Liouville's theorem is obtained in Section 5.2 of [AM].

Chapter 6

Section 2. An excellent historical survey of the Hadamard–Perron Theorem and related issues as well as a wealth of references are contained in Section 4 of [An3]. The literature on the subject both prior to Anosov's book and following it is enormous.

(1) This method goes back to Hadamard [H]. An alternative approach, initiated by Perron [Per2] and described in [An3], is based on the the *theorem of constantly acting perturbations* (see the paper of Alekseyev in [AKK]).

Section 3. **(1)** For the original proofs of Theorem 6.3.1 see [Har2] and [Gr].

Section 4. Anosov introduced the class of systems that now bears his name in [An1]. He called those objects "U-systems". In his classical article [Sm5] Smale introduced the notion of hyperbolic set and presented the basics of the theory. He also started to use the term "Anosov systems" which quickly became standard. In [An3] Anosov developed a number of fundamental tools including the theory of stable and unstable foliations and the closing lemma which are applicable to the more general situation of hyperbolic sets.

(1) As far as we know the proof given here, which avoids both the use of infinite-dimensional spaces and that of stable and unstable manifolds, has never appeared in the literature.

(2) The term "locally maximal" hyperbolic set was introduced by Alekseyev [A2], who called them locally maximal Perron sets. It is not commonly used in the western literature where these sets are usually called "basic". This terminology developed from the original program of Smale who studied systems with global hyperbolic behavior (Axiom A). In particular it is used in a series of Bowen's papers which contains a major part of the theory of hyperbolic sets and some of which we follow in our later exposition. We think that Alekseyev's term describes the nature of these sets more appropriately.

Section 5. **(1)** Structural stability of horseshoes is due to Smale [Sm3].

(2) This was shown by Birkhoff [Bi2].

Section 6. The formal analysis in the linearization problem goes back to Poincaré who considered vector fields rather than maps. Smooth linearization in the resonance-free C^∞ case is due to Sternberg [St] and the extension to nonlinear normal forms to Chen [C]. The theory of normal forms has a vast literature of which we do not attempt to list even the principal sources. The survey by Belitskii [Bel] gives an overview for the smooth case. A major work for the analytic case is that of Brjuno [Br].

(1) The main difference between the C^∞ and the analytic situations is the presence of *small denominators* or near resonances, that is, the existence of multi-indices for which the denominator $\lambda_i - \lambda^k$ in (6.6.2) is nonzero but small.

(2) Our use of the homotopy trick to prove Theorem 6.6.5 is due to a suggestion by Y. Ilyashenko who has used it in similar situations.

Chapter 7

Section 1. **(1)** An early remarkable example of the use of pathological generic properties of the C^0 topology is the classical work [OxU] by Oxtoby and Ulam which shows that volume-preserving homeomorphisms are generically ergodic. Their mechanism of ergodicity is very different from those responsible for ergodicity in smooth systems. Genericity of many other ergodic properties in the same setting was shown in [KS].

Section 2. The Kupka–Smale Theorem was originally proved independently by Kupka for flows [Kup] and Smale [Sm1] for flows and diffeomorphisms. For flows the proof can also be found in [PMe].

There are a number of C^1 genericity results. The most basic one is that periodic points are generically dense in the nonwandering set [Pu]. For Hamiltonian systems the analogous result is in [PuRob]. These results are based on the C^1 Closing Lemma by Pugh [Pu] which states that a nonwandering point can be made periodic by a small C^1-perturbation concentrated in a neighborhood of the point. In this form the Closing Lemma is not true in the C^2 topology [Gu2]. It is still unknown whether a nonlocal C^2 or C^∞ closing lemma holds.

Among other interesting C^1 genericity results are that generically all hyperbolic periodic points of a symplectic map have homoclinic points which are dense both in the stable and in the unstable manifolds [T] and the generic density of hyperbolic points for area-preserving two-dimensional maps [T]. For non-Anosov area-preserving two-dimensional maps density of elliptic points is also C^1 generic [New4].

(1) See [Sm4].

Section 3. There is altogether a vast literature on the subject of bifurcations of which we can only give a minute sample. [ArASI] is a comprehensive survey covering the local and nonlocal theory. The book by Palis and Takens [PT] is a definitive source for a certain class of nonlocal bifurcations related to the appearance of positive entropy from Morse–Smale systems. [Ar5] and [Rue7] contain an introduction to the subject. Local and global bifurcations are also discussed in [GH]. Local normal forms and the homotopy trick are the most useful tools in the theory of local bifurcations. Algebraic geometry and its applications to singularity theory begin to play an important role when multiparametric families are considered. An interesting example of global bifurcations appears in typical families of circle diffeomorphisms such as one-parameter subfamilies of the family $f_{a,b}$ discussed in Exercise 11.1.4. Notice that in the one-parameter family f_{a,b_0} there are uncountably many bifurcation values corresponding to irrational values of the rotation number in addition to the countably many endpoints of the intervals of structural stability.

(1) An interesting feature of period-doubling bifurcations is that they appear in infinite cascades in simple families of maps, such as the quadratic family. Namely, a stable fixed point loses stability while generating a stable period-two orbit, which in turn loses stability and generates a stable period-four orbit, and so on. Meanwhile all of the previous periodic orbits persist as repellers. The first few bifurcations in the quadratic family are the subject of Exercise 1.7.2, they occur for $\lambda = 3$ and $\lambda = 1 + \sqrt{6}$. Via algebra, which is still straightforward in principle, but quite formidable, one can show that for $\lambda = 1 + \sqrt{8}$ a 3-periodic point first appears (this is mentioned, with references both to the original and an elementary proof, in [Str]). By Theorem 15.3.2 the entire periodic-doubling cascade must therefore end somewhat earlier (and, moreover, orbits of all periods must have appeared by then). Indeed, in general these bifurcation values accumulate to a fixed parameter, after which the topological entropy becomes positive. Jacobson [J1] first proved the existence of such cascades in a simple analytic family. Independently Feigenbaum [Fe] numerically found an asymptotic rate of convergence of the bifurcation values as well as self-similarity phenomena associated with this process. This led to an interesting direction in low-dimensional dynamics, referred to as renormalization. Feigenbaum's observations were eventually verified rigorously by Lanford using a computer-assisted proof.

Some aspects of the period-doubling phenomenon appear already in the topological situation, that is, for continuous interval maps. This is evident in the Sharkovsky ordering which appears in Theorem 15.3.2, in the proof of Theorem

15.3.7, and in the results of Section 15.4 concerning the map at the limiting parameter value of such a cascade.

Section 4. This theorem and our proof are due to M. Artin and Mazur [ArM]. It is not known whether one can guarantee that all periodic points are isolated. Yomdin [Yo1] has a C^k version of this theorem with a weaker estimate of the form $P_n(f) \le C^{n^\alpha}$, where α depends on k and the dimension. Kaloshin showed nongenericity [Kal]: There is an open set U of C^k diffeomorphisms such that for any sequence $\{a_n\}$ there is a residual subset of U where $P_n(f)/a_n \to \infty$.

(1) See [N].

Chapter 8

Section 1. Manning [Man2] proves our corollary. [K4] and independently [Bo8] observe that this argument can be modified slightly to prove Theorem 8.1.1.

Section 3. **(1)** See [MisP].

(2) There are several proofs of the inequality $h_{\text{top}}(f) \le \max(\deg P, \deg Q)$. The proof by Gromov [Gro] uses a certain minimality property of holomorphic maps to estimate the volume growth. Alternatively one can use the theory of nonuniformly hyperbolic dynamical systems and similarly to Corollary S.5.11 show that $h_{\text{top}}(f) \le p(f)$. On the other hand it follows immediately that $P_n(f)$ is equal to the number of solutions of an equation of degree $(\deg(f))^n$ plus possibly one for the point ∞ and hence $h_{\text{top}}(f) \le p(f) \le \deg(f)$. Yet another proof was obtained by Lyubich [Ly].

Section 4. The notion of index of an isolated fixed point allows a generalization which is particularly fruitful for dynamics, the *Conley index* (see [Co]).

Section 5. **(1)** See [ShuS].

Section 6. A proof of the Lefschetz Fixed-Point Formula is in [Fr3] (Theorem 5.10). This volume contains a number of applications of the Lefschetz Fixed-Point Formula to dynamical systems as well as Morse inequalities and various other homological constructions in dynamics.

Section 7. The original work by Nielsen is in [Ni1], [Ni2]. The first of these contains the proof of our Theorem 8.7.1. The second set of papers constitutes the main body of his work and contains the estimates of the number of periodic points for maps of compact surfaces of higher genus.

Thurston developed a structural theory of surface diffeomorphisms including constructions of models in homotopy classes. His work is presented in [FLP]. Many of the main results of Nielsen theory can be recovered from Thurston's theory; see [HT2]. A modern source on Nielsen theory is [Ji].

Chapter 9

While we concentrate in this chapter on variational methods that produce an infinite number of orbits there are important results where variational methods are used to produce a bounded number of periodic orbits for a broad class of systems. A representative sample includes papers by Ekeland and Lasry [EL] and Rabinowitz [R] producing periodic orbits for convex Hamiltonians in $\mathbb{R}^{2n}$ on one hand, and results by Conley and Zehnder [CoZ] that C^1 symplectomorphisms of $\mathbb{T}^{2n}$ that are generated by a globally Hamiltonian vector field have at least $2n+1$ fixed points. Both of these directions generated far-reaching developments. In connection with the former let us mention variational constructions of heteroclinic orbits for broad classes of Hamiltonian systems. The latter started a fruitful new approach at the junction between symplectic topology and Hamiltonian dynamics including new cohomological invariants.

Section 1. The first chapter of the book [Mi] presents an excellent exposition of the finite-dimensional Morse theory.

Section 2. **(1)** This can be found in [KT].

(2) This influential example was introduced by Bunimovich. It has strong stochastic properties. See, for example, [Bu] for this and further examples.

Section 3. **(1)** See the notes to Section 13.5.

(2) The proof of Theorem 9.3.7 follows [K6].

(3) Our proof of Theorem 9.3.10 is based on the key Proposition 9.3.12 which is due to Aubry [AuD].

(4) This definition and the results of the pertinent exercises are due to Mather.

Section 6. Proposition 9.6.3 is in [Man3]. Manning also proved the inequality $h_{\mathrm{top}}(g_t) \geq v(M)$. The idea of using entropy to show an abundance of minimal geodesics as in Theorem 9.6.7 comes from [K8].

Section 7. Minimal geodesics on surfaces of genus greater than one were first considered by Morse [Mo].

(1) For compact factors of the hyperbolic plane, that is, compact surfaces of constant negative curvature, an even more precise asymptotic follows from the Selberg trace formula; see also the notes to Section 20.6.

(2) For any metric of area A on a surface of genus $g \geq 2$ [K8] shows $\underline{\lim}_{T\to\infty}(1/T)\log N(T) \geq \sqrt{(4g-4)\pi/A}$; equality implies constant curvature.

Chapter 11

Section 1. The notion of rotation number first appears in Chapter 15 of the third of Poincaré's memoirs in [Po1]. This notion is the basis of several fruitful generalizations. On one hand one can define rotation intervals for noninvertible maps of the circle and rotation sets for maps of the torus. Those notions have turned out to be useful in a number of situations including a very recent proof by Franks and Bangert that there are infinitely many closed geodesics for any Riemannian metric on the two-sphere [B2], [Fr4]. On the other hand for any invariant measure of a flow one can define an asymptotic cycle, an element of the first homology group which determines the average asymptotic homological appearance of orbits [Schw]. This notion is developed and used in Section 14.7. The Poincaré theory presented in this chapter has been reproduced in many sources.

(1) The phenomenon of Proposition 11.1.11 is described by applied mathematicians as "frequency locking".

(2) These regions are sometimes called Arnold tongues because of their shape and since they were studied by Arnold.

Section 2. The Poincaré classification also appeared in Chapter 15 of the third of the memoirs in [Po1]. The topological classification of nontransitive circle homeomorphisms with irrational rotation number is due to Markley [Mark].

Chapter 12

Section 1. **(1)** See [D]. Following [Nit1] we present one of the standard proofs.

Section 2. This example also is in [D]. Again we follow [Nit1].

Section 3. The Arnold Theorem first appeared in [Ar1] and started a major development. Moser (for example, [Mos3]) and Rüssmann obtained analogous results for the case of finite differentiability. Naturally the required degree of differentiability and the allowable size of the perturbation depend on lower bounds for the speed of approximation of the rotation number by rationals. For example, for almost every rotation number Rüssmann only needs to assume $C^{3+\epsilon}$. Arnold conjectured that *any* analytic map with Diophantine rotation number is analytically conjugate to a rotation. For almost every rotation number this conjecture was proved by Herman [Her1] by first reducing to the case of small perturbations. Further developments focused on finding optimal conditions on the rotation number and on the development of new methods not depending on KAM-type procedures. Various aspects of this program were carried out by Herman, Yoccoz [Y], Katznelson and Ornstein [KO], and others. There are examples showing that in most cases the existing conditions are optimal. In particular Yoccoz proved that the Diophantine condition is necessary and sufficient for C^∞ conjugacy [Y], and that the Brjuno condition from [Br] is necessary and sufficient for analytic linearization. On the other hand for rotation numbers of bounded type C^1 conjugacy holds for C^1 diffeomorphisms whose derivative satisfies a condition very similar to that of the Denjoy theory [KO].

Section 5. This example is due to Arnold. See [Ar1].

Section 6. The method used for Theorem 12.6.1 was first introduced by Anosov and Katok in [AnK]. Among its applications were the first constructions of ergodic volume-preserving diffeomorphisms of arbitrary manifolds, constructions of minimal and uniquely ergodic diffeomorphisms of manifolds admitting a locally free S^1-action, and so on. For a Hamiltonian version of the method see [K1]. Later Herman and Yoccoz further developed this method and used it to find in typical perturbations of completely integrable Hamiltonian systems invariant tori with nonstandard properties.

The result of Corollary 12.6.4 was first proved by Furstenberg [Fu1]. The particular construction here is taken from the unpublished manuscript "Constructions in ergodic theory" by Katok and E. A. Robinson.

Section 7. This ergodicity result was first proved for C^2 diffeomorphisms independently by Katok [CFSin, Section 3.6] and Herman [Her1].

Chapter 13

This chapter is dominated by variational methods. The variational approach can be applied to Hamiltonian systems in a very similar way. For a recent exposition see [FM].

Section 2. Theorem 13.2.6 was proved independently by Aubry [Au], a French physicist, and Mather [Mat1]. Mather's method used a variational approach on an infinite-dimensional space, Aubry's method is based on constructing globally minimal states (as in Section 3). Before Aubry's work became known to mathematicians Katok [K6] suggested a simplification of the proof of Mather's result based on rational approximation. In Subsection a we follow [K6] and [K7]. Bangert [B1] shows that Hedlund's classical result [He1] on globally minimal geodesics on the torus comes very close to the construction of the counterpart of Aubry–Mather sets for flows.

(1) See [Her2].

(2) See [Mat3].

(3) Our proof of Theorem 13.2.13 is a simplified version of Fathi's proof [F].

Section 3. This section generally follows Aubry's approach [Au], [AuD] although the presentation is our own. Mather's original proof of the uniqueness of the minimal Aubry–Mather set, Corollary 13.3.11, uses an infinite-dimensional variational problem.

Section 4. The results of this section were obtained jointly by Katok and Mather, but written up separately. Mather's version appeared as [Mat4]. We follow the unpublished account of Katok [K7]. Very close results were obtained by Aubry et al. [AuDA].

Section 5. **(1)** This estimate can be considerably improved. Using a slightly different method Mather shows nonexistence of invariant circles for $\lambda > 4/3$. Using an iterative computer-assisted method MacKay improved this to $\lambda > 63/64$. It is interesting to point out that another computer-assisted proof by de la Llave shows the persistence of uncountably many invariant circles with rotation number near and equal to the golden ratio up to $\lambda = 0.93$. Numerically the critical value appears to be around 0.97.

(2) This is from [GuK].

(3) This is in [Mat2].

Chapter 14

Section 1. This theory is due to Poincaré (the second memoir of [Po1]) and Bendixson [Ben]. In the first half of this century differentiable dynamics was usually called qualitative theory of ordinary differential equations and the study of vector fields in dimension 2 (in particular in the plane and on the torus) was considered one of the centerpieces of the field as it was presented in such classical sources as [CL] and [NS]. Among the highlights of the period were the Denjoy theory in its flow version (see Proposition 14.2.4), the study of structural stability of two-dimensional flows by Andronov and Pontryagin [AP], the construction of the Cherry flow (Section 14.4a), and the classification of orbits of flows on surfaces of higher genus by Maĭer [M]. Later, with better understanding of the hyperbolic paradigm the theory of flows on surfaces became more peripheral.

Section 3. **(1)** See [Sch].

Section 4. **(1)** The Cherry flow was constructed in [Ch].

Section 5. The results about the topological structure of interval exchange transformations first appeared in [Ke] even though they can be extracted from earlier work of Maĭer [M]. Keane also conjectured that almost every irreducible interval exchange transformation is uniquely ergodic. Beyond the elementary level at which we treat the problem there is a fundamental series of results, primarily by Veech [V1], [V2] and Masur [Mas], which assert prevalence of uniquely ergodic interval exchange transformations and describe their metric properties. In particular they solved Keane's conjecture. The main idea is to consider an appropriate space of interval exchanges and introduce a dynamical system on that space in such a way that the properties of an interval exchange are related to asymptotic properties of its orbit under this dynamical system. In Veech's approach this is done via an appropriate inducing construction. Lemma 14.5.7 constitutes the first step. Important contributions to the study of interval exchange transformation using a more direct combinatorial approach were made by Boshernitzan. An interesting general property of interval exchange transformations, also based on Lemma 14.5.7, is that they are never mixing.

(1) The estimate of the number of invariant measures follows an idea of Oseledec who first used it to estimate the spectral multiplicity of Lebesgue measure.

(2) This result and its proof are taken from the unpublished manuscript "Constructions in ergodic theory" by Katok and E. A. Robinson.

Section 6. **(1)** A more general classification without the assumption of existence of a positive invariant measure is contained in [M] for orientable surfaces and has been extended to the nonorientable case by Aranson. The principal difference is the existence of wandering domains and nowhere-dense quasi-minimal sets as in the Cherry example.

(2) Further developments in the study of rational billiards and similar systems are related to the use of powerful tools originating from Teichmüller theory. This idea is due to H. Masur and it first appeared in his study of measured foliations on surfaces and interval exchange transformations [Mas] as an alternative to Veech's approach. A seminal result in that direction is a paper by Kerckhoff, Masur, and Smillie [KMS] who showed that for a set of directions of full measure in a rational billiard system the corresponding flow is ergodic. This was followed by a major tour de force by Masur and Smillie [MS], where very precise information about the size of the set of nonuniquely ergodic systems was obtained.

Section 7. **(1)** By the result of Yoccoz in [Y] the assertion is true without assuming area preservation.

(2) Asymptotic cycles were first introduced by Schwartzman [Schw].

(3) The results of this subsection follow the outline of [K2]. Detailed proofs have never appeared in print.

(4) This estimate is sharp—see [S].

Chapter 15

Section 1. **(1)** This fact was published by Li and Yorke in the early seventies who were unaware of Sharkovsky's general result. The significance of their paper titled "Period three implies chaos" is in launching the word "chaos" as a vague but powerful term for describing complicated behavior of a dynamical system.

Section 2. Theorems 15.2.1 and 15.2.11 are due to Misiurewicz [Mis3] based on prior work with Szlenk [MisS] about piecewise-monotone maps. Our proof is a streamlined version of these two papers.

(1) See [Re].

Section 3. **(1)** Theorem 15.3.2 and the first half of Theorem 15.3.7 were first proved in [Sha]. We follow the best-known exposition [BGMY]. Its strength is the construction of minimal Markov models which also give entropy estimates.

(2) Our proof of Theorem 15.3.7 was influenced by [Nit2].

(3) The square-root trick goes back to Štefan. In fact, maps with $\mathcal{P}(f) = S_{2^n}$, $n \in \mathbb{N}_0$, and $\mathcal{P}(f) = S_\infty$ appear successively in the quadratic family or a similar nice family of unimodal maps as a result of successive period-doubling bifurcations. See the notes to Section 7.3.

Section 4. Theorem 15.4.2 and its proof are from [Mis2]. The results of this section all constitute a description of the topological aspects of the self-similar structure that appears at the limit parameter of a period-doubling cascade.

Sections 5 and 6. The results of these sections are simplified versions of what is known as Milnor–Thurston theory [MiT]. Our presentation is adapted from the presentation of this theory in [Me].

(1) See [PP], [Rue3].

Chapter 16

Section 2. Hyperbolicity as described in Theorem 16.2.1 is one of the leading paradigms in the smooth theory of interval maps. It is an open phenomenon and seems to hold for a dense set of parameters in typical families. This was recently shown for the quadratic family by Graczyk and Świątek [GraSw] who built on work by Sullivan, Świątek and Jacobson, and others. Methods of complex dynamics (Section 17.8) are essential for this work. Nevertheless this set of parameters does not have full Lebesgue measure. Other behavior involves the existence of an ergodic absolutely continuous invariant measure supported on an open set containing critical points. Therefore this set is not hyperbolic but the measure is hyperbolic in the nonuniform sense of Section 4 of our Supplement. A prototype of such maps is $x \mapsto 4x(1-x)$ (Exercises 2.4.2 and 5.1.6). Jacobson proved that in a broad class of families, including the quadratic family, such behavior appears for a set of parameters of positive Lebesgue measure [J2]. Later other proofs were given by Benedicks and Carleson, Rychlik, and others. Conjecturally the union of these two types of behavior exhausts a set of full measure of parameters in typical families, including the quadratic family.

(1) See [Ma2].

Section 4. This section is based on [Me].

Chapter 17

Section 1. Smale gives the general description of a construction of hyperbolic attractors from expanding maps in Section I.9 of [Sm5]. He calls these "DE" for "derived from expanding". We follow the specific description in Katok's article in [AKK]. Williams suggested a more general construction based on branched coverings [Wil] which in particular gives attractors that are not locally homeomorphic to a product of a Cantor set and a manifold.

Attractors that are not hyperbolic sets but exhibit some nonuniformly hyperbolic behavior seem to appear in numerical studies of various models. The most well known of these studies are by Lorenz and Hénon. Such objects are known as strange attractors. Numerous efforts have been made to prove their existence rigorously. A major result in this direction is by Benedicks and Carleson [BC] who considered two-dimensional Hénon maps as a perturbation of the one-dimensional quadratic maps and showed for a set of parameters of positive measure in a family of two-dimensional quadratic maps the existence of an attractor with nonuniformly hyperbolic behavior. This extends results by Jacobson and others discussed in the notes for Section 16.2. We refer to [Rob2] for an introduction and survey of results on the Lorenz and other strange attractors.

Section 2. **(1)** DA maps were introduced in [Sm5]. Our presentation follows [PMe].

(2) Plykin's original construction [Ply] of a hyperbolic attractor on S^2 is direct and does not use the projection from the torus. Later one-dimensional attractors of maps on surfaces were studied extensively by Aranson, Grines, and their students.

Section 3. The construction of Anosov diffeomorphisms on nilmanifolds is due to Smale [Sm5], who acknowledges some contribution by A. Borel.

It is conjectured that every Anosov diffeomorphism of a compact manifold has a lift to a finite cover that is topologically conjugate to an automorphism of a nilmanifold. For the case of manifolds that are finite factors of nilmanifolds this has been proven by Franks and Manning [Fr1], [Fr2], [Man1]. For Anosov diffeomorphisms of tori we prove this in Theorem 18.6.1. Another case of complete topological classification is that of Anosov diffeomorphisms with one-dimensional stable or unstable foliation. Those maps are topologically conjugate to toral automorphisms [Fr2], [New1]. Finally, in recent work of Benoist and Labourie it was shown that an Anosov diffeomorphism with smooth stable and unstable foliations that preserves an affine connection (for example, symplectic ones) is smoothly conjugate to a finite factor of an automorphism of a nilmanifold [BL].

Section 5. An algebraic approach to the study of geodesic flows on manifolds of constant negative curvature was initiated by Gelfand and Fomin who used the theory of infinite-dimensional unitary group representations to obtain countable Lebesgue spectrum, which is a stronger property than mixing [GF].

Section 6. The Riemannian-geometry background for this section can be found in any advanced text on the subject. The book [Kl] by Klingenberg is particularly recommended.

The study of the global behavior of geodesic flows on manifolds of negative sectional curvature was a prime motivation for introducing the notion of global hyperbolic behavior. Ergodicity of the geodesic flows on negatively curved surfaces was proved by E. Hopf [Ho2]; in higher dimensions it is due to Anosov [An3]. In both cases the Hopf argument (Section 5.4) is used, but in the higher-dimensional case the foliations are not always C^1, so it cannot be adopted straightforwardly. The key step is to first show that the foliations are absolutely continuous which then allows one to use Hopf's technique [An3], [AnSin].

The geodesic flow of a Riemannian manifold may be Anosov even if the Riemannian metric is not of negative curvature. Klingenberg found necessary conditions on Riemannian metrics to have Anosov geodesic flow (for example, no conjugate points) and in turn necessary conditions on a smooth manifold to admit such metrics (which are similar to those required for the existence of a negatively curved metric). A survey of some such results and further references can be found at the end of [Kl].

Section 7. The examples discussed in this section have special significance due to the result of Benoist, Foulon, and Labourie [BFL], who showed that any contact Anosov flow with C^∞ stable and unstable foliations has a finite cover that is C^∞ conjugate (not just orbit equivalent) to the geodesic flow on a locally symmetric space. An earlier result in the three-dimensional case is due to Ghys [Gh] and important partial results and tools were developed by Kanai, Katok, and Feres. A comprehensive treatment of symmetric spaces is in [Hel], which in particular contains all necessary background for the discussion in this section.

(1) See [Hel], Chapter 10, Table 5.

(2) See [HT1], [Fri], [Go]. A particularly striking example due to Franks and Williams [FrW] shows that the nonwandering set of an Anosov flow may not be the entire manifold.

Section 8. The classical work of Fatou [Fa] and Julia [Ju] remains the foundation of complex dynamics. After a long period of neglect interest in the subject was revived in the sixties by the works of Brolin, Jacobson, and Guckenheimer, who brought the modern perspectives of hyperbolic dynamics and ergodic theory to the subject. From the seventies the subject experienced explosive growth in quantity and quality due to the refinement of dynamical tools and increased interest from analysts who brought their powerful methods and insights to bear. We make no attempt to survey even just the major sources. The survey article by Blanchard [Bl] may serve as both an introduction and a guide to the state of the field by the mid-eighties. Major contributions to the subject were made by Sullivan, Douady and Hubbard, Herman, Yoccoz, Mañé, Lyubich, Milnor, and others. In particular Douady and Hubbard developed a detailed theory of quadratic maps whose centerpiece is a careful analysis of the Mandelbrot set.

Chapter 18

Sections 1 and 2. The Shadowing Theorem in this general form is due to Anosov [An5]. Its proof was published in Katok's paper in [AKK]. We follow this account for this result and its immediate corollaries.

Section 3. Spectral decomposition was obtained by Smale [Sm5]. Since Smale's original goal was a global topological classification of "typical" or "good" dynamical systems rather than a semilocal analysis his standing assumption was "Axiom A", that the set of nonwandering points is hyperbolic and periodic points are dense in it. In fact hyperbolic sets are more useful for semilocal analysis since they appear in many dynamical systems including some from classical mechanics that do not possess any global hyperbolic structure. Anosov suggested a semilocal version of spectral decomposition [An5] which we basically follow.

In the case of flows there is a basic dichotomy first found by Anosov in the case of Anosov flows: A topologically transitive component of the spectral decomposition is either topologically mixing or has a global section with constant return time, that is, is a suspension of a homeomorphism.

Specification was introduced by Bowen who proved the Specification Theorem in [Bo3].

Section 4. Our exposition of the local product structure follows that of Katok in [AKK].

Section 5. In the proof of Theorem 18.5.5 we follow Bowen's axiomatic approach [Bo6] which appears prominently in Chapter 20.

Section 6. This section is an adaptation of the work of Franks [Fr1], [Fr2] and Manning [Man1] for nilmanifolds (see the notes to Section 17.3). Our version is streamlined; in particular it avoids the use of Markov partitions. The first lemma is proved using an idea due to Anosov.

Section 7. For the original construction of Markov partitions in the discrete-time case see [Sin4] (for Anosov diffeomorphisms) and [Bo2] for compact locally maximal hyperbolic sets. The construction for flows was carried out independently by Bowen [Bo5] and Ratner [Ra]. Markov partitions are a very powerful tool because they allow one to reduce calculations to the symbolic situation where precise tools and results are available. See [Sin3], [Sin5], [Rue5], and [PP] for numerous applications of this method. Our account of existence of Markov partitions closely follows [Bo7].

Chapter 19

Section 1. The first result on Hölder continuity of structures related to a hyperbolic system is Anosov's proof of Hölder continuity of the stable and unstable foliations in [An4]. Hölder continuity of conjugacies for maps and flows seems to have been common knowledge among experts for a long time but we do not know an original reference.

E. Hopf [Ho2] proved that for geodesic flows on surfaces of negative curvature the stable and unstable foliations are C^1. Anosov (Section 24 of [An3]) constructed examples of Anosov systems whose foliations are not differentiable. Hirsch and Pugh [HP] showed that the stable and unstable foliations for a geodesic flow are C^1 if the sectional curvature is strictly between -4 and -1.

This latter assumption corresponds to (19.1.1) with $\alpha > 1$. Hurder and Katok [HK], following an observation by Anosov who found obstructions to C^2 differentiability in the case of codimension-one foliations, found that the precise regularity of the weak stable and unstable foliations for volume-preserving Anosov flows in dimension three and area-preserving Anosov diffeomorphisms in dimension two is $C^{1+O(x \log x)}$, that is, these foliations are differentiable and their derivatives in smooth local coordinates have a modulus of continuity that is $O(x \log x)$. This regularity is exactly the same for all such systems except for those smoothly orbit equivalent to a linear map or a geodesic flow in constant curvature, correspondingly. In general, a bunching assumption guarantees a certain Hölder exponent for the unstable distribution or its derivative in local smooth coordinates and generically this degree of regularity is not exceeded [Has].

Our proofs are kept elementary by using coordinate calculations with conveniently chosen notation. The proof of Hölder continuity of the distributions is adapted accordingly from Anosov's original ideas.

(1) In the proof of Theorem 6.2.8 we showed that the graph transform is a contraction and it is not hard to see that it preserves a Hölder property by arguing similarly to the proof of Theorem 19.1.6. The only difficulty in the proof is that the global extension used in the proof of Theorem 6.2.8 does not provide genuine foliations coming from the hyperbolic set. Thus one can consider a Hölder field of submanifolds inside horizontal cones (such as obtained from the stable distribution under the exponential map) and show that the action of the corresponding graph transform preserves a Hölder condition.

(2) It is not hard to see that for geodesic flows pinching of sectional curvature implies bunching of the geodesic flow. Specifically, if the sectional curvature is between -1 and $-\alpha^2/4$ then the geodesic flow is $\acute{\alpha}$-bunched.

(3) The result of this exercise holds in the smooth category and without the perturbation assumption. To sees this one replaces in the proof the Arnold Theorem 12.3.1 with the global conjugacy result of Herman and Yoccoz ([Her1], [Y]).

Section 2. The Livschitz Theorem appeared originally in [Liv]. It immediately became a major tool in hyperbolic dynamics. Sinai used it to characterize functions that generate the same Gibbs measure [Sin5] (see our Proposition 20.3.10). Another early application [LivSin] shows that for an Anosov diffeomorphism $Jf^n = 1$ for every $x \in \text{Fix}(f^n)$ is a necessary and sufficient condition for existence of an absolutely continuous invariant measure. In addition to the Livschitz Theorem it uses a version of the Hopf argument. This is a stronger assertion than our Theorem 19.2.7.

(1) In fact there are higher regularity results. Livschitz [Liv] proved C^∞ regularity for hyperbolic automorphisms of the torus, Guillemin and Kazhdan [GK] for geodesic flows on negatively curved surfaces, and de la Llave, Marco, and Moriyon [LlMM] in full generality. They also found applications to the smooth classification of various classes of Anosov systems. Other proofs were found by Journé and by Hurder and Katok [HK].

Real-analytic regularity results were obtained by de la Llave. At the other end of the spectrum the Hölder condition can be relaxed to a modulus of continuity whose values over a geometric progression are summable. On the other hand there are counterexamples for continuous cocycles.

Chapter 20

Section 1. Again we follow Bowen's approach [Bo6].

Section 2. Our proof of the Variational Principle follows Walters [W] who in turn uses the approach by Misiurewicz [Mis1].

(1) See [Rue1], [Rue5].

Section 3. We follow Bowen's approach [Bo6] but establish a two-sided estimate in Lemma 20.3.5, which allows us to show in Theorem 20.3.7 that equilibrium states are mixing. In fact equilibrium states satisfy even stronger stochastic properties than mixing. In the framework of Bowen's axiomatic approach (expansiveness, specification, $\varphi \in C^f(X)$) Ledrappier [Led] establishes the K-property. An even stronger Bernoulli property, that is, metric isomorphism to a Bernoulli shift, was established by Bowen in [Bo7] for equilibrium states corresponding to Hölder-continuous functions on a compact locally maximal hyperbolic set.

Section 4. **(1)** This is due to Sinai [Sin5] who used Markov partitions in his proof.

(2) See [LlMM] or [HK]. The same regularity results that are required here are used to show regularity of the solution of the cohomological equation in Livschitz theory (see the notes to Section 19.2).

Section 5. We follow the original paper [Mar] by Margulis.

Section 6. This section reproduces arguments of C. Toll [To]. As we already pointed out, in the discrete-time case the error term in the asymptotic orbit growth is known to be of lower exponential order (Theorem 20.1.6). For geodesic flows in constant negative curvature this is also true due to the Selberg trace formula which has no counterpart in the general case. For geodesic flows on compact manifolds of variable negative curvature whose horosheric foliations are C^1 (in particular surfaces; see the notes to Section 19.1) Dolgopyat [Do,DoP] recently obtained an asymptotic $\mathrm{li}(e^{hT})+O(e^{(h-\epsilon)T})$ as $T \to +\infty$, where $\mathrm{li}(x) = \int_2^x 1/\log u\,du \sim x/\log x$. There are examples due to Parry of special flows over symbolic systems built under a Hölder-continuous function for which the error term cannot be estimated exponentially. This question is intimately related with the behavior of the zeta function near the critical line $\mathrm{Re}(s) = h$ [PP].

Hints and answers to the exercises

1.1.2. Consider $\varphi(x) := d(f^2(x), f(x))/d(f(x), x)$. Example: $f(x) = x - x^3$ on $[-1/2, 1/2]$.

1.1.3. If there is a vector v at p such that $\|Dfv\| \geq \|v\|$ consider a curve tangent to v and use the definition of Df.

1.1.5. Show that it is not surjective by considering two points at maximal distance.

1.2.4. Consider Jordan blocks. $k+1$ is the maximal size of a Jordan block.

1.3.1. If m_n is the number given by the first k digits of 2^n, $\{\cdot\}$ is the fractional part, and $\lg = \log_{10}$ is the logarithm to base 10 then $\lg(m_n/10^{k-1}) \leq \{n \lg 2\} \leq \lg((m_n+1)/10^{k-1})$.

1.3.3. For transitivity it suffices to show 1 is in the orbit closure. Every $g \in \mathbb{Z}_2 \setminus \mathbb{Z}_2^+$ is a limit of odd integers. For m odd and $n \in \mathbb{N}$ there exists $k \in \mathbb{N}$ such that $mk = 1 \pmod{2^n}$.

1.4.1. $C(0)$ is a closed subgroup.

1.4.2. Prove a one-sided version of Lemma 1.4.2 and show that for open $U \neq \varnothing$, $N \in \mathbb{N}$ there exists $n > N$ such that $f^{-n}(U) \cap U \neq \varnothing$.

1.4.4. Use the method of proof of Proposition 1.4.1.

1.6.1. See the proof of Proposition 1.6.4.

1.6.2. Take $f(r,\theta) = \begin{cases} e^{1/(r^2-1)}, & r < 1, \\ 0, & r = 1, \\ e^{-1/(r^2-1)} \sin(1/(r-1) - \theta), & r > 1 \end{cases}$ in polar coordinates (r,θ) on $\mathbb{R}^2$. Use this model to construct a function on S^2.

1.6.3. Such a function has one minimum, one maximum and one multiple saddle. There are $2g+1$ orbits each connecting the minimum (maximum) with the saddle. One way of constructing such functions is to start with the height function on the vertical multiple torus. This function has $2g$ saddles connected in sequence by two orbits. From each of these pairs pick one orbit and on a small neighborhood of the union of these orbits change the function to obtain a single multiple saddle.

1.7.1. $|m^n - 1|$.

1.7.2. (1) 2^n; (2) two fixed points, no other periodic points; (3) divide $f_\lambda(f_\lambda(x)) - x$ by $x - 1$ and $x - x_\lambda$, where x_λ is the nonzero fixed point; (4) show that the period-two orbit is attracting.

1.7.5. Write a string consisting of all possible sequences of zeros and ones without two successive zeros of length 1, 2, 3, … . Insert ones where necessary.

1.7.6. The representation of x in base three can be obtained as follows. Write a string consisting of all possible sequences of zeros and twos of length 1, 2, 3, … . At the end of the collection of blocks of length n insert a block consisting of a one followed by n zeros.

1.8.3. Let p be any number relatively prime to $\det L$. Consider the finite subgroup $p^{-1}\mathbb{Z}^m/\mathbb{Z}^m \subset \mathbb{T}^m$. Show that F_L is invertible on this subgroup, so that all its elements are periodic points.

1.9.9. Let N be the greatest common divisor of lengths of cycles (sequences beginning and ending at the same symbol) and identify two symbols if they are connected by a path whose length is a multiple of N. Let Λ_i be the equivalence classes. For mixing assume without loss of generality that $N = 1$.

1.9.10. Partial sums of ω give (in a nonunique way) sequences that vary within $[0,k]$ up to translation. To disprove mixing consider sequences for which the bound k is attained.

1.9.11. Set $a_{04} = a_{15} = a_{24} = a_{35} = a_{42} = a_{43} = a_{50} = a_{51} = 1$, all other entries zero, define H by $0 \mapsto 00, 1 \mapsto 01, 2 \mapsto 10, 3 \mapsto 11, 4 \mapsto 01, 5 \mapsto 10$. Only the orbit of $\ldots 101010 \ldots$ has more than one preimage.

1.9.12. Show from the representation of Exercise 1.9.10 that S_2^2 is not transitive but B_2 is the union of two transitive invariant sets whose intersection is a period-2 orbit of S_2. Show that this is incompatible with spectral decomposition.

2.1.2. Compare the speed of convergence of orbits to the origin.

2.1.3. (2) Use the method of the previous exercise.

2.1.5. (7) refer to Exercise 1.7.2(3), (4) and use (3).

2.1.7. Use the proof of Proposition 2.1.3.

2.1.8. Find an analytic h such that $h \circ \varphi^1 = \Phi^1 \circ h$ and set $H = \int_0^1 \varphi^{-t} \circ h \circ \varphi^t dt$.

2.1.9. To see impossibility of C^2 linearization expand the conjugacy to second order and find a contradiction. To find a C^1 linearization first find a C^1 function $y = \varphi(x)$ with $\varphi'(0) = 0$ whose graph is invariant under the map. Then take $h(x,y) = (x, y + \varphi(x))$.

2.1.10. Find a fixed point.

2.2.1. Lift the flows and the conjugacy to the universal cover $\mathbb{R}^n$.

2.2.2. In (1) one needs to specify the function only at finitely many points. For (2) one first finds a trigonometric polynomial taking distinct values at the relevant points, and then a positive interpolation polynomial which maps these values to the desired ones. The composition is a trigonometric polynomial.

2.2.3. Take $h(x,t) = \sigma_{f,\psi}^{t+\Phi(x)}(x,t)$.

2.3.1. Take a nonisolated periodic point p of period n and show, using Proposition 1.1.4, that there is a C^1 perturbation of f with only finitely many points of period n near p. For the details see Lemma 7.2.7.

2.3.2. Construct smooth perturbations both with isolated and nonisolated periodic points of period n or $2n$.

2.3.3. Use the argument of Proposition 2.1.7.

2.3.4. First define the conjugacy at the equator.

2.4.1. Repeat the first proof of Theorem 2.4.6.

2.4.2. Take $h(x) = \sin^2(\pi x/2)$.

2.4.4. Use Proposition 2.1.3, Lemma 2.1.4, and Lemma 2.4.10.

2.4.5. Using the notation from the proof of Theorem 2.4.6 take $x = a_p^m$. Then $F^p(x) = m$ and $F^{n+p}(x) = F^n(m) = mk^n$, so $\lim_{n\to\infty} F^{n+p}(x)/k^{n+p} = m/k^p = h(a_p^m)$, that is, $\lim F^n(x)/k^n$ and $h(x)$ coincide for all a_p^m. Being monotone continuous they coincide.

2.4.6. An arbitrary Riemannian metric is given by $ds^2 = g(x)\,dx^2$, where g is a positive periodic function.

2.6.1. Use $E^{\pm}$ instead of eigendirections and linear maps instead of eigenvalues.

2.6.2. Use the construction from the proofs of Theorems 2.6.1 and 2.6.3, replacing the eigendirections of the matrix h by the horizontal and vertical directions in $\mathbb{R}^2$.

2.7.1. Let $\mathcal{F}(f,h) = f \circ h$, $w_{n+1} = -(D_2\mathcal{F}(g,\mathrm{Id}))^{-1}(f_n - g)$, $h_{n+1} = h \circ (\mathrm{Id} + w_{n+1})$, and $f_{n+1} = f \circ h_{n+1}$.

2.8.1. Cover the complement by a countable union of intervals with arbitrarily small sum of lengths.

2.8.3. Construct $\alpha_n \to \alpha$ inductively such that infinitely many coefficients in (2.8.5) are bounded away from 0.

2.9.2. Use Exercises 1.9.11 and 1.9.12. The discontinuities arise at the points where the semiconjugacy of (1.9.6) is not injective.

2.9.3. Any two solutions differ by an invariant function. Show that $\Phi - \varphi$ is not constant, where φ is the solution given by (2.9.5), and use topological transitivity of the shift.

3.1.1. Periodic points correspond to integer translations under the lift. Use the degree to get at least $m^n - 1$ points and the attracting point to get two more.

3.1.2. For $N, m \in \mathbb{N}$ construct a topological Markov chain σ with $p(\sigma) = \log N/m$ from σ_N. Use these numbers to approximate t and use countable unions and compactification to get the desired system.

3.1.3. Use Corollary 1.9.5.

3.1.4. No.

3.1.5. $\log((3+\sqrt{5})/2)$.

3.1.6. Lift the flow to the universal cover and use the fact that nonhomotopic curves on X lift to curves that cannot be close to each other. Deduce that their projections are separated on X. Finally take into account the multiplicity that may appear due to the added short arcs.

3.1.7. If $\{A_i\}$ is a minimal subcover of $\mathfrak{A}$ and $\{B_j\}$ a minimal subcover of $\mathfrak{B}$ then $\{A_i \cap B_j\}$ is a subcover of $\mathfrak{A} \vee \mathfrak{B}$.

3.1.8. Note that $N(\mathfrak{A}) \leq N(f^{-1}(\mathfrak{A}))$; let $a_n = \log N(\mathfrak{A} \vee \cdots \vee f^{-n}(\mathfrak{A}))$ and show that $a_{n+m} \leq a_n + a_m$ to prove convergence.

3.1.9. If $\mathfrak{A}$ is an ϵ-cover with respect to d then $\mathfrak{A}_n = \mathfrak{A} \vee \cdots \vee f^{-n}(\mathfrak{A})$ is a d_n ϵ-cover and $D(f,n,\epsilon) \leq N(\mathfrak{A}_n)$, so $\tilde{h}(f,\epsilon) \leq h(f,\mathfrak{A})$ whence $\tilde{h}(f,\epsilon) \leq \sup_{\mathfrak{A}} h(f,\mathfrak{A})$.

For the other direction let $\mathfrak{A}_\epsilon$ be the covering of X by all ϵ-balls. Then $N(\bigvee_{i=0}^{1-n} f^i(\mathfrak{A}_\epsilon)) \leq S_d(f,\epsilon,n)$ and hence

$$h(f,\mathfrak{A}_\epsilon) \leq h_d(f,\epsilon).$$

On the other hand for a given cover $\mathfrak{A}$ let δ be less than the Lebesgue number of $\mathfrak{A}$. This means that every δ-ball is contained in an element of $\mathfrak{A}$. Then obviously

$$h(f,\mathfrak{A}) \leq h(f,\mathfrak{A}_\delta).$$

These two inequalities imply that $h(f,\mathfrak{A}) \leq h(f)$.

3.2.1. Show that the cardinality of minimal (n,ϵ)-spanning sets grows linearly in $n \in \mathbb{N}$.

3.2.2. Find an (n,ϵ_0)-separated set of preimages of a given point consisting of $|\deg(f^n)|$ points.

3.2.3. Divide S^1 into intervals where the variation of f^k $(k = 1,\dots,n)$ is less than ϵ. The number of such intervals is approximately $|\deg(f^n)|/\epsilon$.

3.2.4. The Markov matrix is $\begin{pmatrix} 0 & 1 & 0 \\ 1 & 0 & 1 \\ 1 & 0 & 0 \end{pmatrix}$, so the entropy is the logarithm of the positive root of $x^3 - x - 1$.

3.2.5. Use the result of Problem 1.9.11 and show that the semiconjugacy H does not decrease entropy by using Proposition 3.2.14.

3.2.6. Zero. Use Proposition 3.1.7 and Corollary 3.2.10.

3.2.7. Take a small disc around the origin in the subspace E^+ for L at the origin (see (1.2.5)) and try to find enough (n, ϵ)-separated points inside that disc.

3.2.8. $\begin{pmatrix} 2 & 1 & 0 & 0 \\ 1 & 1 & 0 & 0 \\ 0 & 0 & 2 & 1 \\ 0 & 0 & 1 & 1 \end{pmatrix}$; for the proof use the previous exercise.

3.2.10. Compare with the proof of Theorem 3.2.9.

3.3.1. Consider a point with a dense positive semiorbit and a tail of zeros.

3.3.2. In Ω_2 define *blocks of rank k* inductvely as follows: A block of rank 0 is the single digit 1. A block of rank k is a string consisting of two copies of a block of rank $k-1$ separated by 2^N zeros, where N is at least the size of the $(k-1)$-blocks used. Now define $A_k \subset \Omega_2$ to be the set of sequences consisting of only zeros except for a single block of rank at most k. Show that $NW(\sigma_{\restriction A_k}) = A_{k-1}$.

3.3.3. Consider the union of examples from the previous exercise and compactify.

3.3.4. Consider the Hausdorff metric $d(\cdot,\cdot)$ (Definition 13.2.1) on closed sets and use Lemma 13.2.3 as follows: If B is closed and invariant let $m(B) = \max\{d(A,B) \mid A \subset B \text{ closed invariant}\}$. Take M such that $m(M) = \min m$. Let us show that M has no proper closed invariant subsets. Otherwise $m_0 := \min m > 0$. Take a closed invariant $M_1 \subset M$ such that $d(M_1, M) = m_0$. By assumption M_1 is not minimal and contains M_2 such that $d(M_2, M_1) \geq m_0$ and hence $d(M_2, M) > m_0$. We continue this process to obtain a sequence M_i such that $d(M_i, M_j) \geq m_0$ contradicting compactness with respect to the Hausdorff metric. (This is different from the proof suggested by Simpson.)

3.3.5. Consider a time change for an irrational linear flow on the torus.

3.3.6. There are points of arbitrarily high period since $\text{Fix}(f^n)$ is always closed. By connectedness $\partial\,\text{Fix}(f^n) \neq \varnothing$ and near every $x \in \partial\,\text{Fix}(f^n)$ there are points of arbitrarily high period. Use this inductively to construct a convergent sequence of points of increasing periods such that an increasing number of iterates stay apart for almost all members of the sequence. Take the limit of this sequence.

4.1.3. Alternate blocks of 0 and 1 of length 1, 2, ...

4.1.7. The averages for almost every (hence some) boundary point converge to $\mu(N)$. But N is nowhere dense, so a residual (hence dense) set of points misses N altogether.

4.1.9. Multiply the linear vector field by a real-analytic function with a single zero of multiplicity at least two. Assume that the new flow has a finite nonatomic Borel probability measure and obtain a contradiction with unique ergodicity of the linear flow.

4.1.10. Consider the set $X_n := \{x \mid k \cdot |T^k x - x| > 1 + \epsilon \text{ for } k = 1, \dots, n\}$ and show that for every interval J of length δ the Lebesgue measure of $J \cap X_n$ is less than $\dfrac{\delta}{(1+\epsilon)n}$.

4.1.12. Consider the absolute value of an eigenfunction, and then consider the ratio of two eigenfunctions for an eigenvalue.

4.1.13. First show that the eigenfunctions form a group, then use the previous exercise.

4.2.1. Use Fourier analysis.

4.2.3. Use Fourier analysis as in Proposition 4.2.2.

4.2.4. Use the fact that A_α commutes with the vertical translation $T_{(0,t)}$ and apply the method of Proposition 4.2.3.

4.2.5. Reduce the question to that of studying the distribution of the second coordinate for an appropriate initial condition in the map A_α. Use the result of the previous exercise and Corollary 4.1.14.

4.2.6. Use the methods of Exercises 4.2.3 and 4.2.4 and induction on m.

4.2.7. Use the method of Exercise 4.2.5 and the result of Exercise 4.2.6.

4.2.9. Use the method of proof of Exercise 2.4.2.

4.2.10. Find an injective Lebesgue-measure-preserving correspondence between $I_{\alpha,\beta}$ and the first-return map (cf. Exercise 4.1.4) induced by a certain rotation on a certain interval $I \subset S^1$.

4.2.11. Use the method of proof of Proposition 4.2.11(2).

4.2.14. Take the intersection of the space of invariant vectors for Π with the simplex σ and show that the extreme points of the resulting simplex correspond to ergodic measures by reducing to the transitive case (Proposition 4.2.14).

4.2.15. Compare with Exercise 1.9.9.

4.2.16. Use the fact that one is the maximal eigenvalue of Π.

4.3.3. Show that for every n the partition ξ^T_{-n} has k^{n+1} elements, all of equal measure.

4.4.1. Take $\xi = \{A, \mathbb{T}^2 \smallsetminus A\}$, where $A = [0,1/2] \times [0,1/2]$, and use minimality and the fact that iterates are isometries to approximate arbitrarily fine partitions.

4.4.2. Generalize the proof of Proposition 4.4.2. Use induction on the rank of cylinders.

4.4.3. They have different entropy.

4.4.4. Use the representation of σ_N as a group automorphism as described in Section 4.2f and the fact that the uniform Bernoulli measure is Haar and consider the action of the associated unitary operator U_{σ_N} on characters.

4.4.5. Modify the method of the previous exercise by orthogonalizing the set of characters with respect to the new measure.

4.4.7. Express the sides of the rectangles of the Markov partition in terms of eigenvalues and eigenvectors of L. Use the fact that the nonzero eigenvalues of L and of the Markov matrix are the same.

4.5.1. Use the method of Exercise 3.1.2.

4.5.2. Take $X = \{a_n\}_{n\in\mathbb{Z}} \cup \{0\} \cup \{1\}$ with $a_n \to 1$ as $n \to \infty$, $a_n \to 0$ as $n \to -\infty$, $f(a_n) = a_{n+1}$, and 0 and 1 fixed.

5.1.2. Use a partition of unity.

5.1.5. Use the previous exercise.

5.1.6. Show that Theorem 5.1.15 applies by using the method of Theorem 5.1.16.

5.2.2. The motion relative to the center of gravity looks like two independent central force problems, so the orbits ar ellipses.

5.2.3. The integral is the momentum with respect to the vertical axis, that is, the third coordinate of the angular momentum as in Subsection d. To describe the motion use spherical (geographic) coordinates with the singularity on the vertical axis.

5.4.2. Find a Möbius transformation that maps p and q to points symmetric with respect to the imaginary axis.

5.4.3. Conjugate by a Möbius transformation so that the axis is the imaginary axis and obtain the answer $\log\left(\left(\operatorname{tr}\begin{pmatrix} a & b \\ c & d \end{pmatrix}\right)^2 - 4\right)$.

5.4.4. The conjugacy has to take the axis of f to that of g, z_1 to z_2, and the direction from z_1 to $f(z_1)$ to that from z_2 to $g(z_2)$.

5.4.5. Map the geodesic to the imaginary axis. The desired curves are lines through the origin. In general the curves, called equidistants, are circular arcs connecting the endpoints of the axis and making equal angles with it.

5.4.6. Map the axis to the imaginary axis and use the fact that Möbius transformations preserve the cross ratio.

5.4.8. Use the fact that any such isometry sends any closed geodesic to a closed geodesic of the same length and the previous exercise together with the fact that any Möbius transformation is uniquely determined by the images of three points on $\partial\mathbb{D}$.

5.4.9. Consider the growth of distances on curves represented by horizontal lines as in Subsection d and use the method of Exercise 3.2.7.

5.5.3. Decompose into blocks $\begin{pmatrix} \lambda & 0 \\ 0 & 1/\lambda \end{pmatrix}$ for real λ, rotations R_α for $\lambda = e^{i\alpha}$, and $\begin{pmatrix} \rho R_\alpha & 0 \\ 0 & \rho^{-1}R_{-\alpha} \end{pmatrix}$ for $\lambda = \rho e^{i\alpha}$.

5.5.4. ω^n is a volume and exterior multiplication on forms induces a multiplicative structure on cohomology, hence the second cohomology of ω is nonzero.

5.5.5. Use the previous exercise.

5.5.6. Use the method of proof of the Moser Theorem 5.1.27 and the Darboux Theorem 5.5.9.

5.5.7. The Hamiltonian is translation invariant. Translations come from flows of Hamiltonians f corresponding to constant force.

5.5.8. Use rotational symmetry. The integral obtained is angular momentum. Independence can be seen by studying how the integral depends on momenta.

5.5.10. To show that $\omega(v, w)$ depends only on the projection of v and w use that the projections are along the flow, hence invariant, and $\omega(X, X_H) = 0$ for every $X \in TM_c$.

5.5.11. For $n = 2$ a geodesic is an oriented great circle and hence identified with an oriented plane which in turn is defined by a unit vector (a positive normal). The space of these is S^2. By rotational symmetry the volume is the standard one. Alternatively take a single great circle together with the unit tangent vectors pointing into one complementary hemisphere as a transversal and compactify by the two tangent directions to again get a sphere.

5.6.2. $\theta = (1/2)\sum_{i=1}^{n}(p_i dq_i - q_i dp_i)$, $v = -(-q, p)$.

5.6.3. Describe v by $dz/dt = iz$.

5.6.5. To preserve θ a vector field has to be Hamiltonian with homogeneous Hamiltonian. To preserve $S\mathbb{T}^n$ the Hamiltonian has to be independent of the configuration coordinate q. Thus these are the Hamiltonians whose action–angle variables coincide with those for the geodesic flow.

6.1.1. See Exercise 1.1.3

6.2.1. Use (1.2.4) and (1.2.5).

6.2.3. (1) Write a differential equation for these functions. (2) Pick $x > 0$ and a C^∞ function φ on $[x', x]$, where $f(x, 0) = (x', 0)$, such that all derivatives of φ and $\varphi \circ f$ coincide at x'. From $\varphi \circ f^n$ and $\varphi \circ (-f^n)$ obtain a C^∞ function. (Smoothness at zero is related to (1).)

6.2.4. Consider $\omega(Df^n v, Df^n w)$ and use invariance of ω and shrinking of v, w to obtain zero. Likewise the stable and unstable subspaces in the general case are isotropic, that is, ω vanishes on them.

6.2.5. Use the time-one map for (1) and a Poincaré map for (2).

6.3.1. By Proposition 1.1.2 there is a fixed point. It is contracting. Use the proof of Theorem 6.3.1 to get a *global* conjugacy.

6.4.2. Let $\langle v,w\rangle'_x = \sum_{n=0}^{\infty}\langle Df^n v, Df^n w\rangle$ for $v,w \in E_x^-$, $\langle v,w\rangle'_x = \sum_{n=0}^{\infty}\langle Df^{-n}v, Df^{-n}w\rangle$ for $v,w \in E_x^+$, and $\langle v,w\rangle'_x = 0$ for $v \in E_x^+, w \in E_x^-$. Prove that $\langle\cdot,\cdot\rangle'_x$ is a continuous scalar product and for the norm $\|v\|'_x := \sqrt{\langle v,v\rangle}$ there is a $\lambda' \in (0,1)$ such that $\|Df^{\mp 1}\|'_x < \lambda'\|v\|'_x$. Now approximate $\langle\cdot,\cdot\rangle'$ by a smooth Riemannian metric.

6.4.3. This is a nonstationary version of Exercise 6.2.4.

6.4.5. Look for an appropriate invariant subset of the horseshoe using the topological conjugacy with the full shift: Consider sequences with at most a single 1.

6.4.7. Use the conjugacy with the full 2-shift by finding, for example, a subshift conjugate to an adding machine (see Exercise 1.3.3).

6.4.8. Use the conjugacy with the full 2-shift and consider an n-step topological Markov chain (see Definition 1.9.10). Compare with Exercise 18.2.2.

6.4.9. Use the Markov partition from Section 2.5d and the previous exercise. Control the identifications in the coding by noting that only some periodic points have no image or preimage in the interior of a rectangle.

6.5.1. Take the box from the proof of Theorem 6.5.5 that contains the horseshoe and carefully choose small rectangular boxes around the iterates of q. Consider the invariant set inside the union.

6.5.2. Using Proposition 6.2.23 produce and use heteroclinic oscillations analogous to the homoclinic oscillations sketched in Subsection b and used in the proof of Theorem 6.5.5.

6.5.3. Consider the time-one map for the gradient flow on the tilted torus (Section 1.6).

6.6.1. Show that there is no invariant C^2 curve tangent to the x-axis.

6.6.2. Describe the resonance terms that cannot be eliminated from the commutation relation.

6.6.3. Note that only the formal part needs any work and that there are no resonances.

6.6.4. Eliminate nonresonance terms and use preservation of area to control resonance terms.

6.6.5. Use Theorem 6.6.5.

7.1.3. Take $\alpha = (1-c)/(1+2(1-c)) < 1/3$ and construct a Cantor set C_α by the same process as the ternary Cantor set but deleting the middle (α^k)th at each stage.

7.1.7. Use Fourier analysis to produce a formal solution. Analyze its properties using the two previous exercises. See also Exercise 14.2.1 concerning time change and Exercise 2.8.3.

7.1.8. First show that for fixed ϵ the complement is open and dense.

7.1.9. Show that for a given interval the set of functions that are not monotone are open and dense. Then consider intervals with rational endpoints.

7.1.10. First show that generically there are no isolated fixed points. Notice that any homeomorphism of $[0,1]$ has at least one fixed point.

7.2.1. Show the contrapositive.

7.2.6. First note that it suffices to consider f on a small neighborhood of 0 in $\mathbb{R}^n$ with 0 as nontransverse fixed point. Perturb in an eigendirection of $Df_{|0}$ corresponding to eigenvalue 1. This does not work in C^2: Consider $f(x) = x + x^2$ on $\mathbb{R}$.

7.2.8. Vectors in the intersection can not expand in either direction.

7.2.9. In Proposition 7.2.9 one needs Kupka–Smale of order 2.

7.2.11. Consider small disks disjoint from the critical points and intersecting the saddle connection. Describe a deformation of the torus that breaks the symmetry around the plane of the saddle connection so as to break the saddle connection. If this deformation is carried out such that no point changes height then we have deformed the metric (the embedding) without changing the height function.

7.3.1. See also Exercise 1.7.2(3).

7.3.2. See the hint for Exercise 1.7.2(3).

7.3.3. Follow the general outline of the proof of Proposition 7.3.3.

7.3.4. Consider the return map to the positive x-axis and use an appropriate version of the analysis for one-dimensional maps.

8.1.1. Notice that the proof of Theorem 8.1.1 works for manifolds with boundary. "Blow up" the points $p_1, \dots, p_n$, that is, construct a manifold N with n boundary components and a continuous map $h: N \to M$ homeomorphic on the interior of N that maps the boundary components to points. Then construct a homeomorphism $F: N \to N$ fixing all boundary points and such that $f \circ H = h \circ F$. Apply Theorem 8.1.1 to F and use the Variational Principle, Theorem 4.5.3.

8.1.2. By the previous exercise it suffices to get positive algebraic entropy on the punctured disk. Without loss of generality p, q are in a small disk D_1. Take a disk D_2 contained in $\mathbb{D}^2$ concentric with D_1 and define f on $\mathbb{D}^2$ to be a diffeomorphism that exchanges p and q by rotating D_1 rigidly by π and leaving $\mathbb{D}^2 \setminus D_2$ fixed. The fundamental group of $\mathbb{D}^2 \setminus \{p,q\}$ is generated by loops a, b based at $x \in \partial\mathbb{D}^2$ such that a, b coincide until they reach the midpoint between p and q and then separate and return on either side of the pair (p, q). With proper labeling we obtain $f_*(a) = b$ and $f_* b = bab^{-1}$. If a_n is the number of a's in the word representing $f_*^n(a)$ and b_n the number of b's (and b^{-1}'s) in $f_*^n(b)$ (after cancellation) then inductively $a_n = b_{n-1}$ and $b_n = b_{n-1} + a_{n-1}$. Then b_n is a Fibonacci sequence and hence grows exponentially. These arguments only depend on the homotopy class of f rel$\{p, q\}$.

8.2.1. Use (3.3.1) and Proposition 3.1.7(2).

8.2.2. Consult the proof of Proposition 3.2.2.

8.3.1. Consider the fixed points. If there is one then it is the only nonwandering point and we are done. Similarly if there are two with derivative not on the unit circle. If there are two with derivative on the unit circle then the map is conjugate to a rotation.

8.3.2. Use the definition of degree via volume and calculate entropy via ϵ-covers using the fact that f is a covering map.

8.3.3. Use Proposition 2.4.9 and the Variational Principle (Theorem 4.5.3).

8.4.1. Use the definition of degree through the number of preimages of a regular value. Reduce the problem to solving a trigonometric equation and this in turn to an algebraic equation.

8.4.2. Reduce the problem to a system of algebraic equations. Use the fact that the number of isolated solutions of such a system does not exceed the product of the degrees of the polynomials involved (cf. the proof of Theorem 7.4.1).

8.4.4. Note that $\operatorname{ind}_{\varphi^\epsilon} x_0$ is independent of such ϵ. Then use the definitions.

8.4.5. Notice that this is a multiple saddle and use the method of the example of the threefold saddle.

8.4.6. Prove that $\|v_P(x,y)\|^2 > C(x^2 + y^2)$ for some $C > 0$ and use this to show that the index is the same as that of the cubic part.

8.6.1. $1 - \deg(f)$.

8.6.2. Use the Lefschetz Fixed-Point Formula (Theorem 8.6.2).

8.6.3. If S admits an Anosov diffeomorphism there is a nonvanishing line field. Then S or a double cover has a nonvanishing vector field. This implies that $\chi(S) = 0$, so S is the torus.

8.6.4. Show that periodic points of any given period all have the same index and use the Lefschetz Fixed-Point Formula.

8.6.5. First show that the conditions of the previous exercise are satisfied. Use Corollary 6.4.19 and Theorem 8.6.2.

8.6.7. Express the trace of a matrix through its eigenvalues by taking an iterate. Reduce the problem to the case where all eigenvalues are either positive real numbers or complex numbers with an argument incommensurable with 2π and use the results of Section 1.4.

8.7.1. Use the fact that $H_k(\mathbb{T}^n, \mathbb{Q}) = \bigoplus_{i_1+\dots+i_n=k} H_{i_1}(S^1, \mathbb{Q}) \otimes \cdots \otimes H_{i_n}(S^1, \mathbb{Q})$ and $(F_A)_{*k}$ is the sum of tensor products of $(F_A)_{*1}$ to get $L(F_A) = \prod_{i=1}^n (1-\lambda_i)$. $p(F_A) = \sum_{|\lambda_i|>1} \log \lambda_i$.

8.7.2. There is an invariant vector v with rational coordinates. Take $\alpha \in \mathbb{R} \smallsetminus \mathbb{Q}$ and $g(x) = Ax + \alpha v \pmod 1$.

8.7.3. Prove first that this condition implies that f_* has an eigenvalue of absolute value greater than one. To that end assume the contrary and show that there are only finitely many possible characteristic polynomials of this kind, hence for the powers of f_*, so all zeros are roots of unity. Now use Theorem 8.7.1.

9.1.2. Argue by contradiction using the gradient flow as in the proof of Proposition 9.1.2.

9.1.3. Use the gradient flow (without assuming nondegeneracy of the extrema). To that end it is helpful to note that level sets are smooth Jordan curves and hence bound smooth disks (a smooth Jordan Curve Theorem).

9.1.4. There have to be extrema, both of index 1; since $\chi(\mathbb{T}^2) = 0$ there must be two other index -1 points. The height function shows that this is attained.

9.1.5. First construct a Morse function f that has one minimum, one maximum, and $\binom{k}{n}$ saddle points of Morse index k, $k = 1, \dots, n-1$ with the extra condition that the values of f are equal at all points of a given index and decrease with k. Then modify f in a neighborhood of its critical values by "collapsing" all points of a given index into one. It is helpful to first make a detailed argument for the case $n = 2$.

9.2.1. Locally replace the boundary by straight lines and argue that this does not change the first-order terms.

9.2.2. Use approximations by smooth strictly differentiably convex curves and compactness arguments.

9.2.3. Choose judicious approximations of the boundary by straight lines and circles to study the linear part. The result is

$$\begin{pmatrix} \frac{\partial S}{\partial s} & \frac{\partial S}{\partial \theta} \\ \frac{\partial \theta'}{\partial s} & \frac{\partial \theta'}{\partial \theta} \end{pmatrix} = \begin{pmatrix} \frac{\partial S}{\partial \theta}\kappa_s - \frac{\sin\theta}{\sin\theta'} & \frac{H}{\sin\theta'} \\ \kappa_{s'}\frac{\partial S}{\partial s} - \kappa_s & \kappa_{s'}\frac{\partial S}{\partial \theta} - 1 \end{pmatrix} = \begin{pmatrix} \frac{\kappa_s H - \sin\theta}{\sin\theta'} & \frac{H}{\sin\theta'} \\ \kappa_{s'}\frac{\kappa_s H - \sin\theta}{\sin\theta'} - \kappa_s & \frac{\kappa_{s'} H}{\sin\theta'} - 1 \end{pmatrix}.$$

9.2.4. It is equal to the radius of curvature at p. Use the previous exercise and let $\theta \to 0$.

9.2.5. To prove hyperbolicity use the explicit form of the differential from Exercise 9.2.3 and the fact that the long diameter exceeds twice the radius of curvature at the points in question. It is convenient to notice that the matrix of the differential coincides at both points. It remains to show that orbits through the foci converge to the long diameter. To that end consider the second iterate of the billiard map and observe that it is monotone on this family.

9.2.6. See the previous exercise.

9.2.7. It may be convenient to replace reflection in the straight side by passage into a reflected copy of the stadium. Take into account the reversed orientation on the mirror image. (See Figure 9.2.6.)

9.3.1. Given a twist map f_1 (which is in particular a positive-tilt map) consider the image S of a vertical line. Along S we have the angle function $\theta_1 > 0$. Now consider the effect of applying a twist map f_2. Since f_2 likewise has a positive angle function, the image of a tangent vector to S at (x, y) under f_2 has angle exceeding $\theta_2(x, y)$ since $\theta_1(f_1^{-1}(x, y)) > 0$ and f_2 preserves orientation. Notice that this proves that the composition of two positive-tilt maps is a positive tilt map.

9.3.2. Define the function $H(s, s')$ as the measure of the region used to define the generating function in the area-preserving case. Replace the use of derivatives by small finite variations. This is carried out in [K5].

9.3.3. To find an orbit of type (p, q) consider the space of nondecreasing sequences satisfying $s_{n+2q} = s_n + 1$ and consider the action functional $L(s) := \sum_{k=0}^{q}[H_1(s_{2kp}, s_{(2k+1)p}) + H_2(s_{(2k+1)p}, s_{(2k+2)p})]$, where H_1 and H_2 are the generating functions for f_1 and f_2, respectively.

9.3.6. Interpolate the generating function $H(s, s')$ of $\tilde{F}$ and the function $s - s'^2/2$ with a slowly changing weight function, keeping (9.3.3).

9.4.1. Use perturbations vanishing outside a small neighborhood of a given point $c(t_0)$ where (9.4.2) fails.

9.5.2. Consider a point on the closed geodesic. Examples: The standard sphere, the standard torus, ...

9.5.3. Minimize length in a nonzero homotopy class.

9.5.4. Use Theorem 9.5.5.

9.5.5. Let $g^{kl} := (g^{-1})_{kl}$ and define the *Christoffel symbols* by

$$\Gamma_{ij}^k = \frac{1}{2}\sum_{l=1}^{n} g^{kl}\left(\frac{\partial g_{il}}{\partial x^j} + \frac{\partial g_{jl}}{\partial x^i} - \frac{\partial g_{ij}}{\partial x^l}\right).$$

9.6.1. Calculate the volume of a ball using polar coordinates around the center. $v(M) = 1$.

9.6.2. Replace a piece of a flat metric by a rotationally symmetric "bulge" that contains more than half of a sphere.

9.7.1. Embed the Möbius strip in such a fashion that the center line forms a "ridge" which is longer than half the edge.

9.7.2. Pass to the orientable double cover and use Theorem 9.7.1.

9.7.3. Start with a flat metric and try to "pinch" it to create very short closed geodesics in the directions of all three generators, at the same time controlling the length of some "diagonals". This construction is due to Hedlund [He1].

9.7.4. The exponential map for any point of the hyperbolic plane is a diffeomorphism.

10.0.1. $f(x) - x$ cannot be positive (negative) for all x.

10.0.2. Show there is a closed interval $J \subset I$ such that $f(J) = [f(x), f^2(x)]$. Note $J \subset f^3(J)$ and use the previous exercise.

10.0.3. Inductively draw the portion of the graph of f^n in the unit square $[0, 1]^2$ to see $P_n(f) \geq 2^n$. To find periodic orbits as desired exploit the similarity to the horseshoe construction (Section 2.5c). Compare with Definition 15.1.10.

10.0.4. Obtain a one-sided 2-shift as a factor of the restriction of f to an invariant subset $\Lambda_1 \subset \Lambda$. See the proof of Theorem 15.1.5.

10.0.5. See the proof of Lemma 5.1.18.

10.0.6. First prove that in coordinates z on $S^2 \setminus \{\infty\}$ and w on $S^2 \setminus \{0\}$ f is holomorphic at every point.

10.0.7. A holomorphic map is not injective at a critical point.

11.1.2. Use Proposition 9.6.4.

11.1.3. Interpret $\frac{F^n(x) - x}{n}$ as the time average of a function on the circle. Then use the Krylov–Bogolubov Theorem 4.1.1 and the Birkhoff Ergodic Theorem 4.1.2.

11.1.5. The intersection is an interval by monotonicity. To show it is nonempty show that $\tau_{0,b} = 0$ and $\tau_{1,b} = 1$ and use continuity. To obtain positive length show that Proposition 11.1.10 applies because $f_{a,b}$ is an entire function.

11.1.6. Yes: If $\{O_1, O_2\}$ is a disjoint open cover of $A_{p/q}$ show that we may assume $\bar{O}_1 \cap \bar{O}_2 = \varnothing$. Use a compactness argument to obtain a contradiction.

11.1.7. Use Propositions 11.1.9 and 11.1.10 as in the proof of Proposition 11.1.11.

11.2.1. Use the Poincaré classification of orbits and the argument for Proposition 2.1.7.

11.2.3. Consider all possible combinations of stable, semistable, and unstable orbits.

11.2.4. f has degree -1. Consider f^2.

11.2.6. Construct a nonatomic f-invariant measure whose support is the set of periodic points. Then map it to Lebesgue measure.

11.2.7. Show that $R_\tau \circ h_1 \circ h_2^{-1} = h_1 \circ h_2^{-1} \circ R_\tau$.

11.2.8. If infinitely many orbits are blown up $f\restriction_{NW(f)}$ is not expansive. Otherwise code by two properly chosen finite unions of intervals with endpoints outside the minimal set (if at most two orbits are blown up then any such interval and its complement will do). Use minimality to show that the coding is injective.

11.2.9. Consider eigenvalues and eigenfunctions of the operator U_{R_α} (see Section 4.1g).

11.2.10. Bound the number of separated orbits using preservation of order.

12.2.1. If the Denjoy set has measure zero the invariant measure is obviously singular. Otherwise note that $f' = 1$ on the Denjoy set, so the restriction of Lebesgue measure to the Denjoy set is invariant.

12.2.2. By defining l_n appropriately arrange for the series $\sum n^{1-1/k}$ at the beginning of the proof to be replaced by the series $\sum 1/n(\log n)^{1+1/k}$.

12.5.1. Use the Baire Category Theorem A.1.22.

12.6.2. Modify the construction by ensuring that on certain sets the derivative of h_n is very close to a prescribed set of values. Then make the successive perturbations so small in measure that this is not destroyed.

12.6.3. Use Exercise 7.1.6.

12.7.1. Use the Poincaré classification.

12.7.2. Take an example from Theorem 12.5.1 or Theorem 12.6.1(1) and let $\mu = h_*\lambda$. Then apply Theorem 12.7.2.

13.1.1. Using uniform continuity of f derive existence of the "uniform twist module", that is, a function β such that for all $y_2 > y_1$ $F(x, y_2) - F(x, y_1) > \beta(y_2 - y_1)$ uniformly in $x \in \mathbb{R}$. This is presented in [K5].

13.1.2. $\omega(t) = \text{const.}\sqrt{t}$.

13.2.1. Use the Variational Principle (Theorem 4.5.3) and Theorem 11.2.9.

13.2.2. The diameters and all orbits that do not intersect the interior of the segment between the foci.

13.2.5. Use Exercises 9.2.3 and 13.1.1.

13.2.6. First prove existence of a Birkhoff periodic orbit of type (p,q) for $\tau_0 < p/q < \tau_1$. See [K5].

13.2.7. Use Proposition 13.2.21 and Exercise 13.2.6.

13.3.2. All ordered orbits are minimal except for the shorter diameter; see Exercise 13.2.2.

13.3.3. Use Corollary 13.3.8 and Proposition 13.2.21 or Exercise 13.2.3.

13.4.1. Let $y(x)$ be the common value of y^+ and y^-. Show that for x_1 and x_2 in a gap the orbits of $(x_1, y(x_1))$ and $(x_2, y(x_2))$ are properly intertwined.

13.5.1. Apply Proposition 13.5.1 in a way similar to the proof of Proposition 13.5.3.

14.1.1. Show that the periods of points are bounded, and then use the Kupka–Smale Theorem 7.2.13.

14.2.1. See Exercise 7.1.7 and show that the time-t map for *every* $t \neq 0$ is topologically transitive (cf. Definition 1.3.1).

14.2.2. Consider a closed transversal and show that the Poincaré map, if defined, is orientation reversing.

14.2.4. Use the construction of Theorem 12.6.1.

14.3.1. Use Proposition 12.2.1.

14.3.2. Use the fact that any $g+1$ disjoint closed curves divide the surface, and the Poincaré–Bendixson Theorem 14.1.1.

14.4.2. Use the fact that any $g+1$ disjoint closed curves divide the surface, and the Poincaré–Bendixson Theorem 14.1.1.

14.4.3. Use minimality of irrational toral translations.

14.4.5. The genus is $[n/2]$. For even n there is one fixed point of index $2-2g=2-n$. For odd n there are two fixed points with index $(1-n)/2$.

14.5.3. It suffices to consider nonatomic measures. Then $\xi(I)$ is a one-sided generator. Alternatively show that the cardinality of the iterated partition ξ^I_{-n} grows linearly.

14.5.4. See Exercise 4.2.10.

14.5.6. Start from $\beta = 2\alpha$ and proceed inductively, adding small pieces as in the proof of Corollary 12.6.4.

14.6.1. First construct an area-preserving flow on the torus with a hole with two half-saddles on the boundary and the rest of the boundary a saddle connection.

14.7.2. Start from two linear flows on the component tori that realize the desired fluxes through the generators. Then make a hole in each torus and modify the flow so that there are two saddle points and the flow is directed from one torus into the other.

14.7.4. See Exercise 14.5.6.

15.1.1. Compare with Theorem 15.1.5 or Theorem 15.1.9.

15.1.2. Adapt (3.2.2) and (3.2.3) for one-sided shifts. Note that the variation of a standard map is the number of arrows in its Markov graph.

15.1.3. Remove "periodic traps". See also Theorem 15.1.9.

15.2.2. Use maps with constant absolute value of slope. Adjust either the number of intervals of monotonicity or the number of fixed points in any way.

15.3.2. Consider, for example, $f(x_i) = x_{i+1}$, $i = 1, \dots, n-1$, and $f(x_n) = x_1$.

15.4.1. Adapt the arguments from Lemmas 15.4.3 and 15.4.5: To get a contradiction assume that μ is nonatomic. First note that $h_{\text{top}}(f) < \log 2/2$ and $h_{\text{top}}(f^2) < \log 2$ imply that neither f nor f^2 have a horseshoe. Thus the bulk of the proofs of Lemmas 15.4.3 and 15.4.5 goes through. To prove Lemma 15.4.3 change the last paragraph as follows: There is a fixed point c such that $(x, c) \cap \operatorname{Fix} f \neq \varnothing$, contradicting the assumption that we have only finitely many fixed points and proving Lemma 15.4.3. To prove Lemma 15.4.5 change the last paragraph thus: There are sequences $w_n < f(w_n) \leq a$ and $w_n < z_n < a < f(z_n)$ such that $w_n \to a$, $z_n \to a$, contradicting piecewise monotonicity and proving Lemma 15.4.5. But this implies that f^2 is not ergodic, hence μ is not nonatomic and by ergodicity $\mu = \delta_x$ for some fixed x.

15.6.1. Use Theorem 15.6.1 and the fact that the semiconjugacy h is injective except on disjoint intervals where all powers of f are monotone.

16.3.1. Use Corollary 16.1.2 followed by Theorem 16.3.1 and the comments at the end of Section 15.2.

16.3.2. Show that C^r perturbations are piecewise monotone.

17.1.1. The intersection K of any decreasing sequence of compact connected sets K_i is connected. To prove this take an open disjoint cover $\{O_1, O_2\}$ of K and note that if $K_i \notin G_j$ for any i, j we have $\varnothing \neq \bigcap_{i \geq N}(K_i \cap (\partial O_1 \bigcup \partial O_2)) \subset K \setminus (O_1 \cup O_2)$ for sufficiently large N.

17.1.2. G is the group of characters of the discrete group of binary rationals $\{m \cdot 2^n \mid m, n \in \mathbb{Z}\}$. We identify Λ with the group of sequences $(\omega_0, \omega_1, \dots)$, where $\omega_0 \in S^1$ and $\omega_{n+1}^2 = \omega_n$ with coordinatewise multiplication.

17.2.4. Consider a surface S obtained by attaching k handles symmetrically around the equator of a round sphere with angles $2\pi/k$. The identification space of orbits of the rotation by $2\pi/k$ around the north–south axis is obviously a torus. Now find an Anosov map on $\mathbb{T}^2$ with two fixed points which can be lifted to S and perform surgery as in the construction of the DA map.

17.3.1. The derivative DF at the identity uniquely determines the automorphism. These derivatives are given by integer matrices $\begin{pmatrix} a & b & 0 \\ c & d & 0 \\ \alpha & \beta & \gamma \end{pmatrix}$, where $ad - cb = \gamma \notin \{0, \pm 1\}$ and $\begin{pmatrix} a & b \\ c & d \end{pmatrix}$ is hyperbolic.

17.3.2. F sends translations to translations and preserves the lattice, hence must preserve (up to orientation) any translation-invariant measure.

17.3.3. To show that $dx\,dy\,dz$ is left-invariant note that left translation of a Heisenberg matrix changes the entries only by a constant.

17.3.4. Exercise 17.3.2 shows that F preserves Haar measure, so by Exercise 17.3.3 it has determinant 1, which by Exercise 17.3.1 implies that $\gamma^2 = 1$.

17.4.2. Consider a parameterized closed curve whose tangent vectors are close to the generator of the flow. Fix a sequence of transverse sections at equally spaced points of the curve and consider the product of the associated Poincaré maps. Introduce adapted coordinates on each transversal and extend the Poincaré maps to the whole Euclidean space preserving hyperbolicity. Then repeat the proof of the Anosov Closing Lemma (Theorem 6.4.15). The unique fixed point of the product of the Poincaré maps corresponds to a periodic orbit for the flow which stays close to the original orbit of the flow after a small reparameterization.

17.5.1. (1) The second horocycle flow $\hat{H}_t(v) = -H_t(-v)$; the tangent vector moves along the horocycle determined by $\gamma(+\infty)$.
(2) Rotation of the vector around its footpoint with unit speed.
(3) The orthogonal geodesic flow: A tangent vector v moves with unit speed along the properly oriented geodesic orthogonal to v.

17.5.2. Use the classification of Möbius transformations into elliptic, parabolic, and hyperbolic ones.

17.6.1. Calculate the volume of a sphere by integrating the volume element generated by orthonormal Jacobi fields. Compare with Exercise 9.6.1.

17.6.2. Use the previous exercise.

17.6.3. Show that $\exp_x: T_x\widetilde{M} \to \widetilde{M}$ is a diffeomorphism.

17.6.4. Use the previous exercise.

17.7.1. Use Theorem 9.6.7 and Exercise 17.6.1 in an argument similar to the proof of Theorem 3.2.9.

17.8.1. Construct an open cover $C = \{C_1, \dots, C_N\}$ of J such that $f(C_i) \cap C_j \neq \varnothing$ implies $C_j \subset f(C_i)$.

17.8.2. Consider "cuts" from the fixed point to ∞ and their preimages for coding.

18.1.2. Take two small disks around the poles and a sufficiently dense collection of orbit segments in between. Then any ϵ-orbit stays sufficiently close to one of these segments while outside the disks. Inside the disks it is trivially close.

18.1.3. Again it suffices to consider disks around the critical points and a sufficiently dense collection of orbit segments outside.

18.1.4. With the notation of the Shadowing Theorem let $f' = f =$ the map in question, $g: \mathbb{T}^2 \to \mathbb{T}^2$ the corresponding time-one map on the tilted torus, and $\alpha = \mathrm{Id}_{\mathbb{T}^2}$. Then $d_{C^0}(f, g) < \epsilon$ but f and g are evidently not conjugate since g does not have a saddle connection.

18.2.1. For a noninvertible map with the appropriate hyperbolicity conditions there may be difficulties in defining unstable directions since preimages are nonunique. Here this is not a problem because by assumption the whole tangent space is expanding.

18.2.2. Use nonuniqueness of unstable manifolds for some perturbations of F which arises from nonuniqueness of preimages.

18.3.1. Show that in $NW(\sigma_A)$ all symbols are essential and equivalent, so that Exercise 1.9.9 can be applied.

18.3.2. On $\mathbb{T}^2$ take two hyperbolic fixed points and a heteroclinic orbit. This a a compact locally maximal hyperbolic set and in the nonwandering set if the diffeomorphism preserves a smooth measure. But its nonwandering set only has two points.

18.3.3. Attach two semicircles to Δ, put a sink (that is, attracting fixed point) in one of them, and extend the map to send this set into itself. To extend from this topological disk to the sphere introduce a source in the complement.

18.3.4. Modify the argument of Theorem 18.3.6 by showing that the closure of the unstable manifold of one of these orbits contains the other orbit entirely.

18.3.6. Use the fact that unstable manifolds are open sets.

18.3.7. Introduce an equivalence relation on periodic points by $p \sim q$ if and only if the weak stable manifold of p intersects the weak unstable manifold of q and vice versa and both intersections are transverse in at least one point. Note that the classes are flow invariant and open.

18.4.1. Obtaining local maximality is easy: A point that stays close for all $n \in \mathbb{Z}$ is by assumption in the intersection of a stable and unstable manifold from the set; hence we have a local product structure and local maximality. Conversely assume local maximality and take a point y as in the statement. Then y is in an ϵ-orbit that coincides with the orbit of y for positive time and is in Λ for negative time. It is shadowed by the orbit of a point $x \in \Lambda$, hence is positively asymptotic to it.

18.4.3. Every closed invariant set in the shift can be approximated arbitrarily well by n-step topological Markov chains. Compare with Exercise 6.4.8.

18.4.4. Use the previous exercise and Exercise 1.9.12 or Exercise 6.4.7.

18.5.1. Use Corollary 1.9.12 and Proposition 3.2.5.

18.5.2. Exercise 8.7.1 yields $P_n(F_L) = \prod_{i=1}^{m} |1 - \lambda_i^n|$. Group the terms according to whether λ_i is on the unit circle (nonhyperbolic) or not. The hyperbolic part gives the right asymptotic, but the nonhyperbolic part, although bounded by 2, has terms arbitrarily close to 0.

18.5.3. Use Theorem 18.5.1 and Exercise 8.7.1.

18.7.1. Use Exercises 18.3.5 and 18.3.9.

18.7.3. Use the previous exercise to show that the semiconjugacy is injective a.e. for every nonatomic measure. Then apply the Variational Principle.

19.1.4. Use the previous exercise as well as the invariant contact structure (Lemma 18.3.7) to show that the strong stable and unstable foliations are C^1. Then use the Hopf argument from Theorem 5.4.16. This result is also shown in Section 20.4.

19.1.5. Perturb a linear Anosov diffeomorphism so that the foliations remain the same but an eigenvalue at some periodic point is changed.

19.1.7. Consider the holonomy map induced by the foliations on an analytic closed curve transverse to the foliations (for example, the projection of the x-axis in $\mathbb{R}^2$ to $\mathbb{T}^2$). Use the Arnold Theorem 12.3.1 to show that it is analytically conjugate to a rotation. Show that the corresponding linear parameter on this transversal is linearly expanded by f. Use this to produce affine coordinates on the universal cover in which the lift F of f is linear.

19.2.1. Show that the difference is cohomologous to the Jacobian of f.

19.2.2. Show that the Fourier coefficients of the solution Φ obtained from Theorem 19.2.1 decay superexponentially.

19.2.3. Using a method like the one used in the proof of Theorem 19.2.5 prove that the function Φ obtained from Theorem 19.2.1 is C^∞ along leaves of the stable and unstable foliations. Then use Fourier analysis to show that this implies smoothness.

20.1.1. Show first that there are infinitely many periodic orbits and then use Theorem 18.5.5.

20.1.2. Note that expansivity is inherited by closed invariant subsets by definition and show that specification is inherited by factors.

20.1.3. By Exercise 1.9.11 it is a sofic system.

20.1.7. Use the previous exercise to show that $\lambda = \lim \mu_n$.

20.1.8. Use the previous exercise and Exercise 18.5.4.

20.3.1. Consider functions of one coordinate only.

20.3.3. Use Lemma 20.3.4 to show that $h_{\mu_\varphi}(f) = 0$ implies that μ_φ is atomic.

20.3.5. Use the previous exercise.

20.4.2. Use Lemma 20.3.4 to show absolute continuity. Find a functional equation for the derivatives and show that its solutions are unique. After that show that (20.4.5) gives a solution.

20.4.3. Instead of area (the smooth measure) use the equilibrium states μ_φ and μ_ψ, where $\psi = \log J^s f$ (the stable Jacobian). Then use the result of the previous exercise.

20.4.4. Take the composition of a linear map whose unstable foliation W is two-dimensional with a volume-preserving diffeomorphism that preserves the leaves of W but changes the eigenvalues of a periodic point.

References

[AM] Abraham, Ralph, and Marsden, Jerrold: *Foundations of mechanics.* Benjamin/ Cummings, New York, 1978.

[ARo] Abraham, Ralph, and Robbin, Joel: *Transversal mappings and flows.* Benjamin, New York, 1967.

[ASm] Abraham, Ralph, and Smale, Stephen: Non-genericity of Ω-stability. In *Global Analysis, Proceedings of Symposia in Pure Mathematics,* vol. 14, pp. 5–8. American Mathematical Society, Providence, RI, 1970.

[AKM] Adler, Roy L.; Konheim, Alan G.; and McAndrew, M. Harry: Topological entropy. *Transactions of the American Mathematical Society,* **114** (1965), 309–319.

[AW] Adler, Roy L., and Weiss, Benjamin: Entropy, a complete metric invariant for automorphisms of the torus. *Proceedings of the National Academy of Sciences,* **57** no. 6 (1967), 1573–1576.

[A1] Alekseyev, Vladimir M.: Invariant Markov subsets for diffeomorphisms. *Uspehi Mat. Nauk,* **23** no. 2 (1968), 209–210.

[A2] ———: Perron sets and topological Markov chains. *Uspehi Mat. Nauk,* **24** no. 5 (1969), 227–229.

[A3] ———: Quasirandom dynamical systems I: Quasirandom diffeomorphisms. *Mathematics of the USSR, Sbornik,* **5** no. 1 (1968), 73–128; II: One-dimensional non-linear oscillations in a field with periodic perturbations. *Mathematics of the USSR, Sbornik,* **6** no. 4 (1968), 505–560; III: Quasirandom oscillations of one-dimensional oscillators. *Mathematics of the USSR, Sbornik,* **7** no. 1 (1969), 1–43.

[A4] ———: *Symbolic dynamics.* 11th Summer mathematical school. Mathematics Institute of the Ukrainian Academy of Sciences, Kiev, 1976.

[AKK] Alekseyev, Vladimir M.; Katok, Anatole B.; and Kušnirenko, Anatole G.: *Three papers in dynamical systems.* Translations of the American Mathematical Society (series 2), vol. 116. American Mathematical Society, Providence, RI, 1981.

[AP] Andronov, Alexander, and Pontrjagin, Lev: Systèmes grossiers. *Comptes Rendus (Doklady) de l'Académie des Sciences de l'URSS,* **14** no. 5 (1937), 247–250.

[An1] Anosov, Dmitriĭ V.: Roughness of geodesic flows on compact Riemannian manifolds of negative curvature. *Soviet Mathematics, Doklady,* **3** (1962), 1068–1070.

[An2] ———: Ergodic properties of geodesic flows on closed Riemannian manifolds of negative curvature. *Soviet Mathematics, Doklady,* **4** (1963), 1153–1156.

[An3] ———: *Geodesic flows on Riemann manifolds with negative curvature.* Proceedings of the Steklov Institute of Mathematics, vol. 90. American Mathematical Society, Providence, RI, 1967.

[An4] ———: Tangent fields of transversal foliations in "$\mathcal{Y}$-systems". *Math. Notes of the Acad. of Sciences, USSR,* **2** no. 5 (1967), 818–823.

[An5] ———: On a class of invariant sets of smooth dynamical systems. In *Proceedings of the Fifth International Conference on Nonlinear Oscillations,* vol. 2, pp. 39–45. Mathematics Institute of the Ukrainian Academy of Sciences, Kiev, 1970.

[AnK] Anosov, Dmitriĭ V., and Katok, Anatole B.: New examples in smooth ergodic theory. Ergodic diffeomorphisms. *Transactions of the Moscow Mathematical Society,* **23** (1970), 1–35.

[AnSin] Anosov, Dmitriĭ V., and Sinai, Yakov: Some smooth ergodic systems. *Russian Mathematical Surveys,* **22** no. 5 (1967), 103–167.

[Ar1] Arnol'd, Vladimir I.: Small denominators I. Mappings of the circle onto itself. *Izvestija Akademiĭ Nauk SSSR Ser. Mat.* **25** (1961), 21–86; English translation: *Translations of the American Mathematical Society (series 2),* **46** (1965), 213–284.

[Ar2] ———: Proof of a theorem of A. N. Kolmogorov on the invariance of quasiperiodic motions under small perturbations of the Hamiltonian. *Russian Mathematical Surveys,* **18** no. 5 (1963), 9–36.

[Ar3] ———: Small denominators and problems of stability of motion in classical and celestial mechanics. *Russian Mathematical Surveys,* **18** no. 6 (1963), 85–193.

[Ar4] ———: *Ordinary differential equations.* MIT Press, Cambridge, MA, 1973.

[Ar5] ———: *Geometric methods in the theory of ordinary differential equations.* Springer Verlag, Berlin, New York, 1983.

[Ar6] ———: *Mathematical methods of classical mechanics.* Springer Verlag, Berlin, New York, 1978, 1989.

[ArAv] Arnol'd, Vladimir I., and Avez, André: *Ergodic problems of classical mechanics.* Addison–Wesley, Amsterdam, 1988.

[ArASI] Arnol'd, Vladimir I.; Afraimovich, Valentin S.; Shilnikov, Leonid P.; and Ilyashenko, Yuli S.: Theory of bifurcations. In *Encyclopedia of Mathematical Sciences, Volume 5, Dynamical Systems V*, pp. 5–218. Springer Verlag, Berlin, New York, 1994.

[Art] Artin, Emil: Ein mechanisches System mit quasiergodischen Bahnen. *Abhandlungen aus dem mathematischen Seminar der hamburgischen Universität*, **3** (1924), 170–175; in *Collected Papers*, pp. 499–504. Springer Verlag, Berlin, New York, 1965.

[ArM] Artin, Michael, and Mazur, Barry: On periodic points. *Annals of Mathematics*, **81** (1965), 82–99.

[Au] Aubry, Serge: The devil's staircase transformation in incommensurate lattices. In *The Riemann problem, complete integrability and arithmetic applications*, Lecture Notes in Mathematics, vol. 925, pp. 221–245. Springer Verlag, Berlin, New York, 1982.

[AuD] Aubry, Serge, and Le Daëron, Pierre-Yves: The discrete Frenkel–Kontorova model and its extensions. I. Exact results for the ground-states. *Physica D*, **8** no. 3 (1983), 381–422.

[AuDA] Aubry, Serge; Le Daëron, Pierre-Yves; and André, Gilles: Classical ground-states of a one-dimensional model for incommensurate structures. Preprint.

[B1] Bangert, Victor: Mather sets for twist maps and geodesics on tori. In *Dynamics reported*, vol. 1, pp. 1–56. Springer Verlag, Berlin, New York, 1988.

[B2] ———: On the existence of closed geodesics on two-spheres. *International Journal of Mathematics*, **4** (1993), 1–10.

[BDT] Beals, Richard; Deift, Percy; and Tomei, Carlos: *Direct and inverse scattering on the real line.* Mathematical Surveys and Monographs, vol. 28. American Mathematical Society, Providence, RI, 1988.

[Be] Beardon, Alan: *Iteration of rational functions: Complex analytic dynamical systems.* Springer Verlag, Berlin, New York, 1991.

[Bel] Belitskiĭ, Grigoriĭ R.: Equivalence and normal forms of germs of smooth mappings. *Russian Mathematical Surveys*, **31** no. 1 (1978), 107–177.

[Ben] Bendixson, Ivar: Sur les courbes définies par les équations différentielles. *Acta Mathematica*, **24** (1901), 1–88.

[BC] Benedicks, Michael, and Carleson, Lennart: Dynamics of the Hénon map. *Annals of Mathematics*, **133** (1991), 73–169.

[BL] Benoist, Yves, and Labourie, François: Sur les difféomorphismes d'Anosov affines à feuilletages stable et instable différentiables. *Inventiones mathematicae*, **111** no. 2 (1993), 285–308.

[BFL] Benoist, Yves; Foulon, Patrick; and Labourie, François: Flots d'Anosov à distributions stable et instable différentiables. *Journal of the American Mathematical Society*, **5** no. 1 (1992), 33–74.

[Bi1] Birkhoff, George D.: Surface transformations and their dynamical applications. *Acta Mathematica*, **43** no. 1–2 (1920), 1–119.

[Bi2] ———: On the periodic motions of dynamical systems. *Acta Mathematica*, **50** (1927), 359–379.

[Bi3] ———: *Dynamical systems.* Colloquium Publications, vol. 9. American Mathematical Society, Providence, RI, 1927.

[Bi4] ———: Proof of the ergodic theorem. *Proceedings of the Academy of Sciences USA* **17** (1931), 656–660.

[BiC] Bishop, Richard L., and Crittenden, Richard J.: *Geometry of manifolds.* Academic Press, New York, 1964.

[Bl] Blanchard, Paul: Complex dynamics on the Riemann sphere. *Bulletin of the American Mathematical Society*, **11** no. 1 (1984), 85–141.

[BGMY] Block, Louis; Guckenheimer, John; Misiurewicz, Michał; and Young, Lai-Sang: Periodic points and topological entropy of one-dimensional maps. In *Global theory of dynamical systems*, edited by Zbigniew Nitecki and Clark Robinson, Lecture Notes in Mathematics, vol. 819, pp. 18–34. Springer Verlag, Berlin, New York, 1980.

[Bo1] Bowen, Rufus: Topological entropy and Axiom A. In *Global Analysis, Proceedings of Symposia in Pure Mathematics*, vol. 14, pp. 23–41. American Mathematical Society, Providence, RI, 1970.

[Bo2] ———: Markov partitions for Axiom A diffeomorphisms. *American Journal of Mathematics*, **92** (1970), 725–747.

[Bo3] ———: Periodic points and measures for Axiom A diffeomorphisms. *Transactions of the American Mathematical Society*, **154** (1971), 377–397.

[Bo4] ———: Periodic orbits for hyperbolic flows. *American Journal of Mathematics*, **94** (1972), 1–30.

[Bo5] ———: Symbolic dynamics for hyperbolic flows. *American Journal of Mathematics*, **95** (1972), 429–459.

[Bo6] ———: Some systems with unique equilibrium states. *Mathematical Systems Theory*, **8** no. 3 (1975), 193–202.

[Bo7] ———: *Equilibrium states and the ergodic theory of Anosov diffeomorphisms*. Lecture Notes in Mathematics, vol. 470. Springer Verlag, Berlin, New York, 1975.

[Bo8] ———: Entropy and the fundamental group. In *The structure of attractors in dynamical systems*, edited by Nelson G. Markley, J. C. Martin, and W. Perrizo, Lecture Notes in Mathematics, vol. 668, pp. 21–29. Springer Verlag, Berlin, New York, 1978.

[BoR] Bowen, Rufus, and Ruelle, David: The ergodic theory of Axiom A flows. *Inventiones mathematicae*, **29** (1975), 181–202.

[Boy] Boyle, Michael: Symbolic dynamics and matrices. In *Combinatorial and graph-theoretical properties in linear algebra*, edited by R. Brualdi, S. Friedland, and V. Klee, IMA Volumes in Mathematics and Its Applications, vol. 50, pp. 1–38. Springer Verlag, Berlin, New York, 1993.

[BP] Brin, Mikhael, and Pesin, Yakov: Partially hyperbolic dynamical systems. *Mathematics of the USSR, Isvestia*, **8** no. 1 (1974), 177–218.

[Br] Brjuno, Alexander D.: The analytical form of differential equations. *Transactions of the Moscow Mathematical Society*, **25** (1971), 131–288; *Transactions of the Moscow Mathematical Society*, **26** (1972), 199–238.

[Bu] Bunimovich, Leonid A.: On the ergodic properties of nowhere dispersing billiards. *Communications in Mathematical Physics*, **65** (1979), 295–312.

[C] Chen, Kuo-Tsai: Equivalence and decomposition of vector fields about an elementary critical point. *American Journal of Mathematics*, **85** (1963), 693–722.

[Ch] Cherry, Thomas M.: Analytic quasi-periodic curves of discontinuous type on a torus. *Proceedings of the London Mathematical Society (second series)*, **44** (1938), 175–215.

[CL] Coddington, Earl A., and Levinson, Norman: *Theory of ordinary differential equations*. McGraw-Hill, New York, 1955.

[CE] Collet, Pierre, and Eckmann, Jean-Pierre: *Iterated maps on the interval as dynamical systems*. Birkhäuser, 1980.

[Co] Conley, Charles: *Isolated invariant sets and the Morse index*. CBMS Regional Conference Series in Mathematics, vol. 38. American Mathematical Society, Providence, RI, 1978.

[CoZ] Conley, Charles, and Zehnder, Eduard: Morse-type index theory for flows and periodic solutions for Hamiltonian equations. *Communications in Pure and Applied Mathematics*, **37** no. 2 (1984), 207–253.

[Con] Connes, Alain: Une classification des facteurs de type III. *Annales scientifiques de l'École Normale Superieure*, **6** no. 2 (1973), 133–252.

[CFSin] Cornfeld, Isaak P.; Fomin, Sergei V.; and Sinai, Yakov G.: *Ergodic theory*. Springer Verlag, Berlin, New York, 1982.

[D] Denjoy, Arnaud: Sur les courbes définies par les équations différentielles à la surface du tore. *Journal de Mathématiques Pures et Appliquées (9. série)*, **11** (1932), 333–375.

[DGS] Denker, Manfred; Grillenberger, Christian; and Sigmund, Karl: *Ergodic theory on compact spaces*. Lecture Notes in Mathematics, vol. 527. Springer Verlag, Berlin, New York, 1976.

[De] Devaney, Robert: *An introduction to chaotic dynamical systems*. Addison–Wesley, Reading, MA, 1989.

[Di] Dinaburg, Efim I.: On the relations among various entropy characteristics of dynamical systems. *Mathematics of the USSR, Isvestia*, **5** (1971), 337–378.

[Do] Dolgopyat, Dmitry: On decay of correlations in Anosov flows. *Annals of Mathematics*, **147** no. 2 (1998), 357–390.

[DoP] Dolgopyat, Dmitry and Pollicott, Mark: Addendum to 'Periodic orbits and dynamical spectra' (by Viviane Baladi). *Ergodic Theory and Dynamical Systems*, **18** no. 2 (1998), 255–292.

[EL] Ekeland, Ivar, and Lasry, Jean-Michel: On the number of periodic trajectories for a Hamiltonian flow on a convex energy surface. *Annals of Mathematics*, **112** no. 2 (1980), 283–319.

[E] Ellis, Robert: *Lectures on topological dynamics*. Benjamin, New York, 1969.

[F] Fathi, Albert: Une interprétation plus topologique de la démonstration du théorème de Birkhoff. *Astérisque*, **103–104** (1983), 39–46.

[FLP] Fathi, Albert; Laudenbach, François; and Poénaru, Valentin: *Travaux de Thurston sur les surfaces*. Astérisque, vol. 66–67. Société Mathématique de France, Paris, 1979.

[Fa] Fatou, Pierre: Sur les équations fonctionelles. *Bulletin de la Société Mathématique de France*, **47** (1919), 161–271; **48** (1920), 33–94; **48** (1920), 208–314.

[Fe] Feigenbaum, Mitchell: Quantitative universality for a class of nonlinear transformations. *Journal of Statistical Physics*, **19** (1978), 25–52.

[FM] Forni, Giovanni, and Mather, John: Action minimizing orbits in Hamiltonian systems. In *Transition to chaos in classical and quantum mechanics*, edited by S. Graffi, Lecture Notes in Mathematics, vol. 1589, pp. 92–186. Springer Verlag, Berlin, New York, 1994.

[Fr1] Franks, John: Anosov diffeomorphisms on tori. *Transactions of the American Mathematical Society*, **145** (1969), 117–124.

[Fr2] ———: Anosov Diffeomorphisms. In *Global Analysis, Proceedings of Symposia in Pure Mathematics*, vol. 14, pp. 61–93. American Mathematical Society, Providence, RI, 1970.

[Fr3] ———: *Homology and dynamical systems*. CBMS Regional Conference Series in Mathematics, vol. 49. American Mathematical Society, Providence, RI, 1982.

[Fr4] ———: Geodesics on S^2 and periodic points of annulus diffeomorphisms. *Inventiones mathematicae*, **108** (1992), 403–418.

[FrW] Franks, John, and Williams, Robert F.: Anomalous Anosov flows. In *Global theory of dynamical systems*, edited by Zbigniew Nitecki and Clark Robinson, Lecture Notes in Mathematics, vol. 819, pp. 158–174. Springer Verlag, Berlin, New York, 1980.

[Fri] Fried, David: Transitive Anosov flows and pseudo-Anosov maps. *Topology*, **22** no. 3 (1983), 299–303.

[Fu1] Furstenberg, Hillel: Strict ergodicity and transformations of the torus. *American Journal of Mathematics*, **83** (1961), 573–601.

[Fu2] ———: The structure of distal flows. *American Journal of Mathematics*, **85** (1963), 477–515.

[Fu3] ———: *Recurrence in ergodic theory and combinatorial number theory*. M. B. Porter Lectures, Rice University. Princeton University Press, Princeton, NJ, 1981.

[G] Gallavotti, Giovanni: *The elements of mechanics*. Springer Verlag, Berlin, New York, 1983.

[Ga] Gantmacher, Feliks R.: *Applications of the theory of matrices*. Interscience Publishers, New York, 1959.

[GF] Gelfand, Israel M., and Fomin, Sergei V.: Geodesic flows on manifolds of constant negative curvature. *Uspehi Mat. Nauk*, **7** no. 1 (1952), 118–137.

[Gh] Ghys, Etienne: Flots d'Anosov dont les feuilletages stables sont différentiables. *Annales scientifiques de l'École Normale Superieure*, **20** no. 2 (1987), 251–270.

[Go] Goodman, Sue E.: Dehn surgery on Anosov Flows. In *Geometric Dynamics*, edited by Jacob Palis, Lecture Notes in Mathematics, vol. 1007, pp. 300–307. Springer Verlag, Berlin, New York, 1983.

[Goo] Goodman, Tim N. T.: Relating topological entropy and measure entropy. *Bulletin of the London Mathematical Society*, **3** (1971), 176–180.

[Gw] Goodwyn, L. Wayne: Topological entropy bounds measure-theoretic entropy. *Proceedings of the American Mathematical Society*, **23** (1969), 679–688.

[GH] Gottschalk, Walter H., and Hedlund, Gustav A.: *Topological dynamics*. American Mathematical Society colloquium publications, vol. 36. American Mathematical Society, Providence, RI, 1955.

[GraSw] Graczyk, Jacek, Świątek, Grzegorz: Generic hyperbolicity in the logistic family. *Annals of Mathematics*, **146** no. 1 (1997), 1–52.

[Gr] Grobman, David M.: Topological classification of neighborhoods of a singularity in n-space. *Mat. Sbornik*, **56** no. 98 (1962), 77–94.

[Gro] Gromov, Mikhael: On the entropy of holomorphic maps. Preprint.

[GH] Guckenheimer, John, and Holmes, Philip: *Nonlinear Oscillations, Dynamical Systems, and Bifurcations of Vector Fields.* Springer Verlag, Berlin, New York, 1983.

[GMN] Guckenheimer, John; Moser, Jürgen; and Newhouse, Sheldon: *Dynamical systems.* CIME Lectures, Bressanone, Italy, June 1978, Progress in Mathematics, vol. 8. Birkhäuser, 1980.

[GK] Guillemin, Victor, and Kazhdan, David: On the cohomology of certain dynamical systems. *Topology*, **19** (1980), 291–299.

[Gu1] Gutierrez, Carlos: Smoothing continuous flows on two-manifolds and recurrence. *Ergodic Theory and Dynamical Systems*, **6** no. 1 (1986), 17–44.

[Gu2] ______: A counter-example to a C^2 closing lemma. *Ergodic Theory and Dynamical Systems*, **7** no. 4 (1987), 509–530.

[GuK] Gutkin, Eugene, and Katok, Anatole: Caustics in inner and outer billiards. *Communications in Mathematical Physics*, **173** (1995), 101-133.

[H] Hadamard, Jaques: Sur l'itération et les solutions asympotiques des équations différentielles. *Bulletin de la Société Mathématique de France*, **29** (1901), 224–228.

[HMO] Hale, Jack K.; Magalhaes, Luis T.; and Oliva, Waldyr M.: *An introduction to infinite-dimensional dynamical systems—geometric theory.* Springer Verlag, Berlin, New York, 1984.

[Ha1] Halmos, Paul: *Measure theory.* Van Nostrand, New York, 1950. Springer Verlag, Berlin, New York, 1974.

[Ha2] ______: *Ergodic theory.* Chelsea, New York, 1956.

[HT1] Handel, Michael, and Thurston, William: Anosov flows on new 3-manifolds. *Inventiones mathematicae*, **59** (1980), 95–103.

[HT2] ______: New proofs of some results of Nielsen. *Advances in Mathematics*, **56** no. 2 (1982), 669–675.

[Har1] Hartman, Philip: *Ordinary differential equations.* Birkhäuser, 1982.

[Har2] ______: On the local linearization of differential equations. *Proceedings of the American Mathematical Society*, **14** (1963), 568–573.

[Has] Hasselblatt, Boris: Regularity of the Anosov splitting and of horospheric foliations. *Ergodic Theory and Dynamical Systems*, **14** no. 4 (1994), 645–666.

[He1] Hedlund, Gustav A.: Geodesics on a two-dimensional Riemannian manifold with periodic coefficients. *Annals of Mathematics*, **33** (1932), 719–739.

[He2] ______: Metric transitivity of the geodesics on closed surface of constant negative curvature. *Annals of Mathematics*, **35** (1934), 787–808.

[Hel] Helgason, Sigurdur: *Differential geometry, Lie groups and Symmetric Spaces.* Academic Press, New York, 1978.

[Hen] Henry, Dan: *Geometric theory of semilinear parabolic equations.* Lecture Notes in Mathematics, vol. 840. Springer Verlag, Berlin, New York, 1981.

[Her1] Herman, Michael R.: Sur la conjugaison différentiable des difféomorphismes du cercle à des rotations. *Publications Mathématiques de l'Institut des Hautes Études Scientifiques*, **49** (1979), 5–234.

[Her2] ______: *Sur les courbes invariants par les difféomorphismes de l'anneau.* Astérisque, vol. 103–104. Société Mathématique de France, Paris, 1983.

[HP] Hirsch, Morris, and Pugh, Charles: Smoothness of horocycle foliations. *Journal of Differential Geometry*, **10** (1975), 225–238.

[HPS] Hirsch, Morris; Pugh, Charles; and Shub, Michael: *Invariant manifolds.* Lecture Notes in Mathematics, vol. 583. Springer Verlag, Berlin, New York, 1977.

[HZ] Hofer, Helmut, and Zehnder, Eduard: *Symplectic invariants and Hamiltonian dynamics.* Birkhäuser, 1994.

[Ho1] Hopf, Eberhard: *Ergodentheorie.* Springer Verlag, Berlin, New York, 1937.

[Ho2] ______: Statistik der geodätischen Linien in Mannigfaltigkeiten negativer Krümmung. *Ber. Verhandlungen der sächsischen Akademie der Wissenschaften zu Leipzig*, **91** (1939), 261–304; Statistik der Lösungen geodätischer Probleme vom unstabilen Typus. II. *Mathematische Annalen*, **117** (1940), 590–608.

[HK] Hurder, Steven, and Katok, Anatole B.: Differentiability, rigidity and Godbillon–Vey classes for Anosov flows. *Publications Mathématiques de l'Institut des Hautes Études Scientifiques*, **72** (1990), 5–61.

[I] Irwin, Michael C.: *Smooth dynamical systems.* Academic Press, London, 1980.

[J1] Jacobson, Michael V.: Properties of the one-parameter family of dynamical systems $x \mapsto Axe^{-x}$. *Uspehi Mat. Nauk*, **31** no. 2 (1976), 239–240.

[J2] ———: Absolutely continuous invariant measures for one-parameter families of one-dimensional maps. *Communications in Mathematical Physics*, **81** no. 1 (1981), 39–88.

[Ji] Jiang, Boju: *Lectures on Nielsen fixed point theory.* Contemporary Mathematics, vol. 14. American Mathematical Society, Providence, RI, 1983.

[Ju] Julia, Gaston: Mémoire sur l'itération des fonctions rationelles. *Journal de Mathématiques Pures et Appliquées*, **4** (1918), 47–245.

[Kal] Kaloshin, Vadim: Exponential growth of number of periodic orbits is not topologically generic. in preparation.

[K1] Katok, Anatole B.: Ergodic perturbations of degenerate integrable Hamiltonian systems. *Mathematics of the USSR, Isvestia*, **7** no. 3 (1973), 535–571.

[K2] ———: Invariant measures for flows on orientable surfaces. *Soviet Mathematics, Doklady*, **14** (1973), 1104–1108.

[K3] ———: Monotone equivalence in ergodic theory. *Mathematics of the USSR, Isvestia*, **11** no. 1 (1977), 99–146.

[K4] ———: A conjecture about entropy. In *Smooth dynamical systems*, edited by Dmitriĭ V. Anosov, pp. 181–203. Mir, Moscow, 1977; English translation: *Translations of the American Mathematical Society (series 2)*, **133** (1986), 91–107.

[K5] ———: Lyapunov exponents, entropy and periodic orbits for diffeomorphisms. *Publications Mathématiques de l'Institut des Hautes Études Scientifiques*, **51** (1980), 137–173.

[K6] ———: Some remarks on Birkhoff and Mather twist map theorems. *Ergodic Theory and Dynamical Systems*, **2** no. 2 (1982), 185–194.

[K7] ———: More about Birkhoff periodic orbits and Mather sets for twist maps. Preprint IHES/M/82/35; Continuation of the preprint "More about Birkhoff periodic orbits and Mather sets for twist maps". University of Maryland preprint.

[K8] ———: Entropy and closed geodesics. *Ergodic Theory and Dynamical Systems*, **2** no. 3–4 (1982), 339–365.

[K9] ———: Nonuniform hyperbolicity and structure of smooth dynamical systems. *Proceedings of the International Congress of Mathematicians, Warszawa*, **2** (1983), 1245–1254.

[KKPW] Katok, Anatole B.; Knieper, Gerhard; Pollicott, Mark; and Weiss, Howard: Differentiability and Analyticity of Topological Entropy for Anosov and Geodesic Flows. *Inventiones mathematicae*, **98** no. 3 (1989), 581–597.

[KSinS] Katok, Anatole B.; Sinai, Yakov G.; and Stepin, Anatole M.: The theory of dynamical systems and general transformation groups with invariant measure. *Journal of Soviet Mathematics*, **7** (1977), 974–1065.

[KS1] Katok, Anatole B., and Spatzier, Ralf J.: First cohomology of Anosov actions of higher rank Abelian groups and applications to rigidity. *Publications Mathématiques de l'Institut des Hautes Études Scientifiques*, **79** (131-156), 1994.

[KS2] Katok, Anatole B., and Spatzier, Ralf J.: Subelliptic estimates of polynomial differential operators and applications to cocycle rigidity. *Mathematical Research Letters*, **1** (1994), 193–202.

[KS] Katok, Anatole B., and Stepin, Anatole M.: Metric properties of measure preserving homeomorphisms. *Russian Mathematical Surveys*, **25** no. 2 (1970), 191–220.

[KSt] Katok, Anatole B., and Strelcyn, Jean-Marie, with the collaboration of François Ledrappier and Feliks Przytycki: *Invariant manifolds, entropy and billiards; smooth maps with singularities.* Lecture Notes in Mathematics, vol. 1222. Springer Verlag, Berlin, New York, 1986.

[Ka] Katok, Svetlana: *Fuchsian groups.* University of Chicago Press, Chicago, 1992.

[KO] Katznelson, Yitzchak, and Ornstein, Donald: The differentiability of the conjugation of certain diffeomorphisms of the circle. *Ergodic Theory and Dynamical Systems*, **9** no. 4 (1989), 643–680.

[Ke] Keane, Michael: Interval exchange transformations. *Mathematische Zeitschrift*, **141** (1975), 25–31.

[KeS] Keane, Michael, and Smorodinsky, Meir: Bernoulli schemes of the same entropy are finitarily isomorphic. *Annals of Mathematics*, **109** (1979), 397–406.

[KMS] Kerckhoff, Stephen; Masur, Howard; and Smillie, John: Ergodicity of billiard flows and quadratic differentials. *Annals of Mathematics*, **124** (1986), 293–311.

[Kh] Khinchin, Alexander Ya.: *Mathematical foundations of information theory*. Dover, New York, 1957.

[Kl] Klingenberg, Wilhelm: *Riemannian Geometry*. de Gruyter, Berlin-New York, 1982.

[KN] Kobayashi, Shoshichi, and Nomizu, Katsumi: *Foundations of differential geometry*. vol. 1. Interscience Publishers, New York, 1963; vol.2. Interscience Publishers, New York, 1969.

[Ko1] Kolmogorov, Andrei N.: On dynamical systems with an integral invariant on the torus. *Doklady Akademiĭ Nauk SSSR*, **93** (1953), 763–766 (in Russian).

[Ko2] ———: Preservation of conditionally periodic movements under a small change in the Hamilton function. *Doklady Akademiĭ Nauk SSSR*, **98** (1954), 527–530 (in Russian); English translation: In *Stochastic behavior in classical and quantum Hamiltonian systems*, Volta Memorial conference, Como, 1977, Lecture Notes in Physics, vol. 93. Springer Verlag, Berlin, New York, 1979.

[Ko3] ———: A new metric invariant of transitive dynamical systems and automorphisms of Lebesgue spaces. *Doklady Akademiĭ Nauk SSSR*, **119** (1958), 861–864 (in Russian); *Proceedings of the Steklov Institute of Mathematics*, **169** no. 4 (1986), 97–102.

[KT] Kozlov, Valeriĭ V., and Treshchëv, Dmitriĭ V.: *Billiards; A genetic introduction to the dynamics of systems with impacts*. Translations of Mathematical Monographs, vol. 89. American Mathematical Society, Providence, RI, 1991.

[Kr] Krengel, Ulrich: *Ergodic theorems*. de Gruyter, Berlin-New York, 1985.

[Kri] Krieger, Wolfgang: On non-singular transformations of a measure space. *Zeitschrift für Wahrscheinlichkeitstheorie und verwandte Gebiete*, **11** no. 2 (1969), 83–97; II. 98–119.

[KB] Kryloff, Nicolas, and Bogoliouboff, Nicolas: La théorie générale de la mesure dans son application à l'étude des systémes dynamiques de la mécanique non linéaire. *Annals of Mathematics*, **38** no. 1 (1937), 65–113.

[KSz] Krzyzewski, Konstantin, and Szlenk, Wiesław: On invariant measures of expanding differentiable mappings. *Studia Mathematicae*, **33** no. 1 (1969), 83–92.

[Ku] Kuksin, Sergei B.: *Nearly integrable infinite-dimensional Hamiltonian systems*. Lecture Notes in Mathematics, vol. 1556. Springer Verlag, Berlin, New York, 1993.

[Kup] Kupka, Ivan: Contribution à la théorie des champs génériques. *Contributions to Differential Equations*, **2** (1963), 457–484.

[LY] Lasota, Andrzej, and Yorke, James A.: On the existence of invariant measures for piecewise monotonic transformations. *Transactions of the American Mathematical Society*, **186** (1973), 481–488.

[L] Lazutkin, Vladimir F.: The existence of caustics for a billiard problem in a convex domain. *Mathematics of the USSR, Isvestia*, **7** (1973), 185–214.

[Led] Ledrappier, François: Mésures d'équilibre d'entropie complèment positive. *Astérisque*, **50** (1977), 251–272.

[LedY] Ledrappier, François, and Young, Lai-Sang: The metric entropy of diffeomorphisms, Part I: Characterization of measures satisfying Pesin's entropy formula. *Annals of Mathematics*, **122** (1985), 509–539; Part II: Relations between entropy, exponents and dimension. *Annals of Mathematics*, **122** (1985), 540–574.

[Le] Levi, Mark: *Qualitative analysis of the periodically forced relaxation oscillations*. Memoirs of the American Mathematical Society, vol. 244. American Mathematical Society, Providence, RI, 1981.

[Lev] Levinson, Norman: A second order differential equation with singular solutions. *Annals of Mathematics*, **50** (1949), 127–153.

[Li] Liao, Shan-Tao: On the stability conjecture. *Chinese Annals of Mathematics*, **1** no. 1 (1980), 9–30.

[LM] Lind, Douglas, and Marcus, Brian: *An introduction to symbolic dynamics and coding*. Cambridge University Press, Cambridge, 1995.

[Liv] Livšic, Alexander: Some homology properties of $\mathcal{Y}$-systems. *Mathematical Notes of the USSR Academy of Sciences*, **10** (1971), 758–763.

[LivSin] Livšic, Alexander, and Sinai, Yakov G.: On invariant measures compatible with the smooth structure for transitive $\mathcal{Y}$-systems. *Soviet Mathematics, Doklady*, **13** no. 6 (1972), 1656–1659.

[LlMM] de la Llave, Rafael; Marco, José Manuel; and Moriyon, Roberto: Canonical perturbation theory of Anosov systems and regularity results for the Livšic cohomology equation. *Annals of Mathematics*, **123** (1986), 537–611; *Bulletin of the American Mathematical Society*, **12** no. 1 (1985), 91–94.

[Ly] Lyubich, Michael Yu.: Entropy of analytic endomorphisms of the Riemannian sphere. *Functional Analysis and its Applications*, **15** no. 4 (1981), 300–302.

[M] Maĭer, Artemiĭ G.: Trajectories on the closed orientable surfaces. *Mathematics of the USSR, Sbornik*, **12** (1943), 71–84.

[Ma1] Mañé, Ricardo: Lyapunov exponents and stable manifolds for compact transformations. In *Geometric Dynamics*, edited by Jacob Palis, Lecture Notes in Mathematics, vol. 1007, pp. 522–577. Springer Verlag, Berlin, New York, 1983.

[Ma2] ———: Hyperbolicity, sinks and measure in one dimensional dynamics. *Communications in Mathematical Physics*, **100** (1985), 495–524; Erratum. *Communications in Mathematical Physics*, **112** (1987), 721–724.

[Ma3] ———: A proof of the C^1 stability conjecture. *Publications Mathématiques de l'Institut des Hautes Études Scientifiques*, **66** (1987), 161–210.

[Ma4] ———: *Ergodic theory and differentiable dynamics.* Springer Verlag, Berlin, New York, 1987.

[Man1] Manning, Anthony: There are no new Anosov diffeomorphisms on tori. *American Journal of Mathematics*, **96** no. 3 (1974), 422–429.

[Man2] ———: Topological entropy and the first homology group. In *Dynamical systems – Warwick 1974*, edited by Anthony Manning, Lecture Notes in Mathematics, vol. 468, pp. 185–190. Springer Verlag, Berlin, New York, 1975.

[Man3] ———: Topological entropy for geodesic flows. *Annals of Mathematics*, **110** (1979), 567–573.

[Mar] Margulis, Grigoriĭ: Certain measures associated with $\mathcal{Y}$-flows on compact manifolds. *Functional Analysis and Its Applications*, **4** (1969), 55–67.

[Mark] Markley, Nelson: Homeomorphisms of the circle without periodic points. *Proceedings of the London Mathematical Society*, **20** (1970), 688–698.

[Mas] Masur, Howard: Interval exchange transformations and measured foliations. *Annals of Mathematics*, **115** (1982), 169–200.

[MS] Masur, Howard, and Smillie, John: Hausdorff dimension of sets of nonergodic measured foliations. *Annals of Mathematics*, **134** (1991), 455–543.

[Mat1] Mather, John: Existence of quasi-periodic orbits for twist homeomorphisms of the annulus. *Topology*, **21** (1982), 457–467.

[Mat2] ———: Glancing billiards. *Ergodic Theory and Dynamical Systems*, **2** no. 3–4 (1982), 397–403.

[Mat3] ———: More Denjoy minimal sets for area-preserving diffeomorphisms. *Commentarii Mathematici Helvetici*, **60** (1985), 508–557.

[Mat4] ———: A criterion for the non-existence of invariant circles. *Publications Mathématiques de l'Institut des Hautes Études Scientifiques*, **63** (1986), 153–204.

[Me] de Melo, Welington: *Lectures on one-dimensional dynamics.* Instituto de Matemática Pura e Aplicada do CNPq, Rio de Janeiro, Brazil, 1988.

[MeS] de Melo, Welington, and van Strien, Sebastian: *One-dimensional dynamics.* Springer Verlag, Berlin, New York, 1993.

[Mi] Milnor, John: *Morse Theory.* Princeton University Press, Princeton, NJ, 1963.

[MiT] Milnor, John, and Thurston, William: On iterated maps of the interval. In *Dynamical Systems, College Park, 1986–1987*, Lecture Notes in Mathematics, vol. 1342, pp. 465–563. Springer Verlag, Berlin, New York, 1988.

[Mis1] Misiurewicz, Michał: A short proof of the variational principle for a $\mathbb{Z}_+^n$ action on a compact space. *Astérisque*, **40** (1976), 147–157.

[Mis2] ———: Invariant measures for continuous transformations of $[0,1]$ with zero topological entropy. In *Ergodic theory, Proceedings, Oberwolfach, Germany 1978*, edited by Manfred Denker and Konrad Jacobs, Lecture Notes in Mathematics, vol. 729, pp. 144–152. Springer Verlag, Berlin, New York, 1979.

[Mis3] ———: Horseshoes for mappings of the interval. *Bull. Acad. Polon. Sci., Ser. Math. Astron. et Phys.* **27** (1979), 167–169.

[MisP] Misiurewicz, Michał, and Przytycki, Feliks: Topological entropy and degree of smooth mappings. *Bull. Acad. Polon. Sci., Ser. Math. Astron. et Phys.* **25** (1977), 573–574.

[MisS] Misiurewicz, Michał, and Szlenk, Wiesław: Entropy of piecewise monotone mappings. *Studia Mathematicae,* **118** (1980), 45–63.

[Mo] Morse, Marston: A fundamental class of geodesics on any closed surface of genus greater than two. *Transactions of the American Mathematical Society,* **26** (1924), 25–60.

[Mos1] Moser, Jürgen K.: On invariant curves of area preserving mappings of an annulus. *Nachrichten der Akademie der Wissenschaften Göttingen, Phys. Kl. IIa,* no. 1 (1962), 1–20.

[Mos2] ______: On the volume element on a manifold. *Transactions of the American Mathematical Society,* **120** (1965), 286–294.

[Mos3] ______: A rapidly convergent iteration method and nonlinear partial differential equations. *Annali della Scuola Norm. Super. de Pisa ser III.* **20** no. 2 (1966), 265–315; no. 3 (1966), 499–535.

[Mos4] ______: Convergent series expansions for quasiperiodic motions. *Mathematische Annalen,* **169** (1967), 136–176.

[Mos5] ______: On a theorem of Anosov. *Journal of Differential Equations,* **5** (1969), 411–440.

[Mos6] ______: *Stable and random motions in dynamical systems (with special emphasis on celestial mechanics).* Princeton University Press, Princeton, NJ, 1973.

[Mu] Munkres, James R.: *Analysis on manifolds.* Addison–Wesley, Reading, MA, 1991.

[N] Nash, John F.: Real algebraic manifolds. *Annals of Mathematics,* **56** (1952), 405–421.

[Ne] Nekhoroshev, Nikolai N.: An exponential estimate of the time of stability of nearly-integrable Hamiltonian systems. *Russian Mathematical Surveys,* **32** (1977), 1–67.

[NS] Nemyckiĭ, Viktor V., and Stepanov, Viacheslav. V.: *Qualitative theory of differential equations.* Princeton University Press, Princeton, NJ, 1960.

[vN] von Neumann, John: Zur Operatorenmethode in der klassischen Mechanik. *Annals of Mathematics,* **33** no. 3 (1932), 587–642; Zusätze zur Arbeit "Zur Operatorenmethode ... ". *Annals of Mathematics,* **33** no. 4 (1932), 789–791.

[New1] Newhouse, Sheldon: On codimension one Anosov diffeomorphisms. *American Journal of Mathematics,* **92** (1970), 761–770.

[New2] ______: Entropy and volume. *Ergodic Theory and Dynamical Systems,* **8*** (Conley Memorial Issue) (1988), 283–300.

[New3] ______: Diffeomorphisms with infinitely many sinks. *Topology,* **13** (1974), 9–18.

[New4] ______: Quasielliptic periodic points in conservative dynamical systems. *American Journal of Mathematics,* **99** (1977), 1061–1087.

[Ni1] Nielsen, Jakob: Über die Minimalzahl der Fixpunkte bei Abbildungstypen der Ringflächen. *Mathematische Annalen,* **82** (1921), 83–93.

[Ni2] ______: Untersuchungen zur Topologie der geschlossenen zweiseitigen Flächen. *Acta Mathematica,* **50** (1927), 189–358; II. **53** (1929), 1–76; III. **58** (1932), 87–167.

[Nit1] Nitecki, Zbigniew: *Differentiable Dynamics.* MIT Press, Cambridge, MA, 1971.

[Nit2] ______: Topological dynamics on the interval. In *Ergodic Theory and Dynamical Systems II, Proceedings Special Year, Maryland 1979–1980,* Progress in Math., vol. 21, pp. 1–73. Birkhäuser, 1982.

[O1] Ornstein, Donald: Bernoulli shifts with the same entropy are isomorphic. *Advances in Mathematics,* **4** no. 3 (1970), 337–352.

[O2] ______: *Ergodic theory, randomness, and dynamical systems.* Yale University Press, New Haven, CT, 1974.

[ORW] Ornstein, Donald; Rudolph, Daniel; and Weiss, Benjamin: *Equivalence of measure preserving transformations.* Memoirs of the American Mathematical Society, vol. 37, no. 262. American Mathematical Society, Providence, RI, 1982.

[Os] Oseledets, Valeriĭ I.: A multiplicative ergodic theorem. Liapunov characteristic numbers for dynamical systems. *Transactions of the Moscow Mathematical Society,* **19** (1968), 197–221.

[Ox] Oxtoby, John: Ergodic sets. *Bulletin of the American Mathematical Society,* **58** (1952), 116–136.

[OxU] Oxtoby, John, and Ulam, Stanislav: Measure preserving homeomorphisms and metrical transitivity. *Annals of Mathematics*, **42** (1941), 874–920.

[PMe] Palis, Jacob, and de Melo, Welington: *Geometric theory of dynamical systems.* Springer Verlag, Berlin, New York, 1982.

[PT] Palis, Jacob, and Takens, Floris: *Hyperbolicity and sensitive chaotic dynamics at homoclinic bifurcations.* Cambridge University Press, Cambridge, 1993.

[P1] Parry, William: Intrinsic Markov chains. *Transactions of the American Mathematical Society*, **112** (1964), 55–56.

[P2] Parry, William: *Entropy and generators in ergodic theory.* W. A. Benjamin, Inc., New York–Amsterdam, 1969.

[PP] Parry, William, and Pollicott, Mark: *Zeta functions and the periodic orbit structure of hyperbolic dynamics.* Astérisque, vol. 187–188. Société Mathématique de France, Paris, 1990.

[Pe1] Peixoto, Maurizio: On structural stability. *Annals of Mathematics*, **69** (1959), 199–222.

[Pe2] ———: Structural stability on two-dimensional manifolds. *Topology*, **1** (1962), 101–120.

[Per1] Perron, Oskar: Zur Theorie der Matrices. *Mathematische Annalen*, **64** (1906), 248–263.

[Per2] ———: Über Stabilität and asymptotisches Verhalten der Integrale von Differentialgleichungssystemen. *Mathematische Zeitschrift*, **29** no. 1 (1928), 129–160.

[Pes1] Pesin, Yakov B.: Families of invariant manifolds corresponding to nonzero characteristic exponents. *Mathematics of the USSR, Isvestia*, **10** no. 6 (1976), 1261–1305.

[Pes2] ———: Characteristic exponents and smooth ergodic theory. *Russian Mathematical Surveys*, **32** no. 4 (1977), 55–114.

[Pes3] ———: Geodesic flows on closed Riemannian manifolds without focal points. *Mathematics of the USSR, Isvestia*, **11** no. 6 (1977), 1195–1228.

[Pet] Petersen, Karl: *Ergodic theory.* Cambridge University Press, Cambridge, 1983.

[Pl] Plante, Joseph: Anosov flows. *American Journal of Mathematics*, **94** (1972), 729–755.

[Ply] Plykin, Roman V.: Sources and sinks for A-diffeomorphisms of surfaces. *Mathematics of the USSR, Sbornik*, **23** (1974), 233–253.

[Po1] Poincaré, Jules Henri: Mémoire sur les courbes définies par les équations différentielles, I. *Journal de Mathématiques Pures et Appliquées, 3. série*, **7** (1881), 375–422; II. **8** (1882), 251–286; III. *4. série*, **1** (1885), 167–244; IV. **2** (1886), 151–217.

[Po2] ———: *Les méthodes nouvelles de la mécanique céleste.* Paris, 1892–1899; English translation: *New methods of celestial mechanics.* Edited by Daniel Goroff. History of Modern Physics and Astronomy, vol. 13. American Institute of Physics, New York, 1993.

[Pol] Pollicott, Mark: *Lectures on ergodic theory and Pesin theory on compact manifolds.* Cambridge University Press, Cambridge, 1993.

[Pu] Pugh, Charles: The closing lemma. *American Journal of Mathematics*, **89** (1967), 956–1009; An improved closing lemma and a general density theorem. *American Journal of Mathematics*, **89** (1967), 1010–1021.

[PuRob] Pugh, Charles, and Robinson, Clark: The C^1 closing lemma, including Hamiltonians. *Ergodic Theory and Dynamical Systems*, **3** (1983), 261–313.

[PuShu] Pugh, Charles, and Shub, Michael: Ergodic attractors. *Transactions of the American Mathematical Society*, **312** no. 1 (1989), 1–54.

[R] Rabinowitz, Paul: Periodic solutions of Hamiltonian systems. *Communications in Pure and Applied Mathematics*, **31** (1978), 157–184.

[Ra] Ratner, Marina: Markov partitions for Anosov flows on n-dimensional manifolds. *Israel Journal of Mathematics*, **15** (1973), 92–114.

[Re] Rees, Mary: A minimal positive entropy homeomorphism of the 2-torus. *Journal of the London Mathematical Society*, **23** (1981), 311–322.

[Ro] Robbin, Joel W.: A structural stability theorem. *Annals of Mathematics*, **94** (1971), 447–493.

[Rob1] Robinson, Clark: Structural stability of C^1 diffeomorphisms. *Journal of Differential Equations*, **22** (1976), 28–73.

[Rob2] ———: *Dynamical systems; stability, symbolic dynamics, and chaos.* CRC Press, Cleveland, OH, 1994.

[Rok1] Rokhlin, Vladimir A.: On the fundamental ideas of measure theory. *Translations of the American Mathematical Society (series 1)*, **10** (1962), 1–54.

[Rok2] ——: New progress in the theory of transformations with invariant measure. *Russian Mathematical Surveys*, **15** no. 4 (1960), 1–22.

[Rok3] ——: Lectures on the entropy theory of measure preserving transformations. *Russian Mathematical Surveys*, **22** (1967), 1–52.

[Ru1] Rudolph, Daniel: A characterization of those processes finitarily isomorphic to a Bernoulli shift. In *Ergodic Theory and Dynamical Systems I, Proceedings Special Year, Maryland 1979–1980*, Progress in Math., vol. 21, pp. 1–64. Birkhäuser, 1981.

[Ru2] ——: *Fundamentals of measurable dynamics: Ergodic theory on Lebesgue spaces.* Oxford University Press, New York, 1990.

[Rue1] Ruelle, David: Statistical mechanics on a compact set with $\mathbb{Z}^\nu$ action satisfying expansiveness and specification. *Transactions of the American Mathematical Society*, **185** (1973), 237–253.

[Rue2] ——: A measure associated with Axiom-A attractors. *American Journal of Mathematics*, **98** no. 3 (1976), 619–654.

[Rue3] ——: Zeta functions for expanding maps and Anosov flows. *Inventiones mathematicae*, **34** no. 3 (1976), 231–242.

[Rue4] ——: An inequality for the entropy of differentiable maps. *Boletim da Sociedade Brasileira Matemática*, **9** (1978), 83–87.

[Rue5] ——: *Thermodynamic formalism.* Addison–Wesley, Reading, MA, 1978.

[Rue6] ——: Ergodic theory of differentiable dynamical systems. *Publications Mathématiques de l'Institut des Hautes Études Scientifiques*, **50** (1979), 27–58.

[Rue7] ——: *Elements of differentiable dynamics and bifurcation theory.* Academic Press, New York, 1989.

[Rü] Rüssmann, Helmut: Über invariante Kurven differenzierbarer Abbildungen eines Kreisringes. *Nachrichten der Akademie der Wissenschaften Göttingen, Math. Phys. Kl.* (1970), 67–105.

[SZ] Salamon, Dietmar, and Zehnder, Eduard: KAM theory in configuration space. *Commentarii Mathematici Helvetici*, **64** no. 1 (1989), 84–132.

[S] Satayev, Evgeni A.: On the number of invariant measures for flows on orientable surfaces. *Mathematics of the USSR, Isvestia*, **9** no. 4 (1975), 813–830.

[Sc] Schmidt, Klaus: *Algebraic ideas in ergodic theory.* CBMS Regional Conference Series in Mathematics, vol. 76. American Mathematical Society, Providence, RI, 1990; *Dynamical systems of algebraic origin.* Progress in Math., vol. 128. Birkhäuser, 1995.

[Sch] Schwartz, Arthur J.: A generalization of a Poincaré–Bendixson theorem to closed two-dimensional manifolds. *American Journal of Mathematics*, **85** (1963), 453–458.

[Schw] Schwartzman, Sol: Asymptotic cycles. *Annals of Mathematics*, **66** (1957), 270–284.

[Sh] Shannon, Claude: The mathematical theory of communication. *Bell Systems Technical Journal*, **27** (1948), 379–423, 623–656; Republished, University of Illinois Press, Urbana, IL, 1963.

[Sha] Sharkovskiĭ, Alexander N.: Coexistence of cycles of a continuous map of the line into itself. *Ukrainskiĭ Matematicheskiĭ Zhurnal*, **16** no. 1 (1964), 61–71; English translation: *International Journal of Bifurcation and Chaos in Applied Sciences and Engineering*, **5** no. 5 (1995), 1263–1273; On cycles and structure of a continuous map. *Ukrainskiĭ Matematicheskiĭ Zhurnal*, **17** no. 3 (1965), 104–111.

[Shi] Shil'nikov, Leonid P.: The existence of a countable set of periodic motions in the neighborhood of a homoclinic curve. *Soviet Mathematics, Doklady*, **8** (1967), 102–106.

[Shu1] Shub, Michael: Endomorphisms of compact differentiable manifolds. *American Journal of Mathematics*, **91** (1969), 175–199.

[Shu2] ——: Dynamical systems, filtrations and entropy. *Bulletin of the American Mathematical Society*, **80** no. 1 (1974), 27–41.

[Shu3] ——: *Global stability of dynamical systems.* Springer Verlag, Berlin, New York, 1987.

[ShuS] Shub, Michael, and Sullivan, Dennis: A remark on the Lefschetz fixed point formula for differentiable maps. *Topology*, **13** (1974), 189–191.

[Si] Siegel, Carl Ludwig: Iteration of analytic functions. *Annals of Mathematics*, **43** (1942), 607–612.

[SiM] Siegel, Carl Ludwig, and Moser, Jürgen K.: *Lectures on celestial mechanics.* Springer Verlag, Berlin, New York, 1971.

[Sin1] Sinai, Yakov G.: On weak isomorphism on transformations with invariant measures. *Mathematics of the USSR, Sbornik*, **63** no. 1 (1964), 23–42.

[Sin2] ———: Dynamical systems with countably-multiple Lebesgue spectrum. II. *Izvestija Akademiĭ Nauk SSSR Ser. Mat.* **30** (1966), 15–68; English translation: *Translations of the American Mathematical Society (series 2)*, **68**, 34–88.

[Sin3] ———: Markov partitions and $\mathcal{Y}$-diffeomorphisms. *Functional Analysis and Its Applications*, **2** no. 1 (1968), 64–89.

[Sin4] ———: The construction of Markovian partitions. *Functional Analysis and Its Applications*, **2** no. 3 (1968), 70–80.

[Sin5] ———: Gibbs measures in ergodic theory. *Russian Mathematical Surveys*, **27** (1972), 21–69.

[Sin6] ———: *Introduction to ergodic theory.* Princeton University Press, Princeton, NJ, 1977.

[Sm1] Smale, Stephen: Stable manifolds for differential equations and diffeomorphisms. *Annali della Scuola Norm. Super. de Pisa ser. III*, **17** (1963), 97–116.

[Sm2] ———: Structurally stable differentiable homeomorphisms with infinitely many periodic points. In *Proceedings of the International Conference on Nonlinear Oscillations*, vol. 2, pp. 365–366. Mathematics Institute of the Ukrainian Academy of Sciences, Kiev, 1963.

[Sm3] ———: Diffeomorphisms with many periodic points. In *Differential and combinatorial topology*, pp. 63–80. Princeton University Press, Princeton, NJ, 1965.

[Sm4] ———: Structurally stable systems are not dense. *American Journal of Mathematics*, **88** no. 2 (1966), 491–496.

[Sm5] ———: Differentiable dynamical systems. *Bulletin of the American Mathematical Society*, **73** (1967), 747–817.

[Sp] Spivak, Michael: *A comprehensive introduction to differential geometry.* Publish or Perish, New York, 1975.

[St] Sternberg, Shlomo: Local contractions and a theorem of Poincaré. *American Journal of Mathematics*, **79** (1957), 809–824; On the structure of local homeomorphisms of Euclidean n-space, II. *American Journal of Mathematics*, **80** (1958), 623–631; The structure of local diffeomorphisms, III. *American Journal of Mathematics*, **81** (1959), 578–604.

[Str] Strogatz, Steven H.: *Nonlinear dynamics and chaos.* Addison–Wesley, Reading, MA, 1994.

[Sz] Szlenk, Wiesław: *An introduction to the theory of smooth dynamical systems.* PWN-Polish Scientific Publishers/Wiley, Warszawa/Chichester, 1984.

[T] Takens, Floris: Homoclinic points in conservative systems. *Inventiones mathematicae*, **18** (1972), 267–292.

[Te] Temam, Roger: *Infinite-dimensional dynamical systems in mechanics and physics.* Springer Verlag, Berlin, New York, 1988.

[To] Toll, Charles: A multiplicative asymptotic for the prime geodesic theorem. Thesis, University of Maryland, 1984.

[V1] Veech, William: Gauss measures for transformations on the space of interval exchange maps. *Annals of Mathematics*, **115** (1982), 201–242.

[V2] Veech, William: The metric theory of interval exchange transformations I: Generic spectral properties. *American Journal of Mathematics*, **106** (1984), 1331–1359; II: Approximation by primitive interval exchanges. 1361–1387; III: The Sah–Arnaud–Fathi invariant. 1389–1422.

[W] Walters, Peter: *An introduction to ergodic theory.* Springer Verlag, Berlin, New York, 1982.

[Wi1] Williams, Robert F.: Classification of one dimensional attractors. In *Global Analysis, Proceedings of Symposia in Pure Mathematics*, vol. 14, pp. 341–361. American Mathematical Society, Providence, RI, 1970.

[Wi2] ———: Classification of subshifts of finite type. *Annals of Mathematics*, **98** (1973), 120–153; Errata. *Annals of Mathematics*, **99** (1974), 380–381.

[Y] Yoccoz, Jean-Christophe: Conjugaison différentiable des difféomorphismes du cercle dont le nombre de rotation vérifie une condition Diophantienne. *Annales scientifiques de l'École Normale Superieure*, **17** (1984), 333–361.

[Yo1] Yomdin, Yosif: A quantitative version of the Kupka–Smale theorem. *Ergodic Theory and Dynamical Systems*, **5** no. 3 (1985), 449–472.

[Yo2] ———: Volume growth and entropy. *Israel Journal of Mathematics*, **57** (1987), 285–300.

[Z] Zimmer, Robert J.: *Ergodic theory and semisimple groups.* Birkhäuser, 1984.

Index

Printed in the United States
By Bookmasters